A Specialist Periodical Report

Spectroscopic Properties of Inorganic and Organometallic Compounds

Volume 5

A Review of the Literature Published during 1971

Senior Reporter
N. N. Greenwood, *School of Chemistry, University of Leeds*

Reporters
D. M. Adams, *University of Leicester*
J. H. Carpenter, *University of Newcastle upon Tyne*
G. Davidson, *University of Nottingham*
M. Goldstein, *North London Polytechnic*
R. Greatrex, *University of Leeds*
B. E. Mann, *University of Leeds*
S. R. Stobart, *Queen's University, Belfast*

The Chemical Society
Burlington House, London, W1V 0BN

ISBN: 0 85186 043 5

Library of Congress Catalog Card No. 74-6662

Organic formulae composed by Wright's Symbolset method

PRINTED IN GREAT BRITAIN BY JOHN WRIGHT AND SONS LTD., AT THE STONEBRIDGE PRESS, BRISTOL

Foreword

This Report provides a comprehensive review of literature published during 1971 on the n.m.r., n.q.r., microwave, vibrational, and Mössbauer spectra of inorganic and organometallic compounds. The coverage and arrangement closely parallel those adopted last year. The Reporters have made strenuous efforts to contain the growing weight of material within the confines of a readable narrative of similar length to those produced in previous years. Extensive use of tabular material at appropriate points has enabled much factual and bibliographic information to be presented without destroying the flow of textual commentary.

It has been our aim not only to record results, but to indicate the wide variety of ways in which spectroscopic information is being used. Each chapter is self-contained and its length, and the number of references it contains, indicate the extent to which each technique is being applied to the study of inorganic and organometallic compounds. Each chapter states the range of work covered and the areas which have been omitted. The references have been obtained by means of the techniques outlined in previous volumes and we have tried to include all significant information published during 1971, though the prolonged United States East Coast dock strike during the crucial time at the end of the year almost frustrated our efforts.

With the appearance of Volume 5, this Specialist Periodical Report on Spectroscopic Properties of Inorganic and Organometallic Compounds has become firmly established as an annual compilation, and its continuing and increasing sales encourage us to believe that the format and overall depth of treatment are generally approved. However, your reporters, both individually and collectively, are far from complacent and would greatly value suggestions from users as to ways in which either the coverage or the presentation of information could be improved.

It is a pleasure to record yet again our thanks to the hard-pressed typists who persevere with our individual and sometimes obscure calligraphy, to the Editoral Staff of the Chemical Society who prepare our manuscript for the printers, and to the production team at the Stonebridge Press for their enthusiastic efforts which have enabled this Report to be published so rapidly.

N. N. G.

Contents

Chapter 5 Characteristic Vibrational Frequencies of Compounds containing Main-group Elements
By S. R. Stobart

Chapter 6 Vibrational Spectra of Transition-element Compounds
By M. Goldstein

Abbreviations

acac	acetylacetone anion
ASTP	tris(*o*-diphenylphosphinophenyl)arsine
ata	$N(CH_2CO_2)_3^{3-}$
azb	azobenzene
bipy	bipyridyl
bn	butylenediamine
CNDO	complete neglect of differential overlap
1,7-CTH	5,7,7,12,14,14-hexamethyl-1,4,8,11-tetrazacyclotetradecane
CyDTA	cyclohexanediaminotetra-acetic acid
DBM	dibenzoylmethane anion
depe	1,2-bisdiethylphosphinoethane
DHMB	Dewar hexamethylbenzene
diars	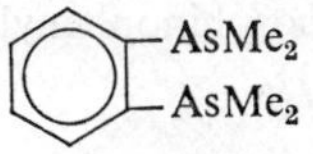
diars′	$Ph_2AsCH_2CH_2AsPh_2$
dien	diethylenetriamine
dimetrien	$H_2NCH_2CH_2NHCH(Me)CH(Me)NHCH_2CH_2NH_2$
diphos, Pf-Pf	$Ph_2PCH_2CH_2PPh_2$
dmaq	8-dimethylarsinoquinoline
DMF	dimethylformamide
dmgH	dimethylglyoxime
dmg	monoanion of dimethylglyoxime
dppa	1,2-bisdiphenylphosphinoacetylene
dpt	dipropylene triamine
ED3A	$(HO_2CCH_2)_2NCH_2CH_2NHCH_2CO_2H$
EDDA	ethylenediaminediacetic acid
EDPA	HO_2CCH_2 \ CH_2CO_2H / NCH_2CH_2N / $HO_2CCH(Me)$ \ $CH(Me)CO_2H$
EDTA	$(HO_2CCH_2)_2NCH_2CH_2N(CH_2CO_2H)_2$
eee	$H_2NCH_2CH_2SCH_2CH_2SCH_2CH_2NH_2$
en	ethylenediamine

ffars	$n = 1, X = Me_2As$
ffos	$n = 1, X = Ph_2P$
$f_6$fos	$n = 2, X = Ph_2P$
$f_8$fos	$n = 4, X = Ph_2P$

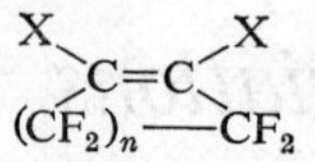

H_5DTPA	$(HO_2CCH_2)_2NCH_2CH_2N(CH_2CO_2H)CH_2CH_2N(CH_2CO_2H)_2$
hfac	hexafluoroacetylacetonato
NTA	$R = CH_2CO_2H$
MIDA	$R = Me$ } $RN(CH_2CO_2H)_2$
IDA	$R = H$
4-mpdpa	4-methyl-2-pyridyldi-(2-pyridyl)amine
OECy	$—SCH_2CH(NH_2)CO_2Et$
PDTA	propylenediaminetetra-acetic acid
Pf=Pf	$Ph_2PCH=CHPPh_2$
phen	1,10-phenanthroline
pic	picoline
γ-picO	γ-picoline *N*-oxide
Pm-Pm	$Me_2PCH_2CH_2PMe_2$
pn	propylenediamine
PTAS	tris(*o*-diphenylarsinophenyl)phosphine
py	pyridine
pyO	pyridine *N*-oxide
QP	tris(*o*-diphenylphosphinophenyl)phosphine
R-en	$RNHCH_2CH_2NH_2$
sal	salicylate anion
salen	*NN'*-ethylenebis(salicylaldiminato)
SBTP	tris(*o*-diphenylphosphinophenyl)stibine
SMCy	$MeSCH_2CH(NH_2)CO_2^-$
sp	*o*-$CH_2=CHC_6H_4PPh_2$
tcne	tetracyanoethylene
terpy	2,2′,2″-terpyridyl
terpyO_3	2,2′,2″-terpyridyl *NN′N″*-trioxide
tetren	tetraethylenepentamine
THF	tetrahydrofuran
TMED	$Me_2NCH_2CH_2NMe_2$
TPP	tetraphenylporphorin
trien	triethylenetetramine
tripyam	tri-(2-pyridyl)amine
TTA	anion of thenoyltrifluoroacetone
TTN	1,1,1-tris(dimethylaminomethyl)ethane
tu	thiourea

Special Ligand Abbreviations

(*used in Chapter 1*)

AA	acetylacetonate (acac)
AMA	2-(2-mercaptophenyl)imino-4-pentanato
AME	2-(2-mercaptoethyl)imino-4-pentanato
ASYM	asym-diethyltetramethyldiamidopyrophosphate
bdpma	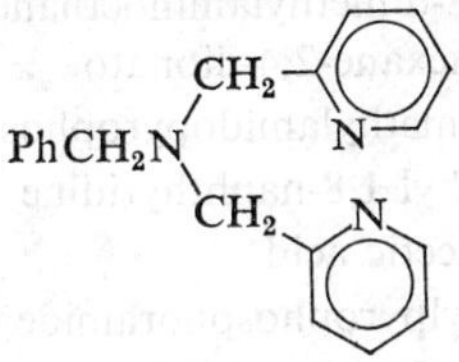
bn	2,3-diaminobutane
C_2 $n = 2$ C_3 $n = 3$	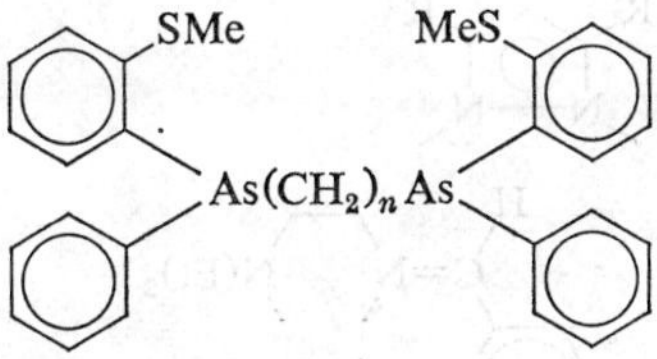
Cpylid	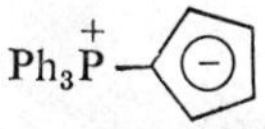
dam	bis(diphenylyarsino)methane
βdc	β-dicarbonyl
$DENB_2$	Schiff's base of benzaldehyde and bis-2-aminoethyl-amine)
β-dik	β-diketonate
1,1-DMA	1,1-dimethylallene
dmdsc	dimethyldiselenocarbamate
dmdtc	dimethyldithiocarbamate
DME	1,2-dimethoxyethane
DMHP	dimethyl hydrogen phosphite
DMMP	dimethyl methyl phosphite
dmphen	dimethyl phenanthrolene
DMSO	dimethylsulphoxide
DNR	2,4-dinitroresorcinol
dpm	$Bu^tCOCHCOBu^{t-}$

$DPNB_2$	Schiff's base of benzaldehyde and bis-(2-amino-propylamine)
DPSO	diphenylsulphoxide
dtc	diethyldithiocarbamato
dtn	diethyldithiophosphinato
dtp	dithiophosphato
Etdtp	*O,O*-diethyldithiophosphato
Et_2en	1-amino-2-diethylaminoethane
ex	ethylxanthogenato
HFA	hexafluoroacetylacetonate
HMPA	$[(CH_3)_2N]_3PO$
im	imidazole
Me_2en	1-amino-2-dimethylaminoethane
mhd	5-methylhexane-2,4-dionato
MONE	ethylhexamethylamidopyrophosphate
2,7-napy	2,7-dimethyl 1,8-naphthyridine
NTA	nitrilotriacetic acid
OMPA	octamethylpyrophosphoramide
pvac	$Bu^tCOCHCOMe^-$
3,5-R_2pz	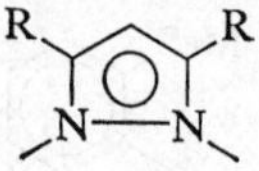
5Cl·SAL·en—NEt_2	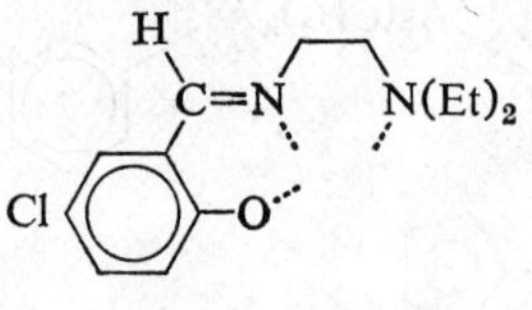
5Cl—SAL—NC_3H	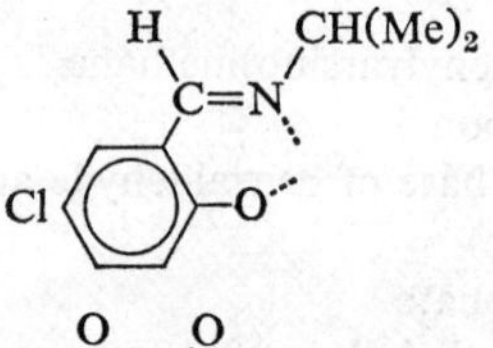
S—DBM	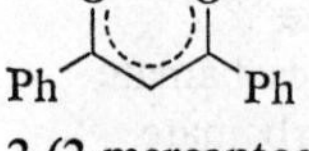
SME	2-(2-mercaptoethyl)salicylaldiminato
SP	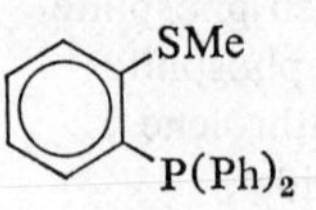 DSP 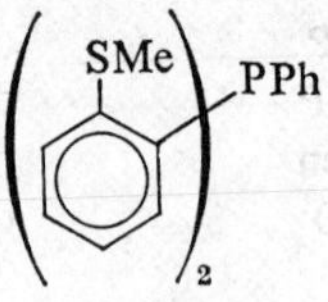
SYM	sym-diethyltetramethyldiamidopyrophosphate

Special Ligand Abbreviations xvii

teta 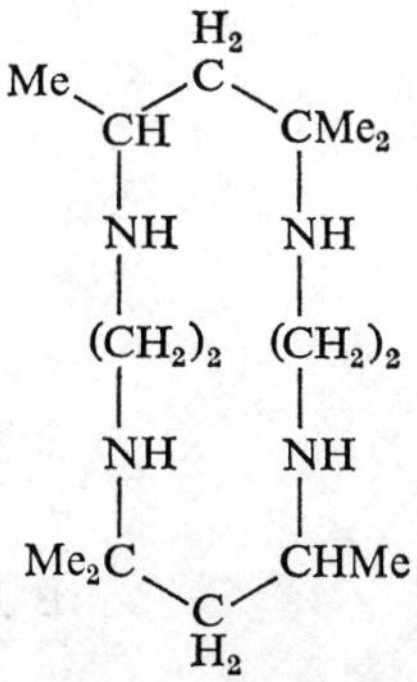

$$\begin{array}{c} \text{Me}\diagdown\text{CH}\diagup\overset{H_2}{C}\diagdown\text{CMe}_2 \\ | \quad\quad\quad | \\ \text{NH} \quad\quad \text{NH} \\ | \quad\quad\quad | \\ (CH_2)_2 \quad (CH_2)_2 \\ | \quad\quad\quad | \\ \text{NH} \quad\quad \text{NH} \\ | \quad\quad\quad | \\ \text{Me}_2\text{C}\diagdown\underset{H_2}{C}\diagup\text{CHMe} \end{array}$$

TFA trifluoroacetylacetonate
tkth tetrakistetrahydroborate
TMA tetramethylallene
tmhd 2,2,6,6-tetramethylheptane-3,5-dionate
TMP $(CH_3O)_3PO$
tn 1,3-diaminopropane

tpma 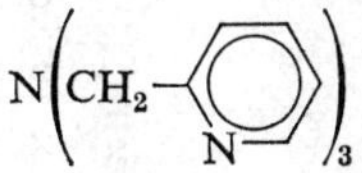

Conversion factors

	cm^{-1}	$J\ mol^{-1}$	eV	$kcal\ mol^{-1}$	$Mc\ s^{-1}$ (MHz)
cm^{-1}	1	11.957	1.2394×10^{-4}	2.8584×10^{-3}	2.9979×10^{4}
$J\ mol^{-1}$	8.3626×10^{-2}	1	1.0364×10^{-5}	2.3904×10^{-4}	2506.2
eV	8068.3	9.6484×10^{4}	1	2.3063	2.4188×10^{8}
$kcal\ mol^{-1}$	349.83	4183.3	4.3359×10^{-2}	1	1.0487×10^{7}
$Mc\ s^{-1}$ (MHz)	3.3356×10^{-5}	3.9903×10^{-4}	4.1344×10^{-9}	9.5345×10^{-8}	1

Mössbauer spectra

For ^{57}Fe ($E\gamma$ = 14.413 keV): 1 mm s^{-1} = $3.879 \times 10^{-4}\ cm^{-1}$ = $4.638 \times 10^{-3}\ J\ mol^{-1}$

= 4.809×10^{-8} eV = $1.109 \times 10^{-6}\ kcal\ mol^{-1}$

= 11.63 Mc s^{-1} (MHz)

For other nuclides multiply the above conversion factors by $E\gamma$ (keV)/14.413.

1

Nuclear Magnetic Resonance Spectroscopy

BY B. E. MANN

1 Introduction

There has been a remarkable increase in the number of papers containing information from n.m.r. spectroscopy of relevance to this report, with an increase of 60% compared to 1970. This is in part due to some 1970 journals arriving late and, for continuity, being included in this volume. In order to accommodate this increase many more papers than before have had to be assigned to the tables at the end of the chapter, and many more papers receive only a brief mention in the text. Following the precedent set in volume 4, references from journals not abstracted for this report, and obtained *via* the Chemical Society's n.m.r. Macroprofile (UKCIS), are listed at the end of the chapter.

As in previous years, ^{1}H n.m.r. has provided the majority of information, but other nuclei have provided useful information. Other nuclei of which direct use has been made this year are: ^{2}H, ^{3}H, ^{6}Li, ^{7}Li, ^{9}Be, ^{11}B, ^{13}C, ^{14}N, ^{17}O, ^{19}F, ^{23}Na, ^{27}Al, ^{29}Si, ^{31}P, ^{35}Cl, ^{39}K, ^{43}Ca, ^{51}V, ^{55}Mn, ^{57}Fe, ^{59}Co, ^{63}Cu, ^{65}Cu, ^{81}Br, ^{85}Rb, ^{87}Rb, ^{93}Nb, ^{117}Sn, ^{119}Sn, ^{127}I, ^{129}Xe, ^{133}Cs, ^{183}W, and ^{199}Hg. Of these nuclei, ^{13}C appears to have attracted the greatest increase in attention with the introduction of commercial spectrometers using Fourier transform.

Several books and reviews have appeared during this year. The series 'Annual Reports on N.M.R. Spectroscopy' (renamed from 'Annual Review on N.M.R. Spectroscopy') has continued with the appearance of volume 4 containing chapters on 'General review of proton magnetic resonance' by G. R. Bedford, 'The investigation of the kinetics of conformational changes by nuclear magnetic resonance' by I. O. Sutherland, 'Nuclear magnetic resonance spectroscopy in pesticide chemistry' by R. Haque and D. R. Buhler, 'A simple guide to the use of iterative computer programmes in the analysis of nuclear magnetic resonance spectra' by C. W. Haigh, 'Nuclear magnetic resonance spectra of polymers' by M. E. A. Cudby and H. A. Willis, and 'Fluorine-19 nuclear magnetic resonance spectroscopy' by K. Jones and E. F. Mooney.[1a] The Chemical Society

[1] (*a*) 'Annual Reports on N.M.R. Spectroscopy', ed. E. F. Mooney, Academic Press, London, vol. 4, 1971; (*b*) 'Nuclear Magnetic Resonance', ed. R. K. Harris (Specialist Periodical Reports), The Chemical Society, London, vol. 1, 1972.

have published volume 1 of 'Nuclear Magnetic Resonance' in their Specialist Periodical Reports series.[1b] Two further volumes have been published in the series 'Progress in Nuclear Magnetic Resonance Spectroscopy'—volume 7 being devoted to 'Fluorine chemical shifts' by J. M. Emsley and L. Phillips[2] while volume 8 is in two parts, 'The narrowing of n.m.r. spectra of solids by high-speed specimen rotation and the resolution of chemical shifts and spin multiplet structures for solids' by F. R. Andrews, and 'Pulsed n.m.r. in solids' by P. Mansfield.[3] A third series 'Nuclear Magnetic Resonance:—Basic Principles and Progress' has been prolific with the appearance of four more volumes: volume 3 contains 'Static quadrupole effects in disordered cubic solids' by O. Kanert and M. Mehring and 'Nuclear magnetic relaxation spectroscopy' by F. Noack;[4] volume 4 contains 'Natural and synthetic high polymers', the main lectures from the seventh colloquium on n.m.r. spectroscopy;[5] volume 5 contains 'Analysis of n.m.r. spectra' by R. A. Hoffman, S. Forsén, and B. Gestblom;[6] and volume 6 contains 'Computer assistance in the analysis of high resolution n.m.r. spectra' by P. Diehl, H. Kellerhals, and E. Lustig.[7]

The book 'Applied Spectroscopy Reviews', volume 4, contains chapters on 'Aluminium-27 and proton n.m.r. of organoaluminiums' by L. Petrakis and F. E. Dickson, and 'Applications of spectroscopy in the study of glassy solids, part II. Infrared, Raman, e.p.r., and n.m.r. spectral studies' by J. Wong and C. A. Angell.[8] Volume 4 of the series 'Determination of Organic Structures by Physical Methods' has been devoted to applications of n.m.r., with chapters on 'Applications of high-field n.m.r. spectroscopy' by W. Naegele, 'Pulsed n.m.r. methods' by N. Boden, 'Nuclear magnetic double resonance spectroscopy' by W. McFarlane, '^{15}N nuclear magnetic resonance' by R. L. Lichter, 'N.m.r. spectra of the heavier elements' by P. R. Wells, '^{13}C nuclear magnetic resonance' by P. S. Pregosin and E. W. Randall, and '^{31}P nuclear magnetic resonance' by J. R. Van Wazer.[9]

Among the books published, some deal with n.m.r. in general, namely 'Nuclear Magnetic Resonance' by W. W. Paudler,[10] 'Introduction to

[2] 'Progress in Nuclear Magnetic Resonance Spectroscopy', ed. J. Emsley, J. Feeney, and L. Sutcliffe, Pergamon Press, Oxford, vol. 7, 1971.
[3] 'Progress in Nuclear Magnetic Resonance Spectroscopy', ed. J. Emsley, J. Feeney, and L. Sutcliffe, Pergamon Press, Oxford, vol. 8, 1971.
[4] 'Nuclear Magnetic Resonance:— Basic Principles and Progress', ed. P. Diehl, E. Fluck, and R. Kosfeld, Springer-Verlag, Berlin, vol. 3, 1971.
[5] 'Nuclear Magnetic Resonance:— Basic Principles and Progress', ed. P. Diehl, E. Fluck, and R. Kosfeld, Springer-Verlag, Berlin, vol. 4, 1971.
[6] 'Nuclear Magnetic Resonance:— Basic Principles and Progress', ed. P. Diehl, E. Fluck, and R. Kosfeld, Springer-Verlag, Berlin, vol. 5, 1971.
[7] 'Nuclear Magnetic Resonance:— Basic Principles and Progress', ed. P. Diehl, E. Fluck, and R. Kosfeld, Springer-Verlag, Berlin, vol. 6, 1971.
[8] 'Applied Spectroscopy Reviews', ed. E. G. Brame, jun., Dekker, New York, vol. 4, 1971.
[9] 'Determination of Organic Structures by Physical Methods', ed. F. C. Nachod and J. J. Zuckerman, Academic Press, New York and London, vol. 4, 1971.
[10] W. W. Paudler, 'Nuclear Magnetic Resonance', Allyn and Bacon, Boston, Mass., 1971.

Magnetic Resonance: Principles and Applications' by R. T. Schumacher,[11] 'Molecular Theories of Nuclear Magnetic Resonance' by G. Mavel,[12] and 'N.m.r. Spectra of Unknowns' by J. A. Moore and D. L. Dalrymple.[13] Other books deal with specific areas of n.m.r., namely 'Relaxation in Magnetic Resonance' by C. P. Poole, jun. and H. A. Farach,[14] 'Spin-temperature and Nuclear Magnetic Resonance in Solids' by M. Goldman,[15] 'Pulse and Fourier Transform n.m.r.' by T. C. Farrer and E. D. Becker,[16] 'The Nuclear Overhauser Effect, Chemical Applications' by J. H. Noggle and R. E. Schirmer,[17] and 'Design of Low-radiofrequency Nuclear Magnetic Resonance Spectrometers' by A. Podcameni.[18] In addition, several books have been published on spectroscopy in general which contain at least one chapter on n.m.r.,[19–23] and the book 'Hydrogen Bonding' by S. N. Vinogradov and R. H. Linnell contains one chapter on the use of n.m.r. to investigate hydrogen bonding.[24]

A number of reviews have also appeared during the year. One review on the use of superconducting magnets in n.m.r. applies their use to the ^{1}H and ^{11}B n.m.r. spectra of decaborane.[25] Applications of nuclear double resonance,[26] and n.m.r. spectra in liquid crystals[27] have also been reviewed. Fourier transform spectroscopy has attracted considerable attention, especially in the context of ^{13}C n.m.r.[28–30] The direct application of n.m.r.

[11] R. T. Schumacher, 'Introduction to Magnetic Resonance: Principles and Applications', Benjamin, New York, 1970.
[12] G. Mavel, 'Molecular Theories of Nuclear Magnetic Resonance', Butterworths, Toronto, 1969.
[13] J. A. Moore and D. L. Dalrymple, 'N.M.R. Spectra of Unknowns', Saunders, Philadelphia, 1971.
[14] C. P. Poole, jun., and H. A. Farach, 'Relaxation in Magnetic Resonance', Academic Press, New York, 1971.
[15] M. Goldman, 'Spin-temperature and Nuclear Magnetic Resonance in Solids', Oxford University Press, Oxford, 1971.
[16] T. C. Farrar and E. D. Becker, 'Pulse and Fourier Transform N.M.R.', Academic Press, New York, 1971.
[17] J. H. Noggle and R. E. Schirmer, 'The Nuclear Overhauser Effect, Chemical Applications', Academic Press, New York and London, 1971.
[18] A. Podcameni, 'Design of Low-radiofrequency Nuclear Magnetic Resonance Spectrometers', Pontificia Univ. Catolica do Rio de Janeiro, Rio de Janeiro, 1969.
[19] D. H. Whiffen, 'Spectroscopy', Wiley, New York, 1971.
[20] E. F. H. Brittain, W. O. George, and C. H. J. Wells, 'Introduction to Molecular Spectroscopy', Academic Press, London, 1970.
[21] 'Physical Methods of Chemistry', ed. A. Weissberger and B. W. Rossiter, Wiley, London, Part 3A, 1971.
[22] 'An Introduction to Spectroscopic Methods for the Identification of Organic Compounds:— N.M.R. and I.R.', ed. F. Scheinmann, Pergamon Press, Oxford, 1970.
[23] R. Chang, 'Basic Principles of Spectroscopy', McGraw-Hill Book Co., New York, 1971.
[24] S. N. Vinogradov and R. H. Linnell, 'Hydrogen Bonding', Van Nostrand Rheinhold Company, New York, 1971.
[25] L. F. Johnson, *Analyt. Chem.*, 1971, **53**, 28A.
[26] W. von Philipsborn, *Angew. Chem. Internat. Edn.*, 1971, **10**, 472.
[27] S. Meiboom and L. C. Snyder, *Accounts Chem. Res.*, 1971, **4**, 81.
[28] F. W. Wehrli, *Chem.-Ztg.*, 1971, **95**, 58.
[29] E. Breitmaier, G. Jung, and W. Voelter, *Angew. Chem. Internat. Edn.*, 1971, **33**, 673.
[30] E. W. Randall, *Chem. in Britain*, 1971, **7**, 371.

to inorganic systems is reviewed with articles on solvent co-ordination numbers of metal ions in solution,[31] organometallic compounds of the lanthanides and actinides,[32] conformational analysis of tris(ethylenediamine) complexes,[33] bond character of β-diketone metal chelates,[34] optical and geometrical isomerization of β-diketone complexes,[35] stereochemistry of bischelate metal(II) complexes,[36] optical activity from asymmetric transition metal atoms,[37] and metal–boron compounds [38] all containing a substantial section on n.m.r. Also, a review entitled 'N.M.R. Spectra of Compounds of the Platinum-group Metals' has appeared.[38a]

Instrumentation and Techniques.—The design of n.m.r. spectrometers [39] and the n.m.r. spectrometers available commercially [40] have been reviewed, and the NV-14, a new polynuclear spectrometer, has been described.[41] Interest continued in the modification of existing n.m.r. spectrometers, with an inexpensive method for obtaining ^{2}H n.m.r. spectra on the XL-100 n.m.r. spectrometer [42] and a modification to a high-resolution Perkin–Elmer R10 spectrometer to record solid-state ^{31}P n.m.r. spectra being described.[43] Variable-temperature n.m.r. received attention with papers on an electrical circuit to balance the probe and provide leakage for low-temperature work,[44] a method to generate a stable stream of cold gas at temperatures down to 100 K without frosting the sample,[45] and a device to overcome lifting of the sample tube.[46] For high-pressure wide-line n.m.r., an apparatus which can withstand pressures of 100 kbar has been described.[47] A single-coil probe damper for pulsed n.m.r. has been described. It improves the time for ring-down following an r.f. pulse by a factor of 2.5.[48] The digitization and processing of n.m.r. spectra by computer and outputing the data ready to feed into LAOCOON 3 were reported.[49] The description of a method to obtain short-term stability of the internal magnetic field–radio frequency lock system to 0.001 Hz offers a method to measure coupling

[31] S. F. Lincoln, *Co-ordination Chem. Rev.*, 1971, **6**, 309.
[32] R. G. Hayes and J. L. Thomas, *Organometallic Chem. Rev.*, 1971, **7**, 1.
[33] J. K. Beattie, *Accounts Chem. Res.*, 1971, **4**, 253.
[34] B. Bock, K. Flatau, H. Junge, M. Kuhr, and H. Musso, *Angew. Chem. Internat. Edn.*, 1971, **10**, 225.
[35] J. J. Fortman and R. E. Sievers, *Co-ordination Chem. Rev.*, 1971, **6**, 331.
[36] R. H. Holm and M. J. O'Connor, *Progr. Inorg. Chem.*, 1971, **14**, 241.
[37] H. Brunner, *Angew. Chem. Internat. Edn.*, 1971, **10**, 249.
[38] G. Schmid, *Angew. Chem. Internat. Edn.*, 1970, **9**, 819.
[38a] R. S. Tobias, *Adv. Chem. Ser.*, 1971, No. 98, p. 98.
[39] D. G. Howery, *J. Chem. Educ.*, 1971, **48**, A327.
[40] D. G. Howery, *J. Chem. Educ.*, 1971, **48**, A389.
[41] F. W. Wehrli, *Chem.-Ztg.*, 1971, **95**, T131.
[42] R. E. Santini, *Analyt. Chem.*, 1971, **43**, 801.
[43] K. B. Dillon and T. C. Waddington, *Spectrochim. Acta*, 1971, **27A**, 1381.
[44] R. H. Geils, *Rev. Sci. Instr.*, 1971, **42**, 265.
[45] L. Silver and R. Rudman, *Rev. Sci. Instr.*, 1971, **42**, 671.
[46] J. Barzilay, *Analyt. Chem.*, 1971, **43**, 160.
[47] R. W. Vaughan, C. F. Lai, and D. D. Elleman, *Rev. Sci. Instr.*, 1971, **42**, 626.
[48] S. B. W. Roeder, N. L. Rhodes, and G. W. Schmidt, *Rev. Sci. Instr.*, 1971, **42**, 1692.
[49] R. E. Rondeau and V. L. Donlan, *Analyt. Chem.*, 1971, **43**, 1699.

constants to ± 0.003 Hz,[50] and an indirect method of measuring $J(^{11}B-^{1}H)$ has been proposed.[51] Perils associated with lock-in detection for n.m.r. lines in solids when the sweep time is much less than T_1 have been examined.[52]

With the appearance of commercial Fourier transformation spectrometers, attention has turned to such spectrometers and problems associated with them. A modification to existing continuous-wave spectrometers to provide a pulse Fourier spectrometer has been described,[53] and a way to obtain Fourier spectra without a digital computer which offers a way to make inexpensive n.m.r. spectrometers was reported.[54] The Overhauser effect, producing intensities not directly related to the number of ^{13}C atoms, can be removed either by the addition of paramagnetic metal ions,[55, 56] or Bu^t_2NO.[57] This effect of addition of paramagnetic ions on ^{13}C n.m.r. spectra has been treated theoretically.[58] On the other hand, a method to retain both the true $^{1}H-^{13}C$ coupling constant and the Overhauser enhancement has been described.[59, 60] Several pulsed n.m.r. experiments have been described which offer a simple method of rapid signal accumulation when $T_2^* < T_2, T_1$, that is, when the resolution is limited by magnetic field inhomogeneity.[61] Alternatively, a pulse technique to detect very weak n.m.r. signals with relaxation times T_1 and T_2 long compared with the decay time T_2^* of the free induction decay has been reported, and applied to detect ^{57}Fe in $Fe(CO)_5$.[62] Aqueous solutions can provide problems for pulsed Fourier transform spectrometers, but a method of improving signal-to-noise ratios for such solutions has been described.[63] When the interval between pulses is short compared to the spin–spin relaxation time, gross discrepancies can occur in ^{13}C signal amplitude. This effect can be overcome by a simple device which randomizes the delay between pulses.[64] The optimization of the pulsed n.m.r. driven equilibrium technique has been discussed.[65] The measurement of specific spin–lattice and spin–spin relaxation times not only for different protons [66] and different

50 A. G. Moritz, *Mol. Phys.*, 1971, **20**, 945.
51 V. S. Bogdanov, A. V. Kessenikh, and V. V. Negrebetsky, *J. Magn. Resonance*, 1971, **5**, 145.
52 D. S. Wollan, *Rev. Sci. Instr.*, 1971, **42**, 682.
53 R. J. Cushley, D. R. Anderson, and S. R. Lipsky, *Analyt. Chem.*, 1971, **43**, 1281.
54 R. R. Ernst, *J. Magn. Resonance*, 1971, **5**, 398.
55 G. N. La Mar, *Chem. Phys. Letters*, 1971, **10**, 230.
56 R. Freeman, K. G. R. Pachler, and G. N. La Mar, *J. Chem. Phys.*, 1971, **55**, 4586.
57 G. N. La Mar, *J. Amer. Chem. Soc.*, 1971, **93**, 1040.
58 D. F. S. Natusch, *J. Amer. Chem. Soc.*, 1971, **93**, 2566.
59 R. Freeman and H. D. W. Hill, *J. Magn. Resonance*, 1971, **5**, 278.
60 O. A. Gansow and W. Schittenhelm, *J. Amer. Chem. Soc.*, 1971, **93**, 4294.
61 P. Waldstein and W. E. Wallace, jun., *Rev. Sci. Instr.*, 1971, **42**, 437.
62 A. Schwenk, *J. Magn. Resonance*, 1971, **5**, 376.
63 A. G. Redfield and R. K. Gupta, *J. Chem. Phys.*, 1971, **54**, 1418.
64 R. Freeman and H. D. W. Hill, *J. Magn. Resonance*, 1971, **4**, 366.
65 W. E. Wallace, jun., *J. Chem. Phys.*, 1971, **54**, 1475.
66 P. Lalanne, A. Andrieux, B. Lemanceau, and C. Lussan, *Ber. Bunsengesellschaft Phys. Chem.*, 1971, **75**, 275.

^{13}C atoms[67, 68] but also for individual lines in high-resolution n.m.r. spectra[69] has been studied. A new method based on Fourier transform spectroscopy for the direct measurement of difference frequencies in n.m.r. has been proposed.[70]

Compounds suitable as internal references have once again received attention. It has been suggested that silicone grease can be used as a convenient internal standard for the quantitative monitoring of reactions by n.m.r. The material is inert to nearly all reagents, thermally stable, non-volatile, moderately soluble in deuteriochloroform and readily removed by passage of the solution through silica gel or alumina.[71] Tetramethylsilane, the reference commonly used for ^{1}H n.m.r., has been suggested as a good reference for ^{13}C and ^{29}Si n.m.r. Solvent effects on these three nuclei in TMS have been investigated and shifts of up to 0.30 p.p.m. for ^{1}H, 0.19 p.p.m. for ^{29}Si and 1.24 p.p.m. for ^{13}C were reported.[72] It has also been proposed that $Me_4N^+ I^-$ should be chosen as the internal reference for ^{14}N and ^{15}N with positive values in the direction of increasing frequency.[73]

A very useful compilation of the chemical shift values for the principal ^{1}H n.m.r. peaks of 50 solvents and other low molecular weight substances dissolved in $DCCl_3$, $(D_3C)_2SO$, C_6H_6 or C_6D_6, C_5H_5N or C_5D_5N, CF_3CO_2H, or D_2O has been published.[74] The use of aluminium, magnesium, barium, or Mischmetal to remove molecular oxygen from common solvents has been investigated by the measurement of T_1, but little difference between the metals was found.[75] The use of ^{3}H n.m.r. has been described.[76]

The Kaplan–Alexander equations for chemical exchange have been generalized to include Heisenberg exchange of high spins ($I > \frac{1}{2}$). General expressions were obtained for the effect of the exchange on the shape and width of lines and on the population of spin eigenstates.[77] The exact solutions for relaxation times T_1 and T_2 in chemically exchanging systems have been obtained, and from these solutions, approximations, involving more rigorous assumptions than have been used previously, can be easily obtained for special conditions.[78] A density matrix description of nuclear magnetic double resonance on chemically exchanging systems has been developed using a double-resonance basis set.[79] The effect of modulation

[67] R. Freeman and H. D. W. Hill, *J. Chem. Phys.*, 1971, **54**, 3367.
[68] R. Freeman and H. D. W. Hill, *J. Chem. Phys.*, 1971, **55**, 1985.
[69] R. Freeman and H. D. W. Hill, *J. Chem. Phys.*, 1971, **54**, 301.
[70] R. R. Ernst, *J. Magn. Resonance*, 1971, **4**, 280.
[71] J. A. Kapecki, *J. Chem. Educ.*, 1971, **48**, 731.
[72] M. Bacon, G. E. Maciel, W. K. Musker, and R. Scholl, *J. Amer. Chem. Soc.*, 1971, **93**, 2537.
[73] E. D. Becker, *J. Magn. Resonance*, 1971, **4**, 142.
[74] R. A. Fletton and J. E. Page, *Analyst*, 1971, **96**, 370.
[75] H. S. Sandhu, *Canad. J. Chem.*, 1971, **49**, 1008.
[76] J. Bloxsidge, J. A. Elvidge, J. R. Jones, and E. A. Evans, *Org. Magn. Resonance*, 1971, **3**, 127.
[77] V. Yu. Zitserman, *Mol. Phys.*, 1971, **20**, 1005.
[78] J. S. Leigh, jun., *J. Magn. Resonance*, 1971, **4**, 308.
[79] P. P. Yang and S. L. Gordon, *J. Chem. Phys.*, 1971, **54**, 1779.

broadening of the shape of unsaturated, Lorentzian dispersion lines has been examined and graphs to determine the linewidth and the number of paramagnetic spins present have been given.[80]

Second-order n.m.r. spectra have received attention. A new computer program, which simplifies the calculation of trial parameters for an AA′BB′ spin system, has been described.[81] The theory of scalar relaxation of the second kind as applied to the $^2A^2B^3X$ and $^2A^3X^3Y$ spin systems has been developed.[82, 83] In the case of the $^2A^2B^3X$ spin system it is predicted and found for 2-bromothiazole that the AB linewidths are field dependent.[84]

Coupling Constants and Chemical Shifts.—In this section papers where the primary consideration is the measurement of coupling constants or chemical shifts are reviewed. The section is thence split into several subsections covering (*a*) coupling constants; (*b*) ^1H and ^{19}F chemical shifts; (*c*) ^{11}B, ^{13}C, ^{29}Si, ^{14}N, and ^{15}N chemical shifts; (*d*) ^{31}P chemical shifts, and (*e*) metal chemical shifts.

Coupling Constants. An attempt has been made to calculate directly-bonded X–H nuclear spin coupling constants from molecular parameters obtained from the extended Hückel model. The agreement is good between experiment and theory, but not good enough to provide structural information.[85] $^1J(^{13}C–^1H)$ has been measured for the compounds (1),[86] where X = H, Me, Ph, Ac, Cl, or $-C_4H_3Fe(CO)_3$, and for (2).[87] It was concluded

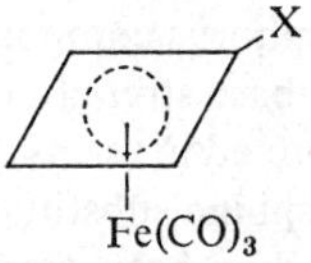

(1)

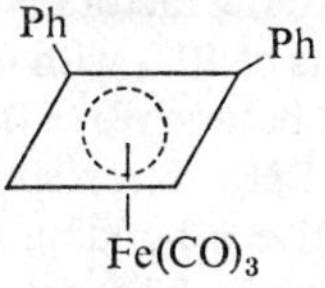

(2)

that the carbon–carbon bonds of the four-membered rings have nearly uniform bond orders. When SnH_4 was treated with FSO_3H at − 78 °C, then a 1 : 3 : 3 : 1 quartet appeared in the ^{119}Sn n.m.r. spectrum, with $^1J(^{119}Sn–^1H)$ of 2960 Hz, due to $[SnH_3]^+$. This value is larger than the 2570 Hz value calculated on the basis of a change from sp^3 hybridization for SnH_4, $^1J(^{119}Sn–^1H)$ = 1933 Hz, to sp^2 hybridization for $[SnH_3]^+$, but the discrepancy is explained as due to the high positive charge on the tin atom.[88]

80 H. A. Buckmaster and J. D. Skirrow, *J. Magn. Resonance*, 1971, **5**, 285.
81 G. Ng and C. R. Dillard, *J. Chem. Educ.*, 1971, **48**, 607.
82 N. C. Pyper, *Mol. Phys.*, 1971, **21**, 961.
83 N. C. Pyper, *Mol. Phys.*, 1971, **21**, 977.
84 R. K. Harris and N. C. Pyper, *Mol. Phys.*, 1971, **20**, 467.
85 J. A. Varga and S. S. Zumdahl, *Theor. Chim. Acta*, 1971, **21**, 211.
86 H. A. Brune, G. Horlbeck, and U.-T. Záhorszky, *Z. Naturforsch.*, 1971, **26b**, 222.
87 H. A. Brune, H. Hütter, and H. Hanebeck, *Z. Naturforsch.*, 1971, **26b**, 570.
88 J. R. Webster and W. L. Jolly, *Inorg. Chem.*, 1971, **10**, 877.

Coupling to ^{11}B has attracted considerable attention. The ^{11}B n.m.r. spectrum of pentaborane(9) (3) (R = H) with all protons decoupled shows $^{1}J(^{11}B-^{11}B)$ of 19.4 ± 0.2 Hz well resolved for the basal boron resonance, but no clear structure was observed in the resonance of the apical boron.[89] Similarly, in (3) (R = Me), coupling has been observed

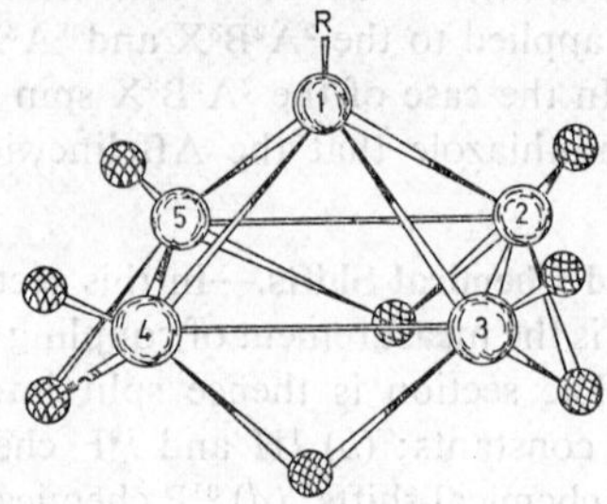

(3)

between the methyl protons and both the apical (*ca.* 6 Hz) and the basal boron atoms (*ca.* 0.8 Hz). Coupling has been observed also between the bridge protons and both the basal terminal protons (*ca.* 5 Hz) and methyl protons on the basal boron atoms (*ca.* 3 Hz).[90] An examination of the ^{19}F n.m.r. spectra of some aromatic amine *N*-oxide adducts of BF_3 has shown $^{1}J(^{11}B-^{19}F)$ to be negative, and variations in chemical shifts were attributed to steric factors.[91] ^{1}H, ^{11}B, ^{19}F, and ^{31}P have been used to examine adducts of BH_3 with tervalent phosphorus compounds. A correlation was found between $^{1}J(^{11}B-^{31}P)$ and the base strength of the phosphine with respect to BH_3. The results were interpreted in terms of a combination of σ-effects, dative π-bonding from the phosphine substituents, and borane hyperconjugation.[92] The same correlation has been found by a second group of workers, but the relationship does not work for a given phosphine with the series BX_3 (X = F, Cl, Br, or H). Relative sign determinations indicated that $^{1}J(^{11}B-^{31}P)$ is positive.[93] The ^{13}C n.m.r. spectrum of 1-methylpentaborane(9) shows $^{1}J(^{11}B-^{13}C)$ of 72.6 ± 0.5 Hz.[94]

With the increase in interest in ^{13}C n.m.r. a number of direct coupling constants have been measured. A number of phosphetan oxides and phosphetanium salts [(4) and (5)] have been investigated by ^{13}C n.m.r. It is found that $^{1}J(^{13}C-^{31}P)$ and $^{3}J(^{13}C-^{31}P)$ are very sensitive to stereochemistry and can be used to determine *cis–trans* isomerism.[95] The first measurements of $^{1}J(^{57}Fe-^{13}C)$ for $Fe(CO)_5$ (23.2 Hz) and $^{1}J(^{57}Fe-^{31}P)$ for

[89] J. D. Odom, P. D. Ellis, and H. C. Walsh, *J. Amer. Chem. Soc.*, 1971, **93**, 3529.
[90] J. B. Leach and T. Onak, *J. Magn. Resonance*, 1971, **4**, 30.
[91] R. S. Stephens, S. D. Lessley, and R. O. Ragsdale, *Inorg. Chem.*, 1971, **10**, 1610.
[92] A. H. Cowley and M. C. Damasco, *J. Amer. Chem. Soc.*, 1971, **93**, 6815.
[93] R. W. Rudolph and C. W. Schultz, *J. Amer. Chem. Soc.*, 1971, **93**, 6821.
[94] P. D. Ellis, J. D. Odom, D. W. Lowman, and A. D. Cardin, *J. Amer. Chem. Soc.*, 1971, **93**, 6704.
[95] G. A. Gray and S. E. Cremer, *Tetrahedron Letters*, 1971, 3061.

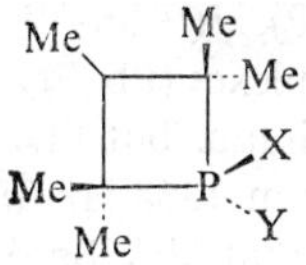

(4)

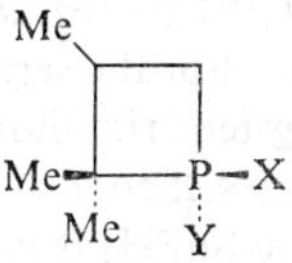

(5)

$Fe(CO)_4(PEt_nPh_{3-n})$ (n = 1—3) (*ca.* 27 Hz) have been reported. It is interesting to note that $Fe(CO)_4(PEt_nPh_{3-n})$ is also fluxional like $Fe(CO)_5$ and only one ^{13}C resonance (CO) is observed.[96] The first measurement of $^1J(^{183}W-^{13}C)$ of 186 Hz has been reported for (durene)$W(CO)_3$,[97] and for a series of compounds $W(CO)_5L$ [L = $(RO)_nPPh_{3-n}$, Ph_3As, Ph_3Sb, Ph_3Bi, or $C_6H_{11}NH_2$] a linear relationship has been reported between the ^{13}C (CO) chemical shift and the force constant for the CO bond.[98] The first measurement of $^1J(^{195}Pt-^{13}C)$ has been reported for *cis*-$PtMe_2(XMe_2Ph)_2$ (X = P or As). On going from X = P to X = As, $^1J(^{195}Pt-^{13}C)$ increases from 594 Hz to 685 Hz, while $^2J(^{195}Pt-X-^{13}C)$ and $^3J(^{195}Pt-^1H)$ decrease. The increase in $^1J(^{195}Pt-^{13}C)$ from X = P to X = As parallels an increase in ν(Pt—C) and could reflect a greater Pt—C bond strength.[99] Subsequently, further measurements of $^1J(^{195}Pt-^{13}C)$ for a number of methylplatinum complexes were reported. $^1J(^{195}Pt-^{13}C)$ is taken as a measure of *trans*-effect and it is found that $-C\langle^{Me}_{OMe}$ > CO ≥ RNC > C_6F_5CN. A plot of $^2J(^{195}Pt-C-^1H)$ against $^1J(^{195}Pt-^{13}C)$ is a straight line passing through the origin, thereby justifying the use of $^2J(^{195}Pt-C-^1H)$ as a measure of *trans*-influence.[100]

The ^{19}F and ^{51}V n.m.r. spectra have been measured for $CsVOF_4$ and V_2O_5 in aqueous and anhydrous hydrofluoric acid at low temperatures. It is found that the apparent coupling $^1J(^{19}F-^{51}V)$ is dependent on the conditions and it is suggested that there is rapid intramolecular exchange between C_{2v} and C_{4v} symmetry.[101] $^1J(^{19}F-^{51}V)$ of 88 Hz is clearly resolved for $[VF_6]^-$.[102]

The relative signs of all the coupling constants involving ^1H, ^{29}Si, and ^{31}P in $(^{28}SiH_3)_2P^{29}SiH_3$ and $(^{28}SiH_3)_2Sb^{29}SiH_3$ have been determined by $^1H-\{^{29}Si\}$ and $^1H-\{^{31}P\}$ 'tickling' experiments. With the exception of $^1K(^{29}Si-^{31}P)$, the reduced coupling constants are probably positive over one and three bonds, and negative over two and four bonds.[103] The first

96 B. E. Mann, *Chem. Comm.*, 1971, 1173.
97 B. E. Mann, *Chem. Comm.*, 1971, 976.
98 O. A. Gansow, B. Y. Kimura, G. R. Dobson, and R. A. Brown, *J. Amer. Chem. Soc.*, 1971, **93**, 5922.
99 A. J. Cheney, B. E. Mann, and B. L. Shaw, *Chem. Comm.*, 1971, 431.
100 M. H. Chisholm, H. C. Clark, L. E. Manzer, and J. B. Stothers, *Chem. Comm.*, 1971, 1627.
101 J. A. S. Howell and K. C. Moss, *J. Chem. Soc. (A)*, 1971, 270.
102 J. A. S. Howell and K. C. Moss, *J. Chem. Soc. (A)*, 1971, 2483.
103 K. D. Crosbie and G. M. Sheldrick, *Mol. Phys.*, 1971, **20**, 317.

measurements of $^1J(^{111}Cd-^{31}P)$, $^1J(^{113}Cd-^{31}P)$, $^1J(^{117}Sn-^{31}P)$, and $^1J(^{119}Sn-^{31}P)$ have been reported. For the series of complexes $(Ph_nEt_{3-n}P)CdI_2$ (n = 0—2), unlike the tungsten, rhodium, or platinum, but like the mercury analogues, as n increases then J decreases from 1368 Hz (n = 0) to 1196 Hz (n = 2).[104] For Me_3SnPH_2 it is reported that $^1J(^{117}Sn-^{31}P)$ is 443 ± 3 Hz and $^1J(^{119}Sn-^{31}P)$ is 463 ± 3 Hz.[105]

Full ^1H, ^{19}F, and ^{31}P n.m.r. data have been reported for $Me_nPH_{3-n}PF_5$ (n = 1—3), including the largest value for $^1J(^{31}P-^{31}P)$ so far measured of + 720 Hz.[106] It is interesting that the two isomers of MePhPPPhMe [(±) and *meso*] have different values of $^1J(^{31}P-^{31}P)$ of − 215 and − 234 Hz.[107] ^1H, ^{11}B, ^{19}F, and ^{31}P n.m.r. spectra have been reported for H_2PPF_2, $(CF_3)_2PPF_2$, and their BH_3 adducts. It is found that $|^1J(^{31}P-^{31}P)|$ increases on co-ordination to BH_3.[108]

The Karplus equation has been discussed. Simple expressions of the angular dependence of vicinal proton coupling constants lead to the prediction of certain invariants, such as $\frac{3}{2}N + \frac{1}{2}L$ in AA′XX′ spectra, which fail to remain constant as changes in the rotamer populations are induced by solvent and/or temperature. This appears to reflect the failure of the Karplus equation to predict accurately the form of the angular dependence in disubstituted ethanes.[109] However, a ratio method has been suggested, because although the magnitudes of the Karplus constants k_1 and k_2 vary, the ratio of k_1 to k_2 is constant.[110] A theoretical treatment of the stereochemical dependence of the vicinal proton–fluorine coupling constants has been given.[111]

An additive relationship for $^2J(^{31}P-N-^{31}P)$ in trimeric phosphonitriles has been proposed, but it predicts that (6) and (7) should have the same value for $^2J(^{31}P-^{31}P)$ although for (6) a value of 46 Hz is found and for (7) a value of 33 Hz. For (8) a value of greater than 100 Hz is predicted, yet a value of 62 Hz is found. The failure of the additive relationship is discussed.[112] The ^1H n.m.r. spectra of the complexes $RuCl_2(CO)_2Q_2$, *trans*-$M^1X(CO)Q_2$, $M^1Cl_3(CO)Q_2$, and $M^2X_2Q_2$ (Q = $PHBu^t_2$; X = halogen; M^1 = Rh or Ir; M^2 = Ni, Pd, or Pt) are of the $[AMX_{18}]_2$ type, and $^2J(^{31}P-^{31}P)$ was derived. The work includes the first measurement of $^2J(^{31}P-^{31}P)$ (*trans*) for Rh^I (*ca.* 330 Hz), Rh^{III} (495 Hz), Ir^I (315 Hz), Ir^{III} (390 Hz), and Ni^{II} (355 Hz). Duplication of ^1H signals for *trans*-MX_2Q_2 (M = Pd or Pt) is attributed to the rotamers (9) and (10).[113]

104 B. E. Mann, *Inorg. Nuclear Chem. Letters*, 1971, **7**, 595.
105 A. D. Norman, *J. Organometallic Chem.*, 1971, **28**, 81.
106 C. W. Schultz and R. W. Rudolph, *J. Amer. Chem. Soc.*, 1971, **93**, 1898.
107 H. C. E. McFarlane and W. McFarlane, *Chem. Comm.*, 1971, 1589.
108 H. W. Schiller and R. W. Rudolph, *Inorg. Chem.*, 1971, **10**, 2500.
109 E. B. Whipple, *J. Magn. Resonance*, 1971, **5**, 163.
110 K. N. Slessor and A. S. Tracey, *Canad. J. Chem.*, 1971, **49**, 2874.
111 G. Govil, *Mol. Phys.*, 1971, **21**, 953.
112 C. W. Allen, *J. Magn. Resonance*, 1971, **5**, 435.
113 A. Bright, B. E. Mann, C. Masters, B. L. Shaw, R. M. Slade, and R. E. Stainbank, *J. Chem. Soc.* (*A*), 1971, 1826.

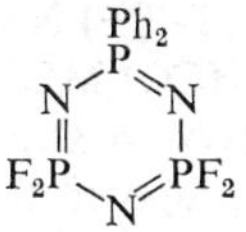
(6)

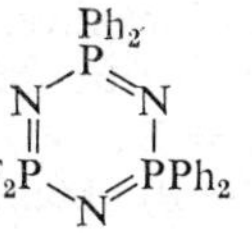
(7)

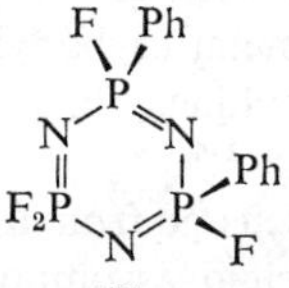
(8)

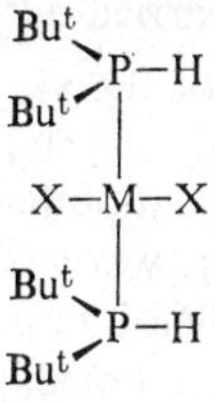
(9)

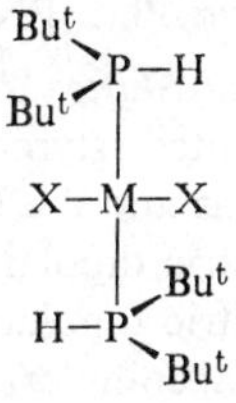
(10)

$^2J(^{31}P-^{31}P)$ has been measured in the ^{31}P n.m.r. spectra of $PtCl_2L^1L^2$ and $[PtClL^1{}_2L^2]^+$ (L^1 and L^2 = tertiary phosphine or phosphite). For these complexes $^2J(^{31}P-^{31}P)$ (*cis*) is 15—20 Hz, but *trans* is *ca.* 470 Hz (L^1 and L^2 are tertiary phosphines) or 715 Hz (L^1 = tertiary phosphine, L^2 = phosphite).[114] A series of complexes *cis*-$PtX_2\{P(OMe)_3\}_2$ (X = Cl, Br, I, NCO, N_3, or Ph) have been prepared and the ^1H n.m.r. spectra analysed as an $X_9AA'X'_9$ spin system; $^3J(^{31}P-^1H)$, $^5J(^{31}P-^1H)$, and $^2J(^{31}P-^{31}P)$ were derived.[115] Similarly, the deuterium-decoupled ^1H n.m.r. spectrum of *cis*-$(CH_3CH_2CD_2CH_2)_2Pt(PPh_3)_2$ has been analysed as an $AA'X_2X'_2$ spin system. However, there are marked differences between these results and those previously reported for *cis*-$Me_2Pt(PEt_3)_2$, *e.g.* $^2J(^{31}P-^{31}P)$ for *cis*-$Bu_2Pt(PPh_3)_2$ is 21 Hz whereas for *cis*-$Me_2Pt(PEt_3)_2$ it is *ca.* 2 Hz.[116] 'Tickling' has been carried out on *cis*-$(PH_3)_2Mo(CO)_4$ and *cis*-$(CCl_3PF_2)_2$-$Mo(CO)_4$, and it was found that $^2J(^{31}P-^{31}P)$ for both complexes is negative. Comparison with data for other Group VIa complexes leads to the conclusion that $^2J(^{31}P-^{31}P)$ is negative, for all complexes of the type *cis*-$L_2M(CO)_4$ (M = Cr, Mo, or W).[117] Consistent with these data is an analysis of the $[AX_n]_3$ spin system, and application to the ^{19}F n.m.r. spectra of *fac*-trisubstituted fluorophosphine complexes. It was shown that $^2J(^{31}P-^{31}P)$ increases as the electronegativity of the groups attached to the phosphorus

114 F. H. Allen and S. N. Sze, *J. Chem. Soc.* (*A*), 1971, 2054.
115 M. J. Church and M. J. Mays, *J. Inorg. Nuclear Chem.*, 1971, **33**, 253.
116 G. M. Whitesides and J. F. Gaasch, *J. Organometallic Chem.*, 1971, **33**, 241.
117 R. M. Lynden-Bell, J. F. Nixon, J. Roberts, J. R. Swain, and W. McFarlane, *Inorg. Nuclear Chem. Letters*, 1971, **7**, 1187.

atom increases and is negative, whereas $^4J(^{19}F-^{19}F)$ is positive.[118] The ^1H, ^{19}F, and ^{31}P n.m.r. spectra of $(CO)_3NiL$, $(CO)_5ML$, *cis*-$(CO)_4ML_2$, and *fac*-$(CO)_3MoL_3$ ($L = Bu^tPF_2$ or Bu^t_2PF; M = Cr, Mo, or W) were analysed and $^2J(^{31}P-^{31}P)$ obtained. It was proposed that $(J_C-J_L)/J_L$ (J_C = P–F or P–H coupling constant of the co-ordinated ligand, J_L = P–F or P–H coupling constant of the free ligand) should be used as a measure of the π-donor strength of the metal carbonyl fragments, $M(CO)_x$. The ^{31}P chemical shifts of the free phosphines, $Bu^t_nPF_{3-n}$ (n = 1—3) were discussed in terms of bond angle changes.[119]

$^2J(^{119}Sn-C-^1H)$ has been discussed from a theoretical basis, and attention was drawn to the fact that rigid assumptions of 5*s* character based on idealized hybrids are misleading.[120] A full analysis of the ^1H n.m.r. spectrum of *cis*- and *trans*-$Ph(S)PNEt_2$ has been carried out.[121]

^1H *and* ^{19}F *Chemical Shifts*. A comparison of the vinyl proton chemical shifts for 4-substituted styrenes, and 1-vinylbicyclo[2,2,2]octanes with CNDO/2 calculations and the Buckingham equation has demonstrated the existence of an electric dipolar field. Previous workers have assumed the existence of an electric dipolar field effect when interpreting the chemical shifts of certain metal complexes.[122] A linear relationship has been reported between τ (CH_2) for symmetric diethyldithiocarbamate complexes and the carbon–nitrogen stretching frequency for a wide range of metal complexes.[123] The correlation found between the ^1H n.m.r. shift of the cyclopentadienyl protons of $RC_5H_4Mn(CO)_3$ and the ^{55}Mn n.q.r. frequency has been explained in terms of π-electron density.[124] For complexes of MeNC to Mn^I and Fe^{II}, the order of ^1H chemical shifts found is $\delta_{ligand} < \delta_{Mn^I} < \delta_{Fe^{II}}$. This order correlates well with the increase in positive charge on the ligand as calculated by the SCCC–MO method.[125]

A reasonably good correlation has been found between the ^1H (methyl) chemical shift of (11) (Y = $SiMe_3$, $GeMe_3$, Me_2P, $AsMe_2$, or SMe) and the Hammett σ-constant; $^1J(^1H-^{13}C)$ was also reported.[126] However, the use of

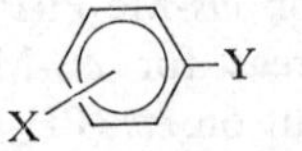

(11)

118 R. K. Harris, J. R. Woplin, and R. Schmutzler, *Ber. Bunsengesellschaft Phys. Chem.*, 1971, **75**, 134.
119 O. Stelzer and R. Schmutzler, *J. Chem. Soc. (A)*, 1971, 2867.
120 P. G. Perkins and D. H. Wall, *J. Chem. Soc. (A)*, 1971, 3620.
121 C. D. Flint, E. H. M. Ibrahim, R. A. Shaw, B. C. Smith, and C. P. Thakur, *J. Chem. Soc. (A)*, 1971, 3513.
122 G. K. Hamer and W. F. Reynolds, *Chem. Comm.*, 1971, 1218.
123 G. St. Nikolov, *Inorg. Nuclear Chem. Letters*, 1971, **7**, 1213.
124 T. B. Brill and G. G. Long, *Inorg. Chem.*, 1971, **10**, 74.
125 P. C. Fantucci, V. Valenti, and F. Cariati, *Inorg. Chim. Acta*, 1971, **5**, 425.
126 R. E. Hess, C. K. Haas, B. A. Kaduk, C. D. Schaeffer, jun., and C. H. Yoder, *Inorg. Chim. Acta*, 1971, **5**, 161.

^{19}F n.m.r. of *m*- and *p*-substituted fluoroaryls to determine substituent effects has been critically examined.[127] By use of this method, the electronic effect of the cyclopentadienyl anion has been determined from the ^{19}F n.m.r. spectra for the lithium salts of the (*m*- and *p*-fluorophenyl)cyclopentadienyl anions and their ferrocenyl derivatives. The values of σ_I and σ_R^0 obtained lead to the conclusion that in the cyclopentadienyl anion a considerable portion of the negative charge is localized in the σ-orbitals.[128]

A simple method of calculating the effects of geminal substituents upon the ^{19}F chemical shift of fluorine nuclei bonded to a sp^3 carbon has been presented,[129] and extended to include fluorine bonded to tin(IV).[130]

^{11}B, ^{13}C, ^{29}Si, ^{14}N, *and* ^{15}N *Chemical Shifts.* ^{11}B N.m.r. has been used extensively to determine the structure of boron compounds and will be discussed further in that section. The ^{11}B chemical shift and $^1J(^{11}B–^{19}F)$ for the boron halides, BXYZ (X, Y, or Z = F, Cl, Br, or I), have been measured. It was found that the values compare reasonably well with those predicted by various MO theories.[131] For the ions $[BWXYZ]^-$ and the CWXYZ molecules, there is a pairwise additive relationship for the substituent effects on ^{11}B and ^{13}C chemical shifts. The pairwise interaction parameters for ^{11}B chemical shifts in tetraco-ordinate boron compounds have been derived.[132]

The interpretation of ^{13}C chemical shifts has received attention. Additive relations have been found for ^{13}C chemical shifts of carbons polysubstituted by halogen, methyl, methoxy, or phenyl groups, after correction for the diamagnetic term. For the series Me_nX, the shift corrected for the diamagnetic term increases across the period of the ligand, but also decreases down the group of the ligand.[133] A semi-empirical MO calculation of the ^{13}C chemical shifts in alkanes has been given.[134] ^{13}C chemical shifts for molecules of the type $X'CX_3$ have been related to the XCX bond angle and z', the number of surrounding electrons at the ^{13}C nucleus. Using this approach, XCX bond angles were calculated from the known ^{13}C chemical shifts and satisfactory agreement with experimental bond angles was found.[135] A deuterium isotope effect of up to 1.36 p.p.m. (C_6H_{12} compared with C_6D_{12}) has been found in the ^{13}C n.m.r. spectra of some deuteriated molecules.[136]

[127] M. J. S. Dewar, R. Golden, and J. M. Harris, *J. Amer. Chem. Soc.*, 1971, **93**, 4187.
[128] A. A. Koridze, S. P. Gubin, A. A. Lubovich, B. A. Kvasov, and N. A. Ogorodnikova, *J. Organometallic Chem.*, 1971, **32**, 273.
[129] L. Phillips and V. Wray, *J. Chem. Soc.* (*B*), 1971, 2068.
[130] L. Phillips and V. Wray, *J. Chem. Soc.* (*B*), 1971, 2074.
[131] M. F. Lappert, M. R. Litzow, J. B. Pedley, and A. Tweedale, *J. Chem. Soc.* (*A*),1971, 2426.
[132] B. F. Spielvogel and J. M. Purser, *J. Amer. Chem. Soc.*, 1971, **93**, 4418.
[133] J. Mason, *J. Chem. Soc.* (*A*), 1971, 1038.
[134] V. N. Solkan, V. M. Mamayev, N. M. Sergeyev, and Yu. A. Ustynyuk, *Org. Magn. Resonance*, 1971, **3**, 567.
[135] D. Purdela, *J. Magn. Resonance*, 1971, **5**, 37.
[136] Yu. K. Grishin, N. M. Sergeyev, and Yu. A. Ustynyuk, *Mol. Phys.*, 1971, **22**, 711.

This year has shown a considerable increase in the application of ^{13}C n.m.r. to transition-metal complexes. A general survey of ^{13}C chemical shifts for carbon attached to transition metals has been carried out for the groups CO, CN^-, CNR, COMe, Me, σ-C_6H_5, C$\langle$ (carbene), π-C_2H_4, π-arene, and π-allyl. For the series (π-Cp)$Fe(CO)_{3-n}X_n$, the ^{13}C chemical shifts are significantly dependent on X. Also the observation of the ^{13}C chemical shift of the central carbon of the carbene ligand at + 362.3 p.p.m. (ref. TMS) for $(CO)_5CrC(OMe)Me$ is consistent with describing the ligand as a carbonium ion.[137] This observation has been endorsed by the observation of the C$\langle^{OMe}_{Me}$ ligand of *trans*-[PtMe{CMe(OMe)}$(AsMe_3)_2]^+$ at + 321.1 p.p.m.[100] The ^{13}C n.m.r. spectra of [PdX(π-allyl)]$_2$ have been reported. It was found the terminal allyl carbon chemical shift (*ca.* + 60 p.p.m.) lies approximately halfway between that of a $=CH_2$ group and that to be expected for a $-CH_2-Pd$ group. It is also interesting that the asymmetry in bonding of the π-allyl group of [PdCl(2-methylallyl)(PPh_3)] shown by an *X*-ray structure determination was also found in the ^{13}C n.m.r. spectra.[138] The application of ^{13}C n.m.r. to $Fe(CO)_5$, $Fe(CO)_4$-(PEt_nPh_{3-n}),[96] (arene)$M(CO)_3$ (M = Cr, Mo, or W),[97] $W(CO)_5L$,[98] and *cis*-$PtMe_2(XMe_2Ph)_2$ (X = P or As)[99] have already been discussed.

A number of new tin alkyls such as (12) have been prepared. As the ^{13}C chemical shift of the unsaturated carbons appear close to those of the free ligand it was concluded that there is no evidence for interaction between the tin and the double bond.[139] ^1H, ^{13}C, and, where relevant, ^{31}P and ^{19}F n.m.r. have been used to investigate some acetylenes containing Si, Ge, Sn, Pb, P, As, or Se. The chemical shifts were interpreted in terms of movement of electronic charge and π-bonding.[140] ^{13}C N.m.r. data have also been

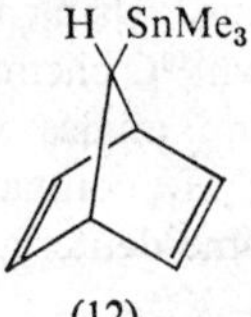

(12)

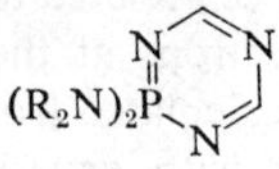

(13)

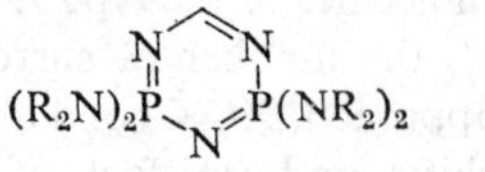

(14)

reported for the phosphorus compounds $(EtO)_2P(O)CH_2X$[141] and phosphorus–nitrogen compounds such as (13) and (14).[142]

It is interesting that $Sb(\overline{CHCH_2CH_2})_n$ (*n* = 3 or 5) show two ^{13}C n.m.r. resonances in the ratio 2 : 1 although the i.r. spectra show the co-ordination

[137] L. F. Farnell, E. W. Randall, and E. Rosenberg, *Chem. Comm.*, 1971, 1078.
[138] B. E. Mann, R. Pietropaulo, and B. L. Shaw, *Chem. Comm.*, 1971, 790.
[139] C. H. W. Jones, R. G. Jones, P. Partington, and R. M. G. Roberts, *J. Organometallic Chem.*, 1971, **32**, 201.
[140] D. Rosenberg and W. Drenth, *Tetrahedron*, 1971, **27**, 3893.
[141] G. A. Gray, *J. Amer. Chem. Soc.*, 1971, **93**, 2132.
[142] A. Schmidpeter and C. Weingand, *Angew Chem. Internat. Edn.*, 1971, **10**, 397.

at the antimony to be a square pyramid. It was therefore concluded that the molecule is fluxional.[143] ^{1}H, ^{13}C, and ^{29}Si chemical shifts of some methylsiloxanes have been measured. It appears that mono-, di, tri-, and tetra-functional silicon atoms in siloxanes exhibit very characteristic sharp and non-overlapping shift ranges, that are useful for both qualitative and quantitative analysis of polysiloxane mixtures.[144]

The ^{13}C chemical shielding anisotropy of solid CS_2, $CaCO_3$,[145] HCN in a liquid crystal,[146] and T_1 and/or T_2 for a number of organic compounds [147–152] have been measured.

Fourier transform spectroscopy has been applied to ^{15}N n.m.r. For the NH_4^+ ion a nuclear Overhauser enhancement of *ca.* – 2.4 has been observed.[153] However, attention has been drawn to potential dangers associated with this *negative* enhancement. For the NH_4^+ ion as the pH is changed, the protons are decoupled from the nitrogen by exchange, thus removing the Overhauser enhancement. Under conditions of exchange which reduces the enhancement to *ca.* – 1, the ^{15}N signal completely vanishes.[154] A further problem associated with ^{15}N n.m.r. is that T_1 is very long (1150 s for PhCN) making saturation of the resonances easy and requiring a long pulse separation for Fourier transform spectroscopy.[155]

^{14}N N.m.r. has been applied to a number of thiocyanate complexes. For some N-bonded complexes it was found that the chemical shifts only cover 30 p.p.m. of the complete range of 1000 p.p.m. found for ^{14}N.[156] However, it has been reported that the N-bonded thiocyanate resonance appears to high field of the free SCN^- ion while the S-bonded $[Hg(SCN)_4]^{2-}$ ion gives a ^{14}N resonance to high field of the free SCN^- ion. This work offers a potential method to discriminate between N- and S-bonded thiocyanate.[157] Only one ^{14}N n.m.r. signal was observed for S_4N_2 which is consistent with (15).[158] ^{14}N N.m.r. chemical shifts of hyponitrous acid and its anion have been reported and discussed.[159]

143 A. H. Cowley, J. L. Mills, T. M. Loehr, and T. V. Long, tert., *J. Amer. Chem. Soc.*, 1971, **93**, 2150.
144 G. Engelhardt, H. Jancke, M. Magi, T. Pehk, and E. Lippmaa, *J. Organometallic Chem.*, 1971, **28**, 293.
145 A. Pines, W.-K. Rhim, and J. S. Waugh, *J. Chem. Phys.*, 1971, **54**, 5438.
146 F. Millet and B. P. Dailey, *J. Chem. Phys.*, 1971, **54**, 5434.
147 T. D. Alger and D. M. Grant, *J. Phys. Chem.*, 1971, **75**, 2538.
148 J. R. Lyerla, jun., D. M. Grant, and C. H. Wang, *J. Chem. Phys.*, 1971, **55**, 4676.
149 H. W. Spiess, D. Schweitzer, U. Haeberlen, and K. H. Hausser, *J. Magn. Resonance*, 1971, **5**, 101.
150 C. F. Schmidt, jun., and S. I. Chan, *J. Chem. Phys.*, 1971, **55**, 4670.
151 O. Olivson and E. Lippmaa, *Chem. Phys. Letters*, 1971, **11**, 241.
152 T. D. Alger, S. W. Collins, and D. M. Grant, *J. Chem. Phys.*, 1971, **54**, 2820.
153 J. M. Briggs, L. F. Farnell, and E. W. Randall, *Chem. Comm.*, 1971, 680.
154 R. L. Lichter and J. D. Roberts, *J. Amer. Chem. Soc.*, 1971, **93**, 3200.
155 E. Lippmaa, T. Saluvere, and S. Laisaar, *Chem. Phys. Letters*, 1971, **11**, 120.
156 K. M. Mackay and S. R. Stobart, *Spectrochim. Acta*, 1971, **27A**, 923.
157 H. Bohland and E. Mühle, *Z. anorg. Chem.*, 1970, **379**, 273.
158 J. Nelson and H. G. Heal, *J. Chem. Soc.* (*A*), 1971, 136.
159 J. Mason and W. van Bronswijk, *J. Chem. Soc.* (*A*), 1971, 791.

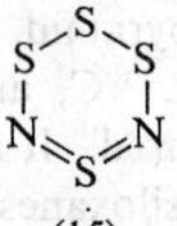

(15)

^{31}P *Chemical Shifts.* The ^{31}P chemical shifts of PX_3 and YPX_3 molecules have been interpreted in terms of the XPX bond angles and the number of surrounding electrons at the ^{31}P nucleus. Satisfactory agreement was found between XPX bond angles calculated from the ^{31}P chemical shift and observed angles.[160] For the ^{31}P chemical shifts of diethyl phenylphosphonates, a linear correlation has been found with the Hammett σ-constants but of opposite sign to that expected on the basis of the electron-withdrawing ability of the substituent.[161] The ^{31}P chemical shifts of $R_xP(CN)_{3-x}$ [162] and 79 phosphonium salts [163] have been reported and discussed.

Linear relationships have been found for the ^{31}P chemical shifts of tertiary phosphines attached to metals. Thus for (π-Cp)Ni(PPh_3)X, a linear relationship was found between the ^{31}P chemical shifts and the wavelength of the first *d–d* transition.[164] For the series of complexes MeL^+, EtL^+, PhL^+, $M^1(CO)_5L$, *cis*-$Ru(CO)_2Cl_2L_2$, *trans*-$M^2Cl(CO)L_2$, *mer*-$M^2Cl_3(CO)L$, *mer*-$M^2Cl_3L_3$, and *cis*- and *trans*-$M^3Cl_2L_2$ (M^1 = Cr, Mo, or W; M^2 = Rh or Ir; M^3 = Pd or Pt; L = tertiary phosphine), a linear relationship between the ^{31}P chemical shift of the free tertiary phosphine and the change in ^{31}P chemical shift on complexation has been reported in a brief note.[165] Full details have been published for the series *trans*-$RhCl(CO)L_2$ [166] and *cis*- and *trans*-$PdCl_2L_2$.[167] ^{31}P N.m.r. spectroscopy has been used to characterize two chlorine-bridged products from the reaction of $RhCl_3,3H_2O$ with PBu^n_3. One compound which shows two equal 1 : 1 doublets is assigned structure (16), while the other shows two 1 : 1 doublets in the ratio 1 : 2 is assigned the structure (17).[168]

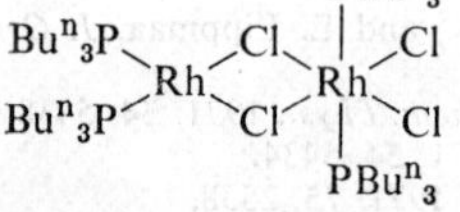

(16)

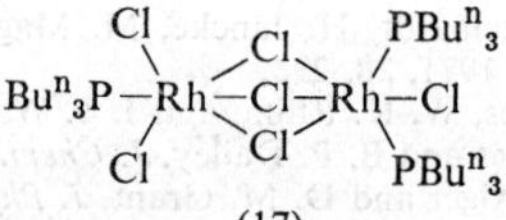

(17)

[160] D. Purdela, *J. Magn. Resonance*, 1971, **5**, 23.
[161] C. C. Mitsch, L. D. Freedman, and C. G. Moreland, *J. Magn. Resonance*, 1971, **5**, 140.
[162] C. E. Jones and K. J. Coskran, *Inorg. Chem.*, 1971, **10**, 1536.
[163] S. O. Grim, E. F. Davidoff, and T. J. Mark, *Z. Naturforsch.*, 1971, **26b**, 184.
[164] J. Thomson, D. Groves, and M. C. Baird, *J. Magn. Resonance*, 1971, **5**, 281.
[165] B. E. Mann, C. Masters, B. L. Shaw, R. M. Slade, and R. E. Stainbank, *Inorg. Nuclear Chem. Letters*, 1971, **7**, 881.
[166] B. E. Mann, C. Masters, and B. L. Shaw, *J. Chem. Soc. (A)*, 1971, 1104.
[167] B. E. Mann, B. L. Shaw, and R. M. Slade, *J. Chem. Soc. (A)*, 1971, 2976.
[168] F. H. Allen and K. M. Gabuji, *Inorg. Nuclear Chem. Letters*, 1971, **7**, 833.

Solvent effects have been investigated for the ^{31}P chemical shift of white phosphorus [169, 170] and P_4S_3 with shifts of up to 35 p.p.m. being observed. In the case of P_4S_3 a linear relationship was found between the chemical shift and $(n_D^2 - 1)/(2n_D^2 + 1)$ for the solvent.[171] Deuteriation of $(MeO)_2P(O)H$ to give $(MeO)_2P(O)D$ produces a ^{31}P chemical shift of up to 1.2 p.p.m. and the ratio of $^1J(^{31}P-^1H) : ^1J(^{31}P-^2H)$ is variable by 1%.[172] Similar deuterium isotope effects of up to 2.39 p.p.m. have been reported for the ^{31}P n.m.r. spectra of compounds such as $PhPH_2$ and $PhPD_2$.[173] T_1 and/or T_2 have been measured for ^{31}P in a number of compounds.[174–177]

Metal Chemical Shifts. In this subsection metal resonances are considered in the order of increasing atomic weight.

The nature of the Mn—Sn bond has been studied by ^{55}Mn n.m.r. and i.r. spectroscopies for the series of compounds $R_{3-x}X_xSn—Mn(CO)_3$, where R = Me or Ph and X = Cl or Br. Molecular orbital considerations have shown that the chemical shift of the ^{55}Mn resonance is a measure of the σ-polarity of the $L—Mn(CO)_5$ bond and that the linewidth is determined by π-bonding. This π-bonding appears to be most pronounced for $Br_3SnMn(CO)_5$.[178] The series of complexes $R_{3-n}X_nSnMn(CO)_5$ has been studied by ^{119}Sn Mössbauer, ^1H n.m.r., and ^{55}Mn n.m.r. spectroscopy; it was concluded that $Mn(CO)_5$ is a better electron donor than methyl. A relationship was reported between the ^{119}Sn isomer shift, the ^1H chemical shift, and $^2J(^{119}Sn-C-^1H)$.[179]

The well-established relationship between λ_{max} for the first *d–d* transition and the ^{59}Co chemical shifts has been tested for a series of complexes of the type $[Co^{III}(NH_3)_x(CN)_yX_z]^{n+}$ $(x + y + z = 6)$, where agreement is found.[180] However, for the series $[Co(NH_3)_{6-n}(C_2O_4)_n]^{3-2n}$ $(n = 0—3)$ the complex $[Co(C_2O_4)_3]^{3-}$ falls off the line, but this complex is paramagnetic (μ_{eff} = 1.35 BM).[181] The pH dependence of the ^{59}Co linewidth has been used to follow the equilibrium:

$$[Co(NH_3)_5(H_2O)]^{3+} \underset{H^+}{\overset{OH^-}{\rightleftharpoons}} [Co(NH_3)_5(OH)]^{2+}$$

The hydroxy-form has a linewidth of *ca.* 0.2 G whereas the aquo-form has a

169 G. Krabbes and G. Grossmann, *Z. Chem.*, 1971, **11**, 270.
170 G. Heckmann and E. Fluck, *Z. Naturforsch.*, 1971, **26b**, 282.
171 G. Heckmann and E. Fluck, *Z. Naturforsch.*, 1971, **26b**, 982.
172 W. J. Stec, N. Goddard, and J. R. Van Wazer, *J. Phys. Chem.*, 1971, **75**, 3547.
173 A. A. Borisenko, N. M. Sergeyev, and Yu. A. Ustynyuk, *Mol. Phys.*, 1971, **22**, 715.
174 S. J. Seymour and J. Jonas, *J. Chem. Phys.*, 1971, **54**, 487.
175 D. W. Sawyer and J. G. Powles, *Mol. Phys.*, 1971, **21**, 83.
176 Nan-I Liu and J. Jonas, *J. Chem. Phys.*, 1971, **55**, 463.
177 S. W. Dale and M. E. Hobbs, *J. Phys. Chem.*, 1971, **75**, 3537.
178 S. Onaka, T. Miyamoto and Y. Sasaki, *Bull. Chem. Soc. Japan*, 1971, **44**, 1851.
179 S. Onaka, Y. Sasaki, and H. Sano, *Bull. Chem. Soc. Japan*, 1971, **44**, 726.
180 N. S. Biradar and M. A. Pujar, *Inorg. Nuclear Chem. Letters*, 1971, **7**, 269.
181 N. S. Biradar and M. A. Pujar, *Z. anorg. Chem.*, 1970, **379**, 88.

linewidth of *ca.* 5 G.[182] T_1 and T_2 measurements have been made for both ^{59}Co and ^{14}N for the complexes $M_3Co(CN)_6$, $Co(NH_3)_6Cl_3$, and $Co(en)_3Cl_3$. The electric field gradient at the cobalt was attributed to ionic association.[183]

^{93}Nb N.m.r. has proved useful to demonstrate the presence of various species in solutions. Thus coupled with ^{19}F n.m.r. it was shown that NbF_5 in EtOH–MeCN contains some $[NbF_4(OEt)(MeCN)]$.[184] Similarly, in HF solution the presence of $[NbOF_5]^{2-}$, $[NbF_6]^-$, and $[NbF_7]^{2-}$ has been demonstrated.[185] The ^{93}Nb n.m.r. spectra of mixtures of $NbCl_5$ and $NbBr_5$ show the presence of all seven species of the type $[NbCl_nBr_{6-n}]^-$ with a statistical distribution. For ^{13}C chemical shifts (for $^{13}CCl_nBr_{4-n}$) and ^{119}Sn chemical shifts (for $^{119}SnCl_nBr_{4-n}$) it has been found that as the number of bromine atoms is increased the signal of the central atom moves to high field. In contrast, for the ^{93}Nb complexes the movement is to low field.[186]

The primary isotope effect has been investigated for the two tin nuclei, ^{117}Sn and ^{119}Sn in the complexes R_nSnCl_{4-n} (R = Me or Et), $(Et_3Sn)_2O$, and $(C_5H_5)_4Sn$; an effect of up to 2.9 p.p.m. was found.[187] ^{119}Sn and ^{13}C n.m.r. spectra of $Me_nSn(SMe)_{4-n}$ have been presented. The shifts appear to be governed by inductive and magnetic anisotropy effects.[188] The ^{119}Sn chemical shifts of $XBu^n_2SnOSnBu^n_2X$ (X = F, Cl, Br, NCS, $OSiMe_4$, OCOMe, OPh, *p*-MeC_6H_4O, *p*-$MeOC_6H_4O$, or *p*-ClC_6H_4O) have been reported. In the cases of X = Cl or Br, two ^{119}Sn signals were observed and the structure (18) was postulated. In all other cases only a broad ^{119}Sn was observed.[189] ^{119}Sn chemical shifts in 40 organotin compounds have been obtained by $^1H\{^{119}Sn\}$ INDOR. The ^{119}Sn shifts for organotin compounds where tin is bonded to hydrogen, lithium, or phosphorus atoms

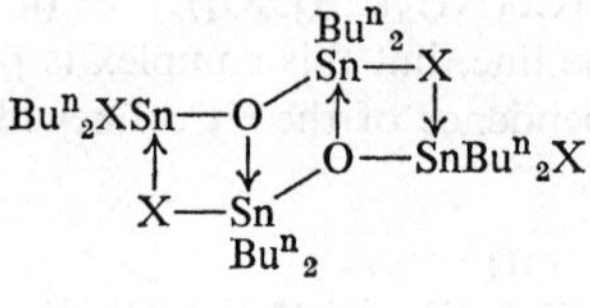

(18)

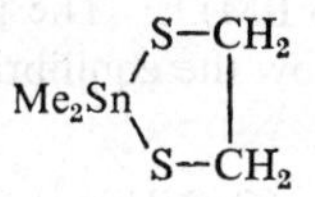

(19)

were reported, along with those for the organotin acetylene, perfluoroalkyl, and perchlorophenyl derivatives for the first time. The most downfield tin shift thus far reported has been measured for (19).[190] Similarly, other tin

182 F. Yajima, A. Yamasaki, and S. Fujiwara, *Inorg. Chem.*, 1971, **10**, 2350.
183 R. Ader and A. Loewenstein, *J. Magn. Resonance*, 1971, **5**, 248.
184 H. Köppel, M. Schönherr, and L. Kolditz, *Z. Chem.*, 1971, **11**, 28.
185 J. A. S. Howell and K. C. Moss, *J. Chem. Soc.* (*A*), 1971, 2481.
186 Yu. A. Buslaev, V. O. Kopanev, and V. P. Tarasov, *Chem. Comm.*, 1971, 1175.
187 A. P. Tupčiauskas, N. M. Sergeyev, and Yu. A. Ustynyuk, *Mol. Phys.*, 1971, **21**, 179.
188 E. V. Van den Berghe and G. P. Van der Kelen, *J. Organometallic Chem.*, 1971, **26**, 207.
189 A. G. Davies, L. Smith, P. J. Smith, and W. McFarlane, *J. Organometallic Chem.*, 1971, **29**, 245.
190 P. G. Harrison, S. E. Ulrich, and J. J. Zuckerman, *J. Amer. Chem. Soc.*, 1971, **93**, 5398.

complexes have been investigated by $^1H\{^{119}Sn\}$ INDOR and the chemical shifts interpreted in terms of the electronegativity of the substituents.[191]

^{183}W and ^{199}Hg chemical shifts have been measured by INDOR. It was found that for *cis*- and *trans*-$W(CO)_4(PBu^n{}_3)_2$ the ^{183}W chemical shifts with respect to WF_6 are + 1969 p.p.m. and + 2021 p.p.m., which are very large.[192] For the complexes $WF_n(OR)_{6-n}$, $WOF_4(OMe)_2$, WOF_4,[$OP(OMe)_2$-Me], WOF_4,[$OS(OMe)_2$], $[WOF_5]^-$, and $[F_4OWFWOF_4]^-$ a qualitative correlation between the ^{183}W chemical shift and ΔE has been reported. The ^{19}F chemical shifts are given for the series $WF_{6-n}(OMe)_n$ by the relationship

$$\delta(F) = 45c + 68t$$

where δ(F) is the ^{19}F chemical shift with respect to WF_6, and c and t are the number of MeO groups *cis* and *trans* to that particular fluorine. $^1J(^{183}W–^{19}F)$ was also discussed.[193] The ^1H and ^{19}F n.m.r. spectra of $(C_2F_3)_2Hg$ and $Me(C_2F_3)Hg$ have been fully analysed and the ^{199}Hg chemical shifts determined. $Me(C_2F_3)Hg$ is at 482 ± 3 p.p.m. and $(C_2F_3)_2Hg$ at 957 p.p.m., both with respect to Me_2Hg.[194]

2 Stereochemistry

This section is subdivided into seven parts, which contain n.m.r. information about magnesium and transition-metal complexes, presented by groups according to the Periodic Table. Within each group categorization is by ligand type. As far as possible cross-references are given at the beginning of each subgroup to compounds discussed elsewhere in this chapter. In this cross-referencing it has not proved possible within the space available to include the many compounds that occur within the sections on dynamic systems, paramagnetic systems, and solid-state n.m.r. Thus many more compounds of relevance to this chapter appear in these sections and in the tables at the end of the chapter.

Complexes of Mg, V, Ti, Zr, Hf, Nb, and Ta.—Information concerning complexes of these elements can be found at the following sources: $(Cp)_2TiPh_2$,[137] $[VOF_4]^-$,[101] $[VF_6]^-$,[102] $[NbF_4(solvent)_2]^+$,[184] $[NbOF_5]^{2-}$, $[NbF_6]^-$, $[NbF_7]^{2-}$,[185] and $[NbCl_nBr_{6-n}]^-$.[186]

^1H N.m.r. has been used to show that $CH_2{=}CH(CH_2)_2MgCl$ rearranges to (20),[195] and that $(Cp)_2Mg,2THF$ does indeed contain THF.[196]

The reaction of uranyl nitrate with acetylacetone in the presence of ammonia or amines has been reinvestigated, and the product reformulated as (21). The bridging hydrogen was found at low field, δ *ca.* 11, and like the

191 A. P. Tupčiauskas, N. M. Sergeyev, and Yu. A. Ustynyuk, *Org. Magn. Resonance*, 1971, **3**, 655.
192 P. J. Green and T. H. Brown, *Inorg. Chem.*, 1971, **10**, 206.
193 W. McFarlane, A. M. Noble, and J. M. Winfield, *J. Chem. Soc.* (*A*), 1971, 948.
194 R. B. Johannesen and R. W. Duerst, *J. Magn. Resonance*, 1971, **5**, 355.
195 A. Maercker and R. Geuss, *Angew. Chem. Internat. Edn.*, 1970, **9**, 909.
196 T. Saito, *Chem. Comm.*, 1971, 1422.

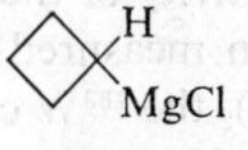

(20)

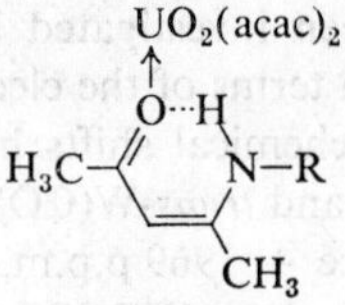

(21)

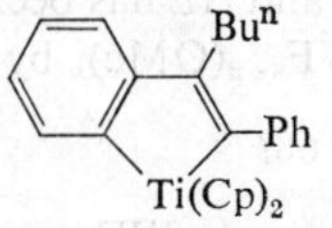

(22)

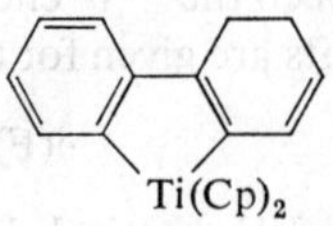

(23)

central proton of the acetylacetonate is exchangeable with D_2O.[197] The ^1H n.m.r. spectra of a series of uranyl β-diketone complexes with pyridine *N*-oxide have been reported.[198]

The ^1H n.m.r. spectrum of the titanium complex (22) shows a unique aromatic proton at τ 4.02 and the α-CH_2 of the n-butyl group is at τ 8.07. These upfield movements were attributed to localized anisotropic fields produced by the π-cyclopentadienyl rings. The aromatic protons in (23) give rise to an ABCD pattern and the cyclopentadienyl protons are equivalent.[199] The ^1H chemical shifts of the cyclopentadienyl protons (*ca.* τ 4.8) in $(\pi\text{-Cp})_2Ti(SR)_2Mo(CO)_4$ compared to those in $(\pi\text{-Cp})_2Ti(SR)_2$ (*ca.* τ 4.2) has been taken as indicating an increase of electron density on the titanium which was attributed to back-bonding from the molybdenum to the titanium. This conclusion was reinforced by i.r. data and an *X*-ray crystal structure.[200] Benzylmagnesium chloride reacts with MCl_4 (M = Ti or Zr) to yield $M(CH_2Ph)_4$ which reacts with HX (X = F, Cl, or Br) to yield $M(CH_2Ph)_3X$, but ethanol yields $Ti(CH_2Ph)_2(OEt)_2$. ^1H N.m.r. was used to show that $Ti(CH_2Ph)_2(OEt)_2$ dimerizes at high concentrations.[201] The cyclopentadienyl resonance of $(Cp)_2MX_2$ (M = Ti, Zr, or Hf; X = Cl or C_6F_5) when X = C_6F_5 consists of a triplet which was explained as arising from long-range ^{19}F–^1H coupling.[202] ^1H N.m.r. has been used to help to characterize the product(s) formed when SO_2 inserts into titanium– or zirconium–alkyl bonds to form compounds such as $(Cp)_2Ti(O_2SMe)_2$. When $(Cp)_2TiMeCl$ is used, two products are obtained which are thought to be the *OO'*- and *O*-sulphinites.[203]

[197] J. M. Haigh and D. A. Thornton, *J. Inorg. Nuclear Chem.*, 1971, **33**, 1787.
[198] M. S. Subramanian and V. K. Manchanda, *J. Inorg. Nuclear Chem.*, 1971, **33**, 3001.
[199] M. R. Rausch and L. P. Klemann, *Chem. Comm.*, 1971, 354.
[200] T. S. Cameron, C. K. Prout, G. V. Rees, M. L. H. Green, K. K. Joshi, G. R. Davies, B. T. Kilbourn, P. S. Braterman, and V. A. Wilson, *Chem. Comm.*, 1971, 14.
[201] U. Zucchini, E. Albizzati, and U. Giannini, *J. Organometallic Chem.*, 1971, **26**, 357.
[202] M. D. Rausch, H. B. Gordon, and E. Samuel, *J. Co-ordination Chem.*, 1971, **1**, 141.
[203] P. C. Wailes, H. Weigold, and A. P. Bell, *J. Organometallic Chem.*, 1971, **33**, 181.

The invariance of the ^{1}H n.m.r. spectrum of (Cp)MX(ox) (M = Ti, Zr, or Hf; X = Cl, Br, or Et; oxH = quinolin-8-ol) over the temperature range + 40 to − 70 °C has been taken as evidence that the cyclopentadienyl ring is π-bonded.[204] ^{1}H N.m.r. has been used to demonstrate that a Zr—H bond in $(Cp)_2ZrHCl$ and $(Cp)_2ZrH_2$ generally adds *cis* across an acetylene. However, some unusual products such as (24) and (25) (R = H or Me)

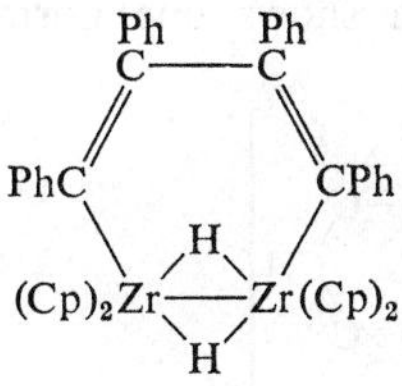

(24)

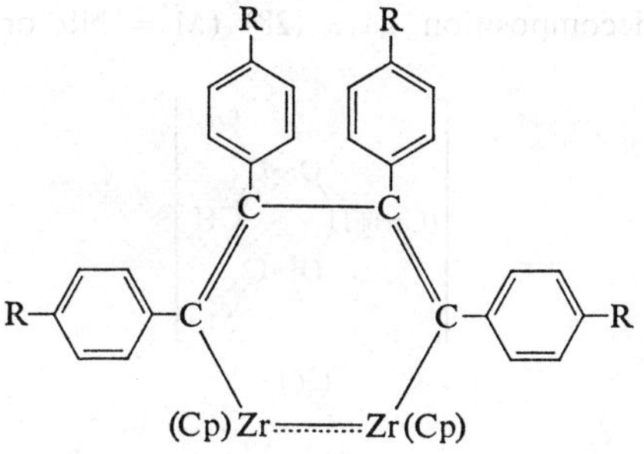

(25)

were obtained.[205] The naphthalene in $\{[(Cp)Zr]_2O\}_2,C_{10}H_8$ cannot be removed by sublimation at 110 °C, yet the ^{1}H n.m.r. spectrum shows the identical spectrum to free naphthalene, with a broad cyclopentadienyl resonance suggesting the possibility that the compound is paramagnetic.[206] The co-ordination of ligands to $MeTiCl_3$ has been investigated by ^{1}H n.m.r., u.v., visible, and i.r. spectroscopy. ^{1}H N.m.r. shows that 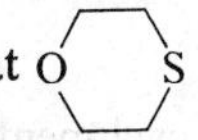

co-ordinates through the sulphur, as the CH_2 next to the sulphur moves 0.31 p.p.m. while that next to oxygen moves only 0.13 p.p.m. on co-ordination.[207]

From the observation that the ^{1}H n.m.r. spectrum of *o*-allylaniline in the presence of $TiCl_4$ over the temperature range − 60 to + 60 °C showed no changes in the *o*-allyl group of the aniline, apart from a small low-field shift of the allyl ^{1}H resonances, compared to the free aniline, it was concluded that the *o*-allyl group is not co-ordinated to the titanium.[208] The ^{1}H n.m.r. of the 2 : 1 adducts of di-isopropyl methylphosphonate (DIMP) with Sn^{II}, Sn^{IV}, and Ti^{IV} halides showed no extra structure compared with free DIMP. This was taken as evidence for the *trans* structure.[209] The so-called five-co-ordinate titanium(IV) acetylacetonates TiX_3(acac) (X = alkoxy or halide) have been shown to be a mixture of well-established

[204] J. Charalambous, M. J. Frazer, and W. E. Newton, *J. Chem. Soc.* (*A*), 1971, 2487.
[205] P. C. Wailes, H. Weigold, and A. P. Bell, *J. Organometallic Chem.*, 1971, **27**, 373.
[206] P. C. Wailes and H. Weigold, *J. Organometallic Chem.*, 1971, **28**, 91.
[207] G. W. A. Fowles, D. A. Rice, and J. D. Wilkins, *J. Chem. Soc.* (*A*), 1971, 1920.
[208] D. A. Baldwin and R. J. H. Clark, *J. Chem. Soc.* (*A*), 1971, 1725.
[209] C. Owens, N. M. Karayannis, L. L. Pytlewski, and M. M. Labes, *J. Phys. Chem.*, 1971, **75**, 637.

four- and six-co-ordinate derivatives:[210]

$$TiX_4 + Ti(acac)_2X_2 \rightleftharpoons Ti(acac)_2X_2TiX_4$$

Some cationic complexes of the $[(Cp)_2Ti(ethylmalonate)]^+$ have been prepared and shown by ^1H n.m.r. to have structure (26) rather than (27).[211]

The ^1H n.m.r. spectrum of $(Cp)_2NbH_3$ shows the presence of two different hydride signals, as has been observed for $(Cp)_2TaH_3$. Thermal decomposition gave (28) (M = Nb or Ta) which shows one normal

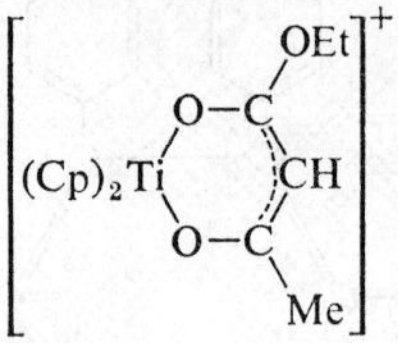

(26)

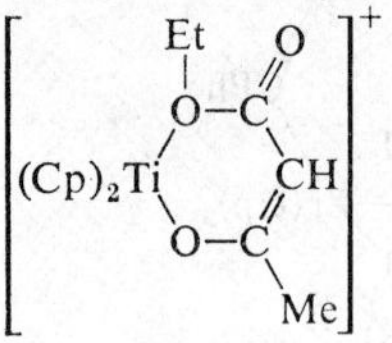

(27)

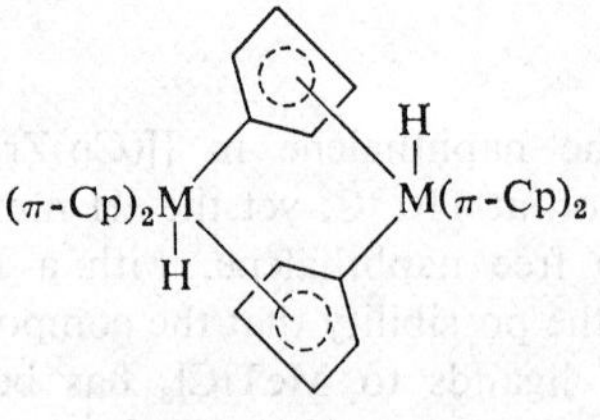

(28)

cyclopentadienyl ^1H n.m.r. resonance and a complex signal.[212] When $(\pi\text{-allyl})_4Ta$ reacts with PF_3 to yield $(\pi\text{-}C_3H_5)Ta(PF_3)_5$, the ^1H n.m.r. shows three signals in the ratio 1 : 2 : 2, confirming the π-allyl structure of the C_3H_5 ligand.[213] The addition of HF to aqueous solutions of $NbCl_5$ or $TaCl_5$ in acetonitrile produces nine species. ^{19}F N.m.r. was used to identify $[MFCl_5]^-$, *cis*- and *trans*-$[MF_2Cl_4]^-$, *fac*- and *mer*-$[MF_3Cl_3]^-$, *cis*- and *trans*-$[MF_4Cl_2]^-$, $[MF_5Cl]^-$, and $[MF_6]^-$.[214] N.m.r. has been used to determine the stability constant between niobium(v) and tantalum(v) halides and various Lewis bases such as Et_2O and nitriles.[215, 216]

Complexes of Cr, Mo, and W.—Information concerning complexes of these elements can be found at the following sources: (arene)$M(CO)_3$, (cycloheptatriene)$M(CO)_3$ (M = Cr, Mo, or W),[97, 137] $(CO)_5ML$, *cis*-$(CO)_4ML_2$,

[210] C. E. Holloway and A. E. Sentek, *Canad. J. Chem.*, 1971, **49**, 519.
[211] D. A. White, *J. Inorg. Nuclear Chem.*, 1971, **33**, 691.
[212] F. N. Tebbe and G. W. Parshall, *J. Amer. Chem. Soc.*, 1971, **93**, 3793.
[213] T. Kruck and H.-U. Hempel, *Angew. Chem. Internat. Edn.*, 1971, **10**, 408.
[214] Yu. A. Buslaev, E. G. Il'in, S. V. Bainova, and M. N. Krutkina, *Doklady Phys. Chem.*, 1971, **196**, 29.
[215] A. Merbach and J. C. Bünzli, *Helv. Chim. Acta*, 1971, **54**, 2536.
[216] A. Merbach and J. C. Bünzli, *Helv. Chim. Acta*, 1971, **54**, 2543.

fac-$(CO)_3ML_3$ (M = Cr, Mo, or W; L = $Bu^t{}_2PF$ or Bu^tPF_2),[119] $M(CO)_5L$ (M = Cr, Mo, or W; L = tertiary phosphine),[165] $Cr(CO)_5CMe(OMe)$, $Mo(CO)_5P(OPr^i)_3$, $(CO)_3(C_5H_5)WMe$,[137] *cis*-$(PH_3)_2Mo(CO)_4$, *cis*-$(CCl_3PF_2)_2Mo(CO)_4$,[117] *fac*-$(RPF_2)_3Mo(CO)_3$, *fac*-$(R_2PF)_3Mo(CO)_3$,[118] $(Cp)_2Ti(SR)_2Mo(CO)_4$,[200] $(Cp)Mo(CO)_2(CHNMe)_2BH_2$,[259] $Mo(CO)_4\{Ph_2AsC_5H_4)_2Fe\}$,[309] $(Cp)Mo(CO)_3CH_2Ph$,[312] $(Cp)Mo(CO)_3CMeCH_2[C(CN)_2]_2CH_2$,[319] $W(CO)_5L$,[98] *cis*- and *trans*-$W(CO)_4(PBu^n{}_3)_2$,[192] $WF_n(OR)_{6-n}$,[193] and $RCH{=}CHW(CO)_2(Cp)(\mu\text{-}CO)Fe(CO)_3$.[302]

The ^1H n.m.r. spectrum (hydride) for $(PEtPh_2)_4MoH_4$ shows a 1 : 4 : 6 : 4 : 1 quintet at τ 12.20; this pattern could be due to fluxional behaviour.[217] $(C_6H_6)_2Mo$ reacts with tertiary phosphines to yield $(C_6H_6)Mo(PR_3)_3$. This compound protonated in acid to give a hydride which gave a quartet at τ 12.3 and integrated against the C_6H_6 as 3 : 1; the ^{31}P n.m.r. spectrum showed a triplet structure. It was therefore concluded that the compound is a dihydride $[(C_6H_6)Mo(PR_3)_3H_2]^{2+}$ and that the compound is fluxional. The ^1H spectrum is still a quartet at − 96 °C.[218, 219] The compound $(C_6H_5CH_3)Mo(PPh_3)_2H_2$ shows a triplet for the ^1H (hydride) n.m.r. spectrum, which was interpreted as due to either a *trans* configuration or a *cis* fluxional configuration.[220] Treatment of $(Cp)_2MH_2$ (M = Mo or W) with AlR_3 produced 1 : 1 complexes $(Cp)MH_2AlR_3$, which show Cp–W–H long-range coupling of 0.5—0.7 Hz.[221] Similarly, treatment of R_2AlH with $HW(CO)_3(Cp)$ yields $R_2AlW(CO)_3(Cp)$. There was only one R and one Cp resonance observed, even at − 65 °C, but the compound was shown to be dimeric by molecular weight measurements; four structures [(29)—(32)] were considered to be possible.[222] The compounds $H_3M_2(CO)_6(OH)_3$ (M = Mo or W) have been reformulated on the basis of ^1H n.m.r. and i.r. evidence as $[(HO)M(CO)_3H_4],4H_2O$.[223] The ^1H n.m.r. spectrum of

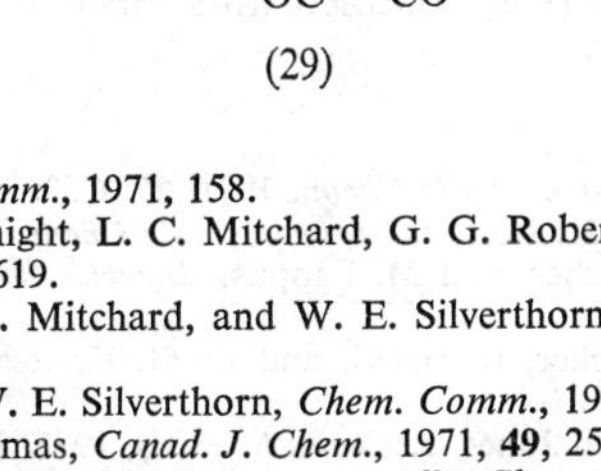

(29)

[217] F. Pennella, *Chem. Comm.*, 1971, 158.
[218] M. L. H. Green, J. Knight, L. C. Mitchard, G. G. Roberts, and W. E. Silverthorn, *Chem. Comm.*, 1971, 1619.
[219] M. L. H. Green, L. C. Mitchard, and W. E. Silverthorn, *J. Chem. Soc.* (*A*), 1971, 2929.
[220] M. L. H. Green and W. E. Silverthorn, *Chem. Comm.*, 1971, 557.
[221] A. Storr and B. S. Thomas, *Canad. J. Chem.*, 1971, **49**, 2504.
[222] R. R. Schrieke and J. D. Smith, *J. Organometallic Chem.*, 1971, **31**, C46.
[223] U. Sartorelli, L. Garlaschelli, G. Ciani, and G. Bonora, *Inorg. Chim. Acta*, **1971**, **5**, 191.

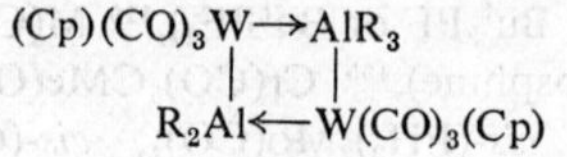

(Cp)(CO)₃W→AlR₃   (Cp)(CO)₂W—CO→AlR₂
    | | | |
R₂Al←W(CO)₃(Cp)   R₂Al←OC—W(CO)₂(Cp)
        (30)   (31)

$$(Cp)(CO)_3W\diagdown_R \diagup^R \diagdown_R \diagup^R$$
$$R\diagup Al \diagdown_R \diagup Al \diagdown W(CO)_3(Cp)$$
(32)

(Cp)W(CO)₃H has been measured in the nematic phase at 220 MHz. The H—W—Cp angle was found to be 53 ± 3°.[224]

Treatment of $[M_2(CO)_{10}]^{2-}$ (M = Cr, Mo, or W) with AgC_4F_7 yields $[M(CO)_5(C_4F_7)]^-$. The ¹⁹F n.m.r. spectrum shows three resonances in the ratio 3:3:1, consistent with (33).[225] Metal–carbene compounds have

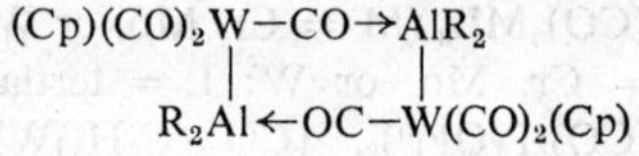

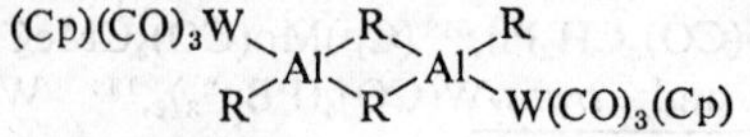

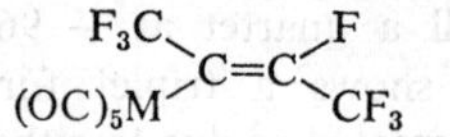

$$F_3C\diagdown\diagup F$$
$$(OC)_5M\diagup C=C\diagdown CF_3$$
(33)

attracted attention: Cr(CO)₅C(α-naphthyl)OMe reacts with (+)-α-phenylethylamine to yield four isomers as shown by the observation of four methyl doublets in the ¹H n.m.r. spectrum. It was suggested that the four isomers arise from restricted rotation about the C—N bond, giving rise to *cis–trans* isomerism, and about the C—naphthyl bond, caused by steric hindrance; structures (34)—(37) were suggested for these isomers.[226] It has been demonstrated that Cr(CO)₅C(OMe)Ph can hydrogen-bond to proton donors, *e.g.* MeOH.[227] The ¹H chemical shifts of (38) have been explained in terms of contributions from (39), increasing in the order Z = O < S < NMe.[228] It has been found that SO₂ reacts with M—CH₂-C≡CMe [M = (Cp)Mo(CO)₃, (Cp)Mo(CO)₂P(OPh)₃, Mn(CO)₅, or (Cp)Fe(CO)₂] to yield (40) (X = lone pair). The CH₂ group shows AB structure in the ¹H n.m.r. spectrum on account of the centre of chirality at sulphur, and long-range coupling to the CH₃ group.[229] However, if SO₃ is used then (40) (X = O) is obtained and the two hydrogens in the CH₂ group are equivalent.[230]

---

[224] M. C. McIvor, *J. Organometallic Chem.*, 1971, **27**, C59.
[225] W. J. Schlientz and J. K. Ruff, *J. Organometallic Chem.*, 1971, **33**, C64.
[226] H. Brunner, E. O. Fischer, and M. Lappus, *Angew. Chem. Internat. Edn.*, 1971, **10**, 924.
[227] H. Werner, E. O. Fischer, B. Heckl, and C. G. Kreiter, *J. Organometallic Chem.*, 1971, **28**, 367.
[228] J. A. Connor and E. M. Jones, *J. Chem. Soc. (A)*, 1971, 1974.
[229] J. E. Thomasson, P. W. Robinson, D. A. Ross, and A. Wojcicki, *Inorg. Chem.*, 1971, **10**, 2130.
[230] D. W. Lichtenberg and A. Wojcicki, *J. Organometallic Chem.*, 1971, **33**, C77.

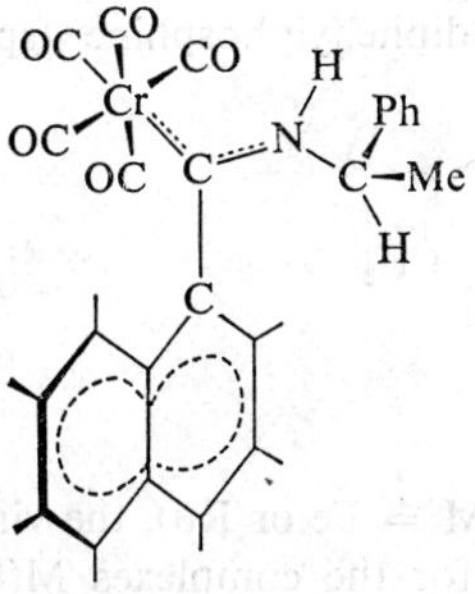

(34)

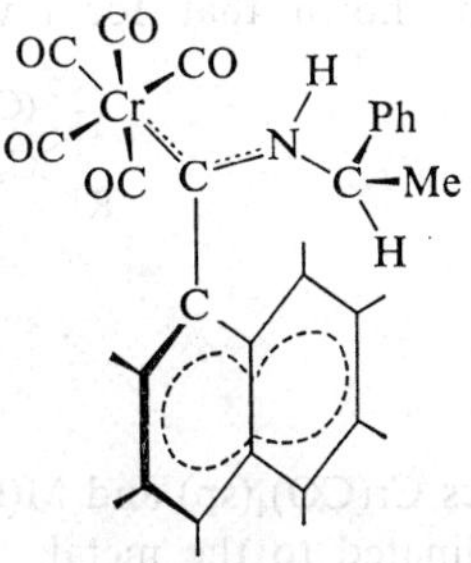

(35)

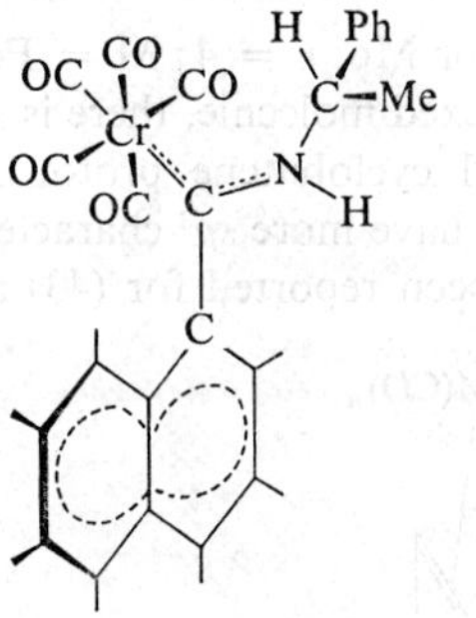

(36)

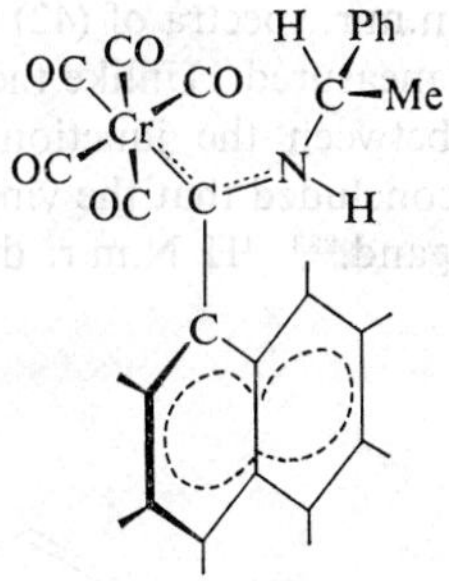

(37)

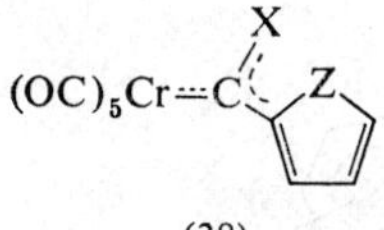

(38)

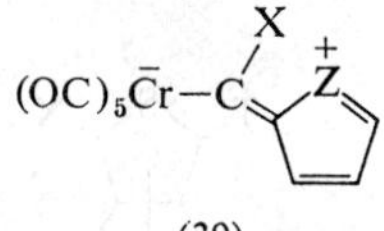

(39)

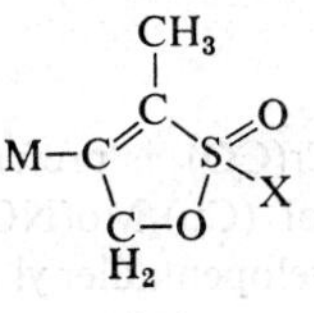

(40)

$Br(CH_2)_2CR^2{=}C{=}CHR^1$ reacts with $[(Cp)Mo(CO)_3]^-$ or $[(Cp)Fe(CO)_2]^-$ to yield (41). The shift in the ^1H n.m.r. signal of the allenic hydrogen was used to demonstrate the co-ordination of the double bond.[231] Similarly, it has been shown that for *o*-vinylphenyldiphenylphosphine (sp) in the

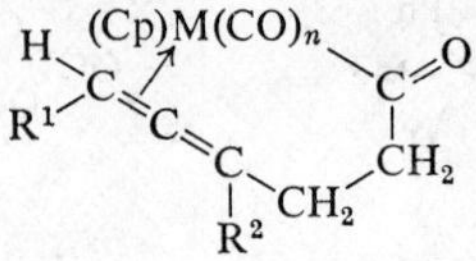

(41)

complexes $Cr(CO)_4(sp)$ and $M(CO)_3(sp)$ (M = Fe or Ru), the vinyl group is co-ordinated to the metal. However, for the complexes $M(CO)_2(sp)_2$ (M = Fe or Ru) one vinyl group is co-ordinated to the metal and one is free. The observation of ^{31}P coupling to the vinyl group was taken as further evidence of co-ordination.[232]

The ^1H n.m.r. spectra of (42) (M = Cr or Mo, n = 4; M = Fe, n = 3) have been measured. Unlike the uncomplexed molecule, there is spin–spin coupling between the junction and vinyl cyclobutene protons. It was therefore concluded that the vinyl protons have more sp^3 character than in the free ligand.[233] ^1H N.m.r. data have been reported for (43) and (44):

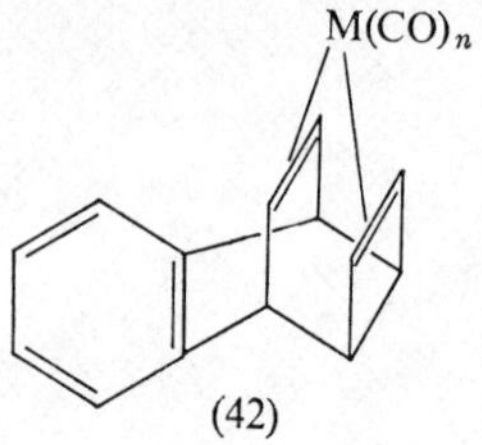

(42)

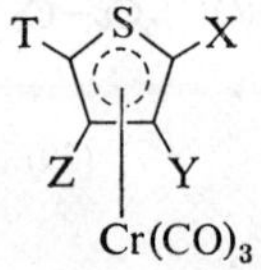

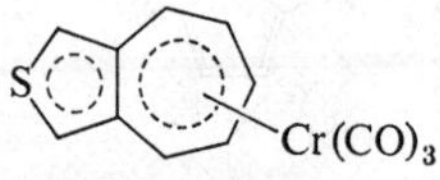

(43) (44)

when X or Y is phenyl, the $Cr(CO)_3$ is preferentially attached to phenyl.[234]

The ^1H n.m.r. spectrum of $(Cp)_2Mo(NO)X$ (X = Me or I), even at -120 °C, shows only one cyclopentadienyl resonance; however, $(Cp)_4Mo$

[231] J. Benaïm, J. Y. Méyour, and J. L. Roustan, *Tetrahedron Letters*, 1971, 983.
[232] M. A. Bennett, G. B. Robertson, I. B. Tomkins, and P. O. Whimp, *Chem. Comm.*, 1971, 341.
[233] Y. Menachem and A. Eisenstadt, *J. Organometallic Chem.*, 1971, **33**, C29.
[234] R. Guilard, J. Tirouflet, and P. Fournari, *J. Organometallic Chem.*, 1971, **33**, 195.

has two cyclopentadienyl signals at room temperature, which is consistent with $(h^5\text{-Cp})_2Mo(h^1\text{-Cp})_2$.[235]

On account of the asymmetric character of $Cr(o\text{-}Pr^iC_6H_4X)(CO)_3$ ($X = NH_2$, NMe_2, or OMe), two methyl signals were observed for the isopropyl group.[236] The ^1H n.m.r. spectra of several *m*- and *p*-substituted anilinechromiumtricarbonyl complexes have been measured. If the chemical shift of the amine proton is plotted against the Hammett σ and σ^- values, a straight line is produced, indicating that resonance interaction between the substituents and the amino-nitrogen is occurring.[237]

When $M(CO)_6$ (M = Cr, Mo, or W) is treated with (*o*-, *m*-, or *p*-$MeC_6H_4)_3P$, then in addition to $(MeC_6H_4)_3PM(CO)_5$ and *cis*- and *trans*-$\{(MeC_6H_4)_3P\}_2M(CO)_4$, (45) is also formed, shown by the movement of signals to high field.[238] Similarly, if $(Ph_2As)_2CH_2$ is used, then in addition to the normal substituted products, (46) (M = Cr or Mo) is also formed.[239]

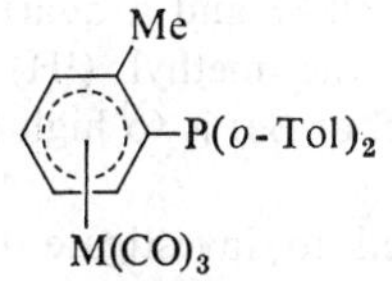

(45)

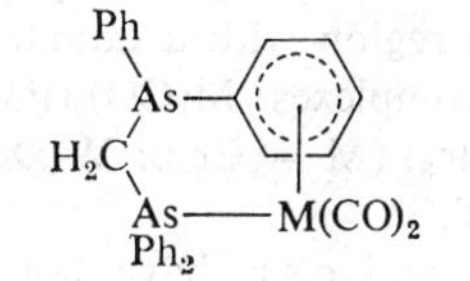

(46)

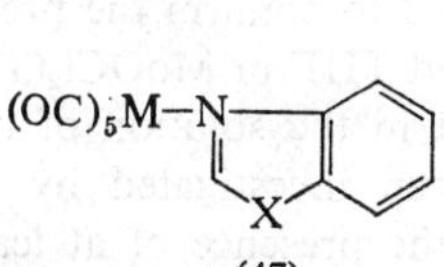

(47)

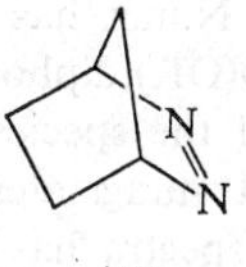

(48)

^{14}N and ^1H n.m.r. have been used to show that for (47) (M = Cr, Mo, or W; X = O, S, or Se) co-ordination is occurring *via* nitrogen rather than X.[240] The observation of two phenyl and t-butyl resonances for $(Cp)Mo(CO)I_2N{=}CPhBu^t$ has been attributed to *cis–trans* isomerism.[241] When (48) is co-ordinated in the complexes $Cr(CO)_5(C_5H_8N_2)$, $Cr(CO)_2$-$(C_5H_8N_2)(C_6H_3Me_3)$, and $Mn(CO)_2(Cp)(C_5H_8N_2)$, the ^1H n.m.r. spectrum is duplicated, from which it was concluded that only one nitrogen atom was co-ordinated.[242]

235 J. L. Calderon and F. A. Cotton, *J. Organometallic Chem.*, 1971, **30**, 377.
236 G. Barbieri and F. Taddei, *Org. Magn. Resonance*, 1971, **3**, 503.
237 A. Wu, E. R. Biehl, and P. C. Reeves, *J. Organometallic Chem.*, 1971, **33**, 53.
238 J. A. Bowden and R. Colton, *Austral. J. Chem.*, 1971, **24**, 2471.
239 R. Colton and C. J. Rix, *Austral. J. Chem.*, 1971, **24**, 2461.
240 W. Beck, J. Chr. Weis, and J. Wieczorek, *J. Organometallic Chem.*, 1971, **30**, 89.
241 M. Kilner and J. N. Pinkney, *J. Chem. Soc. (A)*, 1971, 2887.
242 M. Herberhold and W. Golla, *J. Organometallic Chem.*, 1971, **26**, C27.

On the basis of ^{1}H n.m.r., solvolysis of the iodide has been postulated for (Cp)Mo(NO)I_2L or $[(Cp)Mo(NO)I(PMe_2Ph)_2]^+$ in $(CD_3)_2SO$ or $(CD_3)_2$-CO.[243]

A number of complexes of cyanophosphines and $Cr(CO)_6$ or $Mo(CO)_6$ have been prepared. It is interesting to note that some mixed complexes such as $Cr(CO)_4\{(Me_2N)_2PCN\}P(OMe)_3$ give complex ^{1}H spectra which were interpreted as arising from strong coupling between the phosphorus nuclei, *i.e.* an M_mABY_n spin system.[244] The ^{1}H n.m.r. lineshapes for the complexes $M(CO)_{6-n}[PMeX_2]_n$ (M = Cr, Mo, or W; X = NMe_2 or OMe) have been used to follow changes in $^2J(^{31}P–^{31}P)$ and the results, coupled with i.r. data, have been used to discuss the nature of the metal–phosphorus bond.[245] ^{1}H and ^{31}P n.m.r. data have been reported for $M^1(CO)_5P(M^2Me_3)_3$ (M^1 = Cr, Mo, or W; M^2 = C, Si, Ge, or Sn).[246] When $(CH_2)_2PH$ is complexed to $Mo(CO)_6$, $^1J(^{31}P–^1H)$ increases from 155 to 333 Hz.[247]

The complex *cis*-$Mo(N_2)_2(PMe_2Ph)_4$ gives a complex ^{1}H n.m.r. spectrum in the methyl region with a doublet at τ 8.77 and a quartet at τ 8.58.[248] Unlike the complexes $M(CO)_5(PMe_3)$, the methyl (^{1}H) resonance of $M(CO)_5(XPMe_3)$ (M = Cr or W; X = S or Se) is to high field of that of the free ligand.[249]

^{14}N N.m.r. and e.s.r. have been used to investigate $K_4Mo(CN)_8$ in aqueous solution as a function of temperature, and correlation times have been derived.[250] Treatment of $MoCl_5$ with $LiNMe_2$ yields $Mo(NMe_2)_4$, which is diamagnetic and shows only one methyl resonance from − 90 to + 90 °C.[251] ^{1}H N.m.r. has been used to confirm the presence of alkoxy-groups in $[MoO(OR)(diphos)_2]Cl$ and THF in $MoOCl_2(THF)(diphos)$.[252]

The nature of the species present in the solution of oxoperoxofluoro-molybdates and -tungstates has been investigated by ^{19}F n.m.r. Full analysis of the spectra has shown the presence of at least seven species such as (49).[253] A number of complexes, *e.g.* WF_5NEt_2 and $WF_4(OMe)$-(NEt_2), have been characterized by ^{1}H and ^{19}F n.m.r. spectroscopy. However, the complex $WF_4(OMe)(OPh)$ gives rise to an AA′BC ^{19}F n.m.r. spectrum.[254] The reaction between $WO(OEt)_4$ and HF in ethanol has been

[243] T. A. James and J. A. McCleverty, *J. Chem. Soc.* (*A*), 1971, 1596.
[244] C. E. Jones and K. J. Coskran, *Inorg. Chem.*, 1971, **10**, 1665.
[245] C. E. Jones and K. J. Coskran, *Inorg. Chem.*, 1971, **10**, 55.
[246] H. Schumann, O. Stelzer, J. Kuhlmey, and U. Niederreuther, *Chem. Ber.*, 1971, **104**, 993.
[247] R. Bausch, E. A. V. Ebsworth, and D. W. H. Rankin, *Angew. Chem. Internat. Edn.*, 1971, **10**, 125.
[248] T. A. George and C. D. Seibold, *J. Organometallic Chem.*, 1971, **30**, C13.
[249] E. W. Ainscough, A. M. Brodie, and A. R. Furness, *Chem. Comm.*, 1971, 1357.
[250] R. Poupko, H. Gilboa, B. L. Silver, and A. Loewenstein, *Ber. Bunsengesellschaft. Phys. Chem.*, 1971, **75**, 279.
[251] D. C. Bradley and M. H. Chisholm, *J. Chem. Soc.* (*A*), 1971, 2741.
[252] A. V. Butcher and J. Chatt, *J. Chem. Soc.* (*A*), 1971, 2356.
[253] Yu. A. Buslaev, S. P. Petrosyants, and V. P. Tarasov, *Doklady Phys. Chem.*, 1970, **193**, 548.
[254] A. Majid, R. R. McLean, D. W. A. Sharp, and J. M. Winfield, *Z. anorg. Chem.*, 1971, **385**, 85.

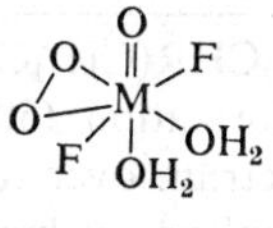

(49)

studied using ^{19}F n.m.r. Signals due to $WOF_2(OEt)_2$, $WOF_3(OEt)(HOEt)$, $[WOF_4(OEt)]^-$, $WOF(OEt)_3$, and $WF_4(OEt)_2$ were detected.[255] Similarly, ^{19}F n.m.r. has been used to show that several species such as *cis*-$[MoO_2(OH)(H_2O)F_2]^-$ exist in molybdenum fluoride solutions.[256]

Complexes of Mn and Re.—Information concerning complexes of these elements can be found at the following sources: $RC_5H_4Mn(CO)_3$,[124] $[Mn(CNMe)_6]^+$,[125] $R_{3-x}X_xSnMn(CO)_5$,[178] $(OC)_5Mn\overline{C{=}CMe{-}SO_2CH_2}$,[229] $(OC)_5Mn\overline{C{=}CMe{-}SO_3CH_2}$,[230] $Mn(CO)_2(Cp)(C_5H_8N_2)$,[242] $Mn(CO)_5CH_2Ph$,[312] $(OC)_5Mn(B_{10}C_2H_{10}Me)$, $(OC)_5Mn(B_8C_2H_8Me)$,[1064] $[(OC)_3Mn(B_9C_6H_{15})]^-$,[1071] $R_{3-n}X_nSnMn(CO)_5$,[179] *cis*-$ReCl_2(PPh_2)(HPPh_2)_3$,[280] and $(OC)_5ReCH{=}CHRFe(CO)_4$.[302]

Treatment of $HMn(CO)_5$ with $(CF_3)_2PX$ gives $XMn(CO)_5$ (X = I, Br, or Cl) or $HMn(CO)_4P(CF_3)_2X$ (X = F, CF_3, or Me). The ^1H and ^{19}F n.m.r. spectra show the latter hydrides to be a mixture of *cis*- and *trans*-isomers. The hydride signal of the *cis*-complex is consistently to low-field of that due to the *trans*-complex, and $|^2J(^{31}P–^1H)|$ (*cis*) is less than 10 Hz, whereas $|^2J(^{31}P–^1H)|$ (*trans*) is between 57 and 72 Hz.[257] The ^1H n.m.r. spectra of $ReH(CO)_n(PPh_3)_{5-n}$ (n = 2 or 3) and $Re(CO)_2(PPh_3)_2S_2H$ have been reported. When $n = 2$, the hydride resonance consists of a doublet (J = 18 Hz) of triplets (J = 30 Hz) demonstrating that the structure is static.[258] The multiplicity of the ^{11}B n.m.r. signal of a wide variety of complexes such as $(Cp)Mo(CO)_2(CHNMe)_2BH_2$, $(C_5H_4Me)Mn(NO)(CHNMe)_2BF_2$, and $(Cp)Fe(CHNMe)_3BH$ has proved useful to determine the number of magnetically active nuclei attached to each boron. The ^1H n.m.r. spectrum is also of interest with the CH proton appearing at $\tau - 1.0$.[259]

Addition of $MeMn(CO)_5$ to $[Mn(CO)_5]^-$ yields a compound with a singlet in the ^1H n.m.r. spectrum at τ 7.19, and it was suggested that $[(OC)_5MnMn(CO)_4(COMe)]^-$ is formed. Subsequent addition of methyl iodide produced not only $MeMn(CO)_5$ with a singlet at τ 10.10 but also a compound thought to be $[Mn(CO)_4(COMe)I]^-$ with a singlet at τ 7.13.[260] The magnitude of the coupling constants obtained from a full analysis of

255 Yu. A. Buslaev, Yu. V. Kokunov, and V. A. Bochkareva, *Russ. J. Inorg. Chem.*, 1971, 1393.
256 Yu. A. Buslaev and S. P. Petrosyants, *Russ. J. Inorg. Chem.*, 1971, **16**, 703.
257 R. C. Dobbie, *J. Chem. Soc.* (*A*), 1971, 230.
258 M. Freni, D. Giusto, and P. Romiti, *J. Inorg. Nuclear Chem.*, 1971, **33**, 4093.
259 P. M. Treichel, J. P. Stenson, and J. J. Benedict, *Inorg. Chem.*, 1971, **10**, 1183.
260 C. P. Casey and R. L. Anderson, *J. Amer. Chem. Soc.*, 1971, **93**, 3554.

the ^{1}H n.m.r. spectrum of $[\overline{CH_2CHR(CH)_4CH}]Mn(CO)_3$ suggested that the compound exists in the conformation (50). However, no change was observed in the ^{1}H n.m.r. spectrum over the temperature range $-$ 90 to + 40 °C.[261] A number of complexes of manganese, rhenium, cobalt, and

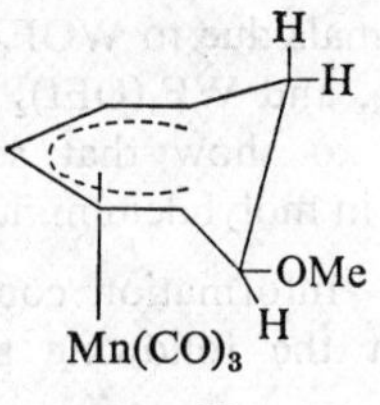

50

rhodium with trimethylsilyl-substituted cyclopentadienyl rings and carbonyls as ligands have been prepared. In the case of the complexes containing the $(Me_3SiC_5H_4)$ ligand, two triplets were observed for the cyclopentadienyl proton resonances, but in the case of the complexes $Me_2Si[(C_5H_4)M]_2$ [M = $Mn(CO)_3$, $Re(CO)_3$, or $Co(CO)_2$] only one singlet cyclopentadienyl resonance was observed, even at low temperatures.[262]

The compound $HM(CO)_4PPh_2(o\text{-}C_2H_3C_6H_4)$ rapidly isomerizes to give a mixture of (51) and (52). ^{1}H N.m.r. was used to demonstrate that when

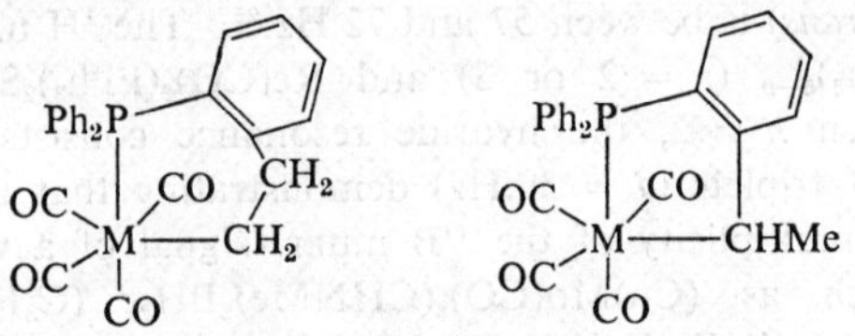

(51) (52)

M = Mn, (52) is the predominant component, and when M = Re, (51) is the predominant component.[263] However, caution is required in the interpretation of ^{1}H n.m.r. data for such compounds as (54). On the basis of ^{1}H n.m.r. data, the product from the reaction of $(o\text{-}C_2H_3C_6H_4)PPh_2$ with $MeMn(CO)_5$ was formulated as (53), whereas an *X*-ray structure determination shows that the structure is really (54). A re-examination of the ^{1}H n.m.r. spectrum showed that the phosphorus only couples to the CH proton, whereas if the product was (53) the phosphorus would couple to all the vinyl protons.[264] The ^{1}H n.m.r. spectrum of (Cp)$Mn(CO)_3$ in the liquid

[261] M. I. Foreman and F. Haque, *J. Chem. Soc. (B)*, 1971, 418.
[262] E. W. Abel and S. Moorhouse, *J. Organometallic Chem.*, 1971, **29**, 227.
[263] M. A. Bennett and R. Watt, *Chem. Comm.*, 1971, 94.
[264] M. A. Bennett, G. B. Robertson, R. Watt, and P. O. Whimp, *Chem. Comm.*, 1971, 752.

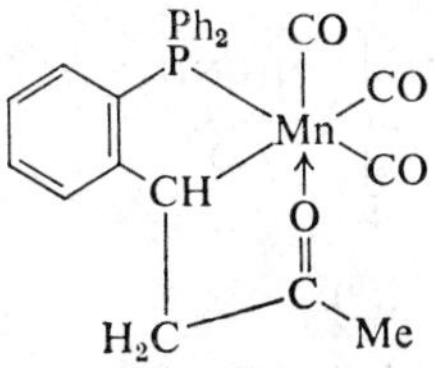

(54)

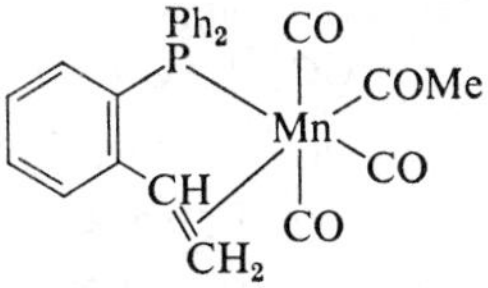

(53)

crystalline nematic phase has been studied, and values of the indirect spin couplings have been determined.[265]

Treatment of (55) (M = Mn) with PMe_2Ph gives (56) with *cis* tertiary phosphines as shown by two 1 : 1 doublets for the methyl resonance. However, if $P(OCH_2)_3CEt$ is used, then (57) (M = Mn or Re) is obtained

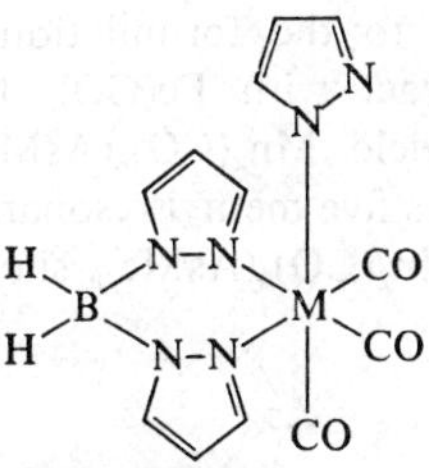

(55)

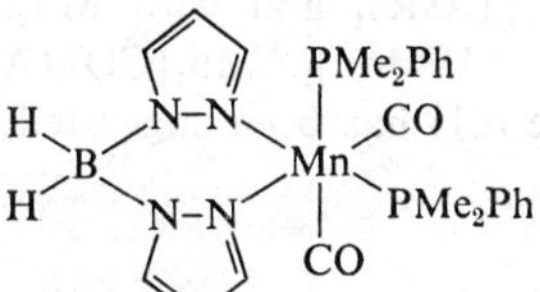

(56)

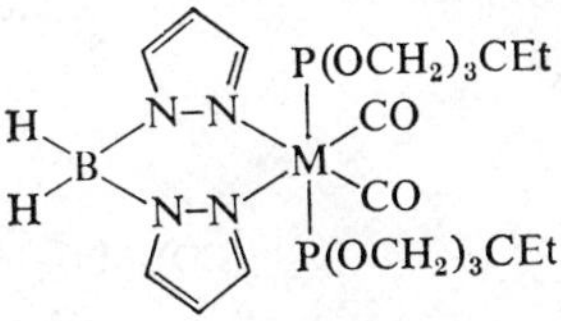

(57)

with a 1 : 2 : 1 triplet for the *trans*-phosphite ligand.[266] This use of the 'virtual coupling' pattern, however, breaks down for $MnBr(CO)_3$-$(PMe_2Ph)_2$ (tertiary phosphines mutually *trans*) when an intermediate coupling pattern has been observed, in contrast to the rhenium analogue which gives a well defined 1 : 2 : 1 triplet for the methyl protons.[267] Similarly (58) shows two methyl doublets on account of the lack of a plane of symmetry through the P—Re—N axis, and (59) (L = $PMePh_2$ or PMe_2Ph) shows five-line multiplets due to the overlap of two methyl triplets. The

[265] G. L. Khetrapal, A. C. Kunwar, and C. R. Kanekar, *Chem. Phys. Letters*, 1971, **9**, 437.
[266] A. Bond and M. Green, *J. Chem. Soc.* (*A*), 1971, 682.
[267] J. T. Moelwyn-Hughes, A. W. B. Garner, and A. S. Howard, *J. Chem. Soc.* (*A*), 1971, 2370.

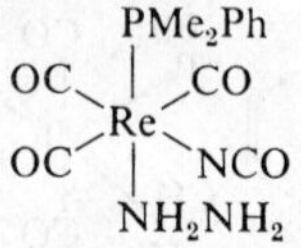

(58)

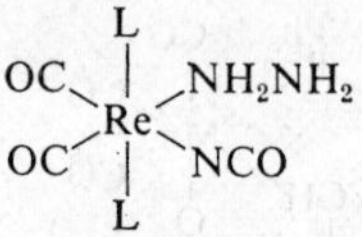

(59)

hydrazine protons were assigned by deuteriation.[268] 'Virtual coupling' has also been used to determine the stereochemistry of the products from $[Re(CO)_5]_2$ and $PMePh_2$. The compound $Re(CO)_5(PMe_2Ph)$, which is paramagnetic, has normal chemical shifts.[269]

Treatment of $(Cp)Mn(CO)_3$ with $P(EMe_3)_3$ (E = C, Si, Ge, or Sn) yields the monosubstituted complexes and the very large complexation shift of the ^{31}P signal of − 161.8 p.p.m. for $(Cp)Mn(CO)_2P(SiMe_3)_3$ merits comment.[270] The observation of two isopropyl methyl resonances in the ratio 6 : 1 when $Ph_3SiMn(CO)_5$ is treated with $(Pr^iO)_3P$ led to the formulation of the product as (60).[271] $(AsR)_5$ and $(AsPh)_6$ react with $Fe(CO)_5$ to yield $[Fe(CO)_3]_2(AsR)_4$ and with $Mn_2(CO)_{10}$ to yield $Mn_2(CO)_8(AsMe)_5$ and $[Mn(CO)_3(AsMe)_4]_2$. $Mn_2(CO)_8(AsMe)_5$ shows five methyl resonances and structure (61) has been suggested, whereas $Mn_2(CO)_6(AsMe)_8$ shows four

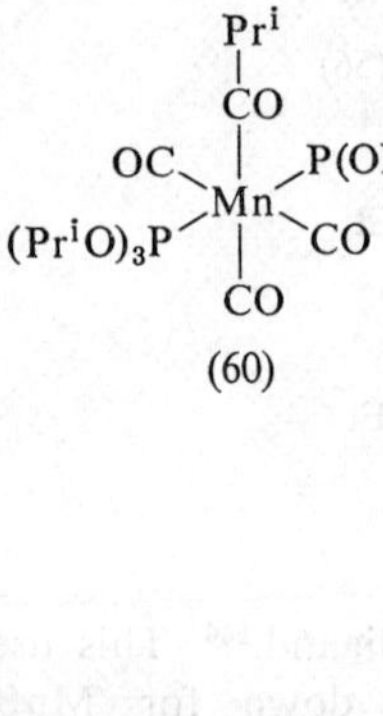

(60)

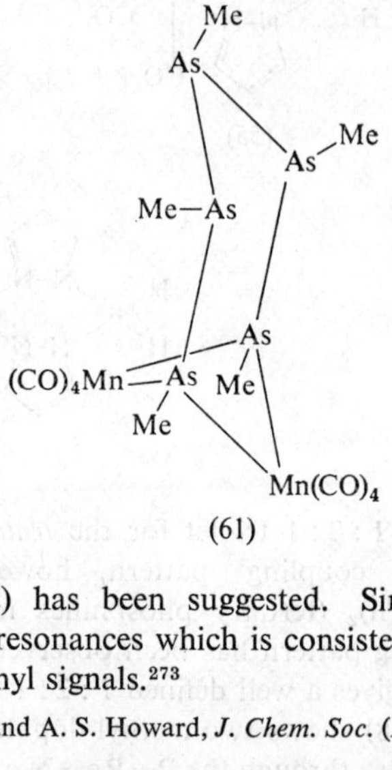

(61)

methyl resonances and structure (62) has been suggested. Similarly, $[Fe(CO)_3]_2(AsMe)_4$ shows two methyl resonances which is consistent with (63);[272] compound (64) shows two methyl signals.[273]

268 J. T. Moelwyn-Hughes, A. W. B. Garner, and A. S. Howard, *J. Chem. Soc.* (*A*), 1971, 2361.
269 J. T. Moelwyn-Hughes, A. W. B. Garner, and N. Gordon, *J. Organometallic Chem.*, 1971, **26**, 373.
270 H. Schumann, O. Stelzer, J. Kuhlmey, and U. Niederreuther, *J. Organometallic Chem.*, 1971, **28**, 105.
271 E. P. Ross, R. T. Jernigan, and G. R. Dobson, *J. Inorg. Nuclear Chem.*, 1971, **33**, 3375.
272 P. S. Elmes and B. O. West, *J. Organometallic Chem.*, 1971, **32**, 365.
273 J. P. Crow, W. R. Cullen, F. L. Hou, and L. Y. Y. Chan, *Chem. Comm.*, 1971, 1229.

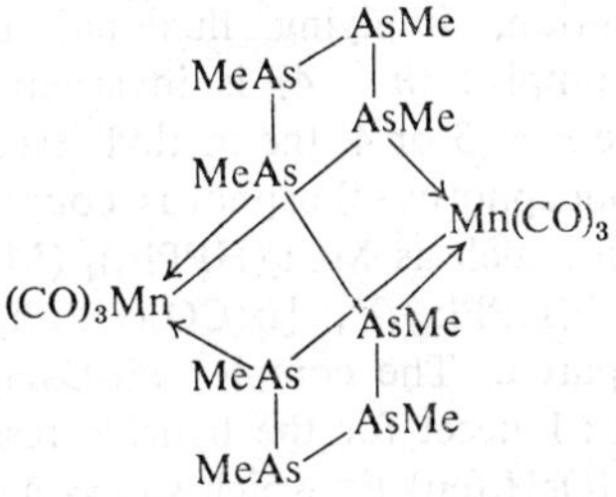

(62)

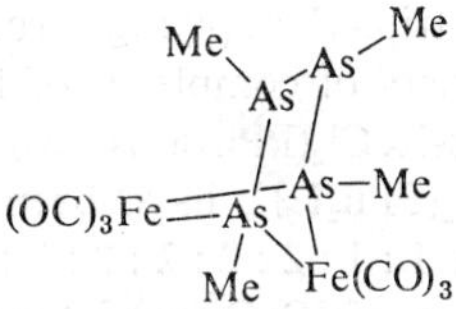

(63)

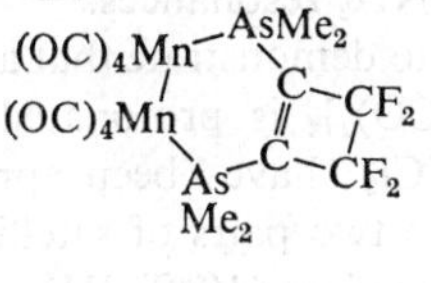

(64)

The ^{31}P n.m.r. spectra of $Et_2PS_2Re(CO)_4$ and $Et_2PS_2Re(CO)_3(PPh_3)$ have been reported.[274] The ^{1}H n.m.r. spectrum of (monohydrogen mesoporphyrin IX dimethyl esterato)tricarbonylrhenium(I) shows an extremely broad resonance at τ 14.9, exchangeable with $DCl–D_2O$, and it was suggested that this lone proton is attached to a porphyrin nitrogen.[275]

Complexes of Fe, Ru, and Os.—Information concerning complexes of these elements can be found at the following sources: $(C_4H_3X)Fe(CO)_3$,[86] $(C_4H_2Ph_2)Fe(CO)_3$,[87] $Fe(CO)_5$, $Fe(CO)_4L$,[96] $(CO)_2(Cp)FeR$,[137, 229, 230, 231] $(o\text{-}C_2H_3C_6H_4PPh_2)M(CO)_3$ (M = Fe or Ru),[232] $(Cp)Fe(CHNMe)_3BH$,[259] $[Fe(CO)_3]$ $(AsR)_4$,[272] $(\pi\text{-}C_2B_4H_6)Fe(CO)_3$, $(\pi\text{-}C_2B_3H_7)Fe(CO)_3$,[1052] $(Cp)Fe(CO)_2(B_{10}C_2H_{10}Me)$, $(Cp)Fe(CO)_2(B_8C_2H_9)$,[1064] $Cs[\{\pi\text{-}(3)\text{-}1,2\text{-}B_9C_2H_{11}\}_2FeH]$,[1081] $(C_{14}H_{12})Fe(CO)_3$,[233] $RuCl_2(CO)_2(PHBu^t{}_2)_2$,[113] and *cis*-$RuCl_2(CO)_2L_2$.[165]

The hydride resonance of $(Me_3P)_4FeH_2$ consists of a doublet of triplets and it was therefore concluded that the stereochemistry is *cis*. In contrast, the hydride resonance of $(Me_3P)_4CoH$ consists of a 1 : 4 : 6 : 4 : 1 quintet and it was suggested that the structure is a fluxional bipyramid.[276] Similarly, the hydride resonance of $FeH_4(PBu^nPh_2)_3$ shows a 1 : 3 : 3 : 1 quartet and is fluxional, like the corresponding osmium complexes.[277] The same behaviour has also been found for $FeH_2(N_2)(PEtPh_2)_3$ which also has a 1 : 3 : 3 : 1 quartet for its hydride resonance.[278] For all the complexes $H_4Ru_3(CO)_{12-n}\{P(OMe)_3\}_n$ (n = 1—4) the hydride resonance consists of an (n + 1)

274 E. Lindner and K.-M. Matejcek, *J. Organometallic Chem.*, 1971, **29**, 283.
275 D. Ostfeld, M. Tsutsui, C. P. Hrung, and D. C. Conway, *J. Amer. Chem. Soc.*, 1971, **93**, 2549.
276 H.-F. Klein, *Angew. Chem. Internat. Edn.*, 1970, **9**, 904.
277 M. Aresta, P. Giannoccaro, M. Rossi, and A. Sacco, *Inorg. Chim. Acta*, 1971, **5**, 115.
278 M. Aresta, P. Giannoccaro, M. Rossi, and A. Sacco, *Inorg. Chim. Acta*, 1971, **5**, 203.

multiplet with binomial distribution, implying fluxional behaviour. However, the spectrum of the complex (n = 4) is invarient down to − 100 °C. Also in the cases where n = 3 or 4, the methyl resonance is a 'virtual triplet' implying strong phosphorus–phosphorus coupling.[279]

A variety of complexes of $PHPh_2$, such as $MCl_2(HPPh_2)_4$ (M = Ru or Os), *cis*-$ReCl_2(PPh_2)(HPPh_2)_3$, $[Rh(HPPh_2)_4]^+$, $[Ir(CO)(HPPh_2)_4]^+$, and *cis*-$OsH_2(PPh_2H)_4$, have been prepared. The complex *cis*-$OsH_2(PPh_2H)_4$ shows a 1 : 1 : 2 : 2 : 2 : 2 : 2 : 2 : 1 : 1 decet for the hydride resonance.[280] Reduction of $[OsO_2(en)_2]^{2+}$ yields $[OsH_2(en)_2]^{2+}$ which shows a resonance 15.4 p.p.m. upfield of t-butanol. The *cis* stereochemistry was demonstrated by the observation of two NH_2 resonances.[281]

^1H N.m.r. has been used to demonstrate that a hydride $[(Cp)Fe(CO)_2]_2H^+$ is formed when $[(Cp)Fe(CO)_2]_2$ is protonated.[282] A number of related complexes containing $SiCl_3$ have been prepared. The compound $(Cl_3Si)_2FeH(CO)(Cp)$ shows two pairs of satellites centred on the hydride resonance which were assigned to $^2J(^{29}Si–^1H)$ and $^1J(^{57}Fe–^1H)$ of 14.5 Hz; however, $^2J(^{13}C—^1H)$ was not detected.[283]

It is interesting that in the ^1H n.m.r. spectrum of $HRu_3(CO)_9(C_{12}H_{15})$, coupling was demonstrated by double resonance between the hydride and the proton at τ 3.81.[284] $HRu_3(CO)_{10}SEt$ dissolves in 98% H_2SO_4 to give a species with a doublet at τ 24.94 and τ 29.39, and the structure (65) was

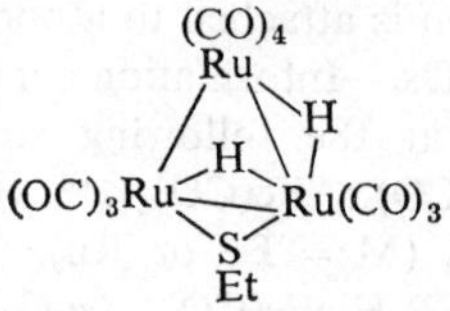

(65)

suggested. If D_2SO_4 is used, then $HDRu_3(CO)_{10}SEt$ is obtained with equal occupancy of the two sites.[285] ^1H N.m.r. has been used to demonstrate the presence of a hydride and two equivalent PPh_3 ligands in (66).[286] Reduction of *mer*-OsY_2LQ_3 yields *mer*-$OsHYLQ_3$ (67) (L = CO or N_2; Q = tertiary phosphine), which shows $^2J(^{31}P–^1H)$ (*trans*) of 90 Hz.[287]

It has been suggested that $(Cp)Fe(CO)_2CH{=}CHCH{=}CHFe(CO)_2(Cp)$ only shows a singlet in the ^1H spectrum of the butadiene fragment on account of the molecule being fluxional *via* a cyclobutadiene intermediate. However,

279 S. A. R. Knox and H. D. Kaesz, *J. Amer. Chem. Soc.*, 1971, **93**, 4594.
280 J. R. Sanders, *J. Chem. Soc.* (*A*), 1971, 2991.
281 J. Malin and H. Taube, *Inorg. Chem.*, 1971, **10**, 2403.
282 D. A. Symon and T. C. Waddington, *J. Chem. Soc.* (*A*), 1971, 953.
283 W. Jetz and W. A. G. Graham, *Inorg. Chem.*, 1971, **10**, 1159.
284 A. Cox and P. Woodward, *J. Chem. Soc.* (*A*), 1971, 3599.
285 A. J. Deeming, R. Ettorre, B. F. G. Johnson, and J. Lewis, *J. Chem. Soc.* (*A*), 1971, 1797.
286 B. R. James and L. D. Markham, *Inorg. Nuclear Chem. Letters*, 1971, **7**, 373.
287 J. Chatt, D. P. Melville, and R. L. Richards, *J. Chem. Soc.* (*A*), 1971, 895.

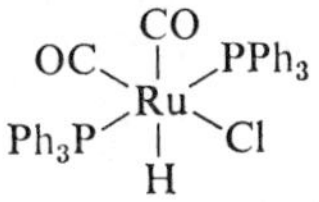

(66)

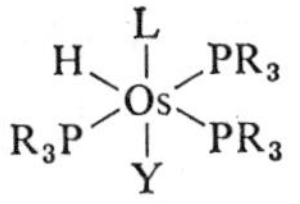

(67)

in benzene or toluene the ^1H spectrum shows an AA′BB′ pattern, even at 116 °C, proving the static nature of the compound.[288] Because of the bulk of the t-butyl and $(Cp)Fe(CO)_2$, both *threo*- and *erythro*-$Bu^tCHDCHDFe(CO)_2(Cp)$ exist principally in one conformer, *e.g.* (68), and thus $^3J(H_aH_b)$ for the *threo*-compound is 4.3 Hz and for the *erythro*-compound is 13.0 Hz. It was thus possible to show that cleavage of the iron–carbon bond with $HgCl_2$ occurs with retention of configuration at the carbon and with Br_2

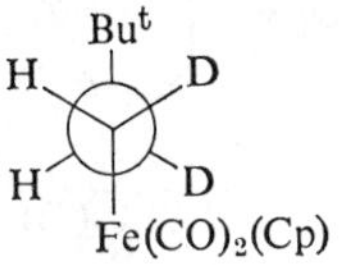

(68)

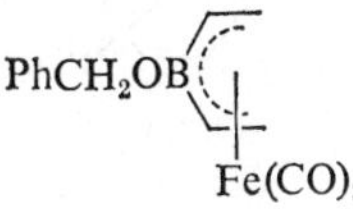

(69)

occurs with inversion of configuration at the carbon.[289] The ^1H n.m.r. spectra of $(Cp)Fe(CO)_2R$ [R = COCHMePh, CHMePh, or S $(O)_2$CHMePh] have been examined in the chiral solvent $(+)$-$C_6H_5CH(CF_3)OH$, but no splitting of the spectrum was observed even at 220 MHz.[290]

Irradiation of $PhCH_2OB(C_2H_3)[(C_2H_3)Fe(CO)_4]$ yields (69); the ^1H n.m.r. spectrum of the vinyl group shows J_{cis} = 12.0 Hz and J_{trans} = 14.5 Hz.[291] Double resonance has been used to assign the ^1H resonances of (70) and related compounds. It is interesting that H_a appears at *ca.* τ 10.[292] The structure (71) has been confirmed by ^1H n.m.r., including double resonance, computer simulation, and change of solvent, at 60, 100, and 220 MHz.[293] However, the observation of signals in the ratio 4 : 1 : 1 : 1 : 1 and the absence of signals in the τ 8—9 region for (72) rules out (73) as the possible structure.[294] The ^1H n.m.r. spectrum of $(C_{14}H_{12})Fe(CO)_3$ shows a doubled structure which was attributed to there being a mixture of isomers, (74) and (75).[295] Similarly, ^1H n.m.r. has been used to demonstrate that when R^1 and R^2 are different, (76) exists as a mixture of two isomers and the relative proportions have been measured.[296]

[288] W. P. Giering, *Chem. Comm.*, 1971, 4.
[289] G. M. Whitesides and D. J. Boschetto, *J. Amer. Chem. Soc.*, 1971, **93**, 1529.
[290] J. J. Alexander and A. Wojcicki, *Inorg. Chim. Acta*, 1971, **5**, 655.
[291] G. E. Herberich and H. Müller, *Angew. Chem. Internat. Edn.*, 1971, **10**, 937.
[292] R. N. Greene, C. H. DePuy, and T. E. Schroer, *J. Chem. Soc.* (*C*), 1971, 3115.
[293] I. I. Kritskaya, G. P. Zol'nikova, I. F. Leshcheva, Yu. A. Ustynyuk, and A. N. Nesmeyanov, *J. Organometallic Chem.*, 1971, **30**, 103.
[294] G. T. Rodeheaver, G. C. Farrant, and D. F. Hunt, *J. Organometallic Chem.*, 1971, **30**, C22.
[295] B. F. G. Johnson, J. Lewis, P. McArdle, and G. L. P. Randall, *Chem. Comm.*, 1971, 177.
[296] H. Müller and G. E. Herberich, *Chem. Ber.*, 1971, **104**, 2772.

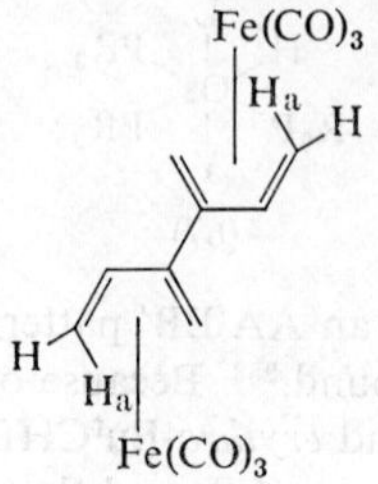

(70)

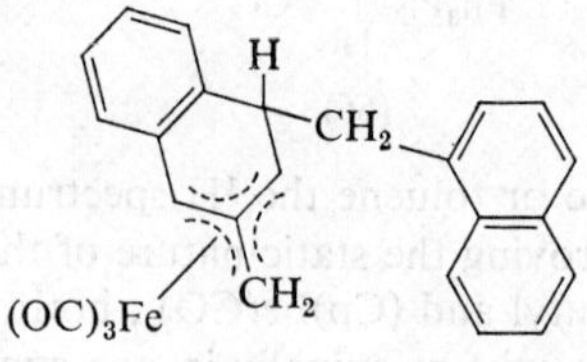

(71)

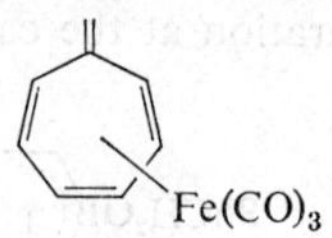

(72)

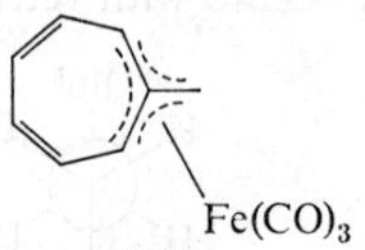

(73)

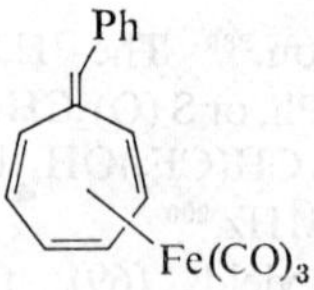

(74)

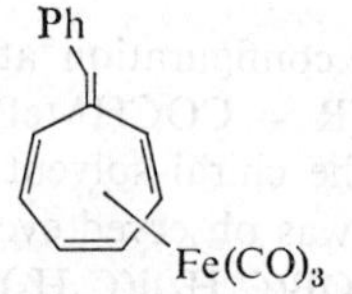

(75)

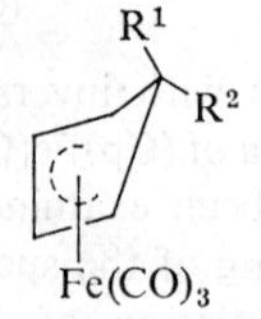

(76)

^1H N.m.r. has been used to demonstrate that cycloheptatrieneirontricarbonyl is exclusively deuteriated in the *exo* position,[297] and the ABCDE spectrum of (77) (M = Fe or Ru) has been analysed.[298]

It has been suggested that the n.m.r. spectra of the compounds (fluoro-olefin)$Fe(CO)_4$ are best explained by there being little π-bonding left in the olefin.[299] From the 'virtual coupling' pattern observed when a fluoro-olefin is added to *trans*-$Os(CO)_3L_2$ it was concluded that when L = $P(OMe)_3$ the phosphites in the product are mutually *cis*, whereas when L = PMe_2Ph the phosphines are mutually *trans*.[300] The products of the reaction of $(C_4Me_4)Fe(CO)_3$ with $(CF_3)_2CO$, C_2F_4, C_3F_6, or $CF_3C{\equiv}CCF_3$ have been characterized by ^{19}F n.m.r. For example, use of C_2F_4 gives three methyl signals in the ratio 1 : 2 : 1 and two ^{19}F multiplets 1 : 1; thus the product was formulated as (78).[301]

[297] H. Maltz and B. A. Kelly, *Chem. Comm.*, 1971, 1390.
[298] A. J. Carty, G. Kan, D. P. Madden, V. Snieckus, M. Stanton, and T. Birchall, *J. Organometallic Chem.*, 1971, **32**, 241.
[299] R. Fields, G. L. Godwin, and R. N. Haszeldine, *J. Organometallic Chem.*, 1971, **26**, C70.
[300] M. Cooke, M. Green, and T. A. Kuc, *J. Chem. Soc.* (*A*), 1971, 1200.
[301] A. Bond and M. Green, *Chem. Comm.*, 1971, 12.

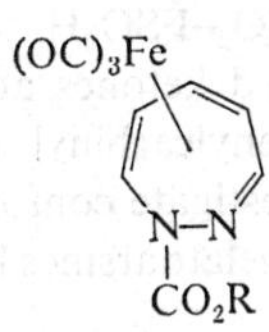

(77)

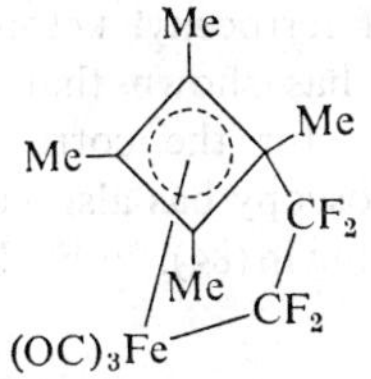

(78)

The compounds RCH=CHFe(CO)$_2$(Cp), RCH=CHW(CO)$_3$(CpH), and RCH=CHRe(CO)$_5$ react with $Fe_2(CO)_9$ to form products such as (79). The co-ordination of the vinyl group was confirmed by ^{1}H n.m.r. with shifts for the vinyl group of up to δ 11.60 being found.[302] The ^{1}H n.m.r. spectrum of $(C_8H_8O)Fe_2(CO)_6$ was interpreted in terms of a static structure such as (80) or (81) rather than a dynamic structure as has been reported

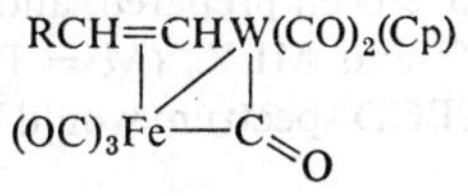

(79)

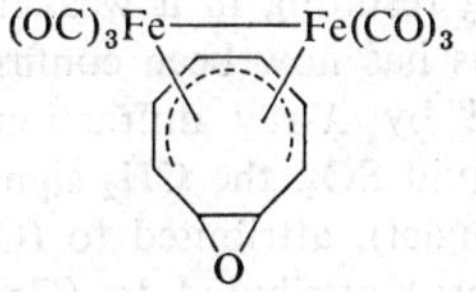

(80)

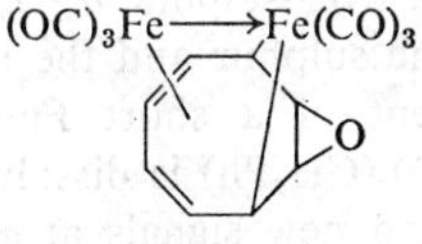

(81)

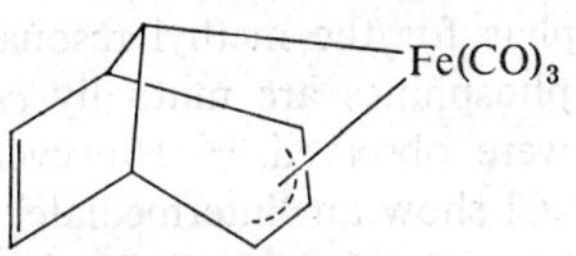

(83)

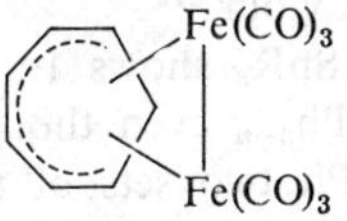

(82)

for $(C_8H_8)Fe_2(CO)_6$.[303] Similarly, compound (82) shows a temperature-independent ^{1}H n.m.r. spectrum consistent with a static structure.[304] Semibullvalene reacts with $Fe_2(CO)_9$ to yield (83); the ^{1}H n.m.r. spectrum consists of five sets of resonances in the ratio 2 : 2 : 1 : 2 : 1 and decoupling is consistent with (83).[305]

302 A. N. Nesmeyanov, M. I. Rybinskaya, L. V. Rybin, V. S. Kaganovich, and P. V. Petrovskii, *J. Organometallic Chem.*, 1971, **31**, 257.
303 H. Maltz and G. Deganello, *J. Organometallic Chem.*, 1971, **27**, 383.
304 F. A. Cotton, B. G. DeBoer, and T. J. Marks, *J. Amer. Chem. Soc.*, 1971, **93**, 5069.
305 R. M. Moriarty, C.-L. Yeh, and K. C. Ramey, *J. Amer. Chem. Soc.*, 1971, **93**, 6709.

A study of ferrocenyl ketones in $CDCl_3$–SO_2–FSO_2H solution at low temperatures has shown that these protonated ketones are excellent ^1H n.m.r. models for the corresponding ferrocenylcarbinyl cations.[306] ^1H N.m.r. spectroscopy has also been used to investigate conformation of the cyclohexane ring in (84).[307, 308] The ferrocenyl chelate arsines $Fe(C_5H_4AsR_2)_2$

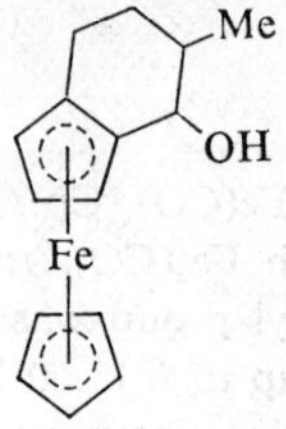

(84)

(= L; R = Me or Ph) have been prepared, and complexed to a number of metals, *e.g.* $Mo(CO)_4L$,[309] and MLX_2 (M = Pd or Pt; X = halogen).[310] For these complexes an ABCD spectrum would be expected, but an AA′BB′ spectrum was found.

The ^{19}F n.m.r. spectrum of $(Cp)Fe(CO)_2(SO_2C_6F_5)$ shows considerable high-field movement relative to $ClSO_2C_6F_5$, and it was suggested that there was strong iron–sulphur d_π–d_π bonding resulting in a weakened π-bond between the sulphur and the ring. This has now been confirmed by the measurement of a short Fe—S bond by *X*-ray diffraction.[311] When $(Cp)Fe(CO)_2(CH_2Ph)$ is dissolved in liquid SO_2, the CH_2 signal at τ 7.31 vanishes and new signals at τ 5.79 (singlet), attributed to $(Cp)Fe(CO)_2$-$S(O_2)CH_2Ph$, and at τ 6.49 (AB pattern) attributed to $(Cp)Fe(CO)_2O$-$S(O)CH_2Ph$, appear; $(Cp)Mo(CO)_3(CH_2Ph)$ and $Mn(CO)_5CH_2Ph$ show similar behaviour.[312]

$(Cp)FeL_2SnR_3$ shows 1 : 2 : 1 triplets for the methyl resonances when L = PMe_nPh_{3-n} even though the phosphines are mutually *cis*, and for L = PMe_2Ph two sets of triplets were observed.[313] However, (Cp)Ru-$(PMePh_2)_2Cl$ and $(Cp)Ru\{P(OMe)_3\}_2Cl$ show an 'intermediate' $AA'X_nX'_n$ coupling pattern.[314] For the complexes $(Cp)Fe(CO)_2SnPh_nCl_{3-n}$ a linear relationship has been reported between τ (Cp) and the i.r. C—H deformation mode of the Cp.[315] The observation of three Cp resonances for $[(Cp)Fe(CO)EPh]_2$ (E = S, Se, or Te) confirms the presence of two

306 C. P. Lillya and R. A. Sahatjian, *J. Organometallic Chem.*, 1971, **32**, 371.
307 B. Gautheron and R. Broussier, *Bull. Soc. chim. France*, 1971, 3636.
308 B. Gautheron and R. Broussier, *Tetrahedron Letters*, 1971, 513.
309 J. J. Bishop and A. Davison, *Inorg. Chem.*, 1971, **10**, 826.
310 J. J. Bishop and A. Davison, *Inorg. Chem.*, 1971, **10**, 832.
311 M. I. Bruce and A. D. Redhouse, *J. Organometallic Chem.*, 1971, **30**, C78.
312 S. E. Jacobson, P. Reich-Rohrwig, and A. Wojcicki, *Chem. Comm.*, 1971, 1526.
313 W. R. Cullen, J. R. Sams, and J. A. J. Thompson, *Inorg. Chem.*, 1971, **10**, 843.
314 T. Blackmore, M. I. Bruce, and F. G. A. Stone, *J. Chem. Soc. (A)*, 1971, 2376.
315 D. S. Field and M. J. Newlands, *J. Organometallic Chem.*, 1971, **27**, 213.

isomers.[316] $EtCOFe(CO)L(Cp)$ (L = phosphorus ligand) shows inequivalence of the CH_2 protons of the ethyl group, confirming the presence of a diastereotopic iron atom.[317] The inequivalence of the cyclopentadienyl groups of $Ph_2GeFe_2(CO)_5(Cp)_2$ has been taken as evidence of *cis–trans* isomerism with the bridging Ph_2Ge and CO groups.[318] $(Cp)Fe(CO)_2CH_2CMe{=}CH_2$ reacts with tcne to yield either $(Cp)Fe(CO)_2\text{-}\overline{CMeCH_2[C(CN)_2]_2CH_2}$ or $(Cp)Fe(CO)_2CH_2\overline{CMe[C(CN)_2]_2CH_2}$, but the former is favoured by the observation of only one AB pattern for the CH_2 groups; the molybdenum analogue reacts similarly.[319]

Although the free ligand (85) has its phenyl protons spread out over the range τ 1.95—2.55, when complexed to $Fe(CO)_4$ all the signals fall between τ 2.0—2.2.[320] The observation in the ^1H n.m.r. spectrum of two groups of three lines, each of equal intensity, with one group being twice the intensity of the second, is consistent with the formula (86).[321] The norbornadiene adduct (87) shows separate ^1H resonances for the protons on each side of the ligand.[322] $[Fe(CNMe)_6]^{2+}$ reacts with $MeNH_2$ to yield (88) which shows

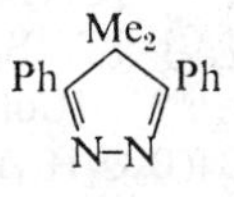

(85)

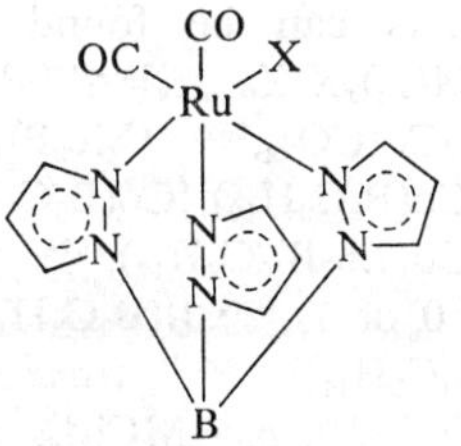

(86)

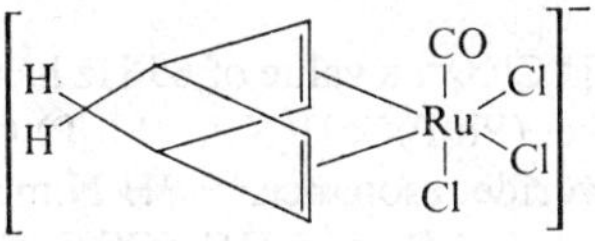

(87)

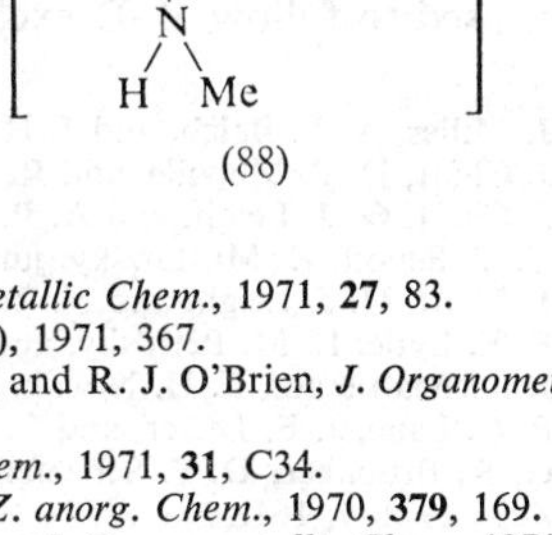

(88)

[316] E. D. Shermer and W. H. Baddley, *J. Organometallic Chem.*, 1971, **27**, 83.
[317] M. Green and D. J. Westlake, *J. Chem. Soc. (A)*, 1971, 367.
[318] A. J. Cleland, S. A. Fieldhouse, B. H. Freeland, and R. J. O'Brien, *J. Organometallic Chem.*, 1971, **32**, C15.
[319] S. R. Su and A. Wojcicki, *J. Organometallic Chem.*, 1971, **31**, C34.
[320] H. Tom Dieck, I. W. Renk, and H.-P. Brehm, *Z. anorg. Chem.*, 1970, **379**, 169.
[321] M. I. Bruce, D. N. Sharrocks, and F. G. A. Stone, *J. Organometallic Chem.*, 1971, **31**, 269.
[322] T. A. Stephenson and E. Switkes, *Inorg. Nuclear Chem. Letters*, 1971, **7**, 805.

a broad resonance at τ 1.62 due to the NH protons and two sets of signals for the isocyanide ligands.[323]

^{31}P N.m.r. data have been reported for some complexes such as $OsCl_2(NO_2)(NO)(PMe_2Ph)_2$,[324] and *trans*-$RuX_2(MeSCH_2CH_2SMe)_2$ shows eight methyl resonances.[325]

A co-ordination shift of the DMSO methyl resonances of 0.58 p.p.m. downfield in $[Ru(NH_3)_5(DMSO)](PF_6)_2$ has been taken as evidence of co-ordination *via* sulphur.[326] For the salts $[Ru(NH_3)_5(NCR)]^{2+}$ there is no correlation of $\nu(C{\equiv}N)$ with the ^1H chemical shifts.[327] However, ^1H n.m.r. has proved useful in identifying polypyridines as complexes with ruthenium(II).[328] The coupling constants of $[M(D_2NCH_2CHMeND_2)_3]^{n+}$ (M = Ru^{II}, Co^{III}, or Pt^{IV}) have been interpreted with the ligands predominantly in one confirmation with the methyl group equatorial.[329] Oxidation of $[(en)_3Ru]^{2+}$ produces $[(en)_2Ru(NH{=}CH{-}CH{=}NH)]^{2+}$ as evidenced by the observation of the CH_2 at δ 2.6 and CH at δ 8.8 with relative intensities 4 : 1.[330]

Complexes of Co, Rh, and Ir.—Information concerning complexes of these elements can be found at the following sources: $(CO)_9Co_3CBr$,[137] $[Co(NH_3)_xX_yZ_{6-x-y}]^{n+}$,[180, 182, 183] $[Co(NH_3)_{6-2n}(C_2O_4)_n]^{3-2n}$,[181] $(Me_3SiCH_2$-$C_5H_4)Co(CO)_2$,[262] $(Me_3P)_4CoH$,[276] $[Co(pn)_3]^{3+}$,[329] $[Co(B_9C_2H_{11})_2]^-$,[1061] $(Cp)Co(B_8C_2H_{10})$, $[Co(B_8C_2H_{10})_2]^-$,[1065] $[Co(B_9C_5H_{13})_2]^-$, $[Co(B_9C_6H_{15})_2]^-$,[1071] $(Cp)Co(7,8$-$B_{10}C_2H_{12})$,[1078] $[(B_9C_2H_{10})_2S_2CHCo]^-$,[1082] $[Co(B_{10}C_2H_{10})_2]^{n+}$ (n = 0 or 1),[1084] $[\{(B_9C_2H_{11})Co(B_8C_2H_{10})\}_2Co]^{3-}$, $[(B_9C_2H_{11})Co(B_8C_2H_{10})$-$Co(B_9C_2H_{11})]^{2-}$,[1086] *trans*-$MX(CO)L_2$, *mer*-$MCl_3(CO)L_2$ (M = Rh or Ir),[113, 165, 166] *mer*-MCl_3L_3 (M = Rh or Ir),[165] $(Me_3SiC_5H_4)Rh(CO)_2$,[262] $Rh_2Cl_4(PBu^n{}_3)_4$, $Rh_2Cl_6(PBu^n{}_3)_3$,[168] $[Rh(HPPh_2)_4]^+$, and $[Ir(CO)$-$(HPPh_2)_4]^+$.[280]

In a review on the complexes of cobalt(III) with flexible quadridentate ligands, the use of ^1H n.m.r. to determine their stereochemistry has been discussed.[331]

For the compound $[CoH_2(dipy)(PR_3)_2]^+ ClO_4^-$, a value of 65 Hz has been reported for $^2J(^{31}P{-}^1H)$ (*cis*).[332] Unlike $(Ph_3P)CoH(N_2)$, $(Ph_3P)_3CoH_2$-(SiR_3) has a linewidth of 40 Hz for the hydride resonance.[333] ^1H N.m.r. has been used to follow H–D exchange between C_3D_6 and $HCo(CO)_4$.[334] The

323 J. Miller, A. L. Balch, and J. H. Enemark, *J. Amer. Chem. Soc.*, 1971, **93**, 4613.
324 J. Chatt, D. P. Melville, and R. L. Richards, *J. Chem. Soc. (A)*, 1971, 1169.
325 J. Chatt, G. J. Leigh, and A. P. Storace, *J. Chem. Soc. (A)*, 1971, 1380.
326 C. V. Senoff, E. Maslowsky, jun., and R. G. Goel, *Canad. J. Chem.*, 1971, **49**, 3585.
327 J. Chatt, G. J. Leigh, and N. Thankarajan, *J. Chem. Soc. (A)*, 1971, 3168.
328 F. E. Lytle, L. M. Petrosky, and L. R. Carlson, *Analyt. Chim. Acta*, 1971, **57**, 239.
329 J. K. Beattie and L. H. Novak, *J. Amer. Chem. Soc.*, 1971, **93**, 620.
330 B. C. Lane, J. E. Lester, and F. Basolo, *Chem. Comm.*, 1971, 1618.
331 G. R. Brubaker, D. P. Schaefer, J. H. Worrell, and J. I. Legg, *Co-ordination Chem. Rev.*, 1971, **7**, 161.
332 G. Mestroni, A. Camus, and C. Cocevar, *J. Organometallic Chem.*, 1971, **29**, C17.
333 N. J. Archer, R. N. Haszeldine, and R. V. Parish, *Chem. Comm.*, 1971, 524.
334 P. Taylor and M. Orchin, *J. Amer. Chem. Soc.*, 1971, **93**, 6504.

observation of the hydride resonance of $HCo(dmg)_2PBu^n_3$ at δ 6.0 is considered to be consistent with the formulation as $H^{\delta+}\cdots Co^{\delta-}$.[335]

$[(Ph_3P)_2(CO)_2Rh]_2$ absorbs hydrogen to give a single resonance at τ 19.34, which was considered to be characteristic of a rhodium–hydrogen bond; however, no coupling to rhodium or phosphorus was reported.[336] The compounds *trans*-$MCl_2H(PRBu^t_2)_2$ (M = Rh or Ir) have been prepared, and when M = Ir the hydride signal is above τ 60, and is the highest-field hydride signal so far reported. The presence of two hydridic hydrogens in $RhClH_2(PBu^t_3)_2$ has been demonstrated by observing the ^{31}P n.m.r. signal as a doublet $^1J(^{103}Rh–^{31}P)$ = 110.3 Hz of triplets when the t-butyl protons were decoupled.[337–339] Similarly, by the observation of a 1 : 5 : 10 : 10 : 5 : 1 sextet for the ^{31}P n.m.r. spectrum of $IrH_n(PEt_3)_2$ with the ethyl protons decoupled, it has been shown that n = 5, not 3 as previously reported. It has also been shown that *mer*-$IrH_3(PR_3)_3$ is *not* fluxional, and the ^1H (hydride) spectrum of *fac*-$IrH_3(PR_3)_3$ has been interpreted.[340] $Ir_2(CO)_6(PPh_3)_2$ reacts with H_2 to yield first $HIr(CO)_3(PPh_3)$, τ 21.8 {J(P–H) = 75 Hz}, and then $H_3Ir(CO)_2(PPh_3)$ which has two doublets at τ 20.6 (J = 17 Hz) and τ 21.5 (J = 120 Hz) in the ratio 2 : 1, consistent with (89).[341]

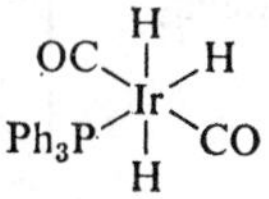

(89)

The compounds $[M(PPh_3)_2(CNR)_3]^+$ (M = Rh or Ir) undergo protonation in glacial acetic acid to yield $[MH(PPh_3)_2(CNR)_3]^{2+}$, as demonstrated by the observation of a ^1H n.m.r. signal in the region τ 20—27.[342] The existence of a bridging hydride in $(Cp)_2M_2Cl_3H$ (M = Rh or Ir) has been demonstrated by the observation of a triplet, $J(^{103}Rh–^1H)$ = 23 Hz, for the hydride resonance when M = Rh.[343]

For compounds of the type R_fCo\{bis(acetylacetone)ethylenedi-imine\}L (L = H_2O or pyridine), the ^1H n.m.r. spectra of the ligands show greater downfield movement for the CF_3 rather than CH_3, implying that there is less negative charge on the ligand. The ^{19}F n.m.r. spectrum shows broadening for the α-fluorine atoms, attributed to coupling to ^{59}Co.[344] ^1H N.m.r. has

[335] G. N. Schrauzer and R. J. Holland, *J. Amer. Chem. Soc.*, 1971, **93**, 1505.
[336] B. L. Booth, M. J. Else, R. Fields, and R. N. Haszeldine, *J. Organometallic Chem.*, 1971, **27**, 119.
[337] C. Masters, W. S. McDonald, G. Raper, and B. L. Shaw, *Chem. Comm.*, 1971, 210.
[338] C. Masters and B. L. Shaw, *J. Chem. Soc. (A)*, 1971, 3679.
[339] C. Masters, B. L. Shaw, and R. E. Stainbank, *Chem. Comm.*, 1971, 209.
[340] B. E. Mann, C. Masters, and B. L. Shaw, *J. Inorg. Nuclear Chem.*, 1971, **33**, 2195.
[341] L. Malatesta, M. Angoletta, and F. Conti, *J. Organometallic Chem.*, 1971, **33**, C43.
[342] J. W. Dart, M. K. Lloyd, J. A. McCleverty, and R. Mason, *Chem. Comm.*, 1971, 1197.
[343] C. White, D. S. Gill, J. W. Kang, H. B. Lee, and P. M. Maitlis, *Chem. Comm.*, 1971, 734.
[344] A. van den Bergen, K. S. Murray, and B. O. West, *J. Organometallic Chem.*, 1971, **33**, 89.

been used to demonstrate that β-bromostyrene reacts with $[Co^{I}(dmg)_2py]^-$ with retention of configuration at the double bond.[345] An AB pattern has been observed for the CH_2 group in $(Cp)RhX(PPh_3)COCH_2Ph$.[346] Similarly, $(Cp)Ir(PPh_3)(CO)CH_2R$ shows a complex resonance for the CH_2 group consistent with AB character.[347]

Metallation of the C(10) atom in benzo[*h*]quinoline in its reaction with $RhCl_3,3H_2O$ has been demonstrated by the disappearance of the signal due to H(10).[348] $(o\text{-}C_2H_3C_6H_4PPh_2)_2RhCl$ shows coupling between the vinyl group and the phosphorus, demonstrating that the vinyl group is co-ordinated to the metal.[349] The methyl resonance of (90) is distorted, implying an AB structure, and ^{31}P n.m.r. 'spin-tickling' has been used to obtain $^2J(^{31}P\text{–}^{31}P)$ of 450 Hz.[350] Previous workers had reported that 1-bromo-2-fluorocyclohexane reacts with $IrCl(CO)(PMe_3)_2$; however, when an attempt was made to follow the reaction in an n.m.r. tube by ^1H and ^{19}F n.m.r., no reaction occurred.[351]

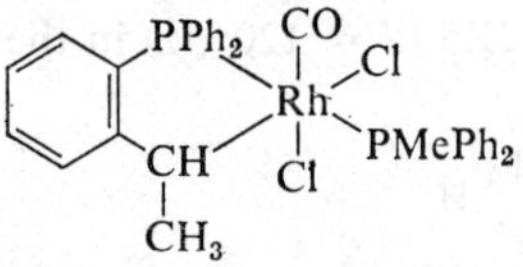

(90)

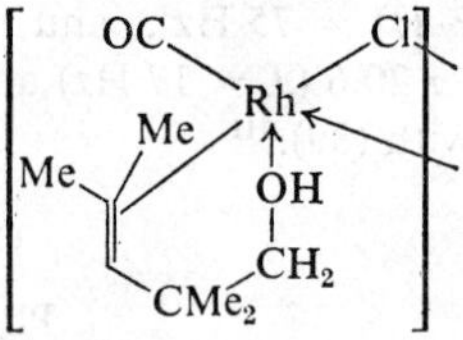

(91)

$Me_2C{=}CHCMe_2OH$ reacts with $PdCl_2$, $PtCl_2$, or $[Rh(CO)_2Cl]_2$ to form complexes such as (91). The signal for the OH proton moves from τ 7.58 in the free ligand to τ 3.76 in the complex, thus providing evidence for the co-ordination of the OH to rhodium.[352] ^1H N.m.r. shifts of quinones, L, move considerably to high field on co-ordination in $M(CO)(PPh_3)_2L$ (M = Co, Rh, or Ir).[353]

^1H N.m.r. has been used to demonstrate that hydrogen and butadiene can react with cobalt(II) and cyanide to give either $[MeCH{=}CHCH_2\text{-}Co(CN)_5]^{3-}$ or $[(1\text{-methylallyl})Co(CN)_4]^{2-}$, depending on the ratio of cyanide to cobalt.[354] Similarly, ^1H n.m.r. has proved useful in demonstrating that $(Ph_3P)_3CoHN_2$ reacts with butadiene to yield (92).[355]

345 M. D. Johnson and B. S. Meeks, *J. Chem. Soc.* (*B*), 1971, 185.
346 A. J. Hart-Davis and W. A. G. Graham, *Inorg. Chem.*, 1971, **10**, 1653.
347 A. J. Oliver and W. A. G. Graham, *Inorg. Chem.*, 1970, **9**, 2653.
348 M. Nonoyama and K. Yamasaki, *Inorg. Nuclear Chem. Letters*, 1971, **7**, 943.
349 M. A. Bennett and E. J. Hann, *J. Organometallic Chem.*, 1971, **29**, C15.
350 M. A. Bennett, S. J. Gruber, E. J. Hann, and R. S. Nyholm, *J. Organometallic Chem.*, 1971, **29**, C12.
351 F. R. Jensen and B. Kruckel, *J. Amer. Chem. Soc.*, 1971, **93**, 6339.
352 L. K. Atkinson and D. C. Smith, *J. Chem. Soc.* (*A*), 1971, 3592.
353 G. La Monica, G. Navazio, and P. Sandrini, *J. Organometallic Chem.*, 1971, **31**, 89.
354 T. Funabiki and K. Tarama, *Chem. Comm.*, 1971, 1177.
355 P. V. Rinze and H. Nöth, *J. Organometallic Chem.*, 1971, **30**, 115.

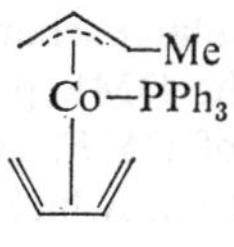

(92)

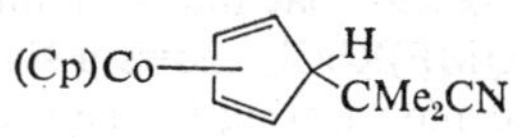

(93)

The radical $\cdot CMe_2(CN)$ adds stereospecifically to $(Cp)_2Co$ to yield (93), as is shown by the presence of the *endo*-hydrogen at τ 7.46.[356] Conversion of $(C_8H_{12})Rh(acac)$ into $[(C_8H_{12})Rh(PR_3)_2]BF_4$ causes the olefin proton to move downfield, which is inconsistent with that expected from consideration of both σ- and π-bonding effects; $[Pd(acac)(PR_3)_2]BF_4$ is similar.[357]

Because $J(^{103}Rh–H)$ and $J(P–H)$ are accidentally equal, the Cp ^1H resonance of $(Cp)Rh(PPh_3)_2$ is a quartet.[358] ^1H N.m.r. has been used to demonstrate the occurrence of base-catalysed deuteriation of the methyl group in compounds such as $[(C_5Me_5)M(OAc)_2(H_2O)]_n$ (M = Rh or Ir),[359] and to demonstrate *cis–trans* isomerism of the bridge methoxides in $[(1,5\text{-cyclo-octadiene})_2Ir_2(\mu\text{-}OMe)_2]$, where two OMe resonances were observed.[360]

There have been a number of papers concerning the magnitude of $^2J(^{31}P–^{31}P)$ either directly or as a result of effects in the ^1H and ^{19}F n.m.r. spectra. Thus $Co(Cp)\{P(OMe)_3\}_2$ shows a 1 : 2 : 1 triplet for the methyl group, implying that $^2J(^{31}P–^{31}P)$ is large.[361] Similarly, $(Me_3P)_3Co(NO)$ shows a 'pseudo-triplet' in the ^1H n.m.r. spectrum.[362] The compounds $[Rh\{P(OMe)_nPh_{3-n}\}_4]^+$ (n = 1 or 2) and *cis*-$Rh[P(OMe)_3]_4HX$ show unusual and inexplicable five-line patterns for the methyl resonance, such as in Figure 1.[363, 364] In the case of fluorophosphines, it has proved possible to calculate $^2J(^{31}P–^{31}P)$ for compounds such as $M(acac)(PF_3)_2$ and $(C_5Me_5)M(PF_3)_2$ (M = Rh or Ir) but the spectra of complexes such as $M_2(PF_4)_8$ have defied analysis.[365, 366] ^{31}P N.m.r. data, including $^2J(^{31}P–^{31}P)$, have been reported for $IrE_2(diphos)_2Cl$ (E = O, S, or Se).[367]

From the observation that $CoCl_2(NO)(PMePh_2)_2$ shows a triplet for the methyl resonance, it has been concluded that the tertiary phosphines are mutually *trans*.[368] $(R_3P)_3RhCl$ reacts with P_4 to yield $(R_3P)_2RhCl(P_4)$; however, only signals due to R_3P were detected in the ^{31}P n.m.r. spectrum

[356] G. E. Herberich and J. Schwarzer, *Angew. Chem. Internat. Edn.*, 1970, **9**, 897.
[357] B. F. G. Johnson, J. Lewis, and D. A. White, *J. Chem. Soc. (A)*, 1971, 2699.
[358] H. Yamazaki and N. Hagihara, *Bull. Chem. Soc. Japan*, 1971, **44**, 2260.
[359] J. W. Kang and P. M. Maitlis, *J. Organometallic Chem.*, 1971, **30**, 127.
[260] M. Green, T. A. Kuc, and S. H. Taylor, *J. Chem. Soc. (A)*, 1971, 2334.
[361] V. Harder, J. Müller, and H. Werner, *Helv. Chim. Acta*, 1971, **54**, 1.
[362] H.-F. Klein, *Angew. Chem. Internat. Edn.*, 1971, **10**, 343.
[363] L. M. Haines, *Inorg. Chem.*, 1971, **10**, 1685.
[364] L. M. Haines, *Inorg. Chem.*, 1971, **10**, 1693.
[365] M. A. Bennett and D. J. Patmore, *Inorg. Chem.*, 1971, **10**, 2387.
[366] R. B. King and A. Efraty, *J. Amer. Chem. Soc.*, 1971, **93**, 5260.
[367] A. P. Ginsberg and W. E. Lindsell, *Chem. Comm.*, 1971, 232.
[368] J. P. Collman, P. Farnham, and G. Dolcetti, *J. Amer. Chem. Soc.*, 1971, **93**, 1788.

and it was suggested that inter- or intra-molecular exchange was occurring.[369] $(py)_3(DMF)RhCl_3$ shows a signal for the DMF protons separate from solvent, demonstrating the co-ordination of DMF. This compound reacts with BH_4^- to yield $(py)_2(DMF)RhCl_2BH_4$, which shows broad

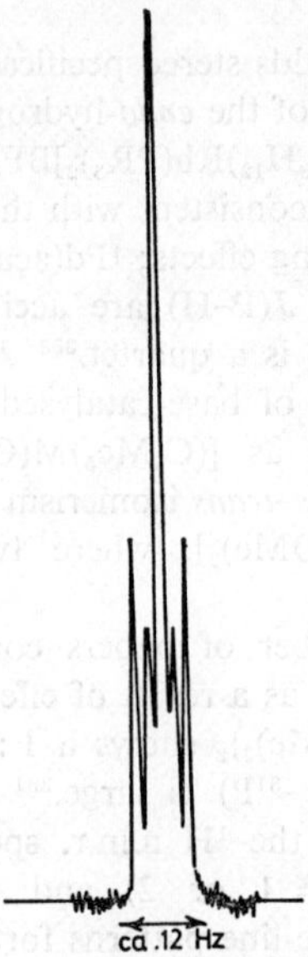

Figure 1 *Methyl resonance pattern observed in the n.m.r. spectra of* $[Rh\{P(OMe)_2Ph\}_4][PF_6]$ *and* $[Rh\{P(OMe)Ph_2\}_4][PF_6]$
(Reproduced by permission from *Inorg. Chem.*, 1971, **10**, 1685)

resonances both in the ^1H and ^{11}B n.m.r. spectra.[370] The ^1H n.m.r. spectrum of $IrCl_2(COEt)(CO)(AsMe_2Ph)$ shows four methyl resonances; thus only two isomers (94) and (95) are possible. When the solution is warmed the methyl resonances collapse, implying arsine exchange.[371] $IrCl_3(SEt_3)_3$ had originally been formulated as the *fac*-isomer, but as two CH_2 signals in the ratio 2 : 1 were observed, the compound has been reformulated as the

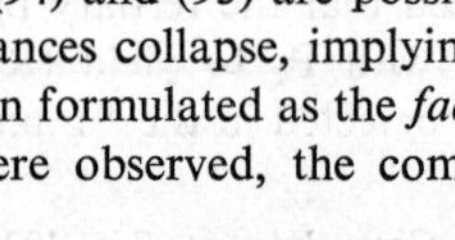

(94)

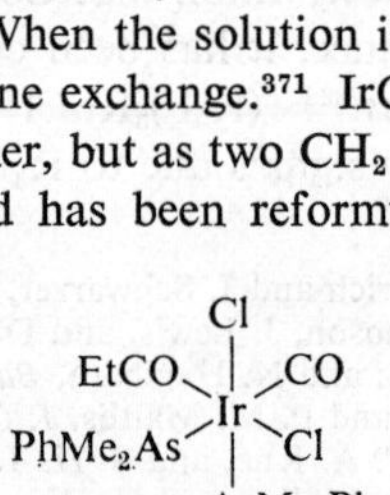

(95)

[369] A. P. Ginsberg and W. E. Lindsell, *J. Amer. Chem. Soc.*, 1971, **93**, 2082.
[370] P. Abley, I. Jardine, and F. J. McQuillin, *J. Chem. Soc.* (*C*), 1971, 840.
[371] R. W. Glyde and R. J. Mawby, *Inorg. Chim. Acta*, 1971, **5**, 317.

mer-isomer.[372] Similarly, $RhCl_3(EtCN)_3$ has been shown to be the *mer*-isomer.[373]

The ^1H n.m.r. spectra of $[Co(phen)_{3-n}(en)_n]^{3+}$ (n = 0—2) have been analysed and the upfield shift of the H(2) and H(9) protons of the phenanthroline ring were explained as due to the ring current of neighbouring rings.[374] The broadness of the ^1H signals of $[Co(en)_3]^{3+}$ compared to the Pt^{IV}, Rh^{III}, Ir^{III}, or Ru^{II} analogues has been attributed to coupling to the ^{59}Co.[375] In a very detailed paper, deuteriation and 'tickling' have been used to assist a complete interpretation of the ^1H n.m.r. spectra of $[Co(pdta)]^-$ and $[Co(edta)]^-$.[376] Similarly, the ^1H n.m.r. spectrum of $[Co(edda)(H_2O)_2]$-ClO_4 has been reinterpreted and is now considered to be consistent with (96) rather than (97).[377]

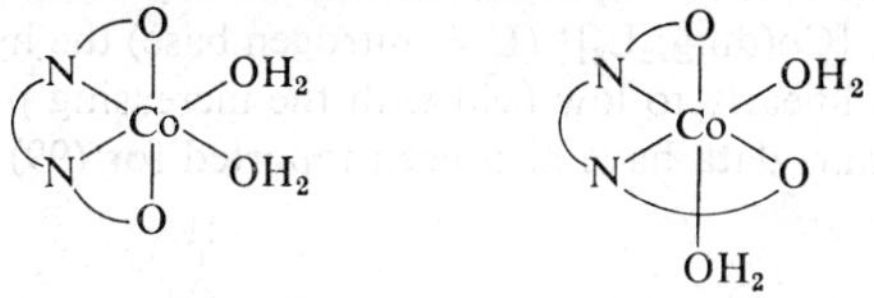

(96) (97)

For complexes of the type $Co(acac)_2NO_2L$ (L = nitrogen ligand) the variations in ^1H n.m.r. chemical shift has been related to inductive and magnetic shielding effects; electric field and *cis* and *trans* effects were found to be small.[378] Similarly, the ^1H chemical shifts for $[Co(en)_2(RCO_2)_2]ClO_4$ have been interpreted in terms of differences in charge distribution.[379] It has been reported that for the complexes $M(acac)_3$, $M(sacac)_3$, and $M(sacsac)_3$ (M = Co, Rh, or Ir) the methine proton moves progressively to low field.[380] The ^1H n.m.r. spectrum of $Co_2(acac)_4(\mu\text{-}OH)$ shows two methyl resonances and one CH resonance, which is consistent with (98).[381] The ^1H n.m.r. spectra of the chelate complexes of β-ketodithioesters with cobalt(III), nickel(II), or zinc(II) have been interpreted as showing π-electron delocalization.[382] The complexity of the methylene resonances of $[Co\{MeC(CH_2NHCH_2CH_2NH_2)_3\}]^{3+}$ has been interpreted as certain conformers being preferred.[383] The two isomers of *trans*-$Co(dmg)_2(py)$-(SCN) (S- and N-bonded thiocyanate) show separate methyl resonances, but in DMSO or DMF an additional common resonance appears which

[372] E. A. Allen, N. P. Johnson, and W. Wilkinson, *Chem. Comm.*, 1971, 804.
[373] B. D. Catsikis and M. L. Good, *Inorg. Chem.*, 1971, **10**, 1522.
[374] H. Ito, J. Fujita, and T. Ito, *Bull. Chem. Soc. Japan*, 1971, **44**, 723.
[375] J. K. Beattie, *Inorg. Chem.*, 1971, **10**, 426.
[376] J. L. Sudmeier, A. J. Senzel, and G. L. Blackmer, *Inorg. Chem.*, 1971, **10**, 90.
[377] K. Kuroda and K. Watanabe, *Bull. Chem. Soc. Japan*, 1971, **44**, 1034.
[378] L. J. Boucher, E. J. Battis, and N. G. Paez, *J. Inorg. Nuclear Chem.*, 1971, **33**, 1373.
[379] A. W. Chester, *J. Inorg. Nuclear Chem.*, 1971, **33**, 3471.
[380] G. A. Heath and R. L. Martin, *Austral. J. Chem.*, 1971, **24**, 2061.
[381] L. J. Boucher and D. R. Herrington, *J. Inorg. Nuclear Chem.*, 1971, **33**, 4349.
[382] G. Dorange and J. E. Guerchais, *Bull. Soc. chim. France*, 1971, 43.
[383] J. E. Sarneski and F. L. Urbach, *J. Amer. Chem. Soc.*, 1971, **93**, 884.

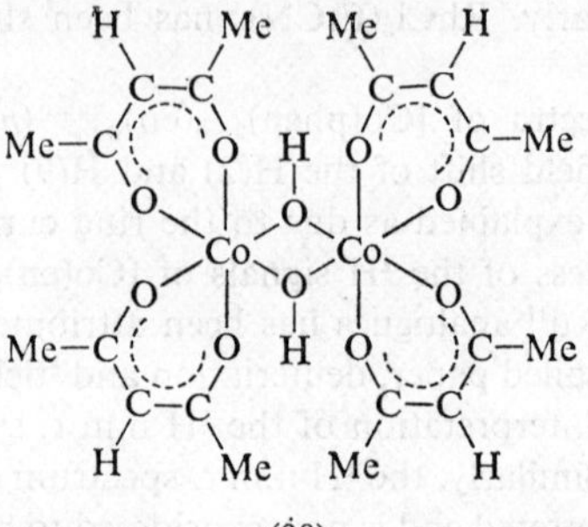

(98)

has been interpreted as the replacement of SCN^- by solvent.[384] It has been reported that for $[Co(dmg)_2L_2]^+$ (L = nitrogen base) the hydrogen-bonded hydrogens move linearly to low field with the increasing pK_a values of the bases.[385] ^1H N.m.r. data have also been reported for (99)[386] and (100).[387]

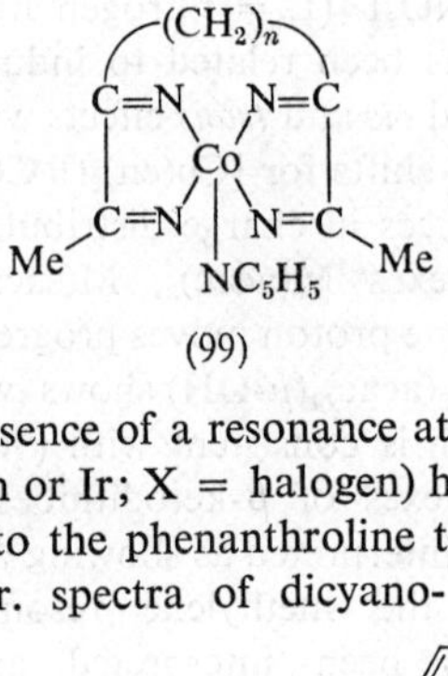

(99)

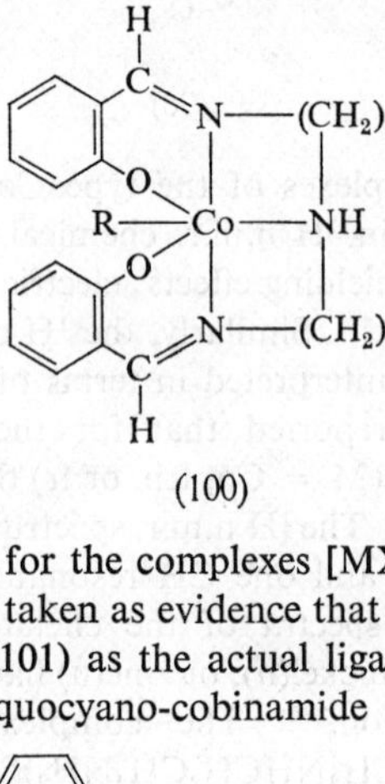

(100)

The absence of a resonance at *ca.* τ 4 for the complexes $[MX_2(phen)_2]^+$ (M = Rh or Ir; X = halogen) has been taken as evidence that water does not add to the phenanthroline to give (101) as the actual ligand.[388] The ^{13}C n.m.r. spectra of dicyano- and aquocyano-cobinamide have been

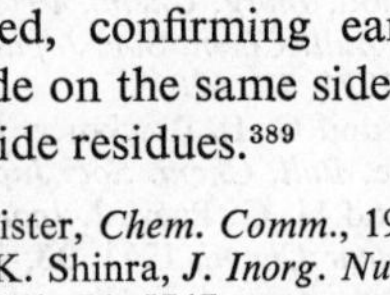

(101)

reported. Although aquocyanocobinamide has only 46 carbon atoms, 60 ^{13}C signals were resolved, confirming earlier suggestions that two isomers exist with the cyanide on the same side of the corrin ring as either the acetamide or propionamide residues.[389]

384 R. L. Hassel and J. E. Burmeister, *Chem. Comm.*, 1971, 568.
385 Y. Yamano, I. Masuda, and K. Shinra, *J. Inorg. Nuclear Chem.*, 1971, **33**, 521.
386 J. Retey, *Helv. Chim. Acta*, 1971, **54**, 5747.
387 W. M. Coleman and L. T. Taylor, *J. Amer. Chem. Soc.*, 1971, **93**, 5446.
388 J. A. Broomhead and W. Grumley, *Inorg. Chem.*, 1971, **10**, 2002.
389 D. Doddrell and A. Allerhand, *Chem. Comm.*, 1971, 728.

Complexes of Ni, Pd, and Pt.—Information concerning complexes of these elements can be found at the following sources: *cis*- and *trans*-$MX_2(PHBu^t_2)_2$ (M = Ni, Pd, or Pt),[113] $(CO)_3Ni(PF_nBu^t_{3-n})$,[119] (Cp)Ni-$(PPh_3)X$,[164] $[Ni(RCOCH_2CS_2Me)_n]^{2+}$,[382] $[Ni(B_{10}C_2H_{10})_2]^{n+}$ (*n* = 0 or 1),[1084] *cis*- and *trans*-$MX_2(PR_3)_2$ (M = Pd or Pt),[165, 167] $Fe(R_2AsC_5H_4)_2MX_2$ (M = Pd or Pt),[310] $MCl(Me_2C{=}CHCMe_2CHOH)_2$ (M = Pd or Pt),[352] $[(C_3H_5)PdCl]_2$, $[C_2H_4PtCl_3]^-$,[137] $[PdX(allyl)]_2$,[138] $[(acac)PdL_2]BF_4$,[357] *cis*-$PtMe_2(EMe_2Ph)_2$ (E = P or As),[99] $Me_2Pt(AsMe_3)_2XY$,[100] $PtCl_2L^1L^2$ (L^1, L^2 are phosphorus ligands),[114] *cis*-$PtX_2\{P(OMe)_3\}_2$,[115] and *cis*-$(EtCD_2CH_2)_2$-$Pt(PPh_3)_2$.[116]

Treatment of a mixture of *cis*-$PtCl_2(PPh_3)_2$ and (102) (X = Ph, Cl, or Br) with N_2H_4 yields *trans*-$PtH(PPh_3)_2$(tetrazolato). However, this product

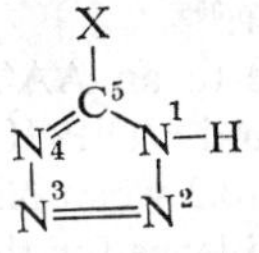

(102)

was shown by the doubling of the hydride resonance to be a mixture of two species. It was suggested that these species were due to the two ways in which the tetrazole ligand could bind, *i.e. via* N(1) or N(2).[390] Contrary to earlier reports, when HCl is added to *trans*-$Pt(PPh_3)_2HCl$, $Pt(PPh_3)_2H_2Cl_2$ is not formed. There is no increase in the size of the hydride resonance, and no other hydride resonances were detected.[391] Treatment of *trans*-$ClPt(PEt_3)_2H$ or *trans*-$ClPt(PEt_3)_2(GeH_2Cl)$ with an excess of GeH_3Cl yields a mixture which shows a complex hydride spectrum. However, the spectrum was interpreted as arising from a mixture of species $HPt(PEt_3)_2$-$(GeH_2Cl)_n(GeHCl_2)_{3-n}$.[392] For a series *trans*-$PtHX(PPh_3)_2$, τ (hydride) has been used to determine the *trans* influence of X, but $^1J(^{195}Pt-^1H)$ was not reported.[393] H_2E (E = S or Se) reacts with $Pt(PPh_3)_3$ to yield a mixture; the ^1H n.m.r. spectrum was interpreted as arising from a mixture of $H_2EPt(PPh_3)_2$ and (103).[394] ^1H N.m.r. data have been reported for a number

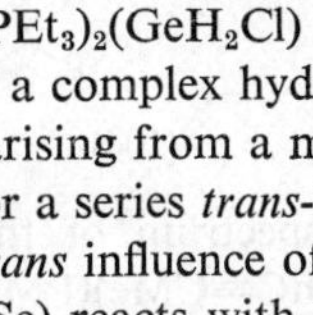

(103)

[390] J. H. Nelson, D. L. Schmitt, R. A. Henry, D. W. Moore, and H. A. Jonassen, *Inorg. Chem.*, 1970, **9**, 2678.
[391] J. T. Dumler and D. M. Roundhill, *J. Organometallic Chem.*, 1971, **30**, C35.
[392] J. E. Bentham and E. A. V. Ebsworth, *J. Chem. Soc.* (*A*), 1971, 2091.
[393] D. M. Roundhill, P. B. Tripathy, and B. W. Renoe, *Inorg. Chem.*, 1971, **10**, 727.
[394] R. Ugo, G. La Monica, S. Cenini, A. Segre, and F. Conti, *J. Chem. Soc.* (*A*), 1971, 522.

of complexes *trans*-$(R_3P)_2MHX$ (M = Ni or Pd; X = halogen or BH_4^-). When *trans*-$(R_3P)_2MX_2$ is mixed with *trans*-$\{(C_6H_{11})_3P\}_2NiH(BH_4)$, new hydride signals due to species such as *trans*-$(R_3P)\{(C_6H_{11})_3P\}MXH$ appear. It is interesting that no coupling was observed between the phosphorus and the hydride for *trans*-$(Bu^n{}_3P)_2PdHCl$.[395] CS_2 inserts into the Pt—H bond of *trans*-$PtHCl(PPh_3)_2$ to yield *trans*-$PtCl(PPh_3)_2SC(S)H$, as shown by a resonance at τ 6.95.[396]

$PtB_3H_7(R_3P)_2$ reacts with PR_3 to yield $Pt(R_3P)_n$ (n = 3 or 4); ^{31}P n.m.r. data were reported, but it was necessary to cool to − 55 °C in order to observe $^1J(^{195}Pt-^{31}P)$. $Pt(PEt_3)_3$ absorbs hydrogen to yield $H_2Pt(PEt_3)_3$, demonstrated by the observation of a quartet with platinum satellites at τ 23.37, and a triplet for the ^{31}P resonance with ethyl protons decoupled.[397] The reaction of $(PPh_3)_2Ni(C_2H_4)$ with Ph_2BBr to yield $[(Ph_3P)_2Ni(\mu-BPh_2)]_2$ has been followed by ^{11}B n.m.r.[398]

cis-$NiMe_2$(diphos) gives rise to an $AA'X_3X'_3$ spectrum for the methyl resonance, where $^3J(^{31}P-^1H)$ and $^3J(^{31}P-^1H')$ are of opposite sign.[399] The observation of a doublet at τ 10.3 when $Ni(acac)_2P(C_6H_{11})_3$ is treated with $[AlMe_3]_2$ has been taken as evidence for the formation of a nickel–methyl bond in $(acac)NiMe\{P(C_6H_{11})_3\}$.[400] When $(p\text{-}MeOC_6F_4CO_2)_2Ni(phen)$ is heated, $(p\text{-}MeOC_6F_4)_2Ni(phen)$ and (104) (R = p-$MeOC_6F_4$ or H) are

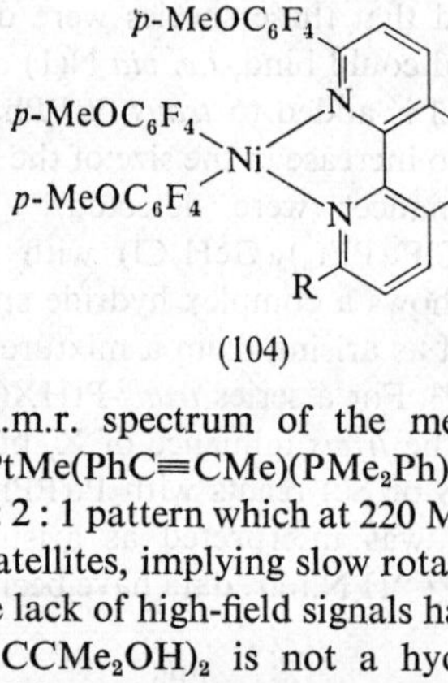

(104)

formed.[401] The ^1H n.m.r. spectrum of the methyl group attached to phosphorus of *trans*-$[PtMe(PhC{\equiv}CMe)(PMe_2Ph)_2]^+$ consists of a 1 : 2 : 2 : 2 : 5 : 8 : 8 : 8 : 5 : 2 : 2 : 2 : 1 pattern which at 220 MHz simplifies to a double triplet with platinum satellites, implying slow rotation about the platinum–acetylene bond.[402] The lack of high-field signals has been taken as evidence that *cis*-$(PPh_3)_2Pt(C{\equiv}CCMe_2OH)_2$ is not a hydride.[403] The ^1H n.m.r. spectra of $[Me_3Pt(NCS)]_4$, $[Me_2Au(NCS)]_2$, $[Me_3Pt(NCS)(C_5H_5N)]_2$, and

[395] M. L. H. Green, H. Munakata, and T. Saito, *J. Chem. Soc.* (*A*), 1971, 469.
[396] A. Palazzi, L. Busetto, and M. Graziani, *J. Organometallic Chem.*, 1971, **30**, 273.
[397] D. H. Gerlach, A. R. Kane, G. W. Parshall, J. P. Jesson, and E. M. Muetterties, *J. Amer. Chem. Soc.*, 1971, **93**, 3543.
[398] C. S. Cundy and H. Nöth, *J. Organometallic Chem.*, 1971, **30**, 135.
[399] M. L. H. Green and M. J. Smith, *J. Chem. Soc.* (*A*), 1971, 639.
[400] P. W. Jolly, K. Jonas, C. Krüger, and Y.-H. Tsay, *J. Organometallic Chem.*, 1971, **33**, 109.
[401] P. G. Cookson and G. B. Deacon, *J. Organometallic Chem.*, 1971, **33**, C38.
[402] M. H. Chisholm and H. C. Clark, *Inorg. Chem.*, 1971, **10**, 2557.
[403] A. Furlani, P. Bicev, M. V. Russo, and P. Carusi, *J. Organometallic Chem.*, 1971, **29**, 321.

$[Me_3Pt(NCS)(EPh_3)]_2$ (E = P or As) have been reported. In the case of $[Me_3Pt(NCS)]_4$, two signals in the ratio 2 : 1 were found, whereas $[Me_3Pt(NCS)(C_5H_5N)]_2$ has two methyl signals. The data were discussed in terms of various isomers.[404]

For the compound *trans*-$PtMe(PMe_2Ph)_2C\langle^{OR^1}_{CH_2R^2}$, when R^1 = H, Me, Et, or Pr^n, the phosphorus methyl protons show a triplet resonance. However, when R^1 = Bu^n or Pe^n, the signal is a double triplet, implying the absence of a plane of symmetry about the Pt—P bond.[405] Similarly, for $[(EtNC)Pt(PMe_2Ph)_2C\langle^{R}_{NHEt}](PF_6)_2$ a complex phosphorus methyl resonance was observed which was interpreted as arising from four inequivalent phosphorus methyl groups.[406] The ^1H n.m.r. spectrum of *trans*-$[PtCl\{C(NHPh)_2\}(PEt_3)_2]ClO_4$ shows two distinct phenyl patterns. This doubling is thought to arise from the carbene ligand being as (105) rather

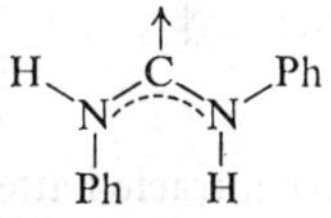

(105)

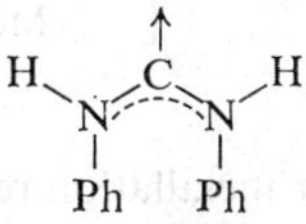

(106)

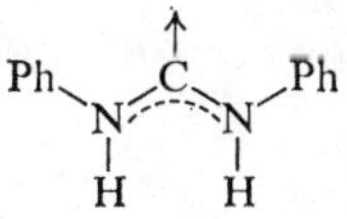

(107)

than as a mixture of (106) and (107) on the basis of molecular models.[407] The ^1H n.m.r. spectrum of (108) shows the α-CH at τ 4.33 with *J*(Pt–H) of 115 Hz. It is interesting that the CH_2 protons of the ethyl group are inequivalent.[408]

For *trans*-$[PtMeL_2(NCR)]^+ X^-$, $^2J(^{195}Pt-^1H)$ of 80 Hz has been taken as evidence that RCN is a weak σ-donor. These compounds add alcohols to

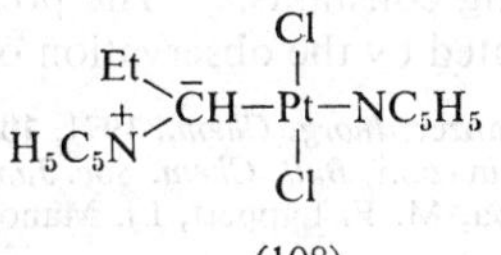

(108)

404 G. C. Stocco and R. S. Tobias, *J. Co-ordination Chem.*, 1971, **1**, 133.
405 M. H. Chisholm and H. C. Clark, *Inorg. Chem.*, 1971, **10**, 1711.
406 H. C. Clark and L. E. Manzer, *J. Organometallic Chem.*, 1971, **30**, C89.
407 E. M. Badley, B. J. L. Kilby, and R. L. Richards, *J. Organometallic Chem.*, 1971, **27**, C37.
408 R. D. Gillard, M. Keeton, R. Mason, M. F. Pilbrow, and D. R. Russell, *J. Organometallic Chem.*, 1971, **33**, 247.

form nitrenes.[409] Treatment of *trans*-$PtIMe(PPh_3)_2$ with $C_6H_{11}NC$ yields *trans*-$[PtMe(PPh_3)_2(CNC_6H_{11})]^+ I^-$, as shown by the observation of the methyl signal at τ 9.95 as a triplet with platinum satellites, $^2J(^{195}Pt-^1H)$ = 60 Hz. However, the methyl group slowly migrates to yield *trans*-$PtI(CMe{=}NC_6H_{11})(PPh_3)_2$.[410] As $^1J(^{195}Pt-^{31}P)$ for $PtCl_2(PEt_3)$-$\overline{CNPh(CH_2)_2N}Ph$ is 2440 Hz, then the PEt_3 is probably *trans* to the carbene rather than the chlorine.[411] Hydrazine reacts with $[Pt(CNMe)_4]^{2+}$ to yield $[Pt(C_4H_9N_4)(MeNC)_2]^{2+}$. A methyl resonance was observed at τ 7.08 split by a hydrogen that can be deuteriated. Thus structure (109) was suggested.[412]

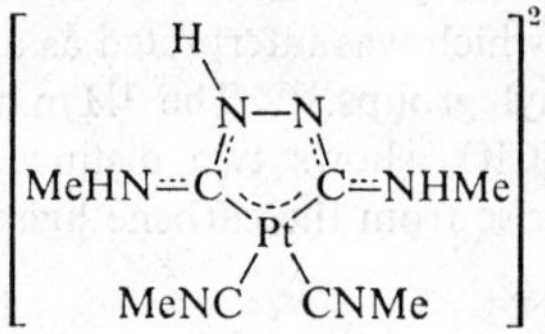

(109)

Intramolecular metallation reactions have attracted attention. For (110) the observation of the CH resonance at δ 7.68 close to that for free PhCH=NPh provides evidence for co-ordination *via* nitrogen rather than the double bond.[413] Compound (111) shows an ABMX spectrum and the ^{19}Fc

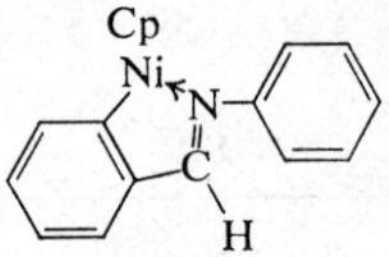

(110)

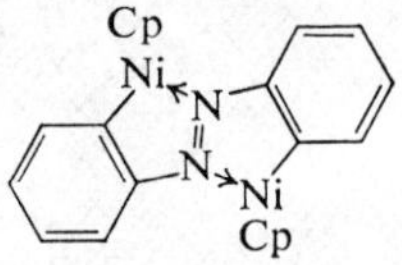

(111)

chemical shift of (112) has been used to derive Taft parameters.[414] The ^{1}H n.m.r. spectrum of (113) shows inequivalence of the NEt and benzylic methylene groups. This behaviour was attributed to the structure being non-planar due to crowding.[415] The stereochemistry of (114), produced by the insertion of norbornadiene into (allyl)Pd(hfac), has been established from the magnitude of coupling constants.[416] The presence of the free olefin in (115) has been demonstrated by the observation of signals at *ca.* τ 4.5.[417]

[409] H. C. Clark and L. E. Manzer, *Inorg. Chem.*, 1971, **10**, 2699.
[410] Y. Yamamota and H. Yamazaki, *Bull. Chem. Soc. Japan*, 1971, **44**, 1873.
[411] D. J. Cardin, B. Cetinkaya, M. F. Lappert, Lj. Manojlović-Muir, and K. W. Muir, *Chem. Comm.*, 1971, 400.
[412] G. Rouchias and B. L. Shaw, *J. Chem. Soc. (A)*, 1971, 2097.
[413] Yu. A. Ustynyuk, V. A. Chertkov, and I. V. Barinov, *J. Organometallic Chem.*, 1971, **29**, C53.
[414] I. V. Barinov, T. I. Voyevodskaya and Yu. A. Ustynyuk, *J. Organometallic Chem.*, 1971, **30**, C28.
[415] S. Trofimenko, *J. Amer. Chem. Soc.*, 1971, **93**, 1808.
[416] R. P. Hughes and J. Powell, *J. Organometallic Chem.*, 1971, **30**, C45.
[417] G. Carturan, M. Graziani, and U. Belluco, *J. Chem. Soc. (A)*, 1971, 2509.

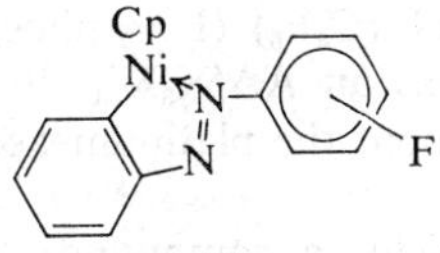

(112)

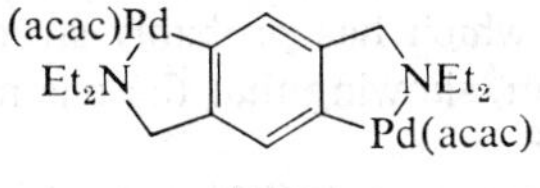

(113)

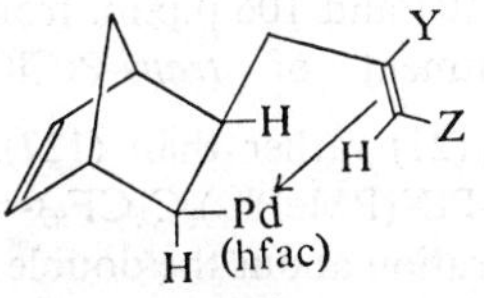

(114)

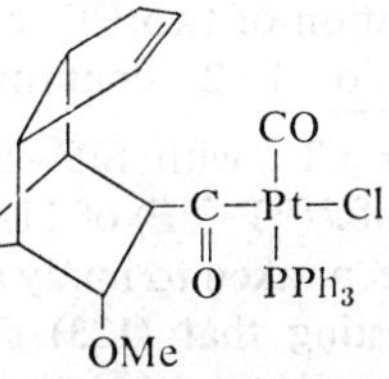

(115)

From the presence of three sharp singlets in the τ 7—9 region with relative intensities 1 : 2 : 3, and from the absence of a signal attributable to a methylene group, it was concluded that the product from heating *cis*-$PtCl_2\{P(OTol)_3\}_2$ is (116).[418] Similarly, *trans*-$PtX_2(PBu^t{}_nR_{3-n})_2$ (n = 1 or 2; R = Ph, *o*-Tol, or Pr^n) undergo internal metallation reactions to yield products such as (117) and (118). The *trans* stereochemistry of the phosphorus nuclei was demonstrated by the observation of an AB pattern with

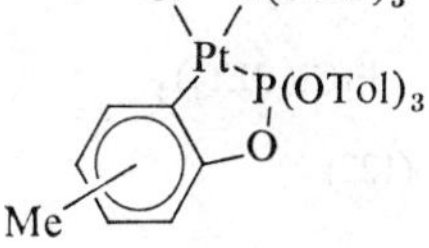

(116)

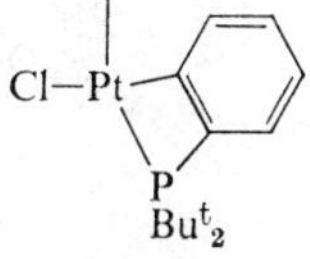

(117)

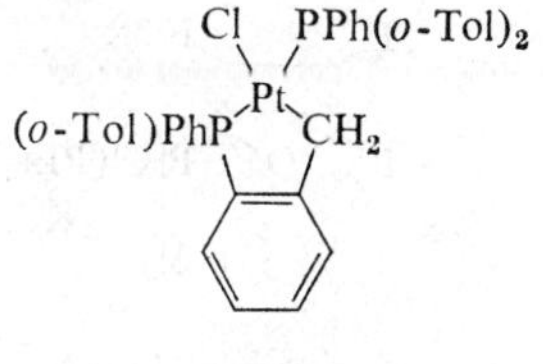

(118)

^{195}Pt satellites in the ^{31}P n.m.r. spectrum, with $^2J(^{31}P-^{31}P)$ of *ca.* 380 Hz. In the case of (117) the ^{31}P n.m.r. signal of the metallated phosphine is *ca.* 90 p.p.m. to high field of the non-metallated one, and in (118) a strongly coupled AB pattern was observed.[419] The values of J(Pt–H), J(H–H), and J(P–H) and the observation of only one ^{31}P n.m.r. signal for each complex have been used to demonstrate that acids, HX, add to acetylene–platinum(0) complexes to yield *trans*-$PtX(CR{=}CHR)(PPh_3)_2$ where the platinum and proton are *cis* with respect to the double bond.[420]

The ^{19}F n.m.r. spectrum of $Pt(C_2F_4)L_2$ is of the $AA'X_2X'_2$ type with platinum satellites. $^2J(^{195}Pt-^{19}F)$ changes from 501 Hz (L_2 = en) to 278 Hz (L = PPh_3), which is consistent with the Chatt–Dewar description of the

[418] E. W. Ainscough and S. D. Robinson, *Chem. Comm.*, 1971, 130.
[419] A. J. Cheney, B. E. Mann, B. L. Shaw, and R. M. Slade, *J. Chem. Soc.* (*A*), 1971, 3833.
[420] B. E. Mann, B. L. Shaw, and N. I. Tucker, *J. Chem. Soc.* (*A*), 1971, 2667.

metal–olefin bond.[421] Similarly, *trans*-$PtXMeL_2(C_4F_6)$ (L = phosphorus ligand) which has probably structure (119) has an $AA'X_3X'_3$ ^{19}F n.m.r. spectrum, showing that there is no rotation about the platinum–acetylene bond.[422]

Treatment of $Pt(PPh_3)_4$ with $(CF_3)_2CN_2$ yields a compound showing three ^{19}F resonances in the ratio 2 : 1 : 1, which is consistent with (120).[423] The observation of two ^{19}F resonances at 104 and 106 p.p.m. from $CFCl_3$ in the ratio 1 : 2 confirm that treatment of *trans*-$PtCl(PEt_3)_2$—$\overline{C{=}CF{-}CF_2CF_2}$ with SiF_4–H_2O yields (121) rather than (122).[424] The magnitude of $J(^{19}F-^{19}F)$ of 11.6 Hz for *cis*-$PtX(PMePh_2)_2C(CF_3){=}C(CF_3)$-HgCl has been taken to imply a *cis* configuration about the double bond.[425] It is interesting that (123) shows two apparent triplets for the OMe resonance, suggesting that $^2J(^{31}P-^{31}P)$ (*cis*) is large.[426]

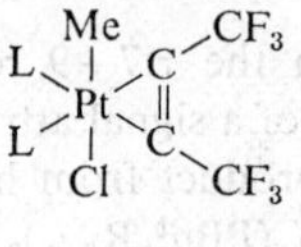

(119)

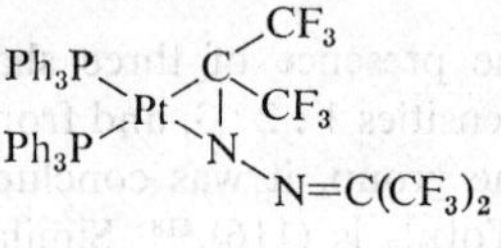

(120)

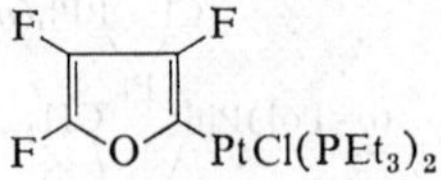

(121)

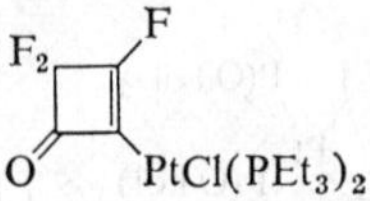

(122)

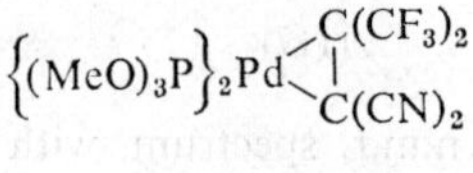

(123)

The ^1H n.m.r. spectra of *trans*-$PtCl_2(CH_2{=}CH'R)(py)$ have been analysed. It was found that the relative sign of $J(H'-H_a)$ (H_a is the α-proton of R) is the same as for the free ligand. The mechanism of olefin rotation was discussed at length.[427] The ^1H n.m.r. spectrum of $(C_2H_4)PtCl_2(MeCN)$ shows no MeCN exchange. If KBr is added, then the signal due to the

[421] R. D. W. Kemmitt and R. D. Moore, *J. Chem. Soc. (A)*, 1971, 2472.
[422] H. C. Clark and R. J. Puddephatt, *Inorg. Chem.*, 1971, **10**, 18.
[423] J. Clemens, R. E. Davis, M. Green, J. D. Oliver, and F. G. A. Stone, *Chem. Comm.*, 1971, 1095.
[424] W. J. Cherwinski and H. C. Clark, *J. Organometallic Chem.*, 1971, **29**, 451.
[425] D. M. Barlex, R. D. W. Kemmitt, and G. W. Littlecott, *Chem. Comm.*, 1971, 199.
[426] H. D. Empsall, M. Green, S. K. Shakshooki, and F. G. A. Stone, *J. Chem. Soc. (A)*, 1971, 3472.
[427] E. Lazzaroni and C. A. Veracini, *J. Organometallic Chem.*, 1971, **33**, 131.

co-ordinated ethylene broadens, ^{195}Pt coupling is lost, and the signal moves to high field; however, the methyl cyanide signal is unaffected.[428] When olefins, *e.g.* $CH_2=CHCN$, are complexed to $Pt(PPh_3)_n$, the olefinic protons undergo a large high-field shift.[429] The ^{1}H n.m.r. spectra of the complexes *trans*-$PdCl_2(R^1N=NR^2)_2$ (R^2 = cyclohexene) were badly resolved at 60 MHz and it was suggested that these complexes merit a detailed investigation at a higher frequency.[430]

The observation of a 1 : 1 : 1 : 1 quartet for the acetylenic proton in $(Ph_3P)_2Pt(HC{\equiv}CR)$ confirms the planar, static structure of the complex. There was not even exchange with a different free acetylene detected,[431] and variation of the temperature over the range − 40 to + 110 °C produces no changes.[432]

^{1}H N.m.r. spectroscopy has been used to show that in solution (124) and (125) have a common intermediate which is believed to be (126).[433] Similarly for (127), the ^{1}H n.m.r. spectrum shows that several isomers exist in

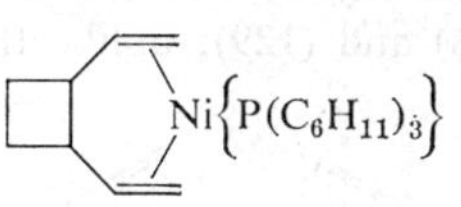

(124)

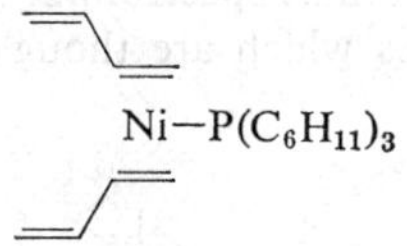

(125)

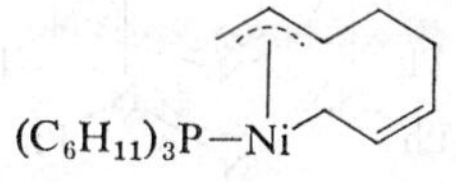

(126)

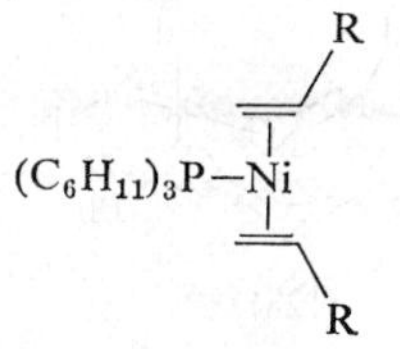

(127)

solution when R = Me; when R = H, a broad singlet is observed at τ 7.22.[434] The co-ordination of cyclohexa-1,4-diene in $Pd_2(C_6H_8)(OAc)_2$,-0.5HOAc has been confirmed by the presence of a multiplet at τ 4.89, whereas free cyclohexene has a multiplet at τ 4.33.[435] However, for the complex $[(PhCH=CH)_2CO]_2Pt$ the ^{1}H n.m.r. spectrum is identical with that of the free ligand and no ^{195}Pt coupling is observed, even at − 66 °C.[436]

[428] T. Weil, L. Spaulding, and M. Orchin, *J. Co-ordination Chem.*, 1971, **1**, 25.
[429] S. Cenini, R. Ugo, and G. La Monica, *J. Chem. Soc.* (*A*), 1971, 410.
[430] L. Caglioti, L. Cattalini, F. Gasparrini, G. Marangoni, and P. A. Vigato, *J. Chem. Soc.* (*A*), 1971, 324.
[431] J. H. Nelson, J. J. R. Reed, and H. B. Jonassen, *J. Organometallic Chem.*, 1971, **29**, 163.
[432] C. D. Cook and K. Y. Wan, *Inorg. Chem.*, 1971, **10**, 2696.
[433] P. W. Jolly, I. Tkatchenko, and G. Wilke, *Angew. Chem. Internat. Edn.*, 1971, **10**, 329.
[434] P. W. Jolly, I. Tkatchenko, and G. Wilke, *Angew. Chem. Internat. Edn.*, 1971, **10**, 328.
[435] J. M. Davidson, *Chem. Comm.*, 1971, 1019.
[436] K. Moseley and P. M. Maitlis, *Chem. Comm.*, 1971, 982.

^{1}H N.m.r. has been used to demonstrate that for $[Pd(CN)(allyl)]_4$ the allyl ligand is symmetrically bonded, but asymmetrically bonded in $Pd(CN)(allyl)(PR_3)$.[437] The ratio of *syn-* to *anti*-isomer concentration in (allyl)(amine)PdCl complexes has been measured by ^{1}H n.m.r. It was found that a bulky substituent on the central carbon atom favours the *anti*-form.[438] The substituent constants of $[2\text{-}(p\text{-}XC_6H_4)C_3H_4]PdL$ and $[1\text{-}(p\text{-}XC_6H_4)C_3H_4]PdL$ (L = cyclopentadienyl or acetoacetonato) have been estimated from their ^{1}H n.m.r. spectra. A plot of the difference in chemical shifts of the *o*- and *m*-protons of the phenyl ring, $\Delta\nu$, against σ_p^+ of X gave a fairly good linear relationship with intercept at $\Delta\nu = 0$.[439] The effect of donor ligands in converting the AM_2X_2 n.m.r. spectrum of $[(C_3H_5)PdCl]_2$ into an AX_4 spectrum has been interpreted as involving $(\pi\text{-}C_3H_5)PdClL$ and $(\sigma\text{-}C_3H_5)PdClL_2$. However, the Raman spectrum has been interpreted as showing that $(\sigma\text{-}C_3H_5)PdL$ is not formed but only $[(\pi\text{-}C_3H_5)PdL_2]^+ Cl^-$.[440]

The ^{1}H n.m.r. spectrum of $[(allyl)(PhNNNPh)Pd]_2$ shows it to consist of two species which are thought to be (128) and (129); unlike the acetate

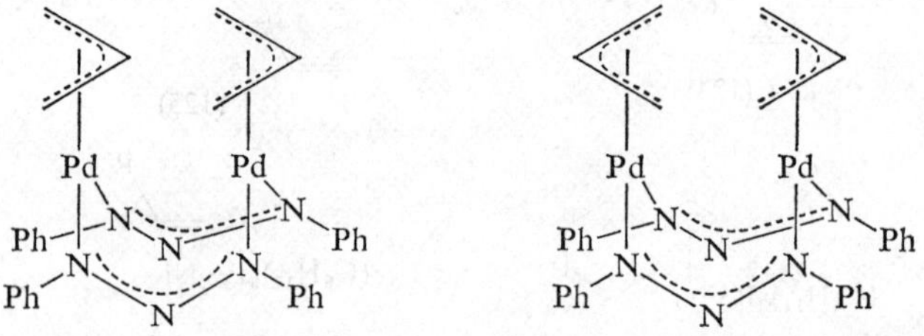

(128) (129)

analogue, the ^{1}H n.m.r. spectrum is invariant with temperature.[441] The observation that the ring protons of $[Me_2\overline{C{-}CH{\cdots}C(COCH_2Me){\cdots}CH}]_2$-$PtCl_6$ are at δ 7.58 is consistent with a delocalized positive charge on the ring.[442]

^{1}H N.m.r. has been used to demonstrate that RC≡CR reacts with $(PhCN)_2PdCl_2$ to give symmetric cyclobutadiene complexes $[(R_4C_4)PdX_2]_2$, which react with alcohols to give allylic complexes (130).[443] A detailed examination of norbornadiene derivatives of palladium and platinum has

[437] B. L. Shaw and G. Shaw, *J. Chem. Soc.* (*A*), 1971, 3533.
[438] J. W. Faller, M. E. Thomsen, and M. J. Mattina, *J. Amer. Chem. Soc.*, 1971, **93**, 2642.
[439] Y. Takahashi, H. Akahori, S. Sakai, and Y. Ishii, *Bull. Chem. Soc. Japan*, 1971, **44**, 2703.
[440] I. A. Leites, V. T. Aleksanyan, S. S. Bukalov, and A. Z. Rubezhov, *Chem. Comm.*, 1971, 265.
[441] T. Jack and J. Powell, *J. Organometallic Chem.*, 1971, **27**, 133.
[442] D. B. Brown and M. J. Strauss, *Chem. Comm.*, 1971, 128.
[443] D. F. Pollock and P. M. Maitlis, *J. Organometallic Chem.*, 1971, **26**, 407.

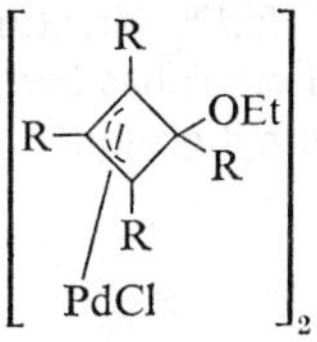

(130)

resolved a controversy over band assignment.[444] For (131) when Y contains an optically active centre, Me_a and $Me_{a'}$, and Me_b and $Me_{b'}$ are no longer equivalent and the spectrum becomes complex with homoallylic coupling between methyl groups. Also the results were interpreted as Me_c being *exo* when Y = H and *endo* when Y = CHMeX (X = Cl, OR, or H).[445]

Treatment of $[(C_6H_{11})_3P]_2NiN{\equiv}NNi[P(C_6H_{11})_3]_2$ with butadiene yields a complex which was shown by ^1H n.m.r. to be (132), a key intermediate in the nickel-catalysed butadiene dimerization.[446]

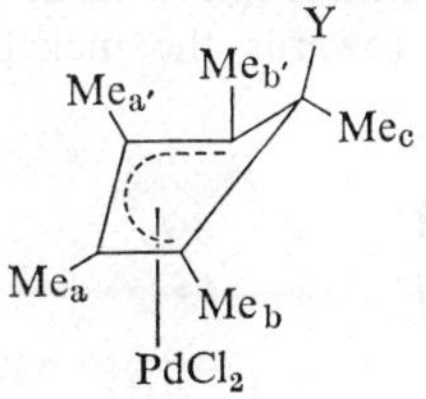

(131)

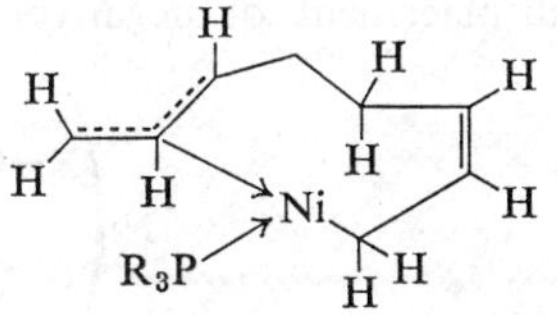

(132)

For $Ni(PMe_3)_4$ both the ^1H and ^{31}P signals are broad, but temperature invariant, and this behaviour has been explained as an $[AX_9]_4$ spin system.[447] The ^{31}P chemical shifts of the complexes $Ni[P(OPh)_{3-n}Cl_n]_4$ and $Ni(PPh_{3-n}Cl_n)_4$ have been interpreted in conjunction with the Faraday effect and susceptibility measurements as indicating a low multiplicity of the nickel–phosphorus bond.[448] Treatment of $Ni[P(O\text{-}o\text{-}C_6H_4Me)_3]_3$ with an alkyl nitrile produces a shift of the ^{31}P n.m.r. signal from − 128.5 to − 130.7 p.p.m.[449]

As the compound $Pt_2Br_2(SMe)_2(AsMe_3)_2$ shows two SMe resonances of relative intensity 1 : 1, the arsines are *cis* as in (133). For the complexes $M_2X_2(SMe)_2(PMe_3)_2$ (M = Pt or Pd) the ^1H (PMe) spectrum is of the $[AX_9]_2$

444 S. J. Betts, A. Harris, R. N. Haszeldine, and R. V. Parish, *J. Chem. Soc.* (*A*), 1971, 3699.
445 P. V. Balakrishnan and P. M. Maitlis, *J. Chem. Soc.* (*A*), 1971, 1721.
446 J. M. Brown, B. T. Golding, and M. J. Smith, *Chem. Comm.*, 1971, 1240.
447 H.-F. Klein and H. Schmidbaur, *Angew. Chem. Internat. Edn.*, 1970, **9**, 903.
448 J. M. Savariault, M. H. Micoud, P. Cassoux, and J. F. Labarre, *Bull. Soc. chim. France*, 1971, 2418.
449 C. A. Tolman, *Inorg. Chem.*, 1971, **10**, 1540.

type and $J(^{31}P-^{31}P)$ is *ca.* 5 Hz.[450] Treatment of nickelmethylxanthate with PPh_3 and HBr in diethyl ether has been shown by ^1H n.m.r. to yield a complex which contains an ethyl group.[451]

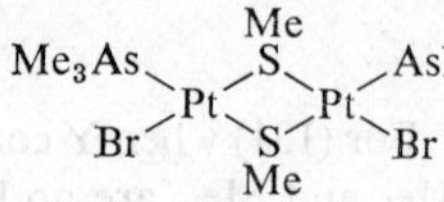

(133)

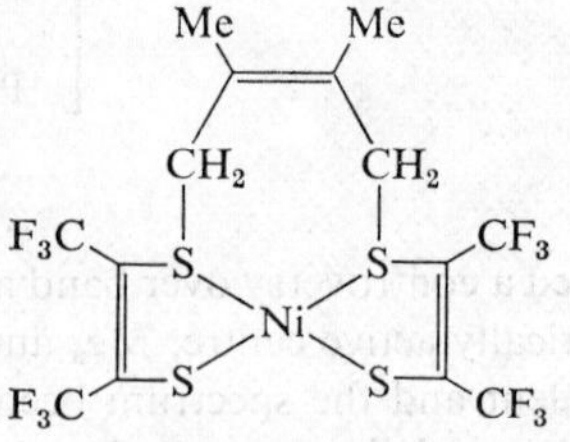

(134)

1,4-Addition of 2,3-dimethylbutadiene to $(CF_3CSCSCF_3)_2Ni$ to yield (134) has been confirmed by the observation of an AB pattern due to the CH_2 groups and one methyl resonance.[452] For the complexes (135) (X = Y = O; X = O, Y = S; or X = Y = S), as oxygen is replaced by sulphur, the resonance of the central proton moves to low field, consistent with the displacement of negative charge towards the nickel atom.[453]

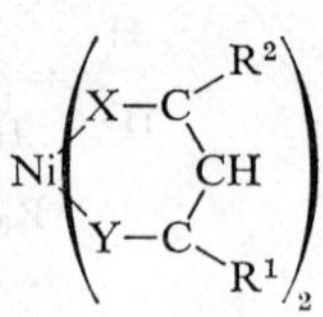

(135)

$(o\text{-}Me_2AsC_6H_4)_3Sb$ (Sbtas), which shows a ^1H resonance at τ 8.80, reacts with $Ni(ClO_4)_2,6H_2O$ to yield the $[Ni(Sbtas)_2]^{2+}$ ion which shows two resonances at τ 8.25 and 8.90 with intensities 1 : 1. This was considered to be evidence for a square-pyramidal rather than trigonal-bipyramidal structure with three $AsMe_2$ units co-ordinated and three free.[454]

For the complexes $[PtL_2(en)]^{n+}$ (L = PPh_3, bipy, SCN^-, *etc.*) the i.r. NH stretching frequency and τ NH have been taken as a measure of the acidity of this proton and hence the *trans* effect of the other ligand.[455] It has been found that the value of $^3J(^{195}Pt-N-C-H)$ in platinum diamine complexes depends chiefly on the oxidation state of the platinum ion, the nature of the ligands *trans* to the chelate ring, and the dihedral angle between the planes of PtNC and NCH; $^4J(Pt-N-C-CH_3)$ is also useful in

450 P. L. Goggin, R. J. Goodfellow, and F. J. S. Reed, *J. Chem. Soc.* (*A*), 1971, 2031.
451 C. Blejean and J. L. Chenot, *J. Inorg. Nuclear Chem.*, 1971, **33**, 3167.
452 J. R. Baker, A. Hermann, and R. M. Wing, *J. Amer. Chem. Soc.*, 1971, **93**, 6486.
453 C. Blejean, *Inorg. Nuclear Chem. Letters*, 1971, **7**, 1011.
454 L. Baracco, M. T. Halfpenny, and C. A. McAuliffe, *Chem. Comm.*, 1971, 1502.
455 G. W. Watt and J. E. Cuddeback, *Inorg. Chem.*, 1971, **10**, 947.

determining conformation.[456] The ^{1}H n.m.r. spectrum of $[Pt(en)_3]^{4+}$ consists of poorly resolved doublets for the CH_2 group with only 0.04 p.p.m. separating the axial and equatorial protons.[457]

A number of complexes (136) (B = CH_2CH_2, *o*-C_6H_4, *etc.*) have been prepared. Protons a, b, and c give rise to an AMX pattern, and although proton d can couple with protons of B, no coupling to protons a, b, or c was observed.[458] Six methyl resonances have been observed for (137)

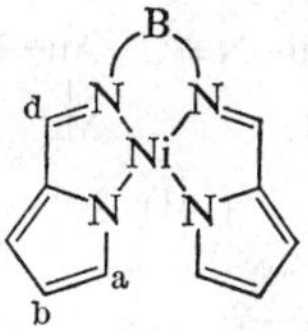

(136)

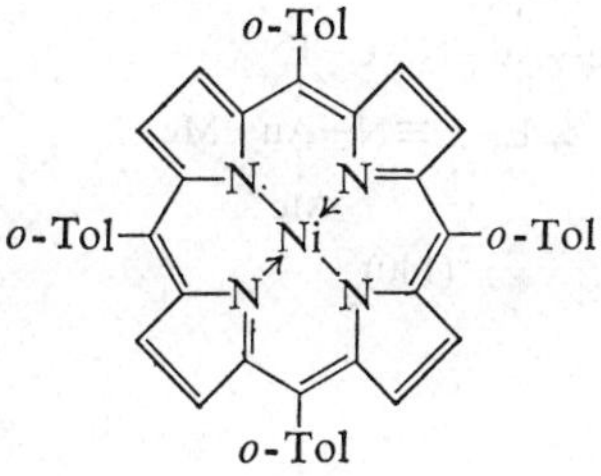

(137)

arising from the four possible isomers due to steric prevention of rotation about the *meso*-carbon *o*-tolyl bond.[459]

Complexes of Cu, Au, Zn, Cd, and Hg.—Information concerning complexes of these elements can be found at the following sources: $[Cu(B_{10}C_2H_{10})_2]^{n+}$ (n = 0 or 1),[1084] $[Me_2Au(NCS)]_2$,[404] $K_2[Zn(NCS)_4]$, $K_4[Cd(NCS)_6]$, $(NH_4)_2[Hg(SCN)_4]$,[157] $[M(B_{10}H_{12})_2]^{2-}$ (M = Zn, Cd, or Hg),[1085] $[Zn(RCOCH_2CS_2Me)_n]^{2+}$,[382] $CdI_2(PR_nPh_{3-n})_2$,[104] $CdBr_2(CH_2{=}CHCH_2CH_2PPh_2)_2$,[1116] $(C_2F_3)_2Hg$, and $(C_2F_3)HgMe$.[194]

The ^{1}H n.m.r. spectrum of (indenyl)$Cu(Bu^tNC)_3$ at room temperature is of the $A_2B_2X_2Y$ type and thus the structure is either (138) or (139) and is

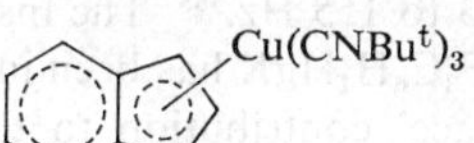

(138)

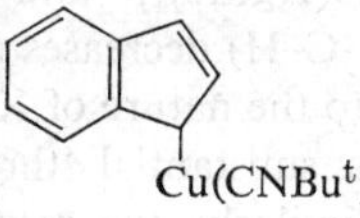

(139)

fluxional.[460] For the compounds $R_2M(CH{=}CH_2)_2$,2CuCl (M = Si or Sn) the resonance of the vinyl group is moved 0.9 p.p.m. upfield, providing evidence for co-ordination of the copper.[461] The diamagnetic complexes $Cu_2(ArN{=}N{-}CR{=}NO)_2$ show well-resolved ^{1}H n.m.r. signals and a structure analogous to $Cu_2(OAc)_4$ has been postulated.[462]

[456] T. G. Appleton and J. R. Hall, *Inorg. Chem.*, 1971, **10**, 1717.
[457] L. H. Novak and J. K. Beattie, *Inorg. Chem.*, 1971, **10**, 2326.
[458] C. J. Jones and J. A. McCleverty, *J. Chem. Soc.* (*A*), 1971, 1052.
[459] F. A. Walker and G. L. Avery, *Tetrahedron Letters*, 1971, 4949.
[460] T. Saegusa, Y. Ita, and S. Tomita, *J. Amer. Chem. Soc.*, 1971, **93**, 5656.
[461] J. W. Fitch, D. P. Flores, and J. E. George, *J. Organometallic Chem.*, 1971, **29**, 263.
[462] S. Gupta, K. C. Kalia, and A. Chakravorty, *Inorg. Chem.*, 1971, **10**, 1534.

When E = O for (140) only one methyl resonance is observed, but when E = S or Se two methyl resonances are observed; for $[Me_2AuCN]_4$ a number of resonances are observed owing to there being four possible isomers such as (141).[463] ^1H N.m.r. has provided evidence that when (142) [464] is treated with methoxide, ring expansion occurs to yield (143), as evidenced

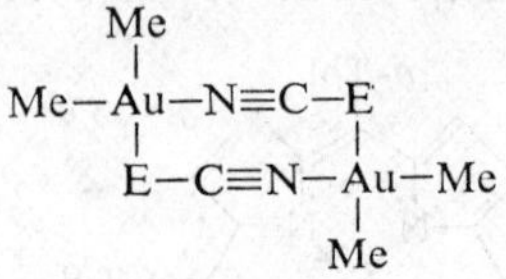

(140)

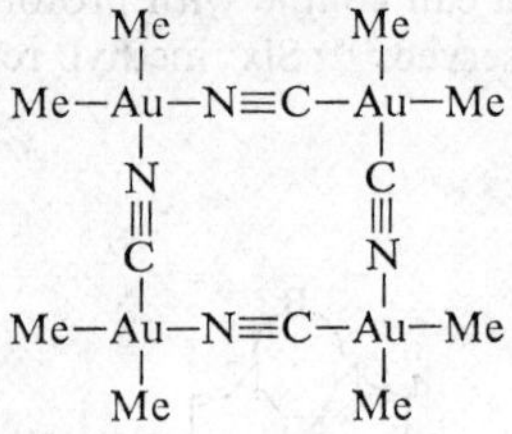

(141)

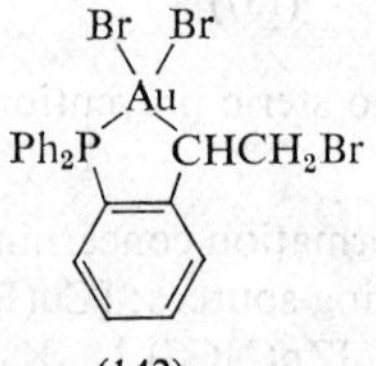

(142)

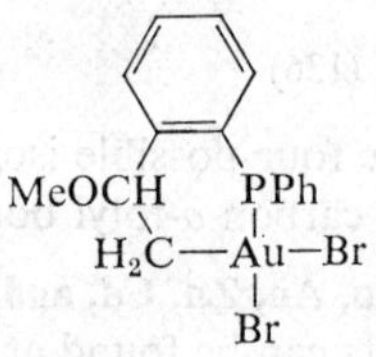

(143)

by the inequivalence of the CH_2 protons.[465] The compound (σ-Cp)Me_2-$AuPPh_3$ shows only one cyclopentadienyl resonance.[466]

The ^1H n.m.r. spectra of $LiAlMe_4$ and Li_2ZnMe_4 are similar, and $^1J(^{13}C-^1H)$ is small (109 Hz).[467]

Although $[N(HgMe)_4]^+$ and $[P(HgMe)_4]^+$ have the same ^1H chemical shift, $^2J(^{199}Hg-C-H)$ decreases from 173 to 155 Hz.[468] The insensitivity of $^4J(^{199}Hg-^{19}F)$ to the nature of X in *o*-$CF_3C_6H_4HgX$ has been interpreted as arising from a substantial 'through-space' contribution to this coupling constant.[469] Similarly, the mercury–*ortho*-hydrogen coupling constant of arylmercuric chlorides is insensitive to methyl substituents on the aryl group, but coupling to the *meta*-hydrogen shows considerable variation, which it was suggested is associated with conformation effects.[470]

463 F. Stocco, G. C. Stocco, W. M. Scovell, and R. S. Tobias, *Inorg. Chem.*, 1971, **10**, 2639.
464 M. A. Bennett, K. Hoskins, W. R. Kneen, R. S. Nyholm, P. B. Hitchcock, R. Mason, G. B. Robertson, and A. D. C. Towl, *J. Amer. Chem. Soc.*, 1971, **93**, 4591.
465 M. A. Bennett, K. Hoskins, W. R. Kneen, R. S. Nyholm, R. Mason, P. B. Hitchcock, G. B. Robertson, and A. D. C. Towl, *J. Amer. Chem. Soc.*, 1971, **93**, 4593.
466 S. W. Krauhs, G. C. Stocco, and R. S. Tobias, *Inorg. Chem.*, 1971, **10**, 1365.
467 J. Yamamoto and C. A. Wilkie, *Inorg. Chem.*, 1971, **10**, 1129.
468 D. Breitinger, K. Geske, and W. Beitelschmidt, *Angew. Chem. Internat. Edn.*, 1971, **10**, 555.
469 W. McFarlane, *Chem. Comm.*, 1971, 609.
470 P. J. Banney and P. R. Wells, *Austral. J. Chem.*, 1971, **24**, 317.

The diuretics $H_2NCONHCH_2CH(OR)CH_2HgCl$ (R = H or Me) do not show ^{199}Hg satellites and integration shows the loss of intensity that would be expected if there was ^{199}Hg coupling.[471] LAME has been used to analyse the ^1H n.m.r. spectra of (144) (X = O or Si; M = Pb or Hg).[472] The ^{19}F n.m.r. spectrum of $\{(CF_3)_3C\}_2Hg$ shows only a singlet with ^{199}Hg satellites; if KF is added a new singlet appears.[473] The decomposition of hydroxy-mercurated propylene in aqueous solution has been followed by ^1H n.m.r.[474]

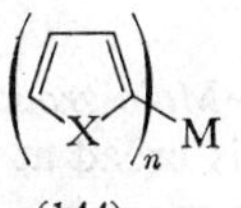
(144)

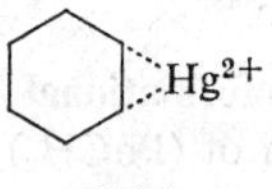
(145)

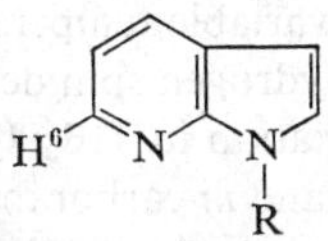
(146)

^1H and ^{13}C n.m.r. have been used to demonstrate that when cyclohexene reacts with $Hg(O_2CCF_3)_2$, (145) is formed.[475] As H(6) in (146) (R = H, HgMe, or HgPh) is unaffected by complexation, it was concluded that the mercury only interacts with the pyrrole nitrogen.[476] A solution of $Hg(CN)_2$ in HF shows only one ^{19}F resonance in the correct position for HF. The ^1H n.m.r. spectrum shows a resonance due to HCN, 4.5 Hz linewidth which sharpens to 0.5 Hz on irradiating ^{14}N at the frequency for NaCN in HF. On this, and i.r. evidence, it was concluded that the ion $[Hg_2(CN)_3]^+$ exists in solution.[477]

3 Dynamic Systems

A number of reviews and papers of general relevance to dynamic n.m.r. spectroscopy have been published. A general review on dynamic organometallic compounds has appeared.[478] Solvation of ions has also attracted attention: in a lengthy review on this subject, a section has been devoted to the use of n.m.r. to determine the solvation number of ions,[479] and the use

[471] G. M. Cree, *J. Organometallic Chem.*, 1971, **27**, 1.
[472] A. P. Ebdon, T. N. Huckerby, and F. G. Thorpe, *Tetrahedron Letters*, 1971, 2921.
[473] B. L. Dyatkin, S. R. Sterlin, B. I. Martynov, and I. L. Knunyants, *Tetrahedron Letters*, 1971, 345.
[474] M. Matsuo and Y. Saito, *Bull. Chem. Soc. Japan*, 1971, **44**, 2889.
[475] G. A. Olah and P. R. Clifford, *J. Amer. Chem. Soc.*, 1971, **93**, 2320.
[476] R. J. Kline, L. F. Sytama, and D. R. Shackle, *Inorg. Nuclear Chem. Letters*, 1971, **7**, 331.
[477] R. J. Gillespie, R. Hulme, and D. A. Humphreys, *J. Chem. Soc. (A)*, 1971, 3574.
[478] K. Vrieze and P. W. N. M. Van Leeuwen, *Progr. Inorg. Chem.*, 1971, **14**, 1.
[479] J. F. Hinton and F. S. Amis, *Chem. Rev.*, 1971, **71**, 627.

of ^{17}O n.m.r. to measure the water-exchange kinetics in labile aquo and substituted aquo transition-metal ions has been reviewed.[480]

A new simpler method to determine activation enthalpies of fast-exchange reactions based on linewidths has been described.[481] Also, a concise matrix formulation of chemical exchange effects on an n.m.r. spectrum using the Bloch equations has been presented. The method accommodates many-site exchange processes, site-dependent relaxation times, differing site populations, and saturation effects in a steady-state first-order spectrum.[482]

Rotational and Conformational Exchange.—*Main-group Compounds.* The ^{1}H n.m.r. spectrum of $(PhCH_2)_2NMeB_3H_7$ is broad at room temperature, sharp at − 80 °C when ^{11}B coupling is removed, but broadens again on further cooling, showing that exchange is occurring with a barrier of *ca.* 6.0 kcal mol^{-1}.[483] Similarly, variable-temperature ^{1}H n.m.r. spectroscopy has revealed efficient boron–hydrogen spin decoupling due to quadrupolar-induced ^{10}B and ^{11}B spin relaxation in TlB_3H_8, $Me_4NB_3H_8$, $(Ph_3P)_2CuBH_4$, $(Ph_3P)_2CuB_3H_8$, *o*-carborane, and *m*-carborane. An additional temperature dependence of the ^{1}H n.m.r. spectrum of $(Ph_3P)_2CuB_3H_8$ provides evidence for slow intermolecular exchange.[484, 485]

KB_4H_9 shows two ^{11}B signals at room temperature, but three signals in the ratio 1 : 2 : 1 at − 45 °C.[486] At 70 °C (147) shows three boron signals in the ratio 1 : 2 : 2, but at 100 °C two signals in the ratio 1 : 4, which was interpreted as being due to movement of the bridge proton.[487, 488] However, treatment of B_5H_{11} with 2 moles of NH_3 at − 112 °C produces $B_5H_{11},2NH_3$.

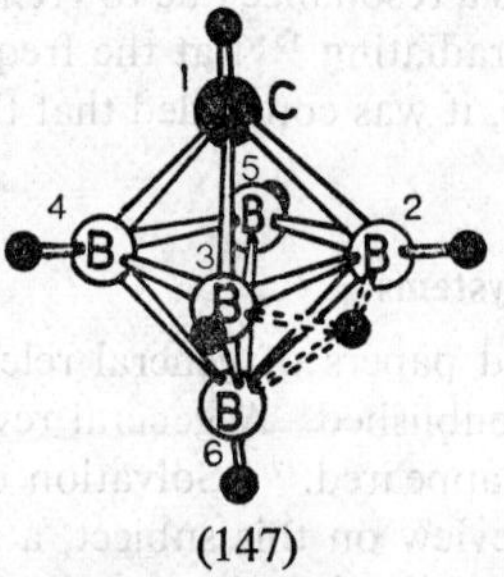

(147)

[480] J. P. Hunt, *Co-ordination Chem. Rev.*, 1971, 7, 1.
[481] F.-H. Marquardt, *J. Chem. Soc.* (*B*), 1971, 366.
[482] L. W. Reeves and K. N. Shaw, *Canad. J. Chem.*, 1970, **48**, 3641.
[483] W. J. Dewkett, H. Beall, and C. H. Bushweller, *Inorg. Nuclear Chem. Letters*, 1971, 7, 633.
[484] H. Beall, C. H. Bushweller, and M. Grace, *Inorg. Nuclear Chem. Letters*, 1971, 7, 641.
[485] C. H. Bushweller, H. Beall, M. Grace, W. J. Dewkett, and H. S. Bilofsky, *J. Amer. Chem. Soc.*, 1971, **93**, 2145.
[486] H. D. Johnson, jun., and S. G. Shore, *J. Amer. Chem. Soc.*, 1970, **92**, 7586.
[487] T. Onak and J. B. Leach, *Chem. Comm.*, 1971, 76.
[488] E. Groszek, J. B. Leach, G. T. F. Wong, C. Ungermann, and T. Onak, *Inorg. Chem.*, 1971, **10**, 2770.

At − 80 °C, the ^{11}B n.m.r. spectrum shows two broad peaks and a well-defined triplet, whereas at − 40 °C this gives two broad signals, and it was suggested that this behaviour was due to the $[B_4H_9]^-$ anion.[489] *X*-Ray diffraction shows that $B_9C_2H_{12}AlMe_2$ has structure (148), but the room-temperature ^{11}B n.m.r. spectrum shows that B(4) and B(7), B(5) and B(11),

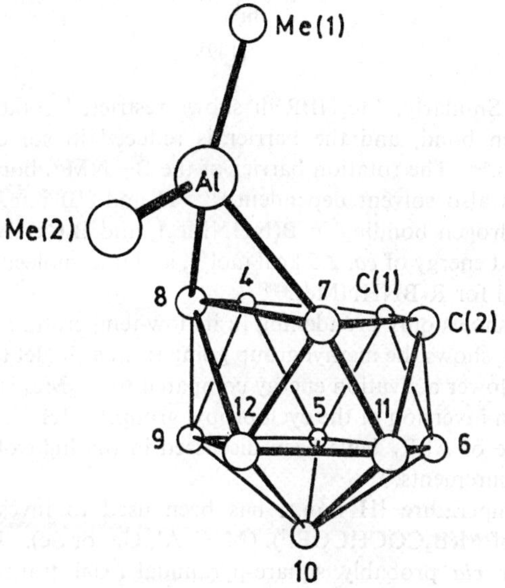

(148)

and B(9) and B(11) are equivalent in pairs but are inequivalent at low temperature. It was therefore suggested that the bridging hydrogen between B(4) and B(5) and the $AlMe_2$ group were interchanging.[490, 491]

$LiBMe_4$ in toluene shows a broad doublet in the ^1H n.m.r. spectrum at 40 °C and a quartet at 100 °C.[492] A number of complexes of type (149) have been prepared,[493] and the potential barrier for free rotation about the boron–carbon (aryl B) measured. As the ΔG values obtained were only slightly dependent on R^2 it was concluded that the barrier to free rotation arose mainly from steric hindrance rather than boron–carbon π-bonding.[494] However, boron–nitrogen π-bonding was invoked for $PhOBHNPr^i_2$ to explain the observation of two isopropyl resonances at 40 °C and only one

489 G. Kodama, J. E. Dunning, and R. W. Parry, *J. Amer. Chem. Soc.*, 1971, **93**, 3372.
490 M. R. Churchill, A. H. Reis, jun., D. A. T. Young, G. R. Willey, and F. M. Hawthorne, *Chem. Comm.*, 1971, 298.
491 D. A. T. Young, R. J. Wiersema, and M. F. Hawthorne, *J. Amer. Chem. Soc.*, 1971, **93**, 5687.
492 D. Groves, W. Rhine, and G. D. Stucky, *J. Amer. Chem. Soc.*, 1971, **93**, 1553.
493 H. A. Staab and B. Meissner, *Annalen*, 1971, **753**, 80.
494 B. Meissner and H. A. Staab, *Annalen*, 1971, **753**, 92.

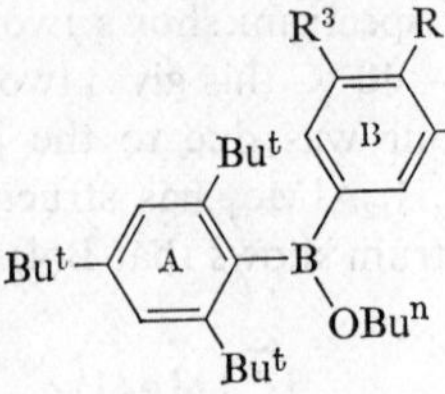

(149)

at 120 °C.[495] Similarly, Me_2NBRPh shows restricted rotation about the boron–nitrogen bond, and the barrier is reduced in some electron-pair donor solvents.[496] The rotation barrier of the B—NMe_2 bond of $(Me_3Si)_2$-$NBClNMe_2$ is also solvent dependent.[497] ^1H and ^{11}B n.m.r. show intramolecular hydrogen bonding in $B(NHNMe_2)_3$ and $B_2(NHNMe_2)_4$, with a hydrogen-bond energy of *ca.* 2.5 kcal mol^{-1}, and intermolecular association was postulated for $R_2BNHNMe_2$.[498]

$Al_2(C_3H_5)_6$ shows only broadening in its low-temperature ^1H n.m.r., but $Al_2Me_4(C_3H_5)_2$ shows the methyl group going from a singlet to a doublet on cooling. The lower activation energy compared to Al_2Me_6 is thought to be due to rotation inversion of the cyclopropyl group.[499] ^1H N.m.r. studies on alkyl exchange of Al_2R_6 have been discussed in the light of new thermochemical measurements.[500]

Variable-temperature ^1H n.m.r. has been used to investigate the rearrangement of $M(B_2COCHCOPr^i)_3$ (M = Al, Ga, or Sc). The reaction is intramolecular *via* probably square-pyramidal–axial transition states.[501] Similarly, the rate of intramolecular interchange of non-equivalent groups has been measured for $Al(acac)_n(hfac)_{3-n}$ [502] and (150) (M = Al or Co; R = C_3H_5 or Pr^i).[503]

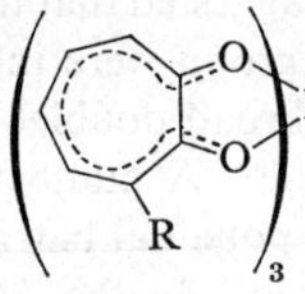

(150)

495 R. A. Kovar, R. Culbertson, and E. C. Ashby, *Inorg. Chem.*, 1971, **10**, 900.
496 T. Totani, K. Tori, J. Murakami, and H. Watanabe, *Org. Magn. Resonance*, 1971, **3**, 627.
497 R. L. Wells, H. L. Paige, and C. G. Moreland, *Inorg. Nuclear Chem. Letters*, 1971, **7**, 177.
498 H. Nöth, *Chem. Ber.*, 1971, **104**, 558.
499 J. W. Moore, D. A. Sanders, P. A. Scherr, M. D. Glick, and J. P. Oliver, *J. Amer. Chem. Soc.*, 1971, **93**, 1035.
500 J. N. Hay, P. G. Hooper, and J. C. Tobb, *J. Organometallic Chem.*, 1971, **28**, 193.
501 J. R. Hutchison, J. G. Gordon, and R. H. Holm, *Inorg. Chem.*, 1971, **10**, 1004.
502 D. A. Case and T. J. Pinnavaia, *Inorg. Chem.*, 1971, **10**, 482.
503 S. S. Eaton and R. H. Holm, *J. Amer. Chem. Soc.*, 1971, **93**, 4913.

$Ph_3As=CRCO_2Me$ is found to exist in solution in two forms:

$$\underset{R}{\overset{Ph_3As}{>}}C-C\underset{OMe}{\overset{O}{<}} \rightleftharpoons \underset{R}{\overset{Ph_3As}{>}}C-C\underset{O}{\overset{OMe}{<}}$$

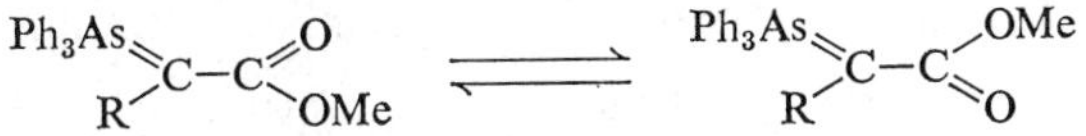

At low temperatures two methyl resonances are observed with the *cis*-isomer being favoured, but at room temperature only one methyl resonance is observed.[504] Restricted rotation about a carbon–nitrogen bond has been demonstrated for (p-$CF_3C_6H_4$)$PhC=NGeMe_3$,[505] $CF_3CONMeSiMe_3$,[506] (Cp)M-(C_5H_4)$CPh=N(p$-Tol) (M = Fe or Ru),[507] (Cp)Fe(C_5H_4)NR^1COR^2,[508] and $Me_2NCSeNH_2$.[509] For $PhCON(MMe_3)CONMe_2$ (M = Si, Ge, or Sn) two methyl resonances were observed at low temperature and an equilibrium was suggested.[510] It is interesting that ΔG for rotation about the carbon–nitrogen bond in $R^1XYCNR^2_2$ (R^1 = Me, $SiMe_3$, or $SnMe_3$; R^2 = Me or Et; X, Y = O, S) is larger when R^1 = $SiMe_3$ or $SnMe_3$ than when R^1 = Me.[511] Similarly, $Me_2NCO_2SiMe_3$ shows restriction about the nitrogen–carbon bond, but while Me_2NCS_2Me shows doubling of the methyl resonance at + 5 °C, even at − 23.5 °C $Me_2NCS_2SiMe_3$ only shows a singlet and it was suggested that this compound is really symmetric as (151).[512]

$$\underset{Me}{\overset{Me}{>}}N-\overset{S}{\underset{S}{\overset{\|}{C}}}\cdots SiMe_3$$

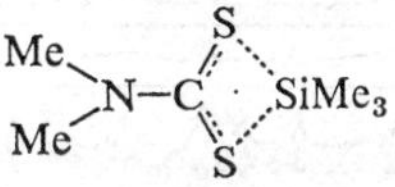

(151)

Restricted rotation about a carbon–nitrogen bond can also arise from steric hindrance. Thus at low temperatures both Bu^tNMe_2 and $Bu^tNMe_2BH_3$ show two methyl resonances in the ratio 2 : 1 for the t-butyl group.[513, 514] $Me_2AlNPhCO_2Me$ shows only two methyl resonances at room temperature, but at − 87 °C four methyl signals are observed, which was attributed to (152) and (153).[515]

The activation energies of chair–chair interconversions of cyclohexane-type rings have been measured for $(CH_2)_5SiMe_2$[516] and 1-Me_3SiX-4-

[504] A. J. Dale and P. Frøyen, *Acta Chem. Scand.*, 1971, **25**, 1452.
[505] R. J. Cook and K. Mislow, *J. Amer. Chem. Soc.*, 1971, **93**, 6703.
[506] L. Birkofer and G. Schmidtberg, *Chem. Ber.*, 1971, **104**, 3831.
[507] R. Damrauer and T. E. Rutledge, *J. Organometallic Chem.*, 1971, **29**, C9.
[508] R. Damrauer, *J. Organometallic Chem.*, 1971, **32**, 121.
[509] L. W. Reeves, R. C. Shaddick, and K. N. Shaw, *J. Phys. Chem.*, 1971, **75**, 3372.
[510] I. Matsuda, K. Itoh, and Y. Ishii, *J. Chem. Soc. (C)*, 1971, 1870.
[511] C. H. Yoder, A. Komoriya, J. E. Kochanowski, and F. H. Suydam, *J. Amer. Chem. Soc.*, 1971, **93**, 6515.
[512] A. E. Lemire and J. C. Thompson, *J. Amer. Chem. Soc.*, 1971, **93**, 1163.
[513] C. H. Bushweller, J. W. O'Neil, and H. S. Bilofsky, *Tetrahedron*, 1971, **27**, 5761.
[514] C. H. Bushweller, W. J. Dewkett, J. W. O'Neil, and H. Beall, *J. Org. Chem.*, 1971, **36**, 3782.
[515] T. Hirabayashi, T. Sakakibara, and Y. Ishii, *J. Organometallic Chem.*, 1971, **32**, C5.
[516] C. H. Bushweller, J. W. O'Neil, and H. S. Bilofsky, *Tetrahedron*, 1971, **27**, 3065.

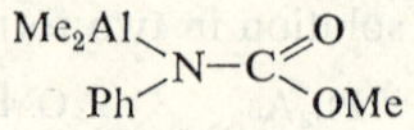

(152)

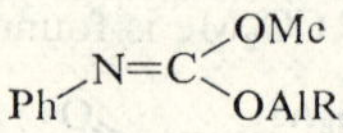

(153)

Rcyclohexane (X = O or NH; R = H, Me, or Bu[t]).[517] For a number of tin alkyls and aryls containing an asymmetric tin atom, *e.g.* $MePhB_2SnBu^i$, coalescence of resonances has been used to measure inversion at the tin. It was found that the rate is concentration-dependent and therefore

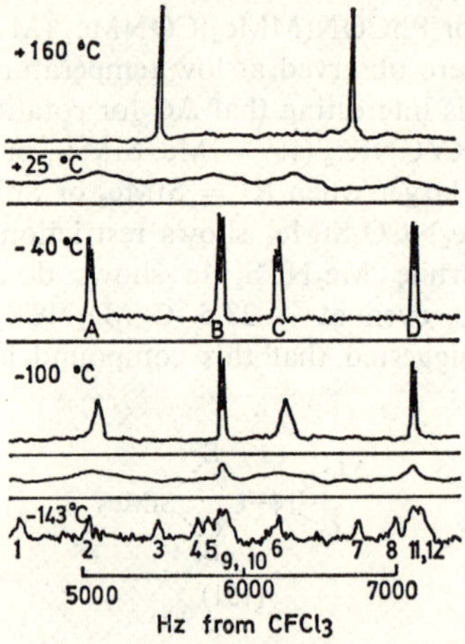

Figure 2 ^{19}F *N.m.r. spectrum of* $F_2PNMeNMePF_2$ *between* + 160 *and* − 145 °C (Reproduced by permission from *J. Amer. Chem. Soc.*, 1971, **93**, 3051)

second-order.[518] The ^1H n.m.r. spectrum of $Ph_2Sn(acac)_2$ had been interpreted as showing the *trans*-structure, but Mössbauer studies suggest the *cis*-configuration. Cooling causes the methyl singlet in the ^1H n.m.r. spectrum to split into a doublet, thus confirming the *cis*-structure.[519] The ^1H n.m.r. spectrum of the $—OCH_n—$ group of a wide range of $Bu^n{}_3SnOR$ compounds has been reported and the temperature at which the Sn–O–CH coupling disappears is taken as a measure of the rate of exchange of the OR group.[520]

Restricted rotation about N—N bond has been observed for $XSiMe_2NMeNMeSiMe_2X$ and ΔG measured.[521] However, for F_2PNMe-$NMePF_2$ the behaviour is more complex, see Figure 2. At − 40 °C,

[517] J. P. Hardy and W. D. Cumming, *J. Amer. Chem. Soc.*, 1971, **93**, 928.
[518] D. V. Stynes and A. L. Allred, *J. Amer. Chem. Soc.*, 1971, **93**, 2666.
[519] N. Serpone and K. A. Hersh, *Inorg. Nuclear Chem. Letters*, 1971, **7**, 115.
[520] A. G. Davies, D. C. Kleinschmidt, P. R. Palan, and S. C. Vasishtha, *J. Chem. Soc.* (*C*), 1971, 3972.
[521] O. J. Scherer and U. Bültjer, *Angew. Chem. Internat. Edn.*, 1971, **10**, 343.

rotation about the N—N bond has become slow and at − 145 °C rotation about the P—N bond is also slow; activation energies were derived.[522] The ^{1}H n.m.r. spectra of $R^1{}_2NS(O)OR^2$ show inequivalence of the CH_2 or CMe_2 group when R^1 is Et or Pr^n at room temperature, but for $R_2NS(O)X$ (X = F, Cl, or Br) cooling to low temperature is necessary to cause inequivalence.[523]

^{1}H N.m.r. has been applied to conformational analysis of phosphacyclohexanes.[524] The ^{1}H n.m.r. coalescence temperature of $Pr^iPhPMMe_3$ (M = Si, Ge, or Sn) has been used to estimate the activation energy of inversion at phosphorus. A linear relationship between the chemical shift of the MMe_3 group and the energy barrier was found.[525] Similarly, the barrier to inversion of acyl phosphines has been measured.[526] However, coalescence of the methyl groups of $Pr^i{}_2$(*p*-Tol)Sb does not occur below 200 °C, thus putting a lower limit of 26 kcal mol^{-1} on the barrier to pyramidal inversion.[527] Variable-temperature ^{1}H n.m.r. spectra of pyridine solutions of $[(HOCH_2)_4P]Cl$ have been interpreted as chemical exchange of the CH_2OH groups of the salt with those of the small amount of $(HOCH_2)_3P$ which is formed at elevated temperatures.[528] The barrier to rotation about the N—P bond in $Pr^i{}_2NPPhClX$ (X = O or S) has been measured,[529] and the failure to observe a ^{19}F n.m.r. signal for the $[SbF_4]^-$ anion in solution has been attributed to exchange and quadrupolar broadening.[530]

The pseudorotation of PF_5 and AsF_5 has been treated theoretically.[531] At 70 °C, PF_4NMeH shows only one fluorine environment and three at − 80 °C. It was therefore suggested that the structure is (154) with slow rotation about the phosphorus–nitrogen bond at low temperature. It is interesting to note that $^2J(F_a–F_{a'})$ is 2.8 Hz and $^2J(F_a–F_e)$ is *ca.* 68 Hz.[532] Compound (155) shows only one type of fluorine at room temperature and two at − 88 °C. However, as δ_F and $^1J(^{31}P–^{19}F)$ are similar for the two fluorines, it was suggested that exchange between axial and equatorial fluorine is still occurring, *i.e.* (155) $\rightleftharpoons$ (156) $\not\rightleftharpoons$ (157).[533] The CH_2 group of $(PhCH_2)_3AsF_2$ is a triplet at low temperature and a singlet at high temperature. It was therefore concluded that intermolecular exchange is occurring.[534]

522 T. T. Bopp, M. D. Havlicek, and J. W. Gilje, *J. Amer. Chem. Soc.*, 1971, **93**, 3051.
523 R. Keat, D. S. Ross, and D. W. A. Sharp, *Spectrochim. Acta*, 1971, **27A**, 2219.
524 J. B. Lambert and W. L. Oliver, jun., *Tetrahedron*, 1971, **27**, 4245.
525 R. D. Baechler and K. Mislow, *J. Amer. Chem. Soc.*, 1971, **93**, 773.
526 W. Egan and K. Mislow, *J. Amer. Chem. Soc.*, 1971, **93**, 1805.
527 J. Jacobus, *Chem. Comm.*, 1971, 1058.
528 S. E. Ellzey, jun., W. J. Connick, jun., and G. J. Boudreaux, *Canad. J. Chem.*, 1971, **49**, 3581.
529 W. B. Jennings, *Chem. Comm.*, 1971, 867.
530 C. J. Adams and A. J. Downs, *J. Chem. Soc. (A)*, 1971, 1534.
531 J. Brickmann, *Ber. Bunsengesellschaft Phys. Chem.*, 1971, **95**, 747.
532 J. S. Harman and D. W. A. Sharp, *Inorg. Chem.*, 1971, **10**, 1538.
533 M. Eisenhut and R. Schmutzler, *Chem. Comm.*, 1971, 1452.
534 C. G. Moreland, R. J. Beam, C. W. Wooten, and S. M. Horner, *Inorg. Nuclear Chem. Letters*, 1971, **7**, 243.

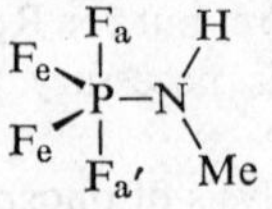

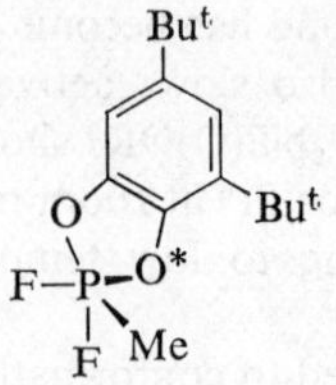

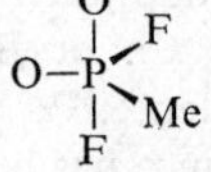

(154) (155) (156)

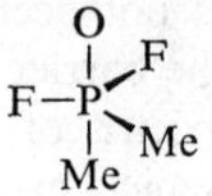

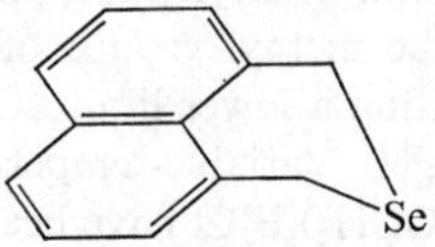

(157) (158)

The barrier to rotation for the S—S bond of BzSSR has been measured.[535] However, (158) shows only one CH_2 signal which broadens at − 80 °C, showing rapid inversion of the five-membered ring.[536] The ^{19}F n.m.r. resonance of FS_4F is temperature dependent.[537] For R_2SeF_2 (R = Me, Et, or Pr^i) coupling between the fluorine and protons or ^{77}Se is only observed at low temperature. It was therefore concluded that fluorine exchange *via* dissociation to $[R_2SeF]^+$ was occurring.[538]

^{17}O N.m.r. measurements on aqueous solutions of $LiIO_3$ have been reported. The iodate resonance having a linewidth $1/T_2 = 995 \pm 20$ Hz lies 206.4 ± 1 p.p.m. downfield from the solvent $H_2{}^{17}O$ line. Relative intensity, as well as shift measurements, indicate that the iodate is present as IO_3^- in the solution; addition of acid results in line broadening. Analysis of the results show that acid-catalysed exchange with water occurs both *via* a mono- and di-protonated iodate.[539]

Transition-metal Carbonyls. The observation that $(Cp)_2Ti(\mu\text{-}SMe)_2M(CO)_4$ (M = Cr, Mo, or W) shows three cyclopentadienyl resonances has been interpreted as the existence of two isomers, (159) and (160); on warming, the resonances coalesce.[540] Restricted rotation about a carbon–oxygen bond for $(CO)_5CrC(C_6H_4X)(OMe)$,[541] and a carbon–nitrogen bond for

535 R. R. Fraser, G. Boussard, J. K. Saunders, J. B. Lambert, and C. E. Mixan, *J. Amer. Chem. Soc.*, 1971, **93**, 3822.
536 A. Biezais-Zirnis and A. Fredga, *Acta Chem. Scand.*, 1971, **25**, 1171.
537 F. Sell, R. Budenz, W. Gombler, and H. Seitter, *Z. anorg. Chem.*, 1971, **380**, 262.
538 K. J. Wynne, *Inorg. Chem.*, 1971, **10**, 1868.
539 R. A. Dwek, Z. Luz, S. Peller, and M. Shporer, *J. Amer. Chem. Soc.*, 1971, **93**, 77.
540 P. S. Braterman, V. A. Wilson, and K. K. Joshi, *J. Chem. Soc.* (*A*), 1971, 191.
541 E. O. Fischer, C. G. Kreiter, H. J. Kollmeier, J. Müller, and R. D. Fischer, *J. Organometallic Chem.*, 1971, **28**, 237.

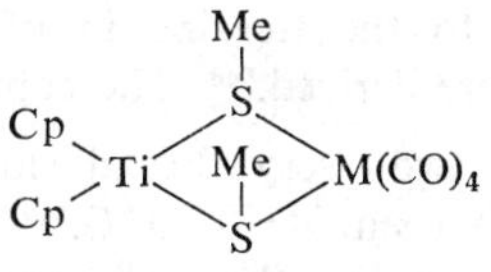

(159)

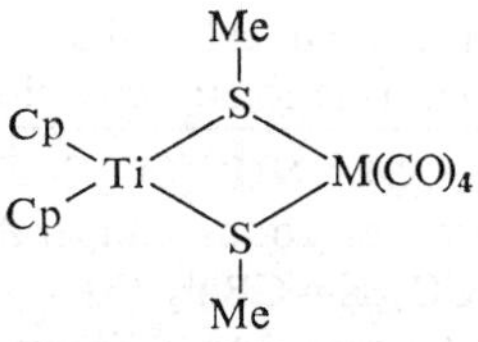

(160)

$(CO)_5CrCEtNMe_2$,[542] have been observed and activation energies derived. Variable-temperature ^1H n.m.r. spectroscopy has been carried out on (arene)$Cr(CO)_3$ (arene = RPh, xylene, or trimethylbenzene). When arene = RCH_2Ph, the shift of the CH_2 group was taken as a measure of the amount of complex in the eclipsed form (161) rather than the staggered form (162) and equilibria data were obtained.[543]

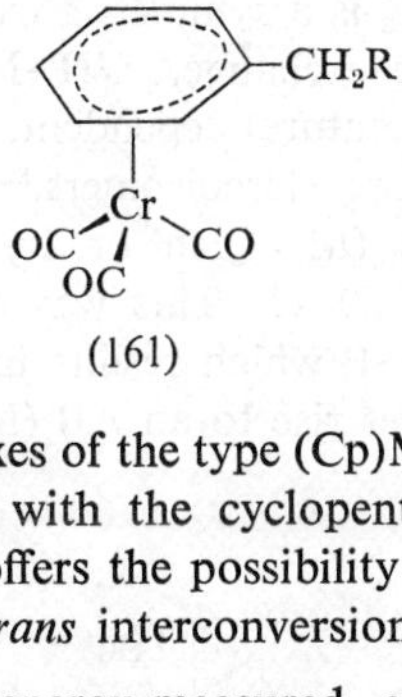

(161)

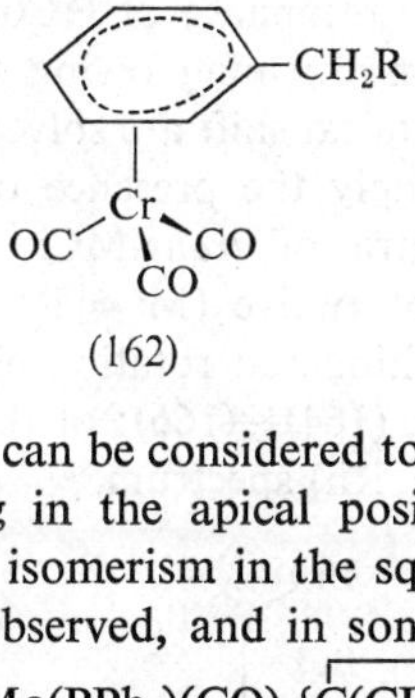

(162)

Complexes of the type (Cp)$Mo(CO)_2LL'$ can be considered to be square-pyramidal with the cyclopentadienyl ring in the apical position. This structure offers the possibility of *cis–trans* isomerism in the square-plane. Such *cis–trans* interconversion has been observed, and in some cases an activation energy measured, *e.g.* for (Cp)$Mo(PPh_3)(CO)_2\{C(CH_2)_3O\}Br$,[544] ($\pi$-indenyl)$Mo(CO)_2LR$ (L = phosphorus ligand or $SbPh_3$, R = H, Cl, Br, I, Me, or Bz),[545] (Cp)$Mo(CO)_2(PMe_nPh_{3-n})AlMe_2$ (n = 1 or 0),[546] (Cp)$Mo(CO)_2(PMePh_2)Cl$, and (Cp)$Mo(CO)(PMePh_2)_2Cl$.[547]

The compounds (163) (M = Mo or W; X = halogen) on cooling first show stoppage of exchange between the chelate and monodentate

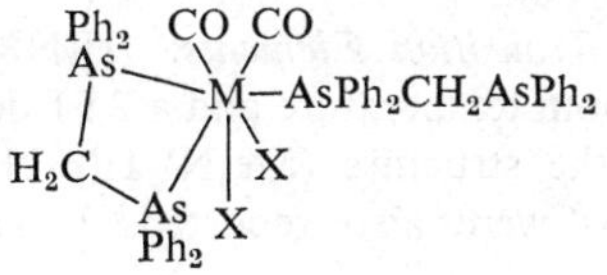

(163)

[542] E. O. Fischer, E. Winkler, C. G. Kreiter, G. Huttner, and B. Krieg, *Angew. Chem. Internat. Edn.*, 1971, **10**, 922.
[543] C. Segard, B. Roques, C. Pommier, and G. Guiochon, *Analyt. Chem.*, 1971, **43**, 1146.
[544] F. A. Cotton and C. M. Lukehart, *J. Amer. Chem. Soc.*, 1971, **31**, 2672.
[545] J. W. Faller, A. S. Anderson, and A. Jakubowski, *J. Organometallic Chem.*, 1971, **27**, C47.
[546] W. R. Kroll and G. B. McVicker, *Chem. Comm.*, 1971, 591.
[547] G. Wright and R. J. Mawby, *J. Organometallic Chem.*, 1971, **29**, C29.

arsine and then extra fine structure due to the freezing in of particular conformations; activation energies were derived.[548] The compound $(Cp)Mo(CO)_2NNHC(CO_2Et)COH$ shows two very broad low-field resonances at room temperature which sharpen at − 80 °C.[549] When $(Cp)M(CO)_2N{=}CBu^t_2$ (M = Mo or W) is cooled, the cyclopentadienyl and t-butyl resonances disappear to be replaced by new ones. It was suggested that only rotation about the N—M bond is consistent with both this and molecular weight, i.r., and u.v. data.[550] At room temperature, $(CO)_3Fe(\mu\text{-}CO)(\mu\text{-}SiMe_2)Fe(CO)_3$ shows one methyl resonance and two at − 72 °C, which was considered to arise from the inequivalence of the $SiMe_2$ methyl groups.[551]

Fluxional behaviour of the ligand $Me_2AsC{=}C(AsMe_2)CF_2CF_2$ (ffars) has been reported for $(ffars)Co_2(CO)_6$ and $(PhC{\equiv}CH)Co_2(CO)_4(ffars)$.[552] The hydride resonance of $HCo(CO)_2(PBu^n_3)_2$ is a symmetric triplet which broadens on warming owing to phosphine exchange. ^2J(P–H) and the hydride chemical shift are solvent (and temperature) dependent, which was taken to imply the presence of two or more stereoisomers.[553] The ^{31}P n.m.r. spectra of *trans*-$MCl(CO)(PBu^t_2R)_2$ (M = Rh or Ir) show six (M = Ir) or twelve (M = Rh) lines at − 60 °C. This was thought to arise from hindered rotation about the M—P which results in the three conformers, (164)—(166); of these (165) gives rise to an AB (M = Ir) or ABX (M = Rh) spectrum.[554]

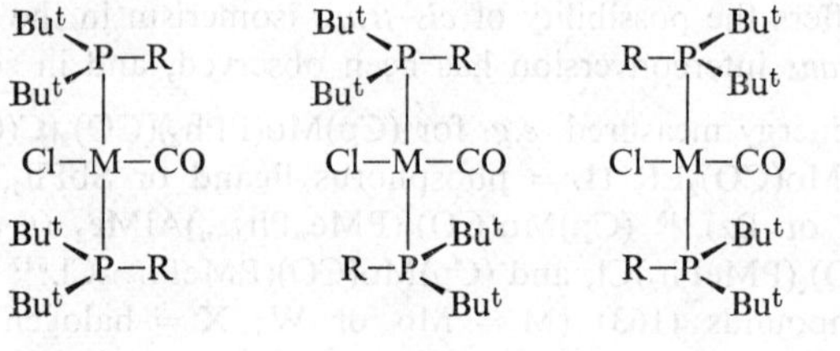

(164) (165) (166)

Other Compounds of Transition Elements. $Ti(NMe_2)_3$ shows a singlet methyl resonance at room temperature and a 2 : 1 doublet at low temperature, consistent with the structure $(Me_2N)_2Ti(\mu\text{-}NMe_2)_2Ti(NMe_2)_2$; data on related compounds were also reported.[555] The rate of molecular

548 M. W. Anker, R. Colton, and C. J. Rix, *Austral. J. Chem.*, 1971, **24**, 1157.
549 M. L. H. Green and J. R. Sanders, *J. Chem. Soc.* (*A*), 1971, 1947.
550 M. Kilner and C. Midcalf, *J. Chem. Soc.* (*A*), 1971, 292.
551 D. Kummer and J. Furrer, *Z. Naturforsch.*, 1971, **26b**, 162.
552 J. P. Crow and W. R. Cullen, *Inorg. Chem.*, 1971, **10**, 2165.
553 G. F. Pregaglia, A. Andreetta, G. F. Ferrari, and R. Ugo, *J. Organometallic Chem.*, 1971, **30**, 387.
554 B. E. Mann, C. Masters, B. L. Shaw, and R. E. Stainbank, *Chem. Comm.*, 1971, 1103.
555 M. F. Lappert and A. R. Sanger, *J. Chem. Soc.* (*A*), 1971, 874.

rearrangement of $Ti(acac)_2X_2$ (X = Pr^iO or o-$Pr^iC_6H_4O$), measured by the coalescence of the methyl signals, increases as the pK_a of the parent phenol decreases.[556] Fluxional behaviour has also been reported for M(Cp)Cl(β-diketonate)$_2$ (M = Ti or Zr).[557] Zr(Cp)(β-diketonate)$_3$ shows two rearrangement processes, (*a*) faster: exchange between A and B in (167), and (*b*) slower: exchange of C and D with B and A, and activation

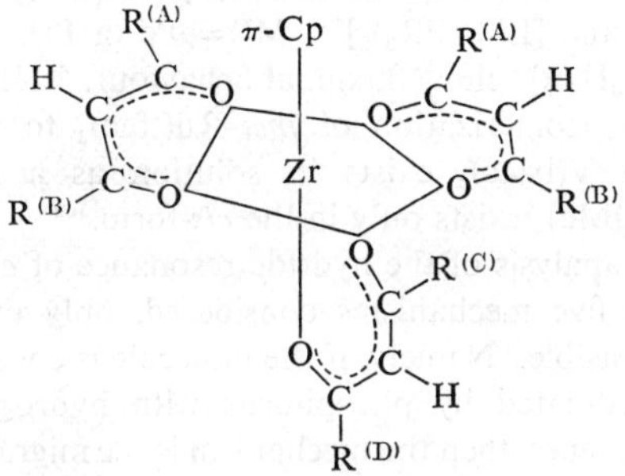

(167)

energies have been derived.[558] (Cp)M(acac)$_2$X (M = Zr or Hf) undergo first-order stereochemical rearrangement processes which interchange the two non-equivalent acetylacetonate ligands.[559] The observation of two cyclopentadienyl resonances at − 50 °C and one at 36 °C for (168) has been attributed to inversion at sulphur.[560]

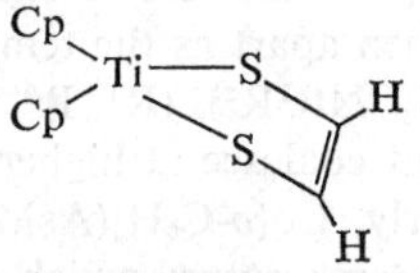

(168)

At room temperature $MoH_4(PMePh_2)_4$ shows a 1 : 4 : 6 : 4 : 1 quartet for the hydride resonance but, on cooling, the spectrum becomes complex suggesting a structure consisting two interpenetrating tetrahedrons of phosphorus and hydrogen atoms.[561] As $MoO_2(acac)_2$ shows only one methyl resonance at room temperature earlier workers had concluded that the compound was *trans*. However, at low temperature two methyl resonances are observed showing the *cis* stereochemistry, and thermodynamic parameters were derived.[562] Similar behaviour was found for

556 J. F. Harrod and K. Taylor, *Chem. Comm.*, 1971, 696.
557 M. J. Frazer and W. E. Newton, *Inorg. Chem.*, 1971, **10**, 2137.
558 J. J. Howe and T. J. Pinnavaia, *J. Amer. Chem. Soc.*, 1970, **92**, 7342.
559 T. J. Pinnavaia and A. L. Lott, tert., *Inorg. Chem.*, 1971, **10**, 1388.
560 H. Köpf, *Angew. Chem. Internat. Edn.*, 1971, **10**, 134.
561 J. P. Jesson, E. L. Muetterties, and P. Meakin, *J. Amer. Chem. Soc.*, 1971, **93**, 5261.
562 B. M. Craven, K. C. Ramey, and W. B. Wise, *Inorg. Chem.*, 1971, **10**, 2626.

$MoO_2(dpm)_2$,[563] and $M(NO)_2(S_2CNMe_2)_2$ (M = Mo or W) shows two methyl resonances which coalesce at 140 °C. $Mo(NO)(S_2CNMe_2)_3$, which shows a 2 : 2 : 1 : 1 methyl pattern, and $Rh(PPh_3)(S_2CNMe_2)_3$, which shows a 2 : 1 : 1 : 1 : 1 methyl pattern at room temperature, also only give one methyl resonance at high temperatures.[564]

The chelate ligand $(Cp)_2M(SMe)_2$ (M = Mo or W) acts as a bidentate ligand (L) to form LRh(1,5-cyclo-octadiene), $[LRh(CO)_2]^+$, $[LRh(C_3H_5)_2]^+$, $LM'Cl_2$, $[L_2M']^{2+}$, and $[LM'(PR_3)_2]^{2+}$ (M' = Pd or Pt). $(Cp)_2M(\mu\text{-}SMe)_2\text{-}M'Cl_2$ and $[RhL(C_3H_5)_2]^+$ show fluxional behaviour.[565] ^1H N.m.r. has been used to follow the isomerization of *mer*-Ru(tfac)$_3$ to the *fac*-isomer at 150 °C. Although V(bzac)$_3$ exists in solution as a *mer–fac* mixture, $V(PhCH_2CSCHCOMe)_3$ exists only in the *cis*-form.[566]

A full lineshape analysis of the hydride resonance of *cis*-$FeH_2\{P(OEt)_3\}_4$ shows that of the five mechanisms considered, only one mechanism of rearrangement is feasible. Namely, if the molecule is considered to be iron, tetrahedrally co-ordinated by phosphorus with hydrogen atoms in the centre of two of the faces, then the mechanism is the migration of one hydrogen across the edge of the tetrahedron on to another face; ^{31}P n.m.r. data were also reported.[567] ^1H N.m.r. data on $MH(NO)L_3$ (M = Ru or Os; L = phosphorus ligand) have been reported. They all show quartets for the hydride resonance at room temperature, but at − 110 °C RuH(NO)-$(PMePh_2)_3$ shows a poorly resolved 1 : 2 : 2 : 2 : 1 signal.[568] Similar fluxional behaviour has been reported for $[HM^1(PF_3)_4]^-$ (M^1 = Ru or Os) and $HM^2(PF_3)_4$ (M^2 = Co, Rh, or Ir),[569] and $HCo\{PPh(OEt)_2\}_4$.[570]

$\{1,2\text{-}(Me_2NCH_2)_2C_5H_3\}(Cp)Fe$ shows two signals for the CH_2 protons, which move together and then apart as the temperature is changed.[571] At room temperature for $Co(S_2CNR^1R^2)_3$ (R^1, R^2 = Et, Bu^i, or Bz) the CH_2 protons are inequivalent, but coalesce at higher temperatures with ΔF *ca.* 17 kcal mol^{-1}.[572, 573] Similarly, $Co\{o\text{-}C_6H_4(AsMe_2)_2\}_2(S_2C_2Ph_2)$ shows two methyl resonances at room temperature which coalesce at 110 °C.[574] The activation energy for the rotation of ethylene in $(\pi\text{-}C_5H_4CN)Rh(C_2H_4)_2$ and $(\pi\text{-}C_5H_4CO_2Me)Rh(C_2H_4)_2$ has been derived from the temperature behaviour of the ^1H n.m.r. spectrum.[575]

563 T. J. Pinnavaia and W. R. Clements, *Inorg. Nuclear Chem. Letters*, 1971, **7**, 1127.
564 R. Davis, M. N. S. Hill, C. E. Holloway, B. F. G. Johnson, and K. H. Al-Obaidi, *J. Chem. Soc.* (*A*), 1971, 994.
565 A. R. Dias and M. L. H. Green, *J. Chem. Soc.* (*A*), 1971, 1951.
566 J. G. Gordon, jun., M. J. O'Connor, and R. H. Holm, *Inorg. Chim. Acta*, 1971, **5**, 381.
567 P. Meakin, E. L. Muetterties, F. N. Tebbe, and J. P. Jesson, *J. Amer. Chem. Soc.*, 1971, **93**, 4701.
568 S. T. Wilson and J. A. Osborn, *J. Amer. Chem. Soc.*, 1971, **93**, 3068.
569 P. Meakin, J. P. Jesson, F. N. Tebbe, and E. L. Muetterties, *J. Amer. Chem. Soc.*, 1971, **92**, 1797.
570 D. D. Titus, A. A. Orio, R. E. Marsh, and H. B. Gray, *Chem. Comm.*, 1971, 322.
571 D. W. Slocum and F. Stonemark, *Tetrahedron Letters*, 1971, 3291.
572 T. H. Siddall, tert., *Inorg. Nuclear Chem. Letters*, 1971, **7**, 545.
573 M. M. Dhingra, G. Govil, and C. R. Kanekar, *Chem. Phys. Letters*, 1971, **12**, 303.
574 J. A. McCleverty and D. G. Orchard, *J. Chem. Soc.* (*A*), 1971, 3784.
575 R. Cramer and J. J. Mrowca, *Inorg. Chim. Acta*, 1971, **5**, 528.

The dynamics of ring inversion for $[Co(en)(NH_3)_4]^{3+}$ in solution are obscured by a relaxation process which dominates the CH_2 magnetic resonance linewidth at low temperatures.[576] However, for $[Rh(D_2NCH_2$-$CH_2ND_2)]^{3+}$ it has been found that the Δ-$\delta\lambda\lambda$ conformer is more stable than the Δ-$\lambda\lambda\lambda$ conformer and the activation energy for flip has been determined.[577]

^1H N.m.r. has been used to examine $[Ni(\textit{meso}\text{-bn})_n]^{2+}$ and $[Ni(\textit{rac}\text{-bn})_n]^{2+}$ (n = 1, 2, or 3; bn = 2,3-diaminobutane), and for n = 2 variable-temperature n.m.r. has been used to obtain ΔG for the octahedral–square-planar equilibrium.[578]

For the compounds $(RSC_2H_4SR)MX_2$ (R = Et, Pr^n, or Bu^n; M = Pd or Pt; X = Cl, Br, or I) the rate of inversion at sulphur increases as the strength of the metal–sulphur bond decreases as a result of the *trans* effect of X, and the platinum compounds are less labile than the corresponding palladium compounds.[579]

It has been suggested that magnetic anisotropy is the main factor influencing ^1H chemical shifts of arylcopper compounds. The temperature dependence of some spectra indicate that there is hindrance to rotation for methyls *ortho* to copper.[580] Similarly, $(m\text{-}CF_3C_6H_4Cu)_8$ shows two broad overlapping CF_3 resonances at − 117 °C in ether which may be due to solvent complexing or restricted $m\text{-}CF_3C_6H_4$ motion.[581]

For (169) (R = H or Me; M = Zn or Cd), coalescence of the isopropyl methyl resonances has been used to obtain activation energies and rates of

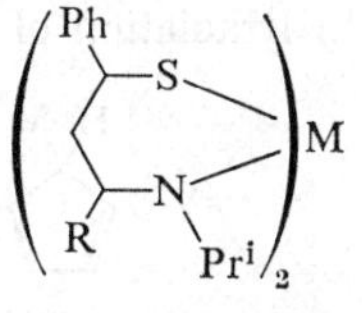

(169)

inversion. It was found that the inversion rates of Zn^{II} complexes are slower than for Cd^{II} and the mechanism is intramolecular. For Cd^{II} the intra- and inter-molecular rates are comparable.[582]

Fluxional Behaviour and Valence Tautomerism.—*Main-group Compounds.* At room temperature ^1H and ^{19}F n.m.r. of compound (169a) show two resonances for R^3 and R^5, but on heating they coalesce on account of rapid movement of $SiMe_3$ between the nitrogens.[583] Similarly, *cis*-$MeO_2CCR{=}C$-

576 R. Bramley and R. N. Johnson, *Chem. Comm.*, 1971, 1309.
577 J. L. Sudmeier and G. L. Blackmer, *Inorg. Chem.*, 1971, **10**, 2010.
578 R. F. Evilia, D. C. Young, and C. N. Reilley, *Inorg. Chem.*, 1971, **10**, 433.
579 R. J. Cross, G. J. Smith, and R. Wardle, *Inorg. Nuclear Chem. Letters*, 1971, **7**, 191.
580 A. Baici, A. Camus, and G. Pellizer, *J. Organometallic Chem.*, 1971, **26**, 431.
581 A. Cairncross and W. A. Sheppard, *J. Amer. Chem. Soc.*, 1971, **93**, 247.
582 S. S. Eaton and R. H. Holm, *Inorg. Chem.*, 1971, **10**, 1446.
583 D. H. O'Brien and C.-P. Hrung, *J. Organometallic Chem.*, 1971, **27**, 185.

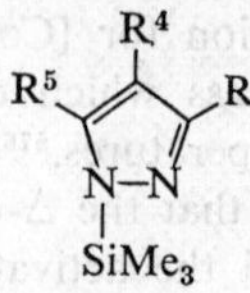

(169a)

(OMe)OSiMe$_3$ shows only one methoxy-resonance, implying rapid tautomerism; however, when R = CO_2Me, cooling into − 40 °C causes the methoxy-resonance to broaden.[584]

σ-Bonded cyclopentadienyls and related systems have received considerable attention. It had been concluded that in the case of $MeCl_2SiCp$, the silicon atom moves around the ring by 1,3-shifts. However, the 'spin-tickling' experiments on which this conclusion was based were wrong and 1,2-shifts actually occur.[585–587] $CpGeH_3$ also shows fluxional behaviour at room temperature but is static at − 17 °C, unlike $(Cp)_2GeH_2$ which is still fluxional at − 50 °C.[588] Similarly $(\sigma\text{-Cp})_2GeFe(CO)_2(\pi\text{-Cp})_2$ shows only a singlet for the σ-Cp group, but at − 50 °C this gives an AA′BB′X spectrum.[589] The spectra of $Sn(MeC_5H_4)_4$ and $Hg(MeC_5H_4)_2$ show fluxional behaviour of the cyclopentadienyl over the temperature range + 20 to − 95 °C, but the variations in J(M–H) were ascribed to changes in population between the three isomers (170)—(172).[590] Compound (173) is also fluxional with interchange of thallium between the rings.[591]

Lineshape analysis of (174) yields an activation energy for $SnPh_3$ migration of 10.1 ± 0.8 kcal mol^{-1}. Irradiation of H(7) causes H(3) and H(4)

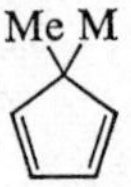

(170)

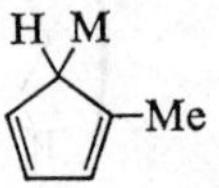

(171)

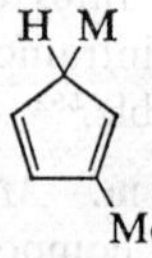

(172)

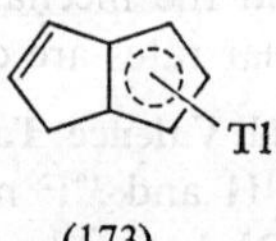

(173)

[584] Y.-N. Kuo, F. Chen, and C. Ainsworth, *Chem. Comm.*, 1971, 137.
[585] F. A. Cotton and T. J. Marks, *Inorg. Chem.*, 1970, **9**, 2804.
[586] N. M. Sergeyev, G. I. Avramenko, A. V. Kisin, V. A. Korenevsky, and Yu. A. Ustynyuk, *J. Organometallic Chem.*, 1971, **32**, 55.
[587] N. M. Sergeyev, G. I. Avramenko, and Yu. A. Ustynyuk, *Inorg. Chem.*, 1971, **10**, 2364.
[588] S. R. Stobart, *J. Organometallic Chem.*, 1971, **33**, C11.
[589] Yu. A. Ustynyuk and A. V. Kisin, *J. Organometallic Chem.*, 1971, **33**, C61.
[590] C. H. Campbell and M. L. H. Green, *J. Chem. Soc. (A)*, 1971, 3282.
[591] H. P. Fritz and F. H. Köhler, *J. Organometallic Chem.*, 1971, **30**, 177.

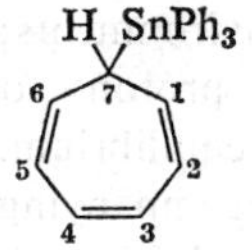

(174)

to decrease in intensity and it was therefore suggested that migration from C(7) to C(4) occurs.[592]

Transition-metal Compounds. [(Norbornyl)Hg]$^{2+}$ shows a static n.m.r. spectrum at − 70 °C but all the peaks collapse at − 30 °C, implying bond migration of the norbornyl unit.[593]

The conversion of [MePt(PMe$_2$Ph)$_2$(allene)]$^+$ into [(2-methylallyl)Pt-(PMe$_2$Ph)$_2$]$^+$ has been followed by ^{1}H n.m.r. If the starting material is cooled to − 20 °C, restricted rotation is observed with the allene perpendicular to the MePt(PMe$_2$Ph)$_2$ plane.[594]

σ-Allyl derivatives of (R$_2$N)$_3$Ti are fluxional.[595] (Cp)M(CO)$_2${(*p*-Tol)$_2$-CNC(*p*-Tol$_2$)} (M = Mo or W) shows four methyl resonances at − 20 °C, two methyl resonances with intensities 1 : 3 at 10 °C and a singlet at 70 °C; thus the structure is (175).[596] Similarly, Ir(C$_3$H$_5$)(CO)(PPh$_3$)$_2$ shows

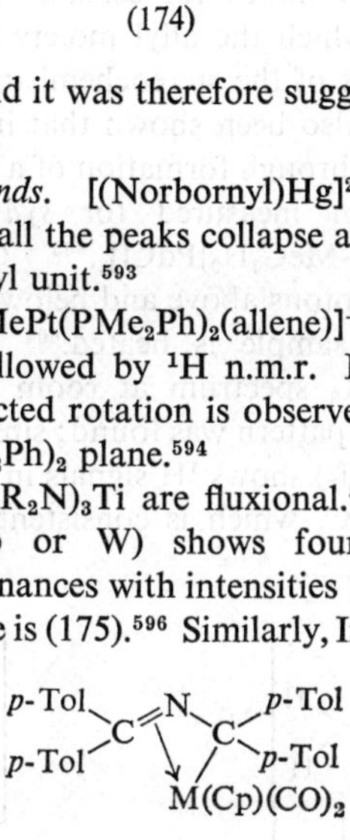

(175)

an AX$_4$ pattern for the allyl group at room temperature, but on cooling this gives first an AM$_2$X$_2$ spectrum at − 20 °C and finally an AGMPX spectrum. Data are also reported for the rhodium analogues, and for Ir(COR)(CO)$_2$(PPh$_3$)$_2$ and IrH$_3$(CO)(PPh$_3$)$_2$.[597] Pd(C$_3$H$_5$)(*p*-O$_2$NC$_6$H$_4$NC)-Cl also shows an AX$_4$ pattern for the allyl group at room temperature.[598] As (2-Pr^{i}C$_3$H$_4$)PdCl(PR$_3$) shows two methyl doublets at room temperature but a broad singlet at + 20 °C, exchange is occurring *via* an intermediate which has a plane of symmetry through the isopropyl–allyl bond.[599]

Ru(C$_3$H$_5$)$_2$(CO)$_2$ has a ^{1}H n.m.r. spectrum which shows that in addition to *syn–anti* exchange at high temperature there is a rapid intramolecular rearrangement at room temperature in which the CH$_2$ groups of the π-allyl

592 R. B. Larrabee, *J. Amer. Chem. Soc.*, 1971, **93**, 1510.
593 G. A. Olah and P. R. Clifford, *J. Amer. Chem. Soc.*, 1971, **93**, 1261.
594 M. H. Chisholm, H. C. Clark, and D. H. Hunter, *Chem. Comm.*, 1971, 809.
595 H.-J. Neese and H. Bürger, *J. Organometallic Chem.*, 1971, **32**, 213.
596 H. R. Keable and M. Kilner, *Chem. Comm.*, 1971, 349.
597 C. K. Brown, W. Mowat, G. Yagupsky, and G. Wilkinson *J. Chem. Soc.* (*A*), 1971, 850.
598 T. Boschi and B. Crociani, *Inorg. Chim. Acta*, 1971, **5**, 477.
599 P. W. N. M. van Leewen, A. P. Pratt, and M. van Diepen, *J. Organometallic Chem.*, 1971, **29**, 433.

groups effectively exchange environments; data were also reported for $Ru(\pi\text{-}C_4H_7)_2L_2$.[600] The CH_2 protons of $[(1\text{-}MeC_3H_4)Ni(CN)_2]^+$ are equivalent owing to a dynamic equilibrium.[601]

Steric factors and their accompanying kinetic and thermodynamic parameters have been determined for certain 1,2,3-h^3-allyl(amine)chloro-palladium complexes in which the allyl moiety is disubstituted. General rules for the establishment of the stereochemistry of allyl complexes have been determined. It has also been shown that interconversion of isomers and epimerization occurs through formation of a σ-bonded intermediate.[602] Similarly ΔG_{73} has been measured for *syn–anti* interconversion of $[\{1\text{-}(H_2C{=}CClCH_2CH_2)\text{-}1\text{-}MeC_3H_3\}PdCl]_2$.[603] Compound (176) shows separate resonances for protons above and below the plane of the ring, but they coalesce when the sample is heated.[604] $(2\text{-}MeC_3H_4)Pd(PMePh_2)(SnCl_3)$ shows an $A_2M_2X_3$ spectrum at room temperature, but at low temperatures an $ABCDX_3$ pattern was found; similar compounds were also examined.[605] $(R_3P)_2Pt(B_3H_7)$ shows ^1H signals in the ratio 3 : 2 : 2 at 25 °C and 2 : 1 : 2 : 2 at − 25 °C, which is consistent with the formulation of the ligand as (177).[606]

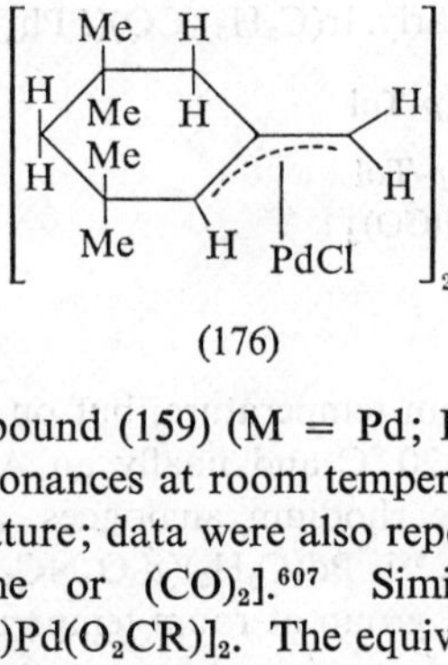

(176)

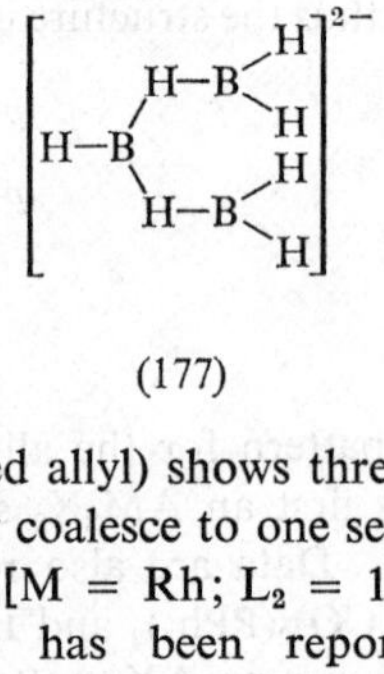

(177)

Compound (159) (M = Pd; L_2 = substituted allyl) shows three sets of allyl resonances at room temperature, but they coalesce to one set at high temperature; data were also reported for (178) [M = Rh; L_2 = 1,5-cyclo-octadiene or $(CO)_2$].[607] Similar behaviour has been reported for $[(\pi\text{-allyl})Pd(O_2CR)]_2$. The equivalence of the π-allylic ligands occurs *via* a rapid intramolecular process, but at higher temperature a bimolecular exchange involving the breakage of palladium–oxygen bonds occurs.[608] The methylene protons of (179) form an AB pattern at − 10 °C which coalesces on warming.[609] $Pt(2\text{-}RC_3H_4)_2$ (R = H or Me) has *cis-* and

600 M. Cooke, R. J. Goodfellow, M. Green, and G. Parker, *J. Chem. Soc.* (*A*), 1971, 16.
601 D. Bingham and M. G. Burnett, *J. Chem. Soc.* (*A*), 1971, 1782.
602 J. W. Faller, M. E. Thomsen, and M. J. Mattina, *J. Amer. Chem. Soc.*, 1971, **93**, 2642.
603 D. J. S. Guthrie, R. Spratt, and S. M. Nelson, *Chem. Comm.*, 1971, 935.
604 C. W. Alexander, W. R. Jackson, and W. B. Jennings, *J. Chem. Soc.* (*B*), 1971, 2241.
605 M. Sakakibara, Y. Takahashi, S. Sakai, and Y. Ishii, *J. Organometallic Chem.*, 1971, **27**, 139.
606 A. R. Kane and E. L. Muetterties, *J. Amer. Chem. Soc.*, 1971, **93**, 1041.
607 S. Trofimenko, *Inorg. Chem.*, 1971, **10**, 1372.
608 J. Powell, *J. Chem. Soc.* (*A*), 1971, 2233.
609 T. Okamoto, *Bull. Chem. Soc. Japan*, 1971, **44**, 1353.

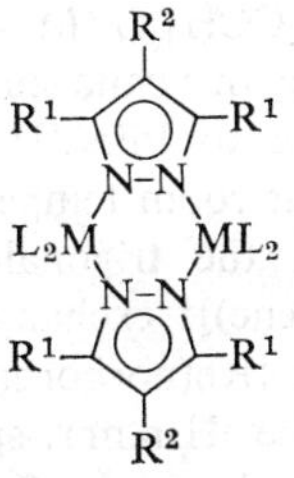

(178)

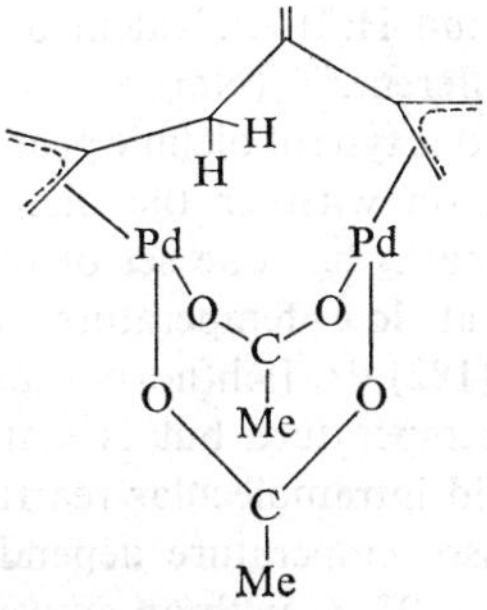

(179)

trans-isomers present in solution; similar isomerism has been also observed for $[PtCl(2\text{-}MeC_3H_4)]_2$ at − 65 °C. At 28 °C, if one platinum satellite is irradiated the other collapses, which was attributed to a short ^{195}Pt relaxation time.[610]

At room temperature, (180) (R = H) shows only one resonance for the α and α′ protons, but two at − 50 °C.[611] $(C_8H_9)(C_8H_8)Co$ is the first cobalt compound to contain a fluxional cyclo-octatetraene ring; related fluxional compounds, *e.g.* $(Cp)(C_8H_8)Co$, have also been prepared.[612] Compound (181)

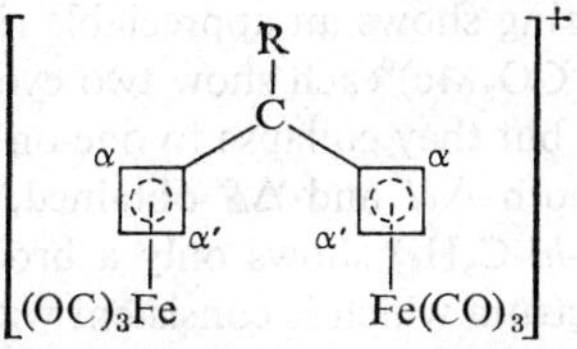

(180)

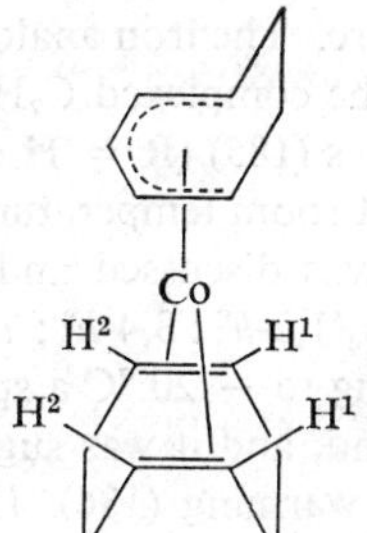

(181)

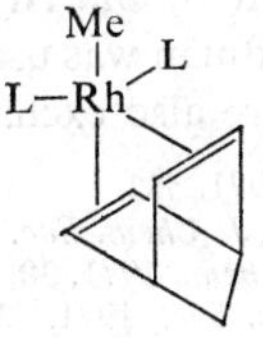

(182)

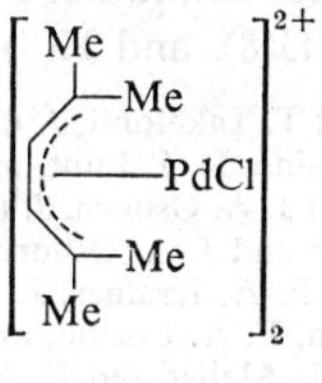

(183)

[610] B. E. Mann, B. L. Shaw, and G. Shaw, *J. Chem. Soc.* (*A*), 1971, 3536.
[611] R. E. Davis, H. D. Simpson, N. Grice, and R. Pettit, *J. Amer. Chem. Soc.*, 1971, **93**, 6688.
[612] A. Greco, M. Green, and F. G. A. Stone, *J. Chem. Soc.* (*A*), 1971, 285.

has H(1) and H(2) equivalent at room temperature, but on cooling they become different.[613] (Norbornadiene)M(PPh_3)(O_2CCH_2X) (M = Rh or Ir) show only one type of olefinic proton, but on cooling this band splits into two depending on whether the olefin is *cis* or *trans* to PPh_3.[614] Similarly, MeRh(diene)L_2 has one set of olefinic protons at room temperature and two sets at low temperature, consistent with the trigonal-bipyramid structure (182).[615] [Rh(norbornadiene)(π-h^5-azulene)]$^+$ exchanges azulene at room temperature but is static at − 80 °C; [Rh(norbornadiene)$_2$L]$^+$ shows rapid intramolecular rearrangement.[616] The ^{1}H n.m.r. spectrum of (183) is also temperature dependent, showing coalescence of the methyl groups at − 25 °C with an activation energy of 13 kcal mol^{-1}.[617]

Although $(Cp)_3TiCl$ shows a single sharp line down to at least − 100 °C, when $(Cp)_4Ti$ is cooled the original single sharp line splits into a doublet and one component broadens, consistent with the formulation (σ-Cp)$_2$-(π-Cp)$_2$Ti.[618] Other workers have also observed this behaviour and have reported data for (Cp)Nb(C_3H_5)$_2$, which shows two cyclopentadienyl resonances at room temperature.[619] Although Mo(Cp)$_2$(NO)(S_2CNMe_2) and [Mo(Cp)$_2$(NO){$S_2C_2(CN)_2$}]$^+$ show only one cyclopentadienyl resonance at room temperature, resonances attributable to a σ-cyclopentadienyl and a π-cyclopentadienyl group are observed at low temperature; there is a detailed discussion of the mechanism.[620] Compound (184) shows only one resonance at 88 °C, but at − 47 °C the spectrum is consistent with this static structure. The iron analogue shows a singlet at δ 6.10 and it was suggested that the complexed C_7H_7 ring shows an appreciable ring current.[621] The complexes (185) (R = H or CO_2Me) each show two cyclopentadienyl resonances at room temperature, but they collapse to one on heating. The mechanism was discussed and both ΔH and ΔS obtained.[622] At 35 °C, PhCCo$_3$(CO)$_6$(1,2-h^2 : 3,4-h^2 : 7,8-h^2-C_8H_8) shows only a broad resonance, but on cooling to − 20 °C a spectrum which is consistent with this formulation is found, and it was suggested that rearrangement occurs *via* a 1,2-shift.[623] On warming (186), H(2) and H(4), and H(6) and H(8) become equivalent with ΔG = 15.9 ± 0.5 kcal mol^{-1}.[624] At − 6 °C, (187) shows a static spectrum, but on warming to 86 °C, H(10) and H(11) coalesce. The double-resonance method of Forsén and Hoffman was used to demonstrate that H(6) and H(8), and H(9) and H(12), are also exchanging and would

[613] S. Otsuka and T. Taketomi, *J. Chem. Soc.* (*A*), 1971, 583.
[614] R. N. Haszeldine, R. J. Lunt, and R. V. Parish, *J. Chem. Soc.* (*A*), 1971, 3696.
[615] D. P. Rice and J. A. Osborn, *J. Organometallic Chem.*, 1971, **30**, C84.
[616] R. R. Schrock and J. A. Osborn, *J. Amer. Chem. Soc.*, 1971, **93**, 3089.
[617] J. Lukas and P. A. Kramer, *J. Organometallic Chem.*, 1971, **31**, 111.
[618] J. L. Calderon, F. A. Cotton, and J. Takats, *J. Amer. Chem. Soc.*, 1971, **93**, 3587.
[619] H. J. de Liefde Meijer and F. Jellinek, *Inorg. Chim. Acta*, 1970, **4**, 651.
[620] W. G. Kita, M. K. Lloyd, and J. A. McCleverty, *Chem. Comm.*, 1971, 420.
[621] T. H. Whitesides and R. A. Budnik, *Chem. Comm.*, 1971, 1514.
[622] M. Rosenblum, W. P. Giering, B. North, and D. Wells, *J. Organometallic Chem.*, 1971, **28**, C17.
[623] B. H. Robinson and J. Spencer, *J. Organometallic Chem.*, 1971, **33**, 97.
[624] R. Aumann, *Angew. Chem. Internat. Edn.*, 1971, **10**, 189.

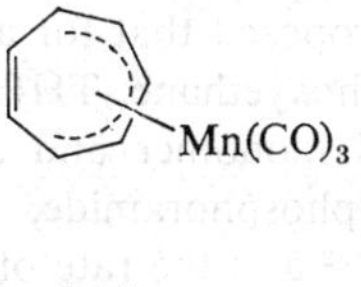

(184)

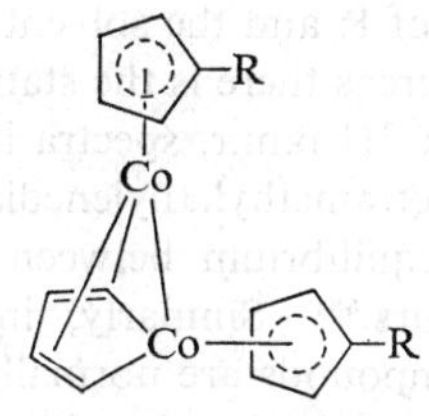

(185)

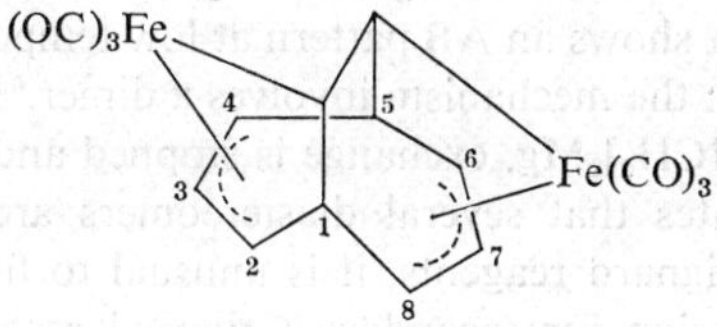

(186)

coalesce above the decomposition temperature.[625] Compound (188) is also fluxional with carboxylate exchange; the norbornadiene analogue was also examined.[626]

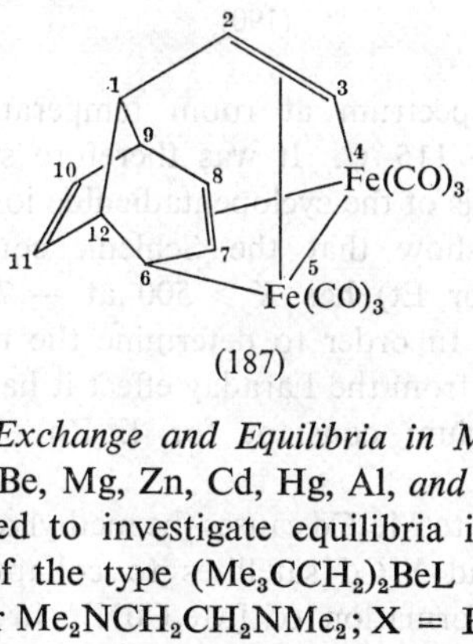

(187)

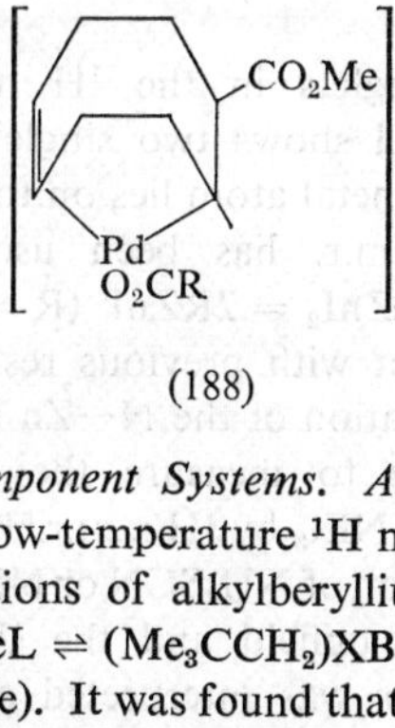

(188)

Ligand Exchange and Equilibria in Multicomponent Systems. Alkyls and aryls of Be, Mg, Zn, Cd, Hg, Al, *and* Ga. Low-temperature ^1H n.m.r. has been used to investigate equilibria in solutions of alkylberyllium complexes of the type $(Me_3CCH_2)_2BeL + X_2BeL \rightleftharpoons (Me_3CCH_2)XBeL$ (L = NMe_3 or $Me_2NCH_2CH_2NMe_2$; X = Br or Me). It was found that the bulk of the neopentyl group drives the equilibrium to the right.[627]

Variable-temperature ^1H n.m.r. of RMgF reveals only a small movement of the methyl resonance and attempts to use ^{19}F n.m.r. have so far proved unsuccessful.[628] ^1H N.m.r. has also been used to investigate the dependence of the position of the Schlenk equilibrium $R_2Mg + MgX_2 \rightleftharpoons 2RMgX$ on

[625] R. Aumann, *Angew. Chem. Internat. Edn.*, 1971, **10**, 560.
[626] M. N. S. Hill, B. F. G. Johnson, and J. Lewis, *J. Chem. Soc. (A)*, 1971, 2341.
[627] G. E. Coates and B. R. Francis, *J. Chem. Soc. (A)*, 1971, 1305.
[628] E. C. Ashby and S. Yu, *J. Organometallic Chem.*, 1971, **29**, 339.

the nature of R and the solvent.[629] Diethyl ether favours the formation of RMgX whereas there is the statistical equilibrium in THF.[630]

From the ^{1}H n.m.r. spectra it has been proposed that for a solution of Et_2Mg in tetramethylethylenediamine, dimethoxyethane, THF, or diethylether, an equilibrium between a disolvated monomer and a disolvated dimer occurs.[631] Similarly, in hexamethylphosphoramide, organomagnesium compounds are normally disolvated,[632] and the rate of cleavage of the carbon–magnesium bond is much smaller than in all solvents previously studied, with the Schlenk equilibrium completely in the RMgX form.[633]

The mechanism of inversion of a Grignard reagent has received attention. Thus $EtCR^1R^2CH_2MgX$ shows an AB pattern at low temperatures and ΔH and ΔS were calculated; the mechanism involves a dimer.[634] If sparteine is added to $[(\pm)\text{-}EtMeCHCH_2]_2Mg$, exchange is stopped and the complexity of the spectrum indicates that several diastereomers are present.[635] In contrast to primary Grignard reagents, it is unusual to find slow (on the n.m.r. time scale) inversion for secondary Grignard reagents. However, although inversion is fast for (189) at 65 °C it is slow at − 8 °C, but all attempts to stop inversion for (190) failed.[636] $(Cp)_2Mg$ and (Cp)MgX

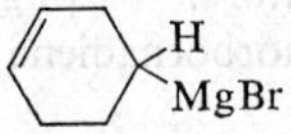

(189)

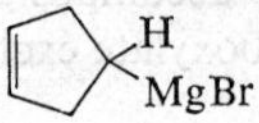

(190)

show singlets in the ^{1}H n.m.r. spectrum at room temperature and (Cp)MgCl shows two singlets at − 115 °C. It was therefore suggested that the metal atom lies on the C_5 axis of the cyclopentadienide ion.[637]

^{1}H N.m.r. has been used to show that the Schlenk equilibrium $R_2Zn + ZnI_2 \rightleftharpoons 2RZnI$ (R = Me or Et) has $K \geqslant 500$ at − 70 °C, in agreement with previous results.[638] In order to determine the magneto-optic rotation of the N—Zn linkage from the Faraday effect it has proved necessary to measure the equilibrium constant for $Et_2Zn + NEt_3 \rightleftharpoons Et_2Zn \leftarrow NEt_3$ by ^{1}H n.m.r.[639]

Addition of NH_4SCN or NH_4OCN to Me_2Cd causes the methyl resonance to move upfield and the ^{111}Cd and ^{113}Cd satellites to collapse. This behaviour was interpreted as the formation of $[Me_2CdX]^-$ in solution

629 D. F. Evans and G. V. Fazakerley, *J. Chem. Soc.* (*A*), 1971, 184.
630 G. E. Parris and E. C. Ashby, *J. Amer. Chem. Soc.*, 1971, **93**, 1206.
631 J. Ducom, *Bull. Soc. chim. France*, 1971, 3529.
632 J. Ducom, *Bull. Soc. chim. France*, 1971, 3523.
633 J. Ducom, *Bull. Soc. chim. France*, 1971, 3518.
634 G. Fraenkel, C. Cottrell, and D. T. Dix, *J. Amer. Chem. Soc.*, 1971, **93**, 1704.
635 G. Fraenkel, C. Cottrell, J. Ray, and J. Russell, *Chem. Comm.*, 1971, 273.
636 A. Maercker and R. Geuss, *Angew. Chem. Internat. Edn.*, 1971, **10**, 270.
637 W. T. Ford, *J. Organometallic Chem.*, 1971, **32**, 27.
638 D. F. Evans and G. V. Fazakerley, *J. Chem. Soc.* (*A*), 1971, 182.
639 G. Levy, J.-P. Laurent, and P. de Loth, *J. chim. Phys.*, 1971, **68**, 621.

which readily exchanges methyl groups with Me_2Cd.[640] If $(CF_3)_2Hg$ and Me_2Cd are mixed, ^1H and ^{19}F n.m.r. show that signals due to $MeHgCF_3$, Me_2Hg, $MeCdCF_3$, and $Cd(CF_3)_2$ rapidly appear.[641] The self-exchange of $CdMe_2$, the exchange of $CdMe_2$ with $GaMe_3$, and of $ZnMe_2$ with $InMe_3$ have been investigated and activation energies derived. Strongly co-ordinating solvents have been shown to enhance the exchange rate between derivatives of Group II metals but decreases the exchange rate when one of the species is from Group III.[642] Similarly, ^1H n.m.r. has been used to follow other organomercury redistribution reactions such as the mixing of $HgPh_2$ with $Hg(CH_2CHO)_2$ to give some $PhHgCH_2CHO$.[643] ^1H N.m.r. studies of the self-exchange of $(Me_3Si)_2Hg$ and $(Me_3Ge)_2Hg$ have shown that these reactions proceed by a second-order process. It was established that the rate of group exchange decreases $(Me_3Sn)_2Hg > (Me_3Ge)_2Hg > (Me_3Si)_2Hg > (MeCl_2Si)_2Hg > Me_2Hg$.[644] Me_3SiHgR and MeHgR were prepared in solution by mixing $(Me_3Si)_2Hg$ or Me_2Hg and R_2Hg. A linear relationship between $^3J(^{199}Hg–Si–C–^1H)$ for $RHgSiMe_3$ and $^2J(^{199}Hg–C–^1H)$ for RHgMe indicates that the mechanism of coupling is basically the same in both series of compounds.[645]

Previous workers have examined the exchange of methyl groups between Al_2Me_6 and $GaMe_3$ by ^1H n.m.r. spectroscopy but have failed to find a fully satisfactory kinetic expression for the observed rates. The results are interpreted in terms of two equilibria:[646]

$$Al_2Me_6 \rightleftharpoons 2AlMe_3$$

$$AlMe_3 + GaMe_3 \rightleftharpoons Me_2AlMe_2GaMe_2$$

The equilibrium mixture of (191), (192), and (193) has been analysed by ^1H n.m.r. over a variety of temperatures and equilibria constants derived;

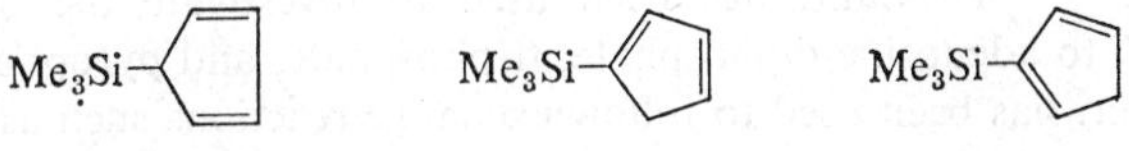

(191) (192) (193)

unlike the carbon analogue, (191) is the most stable.[647] The interaction between $MeAlCl_2$ and $MePr^iNPh$ has been examined and at low temperature the isopropyl group shows two methyl resonances.[648]

[640] N. Röder and K. Dehnicke, *J. Organometallic Chem.*, 1971, **33**, 281.
[641] B. L. Dyatkin, B. I. Martynov, I. L. Knunyants, S. R. Sterlin, L. A. Fedorov, and Z. A. Stumbrevichute, *Tetrahedron Letters*, 1971, 1345.
[642] J. Soulati, K. L. Henold, and J. P. Oliver, *J. Amer. Chem. Soc.*, 1971, **93**, 5694.
[643] F. G. Thorpe, T. N. Huckerby, P. H. Lindsay, and S. W. Breuer, *Tetrahedron Letters*, 1971, 2821.
[644] T. F. Schaaf and J. P. Oliver, *J. Organometallic Chem.*, 1971, **32**, 307.
[645] T. F. Schaaf and J. P. Oliver, *Inorg. Chem.*, 1971, **10**, 1521.
[646] D. S. Matteson, *Inorg. Chem.*, 1971, **10**, 1555.
[647] K. W. Egger and T. L. James, *J. Organometallic Chem.*, 1971, **26**, 335.
[648] Z. Buczkowski, A. Gryff-Keller, and P. Szczeciński, *Tetrahedron Letters*, 1971, 607.

Other compounds of the main-group elements. The interaction of metal ions, especially Ca^{2+}, with sugars has been investigated by ^{1}H n.m.r.[649]

μ-Dimethylaminodiborane and sodium borohydride reversibly form $NaMe_2N(BH_3)_2$ in diglyme. The ^{11}B n.m.r. spectra are temperature dependent: at 60 °C one resonance was observed whereas at 0 °C signals assigned to $[BH_4]^-$, $[B_2H_7]^-$, and $[Me_2N(BH_3)_2]^-$ were detected.[650] The reaction between Me_3PBMe_3 and an excess of Me_3P has been shown to proceed by a dissociative mechanism and activation parameters were derived.[651] ^{11}B N.m.r. has been used to obtain the kinetics of displacement of R_3N from R_3NBH_3 by R_3P.[652] The redistribution equilibria

$$BX_3 + BY_3 \rightleftharpoons BX_2Y + BXY_2 \quad (X \text{ and } Y = \text{halogen})$$

have been measured by ^{11}B n.m.r. and variable temperature has been used to obtain ΔH and ΔS for the system BCl_3–BBr_3.[653]

As the ratio of $PbCl_2$ or Me_2PbCl_2 to $MeAlCl_2$ concentration increases from 0 to 0.5, the methylaluminium ^{1}H resonance moves to low field and then remains constant, implying the formation of 1 : 2 complexes, *i.e.* $PbCl_2,2AlMeCl_2$ and $Me_2PbCl_2,2AlMeCl_2$.[654] Variable-temperature ^{1}H n.m.r. has been used to follow the slow equilibrium between $[(CH_2)_3$-$NAlMe_2]_n$, $n = 2$ or 3, and thermodynamic data obtained. ^{1}H N.m.r. data were also reported for $[(CH_2)_xNMR_2]_n$ ($n = 2$ or 3; $x = 2$—5; R = Me, Et, or Bu^i; M = Al, Ga, or In).[655]

$Cl_nGaBu^n{}_{3-n}$ interacts with Me_3P causing the ^{1}H methyl resonance to move to low field and ^{2}J(P–H) to decrease and then increase; stability constants were derived.[656] In the variable-temperature ^{1}H n.m.r. spectrum of $[Me_2TlOEt]_2$, the methyl resonance of the satellites broadens and then sharpens again. It was suggested that the most likely explanation is that fast monomer–dimer exchange is occurring with thallium–thallium scalar relaxation.[657] ^{31}P N.m.r. has been used to investigate the binding of thallium(I) to adenosine diphosphate, triphosphate, and pyrophosphate.[658]

^{1}H N.m.r. has been used to follow exchange reactions such as [659]

$$GeH_2X_2 + GeH_2Y_2 \rightleftharpoons 2GeH_2XY \quad (X \text{ and } Y = \text{halogen})$$

649 S. J. Angyal and K. P. Davies, *Chem. Comm.*, 1971, 500.
650 P. C. Keller, *Inorg. Chem.*, 1971, **10**, 2256.
651 K. J. Alford, E. O. Bishop, P. R. Carey, and J. D. Smith, *J. Chem. Soc.* (*A*), 1971, 2574.
652 D. E. Walmsley, W. L. Budde, and M. F. Hawthorne, *J. Amer. Chem. Soc.*, 1971, **93**, 3150.
653 M. F. Lappert, M. R. Litzow, J. B. Pedley, R. T. Spalding, and H. Nöth, *J. Chem. Soc.* (*A*), 1971, 383.
654 M. Bolesławski, S. Pasynkiewicz, and H. Pszonka, *J. Organometallic Chem.*, 1971, **28**, C31.
655 A. Storr and B. S. Thomas, *J. Chem. Soc.* (*A*), 1971, 3850.
656 R. Haran and J.-P. Laurent, *Compt. rend.* 1971, **273**, *C*, 1629.
657 A. G. Lee and G. M. Sheldrick, *Trans. Faraday Soc.*, 1971, **67**, 7.
658 J. P. Manners, K. G. Morallee, and R. J. P. Williams, *J. Inorg. Nuclear Chem.*, 1971, **33**, 2085.
659 G. K. Barker and J. F. Drake, *Inorg. Nuclear Chem. Letters*, 1971, **7**, 39.

and halogen exchange in Me_3SnI and Me_3SnBr.[660] R_2SnCl_2 forms 1 : 1 adducts with the potentially quadridentate bis(acetylacetone)ethylenediimine, but ^1H n.m.r. spectra suggest that the adducts are largely dissociated in solution.[661]

Extensive use has been made of ^1H to investigate the 1 : 1 complex between lead(II) or cadmium(II) and diethylenetriaminepenta-acetic acid, and the pH dependence of the complexes.[662] Similar investigations have been carried out on the complex formed between lead(II) and triethylenetetra-aminehexa-acetic acid[663] and tetraethylenepenta-aminehepta-acetic acid.[664] The kinetics of the ligand-exchange reactions of the nitriloacetic acid complex of lead have been measured from the extent of collapse of the ^{207}Pb–^1H spin–spin coupling in the ^1H n.m.r. spectrum of the complex.[665]

Me_3P reacts with PPh_2Cl to yield $[Me_3PPPh_2]^+ Cl^-$, as demonstrated by the observation of coupling from the second phosphorus to the methyl group at low temperature; however, at room temperature this coupling is absent, implying an equilibrium.[666] Similarly, treatment of $Me_2PCH_2CH_2SiMe_3$ or Me_3P with PF_3 causes $|{}^2J(P–H)|$ to decrease, become zero, and then increase. This was interpreted in terms of rapid exchange.[667] The position of the equilibrium between (194) and (195) has been measured by

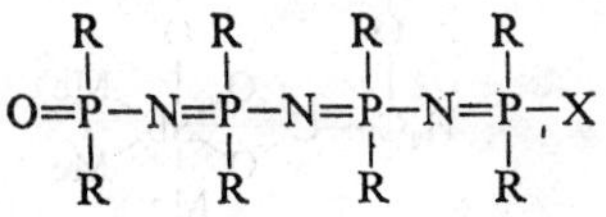

(194)

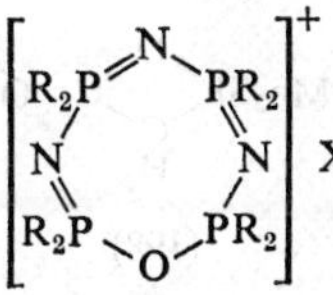

(195)

^{31}P n.m.r. as a function of X.[668] The rate of reaction of salts such as (MeO)(MeS)POSNa with CD_3I has been measured by ^1H n.m.r.,[669] and ^1H n.m.r. has been used to monitor the exchange reaction between Me_2AsD and Et_2AsH.[670] $Me_4As_2S_2$ has been shown by ^1H n.m.r. to exist in solution as an equilibrium mixture of $Me_2As(S)SAsMe_2$ and $Me_2AsSSAsMe_2$ and thermodynamic parameters were derived.[671]

A detailed investigation of SbF_5, AsF_5, and PF_5 in HF has been carried out. At − 140 °C fluorine exchange of $[As_2F_{11}]^-$ in SO_2ClF had stopped

[660] G. Redl and M. Winokur, *J. Organometallic Chem.*, 1971, **26**, C36.
[661] P. J. Smith and D. Dodd, *J. Organometallic Chem.*, 1971, **32**, 195.
[662] P. Letkeman and J. B. Westmore, *Canad. J. Chem.*, 1971, **49**, 2073.
[663] P. Letkeman and J. B. Westmore, *Canad. J. Chem.*, 1971, **49**, 2086.
[664] P. Letkeman and D. T. Sawyer, *Canad. J. Chem.*, 1971, **49**, 2096.
[665] D. L. Rabenstein, *J. Amer. Chem. Soc.*, 1971, **93**, 2869.
[666] F. Ramirez and E. A. Tsolis, *J. Amer. Chem. Soc.*, 1970, **92**, 2553.
[667] J. Grobe and U. Möller, *Z. Naturforsch.*, 1971, **26b**, 639.
[668] A. Schmidpeter and K. Stoll, *Angew. Chem. Internat. Edn.*, 1971, **10**, 131.
[669] C. K. Tseng and J. H.-H. Chan, *Tetrahedron Letters*, 1971, 699.
[670] W. R. Cullen and W. R. Leeder, *Canad. J. Chem.*, 1970, **48**, 3757.
[671] R. A. Zingaro, K. J. Irgolic, D. H. O'Brien and L. J. Edmonson, jun., *J. Amer. Chem. Soc.*, 1971, **93**, 5677.

and three different fluorines were observed, consistent with structure (196).[672] Me_nSiF_{4-n} and SbF_5 in SO_2ClF shows exchange at room temperature, but on cooling the ^1H n.m.r. spectrum shows either a triplet (n = 2) or a quartet (n = 1) which was interpreted as being due to structure (197)

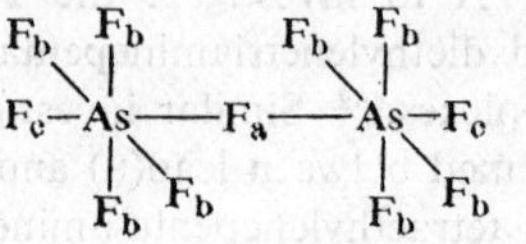

(196)

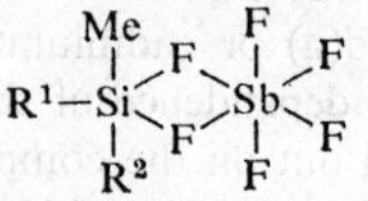

(197)

rather than $[Me_nSiF_{3-n}]^+$.[673] ^{19}F N.m.r. has been used to investigate the systems $SbCl_5$–SbF_5–SO_2ClF, SO_2Cl_2–SbF_5–SO_2ClF, COF_2–SbF_5-SO_2ClF, COF_2–AsF_5–SO_2ClF, and $COClF$–SbF_5–SO_2ClF and species such as chlorofluoroantimony cations were identified.[674] Low-temperature ^{19}F n.m.r. of the system RF–SbF_5–SO_2 has been interpreted as being due to species such as $RSbF_6$, RSb_2F_{11}, and RSb_3F_{21}, with R—F—Sb bridges.[675]

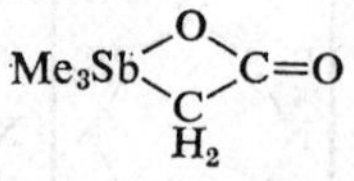

(198)

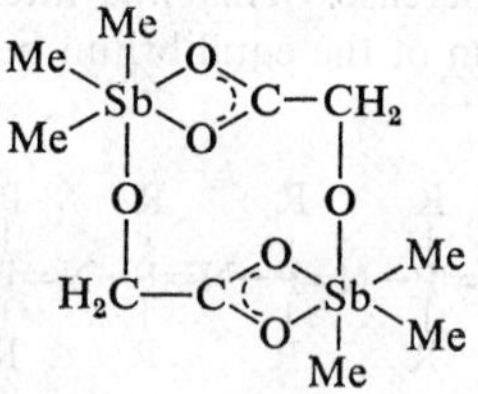

(199)

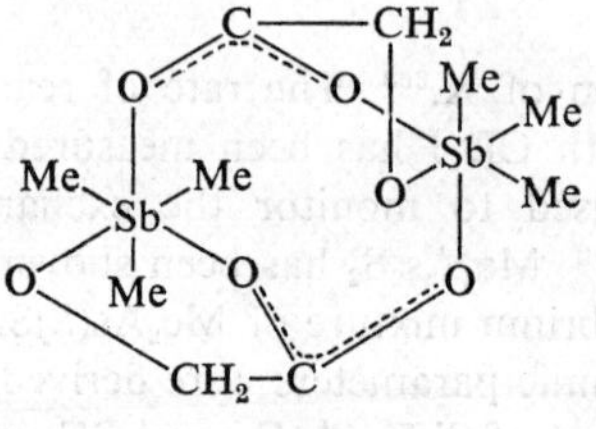

(200)

The ^1H n.m.r. spectrum of $Me_3Sb(OCH_2CO_2)$ shows only one methyl resonance at room temperature but two in the ratio 2 : 1 at − 38 °C. This was interpreted as arising from an equilibrium between structures (198) and (199) or (200).[676] ^1H N.m.r. has been used to establish the stability constant

[672] P. A. W. Dean, R. J. Gillespie, R. Hulme, and D. A. Humphreys, *J. Chem. Soc. (A)*, 1971, 341.
[673] G. A. Olah and Y. K. Mo, *J. Amer. Chem. Soc.*, 1971, **93**, 4942.
[674] J. Bacon, P. A. W. Dean, and R. J. Gillespie, *Canad. J. Chem.*, 1971, **49**, 1276.
[675] J. Bacon and R. J. Gillespie, *J. Amer. Chem. Soc.*, 1971, **93**, 6914.
[676] Y. Matsumura, M. Shindo, and R. Okawara, *J. Organometallic Chem.*, 1971, **27**, 357.

for the equilibrium[677]

$$HCl + X^- \rightleftharpoons XHCl^- \quad (X = \text{halogen})$$

Other compounds of the transition metals. $(Cp)_2Ti$ shows a broad resonance, the position of which is temperature dependent; this behaviour was attributed to an equilibrium between a paramagnetic monomer and dimer; data were also reported for $(C_5Me_5)_2Ti(CO)_2$.[678] The ^1H n.m.r. spectrum of $TiCl_3(PhCOCHCOMe)$ shows two methyl resonances which coalesce on warming and this was interpreted as being due to the equilibrium[679]

$$2\,Cl_3Ti(bzac) \rightleftharpoons Cl_2Ti(bzac)_2 + TiCl_4$$

The ^1H and ^{31}P n.m.r. spectra of the $TiCl_4$–R_3P system have been measured as a function of temperature and the results suggest the presence of fast equilibria involving 1 : 1 and 1 : 2 adducts with the possible existence of species such as $Ti_2Cl_8(PR_3)$. Correlations were found between the ^{31}P chemical shifts for the 1 : 1 adducts and both Taft constants and half-neutralization potentials.[680]

^1H N.m.r. has been used to determine the rate of exchange for the $[Y(hfac)_4]^-$–$[Y(tfac)_4]^-$ system.[681] Ligand exchange rates and stability constants for complexes between nitrilotriacetate and Sc^{3+}, Y^{3+}, La^{3+}, or Lu^{3+},[682, 683] and between cyclohexanediaminetetra-acetate and La^{3+} or Lu^{3+} as a function of pH.[684] A very detailed investigation of the ^1H n.m.r. spectra of 1 : 1 complexes of thorium(IV) and diethylenetriaminepenta-acetic acid as a function of pH has been carried out, including the addition of copper(II) to broaden the uncomplexed groups.[685] The complexes between thorium(IV) and triethylenetetraminehexa-acetic acid has been similarly studied.[686]

Wide-line n.m.r. spectra of ^{17}O and ^{35}Cl for complex oxovanadium(IV) species in hydrochloric acid solution have been carried out over a wide range of concentrations and temperatures. Line broadening attributable to paramagnetic species was interpreted in terms of equilibria between $[VO(H_2O)_4]^{2+}$, $[VO(H_2O)_3Cl]^+$, and $VO(H_2O)_2Cl_2$.[687] Low-temperature ^{19}F n.m.r. spectra of the MF_5–SbF_5–SO_2ClF (M = Nb or Ta) systems appear to indicate ionization of the pentafluorides to form niobium (or tantalum) fluorocations and antimony polyfluoroanions, but the solutions show an

677 R. L. Benoit, A. L. Beauchamp, and R. Domain, *Inorg. Nuclear Chem. Letters*, 1971, **7**, 557.
678 J. E. Bercaw and H. H. Brintzinger, *J. Amer. Chem. Soc.*, 1971, **93**, 2045.
679 D. W. Thompson, R. W. Rosser, and P. B. Barrett, *Inorg. Nuclear Chem. Letters*, 1971, **7**, 931.
680 F. Calderazzo, S. A. Losi, and B. P. Susz, *Helv. Chim. Acta*, 1971, **54**, 1156.
681 N. Serpone and R. Ishayek, *Inorg. Chem.*, 1971, **10**, 2650.
682 A. Merbach and F. Gnaegi, *Helv. Chim. Acta*, 1971, **54**, 691.
683 N. A. Kostromina and N. N. Tananaeva, *Russ. J. Inorg. Chem.*, 1971, **16**, 462.
684 N. A. Kostromina and N. N. Tananaeva, *Russ. J. Inorg. Chem.*, 1971, **16**, 673.
685 A. R. Fried, jun. and A. E. Martell, *J. Amer. Chem. Soc.*, 1971, **93**, 4695.
686 A. R. Fried, jun. and A. E. Martell, *J. Co-ordination Chem.*, 1971, **1**, 47.
687 A. H. Zeltmann and L. O. Morgan, *Inorg. Chem.*, 1971, **10**, 2739.

unusual distribution of species and have very low conductivities. It was concluded that neutral fluxional molecules, MF_4–Sb_nF_{5n+1}, are formed.[688]

Exchange between $[CrO_3F]^-$ and F^- is slow and no exchange was detected between $[CrO_3F]^-$ and Cl^-. However, mixtures of CrO_2F_2 and CrO_2Cl_2 show rapid fluorine exchange.[689]

Both saturation-transfer and variable-temperature n.m.r. have been used to investigate imidazole exchange on ruthenium carbonyl mesoporphyrin-IX dimethyl ester.[690–692] Similarly, lineshape analysis has been used to analyse exchange between free and bound 4-t-butylpyridine in its complex with ruthenium carbonyl tetra-*p*-isopropylphenylporphine.[693] ^1H N.m.r. spectroscopy has been used to measure the equilibrium constant for the aldehyde–hydrate equilibrium of $[Ru(NH_3)_5NC_5H_4CHO]^{2+}$.[694]

The linewidths and chemical shifts of ^1H and ^{31}P of thioamine pyrophosphate in the presence of Co^{2+} and Ni^{2+} indicate that bonding occurs through both the pyrophosphate group and the pyrimidine. Addition of pyruvate provided no evidence of a metal–thiamine pyrophosphate–pyruvate adduct.[695] ^1H N.m.r. has been used to investigate protonation of bis(glycylglycinato)cobalt(III) in acidic aqueous solution,[696] but low temperatures were required to stop ligand exchange for $MeCo(dmg)_2L$ (L = dimer, MeCN, MeNC, or CO).[697]

The exchange reaction between $Rh_2(O_2CMe)_4$ and CF_3CO_2H to yield $Rh_2(O_2CCF_3)_4$ has been followed by ^1H n.m.r.[698] The methyl resonances of $[Rh(norbornadiene)(PMePh_2)_2]^+$ give a broad singlet at 37 °C and a triplet at − 80 °C. Data were also reported for related complexes and for $[RhH_2(PMe_2Ph)_4]^+$, $[Rh(PMePh_2)_3(CO)]^+$, *etc.*[699]

In an extensive paper, ^1H n.m.r. has been used to study metal-ion binding sites, stoicheiometry, stability, and conformation for complex formation between the ligands L-cysteine, *S*-methyl-L-cysteine (and their methyl esters), and DL-methionine and the metal ions Ni^{2+}, Zn^{2+}, Cd^{2+}, Hg^{2+}, Ag^+, and Pb^{2+} as a function of pH.[700] ^{17}O N.m.r. has been used to show that the $[Ni(edta)]^{2-}$ complex at pH 2 exists as $[Ni(H_2O)(Hedta)]^-$, whereas at pH 6–7 there are considerable amounts of $[Ni(H_2O)(edta)]^{2-}$ present, in both cases with one arm of the edta free; the kinetic parameters for

688 P. A. W. Dean and R. J. Gillespie, *Canad. J. Chem.*, 1971, **49**, 1736.
689 A. M. Akena, D. S. Brown, and D. G. Tuck, *Canad. J. Chem.*, 1971, **49**, 1505.
690 M. Tsutsui, D. Ostfeld, and L. M. Hoffman, *J. Amer. Chem. Soc.*, 1971, **93**, 1820.
691 J. W. Faller and J. W. Sibert, *J. Organometallic Chem.*, 1971, **31**, C5.
692 M. Tsutsui, D. Ostfeld, J. N. Francis, and L. M. Hoffman, *J. Co-ordination Chem.*, 1971, **1**, 115.
693 S. S. Eaton, G. R. Eaton, and R. H. Holm, *J. Organometallic Chem.*, 1971, **32**, C52.
694 Z. Zanella and H. Taube, *J. Amer. Chem. Soc.*, 1971, **93**, 7166.
695 W. D. White and R. S. Drago, *Inorg. Chem.*, 1971, **10**, 2727.
696 D. L. Rabenstein, *Canad. J. Chem.*, 1971, **49**, 3767.
697 A. W. Herlinger and T. L. Brown, *J. Amer. Chem. Soc.*, 1971, **93**, 1790.
698 J. L. Bear, J. Kitchens, and M. R. Willcott, tert., *J. Inorg. Nuclear Chem.*, 1971, **33**, 3479.
699 R. R. Schrock and J. A. Osborn, *J. Amer. Chem. Soc.*, 1971, **93**, 2397.
700 D. F. S. Natusch and L. J. Porter, *J. Chem. Soc.* (*A*), 1971, 2527.

H_2O exchange were derived.[701] ^1H N.m.r. has been used to investigate the octahedral complexes of Ni^{2+} with edta, pdta, and Cydta. It was found that the chelate rings are puckered, but the out-of-plane acetate rings of $[Ni(Cydta)]^{2-}$ are the most nearly planar of those studied. The out-of-plane acetate protons exchange 10^4 times faster than in-plane acetate protons and the axial ones three times faster than the equatorial ones.[702] [Ni{2,3-butane(dinitrilo)-*NNN'N'*-tetra-acetate}]$^{2-}$ exhibits slow partial unwrapping followed by nitrogen inversion. Deuterium exchange on a specific out-of-plane acetate ring is stereospecific with the pseudo-axial proton exchanging three times faster than the pseudo-equatorial proton.[703]

^1H N.m.r. line broadening of the methyl groups of methyl pyridines complexes to $Ni(acac)_2$ have been used to obtain rates of ligand exchange and hence ΔH and ΔS.[704] The two complexes (201) (n = 1 or 2) are in equilibrium.[705] When $Ni(dithiopivalate)_2$ and $Ni(perthiopivalate)_2$, or $Ni(dithiotoluate)_2$ and $Ni(perthiotoluate)_2$, are mixed, ^1H n.m.r. shows that some of the corresponding mixed complex is formed. However, in the case

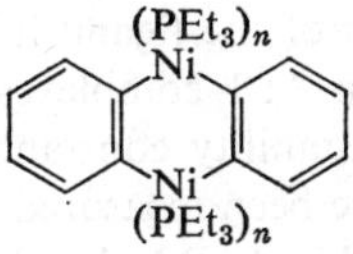

(201)

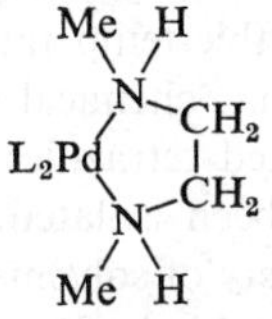

(202)

of the corresponding zinc complexes cooling to − 122 °C is necessary to observe separate resonances for the complexes present in solution.[706] The base-catalysed racemization of the MeNH nitrogen of (202) (L_2 = glycine, dipy, en, *etc.*), which causes coalescence of the methyl resonances, has been examined as a function of pH and rate constants derived.[707] Considerable use of ^1H n.m.r. has been made in the investigation of the method and rate of insertion of butadiene into an allyl–palladium bond.[708–711]

The kinetics of exchange of free and co-ordinated ethylene have been measured by ^1H n.m.r. for the systems $PtCl(acac)(C_2H_4)$ [712] and $(Ph_3P)_2$-$Pt(C_2H_4)$.[713] ^1H N.m.r. has also been used to follow hydrogen–deuterium

701 M. W. Grant, H. W. Dodgen, and J. P. Hunt, *J. Amer. Chem. Soc.*, 1971, **93**, 6828.
702 L. E. Erickson, D. C. Young, F. F.-L. Ho, S. R. Watkins, J. B. Terrill, and C. N. Reilley, *Inorg. Chem.*, 1971, **10**, 441.
703 D. C. Young and C. N. Reilley, *J. Co-ordination Chem.*, 1971, **1**, 95.
704 M. J. Petrin and W. L. Reynolds, *J. Inorg. Nuclear Chem.*, 1971, **33**, 3978.
705 J. E. Dobson, R. G. Miller, and J. P. Wiggen, *J. Amer. Chem. Soc.*, 1971, **93**, 554.
706 A. M. Giuliani, *Inorg. Nuclear Chem. Letters*, 1971, **7**, 1001.
707 T. P. Pitner and R. B. Martin, *J. Amer. Chem. Soc.*, 1971, **93**, 4400.
708 D. Medema and R. van Helden, *Rec. Trav. chim.*, 1971, **90**, 304.
709 R. P. Hughes and J. Powell, *Chem. Comm.*, 1971, 275.
710 D. Medema and R. van Helden, *Rec. Trav. chim.*, 1971, **90**, 324.
711 V. N. Sokolov, G. M. Khvostic, I. Ya. Poddubnyi, and G. P. Kondratenkov, *J. Organometallic Chem.*, 1971, **29**, 313.
712 C. E. Holloway and J. Fogelman, *Canad. J. Chem.*, 1970, **48**, 3802.
713 P.-T. Cheng, C. D. Cook, S. C. Nyburg, and K. Y. Wan, *Inorg. Chem.*, 1971, **10**, 2210

exchange in alkanes catalysed by K_2PtCl_4 in acetic acid.[714] Above − 30 °C $Pt(MePh_2P)_n$ shows rapid phosphine exchange.[715]

On the basis of ^1H n.m.r. line-broadening studies, it has been suggested that histidine forms two complexes with Cu^{2+}, (203) and (204), which are in

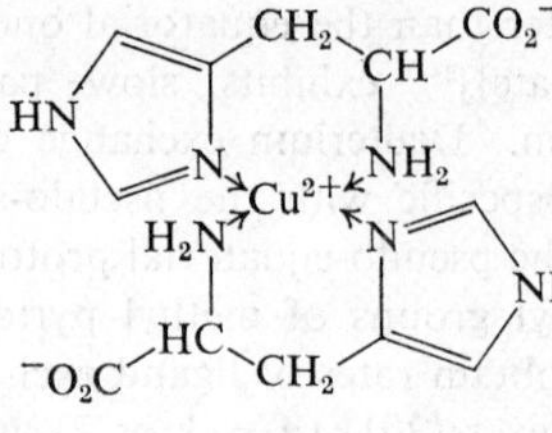

(203)

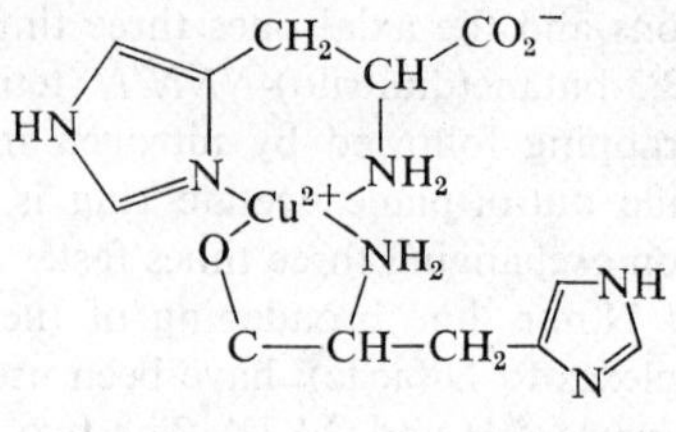

(204)

equilibrium.[716] The rate constant and activation energy for acetic acid (axial) exchange at $Cu_2(O_2CMe)_4$ dissolved in HO_2CCH_3–EtOH have been obtained from variable-temperature ^1H linewidth measurements.[717]

From the ^1H n.m.r. chemical shifts of the ligand it has been concluded that some substituted tetrazoles form 2 : 1 complexes with Ag^+ and one such complex has been isolated.[718] Stability constants for silver(I)–olefin complexes in a variety of solvents have been measured by ^1H n.m.r.[719] The exchange of PR_3 on $MeAuPR_3$ and Me_3AuPR_3 has been followed by ^1H n.m.r.[720–722]

At room temperature, the complex $[PhZn(acac)]_2Zn(acac)_2$ gives only one set of acac resonances, but at − 70 °C this splits into two. This was explained as being due to structure (205). Similar results have been found using other β-diketones.[723] The complexes between Zn^{2+}, Cd^{2+}, Hg^{2+}, and Pb^{2+} and edta, 2-*N*-hydroxyethylethylenediaminetriacetic acid, and ethyleneglycolbis(β-aminoethyl ether)tetra-acetic acid have been examined by i.r. and ^1H n.m.r. as a function of pH. It was found that ligand exchange occurs *via* acid-assisted dissociation pathways and rate constants were derived.[724] Complexes between Zn^{2+} and $\{(H_2OCCH_2)(HONHCOCH_2)NCH_2\}_2$ have also been examined.[725] Stopped-flow ^{35}Cl n.m.r. has been used to study the binding of Hg^{2+} to bovine serum albumin in the presence

714 R. J. Hodges, D. E. Webster, and P. B. Wells, *J. Chem. Soc.* (*A*), 1971, 3230.
715 H. C. Clark and K. Itoh, *Inorg. Chem.*, 1971, **10**, 1707.
716 H. Sigel and D. B. McCormick, *J. Amer. Chem. Soc.*, 1971, **93**, 2041.
717 H. Grasdalen, *Acta Chem. Scand.*, 1971, **25**, 1103.
718 D. M. Bowers, R. H. Erlich, S. Policec, and A. I. Popov, *J. Inorg. Nuclear Chem.*, 1971, **33**, 81.
719 J. Solodar and J. P. Petrovich, *Inorg. Chem.*, 1971, **10**, 395.
720 H. Schmidbaur and A. Shiotani, *Chem. Ber.*, 1971, **104**, 2821.
721 H. Schmidbaur, A. Shiotani, and H.-F. Klein, *Chem. Ber.*, 1971, **104**, 2831.
722 A. Shiotani, H.-F. Klein, and H. Schmidbaur, *J. Amer. Chem. Soc.*, 1971, **93**, 1555.
723 J. Boersma, F. Verbeek, and J. G. Noltes, *J. Organometallic Chem.*, 1971, **33**, C53.
724 G. H. Reed and R. J. Kula, *Inorg. Chem.*, 1971, **10**, 2050.
725 R. J. Motekaitis, I. Murase, and A. E. Martell, *J. Co-ordination Chem.*, 1971, **1**, 77.

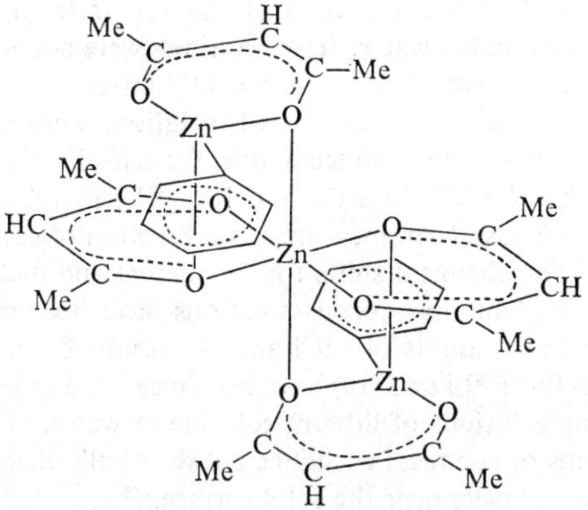

(205)

of a wide selection of ligands, *e.g.* cysteine, thiolactic acid, *o*-penicillamine. Only very approximate formation constants could be determined.[726]

Ionic Solutions. Aqueous electrolyte solutions. In addition to ^1H n.m.r. salt-shifts for a wide range of electrolytes in water at 273 K and 298 K, salt-shifts for alkali-metal chlorides have been determined at 313, 333, and 353 K. Two independent methods were used to assign an absolute shift to the chloride ion. The absolute values so derived were used to set up scales of absolute ionic shifts which were discussed in terms of ion–solvent interactions.[727]

The total effective hydration numbers for a variety of strong electrolytes have been deduced from the temperature dependence of proton shifts, and individual hydration numbers have been suggested. The cations of all the alkali elements appear to have a relatively consistent hydration number of 3 ± 0.5 when the concentration is below 5 mol l^{-1}, but above this concentration ion pairing becomes important. For Mg^{2+} and Ca^{2+}, ion pairing becomes important at concentrations above 2 mol l^{-1}. Based on the assumption that the shift of bound water is not a function of temperature, then

$$h = \frac{55.5}{m}\left[1 - \frac{m_1 + 111.1}{111.1}\left(\frac{d\delta/dT}{d\delta_N/dT}\right)\right]$$

where m = molarity, h = hydration number, m_1 = molarity of exchangeable protons donated by solvent, $d\delta/dT$ is the temperature dependence of the proton shift of the solution, and $d\delta_N/dT$ that for pure water.[728]

The n.m.r. relaxation rate of ^7Li in H_2O and D_2O solutions of LiCl, LiBr, and LiI have been reported. Dipole–dipole interaction was separated

[726] J. L. Sudmeier and J. J. Pesek, *Inorg. Chem.*, 1971, **10**, 860.
[727] J. Davies, S. Ormondroyd, and M. R. C. Symons, *Trans. Faraday Soc.*, 1971, **67**, 3465.
[728] F. J. Vogrin, P. S. Knapp, W. L. Flint, A. Anton, G. Highberger, and E. R. Malinowski, *J. Chem. Phys.*, 1971, **54**, 178.

from the rotational correlation time of the vector from lithium to the proton of the co-ordinated water. The two times were not equal, which was interpreted as due to rotation about the lithium–oxygen bond. ^{1}H and ^{27}Al relaxation rates for $AlCl_3$ and $GaCl_3$ solutions were also reported.[729] By means of the n.m.r. spin-echo technique, the self-diffusion coefficients in aqueous solutions of LiCl, Li_2SiF_6, and $[Me_4X]_2YF_6$ (X = N, P, As, or Sb; Y = Si or Ge) have been measured as a function of concentration and temperature.[730] The chemical shifts for the methyl and methylene protons of diglyme and triglyme in aqueous solutions have been measured in the presence of varying amounts of LiCl and the results discussed.[731] The T_1 of lithium nuclei (both ^{6}Li and ^{7}Li) have been measured in porous porcelain media containing solutions of lithium chloride in water. The results were discussed in terms of chemical exchange between bulk phase ions and ions in electrical double layers near the solid surface.[732]

The pressure dependences of the proton nuclear spin–lattice relaxation times of the water molecules in 4.5 mol l^{-1} aqueous solutions of CsBr, RbBr, KBr, LiCl, $CaCl_2$, and $LaCl_3$ have been measured at 25 °C up to 2.5 kbar. T_1 for aqueous solutions of CsBr, RbBr, and KBr are longer than for pure water and decrease with increasing pressure, but for concentrated solutions of LiCl, $CaCl_2$, and $LaCl_3$ T_1 initially increases and then decreases with increasing pressure. The results were discussed in terms of structure making and breaking by the cations.[733]

Liquid-crystal phases, mainly of the lamellar type, formed in ternary amphiphile–water systems, have been studied by means of ^{2}D and ^{23}Na n.m.r. The ^{2}D resonance of D_2O exhibits quadrupole splittings, partly owing to a partial orientation and partly owing to ^{2}D exchange between water and amphiphile molecules.[734]

NaOH in a KNO_3–$NaNO_3$ melt shows a ^{1}H n.m.r. resonance at + 1.82 p.p.m. (rel. methane) [*cf.* NaOH(aq) at infinite dilution at − 14.9 p.p.m.] and thus there is little hydrogen bonding. However, H_2O in the same melt shows a ^{1}H n.m.r. resonance at − 3.75 p.p.m., close to that of liquid rather than gaseous water at 150 °C, and it was therefore accounted for by ion–water bonding.[735]

It has been concluded that results from nuclear magnetic relaxation rates of the ^{139}La ion in aqueous solutions of halides, nitrate, sulphate, and perchlorate can only be adequately explained in terms of inner-sphere complexes.[736] At low temperatures separate resonances are observed for

729 H. G. Hertz, R. Tutsch, and H. Versmold, *Ber. Bunsengesellschaft. Phys. Chem.*, 1971, **75**, 1177.
730 A. Weiss and K. H. Nothnagel, *Ber. Bunsengesellschaft Phys. Chem.*, 1971, **75**, 216.
731 A. Antony, *J. Inorg. Nuclear Chem.*, 1971, **33**, 2883.
732 L. C. Headley, W. E. Wallace, jun., and P. Waldstein, *J. Magn. Resonance*, 1971, **5**, 168.
733 Y. Lee and J. Jonas, *J. Magn. Resonance*, 1971, **5**, 267.
734 N.-O. Persson and Å. Johansson, *Acta Chem. Scand.*, 1971, **25**, 2118.
735 A. G. Turnbull, *Austral. J. Chem.*, 1971, **24**, 2213.
736 K. Nakamura and K. Kawamura, *Bull. Chem. Soc. Japan*, 1971, **44**, 330.

bulk and bound water for La^{3+}, Zn^{2+}, Ce^{3+}, Er^{3+}, Fe^{3+}, and Ni^{2+}. Integration led to a hydration number of six for all cations except Er^{3+}, which was involved in extensive complex formation.[737] A competitive solvation and complex formation study of $[UO_2]^{2+}$ with H_2O, DMSO, ClO_4^-, NO_3^-, and Cl^- has been carried out. At − 80 °C to − 100 °C the ligand and proton exchange rates are slow enough to permit the direct observation of ^1H n.m.r. signals for water and DMSO molecules in the $[UO_2]^{2+}$ solvation shell. It was found that four ligands complex, the anions complexed being determined by the difference between the number of molecules of water and DMSO complexed and four.[738]

Solvent ^1H n.m.r. line broadening by $[(H_2O)_5CrN_3]^{2+}$ has been used to show the presence of exchange of water protons by both the acid-catalysed and uncatalysed paths.[739] ^{17}O N.m.r. and e.s.r. have been used to determine the rate of exchange and activation energy of H_2O for $[Mn(H_2O)_6]^{2+}$, $[Mn(phen)(H_2O)_4]^{2+}$, and $[Mn(phen)_2(H_2O)_2]^{2+}$.[740] ^{17}O N.m.r. spectra of the water molecules co-ordinated to iron(II) and nickel(II) ions have been observed and it was concluded that both ions are co-ordinated to six water molecules in aqueous solution.[741]

T_1 of aqueous solutions of bovine ferrihaemoglobin and carboxylated haemoglobin have been reported.[742] A detailed investigation of the kinetics of ligand exchange and of transitions between high-spin and low-spin states in the haemin–pyridine–water system has been carried out by n.m.r. spectroscopy.[743]

Kinetic data for water exchange in $[Co(H_2O)_6]^{2+}$, $[Co(NH_3)(H_2O)_5]^{2+}$, $[Co(NH_4)_2(H_2O)_4]^{2+}$, and $[Co(malonate)(H_2O)_4]^{2+}$ have been measured by ^{17}O n.m.r.[744] The temperature dependence of solvent-water proton line broadening by the complex (206) has been measured, and kinetic data for water exchange derived.[745] The Ni^{2+} complex of (207) (R = H or Me) is diamagnetic, but when solvated is paramagnetic. The effect of chemical exchange on the linewidth of the solvent water or DMSO (formyl proton) has been used to obtain thermodynamic rate data of solvent exchange. In all cases a frequency dependence of the proton relaxation time in the first co-ordination sphere of the metal complex was observed. This behaviour was interpreted in terms of a magnetic field dependence of the electron spin relaxation time.[746]

Longitudinal and transverse relaxation times for all three species of nuclei in aqueous solutions of HBF_4 have been measured by pulse

737 A. Fratiello, V. Kubo, S. Peak, B. Sanchez, and R. E. Schuster, *Inorg. Chem.*, 1971, **10**, 2552.
738 A. Fratiello, V. Kubo, and R. E. Schuster, *Inorg. Chem.*, 1971, **10**, 744.
739 R. J. Balahura and R. B. Jordan, *Inorg. Chem.*, 1970, **9**, 2639.
740 M. Grant, H. W. Dodgen, and J. P. Hunt, *Inorg. Chem.*, 1971, **10**, 71.
741 A. M. Chmelnick and D. Fiat, *J. Amer. Chem. Soc.*, 1971, **93**, 2875.
742 R. Kimmich, *Z. Naturforsch.*, 1971, **26b**, 1168.
743 H. A. Degani and D. Fiat, *J. Amer. Chem. Soc.*, 1971, **93**, 4281.
744 P. E. Hoggard, H. W. Dodgen, and J. P. Hunt, *Inorg. Chem.*, 1971, **10**, 959.
745 J. E. Letter, jun. and R. B. Jordan, *J. Amer. Chem. Soc.*, 1971, **93**, 864.
746 L. L. Rusnak and R. B. Jordan, *Inorg. Chem.*, 1971, **10**, 2686.

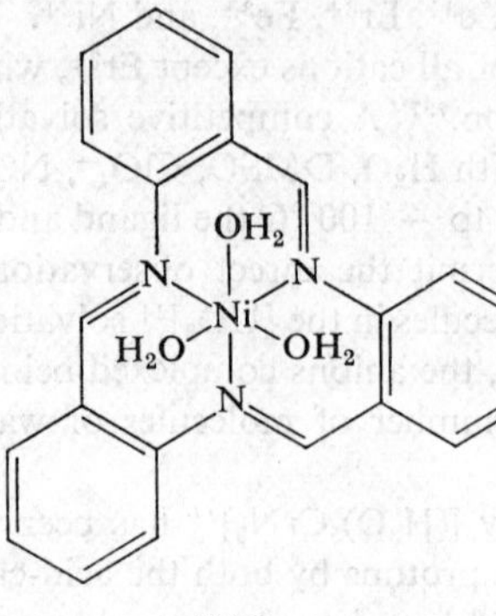

(206)

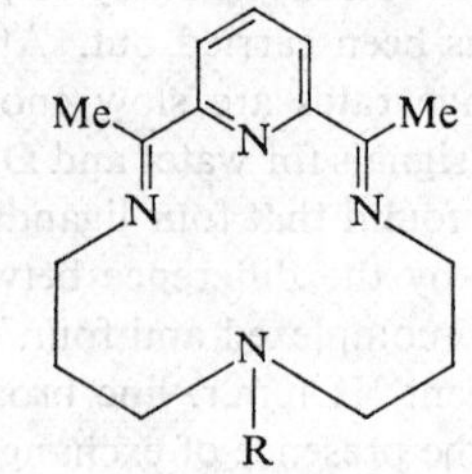

(207)

techniques, and a strong concentration dependence was found. The results demonstrate that the ^{19}F nuclei are participating chemical exchange process which produces a ^{11}B spin multiplet in a collapse.[747]

In a paper concerned with the molar volumes of ions, the n.m.r. method of determining hydration numbers was discussed with respect to ^{27}Al, and it was concluded that for Al^{3+} the hydration number is 5.97 ± 0.07.[748] Results of earlier workers on hydration shifts by Al^{3+} have been recalculated and also produce a hydration number of six for aluminium.[749] Other workers had concluded from n.m.r. measurements that the hydration number of Al^{3+} can be as low as 4.64 and for Ga^{3+} 4.61, but it has been pointed out that these workers, in the preparation of their solutions, passed N_2 through the hot solutions; this would result in the formation of hydroxy-species which would reduce the number of moles of water bound to the metal.[750] The spin–lattice and spin–spin relaxation times of protons and deuterons have been measured as a function of the concentration of the diamagnetic salts $GaCl_3$, $GaBr_3$, and GaI_3 dissolved in H_2O and D_2O. The hydration numbers of the halide ions Br^- (10.3 ± 3 moles) and I^- (19.9 ± 3 moles), and the rotation correlation times of the solvent molecules, as well as the deuterium coupling constant of the molecules hydrating Ga^{3+}, were determined.[751] A hydration and complex formation study of indium halide solutions in water–acetone mixtures has been carried out by proton and indium-115 n.m.r. techniques. At − 100 °C the proton- and ligand-exchange rates in these solutions are slow enough to permit the direct observation of ^1H n.m.r. signals both for bulk water and water molecules in the In^{3+} primary hydration shell. From the relative intensities of these

[747] G. E. Stungis and J. H. Rugheimer, *J. Chem. Phys.*, 1971, **55**, 263.
[748] J. W. Akitt, *J. Chem. Soc.* (*A*), 1971, 2347.
[749] J. W. Akitt, *J. Chem. Soc.* (*A*), 1971, 2865.
[750] J. W. Akitt, N. N. Greenwood, and B. L. Khandelwal, *Spectroscopy Letters*, 1971, **4**, 139.
[751] B. Lemius and S. Domngang, *J. chim. Phys.*, 1971, **68**, 1372.

signals, cation hydration numbers were calculated. These results were correlated with indium-115 n.m.r. measurements which revealed signals attributed to $[In(H_2O)_6]^{2+}$ and $[InI_4]^-$ in iodide solutions and $[InCl_4]^-$ and $[InBr_4]^-$ in the other halide systems. The data were interpreted in terms of indium hexahydrate and tetrahalide complexes, along with additional species at higher halide concentrations.[752]

In contrast to previous results, ^1H chemical shifts for water have been found to move to lower applied fields as the size of the alkyl group in symmetrical tetra-alkylammonium ions is increased for temperatures below 3000 K, although this trend is reversed at higher temperatures.[753] The difference between the chemical shifts of the α- and β-hydrogens of the $[EtNH_3]^+$, $[Et_2NH_2]^+$, $[Et_3NH]^+$, and $[Et_4N]^+$ ions change as the concentration of the solvent sulphuric acid is increased from 0 to 94%. A tentative explanation based on the hydration of ions has been presented.[754] The ^1H n.m.r. linewidths of Me_4AsX (X = Cl, I, or OH) solutions have been analysed on the basis of the ion-pair effect on rotational diffusion quadrupolar relaxation of ^{75}As. It was concluded that the counter ions are separated by 2.5—3.0 Å, with at the most a few water molecules held between the counter ions.[755]

^1H N.m.r. has been used to investigate the anion-exchange resin Dowex AGl. The molal chemical shifts $\delta_{X'}{}^0$ for the anions F^-, Cl^-, Br^-, I^-, NO_3^-, and ClO_4^- were found to be generally lower than for aqueous solutions. The effective hydration numbers have been deduced from the temperature-dependence measurements of the ^1H chemical shifts for the alkali-metal ions and for a few anions in various cross-linkages of Dowex AG5OW and Dowex AGl resins. Hydration decreases as cross-linkage increases.[756]

Non-aqueous electrolyte solutions. The ^1H n.m.r. spectra of the system HCl–H_2O–1,4-dioxan have been measured as a function of HCl and dioxan concentration and the change in $\delta_{H_3O^+}$ with mole fraction of dioxan tabulated.[757]

Three independent methods have been used to estimate the position of the hydroxy-proton resonance of methanol molecules directly interacting with halide ions in electrolyte solutions. The shifts found were Cl^- 1.0, Br^- 3.5, and I^- 3.0.[758]

The solvation of $LiClO_4$ by acetone in nitromethane has been examined by ^1H n.m.r. A plot of the shift of acetone *vs.* the ratio of acetone to $LiClO_4$ shows a discontinuity at 4.3; if LiI is used, the discontinuity occurs at 4.5. It was therefore concluded that Li^+ is solvated by 4 moles of

752 A. Fratiello, D. D. Davis, S. Peak, and R. E. Schuster, *Inorg. Chem.*, 1971, **10**, 1627.
753 J. Davies, S. Ormondroyd, and M. C. R. Symons, *Chem. Comm.*, 1971, 1204.
754 J. T. Edward, *Canad. J. Chem.*, 1971, **49**, 2364.
755 D. W. Larsen, *J. Phys. Chem.*, 1971, **75**, 3880.
756 H. D. Sharma and N. Subramanian, *Canad. J. Chem.*, 1971, **49**, 457.
757 R. Radeglia and A. Weber, *Z. Chem.*, 1971, **11**, 236.
758 S. Ormondroyd, E. A. Phillpot, and M. C. R. Symons, *Trans. Faraday Soc.*, 1971, **67**, 1253.

acetone.[759] Similarly, if dimethylformamide in dioxan is used, the discontinuity occurs for $LiClO_4$ at 3.7—3.8, whereas for LiI it is at 4.3 molecules dimethylformamide per lithium ion.[760]

Lithium dynamic nuclear polarization parameters have been reported for lithium salts in solvents containing OH groups. Observed ^7Li n.m.r. enhancements indicate that the most sensitive experimental variable governing the degree of scalar coupling is the choice of free radical. Of those chosen, nitroxide was by far the best, implying that it can compete effectively with ROH to solvate Li^+.[761]

Addition of polyglycol dimethyl ethers or macrocyclic polyethers to a solution of sodium tetraphenylboron ion-pairs in tetrahydrofuran strongly broadens the ^{23}Na n.m.r. line, indicating that the tetrahydrofuran solvation shell is replaced by the complexing agent.[762, 763] ^{23}Na N.m.r. of lecithin–sodium cholate–water lamellar liquid crystalline phase shows three resonances. This splitting is due to first-order quadrupolar effects, thus indicating that the sodium ions are interacting strongly with the model membrane surfaces.[764]

T_1 for hydrazine containing NaI or KI has been measured and two reorientation activation energies were found.[765] A molten sodium–caesium–rubidium acetate eutectic has been used to study solutions of polyhydric alcohols and phenols between 100 and 150 °C. There are strong solvent–solute interactions and a pronounced downfield shift of the OH proton has been interpreted as indicating hydrogen bonding and exchange.[766] In a detailed study on the solution of KF in glacial acetic acid, it was found that as the concentration of KF increases, the OH protons move to low field and broaden; this broadening was attributed to viscosity effects.[767] Chemical shifts of the ^{23}Na nucleus were measured for sodium tetraphenylborate, perchlorate, iodide, and thiocyanate with reference to aqueous sodium chloride solution at different concentrations in a number of non-aqueous solvents. These shifts were attributed to contact ion pairs and solvation.[768]

From the shift of the THF protons to low field, coupled with molecular weight data, it was concluded that the structure of $(2,6\text{-}Bu^t{}_2\text{-}4\text{-}Me\text{-}C_6H_2O)$-Li,THF is dimeric (208).[769] The concentration- and temperature-dependence of the ^1H and ^7Li n.m.r. spectra of $LiN(SiMe_3)_2$ have been

759 M. K. Wong, W. J. McKinney, and A. I. Popov, *J. Phys. Chem.*, 1971, **75**, 56.
760 C. Lassigne and P. Baine, *J. Phys. Chem.*, 1971, **75**, 3188.
761 J. A. Potenza and J. W. Linowski, *J. Chem. Phys.*, 1971, **54**, 4095.
762 A. M. Grotens, J. Smid, and E. De Boer, *Chem. Comm.*, 1971, 759.
763 E. Shchori, J. Jagur-Grodzinaki, Z. Luz, and M. Shporer, *J. Amer. Chem. Soc.*, 1971, **93**, 7133.
764 G. Lindblom, *Acta Chem. Scand.*, 1971, **25**, 2767.
765 A. A. Borichev, K. P. Mischenko, and V. V. Kushchenko, *J. Gen. Chem.* (*U.S.S.R.*), 1971, **41**, 10.
766 L. L. Burton, S. Sherer, and E. R. Van Artsdalen, *J. Phys. Chem.*, 1971, **75**, 1338.
767 J. Emsley, *J. Chem. Soc.* (*A*), 1971, 2702.
768 R. H. Erlich and A. I. Popov, *J. Amer. Chem. Soc.*, 1971, **93**, 5620.
769 K. Shobatake and K. Nakamoto, *Inorg. Chim. Acta*, 1970, **4**, 485.

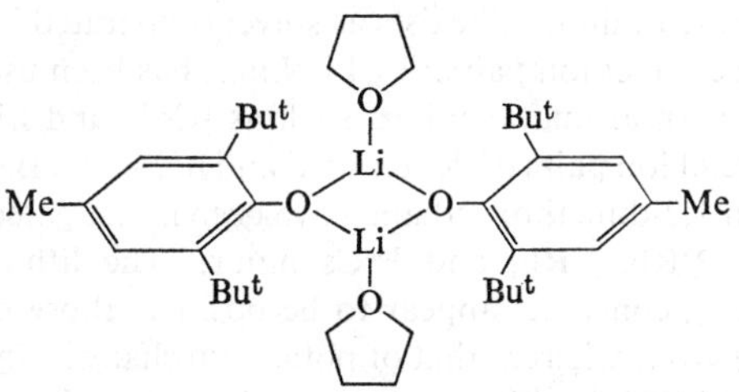

(208)

interpreted in terms of a monomer–dimer equilibrium in THF and a dimer–tetramer equilibrium in hydrocarbon solvents. Thermodynamic parameters were obtained for the monomer–dimer equilibrium.[770]

The ^{7}Li spin–lattice relaxation times in a series of organolithium compounds of known aggregation have been measured in various solvents. The measurements provide lower limits for the ^{7}Li linewidth, which range from 0.05 Hz for symmetrical tetrahedral species to 30 Hz for some larger or smaller aggregates at low temperatures. Factors affecting the spin–lattice relaxation times were discussed.[771]

^{1}H, ^{7}Li, and ^{13}C n.m.r. have been used to investigate the nature of the carbon–lithium bond in 1,1-diphenyl-n-hexyl-lithium (dpb$^-$) and related systems. The solvent dependency of the n.m.r. data was interpreted as evidence that the α-carbon has appreciable sp^3 character.[772] If this compound is examined in a mixture of benzene and THF, then the chemical shifts of the aromatic and aliphatic protons showed minima at a THF to dpb$^-$ ratio of 2. This observation was explained by assuming dimer formation of dpb$^-$ in benzene.[773] From the shifts in the ^{1}H n.m.r. spectra of styryl-lithium, cumylpotassium, and related compounds, it appears that the excess of electronic charge on the aromatic ring is at least 0.5 e^- and that this increases on changing from benzene to THF as solvent or Li^+ to K^+ as counter ion.[774]

A variable-temperature ^{7}Li n.m.r. study of an equimolar mixture of PhLi and *p*-TolLi in ether solution reveals that only two species exist in rapid equilibrium, the kinetics of which are consistent with a bimolecular mechanism. The results are consistent with monomeric aryl-lithium molecules rather than dimeric as has been suggested on the basis of ebullioscopic measurements.[775]

The ^{1}H n.m.r. spectra of several alkali-metal and Grignard derivatives of indole in THF indicate that these species are essentially ionic. The ^{1}H chemical shifts for the alkali-metal salts are in the order Li > K > Na,

[770] B. Y. Kimura and T. L. Brown, *J. Organometallic Chem.*, 1971, **26**, 57.
[771] G. E. Hartwell and A. Allerhand, *J. Amer. Chem. Soc.*, 1971, **93**, 4415.
[772] L. D. McKeever and R. Waack, *J. Organometallic Chem.*, 1971, **28**, 145.
[773] Y. Okamoto and H. Yuki, *J. Organometallic Chem.*, 1971, **32**, 1.
[774] S. Bywater and D. J. Worsfold, *J. Organometallic Chem.*, 1971, **33**, 273.
[775] J. A. Ladd and J. Parker, *J. Organometallic Chem.*, 1971, **28**, 1.

suggesting that the lithium salt exists as solvent-separated ion pairs whereas the last two are contact ion pairs.[776] ^{7}Li N.m.r. has been used to investigate ion pairing between aromatic anions such as $[Cp]^-$ and Li^+.[777]

The alkali-metal ion pairs of deuteriated and non-deuteriated naphthalene radical-anion in 1,2-dimethoxyethane have been investigated by ^{1}H, ^{2}D, ^{6}Li, ^{7}Li, ^{23}Na, ^{39}K, ^{85}Rb, ^{87}Rb, and ^{133}Cs n.m.r. The lithium and sodium hyperfine splitting constants appear to be positive, those of rubidium and caesium are negative, whereas that of potassium changes sign with temperature. The ^{1}H and ^{2}D linewidths were studied as a function of concentration and were found to vary linearly with the reciprocal of the concentration, pointing to the predominance of the Fermi contact interaction. Alkali-metal linewidths were interpreted in terms of intramolecular relaxation processes.[778]

The ^{7}Li chemical shifts of some ion pairs with aromatic dianions have been reported. High-field shifts were found for the $4n + 2$ π-electron systems [cyclo-octatetraene]$^{2-}$, [biphenylene]$^{2-}$, and [acenaphthalene]$^{2-}$, consistent with a diamagnetic ring current, whereas the $4n$ π-electron systems [azulene]$^{2-}$ and [15,16-dimethyldihydropyrene]$^{2-}$ gave low-field shifts consistent with a paramagnetic ring current. Other $4n$ π-electron systems showed a small up-field shift and this was taken to indicate the absence of ring current in these cases.[779]

Low-temperature ^{1}H n.m.r. of $Yb(NO_3)_3\{(Pr^iO)_3PO\}_3$ shows two absorptions, due to free and co-ordinated $(Pr^iO)_3PO$.[780] The ^{19}F n.m.r. shifts of $M_3UO_2F_5$ move to high field in the order M = Cs, Rb, K, NH_4, consistent with the greatest U—F bond covalency in the caesium case; similar conclusions have been drawn from i.r. evidence.[781] ^{1}H N.m.r. has been used to investigate the organic phase from the extraction of zirconium sulphate solutions in sulphuric acid by 0.2 mol l^{-1} di-(2-ethylhexyl) hydrogen phosphate in carbon tetrachloride.[782]

Line-broadening studies show that $[Co(DMSO)_6]^{2+}$ and $[Mn(DMSO)_6]^{2+}$ exchange too fast in DMSO solution to be measured by the n.m.r. technique, whereas $[Fe(DMSO)_6]^{3+}$ and $[Cr(DMSO)_6]^{2+}$ exchange too slowly to be measured. The exchange rate and ΔH and ΔS were determined for $[Cu(DMSO)_6]^{2+}$, $[Ni(DMSO)_6]^{2+}$, and $[Fe(DMSO)_6]^{2+}$. The mechanism of unpaired electron delocalization was discussed.[783] However, the ligand

776 M. G. Reinecke, J. F. Sabastian, H. W. Johnson, jun., and C. Pyun, *J. Org. Chem.*, 1971, **36**, 3091.
777 R. H. Cox, H. W. Terry, jun., and L. W. Harrison, *J. Amer. Chem. Soc.*, 1971, **93**, 3297.
778 B. M. P. Hendriks, G. W. Canters, C. Corvaja, J. W. M. de Boer, and E. de Boer, *Mol. Phys.*, 1971, **20**, 193.
779 R. H. Cox, H. W. Terry, jun., and L. W. Harrison, *Tetrahedron Letters*, 1971, 4815.
780 J. R. McRae and D. G. Karraker, *J. Inorg. Nuclear Chem.*, 1971, **33**, 479.
781 V. I. Sergienko, R. L. Davidovich, V. I. Kostin, and A. A. Matstutsin, *Spectroscopy Letters*, 1971, **4**, 19.
782 T. Sato and T. Nakamura, *J. Inorg. Nuclear Chem.*, 1971, **33**, 1081.
783 G. S. Vigee and P. Ng, *J. Inorg. Nuclear Chem.*, 1971, **33**, 2477.

exchange rates of $[Co(DMSO)_6]^{2+}$ and $[Ni(DMSO)_6]^{2+}$ have been reported.[784]

^{14}N N.m.r. has been used to obtain exchange parameters for $[Fe(MeCN)_6]^{2+}$.[785] T_1 and T_2 for the protons of MeCN in the presence of Ni^{2+} ions have been measured as a function of temperature and frequency by spin–echo techniques. The rotational correlation time of the complex, the electron spin relaxation time, the distance of closest approach of the protons to the nickel ion, the exchange rate, and activation energies were derived. There are significant differences compared with results of other workers:[786] thus ^{14}N relaxation measurements were made and it was suggested that there are four strongly bound molecules of MeCN and two weakly bound.[787]

By measurement of the proton relaxation times, T_1 and T_2, of MeOD containing Co^{2+} ions in dilute solution and the dependence of the observed decay time ($T_2^{\dagger}$) on t_{CP}, the separation between pulses in a Carr–Purcell sequence, the co-ordination number was shown to be six. This has been shown previously by an independent method and thus the use of the dependence of T_2 on t_{CP} to determine co-ordination numbers is justified for solutions of paramagnetic ions.[788]

The temperature dependence of $1/T_{2p}$ of the formyl proton has been used in an investigation of the exchange kinetics of dimethylformamide with Ni^{2+},[789] and the exchange kinetics of 3-picoline *N*-oxide with [hexakis-(3-picoline *N*-oxide)nickel(II)]$^{2+}$ have also been studied.[790]

The solvation number and activation energy for the solvation of Be^{2+} or Al^{3+} by hexamethylphosphoramide, dimethyl methylphosphonate, or trimethyl phosphate have been determined by ^{1}H and ^{31}P n.m.r.[791] From ^{1}H and ^{14}N n.m.r. it has been shown that Zn^{2+} is complexed by four molecules of formamide or *N*-methylformamide.[792]

Two ^{13}C n.m.r. signals, one for the bulk and one for the bound solvent (1.94 p.p.m. upfield), are observed for aqueous dimethyl sulphoxide containing aluminium.[793] ^{27}Al and ^{31}P n.m.r. have been used to measure quantitatively the formation of complexes between aluminium ions and the ionic and molecular species present in phosphoric acid. The results indicate both monomeric and polymeric species in aqueous phosphoric acid. Five ^{31}P lines were observed at low temperatures in solutions of

[784] L. S. Frankel, *Inorg. Chem.*, 1971, **10**, 814.
[785] R. J. West and S. F. Lincoln, *Austral. J. Chem.*, 1971, **24**, 1169.
[786] I. D. Campbell, J. P. Carver, R. A. Dwek, A. J. Nummelin, and R. E. Richards, *Mol. Phys.*, 1971, **20**, 913.
[787] I. D. Campbell, R. A. Dwek, R. E. Richards, and M. N. Wiseman, *Mol. Phys.*, 1971, **20**, 933.
[788] I. D. Campbell, P. E. Nixon, and R. E. Richards, *Mol. Phys.*, 1971, **20**, 923.
[789] L. S. Frankel, *Inorg. Chem.*, 1971, **10**, 2360.
[790] M. L. Yount and S. S. Zumdahl, *Inorg. Chem.*, 1971, **10**, 1212.
[791] J. J. Delpuech, A. Péguy, and M. R. Khaddar, *J. Electroanalyt. Chem. Interfacial Electrochem.*, 1971, **29**, 31.
[792] H. Saitô, Y. Tanaka, and K. Nukada, *J. Amer. Chem. Soc.*, 1971, **93**, 1077.
[793] J. C. Boubel, J. J. Delpuech, M. R. Khaddar, and A. Peguy, *Chem. Comm.*, 1971, 1265.

Al^{3+} in aqueous phosphoric acid. ^{19}F N.m.r. has been used to show that addition of fluoride produces eight new fluoro-phosphato-aluminium complexes containing direct aluminium–fluorine bonds.[794] Separate OH and CH_2 ^1H n.m.r. signals for free and bound ethanol molecules have been observed for solutions of $AlCl_3$ in ethanol at reduced temperatures. It is suggested that the complex formed is $[Al(EtOH)_4]^{3+}$ with some dinuclear and trinuclear complexes.[795]

The low-temperature ^1H n.m.r. spectrum of $GaBr_3$ in MeCN shows that one mole of MeCN is co-ordinated per mole of $GaBr_3$.[796]

4 Paramagnetic Complexes

In this section compounds of the *d*-block transition elements will be considered first and then those of the lanthanide and actinide elements. The solvation of paramagnetic ions has already been dealt with in Section 3 in refs. 678, 685, 687, 695, 701—704, 716, 717, 737, 739—746, and 783—790.

The use of paramagnetic probes in magnetic resonance studies of phosphoryl transfer enzymes has been reviewed.[797]

The significance of non-zero intercepts in the temperature dependence of isotropic shifts for paramagnetic complexes has been investigated. A number of systems have been designed in an attempt to elucidate some of the causes for non-zero intercepts. Ion-pairing interactions, hydrogen-bonding interactions, and temperature-dependent chemical equilibria have been established as principal causes in the systems investigated.[798]

The method of σ-spin delocalization has been re-evaluated by carrying out open-shell INDO/2 calculations. It has been found that (*i*) σ-spin delocalization attenuating with distance is reasonable, (*ii*) alternation in sign of spin round a ring occurs also with σ-spin delocalization and therefore is not diagnostic of π-spin delocalization; (*iii*) the opposite shift for hydrogen and methyl resonances is not diagnostic of π-spin delocalization as this also occurs with σ-spin delocalization.[799]

Compounds of *d*-Block Transition Elements.—In a review on the magnetic resonance methods used in the study of electronic structures of transition metal complexes, the use of n.m.r. for paramagnetic complexes has been reviewed and the information obtainable from n.m.r. compared with that obtainable from e.s.r.[800]

A theoretical interpretation of the effect of spin densities on $[MH_6]^{4-}$ complexes shows that, before covalency parameters can be calculated from shift parameters, it is necessary to determine the spin occupation probabilities of the *d*-electrons.[801] The mechanism by which the unpaired electrons

794 J. W. Akitt, N. N. Greenwood, and G. D. Lester, *J. Chem. Soc.* (*A*), 1971, 2450.
795 H. Grasdalen, *J. Magn. Resonance*, 1971, **5**, 84.
796 C. D. Schmulbach and I. Y. Ahmed, *Inorg. Chem.*, 1971, **10**, 1902.
797 M. Cohn and J. Reuben, *Accounts Chem. Res.*, 1971, **4**, 214.
798 W. D. Perry and R. S. Drago, *J. Amer. Chem. Soc.*, 1971, **93**, 2183.
799 W. DeW. Horrocks, jun., and D. L. Johnston, *Inorg. Chem.*, 1971, **10**, 1835.
800 D. R. Eaton and K. Zaw, *Co-ordination Chem. Rev.*, 1971, **7**, 197.
801 G. N. La Mar, *Inorg. Chem.*, 1971, **10**, 2633.

transfer from the transition-metal *d*-orbitals to the ligand orbitals has been studied on the basis of the method of intergroup configuration interaction.[802]

CNDO molecular orbital calculations have been reported for hexa-aquometal(II) complexes (Ti^{II} to Cu^{II}) and hexa-aquometal(III) complexes (Ti^{III} to Co^{III}). The proton spin transfer coefficients calculated agree very well with spin densities from ^1H n.m.r. shifts if it is assumed that the co-ordinated water molecules are tetrahedral at the oxygen.[803]

INDO calculations have been carried out on a number of acetylacetonate and other complexes. With the aid of these results, ^1H and ^{19}F n.m.r. shift data for complexes of paramagnetic cations have been interpreted in terms of both ligand-to-metal and metal-to-ligand charge transfer.[804] A theoretical treatment of electronic and nuclear relaxation in paramagnetic complexes has been given and used to analyse proton relaxation data in aqueous solutions of Cr^{3+}, Fe^{3+}, and Mn^{2+}.[805]

^1H N.m.r. data have been reported for $V(R^1COCHCOR^1)_n(R^2COCHCOR^2)_{3-n}$ (n = 0—3; R^1, R^2 = Me, CF_3, or Ph). It was found, contrary to expectations, that electron-withdrawing substituents decrease charge transfer to the substituted ligands and increase charge transfer to the remaining ligands.[806] The ^1H and ^2H n.m.r. spectra of $V(acac)_3$ show three methyl resonances when the methyl group is partially deuteriated with shift differences of up to 1.1 p.p.m. being observed.[807] The interaction of VO^{2+} with CH_2Cl_2 has been examined by ^1H n.m.r.[808]

Analysis of the non-Curie temperature dependence of the ligand contact shifts in mixed ligand complexes of Cr^{II} with 4,8-dimethyl-*o*-phenanthroline and 4,4′-dimethyl-αα′-bipyridine has shown that one of the mixed ligand chelates, $ML^1{}_2L^2$, possesses an *A* orbital ground state, whereas $ML^1L^2{}_2$ possesses the *B* orbital ground state in C_2 symmetry.[809] A re-examination of factors causing the broadening of the methyl resonance of $[(4,7\text{-dimethyl-}o\text{-phenanthroline})_3Cr]^{2+}$, including examination of the corresponding ethyl compounds and its nickel analogue, has led to the conclusion that the broadening is electronic.[810]

^1H N.m.r. data have been reported for the oxygen-bridged iron(III) complexes of the type $(Fe^{III}L_2)_2O$ where L is a bidentate Schiff-base ligand. The temperature dependence of the proton Knight shifts in the ligands has been interpreted in terms of the contact hyperfine interaction constant *A* and the exchange integral *J*. The fitting of these shifts represents a novel

802 H. Katô and H. Kato, *Bull. Chem. Soc. Japan*, 1971, **44**, 1734.
803 D. W. Clack and M. S. Farrimond, *J. Chem. Soc.* (*A*), 1971, 299.
804 M. J. Scarlett, A. T. Casey, and R. A. Craig, *Austral. J. Chem.*, 1971, **24**, 31.
805 M. Rubinstein, A. Baram, and Z. Luz, *Mol. Phys.*, 1971, **20**, 67.
806 D. R. Eaton and K. L. Chua, *Canad. J. Chem.*, 1971, **49**, 56.
807 R. R. Horn and G. W. Everett, jun., *J. Amer. Chem. Soc.*, 1971, **93**, 7173.
808 A. J. Leffler, *J. Phys. Chem.*, 1971, **75**, 599.
809 G. N. La Mar and G. R. Van Hecke, *J. Magn. Resonance*, 1971, **4**, 384.
810 G. N. La Mar and G. R. Van Hecke, *Chem. Comm.*, 1971, 274.

method for the evaluation of *J* in such systems.[811] Similarly, *J* has been obtained for μ-oxo-tetraphenylporphineiron(III) and the *p*-tolyl analogue.[812]

Like the analogous *o*-phenanthroline complexes, $[Fe_2(bipy)_4O]Cl_4$ and its methyl-substituted analogues show ^1H n.m.r. shifts consistent with a σ-delocalization mechanism.[813] The ^1H n.m.r. spectra show that [(tetraphenylporphyriniron)$_2$O]$^+$ is paramagnetic, but the shifts are too small for it to be a porphyrin radical and thus the unpaired electron is on the iron.[814]

The complex (209) shows a singlet–triplet spin state equilibrium. There are two intramolecular rearrangements, one showing inversion of molecular

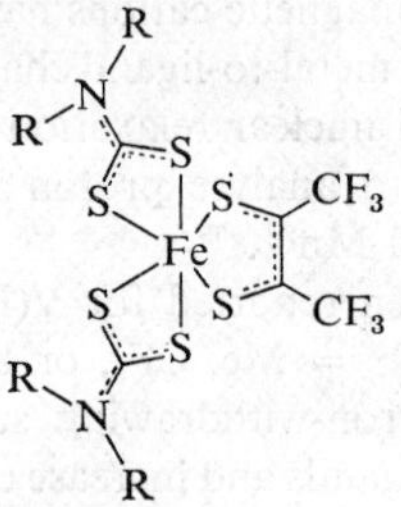

(209)

configuration and the other showing rotation about the carbon–nitrogen bond. It was claimed that this work provides the best available evidence for the operation of a twist mechanism in the rearrangement reactions of chelates. Data were also reported for the ruthenium and cobalt analogues.[815]

Biferrocene $Fe^{II}Fe^{III}$ picrate shows only one very broad resonance, which is consistent with rapid electron exchange.[816] $[(Cp)FeS]_4^+$ is paramagnetic giving a very broad signal at τ 4.23, whereas $[(Cp)FeS]_4^{2+}$ is diamagnetic.[817]

The influence of the paramagnetic compounds $Fe(acac)_3$ and diphenylpicrylhydrazyl on solvents has been investigated by ^{13}C n.m.r. spectroscopy. The shifts were attributed to a pseudo-contact interaction in the collision complex, and hence the warning is given that low or zero scalar coupling has to be interpreted with care.[818] ^1H N.m.r. has been used to demonstrate that ethanol exchange occurs for $[Fe(CO)_2(EtOH)_2]^+$,[819] and to investigate the interaction of amino-acids and peptides with iron(II) ions.[820] The inter-

[811] P. D. W. Boyd and K. S. Murray, *J. Chem. Soc. (A)*, 1971, 2711.
[812] P. D. W. Boyd and T. D. Smith, *Inorg. Chem.*, 1971, **10**, 2041.
[813] M. Wicholas and D. Jayne, *Inorg. Nuclear Chem. Letters*, 1971, 7, 443.
[814] R. H. Felton, G. S. Owen, D. Dolphin, and J. Fajer, *J. Amer. Chem. Soc.*, 1971, **93**, 6332.
[815] L. H. Pignolet, R. A. Lewis, and R. H. Holm, *J. Amer. Chem. Soc.*, 1971, **93**, 360.
[816] D. O. Cowan, R. L. Collins, and F. Kaufman, *J. Chem. Phys.*, 1971, **75**, 2025.
[817] J. A. Ferguson and T. J. Meyer, *Chem. Comm.*, 1971, 623.
[818] O. W. Howarth, *Mol. Phys.*, 1971, **21**, 949.
[819] G. Martini and E. Tiezzi, *Trans. Faraday Soc.*, 1971, **67**, 2538.
[820] A. M. Bowles, W. A. Szarek, and M. C. Baird, *Inorg. Nuclear Chem. Letters*, 1971, **7**, 25.

action of amines, alcohols, aldehydes, and ketones with paramagnetic compounds, namely $Fe(acac)_3$, $FeCl_3$, $FeCl_2$, and $CoCl_2$ can cause proton–proton decoupling.[821, 822] Similarly, the interaction of $CoCl_2$ with phosphites produces loss of phosphorus coupling in the ^1H n.m.r. spectrum.[823]

Dipolar shifts calculated from susceptibility data on single crystals of $Co(acac)_2(py)_2$ are in agreement with ^1H n.m.r. data, but the agreement with ^{13}C shift data is less satisfactory.[824] The n.m.r. contact shifts of $Co(sacsac)_2$ had been previously interpreted in terms of an 'axial field' model, but e.s.r. measurements have now shown a large in-plane anisotropy.[825] Broadline ^1H n.m.r. has been used to investigate Co(salicylaldehyde)$_2$(pyridazine).[826] ^{19}F N.m.r. has been used to demonstrate methyl exchange between cobalt(ethylenediaminebisacetylacetone) {Co(acen)} and methylcobalt(ethylenediaminebistrifluoroacetylacetone) {MeCo(tfen)}:[827]

$$\text{Co(tfen)} + \text{MeCo(acen)} \rightleftharpoons \text{Co(acen)} + \text{MeCo(tfen)}$$

To interpret the temperature dependence of the isotropic shifts of bis(trispyrazoylborate)cobalt(II) it has been necessary to use the McGarvey treatment which takes account of both thermally accessible states and second-order Zeeman interaction.[828] This compound has also been used to investigate the role of substituted anilines as second co-ordination sphere ligands.[829] Similarly, ^{19}F n.m.r. has been used to determine the structure of the ion pair formed between bistripyrazolylmethanecobalt(II) ions and $[PF_6]^-$ ions. It was concluded that the preferred position of the anion lies close to the symmetry axis of the cobalt complex and that a contact ion pair is involved.[830] Restricted rotation about the carbon–nitrogen bond has been found for some tetrahedral cobalt(II) complexes of substituted thioureas.[831]

The ^1H n.m.r. spectra of (4-methylpyridine *N*-oxide)$_6$M$(ClO_4)_2$ (M = Co^{II} or Ni^{II}) have been examined by ^1H n.m.r. in the presence of free ligand. Co^{II} and Ni^{II} give the same ratios of shifts, but this could be fortuitous as this factoring method was considered to be unreliable.[832] The isotropic shifts of (210) (M = Co or Ni) have been attributed to a dominating π-delocalization mechanism and the angle between the two tipped

821 R. Engel, *J. Chem. Soc.* (*C*), 1971, 3554.
822 R. Engel and G. Nathan, *J. Chem. Soc.* (*C*), 1971, 3844.
823 R. Engel and A. Jung, *J. Chem. Soc.* (*C*), 1971, 1761.
824 W. DeW. Horrocks, jun. and D. DeW. Hall, *Inorg. Chem.*, 1971, **10**, 2368.
825 R. J. Fitzgerald and G. R. Brubaker, *Inorg. Chem.*, 1971, **10**, 1324.
826 H. G. Biedermann, P. K. Burkert, and K. E. Schwarzhans, *Z. Naturforsch.*, 1971, **26b**, 968.
827 A. van den Bergen and B. O. West, *Chem. Comm.*, 1971, 52.
828 G. N. La Mar, J. P. Jesson, and P. Meakin, *J. Amer. Chem. Soc.*, 1971, **93**, 1286.
829 D. R. Eaton, H. O. Ohorodnyk, and L. Seville, *Canad. J. Chem.*, 1971, **49**, 1218.
830 D. R. Eaton, L. Seville, and J. P. Jesson, *Canad. J. Chem.*, 1971, **49**, 2751.
831 D. R. Eaton and K. Zaw, *Canad. J. Chem.*, 1971, **49**, 3315.
832 W. D. Perry, R. S. Drago, D. W. Herlocker, G. K. Pagenkopf, and K. Czworniak, *Inorg. Chem.*, 1971, **10**, 1971.

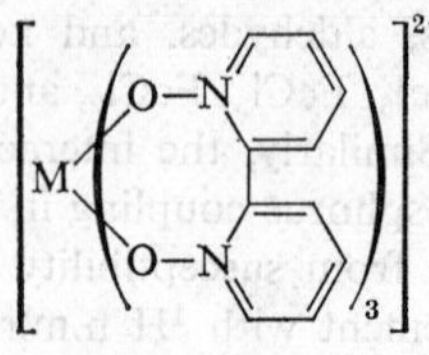

(210)

phenyl rings has been calculated to be 67°.[833] The ^1H n.m.r. of M(4-R-pyridine)$_4X_2$ (M = Co or Ni; R = H or Me; X = halogen or NCS^-) have been reported. When M = Ni, all the CH protons are moved to low field, while the methyl groups are moved to high field. It was therefore concluded that a contact interaction is occurring. However, when M = Co, no such consistency was found and it was therefore concluded that dipolar shifts are also important.[834]

The principal molecular susceptibilities of the pseudotetrahedral dichlorobis(triphenylphosphine)-cobalt(II) and -nickel(II) have been determined from single-crystal measurements. The results were used to evaluate the dipolar contributions to the isotropic ^1H n.m.r. shifts and are consistent with an earlier evaluation based on ^1H n.m.r. data alone.[835]

Spin densities determined by ^1H n.m.r. in cobalt(II) and nickel(II) complexes of guajacole, vanilline, isovanilline, *o*-nitrophenale, and methyl-salicylate have been taken as indicative of reactive ligand sites.[836, 837]

The use of Co(acac)$_2$ and Ni(acac)$_2$ as shift reagents has shown that tropanes are nearly planar [838] and they have also been applied to cyclohexylamine and cycloheptylamine.[839] Similarly, these shift reagents and Cu(ethylacac)$_2$ have been applied to alcohols.[840]

The ^1H n.m.r. chemical shifts of [M(5-R-*o*-phenanthroline)$_3$]$^{2+}$ (M = Co or Ni; R = H or Me) have been reported; all the protons are shifted downfield. Thus the observed isotropic shifts were attributed to both σ- and π-spin delocalization and contributing dipolar shifts in the case of the cobalt species; the ratio method was used.[841] However, in the case of Ni(methyl-substituted *o*-phenanthroline)$_3$Cl$_2$,nH$_2$O it was concluded that σ-spin delocalization is dominant as the π-spin densities do not correlate with Hückel π-spin densities.[842] Some new calculations of relevance to electron

833 I. Bertini, D. Gatteschi, and L. J. Wilson, *Inorg. Chim. Acta*, 1970, **4**, 629.
834 D. Forster, *Inorg. Chim. Acta*, 1971, **5**, 465.
835 W. DeW. Horrocks, jun. and E. S. Greenberg, *Inorg. Chem.*, 1971, **10**, 2190.
836 H. G. Biedermann and K. E. Schwarzhans, *Z. Naturforsch.*, 1971, **26b**, 531.
837 H. G. Biedermann, G. Rossmann and K. E. Schwarzhans, *Z. Naturforsch.*, 1971, **26b**, 78.
838 M. Ohashi, I. Morishima, K. Okada, T. Yonezawa, and T. Nishida, *Chem. Comm.*, 1971, 34.
839 K. Fricke, *Tetrahedron Letters*, 1971, 1237.
840 E. Gillies, W. A. Szarek, and M. C. Baird, *Canad. J. Chem.* 1971, **49**, 211.
841 I. Bertini and L. J. Wilson, *J. Chem. Soc.* (*A*), 1971, 489.
842 M. Wicholas, *Inorg. Chem.*, 1971, **10**, 1086.

delocalization in some tris(*o*-phenanthroline)nickel(II) complex ions have been carried out and the results have been compared with ^1H n.m.r. data.[843]

E.s.r. and ^1H n.m.r. have been used to show that complexes formed by morpholine with $Ni[S_2P(OR)_2]_2$ have probably a five-co-ordinate (approximately C_{4v}) geometry.[844] Similarly, $Ni[S_2P(OR)_2]_2$ forms 1 : 1 adducts with $OP(NMe_2)_3$, but ^{31}P n.m.r. spectra of $OP(NMe_2)_3$ were only detectable in less than 10^{-3} mol l^{-1} solutions in $OP(NMe_2)_3$.[845]

$Ni[S_2P(OPr^i)_2]_2$ is diamagnetic, but interacts with pyridines to give paramagnetic complexes (Evans method). All the pyridine resonances are shifted to low field, which was accounted for in terms of σ-delocalization.[846] The contact shifts observed for the γ-picoline protons in complexes of the type $Ni[S_2P(OEt)_2]_2(\gamma\text{-picoline})_2$ have been explained as spin transfer involving not only the highest bonding orbital of σ-symmetry but also, to some extent, the lowest antibonding and the highest bonding orbitals of π-symmetry.[847]

Examination of the temperature dependence of the ^1H n.m.r. spectrum of the complex (211) (R = H or Me) in various solvents shows an equilibrium between a paramagnetic octahedral structure favoured by low temperature and a diamagnetic square-planar structure favoured by high temperature.[848] The ^1H n.m.r. resonances of $[(R^1R^2C{=}NCH_2CH_2N{=}CR^1R^2)_2Ni]^{2+}$ are rather broad but some resonances could be assigned.[849] As solutions of the compound (212) are concentrated they become paramagnetic (Evans method) and the signals of the CH_2 groups move downfield.[850]

^1H N.m.r. has been used to follow ligand scrambling in paramagnetic complexes

$$NiL^1_2 + NiL^2_2 \rightleftharpoons 2NiL^1L^2$$

where L^1 and L^2 are ligands such as (213) or (214).[851] From the poor agreement between the theoretically calculated π-electron density and those calculated from measurements on $(RCOCHCOR)_2Ni$, it was concluded that the unpaired electrons are delocalized *via* both the σ and π systems.[852]

843 C. L. Honeybourne and S. Morris, *Chem. Phys. Letters*, 1971, **11**, 380.
844 J. R. Angus, G. M. Woltermann, W. R. Vincent, and J. R. Wasson, *J. Inorg. Nuclear Chem.*, 1971, **33**, 3041.
845 J. R. Angus, G. M. Woltermann, and J. R. Wasson, *Inorg. Nuclear Chem. Letters*, 1971, **7**, 837.
846 H. E. Francis, G. L. Tincher, W. F. Wagner, J. R. Wasson, and G. M. Woltermann, *Inorg. Chem.*, 1971, **10**, 2620.
847 M. M. Dhingra, G. Govil, and C. R. Kanekar, *Chem. Phys. Letters*, 1971, **10**, 86.
848 L. Rusnak and R. B. Jordan, *Inorg. Chem.*, 1971, **10**, 2199.
849 D. L. Johnston, I. Bertini, and W. DeW. Horrocks, jun., *Inorg. Chem.*, 1971, **10**, 865.
850 T. I. Benzer, L. Dann, C. R. Schwitzgebel, M. D. Tamburro, and E. P. Dudek, *Inorg. Chem.*, 1971, **10**, 2204.
851 J. C. Lockhart and W. J. Mossop, *Chem. Comm.*, 1971, 61.
852 F. W. Pijpers, H. E. Smeets, and L. B. Beentjes, *Rec. Trav. chim.*, 1971, **90**, 1292.

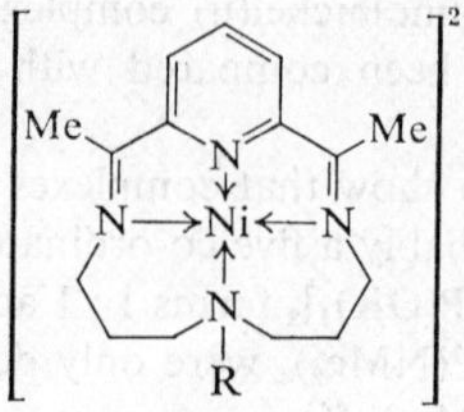

(211)

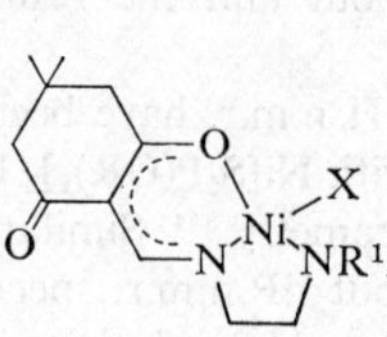

(212)

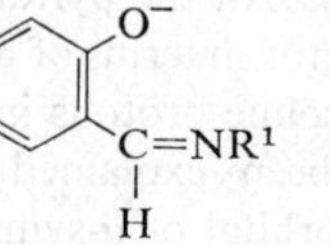

(213)

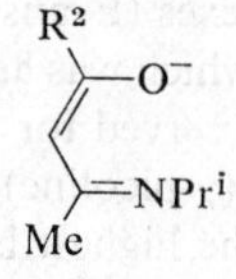

(214)

$[Ni\{(Et_2NCH_2CH_2)_2N(CH_2CH_2PPh_2)\}X]^+$ complexes can exist as high-spin five-co-ordinate or low-spin planar complexes. From the temperature dependence of the ^1H n.m.r. spectrum, the thermodynamic parameters have been calculated.[853] On the basis of ^1H n.m.r. contact shifts of three paramagnetic cations, $[(RC_3H_4)_2Ni]^+$, the presence of a first-order π-delocalization has been suggested.[854]

The use of nickel complexes as shift reagents has also been examined. Thus when $Ni(acac)_2$ is added to anilines, very large ^{14}N chemical shifts have been observed varying from + 2400 p.p.m. for p-$SO_3CF_3C_6H_4NH_2$ to + 1200 p.p.m. for p-$MeC_6H_4NH_2$.[855] If the contact shifts for $Ni(acac)_2$ with various organic compounds, *e.g.* $\overline{HNCH_2CH_2}$, are compared with the corresponding coupling constants, *e.g.* in cyclopropane, there is a good correlation.[856] The shifts induced by $Ni(acac)_2$ in some methylpiperidines have been explained in terms of σ-delocalization of the unpaired electron densities,[857] and the shifts it induces in (215) have also been discussed.[858] $Ni(S_2PR_2)_2$ has been used as a shift reagent for Pr^iNH_2.[859]

The e.s.r. spectra of the compound $Cu(acac)_2(py)$ shows no resolvable ^{14}N hyperfine splitting. ^{14}N N.m.r. was used to show an isotropic ^{14}N coupling constant of $+ 0.70 \pm 0.10$ G, which is too small to be resolved in the e.s.r. spectrum.[860] The binding of Cu^{2+} to inosine has been examined by broadening in the ^1H n.m.r. spectrum.[861] $Cu(hfac)_2$ forms 1 : 1 adducts

853 I. Bertini, P. Dapporto, G. Fallani, and L. Sacconi, *Inorg. Chem.*, 1971, **10**, 1703.
854 H. P. Fritz and F. H. Köhler, *Z. anorg. Chem.*, 1971, **385**, 22.
855 E. I. Berus, Yu. G. Gladkin, V. A. Barkhash, and Yu. N. Molin, *Doklady Phys. Chem.*, 1971, **197**, 355.
856 I. Morishima and T. Yonezawa, *J. Chem. Phys.*, 1971, **54**, 3238.
857 I. Morishima, K. Okada, M. Ohashi, and T. Yonezawa, *Chem. Comm.*, 1971, 33.
858 K. Tori, Y. Yoshimura, and R. Muneyuki, *J. Amer. Chem. Soc.*, 1971, **93**, 6324.
859 C. A. Cabrera, G. M. Wolterman, and J. R. Wasson, *Tetrahedron Letters*, 1971, 4485.
860 B. B. Wayland and M. D. Wisniewski, *Chem. Comm.*, 1971, 1025.
861 N. A. Berger and G. L. Eichhorn, *J. Amer. Chem. Soc.*, 1971, **93**, 7062.

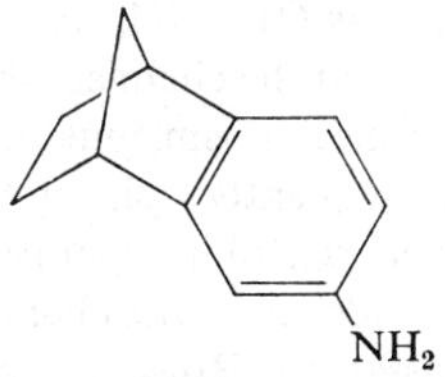

(215)

with olefins and acetylenes; although the ^{1}H n.m.r. spectrum was badly broadened, isotropic shifts were observed.[862]

The Evans method of determining magnetic susceptibility has been applied to show that paramagnetic products are formed during the reaction of $Mn_2(CO)_{10}$ with Ph_3P.[863] Also the susceptibilities of some molybdenum(v) complexes with α-hydroxycarboxylic acids,[864] $(XPR_2NPR_2X)_2M$ (M = Fe, Co, Zn, Ni, Pd, or Pt; R = Me or Ph; X = Se, S, O, or NH),[865, 866] $[Fe(Cp)_2]^+$, $Fe(B_9C_2H_{11})(Cp)$, and related compounds,[867] *mer*-ReX_3L_3, and *cis*- and *trans*-$OsCl_4L_4$ have been measured. Full ^{1}H n.m.r. data have been reported for *trans*-$OsCl_4(PBu^i{}_2Ph)_2$,[868] Co(*N*-methyl-γ-butyrolactam)$_4(ClO_4)_2$,[869] $Co(sacsac)_2EPh_3$ (E = P, As, Sb, or Bi),[870] $(Bu^t{}_2NO)_2CoBr_2$,[871] $CoX_2\{P(NMe_2)F_2\}_2$,[872] *NN'*-bis-(3-isopropylsalicylidene)polymethylenediaminocobalt(II) complexes,[873] $Co(acac)_2(amine)_2$, and $Ni(acac)_2(amine)_2$.[874] Nickel(II) complexes of (216) (X = NH or S)[875] have been measured by the Evans method.

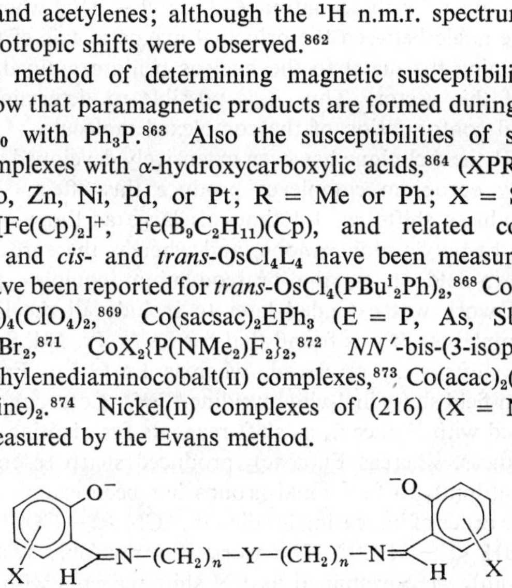

(216)

Compounds of the Lanthanides and Actinides.—Publications dealing with the use of the lanthanides have increased markedly from seven papers last year to 121 papers this year. During 1970, it had been reported that the paramagnetic compound $Eu(dpm)_3$ (dpm = $[Bu^tCOCHCOBu^t]^-$) will

862 R. A. Zelonka and M. C. Baird, *J. Organometallic Chem.*, 1971, **33**, 267.
863 J. R. Miller and D. H. Myers, *Inorg. Chim. Acta*, 1971, **5**, 215.
864 D. H. Brown and J. MacPherson, *J. Inorg. Nuclear Chem.*, 1971, **33**, 4203.
865 A. Davison and E. S. Switkes, *Inorg. Chem.*, 1971, **10**, 837.
866 A. Davison and D. L. Reger, *Inorg. Chem.*, 1971, **10**, 1967.
867 D. N. Hendrickson, Y. S. Sohn, and H. B. Gray, *Inorg. Chem.*, 1971, **10**, 1559.
868 H. P. Gunz and G. J. Leigh, *J. Chem. Soc. (A)*, 1971, 2229.
869 S. K. Madan, *J. Inorg. Nuclear Chem.*, 1971, **33**, 1025.
870 J. F. White and M. F. Farona, *Inorg. Chem.*, 1971, **10**, 1080.
871 D. G. Brown, T. Maier, and R. S. Drago, *Inorg. Chem.*, 1971, **10**, 2804.
872 T. Nowlin and K. Cohn, *Inorg. Chem.*, 1971, **10**, 2801.
873 M. Hariharan and F. L. Urbach, *Inorg. Chem.*, 1971, **10**, 2667.
874 D. A. Fine, *Inorg. Chem.*, 1971, **10**, 1825.
875 W. M. Coleman and L. T. Taylor, *J. Inorg. Nuclear Chem.*, 1971, **33**, 3049.

co-ordinate to functions such as OH, NH_2, *etc.* in organic molecules. The result of this co-ordination is that the chemical shifts of protons are changed *via* a predominantly dipolar mechanism, but that very little paramagnetic broadening occurs and consequently spin–spin coupling information is retained; thus $Eu(dpm)_3$ is very useful in separating overlapping resonances. From the concentration dependence of the chemical shifts it is also possible to determine stability constants. However, perhaps the most exciting possibility arising from the shifts being dipolar in origin is that the magnitude of the shift is given by the equation $K(3\cos^2\theta - 1)/r^3$ where K is a constant, θ is the angle between the principal magnetic axis of the metal and the vector joining the metal to the nucleus being examined, and r is the magnitude of this vector. Thus it is possible to determine the stereochemistry and conformation of the complexed molecule.

During 1971 the technique has been extensively developed and exploited, and not only europium complexes produce this effect. For example, $Pr(dpm)_3$ produces shifts *ca.* 1.4 times as large as those obtained using $Eu(dpm)_3$ in 4-t-butylcyclohexanone, and whereas the europium complex produces down-field shifts, the praseodymium complex produces up-field.[876] This work was extended later to include all the lanthanides as $Ln(dpm)_3$ complexes. It was found that for Ln = Pr, Nd, Sm, Tb, Dy, or Ho, up-field shifts were produced, whereas Ln = Eu, Er, Tm, or Yb produce down-field shifts in 4-vinylpyridine.[877, 878] $Co(acac)_2$ and $Ni(acac)_2$ were compared with $Eu(acac)_3$ as shift reagents for nicotine but produced broad resonances, whereas $Eu(acac)_3$ produced sharp resonances.[879] The binding of $Eu(dpm)_3$ to functional groups has been examined and it was found that the order of increasing binding is $-CN < -CO_2R < -O- < X{=}O < -OH < -NH_2$.[880] The work was not only restricted to ^1H n.m.r.: $Ln(dpm)_3$ were examined as ^{14}N shift reagents with emphasis on pyridine. $Dy(dpm)_3$ was found to be by far the best high-field shift reagent whereas $Yb(dpm)_3$ is the best low-field shift reagent.[881]

In the early investigations, $Eu(dpm)_3$ had only produced down-field shifts. Thus when up-field shifts were observed for the first time this attracted attention. When $Eu(dpm)_3$ is added to 2,6-di-2-propylacetanilide,[882] compound (217),[883] *cis*-3-(1-naphthyl)-1,3,5,5-tetramethylcyclohexan-1-ol,[884] or compound (218) high-field shifts were observed in each

876 P. Bélanger, C. Freppel, D. Tizané, and J. C. Richer, *Chem. Comm.*, 1971, 266.
877 N. Ahmad, N. S. Bhacca, J. Selbin, and J. D. Wander, *J. Amer. Chem. Soc.*, 1971, **93**, 2564.
878 W. DeW. Horrocks, jun. and J. P. Sipe, tert., *J. Amer. Chem. Soc.*, 1971, **93**, 6800.
879 M. Ohashi, I. Morishima, and T. Yonezawa, *Bull. Chem. Soc. Japan*, 1971, **44**, 576.
880 J. K. M. Sanders and D. H. Williams, *J. Amer. Chem. Soc.*, 1971, **93**, 641.
881 M. Witanowski, L. Stefaniak, H. Januszewski, and Z. W. Wolkowski, *Chem. Comm.*, 1971, 1573.
882 T. H. Siddall, tert., *Chem. Comm.*, 1971, 452.
883 P. H. Mazzocchi, H. J. Tamburin, and G. R. Miller, *Tetrahedron Letters*, 1971, 1819.
884 B. L. Shapiro, J. R. Hlubucek, G. R. Sullivan, and L. F. Johnson, *J. Amer. Chem. Soc.*, 1971, **93**, 3283.

case for at least one type of proton.[885] Similarly, when Eu^{3+} is added to $HO_2CCOCH_2CO_2H$, $EtOCOCOCH_2CO_2H$, or $HO_2CCOCH_2CO_2Et$ shifts occur. In the case of $EtOCOCOCH_2CO_2H$, the CH_2 group is shifted to high field whereas in the case of $HO_2CCOCH_2CO_2H$ the CH_2 group is shifted to low field, which was taken as evidence that the stereochemistry of this latter complex is as shown in structure (220) rather than (219).[886]

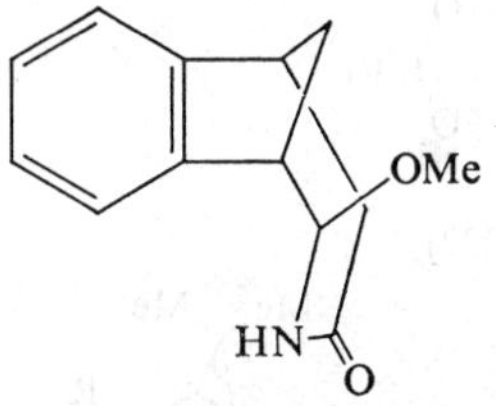

(217)

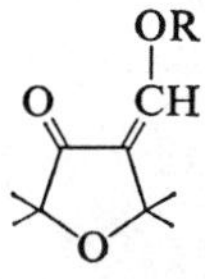

(218)

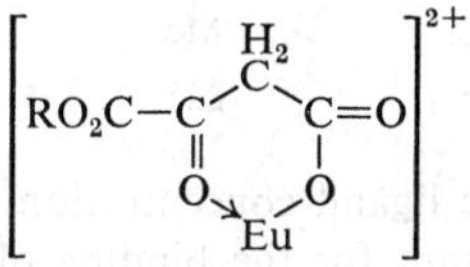

(219)

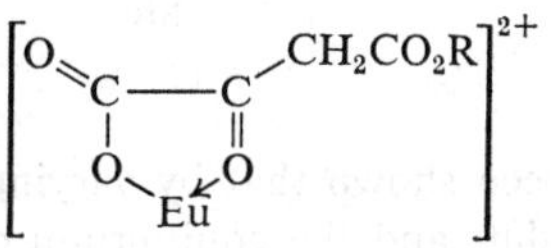

(220)

The technique of using 'shift reagents' is limited to compounds that will complex to these lanthanide compounds. $Eu(dpm)_3$ is not very good at complexing esters, but a new reagent, $Ln(fod)_3$ (Ln = Eu or Pr; fod = $[C_3F_7COCHCOBu^t]^-$), has been introduced which is far better at complexing esters. However, it is more difficult to handle as it readily picks up water and thus should be stored over P_4O_{10}.[887]

The flexibility of these 'shift reagents' was increased by the introduction of optically active 'shift reagents'. The first attempt was not very successful. $Eu(dpm)_3$ had been used to differentiate between the *meso* and racemic forms of $PhSOCH_2SOPh$. The *meso*-form which originally gave a singlet for the CH_2 group produced an AB pattern on the addition of $Eu(dpm)_3$, but the CH_2 group of the racemic form remained a singlet. Thus the chiral shift reagent tris-{3-(t-butylhydroxymethylene)camphorato}europium was tried but it only caused broadening of the CH_2 resonance in the racemic form.[888] However, the addition of the europium complex of 3-heptafluorobutanoyl-(+)-camphor to racemates of alcohols, sulphoxides, an epoxide, or an aldehyde causes separation of the ^1H n.m.r. signals of the enantiomers.[889] More recently two better chiral shift reagents have been

[885] S. B. Tjan and F. R. Visser, *Tetrahedron Letters*, 1971, 2833.
[886] C. Reyes-Zamora and C. S. Tsai, *Chem. Comm.*, 1971, 1047.
[887] R. E. Rondeau and R. E. Sievers, *J. Amer. Chem. Soc.*, 1971, **93**, 1522.
[888] J. L. Greene, jun. and P. B. Shelvin, *Chem. Comm.*, 1971, 1092.
[889] R. R. Fraser, M. A. Petit, and J. K. Saunders, *Chem. Comm.* ,1971, 1450.

introduced, namely tris-{3-(trifluoromethylhydroxymethylene)-(+)-camphorato}europium(III)[890] and compounds (221) and (222) [R^1, R^2 = (223) or (224)].[891]

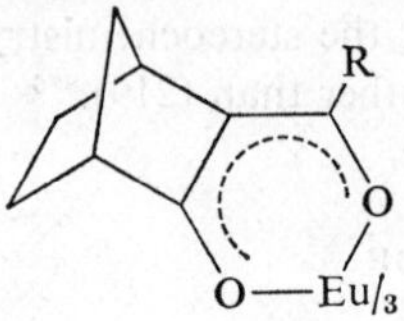

(221)

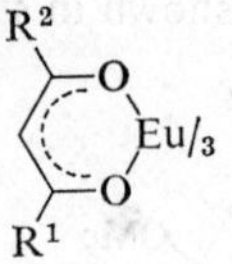

(222)

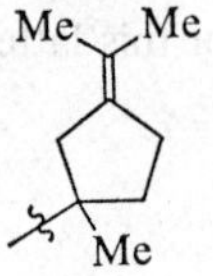

(223)

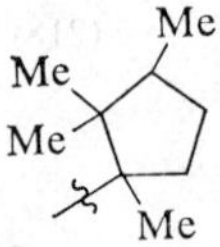

(224)

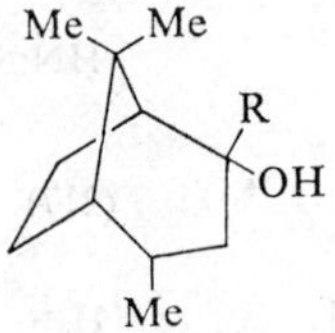

(225)

It has been shown that by varying the ligand concentration the bound chemical shift and the equilibrium constant for the binding of ligand to lanthanide may be determined. The method was applied to show that $Eu(dpm)_3$ is complexed more strongly by n-propylamine than by neopentanol.[892] Deuterium can cause an effect: thus if $Eu(dpm)_3(py)_2$ is added to the partially deuteriated compound (225) (R = H or D) then two sets of ^1H n.m.r. resonances appear, and it was concluded that the deuteriated species is bound more strongly to europium than the non-deuteriated species.[893]

A method based on the shifts and broadening caused by the lanthanides has been developed which offers great potential in both biochemistry (for which it was developed) and inorganic chemistry to determine the full stereochemistry of complexes in solution. The method is based on the fact that the pseudo-contact shift is proportional to $(3\cos^2\theta - 1)/r^3$ whereas the broadening is proportional to $1/r^6$. Thus the observed shifts and broadening can be fed into a computer and a structure produced. The method was applied to the complex formed between adenosine monophosphate and Eu^{3+}, Nd^{3+}, or Ho^{3+}.[894] Other workers have used only the $(3\cos^2\theta - 1)/r^3$ relationship to determine conformation. Thus an

[890] H. L. Goering, J. N. Eikenberry, and G. S. Koermer, *J. Amer. Chem. Soc.*, 1971, **93**, 5913.
[891] G. M. Whitesides and D. W. Lewis, *J. Amer. Chem. Soc.*, 1971, **93**, 5914.
[892] I. Armitage, G. Dunsmore, L. D. Hall, and A. G. Marshall, *Chem. Comm.*, 1971, 1281.
[893] G. V. Smith, W. A. Boyd, and C. C. Hinckley, *J. Amer. Chem. Soc.*, 1971, **93**, 6319.
[894] C. D. Barry, A. C. T. North, J. A. Glasel, R. J. P. Williams, and A. V. Xavier, *Nature*, 1971, **232**, 236.

analytical method of calculation of the lanthanide n.m.r. shift dependence on proton distance has been given,[895] and the equation has been applied to obtain the conformation of adamantan-2-ol, *trans*-4-t-butylcyclohexanol,[896] pyridine and other N-heterocycles,[897] and some rigid bicyclic ethers such as (226).[898] The equation has also been used in conjunction with

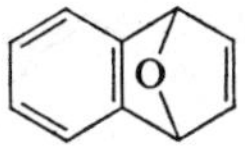

(226)

$Ln(dpm)_3$ (Ln = Pr, Eu, or Tb) to give a complete ^{13}C n.m.r. assignment of borneol.[899] Another method by which the shift reagents can be used to assist assignment of ^{13}C n.m.r. has been reported. In simple molecules the effect of decoupling of individual protons can be used to assign ^{13}C resonances. However, in more complex molecules where proton resonances overlap this was not previously possible. The addition of a lanthanide shift reagent can cause the proton resonances to separate and make this double irradiation method once again applicable. This method was used in the addition of Eu^{3+} to ribose 5-phosphate.[900]

A number of workers have urged caution in the interpretation of shift data. It has been shown that $Pr(dpm)_3$ and $Eu(dpm)_3$ exist in solution as $Pr_2(dpm)_6$ and $Eu_2(dpm)_6$ and allowance has to be made before meaningful stability constant data can be obtained.[901] The paramagnetic-induced proton shifts of alcohols are strongly affected by temperature and by absolute concentrations of both alcohols and the shift reagent [$Pr(dpm)_3$ in this case].[902] When $Eu(dpm)_3$ is added to *cis*-4-t-butyl-1-phenylcyclohexanol, the *m*- and *p*-protons of the phenyl ring fail to move. This is inconsistent with the $(3\cos^2\theta - 1)/r^3$ relationship assuming normal bond lengths and angles.[903] Paramagnetic shifts were obtained in the ^1H n.m.r. spectra of testosterone and 17-α-methyltestosterone by addition of $Eu(dpm)_3(py)_2$. Log–log plots of the shift *vs.* the europium–hydrogen distances were used to separate contact and pseudo-contact shifts. It was concluded that although the shifts are predominantly pseudo-contact, contact interaction is important for protons close to the europium.[904]

[895] A. J. Rafalski, J. Barciszewski, and M. Wiewiòrowski, *Tetrahedron Letters*, 1971, 2829.
[896] S. Farid, A. Ateya, and M. Maggio, *Chem. Comm.*, 1971, 1285.
[897] W. L. F. Armarego, T. J. Batterham, and J. R. Kershaw, *Org. Magn. Resonance*, 1971, **3**, 575.
[898] R. Caple and S. C. Kuo, *Tetrahedron Letters*, 1971, 4413.
[899] J. Briggs, F. A. Hart, G. P. Moss, and E. W. Randall, *Chem. Comm.*, 1971, 364.
[900] B. Birdsall, J. Feeney, J. A. Glasel, R. J. P. Williams, and A. V. Xavier, *Chem. Comm.*, 1971, 1473.
[901] M. K. Archer, D. S. Fell, and R. W. Jotham, *Inorg. Nuclear Chem. Letters*, 1971, **7**, 1135.
[902] L. Tomić, Z. Majerski, M. Tomić, and D. E. Sunko, *Chem. Comm.*, 1971, 719.
[903] N. S. Bhacca and J. D. Wander, *Chem. Comm.*, 1971, 1505.
[904] C. C. Hinckley, M. R. Klotz, and F. Patil, *J. Amer. Chem. Soc.*, 1971, **93**, 2417.

$La(dpm)_3$ and $Eu(dpm)_3$ have been used to assist in assigning the ^1H and ^{13}C n.m.r. spectra of the compound (227). It is interesting to note that the diamagnetic $La(dpm)_3$ complex produces a co-ordination shift in the ^{13}C n.m.r. resonance of the α-carbon of − 3.5 p.p.m. whereas the paramagnetic

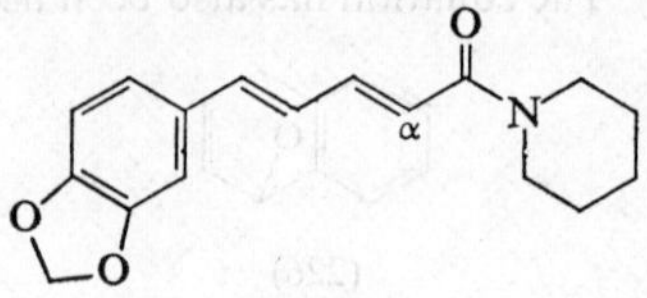

(227)

$Eu(dpm)_3$ complex produces a comparable shift of + 16.5 p.p.m. Thus in addition to the pseudo-contact shift, allowance must be made for a co-ordination shift.[905] Most workers, when using the $(3\cos^2\theta - 1)/r^3$ relationship, have assumed that the principal magnetic axis of the lanthanide is along the lanthanide–ligand bond, *i.e.* along the symmetry axis through the metal. However, an *X*-ray structure shows that for $Ho(dpm)_3$-(picoline)$_2$ the picoline is not coaxial and the NHoN angle is 139.1°. Thus the angle θ cannot safely be related to the lanthanide–ligand bond.[906]

It has been pointed out that the Eu^{3+} ion at room temperature is found as a mixture of electronic states $^7F_0$ and $^7F_1$. The $^7F_0$ state is non-degenerate and has no Zeeman splitting; thus the contact and pseudo-contact mechanisms cannot operate. In the excited state, $^7F_1$, where the crystal-field splittings are much smaller than kT, the pseudo-contact and contact interactions may contribute to the shifts.[907]

There is an almost linear correlation between the basicity of a *p*-substituted aniline and the $Eu(dpm)_3$-induced shift for both the *o*- and *m*-hydrogens, but the linear correlation breaks down for sterically hindered anilines, *e.g.* $PhNMe_2$ or 2,4,6-tri-t-butylaniline.[908] $Yb(dpm)_3$ has been used to induce shifts in aldehydes, ketones, and amines, and the position of the ytterbium was estimated.[909, 910] ^{14}N N.m.r. has been used to examine the interaction of $Yb(dpm)_3$ or $Eu(dpm)_3$ with alkyl amines and pyridines. The magnitude of the shift seems to be related to the basicity of the lone pair of the nitrogen.[911]

The complexes $[M\{N(CH_2CH_2NH_2)_3\}_2](NO_3)_3$ and $[M\{N(CH_2CH_2NH_2)_3\}_2(NO_3)](NO_3)_2$ (M = lanthanide) show signals due to free and

905 E. Wenkert, D. W. Cochran, E. W. Hagaman, R. B. Lewis, and F. M. Schell, *J. Amer. Chem. Soc.*, 1971, **93**, 6271.
906 W. DeW. Horrocks, jun., J. P. Sipe, tert., and J. R. Luber, *J. Amer. Chem. Soc.*, 1971, **93**, 5258.
907 S. I. Weissman, *J. Amer. Chem., Soc.*, 1971, **93**, 4928
908 L. Ernst and A. Mannschreck, *Tetrahedron Letters*, 1971, 3023.
909 Z. W. Wolkowski, *Tetrahedron Letters*, 1971, 821.
910 C. Beauté, Z. W. Wolkowski and N. Thoai, *Tetrahedron Letters*, 1971, 817.
911 M. Witanowski, L. Stefaniak, H. Januszewski, and Z. W. Wolkowski, *Tetrahedron Letters*, 1971, 1653.

co-ordinated ligand; there is slow intermolecular ligand exchange.[912] The ^1H n.m.r. shifts of $(NBu^n_4)_3[MCl_6]$ (M = lanthanide) have been factored into a Fermi contact and dipolar part; the Fermi contact shifts were found to be significant.[913]

The shifts are sufficient in the case of $(Cp)_3PrCNC_6H_{11}$ to permit a conformational analysis on the cyclohexyl ring. There are seven resonances at room temperature and eleven at − 70 °C, and ΔG was determined.[914] Similar shifts have been reported for this compound, $(Cp)_3UOC_6H_{11}$, and $(Cp)_3UO$(cholesteryl).[915]

Only three papers have so far appeared where $Eu(dpm)_3$ or $Eu(fod)_3$ have been applied to organometallic systems, namely the $Fe(CO)_3$ complexes of tropone, cyclohepta-2,4-dien-1-one, 1-methylcyclohepta-2,4-dien-1-ol,[916] and $PhMe_2SnCH_2CH_2CHMeOH$, and a wide range of organometallic compounds, *e.g.* $(Cp)Fe(CO)_2CN$. However, when $Eu(dpm)_3$ was added to $BrMe_2SnCH_2CH_2CHMeOH$ the effect was much smaller and it was suggested that this compound exists in solution as structure (228) which prevents the co-ordination of the europium.[917] Of perhaps

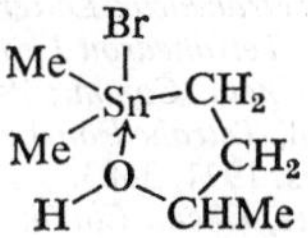

(228)

greater relevance to inorganic chemistry is the observation that $Eu(fod)_3$ will co-ordinate to co-ordinated anions such as fluoride, chloride, azide, or cyanide, *e.g.* in $(Cp)Fe(CO)_2CN$, but not to bromide, iodide, or thiocyanate. It is remarkable that $Eu(fod)_3$ will even co-ordinate to $(Cp)_2WH_2$ and $(Cp)_2Sn$, suggesting the possibility of direct metal–metal interaction.[918]

The effect of $Eu(dpm)_3$ on n.m.r. has been applied to analysis of the following: alcohol mixtures,[919] *cis*- and *trans*-9-hydroxybicyclo[3,3,1]-nonane-*endo*-3-carboxylates,[920] ethoxylated alkylphenol non-ionic surfactants,[921] perhydrophenalenols,[922] alkenols,[923] isoborneol (^{13}C n.m.r.),[924]

912 J. H. Forsberg, T. M. Kubik, T. Moeller, and K. Gucwa, *Inorg. Chem.*, 1971, **10**, 2656.
913 I. M. Walker, L. Rosenthal, and M. S. Quereshi, *Inorg. Chem.*, 1971, **10**, 2463.
914 R. von Ammon, R. D. Fischer, and B. Kanellakopulos, *Chem. Ber.*, 1971, **104**, 1072.
915 R. von Ammon, R. D. Fischer, and B. Kanellakopulos, *Angew. Chem. Internat. Edn.*, 1971, **10**, 819.
916 M. I. Foreman and D. G. Leppard, *J. Organometallic Chem.*, 1971, **31**, C31.
917 M. Gielen, N. Goffin, and J. Topart, *J. Organometallic Chem.*, 1971, **32**, C38.
918 T. J. Marks, J. S. Kristoff, A. Alich, and D. F. Shriver, *J. Organometallic Chem.*, 1971, **33**, C35.
919 D. L. Rabenstein, *Analyt. Chem.*, 1971, **43**, 1599.
920 I. Fleming, S. W. Hanson, and J. K. M. Sanders, *Tetrahedron Letters*, 1971, 3733.
921 G. E. Stolzenberg, R. G. Zaylskie, and P. A. Olson, *Analyt. Chem.*, 1971, **43**, 908.
922 F. A. Carey, *J. Org. Chem.*, 1971, **36**, 2199.
923 H. G. Richey, jun. and F. W. Von Rein, *Tetrahedron Letters*, 1971, 3781.
924 O. A. Gansow, M. R. Willcott, and R. E. Lenkinski, *J. Amer. Chem. Soc.*, 1971, **93**, 4295.

α-monobenzylcyclohexanols,[925] bridged homotropilidenes,[926] 3-methylvinylcyclohexanone,[927] polyethers,[928] $\overline{(CH_2CH_2O)}_n$,[929] cyclic ketones,[930] tertiary amides and thioamides,[931] amides,[932] phosphoryl and amide groups,[933] 2,6-dimethylcyclohexylamines,[934] azetidine derivatives,[935] (4-NH_2-cyclohexyl)$_2CH_2$,[936] R_3PO,[937] *cis*- and *trans*-2,2,3,4,4-pentamethylphosphetan oxides,[938] *syn*- and *anti*-oximes,[939] steroidal alkaloids,[940] chlorinated polycyclodiene pesticide metabolides,[941] triterpenes,[942] the product from tropylium ion and cyclopentadiene,[943] ketals of steroids,[944] 3-substituted bicyclo[3,3,1]nonanes,[945] α-nitrosamino-carbanions,[946] substituted pyrazines,[947] $Pr^iMeNCHS$,[948] 3α,7α-disubstituted bicyclo[3,3,1]nonanes,[949] 5α-cholestan-2α,5-episulphoxides,[950] endrin, dieldrin, and photodieldrin,[951] griseofulvin,[952] bornanes derived from α-pinene,[953] RNHCSOR,[954] some polyfunctional molecules,[955] oximes,[956] sugar oximes,[957] disubstituted

[925] P. Granger, M. M. Claudon, and J. F. Guinet, *Tetrahedron Letters*, 1971, 4167.
[926] G. R. Krow and K. C. Ramey, *Tetrahedron Letters*, 1971, 3141.
[927] C. P. Casey and R. A. Boggs, *Tetrahedron Letters*, 1971, 2455.
[928] H.-D. Scharf and M.-H. Feilen, *Tetrahedron Letters*, 1971, 2745.
[929] J. Dale and P. O. Kristiansen, *Chem. Comm.*, 1971, 670.
[930] P. O. Kristiansen and T. Ledaal, *Tetrahedron Letters*, 1971, 2817.
[931] A. H. Lewin, *Tetrahedron Letters*, 1971, 3583.
[932] L. R. Isbrandt and M. T. Rogers, *Chem. Comm.*, 1971, 1378.
[933] T. M. Ward, I. L. Allcox, and G. H. Wahl, jun., *Tetrahedron Letters*, 1971, 4421.
[934] L. Ernst, *Chem.-Ztg.*, 1971, 95, 325.
[935] T. Okutani, A. Morimoto, T. Kaneko, and K. Masuda, *Tetrahedron Letters*, 1971, 1115.
[936] H. van Brederode and W. G. B. Huysmans, *Tetrahedron Letters*, 1971, 1695.
[937] Y. Kashman and O. Awerbouch, *Tetrahedron*, 1971, **27**, 5593.
[938] J. R. Corfield and S. Trippett, *Chem. Comm.*, 1971, 721.
[939] K. D. Berlin and S. Rengaraju, *J. Org. Chem.*, 1971, **36**, 2912
[940] G. Lukacs, X. Lusinchi, P. Girard, and H. Kagan, *Bull. Soc. chim. France*, 1971, 3200.
[941] J. D. McKinney, L. H. Keith, A. Alford, and C. E. Fletcher, *Canad. J. Chem.*, 1971, **49**, 1993.
[942] D. G. Buckley, G. H. Green, E. Ritchie, and W. C. Taylor, *Chem. and Ind.*, 1971, 298.
[943] S. Itô and I. Itoh, *Tetrahedron Letters*, 1971, 2969.
[944] J. E. Herz, V. M. Rodriguez, and P. Joseph-Nathan, *Tetrahedron Letters*, 1971, 2949.
[945] M. R. Vegar and R. J. Wells, *Tetrahedron Letters*, 1971, 2847.
[946] R. R. Fraser and Y. Y. Wigfield, *Tetrahedron Letters*, 1971, 2515.
[947] A. F. Bramwell, G. Riezebos, and R. D. Wells, *Tetrahedron Letters*, 1971, 2489.
[948] W. Walter, R. F. Becker, and J. Thiem, *Tetrahedron Letters*, 1971, 1971.
[949] J. A. Peters, J. D. Remijnse, A. van der Wiele, and H. van Bekkum, *Tetrahedron Letters*, 1971, 3065.
[950] M. Kishi, K. Tori, and T. Komeno, *Tetrahedron Letters*, 1971, 3525.
[951] L. H. Keith, *Tetrahedron Letters*, 1971, 3.
[952] S. G. Levine and R. E. Hicks, *Tetrahedron Letters*, 1971, 311.
[953] K. Tori, Y. Yoshimura, and R. Muneyuki, *Tetrahedron Letters*, 1971, 333.
[954] R. A. Bauman, *Tetrahedron Letters*, 1971, 419.
[955] H. Hart and G. M. Lowe, *Tetrahedron Letters*, 1971, 625.
[956] Z. W. Wolkowski, *Tetrahedron Letters*, 1971, 825.
[957] J. M. J. Tronchet, F. Barbalat-Ray, and N. Lettong, *Helv. Chim. Acta*, 1971, **54**, 2615.

cyclopropanes,[958] (229),[959] (230),[960] (231) (X = Cl or SO_2Me),[961] (232),[962] (233) and (234),[963] (235),[964] (236),[965] (237) (^{13}C n.m.r.),[966] (238),[967] and (239).[968]

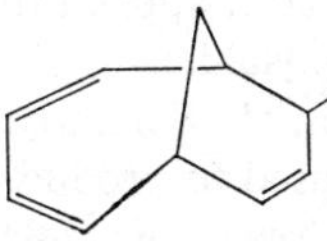

(229)

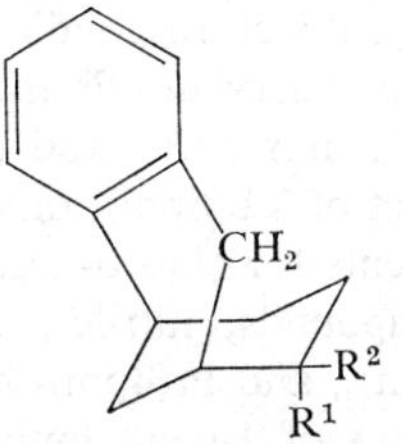

(230) R^1, R^2 = Me, CO_2Me

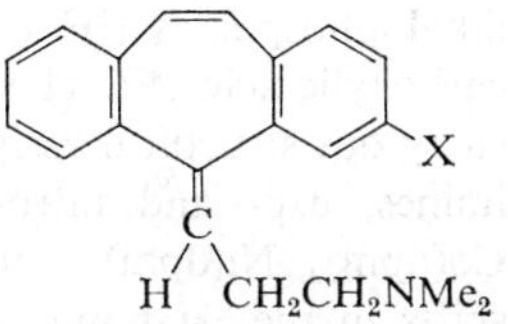

(231)

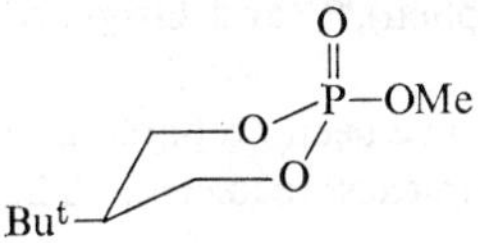

(232)

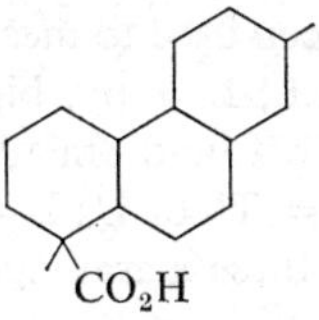

(233)

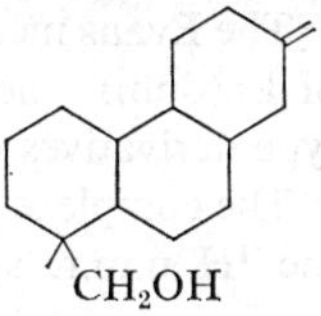

(234)

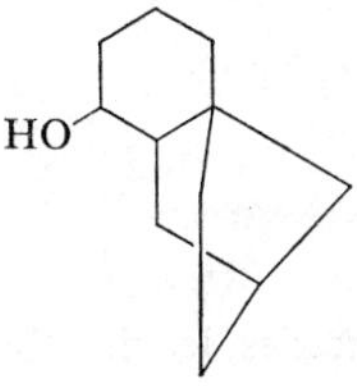

(235)

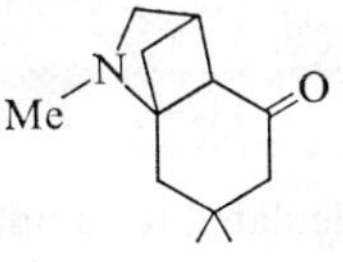

(236)

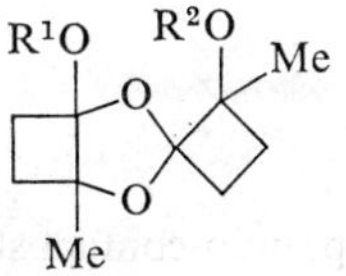

(237)

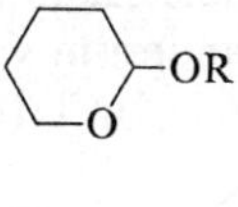

(238)

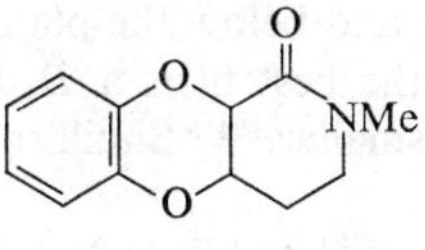

(239)

[958] M. Yoshimoto, T. Hiraoka, H. Kuwano, and Y. Kishida, *Chem. and Pharm. Bull. (Japan)*, 1971, **19**, 849.
[959] M. R. Willcott, J. F. M. Oth, J. Thio, G. Plinke, and G. Schröder, *Tetrahedron Letters*, 1971, 1579.
[960] L. F. Johnson, *Tetrahedron Letters*, 1971, 1703.
[961] D. C. Remy and W. A. Van Saun, jun., *Tetrahedron Letters*, 1971, 2463.
[962] K. C. Yee and W. G. Bentrude, *Tetrahedron Letters*, 1971, 2775.
[963] O. Achmatowicz, jun., A. Ejchart, J. Jurczak, L. Kozerski, and J. St. Pyrek, *Chem. Comm.*, 1971, 98.
[964] L. K. Lala, *J. Org. Chem.*, 1971, **36**, 2560.
[965] Y. Tamura, Y. Kita, H. Ishibashi, and M. Ikeda, *Chem. Comm.*, 1971, 1167.
[966] J. C. Duggan and W. H. Urry, *Tetrahedron Letters*, 1971, 4197.
[967] F. Lafuma and C. Quivoron, *Compt. rend.*, 1971, **272**, *C*, 2020.
[968] J. F. Caputo and A. R. Martin, *Tetrahedron Letters*, 1971, 4547.

The effect of both Eu(dpm)$_3$ and Pr(dpm)$_3$ on n.m.r. has been applied to polyglycol dimethyl ethers,[969] cyclic ketones,[970] some carbohydrates,[971] poly(ethylene oxide) and poly(methyl methacrylate),[972] $[BzR^1R^2PhP]^+$,[973] and $(CH_2)_4$PRO.[974] Similarly, the corresponding M(fod)$_3$ derivatives have been used for poly(methyl methacrylate),[975] the nitrates for $(EtO)_3P$ and alkyl phosphates (^{1}H and ^{31}P n.m.r.),[976] and the perchlorates for some carboxylic acids.[977] The effect of Yb(dpm)$_3$ on n.m.r. has been applied to carry out structural assignments for chloro- and bromo-vinylaldehydes,[978] imines, azo- and nitro-compounds, nitriles, and amides.[979] Similarly, Co(dpm)$_2$, Ni(dpm)$_2$, Eu(dpm)$_3$, and Pr(dpm)$_3$ have been used to produce shifts in the ^{1}H n.m.r. spectra of 1-methylimidazole and seven pyrazole derivatives.[980] Also Eu(dpm)$_3$, Tm(dpm)$_3$, and Pr(dpm)$_3$ have been used to produce shifts in the ^{1}H n.m.r. spectra of some furanoses.[981]

The Evans method has been used to measure the magnetic susceptibilities of Ln(dpm)$_3$ and Ln(dpm)$_3$L, L = py, bipy, or *o*-phen),[982] and Grignard type derivatives of Eu^{2+}, Yb^{2+}, and Sm^{2+}.[983]

The complexes ML$_4$ [M = Th or U; L = (240)] have been prepared and the ^{1}H n.m.r. spectra of these paramagnetic complexes recorded. The

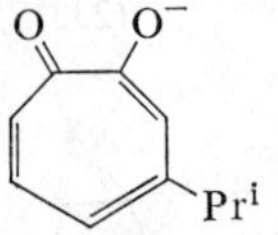

(240)

pseudo-contact shift was calculated for a number of geometries, leaving the rest of the shift attributable to a contact shift. The best fit in the uranium case in CHCl$_3$ was the D_4 square antiprism with alternating ligands distorted 8° above and below the plane perpendicular to the C_4 rotational axis. In DMSO, the best fit is a D_4 square antiprism where the contact shifts are much smaller.[984] Similarly, M(indenyl)$_3$Cl (M = Th or U),

969 A. M. Grotens, J. Smid, and E. de Boer, *Tetrahedron Letters*, 1971, 4863.
970 P .O. Kristiansen and T. Ledaal, *Tetrahedron Letters*, 1971, 4457.
971 R. F. Butterworth, A. G. Pernet, and S. Hanessian, *Canad. J. Chem.*, 1971, **49**, 981.
972 A. R. Katritzky and A. Smith, *Tetrahedron Letters*, 1971, 1765.
973 G. P. Schiemenz and H. Rast, *Tetrahedron Letters*, 1971, 4685.
974 B. D. Cuddy, K. Treon, and B. J. Walker, *Tetrahedron Letters*, 197, 4433.
975 J. E. Guillet, I. R. Peat, and W. F. Reynolds, *Tetrahedron Letters*, 1971, 3493.
976 J. K. M. Sanders and D. H. Williams, *Tetrahedron Letters*, 1971, 2813.
977 F. A. Hart, G. P. Moss, and M. L. Staniforth, *Tetrahedron Letters*, 1971, 3389.
978 C. Beauté, Z. W. Wolkowski, J. P. Merda, and D. Lelandais, *Tetrahedron Letters*, 1971, 2473.
979 C. Beauté, Z. W. Wolkowski, and N. Thoai, *Chem. Comm.*, 1971, 700.
980 R.-M. Claramunt, J. Elguero, and R. Jacquier, *Org. Magn. Resonance*, 1971, **3**, 595.
981 I. Armitage and L. D. Hall, *Canad. J. Chem.*, 1971, **49**, 2770.
982 J. Selbin, N. Ahmad, and N. Bhacca, *Inorg. Chem.*, 1971, **10**, 1383.
983 D. F. Evans, G. V. Fazakerley, and R. F. Phillips, *J. Chem. Soc.* (*A*), 1971, 1931.
984 C. J. Wiedenheft, *Inorg. Nuclear Chem. Letters*, 1971, **7**, 1023.

although it is paramagnetic, gives sharp ^1H n.m.r. spectra.[985] The ^1H n.m.r. spectra of (1,3,5,7-tetramethylcyclo-octatetraene)$_2$M (M = U or Np) have been recorded and compared with that of (cyclo-octatetraene)$_2$M. It was concluded that the shift has both dipolar and contact components with charge transfer from filled ligand π molecular orbitals to vacant f orbitals.[986]

5 Solid State N.M.R.

A number of papers have appeared which are of general interest in solid state n.m.r. A new method for determining the moments of n.m.r. absorption lines from the shape of either the free induction decay or that of the echo has been presented,[987] and a method of measuring second moments in solids by adiabatic rapid-passage n.m.r. has been described.[988] The role of spin symmetry conversion in nuclear relaxation in solids has been discussed.[989] The Debye model of rotational Brownian motion and the rotational random-jump model have been extended to allow for time fluctuations of the rotational diffusion constant and the jump rate constant. This was then applied to n.m.r. lineshapes.[990] A computational method has been devised to interpret the ^1H n.m.r. spectra of polycrystalline paramagnetic hydrates.[991]

Motion in Solids.—The ^7Li n.m.r. spectrum of Li_3P has been measured. The coupling constant for the two sites are 68.5 ± 3.0 kHz and 16.0 kHz (upper limit). These coupling constants imply a degree of covalent bonding. Linewidth changes found between − 60 and 0 °C were attributed to diffusion of the lithium ions.[992]

A ^1H n.m.r. investigation of $Sc_2(SO_4)_3,nH_2O$, $Sc(OH)SO_4,2H_2O$, and $LiSc(SO_4)_2,2H_2O$ has led to the conclusion that the water is in fact water of hydration. Linewidths were used to investigate the water motion with temperature.[993] A minimum in the temperature dependence of the ^{19}F rotating frame spin–lattice relaxation time $T_{1\rho}$ in polycrystalline UF_6 has been observed which allows the determination of the correlation time for slow reorientations of the UF_6 octahedra directly.[994]

^1H, ^2H, ^{35}Cl, and ^{133}Cs n.m.r. have been used to investigate single crystals of $CsMnCl_3,2H_2O$. An activation energy of 8.4 kcal mol^{-1} was computed

[985] P. G. Laubereau, L. Ganguly, J. H. Burns, B. M. Benjamin, J. L. Atwood, and J. Selbin, *Inorg. Chem.*, 1971, **10**, 2274.
[986] A. Streitwieser, jun., D. Dempf, G. N. La Mar, D. G. Karraker, and N. Edelstein, *J. Amer. Chem. Soc.*, 1971, **93**, 7343.
[987] A. J. Dianoux, S. Sýkora, and H. S. Gutowsky, *J. Chem. Phys.*, 1971, **55**, 4768.
[988] W. R. Janzen, *Chem. Phys. Letters*, 1971, **12**, 35.
[989] D. Wallach, *J. Chem. Phys.*, 1971, **54**, 4044.
[990] H. Sillescu, *J. Chem. Phys.*, 1971, **54**, 2110.
[991] V. Haber, *Inorg. Nuclear Chem. Letters*, 1971, **7**, 345.
[992] G. W. Ossman and A. A. Silvidi, *J. Chem. Phys.*, 1971, **54**, 979.
[993] L. N. Komissarova, V. F. Chuvaev, V. M. Shatskii, B. I. Bashkov, and T. A. Zhdanova, *Russ. J. Inorg. Chem.*, 1971, **16**, 666.
[994] R. Blinc, G. Lahajnar, and J. Slivnik, *Chem. Phys. Letters*, 1971, **11**, 344.

for H_2O or D_2O reorientation;[995] $KMnCl_3,2H_2O$ has also been investigated.[996]

Broadline ^1H n.m.r. has been used to examine the solid cyclo-octatetraene complexes (cot)$Fe(CO)_3$, (cot)$Fe_2(CO)_5$, and (cot)$[Fe(CO)_3]_2$. As in solution, (cot)$Fe(CO)_3$ was found to be fluxional with a second moment at room temperature corresponding to complete fluxionality and at low temperature to a static system. Similarly, it was found that (cot)$[Fe(CO)_3]_2$ is rigid at room temperature but (cot)$Fe_2(CO)_5$ is fluxional even at 77 K, in contrast to solution n.m.r. which yields a spectrum of the static form.[997]

^1H and ^2H n.m.r. and ^{35}Cl n.q.r. have been used to study $[H_5O_2]^+[AuCl_4]^-,2H_2O$ from 180 to 300 K. There are phase transitions at 218 K and 290 K in the proton species and at 252 K and above 300 K (not observed) in the deuteriated species. T_1 (^1H) shows a minimum at 273 K which was assigned to an interbond jump motion in the $[H_5O_2]^+$ ion. Proton exchange due to rotation of the hydroxonium ion was also observed and activation energies derived.[998]

^1H N.m.r. has been used to investigate benzene adsorbed on a commercial silica gel. It was concluded that the mean jump-length of the benzene molecules is four times the minimum proton–proton distance of approach.[999] A ^{19}F n.m.r. linewidth investigation of polycrystalline PbF_2 over the temperature range $-$ 50 to + 190 °C has revealed a line-narrowing above $-$ 10 °C which was attributed to vacancy diffusion.[1000]

^2D N.m.r. has been used to investigate deuteriated hydrazine sulphate single crystals, and ^1H n.m.r. for N_2H_5,HC_2O_4. Both sets of investigations led to the conclusion that at room temperature there is rapid rotation about the nitrogen–nitrogen bond; on cooling this rotation stops.[1001, 1002] The ^{31}P spin–lattice relaxation rates have been measured in solid white phosphorus and liquid phosphorus over the temperature range 110—400 K. The results are consistent with large-angle reorientation jumps of the P_4 molecules by 120° about their C_{3v} axes.[1003, 1004] The temperature dependence of the quadrupole splitting of the ^2H magnetic resonance spectra of KD_2AsO_4 and CsD_2AsO_4 have been studied and the electric field gradient tensors at the deuteron sites were determined. At low temperature, the hydrogen-bonded deuteron is off-centre between two oxygens but at high temperature it is exchanging between the two sites.[1005]

995 H. Bug, H. Haas, M. Fleissner, and H. Hartmann, *J. Chem. Phys.*, 1971, **55**, 280.
996 R. D. Spence, W. J. M. de Jonge, and K. V. S. R. Rao, *J. Chem. Phys.*, 1971, **54**, 3438.
997 A. J. Campbell, C. A. Fyfe, and E. Maslowsky, jun., *Chem. Comm.*, 1971, 1032.
998 D. E. O'Reilly, E. M. Peterson, C. E. Scheie, and J. M. Williams, *J. Chem. Phys.*, 1971, **55**, 5629.
999 H. Winkler, M. Nagel, D. Michel, and H. Pfeifer, *Z. phys. Chem. (Leipzig)*, 1971, **248**, 17.
1000 M. Mahajan and B. D. N. Rao, *Chem. Phys. Letters*, 1971, **10**, 29.
1001 T. P. Morton and F. L. Howell, *Mol. Phys.*, 1971, **20**, 1147.
1002 J. Tegenfeldt and I. Olovsson, *Acta Chem. Scand.*, 1971, **25**, 101.
1003 N. Boden and R. Folland, *Chem. Phys. Letters*, 1971, **10**, 167.
1004 N. Boden and R. Folland, *Mol. Phys.*, 1971, **21**, 1123.
1005 R. Blinc, J. Stepišnik, M. Jamšek-Vilfan, and S. Žumer, *J. Chem. Phys.*, 1971, **54**, 187.

The structure of water has been measured over the temperature range − 18 to 178 °C by measuring T_1 (^{2}H) for D_2O, and over the range − 14 to 180 °C by measuring T_1 (^{17}O) for $H_2{}^{17}O$. The results were interpreted in terms of an equilibrium between water forming the lattice and free water. Activation energies, stability constants, ΔH, and ΔS were determined.[1006] The proton spin–lattice relaxation time has been investigated for a single crystal of the ferroelectric $NaH_3(SeO_3)_2$ and polycrystalline $KH_3(SeO_3)_2$. The results show no evidence of hydrogen site rearrangements in the temperature range of the reported dielectric anomaly.[1007]

Structure of Solids.—^{1}H N.m.r. has been used to investigate the state of water in some cation exchange resins. The water signal shows two resonances due to water in the pores of the ionite and in the gel portion.[1008] Similarly, ^{1}H and ^{7}Li n.m.r. have been used to investigate the state of interlayer water and exchange cations of lithium in vermiculite. It was concluded that the Li^+ ions are octahedrally surrounded by oxygen.[1009] The hydrated porosity of macroreticular cation exchange resins have been investigated by means of ^{1}H n.m.r.,[1010] and ^{1}H and ^{23}Na n.m.r. have been used to investigate a synthetic zeolite.[1011]

The low-temperature wideline ^{1}H n.m.r. spectra of ethylene adsorbed on a cation exchange resin in its K^+, Cd^{2+}, and Ag^+ forms has been measured. It was found that the ethylene resonance is broadened increasingly in the order $K^+ < Cd^{2+} < Ag^+$.[1012] Similarly, ^{1}H n.m.r. has been used to investigate the adsorption of ethylene on the sodium and calcium forms of Zeolite Y. At low ethylene concentration on the calcium form a broad resonance occurs, and as the concentration increases the line sharpens and then extra resonances due to polymers occur.[1013]

From ^{1}H and ^{23}Na n.m.r. measurements on Linde 13X and 13Y zeolites it has been concluded that when H_2S is adsorbed it is broken down to H^+ and HS^-.[1014] ^{7}Li, ^{23}Na, and ^{27}Al n.m.r. have been used to study specific atom environment and distribution in γ-faujasite ($Na_{27}Si_{135}Al_{57}O_{384}$,-$n$$H_2O$; *n ca.* 260) and its lithium analogue.[1015]

The ^{1}H n.m.r. chemical shift of H_2O in a single crystal of $K_2C_2O_4,H_2O$ has been carried out as a function of angle and the results are in good agreement with neutron diffraction.[1016] By the use of ^{1}H, ^{23}Na, and ^{81}Br

1006 J. C. Hindman, A. J. Zielen, Z. Svirmickas, and M. Wood, *J. Chem. Phys.*, 1971, **54**, 621.
1007 A. A. Silvidi and D. T. Workman, *J. Chem. Phys.*, 1971, **44**, 4672.
1008 N. I. Nikolaev, G. A. Grigor'eva, N. N. Shapet'ko, and V. A. Arkhipov, *Doklady Phys. Chem.*, 1971, **198**, 408.
1009 A. G. Brekhunets, V. V. Mank, F. D. Ovcharenko, Z. É. Suyunova, Yu. I. Tarasevich, L. M. Shapinskaya, and Yu. V. Shulepov, *Doklady Phys. Chem.*, 1970, **192**, 409.
1010 L. S. Frankel, *Analyt. Chem.*, 1971, **43**, 1506.
1011 H. Lechert and H. J. Henning, *Ber. Bunsengesellschaft Phys. Chem.*, 1971, **75**, 1127.
1012 G. M. Muha, *J. Chem. Phys.*, 1971, 55, 467.
1013 T. A. Egerton and R. D. Green, *Trans. Faraday Soc.*, 1971, **67**, 2699.
1014 H. Lechert and H. J. Hennig, *Z. phys. Chem. (Frankfurt)*, 1971, **76**, 319.
1015 E. E. Genser, *J. Chem. Phys.*, 1971, **54**, 4612.
1016 J.-M. Dereppe, Y. V. Honarker, and M. V. Meerssche, *J. Chem. Phys.*, 1971, **68**, 1615.

n.m.r., the dehydration of polycrystalline $NaBr,2H_2O$ has been studied. As the sample is warmed from room temperature, water is evolved more and more freely, until at 50.8 °C complete dehydration occurs. It was found that the sodium and bromine resonances are sensitive indicators of the anhydrous cubic phase. The results are consistent with the classical mechanism of vacuum dehydration proceeding inwards from the surface of the crystallite.[1017]

The ^{19}F n.m.r. Fourier transform spectrum of CaF_2 powder, NaF single crystal, and C_6F_6 show ^{19}F signals due to CaF_2 and C_6F_6 but none due to NaF, which is too broad. However, if the ^{23}Na nuclei are also pulsed, then the ^{19}F n.m.r. signal of NaF is observed. There was found to be a surprisingly large difference of 114 ± 6 p.p.m. separation from CaF_2 although both are considered to be ionic solids.[1018] In the case of CaF_2, if the ^{43}Ca resonance is irradiated, the ^{19}F linewidth is reduced to 130 Hz.[1019] The ^{19}F chemical shifts of some metal fluorides have been interpreted in terms of the Kondo–Yamashita model. Overlap integrals were calculated and the influence of partial covalency on ^{19}F n.m.r. chemical shifts were discussed.[1020] The measurement of the ^{19}F n.m.r. frequency as a function of orientation for a single crystal of fluorapatite $Ca_5F(PO_4)_3$ about two perpendicular crystal axes has allowed the determination of all the principal components of the nuclear screening tensor, which was found to be axially symmetric with $\sigma_{\parallel} - \sigma_{\perp} = -84$ p.p.m. The negative sign is consistent with the known crystal structure.[1021]

The quadrupole-perturbed n.m.r. spectra of ^9Be in a deuteriated trigycine fluoroberyllate single crystal have been studied as a function of temperature and crystal orientation, and the electric field gradient tensors at the beryllium sites have been determined in the para- and ferro-electric phases.[1022]

The ^{19}F paramagnetic shift has been measured in polycrystalline HoF_3, ErF_3, and YbF_3 as a function of temperature. In all cases two distinct lines are resolved corresponding to two inequivalent fluorine sites.[1023] The ^{17}O n.m.r. spectra of ^{17}O enriched UO_2, $BiUO_4$, and KUO_3 have been observed. It was concluded that the fractional *s*-character in these compounds is small and that there is a negative spin density on the ^{17}O nuclei. Apparently the major contribution to the hyperfine interaction arises from the involvement of excited metal orbitals and/or from exchange polarization of core electrons. Anisotropic interactions in $BiUO_4$ and KUO_3 are predominantly π in character.[1024] The ^{19}F n.m.r. spectra of polycrystalline Na_3PF_6 has been studied. The shift data are consistent with a system

1017 I. D. Campbell and D. J. Mackey, *Austral. J. Chem.*, 1971, **24**, 45.
1018 M. Mehring, A. Pines, W.-K. Rhim, and J. S. Waugh, *J. Chem. Phys.*, 1971, **54**, 3239.
1019 H. E. Bleich and A. G. Redfield, *J. Chem. Phys.*, 1971, **55**, 5405.
1020 V. M. Bouznik and L. M. Avkhutsky, *J. Magn. Resonance*, 1971, **5**, 63
1021 J. L. Carolan, *Chem. Phys. Letters*, 1971, **12**, 389.
1022 R. Blinc, J. Slak, and J. Stepišnik, *J. Chem. Phys.*, 1971, **55**, 4848
1023 S. L. Carr and W. G. Moulton, *J. Magn. Resonance*, 1971, **4**, 400.
1024 M. P. Eastman, H. G. Hecht, and W. B. Lewis, *J. Chem. Phys.*, 1971, **54**, 4141.

containing one 5*f* electron as the unpaired spin. The ^{19}F spectra and the temperature dependence of the field-dependent second moment require the existence of inequivalent ^{19}F nuclei, which is not consistent with the published crystal structure.[1025]

A linear relationship had been found between the deuterium quadrupole coupling constant and the force constant of the X—D bond, but this had not been tested when X is a transition metal. Accordingly the deuterium n.m.r. spectrum of $(Cp)_2MD_2$ (M = Mo or W) has been measured in the solid state, and when the quadrupole coupling constant is plotted against the force constant it does indeed fall on the same straight line.[1026] The ^1H n.m.r. spectrum of solid $Ga(OH)WO_4,3.5H_2O$ shows two resonances, one due to OH and the other due to H_2O.[1027]

Single crystals of a number of manganese carbonyls, $Mn(CO)_5X$ and $(Cp)Mn(CO)_3$, were studied by broadline ^{35}Cl, ^{81}Br, ^{127}I, and ^{55}Mn n.m.r. at room temperature and quadrupole coupling parameters derived. The results were compared with recent MO calculations and considerable deviations were found.[1028] The single-crystal broad-line ^{59}Co n.m.r. spectrum of $Co_2(CO)_8$ has been studied at liquid-nitrogen temperature. The *z*-axis of the electric field gradient tensor was within 6° of the terminal carbonyl opposite the missing bridging carbonyl. This direction agrees well with the value predicted by considering $Co_2(CO)_8$ to have both threefold and fourfold distortions at the cobalt site, but disagrees with a bent bond interpretation for $Co_2(CO)_8$.[1029] High-speed specimen rotation has been used to obtain precise measurement of the relative chemical shifts of ^{63}Cu and ^{65}Cu in polycrystalline CuCl, CuBr, and CuI; no isotope effect was detected.[1030]

The ^{11}B n.m.r. spectra of β-rhombohedral boron and boron carbide have been investigated at room temperature. The results lend support to a model of the boron carbide crystal which assigns a boron atom to the central position of the three-atom chain and a carbon atom to the icosahedron.[1031] Most earlier workers have analysed the ^{11}B n.m.r. measurements in borate powder samples by assuming an axially symmetric electric field. Errors associated with this approach have been analysed.[1032] The ^{11}B n.m.r. spectrum of polycrystalline BF_3 has been fitted by computer simulation with a boron–fluorine interatomic distance of 1.35 ± 0.03 Å.[1033] ^{11}B and ^{27}Al n.m.r. have been used to investigate AlB_2.[1034]

[1025] E. Fukushima and H. G. Hecht, *J. Chem. Phys.*, 1971, **54**, 4341.
[1026] I. Y. Wei and B. M. Fung, *J. Chem. Phys.*, 1971, **55**, 1486.
[1027] M. V. Mokhosoev and A. I. Gruba, *Russ. J. Inorg. Chem.*, 1971, **16**, 508.
[1028] H. W. Spiess and R. K. Sheline, *J. Chem. Phys.*, 1971, **54**, 1099.
[1029] E. S. Mooberry, M. Pupp, J. L. Slater, and R. K. Sheline, *J. Chem. Phys.*, 1971, **55**, 3655.
[1030] E. R. Andrew, J. L. Carolan, and P. J. Randall, *Chem. Phys. Letters*, 1971, **11**, 298.
[1031] T. V. Hynes and M. N. Alexander, *J. Chem. Phys.*, 1971, **54**, 5296.
[1032] H. M. Kriz and P. J. Bray, *J. Magn. Resonance*, 1971, **4**, 69.
[1033] H. M. Kriz and P. C. Taylor, *J. Chem. Phys.*, 1971, **55**, 2601.
[1034] J. P. Kopp and R. G. Barnes, *J. Chem. Phys.*, 1971, **54**, 1840.

^{1}H N.m.r. has been used to investigate the chemisorption of water on to silica,[1035] and the second moment of the ^{1}H n.m.r. lines of some KH_2PO_4-type crystals have been calculated.[1036] From the analysis of the ^{19}F n.m.r. spectra of polycrystalline $KBiF_6$ it has been found that $^1J(^{209}Bi–^{19}F)$ is $\mp$ 2.7 ± 0.3 kHz and the ^{209}Bi quadrupole coupling constant is ± 13.2 ± 12 MHz, the upper signs being favoured.[1037]

From ^{1}H and ^{2}D n.m.r. measurements on $[D_3O]^+[ClO_4]^-$ and $[H_3O]^+[ClO_4]^-$ over the temperature range 100—300 K, it has been concluded that there is a DOD angle of 118.5 ± 0.7° and an OH distance of 1.01 ± 0.02 Å; ^{35}Cl n.m.r. measurements were also reported.[1038] However, other workers have concluded that HOH angle is 102° and the H—H distance is 1.647 Å in $[H_3O]^+[ClO_4]^-$. The barrier to free rotation and the libration frequency were also calculated.[1039]

The ^{19}F n.m.r. spectrum of elemental fluorine has been studied over the temperature range 4.2—105 K.[1040]

6 Group III Compounds

Information concerning complexes of these elements can be found at the following sources: B_5H_9,[89] RB_5H_8,[90, 94] $ArNH_2O,BF_3$,[91] R_3P,BX_3,[92, 93] R_2PPF_2,BH_3,[108] BXYZ,[131] $[BWXYZ]^-$,[132] $(Cp)M(CO)_nBH_2$ and related compounds,[259] $H_2B(N_2C_3H_3)_2Mn(CO)_2(PR_3)_2$,[266] $PhCH_2OB(C_2H_3)_2$,Fe-$(CO)_3$,[291] $B(N_2C_3H_3)_3Ru(CO)_2X$,[321] $[(Ph_3P)_2NiBPh_2]_2$,[398] $[Cl_3PNPCl_3]^+$ $[BCl_4]^-$ and $Cl_3P{=}N{-}PCl_2OBCl_3$,[1203] $[(SCN)_3PNP(NCS)_3][N(NCS)_4]$,[1205] $[BF_4]^-$,[1269] $(Cp)_2MH_2AlR_3$ (M = Mo or W),[221] $R_2AlW(CO)_3(Cp)$,[222] and $LiAlMe_4$.[467]

As was found last year, the majority of the work reported in this section concerns polyborane complexes.

Boron Hydrides and Carboranes.—In this section the ordering is in increasing number of boron atoms in the complex, from 1 to 34.

The ^{11}B n.m.r. spectrum of Bu^n_3B in the presence of $LiBH_4$ or $NaBH_4$ in THF shows considerable temperature dependence for both the Bu^n_3B and BH_4^- resonances, while individually no dependence was found; therefore species such as $H_3B\cdots H\cdots BBu^n_3$ were postulated.[1041] Full ^{1}H n.m.r data have been reported for R_nPH_{3-n},BH_3 (R = Me or Et; n = 1 or 2) and the corresponding BD_3 compounds and these have been used to confirm that there is no hydrogen–deuterium exchange.[1042] Variable-

[1035] J. Demarquay, J. Fraissard, and B. Imelik, *Compt. rend.* 1971, **273**, *C*, 1405.
[1036] G. J. Adriaenssens and J. L. Bjorkstam, *J. Chem. Phys.*, 1971, **55**, 1137.
[1037] E. Fukushima, *J. Chem. Phys.*, 1971, **55**, 2463.
[1038] D. E. O'Reilly, E. M. Peterson, and J. M. Williams, *J. Chem. Phys.*, 1971, **54**, 96,
[1039] M.-H. Cance and A. Potier, *J. chim. Phys.*, 1971, **68**, 941.
[1040] D. E. O'Reilly, E. M. Peterson, D. L. Hogenboom, and C. E. Scheie, *J. Chem. Phys.*, 1971, **54**, 4194.
[1041] V. S. Bogdanov, P. M. Aronovich, A. D. Naumov, and B. M. Mikhailov, *J. Gen. Chem.* (U.S.S.R.), 1971, **41**, 1069.
[1042] J. Davis and J. E. Drake, *J. Chem. Soc.* (*A*), 1971, 2094.

temperature ^1H n.m.r. data on the adduct (241) show that only one conformer exists in solution, as the spectrum is invariant with temperature.[1043] Me_2S,BH_3 is a useful source of BH_3 and reacts with $Ph_2(MeO)P$ to yield $Ph_2(MeO)PBH_3$. Although signals due to Ph and OMe were detected in the

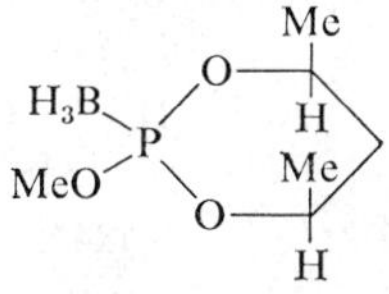

(241)

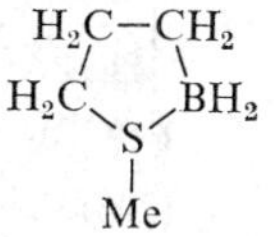

(242)

^1H n.m.r. spectrum, no signals due to BH_3 could be found.[1044] H_3P,BH_3 reacts with $LiBu^n$ to yield $[H_2PBH_3]^- Li^+$. The ^1H n.m.r. spectrum shows not only direct coupling but ^3J(H–H) as well.[1045] The ^1H n.m.r. spectrum of neat $MeS(CH_2)_3BH_2$ has been interpreted as indicating an equilibrium between this and the cyclic internal adduct (242). Data on $MeS(CH_2)_3$-$B(OH)_2$ were also reported.[1046]

The reaction between $NaMe_2N,BH_3,0.5(C_4H_8O_2)$ and diborane in monoglyme was followed by ^{11}B n.m.r., and Me_2NBH_2, $[Me_2N(BH_3)_2]^-$, and $[BH_4]^-$ were identified. Similarly, the reaction between $[Me_2N,BH_3]^-$ and μ-$Me_2NB_2H_5$ also yields $[Me_2N(BH_3)_2]^-$.[1047] ^1H and ^{11}B n.m.r. have been used to show that, unlike the reaction between $LiNH_2$ and B_2H_6 to give $LiBH_4$ and $(NH_2BH_2)_n$, $LiMe_2P$ reacts with B_2H_6 to yield $LiMe_2P$-$(BH_3)_2$.[1048] Me_2NPF_2 and F_2PH form adducts with triborane(7) and diborane(4) and ^1H, ^{11}B, and ^{19}F n.m.r. data were reported. Unlike the adducts of H^-, Me_3N, and THF, where the protons are equivalent, Me_2NPF_2,B_3H_7 and F_2PH,B_3H_7 show a fixity of structure in the symmetrical 2102 mode.[1049] Treatment of H_3P,BH_3 with $[BH_3]^-$ yields $[H_2P(BH_3)_2]^-$, which reacts with $LiBu^n$ and B_2H_6 to give $[HP(BH_3)_3]^{2-}$ and $[P(BH_3)_4]^{3-}$ successively. ^1H N.m.r. shows that the increased negative charge on the complex causes increased shielding of the protons bonded to phosphorus but the BH_3 proton shift alters only slightly.[1050] ^{11}B N.m.r. has been used to show that $[B_3H_8]^-$ and $[B_9H_{14}]^-$ anions are formed when $LiBH_4$ is treated with B_5H_9 at low temperatures.[1051]

The ^{11}B n.m.r. spectrum of $(\pi$-$C_2B_4H_6)Fe(CO)_3$ shows three doublets in the ratio 1 : 2 : 1, which is consistent with structure (243). However, the

[1043] J. Rodgers, D. W. White, and J. G. Verkade, *J. Chem. Soc.* (*A*), 1971, 77.
[1044] J. Beres, A. Dodds, A. J. Morabito, and R. M. Adams, *Inorg. Chem.*, 1971, **10**, 2072.
[1045] E. Mayer, *Inorg. Chem.*, 1971, **10**, 2259.
[1046] R. A. Braun, D. C. Brown, and R. M. Adams, *J. Amer. Chem. Soc.*, 1971, **93**, 2823.
[1047] P. C. Keller, *Inorg. Chem.*, 1971, **10**, 1528.
[1048] L. D. Schwartz and P. C. Keller, *Inorg. Chem.*, 1971, **10**, 645.
[1049] E. R. Lory and D. M. Ritter, *Inorg. Chem.*, 1971, **10**, 939.
[1050] E. Mayer, *Angew. Chem. Internat. Edn.*, 1971, **10**, 416.
[1051] C. G. Savory and M. G. H. Wallbridge, *Inorg. Chem.*, 1971, **10**, 419.

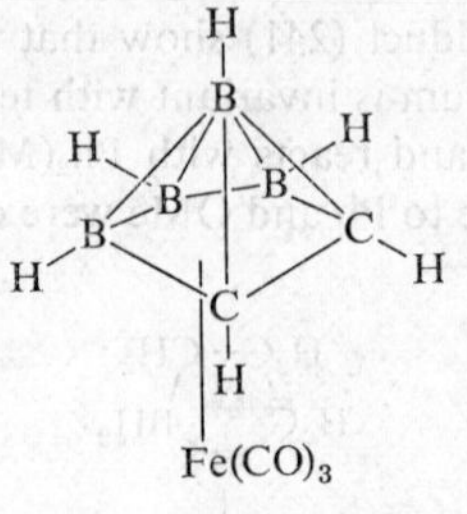

(243)

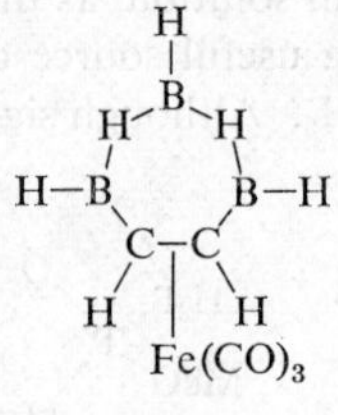

(244)

^{11}B n.m.r. spectrum of $(\pi\text{-}C_2B_3H_7)Fe(CO)_3$ shows only one broad, slightly asymmetric doublet and structure (244) was suggested.[1052] The compound 2,3-μ-$Me_3MC_2Me_2B_4H_5$ (M = Si or Ge) shows four doublets in the ^{11}B n.m.r. spectrum which is consistent with structure (245).[1053] ^{11}B N.m.r.

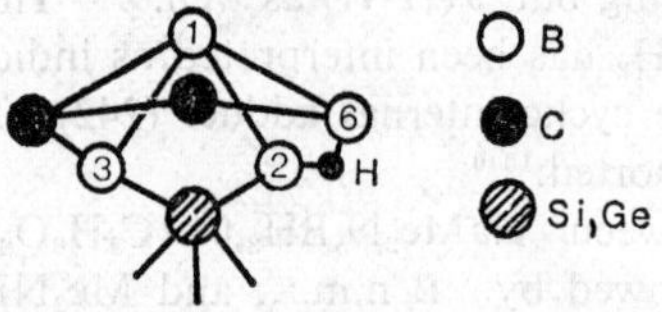

(245)

has been used to show that in the chloro-, bromo-, and iodo-derivatives of the dicarba-*nido*-hexaboranes, $C_2MeB_4H_7$ and $C_2Me_2B_4H_6$ (246), substitution occurs at the B(3) atom.[1054] Treatment of B_4H_{10} with C_2D_2 yields partially deuteriated 2-methyl-, 2,3-dimethyl-, and 2,4-dimethyl-tricarbahexaborane(7) (247). I.r., ^1H n.m.r., and mass spectrometry have been used to show that (*a*) the cage carbon atoms are all deuteriated, (*b*) the methyl groups contain both hydrogen and deuterium, (*c*) all three carboranes contain D_3 and D_4 species, and (*d*) the relative abundance of the D_4 species

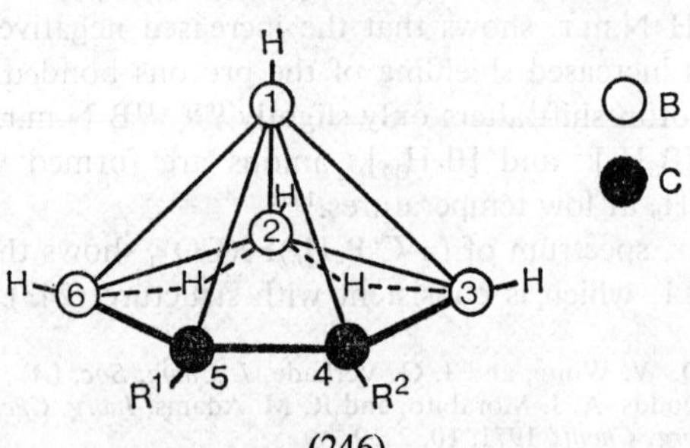

(246)

[1052] R. N. Grimes, *J. Amer. Chem. Soc.*, 1971, **93**, 261.
[1053] C. G. Savory and M. G. H. Wallbridge, *Chem. Comm.*, 1971, 622.
[1054] J. S. McAvoy, C. G. Savory, and M. G. H. Wallbridge, *J. Chem. Soc.* (*A*), 1971, 3038.

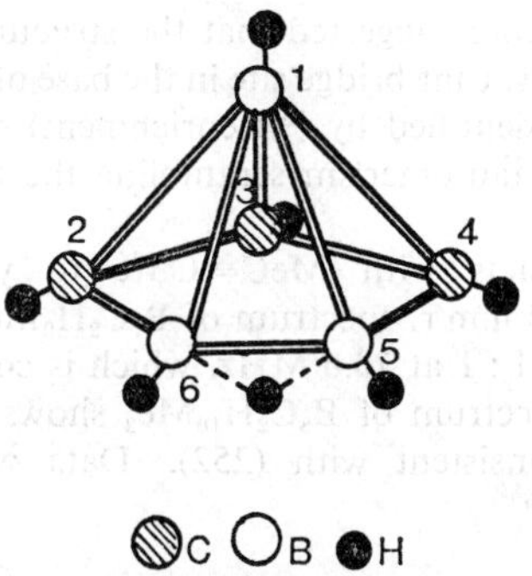

(247)

as compared to the D_3 species is four times as large when a tenfold excess of C_2D_2 rather than equimolar quantities of C_2D_2 and B_4H_{10} are used.[1055]

^{11}B N.m.r. has been used to follow the solvolysis of 1-BrB_5H_8 in various organic bases. It was found that there are competing reactions, one being the reaction of 1-BrB_5H_8 to give degradative products, and the other being rearrangement to 2-BrB_5H_8 and subsequent reaction to yield degradative products.[1056] Pyrolysis of $SiH_3,C_2B_4H_7$ yields $C_3B_5H_7$; this compound shows three ^{11}B signals as doublets in the ratio 2 : 1 : 2 and four ^1H signals in the ratio 2 : 2 : 1 : 2, the first signal being a singlet, *i.e.* CH, while the other signals show ^{11}B coupling. Thus one carbon atom is unsubstituted and it was proposed that structures (248) and (249) were in rapid equilibrium, averaging to (250).[1057] The ^{11}B n.m.r. spectrum of KB_5H_{12} consists of two apparent singlets in the ratio 3.9 : 1 and it was suggested that the structure is a square-pyramidal arrangement of boron atoms on the n.m.r. time scale. The ^{11}B n.m.r. spectrum of LiB_6H_{12} appears to consist of a 2 : 1 : 1 : 2

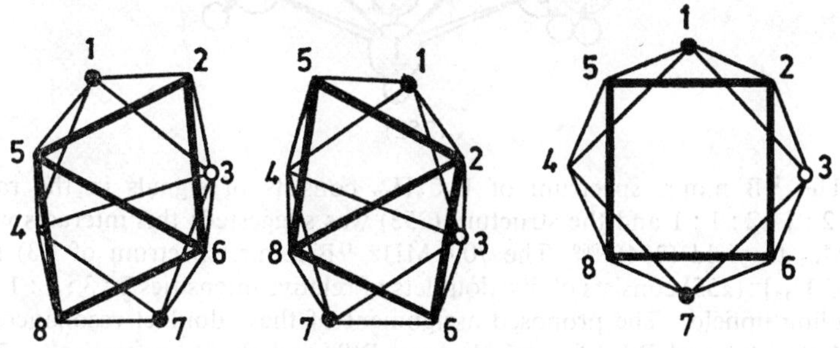

(248) (249) (250)

●, CH groups; ○, unsubstituted carbon atom; the remaining vertices are occupied by BH groups

[1055] D. A. Franz and R. N. Grimes, *J. Amer. Chem. Soc.*, 1971, **93**, 387.
[1056] C. B. Murphy, jun. and R. E. Enroine, *J. Inorg. Nuclear Chem.*, 1971, **33**, 584.
[1057] M. L. Thompson and R. N. Grimes, *J. Amer. Chem. Soc.*, 1971, **93**, 6677.

pattern. It was therefore suggested that the structure is one where a BH_3 group has entered the vacant bridge site in the base of the $[B_5H_8]^-$ unit. The added boron atom (identified by ^{10}B enrichment) shows no ^1H coupling, which suggests rapid tautomerism scrambling the terminal hydrogens on this boron.[1058]

Octaborane(12) reacts with MeC≡CMe to yield $B_7C_2H_9Me_2$ and $B_8C_2H_{10}Me_2$. The ^{11}B n.m.r. spectrum of $B_7C_2H_9Me_2$ shows resonances in the ratio 1 : 1 : 1 : 2 : 1 : 1 at 70.6 MHz, which is consistent with structure (251), whereas the spectrum of $B_8C_2H_{10}Me_2$ shows a 1 : 1 : 1 : 1 : 2 : 1 : 1 pattern, which is consistent with (252). Data were also reported for 4,5-$Me_2B_5C_2H_5Me_2$.[1059]

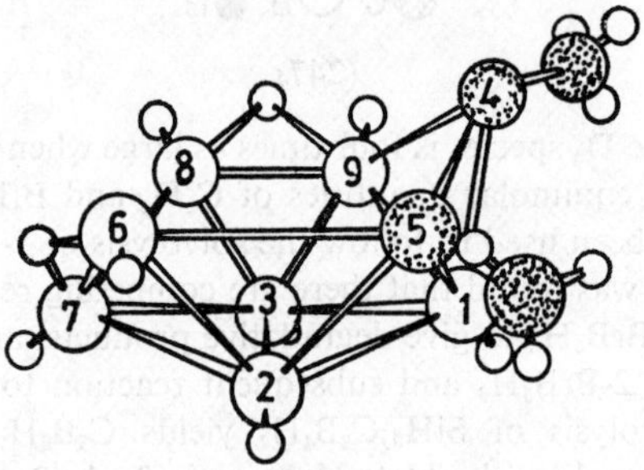

(251)

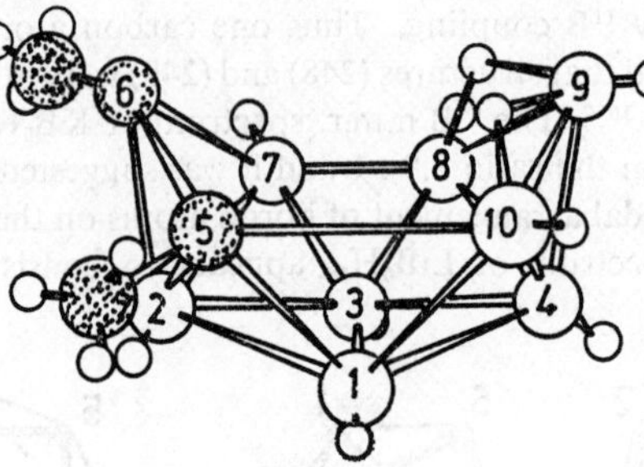

(252)

The ^{11}B n.m.r. spectrum of $B_8C_2H_{12}$ consists of signals in the ratio 1 : 2 : 1 : 2 : 1 : 1 and the structure (253) was suggested; this interacts with NMe_3 to yield (254).[1060] The 70.6 MHz ^{11}B n.m.r. spectrum of [(3)-1,2-$B_9C_2H_{12}]^-$ (255) consists of five doublets of relative intensities 2 : 3 : 2 : 1 : 1, reading upfield. The proposed assignment of these doublet resonances is B(4,7); B(6) and B(9,12); B(5,11); and B(8) and B(10), respectively. The ^{11}B n.m.r. spectrum of [{(3)-1,2-$B_9C_2H_{11}\}_2Co]^-$ contains five doublets of relative intensities 1 : 1 : 4 : 2 : 1, reading upfield. The proposed assign-

[1058] H. D. Johnson, jun., and S. G. Shore, *J. Amer. Chem. Soc.*, 1971, **93**, 3798.
[1059] R. R. Rietz and R. Schaeffer, *J. Amer. Chem. Soc.*, 1971, **93**, 1263.
[1060] P. M. Garrett, G. S. Ditta, and M. F. Hawthorne, *J. Amer. Chem. Soc.*, 1971, **93**, 1265.

ment of these doublets is B(8); B(10); B(9,12) and B(4,7); B(5,11); and B(6), respectively.[1061] $[o\text{-}C_2B_9H_{12}]^-$ is oxidized by Fe^{3+} to yield what is thought to be 5,6-dicarba-*nido*-decaborane (256) on the basis of ^1H and

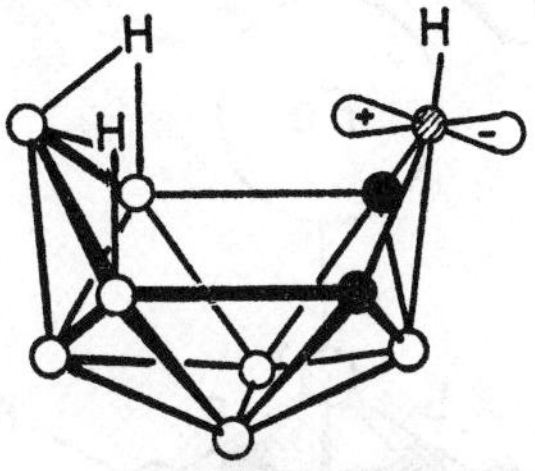

(253)

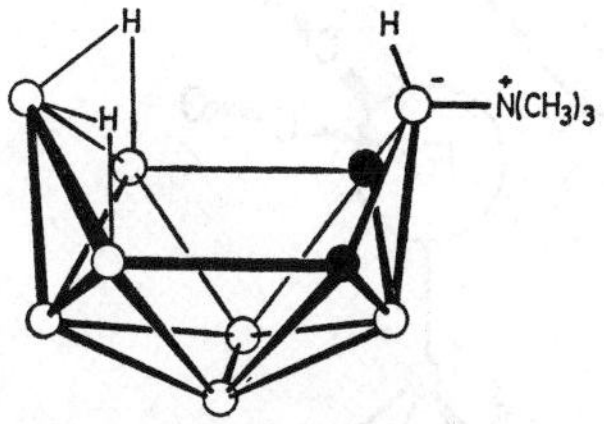

(254)

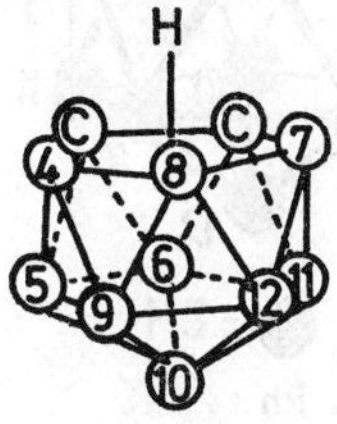

(255)

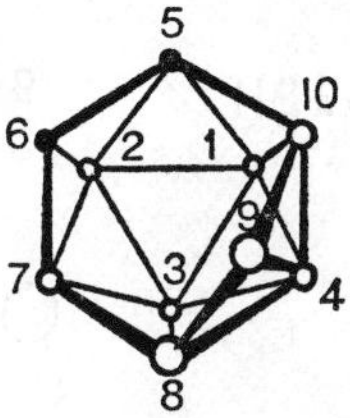

(256)

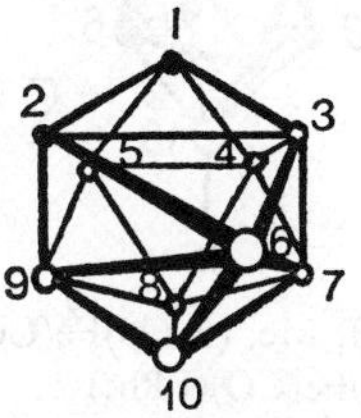

(257)

^{11}B n.m.r. This compound reacts with R_3N,BH_3 to yield 1,2-dicarba-*closo*-decaborane (257).[1062] The ^{11}B n.m.r. spectrum of $Me_2MB_{10}H_{12}$ (M = Ge or Sn) consists of five doublets of equal intensity, which is consistent with (258).[1063] The compounds (259), (260), (261), and (262) have been prepared and characterized by ^1H and ^{11}B n.m.r. For example, in the cases of (259)

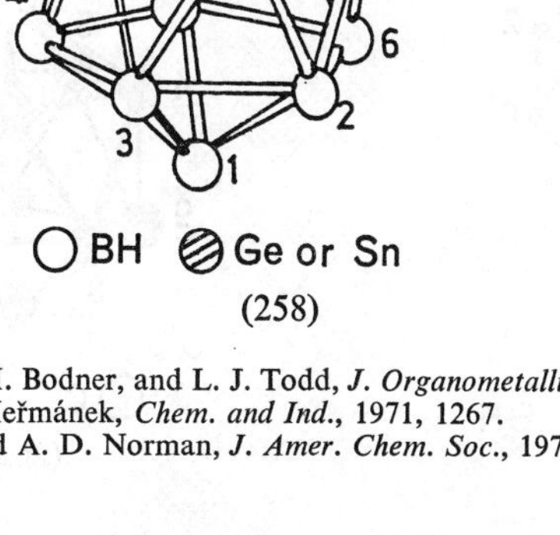

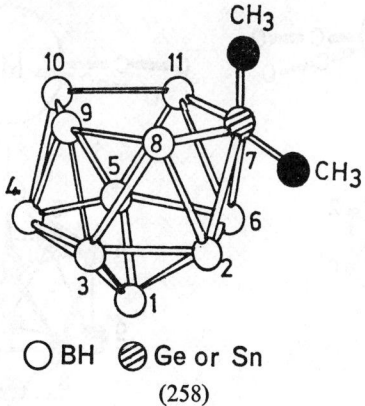

(258)

[1061] A. R. Siedle, G. M. Bodner, and L. J. Todd, *J. Organometallic Chem.*, 1971, **33**, 137.
[1062] J. Plešek and S. Heřmánek, *Chem. and Ind.*, 1971, 1267.
[1063] R. E. Loffredo and A. D. Norman, *J. Amer. Chem. Soc.*, 1971, **93**, 5587.

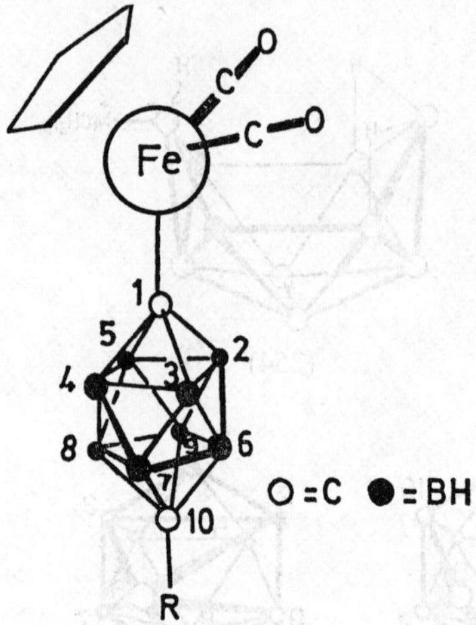

R = H, Me, (π-Cp)Fe(CO)$_2$, or (π-Cp)Fe(CO)(PPh$_3$)

(259)

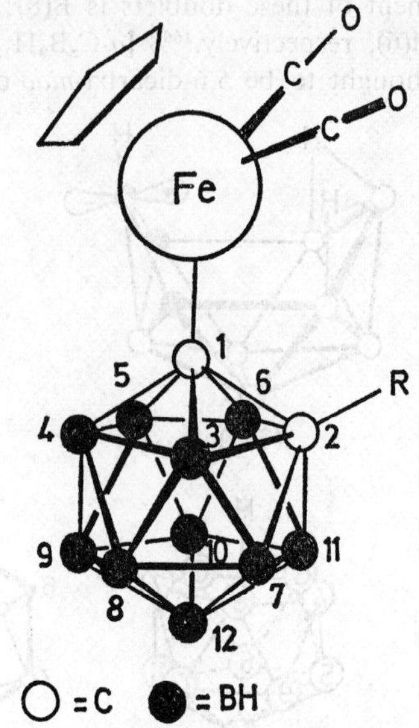

R = Me or Ph

(260)

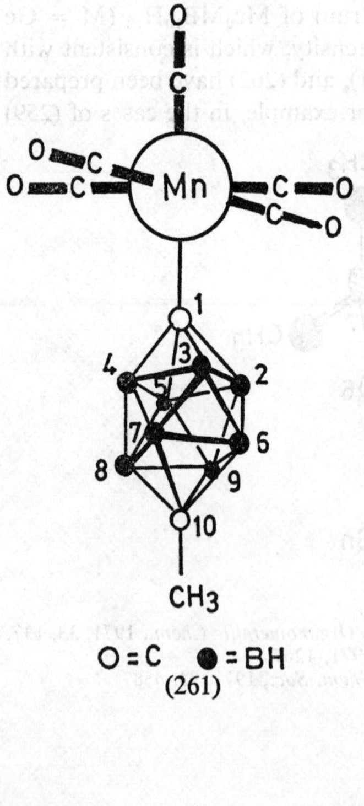

(261)

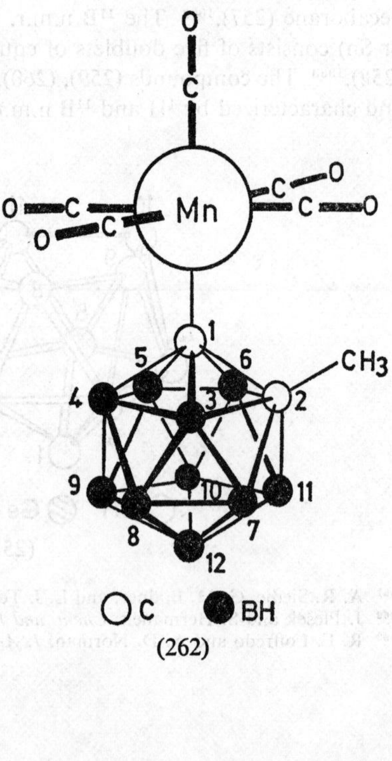

(262)

and (261) only two ^{11}B doublets were observed.[1064] Treatment of 1,6-$B_8C_2H_{10}$ in THF under reflux with two equivalents of $Na^+ [C_{10}H_8]^-$, followed by addition of a fourfold excess of $Na^+ [Cp]^-$ and $CoCl_2$ yielded a mixture of products. $(Cp)Co(B_8C_2H_{10})$ has three doublets in the ^{11}B n.m.r. spectrum in intensity ratio 1 : 2 : 1 and is assigned structure (263).

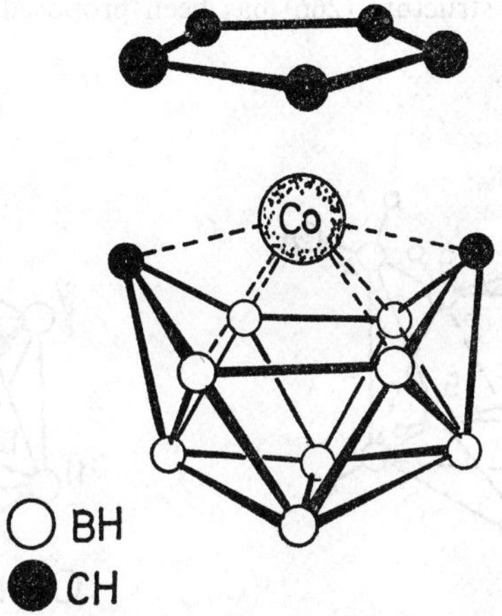

(263)

The other product, $[Co(B_8C_2H_{10})_2]^- Cs^+$, has a similar ^{11}B n.m.r. spectrum.[1065]

The deuterium exchange reactions of $[(3)\text{-}1,2\text{-}B_9C_2H_{12}]^-$, $[(3)\text{-}1,7\text{-}B_9C_2H_{12}]^-$, and $[(3)\text{-}1,2\text{-}B_9C_2H_{13}]^-$ [see (264) for numbering] have been

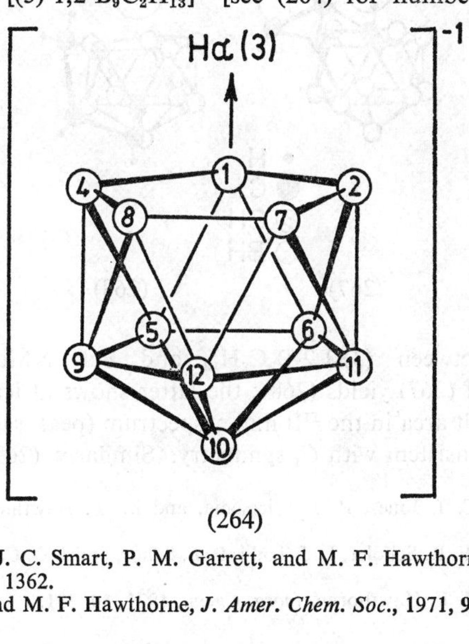

(264)

[1064] D. A. Owen, J. C. Smart, P. M. Garrett, and M. F. Hawthorne, *J. Amer. Chem. Soc.*, 1971, **93**, 1362.

[1065] W. J. Evans and M. F. Hawthorne, *J. Amer. Chem. Soc.*, 1971, **93**, 3063.

studied by ^{11}B and ^{13}C n.m.r. and i.r. spectroscopy, and full assignment of the resonances made.[1066] From an examination of the ^{11}B n.m.r. spectrum of $B_9H_{13}L$ (265), $B_9H_9D_4(SMe_2)$ (1,2,3,7-D), $B_9H_{12}Br(SMe_2)$, and 7-$MeOB_9H_{12}(SMe_2)$ at 32.1, 70.56, and 80.2 MHz it has proved possible to carry out an assignment of many boron resonances.[1067] On the basis of ^{11}B and ^{1}H n.m.r., the structure (266) has been proposed for the product of

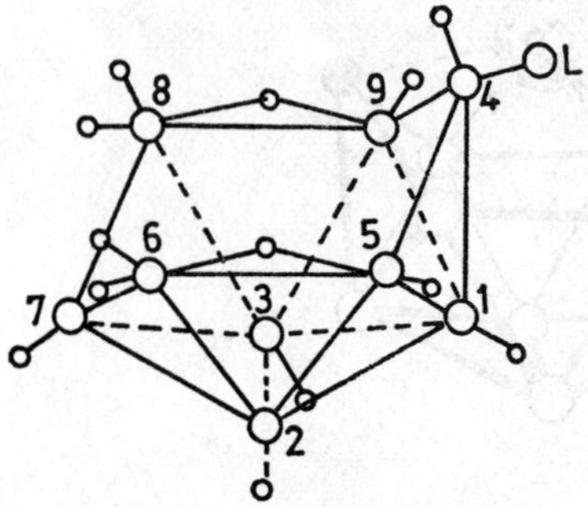

(265)

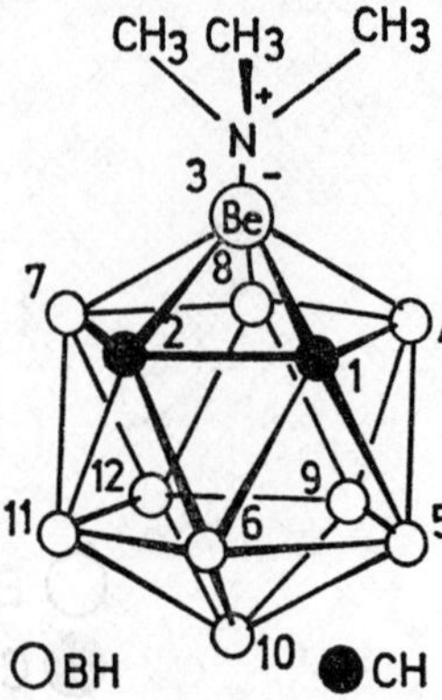

(266)

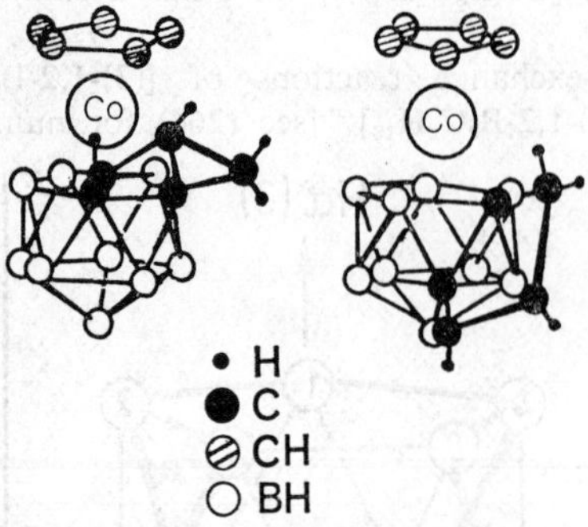

(267) (268)

the reaction between (3)-1,2-$B_9C_2H_{13}$ and BeR_2,NMe_3.[1068] Thermal rearrangement of (267) yields (268); the latter shows at least four distinct resonances of unit area in the ^{11}B n.m.r. spectrum (peak ratios 1 : 3 : 1 : 1 : 1 : 2) which is consistent with C_1 symmetry. Similarly, (269) yields six new

[1066] D. V. Howe, C. J. Jones, R. J. Wiersema, and M. F. Hawthorne, *Inorg. Chem.*, 1971, **10**, 2516.
[1067] G. M. Bodner, F. R. Scholer, L. J. Todd, L. E. Senor, and J. C. Carter, *Inorg. Chem.*, 1971, **10**, 942.
[1068] G. Popp and M. F. Hawthorne, *Inorg. Chem.*, 1971, **10**, 391.

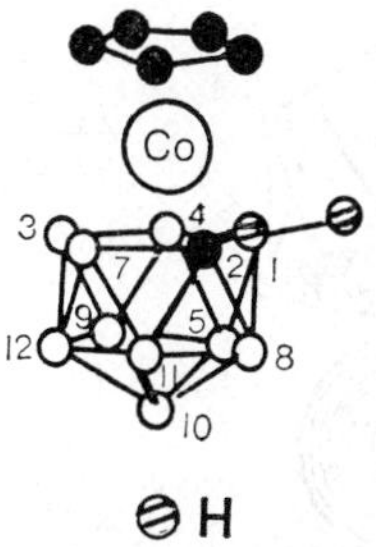

(269)

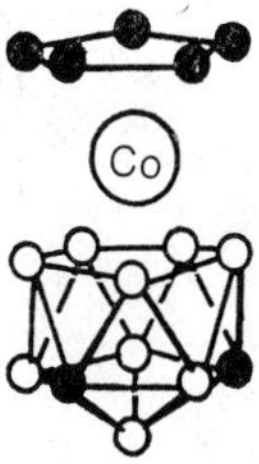

(270)

isomers; one shows six ^{11}B resonances in the ratio 2 : 1 : 2 : 1 : 2 : 1, which is consistent with (270).[1069]

Extensive use has been made of ^{11}B n.m.r. in characterizing carboranes such as (271), (272), (273), (274), and (275).[1070] ^{11}B N.m.r. has been used to establish deuteriation in the 6 and 9 positions in 6,9-$B_{10}H_{12}D_2$, which was

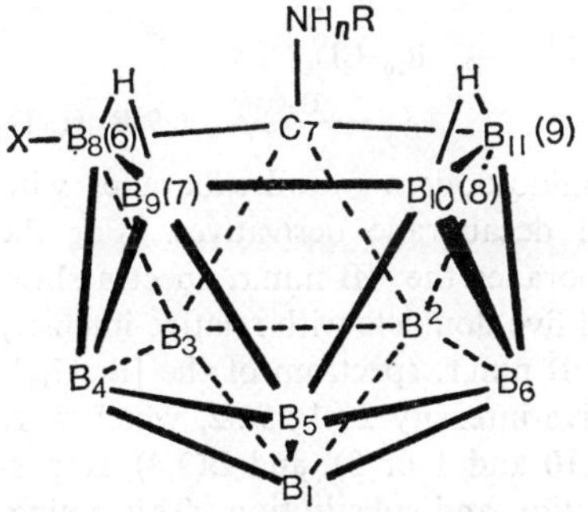

(271)

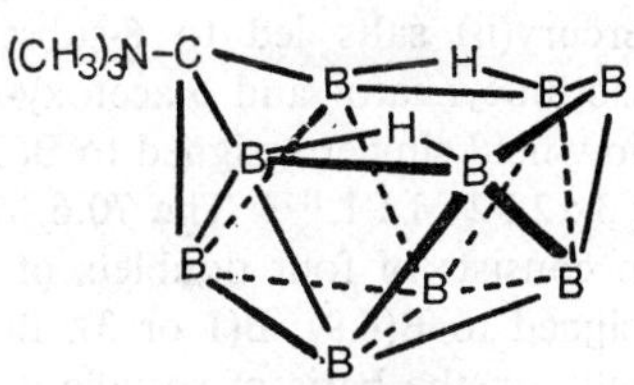

(272)

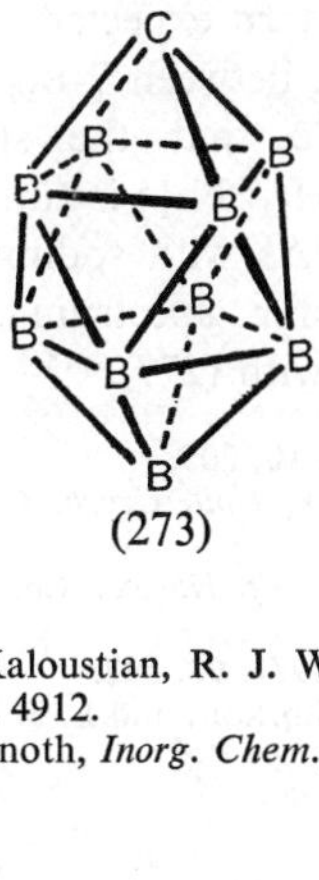

(273)

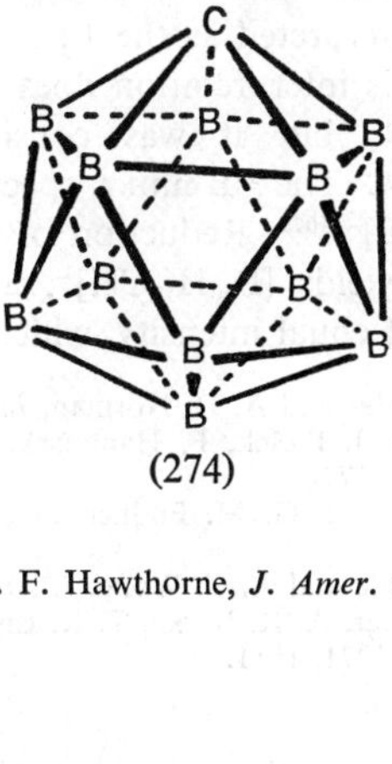

(274)

[1069] M. K. Kaloustian, R. J. Wiersema, and M. F. Hawthorne, *J. Amer. Chem. Soc.*, 1971, **93**, 4912.
[1070] W. H. Knoth, *Inorg. Chem.*, 1971, **10**, 598.

(275)

prepared by the following route:[1071]

$$\mu\text{-}B_{10}H_{10}D_4 \xrightarrow{[BD_4]^-} [B_{10}H_{10}D_5]^- \xrightarrow{D^+} B_{10}H_8D_6 \xrightarrow{H_2O} 6,9\text{-}B_{10}H_{12}D_2$$

The reaction of 6,9-bis(dialkylsulphido)dodecahydrodecaborane with mercury(II) salts led to 6-substituted decaborane derivatives. For the 6-isothiocyanato- and 6-acetoxy-decaboranes the ^{11}B n.m.r. spectra show a low-field singlet assigned to B(6) and five doublets with relative intensity 1 : 3 : 2 : 2 : 1 : 1.[1072] The 70.6 MHz ^{11}B n.m.r. spectrum of the $[B_{10}H_{13}]^-$ ion consists of four doublets of relative intensity 2 : 1 : 5 : 2, which were assigned to B(6,9), B(1 or 3), B(5,7,8,10 and 1 or 3), and B(2,4), respectively, on the basis of specific deuteriation and substitution. This assignment is consistent with structure (276) for the $[B_{10}H_{13}]^-$ ion in solution.[1073] The ^{11}B n.m.r. spectrum of $B_{10}H_{12}PR$ (R = Me or Ph) is badly resolved but was interpreted as the 1 : 1 : 2 : 2 : 2 : 2 pattern expected for C_s symmetry. This interpretation does not distinguish between 7-$B_{10}H_{12}PR$ and 2-$B_{10}H_{12}PR$, but it was considered probable that the structure is 7-$B_{10}H_{12}PR$. The ^{11}B n.m.r. spectrum of $[7\text{-}B_{10}H_{11}PPh]^-$ is similar to that of $[B_{10}H_{11}S]^-$.[1074] Reduction of 1,12-$B_{10}H_{10}CHAs$ with sodium in liquid ammonia yields $[B_{10}H_{12}CH]^-$, and the ^{11}B n.m.r. spectrum shows two doublets of equal intensity, which is consistent with (277).[1075]

[1071] J. A. Slater and A. D. Norman, *Inorg. Chem.*, 1971, **10**, 205.
[1072] B. Štíbr, J. Plešek, F. Hanousek, and S. Heřmánek, *Coll. Czech. Chem. Comm.*, 1971, **36**, 1794.
[1073] A. R. Siedle, G. M. Bodner, and L. J. Todd, *J. Inorg. Nuclear Chem.*, 1971, **33**, 3671.
[1074] J. L. Little and A. C. Wong, *J. Amer. Chem. Soc.*, 1971, **93**, 522.
[1075] D. C. Beer, A. R. Burke, T. R. Engelmann, B. N. Storkoff, and L. J. Todd, *Chem. Comm.*, 1971, 1611.

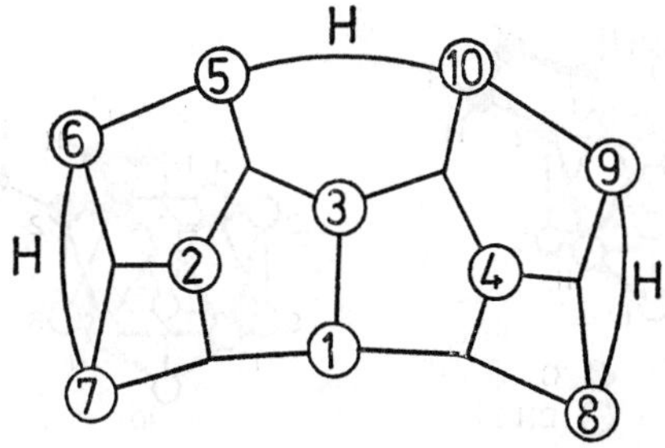

(276)

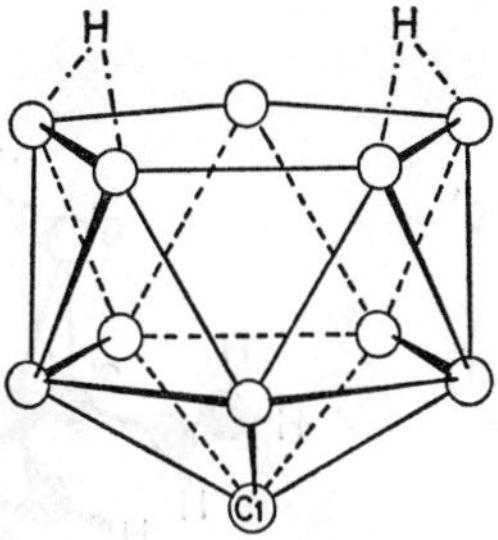

(277)

The ^1H n.m.r. spectra of benzocarborane (278) and related compounds have been reported. The CH resonances of (278) have a chemical shift discrepancy attributable to a ring current of only 0.2 p.p.m., and it was therefore concluded that benzocarborane has little aromatic character.[1076] Extensive use of ^1H and ^{11}B n.m.r. has been made in assigning structures to compounds (279), (280), (281), (282), and (283).[1077] Treatment of $[B_{10}C_2H_{12}]^{2-}$ with $Na^+ [Cp]^-$ and $CoCl_2$ yields a number of complexes including 50% of (Cp)Co(7,8-$B_{10}C_2H_{12}$). The ^{11}B n.m.r. spectrum shows doublets in the intensity ratios 1 : 2 : 1 : 2 : 1 : 2 : 1, consistent with (284). Two other components show ^{11}B n.m.r. doublets in intensity ratios 1 : 1 : 1 : 1 : 1 : 2 : 2 : 1 and ten separate doublets, but the structures are

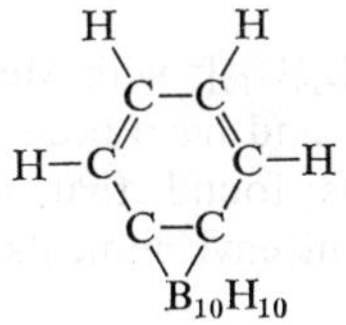

(278)

[1076] D. S. Matteson and N. K. Hota, *J. Amer. Chem. Soc.*, 1971, **93**, 2893.
[1077] D. A. T. Young, T. E. Paxson, and M. F. Hawthorne, *Inorg. Chem.*, 1971, **10**, 786.

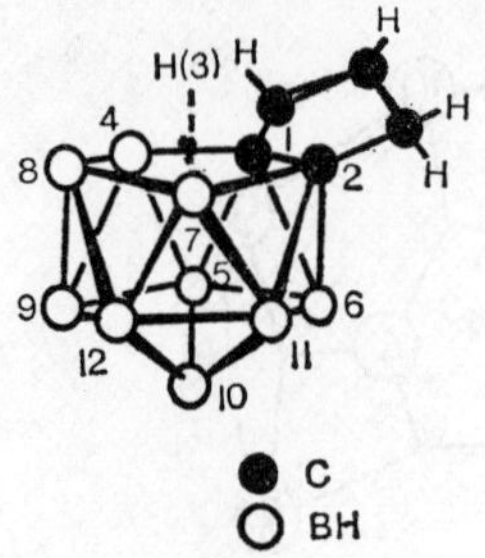

(279)

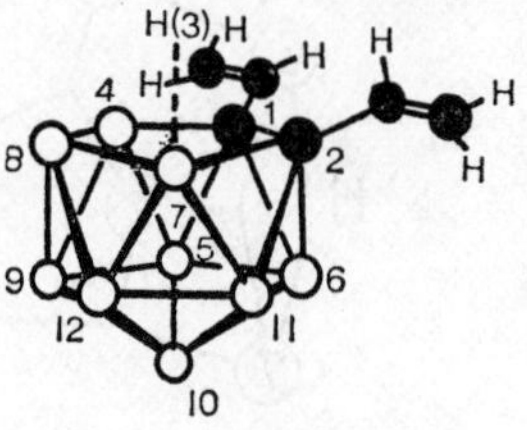

(280)

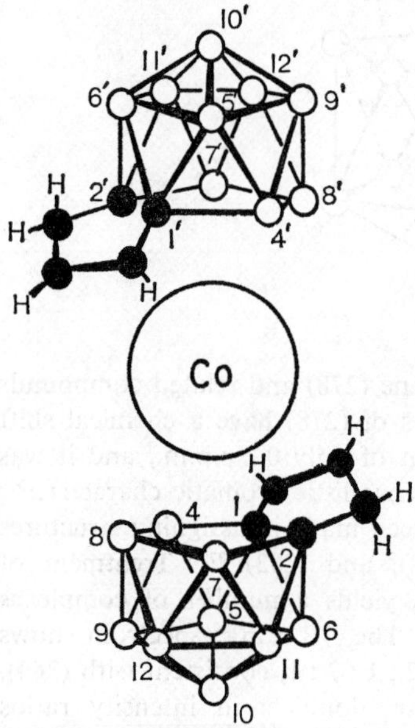

(281)

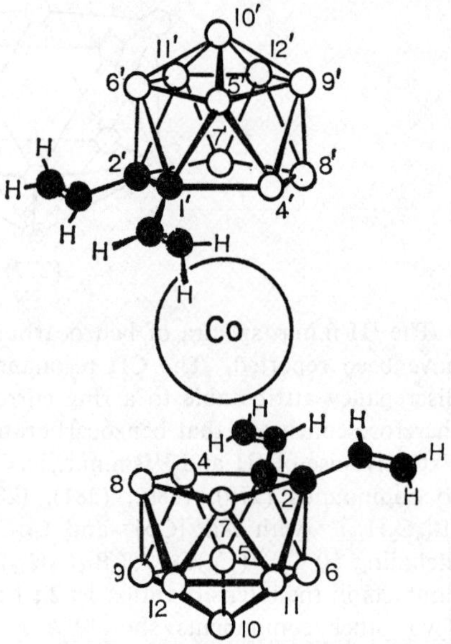

(282)

unknown.[1078] Treatment of $[B_{10}H_{13}]^-$ with Me_3Tl yields $[Me_2Tl]^+ [B_{10}H_{13}]^-$ and $[Me_2Tl]^+ [B_{10}H_{12}TlMe_2]^-$ and have been characterized by ^1H and ^{11}B n.m.r. spectroscopy. It was found that the two methyl groups of $[B_{10}H_{12}TlMe_2]^-$ are in different environments and the structure (285) was proposed.[1079]

When $[B_{11}H_{11}]^{2-}$ is purified, the ^{11}B n.m.r. spectrum shows only a doublet

[1078] G. B. Dunks, M. M. McKown, and M. F. Hawthorne, *J. Amer. Chem. Soc.*, 1971, **93**, 2541.

[1079] N. N. Greenwood, N. F. Travers, and D. W. Waite, *Chem. Comm.*, 1971, 1027.

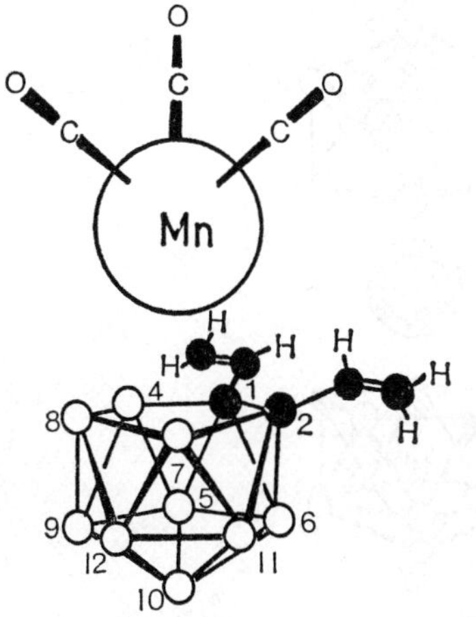

(283)

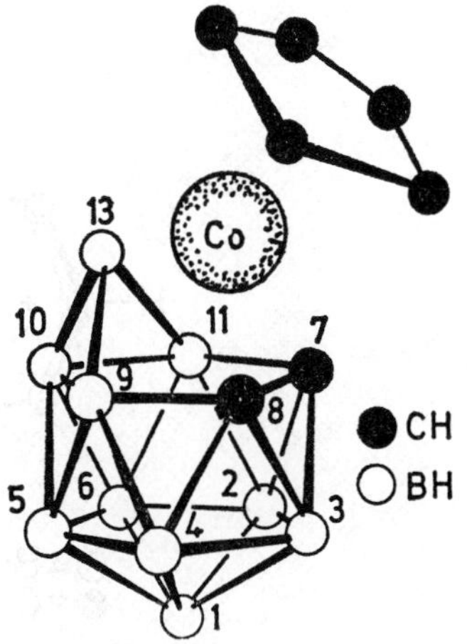

(284)

at 35.0 p.p.m. The weak doublet previously reported at 47.4 p.p.m. is due to a $[B_{10}H_{11}]^{2-}$ impurity.[1080]

When the caesium salt of (286) is dissolved in concentrated hydrochloric acid, $Cs[\{\pi\text{-}(3)\text{-}1,2\text{-}B_9C_2H_{11}\}_2FeH]$ is formed, but extensive overlap of the ^{11}B n.m.r. signals prevent structural assignments.[1081] The ^{11}B n.m.r. spectrum of $Fe(B_9C_2H_{10}SEt_2)_2$ is equally uninformative, apart from a low-field singlet at − 0.2 p.p.m. ^1H and ^{11}B n.m.r. have been used to

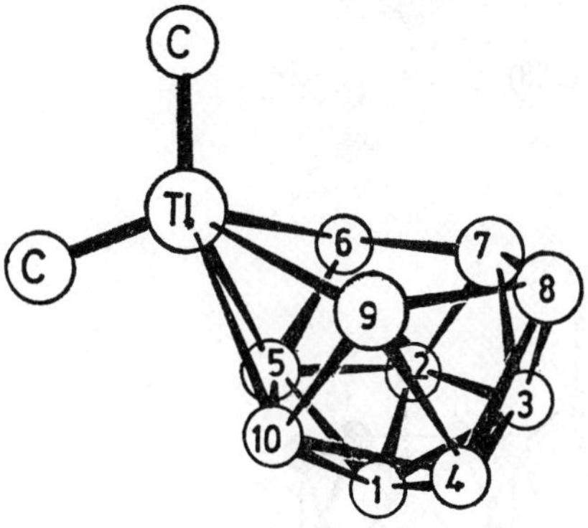

(285)

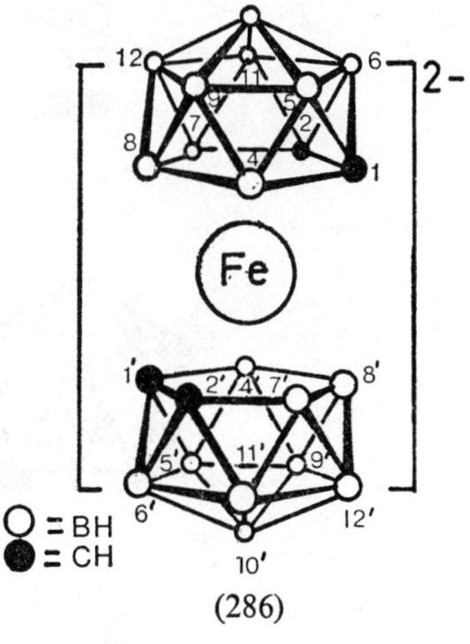

(286)

1080 R. L. Middaugh and R. J. Wiersema, *Inorg. Chem.*, 1971, **10**, 423.
1081 M. F. Hawthorne, L. F. Warren, jun., K. P. Callahan, and N. F. Travers, *J. Amer. Chem. Soc.*, 1971, **93**, 2407.

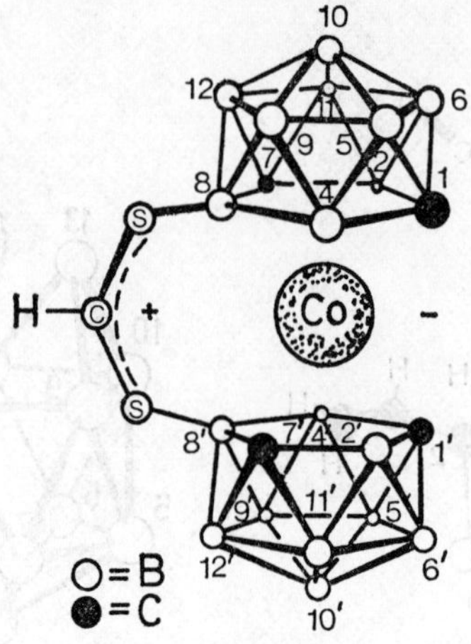
10
12
6
11
7
9
5
2
8
4
1
S
H
C
+
Co
-
S
8′
7′
4
2′
1′
9′
11′
5
12′
6′
10′
=B
=C

(287)

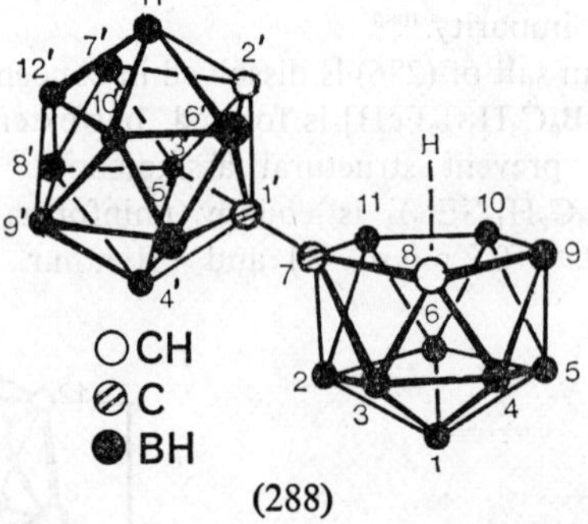
11′
7′
2′
12′
10′
6′
3′
8′
5
1′
9′
4′
H
11
10
8
9
7
6
2
5
3
4
1
CH
C
BH

(288)

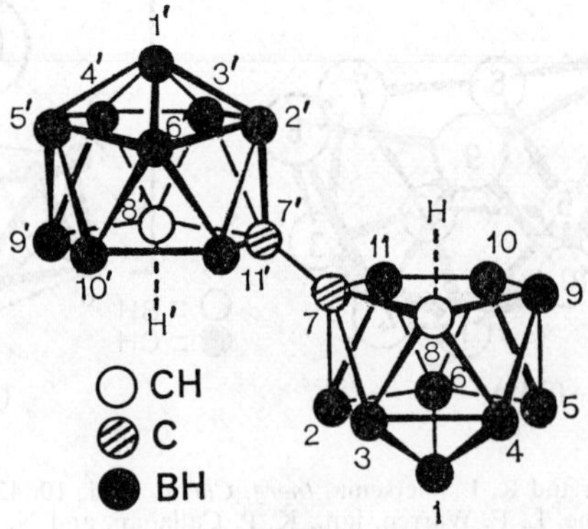
1′
4′
3′
5′
6′
2′
8′
7′
9′
10′
11′
H′
H
11
10
9
7
8
6
5
2
3
4
1
CH
C
BH

(289)

characterize (287) and related compounds.[1082] The ^{11}B n.m.r. spectrum of (288) shows at least ten different types of boron environment and (289) shows at least eight different types of boron.[1083]

1-(1′-1′,2′-$B_{10}C_2H_{11}$)-1,2-$B_{10}C_2H_{11}$ reacts with two mole equivalents of Bu^nLi and then with metal halides to yield (290) (M = Cu^{III}, Cu^{II}, Ni^{II},

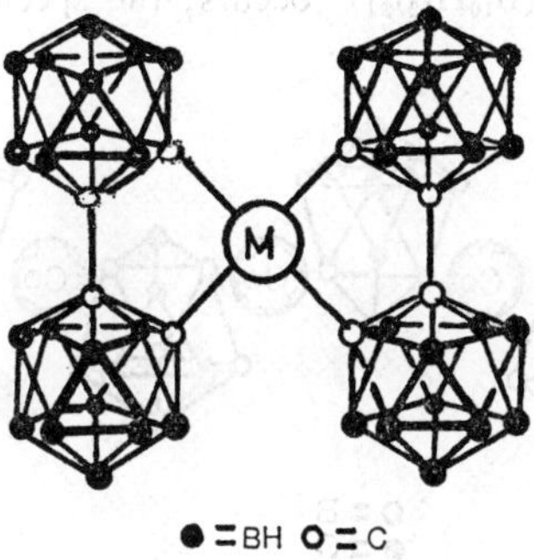

(290)

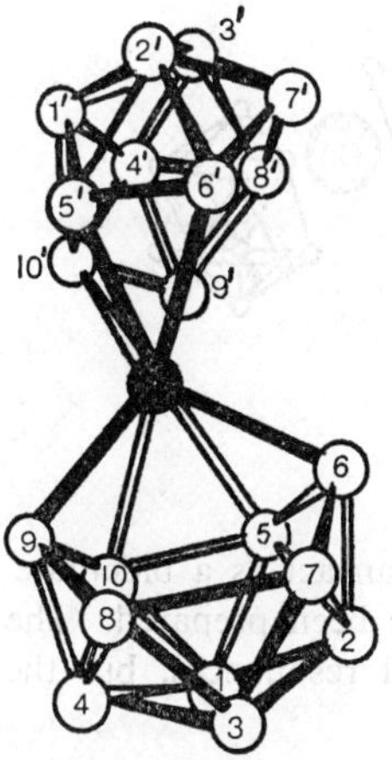

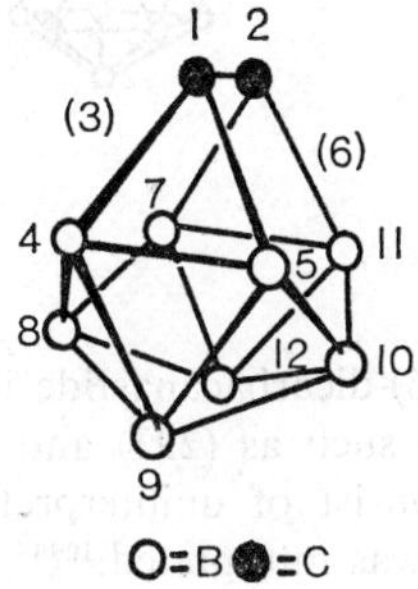

(291) (292)

Ni^{III}, Co^{II}, or Co^{III}), where the stereochemistry at the metal is square-planar or tetrahedral. The cobalt(III) complex shows in the ^{11}B n.m.r. spectrum six sets of doublets in the intensity ratio 2 : 1 : 2 : 1 : 2 : 2. The paramagnetic complexes (M = Cu^{II} or Ni^{III}) show little or no contact shift but no resonances could be detected when M = Co^{II}.[1084]

1082 J. M. Francis and M. F. Hawthorne, *Inorg. Chem.*, 1971, **10**, 594.
1083 M. F. Hawthorne, D. A. Owen, and J. W. Wiggins, *Inorg. Chem.*, 1971, **10**, 1304.
1084 D. A. Owen and M. F. Hawthorne, *J. Amer. Chem. Soc.*, 1971, **93**, 873.

The compounds $[M(B_{10}H_{12})_2]^{2-}$ (M = Zn, Cd, or Hg) and $[M(B_{10}H_{12})]$,-$2Et_2O$ (M = Zn or Cd) have been prepared. At 28.87 MHz, the ^{11}B n.m.r. spectra were poorly resolved, but at 80.53 MHz, the spectra are better resolved. In acetonitrile the ^{11}B n.m.r. spectrum of $[M(B_{10}H_{12})],2Et_2O$ is identical with that of $[M(B_{10}H_{12})]^{2-}$, and thus it was proposed that dissociation to M^{2+} and $[M(B_{10}H_{12})_2]^{2-}$ occurs; the spectra are consistent with (291).[1085]

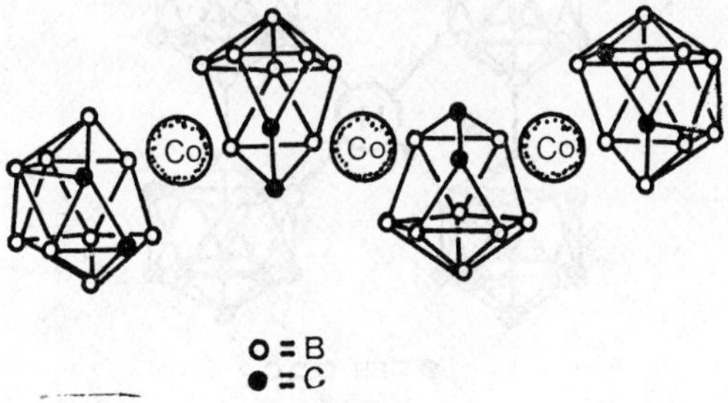

(293)

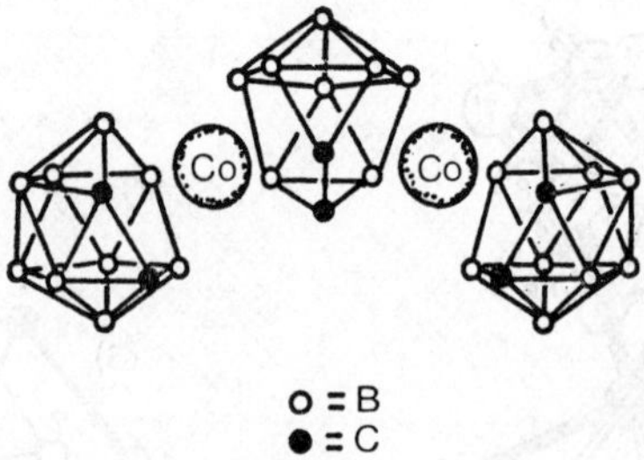

(294)

The (2,6)-dicarbacanastide ion (292) can act as a bidentate ligand and complexes such as (293) and (294) have been prepared. The ^{11}B n.m.r. spectra consist of uninterpretable broad resonances, but the ^1H n.m.r. spectrum was interpreted.[1086]

Other Compounds of Boron.—Bu^{t}Li will deprotonate the heterocyclic organoborane (295) to yield (296) and ^1H n.m.r. is reported.[1087] The reaction of B_2Cl_4 with cyclic olefins has been examined with a view to determining whether *cis*- or *trans*-addition had occurred. In the case of cyclopropene, two products are possible: the *cis*-product (297) would give an AA′BC spectrum whereas the *trans*-product (298) would give an AA′BB′ spectrum. The resonances are broad, but decoupling the boron sharpens

[1085] N. N. Greenwood and N. F. Travers, *J. Chem. Soc. (A)*, 1971, 3257.
[1086] J. N. Francis and M. F. Hawthorne, *Inorg. Chem.*, 1971, **10**, 863.
[1087] R. van Veen and F. Bickelhaupt, *J. Organometallic Chem.*, 1971, **30**, C51.

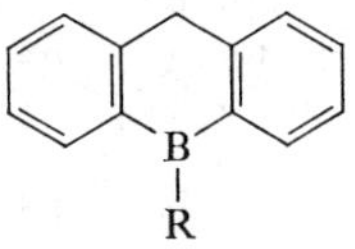

(295)

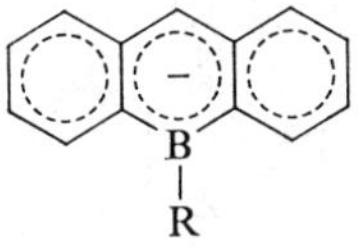

(296)

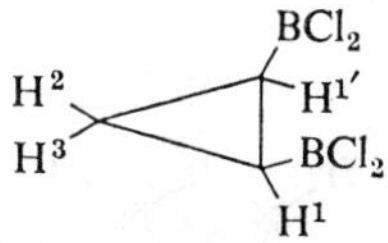

(297)

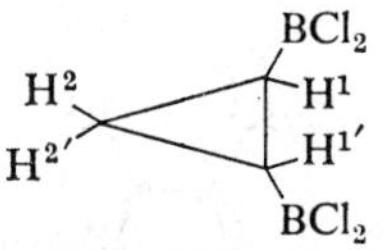

(298)

the resonance and shows the compound to be *cis*.[1088] ^1H N.m.r. has provided evidence for a transient >CHCH$_2$— group when $CH_2{=}CHBr$ reacts with B_2Cl_4, and has been used to follow the reaction of B_2X_4 with halogeno-olefins.[1089]

A mixture of Br_3B,NMe_3 and BF_3 or BCl_3 shows extra resonances, due to redistribution to give species such as Br_2FB,NMe_3. ^{11}B N.m.r. has been used to demonstrate similar reactions for F_3B,NMe_2Ph and Cl_3B,NMe_2Ph.[1090] ^1H and ^{11}B n.m.r. have been used in an attempt to characterize the main product from heating $PhNMe_2,BCl_2Ph$, of empirical formula the same as the initial adduct; it is thought that the boron inserts in the aromatic ring.[1091] RNH_2,BF_3 reacts with PCl_5 to yield $RN(PCl_3)(BCl_2F)$ and $RN(PCl_3)(BCl_3)$, as characterized by ^{11}B, ^{19}F, and ^{31}P n.m.r.[1092] ^{11}B N.m.r. has been reported for $(RNH)_3B$, $(R_2N)_3B$, $(R_2N)_2BCl$, and $(RO)_3B$, and it was concluded that inductive effects and conformation of groups attached to boron have a measurable effect on the ^{11}B chemical shifts.[1093]

The structure of adducts of B_2F_4 and tertiary amines has been studied. $B_2F_4,2NMe_3$ gave only one ^1H n.m.r. and one ^{11}B n.m.r. resonance, consistent with the structure $Me_3NBF_2BF_2NMe_3$. However, the 1 : 1 adduct B_2F_4,NMe_3 has only one ^{11}B resonance, implying exchange of NMe_3 between the two borons.[1094] ^1H N.m.r. spectra have been reported for (299).[1095]

Compounds of the type (300) have very similar chemical shifts to the pyridine analogues, and thus the presence of a ring current has been postulated.[1096] ^{11}B Chemical shifts have been reported for (301).[1097] ^1H

[1088] A. Rosen and M. Zeldin, *J. Organometallic Chem.*, 1971, **31**, 319.
[1089] J. J. Ritter, T. D. Coyle, and J. M. Bellama, *J. Organometallic Chem.*, 1971, **29**, 175.
[1090] S. S. Krishnamurthy and M. F. Lappert, *Inorg. Nuclear Chem. Letters*, 1971, **7**, 919.
[1091] J. R. Blackborow and J. C. Lockhart, *J. Chem. Soc.* (*A*), 1971, 1343.
[1092] H. Binder and E. Fluck, *Z. anorg. Chem.*, 1971, **381**, 116.
[1093] F. A. Davis, I. J. Turchi, and D. N. Greeley, *J. Org. Chem.*, 1971, **36**, 1300.
[1094] B. W. C. Ashcroft and A. K. Holliday, *J. Chem. Soc.* (*A*), 1971, 2581.
[1095] S. Trofimenko, *J. Org. Chem.*, 1971, **36**, 1161.
[1096] S. Gronowitz and A. Maltesson, *Acta Chem. Scand.*, 1971, **25**, 2435.
[1097] D. Nölle and H. Nöth, *Angew. Chem. Internat. Edn.*, 1971, **10**, 126.

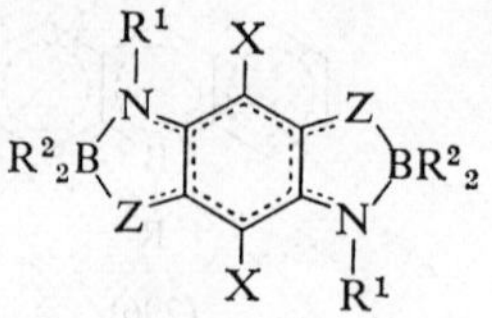

(299)

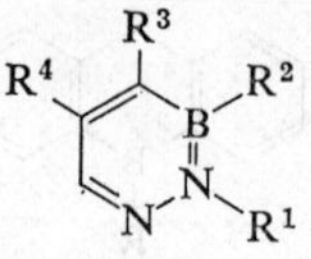

(300)

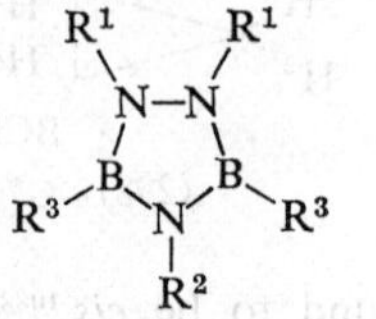

(301)

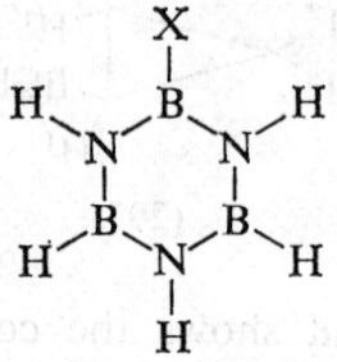

(302)

N.m.r. has been used to analyse the mixture $Me_3N_3B_3Me_{3-n}Ph_n$ formed when $Me_3N_3B_3Me_3$ is treated with PhMgBr or when $Me_3N_3B_3Ph_3$ is treated with MeMgBr.[1098] For a wide range of compounds of the type (302), a linear relationship has been found between the ^{11}B chemical shift and the *para* NH proton chemical shift, but when X = Me, OAr, or SO_3Me there are marked deviations.[1099]

^1H, ^{11}B, ^{19}F, and ^{31}P n.m.r. data on $R^1{}_3PBF_nR^2{}_{3-n}$ have been reported and compared with data on the analogous nitrogen systems.[1100] A study of the ^{19}F n.m.r. spectra of complexes formed by BF_3 with several aliphatic and aromatic ethers shows that the ^{19}F chemical shifts may serve as a measure of the complexing power of the ethers and provide information on the nature of the donor–acceptor bond.[1101] ^1H and ^{11}B n.m.r. have been used to assist characterization of the amine adducts of several amino-acid esters with boranes, BCl_3, and BF_3. No boron–oxygen interactions could be detected, but BCl_3 reacts with glycine to yield $[B(O_2CCH_2NH_3)_3]^{3+}$ $[BCl_4]_3^-$.[1102]

Complexes of Other Group III Elements.—^1H N.m.r. has been used to follow the reaction of aluminium, hydrogen, and secondary amines to yield aminoalanes, *e.g.* H_2AlNEt_2. However, no hydride resonances were found, presumably owing to extensive broadening by the quadrupolar ^{27}Al nucleus.[1103] The reaction of Et_3Al with methyl methacrylate was followed at room temperature in sealed n.m.r. tubes, and it was concluded that both

1098 J. L. Adcock and J. J. Lagowski, *Inorg. Nuclear Chem. Letters*, 1971, **7**, 473.
1099 O. T. Beachley, jun., *J. Amer. Chem. Soc.*, 1971, **93**, 5067.
1100 J.-P. Tuchagues and J.-P. Laurent, *Bull. Soc. chim. France*, 1971, 4246.
1101 S. G. Katal'nikov, A. M. Voloshchuk, M. A. Sokalskii, and T. A. Kozik, *Russ. J. Phys. Chem.*, 1971, **45**, 1028.
1102 E. F. Rothgery and L. F. Hohnstedt, *Inorg. Chem.*, 1971, **10**, 181.
1103 R. A. Kovar and E. C. Ashby, *Inorg. Chem.*, 1971, **10**, 893.

a 1 : 1 and a 1 : 2 complex are formed.[1104] The ^1H n.m.r. spectra of $(PhCH_2)_3Al$ and its etherate in toluene are consistent with a σ-bonded non-fluxional benzyl group, and the invariance from + 25 to − 90 °C is consistent with a monomeric structure.[1105] ^1H N.m.r. has also been used to follow the reaction of alkynes with $[Bu^i_2AlH]_3$, and it was concluded that compounds such as (303) and (304) were formed.[1106]

Treatment of $R^1_3P{=}NSiR^2_2N(AlR^1_3)PR^1_3$ with InR^1_3 or $R^1_3P{=}$-$NSiR^2_2N(InR^1_3)PR^1_3$ with AlR^1_3 yields (305) and ^1H n.m.r. data were

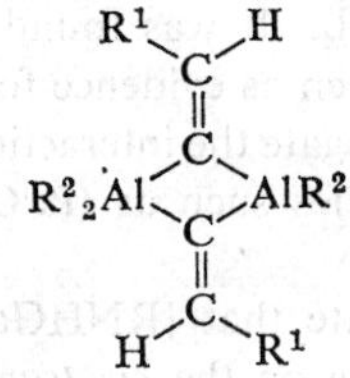

(303)

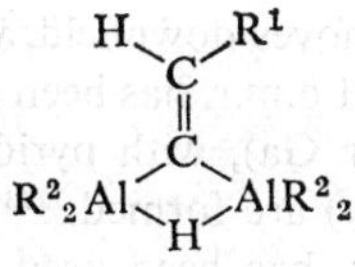

(304)

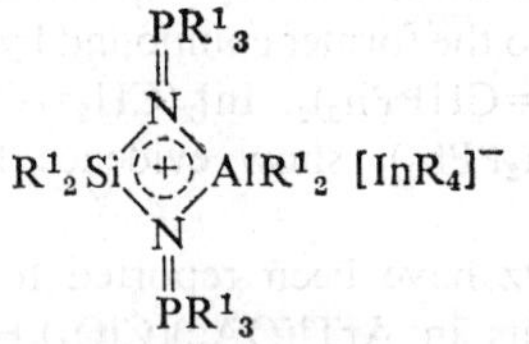

(305)

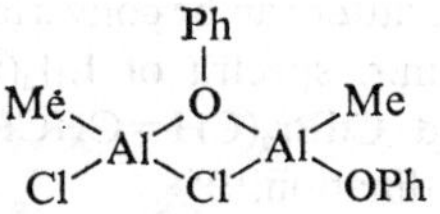

(306)

reported.[1107] The ^1H n.m.r. spectrum of $[Me(PhO)ClAl]_n$ shows the presence of five methyl resonances which are attributed to individual species such as (306).[1108] For complexes of the type $R^1_2Al(R^2COCHCOR^3)$, a linear relationship when R^1 = Me has been found between the chemical shift of the methyl group and the central hydrogen-bonded proton of the free β-diketone in the enol form.[1109] ^1H N.m.r. has been used to examine complex formation between methyl-substituted pyridines and Et_3Al, Et_2AlCl, and $EtAlCl_2$.[1110] $AlCl_3$, PCl_3, and $[MeNH_3]^+ Cl^-$ react to yield various products, including $MeHNPCl_2,NMe,AlCl_3,HCl$ and (307), as characterized by ^1H and ^{31}P n.m.r.[1111] ^1H N.m.r. has been used to investi-

1104 P. E. M. Allen, B. O. Bateup, and B. A. Casey, *J. Organometallic Chem.*, 1971, **29**, 185

1105 J. J. Eisch and J.-M. Biedermann, *J. Organometallic Chem.*, 1971, **30**, 167.

1106 G. M. Clark and G. Zweifel, *J. Amer. Chem. Soc.*, 1971, **93**, 527.

1107 W. Wolfsberger and H. Schmidbaur, *J. Organometallic Chem.*, 1971, **27**, 181.

1108 K. B. Starowieyski, S. Pasynkiewicz, and M. D. Skowronska, *J. Organometallic Chem.*, 1971, **31**, 149.

1109 W. R. Kroll, I. Kuntz, and E. Birnbaum, *J. Organometallic Chem.*, 1971, **26**, 313.

1110 G.-E. Matsubayashi, K. Wakatsuki, and T. Tanaka, *Org. Magn. Resonance*, 1971, **3**, 703.

1111 H. Vollmer and M. Becke-Goehring, *Z. anorg. Chem.*, 1971, **382**, 281.

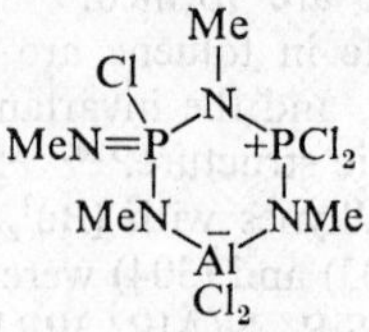

(307)

gate the solution of AlI_3 in Et_2O and CCl_4. It was found that the Et_2O resonance moves downfield, which was taken as evidence for AlI_3OEt_2.[1112] Similarly, ^1H n.m.r. has been used to investigate the interaction of $M(OPr^i)_3$ (M = Al or Ga), with pyridine, and species such as $(Pr^iO)_2Al(\mu\text{-}OPr^i)_2\text{-}Al(OPr^i)_2(py)$ are formed.[1113]

^1H N.m.r. has been used to demonstrate that $[RNHGaH_2]_3$ exists in solution as at least two isomers, depending on the *cis–trans* arrangement of R groups in the cyclohexane-style ring.[1114] $Ga(OEt)X_2$ (X = halogen) is shown by ^1H n.m.r. to be a mixture of two compounds, which were suggested as being $[Ga(OEt)X_2]_n$ (n = 2 or 3) and $X_2Ga(\mu\text{-}X)_2Ga\{(\mu\text{-}OEt)_2\text{-}GaX_2\}_2$; the latter can be converted into the former compound by heating.[1115] The ^1H n.m.r. spectra of $InI_3(CH_2{=}CHPPh_2)_2$, $InI_3(CH_2{=}CHCH_2CH_2\text{-}PPh_2)_2$, and $CdBr_2(CH{=}CHCH_2CH_2PPh_2)_2$ show evidence for double-bond co-ordination.[1116]

Very large values of up to 900 Hz have been reported for thallium–aromatic hydrogen coupling constants in $ArTl(OAc)(ClO_4),H_2O$.[1117] As the ^1H n.m.r. spectrum of $(C_6F_5)_2TlX$ (X = acac, hfac, or PhCOCHCOPh) shows no thallium–hydrogen coupling even at − 60 °C, it was therefore concluded that rapid exchange was occurring.[1118] Both (p-$TolSO_2$)$TlCl_2$ and (p-$TolSO_2$)$_3$Tl show singlets at τ 7.60, consistent with an S-bonded sulphinate structure.[1119] The CH_2 parts of the ethyl groups of (308) and (309) are inequivalent and it was concluded that the inversion of the methyl groups is slow on the n.m.r. time scale;[1120] large coupling constants to thallium were reported.[1121] ^1H N.m.r. has also been reported for (310) (X = H, Tl, or Me_2Tl)[1122] and (311).[1123]

1112 P. J. Ogren, J. P. Cannon, and C. F. Smith, jun., *J. Phys. Chem.*, 1971, **75**, 282.
1113 J. G. Oliver and I. J. Worrall, *J. Inorg. Nuclear Chem.*, 1971, **33**, 1281.
1114 A. Storr and A. D. Penland, *J. Chem. Soc.* (*A*), 1971, 1237.
1115 J. G. Oliver and I. J. Worrall, *J. Chem. Soc.* (*A*), 1971, 2315.
1116 D. M. Roundhill, *J. Inorg. Nuclear Chem.*, 1971, **33**, 3367.
1117 K. Ichikawa, S. Uemura, T. Nakano, and E. Uegaki, *Bull. Chem. Soc. Japan*, 1971, **44**, 545.
1118 G. B. Deacon and V. N. Garg, *Austral. J. Chem.*, 1971, **24**, 2519.
1119 A. G. Lee, *Inorg. Chim. Acta*, 1971, **5**, 346.
1120 R. J. Abraham and K. M. Smith, *Tetrahedron Letters*, 1971, 3335.
1121 K. M. Smith, *Chem. Comm.*, 1971, 540.
1122 A. G. Lee, *J. Chem. Soc.* (*A*), 1971, 880.
1123 J. A. S. Cavaleiro and K. M. Smith, *Chem. Comm.*, 1971, 1384.

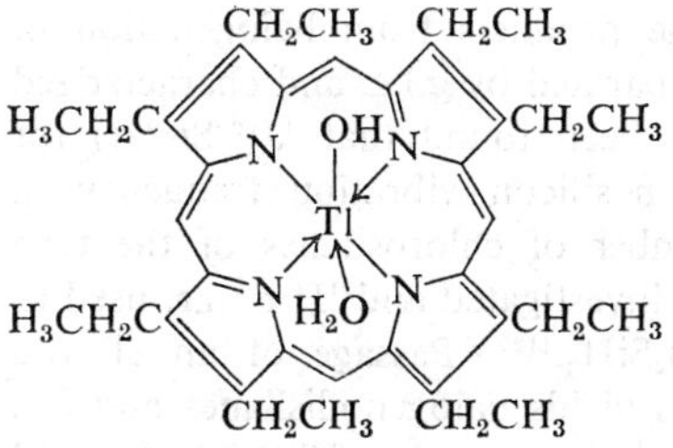

(308)

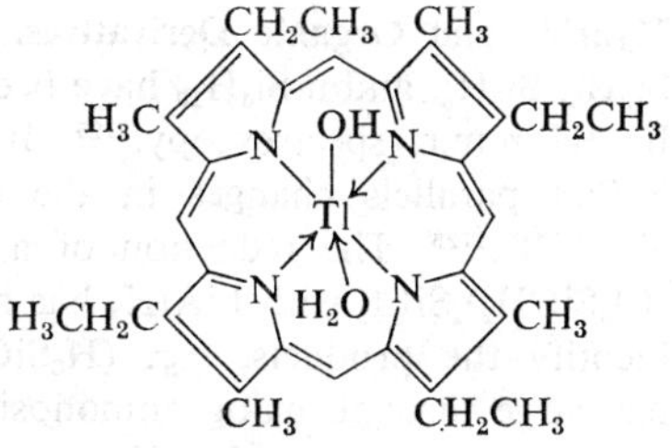

(309)

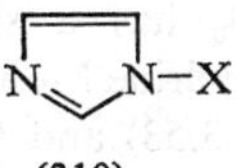

(310)

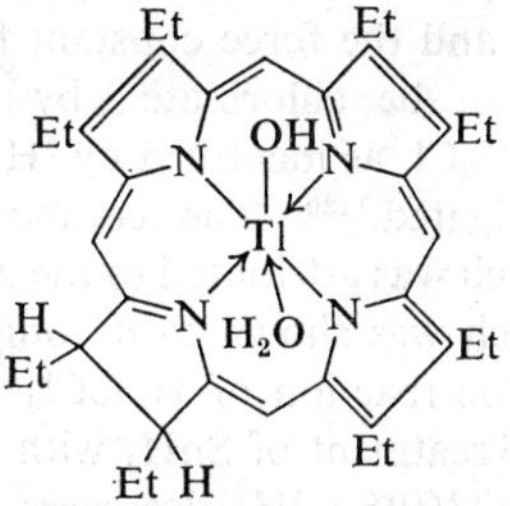

(311)

7 Compounds of Silicon, Germanium, Tin, and Lead

Information concerning complexes of these elements can be found at the following sources: $RC{\equiv}CMR_n$ (M = Si, Ge, Sn, Pb, P, As, or Se),[140] $M^2(CO)_3P(M^1Me_3)_3$ (M^2 = Cr, Mo, or W; M^1 = C, Si, Ge, or Sn),[246] $(Cp)Mn(CO)_2P(MMe_3)_3$ (M = C, Si, Ge, or Sn),[270] $(Me_3C)_2P(MMe_3)$-$NSiEt_3$ (M = Si, Ge, or Sn),[1216] $ArMMe_n$ (M = Si, Ge, P, As, or S),[126] R_3MNCS (M = Si or Ge),[156] $R_2M(CH{=}CH_2)_2,2CuCl$ (M = Si or Sn),[461] 2,3-μ-$Me_3MC_2Me_2B_4H_5$ (M = Si or Ge),[1053] $(H_3MAs)_5$ (M = Si or Ge),[1242] $(SiH_3)_3E$ (E = P or Sb),[103] methylsiloxanes,[144] $(Cl_3Si)_2FeH(CO)(Cp)$,[283] $(Ph_3P)_3CoH_2SiR_3$,[333] $[SiF_6]^{2-}$,[1269] $Me_2MB_{10}H_{12}$ (M = Ge or Sn),[1063] $Ph_2(R)GeFe_2(CO)_3(Cp)_2$,[318] $HPt(PEt_3)_2(GeH_2Cl)_n(GeHCl_2)_{3-n}$,[392] $[SnH_3]^+$,[88] Me_3SnPH_2,[105] $MeSnX_3$,[120] $FSnX_3$,[130] (norbornadienyl)$SnMe_3$,[139] R_nSnX_{4-n},[187] $Me_nSn(SMe)_{4-n}$,[188] $R_{3-n}X_nSnMn(CO)_5$,[179] XBu_2SnOSn-Bu_2X,[189] R_nSnX_{4-n},[190, 191] SnX_n(di-isopropyl methylphosphonate),[209] $(Cp)FeL_2SnR_3$,[313, 315] and $(C_4H_3X)_2Pb$.[472]

A useful listing of n.m.r. data has appeared in the book 'Literature Data for I.R., Raman, N.M.R. Spectroscopy of Si, Ge, Sn, and Pb Organic Compounds' by Licht and Reich.[1124]

[1124] K. Licht and P. Reich, 'Literature Data for I. R., Raman, N. M. R. Spectroscopy of Si, Ge, Sn, and Pb Organic Compounds', Deutscher Verlag der Wissenschaften, Berlin, D.D.R., 1971.

Hydride and Organic Derivatives.—The products from halogenation of Si_2H_6, Si_3H_8, and n-Si_4H_{10} have been separated by g.l.c. and characterized by ^1H n.m.r. spectroscopy.[1125] It has been found that $^1J(^{29}Si–^1H)$ for X_3SiH parallels changes in the silicon–silicon vibration frequency in X_3SiSiX_3.[1126] The reduction of a number of chlorosilanes of the type $(Cl_3SiCH_2)_2SiCl_2$ with $LiAlH_4$ has been investigated and ^1H n.m.r. used to identify the products, *e.g.* $(H_3SiCH_2)_2SiH_2$.[1127] Passage of an electric discharge through halogenomonosilanes yields halogenodisilanes and full ^1H and ^{19}F n.m.r. data have been reported. In the cases of SiH_2FSiH_2F and $SiHF_2,SiHF_2$, the second-order spectra were analysed as $[AX_2]_2$ spectra by use of LAOCOON 2, part 1.[1128] ^1H N.m.r. data have been reported for $(X_3CCO_2)_3SiH$ (X = H or F) and a linear relationship was found between $^1J(^{29}Si–^1H)$ and the force constant for the silicon–hydrogen bond.[1129]

Ge_2H_6 can be chlorinated by $SnCl_4$ to yield among the products $ClGeH_2GeH_2Cl$, as identified by ^1H n.m.r.; similar reactions of Si_3H_8 were also investigated.[1130] The compound $Me_2O,HGeCl_3$ has a singlet at δ 11.23 which was attributed to the hydrogen-bonded hydride $Me_2O\cdots H—GeCl_3$, which was shown to decompose to MeOH (δ 3.53) and $MeGeCl_3$ (δ 1.73). The reaction of $HGeCl_3$ with MeI was also investigated by ^1H n.m.r.[1131] Treatment of SnH_4 with HX at − 78 °C yields SnH_3X. It was found that $^1J(^{119}Sn–^1H)$ decreases from the chloride to iodide, but the hydrogen moves to high field on going from the chloride to iodide.[1132]

Various Me_3Si-substituted ethylenes have been treated theoretically by a combination of the Del–Re method for the σ-system and a revised SCF–LCAO–MO–CI method for the π-system. A good correlation was found between the $=CH_2$ or $=CHX$ proton shift and the corresponding Del–Re parameter.[1133] The $J(Sn–CH_2)$ coupling constants in a series of benzyltin derivatives have been examined as a function of substituents. A plot of J against $\Sigma\sigma^*$ (σ^* is the Taft parameter) shows three equidistant parallel lines, depending on whether there is none, one, or two phenyl groups attached to the tin.[1134] The ^1H n.m.r. chemical shifts of (312) have been taken as evidence that there is no $(3d–2p)\pi$ bonding producing aromaticity in the silicon ring.[1135] ^1H N.m.r. has been used to show that vacuum flow-pyrolysis of Me_3SiCp at 800 °C converts it into a mixture of the

1125 F. Fehér, P. Plichta, and R. Guillery, *Inorg. Chem.*, 1971, **10**, 606.
1126 E. Hengge, *Monatsh.*, 1971, **102**, 734.
1127 G. Fritz and H. Fröhlich, *Z. anorg. Chem.*, 1971, **382**, 9.
1128 J. E. Drake and N. P. C. Westwood, *J. Chem. Soc.* (*A*), 1971, 3300.
1129 E. Henge and E. Starz, *Monatsh.*, 1971, **102**, 741.
1130 J. E. Bentham, S. Cradock, and E. A. V. Ebsworth, *Inorg. Nuclear Chem. Letters*, 1971, **7**, 1077.
1131 T. K. Gar, E. M. Berliner, and V. F. Mironov, *J. Gen. Chem.* (*U.S.S.R.*), 1971, **41**, 343.
1132 J. M. Bellama and R. A. Gsell, *Inorg. Nuclear Chem. Letters*, 1971, **7**, 365.
1133 J. Nagy and M. T. Vandorffy, *J. Organometallic Chem.*, 1971, **31**, 205.
1134 M. Gielen, M. R. Barthels, M. de Clercq, and J. Nasielski, *Bull. Soc. chim. belges*, 1971, **80**, 189.
1135 F. K. Cartledge and P. D. Mollère, *J. Organometallic Chem.*, 1971, **26**, 175.

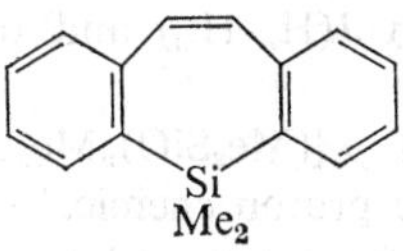

(312)

5-, 1-, and 2-isomers in which the 1- and 2-isomers predominate; on standing, the compound slowly converts into the 5-isomer.[1136] Low-temperature ^1H n.m.r. has been used in an unsuccessful attempt to detect a postulated intermediate Me_4SiO_3.[1137]

The cleavage of Me_4Si by $SbCl_5$ to give Me_3SiCl has been followed by ^1H n.m.r. Hence caution must be used before using Me_4Si as a reference.[1138]

^1H N.m.r. has been used to help to characterize the products of cyclo-addition of silyldiazoalkanes with alkenes and their photolysis products.[1139] The fact that Ph_2C_2 reacts with $Me_3SiSiMe_2H$ in the presence of *trans*-$PtCl_2(Et_3P)_2$ to yield (313) (^1H n.m.r.) has been taken as evidence for the

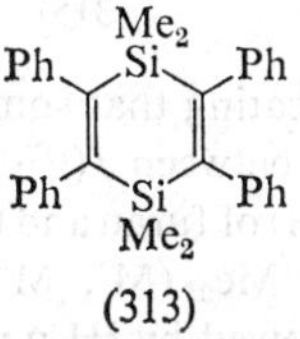

(313)

presence of 'dimethylsilylene' as a reactive intermediate.[1140] The magnetic non-equivalence of the *gem*-dimethyl groups for a series of asymmetric silanes and siloxy-ethers has been shown to be solvent- and temperature-dependent.[1141]

The ^1H n.m.r. spectra of Ph_3MLi (M = C, Si, Ge, Sn, or Pb) have been analysed. The *para*-protons are shielded in the opposite order from that expected on the basis of π-bonding trends of Group IV atoms. From the ^7Li n.m.r. spectra it was concluded that the degree of electron delocalization into the phenyl rings is controlled by the degree of association between Li^+ and the Group IV atom.[1142] The ^1H n.m.r. spectra of Ph_3SiX have been analysed using LAOCOON 3. Variation of substituents on the silicon appears to have little effect on the π-interactions between silicon and the phenyl rings. However, substituents do affect the σ-electron framework as shown

[1136] E. W. Abel and M. O. Dunster, *J. Organometallic Chem.*, 1971, **33**, 161.
[1137] L. Spialter, L. Pazdernik, S. Bernstein, W. A. Swansiger, G. R. Buell, and M. E. Freeburger, *J. Amer. Chem. Soc.*, 1971, **93**, 5682.
[1138] T. J. Pinnavaia and L. J. Matienzo, *J. Inorg. Nuclear Chem.*, 1971, **33**, 3982.
[1139] A. G. Brook and P. F. Jones, *Canad. J. Chem.*, 1971, **49**, 1841.
[1140] K. Yamamoto, H. Okinoshima, and M. Kumada, *J. Organometallic Chem.*, 1971, **27**, C31.
[1141] P. E. Rakita and B. J. Rothschild, *Chem. Comm.*, 1971, 953.
[1142] R. H. Cox, E. G. Janzen, and W. B. Harrison, *J. Magn. Resonance*, 1971, **4**, 274.

by the correlation between $J(H_o–H_m)$ and the electronegativity of the substituent.[1143]

Hammett σ-constants of p-$[(Me_3SiO)_nMe_{3-n}Si]C_6H_4CO_2H$ have been determined from carboxylic proton chemical shifts in pyridine, and compared with those obtained from rate and stability constants. The values were interpreted as showing $(p–d)\pi$ interaction between the silicon and aromatic ring.[1144] The ^1H (methyl) chemical shift and $^1J(^{13}C–^1H)$ (methyl) of 25 substituted phenyltrimethylsilanes have been measured and the results explained in terms of $(p \rightarrow d)\pi$ back-bonding.[1145] For the compounds Me_3MCO_2H (M = C, Si, or Ge) the carboxylic proton chemical shifts changes in accord with the electronegativity C > Si ≈ Ge.[1146]

^1H N.m.r. data have been reported for (314).[1147] The ^1H n.m.r. spectra of (315) and (316) (X = O or S) have been fully analysed. The chemical

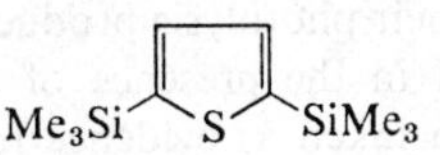

(314)

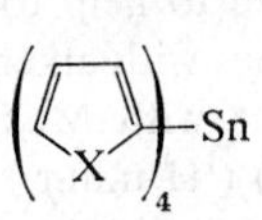

(315)

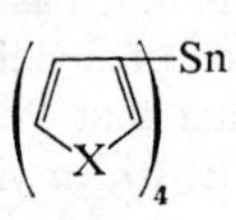

(316)

shifts were interpreted as indicating that some degree of $(d–p)\pi$ bonding is present. A linear correlation between $J(^{117,119}Sn–^1H)$ of these compounds and the corresponding $J(^1H–^1H)$ of furan and thiophen has been observed.[1148]

The reactions of $Me_3M^1M^2Me_3$ (M^1, M^2 = Si, Ge, or Sn) with ICl, HCl, and F_3CI have been followed by ^1H n.m.r.[1149] For the M^1Me_3 group of $Me_3M^1M^2Me_3$, as M^2 is changed from carbon to silicon to germanium to tin, the methyl resonance moves to lower field while the $M^2Me_3Bu^t$ methyl resonance moves to higher field. However, for $Me_3M^1M^2Me_3$ the order is C < Ge < Sn < Si, which was attributed to the electronegativity of M^2.[1150] Full ^1H n.m.r. data have been reported for $(Et_3Si)_3SiX$. It was found that the inductive effect of the ethyl group on the Si_3SiX bonding system is no greater than for the corresponding methyl compounds.[1151] ^1H N.m.r. has shown that the *trans*-isomer of (317) is rigid but the *cis*-isomer is fluxional.[1152]

^1H N.m.r. has been used to assist in the characterization of Diels–Alder adducts of vinylgermanium compounds and dienes.[1153] For the com-

1143 R. H. Cox and W. K. Austin, jun., *J. Organometallic Chem.*, 1971, **26**, 331.
1144 Z. Plzak, F. Mareš, J. Hetflejš, J. Schraml, Z. Papoušková, V. Bažant, E. G. Rochow, and V. Chvalovský, *Coll. Czech. Chem. Comm.*, 1971, **36**, 3115.
1145 M. E. Freeburger and L. Spialter, *J. Amer. Chem. Soc.*, 1971, **93**, 1894.
1146 O. W. Steward, J. E. Dziedzic, and J. S. Johnson, *J. Org. Chem.*, 1971, **36**, 3475.
1147 F. H. Pinkerton and S. F. Thames, *J. Organometallic Chem.*, 1971, **29**, C4.
1148 G. Barbieri and F. Taddei, *J. Chem. Soc. (B)*, 1971, 1903.
1149 C. F. Shaw, tert., and A. L. Allred, *Inorg. Chem.*, 1971, **10**, 1340.
1150 C. F. Shaw, tert., and A. L. Allred, *J. Organometallic Chem.*, 1971, **28**, 53.
1151 H. Bürger and W. Kilian, *J. Organometallic Chem.*, 1971, **26**, 47.
1152 K. Tamao and M. Kumada, *J. Organometallic Chem.*, 1971, **30**, 35.
1153 J. Dubac, P. Mazerolles, A. Laporterie, and P. Lix, *Bull. Soc. chim. France*, 1971, 125.

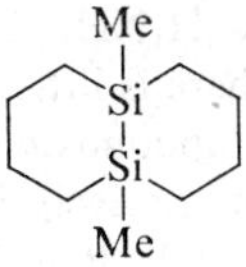
(317)

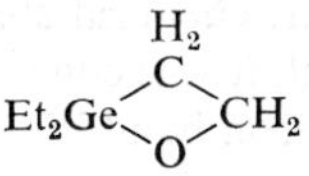
(318)

pounds R_3GeX it has been reported that the α-protons of R = Me or Et move 0.9 p.p.m. to high field on changing from X = Cl to X = Li.[1154] The ^1H n.m.r. spectrum of the ring in (318) is of the AA′XX′ type.[1155] ^1H N.m.r. has been used to help to characterize the products of addition of Me_3SnH, Me_2SnXH, or $MeSnCl_2H$ to norbornadiene,[1156] and to show that *cis*- and *trans*-(2-phenylcyclopropyl)trimethyltin reacts with bromine to yield *cis*- and *trans*-1-bromo-2-phenylcyclopropane with complete retention of configuration.[1157]

Nitrogen Derivatives.—A reinvestigation of the ^1H n.m.r. spectrum of some *NN*-bis(organosilyl)methylhydrazines has resolved previous conflicting interpretations. Variable-temperature n.m.r. data have permitted the unequivocal assignment of the NMe and NH proton absorptions and coupling constants of both *NN*- and *NN′*-bis(organosilyl)methylhydrazines.[1158] The ^1H n.m.r. coupling constants and shifts of $(C_2H_3)Si(NMe_2)_3$, $(C_2H_3)SiMe(NMe_2)_2$, and $(C_2H_3)Me_2Si(NMe_2)$ have been interpreted in terms of $(p\text{–}d)\pi$bonding in the silicon–nitrogen bond.[1159] ^1H N.m.r. data have also been reported for a variety of compounds containing alternating silicon and nitrogen atoms in chains or rings.[1160–1166] Some related systems contain oxygen,[1167] sulphur,[1168] phosphorus,[1169] or germanium[1170, 1171] atoms in the chains or rings. Data were also reported for $Me_3PNSiMe_nX_{3-n}$[1172] and $X_{3-n}R_nSiN{=}S{=}NSiR_nX_{3-n}$.[1173]

1154 E. J. Bulten and J. G. Noltes, *J. Organometallic Chem.*, 1971, **29**, 397.
1155 M. Massol, D. Mesnard, J. Barrau, and J. Satgé, *Compt. rend.* 1971, **272**, *C*, 2081.
1156 H. G. Kuivila, J. D. Kennedy, R. Y. Tien, I. J. Tyminski, F. L. Pelczar, and O. R. Khan, *J. Org. Chem.*, 1971, **36**, 2083.
1157 K. Sisido, K. Ban, T. Isida, and S. Kozime, *J. Organometallic Chem.*, 1971, **29**, C7.
1158 B. Birchlmeir and R. West, *J. Organometallic Chem.*, 1971, **32**, 35.
1159 J. Schraml, Z. Pacl, and V. Chvalovsky, *Coll. Czech. Chem. Comm.*, 1971, **36**, 1578.
1160 D. W. Klein and J. W. Connolly, *J. Organometallic Chem.*, 1971, **33**, 311.
1161 U. Wannagat, E. Bogusch, and P. Geymayer, *Monatsh.*, 1971, **102**, 1825.
1162 U. Wannagat, E. Bogusch, and F. Rabet, *Z. anorg. Chem.*, 1971, **385**, 261.
1163 U. Wannagat, R. Braun, L. Gerschler, and H.-J. Wismar, *J. Organometallic Chem.*, 1971, **26**, 321.
1164 U. Wannagat and L. Gerschler, *Annalen*, 1971, **744**, 111.
1165 L. Gerschler and U. Wannagat, *J. Organometallic Chem.*, 1971, **29**, 217.
1166 U. Wannagat, R. Braun, and L. Gerschler, *Z. anorg. Chem.*, 1971, **381**, 169.
1167 U. Wannagat, F. Rabet, and H.-J. Wismar, *Monatsh.*, 1971, **102**, 1429.
1168 A. D. M. Hailey and G. Nickless, *J. Inorg. Nuclear Chem.*, 1971, **33**, 657.
1169 U. Wannagat, K. Giesen, and F. Rabet, *Z. anorg. Chem.*, 1971, **382**, 195.
1170 U. Wannagat and L. Gerschler, *Z. anorg. Chem.*, 1971, **383**, 249.
1171 F. Rabet and U. Wannagat, *Z. anorg. Chem.*, 1971, **384**, 115.
1172 W. Wolfsberger, H. H. Pickel, and H. Schmidbaur, *Chem. Ber.*, 1971, **104**, 1830.
1173 W. Wolfsberger and H. H. Pickel, *Z. anorg. Chem.*, 1971, **384**, 131.

As the ^{19}F n.m.r. chemical shifts of *p*-$FC_6H_4NHMMe_3$ (M = C or Si) are nearly identical, it was concluded that there is little or no π-interaction in the nitrogen–silicon bond.[1174] For the compounds *p*-$XC_6H_4N(SiMe_3)_2$ it was found that the methyl chemical shift is relatively insensitive to X.[1175] The ^1H n.m.r. spectra of the new compounds $R^1R^2NS(O)R^3CR^4SiMe_3$ show the existence of a centre of chirality.[1176] $(Me_3Si)_2C{=}N{=}N$ reacts with $MeO_2CC{\equiv}CCO_2Me$ to yield a substance which shows two Me_3Si and two MeO resonances and thus must be (319) or (320). On recrystal-

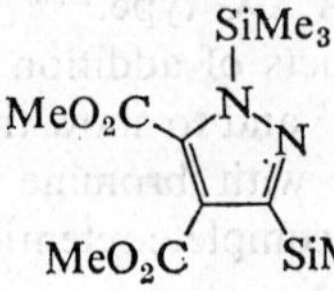

(319)

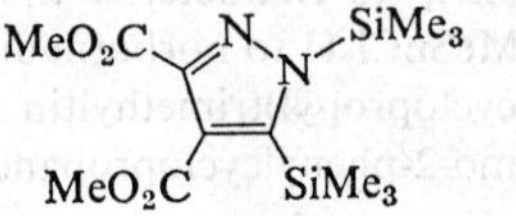

(320)

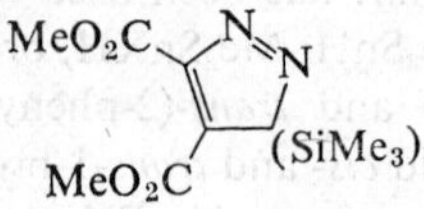

(321)

lization this gives a compound with one $SiMe_3$ and two OMe resonances, which is thought to be (321).[1177] The ^1H n.m.r. spectra of a number of F_3Si— and —SiF_2— amines have been reported and discussed.[1178] Si_2Cl_6 reacts with pyridine to yield (322) as characterized by its AA′LL′X ^1H n.m.r. spectrum.[1179]

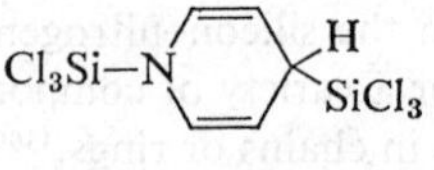

(322)

It has been suggested that instead of using a lanthanide shift reagent, a germanium porphyrin can be used. Thus for (323) (R^1 = H or Ph) the R^2 groups may be put on as a Grignard reagent. For example, when R^1 = octyl separate signals are observed for each proton on account of the ring current effect of the porphyrin; data were also reported for R^1 = $(Cp)Fe(C_5H_4{-})$.[1180]

The ^1H n.m.r. spectra of 1 : 1 adducts of Me_2SnCl_2 with *NN*′-bis-(salicylaldehyde)ethylenedi-imine or *NN*′-bis(benzaldehyde)ethylenedi-imine show considerable changes in $^2J(Sn{-}CH_3)$ as the concentration of di-imine changes, implying co-ordination of nitrogen to tin.[1181] Similarly,

[1174] C. H. Yoder, *Inorg. Nuclear Chem. Letters*, 1971, **7**, 637.
[1175] K. Witke, P. Reich, and H. Kriegsmann, *Z. anorg. Chem.*, 1971, **381**, 280.
[1176] H. Schmidbaur and G. Kammel, *Chem. Ber.*, 1971, **104**, 3252.
[1177] D. Seyferth and T. C. Flood, *J. Organometallic Chem.*, 1971, **29**, C25.
[1178] W. Airey, G. M. Sheldrick, B. J. Aylett, and I. A. Ellis, *Spectrochim. Acta*, 1971, 27A, 1505.
[1179] D. Kummer and H. Köster, *Angew. Chem. Internat. Edn.*, 1971, **10**, 412.
[1180] J. E. Maskasky and M. E. Kenney, *J. Amer. Chem. Soc.*, 1971, **93**, 2060.
[1181] K. Kawakami, M. Miya-Uchi, and T. Tanaka, *J. Inorg. Nuclear Chem.*, 1971, **33** 3773.

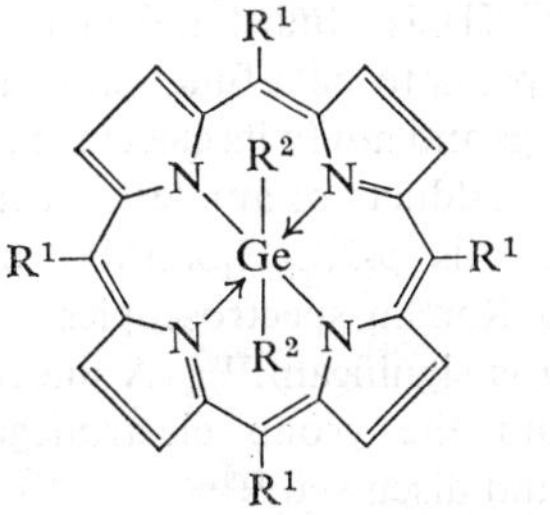

(323)

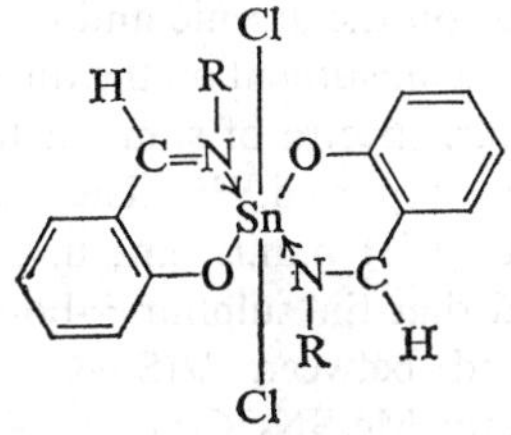

(324)

^1H n.m.r. has been used to assist characterization of (324),[1182] and to study the nature of methanol solutions of a series of bis(acetylacetone)ethylenediimine adducts of organotin(IV) halides.[1183] A ^{19}F n.m.r. study of diadducts of SnF_4 with 19 aromatic amine oxides has been reported. The results indicate that the steric nature of the ligand is probably more important than the base strength in determining the relative chemical shifts. The *trans*-isomer was identified in all of the complexes, whereas the *cis*-isomer is present only with the less bulky ligands.[1184]

Other Complexes.—The n.m.r. spectrum of $C_6F_5CH_2CH_2SiF_3$ is unusual in that no coupling was detected between the CH_2Si and SiF_3 groups.[1185] The use of XeF_2 as a fluorinating agent for organosilicon compounds has been investigated and ^1H and ^{19}F n.m.r. used to assist characterization of the products.[1186] The ^1H n.m.r. spectra for Me_3MXMe (M = C, Si, Ge, or Sn; X = O or S) have been reported and variations in $^1J(^{13}C–H)$ discussed.[1187] The redistribution reactions of Me_2SiX_2 and $MeGeX_3$, and $MeSiX_3$ and Me_2GeX_2, have been followed quantitatively by ^1H n.m.r.[1188]

$^1J(^{31}P–^1H)$ has been measured for $PhPH_2$, $PhPHMMe_3$ (M = C, Si, Ge, or Sn), and $(Me_3Si)_2PH$ and compared with $^1J(^{15}N–^1H)$ for the nitrogen analogues. It was concluded that the changes can be adequately interpreted by considering the redistribution of electrons in the σ-bonding framework, and it was not necessary to invoke π-bonding.[1189] The relative signs of the n.m.r. coupling constants have been measured for $Me_3SnPHPh$ and Me_3SnPPh_2 and the results were briefly discussed.[1190] A number of stannyl arsines of the types Me_3SnAsR_2, $(Me_3Sn)_2AsR$, and $(Me_3Sn)_3As$ have been

1182 N. S. Biradar and V. H. Kulkarni, *J. Inorg. Nuclear Chem.*, 1971, **33**, 2451.
1183 R. Cefalù, L. Pellerito, and R. Barbieri, *J. Organometallic Chem.*, 1971, **32**, 107.
1184 C. E. Michelson and R. O. Ragsdale, *Inorg. Chem.*, 1970, **9**, 2718.
1185 J. M. Birchall, R. N. Haszeldine, M. J. Newlands, P. H. Rolfe, D. L. Scott, A. E. Tipping, and D. Ward, *J. Chem. Soc* (*A*), 1971, 3760.
1186 J. A. Gibson and A. F. Janzen, *Canad. J. Chem.*, 1971, **49**, 2168.
1187 C. H. Yoder and R. Schenck, *J. Inorg. Nuclear Chem.*, 1971, **33**, 2697.
1188 K. Moedritzer, *Z. Naturforsch.*, 1971, **26b**, 517.
1189 P. G. Harrison, S. E. Ulrich, and J. J. Zuckerman, *J. Amer. Chem. Soc.*, 1971, **93**, 2307.
1190 P. G. Harrison, S. E. Ulrich, and J. J. Zuckerman, *Inorg. Nuclear Chem. Letters*, 1971, **7**, 865.

prepared. It was found that ^{2}J(Sn–C–H) is almost independent of the substituent on the arsenic and only appears to be a function of the period to which the substituent on the tin belongs and not to its electronegativity.[1191]

^{1}H N.m.r. spectra of some methyltin adducts of amides and thioamides have been reported.[1192] The compounds p-$XC_6H_4SSnMe_3$ have been examined by ^{1}H n.m.r., i.r., u.v., and Raman spectroscopies, and it was concluded that tin–sulphur π-bonding is significant.[1193] A linear plot has been found between ^{2}J(Sn–C–H) and the group electronegativity of CH_nX_{3-n} for $Me_3SNOC(O)CH_nX_{3-n}$ and discussed.[1194]

8 Compounds of Groups V, VI, VII, and Xenon

Information concerning complexes of these elements can be found at the following sources: S_4N_2,[158] HNO_2, $[NO_2]^-$,[159] phosphetane oxides and phosphetanium salts,[95] $(H_3Si)_3E$ (E = P or Sb),[102] $Me_nH_{3-n}PPF_5$,[106] MePhPPPhMe,[107] R_2P,PF_2 and the BH_3 adducts (R = H or CF_3),[108] $N_3P_3F_{6-n}R_n$,[112] $Ph(S)PNEt_2$,[121] $Me_2EC_6H_4X$ (E = P or As),[126] $RC{\equiv}CEMe_n$ (E = P, As, or Se),[140] $(EtO)_2P(O)CH_2X$,[141] $CHN_3P_2(NR_2)_4$,[142] $R_nP(CN)_{3-n}$,[162] $[R_4P]^+$,[163] $[Fe(CO)_3]_2(AsMe)_4$, $[Mn(CO)_3(AsMe)_4]_2$,[272] $LiMe_2P(BH_3)_2$,[1048] 1,12-$B_{10}H_{10}CHAs$,[1075] $Me_3P{=}NSiMe_2X$,[1172] $Sb(CHCH_2CH_2)_3$,[143] (Cp)Fe(CO)(μ-EPh)$_2$ (E = S, Se, or Te).[316]

The application of chlorine, bromine, and iodine n.m.r. spectroscopy to the study of physico-chemical processes in liquids has been reviewed.[1195]

Phosphorus–Nitrogen Compounds.—Full ^{1}H, ^{19}F, and ^{31}P n.m.r. data, including $^{1}J(^{15}N–^{1}H)$ and $^{2}J(^{15}N–^{19}F)$, have been reported for H_2NPF_2.[1196] Ph_nPF_{5-n} reacts with (325) (X = $SiMe_3$) to yield (325) (X = PPh_nF_{4-n}) and it was found that the equivalence or non-equivalence of the axial or equatorial fluorine atoms is dependent on the position of the methyl group in the piperidyl ring and the position of the ring with reference to the trigonal-bipyramidal structure of the phosphorus.[1197] Treatment of R_2PCl

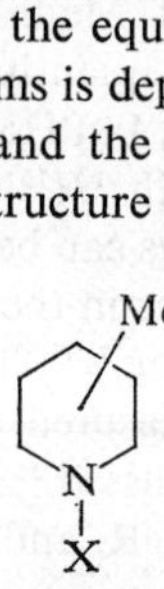

(325)

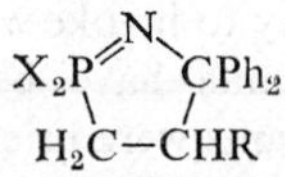

(326)

1191 J. W. Anderson and J. E. Drake, *Canad. J. Chem.*, 1971, **49**, 2524.
1192 G.-E. Matsubayashi, M. Hiroshima, and T. Tanaka, *J. Inorg. Nuclear Chem.*, 1971, **33**, 3787.
1193 T. A. George, *J. Organometallic Chem.*, 1971, **31**, 233.
1194 N. W. G. Debye, D. E. Fenton, S. E. Ulrich, and J. J. Zuckerman, *J. Organometallic Chem.*, 1971, **28**, 339.
1195 C. Hall, *Quart. Rev.*, 1971, **25**, 87.
1196 D. W. H. Rankin, *J. Chem. Soc. (A)*, 1971, 783.
1197 M. J. C. Hewson, S. C. Peake, and R. Schmutzler, *Chem. Comm.*, 1971, 1454.

with $[HNY]^-$ yields R_2PNHY, which is in equilibrium with $R_2PH{=}NY$, and $R_2PPR_2{=}NY$; full ^1H and ^{31}P n.m.r. data were reported including a value of J(P–P) of 250 Hz.[1198] ^1H N.m.r. has been used to demonstrate that when $X_2PN{=}CPh_2$ reacts with $CH_2{=}CHR$, the unsubstituted end of the olefin adds to the phosphorus to yield (326).[1199] $PF_3{=}NPF_2$ has two fluorine resonances and the ^{31}P n.m.r. spectrum consists of a triplet and a quartet consistent with this formulation.[1200] $Me_3SiN{=}PR^2_3$ reacts with $R^1_2PF_3$ or R^1PF_4 to yield $[R^1_nP(N{=}PR^2_3)_{4-n}]^+$ and $[R_nPF_{6-n}]^-$ and full ^1H, ^{19}F, and ^{31}P n.m.r. data were reported.[1201] Treatment of Me_3SiNMe-$CONMeSiMe_3$ with R_nPF_{5-n} yields (327) and (328), as investigated by ^1H and ^{31}P n.m.r.[1202]

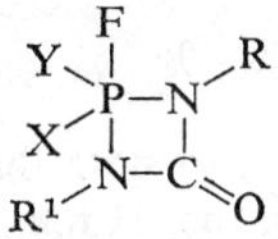

(327)

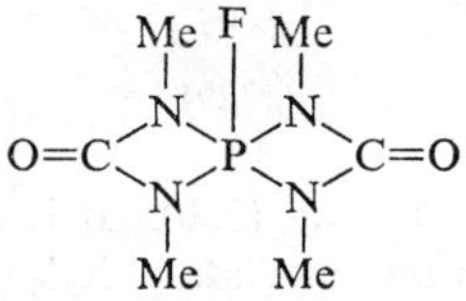

(328)

The mixture of $Cl_3P{=}NPCl_2OBCl_3$ and $Cl_3P{=}NPCl_2O$, formed by the reaction of $Cl_3PNPCl_3^+\,BCl_4^-$ and SO_2, is confirmed by the observation of two AB patterns in the ^{31}P n.m.r. spectrum and a ^{11}B n.m.r. signal typical of a four-co-ordinate boron at − 6.9 p.p.m.[1203] The reaction of $Me_3SiNMePCl_2NSO_2Cl$ with PCl_5 or $PhPCl_2$ to yield $[Cl_3P(\mu NMe)]_2$, $Cl_3P{=}NSO_2Cl$, and $PhPCl_2NSO_2Cl$ has been followed by ^{31}P n.m.r.[1204] Reaction between NH_4NCS and $[Cl_3P{=}NPCl_3]^+\,[BCl_4]^-$ leads to complete substitution of the chlorine atoms by thiocyanate groups. The salt $[(SCN)_3P{=}NP(NCS)_3][B(NCS)_4]$ contains the previously unknown anion $[B(NCS)_4]^-$. The ^{11}B chemical shift of + 17.7 p.p.m. and ^{31}P chemical shift of + 67.0 p.p.m. are taken as evidence of N-bonding by the thiocyanate.[1205]

Treatment of $(Me_2N)_3PO$ with methyl iodide yields $[(Me_2N)_3POP(O)$-$(NMe_2)_2]^+\,I^-$, as characterized by two methyl resonances and two ^{31}P signals.[1206] The ^1H n.m.r. spectra of XPRYPRX show 'virtual coupling' effects.[1207] Full ^1H and ^{31}P n.m.r. data, including J(P–P), have been reported for $(Me_3P{=}N)_3P$, $(Me_3P{=}N)_2PMe$, $[Me_3P{=}N(PMe_2)_2]^+\,Cl^-$, and $[Me_2S(O)N(PMe_2)_2]^+\,Cl^-$.[1208]

1198 A. Schmidpeter and H. Rossknecht, *Z. Naturforsch.*, 1971, **26b**, 81.
1199 A. Schmidpeter and W. Zeiss, *Angew. Chem. Internat. Edn.*, 1971, **10**, 396.
1200 G. E. Graves, D. W. McKennon and M. Lustig, *Inorg. Chem.*, 1971, **10**, 2083.
1201 W. Stadelmann, O. Stelzer, and R. Schmutzler, *Z. anorg. Chem.*, 1971, **385**, 142.
1202 R. E. Dunmur and R. Schmutzler, *J. Chem. Soc.* (*A*), 1971, 1289.
1203 H. Binder and E. Fluck, *Z. anorg. Chem.*, 1971, **381**, 21.
1204 U. Bieller and M. Becke-Goehring, *Z. anorg. Chem.*, 1971, **381**, 209.
1205 H. Binder, *Z. anorg. Chem.*, 1971, **383**, 279.
1206 C. Anselmi, B. Macchia, F. Macchia, and L. Monti, *Chem. Comm.*, 1971, 1152.
1207 H. G. Metzinger, *Org. Magn. Resonance*, 1971, **3**, 485.
1208 W. Wolfsberger, H. H. Pickel, and H. Schmidbaur, *Z. Naturforsch.*, 1971, **26b**, 979.

$[Cl_3PNPCl_3]^+[BCl_4]^-$ reacts with AsF_3 to yield $(NPF_2,PF_5)_3$ and BF_3; the BF_3 was identified as its etherate by ^{11}B and ^{19}F n.m.r. The ^{19}F n.m.r. spectrum of $[NPF_2,PF_5]_3$ in pyridine shows an AM_4X spectrum for the PF_5 unit and an A_2X spectrum for the PF_2 unit, consistent with the structure (329). ^{19}F N.m.r. data were also reported for $(NPF_2)_3$, $(NPF_2)_4$, and py,PF_5.[1209] ^1H and ^{31}P n.m.r. data have been reported for $(R_2NPCl_2N)_2SO_2$,

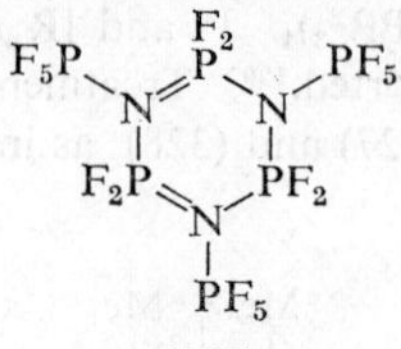

(329)

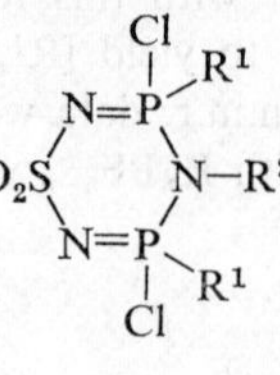

(330)

$(Me_2N)_2PClN_2SO_2$, and (330). It is interesting to note that (330) (R^1 = NMe_2) gives an intermediate $[AX_6]_2$ pattern in the ^1H n.m.r. spectrum.[1210]

The reaction of PCl_5 with NH_4Cl has been studied and the composition of phosphonitrilic chlorides was determined by ^{31}P n.m.r.[1211] Treatment of $N_3P_3Cl_6$ with Ph_2Mg yields (331) (X = Cl) which will react with Me_2NH to give (331) (X = Me_2N), which has two methyl groups in the ^1H n.m.r. spectrum with the same apparent *J*(P–Me), and $N_6P_6Ph_7Cl_5$; the ^{31}P n.m.r. spectrum of this latter compound shows the structure to be (332).[1212]

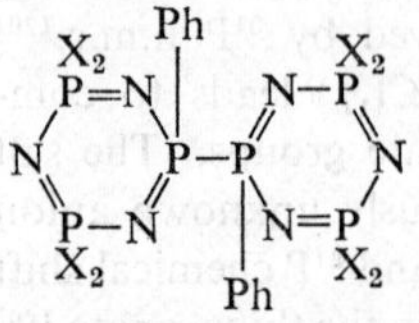

(331)

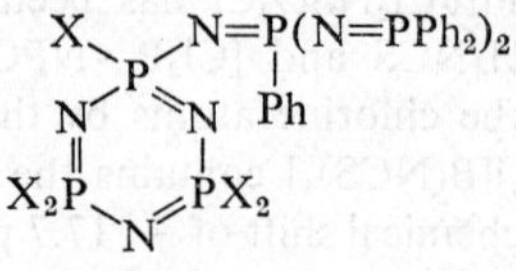

(332)

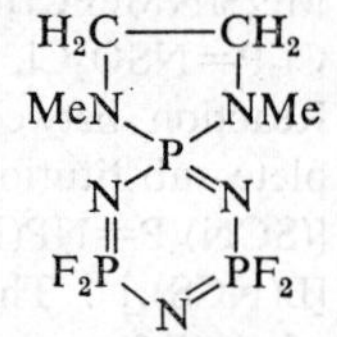

(333)

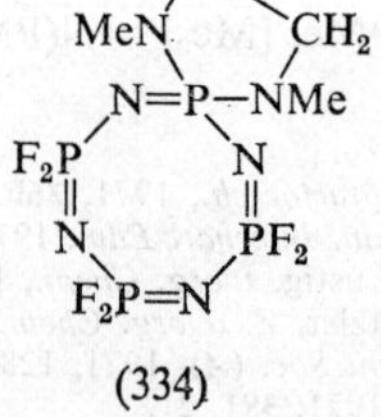

(334)

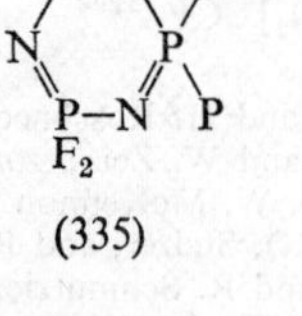

(335)

[1209] H. Binder, *Z. anorg. Chem.*, 1971, **383**, 130.
[1210] U. Klingebiel and O. Glemser, *Chem. Ber.*, 1971, **104**, 3804.
[1211] J. Emsley and P. B. Udy, *J. Chem. Soc. (A)*, 1971, 768.
[1212] M. Biddlestone and R. A. Stone, *J. Chem. Soc. (A)*, 1971, 2715.

$N_3P_3F_6$ or $N_4P_4F_8$ react with $HMeNCH_2CH_2NHMe$ to yield (333) or (334), respectively. The CH_2 groups show only a doublet in the ^1H n.m.r. spectrum, although the protons would be expected to be non-equivalent. Three explanations are possible: (*a*) rapid inversion of the nitrogens, but the spectrum is invarient to − 80 °C; (*b*) planarity of the imidazolidine ring; or (*c*) accidental chemical shift equivalence of the two protons.[1213] $P_3N_3F_6$ reacts with amines to yield products such as (335), and ^1H, ^{19}F, and ^{31}P n.m.r. data were reported.[1214] ^{31}P and ^1H n.m.r. data have been reported for $P_6N_7Cl_9$ and (336).[1215]

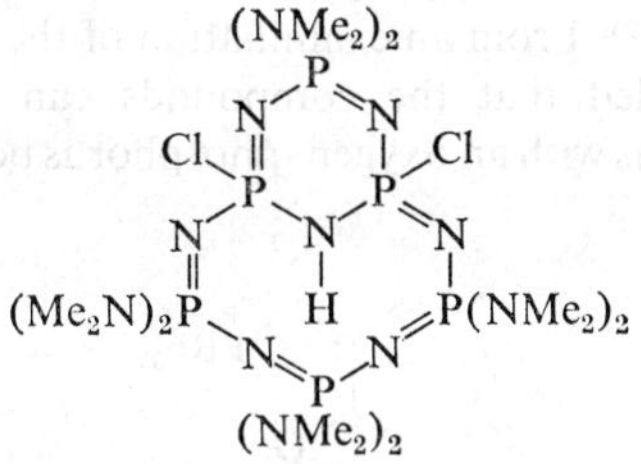

(336)

^1H N.m.r. data have been given for compounds such as $Bu^t_2P(SiMe_3)$-$NSiEt_3$ and have provided evidence for restricted rotation about the silicon–phosphorus bond.[1216]

Other Phosphorus-containing Compounds.—The first stable penta-alkylphosphorus compound, (337), has been prepared and ^1H and ^{31}P n.m.r. data reported.[1217] From a full analysis of the ^1H n.m.r. spectra of trimethyl-allylidene- and -benzylidene-phosphorane, it was concluded that there is delocalization of the ylidic carbanionic charge into the π-system.[1218] The ligands $Ph_2PCH_2CH_2OPR_2$ have been prepared with a view to determining ^2J(P–P) for complexes with metals, and ^1H and ^{31}P n.m.r. data on the free ligands were reported.[1219] The ^1H n.m.r. spectra of (338) (X = N, P, As, or Sb) are all similar, suggesting aromatic character. The α-proton moves to low field, being at τ 1.9 in pyridine and τ 0.7 in (338) (X = Sb).[1220, 1221] The deuteriation of the α-hydrogen of (339) has been followed by ^1H n.m.r.[1222]

The ^1H and ^{31}P n.m.r. data for a series of alkoxyphosphonium hexa-chloroantimonates have been reported and compared with similar data for

1213 T. Chivers and R. Hedgeland, *Inorg. Nuclear Chem. Letters*, 1971, **7**, 767.
1214 E. Niecke, H. Thamm, and G. Flaskerud, *Chem. Ber.*, 1971, **104**, 3729.
1215 W. Harrison, R. T. Oakley, N. L. Paddock, and J. Trotter, *Chem. Comm.*, 1971, 357.
1216 O. J. Scherer and W. Gick, *Chem. Ber.*, 1971, **104**, 1490.
1217 E. W. Turnblom and T. J. Katz, *J. Amer. Chem. Soc.*, 1971, **93**, 4065.
1218 W. Malisch, D. Rankin, and H. Schmidbaur, *Chem. Ber.*, 1971, **104**, 145.
1219 S. O. Grim, A. W. Yankowsky, and W. L. Briggs, *Chem. and Ind.*, 1971, 575.
1220 A. J. Ashe, tert., *J. Amer. Chem. Soc.*, 1971, **93**, 3293.
1221 A. J. Ashe, tert., *J. Amer. Chem. Soc.*, 1971, **93**, 6691.
1222 B. D. Cuddy, J. C. F. Murray, and B. J. Walker, *Tetrahedron Letters*, 1971, 2397.

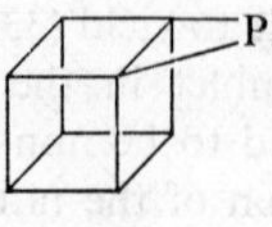

(337)

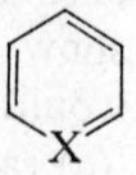

(338)

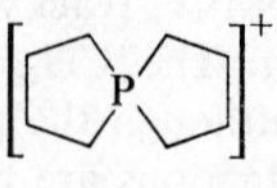

(339)

the corresponding phosphoryl compounds. The ^{31}P chemical shifts cannot be explained by the theoretical treatment of Letcher and Van Wazer, and alternative empirical correlations have been suggested.[1223] The ^{1}H and ^{31}P n.m.r. spectra of $Me_2P(S)-P(S)Ph_2$ has been measured and $^{1}J(P-P)$ was found to be 22.5 Hz.[1224] From an examination of the ^{31}P chemical shifts of (340) it was concluded that the compounds can exist as either five-co-ordinate phosphorus with an oxygen–phosphorus bond and a ^{31}P chemical

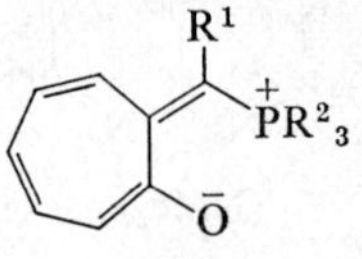

(340)

shift of *ca.* + 50 p.p.m., or four-co-ordinate phosphorus without an oxygen–phosphorus bond and with a ^{31}P chemical shift of *ca.* – 10 p.p.m.[1225]

The course of the reaction of phenyl and methyl polyphosphates with phenol has been followed by ^{31}P n.m.r. spectroscopy.[1226] The ^{31}P n.m.r. spectrum of (341) shows an AA′BB′ pattern, but the ^{1}H n.m.r. spectrum of the CH_2 group is only a singlet.[1227]

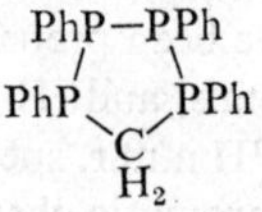

(341)

The ^{19}F n.m.r. spectrum of t-$C_4F_9OPF_4$ shows rapid interconversion of axial and equatorial fluorines in the PF_4 unit. Similarly, (t-$C_4F_9O)_2PF_3$ has only one ^{19}F doublet in the ^{19}F n.m.r. spectrum, but it is not known whether this is due to the t-C_4F_9O groups being axial or whether the molecule is fluxional.[1228] Unlike the sulphur analogue, $(PhO)Ph_2PF_2$

1223 M. Murray, R. Schmutzler, E. Gründemann, and H. Teichmann, *J. Chem. Soc. (B)*, 1971, 1714.
1224 J. Koketsu, M. Okamura, and Y. Ishii, *Inorg. Nuclear Chem. Letters*, 1971, **6**, 15.
1225 I. Kawamoto, T. Hata, Y. Kishida, and C. Tamura, *Tetrahedron Letters*, 1971, 2417.
1226 H. Berger, *Chem. Ber.*, 1971, **104**, 691.
1227 M. Baudler, J. Vesper, P. Junkes, and H. Sandmann, *Angew. Chem. Internat. Edn.*, 1971, **10**, 940.
1228 D. E. Young and W. B. Fox, *Inorg. Nuclear Chem. Letters*, 1971, **7**, 1033.

shows no temperature effects in the ^{19}F n.m.r. spectrum, even at $-80\,°C$. Thus the effects previously observed in the sulphur analogue are now explained by slow phosphorus–sulphur bond rotation. ^{19}F and ^{31}P n.m.r. data were also reported for $(ArO)_{3-n}PR_nF_2$ and $[(ArO)_4P]^+$ (Ar = Ph or C_6F_5).[1229]

The ^1H n.m.r. spectra of 2-chloro-, 2-methoxy-, and 2-phenoxy-1,3,2-dioxaphosphorinanes have been fully analysed by an iterative method. The spectra show that the ring phosphites exist predominantly in one conformation with the P—Cl, P—OMe, and P—OPh groups in the axial position.[1230] ^{19}F and ^{31}P n.m.r. have been used to determine the structure of $[MeNPFCl_2]_2$, $C_6H_4O_2PFCl_2$, $(C_6H_4O_2)_2PCl$, $C_6H_4O_2PF_2Cl$, and $C_6H_4O_2$-PF_3.[1231] Similar compounds, $(C_6H_4O_2)_2PF$, $RP(O_2C_6H_4)_2$, R^1R^2PF-$(O_2C_6H_4)$, and $R_3P(O_2C_6H_4)$ have been examined by other workers and the compounds were characterized by ^1H, ^{19}F, and ^{31}P n.m.r. spectroscopy. Catechol occupies one axial and one equatorial position and thus $(C_6H_4O_2)_2$-PF is unusual with an equatorial fluorine.[1232]

Intramolecular exchange in Me_2PF_3 has been investigated by means of ^1H n.m.r. and double resonance.[1233] The slow disproportionation of P_2I_4 to PI_3 and polymeric P_nI_x species has been examined by ^{31}P n.m.r.[1234]

Other Compounds.—Visible light converts $N_2CR^2COR^1$ into (342). When R^2 = H there is an isomerization shift of *ca.* 3 p.p.m. to high field. This was ascribed to the large diamagnetic anisotropy associated with the strained diazirine ring.[1235] The fluorine–fluorine coupling constant of 16 Hz in $Cs^+[N(SO_2F)COF]^-$ has been taken as evidence for covalent character.[1236] ^{19}F N.m.r. has been used to differentiate between the isomers (343) and (344).[1237]

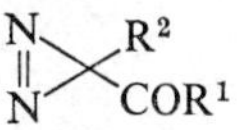

(342)

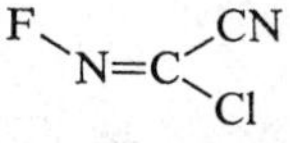

(343)

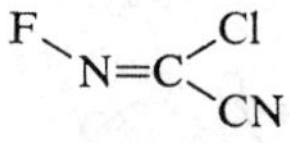

(344)

The ^1H n.m.r. spectra of $[R_nAsPh_{4-n}]^+I^-$ (R = Me or Et) have been discussed. As the positive charge on the arsenic increases, the CH_2 or CH_3 group moves to low field. Thus as the alkyl group is replaced by phenyl, the resonance moves to low field.[1238] The ^1H and ^{19}F n.m.r. spectra of

1229 S. C. Peake, M. Fild, M. J. C. Hewson, and R. Schmutzler, *Inorg. Chem.*, 1971, **10**, 2723.
1230 K. Bergesen and P. Albriktsen, *Acta Chem. Scand.*, 1971, **25**, 2257.
1231 H. Binder, *Z. anorg. Chem.*, 1971, **384**, 193.
1232 G. O. Doak and R. Schmutzler, *J. Chem. Soc.* (*A*), 1971, 1295.
1233 H. Dreeskamp and K. Hildenbrand, *Z. Naturforsch.*, 1971, **26b**, 269.
1234 M. Baudler, P. Junkes, and G. Sadri, *Z. Naturforsch.*, 1971, **26b**, 759.
1235 G. Lowe and J. Parker, *Chem. Comm.*, 1971, 1135.
1236 J. A. Roderiguez and R. E. Noftle, *Inorg. Chem.*, 1971, **10**, 1874.
1237 L. M. Zaborowski and J. M. Shreeve, *Inorg. Chem.*, 1971, **10**, 407.
1238 T. V. Zykova, G. Kh. Kamai, B. D. Chernokal'skii, R. A. Salakhutdinov, and B. E. Abalonin, *J. Gen. Chem.* (*U.S.S.R.*), 1971, **41**, 1049.

(p-$XC_6H_4)_3E$ and (m-$XC_6H_4)_3E$ (X = Cl or F; E = As, Sb, or Bi) and the corresponding oxides have been reported and partially analysed.[1239, 1240] The ^1H n.m.r. spectra of (345) and (346) have been reported; it is interesting that the CH_2 group in (346) gives rise to an AB quartet.[1241]

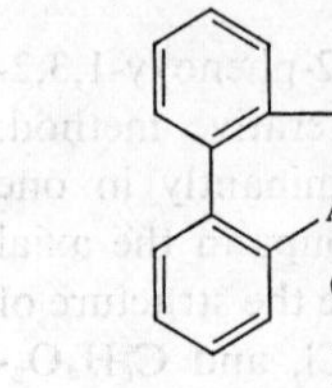

(345)

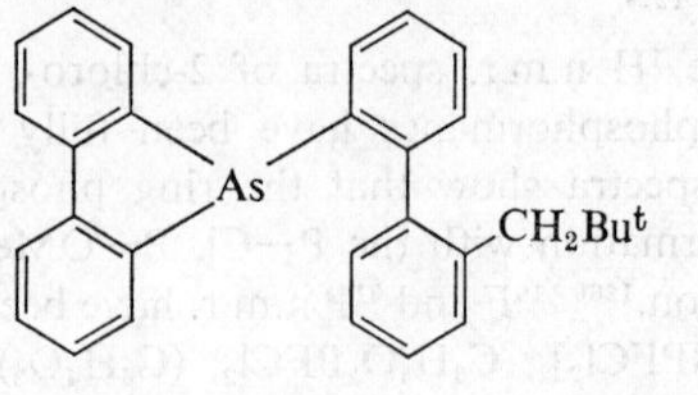

(346)

Both $(H_3MAs)_5$ (M = Si or Ge) show three ^1H n.m.r. resonances, which were accounted for by ring puckering.[1242] Treatment of sulphur with AsF_5 yields a blue solid, $S_8(AsF_6)_2$, which was shown by ^{19}F n.m.r. in HSO_3F to contain the $[AsF_6]^-$ anion.[1243] ^{19}F N.m.r. has been used to show that commercial $HAsF_6$ contains a significant quantity of a dark-green impurity.[1244] The ^1H n.m.r. spectra of $E(S_2CNR_2)_n$ (E = As, Sb, Bi, Se, or Te; R = Me or Et) show no splitting of the resonances and it is thought that this is due to the differences being undetectably small.[1245]

Treatment of $R^1{}_5Sb$ with R^2CO_2H yields $R^1{}_4SbO_2CR^2$. This will form a 1 : 1 adduct with R^3CO_2H. The ^1H n.m.r. spectrum shows that when R^1 is methyl, all the $SbMe_4$ resonances are the same and a low-field signal (τ − 1.50 to − 6.25) due to a hydrogen-bonded hydrogen was observed. It was concluded that the structure is (347).[1246] As only one alkyl resonance was found in the H^1 n.m.r. spectrum of R_2SbCl_3L it was suggested that the

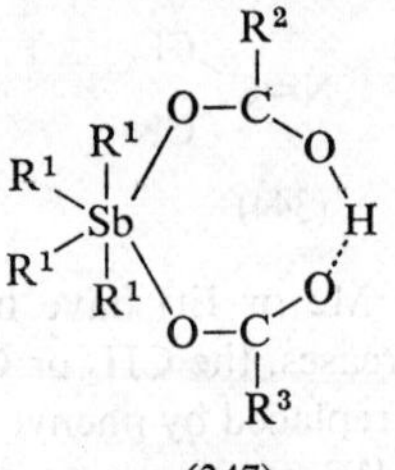

(347)

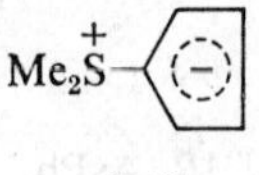

(349)

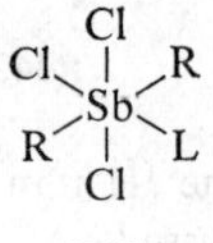

(348)

1239 R. F. de Ketelaere, F. T. Delbeke, and G. P. van der Kelen, *J. Organometallic Chem.*, 1971, **28**, 217.
1240 R. F. De Ketelaere, F. T. Delbeke, and G. P. van der Kelen, *J. Organometallic Chem.*, 1971, **30**, 365.
1241 D. Hellwinkel and B. Knabe, *Chem. Ber.*, 1971, **104**, 1761.
1242 J. W. Anderson and J. E. Drake, *Chem. Comm.*, 1971, 1372.
1243 R. J. Gillespie, J. Passmore, P. K. Ummat, and O. C. Vaidya, *Inorg. Chem.*, 1971, **10**, 1327.
1244 E. W. Lawless, C. J. W. Wiegand, Y. Mizumoto, and C. Weis, *Inorg. Chem.*, 1971, **10**, 1084.
1245 H. C. Brinkhoff and A. M. Grotens, *Rec. Trav. chim.*, 1971, **90**, 252.
1246 H. Schmidbaur and K.-H. Mitschke, *Angew. Chem. Internat. Edn.*, 1971, **10**, 136.

structure is (348).[1247] It has been reported that for the complexes $Ar_{3-n}BiX_n$, separate resonances are found in the ^1H n.m.r. spectrum for the *ortho*-, *meta*-, and *para*-hydrogens.[1248]

The ^1H n.m.r. spectrum of (349) has been analysed and the π-bond orders $p_{r,s}$ estimated from the equation

$$J_{vic} = 7.12 p_{r,s} - 1.18$$

and were found to be within the aromatic range.[1249] ^1H and ^{19}F n.m.r. have been used to characterize trifluoromethylsulphinate esters and trifluoromethylsulphinamides. The compounds $CF_3CH_2OS(O)CF_3$ and CH_3CH_2-$OS(O)CF_3$ show non-equivalent methylene protons.[1250] The compound $N{\equiv}SF_2N{=}SF_2$ shows two triplets in the ^{19}F n.m.r. spectrum.[1251] ^1H and ^{19}F n.m.r. have been used to follow the reactions of the lower fluorides of sulphur with hydrogen sulphide.[1252] The ^{19}F n.m.r. spectrum of SF_4 in the presence of C_6F_5Li at $-$ 40 °C shows the presence of $S(C_6F_5)_4$.[1253] In the

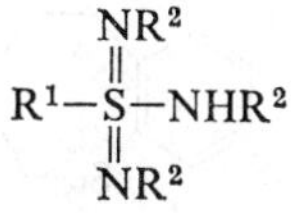

(350)

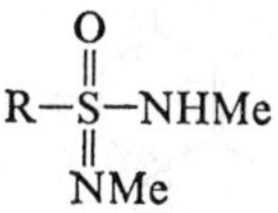

(351)

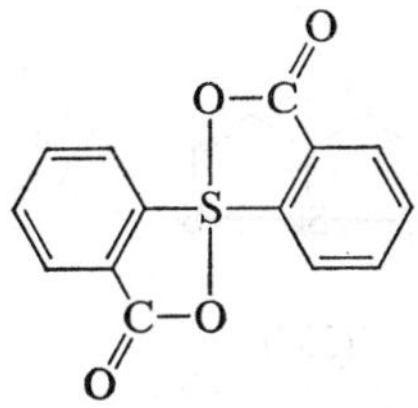

(352)

^{19}F n.m.r. spectrum of $CF_3S(O)F_2CF_2CF_3$, inequivalence of the CF_2 and SF_2 fluorines was found.[1254] ^1H N.m.r. data have also been reported for (350), (351),[1255] and (352).[1256]

It has been reported that (353) has a value of $^2J(^{77}Se\text{–}C\text{–}H)$ of 40 Hz.[1257] ^1H N.m.r. data have been reported for (354) (M = Se or Te), (355),[1258] (356),[1259] (357) (E = S or Se),[1260] compounds of the types (358) and (359),[1261] and the naphthalene analogues,[1262] and of the type (360).[1263]

1247 N. Nishii, Y. Matsumura, and R. Okawara, *J. Organometallic Chem.*, 1971, **30**, 59.
1248 B. C. Smith and C. B. Waller, *J. Organometallic Chem.*, 1971, **32**, C11.
1249 Z. Yoshida, S. Yoneda, and M. Hazama, *Chem. Comm.*, 1971, 716.
1250 D. T. Sauer and J. M. Shreeve, *Inorg. Chem.*, 1971, **10**, 358.
1251 O. Glemser and R. Höfer, *Angew. Chem. Internat. Edn.*, 1971, **10**, 815
1252 B. Meyer, T. V. Oommen, B. Gotthardt, and T. R. Hooper, *Inorg. Chem.*, 1971, **10**, 1632.
1253 W. A. Sheppard, *J. Amer. Chem. Soc.*, 1971, **93**, 5597.
1254 D. T. Sauer and J. M. Shreeve, *Z. anorg. Chem.*, 1971, **385**, 113.
1255 R. Appel and J. Kohnke, *Chem. Ber.*, 1971, **104**, 3875.
1256 I. Kapovits and A. Kálmán, *Chem. Comm.*, 1971, 649.
1257 I. Lalezari, A. Shafiee, and M. Yalpani, *J. Org. Chem.*, 1971, **36**, 2837.
1258 M. Perrier and J. Vialle, *Bull. Soc. chim. France*, 1971, 4591.
1259 E. Ernstbrunner and D. Lloyd, *Annalen*, 1971, **753**, 196.
1260 D. H. Reid, *J. Chem. Soc. (C)*, 1971, 3187.
1261 N. Bellinger, P. Cagniant, D. Cagniant, and M. Renson, *Bull. Soc. chim. France*, 1971, 2689.
1262 P. Cagniant, N. Bellinger, and D. Cagniant, *Bull. Soc. chim. France*, 1971, 2699.
1263 D. Elmaleh, S. Patai, and Z. Rappoport, *J. Chem. Soc. (C)*, 1971, 2637.

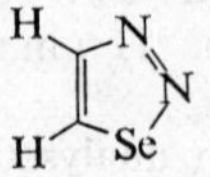

(353)

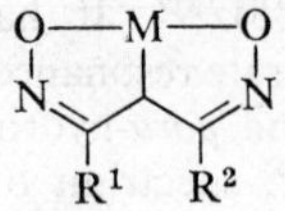

(354)

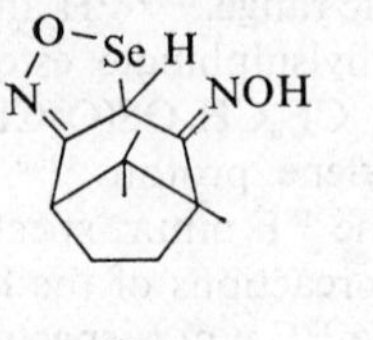

(355)

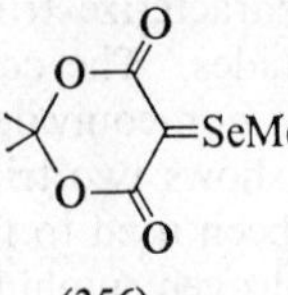

(356)

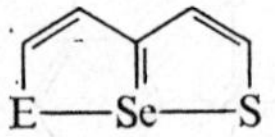

(357)

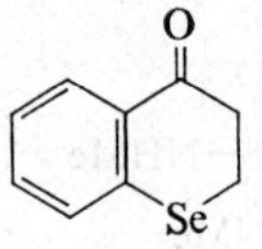

(358)

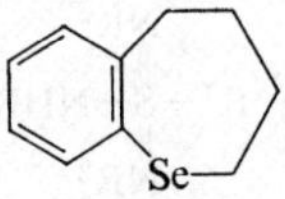

(359)

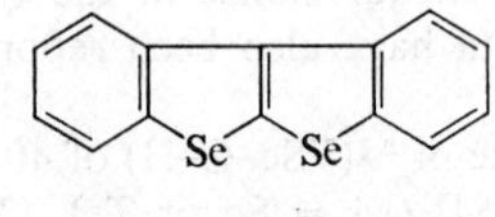

(360)

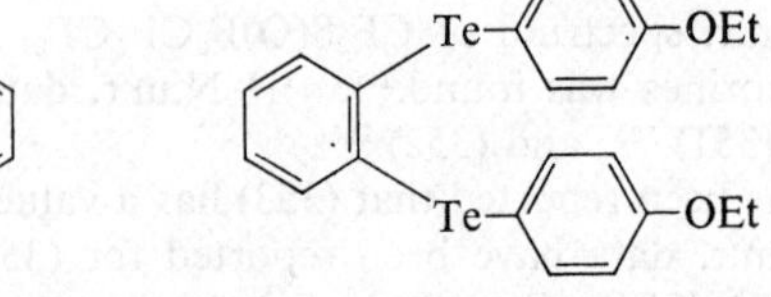

(361)

The ^1H n.m.r. spectrum of the aryltellurium compound (361) has been fully analysed.[1264] The decomposition of Me_2TeI_2 in acetone has been followed by ^1H n.m.r. It was found that signals possibly due to methyl iodide, absorbed water, and $MeTeI_3$ appear. If 50% of methyl or ethyl iodide is added to the reaction mixture, then the decomposition is stopped, but iodobenzene has no effect. No evidence was found for either an ethyl– or a phenyl–tellurium bond.[1265] ^1H N.m.r. data have been reported for the 2 : 1 adducts between olefins and $TeCl_4$,[1266] and full ^1H and ^{19}F n.m.r. data have been reported for some dialkylaminotellurium(VI) fluorides, *e.g.* $TeF_5(NMe_2)$.[1267] ^1H N.m.r. data have also been reported for $Te(dtc)_2$ and $Te(dtc)_4$ (dtc = diethyldithiocarbamate). The ^1H n.m.r. spectrum of

[1264] K. Müllen, *Org. Magn. Resonance*, 1971, **3**, 331.
[1265] K. P. Shrestha and J. S. Thayer, *J. Organometallic Chem.*, 1971, **27**, 79.
[1266] D. Elmaleh, S. Patai, and Z. Rappoport, *J. Chem. Soc.* (*C*), 1971, 3100.
[1267] G. W. Fraser, R. D. Peacock, and P. M. Watkins, *J. Chem. Soc.* (*A*), 1971, 1125.

Te(dtc)$_4$ shows two CH_2 quartets and hence a structure with two ligands unidentate and two didentate was proposed. The broadness of the resonances due to Te(dtc)$_2$ was attributed to paramagnetism.[1268]

The solvent isotope effect on the ^{19}F chemical shifts of the complex fluoro-ions, $[BeF_n]^{2-n}$, $[BF_4]^-$, $[SiF_6]^{2-}$, $[PF_6]^-$, $[FSO_3]^-$, $[F_2PO_4]^-$, $[CF_3CO_2]^-$, and F^- in H_2O and D_2O has been investigated and shifts of up to 2.8 p.p.m. were observed.[1269] The high-resolution ^1H and ^{19}F n.m.r. spectra of the four dihalide ions have been observed in aprotic solvents. The ^1H shielding in each case is 14 p.p.m. less than in the parent molecule. $^1J(^{19}F-^1H)$ in $[FHF]^-$ is 120.5 ± 0.3 Hz.[1270]

Treatment of 2ClF,AsF_5 with OSF_2 yields $[OSClF_2]^+[AsF_6]^-$. In HF solution, the ^{19}F n.m.r. spectrum shows a singlet at 271 p.p.m. from HF, due to $[OSClF_2]^+$, and the lack of a ^{19}F n.m.r. signal attributable to $[AsF_6]^-$ was explained as being due to exchange.[1271] Pulse ^{35}Cl and ^{19}F n.m.r. have been used to investigate ClO_3F over its entire liquid range. It was concluded that the ^{35}Cl relaxation is due to nuclear quadrupole interaction, while the ^{19}F relaxation is dominated by spin–rotation interaction. The results for this quasi-spherical molecule are in accord with rotational diffusion theory, and Hubbard's relationship at the lowest temperatures agrees over the entire range with the extended treatment of McClung.[1272, 1273] The new compounds $HOIOF_4$ and IO_2F_3 have been shown by ^{19}F n.m.r. to exist as a mixture of two components. $HOIOF_4$ exists as the *cis*- and *trans*-isomers whereas IO_2F_3 shows a doublet and triplet for each species, showing that they are (362) and (363) and not (364).[1274]

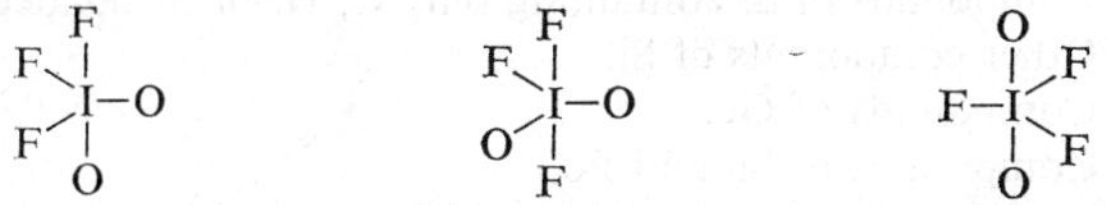

(362) (363) (364)

The density dependence of the ^{129}Xe n.m.r. chemical shifts in O_2 and NO has been investigated and shifts of up to − 1200 Hz at 16.60 MHz recorded. These shifts were attributed to a contact shift which may imply the formation of a true complex with unpaired electron distribution on to the xenon, or may arise by a pseudo-contact mechanism in the collisions.[1275] The ^{19}F n.m.r. spectrum of XeF_4,2SbF_5 has an AB_2 spectrum with ^{129}Xe satellites,

1268 G. St. Nikolov, N. Jordanov, and I. Havezov, *J. Inorg. Nuclear Chem.*, 1971, **33**, 1055.
1269 K. Radley and L. W. Reeves, *J. Chem. Phys.*, 1971, **54**, 4509.
1270 J. S. Martin and F. Y. Fujiwara, *Canad. J. Chem.*, 1971, **49**, 3071.
1271 C. Lau and J. Passmore, *Chem. Comm.*, 1971, 950.
1272 A. A. Maryott, T. C. Farrar, and M. S. Malmberg, *J. Chem. Phys.*, 1971, **54**, 64.
1273 T. C. Farrar, A. A. Maryott, and M. S. Malmberg, *Ber. Bunsengesellschaft Phys. Chem.*, 1971, **75**, 246.
1274 A. Engelbrecht, P. Peterfy, and E. Schandara, *Z. anorg. Chem.*, 1971, **384**, 202.
1275 C. J. Jameson and A. K. Jameson, *Mol. Phys.*, 1971, **20**, 957.

which is consistent with Raman evidence for its existence as $[XeF_3]^+$ $[Sb_2F_{11}]^-$.[1276]

9 Appendix: Compounds not Referred to in Detail

This Appendix contains twenty-one tables, each of which lists compounds which have not been referred to in the main body of the Chapter, but for which n.m.r. data have been reported during the year. The tables list typical compounds from each paper and in this sense are not exhaustive. The tables are organized following the Periodic Table:

Table 1. Compounds of Mg, Sc, Y, Lu, Ti, Zr, Hf, Th, V, Nb, Ta, and Pa.
Table 2. Compounds of Cr, Mo, and W.
Table 3. Compounds of Mn and Re.
Table 4. Derivatives of ferrocene.
Table 5. Compounds of Fe containing only one π-cyclopentadienyl ring attached to any one iron atom.
Table 6. Other compounds of Fe.
Table 7. Compounds of Ru and Os.
Table 8. Compounds of Co.
Table 9. Compounds of Rh.
Table 10. Compounds of Ir.
Table 11. Compounds of Ni.
Table 12. Compounds of Pd and Pt.
Table 13. Compounds of Cu, Ag, Au, Be, Zn, Cd, and Hg.
Table 14. Compounds of B.
Table 15. Compounds of Al, Ga, In, and Tl.
Table 16. Compounds of Si containing only C, H, or Si bonded to Si.
Table 17. Other compounds of Si.
Table 18. Compounds of Ge.
Table 19. Compounds of Sn and Pb.
Table 20. Compounds of P, As, Sb, and Bi.
Table 21. Compounds of S, Se, and Te.

Table 1 *References to further papers containing n.m.r. data on compounds of* Mg, Sc, Y, Lu, Ti, Zr, Hf, Th, V, Nb, Ta, *and* Pa

	Ref.
Chlorophyll	1277
Mg(octaethylporphyrin)(quinoline)$_2$	1278
Sc(octaethylporphyrin)L (L = acac or acetate)	1278
Y(acac)$_3$	1279
M(hfac)$_3$L$_2$ (M = Y or Lu)	1280
$(Et_4N)[Cl_2MeTiCl_3TiCl_2Me]$	1281

1276 R. J. Gillespie, B. Landa, and G. J. Schrobilgen, *Chem. Comm.*, 1971, 1543.
1277 H. Budzikiewicz and K. Taraz, *Tetrahedron*, 1971, **27**, 1447.
1278 J. W. Buchler, G. Eikelmann, L. Puppe, K. Rohbock, H. H. Schneehage, and D. Weck, *Annalen*, 1971, **745**, 135.
1279 J. K. Przystal, W. G. Bos, and I. V. Liss, *J. Inorg. Nuclear Chem.*, 1971, **33**, 679.
1280 M. F. Richardson and R. E. Sievers, *Inorg. Chem.*, 1971, **10**, 498.
1281 R. J. H. Clark and M. Coles, *Chem. Comm.*, 1971, 1587.

Table 1 (*cont.*)

	Ref.
$\{(CF_3)_2CO\}_2TiCl_2$	1282
$(R_2N)_3TiGePh_3$	1283
$Ti(CH_2Ph)_4$	1284
$L_nTiCH_2SiMe_3$	1285
$(Cp)_2Ti(o\text{-}C_6H_4COO)$	1286
$(Cp)_2Ti\{OP(CF_3)_2\}_2$, $(Cp)_2TiO$	1287
$Ti(SnPh_3)_3$, $Ti(NPhCONMe_2)_3$, $Ti(Cp)_2(NPhCONMe_2)$, $Ti(Cp)_2Cl_2$	1288
$Ti(Cp)_2(o\text{-}SC_6H_4COO)$	1289
$Ti(C_7H_7)(C_7H_9)$	1290
$(R_2N)_nTiI_{4-n}$	1291
$(Cp)_2M(N_3)_2$, $[(C_5H_5)_2M(N_3)]_2O$ (M = Ti or Zr)	1292
$(Cp)_2MCl(N{=}CR_2)$, $(Cp)_2M(N{=}CR_2)_2$ (M = Ti, Zr, or Hf)	1293
$M\{(CF_3)_2CHO\}_4$ (M = Ti, Zr, or Hf)	1294
$M(BH_4)_4$ (M = Zr, Hf, or Th)	1295
$(Cp)_2Zr(OR)_nCl_{2-n}$, $Zr(OR)_4$	1296
VO(octaethylporphyrin)	1278
$(Cp)V(CO)_2\{PhP(CH_2CH_2PPh_2)_2\}$	1297
$NbCl(OR^1)_2(S_2PR^2)_2$ (R^1 = Me, Et, or Pr^i; R^2 = C_6H_{11} or MeO)	1298
M(*NN*-diethyldithiocarbamate)$_nX_{5-n}$ (M = Nb, Ta, or Pa)	1299

Table 2 *References to further papers containing n.m.r. data on compounds of* Cr, Mo, *and* W

	Ref.
$HM(Cp)(CO)_3$ (M = Cr, Mo, or W)	1300
$H_3SiM(Cp)(CO)_3$ (M = Cr, Mo, or W)	1301
$[M(CO)_5(C{\equiv}CMe)]^-$ (M = Cr, Mo, or W)	1302

[1282] A. P. Conroy and R. D. Dresdner, *Inorg. Chem.*, 1970, **9**, 2739.
[1283] H. Burger and H.-J. Neese, *J. Organometallic Chem.*, 1971, **32**, 223.
[1284] W. Fruser, K.-H. Thiele, P. Zdunneck, and F. Brune, *J. Organometallic Chem.*, 1971, **32**, 335.
[1285] B. Wozniak, J. D. Ruddick, and G. Wilkinson, *Inorg. Chim. Acta*, 1971, **5**, 3116.
[1286] I. S. Kolomnikov, T. S. Lobeeva, V. V. Gorbachevskaya, G. G. Aleksandrov, Yu. T. Struckhov, and M. E. Vol'pin, *Chem. Comm.*, 1971, 972.
[1287] W. J. Reagan and A. B. Bury, *Inorg. Nuclear Chem. Letters*, 1971, **7**, 741.
[1288] M. F. Lappert and A. R. Sanger, *J. Chem. Soc.* (*A*), 1971, 1314.
[1289] D. N. Sen and U. N. Kantak, *Indian J. Chem.*, 1971, **9**, 254.
[1290] H. O. van Oven and H. J. de Liefde Meijer, *J. Organometallic Chem.*, 1971, **31**, 71.
[1291] H. Bürger, C. Kluess, and H.-J. Neese, *Z. anorg. Chem.*, 1971, **381**, 198
[1292] R. S. P. Coutts and P. C. Wailes, *Austral. J. Chem.*, 1971, **24**, 1075.
[1293] M. R. Collier, M. F. Lappert, and J. McMeeking, *Inorg. Nuclear Chem. Letters*, 1971, **7**, 689.
[1294] K. S. Mazdiyasni, B. J. Schaper, and L. M. Brown, *Inorg. Chem.*, 1971, **10**, 889.
[1295] M. Ehemann and H. Nöth, *Z. anorg. Chem.*, 1971, **386**, 87.
[1296] D. R. Gray and C. H. Brubaker, jun., *Inorg. Chem.*, 1971, **10**, 2143.
[1297] R. B. King, P. N. Kapoor, and R. N. Kapoor, *Inorg. Chem.*, 1971, **10**, 1841.
[1298] D. C. Pantaleo and R. C. Johnson, *Inorg. Chem.*, 1971, **10**, 1298.
[1299] P. R. Heckley, D. G. Holah, and D. Brown, *Canad. J. Chem.*, 1971, **49**, 1151.
[1300] S. A. Keppie and M. F. Lappert, *J. Chem. Soc.* (*A*), 1971, 3216.
[1301] A. P. Hagen, C. R. Higgins, and P. J. Russo, *Inorg. Chem.*, 1971, **10**, 1657.
[1302] W. J. Schlientz and J. K. Ruff, *J. Chem. Soc.* (*A*), 1971, 1139.

Table 2 (*cont.*)

	Ref.
$[(Ph_3P)_2N][RCO_2M(CO)_5]$ (M = Cr, Mo, or W)	1303
$M(CO)_5(SbH_3)$ (M = Cr, Mo, or W)	1304
$(Me_3M^2)_3SbM^1(CO)_5$ (M^1 = Cr, Mo, or W; M^2 = Ge or Sn)	1305
$(CO)_5M^1EMe_2M^2(CO)_5$ (M^1 = Cr, Mo, or W; M^2 = Mn or Re; E = P or As)	1306
$(CO)_5M^1E^1Me_2E^2Me_2M^2(CO)_5$ (M^1, M^2 = Cr, Mo, or W; E^1, E^2 = P or As)	1307
(diphos)$M^2(SMe)_2M^1(CO)_4$ (M^1 = Cr, Mo, or W; M^2 = Pd or Pt)	1308
$\{(Me_2N)_3As\}M(CO)_5$ (M = Cr, Mo, or W)	1309
$M(CO)_4$(*cis*-$AsMe_2CH{=}CHAsMe_2$) (M = Cr, Mo, or W)	1310
(Cp)M(CO)(NO){CPh(OR)}	1311
(Cp)M(CO)(NO)(PPh_3) (M = Cr, Mo, or W)	1312
$Me_nSnPh_{4-n},mM(CO)_3$ (M = Cr or Mo)	1313
$(Me_3Sn)_2C_6H_4Cr(CO)_3$, $Me_3SnCH_2C_6H_5Cr(CO)_3$	1313
$R_2MCr(CO)_5$ (M = Ge or In)	1314
(2-furylNC)$Cr(CO)_5$, $(Me_4N)[Me_3SiCH_2COCr(CO)_5]$, $(OC)_5CrC(OEt)(CH_2SiMe_3)$	1315
$(OC)_5CrCPh(N{=}CHMe)$	1316
$(OC)_5CrC(NH_2)(C_6H_4X)$	1317
$(OC)_5CrCR$(furyl) (R = RS, RCS_2, R_2N, or $PhC{\equiv}C$)	1318
$(OC)_5CrC(OEt)R$ (R = C_6Cl_5 or ferrocenyl)	1319
$(OC)_5CrC(OMe)(CH{=}CH{=}CH{=}CMeOMe)$	1320
trans-$\{(Me_2N)_3As\}_2Cr(CO)_4$	1309
$[(\text{fluorenyl})Cr(CO)_3]^-$	1321
$(C_7H_7\cdot C_6H_2Bu^t{}_2OH)Cr(CO)_3$, $[(C_7H_6\cdot C_6H_2Bu^t{}_2OH)Cr(CO)_3]^+$, $(C_7H_6\cdot C_6H_2Bu^t{}_2O)Cr(CO)_3$, and related compounds	1322
(*N*-alkylpyrrole)$Cr(CO)_3$	1323

[1303] W. J. Schlientz, Y. Lavender, N. Welcman, R. B. King, and J. K. Ruff, *J. Organometallic Chem.*, 1971, **33**, 357.
[1304] E. O. Fischer, W. Bathelt, and J. Müller, *Chem. Ber.*, 1971, **104**, 986.
[1305] H. Schumann and H. J. Breunig, *J. Organometallic Chem.*, 1971, **27**, C28.
[1306] W. Ehrl and H. Vahrenkamp, *Chem. Ber.*, 1971, **104**, 3261.
[1307] H. Vahrenkamp and W. Ehrl, *Angew. Chem. Internat. Edn.*, 1971, **10**, 513.
[1308] P. S. Braterman, V. A. Wilson, and K. K. Joshi, *J. Organometallic Chem.*, 1971, **31**, 123.
[1309] R. B. King and T. F. Korenowski, *Inorg. Chem.*, 1971, **10**, 1189.
[1310] H. G. Metzger and R. D. Feltham, *Inorg. Chem.*, 1971, **10**, 951.
[1311] E. O. Fischer and H.-J. Beck, *Chem. Ber.*, 1971, **104**, 3101.
[1312] A. T. McPhail, G. R. Knox, C. G. Robertson, and G. A. Sim, *J. Chem. Soc. (A)*, 1971, 205.
[1313] T. P. Poeth, P. G. Harrison, T. V. Long, jun., B. R. Willeford, and J. J. Zuckerman, *Inorg. Chem.*, 1971, **10**, 522.
[1314] T. J. Marks, *J. Amer. Chem. Soc.*, 1971, **93**, 7090.
[1315] J. A. Connor and E. M. Jones, *Chem. Comm.*, 1971, 570.
[1316] L. Knauss and E. O. Fischer, *J. Organometallic Chem.*, 1971, **31**, C68.
[1317] E. O. Fischer and H.-J. Kollmeier, *Chem. Ber.*, 1971, **104**, 1339.
[1318] J. A. Connor and E. M. Jones, *J. Chem. Soc. (A)*, 1971, 3368.
[1319] G. A. Moser, E. O. Fischer, and M. D. Rausch, *J. Organometallic Chem.*, 1971, **27**, 379.
[1320] L. Knauss and E. O. Fischer, *J. Organometallic Chem.*, 1971, **31**, C71.
[1321] K. M. Nicholas, R. C. Kerber, and E. I. Stiefel, *Inorg. Chem.*, 1971, **10**, 1519.
[1322] P. J. Pauson, G. R. Proctor, and R. Watson, *J. Chem. Soc. (C)*, 1971, 2399.
[1323] K. Öfele and E. Dotzauer, *J. Organometallic Chem.*, 1971, **30**, 211.

Table 2 (*cont.*)

	Ref.
$(C_{14}H_{14})[Cr(CO)_3]_2$	1324
$Cr(CO)_2(Ph_2AsCH_2AsPh_2)$	1325
$(Cp)Cr(NO)_2(NCO)$	1312
$\{(CF_3)_2CO\}_2CrO_2$	1282
$Mo(CO)_5PH_3$, $[(CO)_5MoPH_2Mo(CO)_5]^-$	1326
$Mo(CO)_5(PPh_2NR_2)$, *cis*-$Mo(CO)_4(PPh_2NR_2)_2$	1327
$Mo(CO)_4(2\text{-}C_5H_4NCH{=}N{-}NH{-}C_5H_4N\text{-}2)$	1328
$C_6F_5HgMo(Cp)(CO)_2L$ [L = CO, PPh_3, or $P(OMe)_3$]	1329
$(C_9H_{11})Mo(CO)_2LCl$ [L = CO, PPh_3, $P(OPh)_3$, or $P(OMe)_3$]	1330
$[(Cp)Mo(CO)_2L_2]^+$, $[(Cp)Mo(CO)_2L]_2$, $(Cp)Mo(CO)_2L^1L^2$ $[L^1 = P(O)(OC_3H_5)_nPh_{2-n}]$	1331
$(Cp)Mo(CO)(PPh_3)_2(NCO)$, $(Cp)Mo(CO)_2(PPh_3)(NCO)$	1312
$L[Mo(CO)_2(COMe)(Cp)]_3$, $[(Cp)MoL]Cl$; $(Cp)MoLCl$ $[L = PhP(CH_2CH_2PPh_2)_2]$	1297
$(Cp)Mo(CO)_2\{P(C_6H_{11})_3\}CMeOEt$	1401*a*
$Mo(CNR)_5X_2$	1332
$[Mo_3(Cp)S_4][SnMe_3Cl_2]$	1333
$(Cp)(CO)M(\mu\text{-}C_5H_4)Mn(CO)_4$ (M = Mo or W)	1334
$[(Me_3SiC_5H_4)M(CO)_3]$ (M = Mo or W)	1335
$HB(N_2C_3HMe_2)_3M(CO)_2SC_6H_4R$ (M = Mo or W)	1336
$(Cp)M(SR)_2$ (M = Mo or W)	1337
$[(Cp)M(SRMe)X]^+$, $[(Cp)_2MLX]^+$ (M = Mo or W)	1338
$[(Cp)_2M(NH_2R)H]^+$ (M = Mo or W)	1339
$CF_3(CH_3)_2SnW(CO)_3(Cp)$	1340
$(Cp)_2WHR$	1341
$(Cp)_2W(C_2H_2R_2)$	1342

[1324] K. Stöckel, F. Sondheimer, T. A. Clarke, M. Guss, and R. Mason, *J. Amer. Chem. Soc.*, 1971, **93**, 2571.
[1325] G. B. Robertson, P. O. Whimp, R. Colton, and C. J. Rix, *Chem. Comm.*, 1971, 573.
[1326] G. Becker and E. A. V. Ebsworth, *Angew. Chem. Internat. Edn.*, 1971, **10**, 186.
[1327] L. K. Atkinson and D. C. Smith, *J. Organometallic Chem.*, 1971, **33**, 189.
[1328] J. G. Dunn and D. A. Edwards, *J. Chem. Soc. (A)*, 1971, 988.
[1329] T. A. George, *J. Organometallic Chem.*, 1971, **33**, C13.
[1330] C. White and R. J. Mawby, *J. Chem. Soc. (A)*, 1971, 940.
[1331] R. J. Haines, A. L. Du Preez, and I. L. Marais, *J. Organometallic Chem.*, 1971, **28**, 97.
[1332] F. Bonati and G. Minghetti, *Inorg. Chem.*, 1970, **9**, 2642.
[1333] P. J. Vergamini, H. Vahrenkamp, and L. F. Dahl, *J. Amer. Chem. Soc.*, 1971, **93**, 6327.
[1334] R. Hoxmeier, B. Deubzer, and H. D. Kaesz, *J. Amer. Chem. Soc.*, 1971, **93**, 536.
[1335] E. W. Abel and S. Moorhouse, *J. Organometallic Chem.*, 1971, **28**, 211.
[1336] S. Trofimenko, *Inorg. Chem.*, 1971, **10**, 504.
[1337] A. R. Dias and M. L. H. Green, *J. Chem. Soc. (A)*, 1971, 2807.
[1338] R. H. Crabtree, A. R. Dias, M. L. H. Green, and P. J. Knowles, *J. Chem. Soc. (A)*, 1971, 1350.
[1339] F. W. S. Benfield and M. L. H. Green, *Chem. Comm.*, 1971, 1274.
[1340] H. C. Clark and B. K. Hunter, *J. Organometallic Chem.*, 1971, **31**, 227.
[1341] M. L. H. Green and P. J. Knowles, *J. Chem. Soc. (A)*, 1971, 1508.
[1342] B. R. Francis, M. L. H. Green, and G. G. Roberts, *Chem. Comm.*, 1971, 1290.

Table 3 *References to further papers containing n.m.r. data on compounds of* Mn *and* Re

	Ref.
$\{RNBMn(CO)_5\}_3$	1712
$C_5X_7Mn(CO)_4L$ (X = H or F; L = CO or PPh_3)	1343
$Me_5Si_2Mn(CO)_5$	1464
$(OC)_5MnCH_2SiMe_3$, $(OC)_5MnCH_2Bu^t$	1285
$(Me_4N)[(Cp)_2Ni_2Mn(CO)_5]$	1572
$CF_3(CH_3)_2SnMn(CO)_5$	1340
$(OC)_4Mn(o\text{-}C_6H_4CH{=}NPh)$	1344
$(OC)_4Mn(\mu\text{-}PMMe_3)_2Mn(CO)_4$ (M = Ge or Sn)	1345
$Mn_2(CO)_8AsMe_2X$	1346
$(OC)_4Mn(PPh_3)L$ [L = $CH_2C{=}CH$, $C(CH_2CO_2R){=}CH_2$, or $C{=}CH{-}S(O)O$]	1347
$MeMn(CO)_3L$, $[(Cp)_2Mn_2(CO)(NO)_2L]^+ [PF_6]^-$ [L = $PhP(CH_2CH_2PPh_2)_2$]	1297
$(C_7H_8X)Mn(CO)_3$	1348
$[(C_6H_3Me_3)Mn(CO)_2(CNR)]^+$	1349
$(Cp)MnH(CO)_2SiCl_3$	1402
$(Cp)Mn(CO)_2AsF_2$	1350
(arene CN)$Mn(CO)_3$, (arene)$Mn(CO)_2(CN)$	1351
$MeCOMn(CO)_3(o\text{-}CH_2{=}CHC_6H_4PPh_2)$, $\{(o\text{-}OCMeCMe)C_6H_4PPh_2\}Mn(CO)_3$, $\{(o\text{-}MeCOCMeC_6H_4)PPh_2\}Mn(CO)_4$	1352
$(Cp)Mn(CO)_2PH_2R$	1353
$(Cp)Mn(CO)_2PR_3$	1354
$(Cp)Mn(CO)_2N_2$	1355
$(C_6F_4 \cdot C_6H_6)Mn(CO)_2(Cp)$	1356
$(Cp)Mn(CO)_2C_2(OMe)_4$	1357
$(Cp)Mn(NO)(S_2CSBu^t)$, $[(Cp)Mn(NO)(SCHMe_2)]_2$	1358
$(CO)_5M^1EMe_2M^2(CO)_5$ (M^1 = Cr, Mo, or W; M^2 = Mn or Re; E = P or As)	1306

1343 R. E. Banks, R. N. Haszeldine, M. Lappin, and A. B. P. Lever, *J. Organometallic Chem.*, 1971, **29**, 427.

1344 M. I. Bruce, B. L. Goodall, M. Z. Iqbal, F. G. A. Stone, R. J. Doedens, and R. G. Little, *Chem. Comm.*, 1971, 1595.

1345 H. Schumann and H.-J. Kroth, *J. Organometallic Chem.*, 1971, **32**, C47

1346 J. Grobe and F. Kober, *J. Organometallic Chem.*, 1971, **29**, 295.

1347 W. D. Bannister, B. L. Booth, R. N. Haszeldine, and P. L. Loader, *J. Chem. Soc.* (*A*), 1971, 930.

1348 F. Haque, J. Miller, P. L. Pauson, and J. B. Pd. Tripathi, *J. Chem. Soc.* (*C*), 1971, 743.

1349 P. J. C. Walker and R. J. Mawby, *J. Chem. Soc.* (*A*), 1971, 3006.

1350 J. Müller and K. Fenderl, *Angew. Chem. Internat. Edn.*, 1971, **10**, 418.

1351 P. J. C. Walker and R. J. Mawby, *Inorg. Chem.*, 1971, **10**, 404.

1352 M. A. Bennett and R. Watt, *Chem. Comm.*, 1971, 95.

1353 M. Höfler and M. Schnitzler, *Chem. Ber.*, 1971, **104**, 3117.

1354 A. J. Hart-Davis and W. A. G. Graham, *J. Amer. Chem. Soc.*, 1971, **93**, 4388.

1355 D. Sellmann, *Angew. Chem. Internat. Edn.*, 1971, **10**, 919.

1356 D. M. Roe and A. G. Massey, *J. Organometallic Chem.*, 1971, **28**, 273.

1357 M. Herberhold and H. Brabetz, *Z. Naturforsch.*, 1971, **26b**, 656.

1358 P. Hydes, J. A. McCleverty, and D. G. Orchard, *J. Chem. Soc.* (*A*), 1971, 3660.

Table 3 (*cont.*)

	Ref.
$(Me_3SiC_5H_4)Mn(CO)_3$	1335
$(C_3H_5)(CO)M(\mu\text{-}C_5H_4)Mn(CO)_4$, $(Cp)HRe(\mu\text{-}C_5H_4)Mn(CO)_4$ (M = Mo or W)	1334
$RSO_3Re(CO)_5$	1359
$ReCl_2(PPh_3)\{N_2C(O)C_6H_4OMe\}$, $ReCl_2\{N_2C(O)Ph\}(PR_3)_3$	1360

Table 4 *References to further papers containing n.m.r. data on derivatives of ferrocene,* $\{Fc = (C_5H_5)FeC_5H_4—\}$

	Ref.
$(HNBFc)_3$	1713
FcX (X = halogen)	1361
$FcPh_2CH$, $FcMeBu^tCOH$, $(FcBu^tCH)_2$, $(Bu^tFcCH)_2$	1362
$Cr(CO)_5C(OEt)Fc$	1319
$[FcPPh_3]^+ [ClO_4]^-$	1363
$FcCR^1{=}CR^2R^3$, $FcCR^1(OR^2)CXCH_2$, $FcCRCH_2CX_2$, $FcC({=}CH_2)CBr{=}CH_2$, $FcCBu^t{=}C{=}CH_2$	1364
trans-$FcCH{=}CHX$	1365
$FcCR^1CHR^2CBr_2$, $FcCH{=}CHR$, $FcCR^1{=}C{=}CR^2H$, $FcCHR{\equiv}CH$, $FcC(C{\equiv}CH){=}C(CH_2)_4CH_2$	1366
$FcCMe_2CH_2CFc{=}CH_2$, $FcCMe_2CH{=}CMeFc$, FcR_3C, and related compounds	1367
$(FcCHMe)_2$	1368
FcNHR	1369
$(FcCOCH_2)_2CH_2$	1370
FcCHROH, $FcCHRNHC_6H_4X$	1371
$[FcCHR]^+$, $(Cp)FeC_5H_3(CH_2)_3CH$	1372
$FcCOCR^1{=}CR^2R^3$, $FcCHOHC_3H_3R_2$, $Fe(C_5H_4COC_3H_5)_2$, and related compounds	1373

1359 E. Lindner and R. Grimmer, *Chem. Ber.*, 1971, **104**, 544.
1360 J. Chatt, J. R. Dilworth, G. J. Leigh, and V. D. Gupta, *J. Chem. Soc* (*A*), 1971, 2631.
1361 F. L. Hedberg and H. Rosenberg, *J. Organometallic Chem.*, 1971, **28**, C14.
1362 M. D. Rausch and C. A. Pryde, *J. Organometallic Chem.*, 1971, **26**, 141.
1363 M. Sato, I. Motoyama, and K. Hata, *Bull. Chem. Soc. Japan*, 1971, **44**, 812.
1364 W. M. Horspool, R. G. Sutherland, and B. J. Thomson, *J. Chem. Soc.* (*C*), 1971, 1563.
1365 R. Asano, I. Moritani, A. Sonoda, Y. Fujiwara, and S. Teranishi, *J. Chem. Soc.* (*C*), 1971, 3691.
1366 W. M. Horspool, R. G. Sutherland, and B. J. Thomson, *J. Chem. Soc.* (*C*), 1971, 1554.
1367 W. M. Horspool, P. Stanley, R. G. Sutherland, and B. J. Thomson, *J. Chem. Soc.* (*C*), 1971, 1365.
1368 C. Baker and W. M. Horspool, *Chem. Comm.*, 1971, 615.
1369 D. W. Slocum, P. S. Shenkin, T. R. Engelmann, and C. R. Ernst, *Tetrahedron Letters*, 1971, 4429.
1370 S. I. Goldberg and J. G. Breland, *J. Org. Chem.*, 1971, **36**, 1499.
1371 G. Marr, B. W. Rockett, and A. Rushworth, *J. Chem. Soc.* (*C*), 1971, 4000.
1372 M. Hisatome and K. Yamakawa, *Tetrahedron*, 1971, **27**, 2101.
1373 W. M. Horspool, R. G. Sutherland, and B. J. Thomson, *J. Chem. Soc.* (*C*), 1971, 1558.

Table 4 (*cont.*)

Ref.

$(FcCH_2)_2C(CO_2H)(CO_2Et)$, $(C_5H_3)FeC_5H_3COC(CO_2R)(CH_2Fc)CH_2$, $(C_5H_4)FeC_3H_4CH_2C(CO_2R)(CH_2Fe)CO$, and related compounds 1374

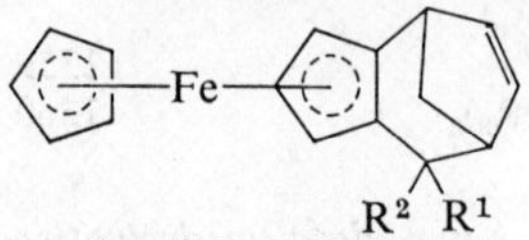

1375

$\{1,3\text{-}(MeCO)_2C_3H_3\}Fe(Cp)$ 1376

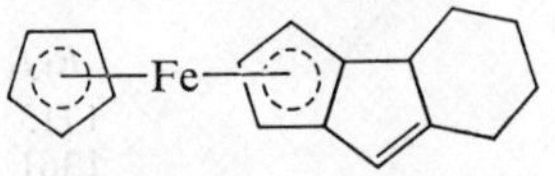

and related compounds 1377

$(Cp)FeC_5H_3COCMePhCH_2CH_2$, $(Cp)FeC_5H_3COCH_2CMePhCH_2$ 1378
$(Cp)FeC_5H_3R^1CR^2R^3OH$ 1379
$(Cp)FeC_5H_3(CH_2NMe_2)\{CH(SCH_2Ph)_2\}$ 1380
$(Cp)FeC_5H_3RCH_2NMe_2$, $(Cp)FeC_5H_3(CH_2OH)(CHPhOH)$ 1381
$(Cp)FeC_5H_2(Ph_2COH)\{2'\text{-}(6'\text{-}Bu^n)C_5H_3N\}\{(CH_2)_4OH\}$ 1382
$(C_5H_4ER_2)_2Fe$ (E = P or As), $(C_5H_4SH)_2Fe$ 1383

$(C_5H_4CH_2CN)_2Fe$, $(C_5H_4CH_2CO_2H)_2Fe$, $C_5H_4FeC_5H_4CH_2COCH_2$,

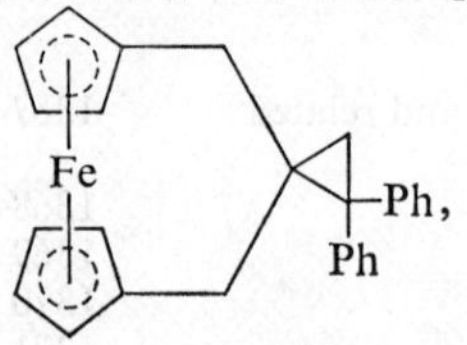

$C_5H_4FeC_5H_4CH_2CH{=}CH$ 1384

$[FcCHR]^+$, $C_5H_4FeC_5H_4CH_2\overset{+}{C}HCH_2CH_2$ 1385
$(C_5H_4R)_2Fe$, $C_5H_4FeC_5H_4(CH_2)_n$ (n = 2—5) 1386

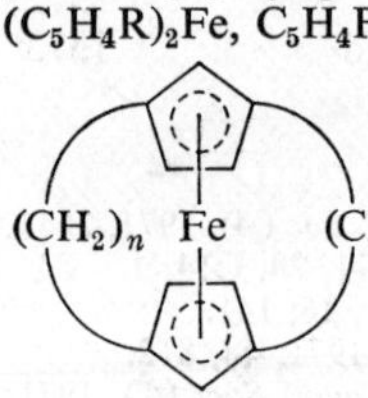

$C_5H_4FeC_5H_3R(CH_2)_n$— 1387

1374 H. Falk and W. Frostl, *Monatsh.*, 1971, **102**, 1259.
1375 T. D. Turbitt and W. E. Watts, *Chem. Comm.*, 1971, 631.
1376 P. Carty and M. F. A. Dove, *J. Organometallic Chem.*, 1971, **28**, 125.
1377 H. Lehner and K. Schlögl, *Monatsh.*, 1971, **102**, 277.
1378 H. d. Abbayes, *Compt. rend.*, 1971, **273**, *C*, 1009.
1379 C. Moïse, D. Sautrey, and J. Tirouflet, *Bull. Soc. chim. France*, 1971, 4562.
1380 M. Brink, *Tetrahedron Letters*, 1971, 2233.
1381 E. B. Moynahan and G. D. Popp, *Canad. J. Chem.*, 1971, **49**, 3565.
1382 D. J. Booth and B. W. Rockett, *J. Chem. Soc.* (*C*), 1971, 3341.
1383 J. J. Bishop, A. Davison, M. L. Katcher, D. W. Lichtenberg, R. E. Merrill, and J. C. Smart, *J. Organometallic Chem.*, 1971, **27**, 241.
1384 M. Sonoda and I. Moritani, *J. Organometallic Chem.*, 1971, **26**, 133.
1385 M. Hisatome and K. Tamakawa, *Tetrahedron Letters*, 1971, 3533.
1386 H. L. Lentzner and W. E. Watts, *Tetrahedron*, 1971, **27**, 4343.
1387 A. D. Brown, jun., and J. A. Winstead, *J. Org. Chem.*, 1971, **36**, 2832.

Table 4 (*cont.*)

	Ref.
$(C_5H_4R)Fe(C_5H_2R^1R^2COCH_2R^3)$	1388
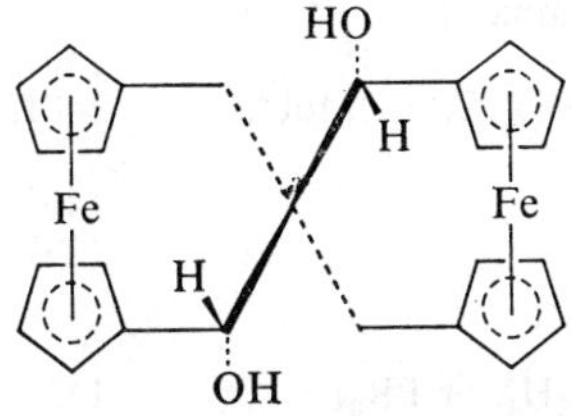	1389

Table 5 *References to further papers containing n.m.r. data on derivatives of iron containing only one π-cyclopentadienyl ring attached to any one iron atom*

	Ref.
$(Cp)Fe(CO)_2CH_2C{=}CMeB_{10}H_{10}$	1390
$(Cp)Fe(CO)_2CMeCH_2C(CN)_2C(CN)_2(CH_2)_n$ (n = 1 or 2)	1391
$[(Cp)Fe(CO)_2COMe_2]^+$	1392
$(Cp)Fe(CO)_2(CH_2)_n(CH{=}CH_2)$ and related compounds	1393
$(Cp)Fe(CO)_2C_5F_7$	1343
$(Cp)Fe(CO)_2SnMe_2CF_3$	1340
$(Cp)Fe(CO)_2HgC_6F_5$	1329
$(Cp)Fe(CO)_2(NCO)$	1312
$[(Cp)Fe(CO)_2]_2Hg$	1394
$(Cp)Fe(CO)_2SnPh\{OS(O)PH\}_2$, $[(Cp)Fe(CO)_2]_2Sn\{OS(O)PH\}_2$	1395
$[(Cp)Fe(CO)_2(C_4H_6)]^+[PF_6]^-$; $(Cp)Fe(CO)_2CH_2CHMe_2$, $(Cp)Fe(CO)CH_2CH_2CH{=}CH_2$ and related compounds	1396
$(Cp)Fe(CO)_2PPh_2Fe(CO)_4$, $(Cp)Fe(CO)(\mu\text{-}CO)(\mu\text{-}PPh_2)Fe(CO)_3$, $(Cp)Fe(CO)_2PPh_2Ni(CO)_3$	1397
$[(Me_3SiC_5H_4)Fe(CO)_2]_2$	1335
$(Cp)Fe(CO)_2CS(OMe)$, $(Cp)Fe(CO)(CS)CO_2Me$	1398
$[(Cp)Fe(CO)_2(diphos)]^+[ClO_4]^-$, $[(Cp)Fe(CO)_2LFe(CO)_2(Cp)]^+[ClO]_4^-$ (L = diphos, $Me_2SCH_2CH_2SMe_2$, or NC_6H_4N)	1399

[1388] J. J. McDonnell and D. J. Pochopien, *J. Org. Chem.*, 1971, **36**, 2092.
[1389] H. Falk, W. Fröstl, and K. Schlögl, *Monatsh.*, 1971, **102**, 1270.
[1390] L. I. Zakharkin, L. V. Orlova, A. I. Kovredov, L. A. Fedorov, and B. V. Lokshin, *J. Organometallic Chem.*, 1971, **27**, 95.
[1391] W. P. Giering and M. Rosenblum, *J. Amer. Chem. Soc.*, 1971, **93**, 5299.
[1392] E. C. Johnson, T. J. Meyer, and N. Winterton, *Inorg. Chem.*, 1971, **10**, 1673.
[1393] J.-Y. Mérour, C. Charrier, J.-L. Roustan, and J. Benaïm, *Compt. rend.* 1971, **273**, *C*, 285.
[1394] S. C. Cohen, S. H. Sage, W. A. Baker, jun., J. M. Burlitch, and R. B. Petersen, *J. Organometallic Chem.*, 1971, **27**, C44.
[1395] R. C. Edmondson, D. S. Field, and M. J. Newlands, *Canad. J. Chem.*, 1971, **49**, 618.
[1396] M. L. H. Green and M. J. Smith, *J. Chem. Soc.* (*A*), 1971, 3220.
[1397] K. Yasufuku and H. Yamazaki, *J. Organometallic Chem.*, 1971, **28**, 415.
[1398] L. Busetto, M. Graziani, and U. Belluco, *Inorg. Chem.*, 1971, **10**, 78.
[1399] M. L. Brown, T. J. Meyer, and N. Winterton, *Chem. Comm.*, 1971, 309.

Table 5 (*cont.*)

	Ref.
$[(Cp)Fe(CO)_2L^1]^+$, $(Cp)Fe(CO)_2L^2$, $(Cp)Fe(CO)L^1Cl$, $(Cp)Fe(CO)L^1L^2$ (L^1 = phosphite; L^2 = phosphonate)	1400
$(Cp)Fe(CO)[P(OR^1)_3]R^2$ (R^2 = SO_2Me or Me)	1401*a*
$(Cp)Fe(CO)\{P(C_6H_{11})_3\}COMe$, $[(Cp)Fe(CO)\{P(C_6H_{11})_3\}C(OR)Me]^+$, $(Cp)Fe(CO)(PPh_3)CHMeOEt$	1401*b*
$(Cl_3Si)_2FeH(CO)(Cp)$	1402
$(Cp)Fe_2(CO)_2\{PhP(CH_2CH_2PPh_2)_2\}$, $[(Cp)Fe\{PhP(CH_2CH_2PPh_2)_2\}]^+ [PF_4]^-$, $(Cp)Fe(CO)\{PhP(CH_2CH_2PPh_2)_2\}COMe$	1297
$[(Cp)Fe\{P(OPh)_3\}_2L]^+ X^-$ (L = RCN, CO, SO_2, C_2H_4, or PR_3)	1403
(Cp)Fe(diphos)H	1404
$(C_5H_4OEt)Fe(CO)(\mu\text{-}CO)_2Fe(CO)(C_5H_4OEt)$, $(C_3H_4OR)Fe(CO)_2R$, $[(C_5H_4OEt)Fe(CO)_2]^+ [BF_4]^-$	1405

Table 6 *References to further papers containing n.m.r. data on derivatives of iron not containing π-cyclopentadienyl rings attached to iron*

	Ref.
$Fe(CO)_4[As(NMe_2)_3]$	1309
$(C_6F_5AsAsC_6F_5)Fe(CO)_4$	1406
$(Me_3Sn)_2Fe(CO)_4$, $[Me_2SnFe(CO)_4]_2$	1407
$Fe(CO)_4[C_2(OMe)_4]$	1357
$Fe(CO)_4(PhCHMeN{=}CHCO_2Et)$	1408
$R_2NPF_2Fe(CO)_4$, $(R_2N)_2PFFe(CO)_4$	1409
$Cl_3SiFeH(CO)_4$	1402
$(Cp)_2Ni_2Fe(CO)_5$	1572
$(Cp)Ni(\mu\text{-}CO)(\mu\text{-}PPh_2)Fe(CO)_3$, $(Cp)CoI(\mu\text{-}PPh_2)Fe(CO)_4$	1397
$Fe(CO)_3(PPh_2C_6H_4CH{=}CH_2)$, $FeX(CO)_3(Ph_2PC_6H_4CHCH_3)$	1410
$(OC)_4Fe\{C_5H_4(OEt)_2\}Fe(CO)_3$, $Fe(CO)_3(C_5H_4O)$	1405
(2-methylcyclohexa-1,3-diene)$Fe(CO)_3$, [(methylcyclohexadienyl)$Fe(CO)_3]^+ [PF_6]^-$, and related compounds	1411
$(R^1CH{=}CHCH{=}CHR^2)Fe(CO)_3$, $(ArCH{=}CHCH{=}CHCH{=}CHCHO)\{Fe(CO)_3\}_2$	1412

1400 R. J. Haines, A. L. Du Preez, and I. L. Marais, *J. Organometallic Chem.*, 1971, **28**, 405.

1401 (*a*) S. R. Su and A. Wojcicki, *J. Organometallic Chem.*, 1971, **27**, 231; (*b*) M. L. H. Green, L. C. Mitchard, and M. G. Swanwick, *J. Chem. Soc.* (*A*), 1971, 794.

1402 W. Jetz and W. A. G. Graham, *Inorg. Chem.*, 1971, **10**, 4.

1403 M. L. H. Green and R. N. Whiteley, *J. Chem. Soc.* (*A*), 1971, 1943.

1404 W. E. Silverthorn, *Chem. Comm.*, 1971, 1310.

1405 A. Eisenstadt, G. Scharf, and B. Fuchs, *Tetrahedron Letters*, 1971, 679.

1406 P. S. Elmes, P. Leverett, and B. O. West, *Chem. Comm.*, 1971, 747.

1407 E. W. Abel and S. Moorhouse, *Inorg. Nuclear Chem. Letters*, 1971, **7**, 905.

1408 J. Y. Chenard, D. Commereuc, and Y. Cauvin, *J. Organometallic Chem.*, 1971, **33**, C69.

1409 W. M. Douglas and J. K. Ruff, *J. Chem. Soc.* (*A*), 1971, 3558.

1410 M. A. Bennett, G. B. Robertson, I. B. Tomkins, and P. O. Whimp, *J. Organometallic Chem.*, 1971, **32**, C19.

1411 A. J. Birch and M. A. Hass, *J. Chem. Soc.* (*C*), 1971, 2465.

1412 H. W. Whitlock, jun., C. Reich, and W. D. Woessner, *J. Amer. Chem. Soc.*, 1971, **93**, 2483.

Table 6 (*cont.*)

	Ref.
$(R^1CH{=}CHCH{=}CHCH{=}CHR^2)Fe(CO)_3$	1413
$[(\text{cyclo-octatrienyl})Fe(CO)_3]^+$	1414
$[(\text{methylcyclo-octatrienyl})Fe(CO)_3]^+$	1415
$[(\text{cycloheptatrienyl})Fe(CO)_3]^+ [PF_6]^-$	1416
$\{C_9H_8(CN)_4\}Fe(CO)_3$	1417
$(CHMe{=}CHCH{=}CHCHMeNHR)Fe(CO)_3$	1418
$(H_2C{=}CCH{=}CHCH{=}CHCH{=}CH)Fe(CO)_3$	1419
$(C_8H_7R)Fe(CO)_3$	1420
(substituted norbornadiene)$Fe(CO)_3$	1421
$[(Me_2CCHCMe_2)Fe(CO)_4]^+ X^-$, $(Me_2C{=}CR{-}CMe{=}CH_2)Fe(CO)_3$	1422
(1,2,3,4,5-pentamethylcyclopentadiene)$Fe(CO)_3$	1423
$[(MeCH{=}CHCH{=}CHCHMe)Fe(CO)_3]^+$, $(MeCH{=}CHCH{=}CHCHMeNHR)Fe(CO)_3$	1424
(tropone)$Fe(CO)_3$, [(cycloheptadienylone)$Fe(CO)_3]^+$, (substituted cycloheptadienone)$Fe(CO)_3$	1425
$[5\text{-}(RC_2B_{10}H_{10})C_5H_4]Fe(CO)_3$, $[R = H \text{ or } -C_5H_4Fe(CO)_3]$	1390
$(C_4Me_4)Fe(CO)_3$, $(C_4Me_3Et)Fe(CO)_3$	1426
$(C_4Me_3Pr^i)Fe(CO)_3$, $(C_4Me_2Pr^i{}_2)Fe(CO)_3$	1427
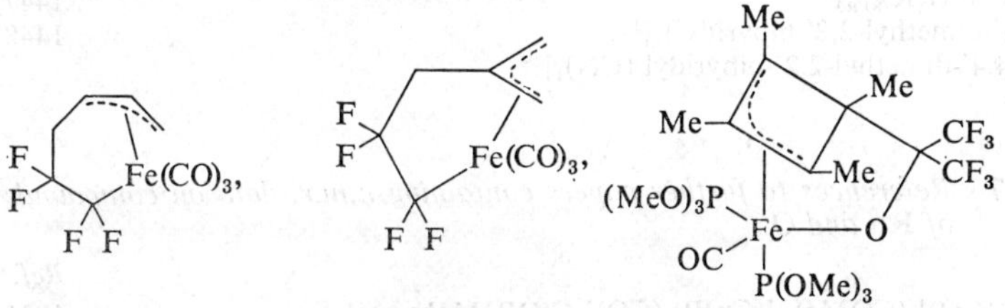	1428
$[SMeFe(CO)_3]_2$, *syn*-$\{SMeFe\}_2(CO)_5PR_3$	1429
$Fe_2(CO)_5L(SR)_2$, $[Fe(CO)_2LSR]_2$, and related compounds (L = tertiary phosphine or arsine, or chelating ditertiary phosphine or arsine, or $SbPh_3$)	1430

[1413] H. W. Whitlock, jun. and R. L. Markezich, *J. Amer. Chem. Soc.*, 1971, **93**, 5290.
[1414] M. Brookhart and E. R. Davis, *J. Amer. Chem. Soc.*, 1970, **92**, 7622.
[1415] M. Brookhart and E. R. Davis, *Tetrahedron Letters*, 1971, 4349.
[1416] G. E. Herberich and H. Müller, *Chem. Ber.*, 1971, **104**, 2781.
[1417] M. Green, S. Tolson, J. Weaver, D. C. Wood, and P. Woodward, *Chem. Comm.*, 1971, 222.
[1418] G. Maglio, A. Musco, R. Palumbo, and A. Sirigu, *Chem. Comm.*, 1971, 100.
[1419] G. T. Rodeheaver, G. C. Farrant, and D. F. Hunt, *J. Organometallic Chem.*, 1971, **30**, C22.
[1420] B. F. G. Johnson, J. Lewis, and G. L. P. Randall, *J. Chem. Soc.* (*A*), 1971, 422.
[1421] J. M. Landesberg and J. Sieczkowski, *J. Amer. Chem. Soc.*, 1971, **93**, 972.
[1422] D. H. Gibson, R. L. Vonnahme, and J. E. McKiernan, *Chem. Comm.*, 1971, 720.
[1423] P. V. Balakrishnan and P. M. Maitlis, *J. Chem. Soc.* (*A*), 1971, 1715.
[1424] G. Maglio, A. Musco, and R. Palumbo, *J. Organometallic Chem.*, 1971, **32**, 127.
[1425] A. Eisenstadt and S. Winstein, *Tetrahedron Letters*, 1971, 613.
[1426] H. A. Brune, H. P. Wolff, and H. Hüther, *Tetrahedron*, 1971, **27**, 3949.
[1427] H. A. Brune, H. P. Wolff, and H. Hüther, *Z. Naturforsch.*, 1971, **26b**, 765.
[1428] A. Bond, M. Green, B. Lewis, and S. F. W. Lowrie, *Chem. Comm.*, 1971, 1230.
[1429] J. P. Crow and W. R. Cullen, *Canad. J. Chem.*, 1971, **49**, 2948.
[1430] J. A. de Beer, R. J. Haines, R. Greatrex, and N. N. Greenwood, *J. Chem. Soc.* (*A*), 1971, 3271.

Table 6 (*cont.*)

	Ref.
$Fe_2Pt(CO)_{10-n}L_n$	1431
$\{CH_2{=}CH(CH_2)CHMeC({=}CH_2)C{=}CCH{=}CH_2\}\{Fe(CO)_3\}_2$, $\{CH_2{=}CHC({=}CH_2)C({=}CH_2)CH{=}CH_2\}\{Fe(CO)_3\}$, and related compounds	1432
$(C_6H_4CH{=}CH_2)\{Fe(CO)_3\}_2$, $(C_6H_4CH{=}CH_2)\{Fe(CO)_3\}_3$	1433
$Fe_2(CO)_6(C_9H_{12})$ (3 isomers)	1434
$(C_{10}H_{10})\{Fe(CO)_3\}_2$ (2 isomers)	1435
$(C_{10}H_{10})\{Fe(CO)_3\}_2$	1436
$(Me_3Si{\cdot}OC{\cdot}CO{\cdot}SiMe_3)\{Fe(CO)_3(SiMe_3)\}_2$	1437
$Me_3SiN\{Fe_3(CO)_{10}\}$	1438
$Fe_2(CO)_5HNEt_2(CONEt_2)_2$	1439
$[Fe_2(CO)_6I(N{=}CR^1R^2)]$	1440
$\{RNC({=}NR)NR\}\{Fe(CO)_3\}_2$	1441
$\{NNC(CO_2Me)_2CHRCH_2\}\{Fe(CO)_3\}_2$	1442
$(RC_6H_3CSHC_6H_4R)\{Fe(CO)_3\}_2$	1443
$(C_{11}H_{12})Fe(CO)_2$ and related compounds, $(C_4H_4)Fe(C_6H_6NCO_2Et)$	1444
$(diene)_2Fe(CO)$	1445
$[(arene)_2Fe]^{2+}$ $[PF_6]^-{}_2$	1446
$[Fe\{P(C_5NH_3R)_3\}_n]^{2+}$	1447
$[Fe(4,4'\text{-dimethyl-}2,2'\text{-bipyridyl})_3]^{2+}$, $[Fe(4,4'\text{-dimethyl-}2,2'\text{-bipyridyl})(CN)_4]^{2-}$	1448

Table 7 *References to further papers containing n.m.r. data on compounds of* Ru *and* Os

	Ref.
$(Cp)Ru(CO)L(COMe)$, $[(Cp)Ru(CO)LC(OR)Me]^+$	1401
$(Cp)Ru(PPh_3)_2C(CF_3){=}CHCF_3$,	1449
$(Cp)Ru(PPh_3)CR{=}CRC(CF_3){=}CH(CF_3)$	1449

1431 M. I. Bruce, G. Shaw, and F. G. A. Stone, *Chem. Comm.*, 1971, 1288.
1432 R. N. Greene, C. H. De Puy, and T. E. Schroer, *J. Chem. Soc.* (*C*), 1971, 3115.
1433 R. Victor, R. Ben-Shoshan, and S. Sarel, *Chem. Comm.*, 1971, 1241.
1434 S. Otsuka, A. Nakamura, and K. Tani, *J. Chem. Soc.* (*A*), 1971, 154.
1435 R. Aumann, *Angew. Chem. Internat. Edn.*, 1971, **10**, 190.
1436 R. Aumann, *Angew. Chem. Internat. Edn.*, 1971, **10**, 188.
1437 M. A. Nasta and A. G. MacDiarmid, *J. Amer. Chem. Soc.*, 1971, **93**, 2813.
1438 E. K. v. Gustorf and R. Wagner, *Angew. Chem. Internat. Edn.*, 1971, **10**, 910.
1439 E. O. Fischer and V. Kiener, *J. Organometallic Chem.*, 1971, **27**, C56.
1440 M. Kilner and C. Midcalf, *Chem. Comm.*, 1971, 944.
1441 N. J. Bremer, A. B. Cutcliffe, M. F. Farona, and W. G. Kofron, *J. Chem. Soc.* (*A*), 1971, 3264.
1442 H. Kisch, *J. Organometallic Chem.*, 1971, **30**, C25.
1443 H. Alper and A. S. K. Chan, *Chem. Comm.*, 1971, 1203.
1444 J. S. Ward and R. Pettit, *J. Amer. Chem. Soc.*, 1971, **93**, 262.
1445 E. Koerner von Gustorf, J. Buchkremer, Z. Pfaifer, and F.-W. Grevels, *Angew. Chem. Internat. Edn.*, 1971, **10**, 260.
1446 J. F. Helling, S. L. Rice, D. M. Braitsch, and T. Mayer, *Chem. Comm.*, 1971, 930.
1447 J. E. Parks, B. E. Wagner, and R. H. Holm, *Inorg. Chem.*, 1971, **10**, 2472.
1448 G. M. Bryant and J. E. Fergusson, *Austral. J. Chem.*, 1971, **24**, 441.
1449 T. Blackmore, M. I. Bruce, F. G. A. Stone, R. E. Davis, and A. Garza, *Chem. Comm.*, 1971, 852.

Table 7 (*cont.*)

	Ref.
$(C_{12}H_{10}N_2)Ru_3(CO)_9$, $(C_{12}H_{10}N_2)Ru_2(CO)_6$	1450
$H_4Ru_4(CO)_{12-n}(PR_3)_n$	1451
$[(C_8H_9)Ru(CO)_3]^+$, $(C_8H_9X)Ru(CO)_3$	1452
$Ru(CO)_3XPPh_2C_6H_4CHMe$	1410
$Ru(CO)_2(PPh_3)(O_2CR)_2$	1453
$Ru(CO)(PPh_3)_2(OCR)Cl$	1454
$RuHX(CO)(CNR)(PPh_3)_2$, $[RuH(CO)_n(CNR)(PPh_3)_{4-n}]^+$	1455
$L_4RuC_6H_4N{=}NPh$	1456
$[RuClL(Me_2AsCH_nCH_nAsMe_2)]Cl_2$ (L = NO or N_2; n = 1 or 2)	1457
$RuCl_2(AsPh_3)_3(O_2)$	1458
$RuCl_3(NO)L_2$, $[RuCl_2(NO)L_3]^+$ (L = PR_3 or AsR_3)	1459
$[Ru(bipy)_2(NO)Cl]^{2+}[PF_6]^-{}_2$	1460
$[Ru(NH_3)_4L_2]^{2+}$	1461
$Me_3M^2M^1(CO)_4X$, $[Me_3M^2M^1(CO)_3X]_2$, $Me_3M^2M^1(CO)_3(PPh_3)X$ $(Me_3M^2)Ru(CO)_3(PPh_3)H$ (M^1 = Ru or Os; M^2 = Si, Ge, or Sn)	1462
$(Me_3Ge)_2M(CO)_4$ and related compounds (M = Ru or Os)	1463
$Me_3SiM(CO)_3(\mu\text{-}SiMe_2)_2MSiMe_3(CO)$ (M = Ru or Os);	1464
$(OC)_3Ru(\mu\text{-}SiMe_2)_3Ru(CO)_3$, $Os_3(CO)_9(\mu\text{-}SiMe_2)_3$	1464
$MH_3(ArNNNAr)(PPh_3)_2$ and related compounds (M = Ru or Os)	1465
$[M(4,4'\text{-dimethyl-}2,2'\text{bipyridyl})_3]^{2+}$ (M = Ru or Os), $[Ru(4,4'\text{dimethyl-}2,2'\text{-bipyridyl})_2L_2]^{n+}$	1448
$OsLX_2Q_3$	1466
$(Ph_2C_2)_2Os_3(CO)_9$	1467

1450 M. I. Bruce, M. Z. Iqbal, and F. G. A. Stone, *J. Organometallic Chem.*, 1971, **31**, 275.
1451 F. Piacenti, M. Bianchi, P. Frediani, and E. Benedetti, *Inorg. Chem.*, 1971, **10**, 2759.
1452 M. Cooke, P. T. Draggett, M. Green, B. F. G. Johnson, J. Lewis, and D. J. Yarrow, *Chem. Comm.*, 1971, 621.
1453 B. F. G. Johnson, R. D. Johnston, J. Lewis, and I. G. Williams, *J. Chem. Soc.* (*A*), 1971, 689.
1454 R. R. Hitch, S. K. Gondal, and C. T. Sears, *Chem. Comm.*, 1971, 777.
1455 D. F. Christian and W. R. Roper, *Chem. Comm.*, 1971, 1271.
1456 M. I. Bruce, M. Z. Iqbal, and F. G. A. Stone, *J. Chem. Soc.* (*A*), 1971, 2820.
1457 P. G. Douglas, R. D. Feltham, and H. G. Metzger, *J. Amer. Chem. Soc.*, 1971, **93**, 84.
1458 M. M. Taqui Khan, R. K. Andal, and P. T. Manoharan, *Chem. Comm.*, 1971, 561.
1459 R. E. Townsend and K. J. Coskran, *Inorg. Chem.*, 1971, **10**, 1661.
1460 J. B. Godwin and T. J. Meyer, *Inorg. Chem.*, 1971, **10**, 471.
1461 R. G. Gaunder and H. Taube, *Inorg. Chem.*, 1970, **9**, 2627.
1462 M. J. Ash, A. Brookes, S. A. R. Knox, and F. G. A. Stone, *J. Chem. Soc.* (*A*), 1971, 458.
1463 S. A. R. Knox and F. G. A. Stone, *J. Chem. Soc.* (*A*), 1971, 2874.
1464 A. Brookes, S. A. R. Knox, and F. G. A. Stone, *J. Chem. Soc.* (*A*), 1971, 3469.
1465 S. D. Robinson and M. F. Uttley, *Chem. Comm.*, 1971, 1315.
1466 P. K. Maples, F. Basolo, and R. G. Pearson, *Inorg. Chem.*, 1971, **10**, 765.
1467 O. Gambino, G. A. Vaglio, R. P. Ferrari, and G. Centini, *J. Organometallic Chem.*, 1970, **30**, 381.

Table 8 *References to further papers containing n.m.r. data on compounds of* Co

	Ref.
$[(Cp)_2Co]^+$	1468
$[(Cp)_2Co]X$, $(Cp)CO(C_5H_5CY^1Y^2COZ)$	1469
$(Me_3SiC_5H_4)Co(CO)_2$	1335
$(Cl_3Si)CoH(CO)(Cp)$	1402
$(Cp)ICo(\mu\text{-}PPh_2)Fe(CO)_4$	1397
$(Cp)NiCo_3(CO)_9$	1572
$(Me_3Sn)_2Co(CO)(Cp)$, $[Me_2SnCo(CO)(Cp)]_2$	1407
$(Cp)CoI(COMe)PPh_3$	1470
$(C_6F_4C_6H_6)Co(Cp)$	1356
$(Cp)Co\{C_7H_8(CH_3)_2(NO)_2\}$ and related compounds	1471
$(C_8H_9)Co(C_8H_{12})$	1472
$[Me_3SiNHMe_2]^+[CoL_4]^-$ ($L = PF_3$ or CO)	1473
$(OC)_3CoCF_2CF_2CH_2CR{=}CH_2$ and related compounds	1474
$F_nH_{3-n}CCOCo(CO)_3PPh_3$	1475
$(ArCH{=}NH)\{CoH(CO)_3\}$	1476*a*
$\{PhCC(P_3N_3F_5)\}\{Co(CO)_3\}_2$	1476*b*
$(Me_3MCCMMe_3)\{Co(CO)_3\}_2$ (M = Si or Sn)	1477
$CH_3CCo_3(CO)_9$	1478
$YCCo_3(CO)_6$(arene)	1479
$RCCo_3(CO)_8L$	1480
RCOCo(salicylaldehyde-ethylenedi-imine)	1481
$L_nCoCH_2MMe_3$ (M = Si or C)	1285
$Ph_4C_4Co(CO)SCF_3$	1482
$(CH_2CHCMeR)Co(CH_2CRCHCH_2)(PPh_3)$	1483
$(MeC_3H_4)Co(C_4H_6)$	1484
$Co(C_8H_{13})(C_8H_{12})$, $Co(CO)_2(C_7H_{10})$	1485
$Co\{P(C_5NH_3CHO)_3\}_2X_2$	1447
$[CoX_2(cis\text{-}Me_2AsCH{=}CHAsMe_2)_2]^+ Y^-$	1310, 1486
$(HNC_6H_4CH{=}NCH_2CH_2N{=}CHC_6H_4NH)CoX(C_5H_5N)$ and related compounds	1487

1468 M. Van den Akker and F. Jellinek, *Rec. Trav. chim.*, 1971, **90**, 1101.
1469 G. E. Herberich and G. Greiss, *J. Organometallic Chem.*, 1971, **27**, 113.
1470 A. J. Hart-Davis, and W. A. G. Graham, *Inorg. Chem.*, 1970, **9**, 2658.
1471 H. Brunner and S. Loskot, *Angew. Chem. Internat. Edn.*, 1971, **10**, 515.
1472 H. Lehmkuhl, W. Leuchte, and E. Janssen, *J. Organometallic Chem.*, 1971, **30**, 407.
1473 R. E. Highsmith, J. R. Bergerud, and A. G. MacDiarmid, *Chem. Comm.*, 1971, 48.
1474 A. Greco, M. Green, and F. G. A. Stone, *J. Chem. Soc.* (*A*), 1971, 3476.
1475 E. Lindner, H. Stirch, K. Geibel, and H. Kranz, *Chem. Ber.*, 1971, **104**, 1524.
1476 (*a*) I. Rhee, M. Ryang, and S. Tsutsumi, *Bull. Chem. Soc., Japan*, 1971, **44**, 2552; (*b*) T. Chivers, *Inorg. Nuclear Chem. Letters*, 1971, **7**, 827.
1477 D. Seyferth and D. L. White, *J. Organometallic Chem.*, 1971, **32**, 317.
1478 D. Seyferth, R. J. Spohn, and J. E. Hallgren, *J. Organometallic Chem.*, 1971, **28**, C34.
1479 B. H. Robinson and J. L. Spencer, *J. Chem. Soc.* (*A*), 1971, 2045.
1480 T. W. Matheson, B. H. Robinson, and W. S. Tham, *J. Chem. Soc.* (*A*), 1971, 1457.
1481 J. Booth, P. J. Craig, B. Dobbs, J. M. Pratt, G. L. P. Randall, and A. G. Williams, *J. Chem. Soc.* (*A*), 1971, 1964.
1482 R. B. King and A. Efraty, *Inorg. Chem.*, 1971, **10**, 1376.
1483 G. Vitulli, L. Porri, and A. L. Segre, *J. Chem. Soc.* (*A*), 1971, 3247.
1484 P. V. Rinze and H. Nöth, *J. Organometallic Chem.*, 1971, **30**, 115.
1485 S. Otsuka and T. Taketomi, *J. Chem. Soc.* (*A*), 1971, 579.
1486 M. A. Bennett and J. D. Wild, *J. Chem. Soc.* (*A*), 1971, 545.
1487 M. Gerloch, B. M. Higson, and E. D. McKenzie, *Chem. Comm.*, 1971, 1149.

Table 8 (*cont.*)

	Ref.
Heptamethyl dicyanocobyrinate and related compounds	1488
$Co\{MeCOC(O_2CCH_3)NO\}_3$	1609
Co(*NN*-dialkyldithiocarbamate)$_n$(*O*-alkylxanthate)$_{3-n}$	1489
(2-pyridyl-CH=CRO)$_3$Co	1490
$(EtNCHCMeNEt)_2Co$	1491
(2-pyrrole-CRO)$_3$Co	1492
[{2-pyridyl-CH=NN=C(NH$_2$)S}$_2$Co]$^+$ and related compounds	1493
Tris{(+)-3-acetylcamphorato}cobalt(III)	1494, 1495
Co(acac)$_2$(NO$_2$)(amine)	1496
Co$_2$(acac)$_4$(N$_3$)$_2$	1497
$[Co(NH_3)_5L]^{n+}$	1498
trans-$Co(NH_3)_4(NO_2)X$	1499
trans-[Co{$\overline{N(CH_2CH_2)}CH_2CH_2NH_2$}$_2$(NO$_2$)$_2$]	1500
[Co(NH$_3$)$_4$L](NO$_3$)$_3$ (L = en or pn)	1501
[Co(en)(diethylenetriamine)X][ZnCl$_4$]	1502
Co(en)$_2$(NH$_3$)(PO$_4$)	1503
[Co(R-pn)$_2$(ClO$_4$)$_2$]$^+$	1504
[CoX(*NN′*-dimethylethylenediamine)$_2$]$^+$ (X = oxalate or CO$_3$)	1505
K[Co(L-valine)$_2$(CO$_3$)]	1506
Dinitrobis(aminoacidato)cobaltate(III)	1507
[Co(NH$_3$)(glycine)(NO$_2$)$_3$]$^-$	1508
Co(CO$_3$)(glycine)(NH$_3$)$_2$	1509
[Co(aminoacidato)$_2$L]$^+$X$^-$ (L = bipyridyl or phenanthroline)	1510
[Co(en)$_2$(L-aspartine)]$^{2+}$, [Co(en)$_2$(L-glu)]$^{2+}$	1511
[(L-hydrogenaspartato)$_2${(−)-pn}Co]$^{3+}$, [(L-aspartato){(−)-pn}$_2$Co]$^{2+}$	1512

[1488] R. Bonnett, J. M. Godfrey, and V. B. Math, *J. Chem. Soc.* (*C*), 1971, 3736.
[1489] H. C. Brinkhoff, *Inorg. Nuclear Chem. Letters*, 1971, **7**, 413.
[1490] C. A. Root, J. E. Rowe, jun., and H. Veening, *Inorg. Chem.*, 1971, 10, 1195.
[1491] R. Bonnett, D. C. Bradley, K. J. Fisher, and I. F. Rendall, *J. Chem. Soc.* (*A*), 1971, 1622.
[1492] C. L. Perry and J. H. Weber, *J. Inorg. Nuclear Chem.*, 1971, **33**, 1031.
[1493] C. J. Jones and J. A. McCleverty, *J. Chem. Soc.* (*A*), 1971, 38.
[1494] R. M. King and G. W. Everett, jun., *Inorg. Chem.*, 1971, **10**, 1237.
[1495] C. S. Springer, jun., R. E. Sievers, and B. Feibush, *Inorg. Chem.*, 1971, **10**, 1242.
[1496] L. J. Boucher and N. G. Paez, *Inorg. Chem.*, 1971, **10**, 1680.
[1497] D. R. Herrington and L. J. Boucher, *Inorg. Nuclear Chem. Letters*, 1971, **7**, 1091.
[1498] R. J. Balahura and R. B. Jordan, *J. Amer. Chem. Soc.*, 1971, **93**, 625.
[1499] C. K. Poon and H. W. Tong, *J. Chem. Soc.* (*A*), 1971, 2151.
[1500] H. P. Fritz and G. Hierl, *Z. Naturforsch.*, 1971, **26b**, 476.
[1501] H. Yoneda, M. Muto, and K. Tamaki, *Bull. Chem. Soc. Japan*, 1971, **44**, 2863.
[1502] A. R. Gainsford and D. A. House, *Inorg. Chim. Acta*, 1971, **5**, 544.
[1503] S. F. Lincoln and A. L. Purnell, *Inorg. Chem.*, 1971, **10**, 1255.
[1504] R. A. Haines and A. J. Smith, *Canad. J. Chem.*, 1971, **49**, 3907.
[1505] S. Yano and Y. Sasaki, *Inorg. Nuclear Chem. Letters*, 1971, **7**, 1229.
[1506] R. D. Gillard and M. G. Price, *J. Chem. Soc.* (*A*), 1971, 2271.
[1507] M. B. Célap, R. G. Denning, D. J. Radanović, and T. J. Janjić, *Inorg. Chim. Acta*, 1971, **5**, 9.
[1508] M. B. Célap, M. J. Malinar, and T. J. Janjić, *Z. anorg. Chem.*, 1971, **383**, 341.
[1509] S. Kanazawa and M. Shibata, *Bull. Chem. Soc. Japan*, 1971, **44**, 2424.
[1510] T. Yasui and B. E. Douglas, *Inorg. Chem.*, 1971, **10**, 97.
[1511] J. I. Legg and J. Steele, *Inorg. Chem.*, 1971, **10**, 2177.
[1512] Y. Kojima and M. Shibata, *Inorg. Chem.*, 1971, **10**, 2382.

Table 8 (*cont.*)

	Ref.
[Co(*RR*- and *SS*-ptnta)]$^-$, [CoCl(*RS*-2,4-ptntaH)]$^-$ {ptnta = [($-O_2CCH_2)_2NCHMe]_2CH_2$}	1513
$(Co(NH_2CH_2CH_2NHCH_2CH_2CH_2NHCH_2CH_2NH_2)(valine)]^{2+}$	1514
[Co{2-pyridyl-$CH_2NH(CH_2)_nNHCH_2$-2-pyridyl)Cl_2]ClO_4	1515
$[Co\{NH_2(CH_2)_2NH(CH_2)_3NH(CH_2)_2NH_2\}(NH_2CHMeCH_2NH_2)]^{3+}$	1516
$[Co(O_2CCHMeNHCH_2CO_2)_2]^-$	1517
Bis(pyridoxylidenevalinato)cobalt(III)	1518
[Co(triethylenetetramine){*N*-methyl-(*S*)-alanine}]$^{2+}$	1519
CoXY (Y = chelating ligands related to edta)	1520
[Co(cyclohexanediaminetetra-acetic acid)]$^-$	1521

Table 9 *References to further papers containing n.m.r. data on compounds of* Rh

	Ref.
$[(C_8H_{11})Rh(Cp)]^+$	1522
$(Cp)Rh(EPh_3)X_2$ (E = P or As; X_2 = Br_2, C_2H_4, $PhCH_2Br$, *etc.*)	1523
$[(Cp)Rh(CO)SnMe_2]_2$, $(Me_3Sn)_2Rh(CO)(Cp)$	1407
$(Cp)RhI(COMe)PPh_3$	1470
$(Cp)Rh(C_6F_4C_6H_6)$	1356
$[(Cp)Rh(arene)]^{2+}$, $[(Cp)Rh(areneH)]^+$	1524
$[(C_5Me_5)Rh(OCOR)_2(H_2O)_n]_m$, $[\{(C_5Me_5)Rh\}_2X_3]^+$	1525
$(C_5Me_5)Rh(CO)_2$, $(C_5Me_5)RhR(CO)X$	1526
$(Cp)Rh(CO)H(SiPh_3)$, $(Cp)(CO)Rh(SiCl_2R)_2$, $(Cp)Rh(CO)I\cdot SiI_3$	1527
$[(Cp)Rh(CO)]_2(CF_3C{\equiv}CCF_3)$	1528
$(Ph_3P)_2RhHCl(GeEt_3)$	1529
$Na_3RhH(CN)_5$	1530
$(PPh_3)_3RhMe$	1531
$RRh(CO)_2(PPh_3)_2$ (R = fluoroaryl)	1532
$Rh(C{\equiv}CR)(CO)(PPh_3)_2$	1533

1513 F. Mizukami, H. Ito, J. Fujita, and K. Saito, *Bull. Chem. Soc. Japan*, 1971, **44**, 3051.
1514 G. R. Brubaker and D. P. Schaeffer, *Inorg. Chem.*, 1971, **10**, 2170.
1515 J. G. Gibson and E. D. McKenzie, *J. Chem. Soc. (A)*, 1971, 1666.
1516 R. H. Lewis and M. D. Alexander, *Inorg. Chim. Acta*, 1971, **5**, 86.
1517 K. Okamoto, J. Hidaka, and Y. Shimura, *Bull. Chem. Soc. Japan*, 1971, **44**, 1601.
1518 E. H. Abbott, *J. Inorg. Nuclear Chem.*, 1971, **33**, 567.
1519 D. A. Buckingham, I. E. Maxwell, and A. M. Sargeson, *Inorg. Chem.*, 1970, **9**, 2663.
1520 B. L. Blackmer and J. L. Sudmeier, *Inorg. Chem.*, 1971, **10**, 2019.
1521 L. J. Zompa and J. M. Shindler, *Chem. Comm.*, 1971, 65.
1522 J. Evans, B. F. G. Johnson, and J. Lewis, *Chem. Comm.*, 1971, 1252.
1523 A. J. Oliver and W. A. G. Graham, *Inorg. Chem.*, 1971, **10**, 1165.
1524 C. White and P. M. Maitlis, *J. Chem. Soc. (A)*, 1971, 3322.
1525 J. W. Kang and P. M. Maitlis, *J. Organometallic Chem.*, 1971, **30**, 127.
1526 J. W. Kang and P. M. Maitlis, *J. Organometallic Chem.*, 1971, **26**, 393.
1527 A. J. Oliver and W. A. G. Graham, *Inorg. Chem.*, 1971, **10**, 1.
1528 R. S. Dickson and H. P. Kirsch, *J. Organometallic Chem.*, 1971, **32**, C13.
1529 F. Glockling and G. C. Hill, *J. Chem. Soc. (A)*, 1971, 2137.
1530 R. A. Jewsbury and J. P. Maher, *J. Chem. Soc. (A)*, 1971, 2847.
1531 M. Michman and M. Balog, *J. Organometallic Chem.*, 1971, **31**, 395.
1532 B. L. Booth, R. N. Haszeldine, and I. Perkins, *J. Chem. Soc. (A)*, 1971, 927.
1533 C. K. Brown, D. Georgiou, and G. Wilkinson, *J. Chem. Soc. (A)*, 1971, 3120.

Table 9 (*cont.*)

	Ref.
LRh(salicylaldimine) [L = $(CO)_2$ or cyclo-octadiene]	1534
$[(C_6H_8O)Rh(CO)Cl]_2$	1535
$Rh(CO)_2Cl$(diazepine)	1536
$Rh(C_{15}H_{20})X$, $Rh_2(C_{15}H_{20})X_2$ (X = C_6F_5, Cl, Cp, $SnCl_3$, or CO)	1537
$[(C_9H_8Me_2)RhCl]_2$	1538
[(bicyclo[3,3,1]nona-2,6-diene)$RhCl]_2$	1539
$(CO)_2RhClL$, $ClRhL_2$ (L = PhCHMeN=$CHCO_2Et$)	1408
bis-(1-substituted tetrazoline-5-thionato)mono(triphenylphosphine)-monochlororhodium(I)	1540
L_nRh(ArNNNAr)	1465
$[Rh(CO)(AsMe_2Ph)_4]^+ [BPh_4]^-$	1541
$Rh(NCS)(CO)(PPh_3)_2(OCMe_2)$	1542
$[RhClL_2]_2$, $RhClL_2(PPh_3)$, $RhClL(PPh_3)_2$ (L = PF_2NEt_2)	1543
$RhCl_3(R^1R^2S)_3$	1544
$Rh_2(O_2CMe)_2(dmg)_2(PPh_3)_2$	1545

Table 10 *References to further papers containing n.m.r. data on compounds of* Ir

	Ref.
$(Cp)IrI_2$, $(Cp)IrIMe(PPh_3)$ and related compounds	1546
$[Ir(Cp)(arene)]^{2+}$, $[Ir(Cp)(areneH)]^+$	1524
$[(C_5Me_5)Ir(OCOR)_2(H_2O)_n]_m$	1525
$(C_5Me_5)Ir(CO)_2$, $(Cp)Ir(CO)RX$	1526
$(C_5Me_5)Ir(CO)RI$	1547
(cyclo-octadiene)$IrHX_2L$	1548
$HIr[P(OPh)_3]_4$, $[H_2Ir\{P(OPh)_3\}_4]^+$, $Ir(NO)\{P(OPh)_3\}_3$	1549
$IrHX_2(CO)L_2$, *trans*-$IrX(CO)L_2$ (L = $PBu^t_nR_{3-n}$)	1550
$[IrHX(C_5H_{10}NH)_4]^+Y^-$	1551
$IrHX_2[P(OR)_3]_3$, $Ir\{C_6H_4OP(OPh)_2\}H\{P(OPh)_3\}$	1552

[1534] R. J. Cozens, K. S. Murray, and B. O. West, *J. Organometallic Chem.*, 1971, **27**, 399.
[1535] R. Rossi, P. Diversi, and L. Porri, *J. Organometallic Chem.*, 1971, **31**, C40.
[1536] R. A. Smith, D. P. Madden, A. J. Carty, and G. J. Palenik, *Chem. Comm.*, 1971, 427.
[1537] R. B. King and P. N. Kapoor, *J. Organometallic Chem.*, 1971, **33**, 383.
[1538] R. Grigg, R. Hayes, and A. Sweeney, *Chem. Comm.*, 1971, 1248.
[1539] J. K. A. Clarke, F. McMahon, J. B. Thomson, and B. Zeeh, *J. Organometallic Chem.*, 1971, **31**, 283.
[1540] U. Agarwala and B. Singh, *J. Inorg. Nuclear Chem.*, 1971, **33**, 598.
[1541] L. M. Haines and E. Singleton, *J. Organometallic Chem.*, 1971, **30**, C81.
[1542] N. J. DeStefano and J. L. Burmeister, *Inorg. Chem.*, 1971, **10**, 998.
[1543] M. A. Bennett, G. B. Robertson, T. W. Turney, and P. O. Whimp, *Chem. Comm.*, 1971, 762.
[1544] J. Chatt, G. J. Leigh, A. P. Storace, D. A. Squire, and B. J. Starkey, *J. Chem. Soc.* (*A*), 1971, 899.
[1545] J. Halpern, E. Kimura, J. Molin-Case, and C. S. Wong, *Chem. Comm.*, 1971, 1207.
[1546] H. Yamazaki, *Bull. Chem. Soc. Japan*, 1971, 44, 582.
[1547] R. B. King and A. Efraty, *J. Organometallic Chem.*, 1971, **27**, 409.
[1548] R. N. Haszeldine, R. J. Lunt, and R. V. Parish, *J. Chem. Soc.* (*A*), 1971, 3711.
[1549] D. Guisto and G. Cova, *Gazzetta*, 1971, **101**, 519.
[1550] B. L. Shaw and R. E. Stainbank, *J. Chem. Soc.* (*A*), 1971, 3716.
[1551] E. R. Birnbaum, *J. Inorg. Nuclear Chem.*, 1971, **33**, 3031.
[1552] E. W. Ainscough, S. D. Robinson, and J. J. Levinson, *J. Chem. Soc.* (*A*), 1971, 3413.

Table 10 (*cont.*)

	Ref.
$IrHCl\{C_6H_4OP(OPh)_2\}\{P(OPh)_3\}_2$	1553
$[Ir(CO)_2(CS)(PPh_3)_2]^+$, $[IrH_2(CO)(CS)(PPh_3)_2]^+$, $Ir(CO_2Me)(CO)(CS)L_2$, $Ir(CO)(CS)(PR_3)_3$, and related compounds	1554
$[IrH_2(NH{=}NAr)(PPh_3)_3][BF_4]$	1555
$IrHCl_2L_3$	1556
$IrCl_2(COCH_2Ph)(PMe_2Ph)_3$	1557
$IrX(CO)R(PR_3)_2$	1558
$Ir(C{\equiv}CR)(CO)(PPh_3)_2$	1533
$Ir(C{\equiv}CR)(CO)(PPh_3)_2$ and its adducts	1559
$Ir(HgMMe_3)(MMe_3)(Et_3P)_2(CO)$, $Ir(HgX)(MMe_3)_2(Et_3P)_2(CO)$ (M = Si or Ge)	1560
$[Ir(CO)L_2(olefin)_2]BPh_4$ and related compounds	1561
$Ir(CO)_2$(salicylaldimine) and related compounds	1534
$[Ir(cyclo\text{-}octadiene)\{S(CH_2CH_2SPh_2)\}]Cl$, $IrCl_3(R^1R^2S)_3$	1544
$[H_2IrCl(CO)\{(Ph_2PCH_2)_4C\}IrCl]_2$	1562
$Ir(CO)(PPh_3)_2(FC_6H_4N_4C_6H_4F)$	1563
$Ir_2(PF_3)_8$	1564
$IrCl_2L^1L^2_3$ (L^1 = anion, tertiary phosphine, amine, *etc.*; $L^2 = PR_3$ or AsR_3)	1565
$L_nIr(ArNNNAr)$	1465
$[IrL_4O_2]^+ [BPh_4]^-$	1541

Table 11 *References to further papers containing n.m.r. data on compounds of* Ni

	Ref.
$[(diphos)_2NiH][PF_6]$	1566
$L_2\overline{NiCF_2CF_2CF_2CF_2}$ and related compounds	1567
$o\text{-}PhC{\equiv}CC_6H_4Ni(C_2Cl_3)(PEt_3)_2$	1568
$(C_8H_{12})_nNi_n\{C_6(CF_3)_6\}$, $L_{2n}Ni_n\{C_6(CF_3)_6\}$ (n = 1 or 2)	1569
$(Cp)Ni(\mu\text{-}PPh_2)_2Ni(Cp)$, $(Cp)Fe(CO)_2(\mu\text{-}PPh_2)Ni(CO)_3$, and related compounds	1397

[1553] M. A. Bennett and R. Charles, *Austral. J. Chem.*, 1971, **24**, 427.
[1554] M. J. Mays and F. P. Stefanini, *J. Chem. Soc. (A)*, 1971, 2747.
[1555] L. Toniolo and R. Eisenberg, *Chem. Comm.*, 1971, 455.
[1556] P. R. Brookes, C. Masters, and B. L. Shaw, *J. Chem. Soc. (A)*, 1971, 3756.
[1557] M. Kubota and D. M. Blake, *J. Amer. Chem. Soc.*, 1971, **93**, 1368.
[1558] A. J. Deeming, B. L. Shaw, and R. E. Stainbank, *J. Chem. Soc. (A)*, 1971, 374.
[1559] C. K. Brown and G. Wilkinson, *Chem. Comm.*, 1971, 70.
[1560] K. A. Hooton, *J. Chem. Soc. (A)*, 1971, 1251.
[1561] A. J. Deeming and B. L. Shaw, *J. Chem. Soc. (A)*, 1971, 376.
[1562] J. Ellermann, R.Gerbeth, and K. Geibel, *Angew. Chem. Internat. Edn.*, 1971, **10**, 826.
[1563] F. W. B. Einstein, A. B. Gilchrist, G. W. Rayner-Canham, and D. Sutton, *J. Amer. Chem. Soc.*, 1971, **93**, 1826.
[1564] T. Kruck, G. Sylvester, and I.-P. Kunau, *Angew. Chem. Internat. Edn.*, 1971, **10**, 725.
[1565] B. L. Shaw and R. M. Slade, *J. Chem. Soc. (A)*, 1971, 1184.
[1566] M. L. H. Green and H. Munakata, *Chem. Comm.*, 1971, 549.
[1567] J. Browning, M. Green, and F. G. A. Stone, *J. Chem. Soc. (A)*, 1971, 453.
[1568] R. G. Miller and D. P. Kuhlman, *J. Organometallic Chem.*, 1971, **26**, 401.
[1569] J. Browning, C. S. Cundy, M. Green, and F. G. A. Stone, *J. Chem. Soc. (A)*, 1971, 448.

Table 11 (*cont.*)

	Ref.
$(C_3H_5)Ni(dihydropentalenylene)Ni(C_3H_5)$	1570
$(\pi\text{-}Cp)(\sigma\text{-}Cp)Ni(Me_2C{=}CCMe_2CHO)$	1571
$(Cp)NiCo_3(CO)_4$, $(Cp)_2Ni_2Fe(CO)_5$, $[Me_4N][(Cp)_2Ni_2Mn(CO)_5]$	1572
$(Cp)Ni(CN)(PBu^n_3)$	1573
$[(Cp)Ni(PBu^n_3)_2]^+ X^-$	1574
$[(Cp)Ni(Ph_2PCH_2PPh_2)]^+ Cl^-$	1575
$(Cp)Ni(PBu^n_3)X$	1576
$NiL(Bu^tNC)_2$ [(L = O_2, $(NC)_2C{=}C(CN)_2$, PhNNPh, *etc.*]	1577
$NiL(Bu^tNC)_2$ (L = olefin, acetylene, or azobenzene)	1578
$(RNC)_2NiC(CF_3)_2OC(CF_3)_2O$ and related compounds	1579
trans-$NiHX(PR_3)_2$	1580
$NiX_2\{P(C_5H_3NCHO)_3\}$ and related compounds	1447
$NiX_2(PRC_6H_4C_6H_4)_2$	1581
$Ni(PPh_2H)_4$	1582
NiX_2(*cis*-$Me_2AsCH{=}CHAsMe_2$) and related compounds	1583
$Ni(CF_3COCHCSCF_3)_2$	1584
$Ni(CF_3COCHCSCH_3)_2$	1585
$Ni(CH_3COCHCSCH_3)_2$	1586
Ni complexes of $(MeCOCH{=}CMeNHCHR)_2$	1587
$[Ni\{(HON{=}CMe)_2C_5NH_3\}_2][ClO_4]_2$	1588
$[Ni(NN\text{-diethyldithiocarbamate})_3]^-$	1489
$Ni(SeSCNEt_2)_2$	1589
$Ni(NCH{=}CHCH{=}C{-}C{=}NCH_2CH_2CH_2)_2$	1492
$Ni_2(NCH{=}CHCH{=}CCH{=}NC_6H_4S)_2$	1493
$[Ni(N{=}CMeCH_2CMe_2NHCH_2N{=}CMeCH_2CMe_2NHCH_2CH_2)]^{2+}$ and related compounds	1590, 1591

1570 A. Miyake and A. Kanai, *Angew. Chem. Internat. Edn.*, 1971, **10**, 801.
1571 M. Sato, K. Ichibori, and F. Sato, *J. Organometallic Chem.*, 1971, **26**, 267.
1572 A. T. T. Hsieh and J. Knight, *J. Organometallic Chem.*, 1971, **26**, 125.
1573 M. Sato, F. Sato, and T. Yoshida, *J. Organometallic Chem.*, 1971, **26**, C49.
1574 M. Sato, F. Sato, and T. Yoshida, *J. Organometallic Chem.*, 1971, **27**, 273.
1575 F. Sato and M. Sato, *J. Organometallic Chem.*, 1971, **33**, C73.
1576 M. Sato, F. Sato, and T. Yoshida, *J. Organometallic Chem.*, 1971, **31**, 415.
1577 S. Otsuka, T. Yoshida, and Y. Tatsuno, *Chem. Comm.*, 1971, 67.
1578 S. Otsuka, T. Yoshida, and Y. Tatsuno, *J. Amer. Chem. Soc.*, 1971, **93**, 6462.
1579 M. Green, S. K. Shakshooki, and F. G. A. Stone, *J. Chem. Soc.* (*A*), 1971, 2828.
1580 M. L. H. Green, Y. Saito, and P. J. Tanfield, *J. Chem. Soc.* (*A*), 1971, 152.
1581 D. W. Allen, F. G. Mann, and I. T. Miller, *J. Chem. Soc.* (*C*), 1971, 3937.
1582 L. Horner and H. Kunz, *Chem. Ber.*, 1971, **104**, 717.
1583 M. A. Bennett and J. D. Wild, *J. Chem. Soc.* (*A*), 1971, 536.
1584 E. Bayer and H.-P. Müller, *Tetrahedron, Letters*, 1971, 533.
1585 O. Siimann and J. Fresco, *J. Chem. Phys.*, 1971, **54**, 740.
1586 O. Siimann and J. Fresco, *J. Chem. Phys.*, 1971, **54**, 734.
1587 E. Larsen and K. Schaumburg, *Acta Chem. Scand.*, 1971, **25**, 962.
1588 E. I. Baucom and R. S. Drago, *J. Amer. Chem. Soc.*, 1971, **93**, 6469.
1589 T. Tanaka and N. Sonoda, *Inorg. Chem.*, 1971, **10**, 2337.
1590 N. F. Curtis, *J. Chem. Soc.* (*A*), 1971, 2834.
1591 V. L. Goedken and D. H. Busch, *Inorg. Chem.*, 1971, **10**, 2679.

Table 11 (*cont.*)

	Ref.
[Ni(NHCMe$_2$CH$_2$CMe=NCH$_2$CH$_2$CH=NCMe$_2$CH$_2$CMe=NCH$_2$CH$_2$CH$_2$]$^{2+}$ and related compounds	1592, 1593
Tetradehydrocorrin-nickel(II) and related compounds	1594—1600

Table 12 *References to further papers containing n.m.r. data on compounds of* Pd *and* Pt

	Ref.
[PdH(diphos)(R$_3$P)][PF$_6$]	1566
trans-{(C$_6$H$_{11}$)$_3$P}$_2$PdHCl	1601
(chlorovinyl)PdCl(PPh$_3$)$_2$	1602
Pd{CH(COMe)$_2$}(acac)L	1603
Pd(C$_6$H$_4$CHMeNMe$_2$)ClL [L = PPh(α-naphthyl)(*o*-tolyl)]	1604
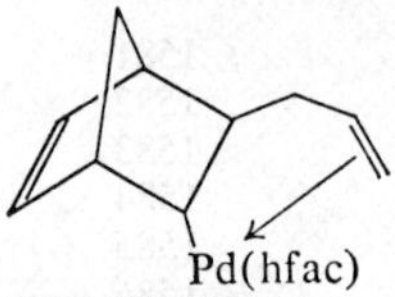 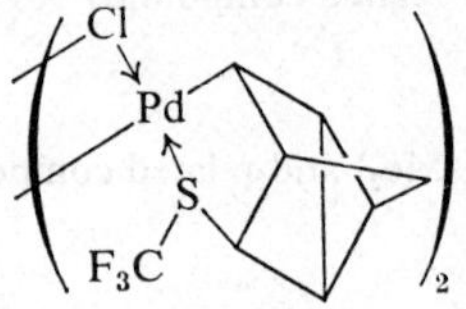	1605
(C$_4$Ph$_4$)Pd(SCF$_3$)$_2$, and related compounds	1482
PdL(Bu^{t}NC)$_2$ (L = olefin, acetylene, or azobenzene)	1578
[(ClCH$_2$CH$_2$C$_3$H$_4$)PdCl]$_2$, [(1-Me-2-ClCH$_2$C$_3$H$_3$)PdCl]$_2$	1535
[{1,3-(MeOCH$_2$)C$_3$H$_3$}PdCl]$_2$ and related compounds	1606
PdX$_2$(Bu^{t}NC)$_2$	1607
PdX$_2$L$_2$, Cl$_2$Pd(PPh$_3$)L (L = PhCHMeN=CHCO$_2$Et)	1408
Pd(CH$_3$CSCHCOCF$_3$)$_2$	1585
Pd(CH$_3$CSCHCOCH$_3$)$_2$	1586
[Pd(Bu^{t}COCHCOBu^t)RS)]$_2$ and related compounds	1608

1592 E. K. Barefield and D. H. Busch, *Inorg. Chem.*, 1971, **10**, 108.
1593 E. K. Barefield and D. H. Busch, *Inorg. Chem.*, 1971, **10**, 1216.
1594 H. H. Inhoffen, J. W. Buchler, L. Puppe, and K. Rohbock, *Annalen*, 1971, **747**, 133.
1595 A. Hamilton and A. W. Johnson, *J. Chem. Soc* (*C*), 1971, 3879.
1596 R. Grigg, A. W. Johnson, and G. Shelton, *Annalen*, 1971, **746**, 32.
1597 R. Grigg, A. P. Johnson, A. W. Johnson, and M. J. Smith, *J. Chem. Soc.* (*C*), 1971, 2457.
1598 A. Hamilton and A. W. Johnson, *Chem. Comm.*, 1971, 523.
1599 A. W. Johnson and W. R. Overend, *Chem. Comm.*, 1971, 710.
1600 I. D. Dicker, R. Grigg, A. W. Johnson, H. Pinnock, K. Richardson, and P. van den Brock, *J. Chem. Soc.* (*C*), 1971, 536.
1601 R. van der Linde and R. O. de Jongh, *Chem. Comm.*, 1971, 563.
1602 I. Moritani, Y. Fujiwara, and S. Danno, *J. Organometallic Chem.*, 1971, **27**, 279.
1603 S. Baba, T. Ogura, and S. Kawaguchi, *Inorg. Nuclear Chem. Letters*, 1971, **7**, 1195.
1604 S. Otsuka, A. Nakamura, T. Kano, and K. Tani, *J. Amer. Chem. Soc.*, 1971, **93**, 4301.
1605 R. P. Hughes and J. Powell, *J. Organometallic Chem.*, 1971, **30**, C45.
1606 K. Tsukiyama, Y. Takahashi, S. Sakai, and Y. Ishii, *J. Chem. Soc.* (*A*), 1971, 3112.
1607 S. Otsuka, Y. Tatsuno, and K. Ataka, *J. Amer. Chem. Soc.*, 1971, **93**, 6705.
1608 D. A. White, *J. Chem. Soc.* (*A*), 1971, 143.

Table 12 (*cont.*)

	Ref.
$[Pd\{MeCOC(CO_2Et)NO\}NHEt]_2$ and related compounds	1609
$[Pd\{(Me_2N)_2BN(PhX_2C)B(NMe_2)_2\}X]_2$	1610
$Pd\{CH(pyrrole)_2\}_2$	1611
Pd(octaethylporphyrin)	1612
Pd(dehydrocorrin)	1613
$M^1XL_2(CH_2M^2Me_3)$ (M^1 = Pd or Pt; M^2 = C or Si)	1285
$M(CH_2C_6H_4PPh_2)_2$ (M = Pd or Pt)	1614
M(cyclo-octadiene)($Bu^tCOCHCOBu^t$),	1615
$M(C_8H_{13})(Bu^tCOCHCOBu^t)$ (M = Pd or Pt)	1615
$[M\{C(CO_2Me){=}C(CO_2Me)C(CO_2Me){=}C(CO_2Me)\}]_n$ (M = Pd or Pt)	1616
$(Ph_3P)_2M(NCO)(CO_2R)$ (M = Pd or Pt)	1617
(bicyclo[3,3,1]nona-2,6-diene)MCl_2 (M = Pd or Pt)	1539
M(DMF)(pentene)Cl_2 (M = Pd or Pt)	1618
$M(SnCl_3)(C_3H_5)(EPh_3)$ (M = Pd or Pt; E = P or As)	1619
$MX(Cp)(PR_3)$, $[M(Cp)(PR_3)_2]^+$ (M = Pd or Pt)	1620
ML_2(ArNNNAr) (M = Pd or Pt)	1465
$MX_2(PhMeAsCH_2CH_2AsMePh)$ (M = Pd or Pt)	1621
$MX_2(PRC_6H_4C_6H_4)_2$ (M = Pd or Pt)	1581
$M(N_3)_2(PPh_3)_2$ (M = Pd or Pt)	1654
$MX_2(R_2S)_2$ (M = Pd or Pt)	1622
(diphos)$M(SMe)_2$, (diphos)$M^1(SMe)_2M^2(CO)_4$ (M^1 = Pd or Pt; M^2 = Cr, Mo, or W)	1308
$M(SSeCNEt_2)_2$	1589
$PtR^1X(PR^2_3)_2$ [X = H, Cl, or R^1; R^1 = $N{=}C(CF_3)_2$]	1623
$PtX(MH_{3-n}X_n)(PEt_3)_2$, $PtHI_2(MH_2X)(PEt_3)_2$ (M = Si or Ge)	1624
(diphos)$Pt(MMe_3)_3$ and related compounds (M = Si, Ge, or Sn)	1625
$[PtX(CO)L_2][BF_4]$, $PtHXL_2$, and related compounds	1626
$Pt(N_2SO_2R)_2(PPh_3)_2$, $Pt(CO_2R)(NCO)(PPh_3)_2$, and related compounds	1627

[1609] D. A. White, *J. Chem. Soc.* (*A*), 1971, 233.
[1610] G. Schmid and L. Weber, *Z. Naturforsch.*, 1971, **26b**, 994.
[1611] F. C. March, D. A. Couch, K. Emerson, J. E. Fergusson, and W. T. Robinson, *J. Chem. Soc.* (*A*), 1971, 440.
[1612] H. Ogashi, N. Masai, Z. Yoshida, J. Takemoto, and K. Nakamoto, *Bull. Chem. Soc. Japan*, 1971, **44**, 49.
[1613] R. Grigg, A. W. Johnson, and G. Shelton, *J. Chem. Soc.* (*C*), 1971, 2287.
[1614] G. Longoni, P. Chini, F. Canziani, and P. Fantucci, *Chem. Comm.*, 1971, 470.
[1615] D. A. White, *J. Chem. Soc.* (*A*), 1971, 145.
[1616] K. Moseley and P. M. Maitlis, *Chem. Comm.*, 1971, 1604.
[1617] W. Beck and K. von Werner, *Chem. Ber.*, 1971, **104**, 2901.
[1618] F. Conti, M. Donati, G. F. Pregaglia, and R. Ugo, *J. Organometallic Chem.*, 1971, **30**, 421.
[1619] J. N. Crosby and R. D. W. Kemmitt, *J. Organometallic Chem.*, 1971, **26**, 277.
[1620] R. J. Cross and R. Wardle, *J. Chem. Soc.* (*A*), 1971, 2000.
[1621] A. J. Cheney and B. L. Shaw, *J. Chem. Soc.* (*A*), 1971, 3549.
[1622] B. E. Aires, J. E. Fergusson, D. T. Howarth, and J. M. Miller, *J. Chem. Soc.* (*A*), 1971, 1144.
[1623] B. Cetinkaya, M. F. Lappert, and J. McMeeking, *Chem. Comm.*, 1971, 215.
[1624] J. E. Bentham, S. Cradock, and E. A. V. Ebsworth, *J. Chem. Soc.* (*A*), 1971, 587.
[1625] A. F. Clemmit and F. Glockling, *J. Chem. Soc.* (*A*), 1971, 1164.
[1626] W. J. Cherwinski and H. C. Clark, *Inorg. Chem.*, 1971, **10**, 2263.
[1627] W. Beck, M. Bauder, G. La Monnica, S. Cenini, and R. Ugo, *J. Chem. Soc.* (*A*), 1971, 113.

Table 12 (*cont.*)

	Ref.
$PtMe_3X\{o\text{-}(AsMe_2)_2C_6H_4\}$	1628
$PtX(CF_2CF_2Me)(PR_3)_2$ and related compounds	1629
trans-$PtMe\{NH{=}CR(OR)\}(PR_3)_2$	1630
cis-$Pt(CN)\{C(CN)_2Me\}(PPh_3)_2$	1631
$PtBr_2(ROCH{=}CHC_6H_4PPh_2)$ and related compounds	1632
$[PtX\{PhP(CH_2)_nCH(OR)CH_2\}]_2$ and related compounds	1633
$Pt(C{\equiv}CCF_3)_2(PPh_3)_2$	1634
trans-$[Pt(CF_3)\{CCH_2R(OMe)\}(PMe_2Ph)_2]^+ [PF_6]^-$ and related compounds	1635
$[Pt(CNR)_2(PPh_3)_2][BF_4]_2$ and related compounds	1636
$Pt(CH_2CHRCH_2)Cl_2(py)_2$	1637
$Fe_2Pt(CO)_{10-n}L_n$, $MPt_2(CO)_5L_3$, $M_2Pt(CO)_7L_3$ (M = Ru or Os)	1431
$Pt(dimethylcyclopropene)(PPh_3)_2$	1638
$PtMeX\{C(CN)_2C(CN)_2\}(PR_3)_2$	1639
$Pt(PPh_3)_2(C_6Me_4O_2)$	1640
$PtCl_2(C_2H_4)(PhCHMeN{=}CHCO_2Et)$	1408
$Pt(PPh_3)_2(MeC{\equiv}CC{\equiv}CMe)$	1641
$Pt(PPh_3)_2\{C{\equiv}C(CH_2)_n\}$, $Pt(O_2CCF_3)\{C{=}CH(CH_2)_n\}(PPh_3)_2$	1642
$[Pt(C_{28}H_{20}OR)Cl]_2$	1643
$(C_5Me_5R)PtCl_2$	1423
$[Pt(N{=}NAr)(PPh_3)_3]^+$	1644
$PtX_2(CNR^1)(PR^2_3)$	1645
$[Pt(S_2PPh_2)(PR_3)_2]^+$, $Pt(S_2PPh_2)_2(PR_3)$	1646
$Pt(O_2CMe)_2(PPh_3)_2$, $Pt(N_2C_6H_3MeSO_2)(PPh_3)_2$	1647
$Pt\{O_3C(CF_3)_2\}(PPh_3)_2$ and related compounds	1648, 1649
$Pt\{SC(NNR)NNPh\}_2$	1650

[1628] A. J. Cheney and B. L. Shaw, *J. Chem. Soc.* (*A*), 1971, 3545.
[1629] H. C. Clark and R. J. Puddephatt, *Inorg. Chem.*, 1970, **9**, 2670.
[1630] H. C. Clark and L. E. Manzer, *Chem. Comm.*, 1971, 387.
[1631] J. L. Burmeister and L. M. Edwards, *J. Chem. Soc.* (*A*), 1971, 1663.
[1632] M. A. Bennett, W. R. Kneen, and R. S. Nyholm, *J. Organometallic Chem.*, 1971, **26**, 293.
[1633] R. N. Haszeldine, R. J. Lunt, and R. V. Parish, *J. Chem. Soc.* (*A*), 1971, 3705.
[1634] W. R. Cullen and F. L. Hou, *Canad. J. Chem.*, 1971, **49**, 3404.
[1635] M. H. Chisholm and H. C. Clark, *Chem. Comm.*, 1971, 1484.
[1636] P. M. Treichel, W. J. Knebel, and R. W. Hess, *J. Amer. Chem. Soc.*, 1971, **93**, 5424.
[1637] K. G. Powell and F. J. McQuillin, *Tetrahedron Letters*, 1971, 3313.
[1638] J. P. Visser, A. J. Schipperijn, L. Lukas, D. Bright, and J. J. de Boer, *Chem. Comm.*, 1971, 1266.
[1639] H. C. Clark and R. J. Puddephatt, *Inorg. Chem.*, 1971, **10**, 416.
[1640] S. Cenini, R. Ugo, and G. La Monica, *J. Chem. Soc.* (*A*), 1971, 416.
[1641] P. B. Tripathy, B. W. Renoe, K. Adzamli, and D. M. Roundhill, *J. Amer. Chem. Soc.*, 1971, **93**, 4406.
[1642] M. A. Bennett, G. B. Robertson, P. O. Whimp, and T. Yoshida, *J. Amer. Chem. Soc.*, 1971, **93**, 3797.
[1643] F. Canziani, P. Chini, A. Quarta, and A. DiMartino, *J. Organometallic Chem.*, 1971, **26**, 285.
[1644] S. Cenini, R. Ugo, and G. La Monica, *J. Chem. Soc.* (*A*), 1971, 3441.
[1645] E. M. Badley, J. Chatt, and R. L. Richards, *J. Chem. Soc.* (*A*), 1971, 21.
[1646] J. M. C. Alison, T. A. Stephenson, and R. O. Gould, *J. Chem. Soc.* (*A*), 1971, 3690.
[1647] T. L. Gilchrist, F. J. Graveling, and C. W. Rees, *J. Chem. Soc.* (*C*), 1971, 977.
[1648] P. J. Hayward and C. J. Nyman, *J. Amer. Chem. Soc.*, 1971, **93**, 617.
[1649] P. J. Hayward, S. J. Saftich, and C. J. Nyman, *Inorg. Chem.*, 1971, **10**, 1311.
[1650] C. E. Forbes, A. Gold, and R. H. Holm, *Inorg. Chem.*, 1971, **10**, 2479.

Table 13 *Reference to further papers containing n.m.r. data on compounds of* Cu, Ag, Au, Be, Zn, Cd, *and* Hg

	Ref.
$KCu(NHCONPrCONH)_2,2H_2O$, $KCu(NHCONPrCONH)_2(NH_2CONHCONHPr)_2$	
$[Ag(NH_2CH_2CH_2NCH_2CH_2)_2]^+$	1500
$Me_2AuX(PR_3)$, $[Me_2Au(PR_3)_2]^+ Cl^-$	1652
$[Me_2AuX]_2$, $[Me_2Au(SMe_2)_2]^+ Cl^-$	1653
$(Ph_3P)AuC{=}NN{=}NNR$	1654
$[R_2Be]_2$	1655
$(PhC{\equiv}C)_2Be(amine)$, $Me(PhC{\equiv}C)Be(amine)$	1656, 1657
$M[C(N_2)CO_2Et]_2$ (M = Zn, Cd, or Hg)	1658
$CaZnBu^n{}_4$	1659
RZn(dimethylmalonate) and related compounds	1660
$[Zn(BH_4)_n]^{2-n}$, $[Zn_3(BH_4)_8]^{2-}$	1661
ZnL_2Cl_2 (L = substituted pyridine)	1662
$Zn(EtNCHCHCHNEt)_2$	1491
$Zn(C_5NH_4CH{=}COR)_2$	1490
$[Zn\{(C_5NH_4CH_2CH_2)_2NCH_2CH_2NH(CH_2CH_2C_5H_4N)\}][ClO_4]_2$	1663
$Zn(C_{14}H_{28}N_6)(ClO_4)_2,0.5H_2O$	1664
$ZnX_2\{P(C_5NH_3CHO)_3\}$, $ZnX_2\{P(C_5NH_3CH{=}NOH)_3\}$	1443
$[M(NH_2CH_2CH_2NHCH_2CH_2)_2]^{2+}$ (M = Zn or Cd)	1500
$M(CH_3CSCHCOCF_3)_2$ (M = Zn or Cd)	1585
$M(CH_3CSCHCOCH_3)_2$ (M = Zn or Cd)	1586
$[M\{(substituted\ pyrrole)_2CH\}_2]^+$ (M = Zn or Cd)	1611
$Hg(CCl{=}CH_2)_2$, $PhHgCCl_2Me$, and related compounds	1665
$(C_6H_3R_3)_2Hg$	1666
$\{CF_2{=}C(CF_3)\}Hg$ and related compounds	1667
$C_6F_5HgSiMe_3$ and related compounds	1668
$C_6F_5HgM(CO)_n(Cp)$ (M = Mo or Fe)	1329
$[(Cp)Fe(CO)_2]_2Hg$	1394
$Ir(HgMMe_3)(MMe_3)_2(Et_3P)_2(CO)$ and related compounds (M = Si or Ge)	1560

1651 J. J. Bour, P. J. M. W. L. Birker, and J. J. Steggerda, *Inorg. Chem.*, 1971, **10**, 1202.
1652 A. Shiotani and H. Schmidbaur, *Chem. Ber.*, 1971, **104**, 2838.
1653 W. M. Scovell, G. C. Stocco, and R. S. Tobias, *Inorg. Chem.*, 1970, **9**, 2682.
1654 W. Beck, K. Burger, and W. P. Fehlhammer, *Chem. Ber.*, 1971, **104**, 1816.
1655 G. E. Coates and B. R. Francis, *J. Chem. Soc. (A)*, 1971, 1308.
1656 G. E. Coates and B. R. Francis, *J. Chem. Soc. (A)*, 1971, 160.
1657 G. E. Coates and B. R. Francis, *J. Chem. Soc. (A)*, 1971, 474
1658 J. Lorberth, *J. Organometallic Chem.*, 1971, **27**, 303.
1659 Y. Kawakami, Y. Yasuda, and T. Tsuruta, *Bull. Chem. Soc. Japan*, 1971, **44**, 1164.
1660 Y. Kawakami and T. Tsuruta, *Bull. Chem. Soc. Japan*, 1971, **44**, 247.
1661 H. Nöth, E. Wiberg, and L. P. Winter, *Z. anorg. Chem.*, 1971, **386**, 73.
1662 J. L. Silver, M. Y. Al-Janabi, R. M. Johnson, and J. L. Burmeister, *Inorg. Chem.*, 1971, **10**, 994.
1663 A. T. Phillip, W. Mazurek, and A. T. Casey, *Austral. J. Chem.*, 1971, **24**, 501.
1664 R. F. Childers and R. A. D. Wentworth, *Inorg. Nuclear Chem. Letters*, 1971, **7**, 519.
1665 D. Seyferth and D. C. Mueller, *J. Organometallic Chem.*, 1971, **28**, 325.
1666 P. G. Cookson and G. B. Deacon, *Austral. J. Chem.*, 1971, **24**, 1599.
1667 B. L. Dyatkin, L. G. Zhuravkova, B. I. Martynov, E. I. Mysov, S. R. Sterlin, and I. L. Knunyants, *J. Organometallic Chem.*, 1971, **31**, C15.
1668 R. Fields, R. N. Haszeldine, and P. J. Palmer, *Tetrahedron Letters*, 1971, 1879.

Table 13 (*cont.*)

	Ref.
$(Ph_3P)_2Hg(\overline{C{=}NN{=}N}R)_2$	1654
$R^1HgCR^2N_2$ (R^1 = Me or Et; R^2 = RHg or CO_2Et)	1658
(3-isobutenyl-5-nortricyclyl)HgCl	1669
$RHgN(SiMe_3)_2$, $(RHg)_2O$, RHgX	1670
$R^1OCR^2R^3CR^4(HgBr)COY$, $BrHgCR^2R^3CR^4(OR^1)COY$	1671
$RCH(OBu^t)CH_2HgX$	1672
Hydroxymercurated but-2-enes, nitromercurated but-2-enes	1673
$N(HgMe)_3$	1674
$Hg(NSOF_2)_2$	1675, 1676
ROHgCl	1677

Table 14 *References to further papers containing n.m.r. data on compounds of* B

	Ref.
$Me_3PP(CF_3)BH_3$, $Me_3PP(CF_3)B_2H_6$	1678
$[B_3H_8]^-$	1679
$2\text{-Br-}1,6\text{-}C_2B_4H_5$, $1,6\text{-}Me_2\text{-}1,6\text{-}C_2B_4H_4$	1680
$2\text{-}Me_3Si\text{-}2,3\text{-}C_2B_4H_7$	1681
$1\text{-}R_3SiB_5H_8$	1682
$2,2'\text{-}(B_5H_8)_2$, $1,2'\text{-}(B_5H_8)_2$	1683
$B_{10}H_{10}C_2MeCX{=}CYZ$	1684
$HCB_{10}H_{10}CSO_nMe$ (n = 1 or 2)	1685
$RC_2B_{10}H_{10}Fe(CO)_2(Cp)$, $RCB_{10}H_{10}C(Cp)Fe(CO)_3$, and related compounds	1390
$[Zn(BH_4)_n]^{n-2}$, $K_2[Zn_3(BH_4)_8]$	1661
$MeS(CH_2)_3BR_2$ (R = H or OH)	1686
$MeCHBrBEt_2$	1687
$C_6H_{13}CH{=}CHB(C_6H_{11})_2$	1688

[1669] E. Vedejs and M. F. Salomon, *Chem. Comm.*, 1971, 1582.
[1670] J. Lorberth and F. Weller, *J. Organometallic Chem.*, 1971, **32**, 145.
[1671] A. J. Bloodworth and R. J. Bunce, *J. Chem. Soc.* (*C*), 1971, 1453.
[1672] D. H. Ballard and A. J. Bloodworth, *J. Chem. Soc.* (*C*), 1971, 945.
[1673] M. Matsuo and Y. Saito, *J. Organometallic Chem.*, 1971, **27**, C41.
[1674] W. Thiel, F. Weller, J. Lorberth, and K. Dehnicke, *Z. anorg. Chem.*, 1971, **381**, 57.
[1675] K. Seppelt and W. Sundermeyer, *Angew. Chem. Internat. Edn.*, 1970, **9**, 905.
[1676] W. Sundermeyer, A. Roland, and K. Seppelt, *Angew. Chem. Internat. Edn.*, 1971, **10**, 419.
[1677] R. Vaidyanathaswamy, D. Devaprabhakara, and V. V. Rao, *Tetrahedron Letters*, 1971, 915.
[1678] A. B. Burg, *J. Inorg. Nuclear Chem.*, 1971, **33**, 1575.
[1679] W. J. Dewkett, M. Grace, and H. Beall, *J. Inorg. Nuclear Chem.*, 1971, **33**, 1279.
[1680] R. R. Olsen and R. N. Grimes, *Inorg. Chem.*, 1971, **10**, 1103.
[1681] W. A. Ledoux and R. N. Grimes, *J. Organometallic Chem.*, 1971, **28**, 37.
[1682] D. F. Gaines and T. V. Iorns, *Inorg. Chem.*, 1971, **10**, 1094.
[1683] D. F. Gaines, T. V. Iorns, and E. N. Clevenger, *Inorg. Chem.*, 1971, **10**, 1096.
[1684] L. I. Zakharkin and V. N. Lebedev, *J. Gen. Chem.* (*U.S.S.R.*), 1971, **41**, 824.
[1685] A. R. Siedle, *J. Inorg. Nuclear Chem.*, 1971, **33**, 3677.
[1686] R. A. Braun, D. C. Brown, and R. M. Adams, *J. Amer. Chem. Soc.*, 1971, **93**, 2823.
[1687] H. C. Brown and Y. Yamamoto, *J. Amer. Chem. Soc.*, 1971, **93**, 2796.
[1688] G. Zweifel, G. M. Clark, and N. L. Polston, *J. Amer. Chem. Soc.*, 1971, **93**, 3395.

Table 14 (*cont.*)

	Ref.
$Et_2B(Cp)$ and related compounds	1689, 1690
$PhB(CH{=}CH)_2CH_2$, $[PhB{=}CHCH{=}CHCH{=}CH]^-$	1691
$(Ph_2B)_{4-n}SnMe_n$, $(diphos)_2Co(BPh_2)_2$	1871
$R_2BCH(BCl_2)CH_2(BCl_2)$	1692
$[Et_3GeBPh_3]Li,4HMPT$	1856
BX_3(hydrazine) (X = Me, F, or Cl)	1693
$(Me_3B)_n\{Me_2N(CH_2)_yNMe_2\}$	1694
$(Me_2B)_2CH_2$, $\{(Me_2N)_2B\}_2CH_2$, $Me_2NBHCH_2BCl(NMe_2BCl_2)_2NMe_2$, and related compounds	1695
$[R_2BCR{=}CHCH{=}CHO]Li$, $RBCR_2CH{=}CHCH_2O$	1696
$[Me_3NBH_2N(CH_2CH_2)_3NBH_2NMe_3]^{2+}$ and related compounds	1697
p-$MeC_6H_4SO_3BX_2NMe_3$ (X = H, Br, or p-$MeC_6H_4SO_3$)	1698
$Ph_2BNMeNMeBPh_2$ and related compounds	1699
$BH_2(NMe_2AlMe_2)_2NMe_2$	1700
$Me(C_6H_{13})BOBu^n$	1701
$R^1{}_2BOCXR^2{}_2$ (R^2 = halogenoalkyl group)	1702
$Et_2BOCR{=}CHR$	1703
$(THF)BCl(CH_2)_3CH_2$	1704
Me_2BSR	1705
$Bu^n(MeO)B(CH_2)_4B(OMe)_2$, $(MeO)_2B(CH_2)_4B(OMe)_2$	1706
$BNH(CH_2)_2N(CH_2)_3NH$, $CH_2CH_2NRBHNH$	1707
$HB(NCMeCHCMeN)_3M(CO)_2SC_6H_4R$ (M = W or Mo)	1336
$[(CF_3)_2CO]_2BPh$	1282
$(BuO)_2B(CH_2)_nX$	1708
$RBO(CH_2)_2O$	1709

1689 H. Grundke and P. I. Paetzold, *Chem. Ber.*, 1971, **104**, 1136.
1690 H. Grundke and P. I. Paetzold, *Angew. Chem. Internat. Edn.*, 1971, **10**, 423.
1691 A. J. Ashe, tert., and P. Shu, *J. Amer. Chem. Soc.*, 1971, **93**, 1804.
1692 A. K. Holliday and R. P. Ottley, *J. Chem. Soc.* (*A*), 1971, 886.
1693 L. K. Peterson and G. L. Wilson, *Canad. J. Chem.*, 1971, **49**, 3171.
1694 A. Storr and B. S. Thomas, *Canad. J. Chem.*, 1970, **48**, 3667.
1695 P. Krohmer and J. Goubeau, *Chem. Ber.*, 1971, **104**, 1347.
1696 A. Suzuki, N. Miyaura, and M. Itoh, *Tetrahedron*, 1971, **27**, 2775.
1697 S. A. Genchur, G. L. Smith, and H. C. Kelley, *Canad. J. Chem.*, 1971, **49**, 3165.
1698 R. Ryschkewitsch, *Inorg. Nuclear Chem. Letters*, 1971, **7**, 99.
1699 H. Nöth, W. Regnet, H. Rihl, and R. Standfest, *Chem. Ber.*, 1971, **104**, 722.
1700 R. E. Hall and E. P. Schram, *Inorg. Chem.*, 1971, **10**, 192.
1701 D. J. Pasto and K. McReynolds, *Tetrahedron Letters*, 1971, 801.
1702 E. W. Abel, N. Giles, D. J. Walker, and J. N. Wingfield, *J. Chem. Soc.* (*A*), 1971, 1991.
1703 W. Fenzl and R. Köster, *Angew. Chem. Internat. Edn.*, 1971, **10**, 750.
1704 H. C. Brown and E. Negishi, *J. Amer. Chem. Soc.*, 1971, **93**, 6682.
1705 H. Vahrenkamp, *J. Organometallic Chem.*, 1971, **28**, 167.
1706 H. C. Brown, E. Negishi, and P. L. Burke, *J. Amer. Chem. Soc.*, 1971, **93**, 3400.
1707 E. F. Rothgery, P. J. Busse, and K. Niedenzu, *Inorg. Chem.*, 1971, **10**, 2343.
1708 L. Miginiac and J. Blais, *J. Organometallic Chem.*, 1971, **29**, 349.
1709 H. C. Brown and S. K. Gupta, *J. Amer. Chem. Soc.*, 1971, **93**, 1819.

Table 14 (*cont.*)

	Ref.
RBOCOCOO	1710
o-$CF_3C_6H_4BF_2$, $[o\text{-}CF_3C_6H_4BX_3]^-$ (X = F or OH)	1711
$[B\{Mn(CO)_3\}NR]_3$	1712
$[B\{(C_5H_4)Fe(Cp)\}NH]_3$	1713
$[MeHNBNMe_3]_3$ and related compounds	1714
$[(MeCO_2)_2BNR_2]_n$	1715
$Cl_2B(MeNPCl_2)_2N$, $Cl_2B(MeN)_2PCl_2$	1716
$B(NHNR_2)_3$ and related compounds	1717
$(Me_3Si)_2NB(NMe_2)_2$, $(Me_3Si)_2NBCl(NMe_2)$	1718
$[Me_2NBCl_3]^-$	1719
$Ph(RS)C{=}N(BCl_2)_2N{=}C(SR)Ph$	1720
$[Me_2NPCl_3][BCl_4]$	1721
$[(CF_3)_3CO]_2BNH_2$, $[(CF_3)_nCF_{3-n}O]_3B$	1722
$OHCC_6H_4NMe_2BF_3$	1723
$ArNH_2BCl_3$ and related compounds	1724
$(Me_2N)_2BN(CX_2Ph)B(NMe_2)_2PdX_2$	1610
$C_5H_3Cl_2NBCl_3$, (thioanisole)BCl_3	1725
$(R_fO)_3B$	1726
$RCHO,BF_3$	1727

Table 15 *References to further papers containing n.m.r. data on compounds of* Al, Ga, In, *and* Tl

	Ref.
$(Me_3M)_n\{Me_2N(CH_2)_mNMe_2\}$ (M = Al, Ga, or In)	1694
R_2MOSCR (M = Al, Ga, or In)	1728
$M[N(SiMe_3)_2]_3$ (M = Al, Ga, or In)	1729
$LiAlH_2F(C_6F_4H)$, $LiAlH_3F$	1730

[1710] P. I. Paetzold, W. Scheibitz, and E. Scholl, *Z. Naturforsch.*, 1971, **26b**, 646.
[1711] T. Chivers, *Canad. J. Chem.*, 1970, **48**, 3856.
[1712] D. T. Haworth and E. S. Matushek, *Inorg. Nuclear Chem. Letters*, 1971, **7**, 261.
[1713] J. C. Kotz and W. J. Painter, *J. Organometallic Chem.*, 1971, **32**, 231.
[1714] M. Pasdeloup, G. Gros, G. Commenges, and J.-P. Laurent, *Bull. Soc. chim. France*, 1971, 754.
[1715] D. T. Haworth and E. S. Matushek, *Chem. and Ind.*, 1971, 130.
[1716] H. Binder, *Z. Naturforsch.*, 1971, **26b**, 616.
[1717] H. Nöth and H. Suchy, *Chem. Ber.*, 1971, **104**, 549.
[1718] H. L. Paige and R. L. Wells, *Inorg. Chem.*, 1971, **10**, 1526.
[1719] R. J. Ronan, J. W. Gilje, and M. J. Biallas, *J. Amer. Chem. Soc.*, 1971, **93**, 6811.
[1720] A. Meller and A. Ossko, *Monatsh.*, 1971, **102**, 131.
[1721] H. Binder and E. Fluck, *Z. anorg. Chem.*, 1971, **381**, 123.
[1722] D. E. Young, L. R. Anderson, and W. B. Fox, *Inorg. Chem.*, 1971, **10**, 2810.
[1723] M. Rabinovitz and A. Grinvald, *Tetrahedron Letters*, 1971, 4325.
[1724] J. R. Blackborow, J. E. Blackmore, and J. C. Lockhart, *J. Chem. Soc.* (*A*), 1971, 49.
[1725] S. Ardjomand and E. A. C. Lucken, *Helv. Chim. Acta*, 1971, **54**, 176.
[1726] D. E. Young, L. R. Anderson, and W. B. Fox, *Chem. Comm.*, 1971, 636.
[1727] M. Rabinovitz and A. Grinvald, *Tetrahedron Letters*, 1971, 641.
[1728] J. Weidlein, *J. Organometallic Chem.*, 1971, **32**, 181.
[1729] H. Bürger, J. Cichon, U. Goetze, U. Wannagat, and H. J. Wismar, *J. Organometallic Chem.*, 1971, **33**, 1.
[1730] R. S. Dickson and G. D. Sutcliffe, *Austral. J. Chem.*, 1971, **24**, 295.

Table 15 (*cont.*)

	Ref.
$H_2B(NMe_2AlMe_2)_2NMe_2$	1700
R_2Al(dimethyl malonate) and related compounds	1660
$Et_2BrAlNHMe_2$, $(EtBrAlNMe_2)_2$	1731
$Al\{OHC(CF_3)_2\}_3$	1294
Et_2Ga(8-oxinate)	1732
$[Et_2GaO_2PX_2]_2$	1733
$(C_6F_5)_3In$	1734
$Et_2In(OSCPh)$	1735
$[Me_2(MeSO_2)In]_2$, $(RSO_2)_3In$	1736
In(tropolonate)$_3$, $In(OCCMeOCHCHCO)_3$	1737
$InX\{SC(CF_3)C(CF_3)S\}$	1738
$R^1{}_2TlNR^2{}_2$	1739
$Tl(Et_2NCOX)$ (X = O or S)	1740

Table 16 *References to further papers containing n.m.r. data on compounds of* Si *containing only* C, H, *or* Si *bonded to* Si

	Ref.
$Me_3SiC{\equiv}CSiMe_3$, $Me_2HSiC{\equiv}CSiMe_2H$ and their $Co_2(CO)_6$ derivatives	1477
$Me_3SiCONHBu^t$	1741
2-Me_3SiCH_2-pyridine and related compounds	1742
p-$Me_3SiC_6H_4CONMe_2$	1743
Me_3SiBu^t, $Me_3SiCH_2SiXMe_2$	1744
$(Me_3Si)_2CHCl$ and related compounds	1745
$Me_3SiC{\equiv}CR$	1746
o-$C_6H_4CH(SiMe_3)CH{=}CHCH(SiPh_3)$	1747
$Me_3SiC_6H_3CH(SiMe_3)CH(SiMe_3)CH(SiMe_3)CH_2$ and related compounds	1748
$PhCH(SiMe_3)CH_2CHO$, $\{Me_2C(SiMe_3)CH_2\}_2CO$,	1749

1731 K. Gosling, A. L. Bhuiyan, and K. R. Mooney, *Inorg. Nuclear Chem. Letters*, 1971, **7**, 913.
1732 B. Sen and G. White, *Inorg. Nuclear Chem. Letters*, 1971, **7**, 79.
1733 J. Weidlein and B. Schaible, *Z. anorg. Chem.*, 1971, **386**, 176.
1734 G. B. Deacon and J. C. Parrott, *Austral. J. Chem.*, 1971, **24**, 1771.
1735 H. Tada and R. Okawara, *J. Organometallic Chem.*, 1971, **28**, 21.
1736 A. T. T. Hsieh, *J. Organometallic Chem.*, 1971, **27**, 293.
1737 D. G. Tuck and M. K. Yang, *J. Chem. Soc.* (*A*), 1971, 3100.
1738 A. F. Berniaz, G. Hunter, and D. G. Tuck, *J. Chem. Soc.* (*A*), 1971, 3254.
1739 B. Walther and K. Thiede, *J. Organometallic Chem.*, 1971, **32**, C7.
1740 R. J. Magee and M. J. O'Connor, *Inorg. Chim. Acta*, 1971, **5**, 554.
1741 P. Jutzi and F. W. Schröder, *Angew. Chem. Internat. Edn.*, 1971, **10**, 339.
1742 W. K. Musker and R. L. Scholl, *J. Organometallic Chem.*, 1971, **27**, 37.
1743 G. P. Schiemenz, *Tetrahedron*, 1971, **27**, 5723.
1744 R. West and G. A. Gornowicz, *J. Organometallic Chem.*, 1971, **28**, 25.
1745 M. A. Cook, C. Eaborn, and D. R. M. Walton, *J. Organometallic Chem.*, 1971, **29**, 389.
1746 R. West and G. A. Gornowicz, *J. Amer. Chem. Soc.*, 1971, **93**, 1720.
1747 F. W. G. Fearon and J. C. Young, *J. Chem. Soc.* (*B*), 1971, 272.
1748 L. Birkofer and N. Ramadan, *Chem. Ber.*, 1971, **104**, 138
1749 R. Calas, J. Dunoguès, and M. Bolourtchian, *J. Organometallic Chem.*, 1971, **26**, 195.

Table 16 (*cont.*)

Compound	*Ref.*
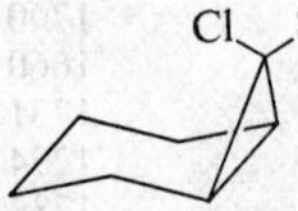 and related compounds	1750
$CH_2CH_2C(SiMe_3)CO_2SiMe_3$	1751
$Me_3SiCR^1{=}C{=}CR^2OCMe_3$ and related compounds	1752
$CH_2CHClCHSiMe_3$	1753
$Ph(Me_3Si)C{=}C{=}C(OMe)SiMe_3$	1754
$Me_3SiC(CONEt_2)CHPh$	1755
$CH_2{=}C(OEt)C(SiMe_3){=}CH_2$, $CH_2BrCH(OEt)C(SiMe_3){=}CH_2$	1756
$Me_3SiCH_2NHC_6H_4OMe$, $Me_3SiCH_2CH_2NHPh$	1757
$CF_2CF_2C(SiMe_3){=}C(SiMe_3)$	1758
Me_3SiR_f	1668
$Me_3SiCH_2ML_n$	1285
$(Me_3SiC_5H_4)M(CO)_n$	1335
$Me_3SiCH_2C(OEt)Cr(CO)_5$	1315
2-Me_3Si-2,3-$C_2B_4H_7$	1681
$R_3Si(CH_2)_nCH{=}CH_2$ (R = Me or Ph)	1759
p-$FC_6H_4SiMe_2Et$	1760
$MeEt_2Si(CH_2)_2OC(OR){=}CH_2$	1761
$Et_3SiCHCHOC_6H_{13}$, $Et_3SiCH{=}CHC_6H_{13}$	1762
(+)-$Ph(Ph_3Si)CHOH$	1763
$Ph_3SiCOMe$	1764
(+)-MePh(1-naphthyl)SiCOPh	1765
(1-naphthyl)$SiMePhCHBr_2$ and its derivatives	1766
$PhSiDMe_2$	1767

1750 D. Seyferth and E. M. Hanson, *J. Organometallic Chem.*, 1971, **27**, 19.
1751 Y.-N. Kuo, F. Chen, C. Ainsworth, and J. J. Bloomfield, *Chem. Comm.*, 1971, 136.
1752 R. Mantione and Y. Leroux, *J. Organometallic Chem.*, 1971, **31**, 5.
1753 D. Seyferth, H. D. Simmons, jun., and H.-M. Shih, *J. Organometallic Chem.*, 1971, **29**, 359.
1754 Y. Leroux and R. Mantione, *J. Organometallic Chem.*, 1971, **30**, 295.
1755 J. Ficini and A. Duréault, *Compt. rend.*, 1971, **273**, *C*, 289
1756 L. L. Shchukovskaya and A. M. Kapustyanakaya, *J. Gen. Chem.* (*U.S.S.R.*), 1971, **41**, 606.
1757 B. E. Cooper and W. J. Owen, *J. Organometallic Chem.*, 1971, **29**, 33.
1758 R. Fields, R. N. Haszeldine, and A. F. Hubbard, *J. Chem. Soc.* (*C*), 1971, 3838.
1759 A. W. P. Jarvie and R. J. Rowley, *J. Chem. Soc.* (*B*), 1971, 2439.
1760 J. Nishimura, J. Furukawa, and N. Kawabata, *J. Organometallic Chem.*, 1971, **29**, 237.
1761 M. F. Shostakovskii, A. S. Atavin, E. P. Vyalykh, G. M. Gavrilova, G. A. Kalabin, and B. A. Trofimov, *J. Gen. Chem.* (*U.S.S.R.*), 1971, **41**, 610.
1762 G. Stork and E. Colvin, *J. Amer. Chem. Soc.*, 1971, **93**, 2080.
1763 M. S. Biernbaum and H. S. Mosher, *J. Amer. Chem. Soc.*, 1971, **93**, 6221.
1764 L. C. Willemsens, *J. Organometallic Chem.*, 1971, **27**, 45.
1765 A. G. Brook and J. D. Pascoe, *J. Amer. Chem. Soc.*, 1971, **93**, 6224.
1766 A. G. Brook, J. M. Duff, and D. G. Anderson, *J. Amer. Chem. Soc.*, 1970, **92**, 7567.
1767 Y. Nagal, H. Matsumoto, M. Hayashi, E. Tajima, M. Ohtsuki, and N. Sekikawa, *J. Organometallic Chem.*, 1971, **29**, 209.

Table 16 (*cont.*)

	Ref.
$Me(ClCH_2)_2SiCH_2SiR_3$	1768
$R_3SiCH_2CHCH_2CClCF_3$	1769
$CH_2{=}CH(CH_2)_nSiR_3$, $SiMe_2CH{=}CHCH_2CH_2$, and related compounds	1770
$R^1{}_3Si(CH_2)_2SiR^2{}_2C_6H_4SiR^2{}_2(CH_2)_2SiR^2{}_2$	1771
$PhCH{=}CHS(CH_2)_2SiR_3$	1772
$(SiH_2CClH)_3$	1773
$(CH_2)_4SiH_2$ and related compounds	1774
$SiMe_2CH{=}CH(CH_2)_3$ and related compounds	1775
$Me_2SiCH_2CHRC({=}CH_2)CH_2$ and related compounds	1776
$Me_2SiCH_2CH{=}CH(CH_2)_2$ and related compounds	1777
$Me_2Si(CH_2)_n$ (n = 3 or 4) and related compounds	1778
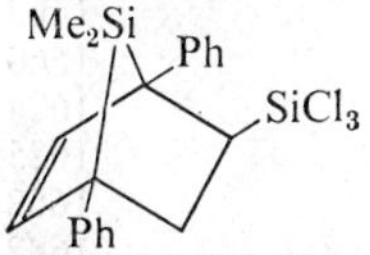	1779
$R^1R^2Si(C_6H_4)_2CH_2$	1780
$R^1R^2Si(C_6H_4)_2C_2H_2$	1781
$MeSiH_2SiMeHSiMeH_2$ and related compounds	1782
Me_6Si_2, $Me_3SiSiMe_3$, Et_6Si_2	1783
$(Me_3Si)_2$ and polysilanes	1784
$Me_3SiSiMe_2C_6F_5$, $(C_6F_3Me_2Si)_2$	1785
$R_3SiSiRCH_2X$	1786
$PhMe_2SiSiMePhCH_2Cl$ and related compounds	1787

[1768] G. Fritz and G. Maass, *Z. anorg. Chem.*, 1971, **386**, 163.
[1769] D. Seyferth and D. C. Mueller, *J. Amer. Chem. Soc.*, 1971, **93**, 3714.
[1770] J. W. Connolly and P. F. Fryer, *J. Organometallic Chem.*, 1971, **30**, 315.
[1771] C. Michel and M. Brini, *Bull. Soc. chim. France*, 1971, 4132.
[1772] F. P. L'vova, N. N. Vlasova, and V. K. Voronov, *J. Gen. Chem.* (*U.S.S.R.*), 1971, **41**, 337.
[1773] G. Fritz and H. Frohlich, *Z. anorg. Chem.*, 1971, **382**, 217.
[1774] T. H. Chao, S. L. Moore, and J. Laane, *J. Organometallic Chem.*, 1971, **33**, 157.
[1775] E. Rosenberg and J. J. Zuckerman, *J. Organometallic Chem.*, 1971, **33**, 321.
[1776] G. Manuel, P. Mazerolles, and J.-C. Florence, *J. Organometallic Chem.*, 1971, **30**, 5.
[1777] S. S. Washburne and R. R. Chawla, *J. Organometallic Chem.*, 1971, **31**, C20.
[1778] D. Seyferth, R. Damrauer, S. B. Andrews, and S. S. Washburne, *J. Amer. Chem. Soc.*, 1971, **93**, 3709.
[1779] A. Laporterie, J. Dubac, P. Mazerolles, and M. Lesbre, *Tetrahedron Letters*, 1971, 4653.
[1780] P. Jutzi, *Chem. Ber.*, 1971, **104**, 1455.
[1781] J. Y. Corey, M. Dueber, and B. Bichlmeir, *J. Organometallic Chem.*, 1971, **26**, 167.
[1782] R. B. Baird, M. D. Sefcik, and M. A. Ring, *Inorg. Chem.*, 1971, **10**, 883.
[1783] M. Gielen, J. Nasielski, and G. Vandendunghen, *Bull. Soc. chim. belges*, 1971, **80**, 175.
[1784] B. K. Nicholson and J. Simpson, *J. Organometallic Chem.*, 1971, **32**, C29.
[1785] M. Weidenbruch, G. Abrotat, and K. John, *Chem. Ber.*, 1971, **104**, 2124.
[1786] K. Tamao and M. Kumada, *J. Organometallic Chem.*, 1971, **30**, 329.
[1787] K. Tamao and M. Kumada, *J. Organometallic Chem.*, 1971, **30**, 349.

Table 16 (*cont.*)

	Ref.
$RSiMe_2SiMe_3$	1788
$(MePh_2Si)_2$ and related compounds	1789
$PhSi\{(CH_2)_4\}_2SiPh$, $Si\{(CH_2)_4\}_3Si$	1790
$[(Me_2CHCH_2)_2Si]_5$	1791
$XMeSi(CH_2)_4SiMeX$	1792
$Me_2Si(CH_2)_4SiMe_2$ and related compounds	1793

Table 17 *References to further papers containing n.m.r. data on other compounds of* Si

	Ref.
$[Me_3SiFe(CO)_3]_2(Me_3SiOCCOSiMe_3)$	1437
$Me_3SiM(CO)_4X$ (M = Ru or Os) and related compounds	1462
$Ir(HgSiMe_3)(SiMe_3)(CO)(Et_3P)_2$ and related compounds	1560
$PtX(SiH_{3-n}X_n)(PEt_3)_2$	1624
$(diphos)PtX(SiMe_3)$	1625
$(Me_3Si)_2Hg$	1668
$Me_3SiB_5H_8$	1682
$(OC)_3M^1(SiMe_3)(SiMe_2)_2M^2(CO)(SiMe_3)$ (M^1 = Ru or Os) and related compounds	1464
$(Cp)(CO)RhSiX_3$	1527
$Cl_3SiMH(CO)_n$ (M = Mn, Fe, or Co)	1402
$(H_3Si)_2NSiH_2N(SiH_3)_2$	1794
$Me_3SiNHCMe{=}CHCN$ and related compounds	1795
$Me_3SiN(CH{=}CH)_2NSiMe_3$	1796
$Me_3SiNMeSiMe_2OSiMe_3$	1797
Me_3SiN_3, $MePhSi(N_3)_2$, $MePhSi(N{=}PPh_3)_2$	1798
$(Me_3Si)_nNH_{3-n}$	1799
$C_7H_6X(NSiMe_3)_2$	1800
$Me_3SiNHPBr_2NSO_2R$	1905
$(Me_3SiNH)_3PS$	1801
$Me_3SiNHPCl_2{=}NPF_2S$ and related compounds	1903
$(Me_3Sn)RNN(SiMe_3)(SnMe_3)$, $RN{=}NSiMe_3$	1802

[1788] H. Sakurai and A. Hosomi, *J. Amer. Chem. Soc.*, 1971, **93**, 1709.
[1789] A. G. Evans, M. A. Hamid, and N. H. Rees, *J. Chem. Soc.* (*B*), 1971, 2165.
[1790] K. Tamao, M. Kumada, and A. Noro, *J. Organometallic Chem.*, 1971, **31**, 169.
[1791] G. R. Husk, R. Wexler, and B. M. Kilcullen, *J. Organometallic Chem.*, 1971, **29**, C49.
[1792] K. Tamao, M. Kumada, and M. Ishikawa, *J. Organometallic Chem.*, 1971, **31**, 17.
[1793] H. Sakurai, Y. Kobayashi, and Y. Nakadaira, *J. Amer. Chem. Soc.*, 1971, **93**, 5272.
[1794] W. M. Scantlin and A. D. Norman, *Chem. Comm.*, 1971, 1246.
[1795] G. A. Gornowicz and R. West, *J. Amer. Chem. Soc.*, 1971, **93**, 1714.
[1796] R. A. Sulzbach and A. F. M. Iqbal, *Angew. Chem. Internat. Edn.*, 1971, **10**, 127.
[1797] P. Boudjouk and R. West, *J. Amer. Chem. Soc.*, 1971, **93**, 5901.
[1798] S. S. Washburne and W. R. Peterson, jun., *J. Organometallic Chem.*, 1971, **33**, 153.
[1799] N. Wiberg and W. Uhlenbrock, *Chem. Ber.*, 1971, **104**, 2643.
[1800] B. Martel and E. Aly, *J. Organometallic Chem.*, 1971, **29**, 61.
[1801] H. H. Falius, J.-P. Giesen, and U. Wannagat, *Inorg. Nuclear Chem. Letters*, 1971, **7**, 281.
[1802] N. Wiberg and M. Veith, *Chem. Ber.*, 1971, **104**, 3191.

Table 17 (*cont.*)

	Ref.
$(Me_3Si)_2NN(SiMe_3)_2$ and related compounds	1803
$(Me_3Si)_2NN{=}CCl_2$	1804
$Me_3SiO_2CN(SiMe_3)N(SiMe_3)CO_2Me$ and related compounds	1805
$[(Me_3Si)_2N]M$ (M = Al, Ga, or In)	1729
$(Me_3Si)_2NB(NMe_2)X$	1718
$[Me_3SiNHMe_2]^+$	1473
$Me_3SiN[Fe_3(CO)_{10}]$	1438
$RHgN(SiMe_3)_2$	1670
$Me_3SiN{=}SF_2O$	1675
$Si(NSOF_2)_4$	1676
N-R_3Si-pyrrole	1806
$Me_2SONSiR_3$, $[Me_2SON]_2SiR_2$	1807
R_3PNSiR_3	1808
$(R_3Si)_2NH$	1809
$R_3SiNHMe$, $R_3SiC{\equiv}CMe(NHMe)CH_2CN$	1810
$R^1R^2R^3{}_2Si$	1811
$R^1R^2R^3SiNCO(CH_2)_2CO$	1812
$(Me_2N)_nSi(CH_2CH_2EMe_2)_{4-n}$ (E = N or P)	1813
$Me_nCl_{3-n}SiNHSiMe_nCl_{3-n}$	1814
$XSiR^1{}_2NR^2SiR^3{}_2X$	1815
$(SiH_nCl_{3-n})_mNR_{3-m}$	1816
F_3SiNR_2 and related compounds	1817
$Cl_2Si(NSiCl_3)_2SiCl_2$, $R^1{}_2Si(NSiR^1{}_3)_2SiR^1{}_2$ ($R^1 = NR^2{}_2$ or OR^3)	1818
$Me_3SiP(SiH_3)_2$, $H_3SiSiH_2P(SiH_3)_2$	1819
$Me_nH_{3-n}SiPH_2$, $MeSiH(PH_2)_2$	1820
MeH_2SiPH_2	1821
$Me_3SiPHPh$	1822
H_3SiOR	1823

[1803] N. Wiberg and M. Veith, *Chem. Ber.*, 1971, **104**, 3176.
[1804] N. Wiberg and W. Uhlenbrock, *Chem. Ber.*, 1971, **104**, 3989.
[1805] N. Wiberg and G. Schwenk, *Chem. Ber.*, 1971, **104**, 3986.
[1806] H. Burger and K. Burczyk, *Z. anorg. Chem.*, 1971, **381**, 176.
[1807] W. Wolfsberger and H. Schmidbaur, *J. Organometallic Chem.*, 1971, **28**, 317.
[1808] W. Wolfsberger, H. H. Pickel, and H. Schmidbaur, *J. Organometallic Chem.*, 1971, **28**, 307.
[1809] L. W. Breed, R. L. Elliot, and J. C. Wiley, jun., *J. Organometallic Chem.*, 1971, **31**, 179.
[1810] F. Ya. Perveev and I. I. Afonina, *J. Gen. Chem.* (*U.S.S.R.*), 1971, **41**, 340.
[1811] S. S. Dua, R. C. Edmondson, and H. Gilman, *J. Organometallic Chem.*, 1971, **27**, 33.
[1812] A. F. Janzen and E. A. Kramer, *Canad. J. Chem.*, 1971, **49**, 3456
[1813] J. Grobe and U. Möller, *J. Organometallic Chem.*, 1971, **33**, 13.
[1814] U. Wannagat, J. Herzig, P. Schmidt, and M. Schulze, *Monatsh.*, 1971, **102**, 1817.
[1815] U. Wannagat and E. Bogusch, *Monatsh.*, 1971, **102**, 1806.
[1816] J. F. Drake and N. P. C. Westwood, *J. Chem. Soc.* (*A*), 1971, 3617.
[1817] W. Airey, G. M. Sheldrick, B. J. Aylett, and I. A. Ellis, *Spectrochim. Acta*, 1971, **27A**, 1505.
[1818] U. Wannagat, H. Moretto, and P. Schmidt, *Z. anorg. Chem.*, 1971, **385**, 164.
[1819] J. W. Anderson and J. E. Drake, *J. Chem. Soc.* (*A*), 1971, 2246.
[1820] G. Fritz and H. Schäfer, *Z. anorg. Chem.*, 1971, **385**, 243.
[1821] J. W. Anderson and J. E. Drake, *J. Chem. Soc.* (*A*), 1971, 1424.
[1822] M. Baudler and A. Zarkadas, *Chem. Ber.*, 1971, **104**, 3519.
[1823] C. Glidewell, *J. Chem. Soc.* (*A*), 1971, 823.

Table 17 (*cont.*)

	Ref.
Me_3SiOR	1824—1831
Et_3SiOR	1832, 1833
Si(*o*-menthyl)ClPh(α-naphthyl)	1834
$Me_2Si(CH_2)_4CH(CMe_2OH)O$ and related compounds	1835
$O\{SiMe(CMe{=}CMe)_2SiMe\}_2O$	1836
$Me_2SiOCMe(PEt)(PEt_2)CMeO$ and related compounds	1837
$(Bu^tO)_3Si(OSiMe_3)$, $(Bu^tO)_2Si(OSiMe_2O)_2Si(OBu^t)_2$	1838
Polysiloxanes	1839
$Si\{OHC(CF_3)_2\}_4$	1294
$Si_2F_5OSiF_3$ and related compounds	1840
$Me_3SiSP(S)F_2$	1841
$R_2SiCl(OSnBu^n{}_2Cl)$	1842
halogenotrisilanes	1843
$CF_3C{=}CHSiF_2SiF_2$, $HC{=}C(CF_3)C(CF_3)CHSiF_2SiF_2$	1844
Ph_2FSiCH_2Ph and related compounds	1845, 1846
SiF_nCl_{4-n}	1847
X_nSiR_{4-n}	1848
$Me_3PC(SiR_nCl_{3-n})_2$	1849
$Cl_3SiSiMe_2CH_2Cl$ and related compounds	1850

[1824] G. R. Van den Berg, D. H. J. M. Platenburg, and H. P. Benschop, *Chem. Comm.*, 1971, 606.
[1825] Y. Yamamoto and D. S. Tarbell, *J. Org. Chem.*, 1971, **36**, 2955.
[1826] R. E. K. Winter and M. L. Honig, *J. Amer. Chem. Soc.*, 1971, **93**, 4616.
[1827] G. S. Burlachenko, T. A. Manukina, and Yu. I. Baukov, *J. Organometallic Chem.*, 1971, **33**, C59.
[1828] P. Bourgeois and G. Merault, *Compt. rend.* 1971, **273**, *C*, 714.
[1829] W. E. Parham and C. S. Roosevelt, *Tetrahedron Letters*, 1971, 923.
[1830] J. P. Picard, R. Calas, J. Dunoguès, and N. Duffaut, *J. Organometallic Chem.*, 1971, **26**, 183.
[1831] K. Rühlmann, H. Seefluth, Th. Kiriakidis, G. Michael, H. Jancke, and H. Kriegsmann, *J. Organometallic Chem.*, 1971, **27**, 327.
[1832] M. Paul and E. Frainnet, *J. Organometallic Chem.*, 1971, **30**, C64.
[1833] R. Bourhis and E. Frainnet, *J. Organometallic Chem.*, 1971, **28**, C11.
[1834] R. J. P. Corriu and G. F. Lanneau, *Tetrahedron Letters*, 1971, 2771.
[1835] A. G. Brook, R. Pearce, and J. B. Pierce, *Canad. J. Chem.*, 1971, **49**, 1622.
[1836] F. Wudl, R. D. Allendoerfer, J. Demirgian, and J. M. Robbins, *J. Amer. Chem. Soc.*, 1971, **93**, 3160.
[1837] J. Satgé, C. Couret, and J. Escudié, *J. Organometallic Chem.*, 1971, **30**, C70.
[1838] I. Kijima, K. Okuda, and Y. Abe, *Bull. Chem. Soc. Japan*, 1971, **44**, 1106.
[1839] F. Métras, J.-C. Lahournère, and J. Valade, *J. Organometallic Chem.*, 1971, **29**, 41.
[1840] K. G. Sharp and J. L. Margrave, *J. Inorg. Nuclear Chem.*, 1971, **33**, 2813.
[1841] D. W. McKennon and M. Lustig, *Inorg. Chem.*, 1971, **10**, 406.
[1842] A. G. Davies and P. G. Harrison, *J. Chem. Soc.* (*C*), 1971, 1769.
[1843] J. E. Drake, N. Goddard, and N. P. C. Westwood, *J. Chem. Soc.* (*A*), 1971, 3305.
[1844] C. S. Liu and J. C. Thompson, *Inorg. Chem.*, 1971, **10**, 1100.
[1845] A. G. Brook and K. H. Pannell, *Canad. J. Chem.*, 1970, **48**, 3679.
[1846] T. J. Hairston and D. H. O'Brien, *J. Organometallic Chem.*, 1971, **29**, 79.
[1847] K. Hamada, G. A. Ozin, and E. A. Robinson, *Bull. Chem. Soc. Japan*, 1971, **44**, 2555.
[1848] J. Grobe and U. Möller, *J. Organometallic Chem.*, 1971, **31**, 157.
[1849] H. Schmidbaur and W. Malisch, *Chem. Ber.*, 1971, **104**, 150.
[1850] K. Tamao and M. Kumada, *J. Organometallic Chem.*, 1971, **30**, 339.

Table 17 (*cont.*)

	Ref.
R_3SiCl	1665
$R^1R^2{}_2SiCl$	1851
$Cl_2Si(CH{=}CH)_2SiCl_2$	1852

Table 18 *References to further papers containing n.m.r. data on compounds of* Ge

	Ref.
$(H_3Ge)_2CH_2$	1853
Me_3GeR	1658, 1741, 1742, 1854, 1855
Et_3GeR	1856
Ph_3GeMe, Ph_3GeSMe	1857
$R_3GeCH_2CH{=}CH_2$	1858
$R_nX_{3-n}GeCH_2$(vinyl)	1859
R_3Ge(vinyl)	1860
(1-naphthyl)$GeMePhCHBr_2$ and related compounds	1766
$R^1R^2Ge(C_6H_4)_2CH_2$	1780
$HPhGe(CH_2)_nCHOH$	1861
Et_3GeR, $Et_2GeCH_2CH(CHCl_2)(CH_2)_n$	1862
$[PhR_2Ge]_2$	1863
$PhGeH_2GeH_2Ph$	1864
Me_6Ge_2, $Et_3SnGeEt_3$, and related compounds	1783
$(R_2N)_3TiGePh_3$	1283
$R_2GeCr(CO)_5$	1314
$[Me_3GeRu(CO)_4]_2$ and related compounds	1462
$(Me_3Ge)_2Os(CO)_4$	1463
$(Ph_3P)_2RhHCl(GeEt_3)$	1529
$Ir(HgGeMe_3)(GeMe_3)_2(Et_3P)_2(CO)$ and related compounds	1560
$PtX(GeH_{3-n}X_n)(PEt_3)_2$	1624
(diphos)$Pt(GeMe_3)X$	1625
$Me_3GeN{=}NR$	1802

[1851] M. A. Kadina and V. A. Ponomarenko, *J. Gen. Chem.* (*U.S.S.R.*), 1971, **41**, 169.

[1852] E. A. Chernyshev, N. G. Komalenkova, and S. A. Bashkirova, *J. Gen. Chem.* (*U.S.S.R.*), 1971, **41**, 1177.

[1853] R. M. Dreyfus and W. L. Jolly, *Inorg. Chem.*, 1971, **10**, 2567.

[1854] S. V. Ponomarev, M. B. Érman, S. A. Lebedev, S. Ya. Pechurina, and I. F. Lutsenko, *J. Gen. Chem.* (*U.S.S.R.*), 1971, **41**, 122.

[1855] D. Seyferth and S. B. Andrews, *J. Organometallic Chem.*, 1971, **30**, 151.

[1856] E. J. Bulten and J. G. Noltes, *J. Organometallic Chem.*, 1971, **29**, 409.

[1857] D. J. Sandman and R. West, *J. Organometallic Chem.*, 1971, **30**, C61.

[1858] M. Massol, J. Barrau, P. Rivière, and J. Satgé, *J. Organometallic Chem.*, 1971, **30**, 27.

[1859] M. Massol, J. Barrau, P. Rivière, and J. Satgé, *J. Organometallic Chem.*, 1971, **30**, 27.

[1860] M. Massol, Y. Cabadi, and J. Satgé, *Bull. Soc. chim. France*, 1971, 3235.

[1861] P. Rivière and J. Satgé, *Angew. Chem. Internat. Edn.*, 1971, **10**, 267.

[1862] D. Seyferth, H.-M. Shih, P. Mazerolles, M. Lesbre, and M. Joanny, *J. Organometallic Chem.*, 1971, **29**, 371.

[1863] P. Rivière and J. Satgé, *Bull. Soc. chim. France*, 1971, 3221.

[1864] F. Feheré and P. Plichta, *Inorg. Chem.*, 1971, **10**, 609.

Table 18 (*cont.*)

	Ref.
$(Me_3Ge)_2NN(GeMe_3)_2$ and related compounds	1803
$Me_nGe(NO_3)_{4-n}$	1865
$MeGeH_2PH_2$	1821
$(OC)_4Mn\{P(GeMe_3)_2\}Mn(CO)_4$	1345
$[Me_2GePPh]_3$	1866
$Me_2GeO\{CMe(PEt_2)\}_2O$ and related compounds	1837
$Me_nGeH_{3-n}F$	1867
$MeGeHX_2$	1868
$MeGeH_nX_{3-n}$	1869
XCH_2GeX_3, $MeGeH_2X$	1870
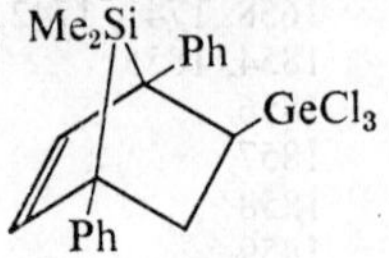	1779

Table 19 *References to further papers containing n.m.r. data on compounds of* Sn *and* Pb

	Ref.
$(Ph_2B)_{4-n}SnMe_n$	1871
Me_3SnR	1313, 1477, 1741, 1742, 1855, 1872
$Et_3SnCH_2CH_2OCOMe$	1873
$Bu^n{}_3SnR$	1874, 1875
$Bu^n{}_2Sn(CH_2CH_2CN)_2$	1966
R_3SnR	1876—1878
$R^1R^2Sn(C_6H_4)_2CH_2$	1780
$Bu^n{}_2Sn(CH{=}CH)_2CH_2$	1691
$Ti(SnPh_3)_3$	1288
$Me_2SnCr(CO)_5$	1314
$CF_3(CH_2)SnM(CO)_n(Cp)_m$	1340
$(Cp)(OC)_2FeSnR_3$	1395

[1865] D. Potts and A. Walker, *Canad. J. Chem.*, 1971, **49**, 2171.
[1866] H. Schumann and H. Benda, *Chem. Ber.*, 1971, **104**, 333.
[1867] C. H. Van Dyke, J. E. Bulkowski and N. Viswanathan, *Inorg. Nuclear Chem. Letters*, 1971, **7**, 1057.
[1868] G. K. Barker, J. E. Drake, R. T. Hemmings, and B. Rapp, *J. Chem. Soc.* (*A*), 1971, 3291.
[1869] J. E. Drake, R. T. Hemmings, and C. Riddle, *J. Chem. Soc.* (*A*), 1971, 600.
[1870] J. M. Bellama and C. J. McCormick, *Inorg. Nuclear Chem. Letters*, 1971, **7**, 533.
[1871] H. Nöth, H. Schäfer, and G. Schmid, *Z. Naturforsch.*, 1971, **26b**, 497.
[1872] J. C. Maire and J. van Rietschoten, *Helv. Chim. Acta*, 1971, **54**, 1054.
[1873] J. Tsurugi, M. Iida, R. Nakao, T. Fukumoto, and N. Murata, *Bull. Chem. Soc. Japan*, 1971, **44**, 777.
[1874] D. J. Peterson, *J. Amer. Chem. Soc.*, 1971, **93**, 4027.
[1875] J.-C. La Hournère and J. Valade, *J. Organometallic Chem.*, 1971, **33**, C4.
[1876] Y. Tanigawa, I. Moritani, and S. Nishida, *J. Organometallic Chem.*, 1971, **28**, 73.
[1877] J.-C. Maire and J. van Rietschoten, *Helv. Chim. Acta*, 1971, **54**, 749.
[1878] D. J. Peterson, *J. Organometallic Chem.*, 1971, **26**, 215.

Table 19 (*cont.*)

	Ref.
$[(Me_3Sn)M(CO)_4X]_2$ (M = Ru or Os)	1462
$Cl_3SnRhC_{15}H_{20}$	1537
$PtSnCl_3(C_3H_5)(PPh_3)$	1619
$(diphos)PtX(SnMe_3)$	1625
$R^1{}_3SnMR^2{}_3$ (M = Si, Ge, or Sn)	1783
Me_3SnNRX	1879
$(Me_3Sn)_2NN(SnMe_3)_2$	1803
$Me_3SnN{=}NR$	1880
$Bu_3SnNR^1R^2$	1881
$R_3SnNN{=}C(C_6H_4R)N{=}N$	1882
$(OC)_4Mn\{P(SnMe_3)_2\}_2Mn(CO)_4$	1345
$(Me_2SnPPh)_3$	1866
$[Bu_2SnO]_n$	1883
$MeSn(O_2SR)_3$	1884
$R_2SiCl(OSnR_2Cl)$	1842
$Me_2SnCl(SSeCNMe_2)$	1589
$Me_nSn(S_2CNR_2)_{4-n}$	1885
$Me_2SnClSP(S)F_2$	1841
M(octaethylporphyrin) (M = Sn or Pb)	1886
Me_3PbCRN_2	1658
R_2PbL (L = 2,2′-methylidynenitrilodiphenol and related terdentate ligands)	1887
$Pb(CF_3COCHCSCH_3)_2$	1585
$Pb(CH_3COCHCSCH_3)_2$	1586

Table 20 *References to further papers containing n.m.r. data on compounds of* P, As, Sb, *and* Bi

	Ref.
$[R_2PH_2]Cl$, R_2Ph, R_2PO_2H	1888, 1889
$[R_2PPR_2H]Cl$	1889
$P_2(CF_3)_4$, $(Cp)_2Ti[OP(CF_3)_2]_2$	1287
$R^1C{=}NNMePR^2NH$	1890
$PHX{=}NC(NR^1R^2){=}NC(NH_2){=}N$	1891

1879 H. W. Roesky and H. Wiezer, *Chem. Ber.*, 1971, **104**, 2258.
1880 N. Wiberg and M. Veith, *Chem. Ber.*, 1971, **104**, 3191.
1881 A. G. Davies and J. D. Kennedy, *J. Chem. Soc.* (*C*), 1971, 68.
1882 K. Sisido, K. Nabika, T. Isida, and S. Kozima, *J. Organometallic Chem.*, 1971, **33**, 337.
1883 D. R. G. Brimage, R. S. Davidson, and P. F. Lambeth, *J. Chem. Soc.* (*C*), 1971, 1241.
1884 R. Lindner, U. Kunze, and J. Koola, *J. Organometallic Chem.*, 1971, **30**, 59.
1885 J. C. May, D. Petridis, and C. Curran, *Inorg. Chim. Acta*, 1971, **5**, 511.
1886 D. G. Whitten, J. C. Yau, and F. A. Carroll, *J. Amer. Chem. Soc.*, 1971, **93**, 2291.
1887 R. Bosco and R. Cefalù, *J. Organometallic Chem.*, 1971, **26**, 225.
1888 F. Seel and K.-D. Velleman, *Chem. Ber.*, 1971, **104**, 2972.
1889 F. Seel and K.-D. Velleman, *Chem. Ber.*, 1971, **104**, 2967.
1890 Y. Charbonnel and J. Barrans, *Compt. rend.*, 1971, **272**, *C*, 1675.
1891 J. Ebeling, M. A. Leva, H. Stary, and A. Schmidpeter, *Z. Naturforsch.*, 1971, **26b**, 650.

Table 20 (*cont.*)

	Ref.
N-organyliminophosphoranes	1892
$Ph_2R^1P{=}NSO_2R^2$	1893
$Me_2S(NPR_2)_2$, $MeS(NPR_2N)_2SMe$, $PhP(NSMe_2N)_3PPh$	1894
$[Ph_2PClNPClPh_2]^+ Cl^-$, $[Ph_2PClNHNHPClPh_2]Cl_2$, $Ph_2PNHNHPPh_2{=}N$	1895
$(RNH)_2PONHNRH$, Cl_3PNMe, $[PO(NHR)NR]_3$, and related compounds	1896
$R^1NHR^2NHPO(NHNH_2)$ and related compounds	1897
$[\{(Me_2N)_3PNHCN_2\}_2]X_2$ and related compounds	1898
$Me_2SPX_2{=}NPF_2O$	1899
$Me_3SnNMePSF_2$, $POF_2NMePSF_2$, and related compounds	1879
$(CF_3)_2PNHMe$, F_2PNHMe	1900
PF_2NMePF_2X, $MeNHPXF_2$	1901
$[Me_2NPCl_3][BCl_4]$	1721
XSO_2NPCl_2OR, XSO_2NRPCl_2O	1902
$SPFXNPCl_2NHSiMe_3$, $SPCl_2NPF_2NPCl_3$	1903
$XSO_2NPCl_2(OR)$, $XSO_2NRPOCl_2$, $FSO_2NRPOClNMe_2$	1904
RSO_2NPBr_3, $RSO_2NPBr_2NHSiMe_3$	1905
$MeN(PF_3)_2NMe$, Me_2NPF_4	1906
$MeN\{P(RNH)_2{=}N\}_2SO_2$, $MeN\{P(N{=}PCl_3)_2{=}N\}_2SOCl$	1907
$(MeNPF_3)_2$, $(MeNPSCl)_2$	1908
$P_3N_3F_{6-n}R_n$	1476*b*; 1909—1912
$N_3P_3Cl_{6-n}(NMe_2)_n$	1913, 1914
$H[PO_2XF]$	1915
$HOPOF_2$, $P_2O_3F_4$	1916

1892 R. Appel and E. Lassmann, *Z. Naturforsch.*, 1971, **26b**, 73.
1893 R. Appel, R. Kleinstück, and K.-D. Ziehn, *Chem. Ber.*, 1971, **104**, 2250.
1894 R. Appel and K.-W. Eichenhofer, *Chem. Ber.*, 1971, **104**, 3859.
1895 W. Haubold, D. Kammel, and M. Becke-Goehring, *Z. anorg. Chem.*, 1971, **380**, 23.
1896 R. J. W. Cremlyn, B. B. Dewhurst, and D. H. Wakeford, *J. Chem. Soc.* (*C*), 1971, 3011.
1897 R. J. W. Cremlyn, B. B. Dewhurst, and D. H. Wakeford, *J. Chem. Soc.* (*C*), 1971, 3011.
1898 R. Appel, B. Blaser, R. Kleinstück, and K.-D. Ziehn, *Chem. Ber.*, 1971, **104**, 1847.
1899 H. W. Roesky and L. F. Grimm, *Chem. Comm.*, 1971, 998.
1900 R. G. Cavell, T. L. Charlton, and W. Sim, *J. Amer. Chem. Soc.*, 1971, **93**, 1130.
1901 J. S. Harman, M. E. McCartney, and D. W. A. Sharp, *J. Chem. Soc.* (*A*), 1971, 1547.
1902 H. W. Roesky and W. G. Böwing, *Chem. Ber.*, 1971, **104**, 3204.
1903 H. W. Roesky, L. F. Grimm, and E. Niecke, *Z. anorg. Chem.*, 1971, **385**, 102.
1904 H. W. Roesky and W. G. Böwing, *Chem. Ber.*, 1971, **104**, 3204.
1905 H. W. Roesky and G. Remmers, *Z. Naturforsch.*, 1971, **26b**, 75.
1906 H. Binder and E. Fluck, *Z. anorg. Chem.*, 1971, **382**, 27.
1907 U. Bieller and M. Becke-Goehring, *Z. anorg. Chem.*, 1971, **380**, 314.
1908 H.-G. Horn, *Chem.-Ztg.*, 1971, **93**, 1971.
1909 H. W. Roesky and W. G. Böwing, *Z. anorg. Chem.*, 1971, **386**, 191.
1910 H. W. Roesky, W. G. Böwing, and E. Niecke, *Chem. Ber.*, 1971, **104**, 653.
1911 H. W. Roesky and E. Janssen, *Z. Naturforsch.*, 1971, **26b**, 679.
1912 E. Niecke, H. Thamm, and O. Glemser, *Z. Naturforsch.*, 1971, **26b**, 366.
1913 B. Green, D. B. Sowerby, and P. Clare, *J. Chem. Soc.* (*A*), 1971, 3487.
1914 B. Green and D. B. Sowerby, *J. Inorg. Nuclear Chem.*, 1971, **33**, 3687.
1915 H. Falius and K.-P. Giesen, *Angew. Chem. Internat. Edn.*, 1971, **10**, 555.
1916 P. A. Bernstein, F. A. Hohorst, M. Eisenberg, and D. D. DesMarteau, *Inorg. Chem.*, 1971, **10**, 1549.

Table 20 (*cont.*)

	Ref.
$(HS_2PMe)_2C_6H_4$ and related compounds	1917
$RSPO_3H_2$	1918
$(CF_3)_2P(S)SP(CF_3)_2$	1919
$[Me_2P(N{=}PR_3)_2]^+ [Me_2PF_4]^-$	1920
H_nPF_{5-n}	1921
$[HFPO_2]^-$, PF_2HO,BF_3, and related compounds	1922
Et_2NPF_4	1923
$M(O_2CMe)_3$ (M = P, As, Sb, or Bi)	1924
$(CF_3)_2EPH_2$ (E = P or As)	1925
o-$(AsMe_nH_{2-n})_2C_6H_4$	1926
$R_2ECH{=}C(ER_2)Ph$ (E = P or As)	1927
$(Ph_2AsCH_2CH_2)_2PPh$ and related compounds	1928
$(o\text{-}Ph_2E^1C_6H_4)_2(o\text{-}Ph_2E^2C_6H_4)E^3$ (E^1, E^2, E^3 = P or As)	1929
$(OC)_5M^1E^1Me_2E^2Me_2M^2(CO)_5$ (M^1, M^2 = Cr, Mo, or W; E^1, E^2, E^3 = P or As)	1307
$Me_2AsCH{=}CHAsMe_2$, $(Me_2As)_2C_6H_4$, and related compounds	1930 1931
$[Ph_3AsCH_2CH(OEt)Me]^+ Br^-$ and related compounds	1932
Me_2AsOR	1933
$(CF_3)_2NOAs(CF_3)_2$	1934
$MeAs(N{=}S{=}N)_2AsMe$	1935
$R_{3-n}As(O)(OH)_n$	1936
RPh_4Sb	1937
R_3Sb, R_3SbBr_2	1304
$M(CO)_5(SbH_3)$ (M = Cr, Mo, or W)	1305
$M^1(CO)_5Sb(M^2Me_3)_3$ (M^1 = Cr, Mo, or W; M^2 = Ge or Sn)	
R_2SbCl_3L	1938

1917 K. Diermert and W. Kuchen, *Angew. Chem. Internat. Edn.*, 1971, **10**, 508.
1918 A. Zwierzak and R. Gramze, *Z. Naturforsch.*, 1971, **26b**, 386.
1919 A. A. Pinkerton and R. G. Cavell, *J. Amer. Chem. Soc.*, 1971, **93**, 2384.
1920 W. Stadelmann, O. Stelzer, and R. Schmutzler, *Chem. Comm.*, 1971, 1456.
1921 F. Seel and K. Velleman, *Z. anorg. Chem.*, 1971, **385**, 123.
1922 L. F. Centofanti and R. W. Parry, *Inorg. Chem.*, 1970, **9**, 2709.
1923 M. Bermann and J. R. van Wazer, *Angew. Chem. Internat. Edn.*, 1971, **10**, 733.
1924 G. Gattow and H. Schwank, *Z. anorg. Chem.*, 1971, **382**, 49.
1925 R. Demuth, J. Grobe, and L. Steiner, *Z. Naturforsch.*, 1971, **26b**, 731.
1926 T. R. Carlton and C. D. Cook, *Inorg. Chem.*, 1971, **10**, 2628.
1927 A. Tzschach and S. Baensch, *J. prakt. Chem.*, 1971, **313**, 254.
1928 R. B. King and P. N. Kapoor, *J. Amer. Chem. Soc.*, 1971, **93**, 4158.
1929 J. W. Dawson and L. M. Venanzi, *J. Chem. Soc.* (*A*), 1971, 2897.
1930 R. D. Feltham and H. G. Metzger, *J. Organometallic Chem.*, 1971, **33**, 347.
1931 S. Trippett and M. A. Walker, *J. Chem. Soc.* (*C*), 1971, 1114.
1932 J. Koketsu and Y. Ishii, *J. Chem. Soc.* (*C*), 1971, 2.
1933 H. G. Ang and K. F. Ho, *J. Organometallic Chem.*, 1971, **27**, 349.
1934 O. J. Scherer and R. Wies, *Angew. Chem. Internat. Edn.*, 1971, **10**, 812.
1935 K. J. Irgolic, L. R. Kallenbach, and R. A. Zingaro, *J. Inorg. Nuclear Chem.*, 1971, **33**, 3177.
1936 G. Doleshall, N. A. Nesmeyanov, and O. A. Reutov, *J. Organometallic Chem.*, 1971, **30**, 369.
1937 A. G. Davies and S. C. W. Hook, *J. Chem. Soc.* (*C*), 1971, 1660.
1938 N. Nishii, Y. Matsumura, and R. Okawara, *J. Organometallic Chem.*, 1971, **30**, 59.

Table 20 (*cont.*)

	Ref.
Me_4SbSR	1939
$Me_3Sb(SR)_2$	1940
$Me_2SbR(NC)X(NMe_2)$	1941
$(Cl_2Sb)_2CH_2$	1942
$Sb(O_2CR)_3$, $Sb(NPhCSNMe_2)_3$	1943
OMO_2CMe (M = Sb or Bi)	1924
$C_6H_4(SbMe_2)(AsMe_2)$	1944
$[Ph_3Bi(CH_2COMe)]X$	1945

Table 21 *References to further papers containing n.m.r. data on compounds of* S, Se, *and* Te

	Ref.
$[(CF_3)_2CO]_2S{=}NR$	1282
$R^1R^2SONR^3$	1946
$(CF_3S)_3N$, $CF_3SN(SCCl_{3-n}F_n)_2$, and related compounds	1947
$Me_2S(NPR_2)_2$, $MeS(NPR_2N)_2SMe$, $[PhP(NSMe_2N)_3PPh]Br_2$	1894
NSF	1948
R_fSFNR	1949
NO_nSO_3F	1950
$CF_3SO_2NHSO_2F$ and related compounds	1951
$N(SOF)(SO_2F)$ and related compounds	1952
$RNSO_2F$	1953—1956
$S_4N_3N(SO_2F)_2$ and related compounds	1957
R_2NSF_3, $R^1N{=}SFR^2$	1958
$PhOSOF_3$, $RN{=}SOF(OPh)$	1959
$SOF_3NC_5H_{10}$, NCNSOFR	1960
$RNSF_2$	1961

[1939] H. Schmidbaur and K.-H. Mitschke, *Chem. Ber.*, 1971, **104**, 1837.
[1940] H. Schmidbaur and K.-H. Mitschke, *Chem. Ber.*, 1971, **104**, 1842.
[1941] J. Koketsu, M. Okamura, and Y. Ishii, *Bull. Chem. Soc. Japan*, 1971, **44**, 1155.
[1942] Y. Matsumura and R. Okawara, *Inorg. Nuclear Chem. Letters*, 1971, **7**, 113.
[1943] J. Koketsu and Y. Ishii, *J. Chem. Soc.* (*C*), 1971, 511.
[1944] B. R. Cook, C. A. McAuliffe, and D. W. Meek, *Inorg. Chem.*, 1971, **10**, 2676.
[1945] R. G. Goel and H. S. Prasad, *J. Chem. Soc.* (*A*), 1971, 562.
[1946] H. Schmidbaur and G. Kammel, *Chem. Ber.*, 1971, **104**, 3234.
[1947] A. Haas and R. Lorenz, *Z. anorg. Chem.*, 1971, **385**, 33.
[1948] R. Appel and E. Lassmann, *Chem. Ber.*, 1971, **104**, 2246.
[1949] R. Mews, G. G. Alange, and O. Glemser, *Inorg. Nuclear Chem. Letters*, 1971, **7**, 627.
[1950] A. M. Qureshi, H. A. Carter, and F. Aubke, *Canad. J. Chem.*, 1971, **49**, 35.
[1951] H. W. Roesky and H. H. Giere, *Inorg. Nuclear Chem. Letters*, 1971, **7**, 171.
[1952] H. W. Roesky and S. Tutkunkardes, *Chem. Ber.*, 1971, **104**, 1655.
[1953] R. Mews and O. Glemser, *Inorg. Nuclear Chem. Letters*, 1971, **7**, 821.
[1954] K. Seppelt and W. Sundermeyer, *Z. Naturforsch.*, 1971, **26b**, 65.
[1955] O. Glemser, H. Saran, and R. Mews, *Chem. Ber.*, 1971, **104**, 696.
[1956] H. W. Roesky, *Angew. Chem. Internat. Edn.*, 1971, **10**, 265.
[1957] H. W. Roesky and M. Dietl, *Z. Naturforsch.*, 1971, **26b**, 977.
[1958] S. P. von Halasz and O. Glemser, *Chem. Ber.*, 1971, **104**, 1247.
[1959] S. P. von Halasz, O. Glemser, and M. F. Feser, *Chem. Ber.*, 1971, **104**, 1242.
[1960] S. P. von Halasz and O. Glemser, *Chem. Ber.*, 1971, **104**, 1256.
[1961] R. Mews and O. Glemser, *Chem. Ber.*, 1971, **104**, 645.

Table 21 (*cont.*)

	Ref.
F_2NCCl_2SF, F_2NCCl_2SOX	1962
$RNSX_2$	1963
$[SF_3]^+[AsF_6]^-$	1964
$HSeCH_2CH_2NH_2$,HCl and related compounds	1965
SeR_2	1966
$S_nCSe_{5-n}Me_2$	1967
$MeSe(CH_2)_2CH(NHZ)CO_2R$	1968
$CH_2SeCHMeCH_2CH_2CHMeX$ (X = S or Se)	1969
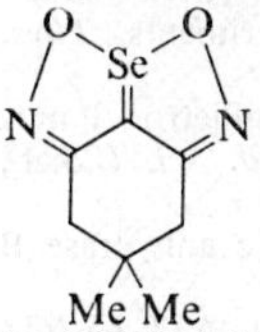	1970
$(Et_2NCSSe)_2M$	1587
$(Me_2N)_2CSSeCl_2$	1971
R_2E (E = Se or Te)	1972
Et_2Te_2, $EtTeCl_3$	1973
$(Ph_2C{=}CH_2)TeCl_2$	1974
$(Me_2N)_2CSTeX_3R$	1975

10 Bibliography

The following is a list of those references obtainable *via* the Chemical Society's n.m.r. Macroprofile (UKCIS) which are in journals not abstracted for the main text. The *Chemical Abstracts* reference number is given in brackets.

Chemical Abstracts, 1971, Volume 74

A. Hirai, J. A. Eaton, and C. W. Searle. ^{57}Fe N.m.r. and some dynamical characteristics of domain walls in α-Fe_2O_3. *Phys. Rev.* (*B*), 1971, **3**, 68 (*CA* 59 097).

H. Hirsch-Kolb, H. J. Kolb, and D. M. Greenberg. N.m.r. studies of manganese binding of rat-liver arginase. *J. Biol. Chem.*, 1971, **246**, 395 (*CA* 61 095).

[1962] L. M. Zaborowski and J. M. Shreeve, *Inorg. Chim. Acta*, 1971, **5**, 311.

[1963] R. Mews and O. Glemser, *Inorg. Nuclear Chem. Letters*, 1971, **7**, 823.

[1964] F. Seel, V. Hartmann, I. Molnar, R. Budenz, and W. Gombler, *Angew. Chem. Internat. Edn.*, 1971, **10**, 186.

[1965] A. Yokoyama, H. Sakurai, and H. Tanaka, *Chem and Pharm. Bull.* (*Japan*), 1971, **19**, 1089.

[1966] G. Ahlgren, B. Äkermark, and M. Nilsson, *J. Organometallic Chem.*, 1971, **30**, 303.

[1967] M. Dräger and G. Gattow, *Chem. Ber.*, 1971, **104**, 1429.

[1968] R. Walter and J. Roy, *J. Org. Chem.*, 1971, **36**, 2561.

[1969] A. Geens and M. Anteunes, *Bull. Soc. chim. belges*, 1971, **80**, 639.

[1970] R. J. S. Beer, J. R. Hatton, E. C. Llanguno, and I. C. Paul, *Chem. Comm.*, 1971, 594.

[1971] K. J. Wynne and P. S. Pearson, *Chem. Comm.*, 1971, 293.

[1972] Y. Okamoto and T. Yano, *J. Organometallic Chem.*, 1971, **29**, 99.

[1973] K. J. Wynne and P. S. Pearson, *Inorg. Chem.*, 1971, **10**, 1871.

[1974] D. Elmaleh, S. Patai, and Z. Rappoport, *J. Chem. Soc.* (*C*), 1971, 3100.

[1975] K. J. Wynne and P. S. Pearson, *Inorg. Chem.*, 1971, **10**, 2735.

M. I. Lobach, V. A. Kormer, I. Yu. Tsereteli, G. P. Kondratenkov, B. D. Babitskii, and V. I. Klepikova. P.m.r. study of bis(π-crotylnickel iodide) reaction with 1,3-butadiene. *J. Polymer Sci., Part B, Polymer Letters*, 1971, **9**, 71 (*CA* 64 540).

J. A. Walter and G. J. Troup. N.m.r. studies of irradiated beryllium oxide. *J. Nuclear Materials*, 1971, **38**, 51 (*CA* 69 957).

A. Lasis. Nature of water bonding in semihydrated gypsum. *Stroit. Materialy*, 1971, **38** (*CA* 69 987).

K. W. Gray and I. Ozier. New technique for the measurement of n.m.r. splittings to very high accuracy. *Phys. Rev. Letters*, 1971, **26**, 161 (*CA* 70 001).

D. G. Hughes and M. R. Smith. N.m.r. detection characteristics of marginal and Robinson oscillators. *J. Phys. (E)*, 1971, **4**, 13 (*CA* 70 007).

D. Freude, D. Mueller, and H. Schmiedel. Lineshape problem of proton resonance spectra of hydroxy-groups on solid surfaces. *Surface Sci.*, 1971, **25**, 289 (*CA* 70 026).

H. M. Gibbs. Nuclear-spin effects are important in spin-exchange experiments. *Phys. Rev. (A)*, 1971, **3**, 500 (*CA* 69 045).

M. Poe, W. D. Phillips, J. D. Glickson, C. C. McDonald, and A. San Pietro. P.m.r. studies of the ferredoxins from spinach and parsley. *Proc. Nat. Acad. Sci. U.S.A.*, 1971, **68**, 68 (*CA* 71 764).

R. W. King and G. C. K. Roberts. N.m.r. studies of human carbonic anhydrase B histidine residues. *Biochemistry*, 1971, **10**, 559 (*CA* 72 015).

E. Fukushima. N.m.r. study of the $NaSBF_6$ structure. *Acta Cryst.*, 1971, **A27**, 65 (*CA* 81 543).

M. I. Lobach, V. A. Kormer, I. Yu. Tsereteli, G. P. Kondratenkov, B. D. Babitskii, and V. I. Klepikova. Reaction of bis(π-crotylnickel iodine) with butadiene studied by the n.m.r. method. *Doklady Akad. Nauk S.S.S.R.*, 1971, **196**, 114 (*CA* 87 092).

G. Deville and A. Landesman. Spin-echo experiments in a flowing liquid. *J. Phys. Radium*, 1971, **32**, 67 (*CA* 89 085).

A. I. Zvyagin, P. S. Kalinin, V. A. Kaplun, and R. S. Shevelevich. Co-ordination of boron in borosilicate glasses studied by an n.m.r. method. *Izvest. Akad. Nauk S.S.S.R., Neorg. Materialy*, 1971, **7**, 350 (*CA* 90 559).

D. O. Shah and R. M. Hamlin, jun. Structure of water in microemulsions: electrical, birefringence, and n.m.r. studies. *Science*, 1971, **171**, 483 (*CA* 91 364).

D. L. Huber. Spin-lattice relaxation near the critical point: rubidium trifluoromanganate(II), manganese difluoride, and europium oxide. *Phys. Rev. (B)*, 1971, **3**, 836 (*CA* 92 713).

G. Englert and F. Wittmann. Water distribution in hydrated tricalcium silicate as a function of the moisture content. *Material. Sci. Eng.*, 1971, **7**, 125 (*CA* 93 277).

T. Minamisono, K. Matuda, A. Mizobuchi, and K. Sugimoto. Quadrupole effects in n.m.r. spectra on short-lived β-radioactive nuclei, ^{12}B and ^{12}N. *J. Phys. Soc. Japan*, 1971, **30**, 311 (*CA* 93 286).

P. I. Rose. Protein–metal-ion binding site: determination with p.m.r. spectroscopy. *Science*, 1971, **171**, 573 (*CA* 93 309).

Yu. A. Buslaev, E. G. Il'in, S. V. Bainova, and M. N. Krutkina. *cis–trans* Isomerism of octahedral complexes of niobium and tantalum fluorochlorides in solution. *Doklady Akad. Nauk S.S.S.R.*, 1971, **196**, 374 (*CA* 93 313).

D. H. Haynes, B. C. Pressman, and A. Kowalsky. N.m.r. study of ^{23}Na complexing by ionophores. *Biochemistry*, 1971, **10**, 852 (*CA* 94 344).

J. D. Glickson, W. D. Phillips, C. C. McDonald, and M. Poe. P.m.r. characterization of alfalfa and soybean ferredoxins; the existence of two ferredoxins in soybean. *Biochem. Biophys. Res. Comm.*, 1971, **42**, 271 (*CA* 94 457).

S. Ogawa and R. G. Shulman. Allosteric transition in haemoglobin. *Biochem. Biophys. Res. Comm.*, 1971, **42**, 9 (*CA* 94 469).

T. J. Swift, T. A. Glassman, C. Cooper, and L. W. Harrison. P.m.r. study of metal-ion–adenine-ring interactions in metal-ion complexes with adenosine triphosphate. *Biochemistry*, 1971, **10**, 843 (*CA* 94 533).

J. D. Brodie and M. Poe. P.m.r. studies of vitamin B_{12}. P.m.r. spectra of some cobalamins and cobinamides. *Biochemistry*, 1971, **10**, 914 (*CA* 94 534).

G. A. Persyn and W. L. Rollwitz. Transient n.m.r. quantitative measurements. *J. Amer. Oil Chemists' Soc.*, 1971, **48**, 67 (*CA* 98 255).

W. L. Rollwitz and G. A. Persyn. On-stream n.m.r. measurements and control. *J. Amer. Oil Chemists' Soc.*, 1971, **48**, 59 (*CA* 98 297).

L. O. Andersson. Varian's new n.m.r. process analyser. *J. Amer. Oil Chemists' Soc.*, 1971, **48**, 47 (*CA* 100 997).

F. Borsa, M. L. Crippa, and B. Derighetti. N.m.r. study of the structural phase transition in lanthanum aluminate. *Phys. Letters* (*A*), 1971, **34**, 5 (*CA* 104 065).

W. G. Proctor. Introduction to water. *J. Amer. Oil Chemists' Soc.*, 1971, **48**, 1 (*CA* 104 817).

Yu. A. Buslaev, E. G. Il'in, and V. D. Kpanev. *trans*-Effect of ylic oxygen in oxofluoroniobates. *Doklady Akad. Nauk S.S.S.R.*, 1971, **196**, 829 (*CA* 105 337).

R. L. Stevenson. Location of the protons in dehydrated γ-faujasite. *J. Catalysis*, 1971, **21**, 133 (*CA* 105 356).

R. L. Streever, T. R. Aucolin, and P. J. Caplan. N.m.r. studies of ^{55}Mn in the manganese-substituted hexagonal ferrite Ba_2Zn_{2-x}:Fe_{12-y}:Mn_{x+y}:O_{22}. *J. Phys. and Chem. Solids*, 1971, **32**, 519 (*CA* 105 364).

B. Pedersen and D. Slotfeldt-Ellingsen. Exploratory deuteron magnetic resonance study of vanadium, niobium, and tantalum deuterides. *J. Less-Common Metals*, 1971, **23**, 223 (*CA* 105 377).

D. L. Hagrman and W. D. Ohlsen. N.m.r. study of the hydrogen nucleus of OH^- in the potassium chloride lattice. *Phys. Rev.* (*B*), 1971, **3**, 1918 (*CA* 105 407).

I. R. McDonald. Free-induction decays in a rotating solid. *Physica*, 1971, **51**, 273 (*CA* 105 440).

G. R. Bulka, S. V. Vedenin, V. M. Vinokurov, T. A. Zakharchenko, N. M. Nizamutdinov, and R. S. Tukhvatullin. *Kristallografiya*, 1971, **16**, 138 (*CA* 104 227).

P. Bhattacharyya, K. N. Pathak, and K. S. Singivi. Many-electron effects on the enhancements of the Korringa constants and spin–lattice relaxation rates in alkali metals. *Phys. Rev.* (*B*), 1971, **3**, 1568 (*CA* 104 336).

A. Lanir and G. Navon. N.m.r. studies of bovine carbonic anhydrase. Binding of sulphonamides to the zinc enzyme. *Biochemistry*, 1971, **10**, 1024 (*CA* 107 612).

B. R. Baker and R. M. Pearson. Reaction of water vapour with anodic aluminium oxide studied by wide-line n.m.r. spectroscopy. *J. Electrochem. Soc.*, 1971, **118**, 353 (*CA* 116 252).

J. Haupt and R. van Steenwinkel. Experimental effects of radiofrequency irradiation on n.m.r. lines in solids. *Z. Naturforsch.*, 1971, **26a**, 260 (*CA* 118 064).

A. I. Nosar. Calculation of n.m.r. line intensities in binary ordered substitution alloys with interstitial impurities. *Fiz. Metall. i Metallov.*, 1971, **31**, 116 (*CA* 118 076).

O. P. Revokatov and V. E. Lyapukhov. Shape of n.m.r. signals in solids during saturation. *Fiz. tverd. Tela*, 1971, **13**, 488 (*CA* 118 088).

H. Schmiedel, D. Freude, and W. Gruender. Measurements of correlation times by means of the temperature dependence of coherent-pulse averaging effects in n.m.r. *Phys. Letters* (*A*), 1971, **34**, 162 (*CA* 118 110).

H. Saji, T. Yamadaya, and M. Asanuma. N.m.r. study of ^{51}V in chromium vanadate(v). *Phys. Letters* (*A*), 1971, **23**, 49 (*CA* 118 112).

D. N. Kravtsov, L. A. Fedorov, A. S. Peregudov, and A. N. Nesmeyanov. Metallotropic transitions in organomercury derivatives of pyrazoles studied by the n.m.r. method. *Doklady Akad. Nauk S.S.S.R.*, 1971, **196**, 110 (*CA* 118 115).

K. B. Utton. N.m.r. study of ferrous fluosilicate. *J. Phys.* (*C*), 1971, **4**, 117 (*CA* 118 121).

M. Hanubusa. Sensitivity enhancement of a pulsed n.m.r. spectrometer. *J. Appl. Phys.*, 1971, **42**, 1077 (*CA* 118 249).

M. I. Emel'yanov, I. N. Nikolaev, and F. M. Samigullin. Self-diffusion of molecules in mixtures of water and nonelectrolytes, studied by a spin-echo method. *Zhur. strukt. Khim.*, 1971, **12**, 161 (*CA* 116 118).

R. E. Watson, A. Misetich, and L. Hodges. Effects of *s–d* hybridization on charge and spin contact densities in noble and transition metals. *J. Phys. and Chem. Solids*, 1971, **32**, 709 (*CA* 117 584).

W. J. Deal, jun., S. G. Mohlman, and M. L. Spring. Conformational equilibria in spin-labelled haemoglobin. *Science*, 1971, **171**, 1147 (*CA* 120 234).

A. S. Mildvan. Nuclear relaxation studies of the role of metals in enzyme-catalysed enolization and elimination reactions. *Adv. Chem. Ser.*, 1971, No. 100, p. 390 (*CA* 120 558).

W. S. Caughey. Structure–function relations in cytochrome *c* oxidase and other haemoproteins. *Adv. Chem. Ser.*, 1971, No. 100, p. 248 (*CA* 120 587).

R. G. Kidd. N.m.r. spectroscopy of organometallic compounds. *Character. Organometallic Compounds*, 1971, **26**, 373 (*CA* 124 352).

V. Niculescu, I. Pop, and N. Rosenberg. N.m.r. and magnetic susceptibilities of the cerium–aluminium intermetallic compounds. *Phys. Letters* (*A*), 1971, **34**, 265 (*CA* 132 170).

K. Siratori. Magnetic resonance of $ZnCr_2Se_4$ with screw spin structure. *J. Phys. Soc. Japan*, 1971, **30**, 709 (*CA* 132 420).

R. L. Streever and P. J. Caplan. N.m.r. studies of ^{151}Eu and ^{153}Eu in europium–iron garnet single crystals. *Phys. Rev.* (*B*), 1971, **3**, 2910 (*CA* 132 447).

A. M. Babeshkin, P. B. Fabrichnyi, and A. N. Nesmeyanov. Study of the transitions on the surface of stannic acid by *X*-ray diffraction, Mössbauer, and p.m.r. spectroscopy. Proceedings of the Conference on Applications of the Mössbauer Effect, 1969, ed. I. Dezsi, 1971, p319 (*CA* 132 779).

E. von Meerwall and T. J. Rowland. Quadrupolar effects in the ^{51}V n.m.r. of vanadium-based transition-metal alloys. *Solid State Comm.*, 1971, **9**, 305 (*CA* 132 792).

N. A. Kostromina and N. N. Tananaeva. Complexes of lanthanum and lutetium with nitrilotriacetate studied by n.m.r. *Zhur. neorg. Khim.*, 1971, **16**, 866 (*CA* 132 793).

A. N. Gil'manov and I. G. Bikchantaev. Palladium–hydrogen system studied by proton magnetic relaxation. *Fiz. Metall. i Metallov.*, 1971, **31**, 510 (*CA* 132 811).

V. M. Ryabikova, B. S. Krumgal'z, I. Yu. Tsereteli, V. I. Derevskaya, V. T. Usacheva, and K. P. Mishchenko. P.m.r. of inorganic salt solutions in DMF and a study of methods for DMF purification. *Izvest. V.U.Z., Khim. khim. Tekhnol.*, 1971, **14**, 45 (*CA* 132 852).

R. Parrot, C. Blanchard, and D. Boulanger. Determination of the spin–lattice coupling coefficients of a paramagnetic ion in tetrahedral symmetry—Mn^{2+} in zinc sulphide. *Phys. Letters* (*A*), 1971, **34**, 109 (*CA* 132 803).

M. Poe, W. D. Phillips, C. C. McDonald, and W. H. Orme-Johnson. P.m.r. and magnetic susceptibility studies on *Clostridium acidi-urici* ferredoxin. *Biochem. Biophys. Res. Comm.*, 1971, **42**, 705 (*CA* 135 197).

J. J. Delpuech, A. Peguy, and M. R. Khaddar. Preferential solvation of diamagnetic cations in water–organic mixtures studied by n.m.r. *J. Electroanalyt. Chem. Interfacial Electrochem.*, 1971, **29**, 31 (*CA* 146 896).

R. Mondelli. Recent progress in applications of n.m.r. spectroscopy. *Chimia e Industria*, 1971, **53**, 54 (*CA* 148 399).

M. R. Zaripov, M. M. Bil'danov, G. M. Kadievskii, and G. R. Enikeeva. Effect of the nature of intramolecular rotary motion on the rate of proton magnetic relaxation in solids. *Doklady Akad. Nauk S.S.S.R.*, 1971, **196**, 136 (*CA* 148 831).

I. V. Matyash, M. A. Piontkovskaya, G. I. Denisenko, and A. M. Kalinichenko. P.m.r. spectra of pyridine adsorbed on various cation-substituted forms of faujasite. *Zhur. strukt. Khim.*, 1971, **12**, 13 (*CA* 148 900).

I. A. Nuretdinov, V. V. Negrebetskii, A. Z. Yankelevich, A. V. Kessenikh, E. I. Loginova, L. K. Nikonorova, and N. P. Grechkin. Proton and ^{31}P n.m.r. spectra at ^{1}H–^{31}P and ^{1}H–^{13}C heteronuclear double resonance of compounds containing the P—N—P group. *Doklady Akad. Nauk S.S.S.R.*, 1971, **196**, 1369 (*CA* 148 903).

Yu. A. Buslaev, V. I. Pakhomov, V. P. Tarasov, and V. N. Zege. ^{19}F Spin–lattice relaxation and *X*-ray study of phase transition in solid K_3ZrF_7 and $(NH_4)_2ZrF_7$. *Phys. Status Solidi* (*B*), 1971, **44**, OK13 (*CA* 148 909).

V. F. Chuvaev, P. Baidala, E. A. Torchenkova, and V. I. Spitsyn. P.m.r. spectra of hydrates of cerium– and thorium–molybdenum heteropoly acids. *Doklady Akad. Nauk S.S.S.R.*, 1971, **196**, 1097 (*CA* 148 911).

A. A. Popel, Z. A. Saprykova, and S. I. Galeeva. Separate determination of paramagnetic ions by the nuclear magnetic relaxation method. Successive titration for the separate determination of paramagnetic ions by the nuclear magnetic relaxation method. *Zhur. analit. Khim.*, 1971, **26**, 259 (*CA* 150 808).

A. A. Popel, Z. A. Saprykova, and S. I. Galeeva. Separate determination of paramagnetic ions by the nuclear magnetic relaxation method. Solutions of diamagnetic ion complexes as titrants for the separate determination of paramagnetic ions. *Zhur. analyt. Khim.*, 1971, **26**, 262 (*CA* 150 809).

A. Oilvson and E. Lippmaa. ^{13}C Spin–lattice relaxation of organic compounds. *Esti NSV Tead. Akad. Toim., Fuus., Mat.*, 1971, **20**, 29 (*CA* 148 833).

Chemical Abstracts, 1971, Volume 75

J. L. Sudmeier and J. J. Pesek. ^{35}Cl N.m.r. studies of metal binding to bovine serum albumin. *Analyt. Biochem.*, 1971, **41**, 39 (*CA* 275).

G. E. Krejcarek, L. Turner, and K. Dus. Investigation of photosynthetic cytochromes *c* by high-resolution n.m.r. spectroscopy. *Biochem. Biophys. Res. Comm.*, 1971, **42**, 983 (*CA* 318).

R. G. Shulman, S. H. Glarum, and M. Karplus. Electronic structure of cyanide complexes of hemes and heme proteins. *J. Mol. Biol.*, 1971, **57**, 93 (*CA* 348).

S. Levai, S. Levai, and M. Catrinescu. Shape of n.m.r. absorption lines. *Rev. fiz. Chim., Ser. A*, 1971, **8**, 62 (*CA* 12 484).

J. A. J. Lourens and E. C. Reynhardt. Second moment of the ^{59}Co n.m.r. line in potassium hexacyanocobaltate. *Phys. Status Solidi* (*B*), 1971, **44**, OK57 (*CA* 13 116).

E. Hengge and F. Hoefler. Vibrational and n.m.r. spectra of tri-iodosilane and mixed trihalogenosilanes. *Z. Naturforsch.*, 1971, **26a**, 768 (*CA* 13 126).

H. J. Keller and H. H. Rupp. Anisotropy of ^{195}Pt n.m.r. shifts in crystalline platinum(II) compounds. *Z. Naturforsch.*, 1971, **26a**, 785 (*CA* 13 129).

A. A. Borichev, K. P. Mishchenko, and V. V. Kushchenko. N.m.r. study of potassium iodide and sodium iodide solutions in hydrazine. *Zhur. obshchei Khim.*, 1971, **41**, 13 (*CA* 13 136).

N. Tanaka. Data processing of n.m.r. spectra by mini-computer. *Enka Biniiru To Porima*, 1971, **11**, 22 (*CA* 13 138).

G. L. Antokol'skii, V. M. Sarnatskii, and V. A. Shutilov. Acoustic nuclear resonance of ^{7}Li in a lithium fluoride crystal. *Akust. Zhur.*, 1971, **17**, 143 (*CA* 13 155).

N. A. Berger and G. L. Eichhorn. Interaction of metal ions with polynucleotides and related compounds. N.m.r. studies of the binding of copper(II) to adenine nucleotides. *Biochemistry*, 1971, **10**, 1847 (*CA* 14 995).

N. A. Berger and G. L. Eichhorn. Interaction of metal ions with polynucleotides and related compounds. N.m.r. studies of the binding of copper(II) to nucleotides and polynucleotides. *Biochemistry*, 1971, **10**, 1857 (*CA* 14 999).

I. P. Beletskaya, V. B. Vol'eva, S. V. Rykov, Q. L. Buchachenko and A. V. Kessenikh. Mechanism of the reactions of organomercury compounds studied by chemically induced dynamic nuclear polarization. *Izvest. Akad. Nauk S.S.S.R., Ser. khim.*, 1971, 44 (*CA* 19 422).

K. Kuroda and N. Ishikawa. Organic fluorine–silicon compounds. Formation of perchlorofluoropolyphenylene from polychlorofluoro(trimethylsilyl)benzenes. *Kogyo Kagaku Zasshi*, 1971, **74**, 495 (*CA* 21 164).

J. D. Bell, R. W. Myatt, and R. E. Richards. P.m.r. evidence of a liquid phase in polycrystalline ice. *Nature Phys. Sci.*, 1971, **230**, 91 (*CA* 26 009).

G. Giessner-Prettre and B. Pullman. Intermolecular nuclear shielding due to the aromatic amino-acids of proteins and porphyrins. *J. Theor. Biol.*, 1971, **31**, 287 (*CA* 27 937).

A. Allerhand and D. Doddrell. Assignments in the ^{13}C n.m.r. spectra of vitamin B_{12}, coenzyme B_{12}, and other corrinoids: application of partially relaxed Fourier-transform spectroscopy. *Proc. Nat. Acad. Sci. U.S.A.*, 1971, **68**, 1083 (*CA* 27 950).

L. V. Nesterov, R. I. Mutalapova, S. G. Salikhov, and E. I. Loginova. Structure of phenylphenoxyphosphonium salts in solution. *Izvest. Akad. Nauk S.S.S.R., Ser. khim.*, 1971, 414 (*CA* 27 964).

H. M. Kriz and P. F. Bray. Distribution of boron sites in glassy boron oxide using ^{11}B n.m.r. *J. Non-Crystalline Solids*, 1971, **6**, 27 (*CA* 27 976).

K. B. Dillon and T. C. Waddington. Structures of some dihalogenotriphenylphosphorus (V) compounds. *Nature Phys. Sci.*, 1971, **230**, 158 (*CA* 28 004).

W. Bremser, H. D. W. Hill, and R. Freeman. Fourier transformation technique in high-resolution n.m.r. spectroscopy. *Messtechnik* (*Brunswick*), 1971, **79**, 14 (*CA* 28 015).

J. Citerne, Y. Crosnier, and R. Gabillard. Measurement of the mean lifetime of a chemical exchange in the earth's magnetic field. *Compt. rend.*, 1971, **272**, *B*, 485 (*CA* 28 026).

B. Lemius, S. Domngang, and J. Wucher. Spin–lattice relaxation times of protons and deuterons in aqueous solutions of gallium chloride. *Compt. rend.*, 1971, **272**, *B*, 489 (*CA* 28 027).

S. Nasu, H. Yasuoka, Y. Nakamura, and Y. Murakami. N.m.r. of ^{59}Co nuclei in precipitated cobalt particles. *Acta Metallurgica*, 1971, **19**, 561 (*CA* 28 033).

E. I. Berus, Yu. G. Gladkii, V. A. Barkhash, and Yu. N. Molin. Effect of substituents on the degree of metal–ligand bond covalence in complexes of substituted anilines with nickel acetylacetonate. *Doklady Akad. Nauk S.S.S.R.*, 1971, **197**, 1362 (*CA* 29 429).

G. Dijkstra. Molecular Spectroscopy. *Chem. Weekblad*, 1971, **67**, OA27 (*CA* 29 505).

V. K. Mukhomorov. Proton magnetic shielding constant in an electric field. *Zhur. strukt. Khim.*, 1971, **12**, 326 (*CA* 25 550).

D. H. Grant. Nuclear magnetic resonance—An introduction. *Canad. Chem. Educ.*, 1971, **6**, 13 (*CA* 29 766).

F. R. N. Gurd, P. J. Lawson, D. W. Cochran, and E. Wenkert. ^{13}C N.m.r. of peptides in the amino-terminal sequence of sperm whale myoglobin. *J. Biol. Chem.*, 1971, **246**, 3725 (*CA* 30 063).

H. H. Strain, B. T. Cope, jun., G. N. McDonald, W. A. Svec, and J. J. Katz. Chlorophylls C1 and C2. *Phytochemistry*, 1971, **10**, 1109 (*CA* 31 463).

P. Rigny. N.m.r. of inorganic solids: applications to some solid fluorides. *Semin. Chim. Etat solide*, 1971, 33 (*CA* 42 459).

M. Zaucer and A. Azman. Calculation of the proton chemical shift in hydrogen-bonded systems. *Rev. Roumaine Chim.*, 1971, **16**, 481 (*CA* 42 824).

M. M. Mirkhidoyatov, M. K. Alyaviya, V. B. Leont'ev, and A. S. Sadykov. Complexing of cadmium iodide with pyridinecarboxylic and aminobenzoic acids, and some of their derivatives studied by n.m.r. *Uzbek. khim. Zhur.*, 1971, **15**, 34 (*CA* 42 825).

N. E. Alekseevskii and E. P. Krasnoperov. N.m.r. in Nb_3Al and $Nb_3AlO,8GeO_2$. *Doklady Akad. Nauk S.S.S.R.*, 1971, **197**, 1048 (*CA* 42 878).

A. Rockenbauer and L. Radics. Theoretical magnetic resonance spectra of AA′A″... XX′X″... systems. Case of deceptive simplicity. *Acta Chim. Acad. Sci. Hung.*, 1971, **68**, 189 (*CA* 42 887).

N. M. Nizamutdinov, S. V. Vedenin, T. A. Zakharchenko, and V. M. Vinokurov. Nuclear and electron paramagnetic resonance in herderite. *Geokhimiya*, 1971, 361 (*CA* 42 893).

R. J. Kurland, R. G. Little, D. G. Davis, and C. Ho. P.m.r. study of high- and low-spin haemin derivatives. *Biochemistry*, 1971, **10**, 2237 (*CA* 44 810).

J. J. Fripiat. Applications of n.m..r to surface chemistry. *Catalysis Rev.*, 1971, **5**, 269 (*CA* 53 467).

A. G. Brekhunets, V. V. Mank, F. D. Ovcharenko, M. A. Piontkovskaya, and G. I. Denisenko. State of water and lithium exchange cations in lithium-substituted type A zeolite studied by an n.m.r. method. *Teor. i eksp. Khim.*, 1971, **7**, 125 (*CA* 53 615).

Yu. A. Buslaev and S. P. Petrosyants. Complexing in molybdenum fluoride oxide solutions studied by ^{19}F n.m.r. *Zhur. neorg. Khim.*, 1971, **16**, 1330 (*CA* 54 099).

N. A. Kostromina and N. N. Tananaeva. Complexes of lanthanum and lutetium with cyclohexanediaminetetra-acetate studied by n.m.r. *Zhur. neorg. Khim.*, 1971, **16**, 1272 (*CA* 54 131).

B. I. Stepanov, A. I. Bokanov, and V. I. Svergun. Spectral properties and structure of tertiary mesityl(ethyl)phosphines. *Zhur. obshchei Khim.*, 1971, **41**, 533 (*CA* 56 049).

C. L. Khetrapal, A. C. Kunwar, C. R. Kanekar, and P. Diehl. N.m.r. investigations on (benzene)chromium tricarbonyl oriented in a nematic phase. *Mol. Crystals Liquid Crystals*, 1971, **12**, 179 (*CA* 56 376).

H. A. Resing and J. K. Thompson. N.m.r. relaxation of water in zeolite 13-X. *Adv. Chem. Ser.*, 1971, No. 101, p. 473 (*CA* 56 406).

N. A. Kostromina and T. V. Ternovaya. Proton resonance spectra of complexes of lanthanum, yttrium, and lutetium with edta. *Teor. i eksp. Khim.*, 1971, **7**, 115 (*CA* 56 421).

M. Yoshimoto, T. Hiraoka, H. Kuwano, and Y. Kishida. Use of a shift reagent in first-order analysis of cyclopropane derivatives in n.m.r. spectroscopy. *Chem. and Pharm. Bull.* (*Japan*), 1971, **19**, 849 (*CA* 56 596).

J. C. Sturman and R. J. Jirberg. Wide-range n.m.r. detector using integrated circuits. *N.A.S.A. Tech. Note*, 1971, 25 (*CA* 56 657).

R. Boving. Magnetic field stabilizer based on proton resonance in flowing water. *Nuclear Instr. Methods*, 1971, **94**, 61 (*CA* 57 729).

R. Z. Sagdeev, Yu. N. Molin, E. V. Dvornikov, V. A. Grigor'ev, and T. A. Luzina. Hyperfine interactions involving ^{13}C nuclei in paramagnetic complexes. *Zhur. strukt. Khim.*, 1971, **12**, 245 (*CA* 56 333).

L. A. Fedorov and E. I. Fedin. Spin–spin coupling constants and some properties of organotin compounds. *Izvest. Akad. Nauk S.S.S.R., Ser. khim.*, 1971, 787 (*CA* 56 351).
H. A. O. Hill, J. M. Pratt, and R. J. P. Williams. Identification and investigation of cobalamins and cobamide coenzymes by n.m.r. and e.s.r. spectroscopy. *Methods Enzymology*, 1971, **18**, 5 (*CA* 58 835).
R. S. Rosenfeld. Trimethylsilyl ethers of some 5β-pregnanes. *Analyt. Biochem.*, 1971, **42**, 382 (*CA* 59 363).
Z. Kecki. N.m.r. contact shift in the study of molecular structure. *Postepy Fiz.*, 1971, **22**, 211 (*CA* 62 657).
J. Sadlej and Z. Kecki. Modified CNDO (complete neglect of differential overlap) method. N.m.r. chemical shifts in the acetonitrile solutions of electrolytes. *Roczniki Chem.*, 1971, **45**, 445 (*CA* 62 916).
H. Yuki and Y. Okamoto. N.m.r. studies of isoprenyl-lithium derived from [3,4-$^2H_5$]1,1-diphenyl-n-butyl-lithium and isoprene. *J. Polymer Sci. Part A*-1, *Polymer Chem.*, 1971, **9**, 1247 (*CA* 64 336).
I. D. Datt, N. V. Rannev, R. P. Ozerov, and V. M. Kuznets. Crystal structure of crystal hydrates of lithium salts. Investigation of the state of water molecules in the lithium hydrochloride deuteriate LiCl,D_2O by electron diffraction and n.m.r. *Kristallografiya*, 1971, **16**, 631 (*CA* 68 526).
L. A. Kutulya, Yu. N. Surov, N. S. Pivnenko, S. V. Tsukerman, and V. F. Lavrushin. Hydrogen bonding of chalcones with trifluoroacetic acid studied by i.r. and n.m.r. spectroscopy. *Zhur. obshchei Khim.*, 1971, **41**, 895 (*CA* 69 096).
E. Kundla. Theory of triple nuclear magnetic resonance. *Eesti NSV Tead. Akad. Toim., Fuus., Mat.*, 1971, **20**, 136 (*CA* 69 272).
H. M. Kriz, M. J. Park, and P. J. Bray. New interpretations of ^{11}B n.m.r. spectra from glassy borates. *Phys. and Chem. Glasses*, 1971, **12**, 45 (*CA* 69 275).
Yu. D. Nechaev, Yu. L. Kleiman, and N. V. Morkovin. Accumulation of n.m.r. spectra using calibrated magnetic-field shifts. *Prib. i Tekh. Eksp.*, 1971, 164 (*CA* 69 294).
A. N. Voronovich, L. S. Lilich, S. V. Petukhov, and M. K. Khripun. Temperature dependence of the proton-relaxation rate in some solutions of 1 : 1 electrolytes. *Doklady Akad. Nauk S.S.S.R.*, 1971, **198**, 865 (*CA* 69 309).
J. Reuben. Gadolinium(III) as a paramagnetic probe for proton-relaxation studies of biological macromolecules. Binding to bovine serum albumin. *Biochemistry*, 1971, **10**, 2834 (*CA* 71 269).
M. Ogawa. Metal–nucleotide complexes. I.r. absorption and n.m.r. spectral studies of 5′-adenylic acid–metal complexes. *Yakugaku Zasshi*, 1971, **91**, 618 (*CA* 71 348).
E. M. Brainina, L. A. Fedorov, and M. K. H. Minacheva. Phenoxy-derivatives of hafnium and zirconium containing cyclopentadienyl and chelate ligands, and their stereochemical structure. *Doklady Akad. Nauk S.S.S.R.*, 1971, **196**, 1085 (*CA* 75 979).
N. I. Nikolaev, G. A. Grigor'eva, N. N. Shapet'ko, and V. A. Arkhipov. Determination of the state of water in sulphonic cation exchangers by the n.m.r. method. *Doklady Akad. Nauk S.S.S.R.*, 1971, **198**, 369 (*CA* 80 632).
P. M. Borodin and Zao. Kim Nguyen. Intermolecular interactions and kinetics of chemical exchange in some lithium fluorosilicate and hydrofluoric acid–silicon dioxide acidic aqueous solutions based on n.m.r. data. *Vestnik Leningrad Univ.* (*Fiz. Khim.*), 1971, 123 (*CA* 80 861).
Yu. A. Lysenko, L. M. Kapkan, A. N. Vedmedskaya, and V. M. Artemova. Ionization of complexes of tin tetrahalides and antimony pentachloride in solvents containing mobile halogen studied by electromigration and n.m.r. methods. *Zhur. obshchei Khim.*, 1971, **41**, 1132 (*CA* 80 913).
D. Feldmann, H. R. Kirchmayr, A. Schmolz, and M. Velicescu. Magnetic materials analyses by nuclear spectrometry. Joint approach to Mössbauer effect and nuclear magnetic resonance. *IEEE, Trans. Magn.*, 1971, **7**, 61 (*CA* 81 885).
B. Maraviglia, F. Weinhaus, S. M. Meyers, and H. Meyer. Nuclear magnetic resonance in deuterium. *U.S. Clearinghouse Fed. Sci. Tech. Inform., AD*, 1971, 34 (*CA* 82 182).
V. M. Mastikhim and E. S. Rudakov. N.m.r. spectra of a complex of propylene with silver(I) in an aqueous solution. *Izvest. Akad. Nauk S.S.S.R., Ser. khim.*, 1971, 1321 (*CA* 82 200).
D. K. Dalling and H. S. Gutowsky. *Z* Dependence of coupling constants between directly bonded nuclei. Value of $J(^{73}Ge–^{13}C)$. *U.S. Clearinghouse Fed. Sci. Tech. Inform., AD*, 1971, 32 (*CA* 82 238).

D. G. Davis and G. Inesi. Proton n.m.r. studies of sarcoplasmic reticulum membranes. Correlation of the temperature-dependent Ca^{2+} efflux with a reversible structural transition. *Biochim. Biophys. Acta*, 1971, **241**, 1 (*CA* 84 306).

R. A. Dwek, R. E. Richards, K. G. Morallee, E. Nieboer, R. J. P. Williams, and A. V. Xavier. Lanthanide cations as probes in biological systems. Proton-relaxation enhancement studies for model systems and lysozyme. *European J. Biochem.*, 1971, **21**, 204 (*CA* 84 312).

O. G. Yarosh, V. K. Voronov, and N. V. Komarov. P.m.r. spectra of some silyl-substituted ethylenes. *Izvest. Akad. Nauk S.S.S.R., Ser. khim.*, 1971, 875 (*CA* 87 723).

E. P. Prokof'ev, Z. H. A. Krasnaya, and V. F. Kucherov. Rapid exchange under equilibrium conditions between *cis*- and *trans*-forms, observed by the p.m.r. method. *Izvest. Akad. Nauk S.S.S.R., Ser. khim.*, 1971, 895 (*CA* 88 005).

V. S. Bogdanov, P. M. Aronovich, A. D. Naumov, and B. M. Mikhailov. Organoboron compounds. Study of the reaction of tri-n-butylborane with lithium and sodium borohydrides using the ^{11}B n.m.r. method. *Zhur. obshchei Khim.*, 1971, **41**, 1063 (*CA* 88 666).

L. Hoff. Application of the $Eu(dpm)_3$ chemical-shift reagent to the determination of the molecular weight of polypropylene glycol by n.m.r. *J. Polymer Sci., Part B, Polymer Letters*, 1971, **9**, 491 (*CA* 89 032).

N. A. Kostromina and N. N. Tananaeva. Mixed complexes of lanthanum and lutetium with edta and nitrilotriacetate studied by an n.m.r. method. *Teor. i eksp. Khim.*, 1971, **7**, 67 (*CA* 91 791).

L. S. Frankel. N.m.r. study of ligand-exchange kinetics of nickel(II) *NN*-dimethylformamide in mixed solvents. *U.S. Clearinghouse Fed. Sci. Tech. Inform., AD*, 1971, 14 (*CA* 92 759).

L. J. Todd, A. R. Siedle, and G. M. Bodner. ^{11}B N.m.r. spectrum of the tridecahydrodecaborate(-1) ion. *U.S. Clearinghouse Fed. Sci. Tech. Inform., AD*, 1971, 20 (*CA* 92 763).

T. V. Zykova, G. Kamai, B. D. Chernokal'skii, R. A. Salakhutdinov, and B. E. Abalonin. P.m.r. spectra and structure of arsonium salts. *Zhur. obshchei Khim.*, 1971, **41**, 1044 (*CA* 92 768).

V. D. Doroshev, N. M. Kovtun, V. N. Seleznev, and V. M. Siryuk. N.m.r. of ^{57}Fe in iron(III) borate single crystals. *Pis'ma Zhur. eksp. teor. Fiz.*, 1971, **13**, 672 (*CA* 92 779).

V. S. Bogdanov, A. V. Kessenikh, and V. V. Negrebetskii. Indirect determination of spin–spin interaction constants $J(^{11}B–H)$ in organic boron compounds. *Izvest. Akad. Nauk S.S.S.R., Ser. khim.*, 1971, 1363 (*CA* 92 782).

F. I. Toth, K. Tompa, and G. Gruener. Frequency-modulated n.m.r. spectrometer for measurement of internal magnetic fields. *KFKI* (*Kozp. Fiz. Kut. Intez.*), Report KFKI-71-29, 1971 (*CA* 92 869).

T. P. Firsova, E. Y. A. Filatov, and V. I. Krlividze. Synthesis of rubidium peroxohydrocarbonate studied by n.m.r. *Izvest. Akad. Nauk S.S.S.R., Ser. khim.*, 1971, 1565 (*CA* 94 118).

A. N. Nesmeyanov, L. V. Rybin, M. I. Rybinskaya, A. V. Arutyunyan, N. T. Gubenko, and P. V. Petrovskii. Complexes of 3,6-diarylpyridazines with iron carbonyls. *Izvest. Akad. Nauk S.S.S.R., Ser. khim.*, 1971, 1574 (*CA* 94 183).

A. A. Popel, Z. A. Saprykova, and S. I. Galeeva. Separate determination of diamagnetic ions by nuclear magnetic relaxation. *Zhur. analit. Khim.*, 1971, **26**, 1211 (*CA* 94 378).

K. Wuethrich. High resolution proton n.m.r. studies of the co-ordination of the haeme iron in cytochrome *c*. 'Probes in Structure and Function of Macromolecular Membranes, Proceedings of the 5th Colloqium Johnson Research Foundation', ed. B. Chance, Academic Press, New York, 1971, vol. 2, p. 465 (*CA* 94 655).

D. Swern and J. P. Wineburg. N.m.r. chemical-shift reagents. Application to the structural determination of lipid derivatives. *J. Amer. Oil. Chemists' Soc.*, 1971, **48**, 371 (*CA* 97 886).

N. Ishikawa, K. Isobe, and K. Kuroda. Organic fluorine–silicon compounds. Preparation and reactions of aryldimethylfluorosilanes containing functional groups in the aryl ring. *Kogyo Kagaku Zasshi*, 1971, **74**, 1680 (*CA* 98 613).

I. Muro, I. Morishima, and T. Yonezawa. Studies of ternary complexes of adrenaline, adenine nucleotide, and metal ion by p.m.r. spectroscopy and its pharmacological significance with respect to the adrenergic mechanism. *Chem.-Biol. Interactions*, 1971, **3**, 213 (*CA* 98 762).

M. Baran, H. Szymczak, and W. Zbieranowski. Study of stresses in crystals by n.m.r. Institute of Nuclear Physics, Cracow, Report, 1971, p. 104 (*CA* 102 189).
W. Vogt and H. Haas. Crystal structure and n.m.r. of $Cs_3Cu_2Cl_7,2H_2O$. *Acta Cryst.*, 1971, **B27**, 1528 (*CA* 102 302).
S. P. Gabuda and Yu. V. Gagarinskii. N.m.r. data on proton positions in solid hydrogen fluoride. *Acta Cryst.*, 1971, **B27**, 1677 (*CA* 102 303).
B. Schnabel, B. Jungnickel, T. Taplick, and K. Heide. P.m.r. determination of the location of water molecules in kieserite, $MgSO_4,H_2O$. *Krist. Tech.*, 1971, **6**, 193 (*CA* 102 370).
J. Chmielewski and W. Antoniak. Developmental trends in n.m.r. equipment. Institute of Nuclear Physics, Cracow, Report, 1971, p. 81 (*CA* 103 066).
G. A. Petsko. P.m.r. studies of 'polywater–water' mixtures. *J. Colloid Interface Sci.*, 1971, **36**, 503 (*CA* 103 339).
M. Punkkinen. Random distribution of paramagnetic centres in the nuclear spin–lattice relaxation. *Phys. kondens. Mater.* 1971, **13**, 79 (*CA* 103 425).
Yu. D. Nechaev, Yu. V. Belov, Yu. L. Kleiman, N. V. Morkovin, and B. I. Ionin. Separation of signals in n.m.r. spectra of organophosphorus compounds by heteronuclear double resonance with accumulation. *Teor. i eksp. Khim.*, 1971, **7**, 424 (*CA* 103 428).
A. G. Kucheryaev, I. M. Ovchinnikov, and N. L. Oliferchuk. Attachment for recording ^{19}F n.m.r. spectra on a proton radiospectrometer. *Prib. i Tekh. Eksp.*, 1971, 230 (*CA* 103 540).
G. Gattow, W. Behrendt, and V. Warzelhan. Chalcogen carbonates. L. Thallium(I) monoalkyl carbonates. *Naturwiss.*, 1971, **58**, 361 (*CA* 104 580).
T. L. Fabry, J. Kim, S. H. Koenig, and W. E. Schillinger. Acidic–alkaline transitions in methaemoglobin and metmyoglobin. 'Probes in Structure and Function of Macromolecular Membranes, Proceedings of the 5th Colloquium Johnson Research Foundation', ed. B. Chance, Academic Press, New York, 1971, vol. 2, p. 311 (*CA* 105 129).
D. G. Davis, T. R. Lindstrom, N. H. Mock, J. J. Baldassare, S. Charache, R. T. Jones, and C. Ho. N.m.r. studies of haemoglobins. Haem proton spectra of human deoxyhaemoglobins and their relevance to the nature of co-operative oxygenation of haemoglobin. *J. Mol. Biol.*, 1971, **60**, 101 (*CA* 105 156).
R. A. Bayne, G. A. Smythe, and W. S. Caughey. E.s.r. and n.m.r. spectra of a haem A–copper complex isolated from bovine heart muscle. 'Probes in Structure and Function of Macromolecular Membranes, Proceedings of the 5th Colloqium Johnson Research Foundation', ed. B. Chance, Academic Press, New York, 1971, vol. 2, p. 613 (*CA* 105 256).
E. Macovshi and F. Mosora. Sodium-ion efflux, from living giant axons of sepia. *Rev. Roumaine Biochim.*, 1971, **8**, 97 (*CA* 106 778).
V. S. Bogdanov, V. P. Litvinov, A. N. Sukiasyan, and Ya. L. Gol'dfarb. Selenides of the thiophen and furan series. P.m.r. spectra and structure of thiophen series selenides. *Zhur. org. Khim.*, 1971, **7**, 1257 (*CA* 109 461).
J. Paasivirta and P. J. Malkonen. Chemical-shift reagents in the study of polycyclic alchols. P.m.r. spectra of methyl-substituted 5-norbornen-2-ols. *Suomen Kem.* (*B*), 1971, **44**, 230 (*CA* 109 464).
C. P. Scholes, R. A. Isaacson, and G. Feher. Determination of the zero-field splitting of Fe^{3+} in haem proteins from the temperature dependence of the spin–lattice relaxation rate. *Biochim. Biophys. Acta*, 1971, **244**, 206 (*CA* 105 122).
K. Jackowski and Z. Kecki. Evaluation of the usefulness of internal standards in high-resolution p.m.r. Institute of Nuclear Physics, Cracow, Report, 1971, p. 159 (*CA* 114 545).
I. A. Kyuntsel, Yu. I. Rozenberg, and A. D. Gordeev. Temperature dependence of the spin–lattice relaxation of ^{121}Sb nuclei in antimony trichloride complexes. *Optika i Spektroskopiya*, 1971, **31**, 67 (*CA* 114 589).
A. J. Dianoux, S. Sykora, and H. S. Gutowsky. Moments of n.m.r. absorption lines from the free induction decay or echo of solids. *U.S. Clearinghouse Fed. Sci. Tech. Inform.*, *AD*, 1971, 14 (*CA* 114 615).
W. van der Lugt, D. I. M. Knottnerus, and W. G. Perdok. N.m.r. investigation of fluoride ions in hydroxyapatite. *Acta Cryst.*, 1971, **B27**, 1509 (*CA* 114 619).
V. V. Mank, V. D. Grebenyuk, N. P. Gnusin, and F. D. Ovcharenko. Effect of iron-ion content in a KU-2 cation exchanger on p.m.r. signal shift. *Ukrain. khim. Zhur.*, 1971, **37**, 542 (*CA* 114 621).

J. W. Hennel and Z. T. Lalowicz. N.m.r. spectra of the ammonium group at helium temperatures. Institute of Nuclear Physics, Cracow, Report, 1971, p. 29 (*CA* 114 654).

D. D. Elleman and S. L. Manatt. N.m.r. spectrometer for the measurement of the moisture content of soils and rocks. 'Geological Problems of Lunar and Planetary Research, Proceedings of AAS/IAP Symposium, 1968', ed. J. Green, Amer. Astronaut. Soc., Tarzana, Calif., 1971, p. 123 (*CA* 115 503).

M. Olczak-Kobza and Z. Kecki. Contact shifts in solutions of transition metals. Institute of Nuclear Physics, Cracow, Report, 1971, p. 144 (*CA* 114 601).

K. A. Andrianov, A. A. Zhdanov, A. M. Evdokimov, and M. M. Levitskii. Reaction of potassium trimethylsilanolate with methanol studied by the ^{1}H n.m.r. method. *Izvest. Akad. Nauk S.S.S.R., Ser. khim.*, 1971, 1360 (*CA* 117 648).

B. M. Moiseev, A. M. Portnov, and L. I. Fedorov. ^{19}F N.m.r. in metamict minerals. *Geokhimiya*, 1971, 749 (*CA* 119 979).

S. G. Katal'nikov, A. M. Voloshchuk, M. A. Sokal'skii, and T. A. Kozik, Boron trifluoride donor–acceptor compounds. ^{19}F N.m.r. of boron trifluoride complexes with some aliphatic and aromatic ethers. *Zhur. fiz. Khim.*, 1971, **45**, 1810 (*CA* 122 899).

A. Johansson and T. Drakenberg. Proton and deuteron magnetic resonance studies of lamellar lyotropic mesophases. *Mol. Crystals Liquid Crystals*, 1971, **14**, 23 (*CA* 123 346).

G. Lindblom and B. Lindman. Ion binding in liquid crystals studied by n.m.r. Cetyltrimethylammonium bromide–hexanol–water system. *Mol. Crystals Liquid Crystals*, 1971, **14**, 49 (*CA* 123 382).

A. Tupciauskas, N. M. Sergeev, and Yu. A. Ustynyuk. Application of heteronuclear magnetic resonance in the determination of shielding constants of the ^{119}Sn nucleus. *Liet. Fiz. Rinkinys*, 1971, **11**, 93 (*CA* 124 794).

O. Bartko and J. Feherova. Determination of the second and fourth moments of e.s.r. and n.m.r. resonance lines. *Cesk. Casopis Fys.*, 1971, **21**, 385 (*CA* 124 825).

J. Faehnrich, B. Sedlak, and E. Repko. Adaptation of high-resolution n.m.r. spectrometers (Tesla) to broad-line measurement. *Cesk. Casopis Fys.*, 1971, **21**, 377 (*CA* 124 905).

R. du Bois. Inexpensive pulsed n.m.r. spectrometer. *Amer. J. Phys.*, 1971, **39**, 1178 (*CA* 126 369).

K. Ochiaia and K. Seto. Determination of acid dissociation constant of ion-exchange resin by n.m.r. *J. Nuclear Sci. Technol.*, 1971. **8**, 377 (*CA* 130 377).

M. M. Gofman, E. L. Rozenberg, and M. E. Dyatkina. Diamagnetic shielding of ^{19}F nuclei in tetrahedral and octahedral fluorides of non-transition elements. *Doklady Akad. Nauk S.S.S.R.*, 1971, **199**, 635 (*CA* 135 499).

N. M. Kovtun, E. E. Solov'ev, A. A. Shemyakov, and V. A. Khokhlov. Nuclear magnetic resonance in some orthoferrites. *Pis'ma Zhur. eksp. teor. Fiz.*, 1971, **14**, 105 (*CA* 135 500).

A. G. Brekhunets, V. V. Mank, F. D. Ovcharenko, M. A. Piontkovskaya, and G. I. Denisenko. N.m.r. in linax zeolite. *Ukrain. khim. Zhur.*, 1971, **37**, 839 (*CA* 135 503).

F. Koch. Detection of the n.m.r. signal of ^{59}Co using a wobbulator. *Studia Univ. Babes-Bolyai, Ser. Phys.*, 1971, **16**, 91 (*CA* 135 505).

Yu. A. Lysenko, L. M. Kapkan, and V. I. Medvedeva. Ionization of esters in solutions of titanium and tin tetrachlorides and antimony pentachloride studied by a p.m.r. method. *Zhur. obshchei Khim.*, 1971, **41**, 1342 (*CA* 135 521).

F. R. McCourt and S. Hess. Nuclear magnetic relaxation in dilute gases of symmetric- and spherical-top molecules. *Z. Naturforsch.*, 1971, **26a**, 1234 (*CA* 135 546).

I. Ozier. Nuclear magnetic relaxation in a dilute gas of spherical rotors. *Z. Naturforsch.*, 1971, **26a**, 1232 (*CA* 135 553).

R. V. Zykova, G. K. H. Kamai, B. D. Chernokal'skii, R. A. Salakhutdinov, and B. E. Abalonin. P.m.r. spectra and structure of ethylarsine derivatives. *Zhur. obshchei Khim.*, 1971, **41**, 1508 (*CA* 135 565).

G. A. Smythe, R. A. Bayne, and W. S. Caughey. Effect of Fe—O—Fe bridging on n.m.r. and e.s.r. spectra of iron(III)–porphyrin systems. 'Probes in Structure and Function of Macromolecular Membranes, Proceedings of the 5th Colloqium Johnson Research Foundation', ed. B. Chance, Academic Press, New York, 1971, vol. 2 p. 553 (*CA* 136 898).

A. N. Nesmeyanov, D. N. Kravtsov, B. A. Kvasov, E. I. Fedin, and T. S. Khazanova. Synthesis and ^{19}F n.m.r. spectra of some fluoroaryl derivatives of tin and lead. *Doklady Akad. Nauk S.S.S.R.*, 1971, **199**, 1078 (*CA* 139 926).

A. N. Nesmeyanov, N. S. Kochetkova, L. A. Fedorov, and R. B. Materikova. Intermolecular exchange of ligands in cyclopentadienylmercury compounds. *Doklady Akad. Nauk S.S.S.R.*, 1971, **199**, 361 (*CA* 139 939).

L. A. Fedorov, M. K. H. Minacheva, and E. M. Brainina. Dynamic transformations of cyclopentadienyl chelate complexes of zirconium and hafnium. *Izvest. Akad. Nauk S.S.S.R., Ser. khim.*, 1971, 1844 (*CA* 140 954).

G. M. Khvostik, I. Y. A. Poddubnyi, V. N. Sokolov, and G. P. Kondratenkov. 'Living' polymers during butadiene polymerization in the presence of π-crotylpalladium chloride. *Doklady Akad. Nauk S.S.S.R.*, 1971, **199**, 893 (*CA* 141 223).

H. Lechert and H. J. Hennig. Dependence of ^{23}Na resonance in the molecular sieve 13X on the sorption of sulphur dioxide. *Z. Naturforsch.*, 1971, **26a**, 1375 (*CA* 144 227).

K. Muthukrishnan and J. Ramakrishna. P.m.r. study of lithium formate monohydrate. *Current Sci.*, 1971, **40**, 486 (*CA* 144 942).

K. Tori. Application of paramagnetic shifts induced by lanthanide complexes in n.m.r. spectroscopy. *Kagaku Kogyo.* 1971, **22**, 1408 (*CA* 145 475).

S. Oshima, K. Asano, T. Nishishita, H. Tsuji, and H. Yokogawa. Data processing of instrumental analysis by small computer. Data processing of high-resolution n.m.r. analysis of phosphates. *Sekiyu Gakkai Shi*, 1971, **14**, 499 (*CA* 145 891).

A. Farkas and E. Tataru. P.m.r. study on light- and heavy-water solutions of beryllium(II) *Studia Univ. Babes-Bolyai, Ser. Phys.*, 1971, **16**, 79 (*CA* 145 908).

G. M. Gusakov, I. I. Evdokimov, and R. K. Mazitov. Effect of ion–ion interaction in solutions on the relaxation of ^{133}Cs nuclei. I. Aqueous solutions of Mn^{2+} ions. *Zhur. strukt. Khim.*, 1971, **12**, 580 (*CA* 145 912).

N. Tanaka and K. Kosaka. Criticisms on methods for measuring time of recovery in high-resolution n.m.r. *Nippon Kagaku Zasshi*, 1971, **92**, 681 (*CA* 145 919).

B. S. Perrett and I. A. Stenhouse. Application of tris(dipivalomethanato)europium(III) as a shift reagent in n.m.r. spectroscopy. Aliphatic and alicyclic alcohols. U.K.A.A., Research Group, Report, 1971, p. 5 (*CA* 145 951).

K. Tsukida and M. Ito. Application of a shift reagent in n.m.r. spectroscopy of esters. Approach to simple identification and simultaneous determination of tocopherols. *Experientia*, 1971, **27**, 1004 (*CA* 146 004).

S. M. Brekhovskikh, T. A. Sidorov, V. S. Grechishkin, and M. L. Zlatogorskii. Arrangement of alkaline ions in silicate and aluminosilicate glasses studied by the n.m.r. method. *Izvest. Akad. Nauk S.S.S.R., neorg. Materialy*, 1971, **7**, 1596 (*CA* 154 522).

P. M. Borodin and M. I. Volodicheva. Chemical shifts of protons in ROH–$SbCl_3$ systems. *Teor. i eksp. Khim.*, 1971, **7**, 472 (*CA* 156 358).

G. Englert and F. Wittmann. Properties of the adsorbed water in hydrated tricalcium silicate. *Zement-Kalk-Gips*, 1971, **24**, 312 (*CA* 156 785).

R. Brazdziunas and V. Lesauskis. Distortion of the shape of a nuclear magnetic resonance line in bridge autodynes. *Liet. Fiz. Rinkinys*, 1971, **11**, 315 (*CA* 156 787).

G. Serfozo and P. Groz. N.m.r. studies on $KNiF_3$ and $RbFEF_3$. *KFKI* (*Kozp. Fiz. Kut. Intez.*), Report KFKI-71-34 1971, (*CA* 156 797).

J. F. Lefelhocz and Brother Columba Curran. I.r., Mössbauer, and n.m.r. studies of some stannous chloride and bromide complexes of thiourea and tetramethylthiourea. Proceedings of the Symposium on Mössbauer Effect Methodology, 1971, vol. 6, p. 55 (*CA* 156 806).

S. K. Novoselov, L. A. Baidakov, and L. P. Strakhov. Chemical shifts in n.m.r. and van Vleck paramagnetism in orthorhombic and cubic modifications of thallium iodide. *Vestnik Leningrad. Univ.* (*Fiz. Khim.*), 1971, 54 (*CA* 156 710).

A. I. Grigor'ev, V. A. Sipachev, and A. V. Novoselova. Structure of beryllium oxyacetates of composition $Be_4O(OCOMe)_5OEt$ and $Be_6O_2(OCOMe)_3$. *Doklady Akad. Nauk S.S.S.R.*, 1971, **199**, 603 (*CA* 156 827).

L. Maija, I. Vevere, L. S. Bresler, I. Yu. Tsereteli, and E. I. Karabanova. Proton n.m.r. spectra of aluminium-β-ketoenolato-alkoxides. *Latvijas PSR Zinatnu Akad. Vestis. Kim. Ser.*, 1971, 439 (*CA* 156 828).

A. M. Smirnov and Yu. S. Konstantinov. Effect of the heterogeneity of a high-frequency magnetic field on the shape of n.m.r. signals. *Vestnik Mosk. Univ., Fiz., Astron.*, 1971, **12**, 473 (*CA* 156 937).

Chemical Abstracts, 1972, Volume 76

P. W. Taylor, J. Feeney, and A. S. V. Burgen. Investigation of the mechanism of ligand binding with cobalt(II) human carbonic anhydrase by ^{1}H and ^{19}F n.m.r. spectroscopy. *Biochemistry*, 1971, **10**, 3866 (*CA* 1259).

J. Reuben and F. J. Kayne. ^{205}Tl n.m.r. study of pyruvate kinase and its substrates. Evidence for a substrate-induced conformational change. *J. Biol. Chem.*, 1971, **246**, 6227 (*CA* 1289).

L. A. Fedorov, D. N. Kravtsov, A. S. Peregudov, E. I. Fedin, and E. M. Rokhlina. Solvation of some trimethyltin compounds studied by p.m.r. *Izvest. Akad. Nauk S.S.S.R., Ser. khim.*, 1971, 1705 (*CA* 2992).

M. K. H. Minacheva, L. A. Fedorov, E. M. Brainina, and R. K. H. Freidline. Synthesis and p.m.r. spectra of mixed cyclopentadienyl chelate compounds of zirconium and hafnium. *Doklady Akad. Nauk S.S.S.R.*, 1971, **200**, 589 (*CA* 3993).

P. S. Kalinin and R. S. Shevelevich. Structural differences in borate glasses and fibres studied by n.m.r. *Izvest. Akad. Nauk S.S.S.R., neorg. Materialy*, 1971, **7**, 1891 (*CA* 6218).

V. F. Chuvaev, V. I. Spitsyn, L. V. Anan'eva, and N. I. Snezkho. State of hydrated water in aluminium ethylenediaminetetra-acetate. *Doklady Akad. Nauk S.S.S.R.*, 1971, **200**, 160 (*CA* 8627).

O. Kanert and D. Kotzur. Recovery of the n.m.r. signal of plastically deformed alkali-metal iodide single crystals. *Z. Naturforsch.*, 1971, **26a**, 1435 (*CA* 8664).

J. Kaufmann, W. Sahm, and A. Schwenk. ^{73}Ge N.m.r. studies. *Z. Naturforsch.*, 1971, **26a**, 1384 (*CA* 8665).

D. Negoui and V. Serban. Complexes of transition metals with tertiary arsines. Splitting reactions of the halogen bridges of the dinuclear complexes of palladium(II) halides with triarylarsines. *Rev. Roumaine Chim.*, 1971, **16**, 1347 (*CA* 9922).

V. V. Negrebetskii, V. S. Bogdanov, and A. V. Kessenikh. Determination of relative signs of spin–spin interaction constants in some ^{15}N-substituted organonitrogen compounds. *Zhur. strukt. Khim.*, 1971, **12**, 716 (*CA* 8620).

R. G. Shulman, K. Wuethrich, and J. Peisach. High-resolution p.m.r. studies of myoglobin. 'Probes in Structure and Function of Macromolecular Membranes', Proceedings of the 5th Colloqium Johnson Research Foundation', ed. B. Chance, Academic Press, New York, 1971, vol. 2, p. 195 (*CA* 11 105).

A. N. Nesmeyanov, V. N. Babin, N. S. Kochetkova, E. I. Mysov, Yu. A. Belousov, and L. A. Fedorov. Isomerism in complexes of iron with benzotriazole. *Doklady Akad. Nauk S.S.S.R.*, 1971, **200**, 1112 (*CA* 13 321).

D. Elmaleh, S. Patai, and Z. Rappoport. N.m.r. investigation of the reaction of 1,1-diarylethylenes with seleninyl chloride and diselenium dichloride. *Israel J. Chem.*, 1971, **9**, 155 (*CA* 13 363).

J. Paasivirta and P. J. Malkonen. Chemical-shift reagents in the study of polycyclic alcohols. P.m.r. spectra of the stereoisomeric 2-methyl-2-norbornanols and 2-methyl-isoborneol. *Suomen Kem.* (*B*), 1971, **44**, 283 (*CA* 13698).

V. A. Drozdov, A. P. Kreshkov, N. D. Rumyantseva, and V. F. Andrianov. Analysis of alkylchlorosilane chlorination products by gas–liquid chromatography and n.m.r. *Plastmassy*, 1971, 65 (*CA* 14 967).

S. O. Grim, H. J. Plastas, C. L. Huheey, and J. E. Huheey. Bond angles, hybridization and phosphorus–metal bonding. *Phosphorus*, 1971, **1**, 61 (*CA* 18 059).

R. M. Pearson. Wide-line n.m.r. studies on transition aluminas. Distribution of protons between surface and bulk phases. *J. Catalysis*, 1971, **23**, 388 (*CA* 18 151).

N. A. Bell and L. A. Nixon. Thermal decomposition of Group IIB metal halide complexes. *p*-Dimethylaminophenyldimethylphosphine complexes. *Thermochim. Acta*, 1971, **3**, 61 (*CA* 18 623).

N. A. Verendyakina, G. B. Seifer, Yu. Ya. Kharitonov, and B. V. Borshagovskii. Cobalt hexa-ammine ferrocyanides. *Zhur. neorg. Khim.*, 1971, **16**, 2714 (*CA* 19 696).

L. N. Komissarova, P. P. Mel'nikov, E. G. Teterin, and V. F. Chuvaev. Scandium phosphates. *Zhur. neorg. Khim.*, 1971, **16**, 2651 (*CA* 19 739).

K. Wuethrich, I. Aviram, and A. Schejter. Structural studies of modified cytochrome *c* by n.m.r. spectroscopy. *Biochim. Biophys. Acta*, 1971, **253**, 98 (*CA* 22 136).

T. R. Lindstrom, J. S. Olson, N. H. Mock, Q. H. Gibson, and C. Ho. N.m.r. studies of haemoglobins. Evidence for preferential ligand binding to β-chains within deoxy-haemoglobins. *Biochem. Biophys. Res. Comm.*, 1971, **45**, 22 (*CA* 22 140).

V. F. Bystrov, N. I. Dubrovina, L. I. Barsukov, and L. D. Bergelson. N.m.r. differentiation of the internal and external phospholipid membrane surfaces using paramagnetic Mn^{2+} and Eu^{3+} ions. *Chem. and Phys. Lipids*, 1971, **6**, 343 (*CA* 22 203).
R. A. Bogatkin. Effect of organosilicon substituents on the properties of the triple bond. *Reakts. Sposobnost Mekh. Reakts. Org. Soedin.*, 1971, 152 (*CA* 24 233).
I. P. Gol'dshtein, E. N. Gur'yanova, L. S. Mel'nichenko, N. N. Zeml'yanskii, T. I. Perepelkova, Yu. K. Maksyutin, and K. A. Kocheshkov. Individuality of compounds of the $RPhSnX_2$ class. *Doklady Akad. Nauk S.S.S.R.*, 1971, **201**, 105 (*CA* 24 523).
V. R. Miller and R. N. Grimes. Carborane formation in alkyne–borane gas-phase systems. Conversion of two-carbon to four-carbon carboranes *via* alkyne insertion. N.m.r. studies of tetracarba-*nido*-hexaboranes. *U.S. Nat. Tech. Inform. Serv., AD. Reps.*, 1971, No. 729 266 (*CA* 25 349).
R. Kosfeld and U. von Mylius. Linewidth and relaxation phenomena in n.m.r. solid-state spectroscopy. *Nuclear Magn. Resonance*, 1971, **4**, 181 (*CA* 25 597).
V. P. Grigor'ev and A. I. Maklakov. Effect of correlation time distributions on the width of an n.m.r. line. *Vysokomol. Soedineniya, Ser. B.* 1971, **13**, 652 (*CA* 25 704).
N. R. Krishna and B. D. N. Rao. Double-resonance study of proton-spin relaxation in a strongly coupled two-spin system. Proceedings of the 15th Nuclear Physics and Solid State Physics Symposium, 1971, vol. 3, p. 323 (*CA* 29 382).
W. N. Lipscomb, R. J. Wiersema, and M. F. Hawthorne. Structural ambiguity of the $[B_{10}H_{14}]^{2-}$ ion. *U.S. Nat. Tech. Inform. Serv., AD. Reps.*, 1971, No. 727 586 (*CA* 29 392).
H. Voorhof. Thermocouple calibration accuracy. *Ingenieurablad*, 1971, **40**, 221 (*CA* 26 682).
F. Conti and M. Paci. Natural-abundance ^{13}C spectra of proteins. Carboxymyoglobin and haemoglobin. *F.E.B.S. Letters*, 1971, **17**, 149 (*CA* 31 435).
J. C. Thompson, W. Haar, W. Maurer, H. Rueterjans, K. Gersonde, and H. Sick. P.m.r. study of the histidine residues of horse myoglobin. *F.E.B.S. Letters*, 1971, **16**, 262 (*CA* 31 436).
R. K. Gupta and S. H. Koenig. The pH and temperature dependence of the n.m.r. spectra of cytochrome *c*. *Biochem. Biophys. Res. Comm.*, 1971, **45**, 1134 (*CA* 31 450).
E. R. Bienbaum. N.m.r. and calorimetric study of the interactions between lanthanide ions and water. *U.S. Nat. Tech. Inform. Serv., PB Rep.*, 1971, No. 202 856 (*CA* 38 039).
F. Noack. Nuclear magnetic relaxation spectroscopy. *Nuclear Magn. Resonance*, 1971, **3**, 83 (*CA* 39 344).
O. Kanert and M. Mehring. Static quadrupole effects in disordered cubic solids. *Nuclear Magn. Resonance*, 1971, **3**, 1 (*CA* 39 345).
J. Haupt. Influence of quantum effects of the methyl-group rotation on the nuclear relaxation in solids. *Z. Naturforsch.*, 1971, **26a**, 1578 (*CA* 39 410).
E. I. Berus, V. F. Anufrienko, Yu. N. Molin, and A. A. Shklyaev. N.m.r. study of adducts of copper–pyridine complexes. *Doklady Akad. Nauk S.S.S.R.*, 1971, **200**, 1129 (*CA* 39 787).
D. Demco and V. Ceausescu. N.m.r. lineshape governed by a time-dependent Hamiltonian, *Inst. Fiz. At.* (*Rom.*) (*Report*) S.R.2, 1971 (*CA* 39 798).
S. P. Ionov and V. S. Lyubimov. Electron structure and n.m.r. spectra. AX6-NYN octahedral systems with σ-bonds. Co^{3+} complexes. *Zhur. fiz. Khim.*, 1971, **45**, 2485 (*CA* 39 799).
V. P. Shatalov, V. S. Glukhovskoi, Yu. A. Litvin, E. S. Kostin, A. R. Samotsvetov, and L. V. Kovtunenko. N.m.r. spectrum of s-butyl-lithium. *Zhur. obshchei Khim.*, 1971, **41**, 1921 (*CA* 39 834).
N. D. Chuvylkin, R. Z. Sagdeev, G. M. Zhidomirov, and Yu. N. Molin. Distribution of spin density in complexes of nickel acetylacetonate with isoquinoline. *Teor. i eksp. Khim.*, 1971, **7**, 612 (*CA* 39 836).
L. C. Gupta and D. L. R. Setty. N.m.r. studies of ^{35}Cl in $K_2CuCl_4,2H_2O$. Proceedings of the 15th Nuclear Physics and Solid State Physics Symposium, 1971, vol. 3, p. 643 (*CA* 39 844).
B. Borcard, E. Hiltbrand, and R. Sechehaye. Free precession in n.m.r. Resonances Magn., Recl. Trav. Sess. Perfect. Inst. Nat. Sci. Appl., Lyon, 1967, publ. 1969, p. 259 (*CA* 39 855).
R. D. B. Waymark and D. R. Bowman. Use of an on-line digital computer for enhancement and integration of nuclear resonance spectra. *Radio Electron. Eng.*, 1971, **41**, 421 (*CA* 39 893).

A. S. Mildvan, N. M. Rumen, and B. Chance. Nuclear relaxation studies of dissolved and crystalline methemoproteins. 'Probes in Structure and Function of Macromolecular Membranes, Proceedings of the 5th Colloqium Johnson Research Foundation', ed. B. Chance, Academic Press, New York, 1971, vol. 2, p. 205 (*CA* 43 053).

R. L. Ward and J. A. Happe. ^{35}Cl N.m.r. studies of the active-site zinc of horse-liver alcohol dehydrogenase. *Biochem. Biophys. Res. Comm.*, 1971, **45**, 1444 (*CA* 43 411).

J. Barciszewski, A. J. Rafalski, and M. Wiewiorowski. Europium shifts in the n.m.r. spectra of lactams. I. Sterochemical assignment of lupanine. *Bull. Acad. polon. Sci., Sér. Sci. chim.*, 1971, **19**, 545 (*CA* 46 365).

D. Zamir and C. Korn. Hydrogen behaviour in metals using n.m.r. *U.S. Nat. Tech. Inform. Serv., AD. Rep.*, 1970, No. 729 690 (*CA* 49 071).

M. Noshiro and Y. Jitsugiri. Determination of silica in glass by ^{19}F n.m.r. *Bunseki Kagaku*, 1971, **20**, 1277 (*CA* 49 191).

P. M. Borodin and Kim Zao. Nguyen. Equilibria in the lithium fluorosilicate–perchloric acid–water system studied by a ^{19}F n.m.r. method. *Zhur. neorg. Khim.*, 1971, **16**, 3248 (*CA* 50 678).

N. A. Kostromina. Complexing of methylglycine with lanthanum and lutetium studied by a proton resonance method. *Zhur. neorg. Khim.*, 1971, **16**, 2966 (*CA* 50 772).

Y. Sugitani, M. Watanabe, and K. Nagashima. Positions of the hydrogen atoms in datolite as determined by n.m.r. *Acta Cryst.*, 1972, **B28**, 326 (*CA* 51 248).

B. Stalinski. N.m.r. studies of molecular motions in solids. *Postepy Fiz.*, 1971, **22**, 249 (*CA* 51 979).

B. D. Flockhart. N.m.r. and e.s.r. methods. *Compr. Analyt. Chem.*, 1971, **2C**, 206 (*CA* 51 990).

H. Barrere and Min Tran. Kim. N.m.r. study of the alpha-phase of the lutetium–hydrogen system. *Compt. rend.*, 1971, **273**, *B*, 823 (*CA* 52 517).

Yu. A. Buslaev, Yu. V. Kokundv, V. A. Bochkareva, and E. M. Shustorovich. *cis*-Effect of multiply bonded oxygen in reactions of molybdenum and tungsten mono-oxo-complexes. *Doklady Akad. Nauk S.S.S.R.*, 1971, **201**, 355 (*CA* 52 519).

I. A. Kyuntsel and Yu. I. Rozenberg. Nuclear relaxation and molecular mobility in crystalline complexes of antimony trihalides with anisole. *Teor. i eksp. Khim.*, 1971, **7**, 565 (*CA* 52 525).

H. Burzynska, J. Dabrowski, and A. Krowczynski. Europium- and praseodymium-induced shifts in the p.m.r. spectra of amines. *Bull. Acad. polon. Sci., Sér. Sci. chim.*, 1971, **19**, 587 (*CA* 52 541).

K. Utvary. N.m.r. spectroscopy as an aid in structure elucidation of inorganic compounds. *Allg. prakt. Chem.*, 1971, **22**, 301 (*CA* 52 568).

J. M. Dereppe, Y. van Honacker, and M. van Meerssche. Proton resonance in hydrates. Potassium oxalate hydrate studied by the ΔH method. *J. Chim. phys.*, 1971, **68**, 1615 (*CA* 52 569).

Z. E. Suyunova, V. V. Mank, Yu. I. Tarasevich, and A. G. Brekhunets. N.m.r. study of the mobility of water in Montmorillonite dispersions. *Ukrain. khim. Zhur.*, 1971, **37**, 1183 (*CA* 52 595).

Yu. A. Buslaev, E. G. Il'in, and M. N. Krutkina. Composition and structure of complexes of tantalum with fluorine, chlorine, and bromine atoms in the internal sphere. *Doklady Akad. Nauk S.S.S.R.*, 1971, **201**, 99 (*CA* 52 623).

S. Albert, H. S. Gutowsky, and J. A. Ripmeester. N.m.r. relaxation studies of solid hexamethylethane and hexamethyldisilane. *U.S. Nat. Tech. Inform. Serv., AD Rep.*, 1971, No. 729 339 (*CA* 52 636).

J. Charvolin and P. Rigny. N.m.r. study of molecular motions in the mesophases of potassium laurate–[$^{2}H_2$]water system. *Mol. Crystals Liquid Crystals*, 1971, **15**, 211 (*CA* 52 644).

V. N. Dubinin, V. V. Kuz'movich, and V. M. Chegoryan. Use of the nuclear alpha-resonance method to study iron and tin in metal–oxygen polymers. *Teor. i eksp. Khim.*, 1971, **7**, 703 (*CA* 52 508).

E. Breitmaier, W. Voelter, and G. Jung. Fourier-transform ^{13}C n.m.r. investigations on metal chelates, carbohydrates, and polypeptides. Proceedings of the 2nd Conference on Applied Physics and Chemistry, 1971, ed. I. Buzas, vol. 1, p. 61 (*CA* 58 411).

T. Schleich, B. J. Blackburn, R. D. Lapper, and I. C. P. Smith. N.m.r. study of the influence of aqueous sodium perchlorate and temperature on the solution conformations of uracil nucleosides and nucleotides. *Biochemistry*, 1972, **11**, 137 (*CA* 59 940).

K. Wuthrich, J. P. Meraldi, A. Tun-Kyi, and R. Schwyzer. Structural studies by n.m.r. of a cation-specific peptide. Proceedings of the 1st European Biophysics Congrees, 1971, ed. E. Broda, vol. 1, p. 93 (*CA* 60 027).

E. Tiezzi. Associations between manganese(II) and peptides studied by n.m.r. and e.p.r. spectroscopy. Proceedings of the 1st European Biophysics Congress, 1971, ed. E. Broda, vol. 1, p. 97 (*CA* 60 032).

M. P. Volarovich, N. I. Gamayunov, and L. Y. U. Vasil'eva. Study of water sorption by the spin-echo method with the use of a pulsed n.m.r. spectrometer. *Kolloid. Zhur.*, 1971, **33**, 922 (*CA* 63 790).

O. D. Kurilenko, A. V. Lenshchenko, V. V. Mank, and R. V. Voitsekhovskii. N.m.r. study of the state of water asdorbed on ion-exchange celluloses. *Ukrain. khim. Zhur.*, 1971, **37**, 960 (*CA* 63 842).

V. V. Mank, V. D. Grebenyuk, and N. P. Gnusin. Shift in the p.m.r. signal of water sorbed by various diamagnetic salt forms of the KU2 cation exchanger. *Ukrain. khim. Zhur.*, 1971, **37**, 956 (*CA* 63 843).

J. Kaerger. Diffusion investigation of water on 13X, 4A, and 5A zeolites using the pulsed field-gradient method. *Z. phys. Chem.* (*Leipzig*), 1971, **248**, 27 (*CA* 63 852).

G. M. Bulgakova, A. N. Shupik, I. P. Skibida, K. I. Zamaraev, and Z. K. Maizus. Proof of the formation of intermediate complexes in the hydroperoxide-catalyst system by the n.m.r. method. *Doklady Akad Nauk S.S.S.R.*, 1971, **199**, 376 (*CA* 63 927).

R. J. R. Logan. Determination of acid–base mechanisms of boron-containing compounds. Applications of ^{11}B n.m.r. *Texas J. Sci.*, 1971, **23**, 107 (*CA* 64 151).

A. G. Lee. Spectroscopic studies of thallium(I) nitroparaffins. *Spectrochim. Acta*, 1972, **28**, *A*, 133 (*CA* 65 888).

R. Van Steenwinkel. Experimental effects of radiofrequency irradiation on n.m.r. lines in solids. Dynamics of the equilibrium establishment. *Z. Naturforsch.*, 1971, **26a**, 1825 (*CA* 65 999).

H. H. Rupp, Anisotropy of the ^{195}Pt n.m.r. shifts of crystalline tetracyanoplatinates(II) and potassium tetracyanoplatinate bromide hydrate. *Z. Naturforsch.*, 1971, **26a**, 1935 (*CA* 66 005).

M. B. Khusidman and V. S. Neshpor. N.m.r. of hexagonal and cubic boron nitride. *Zhur. strukt. Khim.*, 1971, **12**, 1094 (*CA* 66 016).

K. I. Dahlovist and A. B. Hornfeldt. N.m.r. spectrum of oriented selenophen in lyotropic mesophase. *Chemica Scripta*, 1971, **1**, 125 (*CA* 66 018).

2

Nuclear Quadrupole Resonance Spectroscopy

BY J. H. CARPENTER

1 Introduction

Papers dealing with nuclear quadrupole resonance spectroscopy (n.q.r.) continue to be small in number; the number reported here is about the same as last year. Nevertheless, n.q.r. spectra give useful information about the structure and bonding in solids containing quadrupolar nuclei, and complement the gas-phase quadrupole coupling constants available from microwave spectroscopy. Once again, purely organic molecules have not been included.

A review by Schultz,[1] entitled 'Applications of n.q.r. Spectrometry to Analytical Chemistry', is of particular interest to inorganic chemists. After a short discussion of the theory of n.q.r. and the apparatus used, he goes on to discuss selected chemical applications – halogen resonances in Group III trihalides, in Group IV halides, in transition-metal halides, and in hexahalide anions; nitrogen n.q.r.; n.q.r. in minerals; charge-transfer complexes; and quantitative analysis by n.q.r. are considered.

In a series of articles,[2] Smith has also reviewed the subject. He deals with the theory and instrumentation in greater depth than the previous article, and also discusses the chemical applications of n.q.r., including sections on inorganic compounds and metal complexes, ionic lattices, and metals.

This Report includes mention of resonances for sixteen nuclides of thirteen elements: in order of increasing atomic number and mass these are: ^{14}N; ^{23}Na; ^{27}Al; ^{35}Cl, ^{37}Cl; ^{55}Mn; ^{59}Co; ^{63}Cu; ^{75}As; ^{79}Br, ^{81}Br; ^{115}In; ^{121}Sb, ^{123}Sb; ^{127}I; ^{175}Lu. They are arranged in the main text according to the vertical Groups of the Periodic Table of the resonating nuclei, with Main Group elements first, followed by transition metals and lanthanides. This ordering thus places the studies of chemically similar elements together. Once again, papers on chlorine and other halogens predominate: 60% of papers contain references to resonances of ^{35}Cl or ^{37}Cl, and nearly 80% report resonances for one or more of the halogen nuclei.

2 Instrumentation

A versatile wide-range grounded-grid Colpitts super-regenerative oscillator–detector with varactor diode frequency modulation has been described.[3]

[1] H. D. Schultz, *Appl. Spectroscopy*, 1971, **25**, 293.
[2] J. A. S. Smith, *J. Chem. Educ.*, 1971, **48**, 39, 77A, 147A, 243A.
[3] C. T. O'Konski and T. J. Scheffer, *Rev. Sci. Instr.*, 1971, **42**, 1891.

The pulse method is shown to be effective in searching for ^{14}N n.q.r. spectra.[4] The spectrometer described uses a coherent pulse system in the 1—5 MHz range. By using a pulse method, the linewidth is narrowed down to that due to relaxation alone, typically tens of hertz. This gives improved sensitivity.

The use of a commercial frequency synchronizer to control the frequency sweep of a Robinson oscillator has been described.[5] The system was used for obtaining ^{35}Cl n.q.r. signals from powdered specimens, and allowed long time-averaging as well as decreasing the microphonics.

The use of field effect transistors in a n.q.r. spectrometer working in the 3—40 MHz region has been discussed.[6]

By using the temperature variation of the ^{35}Cl resonance line of $NaClO_3$, Larrie *et al.*[7] have made a practical n.q.r. thermometer; an accuracy of about $\pm$ 0.003 K at liquid-nitrogen temperature is claimed.

3 Main Group Elements

Group I (Sodium-23).—The resonance frequency of ^{23}Na in a single crystal of $NaNO_3$ decreases with temperature up to the λ-point at 276 °C, after which the decrease is less gradual.[8] The resonance decreases with pressure at temperatures up to 210 °C; above this it increases, and at the λ transition $(\partial\nu/\partial p)_T$ drops abruptly to a much smaller value. The pressure variation is explained in terms of the anisotropic compression of $NaNO_3$, while the temperature variation above 150 °C is attributed to an increase in the dynamic disorder involving hindered reorientation of the NO_3^- groups.

Group III (Aluminium-27 and Indium-115).—The n.q.r. spectra of some alkyl derivatives of aluminium, AlR_nX_{3-n}, where R = Me, Et, Bu^i, Pr^n, or Ph, X = Cl or Br, and n = 0, 1, 2, or 3, have been observed.[9] The ^{35}Cl resonances, where applicable, are also reported. $AlMe_3$ has previously been shown to be dimeric, and the similarity of the spectra of $AlEt_3$ and $AlBu^i_3$ indicates that these are also dimers in the solid state. In the compounds containing halogens, the asymmetry parameter η for ^{27}Al is much lower than in the trialkyls, suggesting a change in the bridging bond angle. The very low ^{35}Cl resonance frequency implies a high degree of ionic character in the Al—Cl bonds. From the values of e^2Qq and η for ^{27}Al, the AlR_2Cl compounds are considered to have structure (1), whereas the $AlRCl_2$ structures have structure (2).

A moment analysis of the n.q.r. spectrum of ^{115}In in indium metal has been used to measure the isotropic (contact) and anisotropic (classical)

[4] I. A. Safin and D. Y. Osokin, *Pribory i tekhn. Eksperim.*, 1971, **71**, 154.
[5] H. R. Brooker and W. W. Startup, *Rev. Sci. Instr.*, 1971, **42**, 83.
[6] N. Sullivan, *Rev. Sci. Instr.*, 1971, **42**, 462.
[7] R. Larrie, M. Infantes, and J. Vanier, *Rev. Sci. Instr.*, 1971, **42**, 26.
[8] G. J. D'Alessio and T. A. Scott, *J. Magn. Resonance*, 1971, **5**, 416.
[9] M. J. S. Dewar, D. B. Patterson, and W. I. Simpson, *J. Amer. Chem. Soc.*, 1971, **93**, 1030.

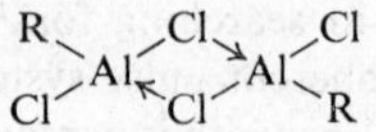

(1)

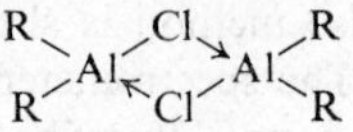

(2)

dipolar interactions.[10] Four transitions at 1.87, 3.76, 5.65, and 7.6 MHz are observed in the metal. The dipolar interactions shift the n.q.r. frequencies and contribute symmetric and asymmetric broadening. Dilute (1% or less) alloys of tin in indium produce broad, asymmetric resonances; the results are in qualitative agreement with n.m.r. data.

Group V (Nitrogen-14, Arsenic-75, Antimony-121, and Antimony-123).— Four resonances were observed at 77 K in the ^{14}N n.q.r. spectrum of $NH_3,\frac{1}{2}H_2O$; these were assigned to two pairs of crystallographically inequivalent nitrogen nuclei, from their Zeeman shifts.[11] The species is shown to be $2NH_3,H_2O$, from the value of e^2Qq; if the ammonia were present as NH_4^+, a near-zero value of e^2Qq would be expected.

From a Zeeman study of ^{14}N in N_2H_4 at 77 K, the principal axes of the field gradient have been deduced.[12] The z-axis is not coincident with the lone pair, but makes an angle of 33° with it (towards the N—N bond). By neglecting the ionicity of the N—N and N—H bonds, bond angles of 90° for θ(H—N—H) and 106° for θ(N—N—H) were obtained (the gas-phase values are 106° and 112° respectively). Some derivatives of hydrazine have also been investigated;[13] values of e^2Qq and η were determined and the ionicity of the N—H and N—C bonds was estimated. Table 1 shows

Table 1 *Quadrupole coupling constants and asymmetry parameters for some derivatives of hydrazine*

Compound	e^2Qq/kHz	η	*Assignment*	ηe^2Qq/MHz
NH_2NH_2	4820.6	0.7843	$-NH_2$	3.78
	4818.5	0.8277	$-NH_2$	3.99
NH_2NH_2,H_2O	4731.2	0.8358	$-NH_2$	3.95
$MeNHNH_2$	4766.3	0.7684	$-NH_2$	3.66
	5480.5	0.5880	$-NHMe$	3.22
Me_2NNH_2	4972.9	0.7613	$-NH_2$	3.79
	4941.1	0.7759	$-NH_2$	3.83
	5943.0	0.3617	$-NMe_2$	2.15
	5931.4	0.3687	$-NMe_2$	2.19
$PhNHNH_2$	5134.3	0.6984	$-NH_2$	3.59
	5430.9	0.6160	$-NHPh$	3.35
$(NH_2NH_3)^+Br^-$	5223.3	0.9339	$-NH_2$	4.88
	2776.2	0.1277	$-NH_3^+$	0.35
$(NH_2NH_3)^+I^-$	5169.5	0.9417	$-NH_2$	4.87
	2666.5	0.2078	$-NH_3^+$	0.55
$(NH_3NH_3)^{2+}2Br^-$	3985.7	0.0559	$-NH_3^+$	0.22

[10] J. P. Palmer and R. R. Hewitt, *Phys. Rev.* (*B*), 1971, **3**, 3676.
[11] S. Eletr and C. T. O'Konski, *J. Chem. Phys.*, 1971, **54**, 4312.
[12] Y. Kamishina, *J. Phys. Soc. Japan*, 1971, **31**, 242.
[13] R. Ikeda, S. Noda, D. Nakamura, and M. Kubo, *J. Magn. Resonance*, 1971, **5**, 54.

the assignments, coupling constants, and other derived parameters. The assignment of the methyl-substituted hydrazines was effected by noting that one resonance varies markedly with methyl substitution, the other being relatively insensitive. For the hydrazinium salts the frequencies with lower values of e^2Qq are assigned to the $-\overset{+}{N}H_3$ nitrogen. The quantity ηe^2Qq is shown to be approximately proportional to the ionicity of nitrogen. The ionic characters of the N—H bonds (in $-NH_2$) and N—C bonds (in methyl-substituted derivatives) are estimated at 42% and 24% respectively.

The Stark effect, on the ^{75}As n.q.r. spectrum of a ferroelectric CsH_2AsO_2 crystal, has been studied using a pulsed electric field.[14] A value of $\partial(\Delta\nu)/\partial E$ = 0.4 kHz kV^{-1} cm was obtained.

Two papers deal with n.q.r. spectra of Group V elements in minerals. The effect on the n.q.r. spectra of small amounts of mercury in antimonites (Sb_4S_6) and realgars (As_4S_4) is similar to that of copper and silver;[15] the linewidths and T_1 of ^{75}As and ^{121}Sb resonances are significantly affected by the amount of the impurity, which is believed to be localized in the interstices of the chain-layer structure of antimonite and in intermediate spaces in realgar.

The detection of n.q.r. signals in minerals is often made difficult by the presence of defects. Franckeite ($Sn_3Pb_5Sb_2S_{14}$, monoclinic) has 284 atoms in a unit cell, making this a severe problem. Nevertheless, resonances have been observed, by cooling to 77 K, for both ^{121}Sb and ^{123}Sb nuclei.[16] For ^{121}Sb, e^2Qq = 345.3 MHz, with $\eta \approx 0$; the value of e^2Qq is similar to that observed in other Sb—S salts, implying a trigonal-pyramidal SbS_3 group with r(Sb—S) = 2.45—2.47 Å and θ(S—Sb—S) = 95—97°.

The temperature dependence of the relaxation time T_1 for ^{121}Sb resonances in $2SbCl_3$,R (R = $C_{10}H_8$, Ph_2CH_2, *p*-xylene, C_6H_6, or PhOMe) and $SbCl_3$,R′ (R′ = PhOEt or PhOMe) does not show any evidence for hindered motion of the methyl group or aromatic ring at high temperatures.[17] Such motions had been indicated by previous studies. In $2SbCl_3,C_6H_6$ a reorientation of the benzene ring about its six-fold axes is deduced from n.m.r. studies; the absence of any effect here indicates that the ^{121}Sb nuclei lie on the six-fold axis and that their thermal displacements from that axis are negligible.

Group VII (Chlorine-35, Chlorine-37, Bromine-79, Bromine-81, and Iodine-127).—Compounds of chlorine and boron are the subject of three investigations. In a study of BCl_3 and the phenylboron chlorides $PhBCl_2$ and Ph_2BCl, the asymmetry parameter η was measured by Zeeman studies on

14 A. A. Boguslavskii and G. K. Semin, *Izvest. Akad. Nauk S.S.S.R., Ser. fiz.*, 1970, **34**, 2519.

15 I. N. Pen'kov, N. A. Ozerova, and N. K. Aidin'yan, *Doklady Akad. Nauk S.S.S.R.*, 1971, **196**, 183.

16 I. N. Pen'kov and I. A. Safin, *Geokhimiya*, 1971, 118.

17 I. A. Kyuntsel, Y. I. Rozenberg, and A. D. Gordeev, *Optics and Spectroscopy*, 1971, **31**, 35.

powdered samples.[18] The value obtained for BCl_3, $\eta = 0.54$, is higher than those for BBr_3 and BI_3, and implies increasing π-bonding on descending the Group. The resonance frequency decreases with increasing substitution of phenyl. A subsequent paper gives the ^{35}Cl n.q.r. frequencies of 23 boron–chlorine compounds;[19] the results are discussed in terms of the *s*-character and charge distribution in the B—Cl bond. An approximate order of donor ability for a ligand L in compounds of the type L,BCl_3 is proposed. Other complexes with the formula L,BCl_3 have been investigated;[20] the majority of these have a tetrahedral configuration of bonds around the boron nucleus [see structure (3)]. The relative order of electron-donor ability of the ligands follows closely that observed for $L,GaCl_3$ complexes. In the complexes of BCl_3 with the nitriles $CH_{3-n}Cl_nCN$, the tetrahedral structure (3) is also suggested for $n = 0$ or 1, but for $n = 2$ or 3 the resonances are best explained by structures (4) or (5), which have previously been postulated.

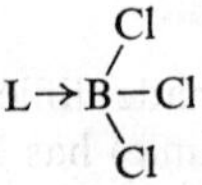
(3)

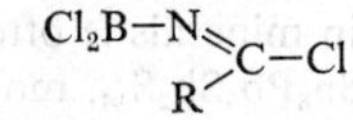
(4)

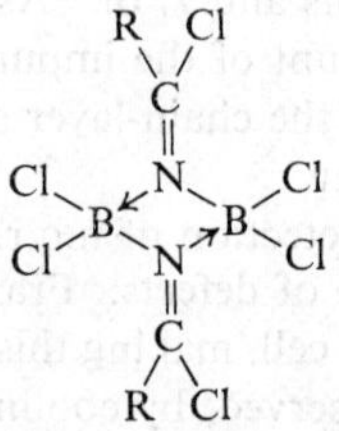
(5)

The n.q.r. and i.r. spectra of a number of compounds containing the C_5Cl_5 entity are reported.[21] In $Tl,C_5Cl_5,2PhMe$, $Bu^n{}_4P,C_5Cl_5$, and C_5H_5NMe,C_5Cl_5 a complex system of closely spaced ^{35}Cl resonances indicates crystallographic inequivalence of chemically equivalent chlorine nuclei in the anion $C_5Cl_5^-$. On the other hand, the complex spectra obtained from $Hg(C_5Cl_5)_2$,C_5Cl_5,HgCl, C_5Cl_5,HgCl,$HgCl_2$, and C_5Cl_5,HgPh clearly show resonances due to allylic and vinylic chlorines in the ratio 1 : 4, indicating a σ-bond between Hg and C_5Cl_5.

In the compounds SiF_mCl_{4-m}, the ^{35}Cl n.q.r. spectra have been reported, in addition to the i.r., n.m.r., and Raman spectra.[22] A correlation of the ^{79}Br, ^{81}Br, and ^{127}I n.q.r. frequencies with inductive (σ_I) and mesomeric (σ_c) Taft parameters in the species $R^1R^2R^3SiX$, where R^1, R^2, R^3 = Et, Ph, Me, Cl, or Br and X = Br or I, has been shown to exist.[23] The number

[18] J. A. S. Smith and D. A. Tong, *J. Chem. Soc.* (*A*), 1971, 173.
[19] J. A. S. Smith and D. A. Tong, *J. Chem. Soc.* (*A*), 1971, 178.
[20] S. Ardjomand and E. A. C. Lucken, *Helv. Chim. Acta*, 1971, **54**, 176.
[21] G. Wulfsberg and R. West, *J. Amer. Chem. Soc.*, 1971, **93**, 4085.
[22] K. Hamada, G. A. Ozin, and E. A. Robinson, *Bull. Chem. Soc. Japan*, 1971, **44**, 2555.
[23] G. K. Semin, E. V. Bryukhova, M. A. Kadina, and G. V. Frolova, *Izvest. Akad. Nauk S.S.S.R., Ser. khim.*, 1971, 1176.

of unbalanced *p*-electrons, $f = (e^2Qq)_{mol}/(e^2Qq)_{atom}$, is related to these parameters:

$$f(\mathrm{Cl}) = 0.314 + 0.045\sigma_I - 0.004\sigma_c$$
$$f(\mathrm{Br}) = 0.348 + 0.071\sigma_I - 0.040\sigma_c$$
$$f(\mathrm{I}) = 0.391 + 0.112\sigma_I - 0.170\sigma_c$$

The signs of the inductive and mesomeric terms as well as the trends in their values are explained in terms of back transfer of electron density into the p_π halide orbitals through $p_\pi \rightarrow d_\pi$ X—Si bonding.

The detection of two ^{35}Cl resonances of unequal intensity in K_2SnCl_4,H_2O confirms the presence of the $Sn^{II}Cl_3^-$ anion,[24] as deduced from *X*-ray and other spectroscopic data. In contrast, there is only one resonance in $SnCl_2,2H_2O$. To explain the value of $(\partial\nu/\partial T)_p$ in K_2SnCl_4,H_2O, a lattice-tilting mode of the $SnCl_3^-$ ion at *ca.* 39 cm^{-1} is postulated.

Several studies have been made of ^{35}Cl resonances in compounds of phosphorus and chlorine. The frequencies observed in $PhPCl_3^+$ Cl^- and $Ph_2PCl_2^+$ Cl^- resemble those of PCl_4^+, supporting the ionic structure.[25] No signal is observed for Ph_4PCl; this is as expected for the PPh_4^+ Cl^- structure. Aromatic cyclic esters, whose structures are given by (6), (7), and (8), have

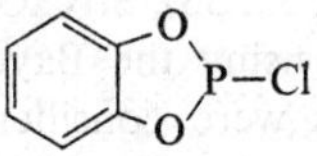

(6)

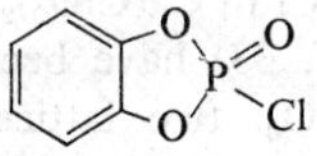

(7)

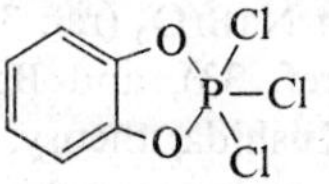

(8)

been studied.[25] In (8), the chlorine resonances indicated that one P—O bond was axial, the other equatorial; hence θ(O—P—O) $\approx$ 90°. In addition, several phosphoryl chlorides XYPOCl and thiophosphoryl chlorides, together with $Cl_3PNPCl_3^+$ PCl_6^-, have been investigated. In another study, the ^{35}Cl n.q.r. spectra of $Ni(PCl_3)_4$ and $Mo(CO)_3(PCl_3)_3$, as well as several $RPCl_2$ and R_2PSCl compounds, where R = Me, Et, or Cl, are reported.[26] In the metal compounds, the complex set of lines appears to originate from chemically equivalent chlorine atoms at crystallographically inequivalent sites.

The temperature variation of the ^{35}Cl n.q.r. spectra of PCl_3, $POCl_3$, $AsCl_3$, and $SbCl_3$ has been investigated up to their melting points.[27] For each compound two resonances have been found (except for $AsCl_3$, which has three resonances), due to chemically different nuclei; this is attributed to intermolecular bonding, and a linear relation between the n.q.r. frequency splitting and the melting point is found. For the phosphorus compounds,

[24] K. Singh and S. Singh, *J. Phys. Soc. Japan*, 1971, **31**, 1069.
[25] R. M. Hart and M. A. Whitehead, *J. Chem. Soc.* (*A*), 1971, 1738.
[26] J. K. B. Bishop, W. R. Cullen, and M. C. L. Gerry, *Canad. J. Chem.*, 1971, **49**, 3910.
[27] M. Hashimoto, T. Morie, and Y. Kato, *Bull. Chem. Soc. Japan*, 1971, **44**, 1455.

the signal fades out somewhat below the melting point; this is considered to be due to free rotation of the molecules.

A limited number of cations form stable salts with $SbBr_6^{3-}$ or $BiBr_6^{3-}$, and of these some did not give a n.q.r. signal. However, strong resonances were observed in salts with the $Et_2NH_2^+$ cation.[28] Since in each case only one resonance occurs, indicating a regular octahedron around the central atom, the ligand–ligand repulsions are more important than the inter-electronic repulsion of the fourteen valence electrons of Sb^{III} or Bi^{III}. Both salts show normal negative $(\partial\nu/\partial T)_p$ at room temperature, but below their phase transitions (265 K for Sb and 255 K for Bi), $(\partial\nu/\partial T)_p$ is positive. The mechanism of destruction of π-bonding is not likely to be operative here, and the cause of the positive temperature coefficient is uncertain, although the change in unit-cell volume at the phase transition may be involved. In the ^{79}Br n.q.r. spectra of $(NH_4)_3Sb_2Br_9$ and $(MeNH_3)_3Sb_2Br_9$, two widely spaced sets of resonances are found;[29] those at 110—130 MHz are assigned to terminal and those at 65—80 MHz to bridging bromine atoms, in the structure indicated by *X*-rays to be composed of two octahedra of bromine atoms sharing a face. Similar spectra are seen for ^{127}I in $(MeNH_3)_3Sb_2I_9$; since only the $\pm\frac{3}{2} \leftrightarrow \pm\frac{1}{2}$ resonances were found, η could not be determined.

The spectra at different temperatures of ^{35}Cl in $NaClO_3$ (ref. 30), of ^{79}Br in $NaBrO_3$ (ref. 31), and of ^{35}Cl in $Ca(ClO_3)_2,2H_2O$ (refs. 32, 33), $Sr(ClO_3)_2$ (ref. 33), and $Ba(ClO_3)_2$ (ref. 33) have been analysed using the Bayer–Kushida theory. For $NaClO_3$, two lattice vibrations were considered necessary for the analysis. In $Ca(ClO_3)_2,2H_2O$, two resonances at *ca.* 30.02 and 29.64 MHz were observed; one paper [32] gives an intensity ratio 3 : 2, suggesting five chlorine nuclei in a unit cell; the other paper [33] describes the peak at lower frequency as 'much less intense'. The latter workers [33] have also analysed their results, using a modification of Brown's method with a quadratic temperature dependence of the lattice frequencies; results give lattice frequencies comparable with those obtained from Raman spectra but, as in $NaBrO_3$, moments of inertia which differ considerably from those expected from the *X*-ray structure are obtained. No signal was observed in the room-temperature ^{35}Cl n.q.r. spectrum of anhydrous $Ba(ClO_3)_2$, but on cooling to 77 K a resonance at 29.704 MHz was observed.[34] This was much weaker than that at 29.902 MHz for $Ba(ClO_3)_2,H_2O$. When the anhydrous salt is formed by heating the monohydrate to 120 °C, no change in unit cell dimensions occurs; simple point-charge and point-dipole

[28] T. B. Brill and G. G. Long, *J. Phys. Chem.*, 1971, **75**, 1898.
[29] T. B. Brill, P. E. Garrou, and G. G. Long, *J. Inorg. Nuclear Chem.*, 1971, **33**, 3285.
[30] D. B. Utton, *J. Chem. Phys.*, 1971, **54**, 5441.
[31] D. D. Early, R. F. Tipsword, and C. D. Williams, *J. Chem. Phys.*, 1971, **55**, 460.
[32] U. Vignaneswara Kumar and B. D. Nageswara Rao, *Phys. Stat. Sol. (B)*, 1971, **44**, 203.
[33] C. V. RamaMohan and J. Sobhanadri, *Mol. Phys.*, 1971, **22**, 575.
[34] C. V. RamaMohan and J. Sobhanadri, *J. Phys. Soc. Japan*, 1971, **31**, 309.

calculations do not give the correct value of the shift in frequency between the two species, which may be due to a change in the space group of the crystal.

No phase transition is observed between 77 and 398 K in the ^{127}I spectrum of powdered H_5IO_6.[35] The coupling constant e^2Qq shows a maximum of 283 MHz at about 300 K, and η decreases with increasing temperature. The large difference of e^2Qq from that in $(NH_4)_3H_2IO_6$ (25.35 MHz) is attributed to different degrees of covalent bonding.

The ^{35}Cl n.q.r. frequencies at 77 K of 59 compounds containing S—Cl bonds are reported;[36] most though not all of these are organic species. The frequencies are interpreted in terms of the sulphur orbital electronegativities, the inductive effects of substituents, and the possible d_π orbitals on sulphur. The $—SO_2—$ group appears to behave like the pseudo-united atom Ge.

Two studies of molecular complexes of the halogens with aromatic molecules suggest that there is little charge transfer, the bonding being due to van der Waals and classical electrostatic forces. Thus in complexes of ICl with pyridine and γ-picoline, the ^{35}Cl and ^{127}I resonances indicate charges of + 0.26, + 0.35, and − 0.61 on the ring, iodine, and chlorine respectively,[37] compared with charges of + 0.32 and − 0.32 on iodine and chlorine in α-ICl. In the complex of Br_2 with benzene, two ^{81}Br resonances due to chemically inequivalent bromine nuclei are observed.[38] A discontinuity in the spectra at *ca.* 60 K implies a phase transition. The ^1H relaxation time T_1 was also studied.

The structure of $ScBr_3$ and ScI_3 is such that each halogen atom bridges two metal atoms. The low resonance frequencies of ^{79}Br and ^{81}Br in $ScBr_3$, and of ^{127}I in ScI_3, indicate a highly ionic lattice.[39] For ^{127}I, e^2Qq = 359.3 MHz and η = 0.243; these values give θ(Sc—I—Sc) = 94°38′, and the occupation numbers of the iodine valence orbitals: $N(s)$ = 1.975, $N(p_x)$ = 1.831, $N(p_y)$ = 1.856, assuming $N(p_z)$ = 2 (x is along the Sc–Sc direction while y is along the I–I direction).

Crystalline titanium tetrabromide shows two ^{81}Br resonances in the cubic form and four resonances in the monoclinic form.[40] A Zeeman study shows that the asymmetry parameters, bond angles (109—111° for Br—Ti—Br), and ionic and double-bond characters are essentially the same in both forms, indicating that in $TiBr_4$ the crystal field has little effect on the molecular structure and bonding.

The ^{35}Cl and ^{79}Br n.q.r. spectra of several square-planar and tetrahedral Ni^{II} complexes have been investigated to determine the amount of π-bonding occurring in the tetrahedral species.[41] The large frequency

[35] K. V. S. Rama Rao and A. Weiss, *Z. Naturforsch.*, 1971, **26a**, 1813.
[36] R. M. Hart and M. A. Whitehead, *Trans. Faraday Soc.*, 1971, **67**, 3451.
[37] H. C. Fleming and M. W. Hanna, *J. Amer. Chem. Soc.*, 1971, **93**, 5030.
[38] P. K. Kadaba, D. E. O'Reilly, E. M. Peterson, and C. E. Scheie, *J. Chem. Phys.*, 1971, **55**, 5289.
[39] P. A. Edwards and R. G. Barnes, *J. Chem. Phys.*, 1971, **55**, 4664.
[40] T. Okuda, Y. Furukawa, and H. Negita, *Bull. Chem. Soc. Japan*, 1971, **44**, 2083.
[41] P. W. Smith and R. Stoessinger, *Chem. Comm.*, 1971, 279.

shifts from the tetrahedral complexes $(Ph_3P)_2NiCl_2$ and $(Ph_3P)_2NiBr_2$ to the square-planar complexes $(Pr^n{}_3P)_2NiCl_2$, $(Pr^n{}_3P)_2NiBr_2$, and $(Bu^n{}_3P)_2NiBr_2$ indicate about 10% π-bonding in the tetrahedral species; this is the first determination of the degree of π-bonding in metal complexes using n.q.r. spectroscopy.

Two resonances of ^{35}Cl and ^{79}Br in $CuCl_2$,X and $CuBr_2$,X, where X is the *N*-oxide of pyridine, γ-picoline, or 2,6-lutidine, are assigned to chemically equivalent halogen nuclei, thus making an earlier suggestion of intermolecular halogen bridges appear unlikely.[42] The ^{35}Cl spectra of $X_2Cu^{II}Cl_4$, where X = $MeNH_3$, $EtNH_3$, or enH_2, show two resonances, as does that of $Co^{III}(NH_3)_6,Cu^{II}Cl_5$, whereas $(NH_4)_2Cu_2Cl_6$ shows three resonances.[43] All have tetragonally six-co-ordinated copper atoms with long (~ 3 Å) bonds to axial chlorine nuclei (except the cobalt compound, in which $CuCl_5$ has a trigonal-bipyramidal structure); the n.q.r. resonances are related to this bonding, and the long-bond interaction is shown to be *ca.* 7—16% of the bonding within the square-planar units.

Several compounds containing the $ZnX_4{}^{2-}$ anion, where X = Cl, Br, or I, have been investigated.[44] In $Na_2ZnCl_4,3H_2O$, two resonances (1 : 3 intensity) of ^{35}Cl are consistent with the C_{3v} site symmetry determined from *X*-ray data. Similarly, in $(NH_4)_3ZnCl_5$ the three resonances (1 : 1 : 2), and the three (2 : 1 : 1) in Cs_2ZnBr_4, agree with *X*-ray data; the latter compound gives only one line at 77 K, indicating a possible phase change on cooling. Three lines (1 : 2 : 1) are observed in Rb_2ZnBr_4 and in $(pyH)_2ZnBr_4$; in $(pyH)_2ZnCl_4$, however, the existence of two lines (1 : 1) indicates C_{2v} symmetry. A phase change occurs in $(pyH)_2ZnBr_4$ between 77 and 298 K. In $(N_2H_5)_2ZnCl_4$, K_2ZnCl_4, and $(NH_4)_2ZnCl_4$ the numbers of resonances are four, six, and twelve respectively. The value of e^2Qq for $ZnCl_4{}^{2-}$ fits in with the trend of other XCl_4 ions and molecules, showing an increase from $ZnCl_4{}^{2-}$ to $AsCl_4{}^+$.

The temperature variation of the ^{35}Cl n.q.r. frequencies in K_2TcCl_6, K_2RuCl_6, and Cs_2RhCl_6 is reported.[45] For all hexahalides except Cs_2MoCl_6 and Cs_2WCl_6, the temperature dependence of compounds with corresponding second- and third-row transition elements is very similar; the linear relation between $d\nu/dT$ and the number of t_{2g} *d*-electrons holds for both rows.

Two differing views of the temperature dependence of the ^{35}Cl n.q.r. frequency in K_2ReCl_6 have been put forward. Armstrong *et al.*[46] invoke a mechanism involving the destruction of π-bonding at high temperatures, and have measured the pressure variation of the resonant frequency.

[42] J. J. R. Fraústo da Silva, L. F. Vilas Boas, and R. Wootton, *J. Inorg. Nuclear Chem.*, 1971, **33**, 2029.
[43] D. E. Scaife, *Austral. J. Chem.*, 1971, **24**, 1993.
[44] D. E. Scaife, *Austral. J. Chem.*, 1971, **24**, 1315.
[45] J. E. Fergusson and D. E. Scaife, *Inorg. Nuclear Chem. Letters*, 1971, **7**, 987.
[46] R. L. Armstrong, G. L. Baker, and H. M. Van-Driel, *Phys. Rev.* (*B*), 1971, **3**, 3072.

O'Leary[47] puts forward arguments against the π-bonding theory, and contends that the validation of a microscopic model requires more than thermodynamic data.

An approximately linear relationship[48] between both ^{35}Cl n.q.r. frequencies and molar composition in solid solutions of K_2IrCl_6 and K_2PtCl_6 parallels similar relations between the unit-cell dimension a_0 or the asymmetric Pt—Cl and Ir—Cl stretching vibrational frequencies and molar composition. The change in n.q.r. frequency is best accounted for in terms of small changes in the bond lengths; vibrational effects, ionic interactions, and π-bonding effects are all ruled out as explanations.

In a predominantly theoretical paper, Armstrong and Jeffrey[49] show that the quadrupolar spin–lattice relaxation time T_1 in R_2MX_6 compounds, determined from n.q.r. measurements, can be related to the rotary-mode frequencies of the MX_6 anions. Calculated results for $PtCl_6^{2-}$ agree well with i.r. and Raman values; rotary-mode frequencies are determined both by this method[50] and from the temperature and pressure variation of the ^{35}Cl n.q.r. frequency[50, 51] in Rb_2PtCl_6, $CsPtCl_6$, and K_2PtCl_6. The lattice frequencies are given in Table 2.

Table 2 *Rotary lattice mode wavenumbers*/cm^{-1} *in compounds containing the* $PtCl_6^{2-}$ *anion*

	K	Rb	Cs
From pressure and temperature variation of ^{35}Cl frequency	58	62	64
From quadrupolar spin–lattice relaxation time	56	65	77
From i.r. and Raman data	63	71	80

The ^{35}Cl resonances of Zeise's salt, $K[PtCl_3(C_2H_4)]$, were investigated in a study of the ground-state properties of compounds showing the kinetic *trans*-effect.[52] In addition, the compounds with ethylene replaced by *cis*-but-2-ene or styrene, and complexes of $PtCl_2$ with cyclo-octa-1,5-diene and norbornadiene, were also studied. In each mono-olefin complex, a pair of resonances at *ca.* 20 MHz was assigned to the inequivalent *cis*-chlorines, whereas a resonance at *ca.* 16 MHz was assigned to the *trans*-chlorine nucleus. The resonances in the *cis*-$PtCl_2$(diene) complexes are at higher frequencies than those of the corresponding amine or phosphine complexes. The ^{81}Br resonances[53] in $K[PtBr_3(C_2H_4)],H_2O$ and (cyclo-octatetraene) $PtBr_2$ support the deduction that the lattice charge distribution is not responsible for the differences of the *cis*- and *trans*-halogen frequencies, and that the *trans*- is significantly more ionic than the *cis*-halogen atoms.

[47] G. P. O'Leary, *Phys. Rev. (B)*, 1971, **3**, 3075.
[48] J. E. Fergusson and D. E. Scaife, *Austral. J. Chem.*, 1971, **24**, 1325.
[49] R. L. Armstrong and K. R. Jeffrey, *Canad. J. Phys.*, 1971, **49**, 49.
[50] D. F. Cooke, and R. L. Armstrong *Canad. J. Phys.*, 1971, **49**, 2381, 2389.
[51] R. L. Armstrong, *J. Chem. Phys.*, 1971, **54**, 813.
[52] J. P. Yesinowski and T. L. Brown, *Inorg. Chem.*, 1971, **10**, 1097.
[53] J. P. Yesinowski and T. L. Brown, *J. Mol. Structure*, 1971, **9**, 474.

In a study of the nature of the gold–halogen bonds in AuX_4^- anions ($X = Cl$ or Br), and the behaviour of the water of crystallization, the compounds $MAuCl_4$ ($M = K$, NH_4, Rb, or Cs), $NaAuCl_4,2H_2O$, $KAuCl_4,2H_2O$, and $NH_4AuCl_4,\frac{2}{3}H_2O$ were studied.[54] The ionic character of the Au—Cl bond is 41% and that of the Au—Br bond is 34%; these are lower than in the equivalent Pt and Pd anions, as expected from the greater electronegativity of gold. The positive $d\nu/dT$ observed for the lowest-frequency line in the hydrated K and Na salts is attributed to weak Cl···H—O bonds between one chlorine atom and the water of crystallization. The anomalous temperature dependence of one line in $CsAuBr_4$ is presumed to be due to torsional oscillations of the $AuBr_4^-$ ion in the crystal involving considerable anharmonicity.

From the temperature dependence of the ^{35}Cl n.q.r. frequencies in $HgCl_2$, the lattice frequency was estimated at *ca.* 30 cm^{-1}, using Brown's procedure.[55] The discrepancy between this value and lattice frequencies of 73, 95, and 124 cm^{-1} is possibly due to volume effects. A Zeeman study[56] of the same resonance in $HgCl_2$ was undertaken to evaluate η, which had a number of differing values in the literature. The values of $\eta = 0.075$ for the lower-frequency and $\eta = 0.09$ for the higher-frequency resonance were in agreement with one previous study, as were the calculated Cl—Hg—Cl angles. From the spectra of $HgCl_2$, Hg_2Cl_2, $HgCl_3^-$, and several other Hg—Cl compounds, an approximately linear relation between the ^{35}Cl n.q.r. frequency and the Hg—Cl stretching frequency or the Hg—Cl bond length was found.[57] The n.q.r. data confirmed that many of the complex Hg—Cl compounds are built up by joining octahedral $HgCl_6$ units. Predictions were made of the structures of some complex chlorides, and the role of bridging in building up the compounds was discussed in terms of the n.q.r. data for bridging chlorine atoms.

The study of HgX_2 donor–acceptor complexes, where $X = {}^{35}Cl$, ^{81}Br, or ^{127}I, and the donor is one of several organic species, showed that whereas the ^{35}Cl n.q.r. frequencies were lower than those in $HgCl_2$, those of ^{81}Br and ^{127}I were higher than in the pure dihalide.[58] The results are interpreted to suggest structure (9) rather than (10), and mercury is thought not to be tetrahedrally co-ordinated.

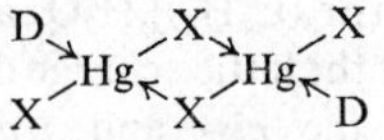

(9)

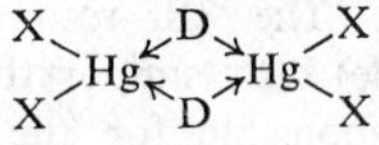

(10)

The compounds Ag_2HgI_4 and Cu_2HgI_4 are thermochromic; the former changes from yellow to red above 323 K and the latter from red to black

[54] A. Sasane, T. Matuo, D. Nakamura, and M. Kubo, *J. Magn. Resonance*, 1971, **4**, 257.
[55] C. V. RamaMohan and J. Sobhanadri, *J. Phys.* (*B*), 1971, **4**, 1393.
[56] G. Chakrapani, G. Satyanandam, and C. R. K. Murty, *J. Phys.* (*C*), 1971, **4**, 1898.
[57] D. E. Scaife, *Austral. J. Chem.*, 1971, **24**, 1753.
[58] T. B. Brill and Z. Z. Hugus, *J. Inorg. Nuclear Chem.*, 1971, **33**, 371.

above 343 K. The cations are ordered below the transition temperature and disordered above it. The temperature variation of the frequency and linewidth of the ^{127}I n.q.r. resonances was studied from 20 K to this temperature.[59] The resonances first narrow with increasing temperature, but then at *ca.* 60 K below the transition temperature the linewidth increases abruptly, causing the disappearance of the signal at *ca.* 40 K below the transition temperature. The increase in linewidth is attributed to thermal diffusion of the Hg^{2+} ions, which are tetrahedrally co-ordinated to iodine at low temperatures, to vacant cation sites, causing 'lifetime broadening'. The ^{63}Cu n.m.r. powder spectrum of Cu_2HgI_4 was also studied.

4 Transition Metals and Lanthanides

Manganese-55.—The effect of ring substitution on the ^{55}Mn n.q.r. frequencies of fourteen monosubstituted and one disubstituted cyclopentadienylmanganese tricarbonyl compounds has been studied.[60] The n.q.r. frequency monitors the variation of π-electron density through interaction of the manganese atom with the π-electrons. Results indicate that n.q.r. is as useful as i.r., u.v., or n.m.r. spectroscopy. The compound $ClHgC_5H_4Mn(CO)_3$ is found to have crystallographically inequivalent Mn atoms but only one ^{35}Cl resonance; a possible chlorine bridge between Hg atoms is proposed.

Cobalt-59.—The temperature variation between 100 and 500 K of the ^{59}Co n.q.r. frequency in a new polytype of $K_3Co(CN)_6$ is reported.[61] This four-layer monoclinic structure has a resonance at 6.06 MHz at 100 K, compared with 7.3 MHz at 4.2 K for a pseudo-orthorhombic twin crystal and 2.7 MHz for a polycrystalline sample. The differences are explained on the basis of a point-dipole model.

Copper-63.—Theoretical results for the solid-echo and echo in local fields in the n.q.r. spectrum of ^{63}Cu agree well with experimental values.[62] The results were obtained with a coherently pulsed spectrometer and are in accordance with those obtained from the conventional spin-echo method.

Lutetium-175.—In a spin-echo experiment, broad resonances at 37, 70, and 110 MHz, closely corresponding to pure n.q.r. frequencies, gave an asymmetry parameter of $\eta = 0.1$ for $^{175}Lu^{3+}$ in lutetium iron garnet.[63]

We thank J. D. Cooper for his generous help in the translation of Russian publications.

[59] P. K. Kadaba and D. E. O'Reilly, *J. Chem. Phys.*, 1971, **55**, 5833.
[60] T. B. Brill and G. G. Long, *Inorg. Chem.*, 1971, **10**, 74.
[61] J. A. J. Lourense and E. C. Reynhardt, *J. Phys. Soc. Japan*, 1971, **30**, 898.
[62] M. A. Korchemkin and Kh. G. Khabibov, *Fiz. tverd. Tela*, 1971, **13**, 1913.
[63] L. D. Khoi and M. Rotter, *Phys. Letters*, 1971, **34A**, 382.

3
Microwave Spectroscopy

BY J. H. CARPENTER

The coverage of topics this year closely follows that of 1970: papers are included which deal with transitions in the radiofrequency, microwave, millimetre-wave, and submillimetre-wave regions of the electromagnetic spectrum, in molecules in the gas phase. This includes molecular-beam electric resonance (m.b.e.r.), molecular-beam magnetic resonance, and molecular-beam masers. Apart from organometallic compounds and a few simple compounds of carbon with oxygen or another hetero-atom, organic species have been excluded.

The information obtained from such studies includes the measurement of bond lengths and angles to a high degree of accuracy; for molecules containing several atoms it is not always possible to obtain the complete structure, but often particular bond lengths or angles, or other structural information, can then be determined. Other structural quantities measured include the dipole moment – including its direction in the molecule and in favourable cases its sign – and coupling constants for quadrupolar nuclei, including sign. All these quantities can be related to theoretical studies of varying degrees of sophistication. In addition, information about molecular force-fields and low-frequency internal motions is often obtained.

As in previous years, the main text is subdivided according to the number of atoms, and in each section the division is by the vertical Group in the Periodic Table of the main atom of interest. In quantitative data, the observed or derived errors are enclosed in parentheses: thus 123.456(78) implies 123.456 ± 0.078.

A review has been published[1] on the Zeeman effect in diamagnetic molecules. The observation of the linear and quadratic Zeeman effects leads to a direct determination of the molecular *g* value, the magnetic-susceptibility anistropy, and the molecular quadrupole moment. Experimentally obtained values of these quantities are tabulated and discussed.

A monograph, 'Dipole Moments in Inorganic Chemistry',[2] which is suitable as an introduction to the subject, discusses techniques of obtaining dipole moments, including m.b.e.r. and microwave spectroscopy. Chapters also deal with the origin of dipole moments and the applications of their measurement in inorganic chemistry.

[1] W. H. Flygare and R. C. Benson, *Mol. Phys.*, 1971, **20**, 225.
[2] G. J. Moody and J. D. R. Thomas, 'Dipole Moments in Inorganic Chemistry', Arnold, London, 1971.

1 Instrumentation and Techniques

A high-resolution bridge spectrometer with superhet detection has been compared with conventional Stark modulated spectrometers and bridge spectrometers without superhet detection.[3] A greater sensitivity under varying conditions than for Stark spectrometers was found.

A cavity spectrometer was used for studies of the rotational spectra of free radicals, including searches for unknown lines.[4] With this instrument, previously observed lines of the OH and NO_2 radicals were observed, in addition to some new lines of NO_2 and some absorptions of the NF_2 radical.

For studies of the relative intensities of microwave lines, the video-detector crystal-current, rather than the microwave power, should be kept constant. An automatic crystal-current leveller for this purpose has been described.[5]

At high temperatures, microwave lines are appreciably Doppler-broadened. To avoid this, a molecular-beam microwave spectrometer has been built;[6] designed to work at 1300—2300 K, the spectrometer utilizes cells with heated sections. A previous paper by the same authors describes four types of microwave absorption cells suitable for high-temperature spectroscopy.[7] Another molecular-beam spectrometer is designed for use with millimetre-waves.[8] The absorption takes place between parallel copper plates, 400 mm long by 100 mm wide; the radiation is fed between them from horns. A half-width of about 5 kHz was obtained for lines of H_2S at 200 GHz.

Double-resonance techniques are being increasingly used in microwave spectroscopy. A spectrometer with the two microwave fields at right-angles has been described, and used in work on propylene oxide.[9] Another paper deals with the technique of radiofrequency–microwave double-resonance;[10] the r.f. field is applied to the electrode of a conventional Stark absorption cell. Various uses for the technique are discussed and illustrated; among the most important advantages are in measurement of the transition frequencies: since there is no Stark voltage, the zero-field frequency is measured. The search for specific transitions and assignment or identification of transitions is also considerably facilitated, and there are many less technical problems than with microwave–microwave double resonance.

The determination of the number of frequency measurements necessary for the microwave identification of a gas in a mixture has been considered.[11]

[3] H. D. Rudolph and D. Schwoch, *Z. angew. Phys.*, 1971, **31**, 197.
[4] L. W. Hrubesh, R. E. Anderson, and E. A. Rinehart, *Rev. Sci. Instr.*, 1971, **42**, 789.
[5] P. Seibt, *Rev. Sci. Instr.*, 1971, **42**, 1895.
[6] J. Hoeft, F. J. Lovas, E. Tiemann, and T. Toerring, *Z. angew. Phys.*, 1971, **31**, 337.
[7] J. Hoeft, F. J. Lovas, E. Tiemann, and T. Toerring, *Z. angew. Phys.*, 1971, **31**, 265.
[8] C. Huiszoon, *Rev. Sci. Instr.*, 1971, **42**, 477.
[9] H. D. Rudolph, H. Dreizler, U. Andresen, *Z. Naturforsch.*, 1971, **26a**, 233; V. Andresen and H. D. Rudolph, *ibid.*, p. 320.
[10] F. J. Wodarczyk and E. B. Wilson, *J. Mol. Spectroscopy*, 1971, **37**, 445.
[11] G. E. Jones and E. T. Beers, *Analyt. Chem.*, 1971, **43**, 656.

Of thirty-three gases taken into account, the number of lines overlapping at 0.2 MHz resolution in the *Q*-band was 45%; if two lines were measured there was still 2% chance of overlap. This indicates that care must be taken when identifying gases from their microwave spectra, and several lines should be measured for positive identification.

2 Diatomic Molecules

In the molecular-beam magnetic-resonance spectrum of HD the electron-coupled spin–spin interaction constant was determined to be 47(7) Hz, its sign being positive.[12] In D_2 the spin–rotational and second-rank tensor interaction constants were determined in the $J = 1$ state.[12] Using calculated values of the magnetic spin–spin interaction constant, the quadrupole coupling, e^2Qq, was $+\ 225.044(24)$ kHz in the $J = 1$ state and $+\ 223.38(18)$ kHz in the $J = 2$ state.

The m.b.e.r. spectrum of CsF has been measured in the $J = 1$ state of the ground state and the first eight excited states.[13] Both the quadrupole coupling constant for ^{133}Cs ($I = 7/2$) and the dipole moment were expanded as power series in $v + \frac{1}{2}$: $e^2Qq/\text{kHz} = 1245.2(1.8) - 16.2(7)(v + \frac{1}{2}) + 0.31(8)(v + \frac{1}{2})^2$, and $\mu/\text{D} = 7.8478(21) + 0.07026(10)(v + \frac{1}{2}) + 0.000195(10)(v + \frac{1}{2})^2$. The value of μ_0 and e^2Qq show good agreement with previous data, but since a greater number of vibrational states were studied, the precision of the coefficients in the series expansion has been improved.

The measurement of the low-lying $J = 0 \rightarrow 1$ transition of AgCl and the $J = 2 \rightarrow 3$ transition of AgBr has resulted in improved values for the halogen quadrupole coupling constants (see Table 1); those for AgBr disagree with previously published results. Temperatures of 650—700 °C were used.[14] Dunham coefficients and equilibrium internuclear distances were obtained for AgBr and also for AgI, whose microwave spectrum had not been previously observed.

By passing a microwave discharge through a quartz tube containing boron trifluoride, BF radicals were formed and the $J = 0 \rightarrow 1$ transitions of ^{10}BF and ^{11}BF at 90—96 GHz were measured.[15] From the B_0 values thus obtained and vibration–rotation constants from optical spectra, r_e was estimated. In addition, the dipole moment and e^2Qq for each boron isotope are reported. Previous *ab initio* calculations of the dipole moment are in good agreement with the observed value, the boron nucleus being at the negative end of the dipole.

Rotational constants and dipole moments have been obtained from the $J = 2 \rightarrow 3$ transition of ^{205}Tl^{79}Br, and in ^{205}Tl^{127}I from the $J = 3 \rightarrow 4$ transition.[16]

[12] R. F. Code and N. F. Ramsey, *Phys. Rev.* (*A*), 1971, **4**, 1945.
[13] H. G. Bennewitz, R. Haerton, O. Klais, and G. Muller, *Chem. Phys. Letters*, 1971, **9**, 19; *Z. Physik*, 1971, **249**, 168.
[14] J. Hoeft, F. J. Lovas, E. Tiemann, and T. Toerring, *Z. Naturforsch.*, 1971, **26a**, 240.
[15] F. J. Lovas and D. R. Johnson, *J. Chem. Phys.*, 1971, **55**, 41.
[16] E. Tiemann, *Z. Naturforsch.*, 1971, **26a**, 1809.

The perturbation of the metastable $a^3\Pi$ ($v = 4$) state of CO by the nearly degenerate $a'^3\Sigma^+$ ($v = 0$) state has been studied in the m.b.e.r. spectrum of the former state.[17] The data obtained are combined with others from optical spectra to obtain the off-diagonal matrix elements of the a–a' interaction as well as the diagonal and off-diagonal elements of the dipole moment operator: $\langle a^3\Pi \| \mu \| a^3\Pi \rangle = 1.3971(3)$ D,

Table 1 *Structural parameters for diatomic molecules*

Molecule	B_0/MHz	r_e/Å	μ_0/D	*Nucleus*	$(e^2Qq)_0$/MHz	*Ref.*
D_2				^2H	0.225044(24)^a	12
CsF			7.8826(21)	^{133}Cs	1.2456(28)^b	13
$Ag^{35}Cl$				^{35}Cl	− 36.50(10)	14
$^{107}Ag^{79}Br$	1943.6420(50)	2.393100(29)		^{79}Br	297.10(15)	14
$^{107}Ag^{127}I$	1345.1105(25)	2.544611(31)		^{127}I	− 1062.17(40)	14
^{10}BF	48022.63(8)			^{10}B	− 9.5(8)	15
^{11}BF	45185.77(6)	1.2625	0.5(2)	^{11}B	− 4.5(4)	15
$^{205}Tl^{79}Br$	1291.9213(30)		4.493(50)			16
$^{205}Tl^{127}I$	813.4699(13)		4.607(70)			16
^{13}CO			1.3747(3)^c			18
CS			0.630(15)^d			19
			1.70(15)^e			19
			1.6(6)^f			19
PN			2.7468	^{14}N	− 5.1728^b	20
$D^{19}F$	325584.98(30)	0.916914				24
$H^{35}Cl$	312989.297(20)	1.2745991		^{35}Cl	− 67.800(95)	24
$H^{37}Cl$	312519.121(20)	1.2745990		^{37}Cl	− 53.436(95)	24
$D^{35}Cl$	161656.238(14)	1.2745990		^{35}Cl	− 67.417(98)	24
$D^{37}Cl$	161183.122(16)	1.2745988		^{37}Cl	− 53.073(113)	24
$H^{79}Br$	250358.51(15)	1.4144691		^{79}Br	535.4(1.4)	24
$H^{81}Br$	250280.58(15)	1.4144705		^{81}Br	447.9(1.4)	24
$D^{79}Br$	127357.639(12)	1.4144698		^{79}Br	530.648(74)	24
$D^{81}Br$	127279.757(17)	1.4144698		^{81}Br	443.363(105)	24
$H^{127}I$	192657.577(19)	1.609018		^{127}I	− 1828.418(200)	24
$D^{127}I$	97537.092(9)	1.609018		^{127}I	− 1823.374(105)	24

^{a} $J = 1$ state. ^{b} Equilibrium value. ^{c} $a^3\Pi$ ($v = 0$) state. ^{d} $A^1\Pi$ ($v = 0$) state. ^{e} $a'^3\Sigma^+$ ($v = 10$) state. ^{f} $k^3\Pi$ ($v = 4$) state.

$\langle a^3\Pi \| \mu \| a'^3\Sigma^+ \rangle = 0.62(2)$ D, the latter being the transition moment of the ($a'^3\Sigma^+$—$a^3\Pi$) Asundi system. Another analysis of the m.b.e.r. spectrum of the metastable $a^3\Pi$ state of ^{13}CO in its four lowest vibrational states [18] yielded the parameters describing the lowest-order magnetic hyperfine interactions of the ^{13}C spin ($I = \frac{1}{2}$) with the spin (**S**) and orbital (**L**) angular momenta of the two unpaired electrons of the $5\sigma\, 2\pi$ open-shell configuration. The **I**.**S** Fermi contact parameter is 1935.1(5) MHz, the **I**.**S** spin–spin dipole-interaction parameters are 106.5(1) MHz (off-diagonal in Λ) and 7.9(1) MHz (diagonal in Λ), and the **I**.**L** spin–orbital parameter is

[17] R. H. Gammon, R. C. Stern, and W. Klemperer, *J. Chem. Phys.*, 1971, **54**, 2151.
[18] R. H. Gammon, R. C. Stern, M. E. Lesk, B. G. Wicke, and W. Klemperer, *J. Chem. Phys.*, 1971, **54**, 2136.

162.5(1) MHz; these are in good agreement with values from restricted Hartree–Fock calculations. The dipole moments for the $v = 0$—3 states show a small but significant drift from the corresponding values for ^{12}CO; this is attributed either to the influence of the ^{13}C hyperfine structure on the $a^3\Pi$ charge distribution or to the breakdown of the Born–Oppenheimer approximation.

By using radiofrequency–optical double-resonance, Field and Bergeman [19] have been able to analyse the perturbations between the $A^1\Pi$ $(v = 0)$ state of CS and the triplet manifold. Dipole moments are obtained for the following states: $A^1\Pi$ $(v = 0)$, $\mu = 0.630(15)$ D, $a'^3\Sigma^+$ $(v = 10)$, $\mu = 1.70(15)$ D, and $k^3\Pi$ $(v = 4)$, $\mu = 1.6(6)$ D. Values calculated using CI show good agreement with observed values, in contrast to those obtained by SCF methods, and support the choice of sign of $^-$C—S$^+$ for the $A^1\Pi$ state.

The values of r_e and ω_e for $^{31}P^{14}N$ are known from optical spectra. A m.b.e.r. study [20] of this species in the $J = 1$ level of the three lowest vibrational states yielded the dipole moment $\mu/D = 2.7511 - 0.0086(v + \frac{1}{2})$, with a calculated value of 3.237 D for $^+$P—N$^-$. From the ^{14}N quadrupole coupling constant: $e^2Qq/\text{MHz} = -5.1728 + 0.0607(v + \frac{1}{2})$, and a calculated values for the field gradient of $eq = -4.839 \times 10^{15}$ esu cm^{-3}, a quadrupole moment for ^{14}N of 1.47×10^{26} cm^2 is obtained, in reasonable agreement with the average of 1.66×10^{26} cm^2 obtained from various other species. The magnetic hyperfine interaction constants C_P and C_N were also obtained.

The use of microwave spectroscopy in the identification of transient species in reacting systems is shown in a paper by Saito and Wentrup.[21] In the pyrolysis of *N*-sulphinylaniline, PhNSO, the $J,K = 2,1 \leftarrow 1,0$ transition of SO in the $^3\Sigma^-$ state was detected. The peak intensity increased with temperature between 650 and 1000 °C, as did that of the $31_{5,27} \leftarrow 30_{6,24}$ peak of SO_2 also observed in the mixtures. These results support the suggestion that the 1-cyanocyclopentadiene obtained from the pyrolysis is derived from phenylnitrene.

In a paper on the gas-phase e.s.r. spectra of IO, SF, SeF, and SeO ($^1\Delta$ state), the dipole moments obtained are compared with those derived from microwave and molecular-beam spectroscopy.[22] The good agreement between the e.s.r. and the inherently more accurate zero-field techniques is noteworthy.

By using submillimetre-wave spectroscopy on a number of transitions in the hydrogen halides, DeLucia *et al.*[23] have been able to obtain more accurate values for rotational and other constants. Using the observed

[19] R. W. Field and T. H. Bergeman, *J. Chem. Phys.*, 1971, **54**, 2936.
[20] J. Raymonda and W. Klemperer, *J. Chem. Phys.*, 1971, **55**, 232.
[21] S. Saito and C. Wentrup, *Helv. Chim. Acta*, 1971, **54**, 273.
[22] C. R. Byfleet, A. Carrington, and D. K. Russell, *Mol. Phys.*, 1971, **20**, 271.
[23] F. C. DeLucia, P. Helminger, and W. Gordy, *Phys. Rev. (A)*, 1971, **3**, 1849.

correction for vibration–rotation interaction and the Dunham correction, and calculated values for the correction for non-spherical electron distribution and the wobble–stretch correction, equilibrium internuclear distances are calculated; values of r_e for the different isotopic species of HCl, HBr, and HI all lie within 10^{-6} Å of each other, in contrast to results obtained by Kaiser,[24] in which r_e for HCl and DCl differed by 5×10^{-5} Å. Values of the spectral constants quoted in Table 1 of this chapter include those listed by DeLucia *et al.*, which are based on earlier measurements. All of the hydrogen halides have now been measured by high-resolution microwave spectroscopy.

3 Triatomic Molecules

By combining i.r. data with new results from microwave work, Winnewisser *et al.*[25] obtained rotation–vibration interaction constants for each vibrational state of HCN and DCN and hence an equilibrium (r_e) structure: r_e(C—H) = 1.06549(24) Å, r_e(C—N) = 1.15321(5) Å, and r_e(N—H) = 2.21870(20) Å. These differ slightly from previous values because of a change in the measured value of α_2. For the analogous HCP and DCP, the equilibrium B_e value is known only for the normal isotopic species; however, a study of the deuteriated molecule in the ground and first-excited bending vibrational states [26] gave values of $\alpha_2 = -32.85(4)$ MHz, and the *l*-type doubling constant $q_l = 44.86(7)$ MHz; only α_1 is now required to evaluate B_e for DCP.

An early microwave study had found rotational lines for SCSe, but a reinvestigation of this compound,[27] in which only lines which bear no relation to those in the earlier work were observed, suggests that the previous study was in error. Transitions were observed in the first and second excited bending vibrational states; these states show first-order Stark effect and hence were more readily observable than the ground-state spectrum, since the dipole moment is low and was determined as $\mu_0 = 0.031(5)$ D. By extrapolation the B_0 values were obtained, giving r_0(C—S) = 1.5530(15) Å and, r_0(C—Se)= 1. 6950(15) Å, these distances being shorter than in OCS and OC Se respectively. From the vibrational frequencies and the *l*-type doubling here observed, the quadratic force field was determined: f(C—Se) = 5.72, f(C—S) = 7.97, f(C—S,C—Se) = 0.591, and $f_\alpha/r_1r_2 = 0.20$, all in mdyn Å^{-1}.

The microwave spectrum of GeF_2 has been observed for the first time.[28] The spectra of molecules containing each of the four zero-spin germanium isotopes (^{70}Ge, ^{72}Ge, ^{74}Ge, and ^{76}Ge) in the ground state and various

[24] E. W. Kaiser, *J. Chem. Phys.*, 1970, **53**, 1686.
[25] G. Winnewisser, A. G. Maki, and D. R. Johnson, *J. Mol. Spectroscopy*, 1971, **39**, 149.
[26] J. W. C. Johns, J. M. R. Stone, and G. Winnewisser, *J. Mol. Spectroscopy*, 1971, **38**, 437.
[27] C. Hirose and R. F. Curl, *J. Chem. Phys.*, 1971, **55**, 5120.
[28] H. Takeo, R. F. Curl, and P. W. Wilson, *J. Mol. Spectroscopy*, 1971, **38**, 464.

excited vibrational states were analysed to give B_e values and an equilibrium structure: r_e(Ge—F) = 1.7321 Å, and θ_e(F—Ge—F) = 97°10′. The α constants and inertia defects were obtained; from the latter, together with vibrational frequencies, the quadratic force field was evaluated: $f_r = 4.08$, $f_\alpha/r_e^2 = 0.316$, $f_{rr} = 0.26$, $f_{r\alpha}/r_e = -0.01$, all in mdyn Å^{-1}. The α constants yielded values of the cubic potential constants: $f_{rrr} = -22.6$, $f_{\alpha\alpha\alpha}/r_e^3 = -0.4$, $f_{rrr'} = -0.3$, $f_{rr\alpha}/r_e = -0.0$, $f_{r\alpha\alpha}/r_e^2 = -0.3$, $f_{rr'\alpha}/r_e = -0.2$, all in mdyn Å^{-2}; the cubic constants are defined as the derivatives of the potential energy with respect to the relevant internal co-ordinates. The dipole moment of the molecule is 2.61(2) D, which fits in with the trend of increasing dipole for the MF_2 species on going down Group IV (see Table 2). This is expected on electronegativity considerations. From Table 2 it

Table 2 *Structural parameters of Group IV difluorides and related molecules*

Molecule	*Parameter*	C	Si	Ge	Sn
MF_2	μ_0/D	0.46	1.23	2.61	
MF_2	θ(F—M—F)/deg	104.9	100.9	97.2	
MF	r(M—F)/Å	1.27	1.60		
MF_2	r(M—F)/Å	1.30	1.60	1.73	
MF_4	r(M—F)/Å	1.32	1.56	1.67	
MH_3Cl	μ_0/D	1.87	1.30	2.10[a]	
MH_3Br	μ_0/D	1.80	1.32	1.98[a]	
MH_3I	μ_0/D	1.65		1.80[a]	
MH_3Cl	$e^2Qq(^{35}Cl)$/MHz	− 74.77	− 40.0	− 46	− 41.62
MH_3Br	$e^2Qq(^{79}Br)$/MHz	577.15	336	380	350
MH_3I	$e^2Qq(^{127}I)$/MHz	−1934	− 1240		− 1273

[a] Mean of dielectric-constant and nonresonant-microwave-absorption values.

can also be seen that the bond angle decreases on descending the Group; such behaviour is also seen in H_2O, H_2S, and H_2Se. The bond length in MF_2 is similar to that in MF for C and Si; the MF_4 bond length progressively decreases on descending the Group.

Two studies have been made of the unstable species NSCl. The structure, ^{35}Cl quadrupole coupling constants, and force field were determined[29] from four isotopic species. The substitution structure is r_s(N—S) = 1.450 Å, r_s(S—Cl) = 2.161 Å, and θ_s(N—S—Cl) = 117°42′. The N—S bond length and the N—S—X angle are identical to those in NSF. The quadrupole coupling constants for ^{35}Cl are: χ_{zz} (along S—Cl bond) = − 43.36 MHz, χ_{xx} (out of plane) = 15.00 MHz, and χ_{yy} = 28.36 MHz. The value of χ_{zz} is lower than is usual for singly bonded chlorine, and using the Townes–Dailey theory and some other assumptions, the structure was found to be 35.5% N=S—Cl, 7.1% −N=S=Cl$^+$, and 57.4% N=S$^+$Cl$^-$. This correlates with the S—Cl bond length, which is greater than the sum of single bond covalent radii, owing to its appreciable ionic character. The dipole moments are found to be:

[29] T. Beppu, E. Hirota, and Y. Morino, *J. Mol. Spectroscopy*, 1970, **36**, 386.

$\mu_a = 0.57(3)$ D, $\mu_b = 1.77(2)$ D, and $\mu = 1.87(2)$ D, these being obtained by Stark-effect measurements on the $1_{01} \leftarrow 0_{00}$ transition.[30]

Numerous transitions in the millimetre-wave spectrum of O_3 have been observed for values of J up to 40 in the 50—320 GHz region.[31] The data have been fitted to a semi-rigid rotor using quartic and sextic centrifugal terms.

Two studies of HDO have analysed the data using terms up to the tenth power of the angular momentum. Several transitions were measured between 8 and 52 GHz.[32] In the other study, transitions were observed in the millimetre-wave and submillimetre-wave regions;[33] the rotational constants were determined with greater precision than in the previous paper; the rotational constant A differed significantly in the two analyses but the other constants were in agreement. The latter paper analysed 55 high-resolution transitions using 22 adjustable parameters. Six microwave transitions were observed in the $v_2 = 1$ state;[32] these were combined with i.r. data to give rotational constants. The constants A and τ_{aaaa} have considerably different values than in the ground state, but the other parameters are less affected. Two transitions in $H_2{}^{17}O$ and one in $H_2{}^{18}O$ have also been observed in the microwave and millimetre-wave regions;[34] these were also combined with i.r. data to give rotational constants.

The observation of 27 new transitions up to 628 GHz in the rotational spectrum of HDS has enabled the rotational constants A, B, and C to be determined from microwave, millimetre-wave, and submillimetre-wave data alone.[35] All the quartic and sextic centrifugal constants were also determined. The study of the $1_{01} \rightarrow 1_{10}$ transition of D_2S at 91.4 GHz using a molecular-beam maser has enabled the complete quadrupole coupling tensor for ^2H to be obtained.[36] In the principal axis system of the quadrupole coupling tensor, the elements are: $\chi_{zz} = 149.0$ kHz, $\chi_{yy} = -59.8$ kHz, and $\chi_{xx} = -89.7$ kHz. The z-axis makes an angle of 1°35′ with the D—S bond (the z-axes for the two deuterium atoms thus meeting at 95°10′ compared with the bond angle of 92°6′), and the x-axis is out of plane.

A microwave study of HOCl has furnished more information than was previously obtained.[37] Values of B and C for four isotopic species, as well as a rough estimate of A for $D^{16}O^{35}Cl$, were combined with i.r. data to give the structure: r(O—Cl) = 1.6895(35) Å, θ(H—O—Cl) = 102°29(27)′. The value of r(O—H) depended on whether I_a^0 and I_b^0 used [r(O—H) =

[30] A. Guarnieri, *Z. Naturforsch.*, 1971, **26a**, 1246.
[31] M. Lichtenstein, J. J. Gallagher, and S. A. Clough, *J. Mol. Spectroscopy*, 1971, **40**, 10.
[32] W. Lafferty, J. Bellet, and G. Steenbeckeliers, *Compt. rend.*, 1971, **273**, *B*, 388.
[33] F. C. de Lucia, R. L. Cook, P. Helminger, and W. Gordy, *J. Chem. Phys.*, 1971, **55**, 5334.
[34] G. Steenbeckeliers and J. Bellet, *Compt. rend.*, 1971, **273**, *B*, 471.
[35] P. Helminger, R. L. Cook, and F. C. de Lucia, *J. Mol. Spectroscopy*, 1971, **40**, 125.
[36] F. C. de Lucia and J. W. Cederberg, *J. Mol. Spectroscopy*, 1971, **40**, 52
[37] A. M. Mirri, F. Scappini, and G. Cazzoli, *J. Mol. Spectroscopy*. 1971, **38**, 218.

0.959(5) Å] or I_b^0 and I_c^0 [0.975(3) Å]. The principal z-axis of the quadrupole coupling tensor was along the O—Cl bond, and the elements were $\chi_{zz} = -121.86(28)$ MHz and χ_{xx} (out of plane) = 59.56(25) MHz. From the asymmetry, the bond is calculated to have 1.6(5)% double-bond character. The high value of χ_{zz} suggests some contribution from $H—\overset{-}{O}\overset{+}{Cl}$, but dipole-moment studies suggest ionization in the opposite, direction.[38] For $HO^{35}Cl$, $\mu_a = 0.367(8)$ D, while for $DO^{35}Cl$, $\mu_a =$ 0.412(15) D. On the assumption that the positive end of the dipole is in the region of the H atom and that it has the same value in the two species, the dipole moment is calculated as 1.3(3) D at an angle of 73(3)° to the a-axis (which is approximately along the O—Cl bond). The μ_b component of the dipole is due only to the O—H group and is similar to that found in MeOH, PhOH, and HNO_3.

4 Tetra-atomic Molecules

The CH_2O molecule has been studied using a molecular-beam maser. In a field of 17 kG the magnetic-susceptibility anisotropy and the molecular quadrupole moment were obtained.[39] From these the anisotropy of the electronic charge distribution was calculated: $\langle a^2 \rangle - \langle b^2 \rangle = 6.19(3)$ Å^2, $\langle b^2 \rangle - \langle c^2 \rangle = 1.72(4)$ Å^2, and $\langle c^2 \rangle - \langle a^2 \rangle = -7.91(5)$ Å^2. In a further experiment with greater accuracy,[40] the hyperfine structure was analysed to give spin–rotational constants in the 1_{10} and 1_{11} states; the spin–spin-interaction strength gives a value for $\mu_n^2 g^2/r^3 = 17.73(12)$ kHz, compared with a calculated value of 17.55(5) kHz, the difference being attributed to a contribution from the electron-coupled spin–spin interaction. The diagonal elements of the spin–rotation tensor were obtained. Another beam-maser study of the $1_{01} \rightarrow 0_{00}$ transition in HCOF and DCOF indicated that the spin–vibration interaction constant C_H was 2 kHz or less.[41] From the value of (e^2Qq) for ^2H (along the D—C bond) of 205.2(4.0) kHz, the field gradient at the D nucleus was calculated as 4.7 mdyn Å^{-1}; this is close to the value of the D—C stretching force-constant, $k = 4.8$ mdyn Å^{-1}. Such agreement has also been observed for DHCO.

The molecule H_2CS is unstable, having a half-life of about six minutes in a copper waveguide. A flow system was used to observe transitions in H_2CS formed by pyrolysis of MeSSMe.[42] The dipole moment, $\mu =$ 1.6474(14) D, and a substitution structure, r_s(S—C) = 1.6108(9) Å, r_s(C—H) = 1.0925(9) Å, and θ_s(HCH) = 116.87(5)°, were derived.

Transitions in the ground state and the fundamentals of both bending vibrational states of linear HCNO were observed[43] in the millimetre-wave

[38] D. G. Lister and D. J. Millen, *Trans. Faraday Soc.*, 1971, **67**, 601.
[39] S. G. Kukolich, *J. Chem. Phys.*, 1971, **54**, 8.
[40] S. G. Kukolich and D. J. Ruben, *J. Mol. Spectroscopy*, 1971, **38**, 130.
[41] S. G. Kukolich, *J. Chem. Phys.*, 1971, **55**, 610.
[42] D. R. Johnson, F. X. Powell, and W. H. Kirchhoff, *J. Mol. Spectroscopy*, 1971, **39**, 136.
[43] M. Winnewisser and B. P. Winnewisser, *Z. Naturforsch.*, 1971, **26a**, 128.

region, for J up to 17. An analysis gave the rotational constants B_v and the centrifugal distortion constants D_v, as well as the l-type doubling constants q_v. A term showing centrifugal effects on q had to be included to fit the data. The isomeric but non-linear HNCO was studied in a molecular-beam maser.[44] From the hyperfine structure of the $1_{01} \rightarrow 0_{00}$ transitions of the normal and deuteriated species, the spin–rotation constants as well as the quadrupole-coupling constants for ^2H and ^{14}N were calculated. In HNCO, $(e^2Qq)_N$ = 2123.0(1.0) kHz, whereas in DNCO it is 2052.7(1.0)kHz. To determine completely the quadrupole coupling tensor for ^{14}N, a measurement on a different rotational state is required. If the deuteron quadrupole coupling tensor is assumed to be cylindrically symmetrical about the D—N bond, then the coupling along the bond is 345(2) kHz. This gives the field gradient eq = 8.26 mdyn Å^{-1}, which is somewhat larger than the corresponding force constant, k = 6.8 mdyn Å^{-1}, in contrast to DCOF.[41]

Two studies [45, 46] at high frequencies have been made of the rotational transitions in ND_3, and the values of B_0, D_J, and D_{JK} obtained are in close agreement. Duxbury and Jones [45] used a Stark field to shift the $J = 4 \rightarrow 5$ transition in $^{14}ND_3$ to coincide with the 1539.7 GHz emission line at a DCN maser. The absorptions observed are forbidden at zero field but become allowed in applied static electric fields. Helminger *et al.*[46] have measured the $J = 1 \rightarrow 2$, $K = 0$ and 1 transitions in $^{14}NH_3$ and $^{15}NH_3$ at *ca.* 615 GHz, and obtained B_0, D_J, and D_{JK} from these and i.r. data. From measurements of the $J = 0 \rightarrow 1$ transitions at *ca.* 572 GHz in $^{14}NH_3$ and $^{15}NH_3$, B_0 values have been derived for these species; the substitution structures obtained from any three isotopic species give r_s(N—H) = 1.0136—1.0138 Å, θ_s(H—N—H) = 107°04′—107°14′; previous substitution structures were based on an erroneous value of B_0 for $^{15}NH_3$.

From Stark-effect measurements on the $1_{01} \leftarrow 0_{00}$ transition of $NH_2{}^{35}Cl$, a value of 1.22(3) D is obtained for the component of the dipole moment along the a-axis.[38]

Although HNO_2 is present at a concentration of less than 1% in a mixture of NO, NO_2, and H_2O, microwave transitions for both *cis*- and *trans*-isomers were observed for eight isotopic species.[47] The substitution structure parameters, dipole moment, and quadrupole coupling tensor components are given in Table 3. The inertia defects indicate that both species are planar. In the *trans*-isomer the dipole moment is approximately along the N—O bond, whereas in the *cis*-isomer it bisects the O—N—O angle. The increased O—N—O angle in the *cis*- relative to the *trans*-isomer corresponds to that in HNO_3. The increase in O—H and N=O bond lengths in the *cis*-form correlates with the stretching-vibration frequencies

[44] S. G. Kukolich, A. C. Nelson, and B. S. Yamanashi, *J. Amer. Chem. Soc.*, 1971, **93**, 6769.
[45] G. Duxbury and R. G. Jones, *Chem. Phys. Letters*, 1971, **8**, 439.
[46] P. Helminger, F. C. de Lucia, and W. Gordy, *J. Mol. Spectroscopy*, 1971, **39**, 94.
[47] A. P. Cox, A. H. Brittain, and D. J. Finnigan, *Trans. Faraday Soc.*, 1971, **67**, 2179.

and force constants. The O—H bond is unusually long and possibly indicates a weak H···Cl interaction. The quadrupole coupling constants for ^{14}N suggest that the main structure is H—O—N=O, although the H^{+} O=N—O^{-} structure is more important in the *cis*-isomer; this is to be expected since it can be stabilized by electrostatic attraction, a view which is supported by charge distributions calculated by MO theory.

Table 3 *Structural parameters of nitrous acid*

Parameter	cis	trans
r_s(O—H)/Å	0.982	0.958
r_s(N—O)/Å	1.392	1.432
r_s(N=O)/Å	1.185	1.170
θ_s(N—O—H)	104.0°	102.1°
θ_s(O—N—O)	113.6°	110.7°
μ_a/D	0.306	1.378
μ_b/D	1.389	1.242
μ/D	1.423(5)	1.855(16)
χ_{aa}/MHz	2.05	1.73
χ_{bb}/MHz	− 5.83	− 5.28
χ_{cc}/MHz	3.78	3.55

The m.b.e.r. spectrum of PH_3 was studied to obtain hyperfine constants and to measure the possible inversion doubling.[48] The components of the spin–rotation interaction constants were found for ^{31}P and ^{1}H, and the dipole moment is 0.57397(20) D. The former quantities were used to obtain the shielding of ^{31}P: $\sigma^{sr} = -366.43$ p.p.m. This combines with a diamagnetic shielding of + 960.834 p.p.m. to give a total shielding of 594.4 p.p.m. A similar calculation for ^{1}H gives σ (total) = 28.28 p.p.m. No effect due to inversion doubling was detected; this must thus be less than 1 kHz, which agrees with calculated barriers in excess of 27 kcal mol^{-1}.

The direct *l*-type doubling transitions of PF_3 for J = 17—24 lie between 8 and 18 GHz. Analysis of their frequencies [49] gave a value of the *l*-type doubling constant, $q_4 = 29.492651(81)$ MHz, while $\zeta_4 = -0.62971(18)$ and the average rotational constant $\bar{C} = 4784.16(25)$ MHz. Two cubic constants were determined: $k_{444} = -14.40$ cm^{-1} and $k_{443} = -26.98$ cm^{-1}.

The structure of PBr_3 was determined from electron-diffraction and microwave data.[50] The gas-phase electron-diffraction data give r_g(P—Br) = 2.2204(30) Å, r_g(Br—Br) = 3.424(6) Å, θ(Br—P—Br) = 101.0(4)°, as well as vibrational amplitudes. B_0 values were determined from the microwave spectra of $P^{79}Br_3$ and $P^{81}Br_3$. Average values r_z and θ_z were found to be related by $\theta_z/\text{deg} = 100.99 - 8.65(r_z/\text{Å} - 2.215)$. From the electron-diffraction value of r^0_α, r_z is estimated to be 2.2155(40) Å, giving

48 P. B. Davies, R. M. Neumann, S. C. Wofsy, and W. Klemperer, *J. Chem. Phys.*, 1971, **55**, 3564.
49 E. Hirota, *J. Mol. Spectroscopy*, 1971, **37**, 20; 1971, **38**, 195.
50 K. Kuchitsu, T. Shibata, A. Yokozeki, and C. Matsumara, *Inorg. Chem.*, 1971, **10**, 2584.

θ_z = 100.95(35)°, which is virtually identical to the θ_α value from electron diffraction. This indicates that θ_α, which is averaged over all vibrational states, is virtually independent of temperature, since θ_z is the average for the ground vibrational state. The observed bond angle is the largest in the phosphorus trihalides.

The observation of the $J = 0 \rightarrow 1$, $1 \rightarrow 2$, and $2 \rightarrow 3$ transitions in AsH_3 and SbH_3 produced an improvement in the values of B_0, and centrifugal, hyperfine, and quadrupole constants could be evaluated.[51] For $^{75}AsH_3$, $e^2Qq = -162.63(3)$ MHz and for $^{75}AsD_3$ $e^2Qq = -164.75(3)$ MHz, whereas for $^{121}SbH_3$, $^{123}SbH_3$, $^{121}SbD_3$, and $^{123}SbD_3$ the values of e^2Qq are 460.31(10), 586.65(11), 465.32(10), and 593.06(11) MHz respectively. In all the species D_{JK} is negative. For SbH_3, r_0 and r_s structures were calculated; r_s(Sb—H) = 1.7032—1.7039 Å, θ_s(H—Sb—H) = 91°29′—91°35′.

The dipole moments of GeH_3X compounds, where X = Cl, Br, or I, have been measured both by dielectric-constant methods and by non-resonant microwave absorption.[52] The values obtained agree to within 3%. The dipole moments of the silyl derivatives are less than those of the methyl compounds, as shown in Table 2, which is not expected on electronegativity considerations, and has been attributed to ($p \rightarrow 3d$) π-bonding between the halogen and silicon. It can be seen from Table 2 that the dipole moments of the germanium compounds do not show similar behaviour to the silyl species, but have dipole moments greater than the methyl compounds in accordance with the lower electronegativity of germanium; for each halide, $\mu_{Ge}/\mu_C \approx 1.13$. Thus either ($p \rightarrow d$) π-bonding does not occur in the germyl halides to any extent, or the atypical behaviour of the silyl compounds is due to other factors.

By using a flow system employing continuous pumping, Krisher *et al.*[53, 54] have obtained the microwave spectrum of the monohalogenostannanes SnH_3X, where X = Cl, Br, or I. Structural parameters obtained are included in Table 4. For SnH_3Cl the bond angle was assumed to be

Table 4 *Structural parameters for the halogenostannanes* SnH_3X

Parameter	Cl	Br	I
r(Sn—X)/Å	2.327(1)[a]	2.4691(3)[a]	2.674(2)
r(Sn—H)/Å	1.70	1.76(7)	
θ(H—Sn—X)	109.5°[b]	105.9(9)°	
e^2Qq/MHz	− 41.62(30)[c]	350(6)[d]	− 1273(8)[e]

[a] Substitution value. [b] Assumed. [c] For ^{35}Cl. [d] For ^{79}Br. [e] For ^{127}I.

[51] P. Helminger, E. L. Beeson, and W. Gordy, *Phys. Rev.* (*A*), 1971, **3**, 122.
[52] J. M. Bellema, S. O. Wandiga, and A. A. Maryott, *Inorg. Nuclear Chem. Letters*, 1971, **7**, 71.
[53] L. C. Krisher, R. A. Gsell, and J. M. Bellema, *J. Chem. Phys.*, 1971, **54**, 2287.
[54] S. N. Wolf, L. C. Krisher, and R. A. Gsell, *J. Chem. Phys.*, 1971, **54**, 4605; 1971, **55**, 2106.

tetrahedral to obtain the Sn—H bond length, whereas in SnH_3I the projection of the Sn—H bonds on to the figure axis was assumed to be equal to that in SnH_3Br to obtain the Sn—I bond length. If the quadrupole coupling constants in the Group IV MH_3X compounds are compared (see Table 2), it is found that for each halogen nuclide $|e^2Qq|$ decreases in the order C $\gg$ Ge $>$ Sn $\gtrsim$ Si. The anomalous behaviour of the silicon compounds parallels that of the dipole moment noted above. The microwave spectra of SO_2ClF for both ^{35}Cl and ^{37}Cl indicate that both molecules are near-prolate rotors.[55] The rotational constants were determined and, by making the assumptions that r(S—O) is between the values found for SO_2F_2 and SO_2Cl_2, and that θ(F—S—Cl) also lies between the values of the corresponding angles in those two molecules, a structure was obtained for SO_2ClF. The bond angles are similar to those on SO_2Cl_2 and SO_2F_2, but the S—F bond length is slightly longer than in SO_2F_2 and S—Cl bond length is shorter than in SO_2Cl_2. This correlates with the stretching-vibrational frequencies and is attributed to the greater positive charge of S in SO_2ClF compared with SO_2Cl_2 enhancing the $(p \rightarrow d)$ π-bonding between S and Cl, though the effect is slight since the asymmetry parameter measured in this work [η = 0.065(6) for ^{35}Cl] is small; the value of χ_{aa} (which is roughly along the S—Cl bond) is slightly smaller than the solid-state value.

5 Molecules Containing Six or More Atoms

Ten isotopic species of planar HC:C·BF_2 have been investigated,[56] and an analysis of the spectra, including quartic centrifugal terms, has been carried out. The preferred substitution structure is: r_s(C—H) = 1.058(3) Å, r_s(C—C) = 1.206(3) Å, r_s(C—B) = 1.513(5) Å, r_s(B—F) = 1.323(5) Å, and θ_s(F—B—F) = 116.5(1.0)°; the molecule has C_{2v} symmetry. The C—C and C—H bond lengths are similar to those in most other ethynyl compounds, but the B—C bond length is the shortest known to date.

The microwave spectra of three isotopic species of Me_3N,BF_3, one of the few donor–acceptor complexes between boron and nitrogen to be studied by this technique, have yielded the B—N bond distance.[57] The bond length, r_s(B—N) = 1.636(4) Å, measured here is somewhat greater than that determined from *X*-ray analysis in the solid state, r(B—N) = 1.585(30) Å, though it fits in well with the trend observed in the other halide complexes in the solid state: Me_3N,BCl_3, r(B—N) = 1.610(6) Å, Me_3N,BBr_3, r(B—N) = 1.603(20) Å, and Me_3N,BI_3, r(B—N) = 1.584(25) Å. The length in Me_3N,BH_3, determined by microwave spectroscopy, is r(B—N) = 1.65(2) Å; the similarity of the B—N bond length

55 C. W. Holt and M. C. L. Gerry, *Chem. Phys. Letters*, 1971, **9**, 621.
56 W. J. Lafferty and J. J. Ritter, *J. Mol. Spectroscopy*, 1971, **38**, 181.
57 P. S. Bryan and R. L. Kuczkowski, *Inorg. Chem.*, 1971, **10**, 200.

to that of Me_3N,BF_3 is as expected from the similar acceptor strengths of BF_3 and BH_3 towards Me_3N.

The molecule HF_2P,BH_3 is shown to have the staggered conformation in the gas phase.[58] The structure, obtained from six isotopic species, is shown in Figure 1. The dipole moment is 2.504(30) D, making an angle of

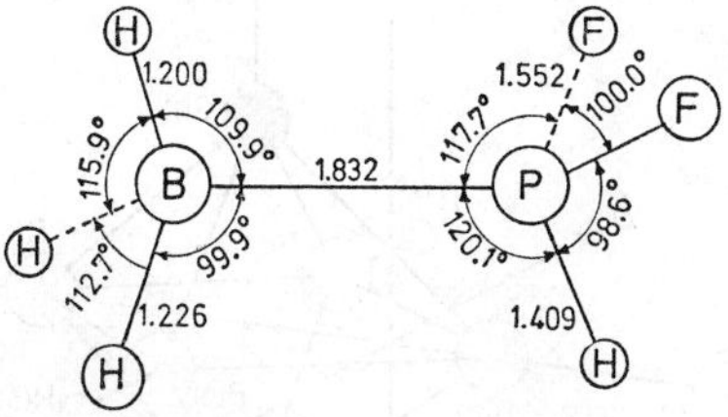

Figure 1 *The structure of difluorophosphine borane*
(Reproduced with permission from *J. Chem. Phys.*, 1971, **54**, 1907)

27.17° to the P—B bond and 32.73° to the P—H bond. The BH_3 group shows a pronounced tilt away from the fluorine atoms, unlike the iso-electronic $HF_2Si—CH_3$. In the boron compound this tilt is explained in terms of the electrostatic interaction between the fluorine atoms, the hydridic hydrogen atoms attached to boron, and the protonic hydrogen attached to phosphorus.

By studying the rotational spectra of eight isotopic species of 2-chloro-1,6-dicarbahexaborane(6), McKown and Beaudet [59] obtained the skeletal structure given in Table 5, where the numbering of atoms is as in Figure 2.

Table 5 *The substitution structure for* 2-*chloro*-1,6-*dicarbahexaborane*(6) (*see Figure* 2 *for atom numbering*)

Atoms	*Bond length*/Å	*Atoms*	*Bond angle*/deg
B(2)—Cl	1.823(10)	B(2)—B(3)—B(4)	87.7(5)
B(2)—B(3)	1.671(10)	B(3)—B(4)—B(5)	91.0(2)
B(3)—B(4)	1.702(5)	B(3)—B(2)—B(5)	93.6(5)
C(6)—B(2)	1.59(4)		
C(6)—B(4)	1.61(4)		

The B—Cl bond length is greater than in BCl_3 and is essentially the sum of the covalent radii. The quadrupole coupling constant $[e^2Qq(^{35}Cl) = 49.7(2.0)\ MHz, \eta = 0]$ is similar to that in m-$B_{10}Cl_{10}C_2H_2$ and higher than in simple B–Cl compounds; this fact, together with the zero asymmetry and long B—Cl bond length, suggests that there is little B–Cl π-bonding in this compound.

The substitution structure [60] for the boron atoms in 2-carbahexa-borane(9) confirms the structure previously proposed by Dunks and

[58] J. P. Pasinski and R. L. Kuczkowski, *J. Chem. Phys.*, 1971, **54**, 1903.
[59] G. L. McKown and R. A. Beaudet, *Inorg. Chem.*, 1971, **10**, 1350.
[60] C. C. S. Cheung and R. A. Beaudet, *Inorg. Chem.*, 1971, **10**, 1144.

Hawthorne.[61] The structure and bond lengths determined in this study are shown in Figure 3; dipole moments for the ^{11}B species were: $\mu_a = 0.65$ D, $\mu_b = 0$, $\mu_c = 1.38$ D, with a total dipole of 1.53(3) D. The bond lengths between the apical and ring boron nuclei are similar to those

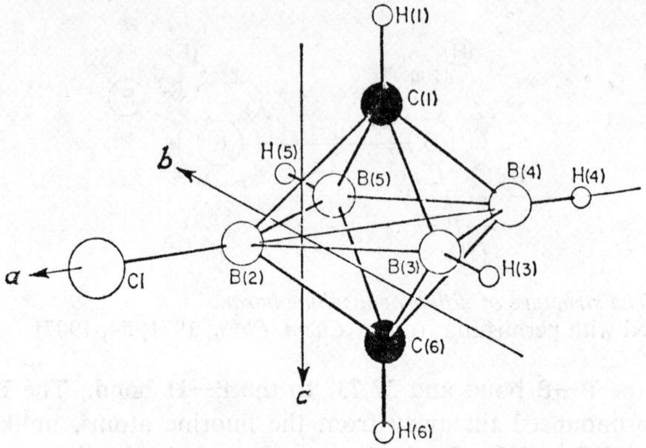

Figure 2 *The structure and labelling convention for* 2-*chloro*-1,6-*dicarbahexaborane*(6)
(Reproduced with permission from *Inorg. Chem.*, 1971, **10**, 1352)

in hexaborane and 2,4-$C_2B_5H_7$, suggesting the sixfold co-ordination of the apical nucleus; other bond lengths also support the given structure. A CNDO/2 calculation gave a position for the carbon atom which closely reproduces the observed moments of inertia.

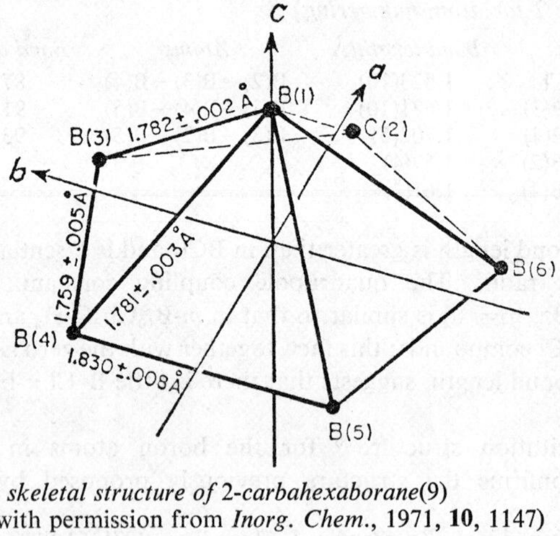

Figure 3 *The skeletal structure of* 2-*carbahexaborane*(9)
(Reproduced with permission from *Inorg. Chem.*, 1971, **10**, 1147)

[61] G. B. Dunks and M. F. Hawthorne, *J. Amer. Chem. Soc.*, 1968, **90**, 7355.

A rotational analysis of the microwave spectrum of $CO(CN)_2$ in the ground state and first two levels of the lowest A_1 bending vibration has been performed.[62] The rotational constants, four planar centrifugal distortion constants, inertia defects, and α for the bending vibration were calculated; the dipole moment is μ = 0.704(7) D. Since only one isotopic species was studied, two structural parameters only could be obtained: on assuming r(C—O) = 1.22 Å, r(C—N) = 1.165 Å, and θ(C—C—N) = 180°, the values r(C—C) = 1.4504 Å and θ(C—C—C) = 115°19′ were obtained, the latter being close to the 114°59′ observed in MeCOCN. The ^{14}N quadrupole structure was analysed to give χ_{aa} = − 2.846(24) MHz, χ_{bb} = 0.045(16) MHz, and χ_{cc} = 2.801(32) MHz; these results are not compatible with a cylindrically symmetrical electric-field gradient around the C—N bond.

Silacyclobutane, $\overline{SiH_2 \cdot CH_2 \cdot CH_2 \cdot CH_2}$, has a low-frequency ring-puckering vibration whose lowest two levels are separated by 75.75 MHz; the next two levels are 7790 MHz apart. The microwave spectra in all four states have been observed.[63] On fitting a quartic potential to the puckering vibration, a barrier height of 442 cm^{-1} was found. Values of the C—H and Si—H bond lengths were assumed, as were the H—C—H and H—Si—H angles, to give r(Si—C) = 1.91 Å, θ(C—C—C) = 103°, and the perpendicular distance of the silicon nucleus from the C—C—C plane was 0.71 Å.

The first example of a planar configuration of bonds around nitrogen in the gas phase has been reported.[64] The microwave spectra of five isotopic species of PF_2NH_2 indicate that the molecule has a plane of symmetry containing all but the fluorine nuclei. This is shown by the almost invariant value of the second moment $P_{bb} = \Sigma_i M_i b_i^2$ (where b_i is the distance of the nucleus of mass m_i along the b-axis direction) and the value μ_b^2 = − 0.009(31) D, indicating zero dipole moment in this direction. In addition, the structures obtained on the assumption of planarity and without this assumption show close agreement. The structure is shown in Figure 4; in addition r_0(P—F) = 1.587 Å, θ(F—P—F) = 94.6°, and θ(F—P—N) = 100.6°. The total dipole moment is 2.576 D, lying almost parallel to the a-axis. The sign is as shown, being obtained from measurements on PF_2NHD (*cis*) as well as the normal species. The planar configuration is in contrast to that obtained from electron-diffraction data,[65] when a pyramidal structure with the HNH plane making an angle of 146° with the P—N bond was found. However, in that study it was assumed that both N—H bonds and both P—N—H angles were equal, which is here shown not to be the case. The planar structure supports a mechanism of (p–d) π-interaction between nitrogen and phosphorus; in NH_2CHO and

[62] R. M. Lees, *Canad. J. Phys.*, 1971, **49**, 367.
[63] W. C. Pringle, *J. Chem. Phys.*, 1971, **54**, 4979.
[64] A. H. Brittain, J. E. Smith, P. L. Lee, K. Cohn, and R. H. Schwendeman, *J. Amer. Chem. Soc.*, 1971, **93**, 6772.
[65] G. C. Holywell, D. W. H. Rankin, B. Beagley, and J. M. Freeman, *J. Chem. Soc.* (*A*), 1971, 785.

$PhNH_2$ the configuration is closer to planar than in most trigonal nitrogen compounds, and in the solid phase Me_2NPF_2 has been found to be planar. Further confirmation of the double-bond character of the P—N bond is to be found in its short length compared with that in the singly bonded NaH_3NPO_3 [r(P—N) = 1.769 Å]; the length in F_2PNH_2, r(P—N) = 1.647 Å, is closer to that of a double bond [*e.g.* in $Ph_2FP{=}NMe$,

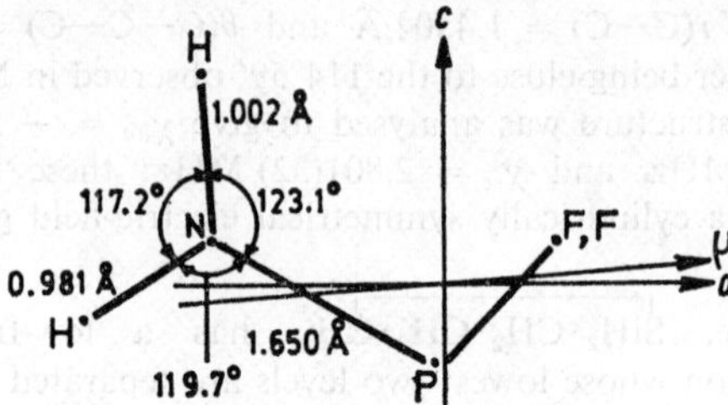

Figure 4 *The projection of* PF_2NH_2 *in its plane of symmetry*
(Based on *J. Amer. Chem. Soc.*, 1971, 93, 6775)

r(P—N) = 1.641 Å]. Such a short bond length is also found in solid Me_2NPF_2. The proton n.m.r. spectrum of F_2PNH_2 shows that, on the n.m.r. time-scale, there is rapid intramolecular conversion of the *cis* and *trans* protons, probably by rotation about the P—N bond. The value of the coupling constant, $J(^{15}N–H)$ = 83.2 Hz, implies *ca.* 30% *s*-character in the nitrogen orbital in the P—N bond; this is close to the 33% *s*-character required for the planar configuration.

In contrast, a *trans* conformation is found for H_2PPF_2 in the gas phase.[66] By assuming r(P—H) = 1.42(1) Å, Kuczkowzki *et al.* obtained the structure: r_0(P—P) = 2.218(38) Å, r_0(P—F) = 1.587(13) Å, θ_0(H—P—H) = 93.2(1.0)°, θ_0(H—P—P) = 90.3(4)°, θ_0(F—P—F) = 98.2(1.2)°, and θ_0(F—P—P) = 97.2(1.6)°. The magnitude and direction of the dipole moment [μ = 1.71(1) D] require a bond moment of 0.5 D for the $F_2\overset{-}{P}{-}\overset{+}{P}H_2$ bond, suggesting an inductive effect of the fluorine atoms. The *trans* conformation has been found for all gaseous diphosphine molecules except P_2H_4; this is explained in terms of steric repulsion and lone-pair interaction.

The potential barrier to internal rotation in CD_3PF_3 has been obtained from measurements of the microwave spectrum in the $|m| = 3$ excited torsional state.[67] Earlier work on the ground state was not able to determine V_6, but here it is calculated as 12.2(3) cal mol^{-1}, with slight variation with

66 R. L. Kuczkowski, H. W. Schiller, and R. W. Rudolph, *Inorg. Chem.*, 1971, **10**, 2505.
67 J. E. Wollrab, E. A. Rinehart, P. B. Reinhart, and P. R. Reed, *J. Chem. Phys.*, 1971, **55**, 1998.

rotational state. The lowering of V_6 compared with CH_3BF_2 (V_6 = 13.77 cal mol^{-1}) parallels a similar decrease in CD_3NO_2 and CH_3NO_2.

The absence of any *b*-type transitions in the microwave spectra of $C_3H_5PH_2$, C_3H_5PHD, and $C_3H_5PD_2$ indicates that the molecules have a plane of symmetry; the conformation is as shown in Figure 5. By assuming

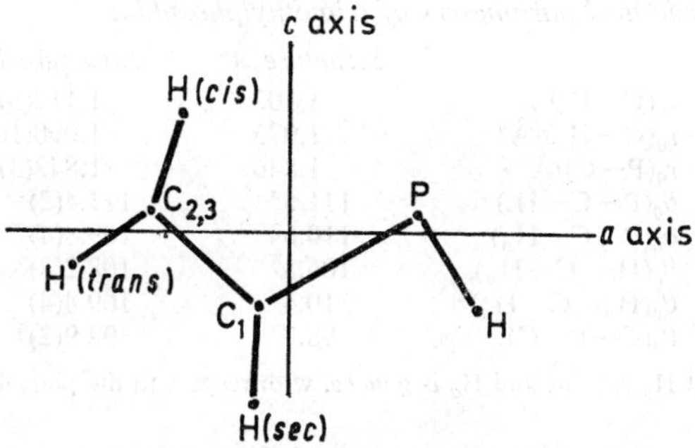

Figure 5 *The projection of cyclopropylphosphine in its plane of symmetry* (Reproduced with permission from *J. Chem. Phys.*, 1971, **54**, 916)

r(C—H) = 1.080 Å, θ(C—C—C) = 60°, and θ(H—C—H) = 116.2°, Dinsmore *et al.*[68] found r(C—C) = 1.5130(1) Å, r(C—P) = 1.8340(2) Å, r(P—H) = 1.4130(2) Å, and θ(C—P—H) = 98.30(2)°, and the angle between the plane of the ring and the C—P bond is 121.10(1)°; the dipole moment is 1.158(4) D. The calculated conformation allows interaction of the phosphorus lone pair with intra-annular electrons; this is also shown by the rather short C—P bond length.

A microwave study of Me_3P, in which one methyl group was partially or totally deuteriated, shows that the methyl groups are tilted by 1.6(5)° relative to the P—C bond.[69] Each methyl group is staggered with respect to the other two P—C bonds. The structure was determined from first and second moment equations, both with (Structure A) and without (Structure B) an effective bond shortening of 0.005 Å on deuteriation; these are given in Table 6. The methyl-group tilt and the greater length of $r(C{-}H_s)$ = 1.112(5) Å compared with $r(C{-}H_a)$ = 1.090(10) Å closely parallel the values found in Me_3N.

The rotational constants B_v for four fundamental vibrational states of $SF_5{}^{35}Cl$ (C_{4v} symmetry) have been obtained from an analysis of the millimetre-wave spectrum.[70]

In tetrahydroselenophen, C_4H_8Se, the very small change of A on substitution of the selenium nucleus implies that it is on the *a*-axis; the nuclear-spin intensity alternation also shows that the molecule has C_2

[68] L. A. Dinsmore, C. O. Britt, and J. E. Boggs, *J. Chem. Phys.*, 1971, **54**, 915.
[69] P. S. Bryan and R. L. Kuczkowski, *J. Chem. Phys.*, 1971, **55**, 3049.
[70] R. Jurek and J. Chanussot, *Compt. rend.*, 1971, **272**, *B*, 941.

symmetry.[71] The dipole moment is 1.93(3) D, and from the rotational constants it is found that the degree of non-planarity is higher than in tetrahydrothiophen.

At dry-ice temperature, BrF_5 has a half-life of about 1.5 minutes in a copper waveguide. The microwave spectra [72] of $^{79}BrF_5$ and $^{81}BrF_5$ yielded

Table 6 *Structural parameters of trimethylphosphine*

	Structure A[a]	*Structure B*[a]
$r_0(C-H_s)/Å^b$	1.102	1.112(5)
$r_0(C-H_a)/Å^b$	1.073	1.090(10)
$r_0(P-C)/Å$	1.846	1.843(3)
$\theta_0(P-C-H_s)$	111.5°	111.4(2)°
$\theta_0(P-C-H_a)$	110.6°	109.8(4)°
$\theta_0(H_s-C-H_a)$	106.8°	108.2(8)°
$\theta_0(H_a-C-H_a)$	110.4°	109.4(4)°
$\theta_0(C-P-C)$	98.7°	98.9(2)°

[a] See text. [b] H_s is *trans* and H_a is *gauche*, with respect to the phosphorus lone pair.

values of the rotational constants B_0 and the quadrupole coupling constants [for ^{79}Br, $e^2Qq = -280.9(3)$ MHz]; the spectra confirm the C_{4v} symmetry. The spectrum of IF_5 has also been observed.[73] The geometrical information from these studies on BrF_5 and IF_5 was compared with that obtained by electron diffraction.[74] The B_α values thus obtained are 4.5 MHz less than the B_0 values. The Br—F bond lengths obtained from the microwave data assuming the electron-diffraction bond angle agree well with those obtained from electron diffraction. For IF_5 the agreement is consistent but not precise; the difference between axial and equatorial bond lengths is poorly determined. The agreement with *X*-ray data is good.

6 Other Studies

Apparatus for measuring linewidths and pressure-induced shifts of microwave lines has been described;[75] various studies of such phenomena are reported.[76–78] In one such study,[78] the molecular quadrupole moment of O_2 is estimated as $Q_{mol} = 1.16(12)$ D Å; this should be an upper limit.

Microwave-i.r. double-resonance studies are continuing; these have been reported for $^{14}NH_3$,[79] for $^{15}NH_3$,[80] and for N_2O.[81]

71 A. K. Mamleev, N. N. Magdesieva, and N. M. Pozdeev, *Zhur. strukt. Khim.*, 1970, **11**, 1124.
72 M. J. Whittle, R. H. Bradley, and P. N. Brier, *Trans. Faraday Soc.*, 1971, **67**, 2505.
73 R. H. Bradley, P. N. Brier, and M. J. Whittle, *Chem. Phys. Letters*, 1971, **11**, 192.
74 A. G. Robiette, R. H. Bradley, and P. N. Brier, *Chem. Comm.*, 1971, 1567.
75 I. C. Story, V. I. Metchnik, and R. W. Parsons, *J. Phys. (B)*, 1971, **4**, 593.
76 I. C. Story, V. I. Metchnik, and R. W. Parsons, *Phys. Letters (A)*, 1971, **34**, 59.
77 P. C. Pandey, K. K. Kirty, and S. L. Srivastava, *J. Phys. (B)*, 1971, **4**, 786.
78 J. S. Murphy and J. E. Boggs, *J. Chem. Phys.*, 1971, **54**, 2443.
79 M. Fourrier, A. van Lerberghe, and M. Redon, *Compt. rend.*, 1971, **273**, *B*, 816.
80 T. Oka and T. Shimizu, *Appl. Phys. Letters*, 1971, **19**, 88.
81 J. Lemaire, J. Houriez, J. Thibault, and B. Maillard, *J. Physique*, 1971, **32**, 35.

4
Vibrational Spectra

BY D. M. ADAMS

1 General Introduction

The organization and format of the chapters dealing with vibrational spectroscopy follow closely those developed by the previous Reporters. The sole innovation is the introduction in Chapter 4 of a section dealing with single-crystal spectroscopy, as this is a developing field with its own jargon and methodology best treated separately.

The year 1971 was a vintage one for books dealing with vibrational spectroscopy. Ferraro's 'Low-frequency Vibrations of Inorganic and Co-ordination Compounds'[1] will find its way into every inorganic laboratory, being in part (organometallic compounds are omitted) the much-needed sequel to this Reporter's 'Metal–Ligand and Related Vibrations'. A particularly valuable work on isotopically labelled compounds by Pinchas and Laulicht[2] contains much of inorganic interest. In Gans' 'Vibrating Molecules'[3] the reader is skilfully piloted through the principles of normal-co-ordinate analysis, with some emphasis on computing methods, whereas the 'Theory of Vibrational Spectroscopy' by Steele[4] promises a challenge for 'Wilson, Decius, and Cross'. It is a particular pleasure to note Jones' treatment of inorganic applications of force-constant analysis.[5] Two new books have improved the selection of introductory symmetry theory available;[6,7] Fackler's book also contains a chapter on molecular vibrations.[8] Important works on 'Lattice Vibrations'[9] and 'Far-infrared Spectroscopy'[10] have appeared. On a more practical level two 'tricks of the trade'

[1] J. R. Ferraro, 'Low-frequency Vibrations of Inorganic and Co-ordination Compounds', Heyden, London, 1971.
[2] S. Pinchas and I. Laulicht, 'Infrared Spectra of Labelled Compounds', Academic Press, London, 1971.
[3] P. Gans, 'Vibrating Molecules: an Introduction to the Interpretation of Infrared and Raman Spectra', Chapman and Hall, London, 1971.
[4] D. Steele, 'Theory of Vibrational Spectroscopy', Saunders, London, 1971.
[5] L. H. Jones, 'Inorganic Vibrational Spectroscopy', Marcel Dekker, New York, 1971, vol. 1.
[6] G. Davidson, 'Introductory Group Theory for Chemists', Elsevier, London, 1971.
[7] M. Orchin and H. H. Jaffé, 'Symmetry, Orbitals and Spectra', Wiley, London, 1971.
[8] J. P. Fackler, 'Symmetry in Co-ordination Chemistry', Academic Press, New York, 1971.
[9] B. Donovan and J. F. Angress, 'Lattice Vibrations', Chapman and Hall, London, 1971.
[10] K. D. Möller and W. G. Rothschild, 'Far-infrared Spectroscopy', Wiley–Interscience, New York, 1971.

books will be useful,[11, 12] as will Tobin's somewhat basic account of 'Laser Raman Spectroscopy'.[13] Other books, not seen personally and some of them of marginal relevance to this field, are given as references.[14–20] In the journals business another publisher has yielded to temptation.[21]

Irish and Chen have reviewed 'Applications of Raman Spectroscopy to Chemical Analysis'[22] including some aspects that may well help stimulate industrial interest, while accounts of the use of i.r. and Raman spectroscopy in the study of glassy solids,[23] factors determining the shapes of i.r. absorption bands in liquids,[24] effects of phase and pressure change on vibrational spectra,[24a] and the effect of electronic relaxation on molecular vibrations[25] all help educate the 'compleat' spectroscopist. Tables for factor-group analysis based upon the correlation method have been presented[26] and should be compared with others published last year.[27] A tentacle of the *ab initio* octopus is reaching steadily into vibrational work and may well be stimulated by Schutte's timely review.[28]

On an historical note, we record that 1971 witnessed publication of the first Raman spectra of matrix-isolated species.[29–31] Used in conjunction with the well-established i.r. equivalent this should result in more convincing assignments than can result from use of the i.r. technique alone.

[11] M. J. de Faubert Maunder, 'Practical Hints on Infrared Spectrometry from a Forensic Analyst', Adam Hilger, London, 1971.
[12] 'Spectroscopic Tricks', ed. L. May, Adam Hilger, London, 1971, vol. 2.
[13] M. C. Tobin, 'Laser Raman Spectroscopy', Wiley–Interscience, London and New York, 1971.
[14] 'Spectroscopy in Inorganic Chemistry', ed. C. N. R. Rao and J. R. Ferraro, Academic Press, New York, 1971, vol. 2.
[15] V. I. Mališek and M. Miler, 'Vibration Spectroscopy', Iliffe, London, 1971.
[16] B. Meyer, 'Low-temperature Spectroscopy: Optical Properties of Molecules in Matrices, Mixed-crystals and Frozen Solutions', Elsevier, Amsterdam, 1971.
[17] J. N. Hodgson, 'Optical Absorption and Dispersion in Solids', Chapman and Hall, London, 1970.
[18] D. P. Craig and S. H. Walmsley, 'Excitons in Molecular Crystals: Theory and Applications', Benjamin, London, 1971.
[19] L. de Galan, 'Analytical Spectrometry', Adam Hilger, London, 1971.
[20] 'Spectroscopic Data Relative to Diatomic Molecules', ed. B. Rosen, Pergamon, Oxford, 1970.
[21] *J. Co-ordination Chem.*, 1971, **1**, no. 1 (August), Gordon and Breach, New York, London, and Paris.
[22] D. E. Irish and H. Chen, *Appl. Spectroscopy*, 1971, **25**, 1.
[23] J. Wong and C. A. Angell, *Appl. Spectroscopy Rev.*, 1971, **4**, 155.
[24] R. P. Young and R. N. Jones, *Chem. Rev.*, 1971, **71**, 219.
[24a] J. E. D. Davies, *J. Mol. Structure*, 1971, **10**, 1.
[25] J. K. Burdett, *Appl. Spectroscopy Rev.*, 1971, **4**, 43.
[26] W. G. Fateley, N. T. McDevitt, and F. F. Bentley, *Appl. Spectroscopy*, 1971, **25**, 155.
[27] D. M. Adams and D. C. Newton, 'Tables for Factor Group and Point Group Analysis', Beckman–RIIC, Croydon, 1970.
[28] C. J. H. Schutte, *Structure and Bonding*, 1971, **9**, 213.
[29] J. S. Shirk and H. H. Claassen, *J. Chem. Phys.*, 1971, **54**, 3237.
[30] H. Huber, G. A. Ozin, and A. Vander Voet, *Nature Phys. Sci.*, 1971, **232**, 166; D. H. Boal and G. A. Ozin, *Spectroscopy Letters*, 1971, **4**, 43.
[31] J. W. Nibbler and D. A. Coe, *J. Chem. Phys.*, 1971, **55**, 5133.

2 Spectra of Small Species

Diatomic Molecules, Ions, and Radicals.—Data for diatomic species continue to accumulate, largely owing to the ever increasing use of matrix isolation. ν(X—X) fundamentals have been observed by i.r. absorption for Ar matrices of fluorine or chlorine: $\nu(^{35}Cl—^{35}Cl)$ 546.3 cm^{-1} and $\nu(^{35}Cl—^{37}Cl)$ 539.4 cm^{-1}. It is not known if the spectra are those of weak complexes: clearly the molecules must be perturbed by some interaction.[32] Raman spectra of nitrogen molecules vibrationally excited by an electric discharge show the lines 2331 ($0 \rightarrow 1$), 2302 ($1 \rightarrow 2$), and 2273 ($2 \rightarrow 3$) cm^{-1}; the development and continued application of this technique will be watched with interest.[33] Matrix-isolated S_2 molecules yield [34] a single Raman line at 716 cm^{-1}. Previous workers using i.r. absorption in similar matrices had reported bands at 660, 680, and 668 cm^{-1}, but these are now believed to be due to non-isolated S_2 molecules. Vapour species over TlX salts have been shown by i.r. matrix-isolation methods to be monomers and dimers, the latter probably being X—Tl—Tl—X:[35] TlF (441), TlCl (261), TlBr (179), TlI (143), Tl_2F_2 (316 and 256), and Tl_2Cl_2 (188.6 and 171.3 cm^{-1}). Force constants and mean square amplitudes have been calculated [36] for MX species where M = Li, Na, K, Rb, Cs, or Al; X = F, Cl, Br, or I. The Raman spectrum of the product of reaction between NSF and AsF_5 or SbF_5 indicates presence of $[AsF_6]^-$; the compounds are formulated as $[NS]^+ [AsF_6]^-$ and a line at 1437 cm^{-1} is assigned to the new thiazyl cation [37] [*cf.* ν(NS) in NSF is at 1372 cm^{-1}]. Simultaneous deposition of OF_2 diluted in Ar with a beam of Li, Na, K, or Mg atoms yields the OF radical, ν = 1028.6 cm^{-1} (independent of metal). ClO (ν = 995 cm^{-1}) is similarly made, but a band at 850 cm^{-1} is also attributed to ClO 'perturbed by another molecule'.[38] ν(OH) in molten NaOH shows weak, broad bands near the fundamental (3619 cm^{-1}), attributed to combinations with quasi-lattice modes.[39]

Between 2100 and 2900 K solid rare-earth oxides other than those of Yb and Eu emit vapours, MO, which have ground-state vibrational frequencies between 808 and 832 cm^{-1} (in Ar and Ne matrices).[40] Matrix-isolated uranium monoxide species show [41] the wavenumbers 776.0 cm^{-1} ($U^{16}O$) and 736.2 cm^{-1} ($U^{18}O$). The vibrational constants of ^{63}CuBi, $^{35}Cl_2$, and $^{35}Cl^{37}Cl$ have been determined.[42, 43]

32 M. R. Clarke and G. Mamantov, *Inorg. Nuclear Chem. Letters*, 1971, **7**, 993.
33 L. Y. Nelson, A. W. Saunders, and A. B. Harvey, *J. Chem. Phys.*, 1971, **55**, 5127.
34 R. E. Barletta, H. H. Claassen, and R. L. McBeth, *J. Chem. Phys.*, 1971, **55**, 5409.
35 J. M. Brom and H. F. Franzen, *J. Chem. Phys.*, 1971, **54**, 2874.
36 S. J. Cyvin, *J. Mol. Structure*, 1971, **8**, 43.
37 O. Glemser and W. Koch, *Angew. Chem. Internat. Edn.*, 1971, **10**, 127.
38 L. Andrews and J. I. Raymond, *J. Chem. Phys.*, 1971, **55**, 3078.
39 A. G. Turnbull, *Austral. J. Chem.*, 1971, **24**, 2213.
40 R. L. DeKock and W. Weltner, *J. Phys. Chem.*, 1971, **75**, 514.
41 S. Abramowitz, N. Acquista, and K. R. Thompson, *J. Chem. Phys.*, 1971, **75**, 2283.
42 Y. Lefebvre and R. Houdart, *Compt. rend.*, 1971, **272**, *B*, 1301.
43 G. Hochenbleicher and H. W. Schrötter, *Appl. Spectroscopy*, 1971, **25**, 360.

Vibrational arcana of the hydrogen halides have been further examined. Hydrogen-bonding in solutions containing hydrogen fluoride has been studied by following changes in ν(HF) in solvents. The largest shift occurred in pyridine, $\nu(HF) = 2713\ cm^{-1}$ (*cf.* $3961\ cm^{-1}$ in the vapour).[44] I.r. absorption due to $(HF)_2$ has been observed at *ca.* $3965\ cm^{-1}$.[45] An extensive i.r. study has been made of matrix-isolated DCl.[46] Hindered rotation occurs in those matrices, offering spherically symmetric substitution sites (Ar, Kr, or SF_6). Bands due to dimers and higher 'multimers' are discussed. I.r. features shown by crystalline HCl and DCl in the region of their second harmonics have been fully assigned and intermolecular force constants deduced.[47] The Raman and i.r. spectra of liquid HBr and of the β, γ, and δ solid phases (with emphasis on δ) in the region of the fundamental (*ca.* $2450\ cm^{-1}$) show that whereas the liquid phase has identical i.r. and Raman wavenumbers, those for the δ-phase are split by $12\ cm^{-1}$. A qualitative explanation is offered based upon vibrational coupling in regions of local order.[48] Some details of the Raman spectrum of the orthorhombic form at liquid helium temperature are due to sample history and crystallite size.[48a] Analysis of vibration–rotation bands of HI and DI yielded many accurate vibrational constants, including ν_0(HI) at $2229.581 \pm 0.006\ cm^{-1}$ $(1 \leftarrow 0)$ and ν_0(DI) at $1599.764 \pm 0.004\ cm^{-1}$ $(1 \leftarrow 0)$.[49] The Raman spectrum of HCl in β-quinol clathrates shows evidence for rotational motion of the guest molecule, which has a stretching frequency close to that of the gas.[50] The same study included clathrates of SO_2, H_2S, and C_2H_2.

Triatomic Molecules, Ions, and Radicals.—Rossotti's lively review of the status of 'polywater' ('Water: how anomalous can it get?') is required reading.[51] The vibrational properties of H_2O, D_2O, and HOD in vapour and liquid states and of 'polywater' have also been reviewed,[52] while a further i.r. study attributes bands at 1400 and $1100\ cm^{-1}$ to impurities rather than to 'polywater'.[53] The ν_2 band of $H_2{}^{18}O(g)$ has been analysed.[54] At pressures up to 7.2 kbar very small shifts of the ν(OH) components in the Raman spectrum of HDO in H_2O are observed, and have been discussed in terms of theories of water.[55] It is claimed that the overtone region shows evidence of the presence of free hydroxy-groups in liquid water.[55a] The i.r.

44 H. Touhara, H. Shimoda, K. Nakanishi, and N. Watanabe, *J. Phys. Chem.*, 1971, **75**, 2222.
45 J. L. Himes and T. A. Wiggins, *J. Mol. Spectroscopy*, 1971, **40**, 420.
46 J. B. Davies and H. E. Hallam, *Trans. Faraday Soc.*, 1971, **67**, 3176.
47 J. Blanchard, L.-C. Brunel, and M. Peyron, *Compt. rend.*, 1971, **272**, *B*, 321, 366.
48 E. L. Pace, *Spectrochim. Acta*, 1971, **27A**, 491.
48a T. S. Sun and A. Anderson, *Spectroscopy Letters*, 1971, **4**, 377.
49 S. C. Hurlock, R. M. Alexander, K. N. Rao, and Sr. N. Dreska, *J. Mol. Spectroscopy*, 1971, **37**, 373.
50 J. E. D. Davies, *Chem. Comm.*, 1970, 270.
51 H. S. Rossotti, *J. Inorg. Nuclear Chem.*, 1971, **33**, 2037.
52 P. Krindel and I. Eliezer, *Co-ordination Chem. Rev.*, 1971, **6**, 217.
53 B. F. Howell and J. Lancaster, *Chem. Comm.*, 1971, 693.
54 J. G. Williamson, K. N. Rao, and L. H. Jones, *J. Mol. Spectroscopy*, 1971, **40**, 372.
55 G. E. Walrafen, *J. Chem. Phys.*, 1971, **55**, 5137.
55a J. J. Péron, C. Bourdéron, and C. Sandorfy, *Canad. J. Chem.*, 1971, **49**, 3901.

absorption band envelopes of liquid H_2O and D_2O in the range 1000—2700 cm^{-1} have been analysed in terms of Cauchy–Gauss product functions, and the effect of temperature (5—70 °C) on the shapes of the intermolecular combination band (2125 cm^{-1}, H_2O; 1555 cm^{-1}, D_2O) and the δ(HOH) band were discussed.[56] Weak i.r. bands at *ca.* 1350 cm^{-1} (H_2O) and 1000 cm^{-1} (D_2O) have been attributed to the first overtones of the respective vibrational modes.[57] The intensity behaviour of the 3540 and 3626 cm^{-1} i.r. bands of water dissolved in acetonitrile upon application of a magnetic field is said to be dependent upon the weather.[58] The products of electrolytically dissociated water vapour have been shown previously by i.r. spectroscopy to include H_2O_3 and H_2O_4; this has now been confirmed by a complementary Raman study.[59] A vibrational analysis of liquid water has been attempted on the basis of a model with mixtures of molecules with different numbers of hydrogen-bonds. For the fully bound species a nine-atom model consisting of one molecule surrounded tetrahedrally by two oxygen atoms and two hydroxy-groups was used.[60] Some of the assignments attributed to various species in liquid water are supported.

Water of hydration in solids continues to attract attention: two survey studies are reported. An i.r. investigation of the vibrational modes of water of crystallization in alkaline-earth halide hydrates has permitted a new means for their structural classification.[61] Spectra of several series of aquohalogeno-complexes have been discussed {$[MX_5,H_2O]^{2-}$, $[CuX_4,$-$2H_2O]^{2-}$, $MX_2,2H_2O$, *etc.*} including some of the water vibrations.[62]

I.r. spectra of $Na_2[Fe(CN)_5NO],2H_2O$ at different degrees of deuteriation were studied [63] as a function of temperature. Doublets were observed for the three fundamentals of isotopically dilute HDO: this was interpreted as evidence of asymmetry of the water molecules. OH Bonds of each water molecule point towards CN groups belonging to different complex ions; as one of these distances is greater than the other, the ν(OH) wavenumbers differ. A typical set is as follows, at 30 °C: ν(OH) = 3603.0 and 3568.0; ν(OD) = 2653.5 and 2629.5; and δ(HOD) = 1427.4 and 1419.0 cm^{-1}. The temperature dependence of 'external' modes of water (below 700 cm^{-1}) was interpreted to show that: (*i*) in $BaCl_2,2H_2O$ and $K_2C_2O_4,H_2O$ the water is 'lattice water', and (*ii*) in $CuCl_2,2H_2O$ at room temperature the water molecules are co-ordinated to copper, but on cooling the state of water approaches that of lattice water.[64] Presence of a single δ(HOH) mode

[56] R. Oder and D. A. I. Goring, *Spectrochim. Acta*, 1971, **27A**, 2149.
[57] R. Oder and D. A. I. Goring, *Spectrochim. Acta*, 1971, **27A**, 2285.
[58] V. I. Klassen, M. A. Orel, M. A. Sarukhanov, I. V. Kagarlitskaya, S. Sh. Rozenfel'd, I. V. Lapatukhin, and L. B. Voloshina, *Doklady Phys. Chem.*, 1971, **197**, 326.
[59] X. Deglise and P. A. Giguére, *Canad. J. Chem.*, 1971, **49**, 2242.
[60] R. A. More O'Ferrall, G. W. Koeppl, and A. J. Kresge, *J. Amer. Chem. Soc.*, 1971, **93**, 1, 9.
[61] H. D. Lutz, H. J. Klüppel, and G. Kho, *Angew., Chem. Internat. Edn.*, 1971, **10**, 183.
[62] D. M. Adams and P. J. Lock, *J. Chem. Soc. (A)*, 1971, 2801.
[63] M. Holzbecher, O. Knop, and M. Falk, *Canad. J. Chem.*, 1971, **49**, 1413.
[64] K. Fukushima, *Bull. Chem. Soc. Japan*, 1971, **44**, 372.

in the i.r. spectrum of $Ba(NO_3)_2,H_2O$ confirms equivalence of the six water molecules in the unit cell, but asymmetric hydrogen-bonds are believed to be involved. The authors therefore suggest that space groups postulated on the basis of *X*-ray work are incorrect and that $P6_1$ or $P6_5$ are more

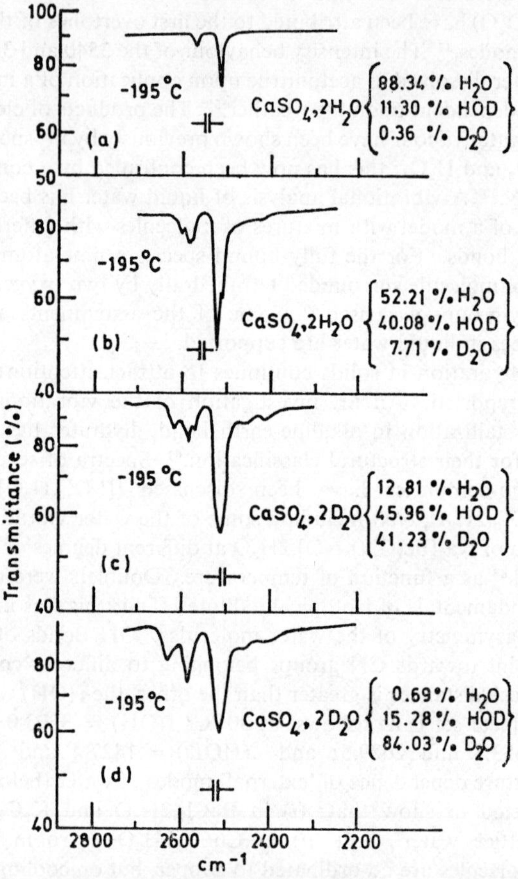

Figure 1 OD *stretching region of deuteriated gypsum films:* (a) 6.01%, (b) 27.75%, (c) 64.41%, (d) 91.67% *deuteriated*
(Reproduced by permission from *J. Chem. Phys.*, 1971, **54**, 5331)

likely.[65] ν(OH) spectra of salt monohydrates such as $BaClO_3.H_2O$ show that water is symmetrically bonded.[65a]

Intermolecular coupling of water molecules in $CuCl_2,2H_2O$ has been studied by i.r. absorption of oriented polycrystalline films using samples in

[65] G. Brink and M. Falk, *Spectrochim. Acta*, 1971, **27A**, 1811.
[65a] L. J. Bellamy, M. J. Blandamer, M. C. R. Symons, and D. Waddington, *Trans. Faraday Soc.*, 1971, **67**, 3435.

which H_2O or D_2O were present in low molar ratios. These uncoupled ν(OH) and δ(HOH) modes are at different frequencies from those given by the normal material in which there is quite strong intermolecular coupling: the difference is used to compute force constants for the coupling. Correlation field splittings are greater for H_2O modes than for D_2O.[66] A similar study of gypsum in the ν(OH) region (see Figure 1) showed that intermolecular coupling is strong; force constants were computed.[67]

Frequencies observed for some triatomic molecules are collected in Table 1. Although the i.r. spectrum of CF_2 has been studied previously,

Table 1 *Vibrational frequencies (expressed as wavenumber*/cm^{-1}) *of some triatomic molecules*

Species	*Fundamentals*			*Ref.*
	ν_1	ν_2	ν_3	
CF_2(g)	1224 ± 1.5	—	1112 ± 1.5	68
NF_2(g)	1074 (pol)	572	—	69
OF_2(l)	925.2 915.7[a]	461.1	821.1	70
OCl_2(l)	634 (pol)	293 (pol)	673	71
XeF_2(g)	514.5	—	555	74
$XeCl_2$(s)	253[a]	—	—	72
ClF_2^+(s)	782	—	—	73
$SnCl_2$(matrix)	341	124	320	75

[a] See text.

both in gas and matrix-isolated form, in this work the technique of rapid-scan spectroscopy was used to detect detail not previously accessible. Successful resolution of rotational fine structure in a spectrum recorded in less than a millisecond suggests a bright future for this technique:[68] $\nu_1 > \nu_3$ for CF_2, NF_2, and OF_2. Polarization information[70] from the Raman spectrum of OF_2 confirms an earlier report of the existence of Fermi resonance between ν_1 and $2\nu_2$. Splitting of bands in the Raman spectrum of solid OCl_2 has been interpreted in terms of a non-centro-symmetric cell containing at least three molecules[71] whilst that of OF_2 contains at least two on C_s or C_1 sites.[71a] $[N{\equiv}N{-}F]^+$ yields frequencies

[66] R. A. Fifer and J. Schiffer, *J. Chem. Phys.*, 1971, **54**, 5097.
[67] R. Kling and J. Schiffer, *J. Chem. Phys.*, 1971, **54**, 5331.
[68] A. S. Lefoha and G. C. Pimentel, *J. Chem. Phys.*, 1971, **55**, 1213.
[69] H. Selig and J. H. Holloway, *J. Inorg. Nuclear Chem.*, 1971, **33**, 3169.
[70] D. J. Gardiner and J. J. Turner, *J. Mol. Spectroscopy*, 1971, **38**, 428.
[71] D. J. Gardiner, *J. Mol. Spectroscopy*, 1971, **38**, 476.
[71a] J. Tremblay and R. Savoie, *Canad. J. Chem.*, 1971, **49**, 3785.
[72] D. H. Boal and G. A. Ozin, *Spectroscopy Letters*, 1971, **4**, 43.
[72a] R. A. Ashby, *J. Mol. Spectroscopy*, 1971, **40**, 639.
[73] T. Surles, H. H. Hyman, L. A. Quarterman, and A. I. Popov, *Inorg. Chem.*, 1971, **10**, 913.
[74] P. Tsao, C. C. Cobb, and H. A. Claassen, *J. Chem. Phys.*, 1971, **54**, 5247.
[75] H. Huber, G. A. Ozin, and A. Vander Voet, *J. Mol. Spectroscopy*, 1971, **40**, 421.

assigned as: ν_1, Σ^+, at 2371 cm^{-1} [ν(N≡N)]; ν_2, Σ^+, at 1057 cm^{-1} [ν(N—F)]; and ν_3, Π, at 391 cm^{-1} (deformation). Assuming 'reasonable' bond lengths the following force constants were estimated:[76] f(N≡N) = 21.23 ± 0.75, f(N—F) = 8.16 ± 0.29 mdyn Å^{-1}. The matrix-isolated product from a 25 : 1 Xe–Cl_2 mixture passed through a microwave discharge shows a single Raman line at 253 cm^{-1}, interpreted as ν_1 of a linear molecule $XeCl_2$. The band was not present when the discharge was not used nor when Kr or Ar was used as matrix.[72] The Raman spectrum of monomeric matrix-isolated $SnCl_2$ has been observed (Table 1). A high-resolution study of HOCl yielded ν_2(HO^{35}Cl) at 1239.9 cm^{-1}, ν_2(HO^{37}Cl) at 1237.7 cm^{-1}.[72a]

The i.r. spectrum of $Ag^+ HF_2^-$ shows a band at 1175 cm^{-1}, assigned as $\delta(HF_2)$; *cf.* 1253 cm^{-1} in $K^+ HF_2^-$ which apparently has stronger H-bonding.[77] The products of passing HBr–Br_2–Ar mixtures through a slow discharge include the hydrogen dibromide radical ($D_{\infty h}$) as revealed by i.r. spectra of the matrix: for HBr_2, ν_3 = 727.4 and ν_1 = 164.7; for DBr_2, ν_3 = 496.1 and ν_1 = 170.0 cm^{-1}. The hydrogen-bonding in HBr_2^- is almost unaffected by removal of the negative charge (giving HBr_2) but ν_3 shows a dramatic increase in quartic character.[78] HBr_2^- in a matrix (formed by co-deposition of HBr–Ar and an atomic beam of alkali metal) has ν_3 = 728 cm^{-1} (H) and 497 cm^{-1} (D). The vibrational potential function includes significant contributions from both cubic and quartic terms.[79] Subjection of Ar–Br_2, Kr–Br_2, or Xe–Br_2 mixtures to microwave discharge allows isolation of the tribromine radical, Br_3, in a matrix. It has Σ_g^+ at 197 (for Ar) or 190 cm^{-1} (for Xe) and (f_r + f_{rr}) = 1.70 mdyn Å^{-1}, compared with 1.23 (for Br_3^-) and 2.45 mdyn Å^{-1} [for Br_2 (f_r)]; Cl_3 could not be made in an analogous manner.[80] Raman spectra of 'PBr_7', *i.e.* $[PBr_4]^+ [Br_3]^-$, and $[PBr_4]^+ [IBrCl]^-$ are reported[81] with ν(I—Cl) = 205 and ν(I—Br) = 150 cm^{-1}. A new molecule, probably LiOF, shows[38] i.r. absorption at 714.0 and 417.5 cm^{-1}. Vibrational spectra of solutions of Cu^I and Ag^I halides in tri-n-butyl phosphate containing added chloride are consistent with the presence of linear ions MX_2^- (Table 2).[82]

Extensive high-resolution i.r. studies of $H^{12}C^{15}N$, $D^{12}C^{15}N$, $H^{13}C^{15}N$, and $^{35}Cl^{12}C^{14}N$ resulted in new, accurate, molecular constants.[83, 84] Quadratic, cubic, and quartic force constants for ClCN have been computed,[85] as have the anharmonic coefficients χ_{13} and χ_{23} for ClCN in Ne and Ar matrices.[86] The values of $\chi_{13,23}$ are close to those for the gas; the

[76] K. O. Christie, R. D. Wilson, and W. Sawodny, *J. Mol. Structure*, 1971, **8**, 245.
[77] A. A. Opalouskii and N. I. Tyuleneva, *J. Struct. Chem.*, 1970, **11**, 22.
[78] V. Bondybey, G. C. Pimentel, and P. N. Noble, *J. Chem. Phys.*, 1971, **55**, 540.
[79] D. E. Milligan and M. E. Jacox, *J. Chem. Phys.*, 1971, **55**, 2550.
[80] D. H. Boal and G. A. Ozin, *J. Chem. Phys.*, 1971, **55**, 3598.
[81] P. Dhamelincourt and M. Crunelle-Cras, *Compt. rend.*, 1971, **272**, *B*, 50.
[82] D. N. Waters and B. Basak, *J. Chem. Soc.* (*A*), 1971, 2733.
[83] B. D. Alpert, A. W. Mantz, and K. N. Rao, *J. Mol. Spectroscopy*, 1971, **39**, 159.
[84] C. B. Murchison and J. Overend, *Spectrochim. Acta*, 1971, **27A**, 2407.
[85] C. B. Murchison and J. Overend, *Spectrochim. Acta*, 1971, **27A**, 1801.
[86] C. B. Murchison and J. Overend, *Spectrochim. Acta*, 1971, **27A**, 1509.

Table 2 *Vibrational frequencies* (cm^{-1}) *of the ions* MX_2^- *in* TBP *solution* [82]

Ion	*Raman* ν_1, Σ_g^+	*I.r.* ν_2, π_u	*I.r.* ν_3, Σ_u^+
$CuCl_2^-$	300	109	405
$CuBr_2^-$	193	81	322
CuI_2^-	148	65	279
$AgCl_2^-$	268	88	333
$AgBr_2^-$	170	61	253
AgI_2^-	132	49	215

matrix does not perturb the anharmonic intramolecular force constants any more than the quadratic ones. Fermi resonance between $\nu_1 + \nu_3$ and $2\nu_2^0 + \nu_3$ in the vibration–rotation spectrum of BrCN has been studied in the region 2770—2880 cm^{-1}.[86a] Raman spectra of liquid and solid ClCN and BrCN, including an oriented single-crystal study of the latter, allowed unambiguous assignment of all lines, including those due to lattice modes.[87] Selected data are in Table 3. The same authors also studied the vibrational spectra of crystalline ICN. In contrast to its congeners it has only one molecule per unit cell (*R*3*m*) and therefore no correlation interaction is possible: nevertheless, the Raman spectra are remarkable in showing high-frequency shoulders on all of the fundamentals. These are attributed to

Table 3 *Single-crystal assignment* [87] *for* ClCN *and* BrCN

ClCN *I.r.* (− 180 °C)	ClCN *Raman* (− 195 °C)	*Assignment*	BrCN *I.r.* (− 180 °C)	BrCN *Raman* (− 195 °C)
—	65	$T_x(B_{2g})$	—	50
—	65	$T_z(A_g)$	—	53
—	90	$T_y(B_{3g})$	—	71
—	97	$R_y(B_{2g})$	—	102
93	—	$R_y(B_{3u})$	104	—
—	130	$R_x(B_{3g})$	—	121
114	—	$R_x(B_{2u})$	121	—
—	405	$\nu_2(B_{2g})$	—	368
—	414	$\nu_2(B_{3g})$	—	377
398	—	$\nu_2(B_{2u} + B_{3u})$	363	—
—	728	$\nu_1(^{37}Cl)$	—	—
—	736	$\nu_1(A_g)$	—	573
734	—	$\nu_1(B_{1u})$	573	—
—	817	$2\nu_2$	740	755
830	839	$2\nu_2$	762	770
2158	2154	$\nu_3(^{13}C)$	2142	2138
2182	2179	$\nu_3(^{15}N)$	2163	2159
—	2210	$\nu_3(A_g)$	—	2190
2209	—	$\nu_3(B_{1u})$	2193	—

[86a] M. Bellouard, *Compt. rend.*, 1971, **273**, *B*, 1099.
[87] M. Pézolet and R. Savoie, *J. Chem. Phys.*, 1971, **54**, 5266.

longitudinal modes (†), which can become active in a non-centrosymmetric space group.[88] Thus: ν_3 [ν(CN)] = 2175† and 2172, ν_1 [ν(C—I)] = 465† and 449, ν_2 (δ) = 330† and 329, and ν_{rot} = 151† and 136 cm^{-1}. The anion in $[Ph_4As][TeCN]$ has ν(CN) at 2081 cm^{-1} in acetonitrile solution.[89]

Intensity measurements of the ν_1 and $2\nu_2$ Raman-active Fermi resonance diads of $^{12}CO_2$ and $^{13}CO_2$ support a previous suggestion[90] that the 02^00 level is above 10^00 for $^{12}CO_2$ but that the reverse order obtains for $^{13}CO_2$.[91] Dipole-moment matrix elements and total band intensities for CO_2 have been deduced from a study of i.r. intensities of the five Σ–Σ bands in the 1.43—1.65 μm region.[92] The high-resolution spectrum of $^{12}C^{18}O_2$ (2200—5100 cm^{-1}) has been analysed[93] and the vibrational levels of nine isotopic variants (^{12,13,14}C, ^{16,17,18}O) of carbon dioxide have been computed using an accurately determined potential function.[94] I.r. intensities for gaseous OCS have been combined with microwave Stark information to establish experimentally that the absolute sign of the dipole-moment derivative for carbonyl stretching is negative.[95] Detailed analysis of high-resolution gas-phase i.r. data for $^{12}CS_2$ and $^{13}CS_2$ in the regions of $\nu_3 - \nu_1$, $\nu_3 - 2\nu_2$, and $\nu_3 + 4\nu_2$ is reported.[96]

The following fundamental frequencies of $S^{18}O_2$ were obtained from a rotational study of *B*-type i.r. bands in the gas phase: $\omega_1 = 1113.9_8$, $\omega_2 = 505.3_0$, and $\omega_3 = 1335.1_8$ cm^{-1}.[97] In combination with $S^{16}O_2$ data, force constants and Coriolis coefficients were deduced. From an investigation of the i.r. spectra of SO_2 with both natural isotopic abundances and ^{18}O enrichment it is shown that bands previously attributed to interaction of internal and lattice modes do in fact arise from isotopic splitting, although there is some intermolecular coupling.[98] A Raman study of SO_2 in a range of solvents shows ν_1 and ν_3 to be lowered on dissolution.[99] Co-deposition of SO_2 in a large excess of Ar with an atomic beam of alkali metal yields prominent new absorptions assigned to the fundamentals of the radical-anion in the charge-transfer complex $M^+ SO_2^-$: interaction between the ions can be neglected.[100] Values are in Table 4.

The previously unobserved ν_2 bending mode of SeO_2 has been found in matrix-isolated specimens by both i.r. and Raman techniques. For $^{80}Se^{16}O_2$,

88 R. Savoie and M. Pézolet, *Canad. J. Chem.*, 1971, **49**, 2459.
89 T. Austad, J. Songstad, and K. Åse, *Acta Chem. Scand.*, 1971, **25**, 331.
90 G. Amat, *Pure Appl. Chem.*, 1969, **18**, 383.
91 H. E. Howard-Lock and B. P. Stoicheff, *J. Mol. Spectroscopy*, 1971, **37**, 321.
92 R. A. Toth, R. H. Hunt, and E. K. Plyler, *J. Mol. Spectroscopy*, 1971, **38**, 107.
93 R. Oberly, K. N. Rao, L. H. Jones, and M. Goldblatt, *J. Mol. Spectroscopy*, 1971, **40**, 356.
94 Z. Cihla and A. Chedin, *J. Mol. Spectroscopy*, 1971, **40**, 337.
95 A. Cunnington and D. H. Whiffen, *Chem. Comm.*, 1971, 981.
96 D. F. Smith, jun., T. Chao, J. Lin, and J. Overend, *Spectrochim. Acta*, 1971, **27A**, 1979.
97 A. Barbe and P. Jouve, *J. Mol. Spectroscopy*, 1971, **38**, 273.
98 A. Barbe, A. Delahaigue, and P. Jouve, *Spectrochim. Acta*, 1971, **27A**, 1439.
99 Y. LeDuff and R. Ouillon, *Compt. rend.*, 1971, **272**, *B*, 757.
100 D. E. Milligan and M. E. Jacox, *J. Chem. Phys.*, 1971, **55**, 1003.

Table 4 *Fundamentals* (cm^{-1}) *for* SO_2^- *radical-anions* [100]

Isotopic content			ν_1	ν_2	ν_3
16	32	16	984.8	495.6	1042.0
16	34	16	976.8	492.8	1029.8
16	32	18	961.2	484.0	1028.4
18	32	16	959.2	484.0	1030.8
18	32	18	946.4	472.0	1006.8

ν_1 = 922.0, ν_2 = 372.5, and ν_3 = 965.6 cm^{-1} (from i.r.),[101] and for SeO_2, ν_1 = 933, ν_2 = 382, and ν_3 = 967 cm^{-1} (from Raman).[102] Bands attributed to dimers and trimers were identified in the Raman experiment: those for the dimer support a centrosymmetric double oxygen-bridged structure with stereochemically active lone pairs on selenium. Assignments: 1002, $\nu(SeO_t)$; 660, $\nu(SeO_b)$; 543, $\nu(SeO_b)$; 363, $\delta(SeO_t)$; and 352 cm^{-1} $\delta(SeO_t)$. Approximate normal modes are in Figure 2.

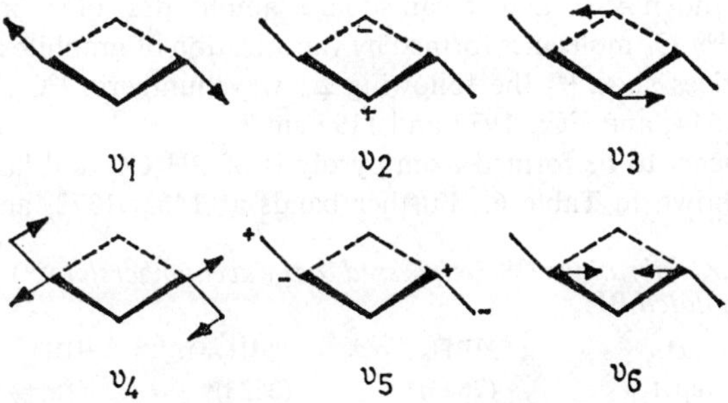

Figure 2 *Schematic representation of the vibrational modes of* $(SeO_2)_2$

Several i.r. studies of matrix-isolated metal oxides have appeared. A particularly valuable contribution gives all three fundamentals for a series M_2O (Table 5). The previously unknown ν_2 bending modes for these species were found *ca.* 200 cm^{-1} *above* predicted energies, indicating, with other evidence, the importance of metal–metal interaction.[103] A further report on Tl_2O is in discord with this work.[104] ν_3 For $U^{16}O_2$ is at 874.3, for

Table 5 *Vibrational frequencies* [103] (cm^{-1}) *of matrix-isolated oxides* M_2O

M	Al	Ga	In	Tl	In, Ga
ν_3	991.7	822.6	734.9	643.6	788.9
ν_1	715.9	595.6	555.1	510.1	550.5
ν_2	503.0	416.5	404.3	381.5	431.4

[101] S. N. Cesaro, M. Spoliti, A. J. Hinchcliffe, and J. S. Ogden, *J. Chem. Phys.*, 1971, **55**, 5834.
[102] D. Boal, G. Briggs, H. Hüber, G. A. Ozin, E. A. Robinson, and A. Vander Voet, *Chem. Comm.*, 1971, 686.
[103] D. M. Makowiecki, P. A. Lynch, and K. D. Carlson, *J. Phys. Chem.*, 1971, **75**, 1963.
[104] J. M. Brom, T. Devore, and H. F. Franzen, *J. Chem. Phys.*, 1971, **54**, 2742.

$U^{18}O_2$ at 826.6, and $U^{16}O^{18}O$ at 854.6 cm^{-1}.[40] Dioxides of Ce, Pr, and Tb each show[39] two i.r. bands in the region 718—756 cm^{-1}. $K^{16}O_2$ and $Rb^{16}O_2$ absorb at 307.5 and 255.0 cm^{-1}, respectively.[105] ν_3 Bands for $Sn^{16}O_2$ (863.1), $Sn^{16}O^{18}O$ (847.2), and $Sn^{18}O_2$ (824.7 cm^{-1}) all show ten components under high resolution, owing to the isotopic distribution of tin. As no other bands could be attributed to these molecules, they are presumed to be linear.[41]

Tetra-atomic Molecules and Ions.—After a lapse of some years, Ketelaar's 'simultaneous transitions' of molecules in CS_2 solution have come to the aid of McDowell in a study of P_4. Spectra of the vapour are known to be consistent with tetrahedral symmetry, but in CS_2 solution there is sufficient intermolecular interaction to allow direct observation in the i.r. spectrum of the A_1 and E modes. In addition, intermolecular combinations such as $[\nu_3(P_4) + \nu_1(CS_2)]$ were found. For P_4 the zeroth-order wavenumbers are $\omega_1(A_1) = 619$, $\omega_2(E) = 373$, and $\omega_3(T_2) = 481$ cm^{-1}. The force field ($f_r = 2.19$ mdyn Å^{-1}) and mean-square amplitudes of vibration were calculated.[106] C_n molecules formed by vaporization of graphite and isolated in Ar matrices show[107] the following Σ_u^+ wavenumbers: $^{12}C_4$, 2164; $^{12}C_5$, 1952 and 1544; and $^{12}C_6$, 1977 and 1197 cm^{-1}.

BH_3 appears to be formed from pyrolysis of BH_3CO and has the wavenumbers shown in Table 6. Further bands at 1403, 1378, and 910 cm^{-1}

Table 6 *I.r. vibrations*[108] *(expressed as wavenumbers*/cm^{-1}*) of matrix-isolated* BH_3

D_{3h}	$^{11}BH_3$	$^{10}BH_3$	$^{11}BD_3$
$\nu_1(A_1')$	$(2623)^a$	$(2623)^a$	$(1856)^a$
$\nu_2(A_2'')$	1125	1132	877
$\nu_3(E')$	2808	2820	2118
$\nu_4(E')$	1604	1610	1172

^{a} ( ) = calculated.

are attributed to boroxine, $H_3B_3O_3$. $H_2B_2O_3$ may be formed if traces of oxygen are present in the gas.[108] Cerf has recorded Raman spectra of binary and ternary mixtures of BCl_3, BBr_3, and BI_3. Using force constants from the parent compounds, it proved possible to assign frequencies to the normal vibrations of all the possible mixed halides (BCl_2Br, $BClBr_2$, BCl_2I, $BClI_2$, $BBrI_2$, BBr_2I, and BClBrI),[109] *cf.* Vol. 3, p. 187. ν_3 Frequencies for the radicals $^{12}CCl_3$ and $^{13}CCl_3$ are at 898 and 869 cm^{-1}, respectively: a planar (D_{3h}) structure is suggested by comparison with BCl_3 and NCl_3 species.[110]

105 L. Andrews, *J. Chem. Phys.*, 1971, **54**, 4935.
106 R. S. McDowell, *Spectrochim. Acta*, 1971, **27A**, 773.
107 K. R. Thompson, R. L. DeKock, and W. Weltner, *J. Amer. Chem. Soc.*, 1971, **93**, 4688.
108 A. Kaldor and R. F. Porter, *J. Amer. Chem. Soc.*, 1971, **93**, 2140.
109 C. Cerf, *Bull. Soc. chim. France*, 1971, 415.
110 R. Steudel, *Z. Naturforsch.*, 1971, **26b**, 475.

I.r. absorption spectra (40—800 cm^{-1}) of matrix-isolated rare-earth trifluorides show interesting differences. For PrF_3 both stretching and bending modes were observed: it is assigned pyramidal symmetry. In contrast, for La, Ce, Sm, and Eu trifluorides only the antisymmetric stretching and bending modes were evident: a planar structure is postulated. The spectrum of NdF_3 could not be interpreted.[111]

Photolysis of CO_2 yields CO_3, which has an i.r. spectrum consistent with C_{3v} symmetry. Values of the wavenumbers for all possible combinations of $^{12,13}C$, $^{16,18}O$ are listed.[112] For $^{12}C^{16}O_3$: $\nu_1(a_1)$ = 1981.1; $\nu_2(a_1)$ = 1073.4; $\nu_3(a_1)$ = 593.2; $\nu_4(b_1)$ = 971.9; and $\nu_5(b_1)$ = 568.2 cm^{-1}. Following a very detailed re-investigation of the i.r. absorption spectra of crystalline NH_3, NH_2D, ND_2H, and ND_3 a full discussion of the origin of the band splittings is given. Coupling between NH_3 molecules is very weak: shifts and band splittings are due to Fermi resonance of ν_1 with the parallel component of the overtone of the deformation vibration.[113] The ν_4 mode of NF_3 has been examined in detail by two groups:[114, 115] ν_0 for this band is placed at 493.43 ± 0.02 cm^{-1}. Integrated intensities of i.r. fundamentals of NF_3 and PF_3 have been compared using semi-empirical CNDO/2 calculations to obtain information on the electronic charge distributions accompanying the normal modes.[116] Re-analysis of high-resolution i.r. data for AsH_3 in the regions of the accidentally near-degenerate pairs of fundamentals $\nu_{1,3}$ and $\nu_{2,4}$ has led to considerably improved values of the vibrational constants.[117] The radical CH_2Br has the following vibrations [118] in Ar matrices: ν_2 [$\delta(CH_2)$] = 1355.7, ν_3 [$\nu(C—^{79}Br)$] = 693.4, and ν_6 (out-of-plane C—H bend) = 367.5 cm^{-1}.

As the author of another one so rightly states,[119a] vibrational studies of the antimony trihalides are 'relativement nombreuses'. This one is justified by the need to sort out minor inconsistencies in the literature and deals with the solid state only. Unfortunately the i.r. work is incomplete and is unlikely to be the last word on the topic. In general, the Raman data are more complete than those of other authors, but the present work does not add to or affect the results of Hooper and James [119b] on SbI_3. Part of the assignment is given in Table 7; a similar assignment of the Raman spectrum of solid $AsBr_3$ is available.[120] The discord between the above work on $SbCl_3$ and that of Ozin and co-workers [121] is distressing and doubtless

111 R. D. Wesley and C. W. DeKock, *J. Chem. Phys.*, 1971, **55**, 3866.
112 M. E. Jacox and D. E. Milligan, *J. Chem. Phys.*, 1971, **54**, 919.
113 H. Wolff, H. G. Rollar, and E. Wolff, *J. Chem. Phys.*, 1971, **55**, 1373.
114 F. N. Masri and W. E. Blass, *J. Mol. Spectroscopy*, 1971, **39**, 98.
115 S. Reichman and S. G. W. Ginn, *J. Mol. Spectroscopy*, 1971, **40**, 27.
116 I. W. Levin and O. W. Adams, *J. Mol. Spectroscopy*, 1971, **39**, 380.
117 K. Sarka, D. Papoušek, and K. N. Rao, *J. Mol. Spectroscopy*, 1971, **37**, 1.
118 D. W. Smith and L. Andrews, *J. Chem. Phys.*, 1971, **55**, 5295.
119 (*a*) E. Chemouni, *J. Inorg. Nuclear Chem.*, 1971, **33**, 2317; (*b*) M. A. Hooper and D. W. James, *Spectrochim. Acta*, 1969, **25A**, 569.
120 J. E. D. Davies and D. A. Long, *J. Chem. Soc.* (*A*), 1971, 1273.
121 E. Denchik, S. C. Nyburg, G. A. Ozin, and J. T. Szymanski, *J. Chem. Soc.* (*A*), 1971, 3157.

Table 7 *Vibrational frequencies*[119a] (cm^{-1}) *for solid* $SbCl_3$ *and* $SbBr_3$

Molecule, C_{3v}	*Unit cell,* D_{2h}	$SbCl_3$	$SbBr_3$
$\nu_1(A_1)$	$A_g + B_{1g}$ (R)	345, —	246, 243
	$B_{2u} + B_{3u}$ (I.r.)	340, 358	255, 242
$\nu_3(E)$	$A_g + B_{1g} + B_{2g} + B_{3g}$ (R)	322, 316, —, —	234, 229, 223,
	$A_u + B_{1u} + B_{2u} + B_{3u}$ (I.r.)	320, 290, —, —	228, —, —, —
$\nu_2(A_1)$	$A_g + B_{1g}$ (R)	176, —	113, —
	$B_{2u} + B_{3u}$ (I.r.)	n.o.[a]	n.o.[a]
$\nu_4(E)$	$A_g + B_{1g} + B_{2g} + B_{3g}$ (R)	154, 145, 141, —	98, 92, 89, —
	$A_u + B_{1u} + B_{2u} + B_{3u}$ (I.r.)	n.o.[a]	n.o.[a]

[a] n.o. = region not observed.

Table 8 *Vibrational frequencies* (cm^{-1}) *for trihalides of antimony and bismuth*

SbF_3*(matrix)*[a]	$SbCl_3$(g)[b]	$BiCl_3$(g)[b]	C_{3v} *assignment*
654	384 (pol)	342 (pol)	ν_1, A_1
624	361	322	ν_3, E
259	154 (pol)	123 (pol)	ν_2, A_1
—	127	107	ν_4, E

[a] I.r., ref. 122; [b] Raman, ref. 121.

Table 9 *Vibrational frequencies* (cm^{-1}) *and assignments for* GeH_3^- *and* GeD_3^- *in liquid ammonia*[123]

GeH_3^-		GeD_3^-		
I.r.	*R*	*I.r.*	*R*	
1740	1739 (pol)	1250	1256 (pol)	ν_1
809	814.5 (pol)	586	588 (pol)	ν_2
886	n.o.[a]	625	n.o.[a]	ν_4

[a] Not observed; ν_3 not observed in either spectrum.

heralds another try. Raman data for $SbCl_3$(g), $BiCl_3$(g), and SbF_3 (matrix) are in Table 8; solid $BiCl_3$ has a complex Raman spectrum consistent with its recently determined crystal structure.[122] Gas-phase Raman data for $BiBr_3$ could not be obtained owing to intense resonance fluorescence from BiBr.

I.r. spectra of liquid ammonia and dimethoxyethane solutions containing GeH_3^- and GeD_3^- complement earlier Raman work (Table 9). Using a simplified force field, the bond angle in GeH_3^- was estimated as 93°. Incomplete data for the silyl anion were also obtained.[123]

$ClOF_3$ in anhydrous HF, BF_3, AsF_5, or SbF_5 forms $ClOF_2^+$ (C_s symmetry) as revealed by Raman spectra of the solutions; wavenumbers for all these

[122] C. J. Adams and A. J. Downs, *J. Chem. Soc. (A)*, 1971, 1534.
[123] T. Birchall and I. Drummond, *J. Chem. Soc. (A)*, 1971, 3162.

are listed and assigned to the appropriate ions, BF_4^-, AsF_6^-, *etc.*[124] Typical are the results for $ClOF_2^+ HF_2^-$ (in cm^{-1}):

1339	a'	ν_1	$\nu(^{35}Cl—O)$	516	a'	ν_3	$\delta_{sym}(OClF)$
1329	a'	ν_1	$\nu(^{37}Cl—O)$	409	a'	ν_4	$\delta_{sym}(FClF)$
746	a'	ν_2	$\nu_{sym}(Cl—F)$	391	a''	ν_6	$\delta_{asym}(OClF)$
729	a''	ν_5	$\nu_{asym}(Cl—F)$				

Molecular force fields have been calculated for $SOCl_2$, $SeOCl_2$, and SO_2Cl_2.[125]

Corrosive materials such as ClF_3 and BrF_3 are readily ampouled for Raman work, but create problems for i.r. work: hence the attractiveness of matrix isolation.[126, 127] Results (Table 10) confirm previous assignments; force constants and thermodynamic functions were calculated.[126]

Table 10 *Vibrational frequencies* (cm^{-1}) *and assignments for* ClF_3 *and* BrF_3

ClF_3			BrF_3		*Assignment,* C_{2v}
Raman (*l*) [73]	*I.r.* (*l*) [73]	*I.r.* (*m*) [127]	*I.r.* (*m*) [127]	*I.r.* (*m*) [126]	
756	758	754	672	672	ν_1, A_1
518	519	523	547	545	ν_2, A_1
324	328	328	235	235	ν_3, A_1
n.o.	703	683.2	597	592	ν_4, B_1
426	416	431	347	346	ν_5, B_1
n.o.	334	332	251.5	250	ν_6, B_2

l = Liquid, *m* = matrix isolated; n.o. = not observed.

I.r. spectra of solid N_2H_2 and N_2D_2 at 77 K each show five bands tentatively assigned as follows (in cm^{-1}), on the basis of a C_{2h} *trans*-di-imide structure (Figure 3):[128]

	N_2H_2	N_2D_2
ν_2	(1552)	(1498)
ν_3	(1496)	(1167)
ν_4	3095	2291
$\nu_5(\nu_6)$	1404	1032
$\nu_6(\nu_5)$	1359	999

( ) = from combinations

Matrix-isolated H_2O_2 (and D_2O_2) show[129] i.r. wavenumbers: $\nu_{1,5}$ = 3204 (2370); ν_2 = 1298.9 (998); ν_6 = 1288.4 (942); and ν_4 = — (598) cm^{-1}. I.r. and Raman spectra of O_2F_2, both as the solid and matrix-isolated, are very similar, implying little intermolecular coupling. Observations[130] of O_2F in some experiments support a decomposition mechanism $O_2F_2 \rightarrow O_2F + F$.

124 R. Bougon, J. Isabey, and P. Plurien, *Compt. rend.*, 1971, **273**, *C*, 415.
125 K. Ramaswamy and S. Jayaraman, *J. Mol. Structure*, 1971, **7**, 474.
126 K. O. Christe, E. C. Curtis, and D. Pilipovich, *Spectrochim. Acta*, 1971, **27A**, 931.
127 R. A. Frey, R. L. Redington, and A. L. Khidir Aljbury, *J. Chem. Phys.*, 1971, **54**, 344.
128 A. Trombetti, *J. Chem. Soc.* (*A*), 1971, 1086.
129 J. A. Lannon, F. D. Verderame, and R. W. Anderson, *J. Chem. Phys.*, 1971, **54**, 2212.
130 D. J. Gardiner, N. J. Lawrence, and J. J. Turner, *J. Chem. Soc.* (*A*), 1971, 400.

$K^{16}O_2K$ and $Rb^{16}O_2K$ are presumed to have a D_{2d} bridged form and show $\nu_5(B_{2u})$ at 433 and 389 cm^{-1}, respectively.[105]

Raman spectra of pure Se_2Cl_2 and Se_2Br_2 and of their solutions in CS_2, CCl_4, and C_6H_6 have been interpreted to show that the chloride has C_2

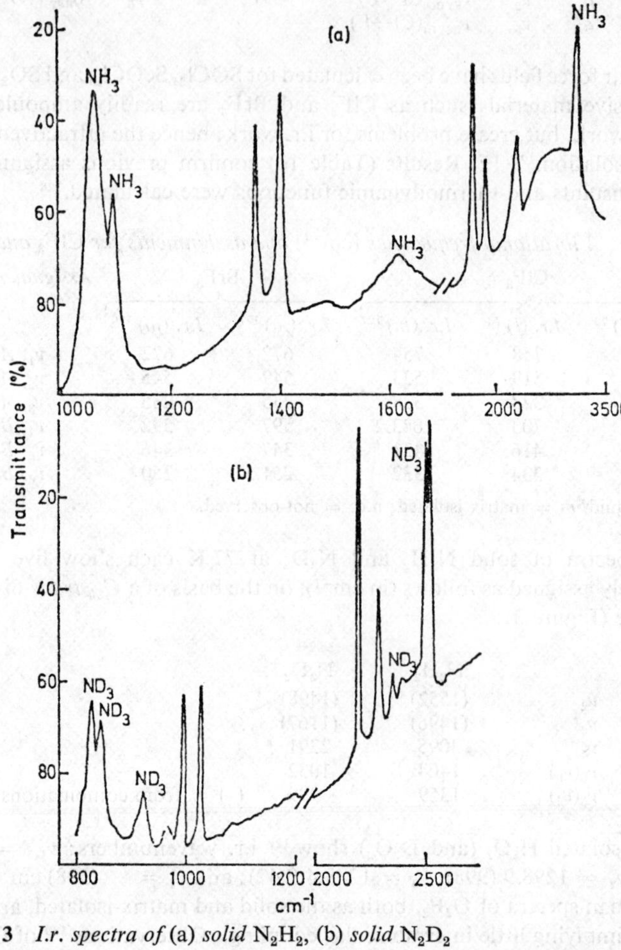

Figure 3 *I.r. spectra of* (a) *solid* N_2H_2, (b) *solid* N_2D_2

symmetry (known previously) but that the bromide is of C_{2v} symmetry. Assignments (cm^{-1}) for the liquid Se_2Cl_2 are:[131] 358 (ν_1, *A*), 289.5 (ν_2, *A*), 133 (ν_3, *A*), 83.5 (ν_4, *A*), 358 (ν_5, *B*), and 148 (ν_6, *B*); ν_1 and ν_5 are separable in solution, *e.g.* in CS_2, $\nu_1 = 366$ and $\nu_5 = 351$ cm^{-1}. Liquid Se_2Br_2 assignments are 254 (ν_1, A_1 and ν_5, B_1), 286 (ν_2, A_1), 105 (ν_3, A_1), 56 (ν_4, A_2),

[131] W. Kiefer, *Spectrochim. Acta*, 1971, **27A**, 1285.

and 120 (ν_6, B_1) cm^{-1}. In each case the lowest-frequency line is the torsional mode. Normal-co-ordinate analyses of S_2Cl_2 and S_2Br_2 are reported.[132]

Another i.r. study (10—400 cm^{-1}; ambient and 90 K) of the mercurous halides, not readily summarized, yielded several weak bands in addition to strong bands previously reported; they are explained in terms of frequencies calculated for the whole unit cell.[133]

New i.r. gas-phase data for fulminic acid (HCNO) and its deuteriate have been assigned as follows:[134] ν_1 [Σ^+, ν(C—H)] = 3336 (2580), ν_2 [Σ^+, ν_{asym}(CNO)] = 2198 (2066), ν_3 [Σ^+, ν_{sym}(CNO)] = 1256 (1254), ν_4 [Π, δ(HCNO)] = 538 (460 and 478), and ν_5 [Π, δ(CNO)] = 331 (—), where all values are those of the *Q*-branch and those in parentheses refer to the deuteriate. High-resolution work on the ν_1 vibration–rotation bands of HCNO [135] and DCNO [136] yielded accurate molecular constants (ν_0 for $D^{12}C^{14}N^{16}O$ = 2620.727 ± 0.002 cm^{-1}), the HCNO study being concerned specifically with hot bands, while normal-co-ordinate analyses for the same molecules [137] and for HNCS, DNCS,[138] and ClCNO [137] provided quantitative descriptions of the vibrations in terms of potential energy distributions, force constants, and Coriolis coupling constants. The i.r. spectrum of ClCNO has been analysed in terms of C_s symmetry and force constants have been computed;[137] some i.r. data for BrCNO are also reported.[139]

Observed (i.r.) and calculated wavenumbers for *cis*- and *trans*-HONO in a matrix are available for several isotopic variants. Typical is the assignment [140] for *cis*-$H^{16}O^{14}N^{16}O$: ν_1 [ν(OH)] = 3142, ν_2 [ν(N=O)] = 1633, ν_3 (HON bend) = 1265, ν_4 [ν(N—O)] = 850, ν_5 (ONO bend) = 610, and ν_6 (OH torsion) = 637 cm^{-1}. ONOF has [141] $\nu_{asym}(NO_2)$ = 1713, $\nu_{sym}(NO_2)$ = 1302, and ν(OF) = 885 cm^{-1}. I.r. data are also listed for $[HCS_2]^-$,[142] $[CS_2Se]^{2-}$, $[CSSe_2]^{2-}$,[143] and $[HCS(SH)]_x$.[144]

Penta-atomic Molecules and Ions.—Tetrahedral and substituted tetrahedral species are back in favour this year. The gas-phase Raman technique is being fruitfully applied, vibrational interactions between tetrahedral ions in solids and the analysis of gas-phase band systems form an apparently bottomless Ph.D. mine and, for chloro-species in particular, distinguishing

132 C. A. Frenzel and K. E. Blick, *J. Chem. Phys.*, 1971, **55**, 2715.
133 T. Osaka, *J. Chem. Phys.*, 1971, **54**, 863.
134 W. Beck, P. Swoboda, K. Feldl, and R. S. Tobias, *Chem. Ber.*, 1971, **104**, 533.
135 B. P. Winnewisser, *J. Mol. Spectroscopy*, 1971, **40**, 164.
136 W. D. Sheasley, C. W. Mathews, E. L. Ferretti, and K. N. Rao, *J. Mol. Spectroscopy*, 1971, **37**, 377.
137 H. H. Eysel and E. Nachbaur, *Z. anorg. Chem.*, 1971, **381**, 71.
138 B. Orel, B. Peterman, M. Obradovič, D. Hadži, and A. Ažman, *Spectroscopy Letters*, 1971, **4**, 39.
139 W. Gottardi, *Angew. Chem. Internat. Edn.*, 1971, **10**, 416.
140 W. A. Guillory and C. E. Hunter, *J. Chem. Phys.*, 1971, **54**, 598.
141 J. E. Sicre and H. J. Schumacher, *Z. anorg. Chem.*, 1971, **385**, 131.
142 G. Gattow, M. Dräger, and R. Engler, *Naturwiss.*, 1971, **58**, 53.
143 G. Gattow and M. Dräger, *Z. anorg. Chem.*, 1971, **384**, 235.
144 G. Gattow and R. Engler, *Naturwiss.*, 1971, **58**, 53.

isotopic structure from other causes of band multiplicity remains a large, pitfall-strewn, field.

Heat capacity measurements indicate that solid sodium and potassium borohydrides undergo order–disorder transitions below room temperature although full structural details are not available. The i.r. and Raman spectra at temperatures below these transitions are consistent with BH_4^- site symmetries T_d (K) and D_{2d} (Na). Space-group changes accompanying the transitions are inferred: $O_h^5 \rightarrow T_d^2$ (K) and $O_h^5 \rightarrow D_{2d}^9$ (Na).[145] $LiBH_4$, which has no heat capacity anomalies, shows spectra consistent with location of BH_4^- on either general positions, two-fold axes, or mirror planes.[145] BF_4^- apparently retains tetrahedral symmetry in $NaBF_4$ and ($NaBF_4$ + NaF) melts, ν_1 decreasing with increasing melt temperature;[146] for solid salts, factor-group splitting is found for all modes other than ν_3.[147] Errors in a force-constant calculation on BF_4^- have been noted.[148] Improved spectra allow the assignment of ν_1(847 cm^{-1}) and ν_2(445 cm^{-1}) for the cation in $NF_4^+AsF_6^-$.[148a] The series $[AlCl_nBr_{4-n}]^-$ ($0 \leqslant n \leqslant 4$) have been studied by i.r. and Raman techniques and a Urey–Bradley field has been found which fits all five members.[149]

Gas-phase Raman *Q*-branch wavenumbers (Table 11) have been reported for MX_4 (X = Cl, Br, or I; M = Si, Ge, Sn, Zr, or Hf), and for CF_4, CCl_4, and SiF_4;[150, 151] details of *O–P* and *R–S* band separations are discussed. Band splitting in the Raman spectra of many of these tetrahalides as solids has been interpreted to show that ν_3 exhibits the most pronounced correlation effects and that ν_1 shows mainly isotopic structure with some hot-band distortion in the case of some of the heavier tetrachlorides.[152a, 152b] Of particular value is a Raman study of ^{35}Cl-enriched solid CCl_4, from which conclusions are drawn that are similar to those of Clark *et al.*,[152] although some evidence for isotopic splitting of ν_2 and ν_4 was found.[153] Gallium dihalides are known to have the composition $Ga^+ [GaX_4]^-$. The first report has appeared of the Raman spectra of the solids, plus i.r. data down to 200 cm^{-1}. Band splittings correlate well with those expected on the basis of site group symmetry (C_2) but are many fewer than predicted by unit-cell analysis of the tetramolecular cell; for the iodide, splittings are even less than required by C_2.[154] General quadratic force-field potential constants have been computed for $SnCl_4$.[155]

145 K. B. Harvey and N. R. McQuaker, *Canad. J. Chem.*, 1971, **5**, 3272.
146 A. S. Quist, J. B. Bates, and G. E. Boyd, *J. Chem. Phys.*, 1971, **54**, 4896.
147 J. B. Bates, A. S. Quist, and G. E. Boyd, *J. Chem. Phys.*, 1971, **54**, 124.
148 M. Fonassier and M. T. Forel, *J. Mol. Spectroscopy*, 1971, **39**, 527.
148a K. O. Christie and D. Pilipovich, *Inorg. Chem.*, 1971, **10**, 2803.
149 R. H. Bradley, P. N. Brier, and D. E. H. Jones, *J. Chem. Soc.* (*A*), 1971, 1397.
150 R. J. H. Clark, B. K. Hunter, and D. M. Rippon, *Chem. and Ind.*, 1971, 787.
151 R. J. H. Clark and D. M. Rippon, *Chem. Comm.*, 1971, 1295.
152 (*a*) R. J. H. Clark and B. K. Hunter, *J. Chem. Soc.* (*A*), 1971, 2999; (*b*) H. F. Shurvell, *J. Mol. Spectroscopy*, 1971, **38**, 431.
153 H. F. Shurvell, *Spectrochim. Acta*, 1971, **27A**, 2375.
154 E. Chemouni, *J. Inorg. Nuclear Chem.*, 1971, **33**, 2325.
155 H. Fujii and M. Kimura, *J. Mol. Spectroscopy*, 1971, **37**, 517.

Large collections of data for substituted tetrahedral species have appeared during the year, but we cannot reproduce them all here. In any case, the interested user of them will need to consult the originals. All i.r.-active fundamentals other than ν_6 of $HCBr_3$ and $DCBr_3$ were found in gas-phase spectra of HCF_3, DCF_3,[163, 164] $HCBr_3$, $DCBr_3$, $HSiBr_3$, and $DSiBr_3$; ν_6 was observed for the liquids.[165] High-resolution studies have appeared for ν_6 and $(\nu_2 + \nu_6)$ of CH_3I,[166] and for ν_4 and $2\nu_4$ of CD_3Cl.[167] A Urey–Bradley

Table 11 *Vibrational frequencies* (cm^{-1}) *of tetrahedral species*

Species	$\nu_1(a_1)$	$\nu_2(e)$	$\nu_3(t_2)$	$\nu_4(t_2)$	*Ref.*
$MgCl_4^{2-}$	252	100	330	142	157
$MgBr_4^{2-}$	150	61	290	90	157
MgI_4^{2-}	107	42	259	60	157
$CdCl_4^{2-}$	265	104	250	104	120
BD_4^-	1621	916, 905	1696, 1677	862, 854	156
BF_4^-	773	357	1080	528	147
$AlCl_4^-$	—	—	480	—	120
CF_4	908.4	434.5	1283.0	631.2	151
CCl_4	460.0	214.2	792, 765	313.5	151
SiF_4	800.8	264.2	1029.6	388.7	151
$SiCl_4$	423.1	145.2	616.5	220.3	151
$SiBr_4$	246.7	84.8	494.0	133.6	151
SiI_4	166.3	57.6	405.0	*ca.* 90	151
$GeCl_4$	396.1	125.0	459.0	171.0	151
$GeBr_4$	235.7	74.7	332.0	111.1	151
GeI_4	156.0	51.6	273.0	77.3	151
$SnCl_4$	369.1	95.2	408.2	126.1	151
$SnBr_4$	222.1	59.4	284.0	85.9	151
SnI_4	147.7	42.4	*ca.* 210	63.0	151
$PbCl_4$	331	90	352	103	158
PBr_4^+	254	116	503, 496	148	159, 162
AsH_4^+	2080	949	2142	818, 813	160
TiI_4	162	51	323	67	161*a*, 161*b*
$ZrCl_4$	377	90	418	113	150
$ZrBr_4$	225.5	60	315	72	150
ZrI_4	158	43	254	55	150
$HfCl_4$	382	101.5	390	112	150
$HfBr_4$	235.5	63	273	71	150
HfI_4	158	55	224	63	150

156 A. M. Heyns, C. J. H. Schutte, and W. S. Chevermann, *J. Mol. Structure*, 1971, **9**, 271.
157 V. A. Maroni, *J. Chem. Phys.*, 1971, **55**, 4789.
158 R. J. H. Clark and B. K. Hunter, *J. Mol. Structure*, 1971, **9**, 354.
159 M. Delahaye, P. Dhamelincourt, and J.-C. Merlin, *Compt. rend.*, 1971, **272**, *B*, 370.
160 J. R. Durig, C. B. Pate, and Y. S. Li, *J. Chem. Phys.*, 1971, **54**, 1033.
161 (*a*) R. J. H. Clark and C. J. Willis, *J. Chem. Soc.* (*A*), 1971, 838; (*b*) H. Bürger, C. Kluess, and H.-J. Neese, *Z. anorg. Chem.*, 1971, **381**, 198.
162 W. Gabes and H. Gerding, *Rec. Trav. chim.*, 1971, **90**, 157.
163 A. Ruoff, H. Bürger, and S. Biedermann, *Spectrochim. Acta*, 1971, **27A**, 1359.
164 A. Ruoff, H. Bürger, and S. Biedermann, *Spectrochim. Acta*, 1971, **27A**, 1377.
165 H. Bürger and J. Cichon, *Spectrochim. Acta*, 1971, **27A**, 2191.
166 H. Matsuura and J. Overend, *Spectrochim. Acta*, 1971, **27A**, 2165.

force-field study of 18 mixed carbon tetrahalides resulted in re-assignment [168] for CF_2I_2: $\nu_1(a_1) = 1064$, $\nu_2(a_1) = 605$, $\nu_3(a_1) = 270$, $\nu_4(a_1) = 112$, $\nu_5(a_2) = 208$, $\nu_6(b_1) = 1110$, $\nu_7(b_1) = 304$, $\nu_8(b_2) = 740$, and $\nu_9(b_2) = 250$ cm^{-1}.

Band splittings (i.r.) shown by SiH_4 matrix-isolated in rare gases, CO, N_2, or CH_4 are interpreted in terms of occupation of at least two types of defect site; there is no evidence of free or nearly free rotation.[169] Raman spectra and assignments are available for SiH_3Br, SiD_3Br (i.r. included),[170] and the series $SiF_{4-n}Cl_n$ ($0 \leqslant n \leqslant 4$),[171, 172] the work of Ozin and co-workers [171] being especially complete. Band contours (i.r.) of $SiHF_3$ and $SiDF_3$ have been analysed.[173] Using available data, Hofler computed force constants for $SiCl_4$ and $SiBr_4$ and used them to make unambiguous assignments for $SiCl_nBr_{4-n}$.[174] A similar service (using some new data) has been performed for $TiCl_nBr_{4-n}$.[175] Vapour-phase Raman band contours provided additional evidence upon which an improved assignment for $Si(Me)_4$ was based.[176] Raman spectra of solid monohalogenostannanes show only two or three lines (cm^{-1}):[177]

	ν(Sn—H)	$\delta(SnH_3)$	*Lattice*
SnH_3Cl	1880	733	38
SnH_3Br	1880	733	54
SnH_3I	1880	—	49

Absence of lines attributable to ν(Sn—halogen) is taken to imply a basically ionic structure containing approximately planar SnH_3^+ ions. The vapour of SnH_3Cl shows i.r. absorption at 1927 [ν(Sn—H)]; 692 [$\delta(SnH_3)$]; 383 and 370 cm^{-1} [*P* and *R*-branches of ν(Sn—Cl)].

Existing vibrational data for $POFCl_2$, POF_2Cl, $PSBr_3$, $PSFBr_2$, and PSF_2Br have been collected and most probable values of frequencies, intensities, and depolarization ratios given. Arguments were based upon an internally consistent set of normal-co-ordinate analyses; the missing Raman lines of POF_2Cl are indicated as 424 and 274 cm^{-1}.[178] The ion PCl_3Br^+ has been demonstrated in compounds of composition $PBCl_7Br$ and $P_2F_6Cl_3Br$.[179] Data and assignment for one of these are in Table 12.

[167] R. W. Peterson and T. H. Edwards, *J. Mol. Spectroscopy*, 1971, **38**, 524.
[168] L. H. Ngai and R. H. Mann, *J. Mol. Spectroscopy*, 1971, **38**, 322.
[169] R. E. Wilde, T. K. K. Srinivasan, R. W. Harral, and S. G. Sankar, *J. Chem. Phys.*, 1971, **55**, 5681.
[170] H. Bürger, J. Cichon, and A. Ruoff, *Z. Naturforsch.*, 1971, **26b**, 1068.
[171] K. Hamada, G. A. Ozin, and E. A. Robinson, *Canad. J. Chem.*, 1971, **49**, 477.
[172] M.-L. Delé-Dubois and F. Wallart, *Compt. rend.*, 1971, **272**, *B*, 1059.
[173] H. Bürger, S. Biedermann, and A. Ruoff, *Spectrochim. Acta*, 1971, **27A**, 1687.
[174] F. Höfler, *Z. Naturforsch.*, 1971, **26a**, 547.
[175] H. Bürger, C. Kluess, K. Wiegel, and F. Höfler, *Z. Naturforsch.*, 1971, **26a**, 550.
[176] S. Sportouch, C. Lacoste, and R. Gaufrès, *J. Mol. Structure*, 1971, **9**, 119.
[177] J. R. Webster, M. M. Millard, and W. L. Jolly, *Inorg. Chem.*, 1971, **10**, 879.
[178] J. S. Ziomek, F. J. Fillwalk, and E. A. Piotrowski, *Appl. Spectroscopy*, 1971, **25**, 212.
[179] F. F. Bentley, A. Finch, P. N. Gates, and F. J. Ryan, *Chem. Comm.*, 1971, 860.

Table 12 *Vibrational data* (cm^{-1}) *and assignment for* PCl_3Br^+ *as its* BCl_4^- *salt* [119a]

I.r.	*Raman*	*Assignment*, C_{3v}
637s	647vw	ν_4, e
577vs	582w	ν_1, a_1
390m	390vs	ν_2, a_1
232m	233s	ν_5, e
213m	213vs	ν_3, a_1
155w	155s	ν_6, e

The vibrational spectra of SbF_4^- in its solid salts M^ISbF_4 (M^I = Na, K, Cs, or NH_4) indicate C_{2v} symmetry.[122]

New Raman data for SF_4 are reported. An assignment is given, which differs from those of other authors in that of the bending modes $\nu_{3,4,5,9}$, and is based upon fine detail previously ignored (Table 13).[180a] Using three

Table 13 *Bending mode assignments for* SF_4

		Ref. 180*a*	*Refs.* 180*b*, *c*
a_1	ν_3	353	464.5
a_1	ν_4	228, 233	228
a_2	ν_5	475	414
b_2	ν_9	228, 233	353

possible assignments, approximate normal co-ordinates were calculated for each and used to obtain i.r. intensities with the aid of dipole derivatives computed by the CNDO/2 method: these were compared with observed i.r. intensities. The calculation proved moderately sensitive to differences in the vibrational assignment, but it was not possible to prove one or the other. However, this is an interesting application of the CNDO/2 method. Further i.r. and Raman data for solid $SeCl_4$ and $TeCl_4$,[181] Raman spectra of $TeCl_2Br_2$(g) (monomer, C_1 symmetry), and Raman and i.r. data for tetraethylammonium salts of $TeCl_2Br_3^-$ (probably square pyramidal) and $TeCl_2Br_4^{2-}$ (? *cis*-octahedral) have appeared.[182]

The first Raman spectra for XeF_4(g) and $XeOF_4$(g) are reported.[74] Together with previous i.r. data, vapour values are now available for all modes of these molecules other than $\nu_7(e_u)$ of XeF_4. Despite appreciable vapour–condensed phase frequency shifts, no changes of assignment were required. Force constants were computed using the OVFF. A particular feature of them is explicit inclusion of the lone pairs, treated as ligands of very small mass present in 'vacant' octahedral positions. The output included very high frequencies for each lone pair (then ignored) while

[180] (*a*) I. W. Levin, *J. Chem. Phys.*, 1971, **55**, 5393; (*b*) K. O. Christie and W. Sawodny, *J. Chem. Phys.*, 1970, **52**, 6320; (*c*) R. A. Frey, R. L. Redington, and A. L. Aljibury, *J. Chem. Phys.*, 1971, **54**, 344.

[181] R. Ponsioen and D. J. Stufkens, *Rec. Trav. chim.*, 1971, **90**, 521.

[182] G. A. Ozin and A. Vander Voet, *Canad. J. Chem.*, 1971, **49**, 704.; *J. Mol. Structure*, 1971, **10**, 397.

retaining the effect of the lone-pair–bond-pair repulsions. Wavenumbers are in Table 14. Both i.r. and Raman data are recorded for $M^{I}XeO_3F$ (M^{I} = K, Rb, or Cs) and $MXeO_3Cl,MCl$ (M = Rb or Cs). $KXeO_3F$ is known to have symmetry C_{2v}^{9} with XeO_3F^- anions linked into chains. More band splitting was observed than indicated below, but was not

Table 14 *Vibrational frequencies* (cm^{-1}) *for* XeF_4 *and* $XeOF_4$ *in the gas phase* [74]

XeF_4, D_{4h}			$XeOF_4$, C_{4v}			
			ν_1	a_1	$\nu(Xe{-}O)$	926.3
ν_1	a_{1g}	554.3	ν_2	a_1	$\nu(XeF_4)$	576.9
ν_2	a_{2u}	291	ν_3	a_1	$\pi(XeF_4)$	285.9
ν_3	b_{1g}	218	ν_4	b_1	$\delta(FXeF)$	225
ν_5	b_{2g}	524	ν_5	b_2	$\nu(XeF_4)$	543
ν_4	b_{1u}	(216)	ν_6	b_2	$\pi(XeF_4)$	(219)
ν_6	e_u	586	ν_7	e	$\nu(XeF_4)$	609
			ν_8	e	$\delta(XeO)$	362
ν_7	e_u	(161)	ν_9	e	$\delta(FXeF)$	161

nearly complete enough to warrant interpretation on any model other than perturbed C_{3v} anion symmetry. $KXeOF_3$: a_1 [$\nu(Xe{-}O)$] = 763, a_1 [$\delta(Xe{-}O)$] = 358, a_1 [$\nu(Xe{-}F)$] = 257, e [$\nu(Xe{-}O)$] = 828, e [$\delta(Xe{-}O)$] = 315 and e [$\delta(OXeF)$] = 222 cm^{-1}.[183]

Using vibrational data for $Os^{16}O_4$ and $Os^{18}O_4$, harmonic frequencies have been obtained (Table 15). With the help of Coriolis constants accurate

Table 15 *Harmonic frequencies* [184] (cm^{-1}) *for* $Os^{16}O_4$ *and* $Os^{18}O_4$

	$Os^{16}O_4$	$Os^{18}O_4$
a_1	974.3	918.5
e	345.2	325.4
t_2	976.9	926.6
t_2	345.0	328.3

force constants were computed, *e.g.* f_r = 8.32 ± 0.06, f_{rr} = 0.21 ± 0.03 mdyn $Å^{-1}$; the value of f_r implies substantial double-bond character.[184] Technetium and rhenium dissolve in molten lithium perchlorate with formation of TcO_4^- and ReO_4^-, as revealed by Raman spectra of the melts. Oxyanion frequencies are comparable with those of aqueous solutions except that ν_1 suffers a red shift of *ca.* 20 cm^{-1}.[185] In molten LiF–NaF–KF Raman vibrations of CrO_4^{2-} are at 880, 840 (pol), 378, and 348 cm^{-1}.[186] Single crystals of KBr doped with M^{2+} CrO_4^{2-} show i.r. absorption in the

183 P. LaBonville, J. R. Ferraro, and T. M. Spittler, *J. Chem. Phys.*, 1971, **55**, 631.
184 R. S. McDowell and M. Goldblatt, *Inorg. Chem.*, 1971, **10**, 625.
185 D. Cohen, S. Fried, and H. Selig, *J. Inorg. Nuclear Chem.*, 1971, **33**, 2687.
186 F. L. Whiting, G. Mamantov, G. M. Begun, and J. P. Young, *Inorg. Chim. Acta*, 1971, **5**, 261.

350—1000 cm^{-1} region, consistent with occupation of three distinct symmetry sites by CrO_4^{2-}. Small M^{2+} dependent shifts (M^{2+} = Mg, Ca, Sr, Ba, or Pb) are taken as evidence of $M^{2+}\cdots CrO_4^{2-}$ ion pairing.[187] Although the oxyanions in the salts $[Ph_4M][MnO_4]$ and $[Ph_4M][ClO_4]$ (M = P or As) are known to be on S_4 sites, their i.r. spectra are as expected for tetrahedral symmetry.[188] Frequencies associated with the PO_4^{3-}, AsO_4^{3-}, and VO_4^{3-} ions in a series of lead apatites have been listed and assigned.[189] I.r. and Raman spectra of M_2TeO_4 (M = K, Rb, or Cs) are interpreted in terms of vibrations of individual TeO_4^{2-} ions. The formation of solid solutions of K_2TeO_4–K_2SO_4 supports this conclusion. The spectra show correlation splitting and are completely different from those of Na_2TeO_4, in which tellurium is octahedrally co-ordinated. Typical of the results are the Raman wavenumbers (in cm^{-1}) for the caesium salt: 803w and 790w (ν_3), 772vs (ν_1), and 313w, 305m, and 263s (deformations).[190] The Br—O bond order of the BrO_4^- ion, calculated from the intensity of the a_1 Raman line, is similar to that of I—O in IO_4^-, in agreement with the known stability of BrO_4^-.[191] Recent results for MO_3X species are in Table 16.[192–194]

able 16 *Vibrational frequencies* (cm^{-1}) *and assignment for* MO_3X *species*

C_{3v}	$MnO_3F(g)$[a]	$[SO_3Cl]^-$ [b]	$[MoO_3S]^{2-}$ [c]	$[WO_3S]^{2-}$ [d]	$[WOS_3]^{2-}$ [e]	*Assignment*
a_1	905.2	1070	905	920 (914)	469	$\nu_{sym}(MO_3)$
a_1	720.7	400	475	450 (459)	875	ν(M—X)
a_1	337.7	640	320	320 (316)	184	$\delta_{sym}(MO_3)$
e	952.5	1250	838	860 (860)	453	$\nu_{asym}(MO_3)$
e	373.9	550	320	320 (316)	184	$\delta_{asym}(MO_3)$
e	264.3	320	235	250 (243)	277	M—X wag

[a] I.r.;[192] [b] i.r. and Raman, as Li^+, Na^+, K^+, and NH_4^+ salts; polarization measurements on ɔlten NH_4SO_3Cl;[193] [c] i.r.,[194a] [d] i.r. wavenumbers with Raman wavenumbers in parentheses,[194a] ⌞aman spectrum; numbering differs from original report; O and X interchanged in mode descrip- ns.[194b]

Full assignments of the spectra of HN=C=NH and DN=C=ND (i.r. in Ar matrix),[195] CF_2=NCl [C_s symmetry, i.r. (g)],[196] cyanogen azide, NCN_3, [i.r. (g) and (matrix), Raman (solutions)],[197] and C_3S_2 (i.r. and

[187] P. J. Miller, G. L. Cessac, and R. K. Khanna, *Spectrochim. Acta*, 1971, **27A**, 2019.
[188] E. J. Baran, *Z. anorg. Chem.*, 1971, **382**, 80.
[189] V. M. Bhatnagar, *Canad. J. Chem.*, 1971, **49**, 662.
[190] P. Tarte and F. Leyder, *Compt. rend.*, 1971, **273**, *C*, 852.
[191] J. D. Witt and R. M. Hammaker, *Inorg. Chem.*, 1971, **10**, 1093.
[192] M. J. Reisfeld, L. B. Asprey, and N. A. Matwiyoff, *Spectrochim. Acta*, 1971, **27A**, 765.
[193] Y. Auger, P. Legrand, E. Puskavic, F. Wallart, and S. Noël, *Spectrochim. Acta*, 1971, **27A**, 1351.
[194] (*a*) M. J. F. Leroy, M. Burgard, and A. Müller, *Bull. Soc. chim. France*, 1971, 1183; (*b*) A. Müller, N. Weinstock, B. Krebs, B. Buss, and A. Ferwanah, *Z. Naturforsch.*, 1971, **26b**, 268.
[195] S. T. King and J. H. Strope, *J. Chem. Phys.*, 1971, **54**, 1289.
[196] R. P. Hirschmann, H. L. Simon, and D. E. Young, *Spectrochim. Acta*, 1971, **27A**, 421.
[197] B. Bak, O. Bang, F. Nicolaisen, and O. Rump, *Spectrochim. Acta*, 1971, **27A**, 1865.

Raman, solid)[198] are available. For C_3S_2 the anharmonicity of ν_7 (deformation about the central carbon atom) apparently changes sign on passing from gas to solid. This very low-frequency skeletal mode (*ca.* 100 cm^{-1} depending upon conditions) is highly anharmonic and is associated with oscillations that cannot be considered infinitesimal (mean vibrational amplitudes are calculated); in other words it behaves as a quasi-linear molecule with little or no barrier to deformation, the subtler effects of this behaviour being seen in intensities and band contours.[198] Unassigned i.r. data for monothioformic acid, HCOSH, are reported.[199] Reaction of hydrogen atoms with HCN yields[200] a matrix-isolated species believed to be $H_2C{=}NH$ with i.r. bands at 3500, 2500, 1600, 1100, and 800 cm^{-1}.

Hexa-atomic Molecules and Ions.—Vibrational spectra of trigonal-bipyramidal molecules continue to attract attention because of the importance of their deformational modes in understanding pseudorotation phenomena. Further Raman work by Miller and Capwell[201a] supports two earlier reports,[201b,c] including location of the long-sought ν_4 at 174(g) and 177(l) cm^{-1}. If ν_4 is important in exchange of axial and equatorial ligands it should show evidence of pronounced anharmonicity, but none was found. It is concluded that the barrier to interchange is at least 2000 cm^{-1} high, but as this is still consistent with rapid ligand interchange the results neither support nor confirm Berry's proposed mechanism. 'Recommended' values for PF_5(g) are: $\nu_1(a_1') = 816$, $\nu_2(a_1') = 648$, $\nu_3(a_2'') = 946.6$, $\nu_4(a_2'') = 575.1$, $\nu_5(e') = 1025$, $\nu_6(e') = 532.5$, $\nu_7(e') = 174$, and $\nu_8(e'') = 520$ cm^{-1}. An electron-diffraction determination of mean vibrational amplitudes in PCl_5 suggests that the correct assignment of the e' modes is that in which the in-plane equatorial bend is at lower frequency than the axial bend.[202] If axial–equatorial ligand exchange is to take place, then the normal co-ordinates associated with the e' vibrational potential must be of the form $(2)^{-\frac{1}{2}}(S_{6a} + S_{7a})$ where S_{6a} and S_{7a} are e' symmetry co-ordinates. From a theoretical analysis of perpendicular band contours, Coriolis constants were extracted and used in defining the e' vibrational potential function. *Two* force fields were found to fit the experimental data; *one* of them is associated with normal co-ordinates of the correct form for ligand interchange.[203] In summary, although these papers all provide additional data, none supplies convincing evidence for the detailed mechanism of pseudorotation.

The previously postulated[204a] triple coincidence of one stretch (ν_4, B_1) and two deformational modes (ν_3, A_1 and ν_8, E) at 480 cm^{-1} in the i.r.

[198] J. B. Bates and W. H. Smith, *Spectrochim. Acta*, 1971, **27A**, 409.
[199] G. Gattow and R. Engler, *Angew. Chem. Internat. Edn.*, 1971, **10**, 415.
[200] P. M. A. Sherwood and J. J. Turner, *J. Chem. Soc. (A)*, 1971, 2474.
[201] (*a*) F. A. Miller and R. J. Capwell, *Spectrochim. Acta*, 1971, **27A**, 125; (*b*) I. W. Levin, *J. Chem. Phys.*, 1969, **50**, 1031; (*c*) I. R. Beattie, K. M. S. Livingstone, and D. J. Reynolds, *J. Chem. Phys.*, 1969, **51**, 1969.
[202] W. J. Adams and L. S. Bartell, *J. Mol. Structure*, 1971, **8**, 23.
[203] L. C. Hoskins and C. N. Perng, *J. Chem. Phys.*, 1971, **55**, 5063.
[204] (*a*) G. M. Begun, W. H. Fletcher, and D. F. Smith, *J. Chem. Phys.*, 1965, **42**, 2236; (*b*) K. O. Christie, *Spectrochim. Acta*, 1971, **27A**, 631.

spectrum of C_{4v} ClF_5 has been confirmed[204b] by a matrix-isolation study: ν_3 and ν_8 are i.r.-active and separate out in the matrix to yield two chlorine-isotope doublets of the correct intensity ratio. I.r. (matrix) data for BrF_5 are reported and found to support a previous assignment;[204a] the close correlation between the vibrational spectra of ClF_3, BrF_3,SF_4, and BrF_5 is emphasized.[127] Force-constant analyses have been computed for ClF_5 and BrF_5 [for both, the axial bond-stretch force constant > equatorial, *viz*, 3.33 and 2.99 (Cl), and 4.24 and 3.42 mdyn $Å^{-1}$ (Br)],[205] and for IF_5 and $XeOF_4$.[206] Raman spectral data have been presented which show the presence of temperature-dependent equilibria in gaseous MF_5 (M = Nb, Ta, or Sb) and demonstrate polymeric species as the main components of the vapours at their boiling points. Previously reported i.r. (matrix) data[207a] are now believed to relate to non-monomeric species and hence the square-pyramidal structure deduced from them is probably erroneous.[207b] The major difficulty in interpretation of the vibrational spectrum of IF_5 is the presence of three strongly polarized bands in the Raman spectrum of the liquid in the ν(I—F) region. It is now shown that in hexafluorobenzene solution there is only a single narrow line (at 700 cm^{-1}); the other components present in the spectrum of IF_5(l) are explained by an association model.[208]

I.r. and Raman spectra of SF_5^- as solids and in solutions strongly support a square-pyramidal structure for this anion;[209] the assignment is in Table 17. The previously claimed VCl_5 has been shown by Griffiths and

Table 17 *Vibrational frequencies*[209] (cm^{-1}) *and assignment for* SF_5^-

Raman		*I.r.*			
(Solid)	*(Solution)*	*(Solid)*	*(Solution)*		
795	794 (pol)	795	797	$\nu_1(a_1)$	$\nu(SF_{ax})$
530	528 (pol)	525	526	$\nu_2(a_1)$	$\nu[SF_{4(eq)}]$
467	465 (pol)	469	467	$\nu_3(a_1)$	$\pi(SF_4)$
434	432	—	—	$\nu_4(b_1)$	$\nu[SF_{4(eq)}]$
262	—	—	—	$\nu_5(b_1)$	$\pi(SF_4)$
348	—	—	—	$\nu_6(b_2)$	$\delta(SF_4)$
596	596 (dpol)	597	595	$\nu_7(e)$	$\nu[SF_{4(eq)}]$
	—	388	—	$\nu_8(e)$	S—F axial wag
242	—	247	241	$\nu_9(e)$	$\delta(SF_4)$

co-workers to be a salt of VCl_5^-. The i.r. spectrum of the compound now formulated as $[PCl_4]^+$ $[VCl_5]^-$ has been tentatively assigned by analogy with PCl_5, $GeCl_5^-$, and $SnCl_5^-$: ν_5 = 420, ν_3 = 320, ν_4 = 192, and ν_6 = 161 cm^{-1}. The novel compounds $[PCl_4]^+$ $[VOCl_4]^-$ and $VBr_3.POBr_3$ have

205 K. Ramaswamy and P. Muthusubramanian, *J. Mol. Structure*, 1971, **7**, 45.
206 E. C. Curtis, *Spectrochim. Acta*, 1971, **27A**, 1989.
207 (*a*) A. L. K. Aljibury and R. L. Redington, *J. Chem. Phys.*, 1970, **52**, 453; (*b*) L. E. Alexander, *Inorg. Nuclear Chem. Letters*, 1971, **7**, 1053.
208 L. E. Alexander and I. R. Beattie, *J. Chem. Soc.* (*A*), 1971, 3091.
209 L. F. Drullinger and J. E. Griffiths, *Spectrochim. Acta*, 1971, **27A**, 1793.

Table 18 *Vibrational frequencies* (cm^{-1}) *of* B_2Cl_4 *in fluid and solid states* [213]

		D_{2d}					D_{2h}		
		Approximate normal modes	*Raman (liquid)*	*I.r. (gas)*			*Approximate normal modes*	*Raman (solid)*	*I.r. (solid)*
ν_1	a_1 (R)	$^{11}B—^{11}B$ stretch	1122	—	ν_1	a_g (R)	$^{11}B—^{11}B$ stretch	1106	—
ν_2	a_1 (R)	B—Cl stretch	401	—	ν_2	a_g (R)	B—Cl stretch	401	—
ν_3	a_1 (R)	Cl—B—Cl deformation	176	—	ν_3	a_g (R)	Cl—B—Cl deformation (in-phase)	187	—
ν_4	b_1 (R)	Torsion	—	—	ν_4	a_u (—)	Torsion	—	—
ν_5	b_2 (I.r., R)	B—Cl stretch (out-of-phase)	—	728	ν_{11}	b_{3u} (I.r.)	B—Cl stretch	—	708
ν_6	b_2 (I.r., R)	Cl—B—Cl deformation (out-of-phase)	289	289	ν_{12}	b_{3u} (I.r.)	Cl—B—Cl deformation (out-of-phase)	—	259
ν_7	e (I.r., R)	B—Cl stretch	915	917	ν_5	b_{1g} (R)	B—Cl stretch	920	—
					ν_9	b_{2u} (I.r.)	B—Cl stretch	—	947
ν_8	e (I.r., R)	Cl—B—Cl out-of-plane wag	—	512	ν_8	b_{2g} (R)	Cl—B—Cl out-of-plane wag	517	—
					ν_7	b_{1u} (I.r.)	Cl—B—Cl out-of-plane wag	—	526
ν_9	e (I.r., R)	Cl—B—Cl in-plane rock	104	104	ν_6	b_{1g} (R)	Cl—B—Cl in-plane rock	253	—
					ν_{10}	b_{2u} (I.r.)	Cl—B—Cl in-plane rock	—	126

also been characterized.[210] The anion in $Cs^+ [VOF_4]^-$ probably has C_{4v} symmetry; it shows[211] ν(V=O) at 1016 and 1023 cm^{-1}, between those of VOF_3 and VOF_5^{2-}. Raman and i.r. data for XeO_3F_2 have been interpreted[212] in terms of D_{3h} symmetry and force constants calculated on the basis of the OVFF; assignment: $a_1' = 806.7$ and 567.4; $a_2'' = 631.7$ and 375.4; $e' = 894$, 318, and 190; and $e'' = 361$ cm^{-1}.

A detailed study of B_2Cl_4 shows that non-planar (D_{2d}) molecules are probably present in the liquid in contrast to planar (D_{2h}) molecules known to be in the solid.[213] The assignment is in Table 18. New data and assignments have appeared for oxalyl fluoride, $(COF)_2$, for both *cis*- and *trans*-isomers;[214] aminodifluorophosphine, H_2NPF_2;[215] $XOClO_3$ (X = Cl or Br);[216] and S_4N_2, for which the structure (1) was deduced,[217] although the

(1)

possible boat, chair, and planar conformations could not be distinguished. I.r. frequencies have been listed for potassium methylamide, $K[CH_3NH]$.[218] Comparison of theoretical and experimental band contours of CF_3CN allowed determination of vibrational constants; assignment of $\nu_3(a_1)$ at 801 cm^{-1} is supported.[219] Medium- and high-resolution i.r. spectra of CH_3HgX(g) {including analysis of the ν_3 [ν(C—H)] and $\rho_r(CH_3)$ modes} led to the assignments of Table 19.[220] An i.r. study of gaseous and matrix-

Table 19 *Vibrational frequencies* (cm^{-1}) *and assignment*[220] *for gaseous* CH_3HgX

	Cl	Br	I	*Band type*	*Assignment*
ν(Hg—C)	504 (pol)	539 (pol)	532 (pol)	$\parallel$	$\nu_3(a_1)$
$\rho_r(CH)_3$	793	792	779	$\perp$	$\nu_7(e)$
$\delta_{sym}(CH_3)$	1199 (pol)	1199 (pol)	1188 (pol)	$\parallel$	$\nu_2(a_1)$
$\delta_{asym}(CH_3)$	1400 ?	1402 ?	1385 ?	$\perp$	$\nu_6(e)$
ν_{sym}(C—H)	2927 (pol)	2927 (pol)	2923 (pol)	$\parallel$	$\nu_1(a_1)$
ν_{asym}(C—H)	3019	3018	3017	$\perp$	$\nu_5(e)$

$\nu_4(a_1)$ and $\nu_8(e)$ out of range of instrument.

[210] I. M. Griffiths, D. Nicholls, and K. R. Seddon, *J. Chem. Soc.* (*A*), 1971, 2513.
[211] J. A. S. Howell and K. C. Moss, *J. Chem. Soc.* (*A*), 1971, 270.
[212] H. H. Claassen and J. L. Huston, *J. Chem. Phys.*, 1971, **55**, 1505.
[213] J. R. Durig, J. E. Saunders, and J. D. Odom, *J. Chem. Phys.*, 1971, **54**, 5285.
[214] D. Smith, D. W. James, and J. P. Derlin, *J. Chem. Phys.*, 1971, **54**, 4437.
[215] D. W. H. Rankin, *J. Chem. Soc.* (*A*), 1971, 783.
[216] K. O. Christe, C. J. Schack, and E. C. Curtis, *Inorg. Chem.*, 1971, **10**, 1589.
[217] J. Nelson and H. G. Heal, *J. Chem. Soc.* (*A*), 1971, 136.
[218] R. C. Makhija and R. A. Stairs, *Canad. J. Chem.*, 1971, **49**, 807.
[219] J. A. Faniran and H. F. Shurvell, *Spectrochim. Acta*, 1971, **27A**, 1945.
[220] Z. Meić and M. Randić, *J. Mol. Spectroscopy*, 1971, **39**, 39.

isolated formamide led to a complete assignment, including the NH_2 'inversion' at 288 cm^{-1} in the vapour.[221]

Hepta-atomic Molecules and Ions.—Few octahedral anions have been studied this year (Table 20). Analysis of the electronic band system of

Table 20 *Vibrational frequencies* (cm^{-1}) *for octahedral species*

	ν_1	ν_2	ν_3	ν_4	ν_5	ν_6	*Ref.*
$Ni[PF_6]_2$	745	572	840	555	465	(402)	222
$K[BrF_6]$ [a]	568	454	*ca.* 400	204, 184	250	138	223
$Rb[BrF_6]$	568	456	*ca.* 400	203, 182	250	144	223
$Cs[BrF_6]$	562	451	*ca.* 400	193, 176	243	156	223
$Cs_2[MnF_6]$ [b]	592	508	620	335	308	—	224
$[PBr_4][TaBr_6]$	236	174	—	—	106	—	159
$Cs_2[OsBr_6]$	—	—	221	—	—	—	225
$Cs_2[OsI_6]$	—	—	165	—	—	—	225
$[Ph_4As][UF_6]$	—	—	525	173	—	—	226
$[Et_4N][UCl_6]$	—	—	310	122	—	—	226
$[Et_4N][UBr_6]$	—	—	214	87	—	—	226
$Cs_2[NpCl_6]$	310	—	265	117	128	—	226*a*

[a] Assigned on basis of D_{3d} symmetry; [b] from electronic spectrum.

ReF_6 near 2 μm at 15 K enabled deduction of the vibrational wavenumbers for the $A\ E_{5/2g} \leftarrow X\ G_{3/2g}$ state: $\nu_1(a_{1g}) = 711$, $\nu_2(e_g) = 661$, $\nu_4(t_{1u}) = 266$, $\nu_5(t_{2g}) = 297$, and $\nu_6(t_{2u}) = 178$ cm^{-1}.[227] Half-bandwidths of the $\nu_3(t_{1u})$ mode of hexachloro-complexes $[MCl_6]^{2-}$ (M = Sn, Pt, or Te) have been computed from far-i.r. reflectance spectra. Values of $\Delta\nu_{\frac{1}{2}}$ for the tellurates are more than twice those for the other salts; this is taken as evidence for the dynamic stereochemical effect of the lone pair.[228] Normal-co-ordinate analyses have been reported for MnF_6^{2-},[228a] for $TaCl_6^-$, $TaBr_6^-$, and for the niobium analogues.[229] T_{1u}-block force constants of improved accuracy have been calculated using data from $^{32}SF_6$ and $^{34}SF_6$.[230]

ν(Os—halogen) frequencies have been assigned[225] for a series of complexes $Cs_2[OsBr_{6-n}I_n]$, where n = 0—6, and fragmentary data reported for $(Et_4N)_2[UOX_5]$ (X = F, Cl, or Br).[226] Rather complete data and

221 S. T. King, *J. Phys. Chem.*, 1971, **75**, 405.
222 H. G. Mayfield and W. E. Bull, *J. Chem. Soc.* (*A*), 1971, 2280.
223 R. Bougon, P. Charpin, and J. Soriano, *Compt. rend.*, 1971, **272**, *C*, 565.
224 C. D. Flint, *J. Mol. Spectroscopy*, 1971, **37**, 414.
225 W. Preetz and H. J. Walter, *J. Inorg. Nuclear Chem.*, 1971, **33**, 3179.
226 J. L. Ryan, *J. Inorg. Nuclear Chem.*, 1971, **33**, 153.
226a B. W. Berringer, J. B. Gruber, T. M. Loehr, and G. P. O'Leary, *J. Chem. Phys.*, 1971, **55**, 4608.
227 J. C. D. Brand, G. L. Goodman, and B. Weinstock, *J. Mol. Spectroscopy*, 1971, **38**, 449.
228 D. M. Adams and M. H. Lloyd, *J. Chem. Soc.* (*A*), 1971, 878.
228a M. L. Mehta, *Spectroscopy Letters*, 1971, **4**, 395.
229 M. N. Auasthi and M. L. Mehta, *J. Mol. Structure*, 1971, **7**, 301.
230 S. N. Thakur, *J. Mol. Structure*, 1971, **7**, 315.

assignments for $ReOF_5$, $OsOF_5$, and IOF_5,[231] $[SbCl_5Br]^-$,[232] $[SiMeF_5]^{2-}$,[233] and the anion $[Xe(OH)O_5]^{3-}$ (believed to be the predominant species in Cs_4XeO_6,nH_2O) [234] have appeared (Table 21). Assignments are proposed for the *cis*-$[MO_2F_4]$ anions in $K_2MO_2F_4,H_2O$ (M = Mo or W).[235]

Table 21 *Vibrational frequencies* (cm^{-1}) *and assignment for* MXY_5 *species*

C_{4v}		$ReOF_5$ [231]	$OsOF_5$ [231]	IOF_5 [231]	$[SbCl_5Br]^-$ [232]	$[Xe(OH)O_5]^{3-}$ [234]
a_1	ν_1	989.8	962.6	927.0	334 (340)[a]	704
	ν_2	738.6	716.4	680.4	290 (290)	685
	ν_3	643	644	640.2	230 (245)	652
	ν_4	309	280.5	362.9	170 (170)	484
b_1	ν_5	652	644	647	310 (305)	—
	ν_6	234	210	307	110 (120)	—
b_2	ν_7	334	332	*ca.* 330	210 —	452
e	ν_8	713.0	700.6	710.3	— (322)	605
	ν_9	260	263	372.2	186 (183)	443
	ν_{10}	365	367	343	—	427
	ν_{11}	125	164	204.8	110 (120)	400

[a] Figures in parentheses are from i.r. spectra, others from Raman spectra.

Larger Molecules and Ions.—I.r. (gas) and Raman (liquid) spectra of $Ni(C^{16}O)_4$ and $Ni(C^{18}O)_4$ have been assigned as shown in Table 22. These

Table 22 *Vibrational frequencies* (cm^{-1}) *of gaseous nickel tetracarbonyl* [236]

	$Ni(C^{18}O)_4$	$Ni(C^{16}O)_4$
ν_1	2084	2131
ν_2	354.4	367.5
ν_3	376	380
ν_4	61	64
ν_5	2010.6	2057.6
ν_6	454.2	458.8
ν_7	417.4	421
ν_8	76	80
ν_9	296	300

data, together with those of $Ni(C^{16}O)_{4-n}(PF_3)_n$, have been used in calculating the general quadratic valence force constants of the tetracarbonyl.[236] The values so obtained are in good agreement with those due to Jones.

Force-constant calculations on the trichlorides, tribromides, and triiodides of Al, Ga, and In suggest that earlier assignments of ν_1 for $GaCl_3$ and InI_3 may be in error as they lead to an irregular trend in $(f_r + 2f_{rr})$ values.[237]

[231] J. H. Holloway, H. Selig, and H. H. Claassen, *J. Chem. Phys.*, 1971, **54**, 4305.
[232] G. Goetz, M. Deneux, and M. J. F. Leroy, *Bull. Soc. chim. France*, 1971, 29.
[233] K. Licht, C. Penker, and C. Dathe, *Z. anorg. Chem.*, 1971, **380**, 293.
[234] G. D. Downey, H. H. Claassen, and E. H. Appelman, *Inorg. Chem.*, 1971, **10**, 1817.
[235] A. Beuter and W. Sawodny, *Z. anorg. Chem.*, 1971, **381**, 1.
[236] G. Bouquet and M. Bigorgne, *Spectrochim. Acta*, 1971, **27A**, 139.
[237] M. N. Auasthi and M. L. Mehta, *J. Mol. Structure*, 1971, **8**, 49.

Several A_2B_7 systems have been studied this year. Raman vibrations for a species $[Cl_3AlClAlCl_3]^-$, assumed to have D_{3h} symmetry and present in KCl–$AlCl_3$ melts, have been assigned as follows: $A' = 425$ and 312; $E' = 435$, 164, and 128; and $E'' = 435$, 164, and 99 cm^{-1}.[238] Both Ga_3Cl_7 and KGa_2Cl_7 yield identical Raman spectra and are believed to contain $[Ga_2Cl_7]^{2-}$, isostructural with $[S_2O_7]^{2-}$. Decomposition according to the scheme $Ga[Ga_2Cl_7] \rightleftharpoons \frac{1}{2}Ga_2Cl_6 + Ga[GaCl_4]$ appears to occur in the melt. Assuming C_{2v} symmetry, the following assignment was made, but no Raman polarization data were obtained and the evidence for it is not good: $A_1 = 420$, 367, 202, 168, 129, and 97; $A_2 = 352$ and 108; $B_1 = 390$ and 146; and $B_2 = 410$, 355, 279, 165, and 154 cm^{-1}.[239] Rather complete i.r. data (above 400 cm^{-1}) and the Raman spectrum [240a] of liquid Cl_2O_7 support a recent assignment.[240b] Raman spectra of Tc_2O_7 show that the vapour- and liquid-phase monomeric structures are similar, but quite different from those of the solid;[241a] there is a close correspondence with Re_2O_7.[241b] Tc_2O_7(s) has a simpler spectrum than Re_2O_7(s), consistent with their different structures.

Raman spectra of solid and liquid $TaBr_5$ show 36 lines; the reader is referred to the paper [242a] for full details. This result (*cf.* only 14 lines were observed by other workers [242b]) represents the difference in instrumental performance. Frequencies for $NbBr_5$ and $TaBr_5$ have also been listed, without assignment.[243]

$Pt_6Br_6Cl_6$ shows [243] i.r. absorption at 358, 310, 231, 196, and 137 cm^{-1}.

3 Single-crystal and other Solid-state Spectroscopy

Papers dealing with single-crystal spectroscopy commonly contain lengthy tables of results and assignments which it would be impracticable to reproduce here. We therefore emphasize assignments for molecules and complex ions, where these have been deduced, and points of particular interest arising from these studies. Also included in this section is work on powder samples interpreted with the aid of unit-cell analyses.

'Simple' Lattice Types.—Single-crystal spectra of the 'simpler' lattice types have been actively investigated for many years, mainly by physicists. A further study of CdI_2 and partial work on PbI_2 are reported.[244] Willemsen's single-crystal Raman results for $PbCl_2$ (preliminary report) [245] are in substantial discord with Ozin's [246] with regard to frequencies and (in several

[238] H. A. Øye, E. Rytter, P. Klaeboe, and S. J. Cyvin, *Acta Chem. Scand.*, 1971, **25**, 559.
[239] E. Chemouni and A. Potier, *J. Inorg. Nuclear Chem.*, 1971, **33**, 2343.
[240] (*a*) A. C. Pavia, J. Roziere, and J. Potier, *Compt. rend.*, 1971, **273**, *C*, 781; (*b*) J. D. Witt and R. M. Hammaker, *Chem. Comm.*, 1970, 667.
[241] (*a*) H. Selig and S. Fried, *Inorg. Nuclear Chem. Letters*, 1971, **7**, 315; (*b*) I. R. Beattie and G. A. Ozin, *J. Chem. Soc.* (*A*), 1969, 2615.
[242] (*a*) J.-C. Merlin and M.-B. Delahaye-Buisset, *Compt. rend.*, 1971, **272**, *B*, 53; (*b*) I. R. Beattie, T. R. Gilson, and G. A. Ozin, *J. Chem. Soc.* (*A*), 1968, 2765.
[243] P. M. Druce and M. F. Lappert, *J. Chem. Soc.* (*A*), 1971, 3595.
[244] C. Carabatos, *Compt. rend.*, 1971, **272**, *B*, 465.
[245] B. Willemsen, *J. Inorg. Nuclear Chem.*, 1971, **33**, 3963.
[246] G. A. Ozin, *Canad. J. Chem.*, 1970, **48**, 2931.

cases) the polarization data. Such disparity is disturbing: one waits the full report with interest. Willemsen's results are in better accord with the selection rules than Ozin's and agree quite closely with those of a third group[247] who found 16 ($PbCl_2$) or 17 ($BaCl_2$) of the predicted 18 Raman lines ($6A_{1g} + 3B_{1g} + 6B_{2g} + 3B_{3g}$ in D_{2h}^{16}). These authors[247] give no experimental details and do not state whether their assignments are based upon single-crystal work, although it is unlikely that they were guessed. A normal-co-ordinate analysis shows a higher bond stretching constant for $BaCl_2$ than for $PbCl_2$. $SrCl_2$ (Cl type) yields a single Raman line at 183 cm^{-1} (T_{2g}).[247] I.r.-active modes of the EO_1 lattice have been determined for several compounds (Table 23), but *LO* modes were reported only for

Table 23 *I.r.-active modes for the EO_1 lattice*

D_{4h}		A_{2u}	E_u	A_{2u}	E_u	*Ref.*
BiOCl		529	283	186	111	248
BiOBr		520	263	125	68	248
LaOCl		510	370	192	135	248
LaOBr		520	328	138	100	248
BaBrF	*TO*	299	201.6	122	97.7	249
	LO	330	272.5	137	116.5	249

BaBrF. The pressure dependence of the i.r.-active *TO* mode in alkaline-earth fluorides has been studied up to 45 kbar. It decreases in the order $CaF_2 > SrF_2 > BaF_2$; in contrast, the Raman-active modes are barely affected by pressure. Gruneisen parameters are considerably lower than those found for AB-type crystals.[250] Raman spectra of rutile and high-pressure orthorhombic TiO_2 were studied up to 40 kbar; the B_{1g} mode of rutile is unusual in showing negative pressure dependence.[251]

Far-i.r. spectra of 12 lanthanide chlorides and 11 bromides (4 structure types studied) are shown to be diagnostic of structure type; numbers of bands observed are in good agreement with those predicted by unit cell analysis.[252]

I.r. absorption by MgO crystals containing OH^- and OD^- has been used to study their mode of incorporation. A band at 2727 cm^{-1} is due to vibration of two OD^- ions coupled to one cation; vibrations at 2444.7 and 2455.5 cm^{-1} are associated with the centres $OD^-\square$ and $OD^-\square\ O^{2-}Mg^{3+}$ respectively ($\square$ = cation vacancy).[253]

Mixed Oxides and Fluorides.—B_{1u} and B_{2u} (D_{2h}) modes of $GdAlO_3$ have been unequivocally assigned using i.r. reflectance (100—1200 cm^{-1}) and

[247] A. Sadoc and R. Guillo, *Compt. rend.*, 1971, **273**, *B*, 203.
[248] A. Rulmont, *Compt. rend.*, 1971, **272**, *B*, 1364.
[249] H. E. Rast, S. A. Miller, and H. H. Caspers, *J. Chem. Phys.*, 1971, **55**, 1484.
[250] J. R. Ferraro, H. Horan, and A. Quattrochi, *J. Chem. Phys.*, 1971, **55**, 664.
[251] M. Nicol and M. Y. Fong, *J. Chem. Phys.*, 1971, **54**, 3167.
[252] L. J. Basile, J. R. Ferraro, D. Gronert, and A. Quattrochi, *J. Chem. Phys.*, 1971, **55**, 3729.
[253] B. Henderson and W. A. Sibly, *J. Chem. Phys.*, 1971, **55**, 1276.

Kramers–Kronig analysis.[254] In a paper primarily concerned with electronic Raman spectra of lanthanides, some single-crystal phonon data are mentioned for YbAlG, TmAlG, ErAlG, DyAlG, TbAlG, EuAlG, and PrF_3 (G = garnet).[255] Cubic perovskites KMF_3 (M = Mg, Mn, Fe, Co, Ni, or Zn) have three i.r.-active modes of T_u symmetry, and are therefore suitable for study as powders. Assignments have been made and force constants calculated for the complete unit cells. I.r. spectra of the orthorhombic perovskite fluorides $NaMF_3$ (M = Mg, Mn, Co, Ni, or Zn) accord well with predictions of factor-group analysis.[256a] Further studies of spinels and related oxides are reported.[265b–f] I.r. frequencies of two forms of $FeNbO_4$ and $Fe_2Nb_2O_6$ have been discussed on the basis of unit-cell analyses.[265g]

I.r. frequencies (59—4000 cm^{-1}) are reported for powder samples of several hexa-, penta-, and tetra-fluorometallates of aluminium, gallium, and iridium with a variety of univalent cations.[257] For compounds of known crystal structure, discussion and assignment are based upon unit-cell analysis. No splittings of ν_3 and ν_4 modes of (MF_6) octahedra were found even in compounds of low symmetry. For pentafluoro-complexes substantially fewer bands than predicted were found, but for tetrafluoro-compounds the numbers of bands are in complete agreement with theory. Thus, for $TlAlF_4$ (space group D_{4h}^1, $Z = 1$) $3A_{2u}$ and $4E_u$ modes are expected (there is an error in the E_g analysis). Seven bands are present: 700, 560, 400, 336, 231, 203, and 105 cm^{-1}, of which the last is mainly due to cation translation (158 cm^{-1} in NH_4^+ salt). The following values (cm^{-1}) are found for hexafluorometallates:

	M_3AlF_6		M_3GaF_6		M_3InF_6			
M =	Na	NH_4	Na	NH_4	Na	K	NH_4	Tl
ν_3	590	570	500	470	467	455	460	410
ν_4	390	385	308	295	250	255	235	200

Sheet and Chain Structures.—Complete or nearly complete observations and assignments have been made for several hexagonal crystals, $AMCl_3$, in which chains of face-sharing (MCl_6) octahedra are found: $AMCl_3$ (A = Rb or Cs; M = Fe, Co, or Ni),[258] $(Me)_4N(MnCl_3)$,[259] and $(Me)_4N(CdCl_3)$.[260]

[254] P. Alain and B. Piriou, *Compt. rend.*, 1971, **272**, *B*, 1185.
[255] J. A. Konigstein and P. Grunberg, *Canad. J. Chem.*, 1971, **49**, 2336.
[256] (*a*) A. P. Lane, D. W. A. Sharp, J. M. Barraclough, D. H. Brown, and D. A. Paterson, *J. Chem. Soc.* (*A*), 1971, 94; (*b*) J. Preudhomme and P. Tarte, *Spectrochim. Acta*, 1971, **27A**, 1817; (*c*) H. D. Lutz and M. Feher, *ibid.*, p. 357; (*d*) J. Preudhomme and P. Tarte, *ibid.*, p. 845; (*e*) *ibid.*, p. 961; (*f*) C. Ricchiccioli-Deltcheff, T. Dupuis, and C. Wadier, *Compt. rend.*, 1971, **273**, *B*, 1020; (*g*) C. Ricchiccioli-Deltcheff, *ibid.*, p. 1095.
[257] P. Bukovec, B. Orel, and J. Siaftar, *Monatsh.*, 1971, **102**, 885.
[258] A. Chadwick, J. T. R. Dunsmuir, I. W. Forrest, A. P. Lane, and S. Fernando, *J. Chem. Soc.* (*A*), 1971, 2794.
[259] D. M. Adams and R. R. Smardzewski, *Inorg. Chem.*, 1971, **10**, 1127.
[260] D. M. Adams and D. C. Newton, *J. Chem. Soc.* (*A*), 1971, 3499.

Both i.r. (absorption and reflectance) and Raman single-crystal methods were employed. Similar work on $CsCuCl_3$ showed that vibrational interaction throughout the entire $[Cu_6Cl_{18}]^{6-}$ repeat of the chain takes place.[260] Detailed assignment for the *cis*-diaquo octahedral chain in $CsMnCl_3,2H_2O$ is also available.[260]

A single-crystal Raman study[261] of red HgI_2 led to the assignment 142 (B_{1g}), 114 (A_{1g}), and 29 and 17 cm^{-1} (E_g) in terms of D_{4h}, all of which were attributed to modes internal to the sheet structure, an assignment subsequently supported by a normal-co-ordinate analysis;[262] powder data for yellow HgI_2 are also reported.[263] The two Raman-active vibrations of the sheet structure of SnF_4 (D_{4h}^{17}) are at 621 (A_{1g}) and 230 cm^{-1} (E_g); calculated i.r.-active modes are in tolerable agreement with those found for the powder.[264] The linear chain structure of α-BiF_5 (C_{4h}^{5}, $Z = 1$) shows i.r. (powder) absorption at 625 and 356 (E_u), and 450 and 250 cm^{-1} (A_u), assignments being based upon comparison with calculated frequencies.[264] SeO_2 has eight formula units in a unit cell of symmetry D_{4h}^{13}; interpretation of single-crystal spectra is therefore difficult. Using a combination of single-crystal Raman data and normal-co-ordinate analysis, an assignment has been proposed.[264] A single-crystal Raman and i.r. reflection study of TeO_2 (D_4, $Z = 4$) yielded the predicted $4A_1 + 5B_2 + 4B_2$ modes in addition to seven of the $9E$ modes (both transverse and longitudinal components) but not all of the A_2 modes were seen owing to experimental limitations. The crystal is piezoelectric and *LO–TO* splitting is expected for the *E* modes, with possible subsequent interaction with A_2 modes; this was found to be the case.[265]

Complex Halides.—Single-crystal Raman and i.r. spectra for $NaBF_4$ show that splitting due to static field effects is greater than those due to the correlation field (Table 24). Complete assignments for ν_3 could not be made because of mixing with $2\nu_4$.[266]

I.r. reflectance spectra of A_2MCl_4 and A_3MCl_5 complexes have been used to investigate the various interactions possible in such crystals.[267] Rather complete factor-group splitting is exhibited by ν_3 of the anion in Cs_2ZnCl_4, for example, and others behave similarly. When tetra-alkylammonium salts are used, splittings are less pronounced and orientational disorder affects the spectra. Instead of a series of sharp bands due to lattice modes (as in Cs salts), broad bands are found which are probably best regarded as density-of-states distributions arising from breakdown of the zone-centre

[261] D. M. Adams and M. A. Hooper, *Austral. J. Chem.*, 1971, **24**, 885.
[262] N. Krauzman, M. Krauzman, and H. Poulet, *Compt. rend.*, 1971, **273**, *B*, 301.
[263] Y. Marqueton, F. Abba, E.-A. Decomps, and M.-A. Nusimovici, *Compt. rend.*, 1971, **272**, *B*, 1014.
[264] I. R. Beattie, N. Cheetham, T. R. Gilson, K. M. S. Livingston, and D. J. Reynolds, *J. Chem. Soc.* (*A*), 1971, 1910.
[265] M. Krauzman and J.-P. Mathieu, *Compt. rend.*, 1971, **273**, *B*, 342.
[266] J. B. Bates, *J. Chem. Phys.*, 1971, **55**, 489.
[267] J. T. R. Dunsmuir and A. P. Lane, *J. Chem. Soc.* (*A*), 1971, 404.

($k = 0$) approximation.[268] Single-crystal i.r. reflectance spectra of NH_4HgCl_3 yielded the predicted $3A_{2u} + 3E_u$ (in D_{4h}) modes. A band at 196 cm^{-1} is attributed to a lattice mode.[268a]

Two concordant i.r. reflectance studies[269, 270] of $K_2CuCl_4,2H_2O$ and some related complexes, and one single-crystal Raman investigation[270]

Table 24 *Assignment*[266] *for single-crystal* $NaBF_4$

	BF_4^-, T_d *Aqueous solution*	*Site*, C_{2v}		*Unit cell*, D_{2h}
ν_1	773 a_1	(782) A_1		785 A_g
				779 B_{1u}
		(369) A_1		368 A_g
				370 B_{1u}
ν_2	357 e		26 cm^{-1}	
		(343) A_2		343 B_{1g}
				Inactive A_u
		(551) A_1		553 A_g
				550 B_{1u}
			26 cm^{-1}	
ν_4	528 f_2	(525) B_1		534 B_{2g}
				516 B_{3u}
			3 cm^{-1}	
		(528) B_2		530 B_{3g}
				526 B_{2u}

have established the assignment rather fully. There is minor disagreement over some mode descriptions within the E_u species but this is largely a matter of semantics that can be resolved by a normal-co-ordinate analysis. The assignment for the *trans*-$[CuCl_2(H_2O)_2]$ molecules in $K_2CuCl_4,2H_2O$ is close to that of Beattie *et al.* for single-crystal $CuCl_2,2H_2O$, emphasizing the environmental similarities.

The single-crystal Raman spectrum of Cs_2HgI_4 reveals considerable correlation splitting of anion modes, all of which was assigned; i.r. and Raman powder data for Ag_2HgI_4 and Cu_2HgI_4 were included.[271]

Square-pyramidal ions occur in K_2SbF_5, $KTeF_5$,[208] and $(Et_4N)_2[InCl_5]$;[272] vibrational assignments have been deduced for all three on the basis of

[268] J. T. R. Dunsmuir and A. P. Lane, *J. Chem. Soc.* (*A*), 1971, 2781.
[268a] J. T. R. Dunsmuir and A. P. Lane, *J. Inorg. Nuclear Chem.*, 1971, **33**, 4361.
[269] J. T. R. Dunsmuir and A. P. Lane, *J. Chem. Soc.* (*A*), 1971, 2724.
[270] D. M. Adams and D. C. Newton, *J. Chem. Soc.* (*A*), 1971, 3507.
[271] N. Krauzman and J.-P. Mathieu, *Compt. rend.*, 1971, **272**, *B*, 955.
[272] D. M. Adams and R. R. Smardzewski, *J. Chem. Soc.* (*A*), 1971, 714.

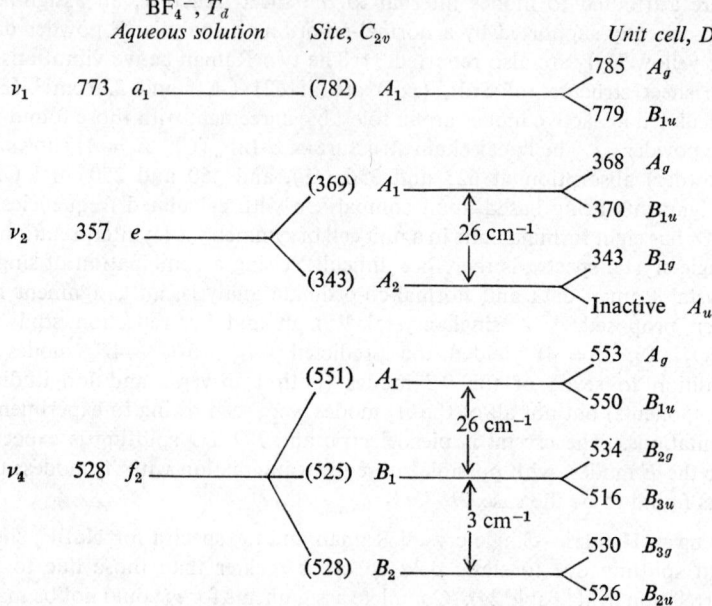

single-crystal Raman spectra plus single-crystal i.r. spectra for the indate. Assignments are summarized in Table 25; ν_7 for the fluorides is surprisingly low in comparison with the other stretching modes.

Table 25 *Assignments for C_{4v} anions*

		$K[TeF_5]$ [208]	$K_2[SbF_5]$ [208]	$(Et_4N)_2[InCl_5]$ [272]
ν_1	a_1	624[a]	557[a]	294
ν_2		517	427	283
ν_3		291	278	140
ν_4	b_1	579	388	287
ν_5		—	—	193
ν_6	b_2	243	220 ?	165
ν_7	e	488	347, 375	274
ν_8		345	307	143
ν_9		—	142 ?	108

[a] Raman wavenumbers listed; i.r. counterparts can be several cm^{-1} removed.

Oxoanion-containing Crystals.—Two Raman studies [273, 274] in good agreement have established the splitting pattern of internal anion modes in K_2CrO_4 and Rb_2CrO_4 (Table 26). Agreement in the lattice mode region was less good, although if full experimental details were published it would be easier to understand the discrepancies, some of which may be due to

Table 26 *Assignment* [273, 274] *of Raman-active anion internal modes in* K_2CrO_4

T_d	Site, C_s	Unit Cell, D_{2h}
884 t_2	$2a'$	919 B_{2g}, 905 A_g, 884 B_{2g}, 868 A_g
	a''	881 B_{1g}, 877 B_{3g}
847 a_1	a'	852 $A_g + B_{2g}$
368 t_2	$2a'$	398 B_{2g}, 397 A_g, 388 B_{2g}, 387 A_g
	a''	393 B_{3g}, 388 B_{1g}
348 e	a'	352 B_{2g}, 347 A_g
	a''	351 B_{3g}, 348 B_{1g}

[273] R. L. Carter and C. E. Bricker, *Spectrochim. Acta*, 1971, **27A**, 569.
[274] D. M. Adams, M. A. Hooper, and M. H. Lloyd, *J. Chem. Soc. (A)*, 1971, 946.

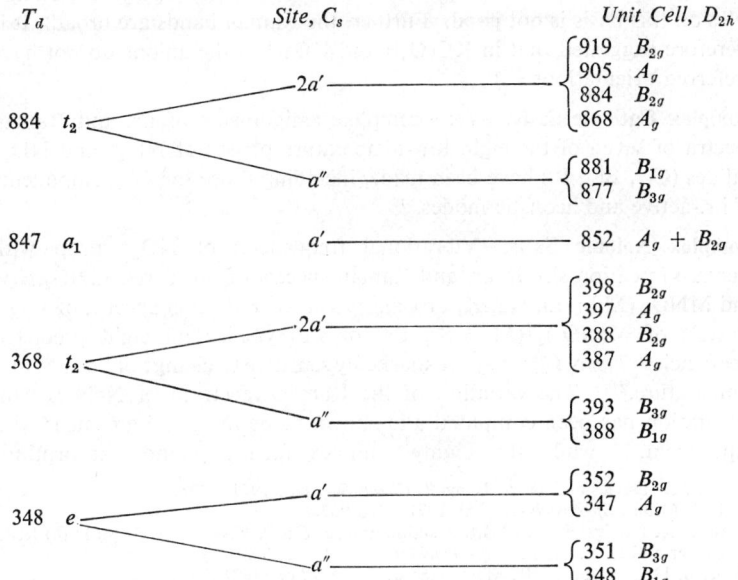

assignment of residuals as fundamentals. With increasing cation size both internal and external modes fall, and splitting out of anion internal modes in the crystals become less complete owing to the greater 'insulating' effect of the larger cations. A detailed discussion of i.r. and Raman powder spectra of several chromates and dichromates has been given in terms of factor-group splitting.[275] In $MgMoO_4$ single-crystal, although the *numbers* of bands due to anion internal modes in the crystal environment are as predicted, those corresponding to ν_1 are *ca.* 50 cm^{-1} higher than in free MoO_4^{2-}, and ν_3 components exhibit pronounced splitting; bending modes are also considerably mixed with lattice modes. These observations are explained by strong Mg–O coupling.[276] $Tb_2(MoO_4)_3$ has symmetry *Pba*2 with $Z = 4$, but the problem is complicated by the presence of three crystallographically independent molybdate groups.[277a] Considerably fewer Raman lines were found than predicted by unit-cell analysis, but the entire spectrum is consistent with an interpretation that assumes an effective site symmetry for MoO_4^{2-} of C_2 as opposed to the true crystallographic site symmetry C_1 (see Figure 4).

The Raman spectrum of $KCrO_3Cl$ has been discussed in terms of the known C_{2h} factor-group symmetry. However, for $KCrO_3F$ and $KOsO_3N$, the O and F or N atoms are *X*-ray-crystallographically indistinguishable, and the space group of these compounds (C_{4h}^6) represents pseudo-symmetry. If the CrO_3F^- and OsO_3N^- ions are taken as specifically oriented, the factor group is C_{2h}, but the agreement of the Raman spectra with selection rules on this basis is not good. Further, the Raman bands are broad. It is therefore suggested that in $KCrO_3F$ and $KOsO_3N$ the anions do not have preferred orientations.[277b]

Complex Cationic Salts.—Very complete assignments of i.r. and Raman spectra of seven of the eight low-temperature phases of NH_4^+ and ND_4^+ halides (Cl^-, Br^-, I^-) have been made, including those for *LO* components of i.r.-active and acoustic modes.[278]

Complex Anionic Salts.—Vibrational frequencies of NO_2^- drop with increase in cation size in i.r. and Raman spectra of powdered $LiNO_2,H_2O$ and MNO_2 (M = Li, Na, K, or Cs).[279] Far-i.r. reflection spectra of single crystals of $M(NO_3)_2$ (M = Sr, Ba, or Pb) yielded all eight predicted frequencies (T_{1u} in T_h^6); they are markedly sensitive to change of both M and temperature.[279a] The variation of the Raman spectrum of $NaNO_3$ with orientation has been computed and shown to be in good agreement with experiment,[280] while the highly complex far-i.r. phonon absorptions

[275] J. E. D. Davies and D. A. Long, *J. Chem. Soc.* (*A*), 1971, 1275.
[276] P. J. Miller, *Spectrochim. Acta*, 1971, **27A**, 957.
[277] (*a*) J. A. Konigstein and J. M. Preudhomme, *J. Chem. Phys.*, 1971, **55**, 461; (*b*) R. L. Carter and C. E. Bricker, *Spectrochim. Acta*, 1971, **27A**, 825.
[278] K. B. Harvey and N. R. McQuaker, *J. Chem. Phys.*, 1971, **55**, 4390.
[279] M. J. Brooker and D. E. Irish, *Canad. J. Chem.*, 1971, **49**, 1289.
[279a] A. M. Bon, *Compt. rend.*, 1971, **272**, *B*, 1297.
[280] A. Sadoc, *Compt. rend.*, 1971, **272**, *B*, 1439.

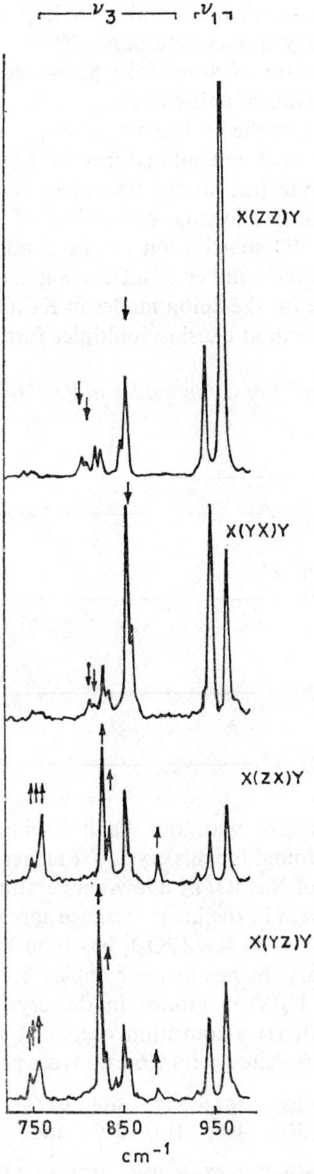

Figure 4 *Polarized Raman spectra of single-crystal* $Tb_2(MoO_4)_3$, T = 77 K, *excited with* 514.5 nm *radiation. Downward pointing arrows indicate Raman lines for which the tensors transforming like species* A_1 *and* A_2 *dominate over those belonging to* B_1 *and* B_2*; for the lines indicated with the arrow up the situation is reversed*

(Reproduced by permission from *J. Chem. Phys.*, 1971, **55**, 461)

(30—170 cm^{-1}) of rare-earth double nitrates $Ln_2(NO_3)_6,M_3(NO_3)_6,24H_2O$ have had their symmetry species determined.[281]

I.r. and Raman spectra of anhydrous powdered Li_2CO_3 and Na_2CO_3 were assigned with the aid of unit-cell analysis. Using reflectance methods the authors showed[282] that the i.r. transmission spectra of these compounds exhibit band maxima that are admixtures of *LO* and *TO* modes. ^{14}C substitution in aragonite (i.r. study) uncouples the ν_2 mode of the isotopically unique anions, allowing estimation of a coupling constant. Satellites on ν_2 after ^{14}C substitution afford a relationship between the coupling constant and the number of interacting anions.[283]

The splitting scheme for the anion modes in $KClO_3$ is as shown in Table 27.[284] The principal method causing multiplet formation in ν_3 is dipole–

Table 27 *Assignment* [284] *of anion modes in* $KClO_3$

C_{3v}	*Site, Cs*	*Unit cell,* C_{2h}
ν_1, a_1	(940) a'	940 A_g; 939 B_u
ν_2, a_1	(620) a'	620 $A_g + B_u$
ν_3, e	(990) a'; (987) a''	1000 B_u; 992 A_u; 982 B_g; 979 A_g
ν_4, e	(490) a'	493 B_u; 488 A_g
	(485) a''	487 B_g; 484 A_u

dipole coupling, consistent with this band having the longest *LO–TO* splitting (*ca.* 30 cm^{-1}) found for this crystal. Further detail has been added to the phonon picture of $NaClO_3$ by a low-temperature Raman study which revealed features not seen in the i.r. reflectance spectrum.[285]

Monoclinic ($P2_1/c$, $Z = 4$) $KH_5(PO_4)_2$ has been treated to a full single-crystal Raman analysis. The results are complex but the authors conclude that the existence of $H_2PO_4^{2-}$ groups in the crystal is probable.[286] In ferroelectric $NH_4H_2PO_4$ (I) a transition occurs at 148 K [to (II)] in the course of which the phosphate-group vibrations are profoundly disturbed:[287]

(I) 340, 400, 467, 537, and 927 cm^{-1}
(II) 300, 360, 400, 415, 550, and 911 cm^{-1}

[281] D. Bloor and J. A. Campbell, *J. Chem. Phys.*, 1971, **54**, 3268.
[282] M. H. Brooker and J. B. Bates, *J. Chem. Phys.*, 1971, **54**, 4788.
[283] W. Sterzel, *Z. anorg. Chem.*, 1971, **385**, 303.
[284] J. B. Bates, *J. Chem. Phys.*, 1971, **55**, 494.
[285] P. J. Miller and R. K. Khanna, *Spectrochim. Acta*, 1971, **27A**, 929.
[286] L. Beys, A. Armengaud, and R. Lafont, *Compt. rend.*, 1971, **273**, *B*, 479.
[287] J.-P. Benoit, *Compt. rend.*, 1971, **273**, *B*, 483.

In a paper which sets a high standard for other practitioners to aim at, Berenblut, Dawson, and Wilkinson analyse the single-crystal Raman spectrum of gypsum ($CaSO_4,2H_2O$) in detail, assigning all the water external modes in addition to anion internal and other lattice modes. They also describe in full the experimental procedure to be followed with a monoclinic crystal.[288] The single-crystal Raman spectrum of NH_4HSO_4 ($P2_1/c$, $Z = 8$) appears to have yielded no information of value in addition to results on the melt.[289] The ν(Cl—O) region in $LiClO_4,3H_2O$ has been reinvestigated in the i.r.[290]

A study [291a] of single-crystal $Cu(HCO_2)_2,4H_2O$ and its deuteriate by Raman spectroscopy supports a previous assignment by Charlton and Harvey.[291b]

With commendable dedication, several single-crystal studies of azides have been completed. Although good crystals of silver azide were obtained and some changes with orientation found, the usual techniques were not followed, so the assignments are not well-proven.[292] The spectra are consistent with the known bimolecular unit cell. The region below 250 cm^{-1} in the Raman spectrum of single-crystal $Ba(N_3)_2$, reinvestigated, shows 7 of 8 A_g and 4 of 5 B_g modes (predicted by unit-cell analysis), but an unequivocal distinction between translatory and rotatory motions was not possible. I.r. (powder) frequencies for $Ba(N_3)_2$ and $Sr(N_3)_2$ are listed.[293] I.r. (down to 400 cm^{-1}) and Raman single-crystal spectra of α-PbN_3 ($Z = 12$) have been discussed.[293a] I.r. and Raman spectra of single-crystal NaNCS yield results in accord with its known space group, D_{2h}^{16}. The splitting of the B_{1u} and B_{2u} components of $\nu_2[\delta(NCS)]$ is less than in KNCS.[294]

Aquo-complexes.—Full details of a single-crystal and i.r. study of $[Mg(H_2O)_6](PO_2H_2)_2$ (D_{4h}^{20}, $Z = 4$ in primitive cell) were not published, but the following assignment for the aquo-cation is given: $\nu_1(a_{1g}) = 363$, $\nu_2(e_g) = 310$, $\nu_3(t_{1u}) = 421$, $\nu_4(t_{1u}) = 215$—200 (4 bands), and $\nu_5(t_{2g}) =$ 285—268 (5 bands) cm^{-1}. The cation is on a site of S_4 symmetry; the numbers of bands found in the ν_4 and ν_5 regions are close to those predicted, but ν_2 should also split.[295] The same group have investigated $MgSO_4,7H_2O$ (D_2^4, $Z = 4$):[296] this is a piezoelectric crystal and various complications were met; some of the longitudinal modes were observed directly. For the

[288] B. J. Berenblut, P. Dawson, and G. R. Wilkinson, *Spectrochim. Acta*, 1971, **27A**, 1849.
[289] P. Dhamelincourt, G. Palauit, and S. Noël, *Bull. Soc. chim. France*, 1971, 2849.
[290] Y. Juncker, M. Krauzman, J.-P. Mathieu, and H. Poulet, *Compt. rend.*, 1971, **272**, *B*, 687.
[291] (*a*) J. Berger, *Compt. rend.*, 1971, **273**, *B*, 927; (*b*) T. L. Charlton and K. B. Harvey, *Canad. J. Chem.*, 1966, **44**, 2717.
[292] J. I. Bryant and R. L. Brooks, *J. Chem. Phys.*, 1971, **54**, 5315.
[293] Z. Iqbal and M. L. Malhotra, *J. Chem. Phys.*, 1971, **55**, 528.
[293a] Z. Iqbal, W. Garrett, C. W. Brown, and S. S. Mitra, *J. Chem. Phys.*, 1971, **55**, 4528.
[294] Z. Iqbal, *J. Mol. Structure*, 1971, **7**, 136.
[295] M. Abenoza, P. Hillaire, and R. Lafont, *Compt. rend.*, 1971, **273**, *B*, 297.
[296] P. Hillaire, M. Abenoza, and R. Lafont, *Compt. rend.*, 1971, **273**, *B*, 255.

cation, correlation to O_h yielded the following assignment, in good agreement with that for the above crystal: $\nu_1 = 374$; $\nu_2 = 318$; $\nu_3 = 418$; $\nu_4 = 210$ and 190; and $\nu_5 = 250$, 240, and 233 cm^{-1}.

Complex Cyanides.—Using single-crystal i.r. reflectance, 24 of the predicted 32 modes in the 40—400 cm^{-1} region have been observed for $K_3Fe(CN)_6$; the assignment deduced [297a] supports that of Nakagawa and Shimanouchi.[297b] A similar study was carried out with $K_3Co(CN)_6$. Further information has also been supplied by Jones *et al.* using $[Co(^{12}C^{14}N)_6]^{3-}$, $[Co(^{13}C^{14}N)_6]^{3-}$, and $[Co(^{12}C^{15}N)_6]^{3-}$ in solution and as powdered solids.[298] A normal-co-ordinate analysis of the unit cell of $Cs_2LiCo(CN)_6$ showed that the force constants were practically unaltered in transferring the complex ion in solution to the lattice environment, but that the low-frequency deformations were inextricably mixed with lattice modes.[299]

Molecular Crystals.—Single-crystal Raman data for ClCN and BrCN are in Table 3. Detailed assignments [300, 301] for $NiCl_2$(thiourea)$_4$ and [Cu(thioacetamide)$_4$]Cl have been made using both i.r. and Raman single-crystal methods. Both crystals have a unimolecular unit cell; the resulting assignments are therefore quite reliable. A single-crystal Raman study of $Mn_2(CO)_{10}$, together with new Raman solution data for $Mn_2(CO)_{10}$ and $Re_2(CO)_{10}$, form the basis of improved assignments for these molecules.[302] The Raman spectrum of solid MoF_5 has been reinvestigated:[303] two of the metal atoms are on C_2 sites and two on C_s sites. Assuming zero or very weak coupling between adjacent corners of the Mo_4F_{20} tetramer, the ν_1 (A_1) molecular mode gives rise to two A_g components in the crystal, one from a C_s site, the other from a C_2 site. Remaining assignments were made similarly, although not all of the expected components were identified. The spectral simplicity is attributed to small site and correlation field splitting of degenerate modes.

Others.—$[Pd(NH_3)_3NO_2]_2[Pd(NH_3)_4](NO_3)_4$ is tetragonal with a unimolecular cell. Single-crystal Raman results [304] assisted assignment of its vibrational spectrum, but the fearsome number of allowed modes makes proof difficult. ν(Pd—NO_2) was attributed to a line at 318 cm^{-1} and ν(Pd—NH_3) in the region 490—520 cm^{-1}.

[297] (*a*) J. T. R. Dunsmuir and A. P. Lane, *J. Chem. Soc.* (*A*), 1971, 776; (*b*) I. Nakagawa and T. Shimanouchi, *Spectrochim. Acta*, 1970, **26A**, 131.
[298] L. H. Jones, M. N. Memering, and B. I. Swanson, *J. Chem. Phys.*, 1971, **54**, 4666.
[299] B. I. Swanson and L. H. Jones, *J. Chem. Phys.*, 1971, **55**, 4174.
[300] D. M. Adams and R. R. Smardzewski, *J. Chem. Soc.* (*A*), 1971, 10.
[301] D. M. Adams and R. R. Smardzewski, *J. Chem. Soc.* (*A*), 1971, 8.
[302] D. M. Adams, M. A. Hooper, and A. Squire, *J. Chem. Soc.* (*A*), 1971, 71.
[303] J. B. Bates, *Spectrochim. Acta*, 1971, **27A**, 1255.
[304] F. P. Boer, V. B. Carter, and J. W. Turley, *Inorg. Chem.*, 1971, **10**, 651.

5

Characteristic Vibrational Frequencies of Compounds containing Main-group Elements

BY S. R. STOBART

In compiling this chapter, the arrangement of material adopted in previous volumes of the series has been adhered to: vibrational data for Main-group compounds, obtained by i.r. or Raman spectroscopy, are divided into eight sections, one for each Main Group of elements in the Periodic Table. Within each section, further subdivision is made where this is considered to be useful by dealing with each element in turn (in order of increasing atomic number) or by collecting together references concerning compounds possessing similar groups of atoms.

1 Group I Elements

The presence of the species SCN^-, LiSCN, and $(LiSCN)_n$ in solutions of LiSCN in polar solvents has been demonstrated using i.r. spectroscopy.[1] An i.r. study of $LiI,3H_2O$ and of partially deuteriated specimens [2] shows that the two hydrogen atoms of H_2O are non-equivalent, the H_2O molecules forming one strong hydrogen-bond $OH\cdots I^-$. Popov and co-workers have investigated acetone solvation of alkali-metal cations;[3] cation–acetone vibrational bands have been detected in the i.r., at 425 (Li^+), 195 (Na^+), and 148 cm^{-1} (K^+). The existence of a frequency dependence on isotopic substitution of solvent (using [$^2H_6$]acetone) and of solute (^6Li) clearly indicates that the observed vibrations involve both cation and solvent. The solvation of Na^+ by THF has also been investigated,[4] in the system THF–$NaAlBu_4$–cyclohexane, by i.r. examination of the 900—1150 cm^{-1} region.

I.r. and Raman spectra for Li_2ZnMe_4 and $LiAlMe_4$ have been reported by Yamamoto and Wilkie; both complexes show i.r. bands at *ca.* 300 cm^{-1}, assigned to Li—C stretching.[5] Corresponding Raman bands could not be observed, suggesting largely ionic bonding. In a similar study [6] of M^IAlMe_4 (M^I = ^6Li, ^7Li, or Na), $\nu(^6Li—C)$ is assigned at 262 cm^{-1}, $\nu(^7Li—C)$ at

[1] M. Chabanel, C. Menard, and G. Guihéneuf, *Compt. rend.*, 1971, **272**, *C*, 253.
[2] G. Brink and M. Falk, *Canad. J. Chem.*, 1971, **49**, 347.
[3] M. K. Wong, W. J. McKinney, and A. I. Popov, *J. Phys. Chem.*, 1971, **75**, 56.
[4] J. A. Olander and M. C. Day, *J. Amer. Chem. Soc.*, 1971, **93**, 3584.
[5] J. Yamamoto and C. A. Wilkie, *Inorg. Chem.*, 1971, **10**, 1129.
[6] C. A. Wilkie, *J. Organometallic Chem.*, 1971, **32**, 161.

256 cm^{-1}, and ν(Na—C) at 132 cm^{-1}. A detailed study has been made of the Raman spectra of $(Bu^tM^I)_4$ (M^I = ^6Li or ^7Li) including a normal-co-ordinate analysis.[6a] I.r. spectra of Li_2SO_4,H_2O and Li_2SO_4,D_2O show bands in the region 360—450 cm^{-1}, attributable to ν(Li···O) and $\nu(Li\cdots OH_2)$,[6b] and $M^I(OBu^t)$ (M^I = Na or K) have been shown[6c] by i.r. and Raman spectroscopy to be cubic tetramers in a number of solvents, $\nu(M^I$—O) appearing at 213 (M^I = Na) and 180 cm^{-1} (M^I = K).

2 Group II Elements

New spectroscopic data for beryllium borohydride, BeB_2H_8, have been obtained[7] in an attempt to elucidate its structure, for which five models have been proposed. Raman and i.r. examination of the four species BeB_2H_8, BeB_2D_8, $Be^{10}B_2D_8$, and BeB_2D_7H shows that, on condensation from the vapour to the solid phase, marked spectral changes occur which are interpreted in terms of the formation of $[BeBH_4]^+[BH_4]^-$. The cation is assigned the C_{2v} structure (1), on the basis of the spectroscopic analysis

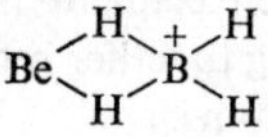

(1)

summarized in Table 1; the BH_4^- ion possesses the expected tetrahedral symmetry. Coates and Francis have reported[8] i.r. wavenumbers for Et_3Be_2H, a band at 1565 cm^{-1} possibly being associated with bridging Be—H—Be. The same authors have also studied various alkynylberyllium complexes:[9, 10] i.r. bands due to ν(C≡C) near 2100 cm^{-1} are strong in

Table 1 *Vibrational bands (wavenumber*/cm^{-1}*) and assignments for* $BeBH_4^+$, *(C_{2v} symmetry)*

(C_{2v} *symmetry*)			
A_1	ν_1	2451	ν_{sym}(BH)
	ν_2	2125	ν_{sym}(BH)
	ν_3	1540	ν_{sym}(BeH)
	ν_4	1128	$\delta(BH_2)$
	ν_5	725	ν(B···Be)
A_2	ν_6	—	BH_2 torsion
B_1	ν_7	2510	ν_{asym}(BH)
	ν_8	970	$\rho(BH_2)$
	ν_9	405	(BeH_2B) bend
B_2	ν_{10}	2075(± 15)	ν_{asym}(BH)
	ν_{11}	1430	ν_{asym}(BeH)
	ν_{12}	1009	$\delta(BH_2)$

[6a] W. M. Scovell, B. Y. Kimura, and T. G. Spiro, *J. Co-ordination Chem.*, 1971, **1**, 107.
[6b] S. Meshitsuka, H. Takahashi, and K. Higasi, *Bull. Chem. Soc. Japan*, 1971, **44**, 3255.
[6c] P. Schmidt, L. Lochmann, and B. Schneide, *J. Mol. Structure*, 1971, **9**, 403.
[7] J. W. Nibler, D. F. Shriver, and T. H. Cook, *J. Chem. Phys.*, 1971, **54**, 5257.
[8] G. E. Coates and B. R. Francis, *J. Chem. Soc. (A)*, 1971, 1308.
[9] G. E. Coates and B. R. Francis, *J. Chem. Soc. (A)*, 1971, 160.
[10] G. E. Coates and B. R. Francis, *J. Chem. Soc. (A)*, 1971, 474.

polymeric species but rather weak in monomers, and though they are in the expected range for the complexes $(PhC{\equiv}C)_2Be,L$ (L = OEt_2, NMe_3, or NEt_3), they are absent in $(PhC{\equiv}C)_2Be,2L$. Vibrations due to bridging and terminal C≡C units have tentatively been distinguished. Low-frequency i.r. and Raman data have been obtained for crystalline formates and acetates of Be and Al.[11] These are analysed in terms of BeO_4 and AlO_6 co-ordination polyhedra; a band at 943 cm^{-1} in $Be(OOCH)_2$, assigned to γ(CH) in earlier work but unshifted on deuteriation, is ascribed to ν_{asym}(Be—O). I.r. data for a beryllium ethoxyacetate have also been given.[11a]

Raman spectra of MgX_2–KX melts (X = Cl, Br, or I) have been measured over a range of Mg^{2+} : X^- mole ratios, and provide evidence for the existence of MgI_4^{2-} species; more complex structures are indicated in the chloride and bromide phases.[12] In the i.r. spectra of MgX_2,6ROH (X = Cl or I; R = Me or Et), bands at 425, 460, and 472 cm^{-1} are assigned[13] to vibrations of an MgO_6 unit. Differing shifts in i.r. frequencies from those observed in the free ligands for β-diketonate complexes of magnesium allow a distinction to be made between chelation of, or bridging between, the metal atoms.[14]

The compounds $M(C{\equiv}CPh)_2$ (M = Ca, Sr, or Ba) have been prepared and their i.r. spectra reported.[15] A decrease of *ca.* 20 cm^{-1} in ν(C≡C) along the series M = Ca—Ba is in accordance with a gradual increase in importance of the ionic contribution $M^+(C{\equiv}CPh)^-$. The only band for which an assignment to ν(M—C) is suggested is that at 324 cm^{-1} in the calcium compound. Some i.r. frequency data for trihydroxyglutarates of Sr and Ba have been given.[16]

3 Group III Elements

Compounds containing B—H Bonds.—An electron diffraction study of $B_2H_4(PF_3)_2$ indicates (2) as the most likely structure. This is further supported by i.r. (Ar matrix) and Raman (gas-phase) spectra which show[17]

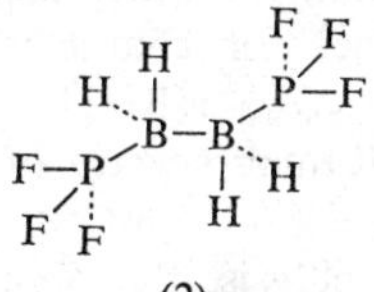

(2)

[11] A. I. Grigor'ev, V. A. Sipachev, and E. G. Pogodilova, *J. Struct. Chem.*, 1970, **11**, 416.
[11a] V. A. Sipachev, A. I. Grigor'ev, and A. V. Novoselova, *Doklady Chem.*, 1971, **196**, 114.
[12] V. A. Maroni, E. J. Hathaway, and E. J. Cairns, *J. Phys. Chem.*, 1971, **75**, 155.
[13] N. Ya. Turova, E. P. Turevskaya, and A. V. Novoselova, *Russ. J. Inorg. Chem.*, 1971, **16**, 482.
[14] D. E. Fenton, *J. Chem. Soc. (A)*, 1971, 3481.
[15] M. A. Coles and F. A. Hart, *J. Organometallic Chem.*, 1971, **32**, 279.
[16] A. I. Mudretsov, *Russ. J. Inorg. Chem.*, 1971, **16**, 636.
[17] E. R. Lory, R. F. Porter, and S. H. Bauer, *Inorg. Chem.*, 1971, **10**, 1072.

no coincidences, consistent with a configuration possessing a centre of symmetry. I.r. and Raman spectra for the alkylphosphineboranes RPH_2BH_3 and R_2PHBH_3 (R = Me or Et) and deuterioborane analogues [18] indicate considerable mixing between vibrational modes. Only in the case of $MePH_2BX_3$ (X = H or D) is a reasonably precise assignment possible, including the assignments shown in Table 2. For all the BH_3 adducts,

Table 2 *Assignment of vibrational wavenumbers*/cm^{-1} *for an unlabelled and a* ^{2}H-*labelled methylphosphineborane*

Assignment	$MePH_2BH_3$	$MePH_2BD_3$
$\nu_{asym}(BX_3)$	2360	1802
$\nu_{sym}(BX_3)$	—	1730
$\delta_{asym}(BX_3)$	1141	849
$\delta_{sym}(BX_3)$	1074	833
$\rho(BX_3)$	670	564
ν(P—B)	581	533

ν(P—B) is close to 557 cm^{-1}, dropping to *ca.* 506 cm^{-1} on deuteriation. An increase of *ca.* 100 cm^{-1} in ν(P—H) from values for the alkylphosphines is observed, consistent with earlier reports of apparently enhanced P—H bonding when adducts are formed by phosphine derivatives. In the co-ordinated boryltosylates RBH_2,NMe_3 and R_2BH,NMe_3 (R = *p*-$Me\cdot C_6H_4SO_3$), ν(B—H) modes occur at 2420, 2320, and at 2520 cm^{-1}, respectively.[19] Monoalkylboranes formulated as RBH_2 have been prepared;[20] where R = Bu^n the product is a clear liquid showing ν(B—H) at 2550 cm^{-1} (terminal) and 1590 cm^{-1} (bridge), the incidence of the latter presumably implying a dimeric structure. For $(R_2BH)_2$ compounds there are no $\nu(B{-}H)_{terminal}$ bands, and $\nu(B{-}H)_{bridge}$ bands are observed strongly in the i.r.,[21] at 1560 cm^{-1} (R = n-butyl), 1580 cm^{-1} (R = isobutyl), and 1590 cm^{-1} (R = s-butyl, cyclopentyl, cyclohexyl, or *exo*-norbornyl).

The vibrational spectra of MeB_2H_5 and $Me_3B_2H_3$ have been assigned.[22] Relationships with earlier data for other methyldiboranes are discussed in terms of frequency ranges for characteristic bands which may be attributed mainly to $\nu(B{-}H)_{terminal}$, $\nu(B{-}H)_{bridge}$, BH_2 wag/rock, ν(B—C), ν(B—B), *etc.* The i.r. and Raman spectra of three closely related *closo*-carboranes $C_2B_3H_5$ [structure (3)], 1,6-$C_2B_4H_6$, and $C_2B_5H_7$ have been measured.[23] In a detailed analysis, the observed data have been fairly fully assigned, and are reproduced for structure (3), *closo*-1,5-dicarbapentaborane(5), in Table 3 to illustrate the wavenumber ranges. Wallbridge

[18] J. Davis and J. E. Drake, *J. Chem. Soc.* (*A*), 1971, 2094.
[19] G. E. Ryschkewitsch, *Inorg. Nuclear Chem. Letters*, 1971, **7**, 99.
[20] H. C. Brown and S. K. Gupta, *J. Amer. Chem. Soc.*, 1971, **93**, 4062.
[21] H. C. Brown and S. K. Gupta, *J. Organometallic Chem.*, 1971, **32**, C1.
[22] J. H. Carpenter, W. J. Jones, R. W. Jotham, and L. H. Long, *Spectrochim. Acta*, 1971, **27A**, 1721.
[23] R. W. Jotham and D. J. Reynolds, *J. Chem. Soc.* (*A*), 1971, 3181.

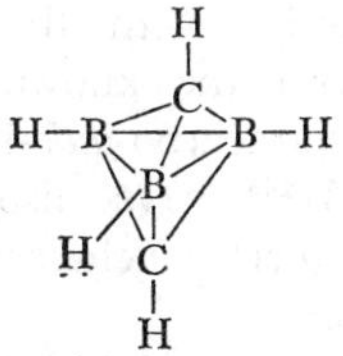

(3)

and co-workers have given[24] unassigned i.r. frequencies for some 3-halogeno-derivatives of the dicarba-*nido*-hexaboranes $C_2B_4H_7I$, 3-X·$C_2(Me)B_4H_6$, and 3-X·$C_2(Me)_2B_4H_5$ (X = Cl, Br, or I). Two new *nido*-carboranes have been obtained from the reaction of acetylene with octaborane(12). These

Table 3 *The vibrational modes of* closo-1,5-*dicarbapentaborane*(5)

Symmetry species	*Approximate description*	*Fundamental frequency* (*wavenumber*/cm^{-1})
A_1'	C—H stretch, sym	3158
	B—H stretch, sym	2631
	B—C stretch, sym	1125
	B—B stretch, sym	842
A_2'	B—H deformation, in-plane, in-phase	(621)
E'	B—H stretch, asym	2618
	C—H deformation, in-phase	1203
	B—B stretch, asym	1100
	B—H deformation, in-plane, out-of-phase	903
	Cage deformation	788
A_1''	—	—
A_2''	C—H stretch, asym	3165
	B—C stretch, asym	1218
	B—H deformation, out-of-plane, in-phase	1119
E''	C—H deformation, out-of-phase	1248
	B—H deformation, out-of-plane, out-of-phase	774
	Cage deformation	546

are[25] *nido*-dicarbanonaborane(11) ($B_7C_2H_{11}$) and *nido*-dicarbadecaborane-(12) ($B_8C_2H_{12}$); i.r. wavenumbers above 1200 cm^{-1} are listed, and the two compounds show respectively thirteen and sixteen weak bands due to cage vibrations in the range 650—1170 cm^{-1}, indicative of very low symmetry. Oxidation of *o*-$C_2B_9H_{12}^-$ by $FeCl_3$ in acidic solution yields a mixture of products, *ca.* 90% being $C_2B_8H_{12}$. I.r. bands[26] in the 1400—1600 cm^{-1} and 1800—2000 cm^{-1} regions, assignable to $\nu(B—H)_{bridge}$, together with n.m.r.

[24] J. S. McAvoy, C. G. Savory, and M. G. H. Wallbridge, *J. Chem. Soc.* (*A*), 1971, 3038.
[25] R. R. Rietz and R. Schaeffer, *J. Amer. Chem. Soc.*, 1971, **93**, 1263.
[26] J. Plešek and S. Heřmánek, *Chem. and Ind.*, 1971, 1267.

and u.v. data, are compatible with the structure 5,6-dicarba-*nido*-decaborane, isoelectronic with the known $B_{10}H_{12}^{2-}$ anion. Partially assigned i.r. data for 2,3,4,5-tetracarbo-*nido*-hexaborane(6) [26a] and (3)-1,2-dicarba-*nido*-undecaborane(13) [26b] have also been reported, and new vibrational studies of tetraborane(10) include results of isotopic substitution involving ^1H, ^2H, ^{10}B, and ^{11}B.[26c]

Brown and Negishi have prepared bis-(3,5-dimethyl)borinan (4) and bis-(3,6-dimethyl)borepan (5) by hydroboration of dienes. The vibration associated with the symmetric stretching of the 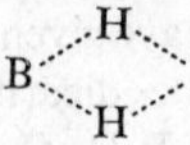

unit has been assigned to i.r. bands at 1565 and 1560 cm^{-1}, respectively.[27]

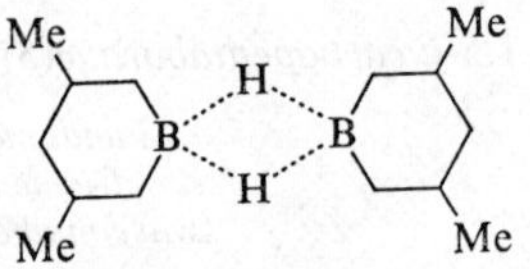

(4)

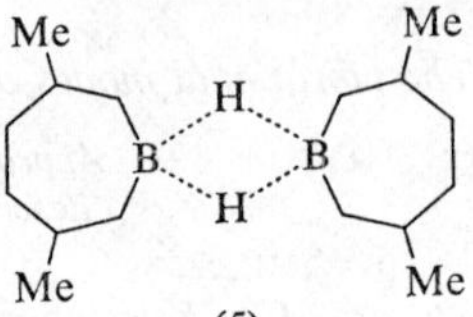

(5)

Kaldor and Porter have carried out a thorough investigation of borazine in Ar and Xe matrices.[28] By using the three species $H_3B_3N_3H_3$, $H_3{}^{10}B_3N_3H_3$, and $D_3B_3N_3H_3$, an almost complete vibrational assignment has been made, agreeing with earlier work apart from the reassignment of the E' mode, ν_{16}, to 1068 cm^{-1}. In addition, two inactive A_2' modes (ν_6, ν_7) are shown from combination bands to lie at 1195 and 782 cm^{-1}, respectively. The reaction of trimethylaluminium with $[Me_2N]_2BH$ yields the novel cyclic compound (6), whose i.r. spectrum includes [29] the following

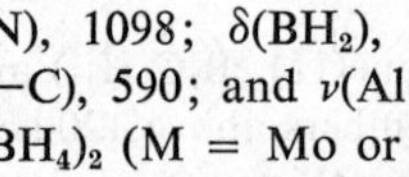

(6)

characteristic absorptions, expressed as wavenumber/cm^{-1}: ν(B—H), 2247; $\delta(BH_2)$, i.p., 1139; ν(B—N), 1098; $\delta(BH_2)$, o.p., 929; δ(AlMe), 769; ν_{asym}(Al—C), 690; ν_{sym}(Al—C), 590; and $\nu(Al—N)_{bridge}$, 515. In the i.r. spectra of $(\pi\text{-}C_5H_5)_2M(O)(BH_4)_2$ (M = Mo or W), bands at ~ 2300 and ~ 1100 cm^{-1} are assigned to terminal B—H stretching and BH_2 deformation

[26a] E. Groszek, J. B. Leach, G. T. F. Wong, C. Ungermann, and T. Onak, *Inorg. Chem.*, 1971, **10**, 2770.
[26b] D. V. Howe, C. J. Jones, R. J. Wiersema, and M. F. Hawthorne, *Inorg. Chem.*, 1971, **10**, 2516.
[26c] A. J. Dahl and R. C. Taylor, *Inorg. Chem.*, 1971, **10**, 2508.
[27] H. C. Brown and E. Negishi, *J. Organometallic Chem.*, 1971, **28**, C1.
[28] A. Kaldor and R. F. Porter, *Inorg. Chem.*, 1971, **10**, 775.
[29] R. E. Hall and E. P. Schram, *Inorg. Chem.*, 1971, **10**, 192.

modes, while those at ~ 2250 and ~ 1350 cm^{-1} are associated with the H-bridges co-ordinating the metal atom.[30]

Two papers [31, 32] by Kuznetsov and Klimchuk state that for the dodecahydroborates $M^I_2B_{12}H_{12}$ (M^I = Na, Rb, or Cs), $M_2{}^IB_{12}H_{12}M^ICl$ (M^I = Rb or Cs), and $M^I_2B_{12}H_{12},xH_2O$ (M^I = Li, x = 5; M^I = Na, x = 4), i.r. bands due to ν(B—H) and ν(B—B)$_{skeletal}$ are found in the 2480—2530 cm^{-1} and 745—780 cm^{-1} regions. I.r. frequency data for $(Me_4N)_2[Hg(B_{10}H_{12})_2]$ have been given; a band at 228 cm^{-1}, together with those at 284 and 244 cm^{-1} in Zn and Cd analogues, is assigned to a skeletal vibration involving the metal atom and the boron clusters.[33]

Compounds containing Al—H or Ga—H Bonds.—Greenwood and Thomas have reported i.r. data for complexes of AlH_3 with quinuclidine, 1,4-dimethylpiperazine, and *NNN'N'*-tetramethyl-*o*-phenylenediamine.[34] Some of their assignments are shown in Table 4. Kovar and Ashby have measured

Table 4 *Vibrational wavenumbers*/cm^{-1} *and assignments for alane–quinuclidine complexes*

$(C_7H_{13}N)_2,AlH_3$	$C_7H_{13}N,AlH_3$	*Assignment*
1685s	1795m	ν_{asym}(Al—H)
—	1760m	ν_{sym}(Al—H)
886m	—	$\delta(AlH_3)$, o.p.
760m	—	$\delta(AlH_3)$, i.p.
—	757s	δ(H—Al—H)
705m	—	δ(N—Al—H)
—	335w	ν(Al—N)
259w	—	$\nu_{asym}(AlN_2)$

ν(Al—H) and δ(Al—H) wavenumbers for dialkylaminoalanes.[35] These are listed in Table 5 and are consistent with formulations as dimeric $[(R_2N)HAlNR_2]_2$, or trimeric $[H_2AlNR_2]_3$ cyclic compounds. In HMgAlEt-H_3, some characteristic i.r. bands that have been observed are given [36] as

Table 5 Al—H *wavenumbers*/cm^{-1} *for some dialkylaminoalanes*

Compound	ν(Al—H)	δ(Al—H)
$HAl(NMe_2)_2$	1824	—
H_2AlNEt_2	1827	734
$HAl(NEt_2)_2$	1822	692
$H_2AlNC_4H_8$	1832	729
$HAl(NC_4H_8)_2$	1824	695
$H_2AlNC_5H_{10}$	1828	725
$HAl(NC_5H_{10})_2$	1825	688

[30] S. P. Anand, R. K. Multani, and B. D. Jain, *J. Organometallic Chem.*, 1971, **26**, 115.
[31] N. T. Kuznetsov and G. S. Klimchuk, *Russ. J. Inorg. Chem.*, 1971, **16**, 619.
[32] N. T. Kuznetsov and G. S. Klimchuk, *Russ. J. Inorg. Chem.*, 1971, **16**, 645.
[33] N. N. Greenwood and N. F. Travers, *J. Chem. Soc.* (*A*), 1971, 3257.
[34] N. N. Greenwood and B. S. Thomas, *J. Chem. Soc.* (*A*), 1971, 814.
[35] R. A. Kovar and E. C. Ashby, *Inorg. Chem.*, 1971, **10**, 893.
[36] S. C. Srivastava and E. C. Ashby, *Inorg. Chem.*, 1971, **10**, 186.

ν(Al—H) (1720 cm^{-1}), δ(Al—H) (750, 790 cm^{-1}), ν(Mg—C) (507 cm^{-1}), and ν(Al—C, bridge) (475 cm^{-1}). Similarly, Al—H stretching vibrations have been distinguished[37] at 1720 cm^{-1} for $LiAlH(NEt_2)_3$ and at 1815, 1770 cm^{-1} in Li_3AlH_6.

Dimeric and trimeric cyclogallazanes $[RNHGaH_2]_n$ represent a new class of compounds; reported i.r. data[38] are given in Table 6. As observed

Table 6 *I.r. spectra of cyclogallazanes in benzene solution* (*wavenumber*/cm^{-1})

$(EtNHGaH_2)_3$	$(EtNHGaD_2)_3$	$(Pr^iNHGaH_2)_2$	$(Pr^iNHGaD_2)_2$	*Assignment*
3338w[a]	3335w	—	—	—
3318m	3310m	3320m	3320w	N—H stretch
3280s	3276s	3280s	3283m	—
1875vs	1350vs	1875vs	1355vs	Ga—H(D) stretch
1820vs, br	1335vs, sh	1820vs, br	1330s, sh	—
745vs, br	502vs	745vs	504vs	Ga—H(D) defn.
—	496vs	—	497vs	—
580s	—	586s	586s	Ring modes
550s, sh	542s	560m, sh	552s	—
510m	522s	490m	536m	—

[a] w = Weak, m = medium, s = strong, vs = very strong, sh = shoulder, br = broad.

previously with gallane derivatives, the strongest absorptions are attributable to ν(Ga—H) modes.

Compounds containing M—C Bonds (M = B, Al, In, or Tl).—Gas-phase Raman spectra for trimethylboron and for trimethylaluminium have been obtained by O'Brien and Ozin.[39] Distinction can be made between monomeric and dimeric trimethylaluminium, the spectrum of the former being very similar to that for Me_3B and consistent with D_{3h} symmetry. Some assignments of vibrational structure in the photoelectron spectra of triethyl- and trivinyl-boron have been given.[40] In the polymeric species $(Ph_2P\cdot S\cdot BMe_2)_n$, the B—C stretch is assigned to a strong i.r. band at 933 cm^{-1}, corresponding to four-co-ordinate boron.[41] I.r. and Raman spectra for the new compounds $(R_2N)_2B\cdot CH_2\cdot B(NR_2)_2$, cyclo-$(R_2N\cdot B\cdot CH_2\cdot)_3$, and cyclo-$(Cl\cdot B\cdot CH_2\cdot)_3$ have been reported and partially assigned,[42] placing ν(B—C) modes within the region 600—820 cm^{-1}, and ν(B—N) at *ca.* 1500 cm^{-1} (asym) and 1370 cm^{-1} (sym), with ν(B—Cl) at 725—760 and 505 cm^{-1}.

An i.r. and Raman study of $K[BAr_4]$ (Ar = *p*-X·C_6H_4—; X = H, OMe, Me, F, Cl, Br, or CF_3) gives particular emphasis to 'X-sensitive' modes in

[37] R. G. Beach and E. C. Ashby, *Inorg. Chem.*, 1971, **10**, 1888.
[38] A. Storr and A. D. Penland, *J. Chem. Soc.* (*A*), 1971, 1237.
[39] R. J. O'Brien and G. A. Ozin, *J. Chem. Soc.* (*A*), 1971, 1136.
[40] A. K. Holliday, W. Reade, R. A. W. Johnstone, and A. F. Neville, *Chem. Comm.*, 1971, 51.
[41] H. Vahrenkamp, *J. Organometallic Chem.*, 1971, **28**, 167.
[42] P. Krohmer and J. Goubeau, *Chem. Ber.*, 1971, **104**, 1347.

the range 650—200 cm^{-1} and includes tentative assignments for vibrations of the BC$_4$ skeleton.[43] For LiAlEt$_4$ (see ref. 6), ν(Al—C) is at 616 cm^{-1}, and some bands at lower wavenumber are assigned to δ(Al—C) modes. Salts of the tetraethynylboron anion B(C≡CH)$_4$$^-$ have been examined [42a] by i.r. spectroscopy. Oxidation at room temperature of InBr with MeBr affords MeInBr$_2$, which shows [44] a strong band (i.r. and Raman) at 523 cm^{-1}, attributed to ν(In—C). In$_2$Br$_4$ also undergoes oxidative addition with methyl halides to form products containing In—C bonds, Raman lines being found [45] at 523 cm^{-1} (Br product) or 518 cm^{-1} (iodide). Observed i.r. data

Table 7 Wavenumber/cm^{-1} of selected bands in the spectra of compounds Et$_2$InX

X	ν_{asym}(In—C)	ν_{sym}(In—C)
Cl	515	460
Br	512	460
I	506	454

for Et$_2$InX (in CH$_2$Cl$_2$ solution) as assigned [46] as shown in Table 7. The compound (7) has been prepared.[47] The only characteristic i.r. absorption due to the cation is ν(P=N) at 1108 cm^{-1}, but the following vibrations of the

$$Me_2Si\underset{N(PMe_3)}{\overset{N(PMe_3)}{(+)}}AlMe_2 \quad InMe_4^-$$

(7)

new InMe$_4$$^-$ anion are assigned: δ_{sym}(Me), 1091 cm^{-1}; ρ(Me), 670 cm^{-1}; ν_{asym}(InC$_4$), 465 cm^{-1}. I.r. frequencies have been listed for (C$_6$F$_5$)$_3$In and (C$_6$F$_5$)$_2$InI, assignments being proposed by comparison with (C$_6$F$_5$)$_2$TlBr and C$_6$F$_5$I. Bands due to 'X-sensitive' modes involving C—In stretching are distinguished at ca. 790 cm^{-1} in each compound.[48]

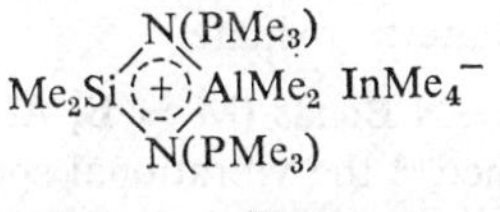

The frequencies of principal i.r. bands (4000—700 cm^{-1}) for (C$_6$F$_5$)$_3$Tl have been given.[49] For the compound (8), (TlC$_2$) modes are assigned [50] as in Table 8. Vibrational frequency data for some Tl^I and Tl^{III} imidazole and triazole complexes appear to have been accidentally omitted from a paper dealing with their properties,[51] but in the case of Me$_2$Tl(C$_3$H$_3$N$_2$) the

---

[42a] A. F. Zhigach, R. A. Svitsyn, E. S. Sobolev, and I. V. Persianova, *Doklady Chem.*, 1971, **196**, 141.
[43] J. T. Vandeberg, C. E. Moore, and F. P. Cassaretto, *Spectrochim. Acta*, 1971, **27A**, 501.
[44] L. Waterworth and I. J. Worrall, *Chem. Comm.*, 1971, 569.
[45] L. Waterworth and I. J. Worrall, *Inorg. Nuclear Chem. Letters*, 1971, **7**, 403.
[46] T. Maeda, H. Tada, K. Yasuda, and R. Okawara, *J. Organometallic Chem.*, 1971, **27**, 13.
[47] W. Wolfsberger and H. Schmidbaur, *J. Organometallic Chem.*, 1971, **27**, 181.
[48] G. B. Deacon and J. C. Parrott, *Austral. J. Chem.*, 1971, **24**, 1771.
[49] G. B. Deacon and J. C. Parrott, *Inorg. Nuclear Chem. Letters*, 1971, **7**, 329.
[50] B. Walther and K. Thiede, *J. Organometallic Chem.*, 1971, **32**, C7.
[51] A. G. Lee, *J. Chem. Soc. (A)*, 1971, 880.

$$\text{Me}_2\text{Tl} \underset{\underset{R_2}{N}}{\overset{\overset{R_2}{N}}{\diamond}} \text{TlMe}_2$$

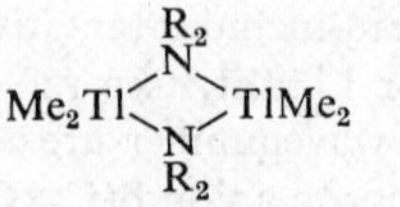

(8)

presence of two bands assignable to $\nu(\text{Tl}-\text{C})$ is taken to imply a non-linear configuration for the C—Tl—C group. A number of alkyl(cyano)-thallium derivatives have been prepared, R(CN)TlY (R = Me, Y = tropolonate, oxinate, OCOEt, OCOPr^i, or OCOCH$_2$Cl; R = Et, Y = tropolonate, oxinate, or OCOMe). A direct relationship between the

Table 8 Wavenumber/cm^{-1} of modes TlC$_2$ in compound (8)

R	$\nu_{\text{asym}}(\text{TlC}_2)$	$\nu_{\text{sym}}(\text{TlC}_2)$
Me	495	461
Et	546	495
C$_5$H$_{10}$N	548	485

frequency of $\nu(\text{C}\equiv\text{N})$ and of $\nu(\text{Tl}-\text{CN})$ [in the wavenumber range 334—375 cm^{-1}] is found,[52] the lowest values for each occurring for those Y groups which interact most strongly with the metal, *i.e.* OCOCH$_2$Cl > OCOC$_n$H$_{2n+1}$ > tropolonate > oxinate.

Compounds containing M—N Bonds (M = B, Al, Ga, In, or Tl).—Shriver and Swanson have assigned[53] the vibrational spectra of Cl$_3$B,NCMe and Br$_3$B,NCMe. Normal-co-ordinate analysis indicates that f_{BN} in these two complexes is significantly greater than in F$_3$B,NCMe. The i.r. spectra of a number of NN'-bis(diphenylboryl) derivatives of hydrazines, Ph$_2$B·NR^1·-NR^2·BPh$_2$ (R^1, R^2 = H or Me), have been recorded and tentatively assigned, $\nu''(\text{BN})''$ being placed[54] in the 1400 cm^{-1} region and $\nu''(\text{NN})''$ at *ca.* 1000 cm^{-1}.

The B—N ring frequencies for dimeric [(MeCO$_2$)$_2$BNR^{1}R^2]$_2$ (R^1, R^2 = H or Me) occur at wavenumbers of *ca.* 930 cm^{-1}, much lower than in the corresponding monomers[55] (*e.g.* 1400 cm^{-1} for R^1, R^2 = Me). For the B-trihalogenoborazines B$_3$X$_3$N$_3$H$_3$ (X = Cl or Br), a very thorough study of the vibrational spectra of isotopically substituted species, together with a normal-co-ordinate analysis, has enabled an almost full assignment to be made.[56] Comparison of out-of-plane force constants with those for benzene indicates that there may be considerable similarity in the two bonding systems. I.r. spectral data indicate that B$_3$X$_3$N$_3$R$_3$ [X = C$_5$H$_5$ or Mn(CO)$_5$; R = H or Me], products from the reactions of NaC$_5$H$_5$ and NaMn(CO)$_5$

---

[52] K. Tanaka, H. Kurosawa, and R. Okawara, *J. Organometallic Chem.*, 1971, **30**, 1.
[53] D. F. Shriver and B. Swanson, *Inorg. Chem.*, 1971, **10**, 1354.
[54] H. Nöth, W. Regnet, H. Rihl, and R. Standfest, *Chem. Ber.*, 1971, **104**, 722.
[55] D. T. Haworth and E. S. Matushek, *Chem. and Ind.*, 1971, 130.
[56] K. E. Blick, K. Niedenzu, W. Sawodny, T. Takasuka, T. Totani, and H. Watanabe, *Inorg. Chem.*, 1971, **10**, 1133.

with *B*-trichloroborazines, contain B—C or B—Mn σ-bonds.[57] Unassigned i.r. data have been listed for (9), named as 1,1,3,3,4,6-hexamethyl-5-phenylcyclo-1,3-disila-5-bor-2-oxa-4,6-diazone.[58] Determination of the crystal structure of the trimeric boron dichloride azide $(BCl_2N_3)_3$ has allowed new aspects of its vibrational spectrum to be discussed: in particular,

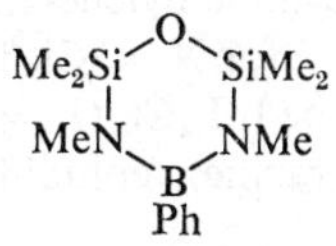

(9)

the observation of three $\nu(N_3)$ fundamentals (at 2219, 2210, and 2160 cm^{-1}) is accounted for by a skew-boat (C_2) rather than a planar (D_{3h}) ring conformation.[59]

I.r. spectra for 6-isothiocyanatodecaborane, $6\text{-NCSB}_{10}H_{13}$, and its acetoxy-analogue, $6\text{-AcOB}_{10}H_{13}$, have been given.[60] In the former, ν(B—N) is found at 1274 cm^{-1} or 1242 cm^{-1}, at a lower wavenumber than for $B(NCS)_3$. Weak bands at *ca.* 335 cm^{-1} in the i.r. spectra of some alane–(N-base) complexes (see ref. 34) are assigned to ν(Al—N). Burger and co-workers have made a full analysis of the vibrational spectra of $M[N(SiMe_3)_2]_3$ (M = Al, Ga, or In), and assign $\nu(M\text{—}NSi_2)$ as shown in Table 9.[61] Several other reports concerning compounds with Al—N bonds

Table 9 *Assignment of* M—N *stretching modes* (*wavenumber*/cm^{-1}) *in the spectra of* $M[N(SiMe_3)_2]_3$ (M = Al, Ga, *or* In)

	M		
Assignment	Al	Ga	In
A_1	390	388	379
E	392	375	360

do not refer directly to ν(Al—N) modes. I.r. spectra for $Et_2BrAl,NHMe_2$ and its thermal decomposition product $(EtBrAlNMe_2)_2$ have been listed but not assigned;[62] i.r. bands due to characteristic vibrations of the pseudohalide group in anionic methyl- and ethyl-aluminium azido-, isocyanato-, and isothiocyanato-complexes and their gallium analogues are observed in the expected ranges;[63] and a band at 1700 cm^{-1} in the i.r. spectrum of the new compound $LiAl(N{=}CBu^t_2)_4$ (whose structure has been determined

[57] D. T. Haworth and E. S. Matushek, *Inorg. Nuclear Chem. Letters*, 1971, **7**, 261.
[58] F. Rabet and U. Wannagat, *Z. anorg. Chem.*, 1971, **384**, 115.
[59] U. Müller, *Z. anorg. Chem.*, 1971, **382**, 110.
[60] B. Štibr, J. Plešek, F. Hanousek, and S. Heřmanek, *Coll. Czech. Chem. Comm.*, 1971, **36**, 1794.
[61] H. Burger, J. Cichon, V. Goetze, U. Wannagat, and H. J. Wismar, *J. Organometallic Chem.*, 1971, **33**, 1.
[62] K. Gosling, A. L. Bhuiyan, and K. R. Mooney, *Inorg. Nuclear Chem. Letters*, 1971, **7**, 913.
[63] F. Weller, I. L. Wilson, and K. Dehnicke, *J. Organometallic Chem.*, 1971, **30**, C1.

crystallographically) is described as a high-frequency azomethine stretching absorption.[64]

I.r. data for a variety of In^{III} complexes have been reported by Tuck and Yang, and include assignments of ν(In—N) to bands in the range 220—285 cm^{-1}, and also ν(In—S) in the 301—371 cm^{-1} region.[65] For $(Bu_4N)_2In(L)_2X$ (L = 1,1-dicyanoethylene-2,2-dithiolate), ν(In—X) are given at 235, 198, and 167 cm^{-1} for X = Cl, Br, or I. Far-i.r. and Raman spectra for complexes of the type InL_3Cl_3 (L = pyrazole, 3-methylpyrazole, pyridine, pyrimidine, or pyrazine) contain bands at 173—230 cm^{-1}, attributed[66] to ν(In—N).

I.r. frequencies for Tl^{III} ethylenediamine complexes have been reported.[67] It is suggested that bands in the region of 500 cm^{-1} and 300 cm^{-1} may possess a contribution from ν(Tl—N); in compounds containing halogeno-substitutents, ν(Tl—X) is more confidently assigned, at *ca.* 280 cm^{-1} (X = Cl), 200 cm^{-1} (X = Br), and 150 cm^{-1} (X = I).

Compounds containing M—O Bonds (M = B, Al, Ga, In, or Tl).—The reaction between Me_3B and O_2 in the gas phase at 125 °C yields small amounts of 3,5-dimethyl-1,2,4-trioxa-3,5-diborolan (10); the i.r. spectrum

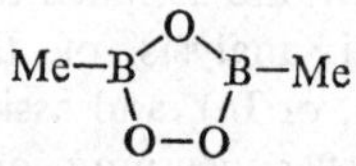

(10)

has been obtained using compensation techniques against known product mixtures, and bands at 1360 and 1167, 928 and 590 cm^{-1} are tentatively assigned to i.p. ring stretches, ν_{asym}(B—Me) and $\delta_{o.p.}$(B—Me), respectively.[68] Perfluoroalkyl borate esters $(R_FO)_3B$ are the products of the reaction between R_FOCl and BCl_3. For $(Bu_F{}^tO)_3B$, the i.r. spectrum of the gas (at 60 °C) shows[69, 69a] absorptions characteristic of B—O, C—F, and C—O bonds at 1422(mw), 1396(s), 1288(vs), 1202(w), 1150(m), 1124(m), 988(s), 952(mw), 730(m), and 534(w) cm^{-1}. Some unassigned i.r. absorption maxima have been given for 3(3*H*)-benzoborepin-3-ol (11).[70] The anionic complex $[(ascorbate)_2B]^{3-}$ has been detected and shows an i.r. band at 972 cm^{-1}, assigned to stretching vibrations of a BO_4 unit.[71] I.r. studies on solid $B(OH)_3$ and B_2O_3 have also been reported.[71a]

[64] H. M. M. Shearer, R. Snaith, J. D. Sowerby, and K. Wade, *Chem. Comm.*, 1971, 1275.
[65] D. G. Tuck and M. K. Yang, *J. Chem. Soc.* (*A*), 1971, 214.
[66] S. A. Cotton and J. F. Gibson, *J. Chem. Soc.* (*A*), 1971, 1696.
[67] F. Ya. Kul'ba, N. G. Yaroslavskii, L. V. Konovalov, A. V. Barsukov, and V. E. Mironov, *Russ. J. Inorg. Chem.*, 1971, **16**, 187.
[68] L. Barton and G. T. Bohn, *Chem. Comm.*, 1971, 77.
[69] D. E. Young, L. R. Anderson, and W. B. Fox, *Chem. Comm.*, 1971, 736.
[69a] D. E. Young, L. R. Anderson, and W. B. Fox, *Inorg. Chem.*, 1971, **10**, 2810.
[70] G. Axelrad and D. Halpern, *Chem. Comm.*, 1971, 291.
[71] E. M. Shvarts, L. I. Korchenenkova, and A. F. Ievin'sh, *Russ. J. Inorg. Chem.*, 1971, **16**, 486.
[71a] P. Broadhead and G. A. Newman, *J. Mol. Structure*, 1971, **10**, 157.

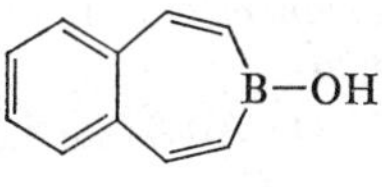

(11)

Raman spectroscopy has been used to augment n.m.r. studies of complex formation between aluminium and phosphate ions in solution.[72] As part of a study of the surface properties of aluminium sulphate, i.r. and Raman spectra have been examined under various stages of compression;[73] observed variations included changes in appearance of bands assigned to H_2O modes, and a splitting of a band at 155 cm^{-1}, assigned to δ(O—Al—O), on progressing to more highly compressed states. Alkoxysiloxy-derivatives of Al and B show [73a] i.r. bands at 1050—1075 cm^{-1} ν(Si—O—Al) and 1320 cm^{-1} ν(B—O—C).

Changes in the i.r. spectrum of $Al(acac)_3$ in nitrobenzene solution following the addition of $AlCl_3$ have been illustrated in a paper dealing with its acylation by n-butyryl chloride.[74] In alkylaluminium–carbonyl complexes of the type $Et_2Al(L)$ [L = dimethylmalonato, dimethyl chloromalonato, dimethyl methylmalonato, or Etacac], a shift of over 100 cm^{-1} to lower wavenumber for ν(C=O) has been taken to indicate chelation by the carbonyl groups.[75] I.r. and Raman examination of aluminium isopropoxide has allowed spectroscopic distinction to be made between previously postulated trimeric and tetrameric allotropic forms.[76]

The monophenoxyaluminium compounds $Me_2Al(OPh)$, Me(PhO)AlCl, and $(PhO)AlCl_2$ are all believed [77] to be trimeric in benzene solution at room temperature, *e.g.* (12), but at higher temperatures the formation of

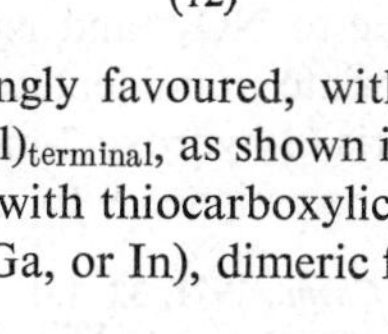

(12)

dimeric species is increasingly favoured, with the resultant appearance of bands due to ν(C—O—Al)$_{\text{terminal}}$, as shown in Table 10. The reaction of trialkyls of Al, Ga, and In with thiocarboxylic acid yields $[R_2MOSCMe]_n$ (R = Me or Et; M = Al, Ga, or In), dimeric for M = Al but monomeric

[72] J. W. Akitt, N. N. Greenwood, and G. D. Lester, *J. Chem. Soc.* (*A*), 1971, 2450.
[73] T. Kawakami, A. Konno, and Y. Ogino, *Bull. Chem. Soc. Japan*, 1971, **44**, 1772.
[73a] I. Kijima, T. Yamamoto, and Y. Abe, *Bull. Chem. Soc. Japan*, 1971, **44**, 3193.
[74] T. Nojiri, M. Motoi, and I. Hashimoto, *Bull. Chem. Soc. Japan*, 1971, **44**, 850.
[75] Y. Kawakami and T. Tsuruta, *Bull. Chem. Soc. Japan*, 1971, **44**, 247.
[76] W. Fieggen and H. Gerding, *Rec. Trav. chim.*, 1971, **90**, 410.
[77] K. B. Starowieyski, S. Pasynkiewicz, and M. D. Skowrońska, *J. Organometallic Chem.*, 1971, **31**, 149.

Table 10 *I.r. bands (wavenumber/cm^{-1}) of dimers of monophenoxy-aluminium compounds*

Compound	ν(C—O—Al)$_{terminal}$	ν(C—O—Al)$_{bridge}$
$Me_2Al(OPh)$	—	1223
MeAl(OPh)Cl	1160	~ 1206
$Al(OPh)Cl_2$	1140	1192

in the case of Ga and In. I.r. and Raman measurements indicate that the Al compounds possess puckered eight-membered $Al_2O_2S_2C_2$ ring skeletons, whereas the Ga and In species presumably have planar four-membered MOSC rings.[78] Frequencies for bands assigned as ring vibrations are shown in Table 11. Similar studies on further systems of this type have also been reported.[78a, 78b]

Table 11 *Ring vibration frequencies for* $[R_2MOSCMe]_n$ *compounds, quoted as wavenumber*/cm^{-1} [*i.r. unless* (*R*) *Raman*]

M	Al (n = 2)		Ga (n = 1)	In (n = 1)
R	Me	Et	Me, Et	Et
ν(M—O)	580	570	550	458
ν(M—S)	346	335	270	222 (R)
δ(OMS)	184 (R)	—	—	+ τ_{Me} 150, 122 (R)

Partial assignments of some i.r. frequencies measured for alkali-metal hydroaluminocarbonates have been given.[79] I.r. spectra (20—4000 cm^{-1}) have been recorded[80] for a series of anhydrous isomorphous complex sulphates $M^IM^{III}(SO_4)_2$ (M^{III} = Al, Ga, In, Cr, or Fe). A detailed consideration of these results indicates that the sulphate ion acts as a triply bridging ligand bound to the tervalent ions.

The vibrational spectra of GaX_3,H_2O (X = Cl or Br) have been interpreted in terms of an O—GaX_3 skeleton, with four-co-ordinate Ga, in contrast to the six-co-ordination for higher hydrates.[81] Bands at *ca.* 410 cm^{-1} are assigned to ν(Ga—O). I.r. bands at 1070 cm^{-1} and 960 cm^{-1} in a gallium tungstate formulated as $Ga(OH)WO_4,3.5H_2O$ are attributed to metal—OH bending modes.[82] Spectral data due to the aminoethanol group have been partially assigned[83] for $Ga(L)_2OC_2H_4NH_2$ complexes, (L = NCS, NO_2, or NO_3). Vibrational bands due to NO_2^+ and NO_3 for the volatile complex $NO_2^+\,Ga(NO_3)_4^-$ are appropriate to unidentate nitrate co-ordination.[84]

In the i.r. spectrum of tris-(3-hydroxy-2-methyl-4-pyranato)indium(III) (13), a band at 482 cm^{-1} can be assigned[85] to ν(In—O), in accordance with

[78] J. Weidlein, *J. Organometallic Chem.*, 1971, **32**, 181.
[78a] J. Weidlein and B. Schaible, *Z. anorg. Chem.*, 1971, **386**, 176.
[78b] J. Weidlein, *Z. anorg. Chem.*, 1971, **386**, 129.
[79] A. S. Berger, N. P. Tomilov, and I. A. Vorsina, *Russ. J. Inorg. Chem.*, 1971, **16**, 42.
[80] P. Rems, B. Orel, and J. Seftar, *Inorg. Chim. Acta*, 1971, **5**, 33.
[81] M. T. Bories, J. Rozière, and A. Potier, *Chem. Comm.*, 1971, 213.
[82] M. V. Mokhosoev and A. I. Gruba, *Russ. J. Inorg. Chem.*, 1971, **16**, 508.
[83] A. M. Golub and Fam Wang Ch'a, *Russ. J. Inorg. Chem.*, 1971, **16**, 185.
[84] D. Bowler and N. Logan, *Chem. Comm.*, 1971, 582.
[85] D. G. Tuck and M. K. Yang, *J. Chem. Soc. (A)*, 1971, 3100.

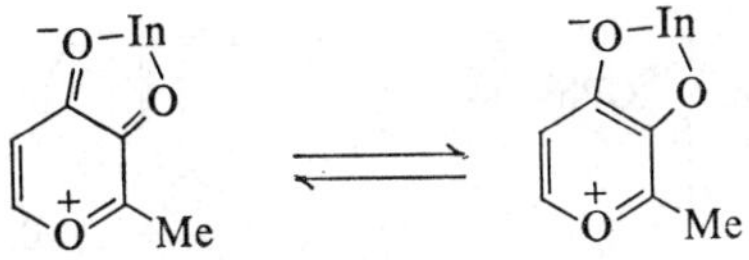

(13)

earlier studies of $In(acac)_3$. Weak bands at 305 and 180 cm^{-1} may be due to motions of the O—In—O grouping in the proposed chelate structure for this complex. A paper [86] devoted to the spectroscopic properties of some aquo-complexes of In^{III} and Fe^{III} includes the assignment of a Raman band at 348 cm^{-1} for $(NH_4)_2[InCl_5(H_2O)]$ to $\nu(In—OH_2)$. Some i.r. data for the alizarinate complex $In(C_{14}H_7O_4)_3$ have been given.[87] Insertion of SO_2 into the In—C bonds of indium trialkyls produces polymeric sulphinates showing a low value for $\nu_{asym}(SO_2)/\nu_{sym}(SO_2)$, consistent with bidentate sulphinate bridging.[88]

I.r. measurements for thallium(I) derivatives have been reported in two instances.[89, 90] Shifts in characteristic ligand frequencies on formation of Tl^I complexes of 8-hydroxyquinolinate and oxalate have been discussed in relation to those observed for alkali-metal(I) analogues, and unassigned i.r. frequencies for the monoalkylcarbonate $Tl(O_2COMe)$ have been listed.

Compounds containing M—S Bonds (M = B, In, or Tl) or Tl—Se Bonds.— Vahrenkamp has reported and partially assigned i.r. and Raman spectra for some boron–sulphur compounds (Table 12).[91] Force-constant

Table 12 *Wavenumbers*/cm^{-1} *for* B—S *stretching vibrations in some boron–sulphur compounds* [*i.r., infrared; R, Raman spectra*]

Compound	ν(B—S)
$B(SMe)_3$	905/930 (asym; i.r.) 430 (sym; R)
$MeB(SMe)_2$	1026/1052 (asym; i.r.) 517 (sym; R)
$PhB(SMe)_2$	908/936 (asym; i.r.) 637 (sym; R)
$Me_2B(SMe)$	574 (R)
Me_2BSH	579 (R)
Me_2BSPh	581 (R)
Ph_2BSMe	885/910 (R)
$(Me_2B)_2S$	594 (asym; i.r.) 543 (sym; R)

calculations suggest an absence of π-bonding in the B—S bond. In diethylindium thiobenzoate [92] (14), an i.r. band at 347 cm^{-1} is assigned to ν(In—S). The complexes (15) and related species containing co-ordinated DMSO, phen, or bipy as well as the dithiolato-ligand have

86 S. A. Cotton and J. F. Gibson, *J. Chem. Soc.* (*A*), 1971, 1693.
87 B. E. Zaitsev, N. P. Vasil'eva, and B. N. Ivanov-Emin, *Russ. J. Inorg. Chem.*, 1971, **16**, 502.
88 A. T. T. Hsieh, *J. Organometallic Chem.*, 1971, **27**, 293.
89 A. G. Lee, *J. Chem. Soc.* (*A*), 1971, 2007.
90 G. Gattow, H. Behrendt, and V. Warzelhan, *Naturwiss.*, 1971, **58**, 361.
91 H. Vahrenkamp, *J. Organometallic Chem.*, 1971, **28**, 181.
92 H. Tada and R. Okawara, *J. Organometallic Chem.*, 1971, **28**, 21.

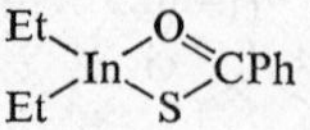

(14)

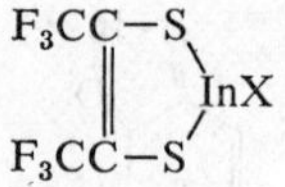

(15) X = Cl, Br, or I

been prepared.[93] Shifts in frequency for two characteristic ligand bands [ν(C$\doteq$C) and ν(R—C$\langle^{C}_{S}$)] confirm an analogy with complexes known to contain a ligand ring-system. Unassigned i.r. frequencies listed for (dicyanoethylene-1,2-dithiolate)- and (toluene-3,4-dithiolate)-complexes of Tl^{III} are qualitatively consistent with four-co-ordination at Tl, by three sulphur atoms and either oxygen or nitrogen.[94]

I.r. frequencies for bands characteristic of the SeCN group have been given for $R_2TlSeCN$ (R = Ph, *o*-tolyl, *m*-tolyl, or *p*-tolyl).[95]

Compounds containing M—Halogen Bonds (M = B, Al, Ga, In, or Tl).—I.r. spectroscopy has been used[96] to study the reaction of BCl_3 and of BF_3 with silica surfaces, ν(B—O) and ν(B—F) or ν(B—Cl) having been identified for chemisorbed species. $NaBF_4$ has a phase transition at 245 °C, distinguished by abrupt loss of the splitting of the ν_2 and ν_4 fundamentals of the anion in the Raman spectrum of the crystalline solid.[97] Likewise, a phase change at 110 K has been observed for NH_4BF_4 through vibrational studies.[97a] Adducts of BCl_3 with aromatic amines[98] (R^1NHR^2,BCl_3) show ν(N—H) at 3100—3200 cm^{-1}; dehydrochlorination yields corresponding anilinochloroboranes R^1NR^2,BCl_2, for which ν(N—H) are at 3300—3400 cm^{-1} for R^2 = H. I.r. frequencies for adducts of (16) with bases such

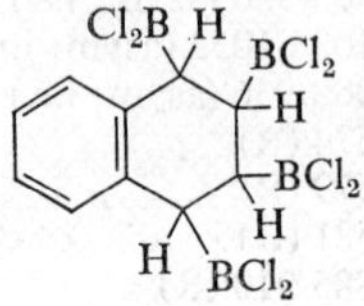

(16)

as Et_3N, Me_3P, or Me_2S have been listed;[99] characteristic 'BCl_2' bands were observed at 800—950 cm^{-1}. An assignment for the vibrational spectrum of $(BFNH)_3$ has been proposed.[99a]

[93] A. F. Berniaz, G. Hunter, and D. G. Tuck, *J. Chem. Soc.* (*A*), 1971, 3254.
[94] G. Hunter and B. C. Williams, *J. Chem. Soc.* (*A*), 1971, 2554.
[95] T. N. Srivastava and K. K. Bajpai, *J. Organometallic Chem.*, 1971, **31**, 1.
[96] B. A. Morrow and A. Devi, *Chem. Comm.*, 1971, 1237.
[97] A. S. Quist, J. B. Bates, and G. E. Boyd, *J. Chem. Phys.*, 1971, **55**, 2836.
[97a] C. J. H. Schutte and D. J. J. van Rensburg, *J. Mol. Structure*, 1971, **10**, 487.
[98] J. R. Blackborrow, J. E. Blackmore, and J. C. Lockhart, *J. Chem. Soc.* (*A*), 1971, 49.
[99] M. Zeblin, *J. Inorg. Nuclear Chem.*, 1971, **33**, 1179.
[99a] K. E. Buck, I. A. Boenig, and K. Niedenzu, *Inorg. Chem.*, 1971, **10**, 1917.

Current evidence for the C_{2v} structure of gaseous $LiAlF_4$ has been assessed.[100] I.r. data for AlF_3 and $NaAlF_4$ have been discussed in terms of distorted AlF_6^- octahedra.[100a] Through study of ligand exchange at $AlCl_3$ for a variety of complexes $AlCl_3$,L by i.r. spectroscopy, Jones and Wood have established a competition order for $AlCl_3$ between a series of donor ligands.[101] The same workers have also obtained i.r. and Raman spectra for $AlCl_3$,1.5L (L = MeCN or HCO_2Me) and $AlCl_3$,2L (L = Me_2O or MeCN).[102] Their results suggest a trigonal-bipyramidal five-co-ordinate structure for $AlCl_3,2Me_2O$, contrasting with an ionic formulation for each of the other complexes, for which bands assignable to vibrations of the species $Al(MeCN)_6^{3+}$, $Al(HCO_2Me)_6^{3+}$, and $AlCl_4^-$ could be distinguished.

There have been a number of reports concerning halogeno—M anions (M = Al, Ga, In, or Tl). From the Raman spectra of molten $AlCl_3$–NaCl, $AlBr_3$–NaBr, and AlI_3–CsI, the assignments in Table 13 have been made.[103]

Table 13 *Raman bands (wavenumbers/*cm^{-1}*) of halogenoaluminium ions*

Ion	$\nu_1(A_1)$	$\nu_2(E)$	$\nu_3(F_2)$	$\nu_4(F_2)$
$AlCl_4^-$	351	121	490	186
$AlBr_4^-$	209	75	409	114
AlI_4^-	146	51	336	82

Compounds of the new anion $[In^IX_3]^{2-}$ (X = Cl, Br, or I) have been obtained;[104] Raman data are assigned on the basis of C_{3v} structures, *e.g.* with ν_1, ν_2, ν_3, and ν_4 at 252, 185, 102, and 97 cm^{-1} for X = Cl. In a review[105] of co-ordination complexes of the Tl^{III} halides, extensive reference is made to vibrational spectroscopy, including summaries of data for a variety of halogenothallium(III) anionic species. Other references[106–108] to MX_4^- and related anions are given in Table 14. I.r. and Raman spectra for solid

Table 14 *References to halogeno–Group III anions characterized by vibrational spectroscopy*

Compound	*Ref.*
$M[AlCl_4]$ (M = Ga or Tl) $Tl[GaCl_4]$	106
$A[InX_4]$ (A = Et_4N, $Pr^n{}_4N$, or Ph_4As; X = Cl, Br, or I) $(Me_4N)_2[InX_5]$ (X = Cl or Br)	107
$Ga_2[Ga_2I_6]$	108

[100] S. J. Cyvin, B. N. Cyvin, D. B. Rao, and A. Snelson, *Z. anorg. Chem.*, 1971, **380**, 212.
[100a] J. Bondam, *Acta Chem. Scand.*, 1971, **25**, 3271.
[101] D. E. H. Jones and J. L. Wood, *J. Chem. Soc.* (*A*), 1971, 3132.
[102] D. E. H. Jones and J. L. Wood, *J. Chem. Soc.* (*A*), 1971, 3135.
[103] G. M. Begun, C. R. Boston, G. Torsi, and G. Mamontov, *Inorg. Chem.*, 1971, **10**, 886.
[104] G. Contreras and D. G. Tuck, *Chem. Comm.*, 1971, 1552.
[105] R. A. Walton, *Co-ordination Chem. Rev.*, 1971, **6**, 1.
[106] F. J. Brinkmann and H. Gerding, *Inorg. Nuclear Chem. Letters*, 1971, **7**, 667.
[107] J. Gislason, M. H. Lloyd, and D. G. Tuck, *Inorg. Chem.*, 1971, **10**, 1907.
[108] W. Lind, L. Waterworth, and I. J. Worrall, *Inorg. Nuclear Chem. Letters*, 1971, **7**, 611.

Ga_2X_6 (X = Cl, Br, or I) have been measured and assigned;[109] in general, the Raman data are in good agreement with those reported previously. Similar spectroscopic results obtained for solid In_2Cl_3 and In_4Cl_7 have not been assigned.[110]

References containing spectroscopic information for adducts of MX_3 (M = Al, Ga, or Tl; X = Cl, Br, or I) with Lewis bases [111–116] are gathered in Table 15.

Table 15 *References containing vibrational data for adducts of Group III halides*

Compound	*Ref.*
$MeCO^+[AlCl_4]^-$ $(MeCO)_2CHCO^+[AlCl_4]^-$	111
$AlCl_3(MeNH_2)_4$ $AlCl_3(MeNH_2)_2$ $AlCl_3,MeNH_3Cl$ $AlCl_3(MeNH_3Cl)_2$ $(Cl_3P{=}N{-}PCl_3)AlCl_4$	112
GaX_3,SbX_3 (X = Cl, Br, or I)	113
$InX_3,2PMePh_2$ $InX_3,2PMe_2Ph$	114
$TlX_3,(bipy)_n$ (X = Cl, Br, or I)	115
$TlX_3,(1,3\text{-propenediamine})_n$	116
(X = Cl, Br, or I; n = 1 or 2.)	

4 Group IV Elements

Attempts to study coals by laser Raman spectroscopy have revealed only bands at *ca.* 1588 and 1360 cm^{-1}, similar to those in activated carbons or graphite, probably as a result of degradation of the sample by carbonization in the laser beam.[117]

The structures of some alkylcarbonium ions have been investigated by i.r. and Raman techniques.[118] In particular, a planar C_3C^+ carbon skeleton has been confirmed for the t-butyl cation Me_3C^+, through spectroscopic comparison with the isoelectronic model Me_3B.

Fujiyama has made a study of the deformation vibrations of the carbon skeletons of $CH_3(CH_2)_nX$ (X = Me, F, Cl, Br, or CN; n = 1, 2, or 3),

109 E. Chemouni, *J. Inorg. Nuclear Chem.*, 1971, **33**, 2333.
110 F. J. Brinkmann and H. Gerding, *Rec. Trav. chim.*, 1971, **90**, 269.
111 A. Germain, A. Commeyras, and A. Casadevall, *Chem. Comm.*, 1971, 633.
112 H. Vollmer and M. Becke-Goehring, *Z. anorg. Chem.*, 1971, **382**, 281.
113 E. Chemouni and A. Potier, *J. Inorg. Nuclear Chem.*, 1971, **33**, 2353.
114 D. M. Roundhill, *J. Inorg. Nuclear Chem.*, 1971, **33**, 3367.
115 F. Ya. Kul'ba, N. G. Yaroslavskii, L. V. Konovalov, A. V. Barsukov, and E. V. Mironov, *Russ. J. Inorg. Chem.*, 1971, **16**, 341.
116 F. Ya. Kul'ba, Yu. A. Makashev, D. M. Markhaeva, and A. V. Barsukov, *Russ. J. Inorg. Chem.*, 1971, **16**, 511.
117 R. A. Friedel and G. L. Carlson, *Chem. and Ind.*, 1971, 1128.
118 G. A. Olah, J. R. DeMember, A. Commeyras, and J. L. Bribes, *J. Amer. Chem. Soc.*, 1971, **93**, 459.

correlating observed frequency variations with the incidence of rotational isomerism.[119]

Barriers to internal rotation in a number of fluorinated ethanes have been determined using spectroscopic data.[120] For the fluorocarbon compounds $CH_3 \cdot CO \cdot OOCF_3$, $CF_3 \cdot CO \cdot OOCF_3$, $(CF_3OO \cdot CO \cdot CF_2)_2CF_2$, $CF_3OO \cdot CO \cdot (CF_2)_3 \cdot CO \cdot F$, and $CF_3 \cdot CF(OF) \cdot OOCF_3$, i.r. spectra measured in the vapour phase have been partially assigned.[121] I.r. frequency data for a number of fluorocarbon and chlorocarbon isothiocyanates include bands assigned[122] to ν(NCS) at wavenumbers of *ca.* 1950—2000 cm^{-1}.

In $Cl_2C{=}C(SiMe_3)Cl$, ν(C=C) is at 1545 cm^{-1}. This very low value has been rationalized in terms of $(p \rightarrow d)\pi$ donation of electron density from the C—C π-bond into the silicon 3*d*-orbitals.[123] $Et_3Si{-}CH{=}C{=}CHMe$ shows[124] a strong i.r. band at 1950 cm^{-1}. For the compounds $R^1R^2C{=}C(OMe)OSiMe_3$ (R^1, R^2 = H, Me, or Ph), ν(C=C) occurs from 1708 cm^{-1} ($R^1 = R^2$ = Me) down to 1630 cm^{-1} ($R^1 = R^2$ = Ph), whereas for $R(MeO_2C)C{=}C(OMe)OSiMe_3$ lower wavenumbers are observed:[125] 1600 cm^{-1} (R = H or Me), 1560 cm^{-1} (R = Ph), or 1530 cm^{-1} (R = CO_2Me).

I.r. spectra for the *sym*-triazine adducts $(C_3H_3N_3)$,L (L = $AlCl_3$, $SbCl_5$, or $TiCl_4$) and $(C_3H_3N_3)$,2L (L = $AlCl_3$, $SbCl_5$, $TiCl_4$, or $SnCl_4$) have been reported and are assigned, assuming a local symmetry of C_{2v} for the *sym*-triazine ligand.[126]

An assignment of the i.r. and Raman spectra of *trans*-oxalyl fluoride, $(COF)_2$, has been published.[126a] Powdered samples of KHC_2O_4 and KDC_2O_4 have been examined by i.r. and Raman spectroscopy,[127] and similar studies of $NaHC_2O_4$, $NaHC_2O_4,H_2O$ and $NaDC_2O_4,D_2O$ have been made.[127a] Among data reported for $Cs_2CO_2F_2$ and $(Me_4N)_2CO_2F_2$ are bands assigned to ν(CO) and ν(CF) vibrations.[128] The i.r. spectra of the solvates Na_2CS_4,L (L = NH_3, $MeNH_2$, $EtNH_2$, or Me_2NH) have been interpreted, frequencies for vibrations of the CS_4^{2-} unit being close to previously reported values.[129, 130] Solid complexes of stoicheiometry $Me_4NX,2CY_4$ and Et_4NX,CY_4 have been isolated.[130a] Estimation of the

[119] T. Fujiyama, *Bull. Chem. Soc. Japan*, 1971, **44**, 1194.
[120] R. A. Pethrick and E. Wyn-Jones, *J. Chem. Soc.* (*A*), 1971, 54.
[121] P. A. Bernstein, F. A. Hohorst, and D. D. DesMarteau, *J. Amer. Chem. Soc.*, 1971, **93**, 3882.
[122] G. Dahms, A. Haas, and W. Klug, *Chem. Ber.*, 1971, **104**, 2732.
[123] J. Dunogues, R. Calas, J. Malzac, N. Duffaut, C. Biran, and P. Lapouyade, *J. Organometallic Chem.*, 1971, **27**, C1.
[124] J. C. Craig and C. D. Beard, *Chem. Comm.*, 1971, 692.
[125] Y. N. Kuo, F. Chen, and C. Ainsworth, *Chem. Comm.*, 1971, 137.
[126] E. Allenstein, V. Beyl, and K. Löhmar, *Z. anorg. Chem.*, 1971, **381**, 40.
[126a] J. R. Durig, S. C. Brown, and S. E. Hannum, *J. Chem. Phys.*, 1971, **54**, 4428.
[127] J. de Villepin and A. Novak, *Spectroscopy Letters*, 1971, **4**, 1.
[127a] J. de Villepin and A. Novak, *Spectrochim. Acta*, 1971, **27A**, 1259.
[128] E. Martineau and J. B. Milne, *Chem. Comm.*, 1971, 1327.
[129] J. Roger, *Compt. rend.*, 1971, **273**, *C*, 1089.
[130] J. Roger, *Compt. rend.*, 1971, **273**, *C*, 1352.
[130a] J. A. Creighton and K. M. Thomas, *J. Mol. Structure*, 1971, **7**, 173.

X—CY_4 stretching force-constants indicate the accepted order for donor–acceptor ability: $I^- > Br^- > Cl^-$; $CI_4 > CBr_4 > CCl_4$.

Compounds containing M—H Bonds (M = Si, Ge, or Sn).—Rotational fine structure of the band arising from the SiH_3 rocking mode, ν_7, of H_3SiCN has been examined in detail;[131] as part of the same work, new vibrational data for Me_3SiCN were obtained. The ν(Si—H) frequencies for the silanes $Et_{3-n}X_nSiH$ (n = 1 or 2; X = F, Cl, Br, Ph, CH_2=CH, MeS, or MeO) have been related to Hammett σ_n constants for X.[132] Sum and difference bands, arising from combination of ring-puckering vibrations with higher frequency fundamentals, have been observed in the vapour-phase i.r. spectra of silacyclopent-3-ene, sila[$^2H_2$]cyclopent-3-ene, silacyclobutane, and germacyclopentane, the results being used to derive ring-puckering potential energy functions.[133]

The first examples of metalloid-substituted polyarsines, $(SiH_3As)_5$ and $(GeH_3As)_5$, have been reported;[134] their vibrational spectra contain bands characteristic of M—H modes (M = Si or Ge). A full assignment of the normal vibrations of vinylgermane (CH_2=$CHGeH_3$ and CH_2=$CHGeD_3$) in terms of the point group C_s has been made by Durig and Turner.[135] For $MeGeHX_2$ and $MeGeDX_2$ (X = F, Cl, Br, or I), i.r. and Raman studies allow assignment of 17 of the 18 fundamentals, assuming C_s symmetry.[136] I.r. frequencies for bands formally assigned to the Ge—H bending vibrations $\nu_7(a')$ and $\nu_{15}(a'')$ are shown in Table 16; normal-co-ordinate calculations

Table 16 *Wavenumbers*/cm^{-1} *for* Ge—H *bending fundamentals in methyldihalogenogermanes*

	$\nu_7(a')$		$\nu_{15}(a'')$	
X	$MeGeHX_2$	$MeGeDX_2$	$MeGeHX_2$	$MeGeDX_2$
F	690	490	700	520
Cl	667	495	707	515
Br	653	482	694	488
I	634	469	656	469

indicate that these modes are described mainly by the valence force-constants f_{CGeH} and f_{XGeH}, contributions from other co-ordinates becoming important only in the case of the difluorogermane. The gas-phase i.r. spectrum of digermylmethane, $(GeH_3)_2CH_2$, has been recorded and

131 H. Bürger and G. Schirawski, *Spectrochim. Acta*, 1971, **27A**, 159.
132 A. N. Egorochkin and N. S. Vyazankin, *Doklady Chem.*, 1971, **196**, 46.
133 J. F. Blanke, T. H. Chao, and J. Laane, *J. Mol. Spectroscopy*, 1971, **38**, 483.
134 J. W. Anderson and J. E. Drake, *Chem. Comm.*, 1971, 1372.
135 J. R. Durig and J. B. Turner, *Spectrochim. Acta*, 1971, **27A**, 1623.
136 G. K. Barker, J. E. Drake, R. T. Hemmings, and B. Rapp, *J. Chem. Soc.* (*A*), 1971, 3291.

partially assigned.[137] Other compounds for which frequencies of Si—H or Ge—H vibrations have been reported are listed in Table 17.[138–147]

Bellama and Gsell have recorded the i.r. spectra of the thermally very unstable monohalogenostannanes H_3SnX (X = Cl, Br, or I) at 77 K.[148]

Table 17 *References to* Si—H *and* Ge—H *compounds characterized by i.r. and/or Raman spectroscopy*

Compound	*Ref.*
$(SiH_3)_2NEt$	138
$(SiH_2Cl)_2NEt$	
$(SiHCl_2)_2NEt$	
$SiHCl_2NHEt$	
$MeSiH_2PH_2$	139
$Me_3SiP(SiH_3)_2$	140
$SiH_3SiH_2P(SiH_3)_2$	
H_3SiOEt	141
H_3SiOPr^i	
H_3SiOBu^t	
$(MeCOO)_3SiH$	142
$(CF_3COO)_3SiH$	
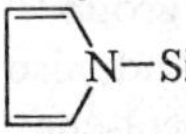	143
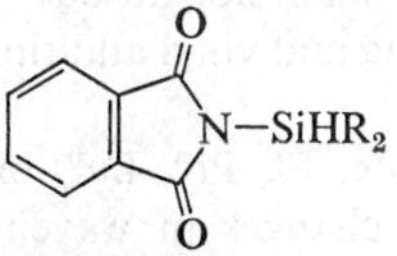 (R = Me_2 or MePh)	144
$MeGeH_2F$	145
Me_2GeHF	
GeH_3CH_2Cl	146
CH_3GeH_2Cl	
GeH_3CH_2Br	
CH_3GeH_2Br	
GeH_3CH_2I	
CH_3GeH_2I	
$HC(OH)(CH_2)_2GePhH$ (cyclic)	147
$HC(OH)CH_2CH(CH)_3GePhH$ (cyclic)	
$HC(OH)(CH_2)_{10}GePhH$ (cyclic)	

[137] R. M. Dreyfuss and W. L. Jolly, *Inorg. Chem.*, 1971, **10**, 2567.
[138] J. E. Drake and N. P. C. Westwood, *J. Chem. Soc.* (*A*), 1971, 3617.
[139] J. W. Anderson and J. E. Drake, *J. Chem. Soc.* (*A*), 1971, 1424.
[140] J. W. Anderson and J. E. Drake, *J. Chem. Soc.* (*A*), 1971, 2246.
[141] C. Glidewell, *J. Chem. Soc.* (*A*), 1971, 823.
[142] E. Hengge and E. Starz, *Monatsh.*, 1971, **102**, 741.
[143] H. Bürger and K. Burczyk, *Z. anorg. Chem.*, 1971, **381**, 176.
[144] A. F. Janzen and E. A. Kramer, *Canad. J. Chem.*, 1971, **49**, 1011.
[145] C. H. Van Dyke, J. E. Bulkowski, and N. Viswanathan, *Inorg. Nuclear Chem. Letters*, 1971, **7**, 1057.
[146] J. M. Bellama and C. J. McCormick, *Inorg. Nuclear Chem. Letters*, 1971, **7**, 533.
[147] P. Rivière and J. Stagé, *Angew. Chem. Internat. Edn.*, 1971, **10**, 267.
[148] J. M. Bellama and R. A. Gsell, *Inorg. Nuclear Chem. Letters*, 1971, **7**, 365.

The bands are complex because of crystalline splitting, but wavenumbers of band-centres and assignments are as shown in Table 18.

Table 18 *I.r. spectral bands (wavenumber*/cm^{-1}) *of* H_3SnX (X = Cl, Br, *or* I)

Compound	ν(Sn—H)	$\delta(SnH_3)$	$\rho(SnH_3)$	ν(Sn—X)
SnH_3Cl	1948	683	566	283
SnH_3Br	1928	670	544	—
SnH_3I	1905	645	493	—

Compounds containing M—C Bonds (M = Si, Ge, Sn, or Pb).—A comprehensive study of the vibrational spectrum of $Si(CH{=}CH_2)_4$ has been made, including an almost complete assignment in terms of D_{2d} geometry; no evidence to suggest donation of π-electron density from the vinyl groups into silicon 3*d* orbitals was found.[149] Unassigned i.r. data for the pentafluorophenylfluorosilanes $Ph_FSiMeFCl$, Ph_FSiMeF_2, $Ph_F(CH_2)_2SiF_3$, and $Ph_F(CH_2)_2SiMeF_2$ have been listed.[150] The i.r. spectra of trimethylsilylmethyl complexes of a variety of metals (Ti, Mn, Co, Pd, Pt, and Au) contain bands assigned to vibrations of the Me_3SiCH_2 group[151] at 1230—1260 cm^{-1} [$\delta_{sym}(CH_3)$], 802—856 cm^{-1} [ν(Si—Me)], and also at 512—560 cm^{-1} [ν(M—C)]. Polymerization of silacyclopent-ene and -ane derivatives by $AlCl_3$ through both ring-opening and vinyl addition has been established by i.r. spectroscopy.[151a]

The vibrational spectra of R_3GeF (R = Me, Et, Pr^n, Bu^n, or Ph) have been almost completely assigned.[152] Large changes in wavenumber for ν(Ge—F) with change of phase, attributed to intermolecular interactions, were noted for these compounds, *e.g.* for gas–liquid–solid, R = Me: 659, 623, 579 cm^{-1}; R = Bu^n: 642, 624, 579 cm^{-1}. A spectroscopic study of Me_nGeI_{4-n} (n = 1, 2, or 3), in which assignments for all the vibrational fundamentals (except torsional modes) are supported by normal-co-ordinate calculations, has been published.[153] Similarly, the vibrational spectra of $MeGeI_3$ and its deuterio-analogue have been studied in detail and assigned, assuming C_{3v} symmetry. An i.r. band observed at 123 cm^{-1} in the spectrum of solid $MeGeI_3$ is attributed to the inactive a_2 torsional mode, leading to a value of 1.1 kcal mol^{-1} for the three-fold barrier to rotation about the Ge—C bond.[154] Detailed interpretations of the i.r. and Raman spectra of liquid $CH_2{=}CH{\cdot}GeX_3$ (X = Cl or Br) are consistent with C_s symmetry for these molecules; the proposed assignments are compared directly with

[149] G. Davidson, *Spectrochim. Acta*, 1971, **27A**, 1161.
[150] J. M. Birchall, R. N. Haszeldine, M. J. Newlands, P. H. Rolfe, D. L. Scott, A. E. Tipping, and D. Ward, *J. Chem. Soc.* (*A*), 1971, 3760.
[151] B. Wozniak, J. D. Ruddick, and G. Wilkinson, *J. Chem. Soc.* (*A*), 1971, 3116.
[151a] T. Araki, D. Terunuma, T. Sato, N. Nagai, M. Furvichi, and S. Nakamura, *Bull. Chem. Soc. Japan*, 1971, **44**, 2725.
[152] K. Licht and P. Koehler, *Z. anorg. Chem.*, 1971, **383**, 174.
[153] J. W. Anderson, G. K. Barker, J. E. Drake, and R. T. Hemmings, *Canad. J. Chem.*, 1971, **49**, 2931.
[154] J. R. Durig, C. F. Jumper, and J. N. Willis, jun., *J. Mol. Spectroscopy*, 1971, **37**, 260.

those for vinylgermane (*q.v.*), reported by the same workers. For $CH_2{=}CHGeCl_3$, a three-fold barrier to rotation of 1.7 kcal mol^{-1} is calculated, from the assignment of a band at 73 cm^{-1} to the torsional vibration.[155]

Strong Raman bands at *ca.* 370 cm^{-1} in the spectra of $C_5H_5GeH_3$ and $(C_5H_5)_2GeH_2$ are assigned to ν(Ge—C) modes.[156] I.r. spectroscopy has been used to show[157] that the products from the reaction of $RGeH_2Cl$ (R = Me, Et, or Ph) are the germacyclic compounds cyclo-R(Cl)Ge-$(—CH_2—)_6$. Incomplete assignments for the i.r. and Raman spectra of two germyl nitrates Me_3GeNO_3 and $Me_2Ge(NO_3)_2$ have been given.[158, 159]

I.r. frequencies have been reported[160] for (neo-$C_5H_{11})_4Sn$ and (neo-$C_5H_{11})_3SnPh$. Strong intermolecular interactions are indicated for triorganotin fluorides R_3SnF (R = Me, Et, Pr^n, Bu^n, or Ph) in the solid state: in the gas phase (i.r. and Raman), ν(Sn—F) are assigned to bands in the 570 cm^{-1} region, dropping to *ca.* 350 cm^{-1} (i.r.) for solid samples; $\Delta\nu$ (solid–vapour) is more than 200 cm^{-1} for four out of the five compounds examined.[161] Bands characteristic of the Ph_3Sn grouping have been assigned in the i.r. spectra of two series of compounds: Ph_3SnX (X = NCS, NCO, OCOMe, OCOPh, or ONO_2) and $(Ph_3Sn)_2Y$ (Y = S, SO_4, SeO_3, or SeO_4);[162] and Ph_3SnX (X = Ph, $SnPh_3$, F, Cl, Br, SPh, $SSnPh_3$, or OOCPh).[163] For the latter series, bands described as ν_{asym}(Sn—Ph) and ν_{sym}(Sn—Ph) are at *ca.* 270 cm^{-1} and *ca.* 230 cm^{-1} respectively, and are almost independent of the nature of X.

The vibrational spectrum of solid Me_2SnF_2 has been measured in the range 800—20 cm^{-1}. All the skeletal modes predicted by factor-group analysis are observed, including the allowed Sn—F vibration ν_{asym}(Sn—F) at 365 cm^{-1}. A band at 218 cm^{-1}, showing variation in intensity with temperature, is assigned as a (Sn—Me) torsion.[164] A drop in wavenumber of nearly 100 cm^{-1} for bands assigned to ν(Sn—Cl) on going from R_2SnCl_2 to R_2SnCl_2,2DMF (R = Ph, *o*-tolyl, or *p*-tolyl) is consistent with an increase in co-ordination at the metal atom.[165] Partially assigned i.r. data for $Bu^t{}_2SnCl_2$ and $Bu^t{}_2Sn(OH)Cl$ have been reported,[166] and some minor reassignments for Me_2SnCl_2 have been proposed following an electron diffraction study.[166a]

[155] J. R. Durig and J. B. Turner, *Spectrochim. Acta*, 1971, **27A**, 395.
[156] S. R. Stobart, *J. Organometalllic Chem.*, 1971, **33**, C11.
[157] K. I. Kobrakov, T. I. Chernysheva, N. S. Nametkin, and A. Ya. Sideridu, *Doklady Chem.*, 1971, **196**, 8.
[158] D. Potts and A. Walker, *Canad. J. Chem.*, 1971, **49**, 2171.
[159] D. Potts and A. Walker, *Canad. J. Chem.*, 1971, **49**, 202.
[160] H. J. Götze, *Chem. Ber.*, 1971, **104**, 3719.
[161] K. Licht, H. Geissler, P. Koehler, K. Hottmann, H. Schnorr, and H. Kriegsmann, *Z. anorg. Chem.*, 1971, **385**, 271.
[162] T. N. Srivastava and S. K. Tandon, *Spectrochim. Acta*, 1971, **27A**, 593.
[163] J. R. May, W. R. McWhinnie, and R. C. Poller, *Spectrochim. Acta*, 1971, **27A**, 969.
[164] M. Goldstein and W. D. Unsworth, *J. Chem. Soc. (A)*, 1971, 2121.
[165] T. N. Srivastava and B. Misra, *J. Organometallic Chem.*, 1971, **32**, 331.
[166] C. K. Chu and J. D. Murray, *J. Chem. Soc. (A)*, 1971, 360.
[166a] H. Fujii and M. Kimura, *Bull. Chem. Soc. Japan*, 1971, **44**, 2643.

For the compounds 4-X·C_6H_4·S·$SnMe_3$ (X = NH_2, F, Cl, Me, Br, H, or NO_2), ν_{asym}(Sn—C) and ν_{sym}(Sn—C) are near 530 and 509 cm^{-1} irrespective of X, while ν(Sn—S) are in the range 326—384 cm^{-1}.[167] Me_2SnPh_2 and $Me_2SnPh[Cr(CO)_3Ph]$ show two bands due to ν(Sn—C), respectively at 530, 536 cm^{-1} (asym) and 519, 522 cm^{-1} (sym); it is suggested that the presence of only one such band (at 530 cm^{-1}) for $Me_2Sn[Cr(CO)_3Ph]_2$ could be due to unusual stereochemistry at the tin atom, possibly a flattened tetrahedral configuration.[168] Unassigned i.r. data have also been given[169] for the following trimethyltin derivatives: $Me_3SnNMeSO_2CF_3$, $Me_3SnNMePSF_2$, $Me_3SnNMePSFCl$, $Me_3SnOPOF_2$, $Me_3SnOPSF_2$, Me_3SnOSO_2F, $Me_3SnOSO_2CF_3$, $Me_3SnSPSF_2$, and $Me_3SnSPSFMe$.

Bis-(*NN*-dimethyldithiocarbamate) salts of the anions $[Me_2SnX_4]^{2-}$ and $[Me_2SnX_3]^-$ (X = Cl, Br, or I) have been prepared; i.r. bands at *ca.* 560 and *ca.* 515 cm^{-1} are assigned to ν_{asym} and ν_{sym}(Sn—C), respectively.[170]

Some characteristic i.r. frequencies for trimethyltin metaborate, $Me_3Sn(BO_2)$, and also for the compounds $MeSn(BPh_2)_3$, $MeSn(BBr_2)_3$,-4diphos, and the related species $Sn(BPh_2)_4$ and $Sn(BBr_2)_4$,4diphos have been reported.[171, 172]

I.r. and Raman data have been obtained for a number of polymeric triorganotin derivatives; references are listed in Table 19.[173–177] In several of these reports the spectroscopic evidence supports proposed five-co-ordination at the tin atoms.

Table 19 *References containing vibrational data for polymeric organotin derivatives*

Compound		*Ref.*
Me_3SnSO_3F		173
Me_3SnSO_3Me		173
$(Me_3Sn)_2EO_4$	(E = S, Se, or Cr;	174
$(Me_3Sn)_2EO_4$,L	L = MeOH, H_2O, py, or DMF)	174
$(Bu^n_3Sn)_2SO_4$		175
$Ph_3SnOCOMe$		176
$Ph_3SnOCOCH_2X$	(X = Cl, Br, or I)	176
$Ph_3SnOCOCHCl_2$		176
$Ph_3SnOCOCF_3$		176
$Ph_3Sn(OOSPh)$		177

167 T. A. George, *J. Organometallic Chem.*, 1971, **31**, 233.
168 T. P. Poeth, P. G. Harrison, T. V. Long, jun., B. R. Willeford, and J. J. Zuckerman, *Inorg. Chem.*, 1971, **10**, 522.
169 H. W. Roesky and H. Wiezer, *Chem. Ber.*, 1971, **104**, 2258.
170 T. Tanaka, K. Tanaka, and T. Yoshimitsu, *Bull. Chem. Soc. Japan*, 1971, **44**, 112.
171 K. Dey, *Z. anorg. Chem.*, 1971, **383**, 338.
172 H. Nöth, H. Schäfer, and G. Schmid, *Z. Naturforsch.*, 1971, **26b**, 497.
173 P. A. Yeats, J. R. Sams, and F. Aubke, *Inorg. Chem.*, 1971, **10**, 1877.
174 B. F. E. Ford, J. R. Sams, R. G. Goel, and D. R. Ridley, *J. Inorg. Nuclear Chem.*, 1971, **33**, 23.
175 R. H. Herber and C. H. Stapfer, *Inorg. Nuclear Chem. Letters*, 1971, **7**, 617.
176 B. F. E. Ford and J. R. Sams, *J. Organometallic Chem.*, 1971, **31**, 47.
177 E. Lindner and U. Kunze, *Inorg. Nuclear Chem. Letters*, 1971, **7**, 573.

I.r. frequency data for some organometallic diazoalkanes prepared by Lorberth[178] include those for the compound $Me_3Pb \cdot C(N_2)CO_2Et$; four bands assigned to Pb—C stretches at 514, 490, 470, and 465 cm^{-1} suggest the structure (17).

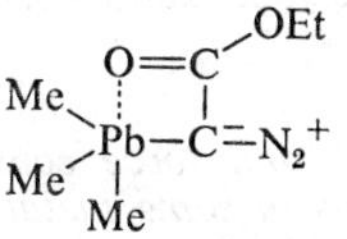

(17)

Compounds containing M—M or M—M′ Bonds (M,M′ = Si, Ge, Sn, or Pb).—Several polysilane derivatives have been investigated by vibrational spectroscopy, leading to a number of assignments for ν(Si—Si) modes. Drake and co-workers have prepared a number of halogenotrisilanes, for which they report[179] the data shown in Table 20. For cyclo-Si_5Me_{10} and cyclo-

Table 20 *Vibrational modes of halogenotrisilanes, expressed as wavenumber*/cm^{-1}. *Results are from i.r. spectra except where indicated as Raman (R)*

Compound	ν(Si—Si)	ν(Si—X)	δ(*skeleton*)
$SiH_3 \cdot SiHCl \cdot SiH_3$	389 (R)	523	150
$SiH_2Cl \cdot SiH_2 \cdot SiH_2Cl$	391 (R)	525	149
1- and 2-$BrSi_3H_7$ (4 : 1 mixture)	379 (R)	339	126

$Si_5Bu^i_{10}$, i.r. bands at 400 and 389 cm^{-1}, respectively, are assigned to ν(Si—Si) vibrations.[180] In the spectra of $(Et_3Si)_3SiX$ (X = H, F, Cl, Br, or I), strong i.r. bands near 480 cm^{-1} have been assigned to $\nu_{asym}(Si_3Si)$, but below 400 cm^{-1} the appearance of several absorptions and the possibility of mixing with other modes does not allow $\nu_{sym}(Si_3Si)$ to be distinguished with any certainty.[181] A correlation between values for f_{SiSi} in disilane derivatives Si_2X_6 and those for f_{SiH} in X_3SiH has been found by Hengge[182] and is discussed in terms of variation in strength of the Si—Si bond with the substituent X. For $Me_3SiSiMe_2(C_6F_5)$ and $Me_2(C_6F_5)SiSi(C_6F_5)Me_2$, and for $SiF_3OSi_2F_5$, i.r. data have been reported,[183, 184] but only above the range at which ν(Si—Si) frequencies are observed.

The vibrational spectra of a number of metal–metal bonded compounds have been examined by Spiro and co-workers;[185, 186] frequencies of bands

[178] J. Lorberth, *J. Organometallic Chem.*, 1971, **27**, 303.
[179] J. E. Drake, N. Goddard, and N. P. C. Westwood, *J. Chem. Soc.* (*A*), 1971, 3305.
[180] G. R. Husk, R. Wexler, and B. M. Kilcullen, *J. Organometallic Chem.*, 1971, **29**, C49.
[181] H. Burger and W. Kilian, *J. Organometallic Chem.*, 1971, **26**, 47.
[182] E. Hengge, *Monatsh.*, 1971, **102**, 734.
[183] M. Weidenbruch, G. Abrotat, and K. John, *Chem. Ber.*, 1971, **104**, 2124.
[184] K. G. Sharp and J. L. Margrave, *J. Inorg. Nuclear Chem.*, 1971, **33**, 2813.
[185] B. Fontal and T. G. Spiro, *Inorg. Chem.*, 1971, **10**, 9.
[186] P. A. Bulliner, C. O. Quicksall, and T. G. Spiro, *Inorg. Chem.*, 1971, **10**, 1.

assigned to ν(M—M) and derived force constants are shown in Table 21. A 16% drop in calculated force constant for f_{SnSn} is observed between Me_6Sn_2 and Ph_6Sn_2, this reduction being much smaller than might be estimated from consideration of the stretching-vibration frequencies alone; by contrast, f_{PbPb} is effectively constant for the two di-lead compounds R_6Pb_2 (R = Me or Ph).

Table 21 *Wavenumbers*/cm^{-1} *and force constants*/mdyn Å^{-1} *for* M—M *stretching vibrations in some metal–metal bonded compounds of the Group IV elements*

Compound	ν(M—M)	f_{MM}
$Me_3SiSiMe_3$	404	1.70
$Me_3GeGeMe_3$	273	1.54
$Me_3SnSnMe_3$	192	1.39
$Me_3PbPbMe_3$	116	0.98
$Ph_3SnSnPh_3$	138	1.17
$Ph_3PbPbPh_3$	113	1.01
$(Ph_3Sn)_4Sn$	159 (asym), 103 (sym)	—

Compounds containing Si—N Bonds.—Electrical discharge reactions between silane–ammonia mixtures result in the deposition of solid films containing Si—N bonds, ν(Si—N) being observed in the 860—930 cm^{-1} region by i.r. spectroscopy.[187]

The i.r. spectrum of Me_3SiNH_2 has been reported[188] and compared with those for $(Me_3Si)_2NH$ and $(Me_3Si)_3N$: the single band assigned to ν(Si—N) occurs at 882 cm^{-1}. Partially assigned i.r. data for $SiCl_3NHMe$, $SiCl_3NHEt$, $SiCl_3NMe_2$, $SiCl_3NEt_2$, $(SiH_2Cl)_2NEt$, $(SiH_3)_2NEt$, and for a mixture of $SiHCl_2NHEt$ and $(SiHCl_2)_2NEt$ have been listed by Drake and Westwood;[138] all of these compounds show bands assignable to ν(Si—N) at *ca.* 940 cm^{-1}. In $(Cl_3Si)_2NH$, ν_{asym}(SiNSi) and ν_{sym}(SiNSi) are at 975 and 766 cm^{-1}, respectively, and corresponding vibrations in some related metallo-derivatives are assigned to bands at the frequencies shown in Table 22.[189] For the compound *p*-$(Me_3Si)_2N\cdot C_6H_4\cdot N(SiMe_3)_2$, $\nu_{sym}(Si_2N)$ modes are at 558 cm^{-1} (i.r.) and 602 cm^{-1} (Raman) whereas only one band attributable to $\nu_{asym}(Si_2N)$ is observed[190] (i.r. 973 cm^{-1}). Spectroscopic

Table 22 *Wavenumbers*/cm^{-1} *of some vibrational bands of metallo-derivatives of* $(Cl_3Si)_2NH$

Compound	ν_{asym}(SiNSi)	ν_{sym}(SiNSi)
$Li^+[N(SiCl_3)_2]^-$	1135	771
$EtZnN(SiCl_3)_2$	1096	820
$Zn[N(SiCl_3)_2]_2$	1084	820
$MeCdN(SiCl_3)_2$	1100	795

187 J. E. Drake and N. P. C. Westwood, *J. Chem. Soc.* (*A*), 1971, 2587.
188 N. Wiberg and W. Uhlenbrock, *Chem. Ber.*, 1971, **104**, 2643.
189 U. Wannagat, H. Moretto, P. Schmidt, and M. Schulze, *Z. anorg. Chem.*, 1971, **381**, 288.
190 K. Witke, P. Reich, and H. Kriegsmann, *Z. anorg. Chem.*, 1971, **380**, 165.

data for p-X·C_6H_4·$N(SiMe_3)_2$ (X = H, Me, Et, MeO, MeCO, F, Cl, Br, I, CN, Ph, or Me_3SiO) have been collected,[190] and in a further publication the frequency of $\nu_{sym}(Si_2N)$ in these compounds is correlated to an empirical equation involving the mass and electronic properties of X.[191]

Seppelt and Eysel report a complete assignment of the vibrational spectrum of tetrakis(trimethylsilyl)hydrazine, on the basis of a D_{2d} configuration for the $Si_2N{-}NSi_2$ skeleton.[192] GVFF Calculations show that f_{NN} is comparable to that for $N_2H_5^+$, and this is taken to indicate interaction between electron pairs on nitrogen and the silicon atoms; ν(N—N) is assigned to a Raman band at 1039 cm^{-1}, with ν_{asym}(SiN) at 925 cm^{-1} and ν_{sym}(SiN) at 579 and 411 cm^{-1}. For $(Me_3Si)_2N{-}N{=}CCl_2$, examined by Wiberg and Uhlenbrock,[193] ν(N—N) is found at 1010 cm^{-1}, with $\nu_{asym}(Si_2N)$ at 943 cm^{-1}.

An i.r. and Raman study of *N*-silylpyrrole and its $-SiCl_3$ and $-SiMe_3$ analogues, together with a normal co-ordinate analysis, indicates that extensive mixing occurs between ν(Si—N) and other vibrations.[143] For seven compounds of type (18), ^{15}N isotopic substitution has been used to

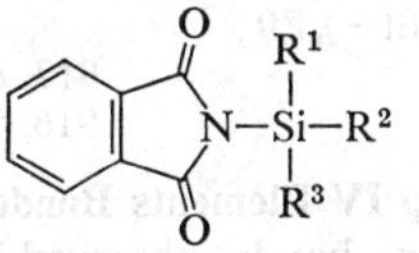

(18) R^1, R^2, R^3 = one of H, Me, Et, $-CH_2CH{=}CH_2$, or Ph

confirm assignment [144, 194] of (Si—N) stretching modes in the range 1065—1083 cm^{-1}.

I.r. spectra of $Me_3Si{\cdot}NMe{\cdot}Cl_2P{=}NSO_2Cl$ and of $[(Me_3Si)NH]_3P{=}O$ and $[(Me_3Si)NH]_2P(O){-}O{-}P(O)[NH(SiMe_3)]_2$ have been listed.[195, 196] All the remaining reports of vibrational data for compounds with Si—N bonds concern long-chain or cyclic species, investigated by Wannagat and co-workers.[197–203b] Of these compounds, the majority for which silicon–nitrogen stretching frequencies have been given are listed in Table 23.

[191] K. Witke, P. Reich, and H. Kriegsmann, *Z. anorg. Chem.*, 1971, **381**, 280.
[192] K. Seppelt and H. H. Eysel, *Z. anorg. Chem.*, 1971, **384**, 147.
[193] N. Wiberg and W. Uhlenbrock, *Chem. Ber.*, 1971, **104**, 3989.
[194] A. F. Jonzen and E. A. Kramer, *Canad. J. Chem.*, 1971, **49**, 3457.
[195] U. Bieller and M. Becke-Goehring, *Z. anorg. Chem.*, 1971, **381**, 209.
[196] H. H. Falius, K.-P. Griesen, and U. Wannagat, *Inorg. Nuclear Chem. Letters*, 1971, **7**, 281.
[197] L. Gerschler and U. Wannagat, *J. Organometallic Chem.*, 1971, **29**, 217.
[198] U. Wannagat, H. Moretto, and P. Schmidt, *Z. anorg. Chem.*, 1971, **385**, 164.
[199] U. Wannagat, R. Braun, and L. Gerschler, *Z. anorg. Chem.*, 1971, **381**, 168.
[200] U. Wannagat, R. Braun, L. Gerschler, and H. J. Wismar, *J. Organometallic Chem.*, 1971, **26**, 321.
[201] U. Wannagat and L. Gerschler, *Z. anorg. Chem.*, 1971, **383**, 249.
[202] U. Wannagat, E. Bognsch, and F. Rabet, *Z. anorg. Chem.*, 1971, **385**, 261.
[203] U. Wannagat, F. Rabet, and H. J. Wismar, *Monatsh.*, 1971, **102**, 1429.
[203a] U. Wannagat, J. Herzig, P. Schmidt, and M. Schulze, *Monatsh.*, 1971, **102**, 1817.
[203b] U. Wannagat, E. Bognsch, and P. Geymayer, *Monatsh.*, 1971, **102**, 1825.

Table 23 *Vibrational data (wavenumbers/*cm^{-1}*) for some* Si—N *compounds*

Compound	ν_{asym}(SiNSi)	ν_{sym}(SiNSi)	*Ref.*
$(Me_3Si\cdot NR\cdot SiMe_2)_2NMe$, R = Me	—	460	197
R = Et	915	451	
R = Pr	948	448	
R = Bu	963	460	
$(Me_3Si\cdot NR\cdot SiMe_2\cdot NMe)_2SiMe_2$, R = Me	—	466	
R = Et	925	460	
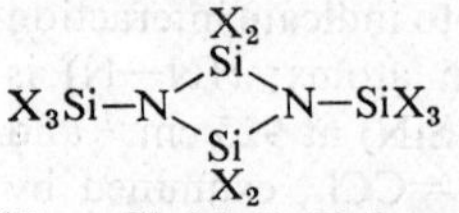	Not assigned		198
(X = Cl, MeO, EtO, or MeNH)			
cyclo-$(-Me_2Si\cdot NH\cdot SiMe_2\cdot NMe-)_2$	947	—	199
cyclo-$(-Me_2SiNMe-)_4$	913	608, 540	200
cyclo-$Me_2Si(-NMe\cdot SiMe_2\cdot NPr-)_2XR^1R^2$			201
X = Si; R^1, R^2 = Me	945, 910	614	
X = Si; R^1 = Me, R^2 = vinyl	945, 906	600	
X = Ge; R^1, R^2 = Cl	938, 890	600	
cyclo-$Me_2Si(-NMe\cdot SiMe_2\cdot NPr-)_2BPh$	916, 900	612	201
cyclo-$Me_2Si(-NMe\cdot SiMe_2\cdot NEt-)_2PR$,			201
R = Et	917, 896	612	
R = Ph	916, 900	610	

Compounds containing Group IV Elements Bonded to P or As.—Anderson and Drake assign weak i.r. bands observed [140] at 400—450 cm^{-1} for $Me_3SiP(SiH_3)_2$ to ν(Si—P) modes; for the phosphine $(Me_3Si)PHPh$, ν(Si—P) is found [204] at 473 cm^{-1}. The i.r. spectra of some cyclic dimers and trimers of (Ph_2MPPh) (M = Si, Ge, or Sn) have been reported. Metal–phosphorus ring-vibrations occur in the 300—450 cm^{-1} region but have not been assigned in detail.[205] The vibrational spectra of the metalloid-substituted polyarsines $(MH_3As)_5$ (M = Si or Ge) (see ref. 134) contain bands at 340—350 cm^{-1} [ν(Si—As)] and 250—265 cm^{-1} [ν(Ge—As) with ν(As—As)].

Compounds containing M—O Bonds (M = Si, Ge, Sn, or Pb).—I.r. bands arising from the absorption of water on to silica gel have been attributed to vibrations of surface silanol groups: deuteriation studies indicate that bands at 870, 620, and 950 cm^{-1} are due to Si—OH bending, Si—OD bending, and Si—O stretching, respectively.[206] I.r. spectroscopy has also been used to study the pyrolysis of silanes chemisorbed on silica,[207] and to investigate, through variation in frequency for ν(O—H), the extent of hydrogen-bonding between trimethylsilanol and a number of different solvents.[208] Some data for solid GeO_2 have also been reported.[209]

[204] M. Baudler and A. Zarkadas, *Chem. Ber.*, 1971, **104**, 3519.
[205] H. Schumann and H. Benda, *Chem. Ber.*, 1971, **104**, 333.
[206] M. Hino and T. Sato, *Bull. Chem. Soc. Japan*, 1971, **44**, 33.
[207] M. J. D. Low, M. Shimizu, and J. C. McManus, *Chem. Comm.*, 1971, 579.
[208] T. Kagiya, Y. Sumida, T. Watanabe, and T. Tachi, *Bull. Chem. Soc. Japan*, 1971, **44**, 923.
[209] G. D. Bagratishvili, M. D. Gogeshvili, R. B. Dzhanelidze, V. A. Chagelishvili, and R. G. Kharati, *Doklady Chem.*, 1971, **196**, 43.

A fairly full investigation of the vibrational spectrum of trimethylsilyl vinyl ether includes i.r. data for both $(CH_3)_3SiOCH{=}CH_2$ and $(CD_3)_3SiOCH{=}CH_2$. A normal-co-ordinate calculation indicates that bands at 730 cm^{-1} (^{1}H-methyl compound) and 785 cm^{-1} (^{2}H-methyl compound) possess mainly ν(Si—O) character; the Si—O force constant is *ca.* 25% lower than that for Me_3SiOMe, this observation perhaps being related to the high reactivity of silyl vinyl ethers.[210] A very extensive compilation of i.r. (and some Raman) data has been made for $R_3GeOGeR_3$ (R = Me or Bu^n), $R^1{}_3GeOR^2$ (15 derivatives), $R^1{}_2Ge(OR^2)_2$ (12 derivatives), and four alkylgermoxacycloalkanes.[211] Characteristic stretching vibrations of the GeC_3, GeC_2, GeOC, and $Ge(OC)_2$ groups are assigned, and in particular for $(Me_3Ge)_2O$ and $(Bu^n{}_3Ge)_2O$, ν_{asym}(GeOGe) are assigned to bands at 794, 841 cm^{-1} respectively, with ν_{sym}(GeOGe) at 467, 400 cm^{-1}.

Structural characteristics of some alkyltin trisulphinates $R^1Sn(O_2SR^2)_3$ (R^1 = Me or Ph; R^2 = Ph or *p*-Me·C_6H_4) and tetrasulphinates $Sn(O_2SR)_4$ (R = Ph or *p*-MeC_6H_4) have been elucidated by i.r. spectroscopy.[212] Frequencies assigned to ν(Sn—O) in some 1 : 1 and 1 : 2 adducts between Ph_2SnCl_2 and sulphoxides[213] are listed in Table 24. The i.r.

Table 24 *I.r. spectral*[a] *wavenumbers*/cm^{-1} *for* ν(Sn—O) *vibrations in sulphoxide adducts of* Ph_2SnCl_2

Adduct	ν(Sn—O)	ν(Sn—Cl)
Ph_2SnCl_2,Me_2SO	421sh, 417, 414sh	328, 322sh
Ph_2SnCl_2,2Me_2SO	418, 408sh	—
Ph_2SnCl_2,Pr_2SO	443, 439sh, 432	328, 321, 317sh
Ph_2SnCl_2,2Pr_2SO	416, 413	—
Ph_2SnCl_2,Bu_2SO	419, 416, 407sh	329sh, 322, 319
Ph_2SnCl_2,2Bu_2SO	419sh, 415, 410	—
Ph_2SnCl_2,$(CH_2)_4SO$	380, 376, 372	336, 331sh, 324
Ph_2SnCl_2,2$(CH_2)_4SO$	378sh, 374, 367	—

[a] sh = shoulder. Samples run as Nujol mulls (CsI plates).

spectra of pyridine *N*-oxide complexes of methyl- and phenyl-tin halides and isothiocyanates contain bands in the range 300—340 cm^{-1}, assigned to Sn—O stretching vibrations.[214] 2 : 1 Adducts between Sn^{II}, Sn^{IV}, and also Ti^{IV} halides and di-isopropyl methylphosphonate have been

[210] A. N. Lazarev, I. S. Ignat'ev, L. L. Schukovskaya, and R. I. Pal'chik, *Spectrochim. Acta*, 1971, **27A**, 2291.
[211] A. Marchand, M.-T. Forel, M. Lebedeff, and J. Valade, *J. Organometallic Chem.*, 1971, **26**, 69.
[212] E. Lindner, U. Kunze, and J. Koola, *J. Organometallic Chem.*, 1971, **31**, 59.
[213] R. S. Randall, R. W. J. Wedd, and J. R. Sams, *J. Organometallic Chem.*, 1971, **30**, C19.
[214] J. H. Holloway, G. P. McQuillan, and D. S. Ross, *J. Chem. Soc. (A)*, 1971, 1935.

characterized by i.r. spectroscopy, a shift of *ca.* 100 cm^{-1} to lower wavenumber in ν(P—O) for the ligand indicating complexing *via* the oxygen atom.[215]

Formation of metallo-polymers by interaction between polysiloxanes and finely divided Fe or Pb has been studied by i.r. spectroscopy.[215a] The i.r. and Raman spectra of a sample of thortveitite, $Sc_2Si_2O_7$, are consistent with D_{3d} symmetry for the $Si_2O_7^{6-}$ ion.[216] Vibrations of the GeO_7 unit have been assigned for some orthogermanates, $MGeO_4$ (M = Zr, Hf, Th, U, or Ge), through a detailed study of their i.r. spectra.[217] A product formulated as $Pb_5O_3(OH)_4$, obtained from treatment of lead(II) acetate solutions with NaOH and ammonia, shows [218] an i.r. band at 490 cm^{-1} which has been attributed to ν(Pb—O).

Compounds containing M—S Bonds (M = Si, Ge, or Sn).—A number of the i.r. bands listed for $Me_3SiSP(S)F_2$ and $Me_2ClSnP(S)F_2$ have been tentatively assigned but these do not include Si—S or Sn—S stretching modes.[219] Sandman and West [220] assign ν(Ge—S) in Ph_3GeSMe to an i.r. band at 395 cm^{-1}. For the $Ge_4S_{10}^{4-}$ ion, which has an adamantane-like structure, the i.r. and Raman frequencies have been listed but not assigned.[221] The compounds 4-X·C_6H_4·S·$SnMe_3$ have ν(Sn—S) at the wavenumbers/cm^{-1} shown [167] in Table 25.

Table 25 *Vibrational spectra of compounds* 4-X·C_6H_4S·$SnMe_3$

X	ν(Sn—S)
NH_2	384
F	371
Cl	358
Me	354
Br	354
H	341
NO_2	326

Compounds containing M—Halogen Bonds (M = Si, Ge, Sn, or Pb).—Gas-phase Raman spectra for the silicon halides SiF_4, SiF_3Cl, SiF_2Cl_2, $SiFCl_3$, and $SiCl_4$ are consistent with the expected symmetry for each molecule.[222] Bromochlorofluoroiodosilane has been prepared by successive exchange reactions from $SiBr_4$. Observed [222a] Si—X stretching fundamentals at 910 cm^{-1} (X = F), 587 cm^{-1} (X = Cl), 486 cm^{-1} (X = Br), and 333 cm^{-1}

[215] C. Owens, N. M. Karayannis, L. L. Pytlewski, and M. M. Labes, *J. Phys. Chem.*, 1971, **75**, 637.
[215a] M. T. Bryk and E. M. Natanson, *Doklady Phys. Chem.*, 1971, **196**, 26.
[216] S. D. Ross, *Spectrochim. Acta*, 1971, **27A**, 1837.
[217] R. Hubin and P. Tarte, *Spectrochim. Acta*, 1971, **27A**, 683.
[218] O. Glemser and Teh-Pei Lin, *Z. anorg. Chem.*, 1971, **382**, 244.
[219] D. W. McKennon and M. Lustig, *Inorg. Chem.*, 1971, **10**, 406.
[220] D. J. Sandman and R. West, *J. Organometallic Chem.*, 1971, **30**, C61.
[221] B. Krebs and S. Pohl, *Z. Naturforsch.*, 1971, **26b**, 853.
[222] K. Hamada, G. A. Ozin, and E. A. Robinson, *Bull. Chem. Soc. Japan*, 1971, **44**, 2555.
[222a] F. Höfler and W. Veigl, *Angew. Chem. Internat. Edn.*, 1971, **10**, 919.

(X = I) give calculated force constants of, respectively, 5.7, 3.0, 2.42, and 1.76 mdyn Å^{-1}. The i.r. spectrum of the 1 : 1 adduct between Me_3N and $HSiCl_3$, thought to be $[HNMe_3]\ SiCl_3$, has been reported,[223] and compared with that of $(HNMe_3)Cl$.

Ill-defined vibrational spectra for a new compound of empirical formula Ge_2F_5 have been obtained; the absence of bands in the 200—300 cm^{-1} region is taken to incidate the absence of Ge—Ge bonding.[224]

The vibrational spectra of some interesting complexes of tin(II) fluoride have been studied by Birchall *et al.*,[225] who conclude that the 1 : 1 adducts SnF_2,B_F (B_F = SbF_5, AsF_5, or BF_3) may be regarded as salts of the fluorine-bridged cation $[Sn—F]_n{}^{n+}$ with $[B_FF]^-$. A characteristic broad band at *ca.* 425—445 cm^{-1}, present in the spectrum of each complex, is tentatively assigned to ν(Sn—F) in the fluorotin cation. Goldstein and Tok have reported [226] i.r. and Raman data for all ten of the binary and mixed trihalogenostannate anions $(SnX_3)^-$, $(SnX_2Y)^-$, and $(SnXYZ)^-$ (X, Y, or Z = Cl, Br, or I). Solution spectra show that all of these species occur as pyramidal anions, this structure being retained in the solid state for Bu_4N^+ salts; however, a lowering of symmetry owing to increased solid-state interactions is apparent for the Ph_4As^+ or Et_4N^+ derivatives. Raman spectra have also been reported [227] for solid $SnCl_2$, $SnCl_2,H_2O$, $KSnCl_3,H_2O$, and $KCl,KSnCl_3,H_2O$. Raman spectra of PbX_2–KX melts (X = Cl or Br) show only one broad band (200—240 cm^{-1} for X = Cl; 140 cm^{-1} for X = Br), attributed [228] to symmetric stretching of Pb—X bonds in the species $[PbX_n]^{(n-2)-}$.

All other reports of vibrational data for Group IV halides [229–236] concern species where the metal possesses a co-ordination number greater than four through adduct formation, mainly by MX_4 derivatives (M = Ge or Sn; X = Cl or Br). In several instances, i.r. and Raman spectroscopy has allowed distinction to be made between possible five- and six-co-ordination

223 M. A. Ring, R. L. Jenkins, R. Zanganeh, and H. C. Brown, *J. Amer. Chem. Soc.*, 1971, **93**, 265.
224 G. P. Adams, J. L. Margrave, and P. W. Wilson, *J. Inorg. Nuclear Chem.*, 1971, **33**, 1301.
225 T. Birchall, P. A. W. Dean, and R. J. Gillespie, *J. Chem. Soc.* (*A*), 1971, 1777.
226 M. Goldstein and G. C. Tok, *J. Chem. Soc.* (*A*), 1971, 2303.
227 J. E. D. Davies and I. R. Tench, *Inorg. Nuclear Chem. Letters*, 1971, **7**, 491.
228 V. A. Moroni, *J. Chem. Phys.*, 1971, **54**, 4126.
229 E. M. Belousova and I. I. Seifullina, *Russ. J. Inorg. Chem.*, 1971, **16**, 80.
230 Yu. K. Shaulov, I. Ya. Markova, Yu. A. Popov, G. K. Selivanov, and Yu. I. Kol'tsov, *Russ. J. Inorg. Chem.*, 1971, **16**, 75.
231 D. Cunningham, M. J. Frazer, and J. D. Donaldson, *J. Chem. Soc.* (*A*), 1971, 2049.
232 M. D. Hobday and T. D. Smith, *J. Chem. Soc.* (*A*), 1971, 1453.
233 L. L. Popova, A. D. Garnovskii, V. I. Minkin, I. D. Sadekov, K. M. Yanasov, and B. S. Lokshin, *Russ. J. Inorg. Chem.*, 1971, **16**, 692.
234 M. Goldstein and W. D. Unsworth, *Spectrochim. Acta*, 1971, **27A**, 1055.
235 G. E. Matsubayashi, M. Hiroshima, and T. Tanaka, *J. Inorg. Nuclear Chem.*, 1971, **33**, 3787.
236 R. Barbieri, R. Cefalù, S. C. Chandra, and R. H. Herber, *J. Organometallic Chem.*, 1971, **32**, 97.

at the metal atom. The presence of only two bands assignable to ν(Sn—X) in the Raman spectra of SnX_4,2B (X = Cl or Br; B = phosphine or arsine) has been taken to indicate *trans*-configurations, with ν_{sym}(Sn—Cl) at 264—283 cm^{-1}, ν_{asym}(Sn—Cl) at 238—252 cm^{-1}, and corresponding ν(Sn—Br) modes *ca.* 100 cm^{-1} lower in wavenumber.[231]

5 Group V Elements

Compounds containing E—H Bonds (E = N, P, or As).—I.r. bands assigned to vibrations of co-ordinated hydrazoic acid in the solid 1 : 1 adduct HN_3,$SbCl_5$ occur at the following wavenumbers:[237] 3280 cm^{-1} (N—H stretch), 2185 and 1308 cm^{-1} (N_3 stretching), 1170 cm^{-1} (NH deformation), and 710, 520 cm^{-1} (N_3 deformations). The i.r. spectrum of difluorophosphorane (H_3PF_2) contains bands at 2500 and 970 cm^{-1}, assigned[238] to ν(P—H) and $\delta(PH_3)$. The stannylphosphine Me_3SnPH_2 has been examined in the vapour phase by i.r. spectroscopy;[239] ν(P—H) is observed at 2285 cm^{-1} with $\delta(PH_2)$ at 1058 cm^{-1}. For several phosphinamines R_2PHNY investigated by i.r. spectroscopy, an equilibrium between two forms (19a) and (19b) is observed:[240]

$$\underset{(19a)}{R_2P{-}NH{-}Y} \rightleftharpoons \underset{(19b)}{R_2PH{=}NY}$$

When R = Me and Y = SO_2CF_3, $SO_2C_6H_4Me$, or $PS(OPh)_2$, or when R = Ph and Y = SO_2CF_3, the isomer (19b) is isolated, with ν(P—H) at 2390, 2350, 2300, and 2350 cm^{-1}, respectively; when R = Ph and Y = $SO_2C_6H_4Me$ or $PS(OPh)_2$, formation of (19a) is favoured, with ν(N—H) at 3230 or 3220 cm^{-1}. Other references which include N—H stretching frequencies are collected in Table 26.

Compounds containing E—C Bonds (E = N, P, As, or Sb).—(Spectroscopic data for a number of organo-Group V derivatives with P—N, P—O, Sb—O, N—S, or P—S bonds, which are dealt with in subsequent sections, may also include reference to characteristic frequencies for N—C, P—C, or Sb—C vibrations.)

I.r. frequencies for F_2NCCl_2SF and $F_2NCCl_2S(O)F$ have been listed.[241] Unassigned frequency data for the adduct Me_3P,PF_5 have been reported.[242] A detailed investigation of the vibrational spectra of $MePCl_2$, $MePOCl_2$, MePOFCl, $MePOF_2$, and $MePSCl_2$ in the range 30—350 cm^{-1} has led to the assignment of bands at the following wavenumbers to methyl torsional modes:[243] 228 cm^{-1} ($MePCl_2$), 264 cm^{-1} ($MePOCl_2$), 240 cm^{-1} ($MePOF_2$),

[237] A. Schmidt, *Z. anorg. Chem.*, 1971, **381**, 31.
[238] F. Seel and K. Velleman, *Z. anorg. Chem.*, 1971, **385**, 123.
[239] A. D. Norman, *J. Organometallic Chem.*, 1971, **28**, 81.
[240] A. Schmidpeter and H. Rossknecht, *Z. Naturforsch.*, 1971, **26b**, 81.
[241] L. M. Zaborowski and J. M. Shreeve, *Inorg. Chim. Acta*, 1971, **5**, 311.
[242] C. W. Schultz and R. W. Rudolph, *J. Amer. Chem. Soc.*, 1971, **93**, 1898.
[243] J. R. Durig and J. M. Casper, *J. Phys. Chem.*, 1971, **75**, 1956.

Table 26 *References to compounds with* N—H *bonds characterized by vibrational spectroscopy*

Compound	*Ref.*
$(RNHGaH_2)_n$, n = 1 or 2, R = Et or Pr^i	38
$Ph_2B\cdot NH\cdot NH\cdot BPh_2$ } $Ph_2B\cdot NH\cdot NMe\cdot BPh_2$	54
R^1R^2NH,BCl_3, various R^1, R^2	98
$SiCl_3NHMe$ } $SiCl_3NHEt$	138
Me_3SiNH_2	188
$(Cl_3Si)_2NH$	189
$[(Me_3Si)NH]_3PO$ } $[(Me_3Si)NH]_2P(O)\cdot O\cdot P(O)[NH(SiMe_3)]_2$	196
cyclo-$(-NHSiMe_2NMeSiMe_2-)_2$	199
MeHNNHMe	256
$MeNHPH_4$	260
$(CF_3)_2P(E)NHMe$ } $F_2P(E)NHMe$ E = O or S	286
S_7NH } $1,3\text{-}S_6N_2H_2$ } $1,4\text{-}S_6N_2H_2$ } $1,5\text{-}S_6H_2H_2$	307

and 221 cm^{-1} ($MePSCl_2$), allowing calculation of the three-fold barrier to internal rotation in each case.

A normal-co-ordinate analysis of $P(C{\equiv}CH)_3$ yields [243a] a value of 3.56 mdyn $Å^{-1}$ for f_{PC} compared to that of *ca.* 3.0 mdyn $Å^{-1}$ in PMe_3, suggesting *d*-orbital participation in the P—C bond. I.r. and Raman spectra for a number of t-butylphosphine derivatives have been partially assigned by Holmes and Fild.[244–246] For Bu^tPX_2, Bu^tPOX_2, and Bu^tPSX, P—C stretching vibrations occur at 615, 643, and 589 cm^{-1} when X = F, and at 587, 645, and 607 cm^{-1} for X = Cl: results for Bu^tPF_4 are consistent [246] with a trigonal-bipyramidal model, the t-butyl group lying in an equatorial position, and ν(P—C) being assigned to a band at 583 cm^{-1}. The vibrational spectra of Ph_4EX (E = P, As, or Sb; X = F, Cl, Br, OH, or ClO_4) indicate that the appearance of the region in which 'X-sensitive' phenyl vibrations occur can provide a criterion for distinguishing between four- and five-co-ordinate species.[247]

Durig and co-workers have given full assignments for vibrations of CH_3AsCl_2, CH_3AsBr_2, CH_3AsI_2, and CD_3AsI_2, assuming C_s symmetry;[248] observed wavenumbers for several fundamentals, including ν(As—C), are listed in Table 27. A fairly full vibrational assignment has also been given [248a]

[243a] W. M. A. Smit and G. Dijkstra, *J. Mol. Structure*, 1971, **7**, 223.
[244] R. R. Holmes and M. Fild, *Spectrochim. Acta*, 1971, **27A**, 1525.
[245] R. R. Holmes and M. Fild, *Spectrochim. Acta*, 1971, **27A**, 1537.
[246] R. R. Holmes and M. Fild, *Inorg. Chem.*, 1971, **10**, 1109.
[247] J. R. Orenberg, M. D. Morris, and T. V. Long, jun., *Inorg. Chem.*, 1971, **10**, 933.
[248] J. R. Durig, C. F. Jumper, and J. N. Willis, jun., *Appl. Spectroscopy*, 1971, **25**, 218.
[248a] C. Woods and G. G. Long, *J. Mol. Spectroscopy*, 1971, **40**, 435.

for Me_3AsX_2 (X = Cl or Br). A re-assignment of the skeletal vibrations in $E(C{\equiv}CH)_3$ by Smit and Dijkstra [249] may be summarized as in Table 28. Vibrational frequencies and assignments for $(p\text{-}X{\cdot}C_6H_4)_3As$, $(m\text{-}X{\cdot}C_6H_4)_3$-As (X = F or Cl), and the corresponding oxides have been listed.[250]

Table 27 *Wavenumbers*/cm^{-1} *for some fundamental vibrations of methyldihalogenoarsines*

Assignment	CH_3AsCl_2	CH_3AsBr_2	CH_3AsI_2	CD_3AsI_2
ν(As—C)	581	575	563	520
ν_{asym}(As—X)	364	269	225	224
ν_{sym}(As—X)	388	269	204	199
$\delta(AsX_2)$ 'wag'	224	189	176	172
$\delta(AsX_2)$ 'scissor'	160	100	87	86

Table 28 *Skeletal vibrations of* $E(C{\equiv}CH)_3$, *expressed as wavenumber*/cm^{-1}

	Mode	E = As	E = Sb
A_1	ν(E—C)	517	474
	$\delta(EC_3)$	100	88
E	ν(E—C)	526	474
	$\delta(EC_3)$	89	81

A GVFF calculation supports the analysis of low-frequency spectra for $(CH_3)_3SbX_2$ and $(CD_3)_3SbX_2$ (X = Cl or Br) on the basis of a D_{3h} model. Fundamentals involving motion of the SbC_3 unit occur at the wavenumbers/cm^{-1} shown in Table 29.[251] Data for some other R_3SbX_2 compounds have also been given, where R = Me or Ph; X = F, Cl, Br, NCS, ONO_2, $MeCO_2$, or CD_3CO_2.[251a] A tentative assignment of the i.r. spectrum of

Table 29 *Wavenumbers*/cm^{-1} *for* $SbCl_3$ *fundamentals of* $(CH_3)_3SbX_2$ *and* $(CD_3)_3SbX_2$ (*i.r. data*)

Assignment	$(CH_3)_3SbCl_2$	$(CD_3)_3SbCl_2$	$(CH_3)_3SbBr_2$	$(CD_3)_3SbBr_2$
$\nu_{sym}(SbC_3)$	522	479	514	473
$\delta(SbC_3)$, out-of-plane	193	174	177	165
$\delta(SbC_3)$, in-plane	164	155	n.o.	135

$(Cl_2Sb)_2CH_2$ has been given.[252] The i.r. and Raman spectra of penta-(cyclopropyl)antimony(v), $(cyclo\text{-}C_3H_5)_5Sb$, have been reported and assigned by comparison with those of cyclo-C_3H_5Br. Bands observed in the 600—200 cm^{-1} region suggest a square-pyramidal (C_{4v}) rather than a trigonal (D_{3h}) molecular configuration, in common with the corresponding

[249] W. M. A. Smit and G. Dijkstra, *J. Mol. Structure*, 1971, **8**, 263.
[250] F. T. Delbeke, R. F. De Ketelaere, and G. P. Van der Kelen, *J. Organometallic Chem.*, 1971, **28**, 225.
[251] C. Woods and G. G. Long, *J. Mol. Spectroscopy*, 1971, **38**, 387.
[251a] R. G. Goel, E. Maslowsky, jun., and C. V. Senoff, *Inorg. Chem.*, 1971, **10**, 2572.
[252] Y. Matsumura and R. Okawara, *Inorg. Nuclear Chem. Letters*, 1971, **7**, 113.

pentaphenyl derivative.[253] Bands at 587, 558, and 548 cm^{-1} in the i.r. spectrum of $Me_3Sb(OCH_2COO)$ are attributed to ν(Sb—C) vibrations,[254] as are [255] i.r. absorptions at 500—530 cm^{-1} in $Me_2Sb(R)\cdot NC(X)\cdot NMe_2$ (R = Me, Et, Bu^t, or Ph; X = O or S). Schmidbaur and co-workers have shown [255a] by i.r. and Raman spectroscopy that whereas Me_4SbX (X = N_3, SCN, or CN) are ionic structures [$Me_4Sb^+ X^-$], $Me_4Sb(O_2CMe)$ is essentially non-ionic.[255b]

Compounds containing N—N or N—P Bonds.—A very detailed spectroscopic study of symmetrical dimethylhydrazine has been made.[256] The results indicate that in the liquid phase this compound exists as an approximately 69 : 31% mixture of *gauche* conformers, each of whose spectra has been distinguished and assigned. Assignment of torsional fundamentals results in a calculated height of 3.3 kcal mol^{-1} for the barrier to three-fold methyl rotation, and of 5.0 kcal mol^{-1} for that to internal rotation about the N—N bond. For the two hydrazine derivatives $(Me_3Si)_2N—N(SiMe_3)_2$ and $(Me_3Si)_2N—N{=}CCl_2$, N—N stretching vibrations are reported to occur at 1039 and 1010 cm^{-1}, respectively.[192, 193] Solid ammonium azide has been studied in the region of internal N_3^- vibrations and lattice modes.[257]

In the series of compounds $Cl_3P{=}NX$ [X = CCl_3, C_2Cl_5, or $CCl(CCl_3)_3$], the P—N and N—C stretching vibrations are strongly coupled, giving rise to two bands described [258] as ν(P=N—C), (in-phase) at 950 cm^{-1} and (out-of-phase) at 1400 cm^{-1}. I.r. and Raman data and SVFF calculations for $XP(NMe_2)_2Cl$ and $XP(NMe_2)Cl_2$ (X = O or S) have been published.[259] The i.r. spectrum of $MeNHPF_4$ has been measured;[260] vibrations of the NPF_4 skeleton have been assigned, including ν(P—N) at 962, 954 cm^{-1} [with $\nu_{asym}(PF_2)$], and $\delta_{o.p.}(NPF_2)$ at 533 cm^{-1}. Gas-phase i.r. spectra for two nitrogen-bridged phosphorus fluorides $PF_2\cdot NMe\cdot P(O)F_2$ and $PF_2\cdot NMe\cdot PF_4$ contain strong bands at 927 and 954 cm^{-1}, respectively, assigned to ν_{asym}(PNP), much weaker absorptions at 664 and 663 cm^{-1} being attributed to the symmetric stretching vibrations.[261] Reaction between $S{=}PF_2\cdot N{=}PF_2Cl$ and methanol yields $Me_2S\cdot PF_2{:}NPF_2{:}O$; an i.r. band at 1380 cm^{-1} is assigned [262] to ν(P=N), with ν(P=O) at 1305 cm^{-1}. The compounds $Me_3P{=}N—Si(Me)_{3-n}F_n$ show i.r. bands attributable to

[253] A. H. Cowley, J. L. Mills, T. M. Loehr, and T. V. Long, *J. Amer. Chem. Soc.*, 1971, **93**, 2150.
[254] Y. Matsumura, M. Shindo, and R. Okawara, *J. Organometallic Chem.*, 1971, **27**, 357.
[255] J. Koketsu, M. Okamura, and Y. Ishii, *Bull. Chem. Soc. Japan*, 1971, **44**, 1155.
[255a] H. Schmidbaur, K.-H. Mitschke, J. Weidlein, and St. Cradock, *Z. anorg. Chem.*, 1971, **386**, 139.
[255a] H. Schmidbaur, K.-H. Mitschke, and J. Weidlein, *Z. anorg. Chem.*, 1971, **386**, 147.
[256] J. R. Durig and W. C. Harris, *J. Chem. Phys.*, 1971, **55**, 1735.
[257] Z. Iqbal and M. L. Malhotra, *Spectrochim. Acta*, 1971, **27A**, 441.
[258] D. P. Khomenko, G. G. Dyadyusha, and E. S. Kozlov, *J. Struct. Chem.*, 1970, **11**, 613.
[259] D. Köttgen, H. Stoll, A. Lentz, R. Pantzer, and J. Goubeau, *Z. anorg. Chem.*, 1971, **385**, 56.
[260] J. S. Harman and D. W. A. Sharp, *Inorg. Chem.*, 1971, **10**, 1538.
[261] J. S. Harman, M. E. McCartney, and D. W. A. Sharp, *J. Chem. Soc.* (*A*), 1971, 1547.
[262] H. W. Roesky and L. F. Grimm, *Chem. Comm.*, 1971, 998.

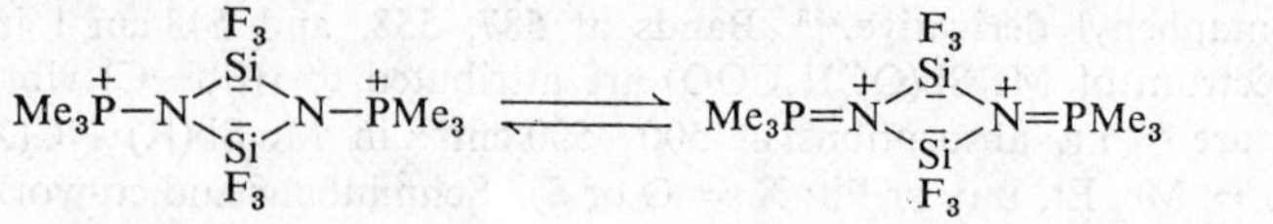

(20a) (20b)

ν(P=N) at 1282, 1286, and 1041 cm^{-1} for n = 1, 2, and 3, respectively.[263] The greatly decreased P=N frequency for the trifluoro-derivative is taken in conjunction with other evidence to indicate a dimeric structure (20) involving penta-co-ordinate silicon. Some other acyclic compounds with P—N bonds for which vibrational data have been given are listed in Table 30.[264, 265]

Table 30 *Compounds with* P—N *bonds characterized by i.r. spectroscopy*

Compound		*Ref.*
$FSO_2N{=}PCl_2OR$	R = Me, Et, Pr, or Bu	264
$ClSO_2N{=}PCl_2OR$		
$FSO_2N(R){-}P(O)Cl_2$		
$ClSO_2N(R){-}P(O)Cl_2$		
$SPF_2N{=}PCl_2NHSiMe_3$		265
$SPFClN{=}PCl_2NHSiMe_3$		
$SPCl_2N{=}PCl_2NHSiMe_3$		
$SPCl_2N{=}PFClNHSiMe_3$		
$SPCl_2N{=}PF_2N{=}PCl_3$		
$SPF_2N{=}PCl_2N{=}PCl_3$		
$SPFClN{=}PCl_2N{=}PCl_3$		
$SPCl_2N{=}PCl_2N{=}PCl_3$		
$SPCl_2N{=}PF_2N{=}PF_3$		
$SPCl_2N{=}PFClN{=}PCl_3$		
$[Ph_3PNClMe]Cl$		265*a*
$[Ph_3PNClEt]Cl$		
$Me_3SO_2N(Me)P(O)Cl_2$		265*b*
$ClSO_2N(Et)P(O)F_2$		

Cyclic phosphonitrilic derivatives have received considerable attention during 1971, a large number of such compounds having been examined by i.r. or Raman spectroscopy. Two isomers of the dimeric species $[PhP(S)NEt]_2$ have been isolated, and structural distinction between *cis*- and *trans*-forms has been achieved by using vibrational spectroscopy.[266] One isomer which shows ten i.r.–Raman coincidences is assumed to possess a *cis*-structure: no coincidences between i.r. and Raman bands are observed for the other, suggesting much higher symmetry, consistent with

[263] W. Wolfsberger, H. H. Pickel, and H. Schmidbaur, *Chem. Ber.*, 1971, **104**, 1830.
[264] H. W. Roesky and W. G. Böwing, *Chem. Ber.*, 1971, **104**, 3204.
[265] H. W. Roesky, L. F. Grimm, and E. Niecke, *Z. anorg. Chem.*, 1971, **385**, 102.
[265a] R. M. Kren and H. H. Sisler, *Inorg. Chem.*, 1971, **10**, 2630.
[265b] H. W. Roesky and W. Grosse Böwing, *Z. anorg. Chem.*, 1971, **386**, 191.
[266] C. D. Flint, E. H. M. Ibrahim, R. A. Shaw, B. C. Smith, and C. P. Thakur, *J. Chem. Soc. (A)*, 1971, 3513.

the *trans*-configuration (21). Sowerby and co-workers have tabulated i.r. frequency data for the following fluorophosphonitriles:[267] *cis*- and *trans*-$P_3N_3F_3(NMe_2)_3$, *cis*- and *trans*-$P_3N_3F_2(NMe_2)_4$, *gem*-$P_3N_3FCl_2(NMe_2)_3$, *gem*-$P_3N_3F_3(NMe_2)_3$, *cis*- and *trans*-$P_3N_3F_2Cl_2(NMe_2)_2$ [5,5-difluoride and 1,3-difluoride], and *cis*- and *trans*-$P_3N_3F_4(NMe_2)_2$. The spectra for these

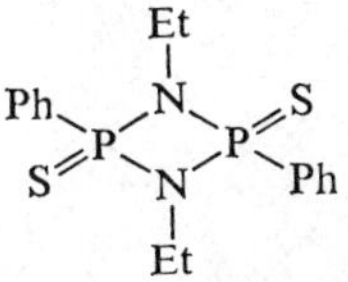

(21) *trans*-form

compounds are assigned in some considerable detail, particular attention being paid to ring-stretching vibrations. For the tris- and tetrakis-amino-species, bands at *ca.* 750 cm^{-1} are attributed to ν_{asym}(P—N), ν_{sym}(P—N) appearing at *ca.* 680 cm^{-1}. Chivers has reported[268] the preparation of some unusual phosphonitrilic ethynyl derivatives from $P_3N_3F_6$. With PhC≡CLi, the compound $P_3N_3F_5$(C≡CPh) is obtained, having i.r. absorptions assigned to ν(C≡C) at 2210 cm^{-1} and ν_{asym}(PNP) at 1275 cm^{-1}. Further reaction yields *gem*-$P_3N_3F_4(C{\equiv}CPh)_2$, with ν(C≡C) at 2200 cm^{-1} and ν_{asym}(PNP) at 1260, 1245 cm^{-1}. The monoethynyl derivative and $Co_2(CO)_8$ react to give (22), which shows ν(CO) at 2035, 2022 cm^{-1} and

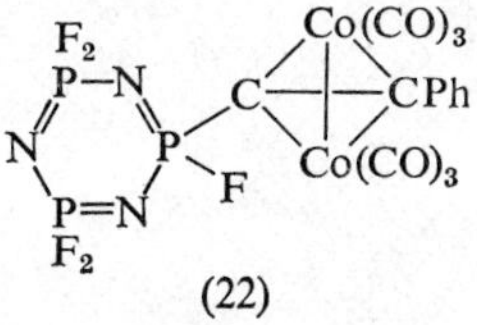

(22)

ν_{asym}(PNP) at 1269 cm^{-1}. In the spectra of the monoaminofluoro-phosphonitriles $P_3N_3F_5NR_2$, $\nu(P{-}N_{amino})$ exhibits considerable frequency variation[269] depending on R, being assigned at 1025, 818, 758, 820, or 698 cm^{-1} for R_2 = H_2, HMe, Me_2, HEt, or Et_2.

Table 31[270–276] lists the remaining phosphorazene compounds examined by i.r. spectroscopy during 1971; in some cases characteristic ν(PN)

[267] B. Green, D. B. Sowerby, and P. Clare, *J. Chem. Soc.* (*A*), 1971, 3487.
[268] T. Chivers, *Inorg. Nuclear Chem. Letters*, 1971, **7**, 827.
[269] E. Niecke, H. Thamm, and G. Flaskemd, *Chem. Ber.*, 1971, **104**, 3729.
[270] E. Niecke, H. Thamm, and O. Glemser, *Z. Naturforsch.*, 1971, **26b**, 366.
[271] H. W. Roesky, W. G. Böwing, and E. Niecke, *Chem. Ber.*, 1971, **104**, 653.
[272] E. Kobayashi, *Bull. Chem. Soc. Japan*, 1971, **44**, 2280.
[272a] E. S. Kozlov, D. P. Khomenko, and G. G. Dyadyusha, *Spectroscopy Letters*, 1971, **4**, 343.
[273] M. Biddlestone and R. A. Shaw, *J. Chem. Soc.* (*A*), 1971, 2715.
[274] H. R. Allcock and E. J. Walsh, *Inorg. Chem.*, 1971, **10**, 1643.
[275] U. Bieller and M. Becke-Goehring, *Z. anorg. Chem.*, 1971, **380**, 314.
[276] T. Chivers and R. Hedgeland, *Inorg. Nuclear Chem. Letters*, 1971, **7**, 767.

frequencies have been given, with occasional more complete assignments of the reported spectra.

Table 31 *References to publications giving vibrational data for cyclic* P—N *compounds*

Compound	*Ref.*
$P_3N_3F_5Me$	270
$P_3N_3F_5CH{=}CH_2$	270
$P_3N_3F_5SMe$	270
$P_3N_3F_5SPh$	270
$P_3N_3F_5OMe$	270
$P_3N_3F_5OEt$	270
$P_3N_3F_5OPh$	270
$P_3N_3F_5NPF_2OMe$	265*b*
$P_3N_3F_5NPF_2OEt$	265*b*
$P_3N_3F_5NPCl_2OEt$	265*b*
$P_3N_3F_5(NPCl_2)_2OEt$	265*b*
$P_3N_3F_5PCl_2NEtPOCl_2$	265*b*
$P_3N_3F_5NPCl_2X$, $P_3N_3F_5NPF_2X$ (X = $NHSiMe_3$, $NPCl_3$, $NPCl_2NHSiMe_3$, or $NPCl_2NPCl_3$)	271
$P_3N_3F_5NPCl_2NMeSiMe_3$	271
$P_3N_3F_5NPCl(NHMe)_2$	271
$P_3N_3(NH_2)_6,HCl$	272
$P_3N_3(NH_2)_6,H_2O$	272
$P_3N_3Cl_6$	272*a*
$P_3N_3Cl_5(NPPh_3)$	273
$P_3N_3PhCl_4(NPPh_3)$	273
gem-$P_3N_3Ph_2Cl_4$	273
$P_6N_6Ph_2Cl_8$	273
$P_6N_6Ph_7Cl_5$	273
$P_5N_5F_9NH_2$	271
$P_5N_5F_9NPCl_3$	271
$P_6N_6F_{11}NH_2$	271
$P_6N_6F_{11}NPCl_3$	271
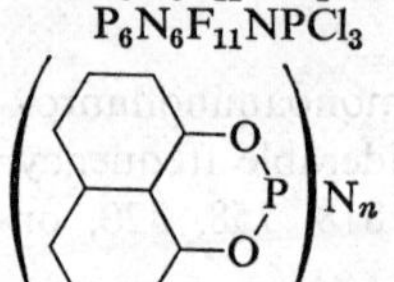 (n = 3 or 4)	274
$MeN(-PR_2\cdot N-)_2SO_2$ (R = NH_2 or NHMe)	275
$MeN(-PRCl\cdot N-)_2SO_2$ (R = NMe_2, NHPh, or Ph)	275
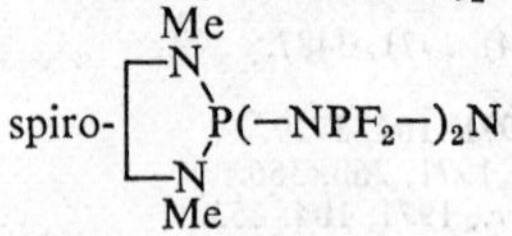	276
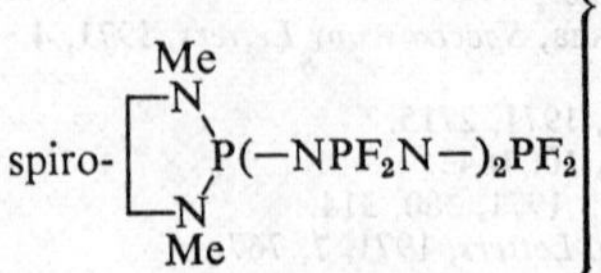	276

Compounds containing As—N, Bi—N, P—P, or As—As Bonds.—A detailed study of $(Me_2N)AsCl_2$ by i.r. and Raman spectroscopy has been reported.[276a] In the i.r. spectra of the compounds $PhAsCl_3N_3$ and cyclo-$As_3N_3Ph_3Cl_3$, in the bands 930—970 cm^{-1} region are assigned as predominantly arising from (As—N) stretching vibrations.[277] The vibrational spectra for the azides Me_2AsN_3 and Me_2BiN_3 can be interpreted in terms of a bent MN_3 skeleton (C_s symmetry); wavenumbers for ν(As—N) and ν(Bi—N) are given as 438 and 352 cm^{-1}, respectively.[278]

I.r. and Raman spectra have been recorded and partially assigned[278a] for H_2PPF_2, $(CF_3)_2PPF_2$, and their BH_3 adducts; Raman bands at 446 and 486 cm^{-1} in the first two compounds are assigned to ν(P—P) modes. A spectroscopic study of tetramethyldiarsine, $Me_2AsAsMe_2$, shows that in the solid state only the *trans*-conformer, for which a full assignment is offered, is present; the liquid phase appears to be a mixture of *trans*- and *gauche*-isomers, with the former predominating.[279] The arsenic–arsenic stretching frequencies occur at 272 cm^{-1} (*trans*-conformer) and 254 cm^{-1} (*gauche*-conformer), both bands being observed in the Raman spectrum at room temperature but that due to the *gauche*-isomer disappearing on cooling (Figure 1). Further intensity measurements in the temperature range 21—95 °C indicate an energy difference of *ca.* 0.6 kcal mol^{-1} between the two conformers, assuming a direct relationship between intensity and concentration. In the polyarsines $(MH_3As)_5$ (M = Si or Ge), ν(As—As) modes are observed[134] at *ca.* 270 cm^{-1}.

Compounds containing E—O Bonds (E = N, P, As, Sb, or Bi).—Co-deposition of NO, N_2O, or NO_2 with alkali metals in an Ar matrix yields a number of products resulting from strong charge-transfer interactions.[280] In particular, the formation of $M_x^+NO^-$ is inferred from bands varying in wavenumber between 1350 and 1375 cm^{-1}, assigned to $\nu(NO^-)$; ion-pairs $M_x^+NO_2^-$ are thought to account for absorptions at 1200—1240 cm^{-1}; and the isolation of an anion $O_2N{=}N^-$ with $\nu_4(b_1)$ at 1205 cm^{-1} is suggested. Nitrosyl difluorophosphate, $NOPO_2F_2$, has been prepared and shows an i.r. absorption at 2273 cm^{-1}, assigned to the NO^+ stretching vibration.[281] In $NO_2[Ga(NO_3)_4]$, symmetric and antisymmetric $\nu(NO_2^+)$ modes are found at 1401 and 2357, 2377 cm^{-1}, respectively.[84] The i.r. spectrum of gaseous $Cl_2C(NO_2)_2$ contains bands at 1333, 1297 cm^{-1} and 1636, 1627 cm^{-1} owing[282] to $\nu_{sym}(NO_2)$ and $\nu_{asym}(NO_2)$. The related salt $K[CCl(NO_2)_2]$ shows similar bands in the solid state at 1210 cm^{-1} and 1448 cm^{-1}.

[276a] J. R. Durig and J. M. Casper, *J. Mol. Structure*, 1971, **10**, 427.
[277] V. Krieg and J. Weidlein, *Angew. Chem. Internat. Edn.*, 1971, **10**, 516.
[278] J. Müller, *Z. anorg. Chem.*, 1971, **381**, 103.
[278a] H. W. Schiller and R. W. Rudolph, *Inorg. Chem.*, 1971, **10**, 2500.
[279] J. R. Durig and J. M. Casper, *J. Chem. Phys.*, 1971, **55**, 198.
[280] D. E. Milligan and M. E. Jacox, *J. Chem. Phys.*, 1971, **55**, 3404.
[281] V. P. Babaeva and V. Ya. Rosolovskii, *Russ. J. Inorg. Chem.*, 1971, **16**, 471.
[282] A. O. Diallo, *Spectrochim. Acta*, 1971, **27A**, 239.

Normal-co-ordinate calculations for P_4O_6 and P_4O_{10} have been reported but not discussed.[283] Lithium salts of two new dihalogeno-oxophosphorus acids have been prepared:[284] these are $Li(PO_2ClF)$ and $Li(PO_2BrF)$, their i.r. absorptions, assigned to $\nu_{asym}(PO_2)$ and $\nu_{sym}(PO_2)$, being observed at 1254 and 1133 cm^{-1} (fluorochloro-species) and 1282 and 1160 cm^{-1} (fluorobromo-species). The i.r. spectrum of gaseous $(CF_3)_2P(O)F$ has been

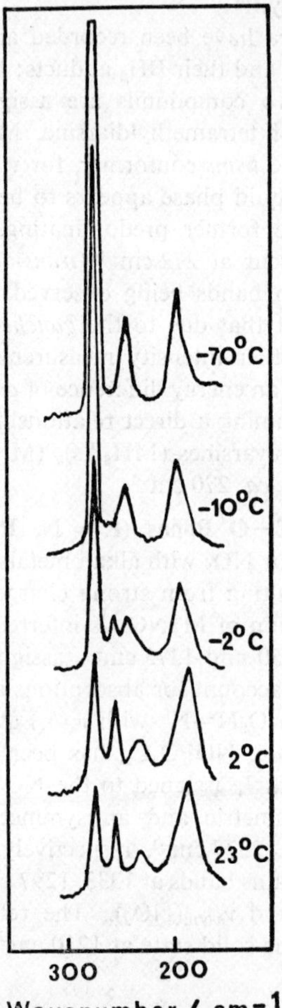

Figure 1 *Effect of cooling the sample on the low-frequency Raman spectrum of tetramethyldiarsine*
(Reproduced by permission from *J. Chem. Phys.*, 1971, **55**, 198)

[283] S. J. Cyvin and B. N. Cyvin, *Z. Naturforsch.*, 1971, **26a**, 901.
[284] H. Falius and K. P. Giesen, *Angew. Chem. Internat. Edn.*, 1971, **10**, 555.

reported by Dobbie.[285] Among wavenumbers listed, ν(P=O) is at 1366 cm^{-1} with δ(P=O) at 589 cm^{-1}, and ν(P—F) is tentatively assigned to a band at 907 cm^{-1}; unassigned i.r. data for the phosphinic anhydride $[(CF_3)_2P(O)]_2O$ are also given. Fairly detailed assignments for the vapour- and solid-phase i.r. spectra of $(CF_3)_2P(E)NHMe$ and $F_2P(E)NHMe$ (E = O or S) have been made. For E = O, bands due to ν(P=O) modes are at 1324 and 1327 cm^{-1}, respectively, corresponding ν(P=S) vibrations appearing at 696, 697 cm^{-1} for the two sulphides.[286]

Goubeau and co-workers have discussed the vibrational spectra of methoxyphosphine oxides and related derivatives in a series of three publications.[287–289] Observed wavenumber ranges are: ν(P=O), 1200—1280 cm^{-1}; ν(P=S), 600—690 cm^{-1}; ν_{sym} and $\nu_{asym}(PO_3)$, *ca.* 750 and 840 cm^{-1}; ν_{sym} and $\nu_{asym}(PS_3)$, *ca.* 450 and 560 cm^{-1}. Compounds for which data have been given include, for E = O or S, $(MeO)_3PE$, $(MeS)_3PE$, $(MeO)_2(MeS)PE$, $(MeO)(MeS)_2PE$, $(MeO)_2PECl$, and $(MeO)PECl_2$. References 290—292 also contain frequency data for several other compounds with P—O bonds.

In the trisacetato-compounds $(MeCO_2)_3E$ (E = P, As, Sb, or Bi), the $MeCO_2$ group is deduced to be unidentate by analysis of the i.r. spectra. Assignments to ν(E—O) are made for bands at 875 cm^{-1} (E = P), *ca.* 800 cm^{-1} (E = As), and 684 cm^{-1} (E = Sb) [not assigned for E = Bi].[293] Vibrational spectra for the complexes $SbCl_3,(Ph_3AsO)_2$ (23), $SbCl_3$,-$(Ph_3PO)_2$ (24), $BiCl_3,(Ph_3AsO)_2$ (25), and $BiCl_3,(Ph_3PO)_2$ (26) have been measured and assigned, and they indicate that $SbCl_3$ is the stronger acid, Ph_3AsO the stronger base. Element–oxygen stretching vibrations are observed [294] to be as shown in Table 32. The i.r. spectrum of cubic $ThAs_2O_7$,

Table 32 *Observed stretching vibrations for complexes* (23)—(26)

Mode	*Wavenumber*/cm^{-1}
ν(PO)	1136, 1129 (24); 1152 (26)
ν(AsO)	824, 808 (23); 840 (25)
ν(SbO)	394, 350 (24)
ν(BiO)	406 (26)

reported by Hubin, suggests a D_{3d} structure with a linear As—O—As bridge for the $As_2O_7^{4-}$ ion.[295] Bands at 798, 880, and 800 cm^{-1} in the i.r.

[285] R. C. Dobbie, *J. Chem. Soc. (A)*, 1971, 2894.
[286] R. G. Cavell, T. L. Charlton, and W. Sim, *J. Amer. Chem. Soc.*, 1971, **93**, 1130.
[287] O. A. Wafa, A. Lentz, and J. Goubeau, *Z. anorg. Chem.*, 1971, **380**, 128.
[288] V. Hornung, O. A. Wafa, A. Lentz, and J. Goubeau, *Z. anorg. Chem.*, 1971, **380**, 137.
[289] J. Goubeau and A. Lentz, *Spectrochim. Acta*, 1971, **27A**, 1703.
[290] S. R. Rafikov, N. D. Kazakova, G. A. D'yachkov, and O. V. Agashkin, *Doklady Chem.*, 1971, **196**, 111.
[291] S. Nakayama, M. Yoshifuji, R. Okazaki, and N. Inamoto, *Chem. Comm.*, 1971, 1186.
[292] A. K. Sheinkman, G. V. Samoilenko, and S. N. Baranov, *Doklady Chem.*, 1971, **196**, 169.
[293] G. Gattow and H. Schwank, *Z. anorg. Chem.*, 1971, **382**, 49.
[294] S. Milićev and D. Hadži, *Inorg. Nuclear Chem. Letters*, 1971, **7**, 745.
[295] R. Hubin, *Spectrochim. Acta*, 1971, **27A**, 311.

spectra of $(CF_3)_2As(CF_3NO)$, $As(CF_3NO)_3$, and $(CF_3NO)_2AsCl$ have been assigned to ν(As—O) vibrations.[296]

In the i.r. spectra of $(R_3SbN_3)_2O$ (R = Me or Ph), bands at 768 and 750 cm^{-1} are attributed to ν(Sb—O—Sb) vibrations, some other characteristic frequencies for these and related organoantimony pseudohalides also being given.[297] The adducts R_2SbCl_3,L (R = Me, Et, or Ph; L = DMSO, hmpa, tppo, or pyO) show characteristic ν(Sb—O) i.r. absorptions (360—440 cm^{-1} region) and/or a marked frequency decrease in ligand ν(X—O) (when X = S, P, or N) on complex formation,[298] and the pattern of bands due to ν(Sb—C) and ν(Sb—Cl) is consistent with the structure (27). Far-i.r.

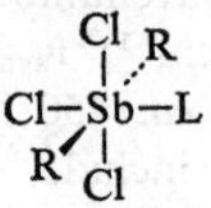

(27)

spectra for the antimonates M^ISbO_3 (M^I = Li, Na, K, or Ag) have been measured but not assigned.[299]

New i.r. data for anhydrous $[Ph_3SbClO_4]_2O$ and its bismuth analogue can be interpreted in terms of non-ionic, five-co-ordinate structures with M—O—M bridges (M = Sb or Bi), and this structure has been confirmed crystallographically.[300] The i.r. spectra of Ph_3BiX (X = CO_3, SO_4, SeO_4, CrO_4, or oxalate) indicate that these compounds have non-ionic polymeric structures with bridging anions, the bismuth atoms being five-co-ordinate. For X = CO_3, ν(Bi—O) is at 300 cm^{-1}; similar bands for the other species are found at *ca.* 250 cm^{-1} or below.[301] I.r. data recorded during the study of bismuth molybdate catalysts include a band at 1290 cm^{-1}, assigned[301a] to ν(Bi=O) of $(BiO)_2MoO_4$.

Compounds containing N—S or P—S Bonds.—For salts of the $N(SCl_2)^+$ cation with $AlCl_4^-$ and $SbCl_6^-$, unassigned Raman frequencies have been given.[302] A novel compound $N{\equiv}S(F_2){\cdot}N{=}SF_2$, containing S—N single, double, and triple bonds, has been prepared: i.r. and Raman measurements allow ν(S≡N) and ν(S=N) to be assigned, at 1470 and 1222 cm^{-1}, respectively.[303] The molecules $XNS(F_2)O$ have i.r. absorptions at 1180 (X = I), 1163 (X = Br), 1122 (X = F), and 1210 cm^{-1} (X = H), ascribed to ν(S=N) vibrations.[304] Among i.r. data for $R^1R^2S(O)NR^3$ compounds

[296] H. G. Ang and K. F. Ho, *J. Organometallic Chem.*, 1971, **27**, 349.
[297] R. G. Goel and D. R. Ridley, *Inorg. Nuclear Chem. Letters*, 1971, **7**, 21.
[298] N. Nishii, Y. Matsumura, and R. Okawara, *J. Organometallic Chem.*, 1971, **30**, 59.
[299] R. Franck and C. Rocchiccioli-Deltcheff, *Compt. rend.*, 1971, **273**, *B*, 128.
[300] G. Ferguson, R. G. Goel, F. C. March, D. R. Ridley, and H. S. Prasad, *Chem. Comm.*, 1971, 1547.
[301] R. G. Goel and H. S. Prasad, *Canad. J. Chem.*, 1971, **49**, 2529.
[301a] M. A. Dalin, N. A. Mangasaryan, B. R. Serebryakov, V. L. Mekhtieva, H. E. Portyanskii, and K. M. Mekhtiev, *Doklady Phys. Chem.*, 1971, **200**, 819.
[302] O. Glemser and J. Wegener, *Inorg. Nuclear Chem. Letters*, 1971, **7**, 623.
[303] O. Glemser and R. Höfer, *Angew. Chem. Internat. Edn.*, 1971, **10**, 815.

(R^1, R^2, R^3 = Me, Et, or Bu) are bands due[305] to ν(S=N) at *ca.* 1130 cm^{-1}, and also to ν(S=O), at *ca.* 1230 cm^{-1}. Three bands in the 1040—1190 cm^{-1} range in the i.r. spectrum of $(MeAs)_2S_2N_4$ have been tentatively assigned to (NSN) vibrations.[306] The vibrational spectra of the sulphur imides S_7NH, 1,3-$S_6N_2H_2$, 1,4-$S_6N_2H_2$, and 1,5-$S_6N_2H_2$ have been discussed, being surprisingly simple although the molecules are of low symmetry. In particular, although ν(S—N) modes can be distinguished (*ca.* 850—650 cm^{-1}), fewer than the predicted number of bands are observed.[307] Mews and Glemser assign bands at 1462 and 1348 cm^{-1} in the gas-phase i.r. spectrum of $ClF_2C\cdot CF_2\cdot NS(O)F_2$ to asymmetric and symmetric stretching of the NSO group of atoms.[308] Some further frequencies for ν(SN) vibrations[309, 310] are shown in Table 33, and references to the large number

Table 33 *Wavenumbers*/cm^{-1} *for* SN *stretching vibrations*

Compound	ν(SN)
$Me_3Si-N=S=N-SiMe_3$	1242, 1140
$ClMe_2Si-N=S=N-SiMe_2Cl$	1250, 1154
$Cl_2MeSi-N=S=N-SiMeCl_2$	1258, 1177
$Cl_2EtSi-N=S=N-SiEtCl_2$	1238, 1176
$(CF_3)Br_2CCBr(CF_3)(NSF_2)$	1403
$(CF_3)BrC:C(CF_3)NSCl_2$	1340
cis- and *trans-*$(CF_3)BrC:C(CF_3)NSF_2$	1410

of other compounds containing S—N bonds for which mainly unassigned vibrational data have been given are collected in Table 34.[311–320]

Pinkerton and Cavell have reported gas-phase i.r. spectra[321] for the sulphur-bridged systems $(CF_3)_2P(S)\cdot S\cdot P(CF_3)_2$ and $(CF_3)_2P(S)\cdot S\cdot S\cdot P(S)$-$(CF_3)_2$. In the first of these, absorptions at 783 cm^{-1} and 500 cm^{-1} are assigned to ν(P=S) and ν(P—S) respectively, whilst in the second ν(P=S) is at 781 cm^{-1}, ν(P—S) at 447 cm^{-1}, and δ(P=S) at 370 cm^{-1}. I.r. and

304 K. Seppelt and W. Sundermeyer, *Z. Naturforsch.*, 1971, **26b**, 65.
305 H. Schmidbaur and G. Kammel, *Chem. Ber.*, 1971, **104**, 3234.
306 O. J. Scherer and R. Wies, *Angew. Chem. Internat. Edn.*, 1971, **10**, 812.
307 J. Nelson, *Spectrochim. Acta*, 1971, **27A**, 1105.
308 R. Mews and O. Glemser, *Inorg. Nuclear Chem. Letters*, 1971, **7**, 821.
309 W. Wolfsberger and H. H. Pickel, *Z. anorg. Chem.*, 1971, **384**, 131.
310 R. Mews and O. Glemser, *Inorg. Nuclear Chem. Letters*, 1971, **7**, 823.
311 H. Bürger, K. Burczyk, A. Blaschette, and H. Safari, *Spectrochim. Acta*, 1971, **27A**, 1073.
312 H. W. Roesky and S. Tutkunkardes, *Chem. Ber.*, 1971, **104**, 1655.
312a A. J. Banister and J. R. House, *J. Inorg. Nuclear Chem.*, 1971, **33**, 4057.
313 S. P. von Halasz and O. Glemser, *Chem. Ber.*, 1971, **104**, 1247.
314 S. P. von Halasz, O. Glemser, and M. F. Feser, *Chem. Ber.*, 1971, **104**, 1242.
315 H. W. Roesky and H. H. Giere, *Inorg. Nuclear Chem. Letters*, 1971, **7**, 171.
316 H. W. Roesky and W. G. Böwing, *Angew. Chem. Internat. Edn.*, 1971, **10**, 344.
317 R. Keat, D. S. Ross, and D. W. A. Sharp, *Spectrochim. Acta*, 1971, **27A**, 2219.
318 A. Haas and R. Lorenz, *Z. anorg. Chem.*, 1971, **385**, 33.
319 U. Klingebiel and O. Glemser, *Chem. Ber.*, 1971, **104**, 3804.
320 H. W. Roesky, *Angew. Chem. Internat. Edn.*, 1971, **10**, 266.
321 A. A. Pinkerton and R. G. Cavell, *J. Amer. Chem. Soc.*, 1971, **93**, 2384.

Table 34 *References to compounds containing* S—N *bonds*

Compound	*Ref.*
Me_2NSO_2F	
Me_2NSO_2Cl	311
Me_2NSO_2Br	
$Me_2S(O){=}NP(O)F_2$	
$Me_2S(O){=}NP(S)F_2$	
$Me_2S(O){=}NSO_2F$	312
$Me_2S(O){=}NS(O)(F){=}NSO_2F$	
$Me_2S(O){=}NC(Me){=}NSO_2F$	
$PhSN{=}S{=}NSPh$	312*a*
$N{\equiv}CN{=}S(F)R$ (R = NMe_2, NEt_2, or NC_5H_{10})	
$CF_3C(O)N{=}S(F)R$ (R = NMe_2, NEt_2, or NC_5H_{10})	
$MeN{=}S(F)NEt_2$	313
$(CF_3)_2C(F)N{=}S(F)NEt_2$	
$N{\equiv}CN{=}S(O)F_2$	
$N{\equiv}CN{=}S(O)(F)NMe_2$	314
$N{\equiv}CN{=}S(O)(F)NEt_2$	
$N{\equiv}CN{=}S(O)(F)OPh$	
$CF_3SO_2NRSO_2F$ (R = H or Me)	
$Ag[CF_3SO_2NSO_2F]$	315
$CF_3SO_2NHP(O)Cl_2$	
$CF_3SO_2NHSO_2Cl$	
$FSO_2NMeP(O)Cl_2$	316
$R^1_2NS(O)F$	
$R^1_2NS(O)Cl$	
$R^1_2N(R^2O)SO$ (R^1, R^2 various)	317
$(R^1_2N)(R^2_2N)SO$	
$(CF_3S)_2NSCF_2Cl$	
$(CF_2ClS)_2NSCF_3$	
$(CF_3S)_2NSCFCl_2$	318
$(CFCl_2S)_2NSCF_3$	
$(CF_3S)_2NSCCl_3$	
$SO_2[NPCl_2NMe_2]_2$	
$SO_2[NPCl_2NEt_2]_2$	
$SO_2[NPCl(NMe_2)_2]_2$	
cyclo-$SO_2[NPCl_2]_2NEt$	319
cyclo-$SO_2[NPClNMe_2]_2NMe$	
cyclo-$SO_2[NPClNMe_2]_2NEt$	
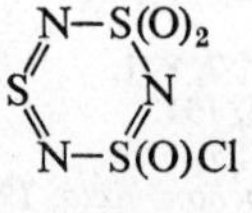	320

Raman frequencies for two isomers (α- and β-) of $P_4S_3I_2$ have been listed, and possess numerous coincidences in each case, indicating that neither isomer is centrosymmetric.[322]

A complete vibrational assignment for $PhSPF_4$ is consistent with an equatorially substituted trigonal-bipyramidal configuration about the phosphorus atom.[323] Doubling of several spectral bands due to vibrational

[322] G. J. Penney and G. M. Sheldrick, *J. Chem. Soc. (A)*, 1971, 1100.
[323] A. H. Norbury, S. Peake, and R. Schmutzler, *Spectrochim. Acta*, 1971, **27A**, 151.

fundamentals of $MeSP(O)Cl_2$ occurs in solution or in the liquid phase; this is attributed to the incidence of geometric isomers as a result of rotation of the MeS— group about the S—P bond.[324] In 2,4,5-$Cl_3C_6H_2O \cdot P(S)(OMe)_2$, ν(P=S) is found[325] at 616 cm^{-1}. I.r. bands at *ca.* 650 and 1230 cm^{-1} are assigned to (S—P=S) stretching vibrations[326] for the compounds $(RO)_2P(S)$—S—N◁ (R = Et, Pr^n, Pr^i, or Bu^i). Frequencies for (P—S) vibrations in $P(SMe)Cl_2$, $P(SMe)_2Cl$, and $P(SMe)_3$ have been reported, along with other data, and are shown in Table 35.[327] Examination of

Table 35 P—S *vibrations of* $P(SMe)Cl_2$, $P(SMe)_2Cl$, *and* $P(SMe)_3$

Compound	*Wavenumber*/cm^{-1}
$P(SMe)Cl_2$	470
$P(SMe)_2Cl$	468, 434
$P(SMe)_3$	499, 476, 452, 437

cyclo-$(CH_2)_3S_2P(S)Me$ by i.r. spectroscopy suggests the occurrence of two conformers in equilibrium.[328] The Ph_4As^+ salts of the new anions CF_3-PS_2OH^- and $CF_3PSO_2H^-$ have been characterized through i.r. spectroscopy.[328a]

Other Compounds containing a Group V Element Bonded to a Group VI Element.—Rauchle *et al.* have studied the i.r. and Raman spectra of $P(NMe_2)_3$ and X=$P(NMe_2)_3$, (X = O, S, Se, or Te). Quite complete assignments are proposed (except for skeletal deformations) and it appears that for the P^V compounds the molecular symmetry is at least C_3, while for $P(NMe_2)_3$ the number of observed bands can only be consistent with lower (probably C_s) symmetry. The ν(P=X) frequencies show an unusual trend: ν(P=O) is at 1210, ν(P=S) at 565, ν(P=Se) at 530, and ν(P=Te) at 519 cm^{-1}; the authors propose that the small change on going from X = S through Te is indicative of increasing stretching force constants, accounted for in terms of more efficient π-orbital-overlap between Se or Te and the phosphorus atom.[329] The Raman spectrum of solid P_4Se_3 has been reported[330] and compared with that of P_4S_3; the related compound As_4S_3 has also been studied,[331] strong i.r. bands at 370 and 340 cm^{-1} being assigned to ν(As—S) vibrations with δ(SAsS) at 176 cm^{-1}; and in As_2Se_3, ν(As—Se) modes occur[331] within the range 200—280 cm^{-1}.

324 R. A. Nyquist, *Spectrochim. Acta*, 1971, **27A**, 697.
325 R. A. Nyquist and W. W. Muelder, *Appl. Spectroscopy*, 1971, **25**, 449.
326 L. Almasi, A. Hantz, and T. Baicu, *Chem. Ber.*, 1971, **104**, 3982.
327 N. Fritzowky, A. Lentz, and J. Goubeau, *Z. anorg. Chem.*, 1971, **386**, 67.
328 E. A. Ishmaeva, O. A. Raevskii, R. A. Cherkasov, F. G. Khalitov, V. V. Ovchinnikov, and A. N. Pudovik, *Doklady Phys. Chem.*, 1971, **197**, 302.
328a A. A. Pinkerton and R. G. Cavell, *Inorg. Chem.*, 1971, **10**, 2720.
329 F. Räuchle, W. Pohl, B. B. Laich, and J. Goubeau, *Ber. Bunsengellschaft phys. Chem.*, 1971, **75**, 66.
330 V. A. Moroni and R. V. Schablaske, *J. Inorg. Nuclear Chem.*, 1971, **33**, 3182.
331 H. J. Whitfield, *Austral. J. Chem.*, 1971, **24**, 697.

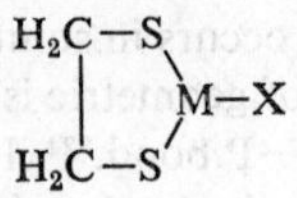

(28) M = As or Sb; X = Cl or Br

The i.r. and Raman spectra of compounds with structure (28) have been reported. Above 400 cm^{-1} the spectra are virtually identical, while at lower frequency MS_2 stretching modes are assigned as shown in Table 36.[332] I.r. and Raman frequencies for dialkylaminotellurium fluorides $(R_2N)TeF_5$ (R = Me or Et), $(NMe_2)_2TeF_4$, and C_4H_8N,TeF_5 have been measured and partially assigned, bands consistently near to 620 cm^{-1} being attributed to ν(Te—N) vibrations.[333]

Table 36 MS_2 *stretching modes of the compounds* (28)

M	X	*Wavenumber*/cm^{-1}
As	Cl	390, 359
As	Br	389, 361
Sb	Cl	357, 324
Sb	Br	355, 315

Compounds containing Group V–Halogen Bonds.—A very large amount of information concerning characteristic frequencies for Group V–halogen bond-stretching vibrations is embodied in the text of the preceding subdivisions of Section 5 (refs. 237—333) (reports for fluoro-derivatives of phosphorus being especially abundant*); no further reference to these data will be made below.

Frequencies for ν(N—Cl) in $BaNClO_3$, R_2NClO_3 (R = alkyl) and $[CO\cdot CMe_2\cdot NH\cdot CO]NCl$ have been given.[333a] Three reports concerning complexes between halogeno-derivatives and nitrogen bases lead to the following assignments:[333b—d] $\nu(N\cdots I)$, *ca.* 90 cm^{-1}; $\nu(N\cdots ICl)$, 140 cm^{-1}; $\nu(N\cdots IBr)$, 133 cm^{-1}; $\nu(N\cdots I_2)$, 93 cm^{-1}; and $\nu(N\cdots Br)$, 88—120 cm^{-1}.

The compounds Bu^tOPF_4 and $(Bu^tO)_2PF_3$ have i.r. spectra[334] containing absorptions in the region 840—980 cm^{-1}, owing to ν(P—F). Evidence has been found for co-ordination of the PF_6^- ion: more i.r. bands are observed for the anion in the complex $[Cu(py)_4(PF_6)_2]$ than would be expected for O_h symmetry, but a reasonable interpretation assuming D_{4h} symmetry (one fluorine atom being co-ordinated to the metal) is possible.[335]

332 P. N. Gates, P. Powell, and D. Steele, *J. Mol. Structure*, 1971, **8**, 477.
333 G. W. Fraser, R. D. Peacock, and P. M. Watkins, *J. Chem. Soc.* (*A*), 1971, 1125.
333a K. Höhne, J. Jander, K. Knuth, and D. Schlegel, *Z. anorg. Chem.*, 1971, **386**, 316.
333b A. Mishra and A. D. E. Pullin, *Austral. J. Chem.*, 1971, **24**, 2493.
333c G. W. Brownsen and J. Yarwood, *J. Mol. Structure*, 1971, **10**, 147.
333d J. D'Houdt and Th. Zeegers-Huyskens, *J. Mol. Structure*, 1971, **10**, 135.
334 D. E. Young and W. B. Fox, *Inorg. Nuclear Chem. Letters*, 1971, 7, 1033.
335 S. A. Bell, J. C. Lancaster, and W. R. McWhinnie, *Inorg. Nuclear Chem. Letters*, 1971, 7, 405.

* See in particular refs. 238, 244, 246, 260, 261, 268, 269, 276, 281, 284–286, and 323.

The Raman spectrum of crystalline SbF_5 at −100 °C has been reported and is comparable with the spectrum of the liquid, the principal bands being at 717 and 668 cm^{-1}. This is indicative of retention of a structure involving *cis*-fluorine-bridged SbF_5 units for both phases, in accordance with ^{19}F n.m.r. evidence. The Raman spectra of mixed TaF_5–SbF_5 phases are also consistent with the presence of *cis*-fluorine-bridged, neutral mixed-metal polymers.[336] The study by Raman spectroscopy of $SbCl_3$ in HCl has given evidence for the formation of $SbCl_4^-$ and $SbCl_6^{3-}$ ions in solution. Assignments of the observed bands for these species in terms of assumed T_d and O_h symmetry respectively are shown in Table 37.[337] Antimony

Table 37 *Assignments of bands under assumed* T_d *and* O_h *symmetries for* $SbCl_4^-$ *and* $SbCl_6^{3-}$ *ions*

$SbCl_4^-$ (T_d)		$SbCl_6^{3-}$ (O_h)	
Assignment	*Wavenumber*/cm^{-1}	*Assignment*	*Wavenumber*/cm^{-1}
$\nu_1(a_1)$	340	$\nu_1(a_{1g})$	329
$\nu_2(e)$	129	$\nu_2(e_g)$	279
$\nu_3(f_2)$	{288, 238}	$\nu_3(f_{2g})$	135
$\nu_4(f_2)$	108		

chloride complexes with the Schiff-base derivatives [M^{II} (sqen)] (M = Co, Ni, or Cu) exhibit i.r. bands at *ca.* 300 cm^{-1}, which have been assigned to $\nu(Sb^{III}—Cl)$ or $\nu(Sb^{V}—Cl)$ vibrations as appropriate.[338] Unassigned i.r. frequencies for $HBiCl_4$,2py and $HBiCl_4$,2aniline have been reported.[339]

6 Group VI Elements

Compounds containing O—H or S—H Bonds.—I.r. spectra have been measured for the seven known hydrates HX,$x$$H_2O$ (X = Cl or Br; x = 1, 2, or 3) and HBr,4H_2O, and also provide evidence for the existence of HCl,4H_2O. For x = 2, the spectra reveal the presence of a very strong central hydrogen-bond in $H_5O_2^+$; indeed, it appears that in these species the $H_5O_2^+$ ion vibrates as a discrete entity.[340] I.r. bands associated with stretching or bending of E—H or E—D bonds in the spectra of $(CF_3)_2$PEY (E = O or S; Y = H or D) are split into doublets; it is suggested that this effect arises from rotational isomerism through intra-molecular hydrogen-bonding.[341]

Compounds containing E—C Bonds (E = S, Se, or Te).—Trimethyl-sulphonium bromide, assumed to possess the structure $Me_3S^+Br^-$, has

[336] P. A. W. Dean and R. J. Gillespie, *Canad. J. Chem.*, 1971, **49**, 1736.
[337] K. I. Petrov, V. E. Plyushchev, V. V. Formichev, and G. V. Zimina, *Russ. J. Inorg. Chem.*, 1971, **16**, 696.
[338] M. D. Hobday and T. D. Smith, *J. Chem. Soc. (A)*, 1971, 3424.
[339] A. G. Galinos and P. B. Issopoulos, *Compt. rend.*, 1971, **272**, *C*, 2157.
[340] A. S. Gilbert and N. Sheppard, *Chem. Comm.*, 1971, 337.
[341] R. C. Dobbie and B. P. Straughan, *Spectrochim. Acta*, 1971, **27A**, 255.

been examined by i.r. and Raman spectroscopy, the vibrations of the Me_3S^+ unit being assigned in terms of C_{3v} symmetry through correlation with those for dimethyl sulphide.[342] Far-i.r. spectra of Me_2E (E = O, S, Se, or Te) at liquid-nitrogen temperature have given the wavenumbers/cm^{-1} for torsional modes and values for derived barrier heights that are shown in Table 38.[343] The vibrational spectrum of dimethyl sulphide has

Table 38 *Far-i.r. spectra of* Me_2E (E = O, S, Se, *or* Te) *and assignment of the bands*

E		O	S	Se	Te
Mode	B_2	268	247	207	185
	A_2	245	210	175	
ΔE/kcal mol^{-1}		3.50	3.22	2.40	2.47

been discussed in three papers,[344–345a] in which are contained data for $(CH_3)_2S$, CH_3SCD_3, and $(CD_3)_2S$ and, in two cases,[345, 345a] valence force field calculations; these studies appear to contain little new information. The synthesis of the compounds S=C(SMe)$_2$, S=C(SeMe)$_2$, S=C(SMe)-(SeMe), and Se=C(SeMe)$_2$ has allowed a number of characteristic E—C frequencies to be measured;[346] data for i.r. absorptions assigned to skeletal stretching vibrations are given in Table 39. In silylated sulphur ylides of the

Table 39 *Sulphur– and selenium–carbon stretching bands* (*wavenumbers*/cm^{-1})

ν(C=S)	*ca.* 1050	ν(C=Se)	856, 883
ν_{asym}(SCS)	859, 874	ν_{sym}(SCS)	511, 494
ν_{asym}(SeCSe)	720, 760	ν_{sym}(SeCSe)	468
ν_{asym}(SCSe)	780, 814	ν_{sym}(SCSe)	457

type $R_2N\underset{\underset{O}{\|}}{S}(R){=}CHSiR_3$, i.r. bands due[346a] to ν(S=C) are rather variable, in the range 900—1050 cm^{-1}.

New spectroscopic results for Me_2SeF_2 and its deuteriated analogue have allowed a more or less complete assignment, based on C_{2v} symmetry, to be made.[347] Selenium–carbon vibrations are at the following wavenumbers: for the hydrido-species, ν_{asym}(Se—C), 621 cm^{-1}, ν_{sym}(Se—C),

[342] J. W. Ypenburg, E. Van der Leij-van Wirdum, and H. Gerding, *Rec. Trav. chim.*, 1971, **90**, 896.
[343] J. R. Durig, C. M. Player, J. Bragin, and Y. S. Li, *J. Chem. Phys.*, 1971, **55**, 2895.
[344] J. W. Ypenburg and H. Gerding, *Rec. Trav. chim.*, 1971, **90**, 885.
[345] M. Tranquille, P. Labarise, M. Fouassier, and M. T. Forel, *J. Mol. Structure*, 1971, **8**, 273.
[345a] G. Geiseler and G. Hanschmann, *J. Mol. Structure*, 1971, **8**, 293.
[346] M. Dräger and G. Gattow, *Chem. Ber.*, 1971, **104**, 1429.
[346a] H. Schmidbaur and G. Kammel, *Chem. Ber.*, 1971, **104**, 3252.
[347] R. H. Larkin, H. D. Stidham, and K. J. Wynne, *Spectrochim. Acta*, 1971, **27A**, 2261.

605 cm^{-1}, and δ(CSeC), 295 cm^{-1}; corresponding values for the deuterio-species are 580, 561, and 271 cm^{-1}, respectively. A comparison has been made between the vibrational spectra of thiourea, $SC(NH_2)_2$, and selenourea, $SeC(NH_2)_2$: normal-co-ordinate calculations indicate that E—C stretching vibrations are of mixed composition.[348] For the sulphur compound, ν(S=C) is contributed to by ν_4 (1414 cm^{-1}), ν_6 (733 cm^{-1}), and ν_7 (487 cm^{-1}), with ν_6 predominating, while in selenourea, Se—C stretching arises mainly from the ν_7 fundamental at 390 cm^{-1}, with some ν_6 character (640 cm^{-1}). An assignment [349] for the normal vibrations of thiophane, $(CH_2)_4S$, places ν(CSC) modes at 795 and 691 cm^{-1}. Similar results for $(CH_2)_4Se$ are consistent with a planar arrangement rather than with C_2 symmetry.[350] For a series of tetramethylthiourea (tmtu) adducts $RTeX_3$,tmtu (R = Me or Et, X = Cl, Br, or I; R = *p*-MeO·C_6H_4, X = Cl) ν(Te—C) modes are [351] at 472—533 cm^{-1}, with ν(Te—S) at 190—228 cm^{-1}.

Compounds containing O—O, S—O, Se—O, or S—S Bonds.—The discovery of a convenient preparative route to trifluoromethyl hydroperoxide, CF_3OOH, has made possible an investigation of some of its properties. I.r. frequencies have been listed and partially assigned, placing ν(O—O) at 862 cm^{-1}. Reaction with acid fluorides yields the trifluoromethyl peroxy-esters $CF_3C(O)OOCF_3$, $MeC(O)OOCF_3$, $[CF_3OOC(O)CF_2]_2CF_2$, $CF_3OOC(O)(CF_2)_3C(O)F$, and $CF_3CF(OF)OOCF_3$, for which i.r. data have also been listed, ν(O—O) not being assigned for these compounds.[121] Partially assigned i.r. data for CF_3OOCl, a stable yellow gas at room temperature, support the formulation of this molecule as a peroxide structure. A band at 813 cm^{-1} is tentatively assigned to the ν(O—O) stretching vibration.[352] For peroxydisulphuryl difluoride, $S_2O_6F_2$, a very strong polarized Raman line at 801 cm^{-1} with an i.r. counterpart at 795 cm^{-1} is assigned [353] to ν(O—O).

Reaction between OSF_2 and $2ClF,AsF_5$ affords the complex $[OSClF_2][AsF_6]$, characterized by i.r. and Raman spectroscopy as well as by other means.[354] Bands attributed to vibrations of the $OSClF_2^+$ cation occur at the following wavenumbers/cm^{-1}: 1470, ν(S—O); 980, ν_{asym}(SF); 925, ν_{sym}(SF); 638, ν(S—Cl); 457, 442, 308, (bending modes). Valence force-field calculations for $(CH_3)_2SO$, $(CD_3)_2SO$, and $(CH_3)_2SO_2$ indicate that ν(SO) mixes appreciably with other modes in these molecules.[345, 345a]

The preparation of a number of iminosulphur oxydifluoride derivatives

348 G. B. Aitken, J. L. Duncan, and G. P. McQuillan, *J. Chem. Soc. (A)*, 1971, 2695.
349 Ya. M. Kimel'fel'd, A. Usmanov, G. N. Zhizhin, and V. P. Litvinov, *J. Struct. Chem.*, 1970, **11**, 799.
350 Ya. M. Kimel'fel'd, A. Usmanov, G. N. Zhizhin, V. A. Volovin, and N. N. Magdesieva, *J. Struct. Chem.*, 1970, **11**, 1056.
351 K. J. Wynne and P. S. Pearson, *Inorg. Chem.*, 1971, **10**, 2735.
352 C. T. Ratcliffe, C. V. Hardin, L. R. Anderson, and W. B. Fox, *J. Amer. Chem. Soc.*, 1971, **93**, 3886.
353 A. M. Qureshi, L. E. Levchuk, and F. Aubke, *Canad. J. Chem.*, 1971, **49**, 2544.
354 C. Lau and J. Passmore, *Chem. Comm.*, 1971, 950.

has been reported, the most interesting being $OS(NSOF_2)_2$. I.r. spectra have been measured, and partially assigned [355] in the way shown in Table 40.

Table 40 *I.r. bands (wavenumber/cm^{-1}) and assignments in some imino-sulphur oxydifluoride derivatives*

Compound	ν(SO)	ν(SN)	ν_{asym}(SF)	ν_{sym}(SF)
$OS(NSOF_2)_2$	{1415, 1234}	1166	865	845
$P(NSOF_2)_3$	1415	1238	840	—
$As(NSOF_2)_3$	1400—1500	1215	820	760
$Si(NSOF_2)_4$	1466—1481	1308	905	866
$MeSi(NSOF_2)_3$	1460	1295	900	852
$Me_2Si(NSOF_2)_2$	1455	1289	880	832
$Me_3Si(NSOF_2)$	1495	1275	852	819

Fairly detailed assignments for the vibrational spectra of $FOSO_2F$, $ClOSO_2F$, and $S_2O_6F_2$ have been made, consistent with C_s symmetry for the first two species and C_2 symmetry for peroxydisulphuryl difluoride.[353] For FSO_2NCl_2, i.r. absorptions at 1458 and 1222 cm^{-1} are attributed [356] to ν_{asym}(S—O) and ν_{sym}(S—O), respectively, with ν(S—F) at 844 cm^{-1}; the i.r. spectrum of perfluoromethylsulphinic acid, $CF_3S(O)OH$, has been measured but not assigned.[357] Sauer and Shreeve have reported gas-phase i.r. data for three bis(fluoroalkyl)sulphur oxydifluorides:[358] for $(CF_3)_2S(O)F_2$, $CF_3S(O)F_2C_2F_5$, and $(C_2F_5)_2S(O)F_2$, ν(SO) modes are at 1328, 1318, and 1316 cm^{-1}, with bands at *ca.* 715 cm^{-1} assigned to ν(SF). In the silylated ylides $R_2NS(R)(=O)=CHSiR_3$, wavenumbers [358a] for ν(S=O) are in the region of 1200 cm^{-1}. Unassigned i.r. data for SO_2 adducts of disubstituted hydrazines have been listed.[359]

Unassigned i.r. spectra have been listed for the following compounds:[360] $Cl_3CSSNCO$, $[Cl_3CSSN(H)]_2CO$, $Cl_3CSSN(H)CO_2Me$, $Cl_3CSSN(H)CO_2$-C_6H_{11}, and $Cl_3CSSN(H)CONHPh$. Gillespie and co-workers have interpreted the Raman spectrum of $S_4(SO_3F)_2$ in terms of an ionic system containing a square-planar cation,[361] as shown in Table 41. I.r. data for arylselenonic acids $RSeO_3H$ (R = Ph, *p*-Cl·C_6H_4, *p*-Br·C_6H_4, or *p*-Me·C_6H_4) include [362] assignment of bands at *ca.* 885 cm^{-1} to $\nu_{sym}(SeO_2)$, and at 940—950 cm^{-1} to $\nu_{asym}(SeO_2)$.

355 W. Sundermeyer, A. Roland, and K. Seppelt, *Angew. Chem. Internat. Edn.*, 1971, **10**, 419.
356 H. W. Roesky, *Angew. Chem. Internat. Edn.*, 1971, **10**, 265.
357 H. W. Roesky, *Angew. Chem. Internat. Edn.*, 1971, **10**, 810.
358 D. T. Sauer and J. M. Shreeve, *Z. anorg. Chem.*, 1971, **385**, 113.
359 J. M. Kanamueller, *J. Inorg. Nuclear Chem.*, 1971, **33**, 4051.
360 H. Bayreuther and A. Haas, *Chem. Ber.*, 1971, **104**, 2588.
361 R. J. Gillespie, J. Passmore, P. K. Ummat, and B. C. Uaidya, *Inorg. Chem.*, 1971, **10**, 1327.
362 K. Dostal, Z. Zak, and M. Cernik, *Chem. Ber.*, 1971, **104**, 2044.

Table 41 *Assignment of Raman spectrum of* $S_4(SO_3F)_2$

Wavenumber/cm^{-1}	S_4^{2+}	SO_3F^-
330	$\nu_3(B_{2g})$	—
382	—	$\nu_6(E)$
460	$\nu_4(E_u)$	—
530	$\nu_2(B_{1g})$	—
564	—	$\nu_3(A_1)$
584	$\nu_1(A_{1g})$	—

Compounds containing Group VI–Halogen Bonds.—Oxohalogen compounds have received some attention during 1971 and vibrational data have featured in several reports. Schack *et al.* have prepared bromine perchlorate, $BrOClO_3$: the i.r. spectrum reported [363] for this molecule is summarized in Table 42. For $BrOSO_2F$ and $Cs[Br(OSO_2F)_2]$, Raman spectra have been

Table 42 *The i.r. spectrum of* $BrOClO_3$ (C_s *symmetry*)

Vapour phase Wavenumber/cm^{-1}	*Ar matrix Wavenumber*/cm^{-1}	*Assignment*	
	1279vs	$\nu_1(A')$	$\nu_{asym}(ClO_3)$
1275vs	1262vs		
	1253m	$\nu_9(A'')$	
1039s	1037s	$\nu_2(A')$	$\nu_{sym}(ClO_3)$
683m	686m	$\nu_3(A')$	ν(O—Br)
648s	651vs, 643m	$\nu_4(A')$	ν(O—Cl)

recorded. C_s symmetry is indicated for the former, and both derivatives show bands assigned to ν(Br—O), at 464 and 437 cm^{-1}, respectively.[364] Two other bands in the spectrum of $BrOSO_2F$, at 884 and 175 cm^{-1}, are attributed to ν(S—OBr) and δ(Br—O); the former may also account for a further band at 1020 cm^{-1} observed for the anionic species.

New interpretations of some i.r. data for metal chlorates, bromates, and iodates have been put forward by Sterzel and Schnee,[365] some wavenumbers/cm^{-1} and assignments being shown in Table 43. The influence of the cation

Table 43 *I.r. spectra of metal chlorates, bromates, and iodates*

Compound	ν_1	ν_2	ν_3	ν_4
$KClO_3$	939	614	971	489
$NaClO_3$	937	620	969	482
$KBrO_3$	810	428	790	361
$NaBrO_3$	820	440	795	363
KIO_3	796	348	745	306
$NaIO_3$	794	355, 380	762, 774	331

363 C. J. Schack, K. O. Christie, D. Pilipovich, and R. D. Wilson, *Inorg. Chem.*, 1971 **10**, 1078.
364 A. M. Qureshi and F. Aubke, *Inorg. Chem.*, 1971, **10**, 1116.
365 W. Sterzel and W.-D. Schnee, *Z. anorg. Chem.*, 1971, **383**, 231.

and of crystal symmetry on the vibrational spectra of seven metal periodates has been discussed.[366] The laser Raman spectrum of crystalline periodic acid, H_5IO_6, indicates somewhat distorted octahedral symmetry; in aqueous solution, a smaller number of bands is observed, suggesting the formation of an undistorted octahedral complex of dimerized periodate anion.[367] The compounds I_2O_5 and HIO_3 have been used as examples in a discussion of the contribution of combination bands (nearly all of which will be i.r.- and Raman-active for solids) to the width of spectral bands; it is shown that by calculating combination band frequencies from a complete spectral analysis, the band-contours observed for solid samples of the aforementioned compounds can be rationalized.[367a] I.r. frequencies for a large number of salts of I_2O_7 with oxides (e.g. $2K_2O,I_2O_7,9H_2O$) have been listed,[368] with some suggested assignments for $\nu(I-O)$.

Characteristic frequencies arising from the vibrations of two fluorosulphur cations have been reported. New products obtained from the reaction between S_2F_2 and BF_4 or AsF_4 appear to be salts of the S_2F^+ ion,[369] with a characteristic i.r. band due to $\nu(S-F)$ at 850 cm^{-1}. The i.r. spectra of the adducts CF_3SF_3,BF_3, CF_3SF_3,AsF_5, and CF_3SF_3,SbF_5 show that they have ionic structures $[CF_3SF_2]^+[MF_{n+1}]^-$ (M = B, As, or Sb; n = 3 or 5). The following wavenumbers/cm^{-1} are attributed [370] to $CF_3SF_2^+$: 1282, 1078 (C—F stretch); 885, 766 (S—F stretch); 678, 472, and 429 (deformations).

For RN=SF$_2$, $\nu(S-F)$ lies in the range 708—760 cm^{-1}, but is in the 630—725 cm^{-1} region [371] for the corresponding compounds $RN=S(F)CF(CF_3)_2$ (R = Me, CF$_3$, CN, or CF$_3$CO). Frequencies for $\nu(S-F)$ vibrations are also contained in a number of earlier references, particularly those relating to (fluoro)sulphur–nitrogen and sulphur–oxygen derivatives.

Stabilization of SeCl$_2$ has been achieved by co-ordination to tetramethylthiourea.[372] I.r bands at 252 cm^{-1} (solid) and at 233 and 248 cm^{-1} (CHCl$_3$ solution) are attributed to axial $\nu(Se-Cl)$ modes of a T-shaped structure (29). Wynne and Pearson have studied the i.r. spectra of the adducts RSeCl$_3$,SbCl$_5$ (R = Me, Et, or Ph) and RTeCl$_3$,SbCl$_5$ (R = Me,

$$\begin{array}{c} \text{Cl} \\ | \\ (Me_2N)_2C-S-Se \\ | \\ \text{Cl} \end{array}$$

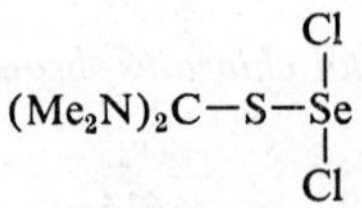

(29)

[366] H. Siebert and G. Weighardt, *Spectrochim. Acta*, 1971, **27A**, 1677.
[367] A. J. Fatiadi, *Chem. and Ind.*, 1971, 64.
[367a] P. M. A. Sherwood, *Spectrochim. Acta*, 1971, **27A**, 1019.
[368] M. Dratovsky, V. Kozisek, and B. Strauch, *Coll. Czech. Chem. Comm.*, 1971, **36**, 3810.
[369] F. Seel, V. Hartmann, I. Molnar, R. Budenz, and W. Gombler, *Angew. Chem. Internat. Edn.*, 1971, **10**, 186.
[370] M. Kramar and L. C. Duncan, *Inorg. Chem.*, 1971, **10**, 647.
[371] R. Mews, G. G. Alange, and O. Glemser, *Inorg. Nuclear Chem. Letters*, 1971, **7**, 627.
[372] K. J. Wynne and P. S. Pearson, *Chem. Comm.*, 1971, 293.

Et, or *p*-MeO·C_6H_4). All are consistent with an ionic formulation $RMCl_2^+SbCl_6^-$; for M = Se in particular, the high frequencies of ν(Se—Cl) (*e.g.* 424 cm^{-1} for R = Me) are not consistent with the alternative, chlorine-bridged structure.[373] Determination of the crystal structure of $TeCl_4,AlCl_3$ has enabled a reassignment of the vibrational spectrum of this adduct to be made.[374]

7 Group VII Elements

I.r. spectroscopy has provided strong evidence for hydrogen-bonding between the fluoride ion and acetic acid[375] in a study of the system KF–AcOH. A compound with high fluorinating capacity which is formed when RbCl is heated at 140 °C in a stream of fluorine has now been shown to be $Rb(ClF_4)$. The Raman spectrum consists of three lines (502, 414, and 287 cm^{-1}), in full accordance with spectra of known ClF_4^- salts.[376]

Raman spectra of BrF_3–HF mixtures have been found to be very complex and have been subjected to curve analysis. Intensities of bands thus distinguished, which could be assigned to ν_1 of the species BrF_2^+ and BrF_4^- (at 625 and 528 cm^{-1}, respectively), have been used to calculate ionic concentrations and hence the equilibrium constant for

$$BrF_3 + HF \rightleftharpoons HF_2^- + BrF_2^+$$

Values for 10^3K ($K = [HF_2^-][BrF_2^+]/[HF][BrF_3]$) in the range 6.14—2.09 were derived[377] for varying mole fractions of BrF_3.

8 Group VIII Elements

The Raman spectrum of the complex $XeF_4,2SbF_5$ is consistent with the ionic formulation $XeF_3^+Sb_2F_{11}^-$. Strong, polarized bands observed at 584 and 643 cm^{-1} are not attributable to any of the species $Sb_2F_{11}^-$, XeF_4, XeF^+, or $Xe_2F_3^+$, and are assigned to the ν_1 and ν_2 vibrations of a T-shaped XeF_3^+ ion,[378] by analogy with data for the analogous neutral species ClF_3 and BrF_3.

Raman spectroscopy has also been used[379] to investigate the 1 : 1 complex $XeF_2,XeOF_4$ (shown crystallographically to be isostructural with XeF_2,IF_5), the results confirming that this adduct is a molecular complex consisting of XeF_2 and $XeOF_4$ molecules bound as a result of the appreciable bond polarity in XeF_2. Observed Raman bands (wavenumbers/cm^{-1}) for the complex are compared with those for free XeF_2 and $XeOF_4$ in Table 44.

[373] K. J. Wynne and P. S. Pearson, *Inorg. Chem.*, 1971, **10**, 1871.
[374] B. Krebs, B. Buss, and D. Altena, *Z. anorg. Chem.*, 1971, **386**, 257.
[375] J. Emsley, *J. Chem. Soc.* (*A*), 1971, 2702.
[376] J. Shamir and N. Parchi, *Spectroscopy Letters*, 1971, **4**, 57.
[377] T. Sorles, H. H. Hyman, L. A. Quartermain, and A. I. Popov, *Inorg. Chem.*, 1971, **10**, 611.
[378] R. J. Gillespie, B. Landa, and G. J. Schrobilgen, *Chem. Comm.*, 1971, 1543.
[379] N. Bartlett and M. Wechsberg, *Z. anorg. Chem.*, 1971, **385**, 1.

Table 44 *Vibrational frequencies (wavenumber/*cm^{-1}*) observed in the Raman spectrum of* $[XeF_2,XeOF_4]$

Complex	*Assignment*	*Free molecule*
903	ν_1 of $XeOF_4$	919
573	ν_2 of $XeOF_4$	562
532	ν_4 of $XeOF_4$	530
494	ν_1 of XeF_2	497
378	ν_8 of $XeOF_4$	364
301	ν_3 of $XeOF_4$	286
254	ν_5 of $XeOF_4$	231
188	ν_9 of $XeOF_4$	—
125	Lattice modes	—

Raman data also suggest that a similar interaction between XeF_2 and XeF_5^+ is occurring in 1 : 1 and 1 : 2 complexes of the former with $[XeF_5^+][AsF_6^-]$.

6

Vibrational Spectra of Transition-element Compounds

BY M. GOLDSTEIN

1 Introduction

The arrangement of material in this chapter follows the pattern of previous years and is the same as in the preceding chapter. Information on vibrations of M—X bonds is presented in the sequence of vertical groups of the transition elements M in the Periodic Table. Within each group the elements X are also arranged according to vertical groups in the Periodic Table. A number of studies reported span more than one transition-metal group, and are conveniently dealt with at this stage, with some cross-referencing as necessary in later sections.

Raman and i.r. (to 33 cm^{-1}) spectra of $M_6O_{19}^{8-}$ (M = Nb or Ta) and $M_6O_{19}^{2-}$ (M = Mo or W) anions, as K$^+$ or $Bu^n_4N^+$ salts, respectively, have been assigned on the basis of O_h symmetry as in Table 1.[1a] A normal-co-ordinate analysis was used to aid assignments and to obtain the force constants and potential-energy distributions in the modes. The ratio of force constants of terminal, bridging, and central metal–oxygen bonds was found to be close to 8 : 4 : 1. A less-detailed study of $W_6O_{19}^{2-}$ has also been made.[1b]

The increase in approximate half-band widths of the $\nu_3(t_2)$ i.r. bands of some crystalline oxyanions (as KBr discs, mainly of K$^+$ salts) with increasing charge of the anion in the following isoelectronic series has been discussed:[2]

$$MnO_4^- < CrO_4^{2-} < VO_4^{3-} < TiO_4^{4-}$$
$$TcO_4^- < MoO_4^{2-}$$
$$(OsO_4 <)ReO_4^- < WO_4^{2-}$$
$$ClO_4^- < SO_4^{2-} < PO_4^{3-}$$

There has been continued growth in the use of vibrational spectroscopy in studies of a series of metal complexes of a particular ligand (or type of ligand). These studies are discussed in the following chapter if the emphasis is on internal vibrations of the ligands, or under the metals concerned where these are in the same periodic group. In many cases, however, such studies

[1] (*a*) R. Mattes, H. Bierbüsse, and J. Fuchs, *Z. anorg. Chem.*, 1971, **385**, 230; (*b*) C. M. Flynn, jun., and M. T. Pope, *Inorg. Chem.*, 1971, **10**, 2524, 2745.
[2] E. J. Baran and A. Müller, *Spectrochim. Acta*, 1971, **27A**, 517.

Table 1 *Assignment* (cm^{-1}) *of vibrational modes of* $M_6O_{19}^{n-}$ *anions*

	$Nb_6O_{19}^{8-}$	$Ta_6O_{19}^{8-}$	$Mo_6O_{19}^{2-}$	$W_6O_{19}^{2-}$
$\nu_1(A_{1g})$	896	871	980	992
$\nu_{12}(F_{1u})$	858	844	957	972
$\nu_4(E_g)$	842	835	951	968
$\nu_5(E_g)$	730	720	809	836
$\nu_{13}(F_{1u})$	710	690	798	812
$\nu_8(F_{2g})$	532	519	—	658
$\nu_{14}(F_{1u})$	530	540	602	586
$\nu_2(A_{1g})$	505	470	580	557
$\nu_6(E_g)$	463	402	—	501
$\nu_{15}(F_{1u})$	411	400	438	444
$\nu_{16}(F_{1u})$	312	334	356	369
$\nu_9(F_{2g})$	283[a]	275	—	—
$\nu_3(A_{1g})$	283[a]	218[a]	278	230
$\nu_{17}(F_{1u})$	234	222	220	226
$\nu_{10}(F_{2g})$	210	175	163	215
$\nu_7(E_g)$	225	218[a]	197	178
$\nu_{18}(F_{1u})$	195	174	192	174
$\nu_{11}(F_{2g})$	100	—	125	122

[a] Coincident bands.

cover a range of metal salts and deal largely with metal–ligand vibrations; they are therefore appropriately included at this stage.

From assignment of ν(MO) and ν(MN) modes in the far-i.r. spectra of pyridine complexes of metal(II) nitrates (M = Ni, Co, Cu, or Zn), several distinct phases $M(py)_x(NO_3)_2$ [x = 2, 3, 4, or 6, depending on M] can be identified.[3] Metal–nitrogen stretching modes have also been assigned to i.r. bands in the range 187—304 cm^{-1} for transition metal(II) solvates containing benzonitriles as ligands.[4] For $[M(N\text{-n-butylimidazole})_6]^{2+}$ salts (M = Cd, Mn, Fe, Co, or Ni), bands in the range 259—161 cm^{-1} are attributed to ν(MN) on the basis of absence in the spectrum of the free ligand, and follow the Irving–Williams order.[5] The i.r. and Raman spectra of $[M(NH_3)_6]^{3+}$ salts (M = Co, Cr, or Rh) have been assigned; ν(MN) modes are in the ranges 523—429 cm^{-1} (Raman) and 493—476 cm^{-1} (i.r.).[6] Partial assignments for some of the $[M'F_6]^{3-}$ anions used in the study (M′ = Al, Ga, In, Sc, V, Cr, or Fe) are also given.

Assignments have been made[7] of the i.r. spectra of *N*-aryl Schiff-base complexes (1) by noting the effects of (*a*) ^{15}N-labelling (for M = Cu, R = Me), (*b*) changing M (Co, Cu, or Zn), and (*c*) varying R (for M = Cu). Two bands (505 and 427 cm^{-1}) are assigned as ν(CuN) for R = Me; they shift 3—4 cm^{-1} on ^{15}N substitution and vary (559—491 and 507—417 cm^{-1}, respectively) according to the σ_p parameter of R.

The effect of isotopic substitution of the metal rather than of the co-ordinating ligand atom has been used further in identification of metal–

[3] R. H. Nuttall, A. F. Cameron, and D. W. Taylor, *J. Chem. Soc. (A)*, 1971, 3103.
[4] C. A. A. Van Driel and W. L. Groeneveld, *Rec. Trav. chim.*, 1971, **90**, 389.
[5] J. Reedijk, *J. Inorg. Nuclear Chem.*, 1971, **33**, 179.
[6] K. Wieghardt and H. Siebert, *J. Mol. Structure*, 1971, 7, 305.
[7] G. C. Percy and D. A. Thornton, *Inorg. Nuclear Chem. Letters*, 1971, **7**, 599.

ligand modes. Thus certain i.r. bands in the 300—180 cm^{-1} region of the spectra of complexes of the 8-quinolinolato (Q) ligand, MQ_2 (M = Ni, Cu, or Zn), $MQ_2,2H_2O$ (M = Fe, Ni, Cu, or Zn) and FeQ_3, shift by 2—6 cm^{-1} on metal isotopic substitution ($^{54,57}Fe$, $^{58,62}Ni$, $^{63,65}Cu$, or $^{64,68}Zn$) and are accordingly assigned as ν(MO) and ν(MN).[8]

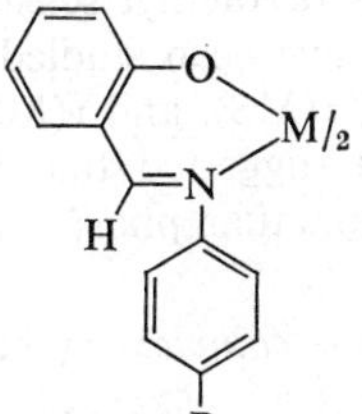

R

(1)

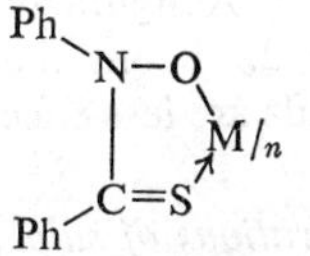

n = 2, M = Pd, Co, Fe, Zn, or Cd;
n = 3, M = Co or Rh

(2)

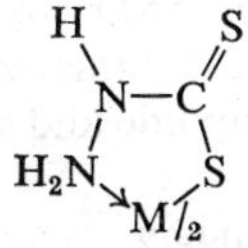

(3)

I.r. spectra of the hypophosphites $M(H_2PO_2)_2$ and $M(H_2PO_2)_2,py_2$ (M = Co or Zn) have been assigned in terms of $\nu(PO_2)$ (*ca.* 1150 cm^{-1}), ν(MO) (355—303 cm^{-1}) and ν(M—py) modes (262—240 cm^{-1}).[9] Metal–oxygen modes have also been assigned in the i.r. spectra of chelates (2) of *N*-thiobenzoyl-*N*-phenylhydroxylamine in the 495—450 cm^{-1} range,[10] in pyridine *N*-oxide complexes of several 3*d* metals (as perchlorate salts)[11] in the region 485—338 cm^{-1}, and in Co^{II}, Ni^{II}, and Cu^{II} complexes of the new ligand 2,3-di-(2-pyridine *N*-oxide)quinoxaline[12] at 392—370 cm^{-1} [ν(NO) at 1254—1188 cm^{-1}].

The i.r. spectra of $[ML_6](ClO_4)_3$ and $[ML_6](ClO_4)_2$ [M^{III} = Al, Cr, or Fe; M^{II} = Mn, Fe, Co, Ni, or Zn; L = $(CH_2)_4SO$] have been partially assigned.[13] The splitting of ν(S=O) in the complexes is interpreted on the basis of S_6 cation symmetry, while bands in the range 499—388 cm^{-1} are attributed to ν(MO) by analogy with Me_2SO complexes.

In the i.r. spectra of the hydrates $CoL_4(H_2O)_2X_2$ (L = 2-picoline or 3-ethylpyridine; X = Br or I), Mn(3-ethylpyridine)$_4(H_2O)_2I_4$, and Mn-$(H_2O)_4Cl_2$, $\rho(H_2O)$ [655—588 cm^{-1}], $\omega(H_2O)$ [625—556 cm^{-1}], and ν(MO) [395—311 cm^{-1}] have generally been assigned.[14]

Complexes of the dithiocarbazato-ligand $H_2NNHCS_2^-$ (L), of compositions $CrL_3,2H_2O$, $NiL_2,2H_2O$, and ML_2 (M = Pd, Pt, Zn, Cd, or Pb) are formulated as in (3), rather than as involving *S,S*-bonding, largely on the basis of i.r. spectral evidence.[15] Bands in the region 399—337 cm^{-1} are attributed to ν(MS). Similarly, Ni, Pd, and Pt complexes of

[8] N. Ohkaku and K. Nakamoto, *Inorg. Chem.*, 1971, **16**, 798.
[9] G. Brun and M. Dumail, *Compt. rend.*, 1971, **272**, *C*, 1867.
[10] R. Dietzel and Ph. Thomas, *Z. anorg. Chem.*, 1971, **381**, 214.
[11] N. M. Karayannis, C. M. Mikulski, M. J. Strocko, L. L. Pytlewski, and M. M. Labes, *J. Inorg. Nuclear Chem.*, 1971, **33**, 3185.
[12] M. Tong and D. G. Brewer, *Canad. J. Chem.*, 1971, **49**, 3425.
[13] C. V. Berney and J. H. Weber, *Inorg. Chim. Acta*, 1971, **5**, 375.
[14] M. Goodgame and P. J. Hayward, *J. Chem. Soc.* (*A*), 1971, 3406.

H_2NNHCS_2Me have $\nu(MS)$ in the 418—315 cm^{-1} range.[15] A somewhat lower range (367—246 cm^{-1}) of $\nu(MS)$ is suggested [16] for tetraethyl- and tetramethyl-dithio-oxamide complexes of the types ML_3A_2 (M = Mn, Fe, Co, Ni, or Cu; A = ClO_4^- or $FeCl_4^-$), $CuLX_2$ (X = Cl or Br), and $CuL_2(ClO_4)_2$.

A large number of complexes of dialkyl sulphides or diethyl selenide with halides of Pd^{II}, Pt^{II}, Ru^{III}, Os^{III}, Rh^{III}, or Ir^{III} have been studied in the far-i.r. spectral region.[17] Assignments for $\nu(MX)$, $\nu(MS)$, and $\nu(MSe)$ are summarized in Table 2. From these data it is suggested that the sulphide and selenide ligands are less effective π-acceptors than phosphines.

Table 2 *Metal–ligand vibrations of sulphide and selenide complexes* (cm^{-1})

Compound[a]	$\nu(MCl)$	$\nu(MBr)$	$\nu(MS)$	$\nu(MSe)$
trans-$[PdX_2L_2]$	362—336	268—255	326—295	223—220
trans-$[PtX_2L_2]$	343—330	240[b]	325—301	215[b]
fac-$[RuX_3L_3]$[c]	324, 316[b]	246[b]	316—289	—
fac-$[OsCl_3(Et_2S)_3]$	320, 296[b]	—	obsc.[b, d]	—
fac-$[RhX_3L_3]$	346—303	265—247	324—290	228—207[e]
fac-$[IrX_3L_3]$	322—296	240, 216[b]	318—270	202[b]

[a] L = R_2S (R = Et, Pr^n, Bu^n, Bu^i, or Bu^s) or Et_2Se; X = Cl or Br. [b] One example only. [c] $[RuCl_3(Et_2S)_2]_2$ has $\nu(RuCl)$ *ca.* 330, $\nu(RuS)$ at 290 cm^{-1}. [d] obsc. = obscured. [e] Data for X = I included.

Identification of $\nu(MX)$ modes in transition-metal halide complexes of heterocyclic ligands is now commonly used as a structural technique. Thus tetrahedral co-ordination of the metal atoms is proposed for the following complexes of ligands (4) and (5): $Co(LE)X_2$, $Co(LS)X_2$, $Ni(LE)Br_2$,

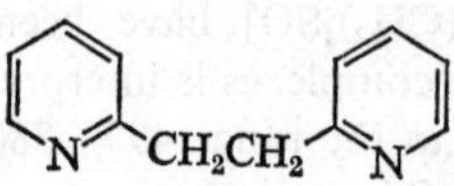

(LE)

(4)

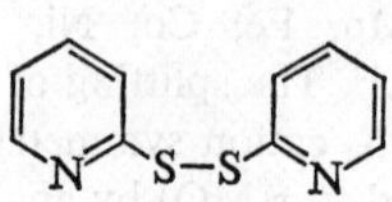

(LS)

(5)

$Ni(LE)I_2$, $Zn(LE)Cl_2$, and $Zn(LS)Cl_2$ (X = Cl, Br, I, or NCS).[18] However, for $Ni(LE)X_2$ (X = Cl or NCS) and $Ni(LS)X_2$ (X = Cl, Br, or NCS), five-co-ordinate structures involving X-bridging are proposed. For example, the assignment of $\nu(NiBr)$ in $Ni(LS)Br_2$ at 212 cm^{-1} is intermediate between values found for tetrahedral and octahedral co-ordination.

Only one $\nu(MX)$ i.r. band is shown by complexes $MX_2(pyrazine)_2$ (M = Co or Ni; X = Cl or I), and in each case this is *ca.* 40—80 cm^{-1}

[15] M. Akbar, S. E. Livingstone, and D. J. Phillips, *Inorg. Chim. Acta*, 1971, **5**, 119.
[16] G. Peyronel, G. C. Pellacani, A. Pignedoli, and G. Benetti, *Inorg. Chim. Acta*, 1971, **5**, 263.
[17] B. E. Aires, J. E. Fergusson, D. T. Howarth, and J. M. Miller, *J. Chem. Soc. (A)*, 1971, 1144.
[18] M. Keeton and A. B. P. Lever, *Inorg. Chem.*, 1971, **10**, 47.

higher than in analogous X-bridged systems [19] (values in cm^{-1}):

	X = Cl	X = Br	X = I
$CoX_2(pyrazine)_2$	260	203	177
$NiX_2(pyrazine)_2$	260	220	197

A sheet structure containing bridging pyrazine and terminal halogen atoms is suggested, and confirmed by a single-crystal *X*-ray study on $CoCl_2$-$(pyrazine)_2$.

Metal–halogen stretching modes (cm^{-1}) have also been identified in the expected regions of the i.r. spectra of complexes of the triphosphine $(Ph_2PCH_2CH_2)_2PPh$ (L) of formula $[LMCl]^+$ (M = Ni, Pd, or Pt), $[LMCl_2]$ (M = Co or Ru), [LRhCl], $[LMCl_3]$ (M = Rh or Ir), $[LOsCl_4]$, and $L_2Re_3Cl_9$;[20] of anions $[MX_3(4\text{-picoline})]^-$ [M = Co or Zn; X = Cl or Br; ν(MN) at 232—220 cm^{-1}];[21] and of pyridiazine and phthalazine complexes of types MX_2L_n [M = Co, Zn, or Cd; X = Cl, Br, or $\frac{1}{2}SO_4$; n = 1 or 2; ν(MN) = 276—200 cm^{-1}].[22]

A number of complexes of stoicheiometries such as MX_2L, $(MX_2)_2L_3$, and MX_2L_2 have been obtained in which L is a heterocyclic ligand of type (6) (E = S or Se; R^1, R^2, or R^3 = various combinations of H, Me, Cl, NO_2, or NH_2) and MX_2 is a dichloride.[23, 24] Various structural possibilities are indicated, based on the appearance or absence of ν(MX) bands in i.r. regions characteristic of terminal or bridging M—X groups.

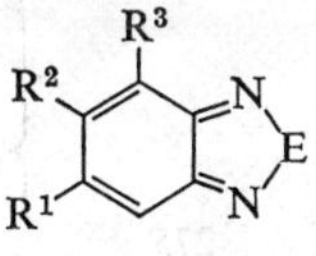

(6)

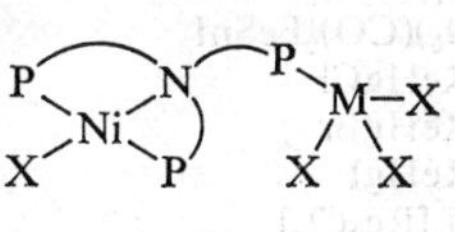

(7)

Dinuclear complexes $NiM(NP_3)X_4$ [M = Mn, Fe, Co, Ni, or Zn; X = Cl, Br, or I; NP_3 = $(Ph_2PCH_2CH_2)_3N$] are formulated as (7), largely on the basis of general observation of three ν(MX) i.r. bands [ν(NiX) probably obscured].[25]

A comprehensive survey has been published which deals with the vibrational spectra of intra- and inter-metal and semimetal bonds.[26] Data are tabulated and discussed for $\nu(M^1—M^2)$ vibrations where M^1 or M^2 = Si, Ge, Sn, Pb, P, As, Sb, Bi, Zn, Cd, Hg, B, Al, Ga, In, Tl, Se, Te, or a transition metal. This useful and timely review includes cluster compounds, and covers the literature up until the end of 1970.

Metal–metal vibrations reported during 1971 are listed in Tables 3—5.

[19] P. W. Carreck, M. Goldstein, E. M. McPartlin, and W. D. Unsworth, *Chem. Comm.*, 1971, 1634.
[20] R. B. King, P. N. Kapoor, and R. N. Kapoor, *Inorg. Chem.*, 1971, **10**, 1841.
[21] M. A. Buhannic and J. E. Guerchais, *Bull. Soc. chim. France*, 1971, 55,
[22] J. R. Allan, G. A. Barnes, and D. H. Brown, *J. Inorg. Nuclear Chem.*, 1971, **33**, 3765.
[23] R. H. Hanson and C. E. Meloan, *Inorg. Nuclear Chem. Letters*, 1971, **7**, 461.
[24] R. H. Hanson and C. E. Meloan, *Inorg. Nuclear Chem. Letters*, 1971, **7**, 467.
[25] M. Bacoi, R. Morassi, and L. Sacconi, *J. Chem. Soc.* (*A*), 1971, 3686.
[26] E. Maslowsky, jun., *Chem. Rev.*, 1971, **71**, 507.

Table 3 *Metal–metal vibrations* (cm^{-1}) *of dinuclear complexes*

Complex	ν (*i.r.*)	ν (*Raman*)	*Ref.*
$(\pi\text{-}C_5H_5)(CO)_3CrGeMe_3$	191		27[a]
$(\pi\text{-}C_5H_5)(CO)_3MoGeMe_3$	178	180	27[a]
$(\pi\text{-}C_5H_5)(CO)_3WGeMe_3$	166	171	27[a]
$(\pi\text{-}C_5H_5)(CO)_3CrSnMe_3$	183	183	27[a]
$(\pi\text{-}C_5H_5)(CO)_3MoSnMe_3$	170	168	27[a]
$(\pi\text{-}C_5H_5)(CO)_3WSnMe_3$	166	165	27[a]
$(\pi\text{-}C_3H_5)(PhNC)PdSnCl_3$	212		28[b]
$(\pi\text{-}C_3H_5)(p\text{-}O_2NC_6H_4NC)PdSnCl_3$	182		28[b]
$(\pi\text{-}C_3H_5)(PhNC)PdGeCl_3$	229		28[b]
$(\pi\text{-}C_3H_5)(p\text{-}O_2NC_6H_4NC)PdGeCl_3$	225		28[b]
$(CO)_4CoSnBr_3$	182		29[c]
$(CO)_4CoSnI_3$	156		29[c]
$(CO)_4CoGeBr_3$	200		29[c]
$(CO)_4CoGeI_3$	161		29[c]
$(CO)_4CoSiMe_3$	295	292	30[b, d]
$(CO)_5MnSnCl_3$	197		31
$(CO)_5MnSnBr_3$	178		31
$(CO)_5MnSnMe_3$	179		31
$(CO)_5MnSnPh_3$	170		31
cis-$[Ru(CNEt)_2(PPh_3)_2(SnCl_3)Cl]$		180	32[b, e]
cis-$[Ru(CNEt)_2(AsPh_3)_2(SnCl_3)Cl]$		179	32[b, e]
cis-$[Ru(CNEt)_2(SbPh_3)_2(SnCl_3)Cl]$		185	32[b, e]
$(\pi\text{-}C_5H_5)(CO)_2FeSnCl_3$	222		33[b, f]
$(\pi\text{-}C_5H_5)(CO)_2FeSnBr_3$	198		33[b, f]
$(\pi\text{-}C_5H_5)(CO)_2FeSnI_3$	164		33[b, f]
$(CO)_5ReHgCl$	143		34[b, f]
$(CO)_5ReHgBr$	116		34[b, f]
$(CO)_5ReHgI$	108		34[b, f]
$[Bu^n_4N]_2[Re_2Cl_8]$		275	35[g]
		274	36
$[Ph_4As]_2[Re_2Cl_8]$		275	36
$[Bu^n_4N]_2[Re_2Br_8]$		275	35[g]
$[Ph_4As]_2[Re_2Br_8]$		278	36
$Re_2(O_2CMe)_4Cl_2$		289	35[g]
$Re_2(O_2CMe)_4Br_2$		289	35[g]
$Mo_2(O_2CMe)_4$		406	35[g]

[27] D. J. Cardin, S. A. Keppie, M. F. Lappert, M. R. Litzow, and T. R. Spalding, *J. Chem. Soc.* (*A*), 1971, 2262.
[28] T. Boschi and B. Crociani, *Inorg. Chim. Acta*, 1971, **5**, 477.
[29] K. L. Watters, W. M. Butler, and W. M. Risen, jun., *Inorg. Chem.*, 1971, **10**, 1970.
[30] J. R. Durig, S. J. Meischen, S. E. Hannum, R. R. Hitch, S. K. Gondal, and C. T. Sears, *Appl. Spectroscopy*, 1971, **25**, 182.
[31] S. Onaka, *Bull. Chem. Soc. Japan*, 1971, **44**, 2135.
[32] B. E. Prater, *Inorg. Nuclear Chem. Letters*, 1971, **7**, 1071.
[33] S. R. A. Bird, J. D. Donaldson, A. F. le C. Holding, B. J. Senior, and M. J. Tricker, *J. Chem. Soc.* (*A*), 1971, 1616.
[34] A. T. T. Hsieh and M. J. Mays, *J. Chem. Soc.* (*A*), 1971, 2648.
[35] W. K. Bratton, F. A. Cotton, M. Debeau, and R. A. Walton, *J. Co-ordination Chem.*, 1971, **1**, 121.
[36] C. Oldham, J. E. D. Davies, and A. P. Ketteringham, *Chem. Comm.*, 1971, 572.

Table 3 (*cont.*)

Complex	*ν (i.r.)*	*ν (Raman)*	*Ref.*
$Re_2(O_2CR)_4X_2$		295—277	36[h]
$Re_2(O_2CMe)_2Cl_4,2H_2O$		274	
$Re_2(O_2CMe)_2Br_4,2H_2O$		288	
$(nbd)_2RhSnCl_3$	165, 146		37[b, i]
$(nbd)_2RhSnBr_3$	159, 144		
$[Et_4N][Cl_2(CO)RhSnCl_3]$	210		
trans-$[(Et_3P)_2ClPtSiH_nCl_{3-n}]$	330	335	38[j]
trans-$[(Et_3P)_2BrPtSiH_2X]$	340—328	335	
trans-$[(Et_3P)_2IPtSiH_2Y]$	340—326	344—335	
trans-$[(Et_3P)_2ClPtGeH_3]$	225		
trans-$[(Et_3P)_2ClPtGeH_2Cl]$	252		
trans-$[(Et_3P)_2BrPtGeH_2Z]$	245—241		
trans-$[(Et_3P)_2IPtGeH_2Z]$	244—242		
$Au_2(S_2CNBu^n_2)_2$		185	39
$[Hg_2]^{2+}$		182[k]	40*a*

[a] Slightly different values are given for some of these compounds in a preliminary report (D. J. Cardin, S. A. Keppie, and M. F. Lappert, *Inorg. Nuclear Chem. Letters*, 1968, **4**, 365). [b] ν(M′X) modes also assigned. [c] Decrease in ν(Co—M) with increasing mass of halogen is paralleled by decrease in Co—M force constant. [d] Data shown are for solid. In the i.r. spectrum of the gas, ν(Co—Si) is at 298 cm^{-1}. [e] The *cis* geometry is inferred from the appearance of two ν(NC) i.r. bands (at *ca.* 2186 and 2165 cm^{-1}). [f] ν(CO) modes listed. [g] These compounds may be considered as containing M≡M bonds. Vibrational analyses were carried out with moderately consistent results, and *f*(MM) found to be in the range 4.5—3.1 mdyn $Å^{-1}$. $K_4Mo_2Cl_8,nH_2O$ (n = 0 or 2) also studied. [h] R = Me, Et, Pr[n], C_6H_{11}, or Ph; X = Cl or Br. Bands also assigned to ν(ReX) and ν(ReO). High frequency and intensity of ν(ReRe) ascribed to the quadruple Re≣Re bond. [i] (nbd) = norbornadiene. [j] I.r. data for C_6H_6 solutions. Raman values not observed in all cases. n = 3—0; X = H, Cl, or Br; Y = H, Cl, Br, or I; Z = Cl or Br. ν(Pt–halogen) observed in several cases at 282—260 (Cl), 175 (Br), or 148 cm^{-1} (I). Two bands generally assigned as ν(Pt—P), at 450—415 and 385—360 cm^{-1}. Assignments also given for ν(M′H) and δ(M′H) (M′ = Si or Ge). [k] Value for aqueous $Hg_2(NO_3)_2,2H_2O$. This corrects the earlier and widely quoted assignment (L. A. Woodward, *Phil. Mag.*, 1934, **18**, 823) of 169 cm^{-1}.

Table 4 *Metal–metal vibrations* (cm^{-1}) *of trinuclear complexes*

Complex	$\nu_{sym}(MM'_2)$ *i.r.*	$\nu_{sym}(MM'_2)$ *Raman*	$\nu_{asym}(MM'_2)$ *i.r.*	$\nu_{asym}(MM'_2)$ *Raman*	*Ref.*
$Cl_2Sn[Fe(CO)_2(\pi\text{-}C_5H_5)]_2$			230		33[a, b]
$Br_2Sn[Fe(CO)_2(\pi\text{-}C_5H_5)]_2$			230		
$I_2Sn[Fe(CO)_2(\pi\text{-}C_5H_5)]_2$			232		
$(SCN)_2Sn[Fe(CO)_2(\pi\text{-}C_5H_5)]_2$			204 or 168		
$(HCO_2)_2Sn[Fe(CO)_2(\pi\text{-}C_5H_5)]_2$			176		
$(CH_3CO_2)_2Sn[Fe(CO)_2(\pi\text{-}C_5H_5)]_2$			174		
$Hg[Co(CO)_4]_2$		161			40*b*
$[Zn\{Fe(CO)_4\}_2{}^{2-}]$		176			
$[Cd\{Fe(CO)_4\}_2{}^{2-}]$		167			
$[Hg\{Fe(CO)_4\}_2{}^{2-}]$		166			

37 J. V. Kingston and G. R. Scollary, *J. Chem. Soc.* (*A*), 1971, 3399.
38 J. E. Bentham, S. Cradock, and E. A. V. Ebsworth, *J. Chem. Soc.* (*A*), 1971, 587.
39 F. J. Farrell and T. G. Spiro, *Inorg. Chem.*, 1971, **10**, 1606.
40 (*a*) C. G. Davies, P. A. W. Dean, R. J. Gillespie, and P. K. Ummat, *Chem. Comm.*, 1971, 782; (*b*) H. Behrens, H.-D. Feilner, E. Lindner, and D. Uhlig, *Z. Naturforsch.*, 1971, **26b**, 990.

Table 4 (*cont.*)

Complex	$\nu_{sym}(MM'_2)$ *i.r.*	$\nu_{sym}(MM'_2)$ *Raman*	$\nu_{asym}(MM'_2)$ *i.r.*	$\nu_{asym}(MM'_2)$ *Raman*	*Ref.*
$Zn[Mn(CO)_5]_2$		167	257		
$Cd[Mn(CO)_5]_2$		168	210		34[c], 41[d]
$Hg[Mn(CO)_5]_2$		165	186		
$Cd[Mn(CO)_5]_2$(diglyme)		165	165	167	
$Cd[Mn(CO)_5]_2$(terpy)	155	158		167	
$Cd[Mn(CO)_5]_2$(phen)	162	162	175	174	41[d]
$Cd[Mn(CO)_5]_2$(bipy)		166	172		
$Cd[Mn(CO)_5]_2py_2$		157	163		
$Cd[Mn(CO)_5]_2$(4-picoline)$_2$		158	163		
cis-$(H_3Ge)_2Fe(CO)_4$		229		217	42[e]
$Zn[Re(CO)_5]_2$		108	247		
$Cd[Re(CO)_5]_2$		110	193		
$Hg[Re(CO)_5]_2$		111	163		34[c]
$Hg[Co(CO)_4]_2$		161	196		
$Hg[W(CO)_3(\pi-C_5H_5)]_2$		106	166		
$[(CO)_5Mn]Hg[Re(CO)_5]$			{180, 165}[f]		34
$[Me_4N]_2[Cl(CO)Rh(SnCl_3)_2]$			208[b]		37[a]
$[Ph_4As][Cl(CO)Pd(SnCl_3)_2(EtOH)_2]$			177[b]		
$[Et_4N][Cl(CO)Pt(SnCl_3)_2]$			204[b]		
$[Ph_4As]_2[$*cis*-$Cl_2Pt(SnCl_3)_2]$			{202, 184}[b]		43[a]
$[Ph_4As]_2[$*trans*-$Cl_2Pt(SnCl_3)_2]$			{209, 184}[b]		
$[Hg_3][AsF_6]_2$		118[g]			40a

[a] ν(SnX) also assigned. [b] No distinction made between ν_{sym} and ν_{asym}. [c] M—M′ force constants calculated. [d] ν(CO) modes listed. [e] The unusual assignment of $\nu_{sym}(MM'_2) > \nu_{asym}(MM'_2)$ is shown to be correct by Raman polarization measurements on the liquid compound. [f] ν(HgMn) and ν(HgRe), undistinguished. [g] Value for SO_2 solution. A Raman line at 113 cm^{-1} shown by a solution of Hg in HSO_3F is similarly attributed to the $[Hg_3]^{2+}$ ion.

Table 5 *Metal–metal vibrations* (cm^{-1}) *of polynuclear complexes*

Complex	ν (cm^{-1})[a]	*Ref.*
$[ClRu(SnCl_3)_5]^{4-}$	*ca.* 210	44
$[Ph_4As]_2[(CO)Rh(SnCl_3)_3]$	181	37[b]
$[Et_4N]_2[ClPd(SnCl_3)_3]$	200	43[b]
$[CdFe(CO)_4]_n$	217	45[c]
$[HgFe(CO)_4]_n$	196	

[a] All data from i.r. spectra. [b] ν(SnCl) modes listed. [c] These compounds are suggested as having infinite zig-zag chain structures. The ν(CO) modes resemble the pattern expected for C_{2v} symmetry about the Fe atoms.

41 A. T. T. Hsieh and M. J. Mays, *J. Chem. Soc.* (*A*), 1971, 729.
42 S. R. Stobart, *Inorg. Nuclear Chem. Letters*, 1971, 7, 219.
43 J. V. Kingtson and G. R. Scollary, *J. Chem. Soc.* (*A*), 1971, 3765.
44 H. Okuno, T. Ishinuri, K. Mizumachi, and H. Ihochi, *Bull. Chem. Soc. Japan*, 1971, **44**, 415.
45 T. Takano and Y. Sasaki, *Bull. Chem. Soc. Japan*, 1971, **44**, 431.

A number of complexes of type $(\pi\text{-}C_3H_5)(PPh_3)_3Pt(SnCl_3)$ have been studied in the i.r., and ν(SnCl) but not ν(PtSn) frequencies given.[46a] From a study of the far-i.r. and Raman spectra of $(cod)_3Pt_3(SnCl_3)_2$, 'cluster modes' have been assigned at 170, 143, 112, and 81 cm^{-1}; assignment of these and other skeletal modes was assisted by simplified force-constant calculations.[46b] Compounds $(Cl_3M)CoL(CO)_3$ and $(Cl_3M)CoL_2(CO)_3$ (M = Si, Ge, or Sn; L = Bu^n_3P or Ph_3P) give similar ν(CO) patterns in the i.r., but there are differences between the two series in the low-frequency region from which it is deduced that the compounds $(Cl_3M)CoL_2(CO)_3$ should be formulated $[CoL_2(CO)_3]^+[SnCl_3]^-$.[47]

2 Scandium and Yttrium

A review of the co-ordination chemistry of scandium[48] includes reference to several recent vibrational spectroscopic studies, *e.g.* of $[Sc(chel)_3]^{3+}$, $[Sc(chel)_2X_2]^+$, and ScX_3L_n (chel = bipy or phen; X = halide; L = a phosphine, NH_3, py, *etc.*) and of various complexes of Sc^{III} with oxo-ligands such as Ph_3PO and Me_2SO.

I.r. bands of complexes $Sc(NCS)_3,x$py (x = 2, 3, or 4) have been listed.[49] The assignment of a band at 465 cm^{-1} to ν(Sc—py) is based on a disproven correlation and is clearly in error.

A number of reports have assigned scandium–oxygen modes, as listed in Table 6.

Table 6 *Assignment of scandium–oxygen modes* (cm^{-1}) *in i.r. spectra*

Compound	ν(Sc—O)	ν(O—Sc—O)	*Ref.*
$Sc_2O(OH)_2(HCO_3)_2,5H_2O$[a]	400	585	50[b]
$KSc(CO_3)_2,2H_2O$	—	676	51[b]
$NH_4Sc(CO_3)_2,2H_2O$	410	675	51[b]
$CsSc(CO_3)_2,H_2O$	405	675	51[b]
$Na_2Sc(CO_3)_4,11H_2O$	407	—	51[b]
$ScPO_4$(c)	565, 660	—	52[c]
$RH[Sc(NCS)_4R_2]$ [d]	425—415	—	53
$[Sc(NCS)_2R_n](NCS)$ [d]	425—415	—	53

[a] Suggested formula of basic scandium carbonate. [b] I.r. spectroscopy used to follow thermal decomposition; data given for various temperatures. [c] PO_4^{3-} ion has D_{2d} symmetry (C_{3v} in amorphous $ScPO_4$). [d] R = antipyrine (n = 3) or pyrimidone (n = 2), with carbonyl co-ordination to Sc in each case.

46 (*a*) J. N. Crosby and R. D. W. Kemmitt, *J. Organometallic Chem.*, 1971, **26**, 277; (*b*) A. Terzis, T. C. Strekas, and T. G. Spiro, *Inorg. Chem.*, 1971, **10**, 2617.
47 K. Ogino and T. L. Brown, *Inorg. Chem.*, 1971, **10**, 517.
48 G. A. Melson and R. W. Stotz, *Co-ordination Chem. Rev.*, 1971, **7**, 133.
49 T. M. Sas, L. N. Komissarova, and N. I. Anatskaya, *Russ. J. Inorg. Chem.*, 1971, **16**, 45.
50 L. N. Komissarova, Z. N. Prozorovskaya, V. F. Chuvaev, and N. M. Kosinova, *Russ. J. Inorg. Chem.*, 1971, **16**, 23.
51 I. V. Arkhangel'skii, V. A. Zhorov, L. N. Komissarova, E. G. Teterin, and V. M. Shatskii, *Russ. J. Inorg. Chem.*, 1971, **16**, 339.
52 A. Muck and F. Petrů, *Z. anorg. Chem.*, 1971, **383**, 104.
53 Yu. G. Eremin and V. S. Katochkina, *Russ. J. Inorg. Chem.*, 1971, **16**, 189.

The i.r. spectrum of $H_3[Y(HPO_3)_3]$ indicates that the HPO_3^{2-} ligands are of C_s symmetry in the complex.[54]

In $[M(NH_3)_6][ScF_6]$ compounds (M = Co, Cr, or Rh), the $[ScF_6]^{3-}$ fundamentals identified are at 481 (ν_1), *ca.* 455 (ν_3), *ca.* 253 (ν_4) and 233 cm^{-1} (ν_5).[6]

3 Titanium, Zirconium, and Hafnium

An isomer of $[(\pi\text{-}C_5H_5)_2TiH]_2$ is thought to be a linear polymer with H—Ti—(H—Ti—)$_x$H chains; an i.r. band at 1140 cm^{-1} is attributed to an asymmetric stretching mode of the bridge.[55]

The Raman spectrum of $Zr(BH_4)_4$ has been assigned in terms of the known tetrahedral $Zr(BH)_4$ arrangement with triple hydrogen bridges between each B and the Zr atom; four lines are polarized in agreement with T_d symmetry predictions ($4A_1 + 5E + 9T_2$).[56] For $Hf(BH_4)_4$, $Th(BH_4)_4$, their fully deuteriated analogues, and $[M(BH_4)_5]^-$ salts (M = Zr, Hf, or Th), less detailed (i.r.) assignments have been given;[57] these are compared with those of $Zr(BH_4)_4$ in Table 7.

Values of i.r. ν(TiN) modes in some new Ti^{III} dialkylamides (*e.g.* $[Ti(NMe_2)_3]_2$, 587 cm^{-1}; $[Ti(NEt_2)_3]_2$, 606 cm^{-1}) are very close to those of similar Ti^{IV} compounds.[58]

In titanium alkoxide derivatives of diethylhydroxylamine, $(Et_2NO)_n$-$Ti(OR)_{4-n}$ (n = 1—4; R = Et, Pr^i, or Bu^t), ν(TiO) assignments are given over the region 492—672 cm^{-1}; an even wider range is suggested in oxime derivatives such as $(Me_2C{=}NO)_nTi(OPr^i)_{4-n}$ (n = 1—4).[59] The value of 1065 cm^{-1} given [60] for ν(TiO) in $(\pi\text{-}C_5H_5)_2Ti\{OP(CF_3)_2\}_2$ is more appropriate for ν(Ti=O).

Raman depolarization ratios and relative intensities have been given for the totally symmetric M—O vibration of several eight-co-ordinate Zr^{IV} and Th^{IV} complexes (Table 8).[61] The linear relationship found between the molecular polarizabilities and the mean Taft σ-constant of R^1 and R^2 (Table 8) is interpreted in terms of M—O bond covalency.

Compounds $ZrO(ClO_4)_2,2H_2O$, $ZrO(OH)ClO_4,H_2O$, and $HfO(ClO_4)_2$,-$3H_2O$ have i.r. spectra indicative of ClO_4^- co-ordination, but no bands characteristic of M=O bonds.[62] The t_2 ν(TiI) mode of TiI_4 is readily located at 323 cm^{-1}, but in compounds $(R_2N)_3TiI$ and $(R_2N)_2TiI_2$

[54] A. Muck and F. Petrů, *Z. Naturforsch.*, 1971, **26b**, 64.
[55] R. H. Marvitch and H. H. Brintzinger, *J. Amer. Chem. Soc.*, 1971, **93**, 1246.
[56] B. E. Smith and B. D. James, *Inorg. Nuclear Chem. Letters*, 1971, **7**, 857.
[57] M. Ehemann and H. Nöth, *Z. anorg. Chem.*, 1971, **386**, 87.
[58] M. F. Lappert and A. R. Sanger, *J. Chem. Soc.* (*A*), 1971, 874.
[59] A. Singh, C. K. Sharma, A. K. Rai, V. D. Gupta, and R. C. Mehrotra, *J. Chem. Soc.* (*A*), 1971, 2440.
[60] W. J. Reagan and A. B. Burg, *Inorg. Nuclear Chem. Letters*, 1971, **7**, 741.
[61] C. J. Wiedenheft, *Inorg. Nuclear Chem. Letters*, 1971, **7**, 439.
[62] E. N. Lebedeva, S. S. Korovin, N. P. Tomilov, and K. I. Petrov, *Russ. J. Inorg. Chem.*, 1971, **16**, 355.

Table 7 *Assignments*[a] (cm^{-1}) *for tetrahydroborates of* Zr, Hf, *and* Th

Assignment[56]	$Zr(BH_4)_4$	$Hf(BH_4)_4$	$Hf(BD_4)_4$	$Th(BH_4)_4$	$Th(BD_4)_4$	$[M(BH_4)_5]^-$ [b]	*Assignment*[57]
ν(BH) terminal	2570p	2581	1925	2530	1880	2470—2430, 2410—2400	ν(BH) terminal
Bridge expansion and stretching	2215, 2125	2195, 2135	1625, 1545	2270, 2200, 2100	1700, 1620, 1550	2270—2220, 2215—2080	$\nu(BH_nM)$
Bridge expansion	2180p						
Bridge stretch, in-phase	1285p	1220	920	1165	890	1175—1115	$\delta(BH_m)$
(Absent in i.r.)	1077dp						
ν(M—B)	549p	487	465	450	490		
δ(M—B) or bridge mode	216dp						

[a] Raman data for C_6H_6 solution of $Zr(BH_4)_4$ (p = polarized, dp = depolarized). Other data from i.r. spectra of solids. [b] R_4N^+ (M = Zr, Hf, or Th) or Li^+ (M = Th) salts.

(R = Me or Et), ν(TiI) is difficult to assign owing to the proximity of $-NR_2$ deformations.[63]

In $TiCl_4$(*o*-allylaniline), the i.r. skeletal modes are very similar to those of the five-co-ordinate $TiCl_4$(aniline), whereas $TiX_4(aniline)_2$ (X = Cl or

Table 8 *Raman data on the* ν_{sym}(MO) *mode of β-diketonates of* Zr^{IV} *and* Th^{IV}, $M(R^1COCH_2COR^2)_4$, *in* $CHCl_3$ *solution*

M	R^1	R^2	Δν (cm^{-1})	ρ
Zr	Me	Me	444	0.08
Zr	CF_3	Me	431	0.15
Th	Me	Me	427	0.10
Th	Ph	Pr^n	405, (432sh)	0.18
Th	Ph	Ph	406	0.20
Th	C_3F_7	Bu^t	485	0.15

Br) contain octahedral titanium.[64] In these compounds ν(TiN) is suggested to be at *ca.* 255 cm^{-1}. Less-detailed i.r. assignments are listed for some acetylurea and benzoylurea complexes of chlorides or bromides of M^{IV} (Ti, Zr, or Sn), M^{III} (Al or In), or Sb^V; for such adducts, ν(M ← O) is placed at 405—360 cm^{-1}.[65]

Some i.r. data have been given (mainly as line-diagrams) for $ZrCl_4(py)_2$, $(pyH)_2ZrCl_6$, and $(pyH)_2ZrOCl_4,nH_2O$ (*n* = 0 or 2).[66]

A variety of complexes $MeTiCl_3$,B and $MeTiCl_3$,2B have been prepared, all of which show ν(TiC) in the 500—450 cm^{-1} i.r. region, with ν(TiCl) between 390 and 360 cm^{-1}.[67] Depending on the mole ratios used, $MeTiCl_3$

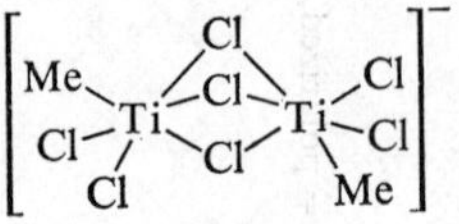

(8)

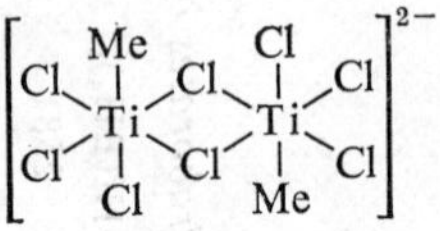

(9)

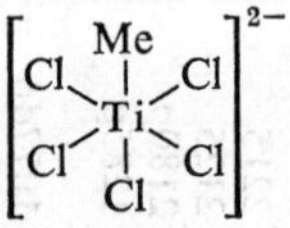

(10)

[63] H. Bürger, C. Kluess, and H.-J. Neese, *Z. anorg. Chem.*, 1971, **381**, 198.
[64] D. A. Baldwin and R. J. H. Clark, *J. Chem. Soc. (A)*, 1971, 1725.
[65] R. C. Paul, S. Sood, and S. L. Chadha, *J. Inorg. Nuclear Chem.*, 1971, **33**, 2703.
[66] G. M. Toptygina, I. B. Barskaya, and I. Z. Babievskaya, *Russ. J. Inorg. Chem.*, 1971, **16**, 686.
[67] G. W. A. Fowles, D. A. Rice, and J. D. Wilkins, *J. Chem. Soc. (A)*, 1971, 1920.

reacts with $[Et_4N]X$ (X = Cl or Br) to afford three series of anions.[68] These are formulated as in (8), (9), or (10), largely by comparison of their i.r. spectra with those of $[TiX_6]^{2-}$ (X = Cl or Br), $[Ti_2Cl_9]^-$, and $[Ti_2Cl_{10}]^{2-}$. General ranges for these types of complex are given as (cm^{-1}):[68]

	$[Me_2Ti_2X_7]^-$	$[Me_2Ti_2X_8]^{2-}$	$[MeTiX_5]^{2-}$
$\nu(TiCl)_t$	376—419	315—373	308—320
$\nu(TiCl)_b$	228—320	255	—
$\nu(TiBr)_t$		240—298	243—245

$TiCl_3$(bzac) and $TiCl_3$(acac) have been ascribed monomeric five-co-ordinate structures since their i.r. ν(TiCl) modes (391 and 399 cm^{-1}, respectively) are *ca.* 15—20 cm^{-1} higher than in corresponding octahedral $TiCl_2(diketonate)_2$ compounds.[69]

Complexes of Ti^{IV} with bidentate Schiff bases of type (11), $TiCl_2L_2$, and with related quadridentate ligands such as (12), $TiCl_2L$, show i.r. bands of

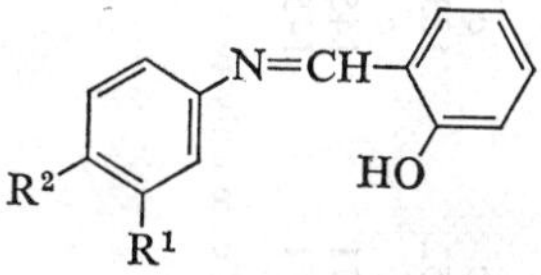

(11)

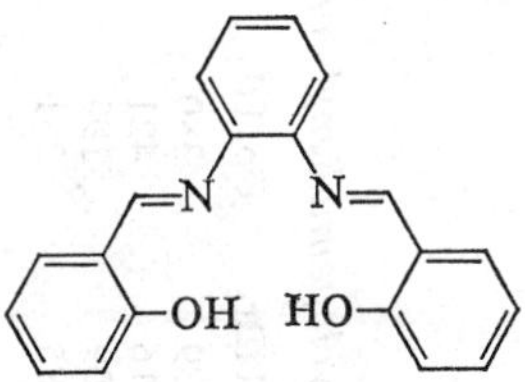

(12)

medium intensity at 475—430 and 565—545 cm^{-1}, which are assigned [70] to ν(TiO) and ν(TiN), respectively, by analogy with other reports.

Cyclopentadienyl derivatives of group IVA metals continue to be of spectroscopic interest. The i.r. (to 33 cm^{-1}) and Raman spectra of some well-known compounds of this type have been studied in detail, with particular attention to low-frequency skeletal modes; assignments (Table 9) were aided by approximate normal-co-ordinate analysis.[71] The force constant calculations indicated that (*a*) relative tilting of the two π-C_5H_5 rings slightly weakens the metal–ring bond, and (*b*) replacement of halogen in MX_4 (M = Ti, Zr, or Hf) by π-C_5H_5 weakens the remaining M—X bonds. The authors argue that, contrary to previous suggestions, the relative weakness of the symmetric ring-breathing mode (*ca.* 1130 cm^{-1}) does not necessarily imply that the metal–ring bonds are considerably ionic.[71]

[68] R. J. H. Clark and M. Coles, *Chem. Comm.*, 1971, 1587.
[69] D. W. Thompson, R. W. Rosser, and P. B. Barrett, *Inorg. Nuclear Chem. Letters*, 1971, 7, 931.
[70] N. S. Biradar and V. H. Kulkarni, *J. Inorg. Nuclear Chem.*, 1971, **33**, 3847.
[71] E. Maslowsky, jun. and K. Nakamoto, *Appl. Spectroscopy*, 1971, **25**, 187.

Table 9 *Metal–ligand frequencies* (cm^{-1}) *observed for π-cyclopentadienyl compounds*[a] *of* Ti, Zr, *and* Hf: Cp_2MX_2

Mode	*Species*	Cp_2TiF_2	Cp_2TiCl_2	Cp_2TiBr_2	Cp_2TiI_2	Cp_2ZrCl_2	Cp_2ZrBr_2	Cp_2ZrI_2	Cp_2HfCl_2
ν_{sym}(M–ring)	a_1	357	359	389	380	358	356	357	360
ν_{asym}(M–ring)	b_2	417	413	416	421	358	356	357	360
ν_{sym}(MX)	a_1	544	400	209	197	333	207	178	310
ν_{asym}(MX)	b_1	568	400	285	231	333	221	194	310
tilt[b], tilt, tilt, tilt	$a_1 + b_1 + a_2 + b_2$	258, 293	245, 300	245, 346	264, 347	266, 310	269, 315	272, 292	264, 284
δ_{sym}(XMX)	a_1	194	143	*c*	*c*	123	85	95	123
δ_{asym}(XMX)	b_1	234	207	152	123	165	128	115	165
δ_{asym}(ring–MX)	a_2	215	166	*c*	*c*	140	109	125	145
δ_{sym}(ring–M–ring)	a_1	234	184	171	145	165	155	155	165
δ_{asym}(ring–M–ring)	b_2	234	184	171	145	154	135	138	165

[a] Cp = (π-C_5H_5). Assignments given also for $CpTiCl_3$. [b] In each case only two ring tilting modes observed, assumed to consist of two accidentally degenerate pairs. [c] Not observed.

The data of Table 9 lend support to the following assignments of some 8-quinolinolato (ox) derivatives (cm^{-1}):[72]

	ν(MX)	ν(M–ring)
$(\pi\text{-}C_5H_5)TiCl(ox)_2$	380	356
$(\pi\text{-}C_5H_5)TiBr(ox)_2$	250	355
$(\pi\text{-}C_5H_5)ZrCl(ox)_2$	321	333
$(\pi\text{-}C_5H_5)ZrBr(ox)_2$	218	332
$(\pi\text{-}C_5H_5)HfCl(ox)_2$	288	296

4 Vanadium, Niobium, and Tantalum

In hydrido-derivatives of niobocene and tantalocene of type $(\pi\text{-}C_5H_5)_2$-M(L)H (L = CO, PMe_3, PEt_3, or C_2H_4),[73] ν(TaH) at 1750—1650 cm^{-1} is higher than ν(NbH) at 1735—1635 cm^{-1}.

I.r. spectra of cyanovanadium complexes have been reported, and assignments included those for ν(V—C) and δ(V—C—N) as follows (cm^{-1}):[74]

	ν(V—C)	δ(V—C—N)
$K_4[V(CN)_7]$	363, 340	477, 448
$Na_3[V(CN)_6]$	338	446
$K_4[V(CN)_6]$	329	406
$K_5[V(CN)_6]$	720	800

The high values for the V^I species are attributed to d_π–p_π effects.[74] In a compound formulated as $K_{3.5}V(CN)_{5.5}(OH)_{0.5}$, an i.r. band at 610 cm^{-1} may be due to ν(V—O) in a V—(OH)—V bridge.[75]

Crystallographic data for $LaVO_4$ have been given, and used to discuss the i.r. and Raman spectra (to 250 cm^{-1}) of this compound *via* factor group analysis (C_{2h}^5).[76] The analysis is in terms of VO_4^{3-} modes, but bands at 305 (Raman) and 295 cm^{-1} (i.r.) are tentatively assigned to ν(LaO). In a further paper,[77] the same authors interpret the i.r. spectra (to 300 cm^{-1}) of the lanthanide vanadates $LnVO_4$ (Ln = Ce, Pr, Nd, Sm, Eu, or Gd) in terms of the VO_4^{3-} D_{2d} sites in the crystal (D_{4h}^{19} space group). The expected splitting (to $B_2 + E$) of ν_3 (in T_d) of the VO_4^{3-} ion is observed (at *ca.* 865 and *ca.* 810 cm^{-1}) but that of ν_4 (*ca.* 443 cm^{-1}) is not. I.r. data (1000—400 cm^{-1}) have also been listed[78,79] for Tl_3VO_4, $Tl_4V_2O_7$, $TlVO_3$, $Tl_3V_5O_{14}$, $Tl_6V_{10}O_{28},4H_2O$, $K_2Zn_2V_{10}O_{28},16H_2O$, and other $V_2O_7^{4-}$ salts and heteropolyvanadates such as $Cs_2Zn_2V_{10}O_{28},14\frac{1}{2}H_2O$. Typically,

[72] J. Charalambous, M. J. Frazer, and W. E. Newton, *J. Chem. Soc. (A)*, 1971, 2487.
[73] F. N. Tebbe and G. W. Parshall, *J. Amer. Chem. Soc.*, 1971, **93**, 3793.
[74] R. Nast and D. Rehder, *Chem. Ber.*, 1971, **104**, 1709.
[75] B. G. Bennett and D. Nicholls, *J. Chem. Soc. (A)*, 1971, 1204.
[76] E. J. Baran and P. J. Aymonino, *Z. anorg. Chem.*, 1971, **383**, 220.
[77] E. J. Baran and P. J. Aymonino, *Z. anorg. Chem.*, 1971, **383**, 226.
[78] L. Žurkova, M. Gregorova, and M. Dillinger, *Coll. Czech. Chem. Comm.*, 1971, **36**, 1906.
[79] (*a*) L. Ulicka and M. Dillinger, *Coll. Czech. Chem. Comm.*, 1971, **36**, 2036; (*b*) P. Schwendt, P. Petrovič, and L. Zůrkova, *ibid.*, p. 3780; (*c*) L. Zůrkova, V. Suchá, and M. Dillinger, *ibid.*, p. 3788.

ν(V=O) is given in the 938—670 cm^{-1} range, with ν(V—O—V) at *ca.* 485 and 670 cm^{-1} in both $Tl_4V_2O_7$ and $TlVO_3$. In peroxy-compounds of vanadium, i.r. bands (cm^{-1}) are assigned as follows:[80]

	ν(V=O)	ν(O—O)	ν_{sym}(V<O—O)	ν_{asym}(V<O—O)
K_3VO_8	—	850	560	605
$NH_4[VO(O_2)(H_2O)(bipicoline)],H_2O$	929	839	570	610
$Ba[VO_2(O_2)(terpyridyl)]_2$	{950, 968	832	570	605

Broad bands in the region 850—700 cm^{-1} of the i.r. spectra of $NH_4[VO_2F_2]$, $Na[VO_2F_2],H_2O$, $[enH_2][VO_2F_2]_2$, and $NaVOF_4,0.7H_2O$ are suggested as arising from ν(V—O—V) modes of polymeric structures.[81, 82] However, in $[2,2'\text{-bipyH}_2][VOF_4]_2$ and [quinolinium]$[VOF_4]$, a band at *ca.* 1000 cm^{-1} is found, probably ν(V=O), indicating monomeric anions in these cases.[81] Similar values for ν(V=O) are quoted for (MeCN)$VOCl_3$ (1016 cm^{-1}) and $(MeCN)_2VOCl_3$ (992 cm^{-1}),[83] and for eight salicylaldehyde thiosemicarbazone complexes of oxovanadium(IV) salts (993—964 cm^{-1}).[84]

In the complexes VOL_2 derived from the Schiff-base anion (13), ν(V=O) occurs at *ca.* 900 cm^{-1}, whereas in VOL,nH_2O [L = (14), n = 0 or 1] the corresponding band is at *ca.* 1000—980 cm^{-1}.[85]

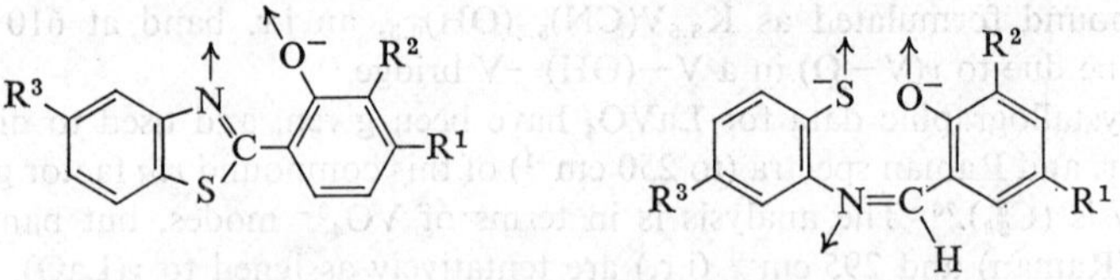

R^1 = H, Cl, Br, or NO_2; R^2 = H or Cl; R^3 = H or Cl

(13) (14)

I.r. spectroscopy has assisted in assigning the formula $Tl_2[VL_3]$ to complexes of V^{IV} prepared from oxo-vanadium salts (L = a catechol derivative), mainly by demonstrating the absence of V=O groups.[86] The product from the action of NH_3 on $VOCl_3$ at 373 K has also been identified by i.r. methods.[87a]

80 F. Offner and J. Dehand, *Compt. rend.*, 1971, **273**, *C*, 50.
81 A. K. Sengupta and B. B. Bhaumik, *Z. anorg. Chem.*, 1971, **384**, 251.
82 A. K. Sengupta and B. B. Bhaumik, *Z. anorg. Chem.*, 1971, **384**, 255.
83 J. P. Brunette, R. Heimburger, and M. J. F. Leroy, *Compt. rend.*, 1971, **272**, *C*, 2147.
84 N. V. Gerbeleu and M. D. Revenko, *Russ. J. Inorg. Chem.*, 1971, **16**, 557.
85 C. C. Lee, A. Syamal, and L. J. Theriot, *Inorg. Chem.*, 1971, **10**, 1669.
86 R. P. Henry, P. C. H. Mitchell, and J. E. Prue, *J. Chem. Soc.* (*A*), 1971, 3392.
87 (*a*) N. I. Vorob'ev, V. V. Pechkovskii, and L. V. Kobets, *Russ. J. Inorg. Chem.*, 1971, **16**, 473; (*b*) L. B. Hubert-Pfalzgraf, J. Guicu, and J. G. Reiss, *Bull. Soc. chim. France*, 1971, 3855.

Complexes $Nb(OMe)_5L$ have ν(NbL) at 170 (L = py), 164 (L = $C_5H_{10}NH$), 138 (L = C_4H_8ONH), 268 (L = Me_3PO), or 270 cm^{-1} [L = $(Me_2N)_3PO$].[87b]

In oxalato-complexes of niobium, the in-phase ν(NbO) mode is assigned in the Raman spectrum at 440—408 cm^{-1}, while the out-of-phase i.r.-active component is at 440—370 cm^{-1}. As expected, these bands in corresponding Ta compounds are shifted *ca.* 30 cm^{-1} to lower frequencies.[88a] Strong bands in the region 296—280 cm^{-1} are assigned as ν(VS) in $V(S_2PX_3)_3$ (X = Me, Ph, OEt, F, or CF_3).[88b]

Metal–halogen i.r. stretching frequencies (cm^{-1}) have been given for $[Me_4N][VCl_3]$ (290 and 250), $RbVCl_3$ (310 and 263), $Cs_2VCl_4(H_2O)_2$ (260 and 285vwsh), $(NH_4)_3VCl_3(H_2O)_3$ (265 and 285sh),[89] $[Et_4N]_2[NbOX_5]$ (330, X = Cl; 236, X = Br),[90] $Nb(dtc)_2X_3$ (335, X = Cl; 260, X = Br), and $Ta(dtc)_2X_3$ (320, X = Cl; 225, X = Br).[91] In the dtc compounds (dtc = *NN*-diethyldithiocarbamate), ν(MS) occurs at *ca.* 360 cm^{-1}; sulphido-complexes $M(dtc)_3S$ have also been obtained, showing ν(M=S) at 502 (M = Nb) or 483 cm^{-1} (M = Ta).[91a] ν(TaF) assignments are available [91b] for hydrazinium salts of TaF_6^-, TaF_7^{2-}, and TaF_8^{3-}.

I.r. spectra (to 130 cm^{-1}) of $[pyH]_2[Ta_6Br_{12}]Cl_6$, $[pyH]_2[Ta_6Cl_{12}]Br_6$, and $[pyH]_2[Ta_6Br_6Cl_6]Br_6$ have been measured in a study of exothermic irreversible transformations of $[pyH]_2[M_6Br_{12}]Cl_6$ salts (M = Nb or Ta) which occur at *ca.* 200 °C.[92]

5 Chromium, Molybdenum, and Tungsten

Metal–hydrogen frequencies reported during 1971 for compounds of these elements are listed in Table 10.

A graphical relationship between the torsional frequencies of sandwich compounds $(\pi\text{-}C_5H_5)_2M$ and $(\pi\text{-}C_6H_6)_2M$, and the corresponding torsional barrier, has been calculated [100] and used to show that previous assignments of the torsion in $(\pi\text{-}C_6H_6)_2Cr^{n+}$ (n = 0 or 1) are in error.

88 (*a*) M. Muller and J. Dehand, *Bull. Soc. chim. France*, 1971, 2843; (*b*) R. G. Cavell, E. D. Day, W. Byers, and P. M. Watkins, *Inorg. Chem.*, 1971, **10**, 2716.
89 L. F. Larkworthy, K. C. Patel, and D. J. Phillips, *J. Chem. Soc.* (*A*), 1971, 1347.
90 D. Brown and C. E. F. Rickard, *J. Chem. Soc.* (*A*), 1971, 81.
91 (*a*) P. R. Heckley, D. G. Holah, and D. Brown, *Canad. J. Chem.*, 1971, **49**, 1151; (*b*) B. Frlec and M. Vilhar, *J. Inorg. Nuclear Chem.*, 1971, **33**, 4069.
92 B. Speckelmeyer, C. Brendel, M. Dartmann, and H. Schäfer, *Z. anorg. Chem.*, 1971, **386**, 15.
93 F. Pennella, *Chem. Comm.*, 1971, 158.
94 M. L. H. Green and W. E. Silverthorn, *Chem. Comm.*, 1971, 557.
95 M. L. H. Green, J. Knight, L. C. Mitchard, G. G. Roberts, and W. E. Silverthorn, *Chem. Comm.*, 1971, 1619.
96 R. H. Crabtree, A. R. Dias, M. L. H. Green, and P. J. Knowles, *J. Chem. Soc.* (*A*), 1971, 1350.
97 M. L. H. Green, L. C. Mitchard, and W. E. Silverthorn, *J. Chem. Soc.* (*A*), 1971, 2929.
98 F. W. S. Benfield and M. L. H. Green, *Chem. Comm.*, 1971, 1274.
99 M. L. H. Green and P. J. Knowles, *J. Chem. Soc.* (*A*), 1971, 1508.
100 J. Laane, *J. Coordination Chem.*, 1971, **1**, 75.

A detailed study of the Raman spectra of the $[Mo(CN)_8]^{4-}$ and $[W(CN)_8]^{4-}$ ions (K^+ salts, dihydrates) indicates that these species retain a dodecahedral structure in aqueous solution, in contrast to the results of earlier investigations which suggested that a D_{4d} structure was adopted on dissolution.[101]

Table 10 *Molybdenum- and tungsten-hydrogen i.r. stretching modes* (cm^{-1})

Compound	ν(MH) *or* ν(MD)	*Ref.*
$(MePh_2P)_4MoH_4$	1800, 1714[a]	93[b]
$(MePh_2P)_4MoD_4$	1293, 1235[c]	
$(\pi\text{-}MeC_6H_5)Mo(PPh_3)_2H_2$	1740, 1720	94
$(\pi\text{-}C_6H_6)Mo(PPh_3)_2H_2$	1775, 1730	
$[(\pi\text{-}C_6H_6)Mo(PPhMe_2)_3H_2][PF_6]_2$	1880	95
$[(\pi\text{-}C_5H_5)Mo(PPh_3)H][PF_6]$	1863	96
$[(\pi\text{-}C_6H_6)Mo(PPhMe_2)_3H][PF_6]$	1915	97
$[(\pi\text{-}C_6H_6)Mo(PPhMe_2)_3D][PF_6]$	1370	
$[(\pi\text{-}C_6H_6)Mo(PPh_2Me)_3H][PF_6]$	1915	
$[(\pi\text{-}C_5H_5)_2Mo(NH_2Me)H][PF_6]$	1890	98
$[(\pi\text{-}C_5H_5)_2Mo(NH_2R)H][PF_6]$[d]	1830	
$[(\pi\text{-}C_5H_5)_2W(NH_2Me)H][PF_6]$	1940	
$[(\pi\text{-}C_5H_5)_2W(Ph)H]$	1913	99[e]
$[(\pi\text{-}C_5H_5)_2W(C_6D_5)H]$	1913	
$[(\pi\text{-}C_5H_5)_2W\{P(OMe)_3\}H][PF_6]$	1958	96
$[(\pi\text{-}C_5H_5)_2W(PPh_3)H][PF_6]$	1942	

[a] δ(MoH) at 775 and 640 cm^{-1}. [b] Ranges 1850—1700 cm^{-1} and 800—600 cm^{-1} quoted for ν(MoH) and δ(MoH), respectively, in $(EtPh_2P)_4MoH_4$ and $(Ph_2PCH_2CH_2PPh_2)_2MoH_4$. [c] δ(MoD) at 558 and 460 cm^{-1}. [d] R = Et, Pr^i, Bu^n, or Bu^s. [e] ν(W–ring) and ν(W–Ar) occur in the region 407—371 cm^{-1}.

All Raman-active vibrations of the CrN_6 group in $[Cr(NH_3)_6]^{3+}$ complexes have been observed and assigned in terms of O_h symmetry;[102] calculated M—N force constants were compared with those derived for $[Co(NH_3)_6]^{3+}$. The vibrational spectrum of $[Cr(NH_3)_6Cl]^{2+}$ has also been measured;[102] by using Cr isotopic shifts ($^{50,53}Cr$) or by making analogies, i.r. bands have been assigned to ν(Cr—N) in the following tris(chelates) (values in cm^{-1} for Cr in natural abundance):[103]

$[Cr(bipy)_3]^{n+}$	n =	3	2	1	0
		385, 349	351, 343	371, 343	383, 308

These small variations within geometrically analogous compounds show that the Cr—N bond strengths are similar, and can be rationalized in terms of σ- and π-bonding contributions.[103]

The i.r. spectrum of $Mo(NMe_2)_4$ has been compared with those of analogous Ti and V compounds; the $\nu_{asym}(MoN_4)$ mode at *ca.* 565 cm^{-1} is split, indicating D_{2d} rather than T_d MoN_4 skeletal symmetry.[104]

[101] T. V. Long and G. A. Vernon, *J. Amer. Chem. Soc.*, 1971, **93**, 1919.
[102] T. V. Long and D. J. B. Penrose, *J. Amer. Chem. Soc.*, 1971, **93**, 632.
[103] J. Takemoto, B. Hutchinson, and K. Nakamoto, *Chem. Comm.*, 1971, 1007.
[104] D. C. Bradley and M. H. Chisholm, *J. Chem. Soc.* (*A*), 1971, 2741.

When WCl_6 is heated under reflux with excess of CCl_3CN, a compound of composition $WCl_6,2(CCl_3CN)$ is obtained, which shows in the i.r. spectrum (a) a band at 2300 cm^{-1} due to co-ordinated CCl_3CN [ν(CN) of free ligand = 2250 cm^{-1}]; (b) all the other bands expected for CCl_3CN and W—Cl groups; and (c) a band at 1286 cm^{-1} attributed to $\nu(W\equiv N)$. Structure (15) is confirmed by a full X-ray study.[105]

$$Cl_3C-C\equiv N \rightarrow WCl_4 \equiv N-C_2Cl_5$$

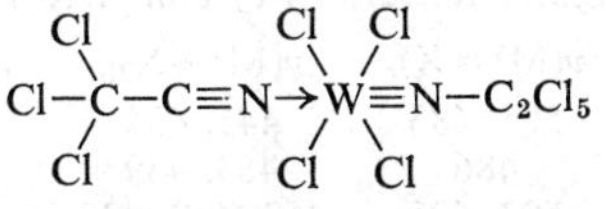

(15)

Detailed i.r. and Raman assignments have been given for $K_2Cr_3O_{10}$, α- and β-$Cs_2Cr_3O_{10}$, $K_2Cr_4O_{13}$, $Cs_2Cr_4O_{13}$, and CrO_3; some data are given in Table 11.[106] Comparison with other oxo-chromium(VI) species shows that

Table 11 *Some vibrational assignments of oxo-chromium compounds* (cm^{-1})

	$K_3Cr_3O_{10}$	$Cs_2Cr_4O_{13}$	CrO_3
$\nu_{asym}(CrO_2)$	—	982 (i.r.)	1003 (R)
$\nu_{sym}(CrO_2)$	960 (i.r.)	958 (i.r.)	979 (calc)
$\nu_{asym}(CrO_3)$	—	935 (i.r.)	—
$\nu_{sym}(CrO_3)$	—	903 (i.r.)	—
$\nu_{asym}(CrOCr)$	$\begin{cases} 827 \text{ (i.r.)} \\ 775 \text{ (i.r.)} \end{cases}$	825 (i.r.)	$\begin{cases} 915 \text{ (calc)} \\ 894 \text{ (i.r.)} \end{cases}$
$\nu_{sym}(CrOCr)$	552 (i.r.)	520 (i.r.)	$\begin{cases} 563 \text{ (R)} \\ 491 \text{ (i.r.)} \end{cases}$
$\delta(CrO_2)$	383 (i.r.)	380 (i.r.)	398 (R)
$\delta(CrO_3)$	357 (i.r.)	335 (i.r.)	—
$\rho(CrO_3)$	230 (R)	247 (R)	—
$\delta(CrOCr)$	—	—	$\begin{cases} 318 \text{ (i.r.)} \\ 100 \text{ (R)} \\ 50 \text{ (calc)} \end{cases}$

the Cr—O force constant is almost linearly dependent upon Cr—O distance. In tetra-alkoxides of Cr^{IV},[107a] $\nu_{asym}(CrO_4)$ (t_2 in T_d) is at ca. 600 cm^{-1}.

I.r. spectroscopy (with X-ray powder diffraction, d.t.a., etc.) has been used to study the tribochemical interactions which occur when WO_3 is ground in a vibrator mill with MgO,[107b] or with CaO or SrO,[108] or with a series of Ba compounds.[109] I.r. curves have also been published for paratungstates of sodium.[110a]

---

[105] M. G. B. Drew, K. C. Moss, and N. Rolfe, *Inorg. Nuclear Chem. Letters*, 1971, **7**, 1219.
[106] R. Mattes, *Z. anorg. Chem.*, 1971, **382**, 163.
[107] (a) E. C. Alyea, J. S. Basi, D. C. Bradley, and M. H. Chisholm, *J. Chem. Soc.* (*A*), 1971, 772; (b) H. Häusler, R. Möbius, and P.-E. Nau, *Z. anorg. Chem.*, 1971, **386**, 270.
[108] R. Albrecht, H. Häusler, and R. Möbius, *Z. anorg. Chem.*, 1971, **382**, 177.
[109] R. Albrecht, H. Häusler, and R. Möbius, *Z. anorg. Chem.*, 1971, **384**, 211.
[110] (a) C. M. Wolff and J. P. Schwing, *Compt. rend.*, 1971, **272**, *C*, 1974; (b) R. G. Cavell, W. Byers, and E. D. Day, *Inorg. Chem.*, 1971, **10**, 2710.

In $Cr(S_2PX_2)_3$ (X = Me, Ph, OEt, CF_3, or F) a strong i.r. band near 315 cm^{-1} is assigned to ν(CrS).[110b]

New metal complexes of the $[MoS_4]^{2-}$, $[WS_4]^{2-}$, and $[WSe_4]^{2-}$ anions, of type $[Ph_4P]_2[M^2(M^1X_4)_2]$ (M^2 = Ni, Co, or Zn), have been studied by i.r. and other techniques (Table 12). Evidence is presented suggesting that the

Table 12 *Far-i.r. assignments* (cm^{-1}) *of complexes* $[M^2(M^1X_4)_2]^{2-}$

Complex	$\nu(M^1{=}X)_t$	$\nu(M^1{-}X)_b$	$\nu(M^2{-}X)$	*Ref.*
$[Ni(MoS_4)_2]^{2-}$	505, 483	445, 434	325	111
[a]$[Fe(MoS_4)_2]^{2-}$	486	463, 439	349	111
$[Zn(MoS_4)_2]^{2-}$	502, 491	452, 440, 430	278	111
$[Ni(WS_4)_2]^{2-}$	490, 487	449, 447	328	112, 113
[a]$[Fe(WS_4)_2]^{2-}$	499, 490	450, 437	318	112
$[Co(WS_4)_2]^{2-}$	500, 491	450, 442	289	112
$[Zn(WS_4)_2]^{2-}$	489	456	280	112
$[Zn(WSe_4)_2]^{2-}$	330, 320	295	—	111

[a] Impure.

complex $[Ni(WS_4)_2]^{2-}$ involves planar co-ordination of nickel in a D_{2h} $[S_2WS_2NiS_2WS_2]^{2-}$ arrangement.[113]

Molybdenum–oxygen vibrations have been assigned in a number of compounds as listed in Table 13. Data on $[Mo_6O_{19}]^{2-}$ and $[W_6O_{19}]^{2-}$ have been referred to in an earlier section.[1] The anions $[VW_5O_{19}]^{3-}$ and $[V_2W_4O_{19}]^{4-}$ have also been studied.[1b]

Table 13 *Molybdenum–oxygen vibrations* (cm^{-1})

Compound	$\nu(Mo{=}O)$	*Ref.*
$H_2[MoOBr_5]$	970	114
$H[MoOBr_4]$	970	114
$[bipyH_2][MoOCl_5]$	985	115
$[enH_2][MoOCl_5]$	980	116
$Na[Mo_2O_4(glutat)(H_2O)],3H_2O$[a]	965; 730 [ν_{asym}(Mo—O—Mo)]; 470 [ν_{sym}(Mo—O—Mo)]	117
$Mo_2O_4Cl_2(bipy)_2$	952; 770, 735, 720 [ν(Mo—O—Mo)]	115
MoO(OH)(TPP)	901, 648, 614	118[b]
MoO(OOH)(TPP)[c]	941, 901, 649, 616	118[b]
$[MoO(OH)(TPP)]_2$	944, 900, 619, 551	118[b]
[MoO(Cl) (TPP)], HCl	990	118[b]

111 A. Müller, E. Ahlborn, and H.-H. Heinsen, *Z. anorg. Chem.*, 1971, **386**, 102.
112 A. Müller, E. Diemann, and H.-H. Heinsen, *Chem. Ber.*, 1971, **104**, 975.
113 A. Müller and E. Diemann, *Chem. Comm.*, 1971, 65.
114 H. K. Saha and A. K. Banerjee, *J. Inorg. Nuclear Chem.*, 1971, **33**, 2989.
115 H. K. Saha and M. C. Halder, *J. Inorg. Nuclear Chem.*, 1971, **33**, 3719.
116 H. K. Saha and M. C. Halder, *Z. anorg. Chem.*, 1971, **380**, 97.
117 T. S. Huang and G. P. Haight, *J. Amer. Chem. Soc.*, 1971, **93**, 611.
118 E. B. Fleischer and T. S. Srivastava, *Inorg. Chim. Acta*, 1971, **5**, 151.

Table 13 (*cont.*)

Compound	ν(Mo≡O)	*Ref.*
$MoOCl_3(bipy)$[d]	980	115
$MoOCl_3$(diphos) (red form)	941	119[e]
$MoOCl_3$(diphos) (brown form)	952	119[e]
$MoOCl_2$(diphos)	976	119[e]
[$MoOCl_2$(diphos)(THF)]	938	119[e]
[MoOX$(diphos)_2$]Y	943—940	119[e]
mer-[$MoOCl_2(PEt_2Ph)_3$] (green)	940	120[f]
mer-[$MoOCl_2(PMe_2Ph)_3$] (green form)	943	120[f]
mer-[$MoOCl_2(PMe_2Ph)_3$] (blue form)	953	120[f]
$MH[Mo_2O_2(edta)],nH_2O$[g]	764—744 [ν(Mo—O—Mo)]	121

[a] glutat = glutathione anion, $^{-}O_2CCH(NH_2)(CH_2)_2CONHCH(CH_2S^{-})CONHCH_2CO_2^{-}$. [b] TPP = tetraphenylporphyrin; no comment was made on the 650—550 cm^{-1} bands. [c] ν(O—O) at 1132 cm^{-1}. [d] Three isomers obtained, all with the same value for ν(Mo=O). [e] diphos = $Ph_2PCH_2CH_2PPh_2$; X = Cl, OMe, or OEt; Y = Cl, NCS, or NCSe. [f] The blue PMe_2Ph isomer and the (green) PEt_2Ph complex have been shown by *X*-ray diffraction each to contain the same geometrical co-ordination environment, but there are notable differences in inter-bond angles which are attributed to ligand repulsion. [g] M = Li, n = 7; M = K, n = 2; M = Na or NH_4, n = 8; M = $\frac{1}{2}$Ba, n = 9. In a preliminary note,[122] the same authors suggest a band at 931 cm^{-1} as characteristic of the

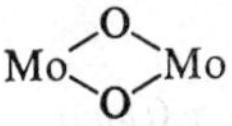

ring deduced to be present.

More Raman bands are found for solid MoF_4 than the two (A_{1g} + E_g) expected from the sheet structure of the type deduced from powder data on NbF_4 (note that MoF_5 and NbF_5 are isostructural).[123] Thus four bands (746m, 722s, 710w, and 690w cm^{-1}) are found in the stretching region, and five more below 300 cm^{-1}. The data were compared with those for MoF_5 and MoF_6, but no firm conclusions could be drawn.

Unassigned i.r. data have been listed for $XeCr_2F_{10}$ (obtained from XeF_2 + CrF_5 at 60 °C),[124] and for (RO)WF_5 (R = Me or Ph).[125] In $(Et_2N)WF_5$ and $(Et_2N)WF_4(OR)$ (R = Me or Ph), strong bands in the 700—500 cm^{-1} range [125, 126] are assigned [126] as ν(WF); these bands obscure assignment of ν(WN). Complexes $(Ph_3PS)WCl_5$ and $(Ph_3PSe)WCl_5$ show ν(WCl) at 329sh, 324sh, and 319vs, and at 331sh, 329sh, and 320vs cm^{-1} respectively.[127] The decreases in ν(PS) (96 cm^{-1}) and ν(PSe) (23 cm^{-1}) on co-ordination in these compounds are amongst the largest ever reported, and suggest bridging S (or Se). These i.r. data, together with conductance and other results, suggest the formulation $[W_2Cl_8L_2]Cl_2$.[127]

119 A. V. Butcher and J. Chatt, *J. Chem. Soc.* (*A*), 1971, 2356.
120 J. Chatt, Lj. Manojlović-Muir, and K. W. Muir, *Chem. Comm.*, 1971, 655.
121 J. Kloubek and J. Podlaha, *J. Inorg. Nuclear Chem.*, 1971, **33**, 2981.
122 J. Kloubek and J. Podlaha, *Inorg. Nuclear Chem. Letters*, 1971, **7**, 67.
123 J. B. Bates, *Inorg. Nuclear Chem. Letters*, 1971, **7**, 957.
124 J. Slivnik and B. Žemva, *Z. anorg. Chem.*, 1971, **385**, 137.
125 A. Majid, R. R. McLean, D. W. A. Sharp, and J. M. Winfield, *Z. anorg. Chem.*, 1971, **385**, 85.
126 A. Majid, R. R. McLean, T. J. Onellette, D. W. A. Sharp, and J. M. Winfield, *Inorg. Nuclear Chem. Letters*, 1971, **7**, 53.
127 P. M. Boorman and K. J. Reimer, *Canad. J. Chem.*, 1971, **49**, 2926.

6 Manganese and Rhenium

Manganese–hydrogen stretching frequencies reported are listed in Table 14.

Table 14 *Manganese–hydrogen stretching frequencies* (cm^{-1})

Compound	ν(MnH) *or* ν(MnD)	*Ref.*
$HMn(CO)_4[P(CF_3)_3]$	1814	128[a]
$HMn(CO)_4[P(CF_3)_2F]$	1798	128[a]
$HMn(CO)_4[P(CF_3)_2Me]$	1804	128[a]
$(\pi\text{-}C_5H_5)Mn(CO)_2H(SiCl_3)$	1889	129[a]
$(\pi\text{-}C_5H_5)Mn(CO)_2D(SiCl_3)$	1360	129[a]
$(\pi\text{-}C_5H_5)Mn(CO)_2H(SiPh_3)$	1900	130[b]
$(\pi\text{-}C_5H_5)Mn(CO)_2D(SiPh_3)$	1355	130[b]

[a] I.r. data. [b] Raman data. Kinetics of the displacement of Ph_3SiH from the Mn—H compound studied by i.r. spectroscopy.

I.r. and Raman measurements on $ClRe(CO)_5$ and $BrRe(CO)_5$ have been extended, and an almost complete set of assignments has been made by analogy with related systems.[131] Detailed assignments (Table 15) have also

Table 15 *Vibrational assignments* (cm^{-1}) *for* $[Mn(CN)_5NO]^{3-}$

		Approx. description[a]	ν *(obs).*	ν *(calc.)*
A_1 modes:				
	ν_1	C—N str. ax.	2124	2124
	ν_2	C—N str. eq.	2129	2129
	ν_3	N—O str.	1706	1705
	ν_4	Mn—N str.	660	660
	ν_5	Mn—C—N bend eq. ip.	461	461
	ν_6	Mn—C str. ax.	370	370
	ν_7	Mn—C str. eq.	315	315
	ν_8	N—Mn—C def.	104[b]	111
E modes:				
	ν_{16}	C—N str. eq.	2101	2101
	ν_{17}	Mn—N—O bend	660	662
	ν_{18}	Mn—C—N bend eq. oop.	407	407
	ν_{19}	Mn—C—N bend eq. ip.	453	453
	ν_{20}	Mn—C str. eq.	493	493
	ν_{21}	Mn—C—N bend ax.	270	270
	ν_{22}	N—Mn—C def.	174	174
	ν_{23}	C—Mn—C def. ax.	130	130
	ν_{24}	C—Mn—C def. eq.	—	86

[a] ip. = in-plane; oop. = out-of-plane. [b] Deduced from the overtone at 208 cm^{-1}.

been given of the i.r. spectrum (5000—180 cm^{-1}) of $K_3[Mn(CN)_5NO]$, using data on the ^{15}NO analogue, in terms of the free ion symmetry C_{4v} since no crystal-field effects were observed.[132] A normal-co-ordinate

[128] R. C. Dobbie, *J. Chem. Soc. (A)*, 1971, 230.
[129] W. Jetz and W. A. G. Graham, *Inorg. Chem.*, 1971, **10**, 4.
[130] A. J. Hart-Davis and W. A. G. Graham, *J. Amer. Chem. Soc.*, 1971, **93**, 4388.
[131] W. A. McAllister and A. L. Marston, *Spectrochim. Acta*, 1971, **27A**, 525.
[132] A. Poletti, A. Santucci, and G. Paliani, *Spectrochim. Acta*, 1971, **27A**, 2061.

analysis enabled the potential-energy contributions to the modes to be calculated.

In a series of dinuclear carbonyl derivatives $M_2(CO)_6X_2(MeCN)_2$ (M = Mn or Re; X = Cl, Br, or I), the position of ν(CN) at *ca.* 2320 cm^{-1} and the assignment of a band in the i.r. region 235—210 cm^{-1} to ν(MN) show that the MeCN is co-ordinated.[133] The ν(CO) pattern in solution is best explained in terms of a *trans* structure (16).

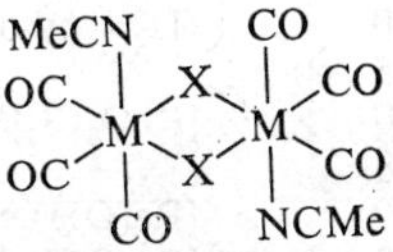

(16)

Partial i.r. assignments have been given for $[Mn(CNMe)_6]I$,[134] and for $[(\pi\text{-}C_5H_5)Mn(NO)(S_2CSR)]$ and $[(\pi\text{-}C_5H_5)Mn(NO)(SR)]_2$ (R = Pr^n, Pr^i, Bu^s, or Bu^t).[135]

In the μ-nitride complexes $[(Et_2PhP)_2X_2Re{\equiv}N \rightarrow A]$ (X = Cl, A = BF_3, BCl_3, BBr_3, or $PtCl_2(PEt_3)_2$; X = Br, A = BCl_3 or BBr_3),[136] $\nu(Re{\equiv}N)$ occurs in the range 1170—1052 cm^{-1} with ν(ReCl) at 297—285 and 265—217 cm^{-1}.

I.r. spectroscopy (2000—400 cm^{-1}) has been used to identify different modifications of MnO_2 and to study their thermal products,[137] the different preparations being characterized by certain values of the ratio of absorbances of bands at 595 ± 20 and 535 ± 20 cm^{-1}.

I.r. data (4000—200 cm^{-1}) have been tabulated for a large number of compounds containing Re—O and Re=O bonds, and some approximate force constant data given.[138] Some of the assignments suggested are shown in Table 16.

Assignments of the i.r. spectra (to 200 cm^{-1}) of $CsReO_4$, $KReO_4$, and ReO_2 have been based on distorted tetrahedral structures, whereas for K_2ReOCl_4,aq, Cs_2ReOCl_5,aq, andCs_2ReOCl_4F, C_{4v} symmetry is assumed.[138] The new series of lanthanide complexes $Ln(ReO_4)_3,n(TMU)$ (TMU = *NNN′N′*-tetramethylurea) show three (n = 5) or one (n = 6) i.r. ν(ReO) bands, indicating respectively co-ordinated or ionic per-rhenate groups.[139]

ν(Re=O) is at 1020 cm^{-1} in $ReOCl_4,OPCl_3$,[140a] 951 cm^{-1} in the $[ReOCl_5]^{2-}$ anion,[140b] and at *ca.* 975 cm^{-1} in some benzoylazo-complexes

[133] J. G. Dunn and D. A. Edwards, *J. Organometallic Chem.*, 1971, **27**, 73.
[134] P. C. Fantucci, V. Valenti, and F. Cariati, *Inorg. Chim. Acta*, 1971, **5**, 425.
[135] P. Hydes, J. A. McCleverty, and D. G. Orchard, *J. Chem. Soc.* (*A*), 1971, 3660.
[136] J. Chatt and B. T. Heaton, *J. Chem. Soc.* (*A*), 1971, 705.
[137] G. A. Kolta, F. M. Abdel Kerim, and A. A. Abdul Azim, *Z. anorg. Chem.*, 1971, **384**, 260.
[138] B. Jeżowska-Trzebiatowska, J. Hanuza, and M. Bałuka, *Spectrochim. Acta*, 1971, **27A**, 1753.
[139] M. Perrier and G. Vincentini, *J. Inorg. Nuclear Chem.*, 1971, **33**, 2497.
[140] (*a*) A. Guest and C. J. L. Lock, *Canad. J. Chem.*, 1971, **49**, 603; (*b*) J. E. Fergusson and J. L. Love, *Austral. J. Chem.*, 1971, **24**, 2685.

Table 16 *Some assignments for rhenium–oxygen compounds*[138]

Mode	*Range* (cm^{-1})	*Types of compound*
ν(Re=O)	995—912	$M^{I}ReO_4$; $M^{I}_2ReOX_5$ (X = F or Cl); $[ReO_2L_4]$, $[ReOCl_3L_2]$, $[Re_2O_3L_4Cl_4]$, $[Re_2O_3(CN)_8]^{6-}$ (L = py or ½en); $[ReO_3Cl_3]^{2-}$; $[Re_2OCl_{10}]^{4-}$
$\nu_{sym}(ReO_2)$	890—880	$[ReO_2(CN)_4]^{3-}$; $[ReO_2L_4]^{+}$ (L = py or NH_3); $[Re_2O_3L_4Cl_4]$ (L = py or ½en); $[Re_2O_3(CN)_8]^{6-}$
$\nu_{asym}(ReO_2)$	835—775	
$\delta(ReO_2)$	270—220	
$\nu_{sym}(ReO_3)$	*ca.* 935	$[ReO_3Cl_3]^{2-}$
$\nu_{asym}(ReO_3)$	*ca.* 890	
$\delta(ReO_3)$	*ca.* 318	
ν(Re—OH)	570—550	$[Re(OH)_2en_2]^{3+}$; $[ReO(OH)en_2]^{2+}$
ν_{asym}(ReORe)	860—720	$[Re_2OCl_{10}]^{4-}$; $[Re_2O_3(CN)_8]^{6-}$; $[Re_2O_3L_4Cl_4]$ (L = py or ½en)
ν_{sym}(ReORe)	205—200	

of rhenium oxyhalides with tertiary phosphines.[141] Two forms of $ReOCl_3$ have been characterized crystallographically,[142] but both show the same value for ν(Re=O) at 1018 cm^{-1}.

Bands in the i.r. spectrum of $Re_2O_3(S_2CNEt_2)_4$ at 960mw and 670vs cm^{-1} are absent from the spectrum of the $Et_2NCS_2^-$ ion.[143] These are attributed to $\nu(ReO)_t$ and $\nu(ReO)_b$, respectively, indicating that the complex contains an O=Re—O—Re=O moiety, shown by a full *X*-ray study to be almost linear. The complex reacts with MeOH to afford $ReO(S_2CNEt_2)_2(OMe)$, for which ν(Re=O) is at 940 cm^{-1}.

Differences between the i.r. spectra (800—200 cm^{-1}) of the anions in $[M(NH_3)_6][MnF_6]$ (M^{III} = Co, Cr, or Rh) and those in K_3MnF_6 and K_2NaMnF_6 have been attributed to Jahn–Teller effects.[144] In the hexa-ammine salts, the spectra are consistent with C_{3i} site symmetry of the $[MnF_6]^{3-}$ anions. Other assignments of ν(Mn–halogen) modes [25] have already been mentioned.

The far-i.r. spectra (to 33 cm^{-1}) of the new complexes $[R_4N][ReCl_5]$ (R = Et or Pr^n) are considered to be inconsistent with dimeric (as Nb_2Cl_{10}) or *trans* chlorine-bridged polymeric structures.[145, 146] As the frequencies and relative intensities of bands attributed to $\nu(ReCl)_t$ resemble those of *cis*-$[ReCl_4(MeCN)_2]$, a *cis*-bridged polymeric structure is proposed. The far-i.r. spectra of derivatives $[R_4N][ReCl_5L]$ (R = alkyl; L = Cl, MeCN, $HCONMe_2$, py, pyrazine, Me_2SO, Ph_3P, or thiourea) are said to be in accord with C_{4v} symmetry.[146]

Other ν(ReCl) i.r. assignments not mentioned above are (cm^{-1}): $[ReCl_3(PR_2Ph)_3]$ (R = Me or Et; *ca.* 310 and *ca.* 260; *mer* geometry

[141] J. Chatt, J. R. Dilworth, G. J. Leigh, and V. D. Gupta, *J. Chem. Soc. (A)*, 1971, 2631.
[142] P. W. Frais, C. J. L. Lock, and A. Guest, *Chem. Comm.*, 1971, 75.
[143] D. G. Tisley, R. A. Walton, and D. L. Wills, *Inorg. Nuclear Chem. Letters*, 1971, **7**, 523.
[144] K. Wieghardt and H. Siebert, *Z. anorg. Chem.*, 1971, **381**, 12.
[145] D. G. Tisley and R. A. Walton, *Inorg. Nuclear Chem. Letters*, 1971, **7**, 537.
[146] D. G. Tisley and R. A. Walton, *J. Chem. Soc. (A)*, 1971, 3409.

suggested);[147] $[ReOCl_5]^{2-}$ (313);[140b] $[Re(CO)_2(NO)Cl_2]_2$ (331, 287, and 252), $[Re(CO)_2(NO)Cl_2(C_4H_8S)]$ (317 and 292), and $[Re(CO)_2(NO)Cl_2py]$ (320 and 295);[148] $[Re(CO)(NO)Cl_2(C_8H_{14})]$ (326, 283, and 250) and $[Re(CO)_2(NO)Cl(acac)]$ (316);[149] and $[\{Ph_2P(CH_2CH_2PPh_2)_2\}_2Re_3Cl_9]$ (358 and 331).[20]

7 Iron, Ruthenium, and Osmium

Vibrational data on ν(MH) modes (M = Fe, Ru, or Os) are collected in Table 17. Identification of i.r. bands due to bridging hydrogen atoms in

Table 17 ν(M—H) *assignments* (cm^{-1}) *for* Fe, Ru, *and* Os *compounds*

Compound	ν(MH) *or* ν(MD)	*Ref.*
$H_2Fe(Ph_2PCH_2CH_2PPh_2)_2$	1875, 1825[a]	150
$HFeCl(Ph_2PCH_2CH_2PPh_2)_2$	1920[b]	150
$H_4Fe(PEtPh_2)_3$	1880[c]	150
$H_4Fe(PBu^nPh_2)_3$	1880[c]	150
$D_4Fe(PEtPh_2)_3$	1380, 1340[b]	150
$D_4Fe(PBu^nPh_2)_3$	1330[b]	150
$H_2Fe(CO)(PEtPh_2)_3$	1870	151[d]
$H_2Fe(N_2)(PMePh_2)_3$	1881	151[d]
$H_2Fe(N_2)(PEtPh_2)_3$	1840	151[d]
$H_2Fe(N_2)(PBu^nPh_2)_3$	1858	151[d]
$HFe(\pi\text{-}C_5H_5)(Me_2PCH_2CH_2PMe_2)$	1830	152
$HFe(\pi\text{-}C_5H_5)(CO)(SiCl_3)_2$	1960	129
$HRu(\pi\text{-}C_5H_5)(PPh_3)_2$	1950	153
$BH_4Ru(\pi\text{-}C_5H_5)(PPh_3)_2$	1996[e]	153
$HRu(NO)(PPh_3)_3$	1965	154[f]
$HRu(NO)(PMePh_2)_3$	1900	154[f]
$HRu(NO)(PPr^iPh_2)_3$	1990	154[f]
$HRu(NO)[P(C_6H_{11})Ph_2]_3$	1995	154[f]
$HOs(NO)(PPh_3)_3$	2050	154[f]
$HOs(NO)(PMePh_2)_3$	1985	154[f]
$HOs(dtt)(CO)(PPh_3)_2$	2002	155[g]
$H_3Os(dpt)(PPh_3)_2$[h]	2150, 2100, 1910	155[g]
$HOsCl(CO)(PR_3)_2$	2012	156[i]
$DOsCl(CO)(PR_3)_2$	1508	156[i]
$HOsCl(CO)(PR_3)_2py$	2040	156[i]

[147] H. P. Gunz and G. J. Leigh, *J. Chem. Soc. (A)*, 1971, 2229.
[148] F. Zingales, A. Trovati, F. Cariati, and P. Uguagliati, *Inorg. Chem.*, 1971, **10**, 507.
[149] A. Trovati, P. Uguagliati, and F. Zingales, *Inorg. Chem.*, 1971, **10**, 851.
[150] M. Aretsa, P. Giannoccaro, M. Rossi, and A. Sacco, *Inorg. Chim. Acta*, 1971, **5**, 115.
[151] M. Aresta, P. Giannoccaro, M. Rossi, and A. Sacco, *Inorg. Chim. Acta*, 1971, **5**, 203.
[152] W. E. Silverthorn, *Chem. Comm.*, 1971, 1310.
[153] T. Blackmore, M. I. Bruce, and F. G. A. Stone, *J. Chem. Soc. (A)*, 1971, 2376.
[154] S. T. Wilson and J. A. Osborn, *J. Amer. Chem. Soc.*, 1971, **93**, 3068.
[155] S. D. Robinson and M. F. Uttley, *Chem. Comm.*, 1971, 1315.
[156] F. G. Moers, *Chem. Comm.*, 1971, 79.

Table 17 (*cont.*)

Compound	ν(MH) *or* ν (MD)	*Ref.*
$HOsCl(CO)(PMe_2Ph)_3$	1929	157*a*[j]
$DOsCl(CO)(PMe_2Ph)_3$	1386	
$H_2OsCl_2(PEt_2Ph)_3$	2188, 1968	
$H_2OsCl_2(PMePh_2)_3$	2171, 2029	
$HOsCl_3(PEt_2Ph)_3$	2216, 2004, 1900	
$HOsCl(N_2)(PEt_2Ph)_3$	1936	
$HOsCl(N_2)(PEtPh_2)_3$	1934	
$HOsCl(N_2)(PMe_2Ph)_3$	1912	
$DOsCl(N_2)(PMe_2Ph)_3$	1374	
$HOsBr(N_2)(PMe_2Ph)_3$	1953, 1918	
$[H_2Os(en)_2][ZnCl_4]$	2150[k]	157*b*
$[D_2Os(en)_2][ZnCl_4]$	1537	

[a] The observation of these two bands (CH_2Cl_2 solution) indicates *cis* octahedral geometry. [b] Data for solid state. [c] Values for solutions; more bands found for the solids, including an unusual feature *ca.* 2400 cm^{-1}. [d] $\nu(N_2)$ at 2065—2058 cm^{-1}. [e] BHRu bridging mode; value given in text is 1992 cm^{-1}. [f] Deuteriation of these compounds shifts ν(NO); *trans* structures suggested for the solids, but all the compounds are stereochemically non-rigid in solution (n.m.r. study). [g] dtt = 1,3-di-*p*-tolyltriazenido; dpt = 1,3-diphenyltriazenido: Ar—N⩸N⩸N—Ar. [h] Probably seven-co-ordinate. [i] R = C_6H_{11}. [j] Data for mulls. Some solution phase ν(MH) values, $\nu(N_2)$, and ν(CO) reported, and used in stereochemical assignments. [k] Two bands probably present in an asymmetric envelope, indicating *cis* geometry.

$(\pi\text{-}C_2B_3H_7)Fe(CO)_3$ but not in $(\pi\text{-}C_2B_4H_6)Fe(CO)_3$ has been used to support proposed structures for these two complexes.[158] Unassigned i.r. data on bis(dicarbollyl)-iron(III) and -iron(II) species, such as $[Ph_3PMe][(B_9C_2H_{11})_2\text{-}Fe^{II}H]$, are available.[159]

A study of the Raman spectrum of $[^2H_{10}]$ferrocene in the polycrystalline and solution (CS_2) states[160] lends support to the recent assignments[161] of ferrocene, although a disputed assignment [that of the e_g ν(CC) mode at 1356 cm^{-1}] could not be unambiguously confirmed. Other cyclopentadienyliron compounds for which partial i.r. data have been reported are ferrocenyltriphenylphosphonium perchlorate (17),[162] cationic species

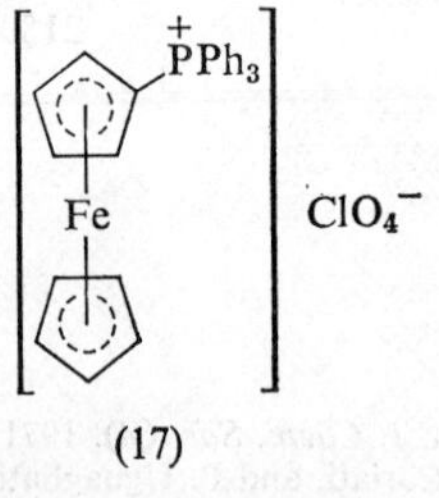

(17)

[157] (*a*) J. Chatt, D. P. Melville, and R. L. Richards, *J. Chem. Soc.* (*A*), 1971, 895; (*b*) J. Malin and H. Taube, *Inorg. Chem.*, 1971, **10**, 2403.
[158] R. N. Grimes, *J. Amer. Chem. Soc.*, 1971, **93**, 261.
[159] M. F. Hawthorne, L. F. Warren, K. P. Callahan, and N. F. Travers, *J. Amer. Chem. Soc.*, 1971, **93**, 2407.
[160] R. T. Bailey, *Spectrochim. Acta*, 1971, **27A**, 199.
[161] T. V. Long and R. F. Huege, *Chem. Comm.*, 1968, 1239.
[162] M. Sato, I. Motoyama, and K. Hata, *Bull. Chem. Soc. Japan*, 1971, **44**, 812.

$[(\pi\text{-}C_5H_5)Fe\{P(OPh)_3\}L]^+$ (L = MeCN, CO, SO_2, or C_2H_4),[163] and salts of the $[(\pi\text{-}C_5H_5)Fe(CO)_2]_2H^+$ ion.[164]

Bands due to the $[HgCl_3]^-$ ion at 282 cm^{-1} (i.r., ν_{asym}) and 289 cm^{-1} (Raman, ν_{sym}) are shown by $[(\pi\text{-}C_5H_5)Ru(PPh_3)_2(MeCN)][HgCl_3]$, whereas the zinc analogue shows ν(ZnCl) i.r. bands at 332, 301, 242, and 239 cm^{-1}, suggesting the formulation $[(\pi\text{-}C_5H_5)Ru(PPh_3)_2(MeCN)][Zn_2Cl_6]$.[153]

I.r. spectra (1300—800 cm^{-1}) have been illustrated (no frequencies given) for $(R)_2Sn(L)_2$, $(R)SnPh(L)_2$, (R)SnPh(L)(OH), and $Ph_2Sn(L)_2$ [R = $(\pi\text{-}C_5H_5)Fe(CO)_2$; L = PhS(O)O; ν(S=O) = 850 cm^{-1}].[165]

Carbonyl derivatives of Fe, Ru, and Os have been studied by matrix isolation. Irradiation of an argon matrix containing $Fe(CO)_5$ and C_2H_4 gives $Fe(CO)_4(C_2H_4)$, for which i.r. frequencies have been listed.[166] Two forms of $Fe_2(CO)_8$ are believed to be formed when matrix-isolated $Fe_2(CO)_9$ at 20 K is subjected to u.v. irradiation; bridged (C_{2v}) and un-bridged (D_{3d}) structures are postulated on the basis of ν(CO) i.r. frequencies.[167] Matrix-isolated $Fe_3(CO)_{12}$ appears from its i.r. spectrum to have the double-bridged C_{2v} structure observed in the solid, and not the solution-phase structure.[168] On the other hand, the spectra of $Ru_3(CO)_{12}$, $Os_3(CO)_{12}$, and $Ru_3(CO)_{10}(NO)_2$ in matrices at 20 K are similar to those of solutions, and indicate geometries somewhat distorted from D_{3h}.

ν(CO) frequencies in conjunction with Mössbauer data have been used in elucidation of the structures of approximately 50 compounds of types such as $[Fe(CO)_2L(SR)]_2$, $[Fe_2(CO)_5L(SR)_2]$, $[\{Fe(CO)_2(SR)\}_2L]$, $[Fe(CO)_3$-$(SR)_2Fe(CO)L]$, and $[Fe_2(CO)_3L_2(SR)_2]$, where L is a tertiary phosphine, arsine, *etc.* or a ditertiary diphosphine or diarsine.[169]

Structures of $[(Me_3Si)_2Fe(CO)_4]_2$ (18) and its reaction product with HCl, $[C_{14}H_{20}Si_2Fe_2O_8]$ (19), have been proposed using i.r. spectral data.[170]

The ν(CO) i.r. bands of $[(\pi\text{-}C_5H_5)Fe(CO)_2]_2$ in heptane solution containing various amounts of $AlBu^i_3$ show three types of concentration

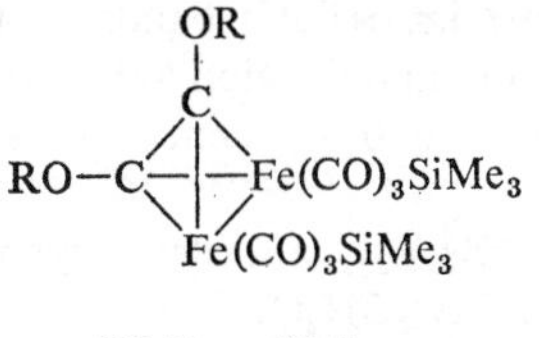

(18) R = $SiMe_3$
(19) R = H

[163] M. L. H. Green and R. N. Whiteley, *J. Chem. Soc.* (*A*), 1971, 1943.
[164] D. A. Symon and T. C. Waddington, *J. Chem. Soc.* (*A*), 1971, 953.
[165] R. C. Edmondson, D. S. Field, and M. J. Newlands, *Canad. J. Chem.*, 1971, **49**, 618.
[166] M. J. Newlands and J. F. Ogilvie, *Canad. J. Chem.*, 1971, **49**, 343.
[167] M. Poliakoff and J. J. Turner, *J. Chem. Soc.* (*A*), 1971, 2403.
[168] M. Poliakoff and J. J. Turner, *J. Chem. Soc.* (*A*), 1971, 654.
[169] J. A. de Beer, R. J. Haines, R. Greatrex, and N. N. Greenwood, *J. Chem. Soc.* (*A*), 1971, 3271.
[170] M. A. Nasta and A. G. MacDiarmid, *J. Amer. Chem. Soc.*, 1971, **93**, 2813.

dependence.[171] Bands are attributed to the original carbonyl (2005, 1962, and 1794 cm^{-1}), a 1 : 1 adduct (2026, 1985, 1828, and 1682 cm^{-1}), and a 1 : 2 adduct (2042, 2004, and 1682 cm^{-1}). Similar studies show that with $AlBu^i_3$ the non-bridged form of $[(\pi\text{-}C_5H_5)Ru(CO)_2]_2$ (20) is converted into carbonyl-bridged 1 : 1 [ν(CO) = 2006, 1988, 1831, and 1680 cm^{-1}] and 1 : 2 [ν(CO) = 2045, 2006, and 1680 cm^{-1}] adducts [(21) and (22)], in

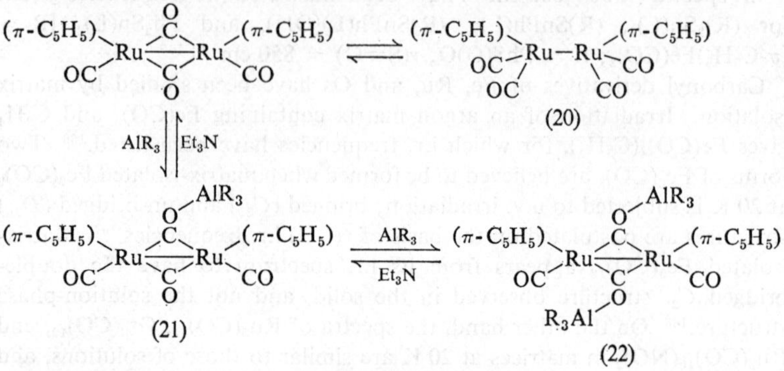

which $AlBu^i_3$ is co-ordinated to the bridging CO oxygen atom(s). The process is reversed on addition of Et_3N.

Unassigned i.r. data have been listed for several phenylazophenyl (azb) complexes, such as $[(azb)Ru(CO)_2Cl]_2$, $[(azb)(L)Ru(CO)_2Cl]$, $[(azb)Ru(CO)_2Cl_2]^-$, and $[(azb)Ru(CO)_2(chel)]$ (L = unidentate ligand; chel = bidentate chelate).[172] Values of ν(CO) and ν(NO) have been tabulated for compounds such as $(R_3M)Fe(CO)_{3-n}(NO)(L)_n$ [n = 0 or 1; R_3M = Ph_mSnCl_{3-m}, Cl_3Ge, Ph_3Pb, *etc.*; L = Ph_3P, $(PhO)_3P$, or $PhEt_2As$].[173]

Cyano- and isocyano-iron(II) complexes, $[Fe(CNR^1)_4(CN)_2]$, $[Fe(CNR^1)_4(CN,BF_3)_2]$, $[Fe(CN,BF_3)_6]^{4-}$, and $[Fe(CNR^2)_6]^{2+}$ (R^1 = H, Me, or Et; R^2 = Me or Et),[174] show i.r. ν(FeC) bands around 420—380 cm^{-1} and various δ(Fe—C≡N) modes in the region 625—470 cm^{-1}. Other authors [134] have quoted the i.r. range 380—300 cm^{-1} for ν(FeC) in $[Fe(CNMe)_6]Cl_2,3H_2O$.

Oxidation of $[Ru(MeNH_2)_6]I_2$ affords a cyano-compound [ν(CN) = 2040 cm^{-1}], probably $Ru(CN)_3,3H_2O$.[175]

Assignment of vibrations involving ^{15}N compared with those in ^{14}N analogues has been used [176] in stereochemical examination of reactions of *trans*-$[RuCl(N_2)(diars)_2][PF_6]$.

[171] Sr. A. Alich, N. J. Nelson, and D. F. Shriver, *Chem. Comm.*, 1971, 254.
[172] M. I. Bruce, M. Z. Iqbal, and F. G. A. Stone, *J. Chem. Soc.* (*A*), 1971, 2820.
[173] M. Casey and A. R. Manning, *J. Chem. Soc.* (*A*), 1971, 256.
[174] D. Hall, J. H. Slater, B. W. Fitzsimmons, and K. Wade, *J. Chem. Soc.* (*A*), 1971, 800.
[175] W. R. McWhinnie, J. D. Miller, J. B. Watts, and D. Y. Waddan, *Chem. Comm.*, 1971, 629.
[176] P. G. Douglas, R. D. Feltham, and H. G. Metzger, *J. Amer. Chem. Soc.*, 1971, 93, 84.

X-Ray photoelectron spectroscopy (with other techniques) has shown [177a] that oxidation of $[Ru(en)_3]^{2+}$ affords a di-imine chelate of Ru^{II} (23) [δ(=N—H) = 1199; δ(=C—H) = 837 cm^{-1}; ν(C=N) obscured], rather than a Ru^{VI} amide as previously believed.

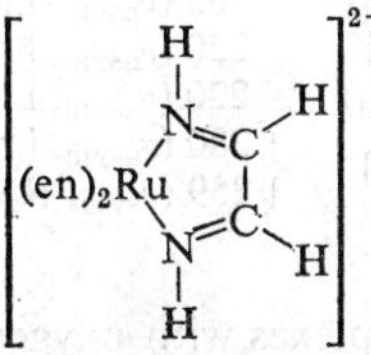

(23)

In $[Ru(NH_3)_5(Me_2SO)][PF_6]_2$ and three deuteriated analogues, two i.r. bands in the region 450—400 cm^{-1} are attributed to ν(RuN); the Me_2SO is said to be S-bonded.[177b]

Thermolysis of iron sulphates has been followed by i.r. spectroscopy.[178]

Coupling of L-histidine methyl ester (or several peptide esters containing L-histidine) to the propionic acid side-chains of ferric protoporphyrin IX chloride affords disubstituted derivatives. These show ν(Fe–ligand) at 385—381 cm^{-1}, but the absence of bands attributable to ν(Fe—O—Fe) at 903 and 840 cm^{-1} indicates that oxygen-carrying dimers are not formed.[179]

Solutions of *Clostridium pasteurianum* nibredoxin show Raman bands at 365 and 314 cm^{-1}, assigned respectively to ν_{asym} and ν_{sym} of a FeS_4 tetrahedron.[180] Unassigned i.r. data have been given for compounds $[Fe(RR'dtc)_2(tfd)]$ containing FeS_6 groups [RR'dtc = an *NN'*-dialkyl dithiocarbamato-ligand; tfd = bis(perfluoromethyl)-1,2-dithieten].[181]

Iron–ligand stretching and other modes have been assigned in the i.r. (500—200 cm^{-1}) and Raman (400—170 cm^{-1}) spectra of complexes $FeL_4(ClO_4)_3$ and FeL_2X_3 (L = Ph_3AsO or Ph_3PO; X = Cl, Br, NO_3, or NCS).[182] The vibrational data together with conductance and e.s.r. measurements, show that the perchlorates are ionic, $[FeL_4](ClO_4)_3$ [showing, *e.g.* ν(Fe—$OAsPh_3$) at 435—400 cm^{-1}], while the nitrates [ν(Fe—ONO_2) at 320 and 280 cm^{-1}] and isothiocyanates [ν(Fe—NCS) at 318 and 280 cm^{-1}] are believed to be five-co-ordinate species. The halogeno-compounds, however, are formulated as [*trans*-FeL_4X_2][FeX_4];[182] analogous structures are proposed by the same authors [183] for compounds $Fe(Me_2SO)_2X_3$ (Table 18).

177 (*a*) B. C. Lane, J. E. Lester, and F. Basolo, *Chem. Comm.*, 1971, 1618; (*b*) C. V. Senoff, E. Maslowski, and R. G. Goel, *Canad. J. Chem.*, 1971, **49**, 3585.
178 E. V. Margulis, M. M. Shokarev, L. A. Savchenko, N. I. Kopylov, and L. I. Beisekeeva, *Russ. J. Inorg. Chem.*, 1971, **16**, 392.
179 A. van der Heijden, H. G. Peer, and A. H. A. van den Oord, *Chem. Comm.*, 1971, 369.
180 T. V. Long, T. M. Loehr, J. R. Allkins, and W. Lovenberg, *J. Amer. Chem. Soc.*, 1971, **93**, 1809.
181 L. H. Pignolet, R. A. Lewis, and R. H. Holm, *J. Amer. Chem. Soc.*, 1971, **93**, 360.
182 S. A. Cotton and J. F. Gibson, *J. Chem. Soc.* (*A*), 1971, 859.

Table 18 ν(FeX) *assignments* (cm^{-1}) *for* FeL_2X_3 *complexes* (L = Ph_3PO, Ph_3AsO, *or* Me_2SO)

Compound	*Cation*	*Anion*
[*trans*-$Fe(Ph_3PO)_4Cl_2$][$FeCl_4$]	288 (ν_{asym}, i.r.)	336 (ν_1, R); 378 (ν_3, i.r.)
[*trans*-$Fe(Ph_3AsO)_4Cl_2$][$FeCl_4$]	285 (ν_{asym}, i.r.)	
[*trans*-$Fe(Ph_3PO)_4Br_2$][$FeBr_4$]	220 (ν_{asym}, i.r.)	206 (ν_1, R); 290 (ν_3, i.r.)
[*trans*-$Fe(Ph_3AsO)_4Br_2$][$FeBr_4$]	220 (ν_{asym}, i.r.)	
[*trans*-$Fe(Me_2SO)_4Cl_2$][$FeCl_4$]	280 (ν_{asym}, i.r.); 259 (ν_{sym}, R)	336 (ν_1, R); 108 (ν_2, R); 380 (ν_3, i.r.); 136 (ν_4, i.r.)

In a series of iron(III) complexes with oxygen-donating ligands [Me_2SO, $CO(NH_2)_2$, $HCONMe_2$, or phenazone], ν(FeO) contributes to bands in the 470—300 cm^{-1} region; for the Me_2SO complexes, the narrower range 470—455 cm^{-1} is quoted.[183] Complexes of 4-substituted pyridine *N*-oxides of types $[FeL_6]^{3+}$ and $[FeL_4Cl_2]^+$ have ν(FeO) in the 448—372 cm^{-1} range.[183b] In iron(III) phosphonates [FeL_3] (L = isopropyl hydrogen methylphosphonate, ethyl hydrogen ethylphosphonate, or methyl hydrogen methylphosphonate), 4—6 i.r. bands in the wide range 587—237 cm^{-1} are described as ν(FeO) modes.[184]

Assignments (i.r. and Raman) for the FeO_6 skeletal vibrations have been suggested for $Fe(O_2PX_2)_3$ (X = F or Cl).[184b] O-Co-ordination of the ligand has been indicated from an i.r. study of $[FeCl_2(HCONH_2)_2]_n$ and confirmed by a single-crystal *X*-ray study.[184c]

The observation of two i.r. ν(Ru—py) bands (256 and 244 cm^{-1}) indicates that $[RuCl_3(py)_3],0.5C_6H_6$ has *fac* geometry.[185]

$TlFeF_6$ is of cubic symmetry and shows only the two expected i.r.-active internal modes ν(FeF) at 448 cm^{-1} and δ(FeF) at 268 cm^{-1}.[186] The spectra of corresponding K, Rb, Cs, Ag, Na, and Li salts are more complex on account of their less-symmetrical structures.

The presence of only one ν(FeX) mode (380 cm^{-1}, X = Cl; 293 cm^{-1}, X = Br) and/or two ν(CO) modes [2063—2045 and 2030—2000 cm^{-1}] in the i.r. spectra of complexes $Fe(CO)_2LX_4$ [L = $MeC(CH_2PPh_2)_3$; X = Cl, Br, or I] has been used to deduce C_{2v} co-ordination geometry about the Fe atom in these compounds.[187] Other data on iron–halogen frequencies [24, 25] have already been mentioned.

Assignments of ruthenium– and osmium–halogen vibrations are listed in Table 19. Far-i.r. data (361, 305, and 245 cm^{-1}) have also been reported

[183] (*a*) S. A. Cotton and J. F. Gibson, *J. Chem. Soc.* (*A*), 1971, 1690; (*b*) N. M. Karayanis, J. T. Cronin, C. M. Mikulski, L. L. Pytlewski, and M. M. Labes, *J. Inorg. Nuclear Chem.*, 1971, **33**, 4344.

[184] (*a*) E. P. Scheide and G. G. Guilbault, *Spectroscopy Letters*, 1971, **4**, 329; (*b*) J. Pebler and K. Dehnicke, *Z. Naturforsch.*, 1971, **26b**, 747; (*c*) G. Constant, J. C. Daran, and Y. Jeanin, *J. Inorg. Nuclear Chem.*, 1971, **33**, 4209.

[185] J. Chatt, G. J. Leigh, and A. P. Storace, *J. Chem. Soc.* (*A*), 1971, 1380.

[186] S. Shearer-Turrell, A. Tressaud, and J. Portier, *J. Mol. Structure*, 1971, **7**, 289.

[187] H. Behrens, E. Lindner, and H.-D. Feilner, *Z. anorg. Chem.*, 1971, **385**, 325.

Table 19 *Ruthenium– and osmium–halogen stretching modes* (cm^{-1})

Compound	ν(MX)	*Ref.*
fac-[$(PhMe_2P)_2RuCl_3(NO)$]	324, 314, 290, 278	188
[$(Me_2SO)_4RuCl_2$][a, b]	345	189
[$(o\text{-}C_6Cl_4O_2)RuCl_2(PPh_3)_2$]	342	190
Cs[$RuCl_4(PhNH_2)_2$][c]	304	191*a*[d]
Cs[$RuBr_4(PhNH_2)_2$][c]	221	191*a*[d]
$RuCl_3(NH_3)_3$	311, 290, 283	191*b*
[$RuCl_3(PhSR)_3$][e]	338—328, 318—312	185[f]
[$RuBr_3(PhSR)_3$][e]	268—260, 238—230	185[f]
[$RuCl_2(dis)_2$]	340—328	185[f]
[$RuBr_2(dis)_2$]	238	185[f]
[$RuCl_2(dis)_2$]+	346	185[f]
[$RuCl_2(CO)_2(dis)$]	344	185[f]
[$RuCl_3\{MeC(CH_2SEt)_3\}$]	318, 282	185[f]
[$RuCl_2(CO)\{MeC(CH_2SEt)_3\}$]	308, 282	185[f]
[$RuCl_2\{PhP(CH_2CH_2PPh_2)_2\}$]	284, 279	20
[$OsCl_4\{PhP(CH_2CH_2PPh_2)_2\}$]	317, 313, 301	20
[$OsCl_2(Y)(PR_3)_3$]	303—240	192[g]
[$OsCl_2(Y)_2(PR_3)_2$]	314—277	192[g]
[$OsBr_2(Y)(PR_3)_3$]	200—173	192[g]
[$OsBr_2(Y)_2(PR_3)_2$]	234—193	192[g]
[$OsCl_2(NO_2)(NO)(PMe_2Ph)_2$]	321, 302	192[g]
[$Os_2Cl_3(PR_3)_6$]+	286—273	192[g]
[$H_2OsCl_2(PEt_2Ph)_3$]	284, 254	157
[$HOsCl(N_2)(PEt_2Ph)_3$]	292	157
[$HOsCl(N_2)(PMe_2Ph)_3$]	288	157
[$HOsBr(N_2)(PMe_2Ph)_3$]	189	157
[$HOsCl(CO)(PMe_2Ph)_3$]	268	157

[a] *trans* geometry inferred from the ν(MX) i.r. data. [b] Mid-i.r. evidence suggests both S- and O-co-ordinated Me_2SO. [c] The observation of only one ν(RuX) band and one ν(RuN) mode (235 cm^{-1}) indicates that the anions are of *trans* geometry. [d] Unassigned far-i.r. data listed for several other complexes of type A[$RuX_4(an)_2$], [$Ru(an)_6$]Br_2, and $Ru(an)_4Br_2$ [A^+ = Cs, Me_4N, $PhNH_3$, or C_5H_5NH; X = Cl, Br, or I; an = substituted aniline], [e] R = Me, Et, Pr^n, or Bu^n. [f] See also ref. 17 (Table 2); (dis) = $RSCH_2CH_2SR$, where R = Me, Et, Pr^n, or Ph. [g] Y = CO, MeNC, PhNC, MeCN, or PhCN; PR_3 = PEt_3, PMe_2Ph, PEt_2Ph, *etc.* (28 complexes in all).

for $Ru(PPh_3)_2(SO_2)Cl_2,2Me_2CO$; these bands presumably include ν(RuCl), but their assignment is not clear.[193a] Unassigned i.r. data in ν(RuCl) and ν(RuS) regions have been listed for $RuCl_3$ complexes of 1,4-thioxan, 1,4-dithian, and 1,2-disubstituted thioethanes.[193b]

188 R. E. Townsend and K. J. Coskran, *Inorg. Chem.*, 1971, **10**, 1661.
189 B. R. James, E. Ochiai, and G. L. Rempel, *Inorg. Nuclear Chem. Letters*, 1971, **7**, 781.
190 A. L. Balch and Y. S. Sohne, *J. Organometallic Chem.*, 1971, **30**, C31.
191 (*a*) D. L. Key, L. F. Larkworthy, and J. E. Salmon, *J. Chem. Soc.* (*A*), 1971, 2583; (*b*) F. Bottomley and S. B. Tong, *Canad. J. Chem.*, 1971, **49**, 3739.
192 J. Chatt, D. P. Melville, and R. L. Richards, *J. Chem. Soc.* (*A*), 1971, 1169.
193 (*a*) S. Cenini, A. Fusi, and G. Capparella, *J. Inorg. Nuclear Chem.*, 1971, **33**, 3577; (*b*) D. A. Rice and C. W. Timewell, *Inorg. Chim. Acta*, 1971, **5**, 683.

8 Cobalt, Rhodium, and Iridium

Metal–hydrogen stretching frequencies involving these elements are collected in Table 20. Compounds $(\pi\text{-}C_5Me_5)_2M_2Cl_3H$ (M = Rh or Ir) are

Table 20 *ν*(MH) *assignments for* Co, Rh, *and* Ir *compounds* (cm^{-1})

Compound	*ν*(MH) *or ν*(MD)	*Ref.*
$HCo(CO)(PEt_2Ph)_3$	1910	194
$H_2Co(PPh_3)_3(SiR_3)$	*ca.* 1945	195
(R = F or OEt)		
$HCo(dmg)_2(PBu^n_3)$	2240	196
$DCo(dmg)_2(PBu^n_3)$	1680	
(dmg = dimethylglyoximato)		
$HCo[PPh(OEt)_2]_4$ [a]	2017	197
$H_3Co(PPh_3)_3$	1940, 1890, 1760	198
$HRhCl_2(PBu^t_3)_2$ [b]	1938	199
$H_2RhCl(PBu^t_3)_2$	2242, 2227	
$HRh(PPh_3)_2Cl_2$	2105	
$HRh(PPh_3)_2(SnBu^n_3)Cl$	2052	
$HRh(PPh_3)_2[SiMe(OSiMe_3)_2]Cl$	2122	
$HRh(PPh_3)_2(GeMe_3)Cl$	2080, 2035	
$HRh(PPh_3)_2(GeEt_3)Cl$	2107, 2062	200
$HRh(AsPh_3)_2(GeMe_3)Cl$	2057, 2025	
$HRh(AsPh_3)_2(GeEt_3)Cl$	2114, 2042	
$HRh(PPh_3)_2(GeCl_3)Cl$	2123, 2098	
$HRh(cod)(GeCl_3)Cl$	2000	
cis-$[H_2Rh\{P(OMe)_3\}_4][BPh_4]$	1976	
cis-$[H_2Rh(PBu^n_3)_4][BPh_4]$	2020	
cis-$[H_2Rh\{P(OPh)_3\}_4][BPh_4]$	1989	201[c]
trans-$[HRh\{P(OMe)_3\}_3Br][BPh_4]$	2072	
trans-$[HRh\{P(OMe)_3\}_3I][BPh_4]$	2081	
$[H_2BH_2Rh(CO)\{P(C_6H_{11})_3\}_2]$	1960[d]	202
$[HRhX_2(PBu^t_2R)]$	1946—1936	203[c]
(X = Cl or Br; R = Me, Et, or Pr^n)		
$[HRhCl_2(PBu^tPr^n_2)(L)]$	2198—2083	
(L = MeCN, MeNC, $P(OMe)_3$, or py)		
$[H_2RhCl(L)_2]$	2212—2066[e]	203
(L = PBu^t_3, PBu^t_2Me, or PBu^tMePh)		
$[H_2Rh(dtt)(PPh_3)_2]$	2080, 2056, 2036	
$[HIr(dpt)(PPh_3)_2]$	2160	155[f]
$[H_2Ir(dpt)(PPh_3)_2]$	2164, 2140	
$[HIr(PMe_2Ph)_3py_2][ClO_4]_2$	2205	
$[HIr(PMe_2Ph)_3(phen)][ClO_4]_2$	2235	204[g]
$[HIr(PMe_2Ph)_3(CO)_2][BPh_4]_2$	1939	
$[HIr(PMe_2Ph)_3I_2]$	2070	
$[HIr(PPh_3)_2(NO_3)_2]$	2274, 2259	205

194 M. Aresta, C. F. Nobile, M. Rossi, and A. Sacco, *Chem. Comm.*, 1971, 781.
195 N. J. Archer, R. N. Haszeldine, and R. V. Parish, *Chem. Comm.*, 1971, 524.
196 G. N. Schrauzer and R. J. Holland, *J. Amer. Chem. Soc.*, 1971, **93**, 1505.
197 D. D. Titus, A. A. Orio, R. E. Marsh, and H. B. Gray, *Chem. Comm.*, 1971, 322.
198 A. Yamamoto, S. Fitazume, L. S. Pu, and S. Ikeda, *J. Amer. Chem. Soc.*, 1971, **93**, 371.
199 C. Masters, W. S. McDonald, G. Raper, and B. L. Shaw, *Chem. Comm.*, 1971, 210.
200 F. Glockling and G. C. Hill, *J. Chem. Soc.* (*A*), 1971, 2137.
201 L. M. Haines, *Inorg. Chem.*, 1971, **10**, 1693.
202 L. Vaska, W. V. Miller, and B. R. Flynn, *Chem. Comm.*, 1971, 1615.
203 C. Masters and B. L. Shaw, *J. Chem. Soc.* (*A*), 1971, 3679.
204 B. L. Shaw and R. M. Slade, *J. Chem. Soc.* (*A*), 1971, 1184.
205 D. N. Cash and R. O. Harris, *Canad. J. Chem.*, 1971, **49**, 867.

Table 20 (*cont.*)

Compound	ν(MH) *or* ν(MD)	*Ref.*
$[HIr(pip)_4I]I,H_2O$	2190	206[h]
$[HIr(pip)_4Br]Br,H_2O$	2195	
$[HIr(pip)_4I][ClO_4]$	2193	
$[HIr(pip)_4Br][ClO_4]$	2194	
$[HIr(pip)_4Cl][ClO_4]$	2199	
$[HIr\{P(OAr)_3\}_3Cl_2]$	2205—2180	207[i]
$[HIr\{P(OAr)_3\}_3Br_2]$	2200—2165	
$[HIr\{P(OAr)_3\}_3I_2]$	2180—2145	
$[HIr\{P(OPh)_3\}_4]$	2055	
$[HIr(PPh_3)_3(CO)]$	2120	
$[HIr\{P(OAr)_3\}_2X(chel)]$	2090—2070	207, 208[i]
$[HIr\{P(OPh)_3\}_3(chel)]$	2060	
$[H_2Ir(CO)(MeCN)(PPh_3)_2]^+$	2140	209
$[HIr(C_5H_5)(PPh_3)_2]I$	2150	210
$[HIr(PBu^t_2R)_2Cl_2]$[j]	*ca.* 2000	211
$[H_3Ir\{PhP(CH_2CH_2PPh_2)_2\}]$	2030, 1975	20
$[H_2Ir(CO)(CS)\{P(C_6H_{11})_3\}_2][ClO_4]$	2151, 2129, 2143	212[k]
$[D_2Ir(CO)(CS)\{P(C_6H_{11})_3\}_2][ClO_4]$	1540, 1490	
$[H_5Ir(PEt_2Ph)_2]$[l]	2240	213
$[H_3Ir(PEt_2Ph)_2(PPh_3)]$	2049, 1751	
$[H_3Ir(PEt_2Ph)_2\{P(OMe)_3\}_3]$	2020, 1764	
$[H_3Ir(PEt_2Ph)_2(AsMe_2Ph)]$	2080, 1727	
$[H_3Ir(PEt_2Ph)_2(SbPh_3)]$	2096, 2008, 1733	
$[H_3Ir(PEt_2Ph)_2(MeNC)]$	2016, 1739	
$[H_3Ir(PEt_3)_2(AsMe_2Ph)]$	2092, 2000, 1721	
$[HIr(CO)X_2(PBu^tR_2)]$	2240—2200	214[c, m]
$[HIr(CO)Cl_2(PBu^t_2R)]$	2270—2230	
$[HIr(cod)Cl_2]_2$	2260	215[n]
$[HIr(cod)X_2(L)_2]$	2257—2215	
$[HIrCl_2(L)_3]$	2251—2213	216[c, o]

[a] Single-crystal *X*-ray study shows Co co-ordination to be approx. trigonal bipyramidal with H atom axial. [b] Only one ν(RhCl) band observed (340 cm^{-1}); a square-pyramidal structure (H axial, *trans* Cl_2) suggested (the crystal structure of the $PBu^tPr^n_2$ analogue is given.) [c] The geometries given have been suggested by the spectroscopic data. [d] $\nu(BH_2Rh)$ bridging mode, obscured in $[H_2BH_2Rh(CO)(PPh_3)_2]$ and in $[H_2BH_2Ir(CO)(PR_3)_2]$ ($R = C_6H_{11}$) [e] Two bands in this range in each case. [f] dtt = 1,3-di-*p*-tolyltriazenido, dpt = 1,3-diphenyltriazenido: Ar—N⩵N⩵N—Ar. [g] Assignments given for other skeletal modes; configurations suggested on the basis of these, n.m.r., and other data. [h] pip = piperidine; data given for KBr discs; solvent sensitivity shows configurations are *trans*. [i] Ar = Ph, *p*-tolyl, *m*-tolyl, *o*-tolyl, or *p*-ClC_6H_4. [j] R = Me, Et, or Pr^n; only one ν(IrCl) band observed (*ca.* 315 cm^{-1}); square-pyramidal structures (H axial, *trans* Cl_2) suggested. [k] In a series of compounds of this type, ν(CS) was at *ca.* 1320 cm^{-1}. [l] Previously thought to be a trihydride. [m] X = Cl or Br; R = Me, Et, Pr^n, or Bu^n. [n] X = Cl, Br, I, or CN; L = PPh_3, $PMePh_2$, or $AsPh_3$ (8 compounds). [o] L = PEt_3, PPr^n_2Ph, or PBu^n_2Ph.

[206] E. R. Birnbaum, *J. Inorg. Nuclear Chem.*, 1971, **33**, 3031.
[207] E. W. Ainscough, S. D. Robinson, and J. J. Levison, *J. Chem. Soc.* (*A*), 1971, 3413.
[208] M. A. Bennett and R. Charles, *Austral. J. Chem.*, 1971, **24**, 427.
[209] G. R. Clark, C. A. Reed, W. R. Roper, B. W. Skelton, and T. N. Waters, *Chem. Comm.*, 1971, 758.
[210] H. Yamazaki, *Bull. Chem. Soc. Japan*, 1971, **44**, 582.
[211] C. Masters, B. L. Shaw, and R. E. Stainbank, *Chem. Comm.*, 1971, 209.
[212] M. J. Mays and F. P. Stefanini, *J. Chem. Soc.* (*A*), 1971, 2747.
[213] B. E. Mann, C. Masters, and B. L. Shaw, *J. Inorg. Nuclear Chem.*, 1971, **33**, 2195.
[214] B. L. Shaw and R. E. Stainbank, *J. Chem. Soc.* (*A*), 1971, 3716.
[215] R. N. Haszeldine, R. J. Lunt, and R. V. Parish, *J. Chem. Soc.* (*A*), 1971, 3711.
[216] P. R. Brookes, C. Masters, and B. L. Shaw, *J. Chem. Soc.* (*A*), 1971, 3756.

$$(\pi\text{-}C_5Me_5)_2ClM\underset{Cl}{\overset{H}{\diagdown\diagup}}MCl(\pi\text{-}C_5Me_5)_2$$

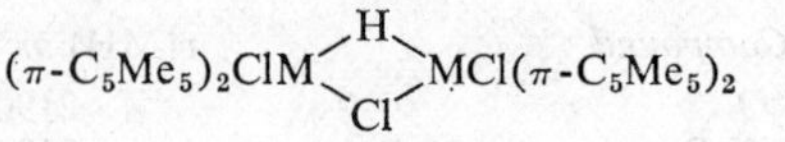

(24)

formulated as in (24); $\nu(MH)_b$ is at 1151 (M = Rh) and 1155 cm^{-1} (M = Ir), shifting to 812 and 819 cm^{-1}, respectively,[217] on deuteriation. Metal–halogen assignments in these compounds are (cm^{-1}):

$\nu(MCl)_t$ = 280 (Rh), 290 (Ir);
$\nu(MCl)_b$ = 255 (Rh), 256 (Ir).

I.r., Raman, and n.m.r. spectroscopy have been used to study the reaction of NaCN with $[Rh(CO)_2Cl]_2$; evidence for three intermediate species in the formation of $[RhH(CN)_5]^{3-}$ salts in solution was obtained, but vibrational assignments were not given.[218]

The existence of the *cis*-isomer of $[Co(CN)_3(NH_3)_3]$ has now been established by X-ray methods.[219] Two $\nu(CN)$ modes are found in both i.r. (2133 and 2127) and Raman (2136 and 2124 cm^{-1}) spectra, in agreement with C_{3v} symmetry. Partial vibrational data have also been given for the 'super-complexes' $[Co(CN)_3(NH_3)_3]MX_n$ $[MX_n = Hg(NO_3)_2, Hg(ClO_4)_2, H_2O,0.75AgNO_3,$ or $AgNO_3]$; these show [219] bands due to the bridging cyano groups (Co—C≡N—M) between 2180 and 2150 cm^{-1}.

Salts of the new anion *cis*-$[Co(SO_3)_2(CN)_4]^{5-}$ have been compared with those of the isomeric *trans*-anion in terms of the number and frequencies of their i.r. and Raman modes.[220] Some of the range (cm^{-1}) quoted for these compounds include:

$\nu(CN)$	2149—2109	$\nu_{asym}(SO_3)$	1100—1057
$\nu_{sym}(SO_3)$	1001— 956	$\delta_{sym}(SO_3)$	654— 621
$\delta(CoN)$ }	559— 504	$\nu(CoC)$	463— 387
$\delta_{asym}(SO_3)$ }		$\nu(CoS)$	271— 187

I.r. spectroscopy has been used to study olefin hydroformylation reactions which produce compounds of the type $[Rh(CO)_2(O_2CR)_2]$,[221] and CO insertion reactions into π-allyl complexes of Rh and Ir.[222]

Spectra of $K_3[Co(CN)_5X]$ (X = NCS, SCN, N$_3$, or NCSe) have been compared.[223]

Variable-temperature i.r. studies (− 77 and + 40 °C) indicate that the acyl complexes $FCH_2\cdot C(O)\cdot Co(CO)_3(PPh_3)$ and $F_2CH\cdot C(O)\cdot Co(CO)_3$-

---

[217] C. White, D. S. Gill, J. W. Kang, H. B. Lee, and P. M. Maitlis, *Chem. Comm.*, 1971, 734.
[218] R. A. Jewsbury and J. P. Maher, *J. Chem. Soc. (A)*, 1971, 2847.
[219] H. Siebert, C. Siebert, and K. Wieghardt, *Z. anorg. Chem.*, 1971, 380, 30.
[220] H. Siebert, C. Siebert, and S. Thym, *Z. anorg. Chem.*, 1971, 383, 165.
[221] B. Heil, L. Markó, and G. Bor, *Chem. Ber.*, 1971, 104, 3418.
[222] C. K. Brown, W. Mowat, G. Yagupsky, and G. Wilkinson, *J. Chem. Soc. (A)*, 1971, 850.
[223] D. F. Gutterman and H. B. Gray, *J. Amer. Chem. Soc.*, 1971, 93, 3364.

(PPh_3) each exist in two isomeric forms arising from restricted rotation about the Co—C(acyl) bonds.[224]

Reaction between [(π-C_5H_5)$_2$Co] and Et_2O–HX solutions gives products identified as [(π-C_5H_5)$_2$Co]$_2$[CoX_4] (X = Cl, Br, or I) on the basis of i.r. and other evidence.[225] In the presence of Ph_3P the products are [(π-C_5H_5)$_2$Co]X, whereas with HF [(π-C_5H_5)$_2$Co][HF_2] is formed.

I.r. frequencies of the BF_4^- anion have been given for some new diene complexes of types [M(diene)$_2$][BF_4] and [M(diene)L_2][BF_4] (M = Rh or Ir; diene = 1,5-C_8H_{12} or bicyclo[2,2,1]heptadiene; L = MeCN, Ph_3P, *etc.*).[226]

In compounds M(PPh_3)$_2$(CO)(NCE) (M = Rh or Ir; E = O, S, or Se), i.r. measurements have shown that the NCE ligands are N-bonded, giving the first examples of NCSe N-bonded to a d^8 metal.[227]

Metal–nitrogen stretching modes have been assigned in [(Me_3Si)$_2$N]$_2$Co (i.r., 362 cm^{-1}),[228] aqueous [Co(NH_3)$_2$(NO_2)$_4$]$^+$ salts (Raman, 480 cm^{-1}),[229] methylcobaloximes of type Co(HON=C(Me)·C(Me)=NO^-)$_2$Me(base) [i.r., 516 ± 4 and 436 ± 8 cm^{-1}; ν(Co—Me) at 330—325 cm^{-1}],[230] $IrCl_4$phen (i.r., 234 and 227 cm^{-1}),[231] and [IrX_3phen]$_2$ (i.r., 280—240 cm^{-1}; X = Cl or Br).[231]

I.r. data have also been given for [CoL_6][ClO_4]$_2$ (L = Me_2SO or $HCONMe_2$),[183] three geometrical isomers of [(CO_3)Co(NH_3)$_2$(glycinato)],[232] and the sulphide Co_3(π-C_5H_5)$_3S_2$ (25).[233]

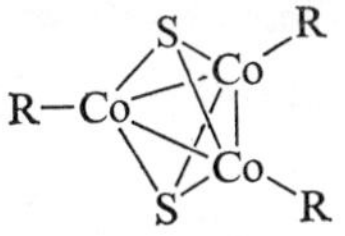

(R = π-C_5H_5)

(25)

Amine oxide complexes [234] [Co(R_3NO)$_2X_2$] (R = Et or Pr^n; X = Cl, Br, I, NCS, or ClO_4) show ν(NO) in the range 982—930 cm^{-1} and bands presumed to be ν(CoO) at 590—545 cm^{-1}.

Two classes of iridium *p*-tolylsulphinates (pts) have been prepared. Compounds (pts)Ir(CO)(PPh_3)$_2$ and (pts)Ir(Me)(CO)(PPh_3)$_2$I are oxygen-

224 E. Lindner, H. Stich, K. Geibel, and H. Kranz, *Chem. Ber.*, 1971, **104**, 1524.
225 M. Van den Akker and F. Jellinek, *Rec. Trav. chim.*, 1971, **90**, 1101.
226 M. Green, T. A. Kuc, and S. H. Taylor, *J. Chem. Soc.* (*A*), 1971, 2334.
227 N. J. Destefano and J. L. Burmeister, *Inorg. Chem.*, 1971, **10**, 998.
228 D. C. Bradley and K. J. Fisher, *J. Amer. Chem. Soc.*, 1971, **93**, 2058.
229 B. Belsot and P. Devrainne, *Compt. rend.*, 1971, **272**, *C*, 1454.
230 D. Benlian and G. Hernandorena, *Compt. rend.*, 1971, **272**, *C*, 2001.
231 J. A. Broomhead and W. Grumley, *Inorg. Chem.*, 1971, **10**, 2002.
232 S. Kanazawa and M. Shibata, *Bull. Chem. Soc. Japan*, 1971, **44**, 2424.
233 M. Sorai, A. Kosaki, H. Suga, S. Seki, T. Yoshida, and S. Otsuka, *Bull. Chem. Soc. Japan*, 1971, **44**, 2364.
234 D. W. Cunningham and M. O. Workman, *J. Inorg. Nuclear Chem.*, 1971, **33**, 3861.

bonded [ν(SO) at *ca.* 1080, ν_{asym}(SOIr) at 870—840 cm^{-1}], whereas S co-ordination is found for (pts)Ir$(CO)_2(PPh_3)_2$, (pts)Ir(O_2)(CO)$(PPh_3)_2$, and (pts)Ir(SO_4)(CO)$(PPh_3)_2$ [$\nu_{asym}(SO_2)$ at 1240—1140, $\nu_{sym}(SO_2)$ at 1060—1010 cm^{-1}].[235] In the dioxygen complex (pts)Ir(O_2)(CO)$(PPh_3)_2$,[235] $\nu(IrO_2)$ is at 835 cm^{-1}.

Metal–sulphur stretching frequencies (cm^{-1}) have been assigned in the i.r. spectra of *trans*-[MCl(CO)$(SO_2)(PPh_3)_2$] (295, M = Rh; 315, M = Ir),[236] and of [(π-$C_5H_5)_2$Mo$(SBu^n)_2$] (293 and 283).[237] In the complexes CoX_2L_2 (X = Cl, Br, or I) of the dithiolen ligand (26),[238] the sulphur atom of the thione group is the co-ordination site with ν(CoS) at *ca.* 310 cm^{-1}.

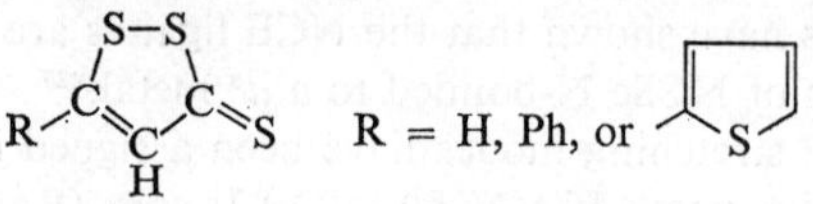

(26)

There have been a large number of reports of metal–halogen modes involving Co, Rh, or Ir; these are summarized in Table 21. References

Table 21 *Some cobalt–, rhodium–, and iridium–halogen vibrational assignments*

Compound	ν (cm^{-1})	*Ref.*
[(triphos)$CoCl_2$]	338, 322	20[a]
[(triphos)$RhCl_3$]	302, 278, 269	
[(triphos)RhCl]	274	
[(triphos)$IrCl_3$]	314	
[$CoCl_2L_2$]	335—324, 307, 285	239[b]
[$CoBr_2L_2$]	277—268, 250—243	
[CoI_2L_2]	*ca.* 245, *ca.* 237	
[(π-$C_5H_5)_2M(SR)_2CoCl_2$]	342—320, 294—290	237[c]
[(π-$C_5H_5)_2M(SR)_2CoBr_2$]	248—243, *ca.* 266	
[(π-$C_5H_5)_2M(SR)_2CoI_2$]	206—205, *ca.* 186	
fac-[$CoCl_3$(trias)]	348, 328, 320	240[d, e]
fac-[$RhCl_3$(trias)]	323, 302, 296	
mer-[$RhCl_3$(trias)]	339, 307, 262	
mer-[$RhBr_3$(trias)]	289, 286, 244	
mer-[$RhCl_2$I(trias)]	338	
mer-[$IrCl_3$(trias)]	321, 287, 246	
[$(Ph_3M)_2$RhHCl(GeR_3)]	334—284[f]	200[g]
[$RhHCl_2(PBu^tPr^n_2)_2$]	355—340[f]	203[e, h]
[$RhHCl_2(PBu^t_2R)_2$]		

[235] C. A. Reed and W. R. Roper, *Chem. Comm.*, 1971, 1556.
[236] L. Vaska, *Inorg. Chim. Acta*, 1971, **5**, 295.
[237] A. R. Dias and M. L. H. Green, *J. Chem. Soc. (A)*, 1971, 2807.
[238] F. Petillon and J. E. Guerchais, *Bull. Soc. chim. France*, 1971, 2455.
[239] E. A. Allen, N. P. Johnson, D. T. Rosevear, and W. Wilkinson, *J. Chem. Soc. (A)*, 1971, 2141.
[240] R. G. Cunninghame, R. S. Nyholm, and M. L. Tobe, *J. Chem. Soc. (A)*, 1971, 227.

Table 21 (*cont.*)

Compound	ν (cm^{-1})	*Ref.*
trans-$[RhCl_2(PBu^t_2R)_2]$	360—352, 348—344	203[h]
$[RhHCl_2(PBu^tPr^n_2)(L)]$	334—326	203[h]
$[RhH_2Cl(PBu^t_3)_2]$	264	203[h]
$[RhH_2Cl(PBu^tMePh)_2]$	244, 277	203[h]
$[RhCl(PBu^t_2H)_2]$	290, 229	203[h]
$[(o\text{-}Ph_2PC_6H_4CHMe)Rh(CO)Cl_2]_2$	*ca.* 230 (*trans* to C), 265 (*trans* to P), 289 (bridging, *trans* to CO)	241
$[(Me_2C{=}CHCMe_2CH_2OH)Rh(CO)Cl]_2$	300, 250[i]	242
$[(Ph_3P)Rh(CO)Cl]_2$	295[j]	243[k]
$[(Ph_3P)Rh(CO)Cl(L)]$	*ca.* 310[e]	243[k]
$[RhCl_2(C_8H_{15}O_2)(L)]_2$ [j, l]	*ca.* 330 (terminal), *ca.* 245 (bridging)	244[m]
$[Rh_2Cl_2(bhq)_4],0.25(CHCl_3)$[l]	233	245
$[Rh_2Cl_2(php)_4],2(CHCl_3)$[l]	218, 211	245
$[RhCl_3(dmp)]^n$	357, 343, 320	246
$[RhCl_3(dmp)(H_2O)],2MeOH$	345, 336, 323, 309	246
$[IrCl_3(dmp)]^n$	339, 320, 314	246
mer-$[RhCl_3(PhSR)_3]$[e]	*ca.* 330, *ca.* 305, *ca.* 295	247[o]
fac-$[RhCl_3(L)_3]$[e]	*ca.* 315, *ca.* 305	247[o]
trans-$[RhCl_2(EtSCH_2CH_2SEt)_2]X$[e]	348	247[o]
fac-$[IrCl_3\{MeC(CH_2SEt)_3\}]$	312, 296	247[o]
fac-$[IrCl_3\{S(CH_2CH_2SPr^n)_2\}]$[e]	320, 300	247[o]
$[IrCl_3\{S(CH_2CH_2SPh)_2\}]$	310	247[o]
$[IrCl_4(phen)]$	364, 339, 330, 317	231
$[IrCl_3(phen)]_2$	345, 333, 321, 298	231
$[IrBr_3(phen)]_2$	234	231
$[IrCl_2(CH_2{=}CMeCH_2)(CO)(PhMe_2P,As)_2]$	*ca.* 308 (*trans* to CO), 261 (*trans* to R)	248
$[IrCl_2(MeCHClCH_2)(CO)(PhMe_2P)_2]$	311 (*trans* to CO), 260 (*trans* to R)	248
$[Ir(CO)(PPh_3)_2(X)(Y)Cl]$[p]	345—329	205
$[(C_8H_{12})\{(allyl)_2O\}IrCl]_2$	305, 275[j]	249
$[Ir(piperidine)_4Cl_2](ClO_4)$	315	206
$[HIr(piperidine)_4Cl](ClO_4)$	241	206
$[HIr(piperidine)_4Br]X$[q]	148—143	206
$[HIr(piperidine)_4I]Y$[q]	114—109	206
$[IrCl_2(X)(PMe_2Ph)_3]$[r]	*ca.* 330, *ca.* 320	204
$[IrCl_2(L)(PR_2Ph)_3]Y$[r, s]	334—320[e]	204
$[IrClI_2(PMe_2Ph)_3]$	277	204

241 M. A. Bennett, S. J. Gruber, E. J. Hann, and R. S. Nyholm, *J. Organometallic Chem.*, 1971, **29**, C13.
242 L. K. Atkinson and D. C. Smith, *J. Chem. Soc. (A)*, 1971, 3592.
243 D. F. Steele and T. A. Stephenson, *Inorg. Nuclear Chem. Letters*, 1971, **7**, 877.
244 J. A. Evans, D. R. Russell, A. Bright, and B. L. Shaw, *Chem. Comm.*, 1971, 841.
245 M. Nonoyama and K. Yamasaki, *Inorg. Nuclear Chem. Letters*, 1971, **7**, 943.
246 G. C. Kulasingam, *Inorg. Chim. Acta*, 1971, **5**, 180.
247 J. Chatt, G. J. Leigh, A. P. Storace, D. A. Squire, and B. J. Starkey, *J. Chem. Soc. (A)*, 1971, 899.
248 A. J. Deeming, B. L. Shaw, and R. E. Stainbank, *J. Chem. Soc. (A)*, 1971, 374.
249 G. Pannetier, R. Bonnaire, and P. Fougeroux, *J. Organometallic Chem.*, 1971, **30**, 411.

Table 21 (*cont.*)

Compound	ν (cm^{-1})	*Ref.*
$[(R^1CO)IrCl(X)(CO)(L)_2]$[t]	330—215[l]	250
$[R^2IrCl_2(CO)(PPh_3)_2]$[t]		
$[IrHCl_2(CO)(PBu^t_nR_{3-n})_2]$	310—298, 265—257	214[e, o]
trans-$[IrCl(CO)(PBu^t_nR_{3-n})_2]$	308—300	
$[HIrCl_2L^1_3]$	275—271, 243—234	216[e, v]
fac-$[IrCl_3(PR^1_3)_2]$	307—244[u]	
fac-$[IrCl_3(PR^2_2Ph)_3]$		
fac-$[IrICl_2(PEt_3)_3]$	278, 257	
fac-$[IrI_2Cl(PEt_3)_3]$	263	
$[IrCl_3(CO)(L^2)_2]$	320—276[v]	
$[IrCl_3(PMe_2Ph)_2(L^3)]$	298—246[v]	
$[(cod)IrCl(PPh_2R^1)]$	330—249[l]	215[e, w]
$[(cod)IrHCl_2]_2$		
$[(cod)IrHCl_2(L^1)]$		
$[(cod)IrX_2Y(L^2)]$		
$[R^2H][(cod)IrCl_4]$		
$[quinH][(cod)IrCl_2Br_2]$		

[a] triphos = $PhP(CH_2CH_2PPh_2)_2$. [b] L = piperidine, morpholine, or (probably N-co-ordinated) thiomorpholine. [c] M = Mo or W; R = alkyl; similarly, ν(FeCl) in $(\pi\text{-}C_5H_5)_2Mo(SMe)_2FeCl_2$ is at 205 cm^{-1}. [d] trias = $(o\text{-}Me_2AsC_6H_4)_2AsMe$. [e] I.r. data used in stereochemical assignments. [f] Two bands in this range. [g] R = Me or Et; M = P or As. [h] R = Me, Et, or Pr^n; L = MeNC, MeCN, py, or $P(OMe)_3$. [i] These values are said to be consistent with bridging Rh—Cl groups *trans* to CO; both the oxygen atom and the olefinic double bond are believed to be co-ordinated to Rh. [j] I.r. data used in support of halogen-bridged dimeric structure. [k] L = $AsPh_3$, py, Me_2S, *etc.* [l] See text. [m] L = py, PPh_3, or $AsMe_2Ph$. [n] dmp = 2,9-dimethyl-1,10-phenanthroline; five-co-ordinate structures suggested. [o] R = Me, Et, Pr^n, or Bu^n; L = PhSEt, PhSPr^n, or $\frac{1}{3}MeC(CH_2SEt)_3$; X = ClO_4 or BF_4. [p] X = NO_3, NCO, NCS, or NO_2; Y = Cl or NO_3. [q] X = ClO_4 or Br,H_2O; Y = ClO_4 or I,H_2O. [r] X = NO_3, OMe, OH, NO_2, or CO_2Me; L = CO, EtOH, py, NH_3, phosphine, *etc.*; Y = ClO_4 or NO_3; R = Me or Et. [s] Two geometrical isomers of $[IrCl_2(CO)(PMe_2Ph)_3](ClO_4)$ characterized; ν(IrCl) in one form is at 332 (*trans* $IrCl_2$), while corresponding bands in the other form are at 309 and 285 cm^{-1} (Cl *trans* to CO and phosphine). [t] R^1 = Ph, $PhCH_2$, aryl, or Me; X = Cl or Br; L = Ph_3As, Ph_3P, PMe_2Ph, *etc.*; R^2 = alkyl or aryl (25 compounds). [u] 2 or 3 bands in these ranges. [v] L^1 = PEt_3, PPr^n_2Ph, or PBu^n_2Ph; L^2 = $AsEt_3$ or PEt_2Ph; L^3 = PEt_3, PBu^n_3, or $AsMe_2Ph$; R^1 = Me, Et, or Bu^n; R^2 = Me, Et, Pr^n, or Bu^n. [w] R^1 = Me or Ph; L^1 = PPh_3, $PMePh_2$, or $AsPh_3$; X or Y = Cl or Br; L^2 = PPh_3, $AsPh_3$, or py; R^2 = py or quin.

17—22, 25, 199, 211, and 217 contain similar data which have already been mentioned.

When $RhCl_3,3H_2O$ is heated under reflux in MeOH–2-methylallyl alcohol (C_3H_7OH), the compound $[RhCl_2(C_8H_{15}O_2)]_n$ can be obtained, which on treatment with a base L (L = py, PPh_3, or $AsMe_2Ph$) affords $[RhCl_2(C_8H_{15}O_2)(L)]_2$.[244] These have ν(RhCl) bands (Table 21) attributable to both bridging and terminal halide groups, and the bis-(4-methylpyridine) analogue has been shown (*X*-ray diffraction) to have the bridged structure (27).[244] The low ν(RhCl) frequencies (Table 21) of $[Rh_2Cl_2(bhq)_4]$ and $[Rh_2Cl_2(php)_4]$ solvates are consistent with structure (28), deduced from mid-i.r. and n.m.r. data.[245]

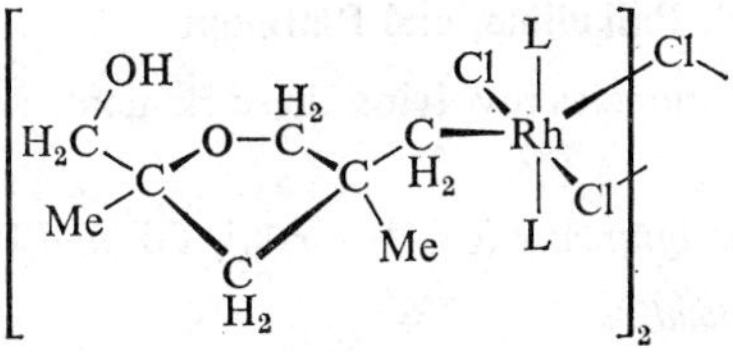

(27) L = 4-methylpyridine

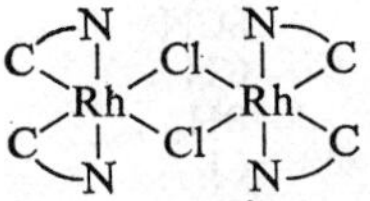

(28)

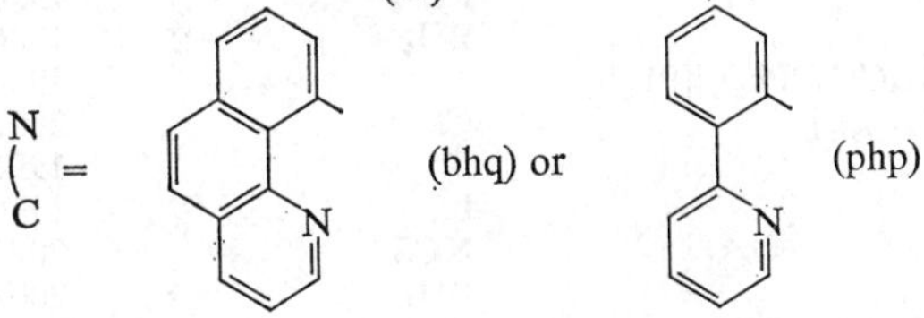

The use of i.r. ν(IrCl) frequencies in making stereochemical assignments is very elegantly illustrated by work on cod,[215] and acyl- and aryl-complexes [250] of Ir (Table 21). Thus ranges of stretching frequencies (cm^{-1}) of Ir—Cl bonds in various stereochemical environments are as follows:

ν(IrCl)	trans-*group*	*Ref.*
330—302	Cl	215, 250
316—294	CO	250
308—260	C=C	215
262—240	R or Ar	250
257—249	H	215
247—215	RCO or ArCO	250

The compound previously described as $[(Bu^n_3P)_4Rh_2Cl_6]$ and formulated as in (29),[251] is now believed to be a mixture of (30), *mer*-$[(Bu^n_3P)_3RhCl_3]$, and $[(Bu^n_3P)Cl_2RhCl_3RhCl(PBu^n_3)_2]$; unassigned i.r. data (500—200 cm^{-1}) for each of these three components have been listed.[252]

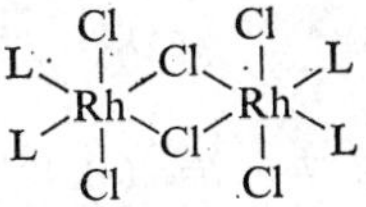

(29)

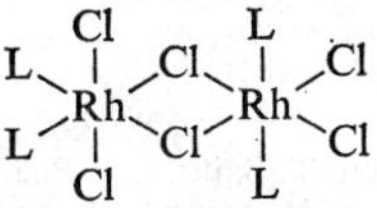

(30)

[250] M. Kubota and D. M. Blake, *J. Amer. Chem. Soc.*, 1971, **93**, 1368.
[251] J. Chatt and B. L. Shaw, *J. Chem. Soc.*, 1964, 2508.
[252] F. H. Allen and K. M. Gabuji, *Inorg. Nuclear Chem. Letters*, 1971, **7**, 833.

9 Nickel, Palladium, and Platinum

Metal–hydrogen frequencies involving these elements are listed in Table 22.

Table 22 *ν(MH) assignments* (cm^{-1}) *for* Ni, Pd, *and* Pt *compounds*

Compound	X	ν(MH)	*Ref.*
$[HNiX\{P(C_6H_{11})_3\}_2]$	Cl	1916	
	Br	1917	
	I	1976	253
	SCN	1928	
	CN	1870	
	BH_4	1920[a]	254
$[HNiX(PPr^i_3)_2]$	Cl	1937	
	Br	1979	253
	I	1990	
	BH_4	1986[a]	254
$[HNi(Ph_2PCH_2CH_2PPh_2)_2][PF_6]$		1930	255
$[HPdX\{P(C_6H_{11})_3\}_2]$	Cl	2002	254[b], 256
	Br	1991[b]	
	I	1966[b]	
	NCS	2022[b]	254
	BH_4	2002[a]	
$[HPdX(PPr^i_3)_2]$	Cl	2010[b]	
	BH_4	2013[a]	
$[HPd(PX_3)(Ph_2PCH_2CH_2PPh_2)][PF_6]$	C_6H_{11}	1895	255
	Pr^n	1959	
trans-$[HPtX(PPh_3)_2]$	CF_3CO_2	2280	
	Br	2215	
	$(CH_2)_2(CO)_2N$—	2190	
	MeCOS—	2140	257[c]
	NCS	2250	
	CN	2080	
	Cl	2220[d]	
	ClO_4	2312	258
trans-$[HPtX(PPh_3)_2](ClO_4)$	Ph_3P	2124	
	Ph_3Sb	2130, 2163	
	C_2H_4	2098, 2089	
	C_3H_6	2082	
	CO	2188, 2087	
	NH_3	2202, 2186	258[c]
	$MeNH_2$	2195	
	$EtNH_2$	2193	
	Me_2NH	2194	
	py	2203	
	$(H_2N)_2CS$	2150	

[253] M. L. H. Green, T. Saito, and P. J. Tanfield, *J. Chem. Soc.* (*A*), 1971, 152.
[254] M. L. H. Green, H. Munakata, and T. Saito, *J. Chem. Soc.* (*A*), 1971, 469.
[255] M. L. H. Green and H. Munakata, *Chem. Comm.*, 1971, 549.
[256] R. van der Linde and R. O. de Jongh, *Chem. Comm.*, 1971, 563.
[257] D. M. Roundhill, P. B. Tripathy, and B. W. Renoe, *Inorg. Chem.*, 1971, **10**, 727.
[258] I. V. Gavrilova, M. I. Gel'fman, N. V. Ivannikova, and V. V. Razumovskii, *Russ. J. Inorg. Chem.*, 1971, **16**, 596.

Table 22 (*cont.*)

Compound	X	ν(MH)	*Ref.*
[HPtX($Ph_2PCH_2CH_2PPh_2$)]	$SiMe_3$	2000	259
	$Cl(SnMe_3)_2$	1960	259
	$(SnMe_3)_3$	1960	259
[HPt($Ph_2PCH_2CH_2PPh_2$)][PF_6]		1978	20

[a] Other assignments include: δ(MH), 750—730; ν(BH), 2363—2260; δ(BH), 1140—1070; $\nu(MH_2B)$, 2060—1980 and 1868—1800 cm^{-1}. [b] δ(PdH) at 753—705 cm^{-1}. [c] Data in general accord with the *trans*-effect series. [d] Given as 2232 cm^{-1} in ref. 258. A different crystalline form of this compound apparently also exists, showing ν(PtH) in the solid at 2277, 2267, and 2232 cm^{-1} (ref. 260).

Reaction of [$Pt(PPh_3)_4$] with HCl (1 : 2 molar proportions) is shown [260] by i.r. and n.m.r. studies to give *trans*-[$(Ph_3P)_2PtHCl$] and not [$(Ph_3P)_2PtH_2Cl_2$] as previously believed.

Adducts formed by reaction of [$Pt(PPh_3)_n$] (n = 2 or 3) with H_2E or PhEH (E = S or Se) are formulated respectively as (31) [ν(PtH) = 2116 (E = S) or 2140 cm^{-1} (E = Se)] and (32) [ν(PtH) = *ca.* 2130 (E = S or Se); δ(PtH) = 807 (E = S) or 795 cm^{-1} (E = Se)].[261]

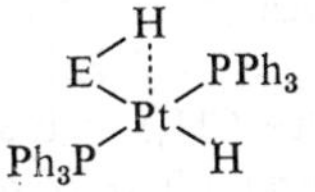

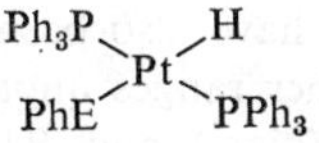

(31) E = S or Se (32)

The i.r. ν(MC) frequency in *trans*-[$(R^1_3P)_2M(C{\equiv}CR^2)_2$] ($R^2$ = Me, $CH{=}CH_2$, Ph, FCH_2, or $HC{\equiv}C$; M = Ni, Pd, or Pt), assigned at 580—540 cm^{-1}, increases with increasing electronegativity of R^2 because of enhanced d_π–p_π interaction between M and $-C{\equiv}CR^2$.[262]

Largely unassigned i.r. data are available for: π-hexakis(trifluoromethyl)benzene (Ar) complexes of nickel such as $(1,5\text{-}C_8H_{12})_nNi_n(Ar)$ (n = 1 or 2) and $L_2Ni(Ar)$ (L = phosphine, *etc.*);[263] fluoroalkyl derivatives $L_2NiCF_2CF_2CF_2CF_2$, $L_2NiCF_2CFHCFHCF_2$, $L_2Ni(CF{=}CF_2)X$, $L_2Ni(CCl{=}CF_2)Cl$, *etc.* (L = arsine; X = Cl or Br);[263] isocyanide derivatives $(Bu^tNC)_2NiC(CF_3)_2{\cdot}X{\cdot}C(CF_3)_2{\cdot}X$ (X = O or N), $(PhNC)_2NiC(CF_3)_2{\cdot}O$, $(Bu^tNC)_2NiCF_2CF_2CF_2CF_2$, *etc.*;[264] and the propene (C_3H_6) and bivinyl (C_4H_6) complexes K[$PtCl_3(Me_2SO)(C_3H_6)$],H_2O, [$PtCl_2(Me_2SO)(C_3H_6)$], K_2[$Pt_2Cl_6(C_4H_6)$], and [$Pt_2Cl_4(Me_2SO)_2(C_4H_6)$].[265]

259 A. F. Clemmit and F. Glockling, *J. Chem. Soc.* (*A*), 1971, 1164.
260 J. T. Dumler and D. M. Roundhill, *J. Organometallic Chem.*, 1971, **30**, C35.
261 R. Ugo, G. La Monica, S. Cenini, A. Segre, and F. Conti, *J. Chem. Soc.* (*A*), 1971, 522.
262 H. Masai, K. Sonogashira, and N. Hagihara, *J. Organometallic Chem.*, 1971, **26**, 271.
263 J. Browning, C. S. Cundy, M. Green, and F. G. A. Stone, *J. Chem. Soc.* (*A*), 1971, 448, 453.
264 M. Green, S. K. Shakshooki, and F. G. A. Stone, *J. Chem. Soc.* (*A*), 1971, 2828.
265 Yu. N. Kukushkin and I. V. Pakhomova, *Russ. J. Inorg. Chem.*, 1971, **16**, 226.

Analysis of the high-resolution photoelectron spectra of C_2H_4 and C_2D_4 has given support [266] to the recent interpretation [267] of the vibrational spectrum of Zeise's salt, $K[PtCl_3(C_2H_4)],H_2O$, in terms of $(C_2H_4)-Pt$ bonding. I.r. bands due to co-ordinated C_2H_4 in $L_2Pd(C_2H_4)$ [L = Ph_3P, $(C_6H_{11})_3P$, or (o-tolyl·O)$_3$P] have been reported.[256]

$Me_3Pt(SCN)$ is tetrameric in benzene;[268] i.r., Raman, and n.m.r. data indicate an acentric D_{2d} structure comprising a tetrahedral array of Pt atoms linked by SCN bridges (three-co-ordinate S atoms). Assignments include (cm^{-1}):

	I.r.	*Raman*
ν(C≡N)	2194, 2184	2182
ν(CS)	749, 729	747
ν(PtC)		592, 582, 566
δ(NCS), ν(PtN)	454	456, 446
δ(PtC_3), ν(PtS), δ(PtNCSPt)	254, 243, 194	263, 238

The i.r. and Raman spectra of derivatives $[Me_3Pt(NCS)(L)]_2$ (L = py, Ph_3P, or Ph_3As) have also been discussed; $\nu(PtC_3)$ is in the region 579—548 cm^{-1}.[268] Other ranges quoted for $\nu(Pt-CH_3)$ include 555—510 cm^{-1} (i.r.) for $PtMe_2$(diars) and $RPtXMe_2$(diars) [diars = o-$Me_2AsC_6H_4$, or racemic or *meso*-$(PhMeAsCH_2)_2$; X = halide; R = Me, MeCO, PhCO, σ-allyl, or σ-2-methylallyl; ν(PtCl) = 254—226 cm^{-1}],[269, 270] and 526—514 cm^{-1} (Raman) for alkoxycarbene complexes $[(L)PtMe(Q)_2](PF_6)$ [(33); L = MeOCMe, $MeOCCH_2Ph$, $\overline{CH_2CH_2COCH_2}$, *etc.*; Q = $PhMe_2P$, $PMe_2(C_6H_{11})$, or Me_3As; ν(CO) of carbene = 1320—1255 cm^{-1} (i.r.)].[271a] In the case of $[\{(C_6H_{11})_2PCH_2CH_2P(C_6H_{11})_2\}PdEt_2]$,[256] ν(PdC) is at 530 cm^{-1}, while in compounds such as *trans*-$[Pd(TeEt_2)(Ar)X]$ (X = halide; Ar = Ph, o-tolyl, *etc.*), ν(PdC) is at *ca.* 515—479 cm^{-1}.[271b]

In the palladiacyclopentadiene complex (34), an i.r. band attributed to ν(C=O) is at 1620 cm^{-1}, compared with 1705 ± 20 cm^{-1} in four-

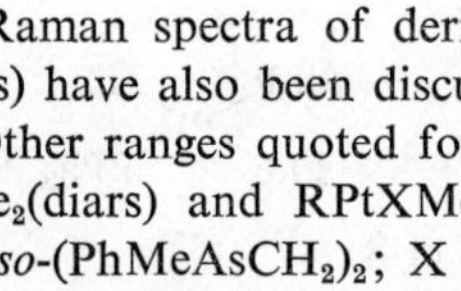

(33)

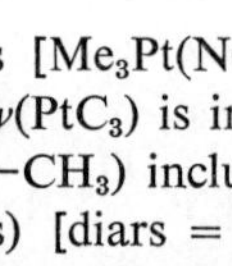

(34) R = $MeCO_2$

[266] C. R. Brundle and D. B. Brown, *Spectrochim. Acta*, 1971, **27A**, 2491.
[267] J. Hiraishi, *Spectrochim. Acta*, 1969, **25A**, 749.
[268] G. C. Stocco and R. S. Tobias, *J. Coordination Chem.*, 1971, **1**, 133.
[269] A. J. Cheney and B. L. Shaw, *J. Chem. Soc.* (*A*), 1971, 3545.
[270] A. J. Cheney and B. L. Shaw, *J. Chem. Soc.* (*A*), 1971, 3549.
[271] (*a*) M. H. Chisholm and H. C. Clark, *Inorg. Chem.*, 1971, **10**, 1711; (*b*) S. Sergi, F. Faracne, L. Silvestro, and R. Pietropaolo, *J. Organometallic Chem.*, 1971, **33**, 403.

co-ordinate adducts with amines and phosphines, suggesting that (34) is associated *via* >C=O···Pd interaction.[272]

The single-crystal *X*-ray structure[273] of the product from reaction of $Ni(CO)_4$ with *o*-$I_2C_6H_4$ (1 : 1) shows that the correct formulation is $\{C_6H_4(CO)_2\}NiI$ and not $(C_6H_4)(CO)NiI$ as pre-supposed.[274] The structure contains alternate iodide and carboxyl bridges and incorporates the unique C-bonded *o*-phthaloyl-C,C′,O grouping (35). I.r. bands at 1796

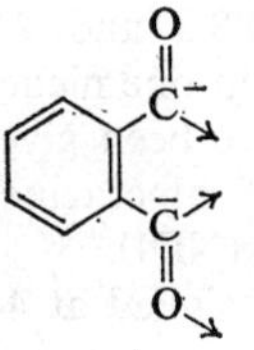

(35)

and 1768 cm^{-1} are thus attributable to the bridging carboxyl[273] rather than to bridging carbonyl.[274]

A complete i.r. assignment has been suggested for $[Pt(CO)Cl_2]en$ (in which the *trans* configuration of en is assumed), including the following skeletal modes (cm^{-1}):[275]

$\nu(PtC)$	536
$\nu(PtN)$	516
$\nu_{asym}(PtCl)$	344
$\nu_{sym}(PtCl)$	318

Similar results have been given for other complexes $[Pt(CO)Cl_2](LL)$, where (LL) is a bidentate ligand.

The change in bands assigned to $\nu(CO)$ has been studied during the formation of uncharacterized polymeric halogenocarbonyl complexes of Pd.[276]

Values of $\nu(Ni-CN)$ for benzene solvates of $[Ni(CN)_4]^{2-}$ salts have been given.[277]

The first species containing only N_2 as a ligand have been identified by i.r. spectroscopy when Ni atoms and pure N_2 ($^{14}N_2$ or $^{15}N_2$) or N_2–Ar mixtures are co-deposited on a KBr window at 26—17 K.[278] By studying the relative intensities of the i.r. bands of the deposits obtained under

272 K. Moseley and P. M. Maitlis, *Chem. Comm.*, 1971, 1604.
273 N. A. Bailey, S. E. Hull, R. W. Jotham, and S. F. A. Kettle, *Chem. Comm.*, 1971, 282.
274 E. W. Gowling, S. F. A. Kettle, and G. Sharples, *Chem. Comm.*, 1968, 21.
275 T. Theopanides and P. C. Kong, *Inorg. Chim. Acta*, 1971, **5**, 485.
276 Yu. A. Kushnikov, A. Z. Beilina, and V. F. Vozdvizhenskii, *Russ. J. Inorg. Chem.*, 1971, **16**, 218.
277 A. Sopková, J. Chomič, and E. Matejčiková, *Monatsh.*, 1971, **102**, 961.
278 J. K. Burdett and J. J. Turner, *Chem. Comm.*, 1971, 885.

different conditions, both $Ni(N_2)$ and $Ni(N_2)_2$ species were definitely identified (cm^{-1}):

	$\nu(N_2)$
$Ni(^{14}N_2)$	2169.4
$Ni(^{15}N_2)$	2096.4
$Ni(^{14}N_2)_2$	2179.8
$(^{14}N_2)Ni(^{15}N_2)$	2177.2, 2108.2
$Ni(^{15}N_2)_2$	2106.2

Two other bands found (2188.3 and 2196.4 cm^{-1}) were attributed to complexes containing more than one nickel atom.

Selected i.r. assignments have been given for complexes $NiCl_2L_2$ (L = *N*-2-aminoethyl derivatives of piperazine, pyrrolidine, or morpholine), including suggestions for ν(NiN).[279] In the new five-co-ordinate $Ni(NCS)_2(PMe_3)_2$, ν(NiN) is assigned at 440 cm^{-1} (Raman) and 412 cm^{-1} (i.r.).[280]

Stereochemical assignments for compounds PdX_2L_2 (X = Cl or Br; L = thiazole, benzothiazole, or their derivatives) have been made from consideration of i.r. ν(PdN) and ν(PdX) frequencies; in all cases the organic ligands are considered to be N-co-ordinated.[281]

Both intramolecular vibrations of the constituent ions and lattice vibrations due to interaction between the ions have been assigned in the i.r. spectra (to 30 cm^{-1}) of the Magnus-type salts $[Pden_2][PtCl_4]$, $[Pden_2][PdCl_4]$, and $[Pten_2][PtCl_4]$.[282] At liquid-nitrogen temperatures the chelate rings are considered to have the $\delta\lambda$ conformation, whereas at room temperature the spectra indicate that either there is a co-existence of two conformers (C_{2h} and D_2), or the cations have a lower site symmetry. Calculated frequencies for C_{2h} and D_2 conformers are given.

I.r. spectroscopy has been used to show[283] that the complex $Pden_2Cl_2,HCl,H_2O$ should be formulated $[Pd(en)(enH)Cl]Cl_2,H_2O$.

Table 23 *Comparison of* ν(NiX) *and* ν(NiP) *between different configurations*

X	*trans*-$[^{58}NiX_2(PR^1_2R^2)_2]$		*cis*-$[^{58}NiX_2(diphos)]$		$[NiX_2(PPh_3)_2]$[a]	
	ν(NiX)	ν(NiP)	ν(NiX)	ν(NiP)	ν(NiX)	ν(NiP)
Cl	401[b]	252[b]	{341, 328}[e]	{379, —}[e]	{341, 305}[f]	{190, 164}[f]
Br	338[c]	265[c]	{290, 266}	{365, 308}	{265, 232}	{193, 184}
I	260[d]	252[d]	{260, 212}	{353, 278}	215	{198, 182}

[a] Tetrahedral. [b] R^1 = Ph, R^2 = Et. [c] R^1 = R^2 = Et. [d] R^1 = Me, R^2 = Ph; Ni in natural abundance. [e] Two forms obtained, assumed to differ only in the conformation of the diphosphine ligand. [f] ^{58}Ni complex.

279 G. Contreras and E. Astegarrabia, *Inorg. Chem. Acta*, 1971, **5**, 54.
280 A. Merle, M. F. Obier, M. Dartiguenave, and Y. Dartiguenave, *Compt. rend.*, 1971, **272**, *C*, 1956.
281 M. N. Hughes and K. J. Rutt, *Spectrochim. Acta*, 1971, **27A**, 924.
282 Y. Omura, I. Nakagawa, and T. Shimanouchi, *Spectrochim. Acta*, 1971, **27A**, 1153.
283 D. A. Johnson and W. H. Delphin, *Inorg. Nuclear Chem. Letters*, 1971, **7**, 717.

I.r. assignments for ν(NiP) and ν(NiX) in [$NiX_2(Ph_2PCH_2CH_2PPh_2)$] (X = Cl, Br, or I) have been made on the basis of ^{58,62}Ni isotopic shifts (Table 23).[284] In these *cis*-complexes, ν(NiP) is always greater than in *trans*-systems such as [$NiX_2(PEt_3)_2$], whereas the reverse is true of ν(NiX). These effects have their origin in the strong *trans* influence of phosphine ligands. In the new five-co-ordinate complex $Ni(NO_2)_2(PMe_3)_2$,[285] ν(NiP) is given as 377 cm^{-1} with ν(NiN) at 415 cm^{-1}. Metal–phosphorus modes have also been assigned in i.r. spectra as follows (cm^{-1}):

	ν(MP)	ν(MCl) *or* ν(MBr)	*Ref.*
[$(Ph_3P)_3NiBr$]	196	252	286
[$(Ph_3P)_2NiBr]_n$	190	276, 270, 245	286
{$(C_6H_{11})_3P$}$_2$Pd	395		256
cis-[$PtClF(PPh_3)_2$]	443, 422	302	287
cis-[$PtBrF(PPh_3)_2$]	444, 422	192	287

In 32 complexes of the type [Ni(acac)$_2$(substituted pyridine)$_2$], i.r. bands at *ca.* 575 and *ca.* 425 cm^{-1} have been assigned as ν(NiO), the actual frequencies depending on the electron-releasing or -attracting characteristics of the substituents in the pyridine ring.[288]

Novel cyclic compounds such as (36) and (37) are formed from peroxobis-(triphenylphosphine)platinum(II) and hexafluoroacetone.[289] These show

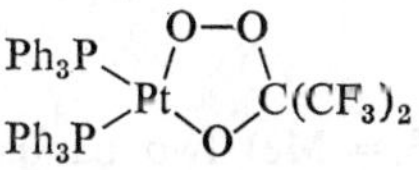

(36)

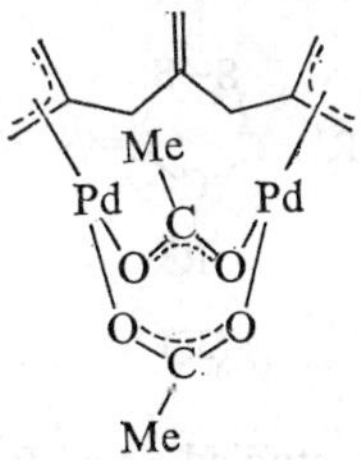

(38)

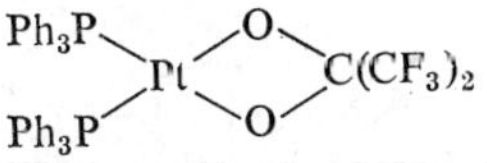

(37)

ν(PtO) in the range 400—300 cm^{-1}, with ν(O—O) of type (36) at 808—780 cm^{-1}. Somewhat different values for ν(PtO) (622, 578, 500, and 455 cm^{-1}) are quoted [242] for [$Me_2C{=}CHCMe_2CH_2OPtCl]_2$. Unassigned i.r. data have been given for the acetato-bridged complex (38) and some of its derivatives.[290]

A UBFF vibrational analysis of a simplified model has been used to assist assignment of the i.r. spectrum (to 250 cm^{-1}) of the bis-(*N*-cyano-dithiocarbimato)nickel(II) anion.[291] Vibrational mixing is less than in

[284] C. Udovich, J. Takemoto, and K. Nakamoto, *J. Coordination Chem.*, 1971, **1**, 89.
[285] A. Merle, M. Dartiguenave, and Y. Dartiguenave, *Compt. rend.*, 1971, **272**, *C*, 2046.
[286] C. S. Cundy and H. Nöth, *J. Organometallic Chem.*, 1971, **30**, 135.
[287] R. D. W. Kemmitt, R. D. Peacock, and J. Stocks, *J. Chem. Soc.* (*A*), 1971, 846.
[288] J. M. Haigh, N. P. Slabbert, and D. A. Thornton, *J. Mol. Structure*, 1971, **7**, 199.
[289] P. J. Hayward and C. J. Nyman, *J. Amer. Chem. Soc.*, 1971, **93**, 617.
[290] T. Okamoto, *Bull. Chem. Soc. Japan*, 1971, **44**, 1353.
[291] Lakshmi, P. Bhaskara Rao, and U. Agarwala, *Inorg. Chim. Acta*, 1971, **5**, 354.

related dialkyldithiocarbamates, and bands at 525, 345, and 335 cm^{-1} are largely associated with Ni—S bond stretching. The reaction of CS_2 with $[Ni(aziridine)_n]^{2+}$ (n = 4 or 6) or $[Ni(2\text{-methylaziridine})_4]^{2+}$ affords insertion products with $\nu(NiS)$ given as[292a] *ca.* 380 cm^{-1}. Bands at 378 and 317 cm^{-1} in the i.r. spectrum of $[PtS(Ph_3P)_2]_2(CHCl_3)_2$ are assigned as $\nu(PtS)$, the low values indicating sulphur-bridging.[261] In the S-bonded Me_2SO complexes[265] $K[PtCl_3(Me_2SO)]$, $[PtCl_2(Me_2SO)(C_3H_6)]$, and $[Pt_2Cl_4(Me_2SO)_2(C_4H_6)]$ (probably *cis*), $\nu(PtS)$ is in the region 445—426 cm^{-1}. Assignments of $\nu(CN)$, $\nu(CS)$, ν(Pd—halogen), and $\nu(PdS)$ [348—312 cm^{-1}] have been listed for Pd^{II} complexes of tetramethyl- and tetraethyl-dithio-oxamide.[292b]

A large number of compounds containing the SMe group bridging two Pd or Pt atoms have been studied by i.r. and Raman (and n.m.r.) spectroscopy, and consistent sets of vibrational assignments made.[293] Of particular interest are the different range for $\nu(S{-}CH_3)$ and $\nu(M{-}S)$, depending on the nature of the *trans*-ligand (Table 24), which show that the *trans* influence of the methylthio-group is much greater than that of chloride or of bromide.

Complexes NiI_2L_2, where L is the thio-ligand (39), have been prepared.[294] When R^1 = Ph and R^2 = H, $\nu(NiI)$ is located at 292 and 272 cm^{-1},

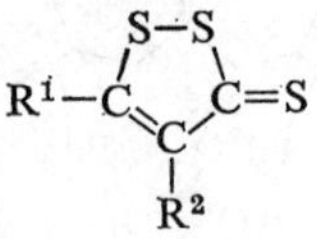

(39)

whereas in other compounds studied (*e.g.* R^1 = R^2 = Me) two bands are found in the 230—210 cm^{-1} region. Together with electronic spectral data, these results are said to indicate planar geometry in the first case and tetrahedral co-ordination in the others.

In $[(Ph_3P)_2NiCl(C_6H_5)]$, $\nu(NiCl)$ is assigned[295] at 348 cm^{-1}. Other references to ν(Ni—halogen) modes[18—20, 25, 284, 286] have been mentioned earlier.

There have been many reports which include values for palladium– and platinum–halogen stretching frequencies. Several of these have already been mentioned in this or previous sections.[17, 20, 269, 270, 275, 281, 282, 287, 292b, 293] The following discussion will serve to illustrate the type of structural

[292] (*a*) B. J. McCormick, R. I. Kaplan, and B. P. Stormer, *Canad. J. Chem.*, 1971, **49**, 699; (*b*) A. C. Fabretti, G. C. Pellaconi, and G. Peyronel, *J. Inorg. Nuclear Chem.*, 1971, **33**, 4247.

[293] P. L. Goggin, R. J. Goodfellow, and F. J. S. Reed, *J. Chem. Soc.* (*A*), 1971, 2031.

[294] F. Y. Petillon and J. E. Guerchais, *Canad. J. Chem.*, 1971, **49**, 2598.

[295] M. Hidai, T. Kashiwagi, T. Ikeuchi, and Y. V. Chida, *J. Organometallic Chem.*, 1971, **30**, 279.

Table 24 *Assignments*[a] *of vibrational spectra of methylthio-bridged complexes*

Assignment	$[M_2X_3(SMe)(YMe_3)_2]$[b,c]	*cis*-$[M_2X_2(SMe)_2(YMe_3)_2]$[b,c]	*trans*-$[Pt_2Cl_2(SMe)_2(PMe_3)_2]$	$[Pt_2(SMe)_2(YMe_3)_4](BF_4)_2$[b]
$\nu(S-CH_3)$	*ca.* 680	*ca.* 710,[d] *ca.* 680[e]	685 ?	*ca.* 680
$\nu_{asym}(MP_2)$ or $\nu_{asym}(M_2P_2)$	*ca.* 380	*ca.* 380—360 (both rows)	*ca.* 380	*ca.* 395, *ca.* 375
$\nu_{sym}(MP_2)$ or $\nu_{sym}(M_2P_2)$	*ca.* 390			
$\nu(M_2S)$ or $\nu(M_2S_2)$	2 bands, *ca.* 385—340	*ca.* 355,[e] *ca.* 300,[d] *ca.* 295[d]	*ca.* 360,[e] *ca.* 300[d]	*ca.* 350, *ca.* 320
$\nu_{asym}(MAs_2)$ or $\nu_{asym}(M_2As_2)$	*ca.* 280	*ca.* 264	—	*ca.* 260 (both rows)
$\nu_{sym}(MAs_2)$ or $\nu_{sym}(M_2As_2)$	*ca.* 285	*ca.* 266	—	
$\nu_{asym}(M_2Cl_2)_t$ or $\nu_{sym}(M_2Cl_2)_t$	*ca.* 310	*ca.* 310	*ca.* 320	—
$\nu(M_2Cl)_b$	*ca.* 260	—	—	—

[a] Approximate values from i.r. and/or Raman data of the series. Assignments for $\nu(YC_3)$ and $\delta(PC_3)$ given in several cases. [b] M = Pd or Pt; X = Cl or Br; Y = P or As. [c] 7 compounds. [d] *trans* to Y. [e] *trans* to X.

information which has been obtained from studies of metal–halogen frequencies of these elements.

The first example of a carbene complex of type [PtX_2(L)(carbene)] with the *trans* configuration (40) has been obtained (confirmed by single-crystal *X*-ray study), showing ν(PtCl) at 340 and 342sh cm^{-1}.[296] Isomerization to the *cis*-isomer [ν(PtCl) = 308 and 277 cm^{-1}] apparently

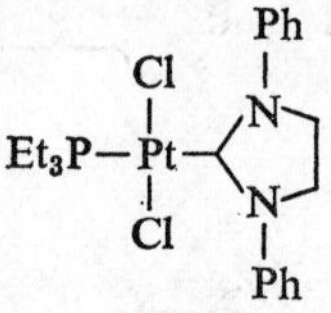

(40)

occurs at 210 °C. Other *cis* carbene complexes of Pt^{II},[297] *cis*-[$PtCl_2${C(X)NHPh}(PEt_3)] (X = MeO, EtO, Pr^iO, PhNH, MeNH, EtNH, or Bu^sNH) and *cis*-[$PtBr_2${C(OEt)NHPh}(PEt_3)], show ν(PtCl) at 306—294 and 284—270 cm^{-1}, and ν(PtBr) at 201 and 192 cm^{-1}.

I.r. spectroscopy (with d.t.a.) has been used to show that heating of *trans*-[$PtCl_2${$(CH_2)_5S$}$_2$] [ν(PtCl) = 345 cm^{-1}], gives rise to the corresponding *cis*-isomer [ν(PtCl) = 326 and 312 cm^{-1}].[298] Similarly, complexes [$PtCl_2$(MeCN)(L)] (L = C_2H_4 or CO) have been assigned *cis* configurations on account of the observation of two ν(PtCl) i.r. modes (357 and 331 cm^{-1}, L = C_2H_4; 353 and 344 cm^{-1}, L = CO).[299]

Complexes [(Ar)PtCl($TeEt_2$)$_2$] are considered to be *trans* since they show low ν(PtCl) modes (271 cm^{-1}, Ar = Ph; 269 cm^{-1}, Ar = mesityl) as expected for Cl *trans* to the strongly activating Ar group.[300] Low values for ν(PtX) [*ca.* 240 cm^{-1}, X = Cl; *ca.* 170 cm^{-1}, X = Br] are also found in the complexes [Me(L)$_2$PtX($CF_3C{\equiv}CCF_3$)] (L = Me_3P, Me_3As, or Me_3Sb; X = Cl or Br; ν(C≡C) = *ca.* 1850 cm^{-1}), in agreement with *trans*-CH_3 as in structure (41).[301]

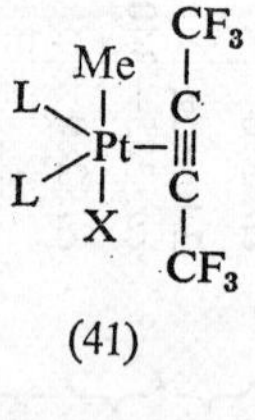

(41)

296 D. J. Cardin, B. Cetinkaya, M. F. Lappert, Lj. Manojlović-Muir, and K. W. Muir, *Chem. Comm.*, 1971, 400.
297 E. M. Badley, J. Chatt, and R. L. Richards, *J. Chem. Soc.* (*A*), 1971, 21.
298 E. A. Allen, N. P. Johnson, D. T. Rosevear, and W. Wilkinson, *Chem. Comm.*, 1971, 171.
299 T. Weil, L. Spaulding, and M. Orchin, *J. Co-ordination Chem.*, 1971, **1**, 25.
300 S. Sergi, F. Faraone, and L. Silvestro, *Inorg. Chem. Nuclear Letters*, 1971, **7**, 869.
301 H. C. Clark and R. J. Puddephatt, *Inorg. Chem.*, 1971, **10**, 18.

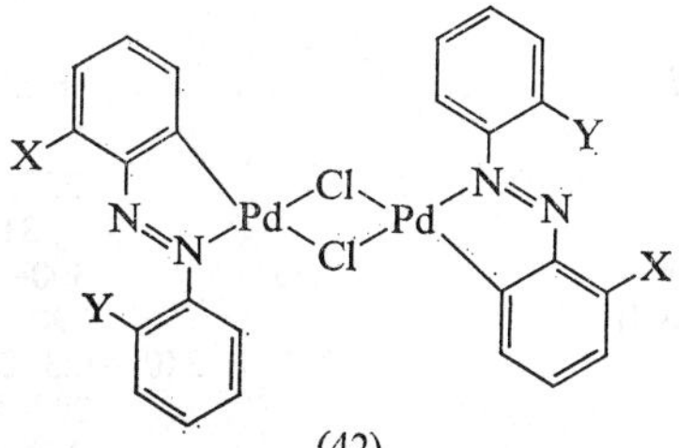

(42)

It has been suggested that in complexes (42), i.r. bands at *ca.* 240 cm^{-1} and *ca.* 270 cm^{-1} arise from ν(PdCl) of the bonds *trans* to C and N, respectively.[302] Pd—Cl and Pt—Cl bridging modes have also been assigned as in Table 25.

Table 25 *Assignments* (cm^{-1}) *of bridging* ν(PdX) *and* ν(PtX) *modes*[a]

Compound	$\nu(MX)_b$	$\nu(MX)_t$	*Ref.*
[(π-allyl)$_2$Pd$_2$Cl$_2$]	258[b]	—	303
[(alkene)$_2$Pt$_2$Cl$_2$][c]	327—310, 296—283	365—348	304
[(Me$_2$C=CHCMe$_2$CH$_2$O)$_2$Pt$_2$Cl$_2$]	274[d]	—	242
[PtCl$_2$Me$_2$]$_n$	234	332	305
[PrBr$_2$Me$_2$]$_n$	202	255	305
[PtI$_2$Me$_2$]$_n$	127	171	305

[a] Refs. 293 and 302 are discussed in text. [b] Band disappears on reaction with Ph$_3$P or Me$_2$SO, new bands appearing at 350—290 cm^{-1}. [c] alkene = C_2H_4, 2-methylpropene, 2-methylbut-2-ene, 2-phenylpropene, cyclohexene, cycloheptene, or cyclo-octene. Undistinguished 'ν(PtCl)' frequencies (288—218 cm^{-1}) listed for [Pt$_2$Cl$_2$(all)$_2$] (all = allyl, 2-methylallyl, 1,2-dimethylallyl, 1,1,2-trimethylallyl, or 1,2,3-trimethylallyl). [d] Band disappears on reaction with *p*-toluidine giving [(Me$_2$C=CHCMe$_2$CH$_2$O)PtCl(*p*-toluidine)] with $\nu(PtCl)_t$ at 253 and 307 cm^{-1}.

Further data on palladium– and platinum–halogen modes are listed in Table 26.

Table 26 *Further data on* ν(PdX) *and* ν(PtX) *frequencies* (cm^{-1})

Compound	ν(MX)	*Ref.*
trans-[PdCl$_2$(PBu$^t_2$R)][a]	354—342	203
[Pd(C$_3$H$_5$—C=N—Ph)(PhNC)Cl]$_2$	317, 274	28
[Pd(C$_5$Me$_5$·CHMeR)Cl$_2$][b]	322—300	306
[Pd(C$_5$Me$_5$H)Cl$_2$]	335, 315, 290	306
cis-[PdCl(PEt$_3$)$_2$py][BF$_4$]	305	307
[PdCl(PPh$_3$)$_3$][BF$_4$]	312	307
cis-[PtCl(PEt$_3$)$_2$py][BF$_4$]	315	307
[PtCl(PPh$_3$)$_3$][BF$_4$]	317	307

[302] D. R. Fahey, *J. Organometallic Chem.*, 1971, **27**, 283.
[303] L. A. Leites, V. T. Aleksanyan, S. S. Bukalov, and A. Z. Rubezhov, *Chem. Comm.*, 1971, 265.
[304] B. E. Mann, B. L. Shaw, and G. Shaw, *J. Chem. Soc.* (*A*), 1971, 3536.
[305] J. R. Hall, and G. A. Swile, *Austral. J. Chem.*, 1971, **24**, 423.
[306] P. V. Balakrishnan and P. M. Maitlis, *J. Chem. Soc.* (*A*), 1971, 1721.
[307] K. R. Dixon and D. J. Hawke, *Canad. J. Chem.*, 1971, **49**, 3252.

Table 26 (*cont.*)

Compound	ν(MX)	*Ref.*
[Pt(C_5Me_5R)Cl_2][c]	337—300	308
[Pt(C_5Me_5H)Br_2]	227, 222	308
[Pt(diene)Cl_2][d]	338, 318, 298	308
[Pt(diphos)(MMe_3)Cl][e]	300—280	259
[Pt(diphos)($SnMe_3$)$_2$HCl]	290—280	259
[Pt(R^1NC)(PR^2_3)Cl_2][f]	340—323, 298—277	297
[Pt(all)(L)Cl][g]	297—289	304
[$PtCl_2$(diene)][h]	326, 305	304
[$PtBr_2$(diene)][h]	215?	304
[(PhC≡CPh)$PtCl_2$(Ph_3P)]	332, 298[i]	309
[(Ph_2MeP)$_2$PtCl{C(CF_3)=(CF_3)HgCl}]	348, 310[j]	309

[a] R = Me, Et, or Pr^n. [b] R = Cl, OMe, OEt, or OPr^i. [c] R = H, Et, or CH = CH_2. [d] diene = hexamethylbicyclo[2,2,0]hexa-2,5-diene. [e] M = Si, Ge, or Sn. [f] R^1 = Ph, *p*-tolyl, Me, or Et; PR^2_3 = PEt_3, PEt_2Ph, *etc.*; 7 complexes. [g] all = allyl or 2-methylallyl; L = isoquinoline, Ph_3P, Ph_3As, or py. [h] diene = 2,5-dimethylhexa-1,5-diene. [i] *cis* geometry suggested by the appearance of these two bands; ν(C≡C) at 1990, 1965 cm^{-1}. [j] Bands described as 'metal–chlorine stretching'; ν(C=C) at 1578 cm^{-1}.

10 Copper, Silver, and Gold

Co-condensation of Cu or Ag atoms with CO at 20 K gives deposits with i.r. bands at 1989 and 1975 cm^{-1} (Cu), or 1968 and 1939 cm^{-1} (Ag), which shift by *ca.* 45 cm^{-1} to lower frequency when $C^{18}O$ is used.[310] The bands are assigned to terminal carbonyl groups co-ordinated to the metal atoms, and their sharpness (*ca.* 5 cm^{-1} wide) indicates that discrete molecular species are present.

A straight-line correlation has been found between the position of the main electronic absorption maxima and the i.r. ν(CuN) frequencies of complexes Cu(N—N)$_2X_2$ [(N—N) = en, 1,3-propylenediamine, *etc.*; X = Cl, Br, I, NCS, ClO_4, BF_4, or NO_3].[311] Similar correlations involving ν(CuN), ν(CuX), and electronic spectra have been noted for the corresponding 1 : 1 complexes Cu(N—N)X_2 (X = Cl, Br, NO_3, or NCS), and are interpreted in terms of increasing tetragonal distortion and increasing in-plane bond strength.[312] The nature of the Cu—X bond is thought to affect the ν(CuN) frequency *via* a *trans* influence.

The variation in ν(CuN) with temperature in complexes [Cu(*asym*-Et_2en_2)$_2$]X_2 [*e.g.* for X = BF_4, ν(CuN) = 403 (+78°C), 407 (0 °C), or 410 cm^{-1} (−196 °C)] has been related to increasing tetragonal distortion of the cation at lower temperatures.[312b]

An *X*-ray crystallographic study has demonstrated regular tetrahedral co-ordination in the cation of [Cu^I(py)$_4$](ClO_4).[313] In this and the 4-picoline

308 P. V. Balakrishnan and P. M. Maitlis, *J. Chem. Soc.* (*A*), 1971, 1715.
309 D. M. Barlex, R. D. W. Kemmitt, and G. W. Littlecott, *Chem. Comm.*, 1971, 199.
310 J. S. Ogden, *Chem. Comm.*, 1971, 978.
311 A. B. P. Lever and E. Mantovani, *Inorg. Chem.*, 1971, **10**, 817.
312 (*a*) A. B. P. Lever and E. Mantovani, *Inorg. Chim. Acta*, 1971, **5**, 429; (*b*) A. B. P. Lever, E. Mantovani, and J. C. Donin, *Inorg. Chem.*, 1971, **10**, 2424.
313 A. H. Lewin, R. J. Michl, P. Ganis, U. Lepore, and G. Avitabile, *Chem. Comm.*, 1971, 1400.

and quinoline analogues, i.r. bands of the ligands showed the usual shifts to higher frequencies, but no ν(CuN) modes were located.

Complexes of the macrocyclic ligands (43) and (44), of types [Cu(L)]-$(ClO_4)_2$,$n$$H_2O$ and [Cu(L)X](ClO_4),$n$$H_2O$ (X = Cl, Br, or I; n = 1 or 0)

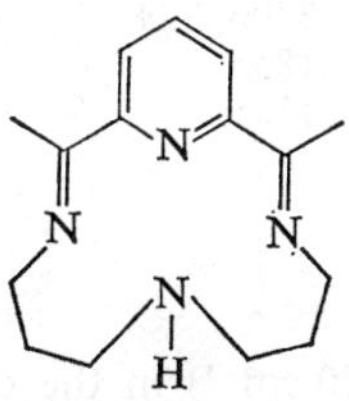

(43)

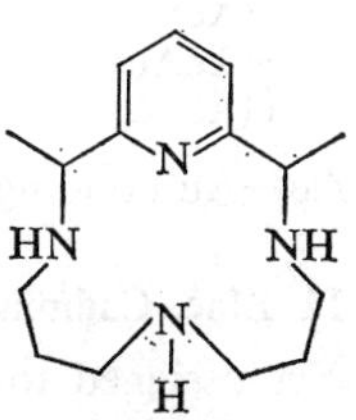

(44)

have been studied, and i.r. modes due to the ClO_4^-, H_2O, and N—H groups listed.[314] It is believed that the ClO_4^- groups are unco-ordinated except possibly in [Cu(L)]$(ClO_4)_2$ when (L) = (44).

I.r. spectroscopy has been used to study reactions in which EtOH or H_2O add to the azomethine group of complexes of Cu^{II} with *NN'*-bis-(2'-pyridylmethylene)ethane-1,2-diamine.[315]

In complexes (generally 1 : 1) of $CuCl_2$ with ligands of type (6), ν(CuCl) bands in the range 316—275 cm^{-1} are said to be consistent with bridging chloride groups.[23]

I.r. data for several organogold–phosphine complexes have been reported; selected assignments are shown in Table 27.[316, 317]

Table 27 *Some assignments* (cm^{-1}) *of phosphine-gold complexes*[a]

Compound	ν_{asym} (PC_3)	ν_{sym} (PC_3)	ρ (MeAu)	ν (AuC)	ν (AuP)	ν (AuCl)
Me_3PAuMe	742	680	696	531, 534[b]	356, 360[b]	—
Me_3PAuCl	750, 725	689	—	—	379	309
Me_3PAuMe_3	742	677	765	538	391	—
Et_3PAuMe	770, 753	708	740	529	386	—
Et_3PAuMe_3	763, 755	730	739	535	369	—
Me_3PAuMe_2Cl	749	675	820	557	426[c]	284
Me_3PAuMe_2Br	752	676	822	548	420[c]	—
Me_3PAuMe_2I	747	674	814	535	404[c]	—
$[(Me_3P)_2AuMe_2]Cl$	749	674	800	516	400[c]	—

[a] I.r. unless stated otherwise. [b] Raman. [c] Tentative assignments of weak, broad bands.

[314] L. F. Lindog, N. E. Tokel, L. B. Anderson, and D. H. Busch, *J. Co-ordination Chem.*, 1971, **1**, 7.
[315] M. Cressey, E. D. McKenzie, and S. Yates, *J. Chem. Soc.* (*A*), 1971, 2677.
[316] H. Schmidbaur and A. Shiotani, *Chem. Ber.*, 1971, **104**, 2821.
[317] A. Shiotani and H. Schmidbaur, *Chem. Ber.*, 1971, **104**, 2838.

For $Au_2(S_2CNBu^n_2)_2$, assignments (cm^{-1}) include the following:[39]

	I.r. (*mull*)	*Raman* ($CHCl_3$)
ν(CS)	825	813p, 762
δ(SCS)	440	440p
ν(AuS)	375, 337	340p, 328
ν(AuAu)		185p
δ(AuSC)		138s

Significant Au—Au bonding is indicated.

11 Zinc, Cadmium, and Mercury

I.r. frequencies assigned to ν(BH) (2440—2020 cm^{-1}) in the compounds $Zn(BH_4)_2(C_4H_8O)_2$, $LiZn(BH_4)_3$ (in Et_2O or C_4H_8O), and $NaZn(BH_4)_3$ (in Et_2O) have been listed.[318a] K_2ZnH_4 shows strong i.r. bands at 1400 and 650 cm^{-1}, consistent with octahedrally co-ordinated zinc.[318b]

ν(CN) frequencies have been reported for $[Hg_2(CN)_3]^+$ (2260 and 2245 cm^{-1}),[319] and for $[Hg(CN)_3]^-$ and $[Hg(CN)_4]^{2-}$ ions.[320a]

When $[Et_4N]X$ (X = OCN or SCN) is dissolved in $CdMe_2$, the linear C—Cd—C skeleton becomes bent, as evinced from the appearance of two i.r. ν(CdC) bands; this is ascribed to formation of complexes $[Me_2CdX]$.[320b]

Reasonable Coriolis coupling constants for ν_8 [ν(CH)] of $HgMe_2$ can only be obtained if free rotation of the methyl groups is assumed.[321] An i.r. and Raman study has led to the conclusion that $HgEt_2$ and $Hg(CH_2CD_3)_2$ are centrosymmetric.[322] Invariance of the Raman spectrum over the temperature range + 80 to − 196 °C is taken as evidence for free or 'quasi-free' internal rotation of the ethyl groups. Detailed assignments made on the basis of C_{2h} symmetry (considered to be more likely than D_{2h}) have been given.[322]

A number of methylmercury compounds have been studied. In Table 28 some assignments for MeHgF[323] and MeHgNCO[324] are compared. The isocyanate is 'bent' at the nitrogen atom, giving rise to two components of δ(NCO) [625 (i.r.), 635 (Raman), and 610 cm^{-1} (i.r. and Raman)]. The i.r. ν_{sym}(NCO) band is split (1285 and 1195 cm^{-1}) by Fermi resonance with δ(NCO) at 625 cm^{-1}. The ν(HgN) and ν(HgC) values have been compared[324] with those of $MeHgN_3$.

Similar assignments to those in Table 28 have been given for common modes in $(MeHg)_2NCN$ and $MeHgC_5H_5$ [$\nu(Hg—C_5H_5)$ = 338 cm^{-1}], together with data on $(EtHg)_2NCN$ and $EtHgC_5H_5$ [$\nu(Hg—C_5H_5)$ = *ca.*

318 (*a*) H. Nöth, E. Wiberg, and L. P. Winter, *Z. anorg. Chem.*, 1971, **386**, 73; (*b*) E. C. Ashby and R. G. Beach, *Inorg. Chem.*, 1971, **10**, 2486.
319 R. J. Gillespie, R. Hulme, and D. A. Humphreys, *J. Chem. Soc.* (*A*), 1971, 3574.
320 (*a*) K. G. Ashurst, N. P. Finkelstein, and L. A. Goold, *J. Chem. Soc.* (*A*), 1971, 1899; (*b*) N. Röder and K. Dehnicke, *J. Organometallic Chem.*, 1971, **33**, 281.
321 J. Mink and L. Nemes, *J. Organometallic Chem.*, 1971, **28**, C39.
322 J. L. Bribes and R. Gaufrés, *Spectrochim. Acta*, 1971, **27A**, 2133.
323 D. Breitinger, A. Zober, and M. Neubauer, *J. Organometallic Chem.*, 1971, **30**, C49.
324 H. Leimeister and K. Dehnicke, *J. Organometallic Chem.*, 1971, **31**, C3.

Table 28 *Assignments* (cm^{-1}) *for* MeHgX (X = F *or* NCO)

	MeHgF [323]		MeHgNCO [324]	
	I.r.	*Raman*	*I.r.*	*Raman*
$\nu_{asym}(CH_3)$	2990	2985	2910	2912
$\nu_{sym}(CH_3)$	2930	2933	—	2930
$\delta_{asym}(CH_3)$	—	1420	—	1358
$\delta_{sym}(CH_3)$	—	1209	—	1213
$\rho(CH_3)$	790	785	780	—
ν(HgC)	561	573	565	573
ν(HgX)	482[a]	414[a]	362	356
δ(CHgX)	—	170	—	160

[a] The large difference between these values was not commented upon and appears to be a typographical error.

336 cm^{-1}].[325] The vibrational spectra of the organomercury azides $RHgN_3$ (R = cyclopropyl, cyclopentyl, or cyclohexyl) have also been discussed.[326] In $(C_3H_5)HgN_3$, each of $\nu_{sym}(N_3)$, ν(HgC), and ν(HgN) are split into two components. This is believed to be due to the presence of two isomers, (45) and (46), although the spectral data were obtained for the solid phase, and correlation splitting cannot be ruled out.

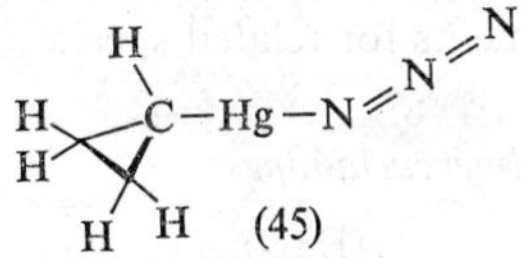
(45)

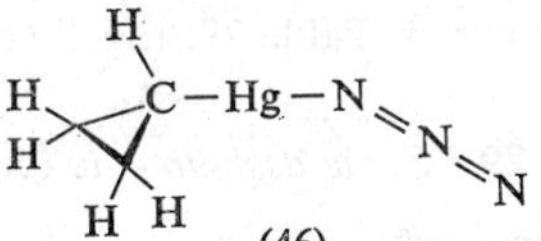
(46)

I.r. frequencies have been listed for $Hg(C_5Cl_5)_2$ [327] and $PhHgCClXCF_3$ (X = Cl or Br).[328]

In the two-co-ordinate $\{(Me_3Si)_2N\}_2Zn$, $\nu_{asym}(ZnN_2)$ is at 436 cm^{-1}, compared with 362 cm^{-1} in the Co analogue, suggesting stronger metal–nitrogen bonding in the zinc compound.[228]

For the (explosive) deprotonated *N*-sulphinylphenylhydrazine complex $Hg(PhNNSO)_2$, i.r. bands have been assigned as follows:[329] ν_{asym}(NSO), 1288; ν_{sym}(NSO), 1000; ν(HgN), 280 and 266 cm^{-1}.

Metal isotopes ($^{64,68}Zn$) have now been used to assist in the assignment of Raman spectra, the first example being for $[Zn(NH_3)_4]I_2$.[330] Comparison was also made with the spectrum of the fully deuteriated analogue, and a normal-co-ordinate analysis on the cation assisted the assignments. The ν(ZnN) modes were thus shown to be 432 cm^{-1} (ν_{sym}, a_1) and 412 cm^{-1} (ν_{asym}, t_2) in the ^{64}Zn compound. This study confirms the assignment of ν(ZnN) in $[Zn(NH_3)_4][PtCl_4]$ to an i.r. band at 430 cm^{-1}, and supports the

[325] J. Lorberth and F. Weller, *J. Organometallic Chem.*, 1971, **32**, 145.
[326] A. F. Shihada and K. Dehnicke, *J. Organometallic Chem.*, 1971, **26**, 157.
[327] G. Wulfsberg and R. West, *J. Amer. Chem. Soc.*, 1971, **93**, 4085.
[328] D. Seyferth and D. C. Mueller, *J. Amer. Chem. Soc.*, 1971, **93**, 3714.
[329] W. K. Glass and J. O. McBreen, *Inorg. Nuclear Chem. Letters*, 1971, **7**, 733.
[330] K. Nakamoto, J. Takemoto, and T. L. Chow, *Appl. Spectroscopy*, 1971, **25**, 352.

value of 390 cm^{-1} for ν(CdN) in the Cd^{II} analogue.[331a] The latter assignment is comparable with that of 370 cm^{-1} for ν(CdN) in $CdX_2(NH_3)_2$ (X = Cl, Br, or I).[331b]

Similar values (*ca.* 450 cm^{-1}) of ν(ZnN) have been suggested in the Raman spectra of a series of zinc complexes with multidentate amine ligands.[332] The actual observed frequency ranges (in cm^{-1}) apparently correlate with the metal co-ordination number:

Co-ordination number	ν(ZnN)
4	475—460
5	440—430
6	425—415

Somewhat lower values (*ca.* 240 and 180 cm^{-1}) are suggested[333a] for ν(ZnN) in complexes $[Zn(bipy)_2(phen)]X_2$ on the grounds that metal isotope shifts have been observed for $[Zn(bipy)_3]^{2+}$ in similar regions.

Detailed i.r. assignments, supported by force-constant calculations, have been given for $[HgNH_2X]_n$ (X = Cl or Br).[333b]

The i.r. and Raman spectrum of $N(HgMe)_3$ indicate C_1 symmetry in the solid, but C_{3v} in solution.[334] The solution-phase data (Raman, cm^{-1}) are shown in Table 29, together with assignments for related species.[325, 335]

Table 29 *Some assignments* (cm^{-1}) *for alkylmercuriamines*

Compound[a]	ν(HgMe)	ν(HgN)	*Ref.*
$N(HgMe)_3$	535 (A_1), 528 (E)	465 (E), 407 (A_1)	334[b]
$[H_3N(HgMe)]^+$	571	585	335[c]
$[H_2N(HgMe)_2]^+$	541	575 (asym), 515 (sym)	335[c]
$[HN(HgMe)_3]^+$	540	585 (asym), 420 (sym)	335[c]
$[N(HgMe)_4]^+$	560	590 (asym), 141 (sym)	335[c]
$(MeHg)_2NCN$	552, 546 (i.r.); 559, 552 (Raman)	427 (i.r.)	325
$(EtHg)_2NCN$	531 (i.r.); 530, 527 (Raman)	426 (i.r.); 461 (Raman)	325

[a] Cations as ClO_4^- salts. [b] Solution-phase Raman data; δ(CHgN) = 294 cm^{-1} (A_1). [c] *N*-deuterio-species also studied, and a simplified normal-co-ordinate analysis performed.

The isoelectronic ion $[O(HgMe)_3]^+$, on the other hand, is trigonal planar (D_{3h}) in the crystalline azide salt.[334] Assignments (i.r.) for $[(RHg)O(HgMe)_2](N_3)$ [R = Et, Pr^n, or Ph; ν(HgO) = *ca.* 550 and *ca.* 531 cm^{-1}] and re-assignments (i.r. and Raman) for $(MeHg)_2O$ have also been given.[334]

[331] (*a*) E. J. Baran, *Monatsh.*, 1971, **102**, 79; (*b*) K. C. Patil and E. A. Secco, *Canad. J. Chem.*, 1971, **49**, 3831.
[332] G. R. Cayley and D. N. Hague, *Trans. Faraday Soc.*, 1971, **67**, 2896.
[333a] S. N. Ghosh, *J. Inorg. Nuclear Chem.*, 1971, **33**, 3200; (*b*) K. Niwa, H. Takahashi, K. Higasi, and T. Kajiura, *Bull. Chem. Soc. Japan*, 1971, **44**, 3010.
[334] W. Thiel, F. Weller, J. Lorberth, and K. Dehnicke, *Z. anorg. Chem.*, 1971, **381**, 57.
[335] Nguyen Quy Dao and D. Breitinger, *Spectrochim. Acta*, 1971, **27A**, 905.

The Raman spectra of salts of the $[P(HgMe)_4]^+$ ion have been assigned on the basis of T_d symmetry as follows: ν_1 (a_1), 119; ν_3 (t_2), 354; ν_4 (t_2), 54 cm^{-1}.[336]

In crystalline hydrated zinc nitrate, the two sets of non-equivalent water molecules give different Raman frequencies in the δ(HOH) region.[337] $\nu(Zn-OH_2)$ is placed at 376 cm^{-1}, with δ(OZnO) at 241 cm^{-1}; corresponding values for the deuteriate are 358 and 226 cm^{-1}. An i.r. and Raman spectral study of aqueous $Zn(NO_3)_2$ has also been made.[338]

In a series of hydroxycadmium halides of types Cd(OH)X, $Cd_3(OH)_4X_2$, $Cd_2(OH)_3X$, $Cd_5(OH)_7Br_3$, and $Cd_7(OH)_{10}I_4$, ν(CdO) is assigned [339] to up to three bands in the region 360—280 cm^{-1}.

Largely on the basis of i.r. and Raman data [*e.g.* ν(HgS) at 299; ν(CN) at 2160 and 2140; δ(SCN) at 462 and 433 cm^{-1}], the species $[NCS(HgOH)_n$-HgSCN] and [HOHg···(NCS)Hg(SCN)···HgOH] are believed to be formed when $Hg(SCN)_2$ and $Hg(NO_3)_2$ solutions are mixed.[340a]

In $[HgCl_2(Bu_2{}^tS)]$,[340b] ν(HgS) is at 288 cm^{-1}.

The Raman spectra of polycrystalline solid solutions of $Zn_xCd_{1-x}S$ have been studied.[341] The LO mode of CdS (306 cm^{-1}) shifts linearly to higher frequency as the concentration of Zn^{2+} is increased, whereas the ZnS LO mode disappears on addition of Cd^{2+}. The study has important bearing on the use of Raman spectroscopy in solid-solution content analysis.

Reports of metal–halogen modes of some zinc,[18, 21, 22, 25] cadmium,[22, 23] and mercury [23] complexes have been mentioned in previous sections. Data have also been reported for complexes MX_2L_2 and MX_2L (M = Zn, Cd, or Hg; X = halide; L = thiomorpholin-3-one or thiazolidine-2-thione),[342] and for $Zn(MeCONHNH_2)_2Cl_2$ and $Zn(MeCONDND_2)_2Cl_2$.[343]

In $HgX_2(PPh_3)$ and $HgX_2(SbMePh_2)$, ν(HgX) modes have been assigned as follows (cm^{-1}):[344]

ν(HgCl)	*ca.* 280
ν(HgBr)	*ca.* 190
ν(HgI)	*ca.* 160

These data have been used to suggest halide-bridged dimeric structures for these compounds; ν(HgSb) is given as *ca.* 54 cm^{-1}.

I.r. and Raman spectral assignments have been given for halogenomercurates(II), $[HgX_3]^-$ and $[HgX_4]^{2-}$ (X = Cl, Br, or I), in the solid state

336 D. Breitinger, K. Geske, and W. Beitelschmidt, *Angew. Chem. Internat. Edn.*, 1971, **10**, 555.
337 M. H. Brooker and D. E. Irish, *Canad. J. Chem.*, 1971, **49**, 1510.
338 D. E. Irish and M. H. Brooker, *Trans. Faraday Soc.*, 1971, **67**, 1916, 1923.
339 L. Walter-Levy and D. Groult, *Bull. Soc. chim. France*, 1971, 1221.
340 (*a*) R. P. J. Cooney and J. R. Hall, *Inorg. Nuclear Chem. Letters*, 1971, **7**, 1017; (*b*) P. Biscarini, L. Fusina, and G. D. Nivellini, *Inorg. Chem.*, 1971, **10**, 2564.
340 R. P. J. Cooney and J. R. Hall, *Inorg. Nuclear Chem. Letters*, 1971, **7**, 1017.
341 J. Shamir and S. Larach, *Spectrochim. Acta*, 1971, **27A**, 2105.
342 D. De Filippo, F. Devillanova, C. Preti, and G. Verani, *J. Chem. Soc.* (*A*), 1971, 1465.
343 Yu. Ya. Kharitonov and R. I. Machkhoshvili, *Russ. J. Inorg. Chem.*, 1971, **16**, 638.
344 K. Brodersen, R. Palmer, and D. Breitinger, *Chem. Ber.*, 1971, **104**, 360.

and in solution.[345] Compounds of composition $A[Hg_2X_5]$ and $A_2[Hg_3X_8]$, except $A[Hg_2I_5]$ (A = cation), are extensively dissociated to $[HgX_3]^-$ in solution.

Reaction between Hg and AsF_5 in SO_2 affords $[Hg_3]^{2+}[AsF_6^-]_n$. This shows i.r. and Raman bands due to $[AsF_6]^-$, and a single polarized Raman line in SO_2 solution at 118 cm^{-1} which is assigned as ν(Hg—Hg) of a linear $[Hg_3]^+$ cation.[40] The compound $[Hg_{3n}]^{n+}[AsF_6^-]_n$ has also been obtained (ν_3 of $[AsF_6]^-$ = 699 cm^{-1}).[346]

12 Lanthanides

I.r. data have been given for the following double sulphates, and assignments made for vibrations of the SO_4^{2-} groups:[347]

$Ln_2(SO_4)_3,K_2SO_4,2H_2O$	(Ln = La or Ho)
$2Ln_2(SO_4)_3,3K_2SO_4,8H_2O$	(Ln = Ce or Gd)
$Ln_2(SO_4)_3,5K_2SO_4$	(Ln = La or Sm)

Vibrational spectroscopic data for several lanthanide vanadates $LnVO_4$ have been reported,[76, 77] but only for $LaVO_4$ is a detailed interpretation given. Bands at 305 cm^{-1} (Raman) and 295 cm^{-1} (i.r.) are tentatively assigned to ν(LaO).[76]

The i.r. spectrum of $EuCO_3$ supports the suggestion that this compound is isomorphous with the aragonite modification of $CaCO_3$; thus C_s symmetry of the CO_3^{2-} ion is indicated by appearance of ν_1 and splitting of ν_3 and ν_4.[348]

I.r. data have been given for various lanthanide complexes of edta, H[Ln(edta)],nH$_2$O (n = 5 or 7),[349] and of 1,1-hydrazine diacetates.[350]

The phosphorus dichloridate and dibromidate complexes $Nd(O_2PX_2)_3$ (X = Cl or Br) are believed to be polymeric with trigonal (D_3) local symmetry about Nd, on the basis of an i.r. study.[351] ν(NdO) is assigned at 410 and 385 cm^{-1} (X = Cl) and 415 and 370 cm^{-1} (X = Br). D_3 Symmetry has also been suggested for 15 lanthanide β-diketonates [LnL$_3$] (Ln = La, Pr, Nd, or Sm; HL = acac, bzac, bzbz, or ttfa).[352a] Assignments of ν(MO) are, *e.g.* for M = La; 650 (B_2), 525 (A_1), and 405 (A_2).

I.r. bands described as ν(MO) have also been assigned[352b] for several lanthanide 8-quinolinolates at 490—483 cm^{-1}, and for complexes Sm(OR)$_2$-(chel) [R = Pr^i or Bu^t; (chel) = anion derived from ethyl 1-methylacetoacetate or ethyl benzolylacetate][352c] at 635—600 and 350—285 cm^{-1}.

345 M. A. Hooper and D. W. James, *Austral. J. Chem.*, 1971, **24**, 1331, 1345.
346 R. J. Gillespie and P. K. Ummat, *Chem. Comm.*, 1971, 1168.
347 K. I. Petrov, G. N. Voronskaya, L. D. Iskhakova, and V. E. Plyushchev, *Russ. J. Inorg. Chem.*, 1971, **16**, 675.
348 W. Sterzel and W-D. Schnee, *Z. Naturforsch.*, 1971, **26b**, 615.
349 J. L. Mackey, D. E. Goodney, and J. R. Cast, *J. Inorg. Nuclear Chem.*, 1971, **33**, 3699.
350 V. P. Khramov and G. A. Aliev, *Russ. J. Inorg. Chem.*, 1971, **16**, 517.
351 E. J. Schimitschek, J. A. Trias, and C. Y. Liang, *Spectrochim. Acta*, 1971, **27A**, 2141.
352 (*a*) P. C. Mehta and S. P. Tandon, *Z. Naturforsch.*, 1971, **26a**, 759; (*b*) H. F. Aly, F. M. Abdel Kerim, and A. T. Kandil, *J. Inorg. Nuclear Chem.*, 1971, **33**, 4340; (*c*) P. C. Mehta, S. S. L. Surana, and S. P. Tandon, *Spectroscopy Letters*, 1971, **4**, 349.

Tentative assignments have been proposed for the i.r. spectra of lanthanide oxide halides MOX (X = Cl or Br).[353] Raman data are also listed, and the numbers of bands observed by each technique approximate to group-theoretical predictions.

13 Actinides

Data on $Th(BH_4)_4$, $Th(BD_4)_4$, and $[Th(BH_4)_5]^-$ have been given in Table 7.[57]

Matrix-isolation methods have been used to show formation of carbonyl complexes of uranium, $[U(CO)_n]$, showing ν(CO) at 1961 (n = 6), 1938 (n = 5), 1919 (n = 4), 1893 (n = 3), 1855 and 1846 (n = 2), and 1832 and 1817 cm^{-1} (n = 1).[354]

I.r. bands in the spectrum of $(\pi\text{-}C_6H_6)U(AlCl_4)_3$ at 3075, 1020, 740, 682, and 667 cm^{-1} are due to the co-ordinated benzene, while those at 480 and 550 cm^{-1} are associated with motions of chloride bridges.[355]

The i.r. spectrum of α-UO_3 (1100—200 cm^{-1}) has been re-interpreted.[356] Bands at 980 and 890 cm^{-1} are now believed not to be combination bands but to arise from stretching of uranyl-type bonds with U—O lengths near 1.75 Å. *X*-Ray crystallographic evidence supports the view that the structure is more complex than the generally accepted simple hexagonal cell.

X-Ray powder data have been used to assign space group D_{3d}^1 to Cr_2UO_6, although this is not a unique choice. Unit-cell analysis on this basis predicts $3A_{2u} + 4E_u$ i.r. modes, in good agreement with the observed spectrum, *viz*, 670, 570, 487, 455, 326, 234, and 205 cm^{-1}.[357]

A normal-co-ordinate analysis has been made of UO_2F_2, treating it as an infinite sheet structure.[358] Although only three of the eight optically active modes have been reported, the preliminary data are significant. The U—F stretching force constant (0.41 mdyn $Å^{-1}$) is very small (*cf.* 3.78 mdyn $Å^{-1}$ for UF_6), indicating highly ionic bond character.

Two studies on the $[UO_2F_5]^{3-}$ ion have been reported. Raman, i.r., fluorescence, and electronic absorption spectra have been used to establish the energy levels of the NH_4^+ salt.[359] From the vibrational frequencies the U—O force constant and bond distance (1.748 Å) have been calculated. The i.r. assignments include (cm^{-1}): ν(UF), 434; $\nu_{sym}(UO_2)$, 825; $\nu_{asym}(UO_2)$, 890; and $\delta_{sym}(UO_2)$, 280 and 205.[359] The $\nu_{sym}(UO_2)$ (842—791 cm^{-1}), $\nu_{asym}(UO_2)$ (910—848 cm^{-1}), and ν(UF) (409—348 cm^{-1}) modes of

353 L. J. Basile, J. R. Ferraro, and D. Gronert, *J. Inorg. Nuclear Chem.*, 1971, **33**, 1047.
354 J. L. Slater, R. K. Sheline, K. C. Lin, and W. Weltner, *J. Chem. Phys.*, 1971, **55**, 5129.
355 M. Cesari, U. Pedretti, A. Zazzetta, G. Lugli, and W. Marconi, *Inorg. Chim. Acta*, 1971, **5**, 439.
356 S. Siegel and H. R. Hoekstra, *Inorg. Nuclear Chem. Letters*, 1971, **7**, 497.
357 H. R. Hoekstra and S. Siegel, *J. Inorg. Nuclear Chem.*, 1971, **33**, 2867.
358 K. Ohwada, *J. Inorg. Nuclear Chem.*, 1971, **33**, 1615.
359 C. L. Garg and K. V. Narasimham, *Spectrochim. Acta*, 1971, **27A**, 863.

$[UO_2F_5]^{3-}$ salts are influenced by the cation *via* cation···F and cation···O interactions.[360]

Further data on $\nu(UO_2)$ vibrations in uranyl(VI) complexes are listed in Table 30. In $[UO_2\{OP(NMe_2)_3\}_2X_2]$ (X = Cl, Br, NO_3, or NCS),

Table 30 *Further data on* $\nu_{asym}(UO_2)$ *modes* (cm^{-1}) *in uranyl*(VI) *complexes*

Complex	$\nu_{asym}(UO_2)$	*Ref.*
$[UO_2(dithiolato)_2]^{2-}$	910 ± 5	361[a]
$[UO_2(dithiolato)_2(LO)]^{2-}$	909—900	362[a]
$[UO_2(O_2)(C_2O_4)(H_2O)]^{2-}$	900—873	363[b]
$[UO_2(O_2)(H_2O)_2]$	925	363[b]
$[UO_2(acac)(O{=}CMeCH{=}CMeNR^1R^2)]$[c]	912—897	364[f]
$[UO_2(acac)(NR^1R^2{}_2)]$[d]	906—900	364[f]
$[UO_2(chel)(NH_2R)]$[e]	909—892	364[f]
$[UO_2(\beta\text{-diketonate})_2(L)]$[g]	940—895	365
$[UO_2(ONNO)(EtOH)]$[h]	910	366

[a] (dithiolato) = maleonitriledithiolato $[NC(S^-)C{=}C(S^-)CN]$ or isomaleonitriledithiolato; (LO) = pyO, Ph_3PO, or Ph_3AsO. [b] (O_2) = peroxo-ligand. [c] R^1 = H or R^2; R^2 = Ph, H, or alkyl. [d] R^2 = H or R^1; R^1 = Me or Et. [e] (chel) = dibenzoylmethane or di(*p*-methoxybenzoyl)methane; R = Ph, H, or alkyl. [f] Tentative assignments given of ν(U—O) at 526—513 cm^{-1} and of ν(U—N) of amine adducts at 568—516 cm^{-1}. [g] (β-diketonate) = thenoyl- or benzoyl-trifluoracetone, benzoylacetone, dibenzoylmethane, or trifluoro- or hexafluoro-acetylacetone; (L) = pyO or 4-substituted pyridine-*N*-oxide; ν(U—O) given as 485—366 cm^{-1}. [h] (ONNO) = *NN'*-*o*-phenylenebis(salicylideneiminato); crystal structure reported.

$[UO_2\{OP(NMe_2)_3\}_3(NCS)_2]$, and $[UO_2\{OP(NMe_2)_3\}_5](ClO_4)_2$, the $UO_2{}^{2+}$ grouping is apparently linear, since $\nu_{sym}(UO_2)$ could not be observed in the i.r. spectra.[367]

Actinide–halogen wavenumbers (cm^{-1}) have been assigned as follows: $[PaOCl_3(Ph_3PO)_2]$, 261 and 254;[90] $[Pa(dtc)_4Cl]$, 317 (dtc = *NN*-diethyldithiocarbamato);[91] and $[MCl_4(\text{substituted urea})_n]$, 285—242 (M = U or Th; n = 2, 3, 4, or 6).[368] Data on UOF_2 [358] and $[UO_2F_5]^{3-}$ salts [359, 360] have already been mentioned. For solid UCl_6 all three Raman-active modes predicted for O_h symmetry [ν_1 369; ν_2 325; ν_5 126 cm^{-1}] have been observed, together with lines at *ca.* 100 cm^{-1} which are probably lattice modes.[369]

360 V. I. Sergienko, R. L. Davidovich, V. I. Kostin, and A. A. Matstutsin, *Spectroscopy Letters*, 1971, **4**, 19.
361 L. Zimmer and K. H. Leiser, *Inorg. Nuclear Chem. Letters*, 1971, **7**, 563.
362 L. Zimmer and K. H. Leiser, *Inorg. Nuclear Chem. Letters*, 1971, **7**, 1163.
363 R. N. Shchelokov, V. I. Belomestnykh, and R. M. Somova, *Doklady Chem.*, 1971, **196**, 38.
364 J. M. Haigh and D. A. Thornton, *J. Inorg. Nuclear Chem.*, 1971, **33**, 1787.
365 M. S. Subramanian and V. K. Manchanda, *J. Inorg. Nuclear Chem.*, 1971, **33**, 3001.
366 G. Bandoli, D. A. Clemente, U. Croatto, M. Vidali, and P. A. Vigato, *Chem. Comm.*, 1971, 1330.
367 A. K. Majumdar, R. G. Bhattacharyya, and D. C. Bera, *Chem. and Ind.*, 1971, 730.
368 K. W. Bagnell, J. G. H. du Preez, and M. L. Gibson, *J. Chem. Soc.* (*A*), 1971, 2124.
369 J. Shamir and A. Silberstein, *Spectroscopy Letters*, 1971, **4**, 341.

7

Vibrational Spectra of Some Co-ordinated Ligands

BY G. DAVIDSON

The ligands have been subdivided according to the position of the donor atom in the Periodic Table, except for some which contain more than one possible donor atom (*e.g.* $-NO_2$, $-ONO$); these have been treated separately. Each paper is referred to only once in this chapter, and it will be necessary for a reader interested in a complex containing several different ligands to check through all of the possible sections in which it might be mentioned.

1 Carbon Donors

This section will include references to both localized (σ-) M—C bonded and delocalized (π-) M—(C_n) bonded systems.

The i.r. spectra of solutions of C_5H_5Li, C_5H_5Na, C_5H_5MgBr, and $(C_5H_5)_2Mg$ in THF were all consistent with the presence of a symmetric, and presumably largely ionically bound, C_5H_5 unit.[1] It should be noted, however, that very incomplete spectra were obtained because of the presence of large regions of solvent absorption.

A straightforward σ-bonded structure has been proposed for $Ti(CH_2Ph)_4$ on the basis of its i.r. spectrum.[2] $\nu(CH_2)_2$ bands at 2863 and 2930 cm^{-1} are typical of sp^3-hybridized $\rangle CH_2$ units.

As part of a study of low-oxidation-state cyclopentadienyl complexes of Ti and Zr, Wailes and Weigold[3] prepared complexes with formulae $[(C_5H_5)_2ZrCl]_2,C_6H_6$ and $[(C_5H_5)_2ZrCl]_n$. The former gave an i.r. spectrum characteristic of h^5-C_5H_5 species, whereas the latter appeared to contain h^1- and h^5-C_5H_5.

The reaction of CO_2 with $(h^5\text{-}C_5H_5)_2TiPh_2$ gives the complex (1), in which there has been carboxylation of the phenyl ring (confirmed by a single-crystal *X*-ray study). Characteristic i.r. bands of h^5-C_5H_5 and of the arene fragment are found, with further (unassigned) bands at 880m, 1130s, 1280vs, 1620s, and 1660vs (all cm^{-1}).[4]

1 W. T. Ford, *J. Organometallic Chem.*, 1971, **32**, 27.
2 W. Bruser, K. H. Thiele, P. Zdunneck, and F. Brune, *J. Organometallic Chem.*, 1971, **32**, 335.
3 P. C. Wailes and H. Weigold, *J. Organometallic Chem.*, 1971, **28**, 91.
4 I. S. Kolomnikov, T. S. Lobeeva, V. V. Gorbachevskaya, G. G. Aleksandrov, Yu. T. Struchkov, and M. E. Vol'pin, *Chem. Comm.*, 1971, 972.

I.r. evidence has been obtained for hindered rotation about the metal–carbon (of carbene) bond in a number of transition-metal carbene complexes. In $Ar(CO)_2Cr{-}C(OMe)Ph$ (Ar = various arene ligands), 3 or 4 ν(C—O) bands are seen (in C_6H_{12} solution). This can only be explained on the basis of the presence of two isomers, (2a) and (2b).[5]

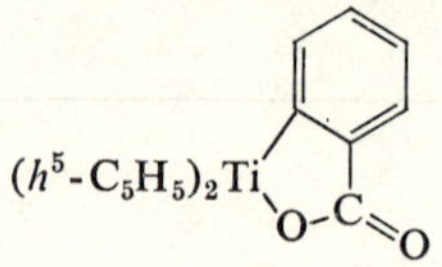

(1)

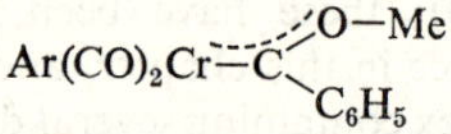

(2a) (*cis-*)

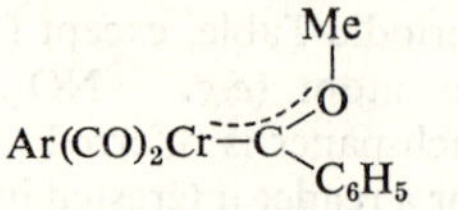

(2b) (*trans-*)

In connection with a mechanistic study, a very thorough investigation of the ν(CO) region of the carbene complexes $M(CO)_5L$ [M = Cr or W; L = $C(OEt)CH_2Ph$ or C(OEt)Ph] was made, including some ^{13}CO enrichment experiments. Observed and calculated frequencies were given for all possible ^{13}CO derivatives of $W(CO)_5C(OEt)Ph$. Bandshape analysis and measurement of integrated intensities were used in deriving dipole-moment derivatives. The results indicated that the carbene ligand is bound in a similar manner to, for example, phosphines, but in a quite different manner from the binding of simple σ-donors (*e.g.* amines).[6a]

Anionic pentacarbonyl-heptafluoro-1-methylpropenyl complexes $[(CO)_5{-}M{-}C(CF_3){=}CF(CF_3)]^-$ (M = Cr, Mo, or W) have been prepared. All show three carbonyl stretching bands, consistent with effective C_{4v} symmetry. The A_1 bands are at 2048, 1872 (Cr); 2059, 1872 (Mo); 2061, 1863 cm^{-1} (W); with the E bands at 1920 (Cr), 1923 (Mo), 1913 cm^{-1} (W).[6b]

Anionic acetylene complexes $[M(CO)_5(C{\equiv}CR)]^-$ (R = Me or Ph; M = Cr, Mo, or W) have been described,[7] which show ν(C≡C) in the region 2065—2110 cm^{-1}. This confirms the presence of σ-bonded acetylide groups, by contrast with π-bonded acetylene complexes, which show a lowering of up to 400 cm^{-1} in ν(C≡C).

Schäfer *et al.*[8–10] have obtained a reasonably complete Raman spectrum of $Cr(C_6H_6)_2$. Using these data, and data from $M(C_6H_6)_2$ (M = Mo, W,

[5] H. J. Beck, E. O. Fischer, and C. G. Kreiter, *J. Organometallic Chem.*, 1971, **26**, C41.
[6] (*a*) J. J. Darensbourg and M. Y. Darensbourg, *Inorg. Chim. Acta*, 1971, **5**, 247; (*b*) W. J. Schlientz and J. K. Ruff, *J. Organometallic Chem.*, 1971, **33**, C64.
[7] W. J. Schlientz and J. K. Ruff, *J. Chem. Soc.* (*A*), 1971, 1139.
[8] L. Schäfer, J. F. Southern, and S. J. Cyvin, *Spectrochim. Acta*, 1971, **27A**, 1083.
[9] J. Brunvoll, J. F. Southern, and S. J. Cyvin, *J. Organometallic Chem.*, 1971, **27**, 69.
[10] S. J. Cyvin, J. Brunvoll, and L. Schäfer, *J. Chem. Phys.*, 1971, **54**, 1517.

or V), $(C_6H_6)M(CO)_3$ (M = Cr or Mo), $(C_6H_6)M(C_5H_5)$ (M = Cr or Mn), and $(C_6H_6)_2Cr^+$, a normal-co-ordinate analysis of complexed benzene was performed. This appears to show that the force field for benzene in these systems is very similar to that in free benzene, with the observed frequency shifts on complex formation being due to kinematic coupling. Since the force field is so similar to that of free benzene, this provides further evidence that the benzene ring maintains a six-fold symmetry, and does not suffer distortion in these complexes.

A study of the i.r. and Raman spectra of the compounds $(Ar)Cr(CO)_3$ (Ar = C_6H_6, C_6H_5F,C_6H_5Me,*o*-$Me_2C_6H_4$, or 1,2,3-$Me_3C_6H_3$) in solution has confirmed that the method of 'local symmetry' is of very restricted validity as far as ν(CO) modes are concerned. Evidence is also presented for assignment of low-frequency modes previously reported only from solid-state spectra. A detailed assignment of the vibrations of (C_6H_5Me)-$Cr(CO)_3$ is proposed.[11]

The vibrations of pyrrole, C_4H_4NH, are shifted only slightly on co-ordination to the $Cr(CO)_3$ unit.[12a] The main difference lies in the higher wavenumber of the out-of-plane CH deformations [768, 736 cm^{-1} in C_4H_4NH; 798, 766 cm^{-1} in $(C_4H_4NH)Cr(CO)_3$]. ν(CO) bands are found at 1927, 1845, and 1801 cm^{-1}, with δ(Cr—C—O) at 682, 646 cm^{-1}.

A series of transition-metal–vinylsultone complexes 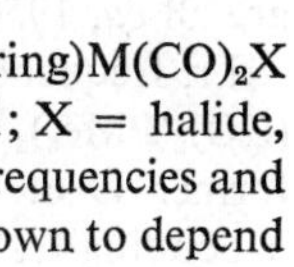

M = $Mn(CO)_5$, $(\pi$-$C_5H_5)Mo(CO)_3$, or $(\pi$-$C_5H_5)Fe(CO)_2$; R = Me or Ph] has been prepared.[12b] All show two characteristic bands in the ν(SO) region, at 1301—1322 cm^{-1} and 1164—1186 cm^{-1}.

Metal–ligand and ligand vibrations in the complexes $(\pi$-ring$)M(CO)_2X$ [M = Cr, Mn, or Fe; ring = C_5H_5, C_5H_4Me, C_6H_6, C_6H_5Cl; X = halide, $Sn(hal)_3$, or Hg(hal)] have been studied.[13] In particular, the frequencies and intensities of bands due mainly to δ(M—C—O) have been shown to depend upon the nature of the ligand X.

Two isomeric compounds have been isolated from the reaction between $MeMn(CO)_5$ and *o*-$CH_2{=}CHC_6H_4PPh_2$ (sp). One is the acyl derivative, $(MeCO)Mn(CO)_3(sp)$, containing bidentate sp [ν(C=O) at 1644 cm^{-1}]. The other is believed to contain a delocalized π-1,2-dimethyloxapropenyl–metal bond, (3) [no ν(C=O) observed]. The latter reacts with CO to give (4) [ν(C=O) at 1622 cm^{-1}]. ν(C≡O) bands were also reported for all three compounds.[14]

[11] G. Davidson and E. M. Riley, *Spectrochim. Acta*, 1971, **27A**, 1649.
[12] (*a*) K. Öfele and E. Dotzaner, *J. Organometallic Chem.*, 1971, **30**, 211; (*b*) D. W. Lichtenberg and A. Wojcicki, *ibid.*, 1971, **33**, C77.
[13] A. R. Manning, *J. Chem. Soc.* (*A*), 1971, 106.
[14] M. A. Bennett and R. Watt, *Chem. Comm.*, 1971, 95.

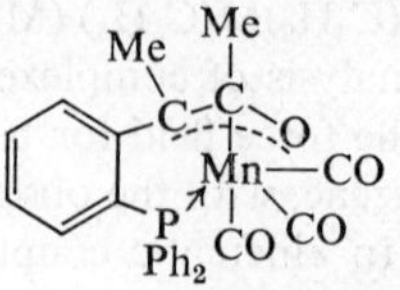

(3)

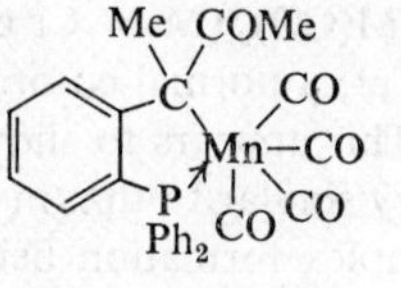

(4)

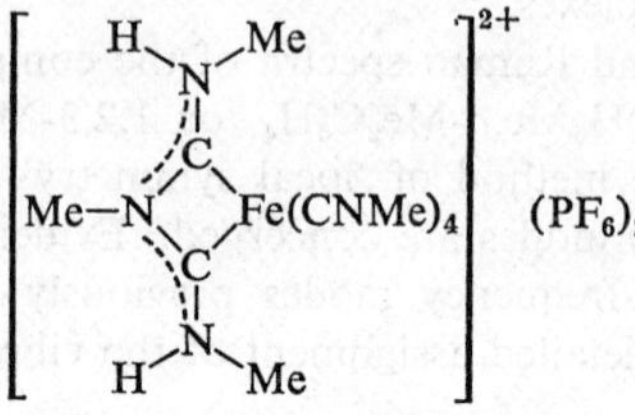

(5)

Methylamine can be added to two of the isocyanide ligands in [Fe(CNMe)$_6$](HSO$_4$)$_2$ to give the complex (5). Assignments of i.r. frequencies to ν(N—H), ν(C≡N), ν(N⋯C⋯N), and δ(NH) in this new complex have been given.[15]

Assignments of ν(C=O) and ν(C—O—R) vibrations have been made for the metal–carbene complexes $\left[\text{M}=\text{C}\begin{smallmatrix}\text{OR}\\\text{Me}\end{smallmatrix} \right]^+$ {M = (π-C$_5$H$_5$)Fe(CO)-[P(hex)$_3$], (π-C$_5$H$_5$)Ru(CO)$_2$, (π-C$_5$H$_5$)Ru(CO)(PPh$_3$), or (π-C$_5$H$_5$)Ru(CO)-[P(hex)$_3$]; hex = cyclohexyl; R = Me or Et}.[16] The values of the former are all ca. 1950 cm^{-1}, the latter ca. 1250 cm^{-1}.

I.r. frequencies have been noted and partially assigned for FeI(COMe)-(CO)$_2$(PMe$_3$)$_2$ and FeI(Me)(CO)$_2$(PMe$_3$)$_2$.[17a] The ν(C=O) band in the former is at 1590 cm^{-1}; this low wavenumber is explained by partial π-bonding from the metal to the acetyl group.

In the complex (h^5-C$_5$H$_5$)Fe(CO)(SnPh$_3$)(PhC≡CPh), ν(C≡C) has been assigned[17b] to an i.r. band at 1870 cm^{-1}, with ν(C≡O) at 1945 cm^{-1}.

ν(C=C) has been assigned in a number of (fluoro-olefin)Fe(CO)$_4$ complexes to a band in the region of 1460 cm^{-1} (CH$_2$=CF$_2$) to 1333 cm^{-1} (CBr$_2$=CF$_2$).[18a] The authors suggest that this is good evidence for a σ-bonded structure $\begin{smallmatrix}\text{C}-\text{C}\\\text{Fe}\end{smallmatrix}$ for the metal–olefin bond. It is unsafe

---

[15] J. Miller, A. L. Balch, and J. H. Enemark, *J. Amer. Chem. Soc.*, 1971, **93**, 4613.
[16] M. L. H. Green, L. C. Mitchard, and M. G. Swanwick, *J. Chem. Soc. (A)*, 1971, 794.
[17] (a) M. Pankowski and M. Bigorgne, *J. Organometallic Chem.*, 1971, **30**, 227; (b) A. N. Nesmeyanov, N. E. Kolobova, V. V. Skripkin, and K. N. Anisimov, *Doklady Chem.*, 1971, **196**, 98.
[18] (a) R. Fields, G. L. Godwin, and R. N. Haszeldine, *J. Organometallic Chem.*, 1971, **26**, C70; (b) J. Y. Chenard, D. Commereuc, and Y. Chauvin, *ibid.*, 1971, **33**, C69.

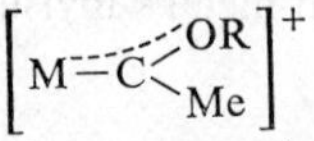

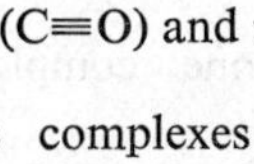

to place great reliance upon the assignment of one band in the spectrum, however, especially as appreciable mixing of normal modes may occur.

Complexes of the 1,α-methylbenzylimine of ethyl glyoxylate (5a) with $Fe(CO)_4$, $Rh(CO)_2Cl$, and $Pt(C_2H_4)Cl_2$ have been prepared.[18b] The first

$$\begin{array}{c} \text{Me} \\ | \\ \text{HC}-\text{N}=\text{CH}-\text{C}-\text{OEt} \\ | \quad\quad\quad \| \\ \text{Ph} \quad\quad\quad \text{O} \end{array}$$

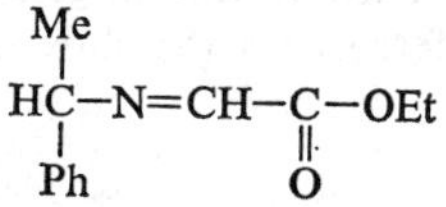

(5a)

two co-ordinate via the C=N double bond, but the Pt complex involves a Pt—N σ-bond. These conclusions are based mainly upon n.m.r. evidence, since the $\nu(C=O)$ and $\nu(C=N)$ frequencies for all are very similar.

A partial assignment of π-allyl ligand vibrations in $(\pi\text{-}C_3H_5)Fe(acac)_2$ has been made:[19] $\nu_{as}(C-C-C)$ 1470 cm^{-1}; $\nu_s(C-C-C)$ ca. 1035 cm^{-1}; $\delta(C-C-C)$ ca. 495 cm^{-1}.

A new, photochemical preparation of bis(butadiene)iron carbonyl, $(C_4H_6)_2Fe(CO)$, has been reported.[20] No vibrational data were given for the co-ordinated butadiene molecules, but $\nu(CO)$ was found at 1984 cm^{-1}.

Following the recent definitive work on the vibrational assignments of ferrocene, Brunvoll et al. have carried out a normal-co-ordinate analysis on that molecule.[21] Few details were given in the published report, but it appeared that the force field for the cyclopentadienyl ring in ferrocene is similar to that in the free $C_5H_5^-$ ion. Differences in vibrational frequencies between the free and complexed C_5H_5 units are derived, it is suggested, from kinematic coupling effects.

The low-frequency vibrations of π-indenyl complexes of Fe and Ru, and σ-indenyl complexes of Hg, have been assigned, by analogy with the homologous cyclopentadienyl derivatives.[22] The following general conclusions were reached: (a) ν(M—ring) in the π-complexes depends upon the electron density associated with the ring, and not its reduced mass. The observed order of frequencies is $(H_4\text{ind})_2Ru > Cp_2Ru > (\text{ind})_2Ru$; (b) ν(Hg—C) in the σ-complexes shows a reverse trend, $(\text{ind})_2Hg > Cp_2Hg$.

A comparison of $\nu(C=O)$ and ν_{as}(Fe—ring) in a series of acylferrocenes and -ferricenium cations shows that the former mode increases in frequency and the latter decreases in frequency on oxidation. Thus, the metal–ring bonding is weaker in the oxidized species, leading to stronger bonding conditions in the ligand moiety.[23] Thus in $(C_5H_5)Fe(C_5H_4COR)$, $\nu(C=O) = 1621$ cm^{-1}, ν_{as}(Fe—ring) = 484 cm^{-1} (R = Ph); 1656, 483 cm^{-1}

---

[19] G. A. Razuvaev, G. A. Domrachev, O. N. Suvorova, and L. G. Abakumova, *J. Organometallic Chem.*, 1971, **32**, 113.
[20] E. Koerner von Gustorf, Z. Pfajfer, and F.-W. Grevels, *Z. Naturforsch.*, 1971, **26b**, 66.
[21] J. Brunvoll, S. J. Cyvin, and L. Schäfer, *J. Organometallic Chem.*, 1971, **27**, 107.
[22] E. Samuel and M. Bigorgne, *J. Organometallic Chem.*, 1971, **30**, 235.
[23] P. Carty and M. F. A. Dove, *J. Organometallic Chem.*, 1971, **28**, 125.

(R = Me); for the cations $(C_5H_5)Fe(C_5H_4OR)^+$, however, the corresponding pairs of figures are 1655, 383 cm^{-1} (R = Ph); 1690, 419 cm^{-1} (R = Me).

ν(C=O) vibrations have also been assigned [24] in the substituted acylferrocenes (6; R^1, R^2, R^3 = H or Me) and (7; R = H, Et, CH_2Ph, *etc.*). They occur, for both series, within the range 1663—1670 cm^{-1}. A closely related series of hydroxyferrocenes (8) showed interesting variations in

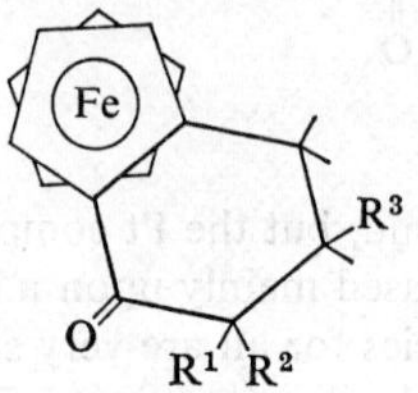

(6)

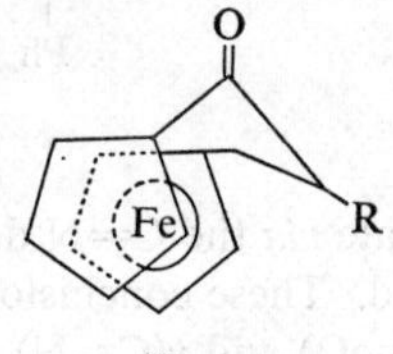

(7)

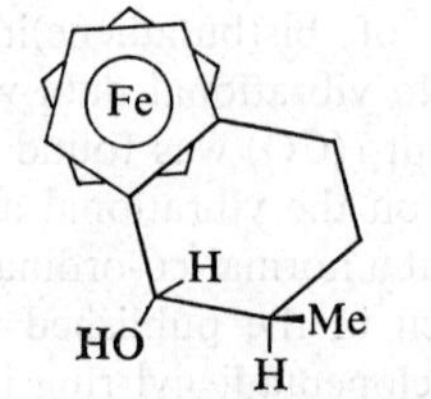

(8a)

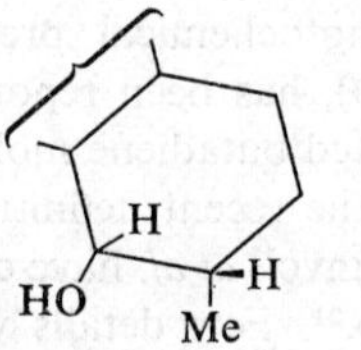

(8b)

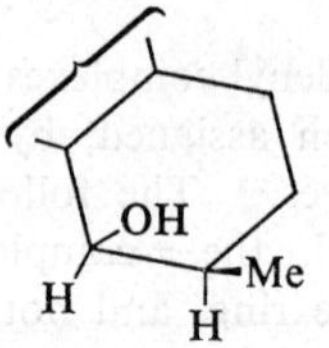

(8c)

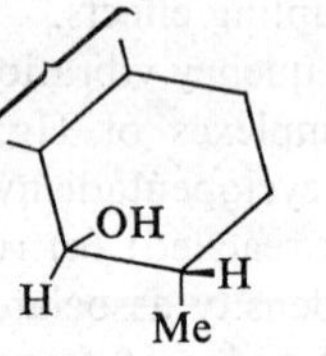

(8d)

ν(O—H).[25] In the *endo*-isomers [(8a) and (8c)], ν(OH) is at *ca.* 3550 cm^{-1}, whereas in the *exo*-isomers [(8b) and (8d)] it lies at *ca.* 3615 cm^{-1}. This suggests that hydrogen-bonding between the hydroxy-group and the iron atom is only possible in the former.

I.r. spectra have been listed, but with no assignment, for a number of symmetrically disubstituted ferrocenes, *e.g.* 1,1′-bis(diphenylarsine)ferrocene.[26]

[24] B. Gautheron and J. C. Leblanc, *Bull. Soc. chim. France*, 1971, 3629.
[25] B. Gautheron and R. Broussier, *Bull. Soc. chim. France*, 1971, 3636.
[26] J. J. Bishop, A. Davison, M. L. Katcher, D. W. Lichenberg, R. E. Merrill, and J. C. Smart, *J. Organometallic Chem.*, 1971, **27**, 241.

Several unassigned i.r. bands have also been reported for $Fe(C_6H_6)_2^{2+}$ (3116s, 1627m, 1456s cm^{-1}).[27]

Tricarbonylcycloheptatrieneiron, $C_7H_8Fe(CO)_3$ (9), which shows a ν(C—H_{exo}) band at 2795 cm^{-1}, is reduced by BuLi in THF to $(C_7H_7)Fe(CO)_3^-$, which shows only two ν(CO) bands in THF solution (1942 and 1868 cm^{-1}).[28]

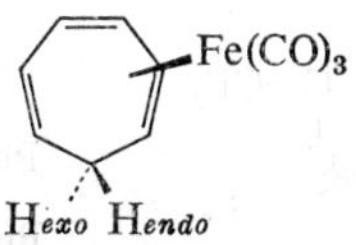

(9)

A single-crystal *X*-ray diffraction study has established that tetracyanoethylene undergoes a 1,3-addition reaction with tricarbonylcycloheptatrieneiron, with formation of a novel 2,3,4,6-*tetrahapto*-bonded system [ν(CO) at 2073, 2011 cm^{-1}]. A similar 1,3-addition compound [ν(CO) at 2086, 2026 cm^{-1}] is formed from the reaction of tricarbonyl-*N*-methoxycarbonylazepineruthenium with tetracyanoethylene.[29]

Reaction between thiobenzophenone ($Ar_2C{=}S$) and $Fe_2(CO)_9$ affords *ortho*-metallated compounds.[30] For example, (10) has ν(CO) at 2069, 2033, 1995, and 1976 cm^{-1}, showing that no bridging CO groups are present.

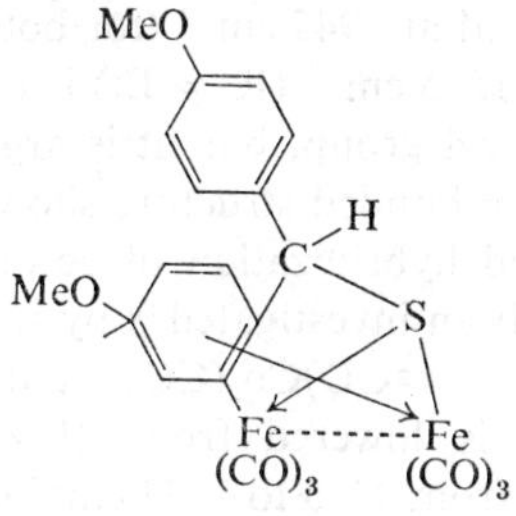

(10)

The complex (11) shows bands which must be assigned[31] to ν(C=C) and ν(C=O) at 1467 and 1425 cm^{-1}. No absorptions above 1500 cm^{-1} which could be so assigned were seen, and thus both the C=C and C=O bonds must be forming bonds to the $Fe(CO)_3$ fragment.

27 J. F. Helling, S. L. Rice, D. M. Braitsch, and T. Mayer, *Chem. Comm.*, 1971, 930.
28 H. Maltz and B. A. Kelly, *Chem. Comm.*, 1971, 1390.
29 M. Green, S. Tolson, J. Weaver, D. C. Wood, and P. Woodward, *Chem. Comm.*, 1971, 222.
30 H. Alper and A. S. K. Chan, *Chem. Comm.*, 1971, 1203.
31 A. N. Nesmeyanov, M. I. Rybinskaya, L. V. Rybin, V. S. Kaganovich, and P. V. Petrovskii, *J. Organometallic Chem.*, 1971, **31**, 257.

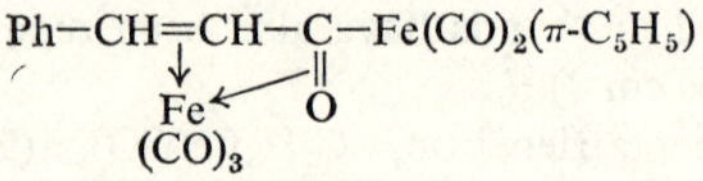

(11)

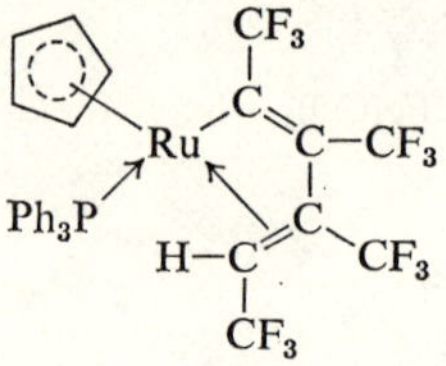

(12)

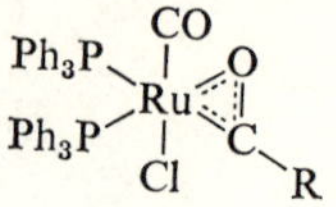

(13)

The spectra of isomers of the (allene-trimer) complex $Fe_2(CO)_6(C_9H_{12})$, which do not contain iron–iron bonds, differ markedly from those of iron–iron-bonded complexes in the carbonyl stretching region, those of the former being much more complex.[32]

The non-co-ordinated C=C bond in the ruthenium complex (12) gives a stretching band[33] at 1640 cm^{-1}.

Reaction of $Ru(PPh_3)_3(H)Cl$ or $Ru(PPh_3)_3(CO)(H)Cl$ with EtCHO or MeCHO affords the π-bonded acyl complexes, formulated as (13; R = Et or Me).[34] This structure is based upon n.m.r., mass spectral, and i.r. data and it is in keeping with the chemical properties of the complex. Thus ν(C≡O) gives a single band at 1945 cm^{-1} (for both complexes); a band at 1510 cm^{-1} (R = Me) or 1505 cm^{-1} (R = Et) is too low to be assigned to ν(C=O) of a σ-bonded acyl group, but (it is argued) is reasonable for a ν(C⋯O) vibration of the π-bonded structure shown.

The nature of the bond hybridization of co-ordinated MeC≡CH and MeC≡CMe ligands has been investigated[35] by studying the i.r. spectra of $(MeC{\equiv}CH)Co_2(CO)_8$, $(MeC{\equiv}CD)Co_2(CO)_8$, and $(MeC{\equiv}CMe)Co_2(CO)_8$. In MeC≡CH, ν(C≡C) is lowered from 2142 to 1550 cm^{-1}, and in MeC≡CMe it is lowered from 2313 to 1633 cm^{-1} upon co-ordination. The data show that the 'acetylenic' carbons are, in fact, sp^2-hybridized in the complexes.

The acetylenic vibrations of a series of cobalt carbonyl complexes have also been studied by Robinson and Spencer.[36] ν(C≡C) (co-ord.) is assigned to the band at 1466 cm^{-1} in $Co_5(CO)_{15}C_3H$ (14a) and to 1527 cm^{-1} in $Co_8(CO)_{24}C_6$ (14b) [X = $C(CO)_9Co_3$]. ν(C≡C) (free) in the latter complex is found at 2070 cm^{-1}.

[32] S. Otsuka, A. Nakamura, and K. Tani, *J. Chem. Soc.* (*A*), 1971, 154.
[33] T. Blackmore, M. I. Bruce, F. G. A. Stone, R. E. Davis, and A. Garza, *Chem. Comm.*, 1971, 852.
[34] R. R. Hitch, S. K. Gondal, and C. T. Sears, *Chem. Comm.*, 1971, 777.
[35] Y. Iwashita, A. Ishikawa, and M. Kainosho, *Spectrochim. Acta*, 1971, **27A**, 271.
[36] B. H. Robinson and J. L. Spencer, *J. Organometallic Chem.*, 1971, **30**, 267.

I.r. frequencies for ν(CO) and ν(C≡C) have been listed for a large number of carbonyltris(triphenylphosphine) complexes of Rh^I and Ir^I with alk-1-ynes. All bands assignable to alkynyl stretching modes fall into the region 2097—2159 cm^{-1}. This suggests a relatively slight perturbation of the triple bond, and terminal σ-bonding to the metal.[37] In the Rh complexes $\{Rh[P(OMe)_3]_4[C_2(CF_3)_2]\}BPh_4$ and $\{Rh[P(OMe)_3]_4[C_2(CO_2Me)_2]\}BPh_4$, however, bands assigned to ν(C≡C) are found at 1834 and 1805 cm^{-1}, respectively, and these must be formulated as π-complexes.[38]

ν(C=O) vibrations have been assigned [38b] in a number of acyl–Rh complexes containing Rh—C σ-bonds, e.g. (15a).

$$X-C\equiv C-H \qquad\qquad X-C\equiv C-C\equiv C-X$$
$$\quad\ \ |\qquad\qquad\qquad\qquad\qquad\ \ |$$
$$Co_2(CO)_6 \qquad\qquad\qquad Co_2(CO)_6$$

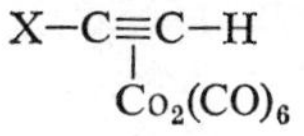

(14a)　　　　　　　　　(14b)

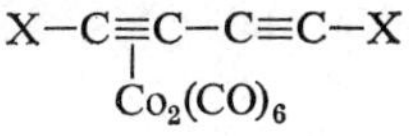

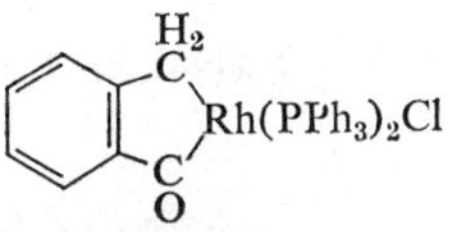

(15a)　　　　　　　　　(15b)

Largely unassigned data have been listed for the rhodium complexes of 1,2,5,6,8-pentamethylenecyclodecane (allene pentamer, $C_{15}H_{20}$) (15b), where X = σ-C_6F_5 or CO.[38c]

Unassigned i.r. data have been listed (in the region 1500—600 cm^{-1}) for $[IrCl(cyclo-octene)_2]_2$, which is a Cl-bridged species.[39]

I.r. spectra in the ν(CH) region for some (π-cyclohexadienyl)–Rh and –Ir complexes have been listed by White and Maitlis.[40]

Nelson and Jonassen have reviewed[41] the co-ordination of olefins and acetylenes to Ni, Pd, and Pt. Shifts in ν(C=C) and ν(C≡C) upon co-ordination are discussed in terms of the strength of the metal–ligand interaction.

Sato et al.[42] have prepared the first complex containing a four-membered lactone ring co-ordinated to a metal via its π-electrons: $(h^1$-$C_5H_5)(h^5$-$C_5H_5)$-(π-2,2,4-trimethyl-3-hydroxypent-3-enoic acid lactone)nickel (16). The

---

[37] C. K. Brown, D. Georgian, and G. Wilkinson, *J. Chem. Soc.* (A), 1971, 3120.
[38] (a) L. M. Haines, *Inorg. Chem.*, 1971, **10**, 1693; (b) K. G. Powell and F. J. McQuillin, *Chem. Comm.*, 1971, 931; (c) R. B. King and P. N. Kapoor, *J. Organometallic Chem.*, 1971, **33**, 383.
[39] J. L. Herdé and C. V. Senoff, *Inorg. Nuclear Chem. Letters*, 1971, **7**, 1029.
[40] C. White and P. M. Maitlis, *J. Chem. Soc.* (A), 1971, 3322.
[41] J. H. Nelson and H. B. Jonassen, *Co-ordination Chem. Rev.*, 1971, **6**, 27.
[42] M. Sato, K. Ichibori, and F. Sato, *J. Organometallic Chem.*, 1971, **26**, 267.

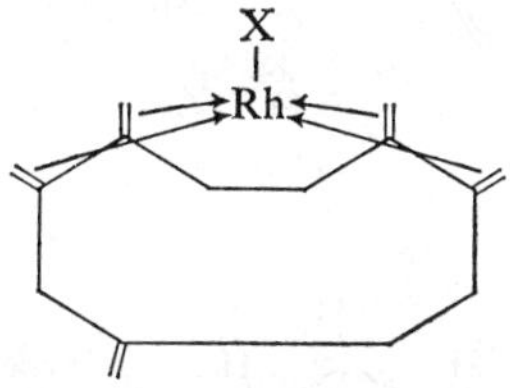

complex possesses an i.r. spectrum in the region 1600—1800 cm^{-1} which is very similar to that of the lactone dimer of dimethylketen, except that all of the wavenumbers are lower.

In the Ni complex (17a), a band at 1735 cm^{-1} has been assigned to a CO_2 stretching vibration.[43]

I.r. (and n.m.r.) spectroscopy has been used to show that, contrary to a previous report (P. W. Jolly, I. Tkatchenko, and G. Wilke, *Angew. Chem. Internat. Edn.*, 1971, **10**, 328, 329), the complex 1-h^1-6,7,8-h^3-*cis*-2-*trans*-6-octadienediyl(tricyclohexylphosphine)nickel(II) (and its dodecadeuterio-analogue) retains the same structure (17b) in the solid state and in solution.[44]

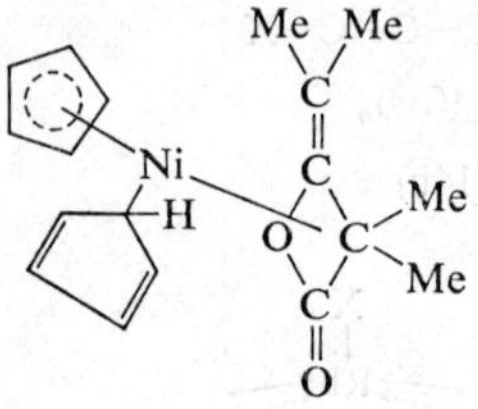

(16)

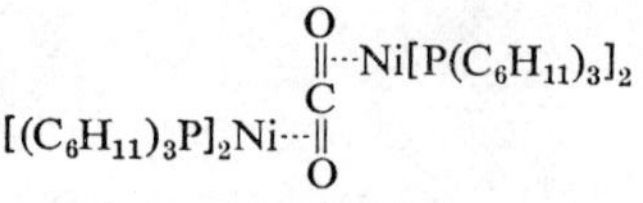

(17a)

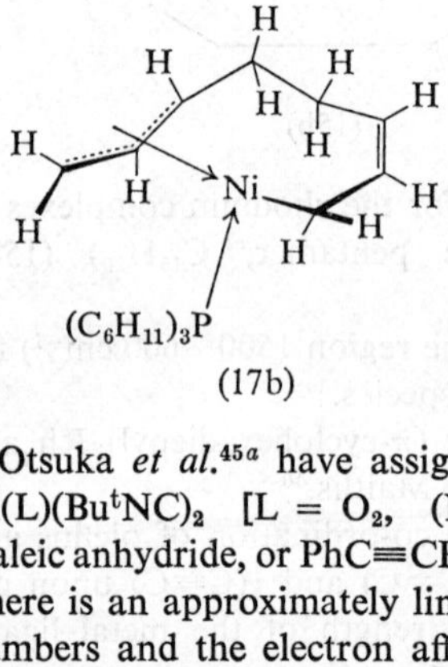

(17b)

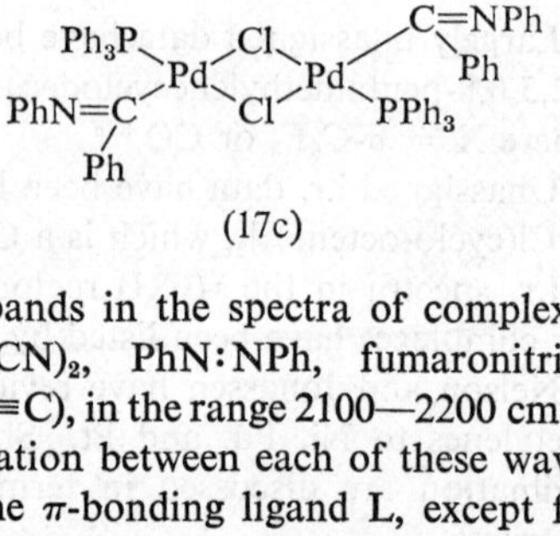

(17c)

Otsuka *et al.*[45a] have assigned two bands in the spectra of complexes $Ni(L)(Bu^tNC)_2$ [L = O_2, $(NC)_2C{:}C(CN)_2$, PhN:NPh, fumaronitrile, maleic anhydride, or PhC≡CPh] to ν(N≡C), in the range 2100—2200 cm^{-1}. There is an approximately linear correlation between each of these wavenumbers and the electron affinity of the π-bonding ligand L, except for L = PhC≡CPh.

The phenylamino-complex (17c) has been prepared.[45b] It contains a Pd—C σ-bond, and it gives rise to ν(C=N) bands at 1584 and 1609 cm^{-1}. This complex undergoes bridge-splitting reactions to give *e.g.* PhN=C(Ph)-$Pd(PPh_3)_2Cl$ [ν(C=N) 1568 cm^{-1}].

[43] P. W. Jolly, K. Jonas, C. Krüger, and Y.-H. Tsay, *J. Organometallic Chem.*, 1971, **33**, 109.

[44] J. M. Brown, B. T. Golding, and M. J. Smith, *Chem. Comm.*, 1971, 1240.

[45] (*a*) S. Otsuka, T. Yoshida, and Y. Tatsumo, *Chem. Comm.*, 1971, 67; (*b*) B. Crociani, M. Nicolini, and T. Boschi, *J. Organometallic Chem.*, 1971, **33**, C81; (*c*) J. Clemens, R. E. Davis, M. Green, J. D. Oliver, and F. G. A. Stone, *Chem. Comm.*, 1971, 1095.

Bis(trifluoromethyl)diazomethane reacts with zerovalent Ni, Pd, or Pt complexes to give $L_2M[(CF_3)_2C{=}N{-}N{=}C(CF_3)_2]$ (M = Ni, Pd, or Pt; L = PPh_3 or Bu^tNC). One of the C=N bonds is π-bonded to the metal, and one is unco-ordinated [ν(CN) at 1560 cm^{-1}].[45c]

The i.r. spectra of $[Cl_2M(olefin)]_2$ (M = Pt or Pd; olefin = vinyl alcohol, vinyl ethers, or their derivatives) have been studied.[46] A normal-co-ordinate analysis of the co-ordinated vinyl alcohol indicated that the lone-pair electrons of oxygen were conjugated with π-electrons of the double bond to some extent, even in the co-ordinated state. Wavenumbers of metal–olefin stretching vibrations (400—440 cm^{-1} region for Pt, 375—425 cm^{-1} region for Pd) for a variety of olefin derivatives have also been listed, and interpreted in terms of M–olefin bond strengths.

Unassigned i.r. spectra have been listed for the following tetracyano-ethylene (tcne) complexes: $Pd(PhNC)_2$(tcne), $Pd(PhNC)(PPh_3)$(tcne), Pd(diphos)(tcne), $Rh(PhNC)_2(PPh_3)$(tcne)Cl, and $Rh(PhNC)_3$(tcne)Cl.[47]

Complexes $M(dba)_2$ (M = Pd or Pt; dba = dibenzylideneacetone, PhCH=CHCOCH=CHPh) have been studied.[48a] In the i.r. spectra, ν(C=C) is found to be in essentially the same position (and with similar intensity) as in the free ligand (1621 cm^{-1} for free dba, 1624, 1613 cm^{-1} in the Pt, Pd complexes, all in $CHCl_3$ solution): ν(C=O), however, is very much changed (new bands appear at 1527 cm^{-1} for M = Pt; 1544 cm^{-1} for M = Pd). These data, together with u.v.—visible spectroscopy results and steric considerations, indicate the structure (18) for the complexes.

The position of ν(CN) for several cyano-allylic complexes of Pd^{II} is found to be in the range 2110—2151 cm^{-1}.[48b]

Spectra in the ν(C—H) and/or ν(C—D) ranges are reported for *cis*-$PtL_2(PPh_3)_2$ (L = CH_3, CH_2D, CHD_2, or CD_3). The bands assigned to CH_3 and CD_3 agree closely with those reported by Adams (*J. Chem. Soc.*, 1962, 1220) for analogous PMe_3 compounds, but use of the average rule (W. J. Lehmann, *J. Mol. Spectroscopy*, 1961, **7**, 261) and data for the partially deuteriated groups suggests a re-assignment as follows: CH_3: 2934 (ν_{as}), 2878 (2856)* (ν_s), 2806 (overtone); CD_3: 2195 (ν_{as}), 2099 (2077)* (ν_s), 2050 (overtone) (all in cm^{-1}; * value corrected for Fermi resonance with the overtone).[49]

I.r. wavenumbers assigned to ligand vibrations corresponding to ν(NH), $\nu(CH_3)$, and ν(C=N), the last in the 1540—1600 cm^{-1} region, have been reported for the cyclic complex (19) of Pt with 3,6-σ-2,4,5,7-tetra-azaoctene-3,3,6-triyl.[50]

[46] Y. Wakatsuki, S. Nozakura, and S. Murahashi, *Bull. Chem. Soc. Japan*, 1971, **44**, 786.
[47] T. Boschi, P. Uguagliati, and B. Crociani, *J. Organometallic Chem.*, 1971, **30**, 283.
[48] (*a*) K. Moseley and P. M. Maitlis, *Chem. Comm.*, 1971, 982; (*b*) B. L. Shaw and G. Shaw, *J. Chem. Soc.* (*A*), 1971, 3533.
[49] B. A. Morrow and Y. Beauchamp, *Canad. J. Chem.*, 1971, **49**, 2921.
[50] G. Rouschias and B. L. Shaw, *J. Chem. Soc.* (*A*), 1971, 2097.

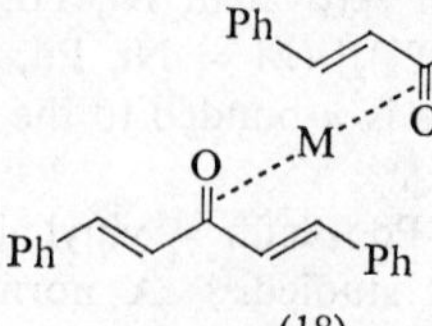

(18)

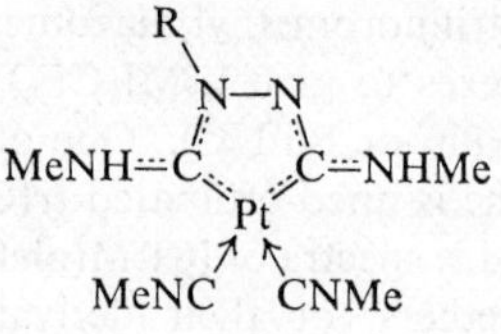

(19)

On heating *cis*-$PtX_2[P(OAr)_3]_2$ (X = Cl, Br, or I; Ar = Ph or *o*-tolyl) in boiling decalin, HX is eliminated and the metal–*ortho*-carbon-bonded complexes (20; Y = H or Me) were isolated, having the expected i.r. bands (1100 and 800 cm^{-1} for Y = H; 1120 and 750 cm^{-1} for Y = Me) typical of such groupings.[51]

The ν(C—OR) vibration in a series of complexes *trans*-$[PtCH_3(L)Q_2]^+$-PF_6^- [Q = PMe_2Ph or $AsMe_3$; L = C(OMe)Me, C(OEt)Me, C(OMe)Et, or C(OMe)Pr^n] has been assigned[52] to a strong i.r. band at *ca.* 1300 cm^{-1}.

Reactions of phosphine–platinum complexes with hexa-2,4-diyne and with 1-phenylbut-1-yn-3-one have yielded complexes of the type (21), with

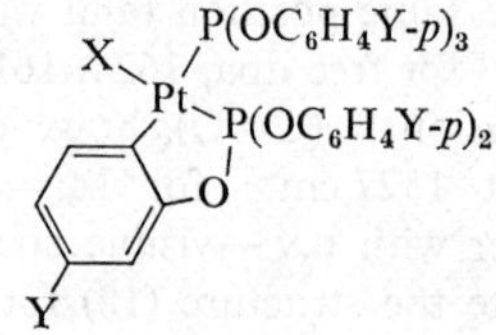

(20)

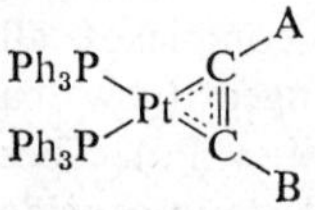

(21)

A = Me, B = C≡CMe and A = Ph, B = C(O)Me, respectively. The former shows two bands, assigned to ν(C≡C), at 1760 cm^{-1} (co-ordinated) and at 2205 cm^{-1} (free); the latter shows only one ν(C≡C) band, at 1720 cm^{-1}, while a Rh analogue[53a] has ν(C≡C) at 1860 cm^{-1}.

The i.r. and Raman spectra of the complexes (21a; L = PMe_3, $AsMe_3$, or $SbMe_3$; X = Cl or Br) and (21b) have been reported.[53b] In (21a), ν(C≡C) was found at *ca.* 1850 cm^{-1}, but in (21b) no ν(C=C) band could be identified.

An interesting pair of Pt–acetylene complexes have been characterized by Furlani *et al.*[54] In (21c), the ν(C≡C) band is assigned to a feature at

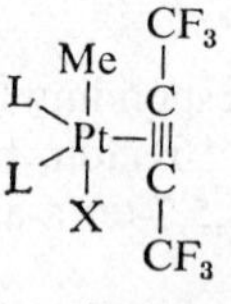

(21a)

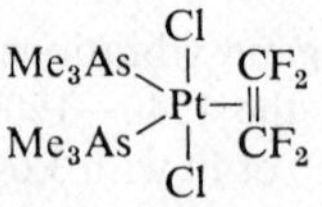

(21b)

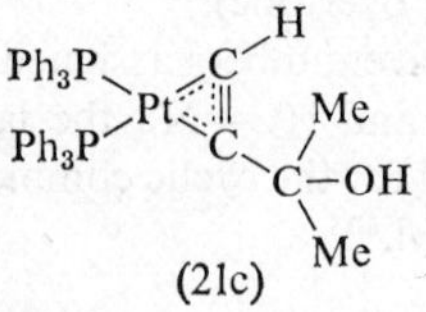

(21c)

[51] E. W. Ainscough and S. D. Robinson, *Chem. Comm.*, 1971, 130.
[52] M. H. Chisholm and H. C. Clark, *Inorg. Chem.*, 1971, **10**, 1711.
[53] (*a*) P. B. Tripathy, B. W. Renoe, K. Adzamli, and D. M. Roundhill, *J. Amer. Chem. Soc.*, 1971, **93**, 4406; (*b*) H. C. Clark and R. J. Puddephatt, *Inorg. Chem.*, 1971, **10**, 18.
[54] A. Furlani, P. Bicev, M. V. Russo, and P. Carusi, *J. Organometallic Chem.*, 1971, **29**, 321.

1680 cm^{-1}, in agreement with that formulation. In $(Ph_3P)_2PtCl$-{C≡C—$CMe_2(OH)$}, however, a closely spaced doublet at 2095, 2088 cm^{-1} is assigned to ν(C≡C), and thus this complex must be a σ-bonded acetylide.

The ν(C≡C) wavenumbers of the dicyanoacetylene complexes $Pt(PPh_3)_2$-(C_4N_2) (A), $Pd(PPh_3)_2(C_4N_2)$ (B), $RhCl(CO)(PPh_3)_2(C_4N_2)$ (C), and $IrCl(CO)(PPh_3)_2(C_4N_2)$ (D) have been reported; they are as follows: (A) 1683; (B) 1751; (C) 1775; (D) 1725 cm^{-1}. These results illustrate the increasing strength of the metal–acetylene interaction on increasing the atomic number of the metal.[55] Bands attributed to ν(C≡C) in the region 1760—1770 cm^{-1} have been observed in the i.r. spectra of several platinum complexes of small-ring acetylenes.[56] In platinum–acetylene–σ-alkenyl and –σ-allenyl complexes, bands assigned to ν(C≡C) (1650—1800 cm^{-1}), ν(C=C) (*ca.* 1550 cm^{-1}), and ν(C=C=C) (1910, 1920 cm^{-1}) have been listed.[57]

[Pt(pent-1-ene)(DMF)Cl_2] gives a band assigned[58] to ν(C=C) at 1503 cm^{-1}.

I.r. and Raman spectra of platinum–tetracyanoethylene complexes (22; X = halide, L = phosphine ligand) indicate that the bonding is

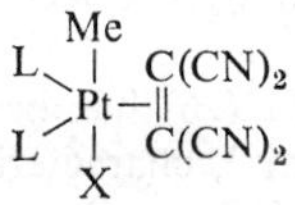

(22a)

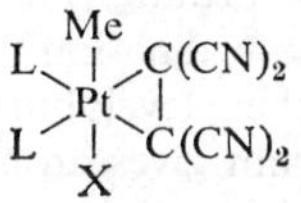

(22b)

intermediate in type between the two forms shown. ν(C≡N), when X = Cl and L = PMe_3, is found at 2230 cm^{-1}, whereas ν(C=C) is assigned[59] to 1201 cm^{-1}.

Absorption bands (i.r.) appear in some Pt^{II} complexes containing two σ- and two π-type bonds, as follows:[60] $Pt(C_9H_9)_2$ (23) 1532, 1262 cm^{-1} [ν(C=C) and $\delta(CH_2)$(scissors), presumably strongly coupled]; $(Ph_3P)_2Pt$-$(C_9H_9)_2$ (24) 1639, 1256 cm^{-1} [more accurately described as ν(C=C), $\delta(CH)_2$ independently]; $Cl_2Pt(C_9H_9)_2$ (25) 1492 cm^{-1} only [?ν(C=C)].

A study[61] of the Pt^{II} complexes of the new ligands tri(*o*-vinylphenyl)-phosphine and its arsenic and antimony analogues by Hall and Nyholm, using i.r. and Raman spectroscopy, indicates two types of aliphatic C=C bonds. This leads to the conclusion that, in the solid state, co-ordination is *via* one double bond, the other two being free.

55 G. L. McClure and W. H. Baddley, *J. Organometallic Chem.*, 1971, **27**, 155.
56 M. A. Bennett, G. B. Robertson, P. O. Whimp, and T. Yoshida, *J. Amer. Chem. Soc.*, 1971, **93**, 3797.
57 B. E. Mann, B. L. Shaw, and N. I. Tucker, *J. Chem. Soc.* (*A*), 1971, 2667.
58 F. Conti, M. Donati, G. F. Pregaglia, and R. Ugo, *J. Organometallic Chem.*, 1971, **30**, 421.
59 H. C. Clark and R. J. Puddephatt, *Inorg. Chem.*, 1971, **10**, 416.
60 M. Aresta and R. S. Nyholm, *Chem. Comm.*, 1971, 1459.
61 D. I. Hall and R. S. Nyholm, *J. Chem. Soc.* (*A*), 1971, 1491.

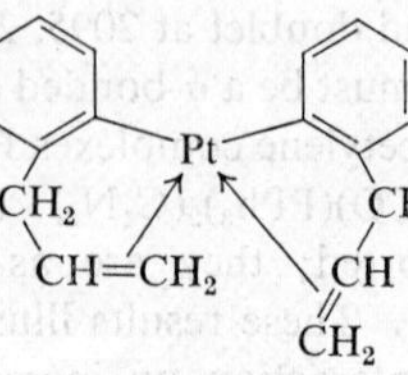

(23)

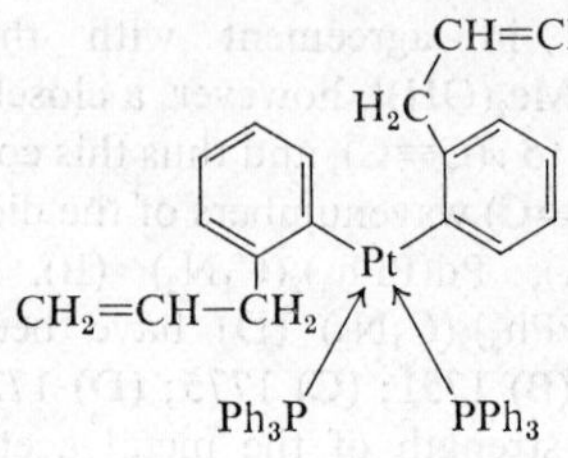

(24)

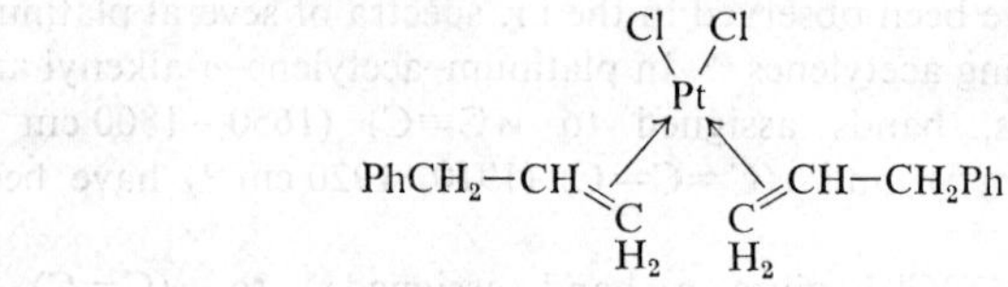

(25)

Small shifts in ν(C=O) for various olefinic ligands containing a carbonyl group (L) on forming $(Ph_3P)_2Pt(L)$ indicate that the $\rangle$C=O group is not involved in co-ordination.[62]

The tetraphenylcyclobutadiene complex of Pt (26) has been prepared. Its i.r. spectrum gives an absorption at 1370 cm^{-1}, characteristic of a co-ordinated tetraphenylcyclobutadiene ring.[63] The compound $[Pt(CF_3)(C_4Me_4)(PMe_2Ph)_2]^+ PF_6^-$ shows i.r. bands (KBr disc) at 1540m, 1445vs,

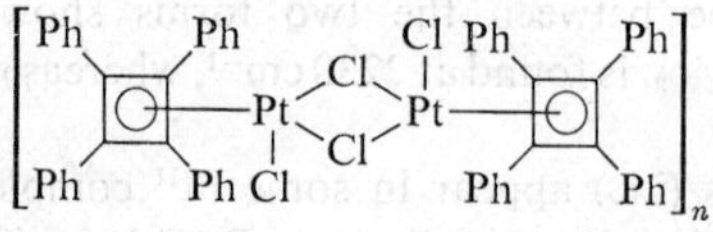

(26)

1360m, 1070s, and 1000vs (all values are cm^{-1}), due to tetramethylcyclobutadiene ligand[64] (*cf.* $[(C_4Me_4)NiCl_2]_2$ – H. P. Fritz, *Adv. Organometallic Chem.*, 1964, **1**, 260).

Contrary to some earlier reports, the i.r. and Raman spectra of $PtMe_3(C_5H_5)$ are consistent with the presence of a h^5-C_5H_5 ring, rather than a σ-bonded h^1-ring.[65] A full assignment of the vibrations of the molecule has been proposed, which is summarized in Table 1.

I.r. absorptions assignable to π-cyclopentadienyl vibrations in some (h^5-C_5H_5)–Pt and –Pd complexes have been collected by Cross and Wardle,[66] and qualitatively interpreted in terms of the expected C_{5v} local symmetry.

[62] S. Cenini, R. Ugo, and C. LaMonica, *J. Chem. Soc.* (*A*), 1971, 409.
[63] F. Canziani, P. Chini, A. Quarta, and A. Di Martino, *J. Organometallic Chem.*, 1971, **26**, 285.
[64] M. H. Chisholm and H. C. Clark, *Chem. Comm.*, 1971, 1484.
[65] J. R. Hall and B. E. Smith, *Austral. J. Chem.*, 1971, **24**, 911.
[66] R. J. Cross and R. Wardle, *J. Chem. Soc.* (*A*), 1971, 2000.

Table 1 *Vibrational assignments for cyclopentadienyltrimethylplatinum*(IV)

Wavenumber/cm^{-1}	*Assignments to* $PtMe_3$ *modes*		*Assignments to* $Pt-(C_5H_5)$ *modes*		
3117	—		$\nu(CH)$		A_1
3102	—		$\nu(CH)$		E_1
2972	$\nu(CH)$	E	—		
2960	$\nu(CH)$	E	—		
2897	$\nu(CH)$	A_1	—		
1425	$\delta(CH_3)$	E	$\nu(CC)$		E_1
1346	$\delta(CH_3)$	?	$\nu(CC)$		E_2
1258	$\delta(CH_3)$	A_1	—		
1218	$\delta(CH_3)$	E	—		
1211	$\delta(CH_3)$	?	—		
1178	—		$\delta(CH)$	∥	E_2
1106	—		$\nu(CC)$		A_1
1056	—		$\delta(CH)$	⊥	E_2
1002	—		$\delta(CH)$	∥	E_1
901	—		$\delta(CCC)$	∥	E_2
865	$\rho(CH_3)$	A_1, E	—		
803	—		$\delta(CH)$	⊥	A_1
789	—		$\delta(CH)$	⊥	E_1
590	$\nu(Pt-C)$	E	—		
559	$\nu(Pt-C)$	A_1	—		
554	—		$\delta(CCC)$	⊥	E_2
316	—		$\rho(Pt-Cp)$		E_1
272	$\delta(PtC_3)$	A_1?	—		
263	—		$\nu(Pt-Cp)$		A_1
182 }	skeletal modes		—		
169 }			—		

In a study of reactions of $Me_2N{=}CH_2^+Br^-$ with CuCl and Cu(CO)Cl, i.r. data on the following isolated compounds have been reported: $[Me_2N{=}CH_2]^+[Cu_2Cl_2Br]^+$ [$\nu(C{=}N)$ 1690 cm^{-1}, *cf.* 1680 cm^{-1} for $Me_2N{=}CH_2^+Br^-$]; $[Cu(CO)(Cl)(CH_2{=}NMe_2)]^+Br^-$ (27) [$\nu(CO)$ 2080 cm^{-1}, $\nu(C{=}N)$ at 1640, 1610 cm^{-1} (at 213 K, in a THF mull)].[67]

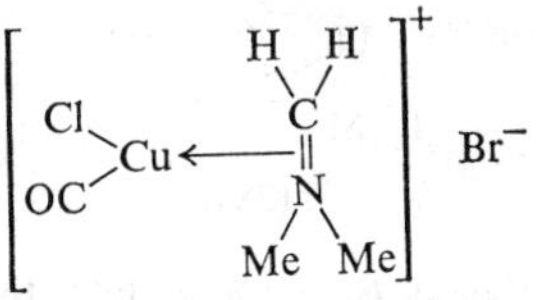

(27)

Several new (olefin)–CuCl complexes have been prepared,[68] which show appreciable decreases in $\nu(C{=}C)$ on complex formation, *e.g.* $Me_3SiCH{=}CH_2$ 1592 cm^{-1}, $(Me_3SiCH{=}CH_2)CuCl$ 1508 cm^{-1}, $Me_3SnCH{=}CH_2$ 1578 cm^{-1}, $(Me_3SnCH{=}CH_2)CuCl$ 1497 cm^{-1}.

[67] R. Mason and G. Rucci, *Chem. Comm.*, 1971, 1132.
[68] J. W. Fitch, D. P. Flores, and J. E. George, *J. Organometallic Chem.*, 1971, **29**, 263.

Cyclo-octa-1,4-diene (1,4-cod) reacts with CuCl to form a complex $[CuCl(1,4\text{-cod})]_2$, which shows two bands assignable to ν(C=C), at 1615 and 1495 cm^{-1}, thus indicating that one C=C interacts much more strongly with Cu than does the other,[69] *cf.* $[Cu(1,5\text{-cod})Cl]_2$, which shows only one ν(C=C), markedly shifted from that of the free ligand. With $HgNO_3$, 1,4-cod forms a bridged dimeric complex $[Ag(1,4\text{-cod})Ag](NO_3)_2$ which shows one ν(C=C) band at 1600 cm^{-1}, indicative of weak interaction of both double bonds of the ligand.

I.r. and Raman spectra of $(C_5H_5)(CH_3)_2Au(PPh_3)$ indicate [70] that in this complex the C_5H_5 group is σ-bonded (h^1-) to the Au atom. Unassigned i.r. bands (1700—950 cm^{-1}) due to the C_6F_6 group are listed [71] for $(C_6F_6)_2$ $Au(PPh_3)X$ (X = Br or I).

Characteristic i.r. bands due to the ligand in tetrazolato-complexes (28; M = Au, Pd, Pt, or Hg) have been listed for R = Me, Pr^i, cyclohexyl, CH_2Ph, Ph, and *p*-MeO·C_6H_4.[72]

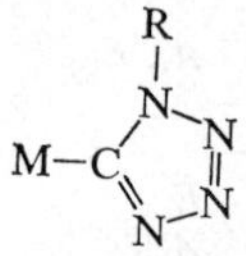

(28)

$(C_5H_5)_4Ce$ has been prepared by Kalostra *et al.*[73] Its i.r. spectrum appears to suggest σ-bound (h^1-)C_5H_5 groups.

I.r. absorptions have been listed for dicycloheptatrienylcerium(IV) dichloride, $(C_7H_7)_2CeCl_2$, and it has been suggested that $C_7H_7^-$ is present.[74]

A series of mono(cyclo-octatetraene)lanthanide chlorides Ln(cot)Cl,2THF (Ln = Ce, Pr, Nd, or Sm) all show i.r. spectra closely similar to K^+ $Ln(cot)_2^-$ complexes, indicating a symmetrical bonding arrangement for the cyclo-octatetraene (a single-crystal *X*-ray study confirmed this).[75]

The bands present in the i.r. spectrum of $(C_5H_5)_2UCl_2$ have been listed,[76a] but no significant attempt was made to assign the observed frequencies.

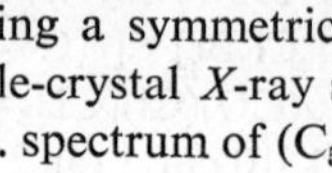

(28a)

[69] H. A. Tayim and M. Kharboush, *Inorg. Chem.*, 1971, **10**, 1827.
[70] S. W. Krauhs, G. C. Stocco, and R. S. Tobias, *Inorg. Chem.*, 1971, **10**, 1365.
[71] R. Usón, P. Royo, and A. Laguna, *Inorg. Nuclear Chem. Letters*, 1971, **7**, 1037.
[72] W. Beck, K. Burger, and W. P. Fehlhammer, *Chem. Ber.*, 1971, **104**, 1816.
[73] B. L. Kalsotra, S. P. Anand, R. K. Multani, and B. D. Jain, *J. Organometallic Chem.*, 1971, **28**, 87.
[74] B. L. Kalsotra, R. K. Multani, and B. D. Jain, *J. Organometallic Chem.*, 1971, **31**, 67.
[75] F. Mares, K. O. Hodgson, and A. Streitweiser, jun., *J. Organometallic Chem.*, 1971, **28**, C24.
[76] (*a*) P. Zanella, S. Faleschini, L. Doretti, and G. Faraglia, *J. Organometallic Chem.*, 1971, **26**, 353; (*b*) J. Jappy and P. N. Preston, *Inorg. Nuclear Chem. Letters*, 1971, **7**, 181.

ν(C=N) modes have been reported for the organometallic imines (28a; R^1_3M = Ph_3Ge, Me_3Sn, Ph_3Sn, or Ph_3Pb; R^2 = Ph).[76b] The wavenumbers in the range 1592—1570 cm^{-1} were shifted by *ca.* 50 cm^{-1} to lower energies compared with those of the purely organic analogues, *e.g.* PhMeC=NPh.

2 Carbonyls

The new species $V(CO)_5(MeTHF)^-$ (MeTHF = 2-methyltetrahydrofuran) has been produced by photolysis of $V(CO)_5^-$ (as the sodium diglyme salt) in 2-methyltetrahydrofuran.[77] ν(CO) bands due to this compound were found as follows (cm^{-1}): 1965vw, A_1; 1820, B_2, correlated with E of $V(CO)_6^-$; 1795vs, E; 1758ms, A_1.

Raman intensity data for $M(CO)_6$ (M = Cr, Mo, or W) support the well-established contribution of M—C π-bonding in such systems, and also indicate that M—C bond orders decrease in the sequence W—C > Cr—C $\gtrsim$ Mo—C. For the C—O bonds in these three molecules, no meaningful bond orders could be calculated because of extensive contributions from the M—C stretching in the C—O stretching eigenvectors.[78]

Distinction between axial and equatorial ^{13}CO stretching vibrations in ^{13}C-enriched, substituted Group VI metal carbonyls has facilitated a discussion of the nature of the intermediate formed during certain substitution reactions of $M(CO)_5$–amine complexes.[79]

I.r. studies of the reaction $M(CO)_6 \rightarrow M(CO)_5 + CO$ (M = Cr, Mo, or W), in inert-gas matrices, at 20 K, have been reported,[80] with a detailed assignment of the carbonyl stretching bands of the $M(CO)_5$ molecules (see Table 2), on the basis of C_{4v} symmetry. Molecular distortions in the $M(CO)_6$ species are also considered.

The i.r. spectra of the compounds $M(CO)_5L$ (M = Cr or W; L = Me_3PS, Me_2PhPS, Ph_3PS, or Me_2PhPSe) show that ν(P=S), ν(P=Se) decrease by 10—25 cm^{-1} on co-ordination.[81] All of the compounds show five ν(CO) bands, and the following figures were quoted for $Cr(CO)_5(Me_3PS)$: 2062m, 1984m, 1944s, 1927s, 1909s (all cm^{-1}).

A general theoretical study[82] has shown that in $M(CO)_5L$ and *fac*-$M(CO)_3L_3$ complexes the i.r. intensity of the ν(CO) band of E symmetry is dependent upon the net charge on L, regardless of apportionment between σ- and π-bonding. The i.r. intensity of the ν(CO) mode of A_1 symmetry, however, increases as the π-acceptor ability of L increases. Thus, the ratio $I(E)/I(A_1)$ is a sensitive indicator of π-bonding in this type of compound.

77 P. S. Braterman and A. Fullarton, *J. Organometallic Chem.*, 1971, **31**, C27.
78 A. Terzis and T. G. Spiro, *Inorg. Chem.*, 1971, **10**, 643.
79 D. J. Darensbourg, M. Y. Darensbourg, and R. J. Dennenberg, *J. Amer. Chem. Soc.*, 1971, **93**, 2807.
80 M. A. Graham, M. Poliakoff, and J. J. Turner, *J. Chem. Soc. (A)*, 1971, 2939.
81 E. W. Ainscough, A. M. Brodie, and A. R. Furness, *Chem. Comm.*, 1971, 1357.
82 W. P. Anderson and T. L. Brown, *J. Organometallic Chem.*, 1971, **32**, 343.

Table 2 *Assignments of ν*(CO) *bands in* $M(CO)_5$ (M = Cr, Mo, W) *in* C_{4v} *symmetry*

Metal	*Assignment*	*Wavenumber*/cm^{-1}	*Band strength*
Cr	A_1	2093.4	vvw
	E	1965.4	s
	A_1	1936.1	m
$Cr(CO)_4(^{13}CO)$		1933.5	vw
Mo	A_1	2098.0	vvw
	E	1972.7	s
	A_1	1932.6	m
$Mo(CO)_4(^{13}CO)$		1929.5	vw
W	A_1	2097.3	vvw
	E	1963.3	s
	A_1	1932.2	m
$W(CO)_4(^{13}CO)$		1928.9	vw

Examples used to illustrate these findings were taken from $Cr(CO)_5L$ and (arene)$Cr(CO)_3$ systems.

The relative Raman intensities and depolarization behaviour of the A_1 and E ν(CO) modes in $(C_6H_6)Cr(CO)_3$ and $(C_5H_5)Mn(CO)_3$ differ considerably (see Figure 1).[83] This observation is interpreted in terms of the bond-derived polarizability tensor elements in the two systems.

A detailed discussion[84] of the solid-state i.r. and Raman spectra of the following arene–metal carbonyl complexes and related systems has been presented: *cis*-diethylenetriaminetricarbonyl-chromium, -molybdenum, and -tungsten; 1,2,3-tri-, 1,2,4,5-tetra-, and 1,3,5-tri-methylbenzenetricarbonyl-chromium; and cyclopentadienyltricarbonylmanganese. Particular attention was paid to the assignment of carbonyl vibrations with respect to factor-group analysis and crystal structure of the complexes.

The i.r.-active ν(CO) modes of phosphine- and arsine-substituted halogenocarbonyls of Mo^{II} and W^{II} have been reviewed.[85]

A large number of molybdenum dicarbonyl complexes $Mo(CO)_2$-$(PR_3)_2L_2$ and $Mo(CO)_2(PR_3)L_3$ have been prepared.[86] All give the expected two C≡O stretches, and the positions of these bands can be correlated quite well with the expected electron-donor and -acceptor characteristics of L.

Carbonyl and nitrosyl stretching frequencies have been listed for $CpM(CO)_2(NO)$, CpM(CO)(NO)[C(OMe)Ph] (Cp = cyclopentadienyl; M = Cr, Mo, or W), and for CpMo(CO)(NO)R {R = [(COEt)Ph] or [$C(NMe_2)Ph$]}.[87]

[83] S. F. A. Kettle, I. Paul, and P. J. Stamper, *Chem. Comm.*, 1971, 235.
[84] H. J. Buttery, S. F. A. Kettle, G. Keeling, P. J. Stamper, and I. Paul, *J. Chem. Soc.* (*A*), 1971, 3148.
[85] R. Colton, *Co-ordination Chem. Rev.*, 1971, **6**, 269.
[86] H. Friedel, I. W. Renk, and H. tom Dieck, *J. Organometallic Chem.*, 1971, **26**, 247.
[87] E. O. Fischer and H. J. Beck, *Chem. Ber.*, 1971, **104**, 3101.

The compounds $[R_2AlW(CO)_3(C_5H_5)]$ (R = Me or Et) have been prepared.[88] When R = Me, carbonyl stretching bands are found at 2014 and 1926 cm^{-1}, but when R = Et, strong peaks occur at 1968, 1692, and

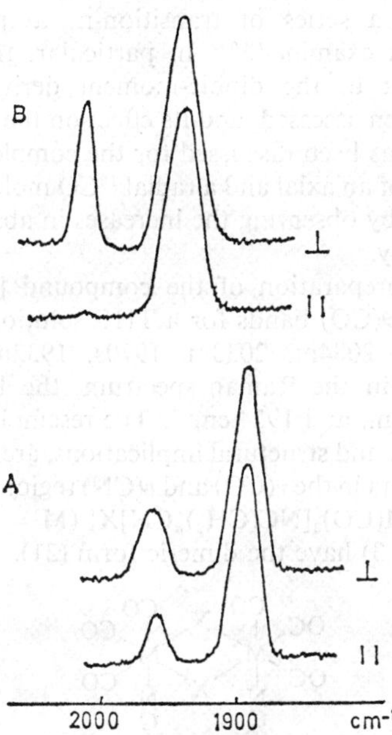

Figure 1 *Solution Raman spectra.* A: $C_6H_6Cr(CO)_3$ *in* CH_2Cl_2, *peaks at* 2016 *and* 1934 cm^{-1}: B: $C_5H_5Mn(CO)_3$ *in* CH_2Cl_2, *peaks at* 1970 *and* 1893 cm^{-1}.

1659 cm^{-1}, indicating the presence of bridging CO groups. Tentative proposals for the structures (29; R = Me) and (30; R = Et) are given.

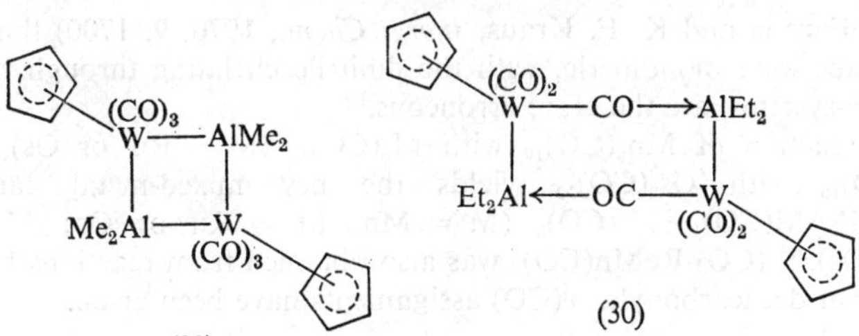

(29)

(30)

In the previous paper, the nature of the bridging carbonyl was not stated explicitly, but it presumably involved, as usual, co-ordination *via* the C atom only. A proposal that the more unusual M—C—O—Al bridging is present in $[Al(\pi\text{-}C_5H_5)W(CO)_3]_3(THF)_3$ has, however, been made by

[88] R. R. Schrieke and J. D. Smith, *J. Organometallic Chem.*, 1971, **31**, C46.

Petersen *et al.*[89] The proposal was made on the basis of some extremely low-frequency (wavenumber < 1600 cm^{-1}) carbonyl stretching vibrations n that complex.

Some factors influencing the absolute i.r. intensity of the carbonyl-stretching modes of a series of transition-metal pentacarbonyl halide derivatives have been examined.[90a] In particular, the contribution of a transverse component to the dipole-moment derivative of the metal–carbonyl bond has been assessed, and its effect on the intensity of the high-energy A_1 vibration has been discussed for the complexes investigated.

The incorporation of an axial and a radial ^{13}CO molecule into $HMn(CO)_5$ has been followed[90b] by observing the increases in absorbance at 1965 and 1982 cm^{-1}, respectively.

A photochemical preparation of the compound $[Et_4N][H_2Mn_3(CO)_{12}]$ has been reported.[91] $\nu(CO)$ bands for a THF solution of the complex are found in the i.r. at 2084m, 2033m, 1970s, 1932m, and 1884wm (all cm^{-1}); for the solid, in the Raman spectrum, the bands are at 2078w, 2027w, 1877m, 1947wm, and 1924 cm^{-1}. The resemblance with the known $[H_2Re_3(CO)_{12}]^-$ anion, and structural implications, are discussed.

I.r. spectroscopic data in the $\nu(CO)$ and $\nu(CN)$ regions have been presented which suggest that $\{M(CO)_3[NC(CH_2)_nCN]X\}$ (M = Mn or Re; X = Cl, Br, or I; n = 1, 2, or 3) have the dimeric form (31). Previous suggestions

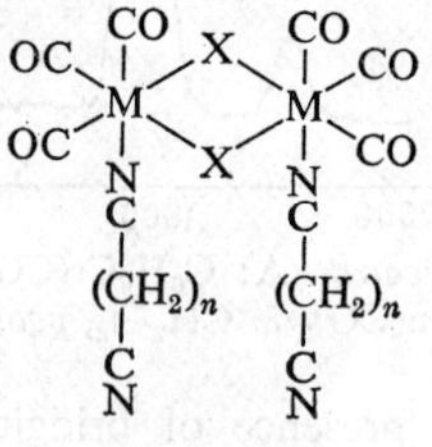

(31)

(M. F. Farona and K. F. Kraus, *Inorg. Chem.*, 1970, **9**, 1700) that these complexes were monomeric, with the dinitrile chelating through the two nitrile π-systems, are therefore erroneous.[92]

The reaction of $Mn_2(CO)_{10}$ with $M_3(CO)_{12}$ (M = Ru or Os), or of $Re_2(CO)_{10}$ with $Os_3(CO)_{12}$, yields the new mixed-metal carbonyls $(CO)_5M'{-}M(CO)_4{-}M'(CO)_5$ (M = Mn, M′ = Ru or Os; M = Re, M′ = Os).[93] $(CO)_5ReMn(CO)_5$ was also obtained from reactions between the parent decacarbonyls. $\nu(CO)$ assignments have been given.

[89] R. B. Petersen, J. J. Stezowski, Ch'eng Wan, J. M. Burlitch, and R. E. Hughes, *J. Amer. Chem. Soc.*, 1971, **93**, 3533.
[90] (*a*) G. Keeling, S. F. A. Kettle, and I. Paul, *J. Chem. Soc.* (*A*), 1971, 3143; (*b*) A. Berry and T. L. Brown, *J. Organometallic Chem.*, 1971, **33**, C67.
[91] G. O. Evans, J. Slater, D. Giusto, and R. K. Sheline, *Inorg. Nuclear Chem. Letters*, 1971, **7**, 771.
[92] J. G. Dunn and D. A. Edwards, *Chem. Comm.*, 1971, 482.
[93] E. W. Abel, R. A. N. McLean, and S. Moorhouse, *Inorg. Nuclear Chem. Letters*, 1971, **7**, 587.

I.r. spectroscopy in the ν(CO) region has been used to show that the metal–metal-bonded complexes $X_nIn[Mn(CO)_5]_{3-n}$ (X = Cl or Br; n = 0, 1, or 2) dissociate into $Mn(CO)_5^-$ ions as the only carbonyl species.[94] However, in the MeCN solution, the evidence suggests that the $In[Mn(CO)_5]_2^+$ cation is also formed. Some crystalline complexes of the latter ion have been isolated.

A new assignment of the vibrations of $Fe(CO)_5$, based on detailed solid-phase i.r. data, has been proposed by Cataliotti *et al.*[95] These assignments are summarized in Table 3.

Table 3 *Revised vibrational assignments for* $Fe(CO)_5$

Assignment		*Wavenumber*/cm^{-1}
A_1'	ν_1	2115
	ν_2	2033
	ν_3	410
	ν_4	440
A_2'	ν_5	593
A_2''	ν_6	2003
	ν_7	623
	ν_8	433
	ν_9	not obs.
E'	ν_{10}	1980
	ν_{11}	643
	ν_{12}	558
	ν_{13}	480
	ν_{14}	not obs.
	ν_{15}	not obs.
E''	ν_{16}	487
	ν_{17}	614
	ν_{18}	not obs.

A reinvestigation[96] of the reactions between N_4S_4 and $Fe(CO)_5$ or $Co_2(CO)_8$ showed that the products are the explosive $Fe(CO)(N_4S_4)$ or $Co_2(CO)(N_4S_4)$, respectively. These have carbonyl stretches at the high values of *ca.* 2180 cm^{-1}, indicating weak M—C bonding. This result is in harmony with the mass-spectroscopic findings of preferential CO loss.

When the i.r. spectrum of $(\pi\text{-}C_5H_5)Fe(CO)_2CH_2Ph$ in SO_2 is recorded soon after its preparation at 243 K, two intense absorptions, at 2062 and 2012 cm^{-1}, are observed. These are very close to (but not identical with) those found for the S-bonded compound $(\pi\text{-}C_5H_5)Fe(CO)_2S(O_2)CH_2Ph$, and they are assigned to the intermediate in the reaction

$$(\pi\text{-}C_5H_5)Fe(CO)_2CH_2Ph + SO_2 \rightarrow \text{Intermediate}$$
$$\downarrow$$
$$(\pi\text{-}C_5H_5)Fe(CO)S(O_2)(CH_2Ph)$$

It is believed[97] that the intermediate is an *O*-sulphinato-complex.

[94] A. T. T. Hsieh and M. J. Hays, *Chem. Comm.*, 1971, 1234.
[95] R. Cataliotti, A. Foffani, and L. Marchetti, *Inorg. Chem.*, 1971, **10**, 1594.
[96] D. A. Brown and F. Frummel, *Chem. Comm.*, 1971, 1579.
[97] S. E. Jacobson, P. Reich-Rohrwig, and A. Wojcicki, *Chem. Comm.*, 1971, 1526.

The i.r. spectra of the series of complexes (π-C_5H_5)Fe(CO)$_2$SnPh$_n$Cl$_{3-n}$ (n = 0—3) have been recorded in the region 4000—250 cm^{-1}. The ν(CO) and δ(CH) (of the π-cyclopentadienyl group) *both* increase on replacing Ph by Cl, and presumably therefore the C—H and C≡O bonds are subjected to similar electronic influences. Changes in the σ- and π-bonding ability of the Sn atom cannot be separated in these molecules.[98]

The heptafulvene complexes (32; R^1 = H, R^2 = Ph; R^1 = R^2 = Ph; R^1 = R^2 = Me) of iron tricarbonyl each show three sharp carbonyl-stretching bands in their i.r. spectra in hexane solution. The frequencies

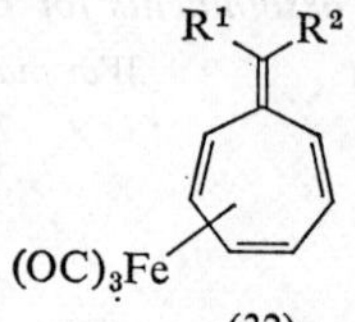

(32)

are not quoted, but are stated to be similar to those normally observed for dieneiron tricarbonyl complexes.[99]

The new Ru^{II} complexes Ru(Cl)H(CO$_2$)(PPh$_3$)$_2$ and *cis*- and *trans*-[RuCl$_2$(CO)$_2$(PPh$_3$)$_2$] have been prepared.[100] The structures (33a—c) are

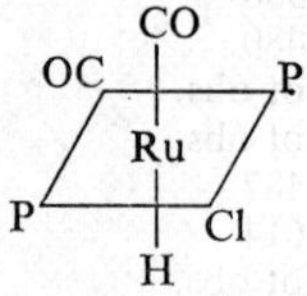

(33a)

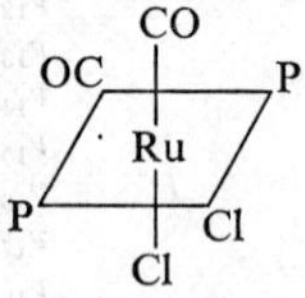

(33b)

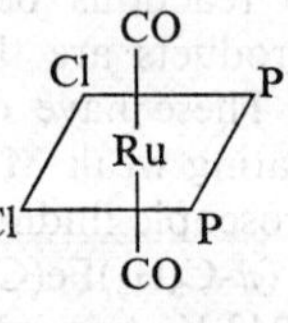

(33c)

assigned to them on the basis of the observation of two ν(CO) modes for the first two systems (in the solid and in C_6H_6 and CH_2Cl_2 solutions), and only one ν(CO) mode in the last complex. The assignment was also assisted by comparison with the ν(CO) bands of other previously well-characterized complexes of similar type.

Carbonyl-stretching bands for the new polynuclear oxocarbonyl $Os_6O_6(CO)_{16}$ occur at 2106s, 2092sh, 2022s, 1993s, 1976sh, 1964sh, 1926s, and

[98] D. S. Field and M. J. Newlands, *J. Organometallic Chem.*, 1971, **27**, 213.
[99] B. F. G. Johnson, J. Lewis, P. McArdle, and G. L. P. Randall, *Chem. Comm.*, 1971, 177.
[100] B. R. James and L. D. Markham, *Inorg. Nuclear Chem. Letters*, 1971, **7**, 373.

1890w (all cm^{-1}). In the related $Os_4O_4(CO)_{12}$, $\nu(CO)$ bands are assigned to 2089s, 2079sh, 2019sh, 1984s, and 1953m (all cm^{-1}).[101]

$\nu(CO)$ and $\nu(NO)$ for the complexes $Co(NO)(CO)_2X$ ($X = I^-$, Br^-, Cl^-, CN^-, or CO) show that all of the species X (except for X = CO) listed give approximately the same transfer of negative charge to the metal.[102] The wavenumbers (all relating to diglyme solutions of the complexes) are given in Table 4.

Table 4 *Carbonyl and nitrosyl stretching wavenumbers in the complexes* $Co(NO)(CO)_2X$

Complex	$\nu(CO)$, *Wavenumber*/cm^{-1}	$\nu(NO)$, *Wavenumber*/cm^{-1}
$Co(NO)(CO)_3$	2104, 2035	1800
$Co(NO)(CO)_2I^-$	2011, 1943	1712
$Co(NO)(CO)_2Br^-$	2013, 1940	1707
$Co(NO)(CO)_2Cl^-$	2012, 1938	1703
$Co(NO)(CO)_2CN^-$	2019, 1952	1726

$\nu(C^{18}O)$ and $\nu(^{13}CO)$ frequencies have been listed[103] for $CoCl_2(CO)(PEt_3)_2$. Some approximate force-constant calculations were carried out.

The complex $Co_3[Fe(CN)_5CO]_2$ has been prepared, and the presence of a linear bridging CO group has been postulated (*i.e.* co-ordination *via* the C *and* the O atoms). Thus, $\nu(CO)$ is observed[104] at 1950 cm^{-1}, some 90 cm^{-1} lower than in $Na_3Fe(CN)_5CO$, which contains a normal terminal CO unit.

The anions $In[Co(CO)_4]_4^-$ and $Tl[Co(CO)_4]_4^-$ have been prepared.[105] They possess a tetrahedral MCo_4 framework; thus, only three carbonyl-stretching bands are observed in the i.r. (2047, 1980, and 1950 cm^{-1} for the In complex), and only one M—Co stretch in the far-i.r. (128 cm^{-1} for In—Co, 107 cm^{-1} for Tl—Co).

A new dinuclear disulphide-bridged derivative of cobalt carbonyl, $Co_2(CO)_6(C_6F_5SSC_6F_5)$, has been characterized.[106] The carbonyl-stretching wavenumbers in this compound are 20—40 cm^{-1} higher than those in $Fe_2(CO)_6(SC_6F_5)_2$, where the S atoms are acting as three-electron donors, and it is suggested that the cobalt complex contains an S—S bond, with the S atoms acting as two-electron donors to the Co [structure (34)].

$(C_6H_6)Co_4(CO)_9$ and $(C_6H_5Me)Co_4(CO)_9$ show strong bridging and terminal $\nu(CO)$ absorptions in the i.r., but no actual frequencies were reported.[107] Structure (35) was suggested for these complexes.

[101] C. W. Bradford and R. S. Nyholm, *J. Chem. Soc.* (*A*), 1971, 2038.
[102] M. Foà and L. Cassar, *J. Organometallic Chem.*, 1971, **30**, 123.
[103] G. Bor, B. F. G. Johnson, J. Lewis, and P. W. Robinson, *J. Chem. Soc.* (*A*), 1971, 696.
[104] E. L. Brown and D. B. Brown, *Chem. Comm.*, 1971, 67.
[105] W. R. Robinson and D. P. Schussler, *J. Organometallic Chem.*, 1971, **30**, C5.
[106] G. Bor and G. Natile, *J. Organometallic Chem.*, 1971, **26**, C33.
[107] I. V. Khand, G. R. Knox, P. L. Pauson, and W. E. Watts, *Chem. Comm.*, 1971, 36.

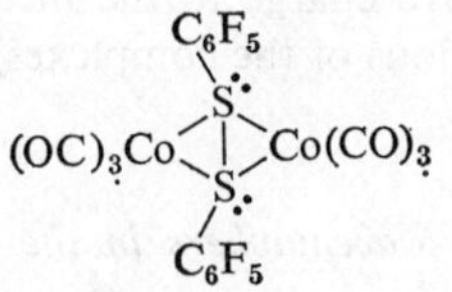

(34)

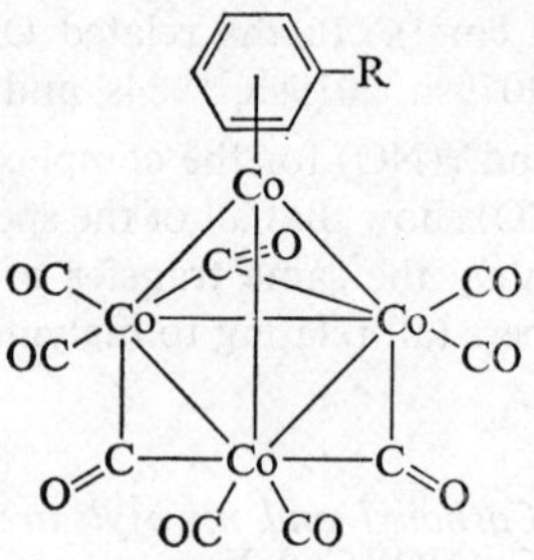

(35)

A variety of new derivatives *trans*-[MA(CO)(PPh$_3$)$_2$] have been prepared (M = Ir or Rh; A = univalent anionic ligand such as halide, pseudohalide, OCOR, *etc.*).[108] A correlation is suggested between ν(CO) and the 'total electronegativity', $\chi_{A(T)}$, or acceptor strength, of the ligand (A) *trans* to CO:

$$\chi_{A(T)} = \chi_F(\Delta\nu^2_{CO})_F/(\Delta\nu^2_{CO})_A = \chi_A + \chi_{A(\pi)},$$

where $\Delta\nu^2_{CO} = [\nu_{CO}(\text{gas})]^2 - [\nu_{CO}(\text{complex in } CHCl_3)]^2$; χ_A = classical 'σ-electronegativity' of A; and $\chi_{A(\pi)}$ = the 'π-electronegativity' or 'π-acidity' of A. The value of $\chi_{A(\pi)}$ for A = F$^-$ is taken to be zero. It is assumed that the magnitude of $\Delta\nu_{CO}$ is a direct measure of the total acceptor strength of A in the particular isostructural series studied. The computed order of π-acidity from this relationship is I > SePh > Br > SPh > CN > Cl > NO$_2$ > NCS ~ N$_3$ > ONO$_2$ > NCO > OCOMe > OCOPh > OPh > OH = F. The order is virtually the same for both the Ir and the Rh series. The most significant support for the above equation comes from the fact that it gives $\chi_{OH(\pi)} = 0$, in accord with the common notion that OH is not a π-acceptor.

The carbonyl-stretching wavenumbers have been listed for the complexes *trans*-[IrCl(CO)(R$_3$M)$_2$] (M = P, R = C$_6$H$_{11}$, *p*-C$_6$H$_4$·Cl, C$_6$H$_5$, *p*-tolyl, Bu^n, Et, or *p*-C$_6$H$_5$·OMe; M = As, R = Ph; M = P, R$_3$ = EtPh$_2$). Correlations of these wavenumbers (1934—1966 cm^{-1}) with (*a*) the rates of oxygen addition to the complexes, and (*b*) the stabilities of the resulting dioxygen adducts (O$_2$)IrCl(CO)(R$_3$M)$_2$ are discussed.[109]

Elemental analysis, n.m.r. and i.r. spectroscopy (C≡O stretching region only), and mass spectrometry have been used to show that Me$_2$Sn[M(CO)-(π-C$_5$H$_5$)]$_2$ (M = Co or Rh), produced as by-products in the reactions of Sn$_2$Me$_6$ with (π-C$_5$H$_5$)M(CO)$_2$, possess the structure (36). The number of ν(CO) bands observed suggested the presence of two isomers, but they could not be separated.[110]

108 L. Vaska and J. Peone, jun., *Chem. Comm.*, 1971, 418.
109 L. Vaska and L. S. Chen, *Chem. Comm.*, 1971, 1080.
110 E. W. Abel and S. Moorhouse, *Inorg. Nuclear Chem. Letters*, 1971, 7, 905.

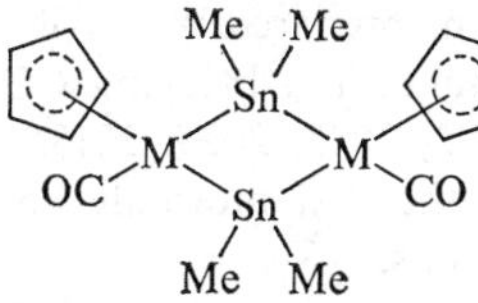

(36)

The Rh—C—O bending frequency has been studied in the i.r. spectra of a number of carbonyl–rhodium complexes.[111] The results show a decrease in the frequency of δ(RhCO) along the series *trans*-$[RhX(CO)(PPh_3)_2]$ (X = F, Cl, Br, or I), corresponding to an increase in the π-acceptor power of the *trans*-ligand.

All four nickel carbonyls $Ni(CO)_n$ (n = 1—4), and probably all six tantalum carbonyls $Ta(CO)_n$ (n = 1—6), have been detected in matrix-isolation experiments.[112] The following wavenumbers were assigned to the nickel species: Ni(CO) 1996 cm^{-1}; $Ni(CO)_2$ 1967 cm^{-1}; $Ni(CO)_3$ 2017 cm^{-1}; $Ni(CO)_4$ 2052 cm^{-1}. These figures seem to suggest an anomalous sequence of C—O bond strengths.

A new technique has been used[113] to observe i.r. spectra of CO chemisorbed on to Ni particles in an Ar matrix at 44 K, the metal being evaporated from a heated filament and condensed into the matrix at 20 K. Bands are found at 2070 and 1890 cm^{-1}.

The reactions between carbon monoxide and Ni^0 phosphino-compounds, such as $Ni(PPh_3)_4$ and $Ni[Ph_2P(CH)_2)_nPPh_2]_2$ (n = 2, 3, or 4), have been followed by studying i.r. spectra in the ν(CO) regions. Compounds $Ni(CO)_3L$, $Ni(CO)_2L_2$, Ni(CO)(L—L)L′, and $Ni(CO)_2$(L—L) (L = PPh_3, L′ = other unidentate phosphine, L—L = chelating diphosphine) were reported, and ν(CO) bands were listed.[114]

New polynuclear carbonyl complexes $(\pi\text{-}C_5H_5)NiCo_3(CO)_9$ (37), $(\pi\text{-}C_5H_5)_2Ni_2Fe(CO)_5$ (38; M = Fe), and $[Me_4N]^+[(\pi\text{-}C_5H_5)_2Ni_2Mn(CO)_5]^-$

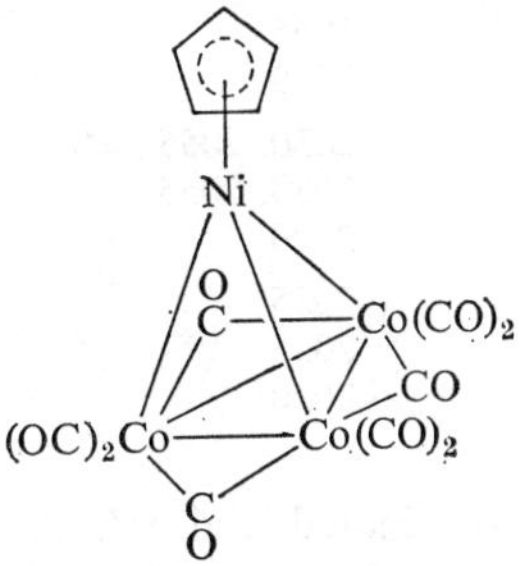

(37)

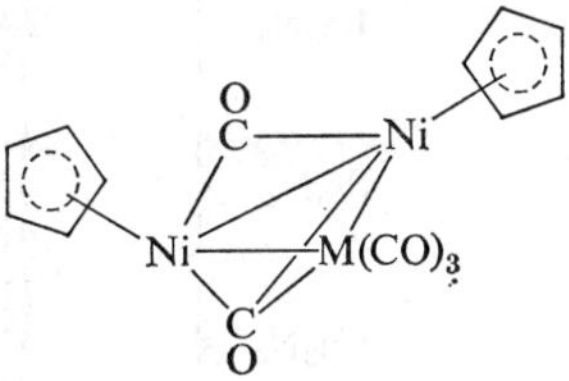

(38)

[111] Yu. S. Varshavskii, M. M. Singh, and N. A. Buzina, *Russ. J. Inorg. Chem.*, 1971, **16**, 725.
[112] R. L. DeKock, *Inorg. Chem.*, 1971, **10**, 1205.
[113] G. Blyholder, M. Tanaka, and J. D. Richardson, *Chem. Comm.*, 1971, 499.
[114] B. Corain, M. Bressan, and G. Favero, *Inorg. Nuclear Chem. Letters*, 1971, **7**, 197.

(38; M = Mn) have been prepared.[115a] All have ν(CO) absorption characteristics of *both* terminal and bridging CO groups. Thus, the first complex gave bands at 2082, 2043, 2025, 2013, and 1850 cm^{-1}. The last complex is believed to be the first molecule known which contains Mn bonded to bridging CO groups.

Carbonyl-stretching frequencies have been reported[115b] for a number of palladium–triphenylphosphine carbonyl complexes. Thus ν(CO) in $Pd(CO)(PPh_3)_3$ is found at 1955 cm^{-1}, but the cluster compounds $[Pd_3(CO)_3(PPh_3)_3]$ (1850 cm^{-1}) and $[Pd_3(CO)_3(PPh_3)_4]$ (1845, 1820 cm^{-1}) show only absorptions due to bridging carbonyl groups. The presence of only one band in the second complex leads to the suggestion that this is a highly symmetrical (D_{3h}) system.

Carbonyl-stretching frequencies in some cationic platinum complexes $[(Ph_3P)Pt(CO)L]^+ BF_4^-$ (L = CO_2Me, CO_2Et, NCO, or Cl) were all found at > 2100 cm^{-1} (L = CO_2Me, 2131; CO_2Et, 2119; NCO, 2131; Cl, 2118 cm^{-1}), and ν(C=O) and ν(C—OR) in these complexes occurred in the expected regions.[116]

3 Nitrogen Donors

Molecular Nitrogen, Azido, and Related Complexes.—Partial solvolysis of alkaline-earth pernitrides M_3N_4 (M = Ca, Sr, or Ba) with H_2O, anhydrous acetic acid, or absolute hydrazine yields products formulated as alkaline-earth–nitrogen complexes, mainly on the basis of their i.r. spectra. Results are shown in Table 5.

Table 5 *Assignment of i.r. bands of nitrogen complexes of alkaline-earths, prepared by solvolysis of alkaline-earth pernitrides*

Pernitride	*Reagent* (*mol. equiv.*)	$\nu(N{\equiv}N)/cm^{-1}$
Ba_3N_4	$\frac{1}{2}H_2O$	1965
	$\frac{1}{2}MeCO_2H$	1965
	$4N_2H_4$	1955
Sr_3N_4	$\frac{1}{2}H_2O$	2030, 2055
	$\frac{1}{2}MeCO_2H$	2030, 2055
	$4N_2H_4$	2020
Ca_3N_4	$\frac{1}{2}H_2O$	2035
	$\frac{1}{2}MeCO_2H$	2035
	$4N_2H_4$	2020

The N_2 molecules are considered to be co-ordinated 'end on'. Approximate values for the N—N distances were calculated as 1.22, 1.20, and 1.20 Å for the Ba, Sr, and Ca compounds, respectively.[117]

[115] (*a*) A. T. T. Hsieh and J. Knight, *J. Organometallic Chem.*, 1971, **26**, 125; (*b*) K. Kudo, M. Hidai, and Y. Uchida, *ibid.*, 1971, **33**, 393.
[116] W. Beck and K. V. Werner, *Chem. Ber.*, 1971, **104**, 2901.
[117] K.-H. Linke and R. Taubert, *Z. anorg. Chem.*, 1971, **383**, 74.

Assignments have been proposed as follows for the co-ordinated azido-groups in some cyclopentadienyl-titanium- and -zirconium-azide complexes:[118] $(\pi\text{-}C_5H_5)_2Ti(N_3)_2$; 2070, 2040 cm^{-1} ($\nu_{as}$), 1332 cm^{-1} ($\nu_s$): $[(\pi\text{-}C_5H_5)_2Ti(N_3)]_2O$; 2060 cm^{-1} ($\nu_{as}$), 1325 cm^{-1} ($\nu_s$): $(\pi\text{-}C_5H_5)_2Zr(N_3)_2$; 2075 cm^{-1} ($\nu_{as}$), 1350 cm^{-1} ($\nu_s$): $[(\pi\text{-}C_5H_5)_2Zr(N_3)]_2O$; 2070 cm^{-1} ($\nu_{as}$), 1338, 1328 cm^{-1} ($\nu_s$).

The complexes of molecular nitrogen *trans*-$[Mo(N_2)_2(diphos)_2]$ {ν(N≡N) at 2035vw, 1985vs cm^{-1} in Nujol; 1973 cm^{-1} in $CHCl_3$ solution} and $[MOCl(N_2)(diphos)_2]$ {ν(N≡N) at 2030vw, 1970vs, 1950sh cm^{-1} in Nujol; 1975 cm^{-1} in dichloromethane solution} have been obtained from novel precursors (diphos = $Ph_2PCH_2CH_2PPh_2$).[119] When the chloro-complex is treated with $MoCl_4(THF)_2$ in 1 : 1 ratio in THF, a blue solid is isolated which has a greatly reduced value for the nitrogen–nitrogen stretching wavenumber (1770, 1720 cm^{-1} in Nujol). It is, therefore, formulated as a dinitrogen-bridge dinuclear complex.

Another series of molybdenum–phosphine–dinitrogen complexes has been prepared by George and Seibold.[120a] The ν(N≡N) stretching wavenumbers have been assigned as follows: *trans*-$[Mo(N_2)_2(diphos)_2]$ 1979 cm^{-1}; *trans*-$[Mo(N_2)_2(PPh_2Me)_4]$ 1926 cm^{-1}; *cis*-$[Mo(N_2)_2(PMe_2Ph)_4]$ 2010 and 1937 cm^{-1}; $Mo(N_2)_2(PBu^n{}_3)_4$ (stereochemistry not stated; said to be impure) 2065, 1980, and 1940 cm^{-1}. These stretching wavenumbers are higher than in the analogous tungsten complexes, thus suggesting that the $Mo-N_2$ bond is weaker than the $W-N_2$ bond (and thus paralleling the known behaviour of Group VI carbonyls).

ν(NN) in the complex (38a) gave no i.r. absorption, but produced a strong Raman scattering line[120b] at 1910 ± 5 cm^{-1}.

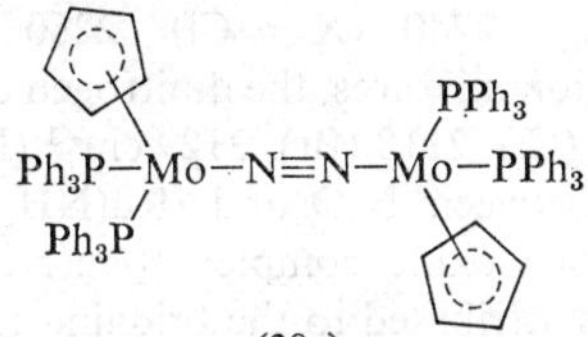

(38a)

The observed nitrogen–nitrogen stretching vibration[121a] in $[(PMe_2Ph)_4\text{-}ClReN_2CrCl_3(THF)_2]$ is at 1875 cm^{-1}. This assignment is based upon ^{15}N substitution experiments (the 1875 cm^{-1} band shifts to 1815 cm^{-1} upon isotopic substitution). This is in the region which would be expected for a bridging dinitrogen group. The lowering of $\nu(N_2)$ in Re⋯N≡N⋯Cr can be understood in terms of population of the π^* molecular orbital of N_2.

118 R. S. P. Coutts and P. C. Wailes, *Austral. J. Chem.*, 1971, **24**, 1075.
119 L. K. Atkinson, A. H. Mawby, and D. C. Smith, *Chem. Comm.*, 1971, 157.
120 (*a*) T. A. George and C. D. Seibold, *J. Organometallic Chem.*, 1971, **30**, C13; (*b*) M. L. H. Green and W. E. Silverthorn, *Chem. Comm.*, 1971, 557.

ν(NN) wavenumbers have been listed[121b] for the complexes $FeH_2(N_2)L_3$, where L = $PMePh_2$ (2058 cm^{-1}), $PEtPh_2$ (2058 cm^{-1}), and $PBuPh_2$ (2065 cm^{-1}).

$\{[(h^5\text{-}C_5H_5)Fe(Me_2PCH_2CH_2PMe_2)]_2N_2\}^{2+}(BF_4^-)_2,2H_2O$ shows a Raman band (but no i.r. band) due to ν(NN) at 2054 cm^{-1}; the cation, therefore, must possess a centrosymmetric structure.[121c]

A dinuclear molecular nitrogen complex has been isolated from the system $(Ph_3P)_2FeCl_3$ + Pr^iMgCl + N_2. $\nu(N_2)$ is assigned to a band at 1761 cm^{-1}, which shifts to 1704 cm^{-1} when $^{15}N_2$ is used. The low intensity and low frequency of this band (by comparison with mononuclear N_2 complexes) is in agreement with the dinuclear formulation. A second $\nu(N_2)$ mode (very weak), at 2016 cm^{-1} (shifts to 1950 cm^{-1} when $^{15}N_2$ is used), is attributed to small amounts of a mononuclear complex.[122]

Solid salts of the type $[Ru(NH_3)_5(X{\equiv}Y)]^{2+}$ (X≡Y = N_2, CO, MeCN, CF_3CN, PhCN, Bu^tCN, or MeNC) have been shown to possess an X≡Y stretching frequency which rises with increasing counter-ion radius. This is interpreted in terms of interaction of the anions with the ammine ligands; smaller anions produce greater polarization of the ammine hydrogen atoms, allowing more electron release on to the Ru, and thence into π^* orbitals of the X≡Y ligand unit.[123]

A band in the i.r. spectrum of each of $[Ru(N_2Ar)(PPh_3)_2Cl](BF_4)$ (Ar = *p*-Me·C_6H_4 or *p*-MeO·C_6H_4) at *ca.* 1700 cm^{-1} is assigned to ν(NN) of the co-ordinated arylazo-group.[124]

From reactions of $[Ru^{II}(NH_3)_5NO]^{3+}$ with hydroxylamine, hydrazine, or ammonia, salts $[Ru^{II}(NH_3)_5(N_2O)]X_2$ have been obtained.[125] The values of ν_1 and ν_3 of the co-ordinated N_2O molecule (observed in the i.r.) were found to vary with X as follows: ν_1: 1150 (X = Cl), 1160 (X = Br), 1175 cm^{-1} (X = I); ν_3: 2240 (X = Cl), 2260 (X = Br), 2250 cm^{-1} (X = I). In some reaction mixtures, the dinitrogen complexes $[Ru^{II}(NH_3)_5(N_2)]X_2$ [$\nu(N_2)$ = 2105 (Cl), 2118 (Br), 2129 cm^{-1} (I)] were also found.

From the reaction between N_2O and $[Ru(NH_3)_5(H_2O)]^{2+}$, the closely related, dinuclear nitrous oxide complex $\{[Ru(NH_3)_5]_2(N_2O)\}(BF_4)_4$ has been isolated.[126] Bands attributed to the bridging N_2O group are found at 2105 and 1155 cm^{-1}. An analogue, probably the bromide, gave similar bands at 2110 and 1160 cm^{-1}.

The kinetics of the displacement of the N_2 ligand in $Os(N_2)Cl_2(PEt_2Ph)_3$ by a number of different phosphines and phosphite ligands have been

[121] (*a*) J. Chatt, R. C. Foy, and R. L. Richards, *J. Chem. Soc.* (*A*), 1971, 702; (*b*) M. Aresta, P. Giannocaro, M. Rossi, and A. Sacco, *Inorg. Chim. Acta*, 1971, **5**, 203; (*c*) W. E. Silverthorn, *Chem. Comm.*, 1971, 1310.

[122] Yu. G. Borodko, M. O. Broitman, L. M. Kachapina, A. E. Slutov, and L. Yu. Ukhin, *Chem. Comm.*, 1971, 1185.

[123] J. Chatt, G. J. Leigh, and N. Thankarajan, *J. Chem. Soc.* (*A*), 1971, 3168.

[124] J. A. McCleverty and R. N. Whiteley, *Chem. Comm.*, 1971, 1159.

[125] F. Bottomley and J. R. Crawford, *Chem. Comm.*, 1971, 200.

[126] J. N. Armor and H. Taube, *Chem. Comm.*, 1971, 287.

followed[127] by monitoring intensity changes of the $\nu(N_2)$ band at 2064 cm^{-1}.

The first Co^{III} complex containing bridging azido-groups has been reported, *i.e.* $(acac)_2Co(N_3)_2Co(acac)_2$. The antisymmetric azide stretch is found at 2080 cm^{-1}, 50 cm^{-1} higher than in the anion $[Co(acac)_2(N_3)_2]^-$, and this is assigned to the bridging azido-group (this band is seen in the solid and also in solution in benzene, $CHCl_3$, or CCl_4, indicating no dissociation; in DMSO solution, however, only a band at 2030 cm^{-1} is seen, showing complex formation of the solvated monomer).[128a] Other assignments proposed for this complex are as follows: $\delta(N_3)$ 560; $\nu_s(N_3)$ 1280; ν(CC) (of acac) 1520; ν(CO) (of acac) 1575 cm^{-1}.

The complexes $CoH(N_2)(PPh_3)_3$ and some derivatives have been studied spectroscopically.[128b] No Co—H stretching vibration was observed for the parent compound, but the ν(NN) vibration was found at 2088 cm^{-1}. The latter vibration is sensitive to variation in the nature of the phosphine ligand, more electron-releasing groups in the phosphine giving a shift to lower wavenumber (2048 cm^{-1} for PBu_3). The anion $Co(N_2)(PPh_3)_3^-$, however, gives ν(NN) at 1845 cm^{-1}.[128c]

The i.r. spectra of $Rh(N_2)(PPh_3)_2Cl$ and $Ir(N_2)(PPh_3)_2Cl$ have been recorded.[129] Comparison of ν(M—Cl) frequencies in these complexes with those of analogous CO complexes suggests that N_2 is a weaker σ-donor and π-acceptor than CO. Also, the $\nu(N_2)$ wavenumber in the Rh complex (2152 cm^{-1}) is higher than that in the Ir complex (2095 cm^{-1}), showing that the M—N_2 bond is stronger for the third-row than for the second-row transition-metal.

The conclusions of the Russian workers concerning the relative π-acceptor abilities of N_2 and CO were confirmed by Darensbourg and Hyde.[130] Measurements of integrated i.r. absorption intensities of ν(CO) in $Ir(CO)(PPh_3)_2Cl$ and of $\nu(N_2)$ in $Ir(N_2)(PPh_3)_2Cl$ indicate that N_2 is a weaker π-acceptor than CO. This is in conflict with conclusions based on frequency shifts, but consistent with the known energies of the π^* orbitals in the two ligands.

When $(Ph_3P)_4M$ (M = Pd or Pt) reacts with MeN_3 or EtN_3, compounds $(Ph_3P)_2M(N_3)_2$ are formed. For M = Pt, $\nu_{as}(N_3)$ is found at 2040 cm^{-1} (in a Nujol mull), but the stereochemistry of the complex is not known. For M = Pd, the initial product gave a single $\nu_{as}(N_3)$ band at 2040 cm^{-1} (in CH_2Cl_2 solution), but chromatography of this product on alumina gave rise to a new band of equal intensity at 2075 cm^{-1}. Thus, the original

[127] P. K. Maples, F. Basolo, and R. G. Pearson, *Inorg. Chem.*, 1971, **10**, 765.
[128] (*a*) D. R. Herrington and L. J. Baucher, *Inorg. Nuclear Chem. Letters*, 1971, **7**, 1091; (*b*) A. Yamamoto, S. Fitazume, L. S. Pu, and S. Ikeda, *J. Amer. Chem. Soc.*, 1971, **93**, 371; (*c*) M. Aresta, C. F. Nobile, M. Rossi, and A. Sacco, *Chem. Comm.*, 1971, 781.
[129] Yu. G. Borod'ko, S. M. Vinogradova, Yu. P. Myagkov, and D. D. Mozzhukhin, *J. Struct. Chem.* (*U.S.S.R.*), 1970, **11**, 251.
[130] D. J. Darensbourg and C. L. Hyde, *Inorg. Chem.*, 1971, **10**, 431.

compound was probably *trans*-, isomerizing to the *cis*-isomer on the chromatography column.[131]

Characteristic bands in the i.r. spectra of the cationic azide-bridged complexes $[(Ph_3P)_2M(N_3)_2M(PPh_3)_2]X_2$ (M = Pd or Pt; X = ClO_4, BF_4, PF_6, or $EtOSO_3$) have been identified.[132] For M = Pd, the antisymmetric azide stretch is at 2079 cm^{-1} (for all cations), with the symmetric stretch at 1256 (ClO_4^-), 1260 (BF_4^-), 1259 (PF_6^-), and 1279 cm^{-1} ($EtOSO_3^-$). When M = Pt, $\nu_{as}(N_3)$ is at 2092 (ClO_4^-), 2102 cm^{-1} (BF_4^-); $\nu_s(N_3)$ is at 1235 (ClO_4^-), 1237 cm^{-1} (BF_4^-).

Palladium(II) complexes with R-azo-cyclohexene derivatives as ligands have been prepared, *e.g.* *trans*-PdL_2Cl_2 (L = $PhN{:}NC_6H_9$ or $MeN{:}NC_6H_9$), and the frequencies of ν(C=C) and ν(N=N) have been compared with those in the free ligands.[133] Shifting of the latter (and not of the former) upon co-ordination indicates that bonding to the metal takes place *via* a metal–nitrogen σ-bond.

In the i.r. spectra of Pt complexes of the type $[(Ph_3P)_3Pt{\cdot}N{:}NC_6H_4{\cdot}$-R-*p*$]^+X^-$ and $[(Ph_3P)_2Pt(X)(N{:}NC_6H_4{\cdot}$R-*p*)], bands are found, all very close to 1580 cm^{-1}, which are assigned to ν(N=N).[134] In $[(Ph_3P)_3Pt{\cdot}NH{:}$-$NC_6H_4{\cdot}$R-*p*$]^{2+}(X)_2^{2-}$, on the other hand, $\nu(\overset{+}{N}H{=}N)$ gives rise to bands near 1160 cm^{-1}.

A complex of methyldiazene, MeN:NH, with cuprous chloride, having the formula (MeN:NH)CuCl, has been prepared.[135] It is almost certainly polymeric, with each Cu bound to two nitrogen atoms from different ligands. ν(NH) is found at 3150 cm^{-1} (*cf.* 3130 cm^{-1} in the free ligand) and ν(N=N) is found at 1520 cm^{-1} (1575 cm^{-1} in the free ligand).

Amines and Related Ligands.—A study of the Raman spectrum of pyridine adsorbed on a variety of metal oxide surfaces has been made, mainly in the 950—1050 cm^{-1} region, to investigate the nature of the bonding of pyridine to these surfaces, and to distinguish between physical and chemical adsorption.[136]

The use of vibrational spectroscopy in assigning *cis* or *trans* stereochemistry to compounds $MX_2(bipy)_2$ and $MX_2(o\text{-}phen)_2$ (M = transition metal; X = halogen, CN, NCS, *etc.*) has been critically discussed.[137]

The structures of transition-metal complexes of the new ligands (39) and (40) were shown by i.r. and other evidence to include both bidentate (N,P) and unidentate (P only) co-ordination behaviour. The i.r. criterion used was the behaviour of the pyridine 8*a* and 8*b* vibrations upon co-ordination

[131] B. Hessett, J. H. Morris, and P. G. Perkins, *Inorg. Nuclear Chem. Letters*, 1971, **7**, 1149.
[132] W. Beck, P. Kreutzer, and K. V. Werner, *Chem. Ber.*, 1971, **104**, 528.
[133] L. Cagliotti, L. Cattalini, F. Gasporini, G. Maragnani, and P. A. Vigato, *J. Chem. Soc.* (*A*), 1971, 324.
[134] S. Cenini, R. Ugo, and C. La Monica, *J. Chem. Soc.* (*A*), 1971, 3441.
[135] M. N. Ackermann, *Inorg. Chem.*, 1971, **10**, 272.
[136] P. J. Hendra, J. R. Horder, and E. J. Loader, *J. Chem. Soc.* (*A*), 1971, 1766.
[137] E. D. McKenzie, *Co-ordination Chem. Rev.*, 1971, **6**, 187.

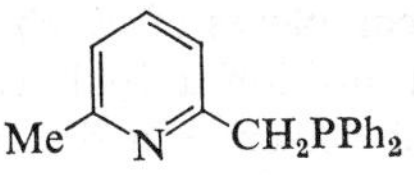

(39)

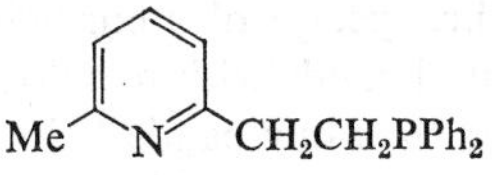

(40)

(behaviour which is well understood from earlier work). These vibrations are at 1586, 1572 cm^{-1} for (39) and 1588, 1576 cm^{-1} for (40), but suffer the same characteristic shifts and intensity changes upon co-ordination as are found for those of pyridine itself.[138]

Bands due to ν(NH) and ν(CH) in some metal(II) *NN'*-di-(3-amino-propyl)piperazine complexes have been listed.[139a]

I.r. frequencies have been listed for new mixed ligand complexes of 3,3′-diaminopropylamine and ethylenediamines, but no discussion was given.[139b]

I.r. and n.m.r. spectra in D_2O solution have been reported[139c] for edda-dx complexes of Cu^{II}, Ni^{II}, Co^{II}, Zn^{II}, Mg^{II}, and Fe^{II}, (bands in the 1600—1700 cm^{-1} region only), at various ratios (moles added base) : (moles ligand present), where edda-dx = *NN'*-ethylenediaminediacetic-*NN'*-diacethydroxamic acid.

The three ν(N—H) bands present in solid di-(4-aminophenyl)methane (mda) (41), at 3450, 3340, and 3240 cm^{-1}, are replaced by two (at 3380 and

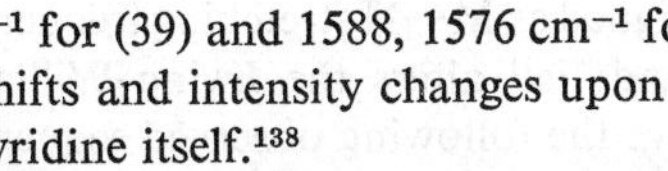

(41)

3280 cm^{-1}) in the solid complex [(mda)$_3$NaCl], suggesting that the hydrogen atoms on the amino-groups are taking part in hydrogen-bonding.[140]

I.r. data, largely unassigned, are given[141] for some pyrrolyl–titanium and related compounds containing M—N or M—C bonds. A pyrrole ring deformation (647 cm^{-1} in the free ligand, 671 cm^{-1} in the Li$^+$ salt) appears as follows: Ti(pyrr)$_4$,2THF 651 cm^{-1}; (π-C$_5$H$_5$)Ti(pyrr)$_2$ 645 cm^{-1}; Li$_3$Cr(pyrr)$_6$,3THF 661 cm^{-1}; Cr(pyrr)$_2$(dioxan)$_2$ 657 cm^{-1}; Li$_2$Ni(pyrr)$_4$ 663 cm^{-1}; (π-C$_5$H$_5$)Ni(pyrr)(PPh$_3$) 644 cm^{-1}.

Evidence for the configuration of *trans*-[Cr(propylenediamine)Br$_2$]$^+$ includes the observation[142] of a single NH$_2$ antisymmetric deformation vibration at 1680 cm^{-1}.

138 W. V. Dahloff, T. R. Dick, G. H. Ford, and S. M. Nelson, *J. Inorg. Nuclear Chem.*, 1971, **33**, 1799.

139 (*a*) J. G. Gibson and E. D. McKenzie, *J. Chem. Soc.* (*A*), 1971, 1029; (*b*) G. Ponticelli, *Inorg. Chim. Acta*, 1971, **5**, 461; (*c*) R. J. Motekaitis, I. Murase, and A. E. Martell, *J. Co-ordination Chem.*, 1971, **1**, 77.

140 I. Goodman, S. J. Kettle, and P. G. Owston, *Chem. and Ind.*, 1971, 1300.

141 D. Tille, *Z. anorg. Chem.*, 1971, **384**, 136.

142 J. A. McLean, jun. and R. I. Goorman, *Inorg. Nuclear Chem. Letters*, 1971, **7**, 9.

The i.r. spectra of a considerable number of complexes of *cis*-chelating bidentate ligands with metal(II) ions of the first transition series have been reported.[143] Four bands below a wavenumber of 600 cm^{-1} in the spectra of the complexes $[M(bipy)_3](ClO_4)_2$ and $[M(phen)_3](ClO_4)_2$ (M = Cr, Mn, Fe, Co, Ni, Cu, or Zn) exhibit a frequency variation which parallels the variation of crystal-field stabilization energy with *d*-orbital configuration. They are therefore assigned as M—N stretching frequencies. The occurrence of spin-paired Cr^{II} and Fe^{II} alters the Irving–Williams series (high-spin complexes only) to give the following observed sequence of frequencies for ν(M—N): Cr > Mn < Fe > Co < Ni > Cu > Zn.

Shifts in some characteristic vibrations of several pyridinemonocarboxylic acid amides on complexing with MoO_2Br_2 (*via* the nitrogen atom of the pyridine ring) have been measured.[144]

Compounds of formula MX_2L (M = Mn, Fe, Co, Zn, Pd, or Pt; X = Cl or Br; L = 2,2′-bipyridyl or 2,2′-[$^2H_8$]bipyridyl) have been prepared and their i.r. spectra investigated (4000—40 cm^{-1}).[145] A normal-co-ordinate analysis was performed on the chloro-complexes, employing a Urey–Bradley force field with resonance parameters and suitable general force-field (GVFF) terms about the metal site. Force constants and frequency positions indicate stronger M—L bonding in the Fe, Pd, and Pt complexes than in those of Mn, Co, and Zn. All observed bands were assigned, both internal modes of bipyridyl and skeletal vibrations.

Complexes of Fe^{II}, Co^{II}, and Ni^{II} with 2,6-bis-(2-diphenylphosphinoethyl)pyridine (42) have been prepared during a series of magnetic studies on

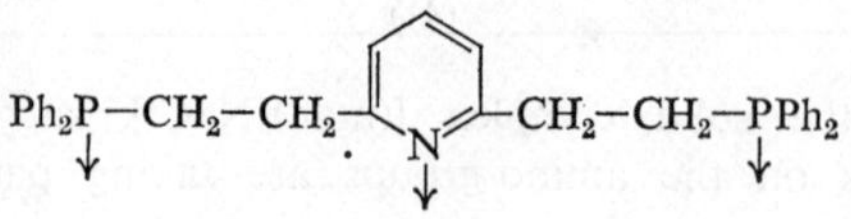

(42)

transition-metal complexes. The pyridine ring deformations in the complexes occur[146] near 1570 and 1600 cm^{-1}.

ν(OH), ν(C=N), ν(C=C), and δ(CH) frequencies have been listed for the following ruthenium complexes: $[RuO_2(OH)_2(bipy)]$, $[RuO_2(bipy)_2]$, $[Ru(OH)_3(phen)]_2O$, and $[RuO_2(phen)_2]$.[147]

The i.r. spectrum of $Co(NH_3)_2(NO_3)_2$ has been reported and partially assigned:[148] ν(Co—N) 460; ν(Co—O) 295: ν(NH) 3350, 3275, 3210, 3184; $\delta(NH_3)$ 1610; and $\rho(NH_3)$ 675, 620 cm^{-1}, together with the vibrations of the

143 G. C. Percy and D. A. Thornton, *J. Mol. Structure*, 1971, **10**, 39.
144 M. K. Alyaviya, M. K. Kasymov, and A. L. Kats, *Russ. J. Inorg. Chem.*, 1971, **16**, 701.
145 J. S. Strukl and J. L. Walter, *Spectrochim. Acta*, 1971, **27A**, 223.
146 W. S. J. Kelly, G. H. Ford, and S. M. Nelson, *J. Chem. Soc.* (*A*), 1971, 388.
147 T. Ishiyama, *Bull. Chem. Soc. Japan*, 1971, **44**, 1571.
148 G. L. McPherson, J. A. Weil, and J. K. Kinnard, *Inorg. Chem.*, 1971, **10**, 1574.

co-ordinated NO_3^- ion: ν_1 1515, 1475; ν_2 1024; ν_3 744; ν_4 1347, 1292, ν_5 812; ν_6 713 cm^{-1}.

A figure illustrating changes in the i.r. spectrum on isomerizing *cis,trans*-$[CoCl_2(NH_3)_2(en)]ClO_4,H_2O$ to *trans,cis*-$[CoCl_2(NH_3)_2(en)]ClO_4$ has been given in a paper on the thermal isomerization of this complex.[149a]

The i.r. spectra of *cis*- and *trans*-isomers of $[Co(cyclam)A_2]^+$, where cyclam = quadridentate 1,4,8,11-tetra-azacyclotetradecane, show differences in the region 800—910 cm^{-1} which are useful in differentiating them.[149b] The bands concerned are believed to be associated with secondary amine and methylene groups of the ligand.

Complexes of Co^{III} with flexible quadridentate ligands have been reviewed.[149c] The review includes a short section on the use of i.r. spectroscopy in investigating their stereochemistry.

A new series of anions $MX_3(\gamma\text{-picoline})^-$ (M = Co^{2+} or Zn^{2+}; X = Cl or Br) has been prepared.[149d] The γ-picoline spectrum is consistent with previous work on tetrahedrally co-ordinated γ-picoline complexes.

Unassigned i.r. data for Co^{II}, Ni^{II}, and Cu^{II} halide complexes of the ligands 6-methyl-2,3-di-(2-pyridyl)quinoxaline (43; R^1 = Me, R^2 = H) and 6,7-dimethyl-2,3-di-(2-pyridyl)quinoxaline (43; R^1 = R^2 = Me) have been given.[150]

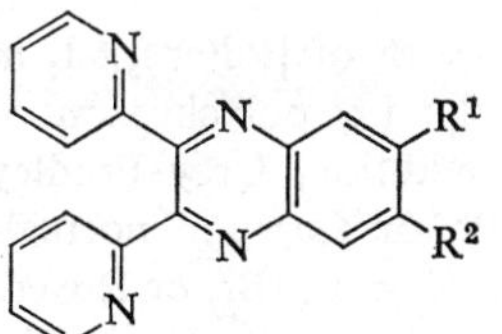

(43)

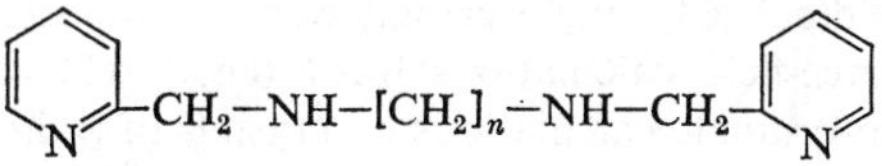

(44)

I.r. wavenumbers for some characteristic ligand vibrations in Co^{III}, Ni^{II}, Cu^{II}, and Pd^{II} complexes with tetra-amines containing terminal pyridyl residues [1,6-bis-(2′-pyridyl)-2,5-diazahexane (44; n = 2) and 1,7-bis-(2′-pyridyl)-2,6-diazaheptane (44; n = 3)] have been listed.[151] ν(NH) occurs between 2400 and 3350 cm^{-1}, but chiefly *ca.* 3330 cm^{-1}; characteristic pyridyl ring vibrations appear at *ca.* 1600 and 1570 cm^{-1}.

Allylamine forms complexes with Rh^{III} as follows: $RhCl_3(C_3H_5NH_2)(H_2O)$; $RhCl_3(C_3H_5NH_2)_2$ (these contain bridging Cl to give a six-co-ordinate polymer); $RhCl_3(C_3H_5NH_2)_3$.[152] In each case, ν(C=C) is found at 1648 cm^{-1}; therefore the C=C bond does not participate in bonding to the metal.

149 (*a*) R. Tsuchiya, Y. Nakata, and E. Kyono, *Bull. Chem. Soc. Japan*, 1971, **44**, 705; (*b*) C. K. Poon, *Inorg. Chim. Acta*, 1971, **5**, 322; (*c*) G. R. Brubaker, D. P. Schaefer, and J. H. Worrell, *Co-ordination Chem. Rev.*, 1971, **6**, 161; (*d*) M. A. Buhannic and J. E. Guerchais, *Bull. Soc. chim. France*, 1971, 55.

150 D. F. Colton and W. J. Geary, *J. Chem. Soc.* (*A*), 1971, 2457.

151 J. G. Gibson and E. D. McKenzie, *J. Chem. Soc.* (*A*), 1971, 1666.

152 H. Sawai and H. Hirai, *Inorg. Chem.*, 1971, **10**, 2068.

A series of 1 : 1 complexes of Ni^{II} halides with the substituted 2-(aminoethyl)- and 2-(aminomethyl)-pyridines 2-[$R^1R^2N(CH_2)_nC_5H_4N$ (R^1, R^2 = H, Me, Et, Pr^i, *etc.*; n = 1 or 2) have been made.[153] On the basis of visible and i.r. spectra, and magnetic behaviour, the compounds are ascribed either distorted tetrahedral or dimeric, five-co-ordinated structures. Both the pyridine ring mode ν_{8a} and the ν(NH) mode of the substituent show shifts to lower wavenumbers on co-ordination, showing that the amine ligands are bidentate.

Square-planar Cu^{II} and Ni^{II} complexes *trans*-$[ML_2(RNH_2)_2]$, where L = succinimide or phthalimide, show a dependence of carbonyl-stretching frequency upon R.[154] This dependence has been correlated with the Taft polar substituent parameter σ^*, showing that the electronic effects of the amine substituents are transmitted so as to affect the electron density in the imide ring.

I.r. wavenumbers to *ca.* 200 cm^{-1} have been listed for $NiL_4(ClO_4)$, $NiL_4(H_2O)_2(ClO_4)_2$, and for $NiL_6(ClO_4)_2$ (L = γ-picoline). Assignments have been proposed, on not very definite evidence, to water, ClO_4^-, and γ-picoline modes.[155]

Some very detailed studies of a number of ethylenediamine complexes of Pd^{II}, Pt^{II}, and Cu^{II} have been made.[156] Thus, the i.r. spectra (to 30 cm^{-1}) of $[M(en)_2]X_2$ (M = Pd; X = Cl, Br, or I; M = Pt; X = Cl or I: M = Cu; X = SCN or $\frac{1}{2}PtCl_4^{2-}$), the Raman spectra of $[Pd(en)_2]Cl_2$ in the solid phase and in aqueous solution, and the i.r. polychroism of $[Pd(en)_2]Cl_2$ single crystal were all reported. In addition, Urey–Bradley force-field calculations for $M(en)_2^{2+}$ (M = Pd, Pt, or Cu) and normal-co-ordinate calculations on crystals of $[Pd(en)_2]X_2$ (X = Cl, Br, or I) were carried out. In this way a good understanding of the spectra emerged. The main points worthy of notice are: (*a*) that modes which are principally M—N stretching, *e.g.* in $Pd(en)_2^{2+}$ at 525 (i.r.), 456 (i.r.), 514 (R), 406 cm^{-1} (R), correlate with the strengths of the co-ordinate bonds, Pt > Pd > Cu, and (*b*) that chelate-ring out-of-plane deformation vibrations will couple strongly with interionic vibrations when anion–cation interaction is considerable. This is borne out experimentally, since in $[Pd(en)_2]X_2$ the i.r. band at 282 cm^{-1} (X = Cl) shifts to 267 (Br), and 250 cm^{-1} (I). Also, the Raman line in $[Pd(en)_2]Cl_2$ crystal at 214 cm^{-1} shifts to 169 cm^{-1} in aqueous solution. The order of anion–cation interaction is shown to be $Cl^- > Br^- > I^- > PtCl_4^{2-}$.

The i.r. spectra of several Pd^{II} complexes with 2,2′-diaminobiphenyl have been reported;[157] some bands have been assigned to specific ligand vibrations, and bands at about 410 cm^{-1} are ascribed to ν(Pd—N).

[153] E. Uhlig, E. Under, and V. Dinjus, *Z. anorg. Chem.*, 1971, **380**, 181.
[154] N. P. Slabbert and D. A. Thornton, *J. Inorg. Nuclear Chem.*, 1971, **33**, 2933.
[155] M. S. Sun and D. G. Brewer, *Canad. J. Chem.*, 1971, **49**, 1502.
[156] Y. Omura, I. Nakagawa, and T. Shimanouchi, *Spectrochim. Acta*, 1971, **27A**, 2227.
[157] S. Onaka, *Bull. Chem. Soc. Japan*, 1971, **44**, 2154.

Experimental frequencies from the i.r. spectra of *trans*-$[PtCl_2(NX_2CX_3)_2]$ (X = H or D) have been used[158] to analyse the normal vibrations for the model $Pt(NX_2CX_3)$ grouping.

A normal-co-ordinate treatment has facilitated a full assignment of the vibrations of co-ordinated *O*-methylhydroxylamine in $[Pt(NH_2OMe)_4]$-$[PtCl_4]$.[159]

A large number of unassigned i.r. frequencies have been reported[160] for some benzoylhydrazine complexes of Pt^{II}.

I.r. frequencies for the anionic species in CuL_2A_2 [L = di-(2-pyridyl)-amine] have been obtained[161] for A = ClO_4^-, PF_6^-, NCS^-, NO_3^-, or NO_2^-.

Using previously established values of the shifts of the ligand bands upon co-ordination to metals, evidence from i.r. spectra has been produced[162] in support of tetrakis-complex formation between Cu^I and the monofunctional amines quinoline, isoquinoline, piperidine, and *N*-benzylidene-methylamine.

The adducts $[Cu(picolinate)_2]$,KSCN and $[Cu(picolinate)_2]$,KOCN show bands in their i.r. spectra very close to those in KSCN and KOCN, respectively.[163]

I.r. wavenumbers down to approximately 350 cm^{-1} have been given for trimethylamine complexed to various salts of transition metals and some other metals, *e.g.* Zn, Cd, and Hg. No structural information was deduced.[164]

The i.r. spectra of a series of $(C_6F_5)_2HgL$ complexes have been presented (L = substituted bipyridyl, *o*-phenanthroline, *etc.*).[165] The ring-stretching and out-of-plane ring deformation modes of the ligands L are displaced to higher wavenumber upon complex formation.

I.r. spectra have been partially assigned[166] for the complexes Hg(morpholine)$(CN)_2$ (45) and Hg(morpholine)$(SCN)_2$ (46). Thus, in (45) ν(NH) is found at 3252 cm^{-1}, ν(CN) at 2185 cm^{-1}, ν(C—O—C) at 1031 cm^{-1}, and

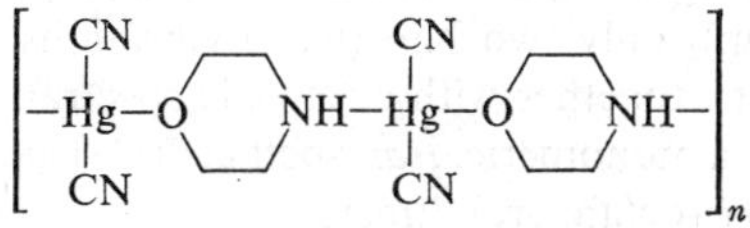

(45)

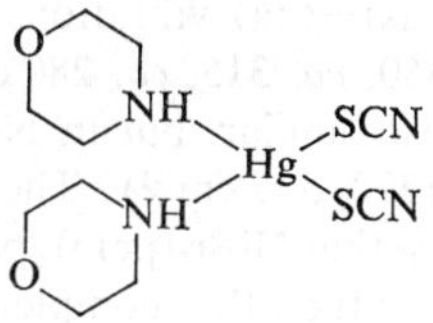

(46)

[158] Yu. Ya. Kharitonov and I. K. Kireeva, *Russ. J. Inorg. Chem.*, 1971, **16**, 733.
[159] M. A. Sarukhanov and A. I. Stetsenko, *Russ. J. Inorg. Chem.*, 1971, **16**, 220.
[160] Yu. Ya. Kharitonov and R. I. Machkhovshvili, *Russ. J. Inorg. Chem.*, 1971, **16**, 404.
[161] J. C. Lancaster, W. R. McWhinnie, and P. L. Welham, *J. Chem. Soc. (A)*, 1971, 1742.
[162] Kuang-Ling H. Chen and R. T. Iwamoto, *Inorg. Chim. Acta*, 1971, **5**, 97.
[163] R. D. Gillard and S. H. Laurie, *J. Inorg. Nuclear Chem.*, 1971, **33**, 947.
[164] C. G. Guilbault and S. M. Billedean, *J. Inorg. Nuclear Chem.*, 1971, **33**, 1411.
[165] A. J. Canty and G. B. Deacon, *Austral. J. Chem.*, 1971, **24**, 489.
[166] I. S. Ahuja and A. Garg, *Inorg. Nuclear Chem. Letters*, 1971, **7**, 937.

ν(Hg—CN) at 420 cm^{-1}. In (46) the following assignments are proposed: ν(NH) 3230; ν(CN) 2110; ν(C—O—C) 1095; ν(CS) 710; δ(SCN) 451, 431; ν(Hg—N) 338 (?); ν(Hg—S) 323 cm^{-1} (?).

I.r. spectra (in the 4000—650 cm^{-1} range) of the quinoline adducts $SnCl_4$,2L, $SnBr_4$,2L, $TiCl_4$,2L, $BiCl_3$,L, $SbCl_5$,L, and $SbCl_3$,L have been assigned in terms of the quinoline fundamentals.[167]

Four internal modes of ethylenediamine have been listed for a series of complexes Sn(diaryl)(en).[168]

In $Yb(NH_2)_2$, the following assignments have been proposed [169] for the NH_2 unit: ν_1 3280; ν_2 3340; ν_3 1510 cm^{-1}.

Oximes.—Some of the i.r. bands due to ligand vibrations in dimethylglyoximato Co^{III} and Fe^{II} pyridine-base complexes have been tabulated.[170a]

Empirical assignments have been made [170b] for most of the observed i.r. absorption bands (4000—200 cm^{-1}) of Ni, Pd, and Pt bis-glyoximates and their *O*-deuteriated derivatives and of Co^{III} chloro-ammino-bis-glyoximate. The assignments are clarified by normal-co-ordinate analyses of the Ni, Pd, and Pt compounds, taking all 19 atoms into account, based on D_{2h} symmetry for the Ni compound and C_{2h} for the others. Bands at 1513 (Ni), 1516 (Pd), and 1517 cm^{-1} (Pt) are largely ν(C=N) (72—76% contribution), whereas those at 444, 351 (Ni); 484, 407 (Pd); and 505, 391 cm^{-1} (Pt) are mainly ν(MN) (~ 50% contribution).

The crystal structure of bis(dihydroxo-boroxalene-diamide-dioximato)-nickel(II) tetrahydrate, $(C_4H_{12}N_8O_8B_2)Ni,4H_2O$, has been determined; the ν(NH) modes are at 3400 and 3380 cm^{-1} (2510 and 2250 cm^{-1} in the N-deuteriated derivative).[170c]

In the far-i.r. spectra of some nickel(II) complexes of aromatic aldehyde oximes, ligand vibrations in NiX_2L_2 [L = pyridine-2-carbaldehyde oxime (pa) or pyridine-2-carbaldehyde *O*-methyl oxime (pam)] have been separated from those due to Ni—X (X = Cl or Br), and the assignments have been used to assist the drawing of stereochemical conclusions.[171] Thus Ni—(pa) vibrations in both of the pa complexes are very similar (*ca.* 350, *ca.* 315, *ca.* 280 cm^{-1}); the Ni—(pam) vibrations in $NiCl_2(pam)_2$ are also similar, but in $NiBr_2(pam)_2$ only two Ni—(pam) vibrations were seen (292, 335 cm^{-1}). These results, together with electronic spectral data, suggest that $NiBr_2(pam)_2$ possesses a monomeric, *trans*-octahedral structure, whereas the other complexes are *cis*-octahedral dimers.

I.r. spectra of some Cu^{II} complexes of pyridine-2-carbaldehyde oxime (pa) have also been reported.[172] Examination of bands due to ν(C=N), ν(NO),

[167] S. S. Singh, *Z. anorg. Chem.*, 1971, **384**, 81.
[168] T. N. Srivastava and K. L. Saxena, *J. Inorg. Nuclear Chem.*, 1971, **33**, 3996.
[169] J. C. Warf and V. Gutmann, *J. Inorg. Nuclear Chem.*, 1971, **33**, 1583.
[170] (*a*) Y. Yamano, I. Masuda, and K. Shinra, *Bull. Chem. Soc. Japan*, 1971, **44**, 1581; (*b*) A. Bigotto, V. Galasso, and G. De Alti, *Spectrochim. Acta*, 1971, **27A**, 1659; (*c*) W. Fedder, H. G. van Schmering, and F. Umland, *Z. anorg. Chem.*, 1971, **382**, 123.
[171] G. Lees, F. Holmes, A. E. Underhill, and D. B. Powell, *J. Chem. Soc.* (*A*), 1971, 337.
[172] F. Holmes, G. Lees, and A. E. Underhill, *J. Chem. Soc.* (*A*), 1971, 999.

and several of the ring vibrations for this ligand supports the conclusion that complexes Cu(pa)(paH)X (X = Cl, Br, I, NO_3, or HSO_4) contain a square-planar $[Cu(pa)(paH)]^+$ group (47), and also indicates that the anion in the nitrato-complex is only weakly co-ordinated.

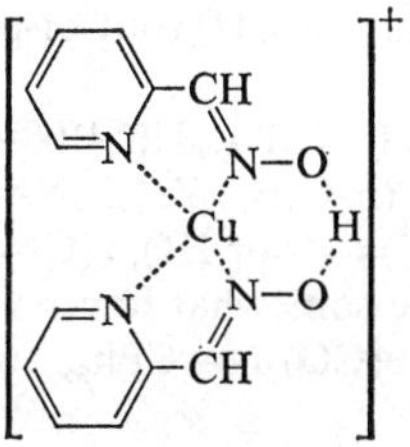

(47)

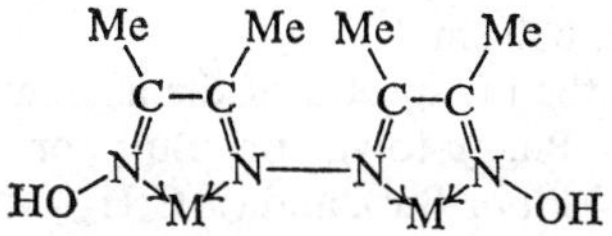

(48)

Spectra of Cu^{II} and Ni^{II} complexes of diacetylazine-dioxime suggest that it binds in a bi-bidentate manner (48). A tentative assignment is given for the ligand spectrum.[173]

Ligands containing ⟩C=N— Groups.—We will consider only those complexes of such ligands which co-ordinate *via* a σ-bond to the nitrogen; derivatives of pyridine rings have been classified under the amines and mentioned in a preceding section.

The i.r. and Raman spectra of some 3(5)-methylpyrazole (mpz) complexes of the general formula $M^{2+}(mpz)_n(anion^-)_2$ have been investigated [174] for a number of bivalent transition-metal ions. In particular, the N—H vibrations of co-ordinated mpz (49) appear to be strongly dependent upon the hydrogen-bonding properties of the anions present in the complexes.

I.r. frequencies have been listed,[175] but not discussed, for several transition-metal complexes of isothiazole (50).

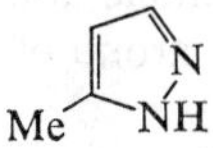

(49)

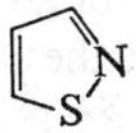

(50)

The i.r. spectra of metallo-octaethylporphyrins (M = Mg, Zn, Cu, Co, Ni, Pd, or $GeCl_2$) have been measured and fairly fully assigned by Nakamoto and his co-workers.[176] Bands assigned mainly to metal–nitrogen stretching vibrations are in the region 330—360 cm^{-1}. 'Metal-sensitive' bands are also distinguished, such a band (in the region of

[173] S. Satpathy and B. Sahoo, *J. Inorg. Nuclear Chem.*, 1971, **33**, 1313.
[174] J. Reedijk, *Rec. Trav. chim.*, 1971, **90**, 117.
[175] M. E. Peach and K. K. Ramaswamy, *Inorg. Chim. Acta*, 1971, **5**, 445.
[176] H. Ogoshi, N. Masai, Z. Yoshida, J. Takemoto, and K. Nakamoto, *Bull. Chem. Soc. Japan*, 1971, **44**, 49.

980 cm^{-1}) possibly being associated with a ring vibration involving a motion of the metal atom.

I.r. bands due to the vibrations of the aromatic *meso*-protons of some metalloporphyrins (Mg^{II}, Zn^{II}, Cu^{II}, Co^{II}, Ni^{II}, and Pd^{II}) have been identified[177] by deuteriation studies; ν(CH) is in the expected 3000 cm^{-1} region, δ(CH) (in-plane) occurs at *ca.* 1220 cm^{-1}, and π(CH) (out-of-plane) at *ca.* 830 cm^{-1}.

In the i.r. spectra of the azomethine derivatives $(\pi\text{-}C_5H_5)_2Ti(Cl)N{=}CR_2$ (R = Ph, *p*-tolyl, or Bu^t; or R_2 = MePh), $(\pi\text{-}C_5H_5)_2Zr(Cl)N{=}CR_2$ (R = Ph or Bu^t), and $(\pi\text{-}C_5H_5)_2M(N{=}CPh_2)_2$ (M = Zr or Hf), ν(C=N) is found at 1640 ± 20 cm^{-1}. This contrasts with the somewhat lower values for $(Me_2PhP)_2Pt(Cl)N{=}CPh_2$ and $(\pi\text{-}C_5H_5)Mo(CO)_2N{=}CPh_2$, where $d_\pi \rightarrow \pi^*$ bonding is probably more significant.[178a]

Complexes of Mn, Fe, Co, Ni, Cu, Zn, and Cd co-ordinated to the ligand *N*-n-butylimidazole have been prepared and examined by i.r. and Raman spectroscopy.[178b]

The novel ketimides $(\pi\text{-}C_5H_5)_2Ti(R)Cl$, $(Ph_3P)_3RhR$, *cis*-$(Ph_3P)_2Pt(R)Cl$, *trans*-$(Ph_3P)Pt(R)H$ [which gives ν(Pt—H) at 2146 cm^{-1}], *cis*-$(Me_2PhP)_2$-$Pt(R)Cl$, *cis*-$(Me_2PhP)_2PtR_2$, and *trans*-$(Me_2PhP)_2Pt(R)H$ [ν(Pt—H) at 2115 cm^{-1}] have been prepared,[179] where R = $(CF_3)_2C{=}N^-$. Trends in ν(C=N) (1715—1650 cm^{-1}) point to the possibility of $(d\pi\text{–}\pi^*)$ π-bonding in the d^8 complexes. The R^- ligand is believed to exert a relatively high *trans*-influence.

Treatment of some bifunctional and terdentate Schiff bases with Ti^{IV} and Zr^{IV} isopropoxides in proportions 1 : 1 or 1 : 2 affords the complexes $M(SB)_x(OPr^i)_{4-2x}$ (M = Ti or Zr; SB^{2-} = anion of the Schiff base SBH_2; x = 1 or 2). The i.r. spectra (4000—400 cm^{-1}) show no bands due to ν(OH), in the 3600—3100 cm^{-1} region, while bands in the 620—520 cm^{-1} range either appear, or are enhanced in intensity compared with those in the spectra of the Schiff bases themselves. ν(C=N) remains almost unaffected, appearing at 1640—1601 cm^{-1}. It is concluded that chelate formation takes place through the (phenolic and alcoholic) oxygen atoms and the nitrogen atom of the >C=N— group of the ligands (51).[180]

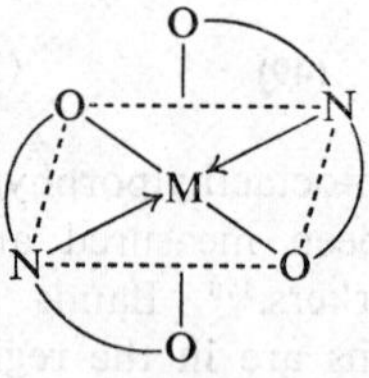

(51)

[177] H. Ogoshi and Z. Yoshida, *Bull. Chem. Soc. Japan*, 1971, **44**, 1722.
[178] (*a*) M. R. Collier, M. F. Lappert, and J. McMeeking, *Inorg. Nuclear Chem. Letters*, 1971, **7**, 689; (*b*) J. Reedijk, *Rec. Trav. chim.*, 1971, **90**, 1249.
[179] B. Cetinkaya, M. F. Lappert, and J. McMeeking, *Chem. Comm.*, 1971, 215.
[180] P. Prashar and J. P. Tandon, *Z. anorg. Chem.*, 1971, **383**, 81.

Bands assigned [181] to $\nu(C=N)$ and $\nu(Ph-O)$ in the i.r. spectra of complexes of Schiff bases of salicylaldehyde and benzaldehyde with $NbCl_5$ occur in the ranges 1610—1660 and 1265—1290 cm^{-1}, respectively. The nitrogen of the azomethine group acts as the donor of electrons. I.r. spectra in the NaCl transmission region have been reported for the complexes MCl_2L_3, $M(SO_4)L_3$, $M(NO_3)_2L_4$ {M = Cr^{III}, Mn^{II}, Fe^{II}, Co^{II}, Cu^{II}, Zn^{II}, Pd^{II}, or Pb^{II}; L = $MeC(:O)NHNH_2$, $PhC(:O)NHNH_2$, or their [$^2H_3$]-analogues}.[182] Assignments were made (for which the evidence was rather poor), and it was suggested that co-ordination was more accurately represented by (52a) than by (52b).

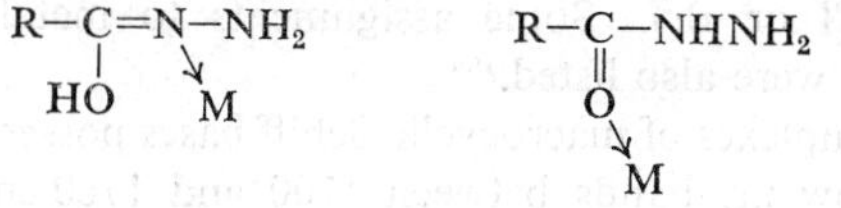

$$\begin{array}{cc} R-C=N-NH_2 & R-C-NHNH_2 \\ | \searrow & \| \\ HO M & O \\ & \searrow \\ & M \\ (a) & (b) \end{array}$$

(52)

Bands due to N—H vibrations in a number of tetrahedral and octahedral imidazolo-complexes of Co^{II} have been reported.[183] In the i.r. spectra of some 1,2-dimethylimidazole complexes of Co^{II}, Ni^{II}, Cu^{II}, and Zn^{II}, bands appearing in the 250 cm^{-1} region for tetrahedral complexes and 380 cm^{-1} for square-planar complexes have been assigned to metal–ligand stretching vibrations.[184]

The i.r. spectra of some cobaloximes (53) have been recorded. Bands assigned to $\nu(C=N)$ are in the 1550 cm^{-1} region, and to $\nu(N-O)$ are at *ca.* 1240 and *ca.* 1090 cm^{-1}. $\nu(Co-N)$ modes are assigned to bands close to 515 cm^{-1} in a variety of similar complexes.[185]

Some of the bands due to ligand vibrations have been assigned in the i.r. spectra of a number of bis-(NN'-diethylbutane-1,3-di-iminato)-complexes of Co^{II} and Zn^{II} (53a).[186]

I.r. spectra have been reported in the 600—200 cm^{-1} region for L, CoL_nX_2 (n = 1 or 2; X = Cl, Br, or I), and $CoL_6(ClO_4)_2$, where L = isothiazole.[187] It was concluded that L was always N-bonded, but no vibrational assignments were made.

Frequencies have been quoted for i.r. bands assigned to $\nu(C=N)$ and $\nu(C=C)$ in the spectra of some 2,4-dimethyl-1,5-benzodiazepinium halogenometallates $M(C_{11}H_{13}N_2)_xX_y$ (M = Co^{II}, Mn^{II}, Zn^{II}, Cd^{II}, Sb^{III}, Cu^{II},

---

[181] L. V. Surpina, O. A. Osipov, and V. A. Kogan, *Russ. J. Inorg. Chem.*, 1971, **16**, 366.
[182] A. D. Ahmed and N. R. Chaudhuri, *J. Inorg. Nuclear Chem.*, 1971, **33**, 189.
[183] W. J. Davis and J. Smith, *J. Chem. Soc. (A)*, 1971, 317.
[184] D. M. L. Goodgame, M. Goodgame, and G. W. R. Canham, *J. Chem. Soc. (A)*, 1971 1923.
[185] N. Yamazaki and Y. Hohokabe, *Bull. Chem. Soc. Japan*, 1971, **44**, 63.
[186] R. Bennett, D. C. Bradley, K. J. Fisher, and I. F. Rendall, *J. Chem. Soc. (A)*, 1971, 1622.
[187] R. Rivest and A. Weisz, *Canad. J. Chem.*, 1971, **49**, 1750.

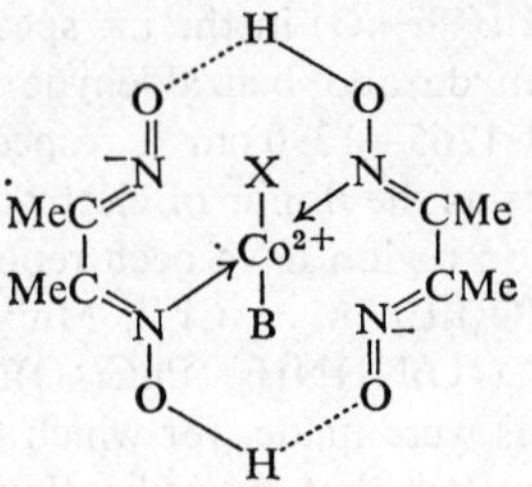

(53)

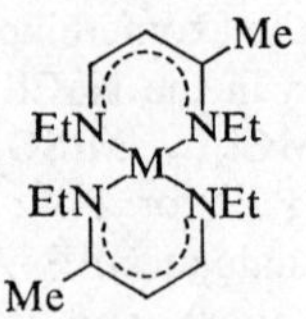

(53a)

or Cu^{I}; X = Cl or Br). Some assignments to metal–ligand stretches (200—300 cm^{-1}) were also listed.[188]

Cobalt(III) complexes of macrocyclic Schiff bases possessing co-ordinated acyl ligands show i.r. bands between 1700 and 1760 cm^{-1} assignable to ν(C=O); differences in the positions of these bands between solution and solid-state spectra, and also on varying the co-ordinating Schiff base, have been interpreted in terms of the existence of more than one configuration for the complexes, *e.g.* for Co(*NN'*-disalicylidene-ethylenediamine) (54).[189]

Ligand vibrations (for C=N and >C=C< in Cu^{II} and Ni^{II} complexes (55) of some co-ordinated anils (derivatives of aromatic amines and nitromalondialdehyde) have been given.[190] Thus, ν(C=N) occurs at *ca.* 1580 cm^{-1} and ν(C=C) at *ca.* 1500 cm^{-1}.

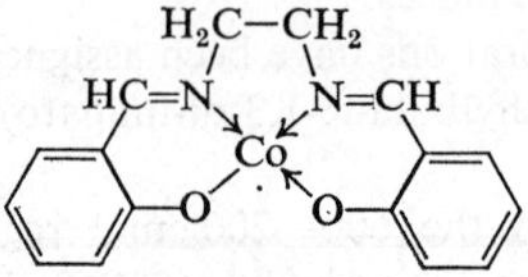

(54)

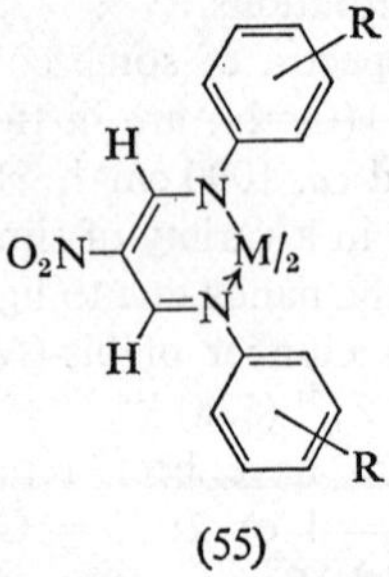

(55)

Some information on ν(C=N) frequencies of the complexes $Ni(ttet)^{2+}$ (56) and $Ni(dtet)^{2+}$ (57), where ttet = 2,4,4-trimethyl- and dtet = 3,3-dimethyl-1,5,8,11-tetra-azacyclotridec-1-ene, has been obtained.[191] Analogous complexes which also contain co-ordinated NCS^- give bands characteristic of N-bonded NCS^-.

[188] A. Ouchi, T. Takeuchi, M. Nakatani, and Y. Takahashi, *Bull. Chem. Soc. Japan*, 1971, **44**, 434.
[189] J. Booth, P. J. Craig, B. Dobbs, J. M. Pratt, G. L. P. Randall, and A. G. Williams, *J. Chem. Soc.* (*A*), 1971, 1964.
[190] S. V. Serina, V. P. Kurbatov, and O. A. Osipov, *Russ. J. Inorg. Chem.*, 1971, **16**, 578.
[191] N. F. Curtis and G. W. Reader, *J. Chem. Soc.* (*A*), 1971, 1771.

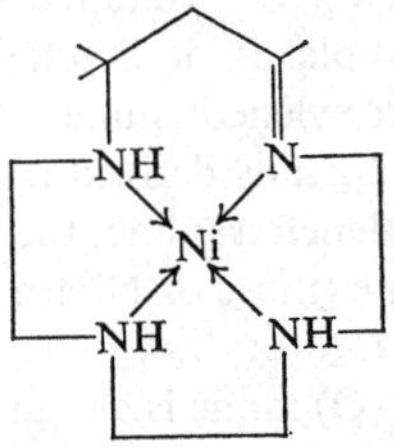

(56)

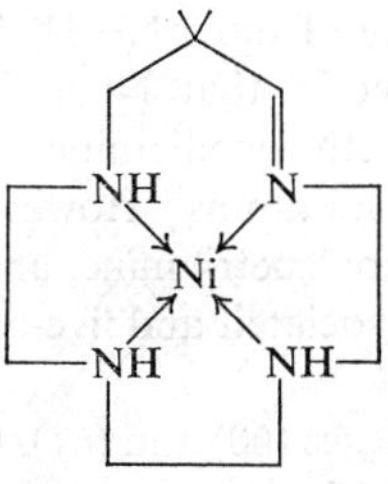

(57)

Some novel macrocyclic complexes have been prepared that contain two metal atoms.[192] They contain the ligands 6,7,8,9,12,19,20,21,22,25-decahydro-8,8,10,21,21,23-hexamethyl-5,26:13,18-bis(azo)dibenz[*i*,*t*][1,2,6,-7,12,13,17,18]octa-azacyclodocosine (taph) and 4-[2-(4-hydrazino-1-phthalazinyl)hydrazino]-4-methyl-2-pentanone (daph), and are formulated as $Ni_2(taph)^{4+}$ and $Ni_2(daph)^{4+}$. There is no evidence (i.r.) for co-ordination of anions (BF_4^-,ClO_4^-), and they are believed to possess the structures (58) and (59) respectively. The imine stretch is found at *ca.* 1640 cm^{-1}, and a

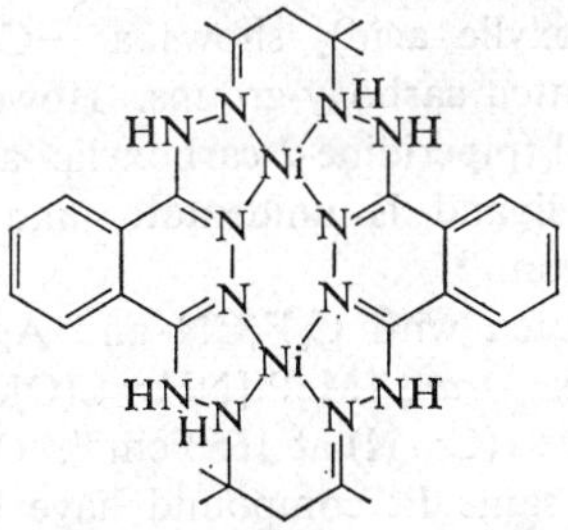

(58)

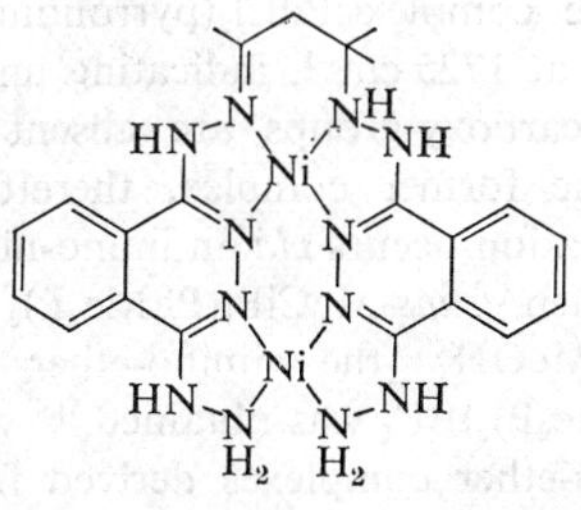

(59)

strong band at 1510 cm^{-1} is believed to be characteristic of the phthalazine ring.

In the i.r. spectra of some Ni^{II} complexes with macrocyclic tetra-imino-ligands, strong bands near 1660 cm^{-1} have been assigned to ν(C=N) vibrations, while bands in the region 3092—3050 cm^{-1} are due to ν(>CH) of the imine groups.[193]

I.r. spectra (666—222 cm^{-1}) have been reported[194] for various 4-methyl-thiazole (met) complexes. All contain N-bonded ligands, and partial assignments were made, *e.g.* for $Ni(met)_2Br_2$, ν(M—Br) 266 cm^{-1}, ν(M—L) 245 cm^{-1}.

[192] W. Rosen, *Inorg. Chem.*, 1971, **10**, 1832.
[193] N. F. Curtis, *J. Chem. Soc.* (*A*), 1971, 2834.
[194] M. N. Hughes and K. J. Rutt, *Inorg. Chem.*, 1971, **10**, 414.

From a study of the ν(N—H) bands of the co-ordinated ligands, Krebs *et al.* concluded[195] that Ni and Cu complexes of Schiff bases of *NN*-disubstituted ethylenediamine and propylenediamine contain four-co-ordinate metal atoms. However, with ligands derived from diethylenetriamine, dipropylenetriamine, and triethylenetetramine, the Cu complexes are certainly associated and five-co-ordinate (those of Ni are only probably so).

In the complexes (60) and (61), the ν(C=O) mode is assigned[196] to bands at 1684 and 1692 cm^{-1}, respectively. The values for the free ligand are 1658 and 1653 cm^{-1}.

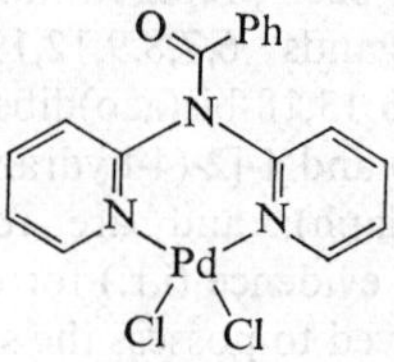
(60)

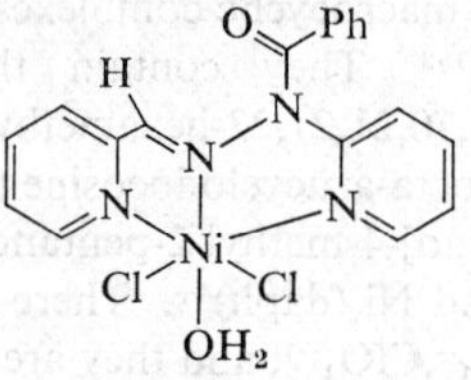
(61)

The complex $PdCl_2$(pyrrolidine-2-carboxylic acid)$_2$ shows a $-CO_2H$ band at 1725 cm^{-1}, indicating unco-ordinated carboxy-groups. However, free carboxy-groups are absent in $PdCl_2$(piperidine-2-carboxylic acid). In the former complex, therefore, the ligand is unidentate, and co-ordination occurs *via* an imino-nitrogen atom.[197]

When *trans*-$MeClPt(PhMe_2P)_2$ was treated with C_6F_5CN and $AgBF_4$ in MeOH, the imino-ether complex *trans*-{MePt[NH=C(OMe)]-$(PhMe_2P)_2$}BF_4 was obtained,[198] which gave ν(C=N) at 1653 cm^{-1}. Other imino-ether complexes derived from the same Pt compound have been obtained using *p*-$(CN)_2C_6F_4$ and MeOH, EtOH, or Pr^iOH. All gave ν(C=N) in the region 1630—1660 cm^{-1} and ν(NH) at 3300—3400 cm^{-1}.

I.r. spectra of pyrazole and imidazole complexes of Cu^{II} have been reported[199a] and comparisons with the free-ligand frequencies have been made. ν(NH) is found to vary with the anions in the same way as for the ligand field and *g*-values. In similar complexes, the copper–ligand vibrations appear to be closely related to the mass of the ligand.

$CuCl_2$ reacts with 2,2′-bisbenzimidazole to yield two compounds of unknown structure whose i.r. frequencies (rock-salt transmission range) have been listed.[199b]

195 H. Krebs, H. Stark, and K. Pachali, *Z. anorg. Chem.*, 1971, **381**, 80.
196 D. St. C. Black and R. C. Srivastava, *Austral. J. Chem.*, 1971, **24**, 287.
197 K. Freund and H. Frye, *Inorg. Nuclear Chem. Letters*, 1971, **7**, 107.
198 H. C. Clark and L. E. Manzer, *Chem. Comm.*, 1971, 387.
199 (*a*) J. Reedijk, J. C. A. Windhirst, N. H. M. van Ham, and W. L. Groeneveld, *Rec. Trav. chim.*, 1971, **90**, 234; (*b*) J. E. Brady and M. A. C. Fogarty, *J. Inorg. Nuclear Chem.*, 1971, **33**, 2706.

From the reaction of glyoxal with the bis-complex of *cis,cis*-1,3,5-triaminocyclohexane and Zn^{II} in aqueous solution, a mono-glyoxaldi-imino-condensate $Zn(C_{14}H_{28}N_6)(ClO_4)_2,\frac{1}{2}H_2O$ and an anhydrous di-condensate have been obtained. These have ν(C=N) at 1680 and 1670 cm^{-1}, respectively, and a band characteristic of the amine group at 1600—1610 cm^{-1}. Trigonal-prismatic co-ordination is suggested from n.m.r. data, *i.e.* containing the unit (62).[200]

Complexes containing the hexafluoroisopropylideneimino ligand $[N{=}C(CF_3)_2]^-$ co-ordinated to B, Si, P, As, or S show a ν(C=N) band [201] in the i.r. between 1650 and 1750 cm^{-1}.

A few ligand bands have been listed for complexes of Sn^{IV} with multi-dentate Schiff bases.[202] 1 : 1 Complexes of lead(IV) with three Schiff bases [derived from condensation of salicylaldehyde with ethylenediamine (63),

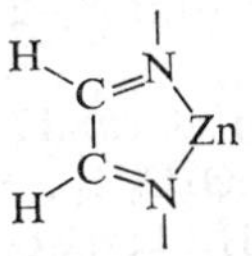

(62)

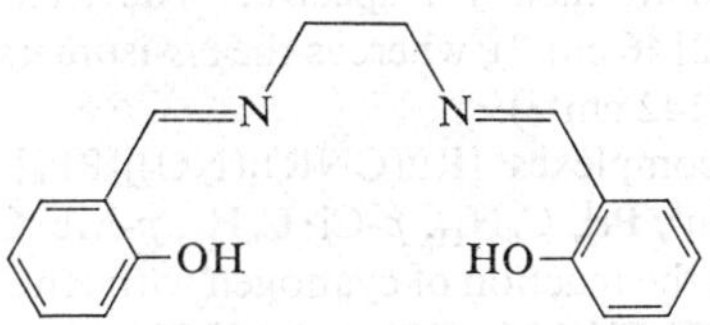

(63)

with *o*-phenylenediamine, and with benzidine] have been obtained.[203] As well as the expected absence of hydrogen-bonded ν(OH) bands, and shifts to longer wavelength of the phenolic C—O stretching modes, the ν(C=N) band shows a shift, suggesting co-ordination of the azomethine nitrogen atom.

Cyanides and Isocyanides.—Connor and Jones [204] have reported ν(CN) (2234 cm^{-1}) for (2-furylCN)$Cr(CO)_5$.

Although a series of ν(CN) frequencies has been reported for two series of $[Fe(CN)_6]^{n-}$ (n = 3 or 4) salts,[205] no attempt was made to correlate these figures with any other properties of the complexes.

An assignment has been proposed [206] for the vibrations of $Fe(CN)_5(NO)^{2-}$, based on solid-phase i.r. and Raman data. A normal-co-ordinate analysis was carried out, but the results are of doubtful validity because of the excessively oversimplified force-field employed.

The crystal structure of $Mn_2[Ru(CN)_6],8H_2O$ shows that the Ru is co-ordinated (approximately octahedrally) to six CN^- ions, each of which is bound through the nitrogen to Mn. Each Mn is co-ordinated to three

200 R. F. Childers and R. A. D. Wentworth, *Inorg. Nuclear Chem. Letters*, 1971, **7**, 519.
201 R. F. Swindell, T. J. Ouellette, D. P. Babb, and J. M. Shreeve, *Inorg. Nuclear Chem. Letters*, 1971, **7**, 239.
202 N. S. Biradar and V. H. Kulkarni, *J. Inorg. Nuclear Chem.*, 1971, **33**, 3781.
203 N. S. Biradar and V. H. Kulkarni, *Z. anorg. Chem.*, 1971, **381**, 312.
204 J. A. Connor and E. M. Jones, *Chem. Comm.*, 1971, 570.
205 J. B. Ayers and W. H. Waggoner, *J. Inorg. Nuclear Chem.*, 1971, **33**, 721.
206 G. Paliani, A. Poletti, and A. Santucci, *J. Mol. Structure*, 1971, **8**, 63.

nitrogen atoms and three H_2O molecules, and two such distorted octahedra are linked, *via* bridging water molecules, to give the unit $Mn_2N_6(H_2O)_6$. The following vibrational assignments are proposed[207] for this system: ν(CN) 2077 cm^{-1}; ν(RuC) 552 cm^{-1}; ν(MnO) 450 cm^{-1}; bound H_2O (rock) 865 cm^{-1}, (wag) 720 cm^{-1}; H_2O bending 1620 cm^{-1}; H—O stretching 3150, 3600 cm^{-1}.

Values of ν(CN) in $[Ru(NH_3)_5(BH_3CN)]^{n+}$ (n = 1 or 2) show the same variation as in corresponding CH_3CN complexes,[208] illustrating the similar ligand properties of BH_3CN^- and CH_3CN. The ν(CN) mode of the N-co-ordinated hydrogen cyanide complex $[Ru(NH_3)_5(NCH)]^{2+}$ occurs at 1960 cm^{-1}.

cis- and *trans*-$[Ru(CNEt)_2(EPh_3)_2X_2]$ (E = P, As, or Sb; X = Cl or Br) have been isolated and characterized[209] by the number of ν(NC) bands observed in their i.r. spectra. The *trans*-isomers give one strong band (2135—2146 cm^{-1}), whereas the *cis*-isomers give rise to two (2168—2187 and 2121—2142 cm^{-1}).

The complexes $[Ru(CNR)_4(NO)][PF_6]_2$ have ν(NO) at *ca.* 1770 cm^{-1} (R = Bu^t, Pr^i, C_6H_{11}, p-Cl·C_6H_4, p-Me·C_6H_4, or p-MeO·C_6H_4).[210]

From the reaction of cyanogen with $RhCl(PPh_3)_3$ in CH_2Cl_2, the cyanogen adduct $RhCl(C_2N_2)(PPh_3)_2$,0.125CH_2Cl_2 has been isolated.[211a] This shows ν(CN) at 2090 and 2240 cm^{-1}, due to co-ordinated C_2N_2. The compound slowly decomposes at ambient temperatures, evolving C_2N_2, the 2090 and 2240 cm^{-1} bands decreasing in intensity.

In the complexes $[Rh(NCBH_3)(CO)(PPh_3)_2]$ and $\{Ir(NCBH_3)(CO)[P(C_6H_{11})_3]\}$, the BH_3CN^- is co-ordinated *via* the nitrogen atom. I.r. evidence supports this conclusion, since ν(BH) is scarcely shifted from the value for the free ion, whereas ν(CN) is increased slightly.[211b]

The C≡N stretch in $[Ir(CO)(MeCN)(PPh_3)_2]^+$ occurs[211c] at 2290 cm^{-1}.

The zerovalent Pt isocyanide complexes $(Ph_3P)_2(CNBu^t)_2$ and $(Ph_3P)_2Pt(CNBu^t)(CO)$ have been characterized.[212] The bis-isocyanide complex gives a band assigned to ν(NC) at 2030 cm^{-1}; this is 100 cm^{-1} less than in the free ligand, and therefore appreciable back-donation Pt → $CNBu^t$ must have occurred. In the carbonyl complex, however, ν(NC) is found at 2110 cm^{-1} and ν(CO) at 1910 cm^{-1}. This shows that CO is a much more effective π-acceptor than Bu^tNC. The bis-isocyanide complex reacts with 1 mole of I_2, CH_3I, or CF_3I to give $[(Ph_3P)_2Pt(CNBu^t)_2R]^+I^-$ (R = I, CH_3, or CF_3), all of which have ν(NC) in the range 2225—2190 cm^{-1}, which is typical for Pt^{II}–isocyanide complexes.

207 M. Rüegg, A. Ludi, and K. Rieder, *Inorg. Chem.*, 1971, **10**, 1773.
208 P. C. Ford, *Chem. Comm.*, 1971, 7.
209 B. E. Prater, *J. Organometallic Chem.*, 1971, **27**, C17.
210 J. W. Dart, M. K. Lloyd, J. A. McCleverty, and R. Mason, *Chem. Comm.*, 1971, 1197.
211 (*a*) M. Bressan, G. Favero, B. Corain, and A. Turco, *Inorg. Nuclear Chem. Letters*, 1971, **7**, 203; (*b*) L. Vaska, W. V. Miller, and B. R. Flynn, *Chem. Comm.*, 1971, 1615; (*c*) G. R. Clark, C. A. Reed, W. R. Roper, B. W. Skelton, and T. N. Waters, *ibid.*, 1971, 758.
212 G. A. Larkin, R. Mason, and M. G. H. Wallbridge, *Chem. Comm.*, 1971, 1054.

ν(CN) wavenumbers have been reported[213] for the new cluster compounds $Au_{11}L_7X_3$ [X = CN or SCN; L = tris-(*p*-substituted phenyl)-phosphine]. All show a single band in solution, close to 2100 cm^{-1}.

Isocyanide complexes of thorium and uranium halides, *e.g.* $(RNC)_4UI_4$, show ν(NC) at higher frequencies than in the corresponding zerovalent metal complexes, suggesting considerable σ-donor character for the uranium– or thorium–carbon bonds.[214]

Nitrosyls.—A comparison between i.r. data, obtained for ν(CN) and ν(NO), and the results of e.s.r. studies of the complexes $[Cr(CN)_5NO]^{3-}$, $[Mn(CN)_5(NO)]^{3-}$, and $[Fe(CN)_5NO]^{2-}$ before and after γ-irradiation suggests formulation[215] of species resembling $[Cr(CN)_5NO]^{4-\text{ or }5-}$ and $[Mn(CN)_5NO]^{2-\text{ and }4-}$.

Nitrosyl-stretching wavenumbers have been reported, in the 1650 cm^{-1} region, for molybdenum complexes $[(\pi\text{-}C_5H_5)Mo(NO)X_2]_2$, $[(\pi\text{-}C_5H_5)\text{-}Mo(NO)X]_2$, and $[(\pi\text{-}C_5H_5)Mo(NO)I(SCH_2Ph)]_2$ (X = I, SCH_2Ph, or SPh), and have assisted tentative structural assignments for these species.[216a]

Nitrosyl-stretching wavenumbers for a variety of tertiary-phosphine and -arsine and isonitrile π-cyclopentadienyl nitrosyl complexes of molybdenum fall into the typical range for a terminal NO group (1650—1690 cm^{-1}).[216b]

The i.r. spectra of $K_3[Mn(CN)_5NO],2H_2O$, $K_3[Mn(CN)_5NO],2D_2O$, $K_3[Mn(CN)_5NO]$, $Ag_2[Mn(CN)_5NO]$, and their ^{15}NO analogues, have been published.[217] Vibrations due to motion of the Mn—N—O grouping could be assigned clearly; see Table 6 for the data on the anhydrous potassium salt.

ν(NO) [and also ν(CO)] have been listed for some new complexes containing gold–iron bonds, $[LAuFe(CO)_3(NO)]$ [L = $(MeO)_3P$, Me_3P, $Ph(C_6H_{11})_2P$, $Ph_2(C_6H_{11})P$, Ph_3P, $(p\text{-}Cl\cdot C_6H_4)_3P$, or $(p\text{-}Me\cdot C_6H_4)_3P$] and $[(Ph_3P)AuFe(CO)_2(NO)L]$ [L = $(PhO)_3P$, Ph_3P, or $PhEt_2As$].[218]

Dinitrosyl complexes of Fe and Co containing the anions of amino-acids and aldoximes as chelating ligands have been studied.[219] Two ν(NO) bands are found (see Table 7). Some mono-nitrosyl compounds of Ni have also been obtained.

Nitrosyl-stretching frequencies in three cationic Ru–nitrosyl complexes {*e.g.* in $[Ru(NO)_2Cl(PPh_3)_2]^+$, ν(NO) is at 1687 and 1845 cm^{-1}} clearly illustrate the danger associated with correlating ν(NO) with the M—N—O bond angle, and underline the notion that ν(NO) is sensitive to the

[213] F. Cariati and L. Naldini, *Inorg. Chim. Acta*, 1971, **5**, 172.
[214] F. Lux and U. E. Bufe, *Angew. Chem. Internat. Edn.*, 1971, **10**, 274.
[215] M. B. D. Bloom, J. B. Rayner, K. D. J. Root, and M. C. R. Symons, *J. Chem. Soc.* (*A*), 1971, 3212.
[216] (*a*) T. A. James and J. A. McCleverty, *J. Chem. Soc.* (*A*), 1971, 1068; (*b*) T. A. James and J. A. McCleverty, *ibid.*, 1971, 1596.
[217] E. Miki, S. Kubo, K. Mizumachi, T. Ishimori, and H. Okuno, *Bull. Chem. Soc. Japan*, 1971, **44**, 1024.
[218] M. Casey and A. R. Manning, *J. Chem. Soc.* (*A*), 1971, 2989.
[219] W. Hieber and H. Führling, *Z. anorg. Chem.*, 1971, **381**, 235.

Table 6 *I.r. spectra and assignments for* $K_3[Mn(CN)_5NO]$

^{14}NO *complex* Wavenumber/cm^{-1}	Strength	^{15}NO *complex* Wavenumber/cm^{-1}	Strength	*Assignment*
2129	w	2125	w	ν(CN)
2103	vs	2103	vs	
2065	sh	2064	sh	
1711	vs	1675	vs	ν(NO)
1695	vs	1663	vs	
659	s	—		$\nu(Mn-^{14}N) + \delta(Mn-^{14}N-O)$
—		661	w	$\nu(Mn-^{15}N)$
—		648	s	$\delta(Mn-^{15}N-O)$
490	w	487	w	$\nu(Mn-C) + \delta(Mn-C-N)$
455	m	454	m	
447	m	446	m	
405	s	404	s	
368	s	368	s	
315	w	314	w	

composition of filled bonding MO's, regardless of the formal assignment of metal and ligand charges.[220]

Values have been assigned to ν(NO) in the following series of Ru–nitrosyl complexes: $[Ru(bipy)_2(NO)L]^{n+}$: L = py, $n = 3$, 1950 cm^{-1}; L = NO_2^-, $n = 2$, 1942 cm^{-1}; L = Cl^-, $n = 2$, 1931 cm^{-1}; L = Br^-, $n = 2$, 1930 cm^{-1}; and $[Ru(phen)_2(NO)L]^{n+}$: L = py, $n = 3$, 1951 cm^{-1}; L = NO_2^-, $n = 2$, 1943 cm^{-1}; L = Cl^-, $n = 2$, 1932 cm^{-1}; L = Br^-, $n = 2$,

Table 7 ν(NO) *modes of* Fe, Co, *and* Ni *complexes containing anions of amino-acids or aldoximes*

Compound	ν(NO), *Wavenumber*/cm^{-1}
$(ON)_2Fe(NH_2CH_2COO)$	1743, 1702
$(ON)_2Fe(NHMeCH_2COO)$	1745, 1699
$Na(ON)_2Fe(NH_2CH_2COO)$	1672, 1712
$(ON)_2Fe(NH_2CHMeCOO)$	1759, 1710
$(ON)_2Fe(NH_2CHPhCOO)$	1762, 1709
$(ON)_2Fe(NH_2CHPr^iCOO)$	1751, 1700*
$(ON)_2Fe(NOHCHC_6H_4O)$	1772, 1711
$(ON)_2Fe(NOHCMeC_6H_4O)$	1788, 1734
$(ON)_2Co(NH_2CH_2COO)$	1821, 1761
$(ON)_2Co(NHMeCH_2COO)$	1824, 1752
$(ON)_2Co(NH_2CHPhCOO)$	1839, 1740*
$(ON)_2Co(NH_2CHMeCOO)$	1845, 1764*
$(ON)_2Co(NH_2CHPr^iCOO)$	1808, 1757
$(ON)[P(C_6H_{11})_3]Ni(NH_2CH_2COO)$	1749*
$(ON)[P(C_6H_{11})_3]Ni(NHMeCH_2COO)$	1760
$(ON)[P(C_6H_{11})_3]Ni(MeCOCHCOMe)$	1779

* In THF solution.

220 C. G. Pierpont, A. Pucci, and R. Eisenberg, *J. Amer. Chem. Soc.*, 1971, **93**, 3050.

1931 cm^{-1}. In the NO_2^- complexes, bands due to the N-bonded group are found at 1450, 1321, 817, 612 cm^{-1} (bipy complex); 1428, 1319, 810, 612 cm^{-1} (phen complex). The relatively high values for the nitrosyl stretches suggest a fairly large positive charge on the complexed NO fragment.[221]

Polarographic half-wave potentials for a series of complexes derived from $Co(NO)(CO)_3$ by replacement of one or two carbonyl groups with neutral donor ligands are correlated with N—O stretching frequencies.[222] They bear a linear relationship with one another. A full discussion of $E_{\frac{1}{2}}$ *vs.* ν(NO) relations has been given by Mašek in *Inorganica Chimica Acta, Reviews*, 1969, **3**, 99.

The cobalt nitrosyl complexes (64) gave[223] the following ν(NO) wavenumbers, all of which are characteristic of bridging nitrosyl groups: X = SPh, 1500 cm^{-1}; X = SePh, 1498 cm^{-1}; X = PPh_2, 1509 cm^{-1}.

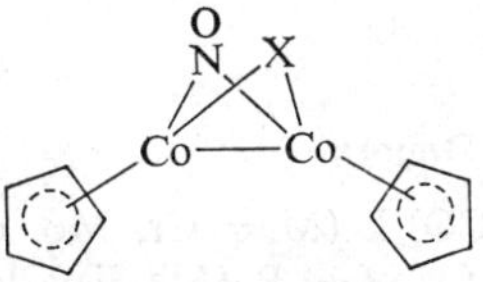

(64)

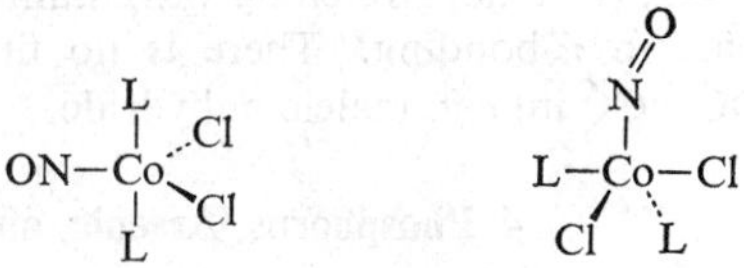

(a) (b)

(65)

The appearance of two bands in the ν(NO) region (1650—1800 cm^{-1}) for the complexes $(R_3P)_2CoCl_2(NO)$ (R = Et, Bu, Ph, or MeC_6H_4; or R_3 = $MePh_2$) has been ascribed by Collman *et al.* to the presence of two isomers (65a) and (65b).[224]

As well as two strong i.r. bands, assigned to ν(NO), at 1850, 1796 cm^{-1}, in the complex $Co_4(NO)_8(NO_2)_2(N_2O_2)$, further bands at 1377, 1327, 1098, and 828 cm^{-1} are probably associated with NO vibrations. The compound in question has been shown crystallographically to contain a quadridentate bridging hyponitrite group, and its vibrations may give rise to some of the additional i.r. frequencies.[225]

Values of ν(NO) have been recorded for the rhodium complexes $RhX_2(NO)(AR_3)_2$ (X = Cl, Br, or I; A = P or As; R = alkyl, aryl, or mixed alkyl and aryl).[226] They span the range 1616—1671 cm^{-1} and show a marked dependence on the nature of the phosphine or arsine ligand present. Similar Ir complexes show ν(NO) in the range 1553—1563 cm^{-1}.

ν(NO) wavenumbers for a series of $M(NO)(PPh_3)L$ complexes (M = Co, Rh, or Ir) are listed in Table 8.[227] The (quinone) C=O stretching and

[221] J. B. Godwin and T. J. Meyer, *Inorg. Chem.*, 1971, **10**, 471.
[222] P. Přibil, J. Mašek, and A. A. Vlček, *Inorg. Chim. Acta*, 1971, **5**, 57.
[223] H. Brunner and S. Loskot, *Z. Naturforsch.*, 1971, **26b**, 757.
[224] J. P. Collman, P. Farnham, and G. Dolcetti, *J. Amer. Chem. Soc.*, 1971, **93**, 1789.
[225] R. Ban, I. H. Sabherwal, and A. B. Burg, *J. Amer. Chem. Soc.*, 1971, **93**, 4926.
[226] S. D. Robinson and M. F. Uttley, *J. Chem. Soc. (A)*, 1971, 1254.
[227] G. La Monica, G. Navazio, P. Sandrini, and S. Cenini, *J. Organometallic Chem.*, 1971, **31**, 89.

Table 8 ν(NO) *for the complexes* $M(NO)(PPh_3)_2L$ (M = Co, Rh, *or* Ir)

M	L	ν(NO), *Wavenumber*/cm^{-1}
Rh	1,4-benzoquinone	1645
	1,4-naphthaquinone	1667
	1,2-naphthaquinone	1623
	maleic anhydride	1630
	tetracyanoethylene	1650
	CO	1670
Co	1,4-naphthaquinone	1705
	maleic anhydride	1700
	fumaronitrile	1715
Ir	3,4,5,6-tetrachloro-1,2-benzoquinone	1592
	maleic anhydride	1672

C≡N (of tcne) stretching wavenumbers suggest that these groups take no part in π-bonding. There is no firm evidence about the behaviour of $\rangle$C=C$\langle$ in, *e.g.* maleic anhydride.

4 Phosphorus, Arsenic, and Antimony Donors

Chromium-group-metal carbonyl complexes $M(CO)_5L$ (M = Cr, Mo, or W) have been reported,[228] where L = $(Me_3Si)_3P$, $(Me_3Ge)_3P$, $(Me_3Si)PPh_2$, $(Me_3Si)_2PPh$, and $(Me_3Sn)_2PPh$. I.r. frequencies assigned to P—M (M = Si, Ge, or Sn) stretching modes, as well as ν(CO) frequencies, have been listed for the complexes. Observed wavenumber shifts for these ν(M—P) from those in the unco-ordinated molecule $(MMe_3)_3P$ are small (< 15 cm^{-1} in all the cases studied).

Stibine complexes of Cr, Mo, and W, *e.g.* $[M(CO)_5SbH_3]$, have been prepared,[229] and show i.r. bands at 775 cm^{-1} due to $\delta(SbH_3)$. Carbonyl frequencies are similar to those in related PH_3 and AsH_3 complexes.

AsF_3 has made its appearance as a ligand in metal carbonyl chemistry. The complex $(\pi\text{-}C_5H_5)Mn(CO)_2AsF_3$ has carbonyl frequencies which, when compared with the PF_3 analogue, indicate that AsF_3 is an even better electron-acceptor than PF_3.[230]

New phosphine-substituted complexes of Re, Ru, Os, Rh, and Ir, containing the ligand PPh_2H, have been prepared. I.r. bands assigned to ν(P—H) occur in the 2300—2400 cm^{-1} region; these are tabulated for a range of neutral and ionic complexes, together with values for ν(M—H) (1900—2000 cm^{-1} region) where appropriate.[231]

The product from the reaction of Me_3SiNMe_2 with a slight excess of $HCo(PF_3)_4$ below 273 K is formulated as $[Me_3SiNHMe_2]^+[Co(PF_3)_4]^-$, ν(PF) appearing at 800 cm^{-1}. The 1 : 1 adduct from Me_2NH and $HCo(PF_3)_4$ is similarly formulated as $[Me_2NH_2]^+[Co(PF_3)_4]^-$. Likewise, the product

[228] H. Schumann, O. Stelzer, J. Kuhlmey, and U. Neiderreuther, *Chem. Ber.*, 1971, **104**, 993.
[229] E. O. Fischer, W. Bathelt, and J. Müller, *Chem. Ber.*, 1971, **104**, 986.
[230] J. Müller and K. Fender, *Angew. Chem. Internat. Edn.*, 1971, **10**, 418.
[231] J. R. Sanders, *J. Chem. Soc.* (*A*), 1971, 2991.

from Me_3SiNMe_2 and $HCo(CO)_4$ contains the $Co(CO)_4^-$ ion, [ν(CO) at 1880 cm^{-1}].[232]

The zwitterionic complex $[CoBr_3^- \leftarrow L{-}L^+]$ [$L{-}L^+ = Ph_2PC{\equiv}\overset{+}{C}PPh_2$-$(CH_2Ph)$] has been obtained [233] [as well as the analogue with $L{-}L^+$ as $Ph_2PCH_2CH_2\overset{+}{P}Ph_2(CH_2Ph)$]. The $\nu(C{\equiv}C)$ wavenumber (2132 cm^{-1}) was compared with that in $[Et_4N]^+[CoBr_3^- \leftarrow PPh_2C{\equiv}CPPh_2]$ (2126 cm^{-1}, all i.r. data from KBr disc spectra). The shift to higher wavenumber for the zwitterionic complex is expected, but the smallness of the shift indicates that the presence of the quaternary phosphorus atom has little effect on the donor properties of the other P. This conclusion follows from the dependence of $\nu(C{\equiv}C)$ upon the $M(d_\pi) \rightarrow P(d_\pi)$ bonding, which controls the amount of π-electron drift from the triple bond into the d_π-orbitals of phosphorus.

Rhodium complexes $RhCl(P_4)(MR_3)_2$ [MR_3 = PPh_3, $P(m\text{-tol})_3$, $P(p\text{-tol})_3$, or $AsPh_3$], which contain the molecular P_4 unit as a ligand, have been prepared by Ginsberg and Lindsell.[234] Their vibrational spectra were compared with that of the free P_4 molecule. Assignment in terms of a bound P_4 molecule of C_{3v} or C_s symmetry is proposed, and the P_4 vibrations are 15—90 cm^{-1} lower in the complexes than in the free ligand.

The new dinuclear trifluorophosphine complex $(PF_3)_4Ir{-}Ir(PF_3)_4$ shows typical ν(PF) bands at 939w and 900m cm^{-1} in the vapour-phase i.r. spectrum.[235]

Characteristic bands in the i.r. spectra of 1,2-bis(dimethylarsino)-ethylene complexes of Ni^{II}, Ni^{III}, and Co^{III} have been reported [236] and used to assist the identification of the ligand complexed in the *cis*- or *trans*-forms (66a) and (66b).

Cu^I nitrate complexes of the types $L_2Cu(NO_3)$, $L_3Cu(NO_3)$, and $L_4Cu(NO_3)$ (L = tertiary phosphine) have been prepared. NO_3^- vibrations indicate the presence of covalent, bidentate; covalent, unidentate; and ionic nitrate in the three series, respectively.[237]

The i.r. spectra of nine complexes HgX_2L [L = Cl, Br, or I; X = (67; R = Me, Ph, or *o*-tolyl)] have been obtained,[238] and $\nu(C{=}O)$ and $\nu(C{-}O)$ were recorded and shown to be consistent with bidentate co-ordination of X *via* As and C=O.

(a) (b)

(66)

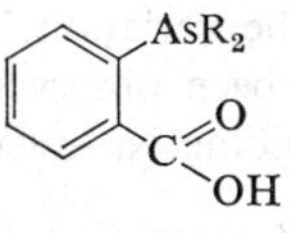

(67)

232 R. E. Highsmith, J. R. Bergerud, and A. G. MacDiarmid, *Chem. Comm.*, 1971, 48.
233 R. C. Taylor and R. A. Kolodny, *Inorg. Nuclear Chem. Letters*, 1971, **7**, 1063.
234 A. P. Ginsberg and W. E. Lindsell, *J. Amer. Chem. Soc.*, 1971, **93**, 2082.
235 T. Kruck, G. Sylvester, and I. P. Kunan, *Angew. Chem. Internat. Edn.*, 1971, **10**, 225.
236 M. A. Bennett and J. D. Wild, *J. Chem. Soc.* (*A*), 1971, 536, 545.
237 W. A. Andersen, A. J. Carty, G. J. Palenik, and G. Schreiber, *Canad. J. Chem.*, 1971, **49**, 761.
238 S. S. Sandhu and H. Singh, *J. Inorg. Nuclear Chem.*, 1971, **33**, 97.

5 Oxygen Donors

Molecular Oxygen, Peroxo-, and Hydroxy-complexes.—The use of vibrational spectroscopy in determining the co-ordination mode of the hydroxy-ligand has been reviewed.[239]

The external molecular vibrations $\nu_{\perp}$ and $\nu_{\parallel}$ of water sorbed on the Ca^{2+} ions in type 5A zeolite have been calculated by a semi-empirical potential function to occur at 433 and 79 cm^{-1}, and have been observed in the far-i.r. absorption at 444 and 71 cm^{-1}, respectively.[240]

The i.r. spectra of some peroxo-vanadate complexes $M^I[V(O{-}O)_3(AA)],nH_2O$ and $M^I_3[V(O{-}O)_3(AA)'],H_2O$ [M^I = NH_4, K, or Na; (AA) = phen or bipy; (AA)′ = $C_2O_4^{2-}$] have been examined.[241] Bands assigned to O—O stretching appear between 850 and 950 cm^{-1}, while absorptions in the 600—650 cm^{-1} region and 950 cm^{-1} region are assigned to O—V—O vibrations.

Reactions of hexacarbonyls with methanolic KOH yield $K_3[M_2(CO)_6(OH)_3],2H_2O$ (M = Mo or W), in which the three OH groups bridge the metal atoms. Further reactions yield $[Mo(CO)_3H(OH)]_4,4H_2O$, $[Mo(CO)_3H(OH)]_4,4Ph_3PO$, $[Mo(CO)(NO)(OH)]_4,4H_2O$, $[Mo(CO)_2(NO)(OH)]_4,4Ph_3PO$, and the analogous W compounds. I.r. frequencies for the H_2O, OH, CO, NO, and P=O groups are listed, but no frequency attributable to ν(M—H) was observed.[242]

The complex $RuCl_2(Ph_3As)_3(O_2)$ has ν(O—O) at 880 cm^{-1} due to the triangular RuO_2 group.[243] Treatment with H_2 causes the 880 cm^{-1} band to disappear, being replaced by a new feature at 1960 cm^{-1} [ν(MH)], while CO gives a dicarbonyl species [ν(CO) at 1950 and 2100 cm^{-1}] with evolution of O_2.

Deuteriation of complexes $M_2[Pt(OH)_6]$ (M = Na or K) resulted in i.r. bands of wavenumbers 685, 1075, and 3400—3500 cm^{-1} being shifted to 530, 900, and 2500 cm^{-1}, respectively. This confirms the assignment of these bands to vibrations of co-ordinated OH.[244]

The i.r. spectra of a long series of uranyl complexes have been reported.[245a] A linear relationship between ν_3 and ν_1 of UO_2 and the paramagnetic susceptibilities has been demonstrated.

The variation in ν(OH) attributable to hydrogen-bonding O—H···Cl has been observed in the i.r. spectra of a number of inorganic halides, including $NaAuCl_4,2H_2O$ and $NaICl_4,2H_2O$.[245b]

[239] V. Baran, *Co-ordination Chem. Rev.*, 1971, **6**, 65.
[240] K. Möller, D. Kunath, and H.-J. Spangenberg, *Spectrochim. Acta*, 1971, **27A**, 353.
[241] J. Sala-Pala and J. E. Guerchais, *J. Chem. Soc.* (*A*), 1971, 1132.
[242] U. Sartorelli, L. Garlaschelli, G. Ciani, and G. Bonora, *Inorg. Chim. Acta*, 1971, **5**, 191.
[243] M. M. Taqui Khan, R. K. Andal, and P. T. Manoharan, *Chem. Comm.*, 1971, 561.
[244] B. N. Ivanov-Emin, L. D. Borzova, N. V. Venskovskii, B. E. Zaitsev, and V. I. Portil'yn, *Russ. J. Inorg. Chem.*, 1971, **16**, 723.
[245] (*a*) M. P. Sahakari and A. J. Mukhedkar, *J. Inorg. Nuclear Chem.*, 1971, **33**, 888; (*b*) K. Ichida, Y. Kuroda, D. Nakamura, and M. Kudo, *Bull. Chem. Soc. Japan*, 1971, **44**, 1996.

Acetylacetonates and Related Complexes.—Published i.r. and Raman data of some β-diketonate complexes have been cited in the context of optical and geometrical isomerization.[246]

Metal-insensitive and metal-sensitive wavenumbers have been listed (from the i.r. spectra, down to *ca.* 350 cm^{-1})[247] for bis-(9,10-anthracenedion-1-olato)metal(II) complexes (metal = Be, Cu, VO, Ni, Mg, Cd, Fe, Co, Mn, Zn, Ca, Cr, Ba, or Pb) (68). The proposed assignments are

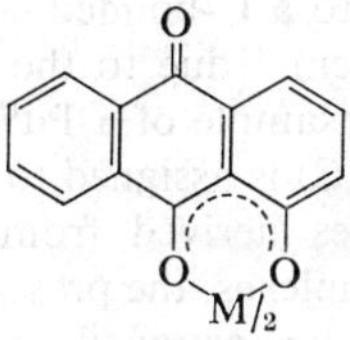

(68)

rather lacking in precision, however: 1634—1610 cm^{-1}, ν(C=O) (uncomplexed); 1585—1558 cm^{-1}, ν(C=C); 1397—1355 cm^{-1} and 1248—1222 cm^{-1}, ν(C=O) and ν(C=C); 671—573 cm^{-1}, ν(M—O).

I.r. wavenumbers have been listed to 400 cm^{-1} and roughly assigned for a series of complexes $TiX_n(acac)_{4-n}$ (n = 2, 3, or 4). These compounds have been postulated as five-co-ordinate, but they were shown[248] to be mixtures of well-established four- and six-co-ordinate compounds.

The i.r. spectra (4000—400 cm^{-1}) of Cr^{III} acetylacetonate and Cr^{III} malondialdehyde in the solid state and in CCl_4 solution have been assigned by comparison with the spectra of other metal derivatives involving these ligands.[249]

The ν(C=O) and ν(C=C) stretching bands of a new 1 : 1 complex, $MnBr_2$(acetylacetone), at 1627 and 1564 cm^{-1}, respectively, are similar to those of the enolic form of acetylacetone. This novel form of linkage is confirmed[250] by a single-crystal *X*-ray analysis, showing a structure of infinite planar $[MnBr_2]_n$ chains with unidentate, O-co-ordinated acetylacetone ligands completing octahedral co-ordination of Mn.

The ν(C=O) frequencies of $M(CF_3COCH_2COR)_2$ (M = Co, Ni, Cu, or Zn; R = Et, Pr^n, Pr^i, Bu^t, Ph, CH_2Ph, or CF_3) show[251] some shifts with change of R. These are attributed to inductive and resonance rather than mass effects.

The reaction between $[Ni(acac)_2]_3$ and dry HBr affords $Ni_3Br(acac)_5$. This in turn, with 7 mole equivalents of pyridine, gives $NiBr(acac)(py)_3$

[246] J. J. Fortman and R. E. Sievers, *Co-ordination Chem. Rev.*, 1971, **6**, 331.
[247] R. A. Walker, *Spectrochim. Acta*, 1971, **27A**, 1785.
[248] C. E. Holloway and A. E. Sentek, *Canad. J. Chem.*, 1971, **49**, 579.
[249] W. O. George, *Spectrochim. Acta*, 1971, **27A**, 265.
[250] S. Koda, S. Ooi, H. Kuroya, Y. Nakamura, and S. Kawaguchi, *Chem. Comm.*, 1971, 280.
[251] M. Yaqub, R. D. Koob, and M. L. Morris, *J. Inorg. Nuclear Chem.*, 1971, **33**, 1946.

and $Ni(acac)_2(py)_2$. I.r. spectra (700—300 cm^{-1}) have been presented in diagrammatic form (with no frequencies or assignments given).[252]

The complex $NiBr_2(C_5H_8O_2)_2$ has been obtained, where $C_5H_8O_2$ is acetylacetone (neutral ligand).[253] The i.r. spectrum showed ν(C=O) at 1693 cm^{-1} (vs), indicating ketonic acetylacetone molecular co-ordination, confirmed by the conclusions drawn from single-crystal *X*-ray study.

Treatment of $Pd(acac)_2$ with PPh_3 (1 : 1) in benzene at room temperature yields $Pd(acac)_2(Ph_3P)$.[254] This has ν(CO) at 1665 cm^{-1} (vs) and 1632 cm^{-1} (s) (i.r. spectra), attributed to a C-bonded $acac^-$ group, with other bands at 1572, 1550s, and 1523vs cm^{-1} due to the chelated enolate group. The compound is thus the first example of a Pd^{II} compound with a C-bonded β-diketonate ligand. ν(Pd—C) is assigned to a band at 540 cm^{-1} (vs).

In Pd^{II} chelate complexes derived from 3-hydroxyiminopentane-2,4-dione, and some related complexes, the presence or absence of a bond near 1520 cm^{-1} in the i.r. spectrum, assigned to carbonyl stretching in a co-ordinated acyl group, has been used to determine whether co-ordination of this type is occurring.[255]

I.r. spectra for a variety of complexes of bis(hexafluoroacetylacetonato)-copper(II), of the type $(hfac)_2CuL$ and $(hfac)_2CuL_2$, have been reported,[256] where L = nitrogen-containing chelate or ethylene glycol. Spectral shifts occurring on co-ordination are discussed; in particular, for 1 : 1 complexes with bases containing $\rangle$N—Me groups, bands in the 2760—2830 cm^{-1} region disappear on chelation to a metal atom.

A novel trinuclear zinc complex has been prepared by Boersma *et al.*[257] The presence of chelating acetylacetonate groups only is supported by the observation that ν(C⋯O) only gives rise to a band at 1579 cm^{-1}, and ν(C⋯C) only at 1521 cm^{-1}. The structure proposed is (69), with one octahedral and two approximately tetrahedral Zn atoms.

Dutt and Sur have made [258] the first mixed-ligand tetrakis-complexes of La, Pr, Nd, and Sm. I.r. spectra (NaCl transmission region) have been reported for the three salts $Na[Pr(acac)_3L]$ (L = propionylacetone, benzoylacetone, or dibenzoylmethane). The shifts in frequencies assigned to modes of L, by comparison with the free ligand, are consistent with the above formulation.

I.r. frequencies (in the NaCl region) have been listed in order to characterize the complexes of lanthanides with various β-diketones and β-ketoesters (*e.g.* ethyl acetoacetate, ethyl benzoylacetate, 3-methyl ethyl acetoacetate, 3-ethyl ethyl acetoacetate), but no discussion was given.[259]

[252] K. Isobe, Y. Nakamura, and S. Kawaguchi, *Inorg. Nuclear Chem. Letters*, 1971, **7**, 929.
[253] S. Koda, S. Ooi, H. Kuroya, K. Isobe, Y. Nakamura, and S. Kawaguchi, *Chem. Comm.*, 1971, 1321.
[254] S. Baba, T. Ogura, and S. Kawaguchi, *Inorg. Nuclear Chem. Letters*, 1971, **7**, 1195.
[255] D. A. White, *J. Chem. Soc.* (*A*), 1971, 233.
[256] D. E. Fenton, R. S. Nyholm, and M. R. Truter, *J. Chem. Soc.* (*A*), 1971, 1577.
[257] J. Boersma, F. Verbeek, and J. G. Noltes, *J. Organometallic Chem.*, 1971, **33**, C53.
[258] N. K. Dutt and S. Sur, *J. Inorg. Nuclear Chem.*, 1971, **33**, 115.
[259] N. K. Dutt, S. Sur, and S. Rahut, *J. Inorg. Nuclear Chem.*, 1971, **33**, 1717, 1725.

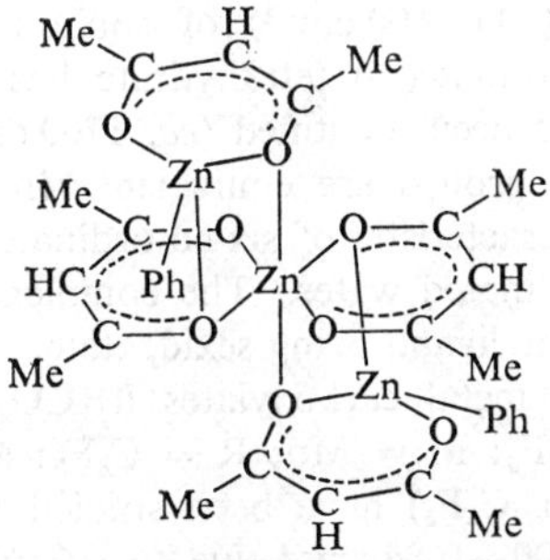

(69)

Carboxylates.—The concentrations of Ba, Sr, and Ca carbonates can be determined, whether they occur singly or together in a sample, by quantitative measurements of i.r. spectra over the range 650—750 cm^{-1}, using KBr discs.[260] There is no interference from nitrates or sulphates. Spectra over this range for the three compounds have been published.

In the i.r. spectra of Ca, Sr, Ba, and Cu^{II} trichloroacetate monohydrates, shifts of $\nu_{as}(CO_2^-)$ and $\nu_s(CO_2^-)$, compared with the free ion, indicate a different type of co-ordination of the CO_2^- group to that found in monochloroacetates.[261] Thus, $\nu_{as}(CO_2^-)$ is at lower wavenumber than in the free ion, the shifts being in the sequence Cu^{II} > Ca > Sr > Ba. Also, the difference between ν_{as} and ν_s of CO_2^- follows the order Cu^{II} < Ca < Sr < Ba. With corresponding monochloroacetates, these sequences are reversed, and ν_{as} is at a higher wavenumber than in the free ion. These observations are interpreted in terms of bidentate co-ordination of the CO_2^- groups in the trichloroacetates, with unidentate bonding for the monochloroacetates. Some relevant assignments are listed in Table 9.

Table 9 *Some vibrational assignments for the trichloroacetate ion and its metal complexes (wavenumbers*/cm^{-1})

Vibration	*Free ion*	Cu	Ca	Sr	Ba
$\nu_{as}(CO_2^-)$	1677	1629	1649	1660	1670
$\nu_s(CO_2^-)$	1353	1379	1373	1379	1353
$\nu_{as}(CCl_3)$	849, 746	843, 762	843, 762	843, 762	843, 749
$\nu_s(CCl_3)$, $\delta(CO_2^-)$	685	685	681	681	681

Partial assignments for ligand vibrations have been proposed for the vanadyl complexes [VO(AA)(BB)],nH$_2$O (AA = malonato or maleato; BB = *o*-phenanthroline or bipyridyl). The results are consistent with square-pyramidal co-ordination at the vanadium atom.[262]

260 D. E. Chasan and G. Norwitz, *Appl. Spectroscopy*, 1971, **25**, 226.
261 A. V. R. Warrier and R. S. Krishnan, *Spectrochim. Acta*, 1971, **27A**, 1243.
262 J. Sala-Pala and J. E. Guerchais, *Bull. Soc. chim. France*, 1971, 2444.

The i.r. spectrum (500—600 cm^{-1}) of sodium *trans*-1,2-cyclohexanediaminetetra-acetatochromate(III) tetrahydrate has been listed, but only one ν(C=O) mode has been identified (*ca.* 1700 cm^{-1}).[263] This indicates that all four carbonyl groups are equivalent. Invariance of the visible spectrum (which is characteristic of six-co-ordinate Cr^{III}) with pH shows that there is no co-ordinated water. The complex is thus formulated as Na[Cr(cydta)],$4H_2O$, the ligand being sexidentate.

A series of anionic metal carboxylates $[(RCO_2)M(CO)_5]^-$ (M = Cr; R = CF_3, C_2F_5, or C_6F_5: M = Mo; R = C_6F_5: M = W; R = H, Me, Et, Ph, CF_3, C_2F_5, or C_6F_5) have been studied.[264] All showed an i.r. band in the region 1600—1684 cm^{-1} due to ν(C=O) of the carboxylate group. The wavenumber of this mode was dependent upon R, but not significantly upon M [*e.g.* all of the C_6F_5 derivatives gave ν(C=O) in the range 1640—1646 cm^{-1}]. Three ν(C≡O) bands were found, in the expected regions, in all cases.

In $(MeCO_2)Re(CO)_5$ and $(PhCO_2)Re(CO)_5$, the (unidentate) carboxylate gives ν(C=O) at *ca.* 1620 cm^{-1}, ν_{as}(C—O—Re) at *ca.* 1340 cm^{-1}. The dimeric compounds $[(MeCO_2)Re(CO)_3]_2$ (and the phenyl analogue) presumably contain bridging carboxylate groups [$\nu(CO_2)$ at *ca.* 1670, *ca.* 1280 cm^{-1}].[265]

When formic acid is adsorbed onto a surface of Fe_2O_3 (preheated to 573 K in a stream of He), an i.r. band at 3670 cm^{-1} [ν(OH, on the surface)] disappeared, and bands due to $\nu(CO_2)$ were observed at *ca.* 1550 and *ca.* 1360 cm^{-1}, together with ν(CH) at 2940 and 2865 cm^{-1}. The i.r. reflectance spectrum of ferric formate shows bands at *ca.* 2950, 2870, 1560, and 1570 cm^{-1}, and therefore this compound must be formed on the Fe_2O_3 surface.[266]

Bands at 1680 and 1600 cm^{-1} in the i.r. spectra of iron(III) atrolactate [(70a or b); R^1 = Ph, R^2 = Me] and iron(III) 2-hydroxyisobutyrate [(70a or b); R^1 = R^2 = Me], respectively, are assigned [267] to co-ordinated carboxylate. In the unco-ordinated ligands, the ν(C—O) mode of the tertiary alcoholic group is at *ca.* 1180 cm^{-1}; this shifts to 1120 and 1100 cm^{-1},

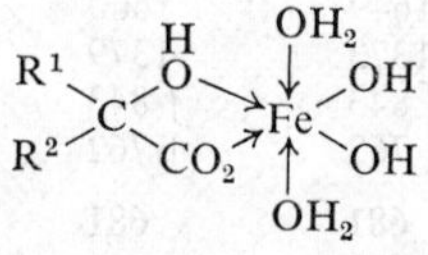

(a)

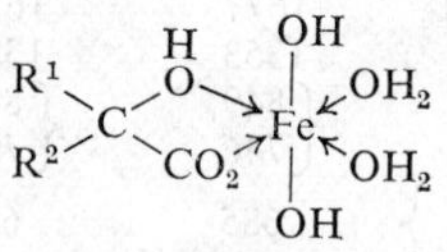

(b)

(70)

[263] N. Tanaka, N. Kanno, T. Tomita, and A. Yamada, *Inorg. Nuclear Chem. Letters*, 1971, **7**, 953.
[264] W. J. Schlientz, Y. Lavender, N. Welcman, R. B. King, and J. K. Ruff, *J. Organometallic Chem.*, 1971, **33**, 357.
[265] E. Linder and R. Grimmer, *J. Organometallic Chem.*, 1971, **31**, 249.
[266] N. Tahezawa, *Chem. Comm.*, 1971, 1451.
[267] K. K. Sen Gupta and A. K. Chatterjee, *Z. anorg. Chem.*, 1971, **384**, 280.

respectively, for the two complexes, and bands at 870 and 820 cm^{-1}, respectively, are attributed to co-ordinated water.

Positions of $\nu_s(CO_2)$ (1267—1431 cm^{-1}) and $\nu_{as}(CO_2)$ (1567—1652 cm^{-1}) have been listed for a number of carboxylate derivatives of ruthenium carbonyl.[268]

In oxalato-aniline complexes of Ru^{III}, *e.g.* $A^+[Ru(oxalate)_2L_2],xH_2O$ (A = K, NH_4, or $PhNH_3$; L = substituted aniline), the presence of only two bands in the ν(C=O) region (1650—1750 cm^{-1}) has been taken as evidence for a *trans*-octahedral structure for these anions.[269]

I.r. frequencies for $K_3[Co(malonate)_3],4H_2O$ have been assigned[270] by analogy with those for other trismalonato-complexes.

An assignment has been proposed for all the i.r. wavenumbers (down to 400 cm^{-1}) of maleic acid, its Na^+ and K^+ salts (hydrated), and its complexes $M(mal),xH_2O$ and $H_2[M(mal)_2],4H_2O$ (M = Co^{2+}, Ni^{2+}, Cu^{2+}, or Zn^{2+}).[271] The positions of the ν(C=O) modes are interpreted as indicating that the M—O bonds are less covalent than those in related oxalato- and malonato-complexes. M—O stretching is stated to contribute to bands at *ca.* 500, *ca.* 750, and *ca.* 400 cm^{-1}, but there is little evidence to support this statement.

Bands due to ν(C=O) in nickel(II)– and cobalt(II)–(*o*-carboxyphenyl tertiary arsine) complexes have been identified. The presence of $\nu_s(CO_2)$ (*ca.* 1390 cm^{-1}) and $\nu_{as}(CO_2)$ (*ca.* 1600 cm^{-1}) suggests co-ordination of the carboxylate ion to the metal.[272]

I.r. data for carboxylato-complexes $Rh(OCOR)(PPh_3)_3$ and $Rh(OCOR)(CO)(PPh_3)_2$ have been reported.[273a] Bands in the regions of 1600—1700 cm^{-1} and 1350—1420 cm^{-1} have been assigned to $\nu_{as}(CO_2)$ and $\nu_s(CO_2)$, respectively, for the co-ordinated carboxylato-group.

Carboxylate CO stretching frequencies have been used[273b] to confirm structural assignments for carboxylato–diene complexes fo Rh^I and Ir^I. The increase in separation between the two (OCO) stretching modes on going from a bridging to a unidentate configuration is clear from the results.

The two carboxylate C—O stretching bands were studied in a series of benzoate complexes.[274] Comparison of spectra of hydrated and anhydrous complexes indicated that the hydrated examples $[Ni(PhCO_2)_2,3H_2O$ and $Cu(PhCO_2)_2,3H_2O]$ contain both ionic and co-ordinated benzoate.

$Pd_2(OAc)_2$(cyclohexa-1,4-diene),$\frac{1}{2}$(HOAc) shows bands due to the HOAc at 1240 and 1720 cm^{-1}. The ν(C—O) bands of the OAc ligands

[268] B. F. G. Johnson, R. D. Johnston, J. Lewis, and I. G. Williams, *J. Chem. Soc.* (*A*), 1971, 689.
[269] D. L. Key, L. F. Larkworthy, and J. E. Salmon, *J. Chem. Soc.* (*A*), 1971, 371.
[270] N. R. Kreten and S. J. Spees, *J. Inorg. Nuclear Chem.*, 1971, **33**, 2437.
[271] D. N. Sathyanarayana and V. V. Savant, *Z. anorg. Chem.*, 1971, **385**, 329.
[272] S. S. Sandhu and S. S. Parmar, *J. Chem. Soc.* (*A*), 1971, 111.
[273] (*a*) R. W. Mitchell, J. D. Ruddick, and G. Wilkinson, *J. Chem. Soc.* (*A*), 1971, 3224; (*b*) R. N. Haszeldine, R. J. Lunt, and R. V. Parish, *ibid.*, p. 3696.
[274] S. F. Pavǩovic, *J. Inorg. Nuclear Chem.*, 1971, **33**, 1475.

(1385 and 1535 cm^{-1}) have a small separation, possibly indicating a symmetrically bound acetate ligand.[275]

When sodium stearate and zinc distearate are heated together in 2 : 1 molar proportions, the product obtained has an i.r. spectrum similar to that of a zinc stearate complex prepared from sodium stearate and zinc thiocyanate. Three $\nu_{as}(CO_2^-)$ bands are observed, at 1622, 1609, and 1595 cm^{-1}. They are attributed to different structural groupings within the sample. The $\nu_s(CO_2^-)$ band is very weak in a mull, but stronger in a KBr disc; an explanation of this is presented.[276]

Assignments for i.r. bands due to vibrations of the dihydroxymalonato-group, for rare-earth complexes of the general form $Ln_2(C_3H_2O_6)_3,7H_2O$, have been given. The ligand is present in the form (71).[277a]

$$\begin{array}{c} CO_2^- \\ | \\ C(OH)_2 \\ | \\ CO_2^- \end{array}$$

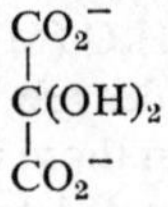

(71)

In Ge^{IV} complexes of *trans*-1,2-cyclohexanediaminetetra-acetic acid (of imprecise stoicheiometry), the i.r. spectra show at least three overlapping bands between 1710 and 1740 cm^{-1}, indicating that the tetra-acetato-ligand is probably quadridentate.[277b]

The compound PhCH(F)CH$_2$CH$_2$Pb(OAc)$_2$F has i.r. bands [277c] due to acetate groups at 1740 and 1245 cm^{-1}.

Keto-, Alkoxy-, Phenoxy-, and Ether ligands.—I.r. spectra (to *ca.* 300 cm^{-1}) of tropolonates (72) of first-row transition-metal + 2 and + 3 ions have

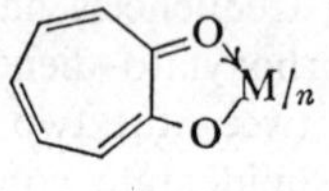

(72)

been discussed,[278] (M = Ca^{2+}, Mn^{2+}, Co^{2+}, Ni^{2+}, Cu^{2+}, Zn^{2+}, Sc^{3+}, Ti^{3+}, V^{3+}, Cr^{3+}, Mn^{3+}, Fe^{3+}, Co^{3+}, or Ga^{3+}). 'M—O stretching modes' are assigned by using the previously established relationship between ν(M—ligand) and the crystal-field stabilization energy (J. M. Haigh, R. D. Hancock, L. G. Hulett, and D. A. Thornton, *J. Mol. Structure*, 1969, **4**, 369, and references therein). ^{18}O Labelling of the Cu^{II} compound has also been used. The general assignments proposed are as follows: 1589—1601 ν(C—C); 1517—1527 ν(C—O) + ν(C—C); 1423—1445 ν(C—C); 1412—

---

[275] J. M. Davidson, *Chem. Comm.*, 1971, 1019.
[276] S. Toda, A. Sakai, and T. Kojma, *Spectrochim. Acta*, 1971, **27A**, 581.
[277] (*a*) A. V. Lapitskaya and S. B. Pirkes, *Russ. J. Inorg. Chem.*, 1971, **16**, 192; (*b*) S. K. Dhar and W. E. Kurez, *Inorg. Nuclear Chem. Letters*, 1971, **7**, 551; (*c*) J. Bornstein and L. Skarlas, *Chem. Comm.*, 1971, 796.
[278] L. G. Hulett and D. A. Thornton, *Spectrochim. Acta*, 1971, **27A**, 2089.

1419 δ(CH); 1306—1347 ν(C—O) (downward shift of 16 cm^{-1} on ^{18}O substitution in the Cu^{II} complex); 1245—1272, 1225—1260, 1210—1225 δ(CH); 1081—827 (up to 8 bands) ν(C—C) + δ(C—H); 753—767, 726—740 possibly ring deformation; 732—310 (about 5 bands) ν(M—O) (downward shifts of 5—26 cm^{-1} in the Cu^{II} complex); all cm^{-1}.

Metal(II) acetaldehyde, propionaldehyde, and benzaldehyde solvates $M(aldehyde)_6[FeCl_4]_2$ and $M(aldehyde)_6[InCl_4]_2$ (M = Mg, Mn, Fe, Co, Ni, or Zn) give decreases in ν(C=O) on complexing.[279]

The i.r. spectra of complexes of the anion derived from benzoin by loss of the hydroxyl proton (ben⁻), of formula $M^{III}Cl_2(ben)$ (M = Fe or Cr), are characteristic of bidentate bonding (73) of the organic ligand. Thus,

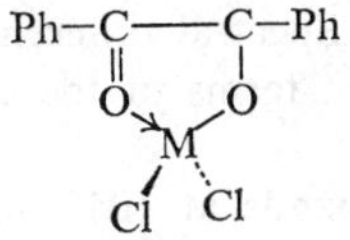

(73)

ν(C=O) is lowered in frequency, and there is no evidence for the presence of an O—H bond.[280]

The ν(C=O) bands of a number of ketones (*e.g.* butanone, acetophenone, and chloroacetone) have been shown[281] to shift to lower frequency in a characteristic manner upon complex formation, in the order Mn < Fe < Co < Ni > Zn.

The positions of some characteristic bands of the polyether ligand [(74) = L] have been found[282] to shift on complex formation. Thus, in

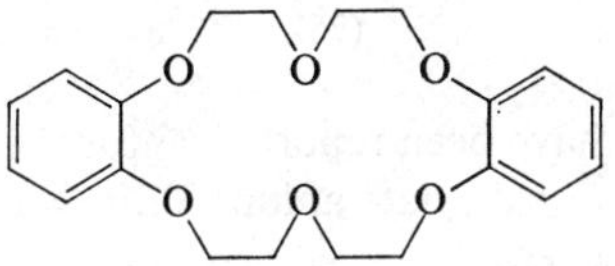

(74)

$K_4CoCl_6L_2$, $K_4CoI_6L_4$, and $K_2Co(NCS)_4L_2$, bands are found at *ca.* 956 cm^{-1} (996 cm^{-1} in the free ligand) and *ca.* 940 cm^{-1} (943 cm^{-1} in the free ligand). Since u.v. and visible spectra show that CoX_4^{2-} ions are present, the complexes are formulated as $2(KL)^+ + (CoCl_4)^{2+} + 2KCl$; $4(KL)^+ + CoI_4^{2-}$; and $2(KL)^+ + Co(NCS)_4^{2-}$, respectively.

279 W. L. Driessen and W. L. Groeneveld, *Rec. Trav. chim.*, 1971, **90**, 87.
280 R. C. Paul, J. Singh, R. D. Sharma, S. S. Parmar, and K. C. Malhotra, *Inorg. Nuclear Chem. Letters*, 1971, **7**, 43.
281 W. L. Driessen and W. L. Groeneveld, *Rec. Trav. chim.*, 1971, **90**, 258.
282 P. C. L. Birkbeck, D. S. B. Grace, and T. M. Shepherd, *Inorg. Nuclear Chem. Letters*, 1971, **7**, 801.

The carbonyl-stretching wavenumber of methyl salicylate decreases by *ca.* 30 cm^{-1} on complex formation with Co^{II}, in *e.g.* bis(methyl salicylato)-diaquocobalt(II), showing that complexing occurs *via* the ester C=O group.[283] In similar complexes with *o*-nitrophenol as ligand, ν(OH) disappears on complexing, and ν_s and $\nu_{as}(NO_2)$ are at lower frequency, suggesting a weakening of the N—O bonds.

I.r. wavenumbers to 250 cm^{-1} of the new dioxan complexes Hg(dioxan)-$(CN)_2$ and Hg(dioxan)$(SCN)_2$ have been listed and partially assigned. Their similarity to those of bands of Hg(dioxan)Cl_2 suggests that the dioxan acts as a bridging ligand and retains its chair form.[284a]

In a study of the conformation of dialkoxyethanes, i.r. spectra of these, of solutions of $HgCl_2$ in them, and of crystalline complexes derived from them, have been reported. Spectral differences in the 1150—1050 cm^{-1} range have been discussed in terms of the conformation of the organic group.[284b]

In $CeCl_3(OMe),2Ph_3PO$, bands at 1045 and 1067 cm^{-1} are associated with Ce—O—C modes, with another band at 1187 cm^{-1} derived from ν(P=O).[285]

I.r. spectra to 400 cm^{-1} have been listed for rare-earth chelate complexes of 2-hydroxy-5-methylacetophenone, of formulae (75). Quantitative assignments have been given for all peaks, on little sound evidence.[286]

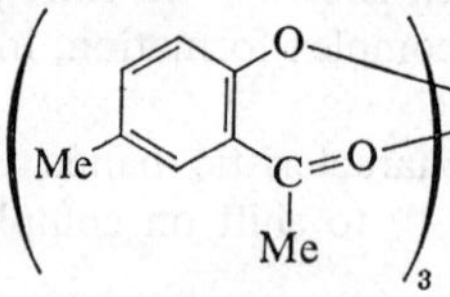

(75)

Unassigned i.r. data have been reported [287] for the lanthanide complexes $M(Hsal)_3(TBP)_2$ (Hsal = salicylate anion; TBP = tri-n-butyl phosphate).

O-Bonded Amides and Ureas.—ν(N—H), ν(C=O), ν(C—N), and the amide I and II band frequencies have been listed for some Ni^{II}, Co^{II}, Co^{III}, and Cu^{II} complexes of *N*-(2-picolyl)- and *N*-2-(2′-pyridyl)ethyl-picolinamides, (76) and (77). Co-ordination is *via* the amide oxygen.[288]

I.r. frequencies (in the rock-salt transmission region) have been reported for complexes of formulae $LnBr_3,5DMA$ and $LnBr_3,5DMA,3H_2O$, where

[283] H. G. Biedermann, G. Rossmann, and K. E. Schwarzhans, *Z. Naturforsch.*, 1971, **26b**, 78.
[284] (*a*) I. S. Ahuja and A. Garg, *J. Inorg. Nuclear Chem.*, 1971, **33**, 1515; (*b*) R. Iwamoto, *Spectrochim. Acta*, 1971, **27A**, 2385.
[285] F. Březma, *Coll. Czech. Chem. Comm.*, 1971, **36**, 2889.
[286] C. L. Garg, K. V. Narasimham, and B. N. Tripathi, *J. Inorg. Nuclear Chem.*, 1971, **33**, 387.
[287] S. P. Sinha, *J. Inorg. Nuclear Chem.*, 1971, **33**, 2205.
[288] M. Nonoyama and K. Yamasaki, *Inorg. Chim. Acta*, 1971, **5**, 124.

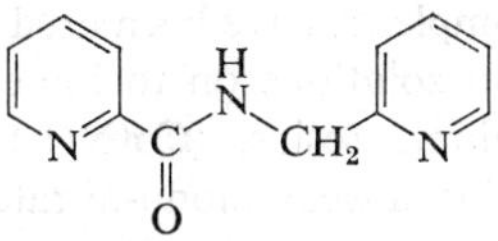

(76)

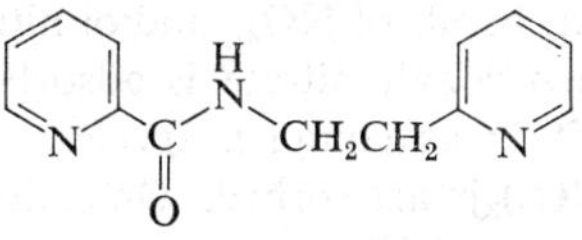

(77)

Ln = lanthanide and DMA = *NN*-dimethylacetamide. The lowered ν(C=O) frequency indicates co-ordination *via* the amide oxygen.[289]

Cyclopropylene urea forms 8 : 1 complexes with all of the lanthanide(III) perchlorates (Pm was not studied; with Lu, anhydrous materials were not obtained). Similar compounds were obtained with a number of lanthanide(III) nitrates. Their i.r. spectra (obtained most satisfactorily by a multiple-internal-reflectance technique) show a shift in the band primarily due to ν(C=O) from 1670 cm^{-1} (in the free ligand) to *ca.* 1635 cm^{-1} in the $Ln(L_8)X_3$ complexes, clearly indicating that the ligand is co-ordinated *via* the carbonyl group.[290]

Nitrates and Nitrato-complexes.—The vibrations of NO_3^- and CO_3^{2-} in bridging situations have been studied by normal-co-ordinate analysis using a hypothetical model (78), in which X is given mass 13 and M mass 60.

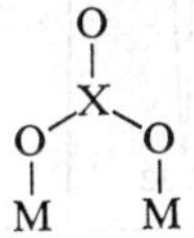

(78)

Variations of M—O and X—O force constants lead to relatively small changes in the XO_3^{n-} spectrum. On the basis of the low-frequency part of their calculated spectra, the authors suggest [291] that such a region may be of use in distinguishing bridging from unidentate groups.

Raman and i.r. spectra have been reported [292] for $VO(NO_3)_3$,2MeCN. *X*-Ray crystallography indicates the presence of two bidentate and one unidentate nitrate groups. Vibrational assignments have been made using isotopic substitution (^2H and ^{15}N) of the MeCN, and by comparison with the spectra of $VO(NO_3)_3$: for the nitrate moiety: ν_{as}(uni-NO_2) 1566 (depol.); ν_s(uni-NO_2) 1260 (pol.); ν_{as}(bi-NO_2) 1215—1230; ν_s(bi-NO_2) 1602—1637; ν(V=O) 1010 (possibly containing some nitrate bands); δ(bi-NO_2) 677, 689; $\delta(\text{uni-}NO_2)_s$ 745; $\delta(\text{uni-}NO_2)_{as}$ 890 (tentative); skeletal stretches and deformations; (uni), 407, 265, 203, 171, and 119; (bi) 436, 351, 300, 235, and 156 (all cm^{-1}); for the MeCN moiety: ν(V—NCMe) 407 [possibly including δ(C—C—N)]; ν(C—C) 945; ν(C≡N) 2290 cm^{-1}.

[289] G. Vicentini and C. Airoldi, *J. Inorg. Nuclear Chem.*, 1971, **33**, 1733.
[290] C. M. Burgess and G. E. Toogood, *Inorg. Nuclear Chem. Letters*, 1971, **7**, 761.
[291] B. Taravel, G. Chauvet, P. Quintard, and P. Delarme, *Compt. rend.*, 1971, **273**, *B*, 85.
[292] F. W. B. Einstein, E. Enwall, D. M. Morris, and D. Sutton, *Inorg. Chem.*, 1971, **10**, 678.

The i.r. bands of NO_3^- and of nitrato-complexes have been used to show that when cobalt nitrate is adsorbed from solutions on to ion-exchange resins, it is as complex species. On anionic resins, $[Co(NO_3)_4]^{2-}$ and $[CoX(NO_3)_3]^-$ are sorbed. On cationic resins, a weak mono-nitrate species may be present.[293]

Following up an earlier report (N. F. Curtis and Y. M. Curtis, *Inorg. Chem.*, 1965, **4**, 804) based upon limited data, it has been confirmed that

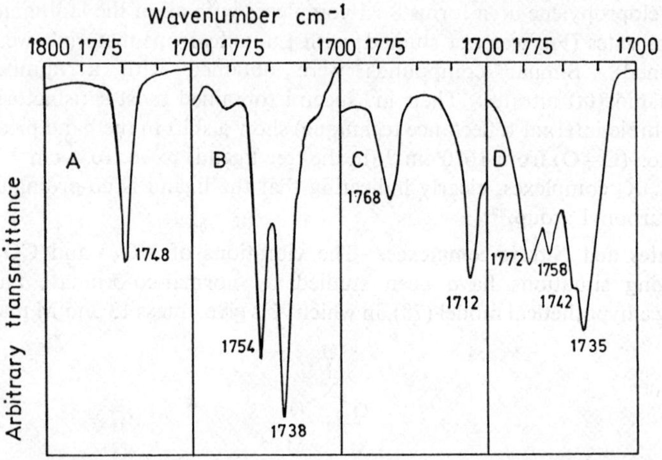

Figure 2 *Combination wavenumbers exhibited by* A: Cu(*asym*-Et_2en)$_2$(NO_3)$_2$ (*red isomer*), B: Co(*asym*-Me_2en)$_2$(NO_3)$_2$, C: Ni(py)$_2$(NO_3)$_2$, *and* D: Co(3-pic)$_3$(NO_3)$_2$
(Reproduced by permission from *Canad. J. Chem.*, 1971, **49**, 1957)

i.r. combination wavenumbers of *ca.* 1750 cm^{-1} may be used as a means of distinguishing between the various modes of co-ordination of nitrate groups.[294] Surprisingly, the region used seems to be free of complications from inter- and intra-molecular coupling of nitrate vibrations. The magnitude of the splitting of the combination bands depends on the strength of the M—NO_3 interaction and, for a given metal and oxidation state, the splitting is generally larger for bidentate than for unidentate co-ordination. The separation between the two combination bands for unidentate co-ordination is in the range 5—26 cm^{-1}, whilst that for bidentate co-ordination is 20—66 cm^{-1} (see Figure 2).

N—O stretching frequencies in the NO_3 group for some Cu^I nitrato-complexes and nitrates $Cu(NO_3)(ZPh_3)_3$ (Z = P, As, or Sb), $Cu(NO_3)$-(ZPh_3)(L), and $[Cu(PPh_3)_2L](NO_3)$ (L = 1,10-phenanthroline or 2,2′-biquinolyl) have been listed.[295]

293 C. Heitner-Wirgun and N. Ben-Zvi, *J. Inorg. Nuclear Chem.*, 1971, **33**, 1493.
294 A. B. P. Lever, E. Mantovani, and B. S. Ramaswamy, *Canad. J. Chem.*, 1971, **49**, 1957.
295 F. H. Jardine, A. G. Vohra, and F. J. Young, *J. Inorg. Nuclear Chem.*, 1971, **33**, 2941.

Bands due to co-ordinated nitrate groups have been assigned from the i.r. and Raman spectra of some tri-isopropyl phosphate (L) complexes $M(NO_3)_3,3L$ (M = Nd, Yb, or Lu).[296]

In tetranitratobis(triphenylphosphine oxide)cerium(IV) $[Ce(NO_3)_4(OPPh_3)_2]$ and its thorium(IV) analogue, NO_3 vibrations are reported,[297] as follows: (Ce); 1262 (ν_1), 1023 (ν_2), 1538 (ν_4), 803 (ν_6) (all cm^{-1}); (Th); 1265 (ν_1), 1017 (ν_2), 1540 (ν_4), 806 (ν_6) (all cm^{-1}). The splitting (into ν_1 and ν_4) of the E' vibration of NO_3^- (D_{3h}) is greater than that observed in lanthanide(III) complexes. There is, however, little distortion of the co-ordinated nitrato-group in (79), so that the splitting does not arise mainly from differences in the N—O′ and N—O″ bond strengths.

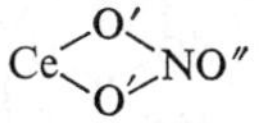

(79)

Ligands containing O—N, O—P, or O—As Bonds.—Drawing data almost exclusively from previous literature, Karayannis *et al.* have considered i.r. spectroscopic evidence for co-ordination of perchlorate in five-co-ordinate complexes $ML_4(ClO_4)_n$, where L is a phosphine oxide or arsine oxide. They conclude that the i.r. evidence *per se* does not provide conclusive information in favour of any one structural possibility.[298]

Metal solvates of phenylbis(dimethylamino)phosphine oxide (papo), of the type $M(papo)_4^{2+}$ $(anion)_2^{2-}$, have been examined by i.r. spectroscopy.[299] A comparison of the spectra with i.r. and Raman spectra of the free ligand indicates co-ordination *via* the oxygen atom. Characteristic i.r. and Raman bands also indicate[300] co-ordination *via* the P atoms of P—O in adducts of hexamethylphosphoramide (HMPA), *e.g.* $M(HMPA)_4^{2+}$ $(anion)_2^{2-}$.

I.r. frequencies have been listed and partially assigned for alkali-metal isopropoxy hydrogen methylphosphonates and pyromethylphosphonates.[301]

Complexes with the general formula MX_4L_2 [M = Sn, Ti, or Zr; X = Cl, Br, I, or NCS; L = $O{:}P(NMe_2)_3$ (HMPA), $O{:}P(NMe_2)_2Cl$ (DMAP), $O{:}P(NMe_2)Cl_2$ (DDP), or $S{:}P(NMe_2)_3$ (HMTP)] have been prepared.[302] From the i.r. spectra, a *trans*-arrangement of HMPA ligands was proposed, but a *cis*-arrangement for the others. ν(P=O) was shifted to lower

[296] J. R. McCrae and D. G. Karraker, *J. Inorg. Nuclear Chem.*, 1971, **33**, 479.
[297] Mazhar-ul-Haque, C. N. Caughlan, F. A. Hart, and R. Vannice, *Inorg. Chem.*, 1971, **10**, 115.
[298] N. M. Karayannis, C. M. Mikulski, M. J. Strocko, L. L. Pytlewski, and M. M. Labes, *J. Inorg. Nuclear Chem.*, 1971, **33**, 2691.
[299] M. W. G. de Bolster and W. L. Groeneveld, *Rec. Trav. chim.*, 1971, **90**, 1153.
[300] M. W. G. de Bolster and W. L. Groeneveld, *Rec. Trav. chim.*, 1971, **90**, 477.
[301] N. M. Karayannis, C. M. Mikulski, M. J. Strocko, L. L. Pytlewski, and M. M. Labes, *Inorg. Chim. Acta*, 1971, **5**, 357.
[302] E. Le Coz and J. E. Guerchais, *Bull. Soc. chim. France*, 1971, 80.

frequencies on co-ordination (more for the *trans*-complexes than for the *cis*-), but the frequency of ν(P=S) was increased.

The i.r. spectra of di-n-butyl phosphates of zirconium have been investigated, with a number of the ligand vibrations being assigned.[303]

I.r. spectra were used to characterize the complexes ML_2, $ML(SO_4)$, and $MLCl_2$, where M = Zr or Hf and L = di-n-butylmethylenephosphonic acid (without discussion).[304]

Partial assignments[305] of the i.r. spectra (to *ca.* 250 cm^{-1}) have been made for the following phosphonato-complexes: $VO(imp)_2$, $Fe(imp)_2$, Cu(mpH), Cu(mp), and M_2(imp)(pmp) (M = Ca, Mn, Ni, or Co), where imp = isopropyl methylphosphonato (80); mp = methylphosphonato

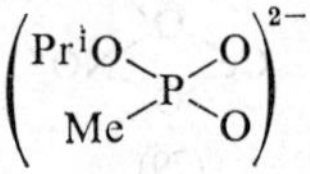

(80)

$(MePO_3)^{2-}$; mpH = hydrogen methylphosphonato $MePO_2(OH)^-$; pmp = pyromethylphosphonato $[O_2P(Me)-O-P(Me)O_2]^{2-}$. Some of the assignments are as follows: ν(PC) 719—787; $\nu(P-CH_3)$ 908—900 (imp complexes), 760—719 (mp and mpH complexes); ν(P—O—C) 101—999; $\nu_s(PO_2)$ 1066—1061; $\nu_{as}(PO_2)$ 1126—1117; $\delta(CH_3)$ 1420—1300; ν_s(P—O—P) 631—662 (pmp complexes) (all cm^{-1}).

PO_3 deformations occur in the regions 560—522 and 529—490 cm^{-1} in the complexes $M(H_2O)(PhPO_3)$ (M = Mn, Co, Ni, Cu, or Zn). Other wavenumbers were also listed.[306a]

In complexes of $Bu^n{}_3P{:}O$ (L) with Re, *e.g.* $(H_3O)_2[L_3ReCl_6]$ and $(H_3O)_2[L_3ReBr_6]$, ν(P=O) is found[306b] at *ca.* 1100 cm^{-1}. This shift from the value in free $Bu^n{}_3P{:}O$ (1160 cm^{-1}) shows that it is co-ordinated to the Re.

Potassium pentacyanonitrosylcobaltate can only be obtained from the black (and not from the red) isomers of $Co(NH_3)_5NO^{2+}$ salts. A study of its i.r. spectrum has been made (4000—200 cm^{-1}).[307] Bands in the region of 1130, 1050, and 980 cm^{-1} have been assigned to the stretching vibrations of the hyponitrite ligand [ν(N=N), ν(N—O), ν(N—O), respectively], consistent with the dimeric formulation $K_6[(NC)_5Co(N_2O_2)Co(CN)_5],4H_2O$.

ν(P—O) frequencies have been reported[308] for the tribenzylphosphine

303 E. G. Tetenin, N. N. Shesterikov, P. G. Krutikov, and A. S. Solovkin, *Russ. J. Inorg. Chem.*, 1971, **16**, 77.
304 H. Meider-Goričan, *J. Inorg. Nuclear Chem.*, 1971, **33**, 1919.
305 N. M. Karayannis, C. M. Mikulski, M. J. Strocko, and M. M. Labes, *Z. anorg. Chem.*, 1971, **384**, 267.
306 (*a*) W. C. Grinanneau, P. L. Chapman, A. G. Menke, and F. Walmsley, *J. Inorg. Nuclear Chem.*, 1971, **33**, 3011; (*b*) K. A. Bol'shakov, N. M. Sinitsyn, V. F. Travkin, and L. N. Antimonova, *Doklady Phys. Chem.*, 1971, **198**, 433.
307 H. Toyuki, *Spectrochim. Acta*, 1971, **27A**, 985.
308 O. A. Serra and A. C. Massabin, *Inorg. Nuclear Chem. Letters*, 1971, **7**, 275.

oxide (tbpo) complexes $CoCl_2(tbpo)_2$, $[Co(tbpo)_4](ClO_4)_2$, and $Co(NO_3)_2$-$(tbpo)_2$.

The AsO_4 vibrations in $[Co(NH_3)_6]AsO_4$ (*a*), $[Co(NH_3)_5(AsO_4)]$ (*b*), and $[Co(NH_3)_4(AsO_4)]$ (*c*) have been assigned from the i.r. spectra of the complexes and their ND_3 analogues.[309] The observed spectra are consistent with T_d, C_{3v}, and C_{2v} symmetry, respectively, for the AsO_4 group. The assignments were: (*a*): ν_3 800, ν_4 406 cm^{-1}; (*b*): ν_1 800, ν_3 851/817, ν_4 492/392 cm^{-1}; (*c*): ν_1 769, ν_2 370, ν_3 856, ν_4 561/505 cm^{-1}.

An intense band at 1242—1255 cm^{-1} in the i.r. spectra of *trans*-dioximato-compounds of Co^{III} with pyridine *N*-oxide has been assigned to ν(N—O).[310] Co-ordination to the metal takes place *via* the oxygen atom of the pyridine oxide.

Adducts $LnL_3(NO_3)_6$ [L = $(RO)_2P(:O)(CH_2)_xP(:O)(OR)_2$; x = 1, 2, or 4; Ln = lanthanide) and $LaL_2(NO_3)_3$ (x = 2 in L) have both L and NO_3 co-ordinated to the metal; ν(P=O) and nitrate frequencies were listed.[311]

ν(P=O) and nitrate bands indicate that in the compounds $Ln(NO_3)_3$,3L [Ln = lanthanide; L = *NN*-(dimethyl)diphenylphosphinamide] the nitrate is co-ordinated, and L is attached to the metals *via* its phosphoryl group.[312]

Compounds $Ln(ClO_4)_3$,6dppa (dppa = diphenylphosphinamide) show lowered ν(P=O) frequencies, indicative of co-ordination *via* oxygen, and splitting of the perchlorate ν_3 band, due to 'solid-state effects'.[313]

When triphenylphosphine oxide (tppo) forms the five-co-ordinate complexes $[UO_2(MeCOS)_2(tppo)]$, $[UO_2(PhCS_2)_2(tppo)]$, $[UO_2(MeCO_2)_2(tppo)]$, and $[UO_2(Et_2NCS_2)_2(tppo)]$, ν(P=O) is lowered by *ca.* 55 cm^{-1}. In the six-co-ordinate systems $[UO_2(MeCO_2)_2(tppo)_2]$ and $[UO_2(NO_3)_2(tppo)_2]$, however, $\Delta\nu$(P=O) is *ca.* − 45 cm^{-1}. In these compounds, $\nu(UO_2)$ is found[314] at 907—920 cm^{-1} and ν(PC) at 439—463 cm^{-1}.

Uranyl–butylphosphato-compounds have been studied by i.r. spectroscopy.[315] Assignment of some of phosphato-bands led to the proposal that phosphate bridging groups were present, as shown in (81).

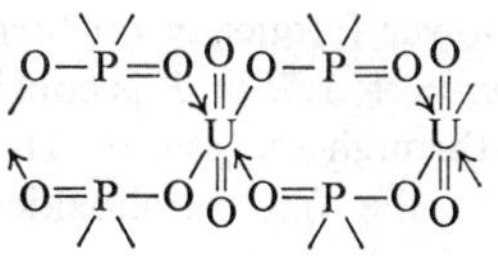

(81)

309 T. A. Beech and S. F. Lincoln, *Austral. J. Chem.*, 1971, **24**, 1065.
310 A. V. Ablov, D. G. Batyr, and M. P. Starysh, *Russ. J. Inorg. Chem.*, 1971, **16**, 368.
311 W. E. Stewart and T. H. Siddall, *J. Inorg. Nuclear Chem.*, 1971, **33**, 2965.
312 G. Vicentini and L. S. P. Braga, *J. Inorg. Nuclear Chem.*, 1971, **33**, 2959.
313 G. Vicentini and P. O. Dunstan, *J. Inorg. Nuclear Chem.*, 1971, **33**, 1749.
314 G. Bandoli, G. Bartolozzo, D. A. Clemente, U. Croatto, and C. Pannatini, *Inorg. Nuclear Chem. Letters*, 1971, **7**, 401.
315 E. G. Teterin, N. N. Shesterikov, P. G. Krutikov, and A. S. Solovkin, *Russ. J. Inorg. Chem.*, 1971, **16**, 146.

I.r. spectra of the complexes $[U(dehp)_4]_n$ and $[U(dehp)_3(NO_3)]_n$ [dehp = di-(2-ethylhexyl)orthophosphoric acid] have been partially assigned.[316]

Measurements of the intensity of $\nu(P=O)$ in the i.r. spectra of $(EtO)_3PO-AsCl_3$ or $(Bu^nO)_3PO-AsCl_3$ mixtures in CS_2, and the application of Job's method of continuous variations, have been used to study co-ordination of the $\rightarrow P=O$ group in these systems.[317] With $(EtO)_3PO$, a 1:1 complex was identified, whereas with $(Bu^nO)_3PO$ both 1:1 (P:As) species were formed. The 2:1 complex could only be identified in concentrated solutions. Stability constants for the adducts were estimated.

Ligands containing O—S or O—Se Bonds.—I.r. spectra show the sulphate to be bridging two metal atoms in a series of transition-metal complexes MX_2L_2, where $X = Cl$, NCS, or SO_4 and $L = o$-, m-, or p-toluidine, anisidine, or phenetidine.[318]

The i.r. spectra of salts of the cations $[Co_2\{(XO_4)_2(OH)\}(NH_3)_6]^+$, where $X = S$ or Se, show that they should be formulated as di-μ-sulphato-μ-hydroxo-complexes (82).[319] Assignments were made to the following

$$\left[(H_3N)_3Co\underset{XO_4}{\overset{XO_4}{\diagdown\!\!\!\!\diagup}}OH\underset{}{\diagdown\!\!\!\!\diagup}Co(NH_3)_3\right]^+$$

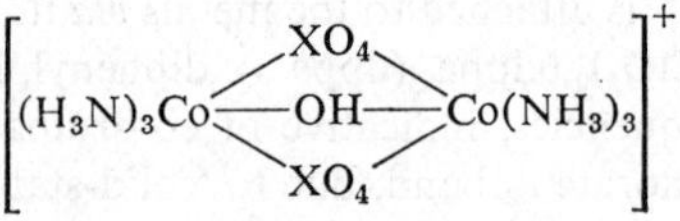

(82)

general regions: $\nu_{as}(S=O)$ 1165—1217; $\nu_s(S=O)$ 1101—1130; $\nu_{as}(S-O)$ 1040—1054; $\nu_s(S-O)$ 964—995; $\delta(SO_4)$ 645—652 and 598—600; $\nu_{as}(Se=O)$ 908—932; $\nu_s(Se=O)$ 860—872; $\nu_{as}(Se-O)$ 822—830; $\nu_s(Se-O)$ 796—808; $\delta(SeO_4)$ 542—550, 504—512, and 392—396 cm^{-1}.

Bands in the i.r. spectra of $CoSO_3$ and $NiSO_3$ (465, 532, 649 ± 1, 949 ± 1, 1020, 1198 ± 3 cm^{-1}) suggest [320] that these are essentially covalent compounds, as the spectra are not typical of ionic sulphites. Similar data have been reported [321] for a number of alcoholates of $CoSO_3$.

The shift of $\nu(SO)$ to lower frequency on forming adducts of the type $R^1R^2SO,mHgCl_2$, where $m = 1, 1.5$, or 2, is consistent with bonding of the sulphoxide to the metal through oxygen.[322] Thus, $\nu(SO)$ is found in the range 1000—1070 cm^{-1} in the free sulphoxides, but between 940 and 1010 cm^{-1} in the adducts.

I.r. spectra of $Ln(ClO_4)_3,8(TMSO)$ (Ln = La, Ce, Pr, Nd, Sm, Eu, or Gd; TMSO = tetramethylene sulphoxide), $Ln(ClO_4)_3,7.5(TMSO)$ (Ln = Tb, Dy, Ho, or Er), and $Ln(ClO_4)_3,7(TMSO)$ (Ln = Tm, Yb, Lu, or Y) all show

---

[316] E. R. Schmid and W. Pfannhauser, *Monatsh.*, 1971, **102**, 1317.
[317] G. Roland and G. Duyckaerts, *Spectrochim. Acta*, 1971, **27A**, 975.
[318] S. N. Das, S. N. Moharana, and K. C. Dash, *J. Inorg. Nuclear Chem.*, 1971, **33**, 3739.
[319] K. Wieghardt and J. Eckert, *Z. anorg. Chem.*, 1971, **383**, 240.
[320] R. Maylor, J. B. Gill, and D. C. Goodall, *Chem. Comm.*, 1971, 671.
[321] R. Maylor, J. B. Gill, and D. C. Goodall, *J. Inorg. Nuclear Chem.*, 1971, **33**, 1975.
[322] P. Biscarini, L. Fusma, and G. D. Nivellini, *J. Chem. Soc. (A)*, 1971, 1128.

ν(S=O) at *ca.* 956—970 cm^{-1} (a downward shift of *ca.* 50 cm^{-1} from the band position for free TMSO.[323] The ν_3 (1080 cm^{-1}) and ν_4 (620 cm^{-1}) bands of ClO_4^- indicate that the perchlorate is not co-ordinated.

Vibrational (i.r. and Raman) spectra of complexes of Me_2SeO (L), *e.g.* $ML_8(ClO_4)_3$ where M = Ce, Pr, Nd, Sm, Eu, Gd, Tb, Dy, Ho, Er, or Yb, indicate co-ordination of L *via* the oxygen atom. Thus ν(SeO) is at 795—799 cm^{-1}, a decrease of wavenumber of 26—21 cm^{-1} compared with the free ligand.[324] The presence of three ν(SeO) bands in the i.r. (together with other data) suggests a symmetry of D_4 for the ML_8 unit. Some other general assignments were made, *e.g.* $\nu_{as}(SeC_2)$ 593—596; $\nu_s(SeC_2)$ 580—589 cm^{-1}.

6 Sulphur and Selenium Donors

Monothio-β-diketones and their metal complexes have been reviewed, including vibrational spectroscopic aspects.[325] General ranges for assignments in the complexes were given as: ν(C$\cdots$C) 1590—1525; ν(C$\cdots$O) 1542—1458; ν(C$\cdots$S) 1270—1220; ν(C$\cdots$S) (possibly coupled with a C—H deformation) 820—790; ν(M—O) 499—437; ν(M—S) 399—376 cm^{-1}. It was pointed out that in the Cu^{I}, Ag^{I}, Cd^{II}, and Hg^{II} complexes, ν(C=O) is at 1640—1605 cm^{-1}, with no absorption in the ν(C$\cdots$O) or ν(M—O) regions, and thus they are S-bonded only. Similarly, As(PhCSCHCOPh)$_3$ gives rise to ν(C=O) at 1640 cm^{-1}, indicating an unco-ordinated C=O group.

It is found that in the complexes containing xanthate groups (83), the following trends are observed[326] when going from four- to five- or six-co-ordinated systems: (i) the ν(M—S) frequency decreases slightly; (ii) the

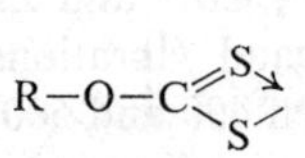

(83)

ν(C—S) frequency increases slightly, and (iii) the ν(C—O) frequency decreases markedly. Thus, if exan = ethyl xanthate (83; R = Et), Zn(exan)$_2$ gives ν(CO) at 1195 cm^{-1}, ν(CS) at 1035 cm^{-1}, and ν(Zn—S) at 360, 345 cm^{-1}, while Zn(exan)$_2$(py) gives ν(CO) at 1150 cm^{-1}, ν(CS) at 1045 cm^{-1}, and ν(Zn—S) at 340, 335 cm^{-1}.

A correlation between ^{1}H n.m.r. chemical shifts of CH_2 and the ν(CN) frequencies of symmetric diethyl dithiocarbamate (*i.e.* bidentate, with equal C—S bonds) has been noted.[327] The correlation, which does not exist for unsymmetric dithiocarbamate complexes (*e.g.* unidentate ones), is discussed

[323] L. B. Zunner and G. Vicentini, *Inorg. Nuclear Chem. Letters*, 1971, **7**, 967.
[324] R. Paetzold and G. Bochmann, *Z. anorg. Chem.*, 1971, **385**, 256.
[325] S. E. Livingstone, *Co-ordination Chem. Rev.*, 1971, **7**, 59; M. Cox and J. Darhen, *ibid.*, p. 29.
[326] M. R. Hunt, A. G. Krüger, L. Smith, and G. Winter, *Austral. J. Chem.*, 1971, **24**, 53.
[327] G. St. Nikolov, *Inorg. Nuclear Chem. Letters*, 1971, **7**, 1213.

in terms of the effects of electron charge densities (see also G. St. Nikolov and N. Tyutyulikov, *Inorg. Nuclear Chem. Letters*, 1971, **7**, 1209).

A few frequencies are listed for the ligand vibrations of some ethylenedithioacetic acid and diethylenetrithioacetic acid complexes.[328] I.r. spectra of metal complexes of dithiocarbazoic acid have been briefly discussed in terms of possible structures.[329]

Assignments [330] of certain i.r. bands in the spectra of some metal chelates of the diphenylthioylthiourea anion include ν(P=S) *ca.* 580 cm^{-1}; ν_{I}(M—S) *ca.* 400 cm^{-1}; and ν_{II}(M—S) *ca.* 330 cm^{-1}.

Chelate complexes of the diphenyldithiophosphinate ion have been prepared for a variety of metals,[331] and characteristic ligand bands in the i.r. spectra have been given. Bands assigned to ν(M—S), in the range 329, 345 cm^{-1} (for Ni and As, respectively) down to 233 cm^{-1} (for Tl), were also listed. The metals studied were Co, Rh, Ir, Ni, Pd, In, Tl, Sn, As, Sb, and Bi.

I.r. spectral data for metal thio-oxinates have been listed and metal-sensitive bands have been distinguished.[332]

Morpholine-4-carbodithioate (mdtc) complexes of a variety of metals have been prepared and ν(M—S) vibrations assigned,[333a] *e.g.* in Zn(mdtc)$_2$, ν(M—S) is found at 356 cm^{-1}.

Transition-metal complexes of *OO'*-diethyl diselenophosphate have been prepared and studied.[333b] $\nu_{as}(PSe_2)$ occurs in the region 515—570 cm^{-1}, with $\nu_{s}(PSe_2)$ at 435—515 cm^{-1}; both of these values are lower wavenumbers than those of vibrations in the free ion, indicating that both selenium atoms are co-ordinated.

Complexes of diphenyldithioarsinate have been prepared,[333c] *e.g.* $M[S_2AsPh_2]_n$, for $n = 2$ (Ni^{II}, Co^{II}, and Zn^{II}) or $n = 3$ (In^{III}, Cr^{III}, and V^{III}). I.r. data for some ligand vibrations are given, in particular for ν(As—S), which occur between 400 and 500 cm^{-1}.

Tetrakis(dithiocarboxylato)vanadium(IV) compounds $(RCS_2)_4V$, where R = Ph, MeC_6H_4, or $PhCH_2$, are believed to be eight-co-ordinate. The ν_s and ν_{as} bands of the CS_2 group in $MeCS_2H$ (581 and 1216 cm^{-1}, respectively) are replaced in $(MeCS_2)_4V$ by bands at 860 cm^{-1} and 1150/1175 cm^{-1}. Similarly, $(PhCS_2)_4V$ gives ν_s and ν_{as} of the CS_2 group at 1015 and 1260 cm^{-1}. These results are consistent with the presence of chelating ligands.[334]

328 J. Podlaha and J. Podlahová, *Inorg. Chim. Acta*, 1971, **5**, 413, 420.
329 C. Battistoni, G. Mattogno, A. Monacci, and F. Toli, *J. Inorg. Nuclear Chem.*, 1971, **33**, 3815.
330 I. Ojuma, T. Ouishi, T. Iwamoto, N. Inamoto, and K. Tamura, *Bull. Chem. Soc. Japan*, 1971, **44**, 2150.
331 A. Muller, V. V. K. Rao, and G. Klinksiek, *Chem. Ber.*, 1971, **104**, 1892.
332 Y. Mido and E. Sekido, *Bull. Chem. Soc. Japan*, 1971, **44**, 2130.
333 (*a*) G. Aravamudan, D. H. Brown, and D. Venkappayya, *J. Chem. Soc.* (*A*), 1971, 2744; (*b*) V. Krishnan and R. A. Zingaro, *J. Co-ordination Chem.*, 1971, **1**, 1; (*c*) A. Müller and P. Werle, *Chem. Ber.*, 1971, **104**, 3782.
334 O. Piovesana and C. Furlani, *Chem. Comm.*, 1971, 256.

The i.r. and Raman spectra of diphenyldithiophosphinates (Ph_2PS_2) of MoO_2, VO, and V have been reported.[335]

The following assignments have been given[336] for (*OO'*-diethylthiophosphato)metal(II) complexes: for VL_3: ν(PO) 1170, ν(PS) 583 cm^{-1}; CrL_3 1157, 560 cm^{-1}; FeL_3 1165, 553 cm^{-1}.

'C—S' and 'C—N' bands in the i.r. spectra of a series of Cr^{III} complexes with tetra-alkylthiurams (84a—c), $(CrL_2Cl_2)^+Cl^-$, have been listed and discussed.[337]

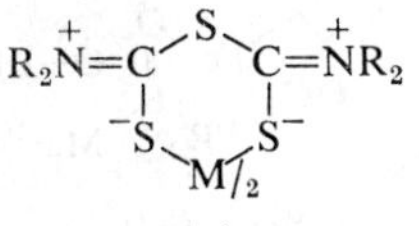

(a)

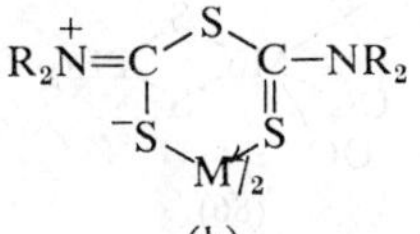

(b)

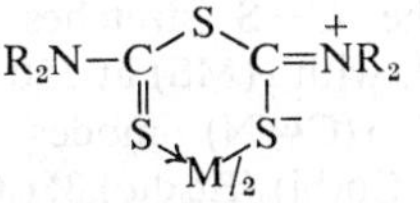

(c)

(84)

Cr^{III} and Co^{III} complexes of dithiocarbazato-anions $R^1R^2NNR^3CS_2^-$ show a strong band at *ca.* 1500 cm^{-1}, attributed to ν(C$\cdots$N), as required by SS-chelation, when R^1 or R^2 is not H, as shown in (85). In $Co(NH_2NHCS_2)_3$ and $Co(NH_2NMeCS_2)_3$, the 1500 cm^{-1} band is missing, indicating NS-chelation.[338]

The i.r. spectrum of (86) shows[339] the characteristic bands of h^5-C_5H_5 and h^1-C_2H_5, together with ν(CO) at 1958, 1843 cm^{-1} (in CH_2Cl_2 solution); $\nu_s(PS_2)$ at 503 cm^{-1}; and $\nu_{as}(PS_2)$ at 598 cm^{-1} and, possibly, 618 cm^{-1}.

For Mn(diethyldithiocarbamate)$_n$, where n = 2 (Mn^{II}) or 3 (Mn^{III}), i.r. spectra have been published,[340] in diagram form, without frequencies or assignments.

1 : 1 Addition compounds of BF_3 with Fe(CO)(h^5-C_5H_5)(PR_3)(SO_2Me) (R = n-C_4H_9 or Ph) yield i.r. absorption spectra that are consistent with the structure (87).[341]

An assignment of vibrational frequencies has been proposed for the ligand vibrations in M(sacsac)$_3$, where M = Co, Rh, or Ir and sacsac =

335 A. Müller, V. V. K. Rao, and E. Diemann, *Chem. Ber.*, 1971, **104**, 461.
336 C. M. Mikulski, N. M. Karayannis, and L. L. Pytlewski, *Inorg. Nuclear Chem. Letters*, 1971, **7**, 785.
337 G. Cantreras and H. Cortes, *J. Inorg. Nuclear Chem.*, 1971, **33**, 1337.
338 C. Battistoni, G. Mattogno, A. Monaci, and F. Tarli, *Inorg. Nuclear Chem. Letters*, 1971, **7**, 1081.
339 E. Lindner and K.-M. Matejcek, *Z. Naturforsch.*, 1971, **26b**, 854.
340 S. Lahiry and V. K. Anand, *Chem. Comm.*, 1971, 1111.
341 D. A. Ross and A. Wojcicki, *Inorg. Chim. Acta*, 1971, **5**, 6.

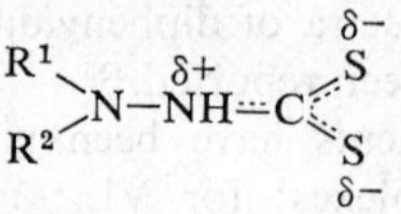

(85)

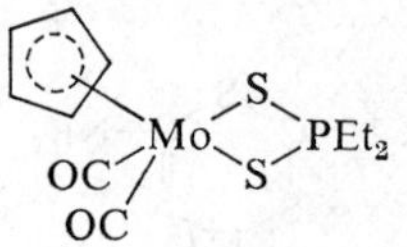

(86)

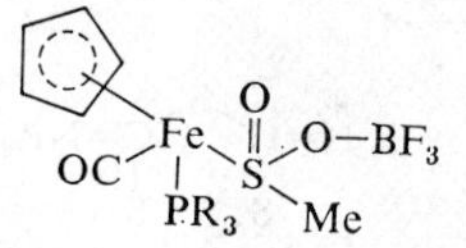

(87)

dithioacetylacetonate. The C—S stretches are assigned[342] to bands between 700 and 750 cm^{-1}, with ν(MS) at 350—360 cm^{-1}.

The wavenumbers of ν(C⋯N) modes of diethyldithiocarbamato (Et_2dtc^-) complexes $K_2Co(bi)_2(Et_2dtc),3H_2O$, $KCo(bi)(Et_2dtc)_2$, and $Co(Et_2dtc)_3$ (bi = biuretato) are all very similar (1490—1495 cm^{-1}).[343] This indicates that the electron-donating capacity of the biuretato-ligand is roughly the same as that of the Et_2dtc^- ligand. ν(CO) of the *O*-ethyl-xanthato-ligand in $Co(exan)(Et_2dtc)_2$ and $Co(exan)_3$ occur at 1230 and 1250 cm^{-1}, respectively, the difference arising from the stronger electron-releasing effect of the NR_2 group compared with OR.

Bands assigned to ν(P—S) of the ligand in bis(diethyldithiophosphinato)-complexes of Ni^{II}, *e.g.* $(Et_2PS_2)_2Ni(py)_2$, and Co^{II}, *e.g.* $(Et_2PS_2)_2Co$, are in the region of 600 cm^{-1} (ν_{as}) and 500 cm^{-1} (ν_s).[344]

The anion of diphenyldiselenophosphinic acid, acting as a bidentate ligand, forms complexes of the type $[Ph_2PSe_2]_nM$, where M = Co^{II}, Ni^{II}, Cd^{II}, or Zn^{II}. I.r. bands in the region 400—600 cm^{-1} are tabulated, those at *ca.* 500 cm^{-1} being tentatively assigned to ν(PSe) modes.[345]

I.r. spectra have been recorded[346] for a series of trimethylstibine sulphide (tmss) complexes with $Co(NO_3)_2$, $Zn(NO_3)_2$, and $Cd(NO_3)_2$, *viz.* $(tmss)_2Co(NO_3)_2$, $(tmss)_4Co(NO_3)_2$, $(tmss)_4Zn(NO_3)_2$, and $(tmss)_4Cd(NO_3)_2$. ν(Sb—S) is found in the range 390—405 cm^{-1} for all of them (*cf.* 433 cm^{-1} in the free ligand), indicating co-ordination *via* the sulphur atom. In solution, all of the 4 : 1 complexes dissociate to 2 : 1 complexes.

The new polysulphide chelate $(NH_4)_3Rh(S_5)_3$ has been prepared[347] from $RhCl_3$ and $(NH_4)_2S_5$. The i.r. spectrum (above 300 cm^{-1} only) shows only two weak bands [due to the $Rh(S_5)_3$ system], at 482 and 445 cm^{-1}. Both were assigned to ν(S—S).

342 G. A. Heath and R. L. Martin, *Austral. J. Chem.*, 1971, **24**, 2061.
343 H. C. Brinkhoff, *Inorg. Nuclear Chem. Letters*, 1971, 7, 413.
344 K. Demert and W. Kuchen, *Chem. Ber.*, 1971, **104**, 2592.
345 A. Müller, P. Christophliemk, and V. V. K. Rao, *Chem. Ber.*, 1971, **104**, 1905.
346 J. Otera, T. Osaki, and R. Okawara, *Inorg. Chem.*, 1971, **10**, 402.
347 R. A. Krause, *Inorg. Nuclear Chem. Letters*, 1971, 7, 973.

Complexes of disulphur or diselenium, $[Ir(S_2)(dppe)_2]Cl$, $[Ir(Se_2)(dppe)_2]Cl$, $[Rh(S_2)(dmpe)_2]Cl$, and $[Rh(S_2)(dmpe)_2]PF_6$, have been prepared [dppe = $Ph_2P(CH_2)_2PPh_2$; dmpe = $Me_2P(CH_2)_2PMe_2$].[348] ν(S—S) was found in the i.r. at 523 cm^{-1}, whereas ν(Se—Se) was identified with a Raman shift at 310 cm^{-1}. These wavenumbers give no indication of the S–S or Se–Se bond orders.

Several new Ni^{II} derivatives of dithio-β-diketones have been prepared,[349] *i.e.* (88) where $R^1 = R^2$ = Me; $R^1 = R^2 = Bu^t$; R^1 = Me, R^2 = Ph;

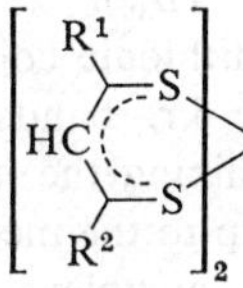

(88)

R^1 = Me, $R^2 = C_8H_{17}$; or R^1 = Ph, $R^2 = CF_3$. No ν(Ni—S) bands are seen at *ca.* 350—400 cm^{-1}, but in the mid-i.r., characteristic bands were seen, as follows: 1496—1489 and 1305—1316 cm^{-1} [ν(C⋯C)]; 1265—1152 and 860—824 cm^{-1} [ν(C⋯S)].

ν(C=C) and ν(CN) modes have been reported[350] for the dithiolen ligands of the tetra-substituted cyclobutadiene complexes $(Me_4C_4)Ni[S_2C_2(CN)_2]$ (1457 and 2206 cm^{-1}) and $(Ph_4C_4)Pd[S_2C_2(CN)_2]$ (1375 and 2198 cm^{-1}). The low values of ν(C=C) are believed to indicate that the sulphur ligand has considerable dithioketone character, as in other dithiolen complexes.

A normal-co-ordinate analysis of the in-plane modes of the bis(maleonitriledithiolato)nickel(II) anion has been performed.[351] The results show that in this highly delocalized system there is a large amount of vibrational mixing. For example, M—S stretching contributes significantly to bands at 498, 454 and (to a lesser extent) at 517, 390, and 357 cm^{-1}. The results also indicate that C—S stretching contributes to bands in the range 1500—450 cm^{-1}. The C—S stretching force-constant calculated suggests a substantial amount of double character in this bond.

A limited amount of i.r. data has been published for some Ni^{II} and Fe^{II} β-mercaptoethylamine complexes.[352] The Ni complexes involve bonding by ligand nitrogen and sulphur atoms, whereas those of Fe showed co-ordination *via* sulphur only.

The i.r. spectra of a number of metal dithiene complexes (89) have been studied, where M = Ni, Pd, or Pt and R = H, Me, or Ph.[353] A normal-co-ordinate analysis helped in the assignment of in-plane vibrations. Thus,

[348] A. P. Ginsberg and W. E. Lindsell, *Chem. Comm.*, 1971, 232.
[349] C. Blejean, *Inorg. Nuclear Chem. Letters*, 1971, **7**, 1011.
[350] E. J. Wharten, *Inorg. Nuclear Chem. Letters*, 1971, **7**, 307.
[351] Lakshmi, P. B. Rao, and U. Agarwala, *Appl. Spectroscopy*, 1971, **25**, 207.
[352] S. A. Grachev, L. I. Shchelkunova, Yu. A. Makashev, and T. Ya. Kul'ba, *Russ. J. Inorg. Chem.*, 1971, **16**, 103.
[353] O. Siimann and J. Fresco, *Inorg. Chem.*, 1971, **10**, 297.

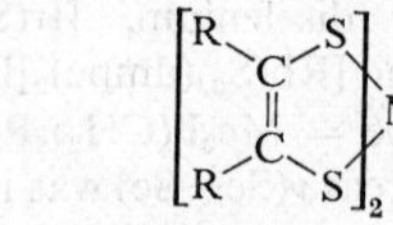

(89)

ν(C⋯C) was found at 1320—1360 cm^{-1}, with ν(C⋯S) at *ca.* 565 cm^{-1}. Ni—S stretching modes were assigned to peaks in the regions 420—465 cm^{-1} (B_{2u}) and 300—330 cm^{-1} (B_{3u}).

Reactions of some phosphinodithioate complexes of Pd^{II} and Pt^{II} have been studied, and characteristic i.r. bands were distinguished.[354] The latter have been used for determining the mode of co-ordination of the diphenylphosphinodithioate group to the metal ions.

Ligand vibrations for thiourea complexes of Pd^{II} and Cd^{II} have been partially assigned.[355]

I.r. spectra (1700—300 cm^{-1}) of M(monothioacetylacetonate)$_2$ complexes and of M(monothiotrifluoroacetylacetonate)$_2$ complexes (M = Cu, Ni, Zn, Cd, Pd, or Pb) have been reported.[356] Assignments were made with the aid of approximate normal-co-ordinate analyses. ν(M—O) modes were extensively coupled with other modes, and ν(M—S) modes were in the range 265—285 cm^{-1}. The M—O and M—S stretching force-constants decrease in the order Pb > Pd > Cd > Zn ~ Cu > Ni.

A study of the conformations of co-ordinated 1,4-dithian (L), 1,4-thioxan (L′), and 2,5-dithiahexane (L″) has been carried out by measuring Raman (and some i.r.) spectra of several complexes of these ligands.[357] Results may be summarized as follows: (*a*) in MX_2L (M = Hg or Cu; X = Cl or Br), $Ag(NO_3)L$, and $[Ag(ClO_4)]_2L_3$, the i.r. spectra (1350—400 cm^{-1}) closely resemble that of L (apart from NO_3 or ClO_4 modes). The lack of i.r. and Raman coincidences (above 400 cm^{-1}) shows that the L group is centrosymmetric. Thus there is little doubt that the (centrosymmetric) chair conformation of L persists in these complexes, and that L is bridging. Assignments were discussed also for ν(MX) and ν(MS). (*b*) In the compounds PtX_2L (X = Cl or Br), the i.r. and Raman (X = Br only) spectra show more bands than the Hg or Cu compounds. Also, some i.r.–Raman coincidences were found, and therefore L is non-centrosymmetric, and the structure (90) was suggested, with L in the boat conformation. (*c*) For $MX_2(L')_2$ and $RhX_3(L')_3$ (M = Cu, Pt, or Pd; X = Cl

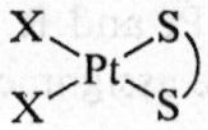

(90)

[354] J. M. C. Alison, T. A. Stephenson, and R. O. Gould, *J. Chem. Soc.* (*A*), 1971, 3690.
[355] Yu. Ya. Kharitonov, V. D. Brega, A. V. Ablov, and N. N. Proskma, *Russ. J. Inorg. Chem.*, 1971, **16**, 307.
[356] O. Siimann and J. Fresco, *J. Chem. Phys.*, 1971, **54**, 734, 740.
[357] D. A. Rice and R. A. Walton, *Spectrochim. Acta*, 1971, **27A**, 279.

or Br), the spectra (to 200 cm^{-1}) show that in all cases L′ is S-bonded. (*d*) In $HgX_2(L'')$, $PtCl_2(L'')$, and $[trans\text{-}RhX_2(L'')_2]^+$ (X = Cl or Br), Raman spectra characteristic of *gauche*, chelating L″ were obtained. (*e*) In $Ag(NO_3)(L'')$, no i.r.–Raman coincidences were found, and therefore *trans*-L″ is present.

I.r. (to 200 cm^{-1}) and Raman spectra of trifluorodithioacetates of Ag^I and Pb^{II}, *viz.* $(CF_3CS_2)Ag$ and $(CF_3CS_2)_2Pb$, have been assigned; the assignments are summarized in Table 10.[358]

Table 10 *Vibrational assignments in* $Ag(S_2CCF_3)$ *and* $Pb(S_2CCF_3)_2$ (*wavenumbers*/cm^{-1})

	$(CF_3CS_2)_2Ag$		$(CF_3CS_2)_2Pb$	
	I.r.	*Raman*	*I.r.*	*Raman*
$\nu_s(CF_3)$	1297	1306	1293	—
$\nu_{as}(CS_2)$	1242	1241	1243	1244
$\nu_s(FCF_2)$	1172	1141	1170	1126
$\nu_{as}(FCF_2)$	1099	1090	1093	1097
$\nu_s(CF_2)$	670	647	662	649

Bands in the i.r. spectra of bis-(*NN*-dialkylthiocarbamato)gold(III) cationic complexes, which have been assigned to ν(C—N), have been listed.[359]

7 Potentially Ambident Ligands

Cyanate, Thiocyanate Complexes, *etc.*, and Iso-analogues.—An extensive and comprehensive review has been published on the i.r. spectra of complexes of the thiocyanate and related ions.[360] Data were tabulated for numerous complexes of the following types: (*a*) $[M(SCN)_z]^{m-}$ [ν(CN, 2 × δ(NCS), ν(CS), and δ(NCS)]; (*b*) $[ML_x(NCS \text{ or } SCN)_y]^{0,\pm z}$ [ν(CN), ν(CS), δ(NCS)]; (*c*) bridged systems, *e.g.* $M(NCS)_nM'$ (including M = Co, M′ = Hg, *n* = 4) [ν(CN), ν(CS), δ(NCS), ν(ML)]; (*d*) $SeCN^-$ complexes [ν(CN), ν(CSe), δ(NCSe), ν(ML)]; (*e*) NCO^- complexes [ν(CN), ν(CO), δ(NCO), ν(ML)]; (*f*) CNO^- complexes [2 × ν(NO), ν(CN), ν(NO), δ(NCO)]. In addition, a large amount of intensity data on the ν(CN) mode of NCS^-, SCN^-, NCO^-, OCN^-, $NCSe^-$, and $SeCN^-$ was given, and a section on far-i.r. spectra was included. A useful discussion of the use of such data in determining the bonding modes of these ligands was given (*N.B.*, some new data were included in this review).

One of the spectroscopic tests for distinguishing the mode of bonding of the thiocyanate ion to metals is measurement of the integrated intensity of the C≡N stretching mode in solutions. Values below *ca.* $3 \times 10^4\ l\ mol^{-1}\ cm^{-2}$ are indicative of S-bonding, while those above

[358] E. Lindner and U. Kunze, *Z. anorg. Chem.*, 1971, **383**, 255.
[359] J. G. M. van der Linden, *Rec. Trav. chim.*, 1971, **90**, 1027.
[360] R. A. Bailey, S. L. Kozak, T. W. Michelson, and W. N. Mills, *Co-ordination Chem. Rev.*, 1971, **6**, 407.

8×10^4 l mol^{-1} cm^{-2} suggest N-bonding. As this approach is restricted by solubility problems, the correlation has been extended to solids.[361] Using an internal intensity standard, it appears to work well, indeed better than criteria involving the actual frequencies of the mode.

ν(NH), $\delta(NH_2)$, $\gamma(NH_2)$, ν(CN), ν(CS), and δ(NCS) modes have been assigned[362] in [picolineH]$^+$[Cr(NCS)$_4$(*m*- or *p*-xylidene)$_2$]$^-$. The following wavenumbers were quoted for ν(CN), ν(CS), and δ(NCS), respectively: 2060, 765, 480 cm^{-1} (*m*-isomer); 2088, 765, 480 cm^{-1} (*p*-isomer).

Nitrosyl- and isocyanato-stretching frequencies, ν(NO) and ν(NCO), have been reported[363] for π-cyclopentadienyl metal carbonyl derivatives containing these ligands (metal = Cr, Mo, W, or Fe).

For the complexes M(NCS)$_2$(DMSO)$_4$ (M = Mn, Ni, or Zn) and Cd(NCS)$_2$(DMSO)$_2$, assignments for vibrations of both the NCS$^-$ and DMSO ligands have been given.[364]

I.r. absorptions associated with the ligand vibrations in some isocyanato-complexes of Mn and Re have been listed (together with data on analogous hydrazine, amine, nitrido-, and azido-complexes.[365] ν_{as}(NCO) is found at *ca.* 2240 cm^{-1}, ν_s(NCO) *ca.* 1320 cm^{-1}, and ν(M—N) *ca.* 340 cm^{-1}.

The following isocyanato- and azido-complexes have been isolated:[366] [Et$_4$N][Mn$_2$(CO)$_6$(N$_3$)$_3$], Mn(CO)$_3$(PPh$_3$)$_2$(NCO), [Et$_4$N][Mn$_2$(CO)$_6$(N$_3$)$_x$(NCO)$_{3-x}$], [Et$_4$N][Re$_2$(CO)$_6$(N$_3$)$_2$(NCO)$_2$], [Ph$_3$PNPPh$_3$][Mn$_2$(CO)$_6$(NCO)$_3$], and Mn(CO)$_3$(PPh$_3$)$_2$(N$_3$). The first three azido-complexes show unusually high $\nu_{as}(N_3)$ frequencies, suggesting that azide bridges are present.

The splitting of both ν(CN) (at *ca.* 2085 cm^{-1}) and ν(CS) (785 cm^{-1}) in [Co(NCS)$_2$A$_2$],H$_2$O (A = acetylhydrazine) indicates *cis*-co-ordination by thiocyanato-groups.[367]

Both linkage isomers of *trans*-[Co(dmg)$_2$(py)(SCN)] have been obtained[368] (dmg = dimethylglyoximato); these have ν(CN) at 2118 cm^{-1} (S-bonded) and 2128 cm^{-1} (N-bonded). These bonding modes, in solvents of high or low dielectric constant, are shown to be kinetically (rather than thermodynamically) controlled.

The complexes [M(Et$_4$dien)(NCX)][BPh$_4$] [M = Ni^{II} or Cu^{II}; X = O, S, or Se; Et$_4$dien = 1,1,7,7-tetraethyl(diethylenetriamine)] are all stable, square-planar, N-bonded systems (*cf.* the Pd analogues, which undergo

[361] R. A. Bailey, T. W. Michelson, and W. N. Mills, *J. Inorg. Nuclear Chem.*, 1971, **33**, 3206.
[362] I. Gănescu, A. Popescu, and M. Proteasca, *Coll. Czech. Chem. Comm.*, 1971, **36**, 3088.
[363] A. T. McPhail, G. R. Knox, C. G. Robertson, and G. A. Sim, *J. Chem. Soc.* (*A*), 1971, 205.
[364] G. V. Tsintsadze, *Russ. J. Inorg. Chem.*, 1971, **16**, 614.
[365] J. T. Moelwyn-Hughes, A. W. B. Garner, and A. S. Howard, *J. Chem. Soc.* (*A*), 1971, 2361, 2370.
[366] R. Mason, G. A. Rusholme, W. Beck, H. Engelmann, K. Joos, B. Lindenberg, and H. S. Smedal, *Chem. Comm.*, 1971, 496.
[367] Yu. Ya. Kharitonov and R. J. Machkhovshvili, *Russ. J. Inorg. Chem.*, 1971, **16**, 492.
[368] R. L. Hassel and J. L. Burmeister, *Chem. Comm.*, 1971, 568.

rapid isomerization to the X-bonded forms, when X = S or Se). All give increased ν(CX) frequencies and ν(CN) intensities with respect to the free ion.[369] The five-co-ordinate Ni(Et$_4$dien)(NCO)$_2$ was also formed; this was N-bonded, but the analogous NCS complex is dimeric and six-co-ordinate, with bridging NCS ligands in the solid state. Two high-frequency ν(CN) modes indicate the presence of a double bridge, while the splitting of lower-frequency ν(CN) bands is diagnostic of the presence of *cis*-terminal NCS groups (91).

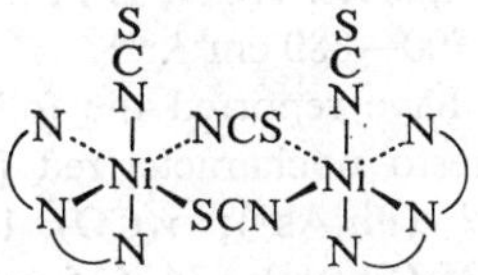

(91)

Bands due to the NCS group in a wide range of complexes of the type Ni(aniline)$_n$(NCS)$_2$, where n = 2, 4, or 6; aniline = variety of substituted anilines, have been reported,[370] with ν(CN) at 2065—2141 cm^{-1}, ν(CS) at 749—816 cm^{-1}, and δ(NCS) at 465—487 and 451—473 cm^{-1}.

The complexes of the bidentate ligand *trans*-[2-(ethylthio)cyclohexyl]-phenylphosphine (L), [NiL$_2$(NCS)](NCS) and [NiL$_2$(NCS)](BPh$_4$), contain five-co-ordinate nickel, whereas [CoL$_2$(NCS)](NCS) is tetrahedrally co-ordinated (one of the L groups remaining unidentate).[371] The i.r. spectra show ν(CN) bands typical of N-co-ordinated thiocyanato-groups, at 2085, 2090, and 2080 cm^{-1}, respectively.

Burmeister *et al.* have shown that the mode of metal to NCX$^-$ (X = S or Se) bonding in a number of Pd–NCX$^-$ and Pt–NCX$^-$ complexes is solvent dependent.[372] From the position of ν(CN), for example, Pt(SbPh$_3$)$_2$(NCS)$_2$ co-ordinates *via* S in MeCN solution, but partially *via* N (and also as the bridge Pt—NCS—Pt) in CH$_2$Cl$_2$. Likewise, Pd(PPh$_3$)$_2$(SeCN)$_2$ bonds *via* Se in DMF and DMSO, but *via* Se *and* N in CHCl$_3$ or CH$_2$Cl$_2$ solution; thus, ν(CN) is at 2120 cm^{-1} in the former pair (Se-bonded), whereas it appears at 2100 cm^{-1} (N-bonded) and at 2125 cm^{-1} (Se-bonded) in the latter pair.

In Pd(dmp)SO$_4$ (dmp = 2,9-dimethyl-1,10-phenanthroline), the pattern of SO$_4$ vibrations is the same as that in the Cu analogue, which is believed to be characteristic of chelating SO$_4$ groups. In the complex Pd(dmp)(SCN)$_2$, ν_3 of SCN appears at 2104 and 2146 cm^{-1}, and ν_1 at 725 and 830 cm^{-1}, perhaps suggesting the presence of one Pd—NCS and one Pd—SCN bond in this complex.[373]

369 J. L. Burmeister, T. P. O'Sullivan, and K. H. Johnson, *Inorg. Chem.*, 1971, **10**, 1803.
370 A. V. Butcher, D. J. Phillips, and J. P. Redfern, *J. Chem. Soc.* (*A*), 1971, 1640.
371 E. Wenschuh and K.-P. Rudolph, *Z. anorg. Chem.*, 1971, **380**, 7.
372 J. L. Burmeister, R. L. Hassel, and R. J. Phelan, *Inorg. Chem.*, 1971, **10**, 2032.
373 R. A. Plowman and L. F. Power, *Austral. J. Chem.*, 1971, **24**, 309.

Ligand vibrations, including ν_{as}(NCO) and $\nu_{as}(N_3)$, have been assigned[374] in a number of platinum isocyanato- and azido-complexes (typical positions: ν_{as}(NCO) ~ 2220 cm^{-1}; $\nu_{as}(N_3)$ ~ 2050 cm^{-1}). In addition, $\nu_{as}(SO_2)$ (*ca.* 1300 cm^{-1}) and $\nu_s(SO_2)$ (*ca.* 1150 cm^{-1}) in $(NHSO_2R)_2$Pt-$(PPh_3)_2$ complexes have been assigned. In the platinum fulminato-complex $Pt(PPh_3)_2(CNO)_2$, the following assignments (with no discussion) were quoted:[375] 2 × ν_s(CNO) 2325 cm^{-1}, ν_s(CNO) 1153 cm^{-1}, and ν_{as}(CNO) 2190 cm^{-1}.

In Cu^{II} thiocyanato-complexes, ν(CN) is in the range 2055—2120 cm^{-1} and ν(CS) is in the range 700—880 cm^{-1}.[376]

Norbury and Sinha[377] have reported the following wavenumbers and assignments for the hitherto uncharacterized $[Ag(NCO)_2]^-$ ion: ν(CN) 2200 (Et_4N^+), 2202 cm^{-1} (Ph_4As^+); ν(CO) 1320 (Et_4N^+), 1335 cm^{-1} (Ph_4As^+); δ(NCO) 632, 635 (Et_4N^+), 624, 615 cm^{-1} (Ph_4As^+).

The structures of two isomers of the $[Au(CN)_2(SCN)_2]^-$ ion (as K^+ or Et_4N^+ salts) have been deduced,[378] mainly by i.r. spectroscopy: S-co-ordinated: ν(CS) 705 cm^{-1}; ν(CN) (of SCN) 2135 cm^{-1}; ν(CN) (of CN) 2150 cm^{-1}: N-co-ordinated: ν(CS) 800 cm^{-1}; ν(CN) (of SCN) 2075 cm^{-1}; ν(CN) (of CN) 2145 cm^{-1}.

The use of i.r. spectra in the 410—500 cm^{-1} region to diagnose the mode of co-ordination of SCN^- and $SeCN^-$ ligands (X) in the complexes ZnX_2L_2 (L = py, picoline, quinoline, or isoquinoline) has been discussed.[379] Data for higher frequency regions were also given. Representative compounds for the different modes of bonding, with useful diagnostic bands, are given in Table 11.

Table 11 *Modes of bonding of* NCS *and* NCSe *in* Zn *complexes*

Type of bonding	*Example*	*Characteristic bands, wavenumbers/cm^{-1}* General range	This compound
M—NCS	$Zn(quin)_2(NCS)_2$	460—500	488—480
M—NCSe	$Zn(isoquin)_2(NCSe)_2$	460—500	488
M—SeCN	$Zn(py)_2(SeCN)_2$	410—440	425
M—NCS—M	$Zn(\gamma\text{-pic})_2(SCN)_2$	460—500	480
		410—440	412
M—NCSe—M	$Zn(\gamma\text{-pic})_2(SeCN)_2$	460—500	499
		410—440	432

The i.r. spectra of several NCS complexes of Al, Ga, and In have been reported.[380] The following assignment is representative: $Al(NCS)_3(pic)_2$ (pic = 4-methylpyridine): ν(C≡N) 2052, 2110 cm^{-1}; δ(NCS) 471 cm^{-1};

374 W. Beck, M. Bander, G. La Monica, S. Cenini, and R. Ugo, *J. Chem. Soc. (A)*, 1971, 113.
375 W. Beck, K. Scharpp, and F. Kern, *Angew. Chem. Internat. Edn.*, 1971, **10**, 66.
376 R. Barbucci, P. Paoletti, and G. Ponticelli, *J. Chem. Soc. (A)*, 1971, 1637.
377 A. H. Norbury and A. I. P. Sinha, *J. Inorg. Nuclear Chem.*, 1971, **33**, 2683.
378 D. Negoiu and L. M. Băloiu, *Z. anorg. Chem.*, 1971, **382**, 92.
379 M. Aslam and W. H. S. Massie, *Inorg. Nuclear Chem. Letters*, 1971, **7**, 961.
380 S. J. Patel, *J. Inorg. Nuclear Chem.*, 1971, **33**, 17.

ν(M—N) 320, 340 cm^{-1}. These are indicative of bonding to the metal *via* nitrogen.

Nitrito- and Nitro-complexes.—Siebert *et al.*[381] have described the preparation and properties of seven salts of the cation $[Co_2\{NO_2,OH\}(NH_3)_8]^{4+}$. Bridging by NO_2 is shown by the appearance of two strong i.r. bands, at 1175—1195 and 1485—1500 cm^{-1}, due to the grouping (92). An aqueous

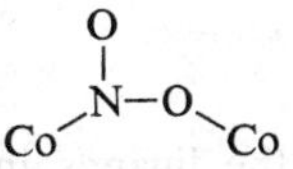

(92)

solution of the chloride salt shows corresponding Raman shifts at 1164 cm^{-1} (weak, polarized) and 1497 cm^{-1} (medium, polarized).

$Co(NO)_2(NO_2)$ possesses a chain structure in the solid phase, with $Co(NO)_2$ groups bridged by bidentate NO_2 ligands. In CCl_4 solution, a dimeric structure was suggested.[382] The i.r. data are summarized in Table 12.

Table 12 *I.r. wavenumbers*/cm^{-1} *and assignments for* $Co(NO)_2(NO_2)$

	$Co(NO)_2(NO_2)$		$Co(^{15}N^{18}O)_2(^{15}N^{18}O_2)$	
	CCl_4 *soln.*	*Solid*	CCl_4 *soln.*	*Solid*
Terminal nitrosyl [ν(NO)]	1867	1878	1789	1800
	1802	1783	1727	1700
Bridging nitrite [$\nu(NO_2)$]	1481	1442	1418	*ca.* 1385
	1164	1192	1112	1135

NO reacts[383] with $Co(PPh_3)_3(NO)$ to give $Co(NO_2)(PPh_3)(NO)_2$, with loss of N_2. ν(NO) bands in the i.r. are found at 1830 and 1760 cm^{-1}, whilst the presence of an O-bonded nitrito-group is supported by the presence of i.r. bands at 1404 cm^{-1} [$\nu_{as}(NO_2)$] and 1095 cm^{-1} [$\nu_s(NO_2)$].

A new class of co-ordination compounds of Co^{III} has been described,[384] *viz.* $M[Co(NH_3)(am)(NO_2)_3],nH_2O$ (am = an amino-acid anion derived from glycine, alanine, α-aminobutyric acid, or norvaline; M = Ag^+ or K^+; n = 0, $\frac{1}{2}$, or 1). Assignments made included the following: $\nu_{as}(CO_2^-)$ 1610—1655 cm^{-1}, NH_2 bend 1575—1580 cm^{-1}, $\nu_{as}(NO_2^-)$ 1425—1440 cm^{-1}, $\nu_s(NO_2^-)$ 1320 cm^{-1} (these show that the NO_2 is present as a nitro-group), NH_3 rock 835 cm^{-1}, NO_2^- bend 827—829 cm^{-1}.

Frequencies due to ν(NO) and $\nu(NO_2)$ in some bridged nitrito-complexes of Ni^{II} have been reported.[385] Wavenumbers of the latter mode are in the

381 H. Siebert, G. Tremmel, and U. Kramer, *Z. anorg. Chem.*, 1971, **383**, 158.
382 C. E. Strause and B. I. Swanson, *Chem. Comm.*, 1971, 55.
383 M. Rossi and A. Sacco, *Chem. Comm.*, 1971, 694.
384 M. B. Celap, M. J. Malinar, and R. J. Janjić, *Z. anorg. Chem.*, 1971, **383**, 341.
385 D. M. L. Goodgame, M. A. Hitchman, and D. F. Marsham, *J. Chem. Soc.* (*A*), 1971, 259.

810—875 cm^{-1} region. Distinction between complexes containing bridging NO_2 groups of types (93) and (94) has been assisted by the more complex appearance of the ν(NO) bands in the spectra of the latter type.

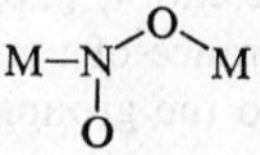

(93)

(94)

The intense vibrations of the ligands in $[Pd(NH_3)_3(NO_2)][Pd(NH_3)_4](NO_3)_4$ have been assigned as follows:[386] ν(NH) 3198 (i.r.), 3209 (R), 3260 (i.r.), 3271 (R), 3295 (R), 3312 (i.r.), 3326 cm^{-1} (R); $\delta_{as}(NH_3)$ 1628 (i.r.), ? 1550 cm^{-1} (i.r.); $\delta_s(NH_3)$ 1243, 1267 (R), 1243, 1255 cm^{-1} (i.r.); $\rho(NH_3)$ 790 cm^{-1} (i.r.); $\nu_s(NO_2)$ 1341 (R), 1341 cm^{-1} (i.r.); $\nu_{as}(NO_2)$ not observed (masked by NO_3^- vibrations); $\delta_s(NO_2)$ 827 (R), 824 cm^{-1} (i.r.); $\rho_W(NO_2)$? 555 cm^{-1} (i.r.).

Ligand vibrational frequencies have been reported[387] for a range of nitro-, nitrato-, and sulphato-complexes of Pd, Pt, Rh, and Ir. They support the presence of (*a*) *N*-bonded NO_2^-, (*b*) unidentate NO_3^-, and (*c*) bidentate SO_4^{2-} groups in these complexes.

The i.r. and Raman spectra of some alkali-metal hexanitro-lanthanates $M_2M'Ln(NO_2)_6$ have been reported[388] for M = Cs or Rb; M′ = Na; and Ln = La, Pr, Nd, or Ho. The spectra are very similar for all of them, and the similarity of the NO_2 bands to those found in $NaNO_2$ and the absence of ν(Ln—N) are consistent with very weak Ln—N bonding.

Other Ligands containing O and N Donor Atoms.—The i.r. spectra of some metal acetylhydrazine complexes (95) have been analysed, bands being assigned to motion of one or more groups of atoms.[389]

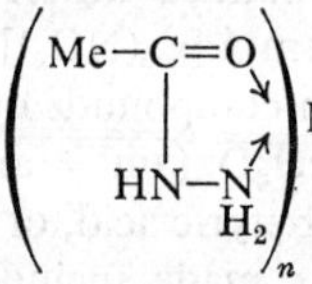

(95)

I.r. spectra of 2,4-dinitrosoresorcinol and seven metal complexes of it have been reported.[390] The spectra of the co-ordinated ligand do not resemble that of the free ligand, but the authors were unable to interpret the spectra.

[386] F. P. Boer, V. B. Carter, and J. W. Turley, *Inorg. Chem.*, 1971, **10**, 651.
[387] J. J. Levison and S. D. Robinson, *J. Chem. Soc.* (*A*), 1971, 762.
[388] J. C. Barnes and R. D. Peacock, *J. Chem. Soc.* (*A*), 1971, 558.
[389] Yu. Ya. Kharitonov and R. I. Machkhovshvili, *Russ. J. Inorg. Chem.*, 1971, **16**, 147.
[390] R. S. Bottei and C. P. McEachern, *J. Inorg. Nuclear Chem.*, 1971, **33**, 9.

Some i.r. spectral data for Cr^{III} complexes of some naturally occurring amino-acids, of the type CrL_3,xH_2O, and related structures have been given, with partial assignment of some of the ligand vibrations.[391]

Deuteriation of (+)-K Co(L-valine)$_2$CO$_3$ has confirmed the assignment of several i.r. bands to NH_2 stretching frequencies.[392]

The i.r. spectrum of (*cis,cis*-1,3,5-triaminocyclohexane-*NN′N″*-tri-acetato)cobalt(III) hydrate shows [393] a single absorption at 1639 cm^{-1}, very close to the frequency reported for $\nu_{as}(CO_2^-)$ in β-tris(glycinato)cobalt(III). It appears that all six donor groups of the ligand are co-ordinated, and structure (96) has been suggested for the complex.

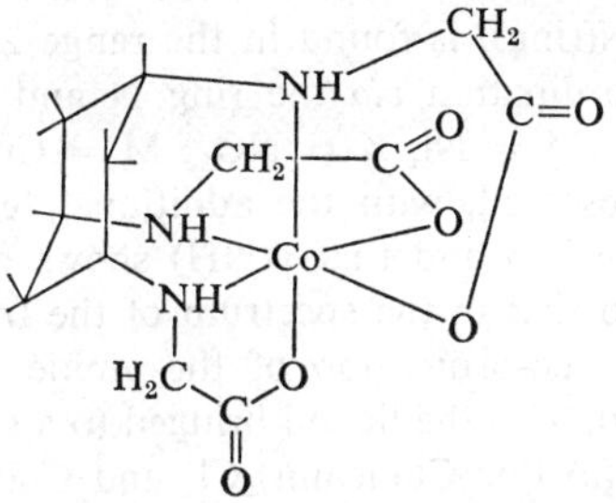

(96)

Complexes prepared from $RhCl_3,4H_2O$ and substituted glutamic acids have been formulated as $[Rh(L)Cl(H_2O)_2]$, where L = glutamato, β-hydroxyglutamato, *N*-benzoylglutamato, or *p*-nitro-*N*-benzoylglutamato, on the basis of elemental analyses, shifts to higher frequencies in the ν(NH) region, shifts in $\nu_s(CO_2^-)$ and $\nu_{as}(CO_2^-)$, and the absence of bands attributable to non-ionized CO_2H groups.[394]

A few tentative assignments of i.r. frequencies have been made [395] for Ni^{II} and Cu^{II} complexes of substituted aroyl-hydrazones R^{1}CH=N-—NHCOR^2. It is suggested that $NiCl_2$ gives a complex derived from the keto-form (97), whereas $Ni(OAc)_2$ and $Cu(OAc)_2$ form complexes derived from the enol form (98).

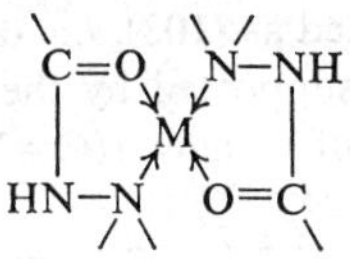

(97)

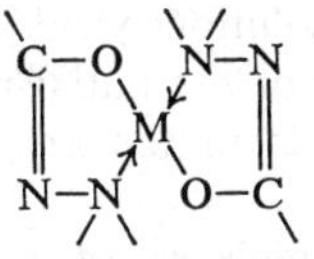

(98)

[391] H. Mizuochi, A. Uehara, E. Kyuno, and R. Tsuchiya, *Bull. Chem. Soc. Japan*, 1971, **44**, 1555.
[392] R. D. Gillard and M. G. Price, *J. Chem. Soc.* (*A*), 1971, 2271.
[393] L. J. Zompa and J. M. Shindler, *Chem. Comm.*, 1971, 65.
[394] H. Kalberer and H. Frye, *Inorg. Chem. Nuclear Letters*, 1971, **7**, 215.
[395] L. El-Sayed and M. F. Iskander, *J. Inorg. Nuclear Chem.*, 1971, **33**, 435.

Complexes of Ni^{II} and Cu^{II} salts with 2-acetamidothiazole (acamt, 99) and 2-acetamidobenzothiazole (acambt, 100) fall into three groups:[396] (*a*) $Ni(acambt)_2X_2$ (X = Cl, Br, or I), $Ni(acamt)_2Cl_2,H_2O$, $Ni(acamt)_2Br_2,2H_2O$, and $Cu(acamt)_2Br_2$ are octahedral (from the electronic spectra

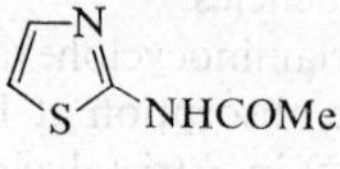
(99)

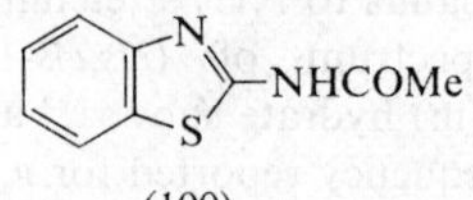
(100)

and magnetic properties), ν(CO) ('amide I') is lowered by *ca.* 35 cm^{-1}, ν[M—O(carbonyl)] is found at *ca.* 350 cm^{-1} (acambt) or *ca.* 400 cm^{-1} (acamt), and ν[M—N(ring)] is found in the range 230—280 cm^{-1}. These data all suggest co-ordination *via* the ring N and the amide O atoms. (*b*) For $M(acamt)_2X_2$ (M = Ni, X = NO_3; M = Cu, X = NO_3 or ClO_4) similar effects are observed, with the additional features that X is not bound (deduced from i.r.) and that ν(NH) shows a ten-fold increase in intensity compared to that in the spectrum of the free ligand. The latter effect is attributed to co-ordination of the amide NH group; thus the structures are as in (*a*), with the ligand bridged to a second metal atom *via* the amide N atom. (*c*) For $Cu(acamt)_2Cl_2$ and $Cu(acambt)_2Cl_2$, no shift occurs in ν(CO), and no ν(Cu—O) band was found. These structures are believed to involve co-ordinated halide, and bridging ligands co-ordinated *via* the ring and amide nitrogens.

Observation of the shifts in the frequency of the 'amide I' band in the i.r. spectrum of poly-L-lysine on co-ordination to Ni^{II} and Pd^{II} has assisted in elucidating the structures of poly-L-lysine–metal complexes in solution.[397]

I.r. spectra of the ligand vibrational modes in four metal–DL-α-serine complexes (with Pt^{II}, Pd^{II}, Ni^{II}, and Cu^{II}) have been assigned[398] sufficiently fully for a normal-co-ordinate analysis to have been carried out. Assignments of interest included the following: NH_2 scissor *ca.* 1590 cm^{-1}; NH_2 twist *ca.* 1100 cm^{-1}; NH_2 rock *ca.* 650 cm^{-1}; ν(M—N) 350(Ni)—400(Pd, Pt) cm^{-1}; ν(M—O) 217, 218 cm^{-1} in Pd^{II}, Pt^{II} complexes, respectively. The complexes possess structures of the type (101).

The interaction of 2,2′-bis-(1,2-dihydro-4-oxo-3,1-benzoxazine) (102) with Cu^{2+} solutions yields a complex formulated as (103), *i.e.* derived from a tautomer of the initial reagent.[399] This is supported by the absence of ν(NH) bands in the i.r. and the presence of a new ν(C=N) band at 1613 cm^{-1}.

1 : 1 Complexes of Cu^{II} and Ni^{II} with 1,2-bis-(3-formyl-5-methylsalicylideneamino)ethane or 1,3-bis-(3-formyl-5-methylsalicylideneamino)-

[396] M. N. Hughes and K. J. Rutt, *Inorg. Nuclear Chem. Letters*, 1971, **7**, 1049.
[397] M. Nakai, M. Yoneyama, and M. Hatano, *Bull. Chem. Soc. Japan*, 1971, **44**, 874.
[398] Y. Inomata, T. Inomata, and T. Mariwaki, *Bull. Chem. Soc. Japan*, 1971, **44**, 365.
[399] D. S. Mittal, N. K. Mathur, and P. N. Mathur, *Inorg. Nuclear Chem. Letters*, 1971, **7**, 737.

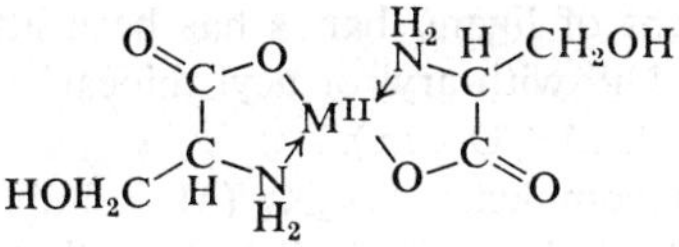

(101)

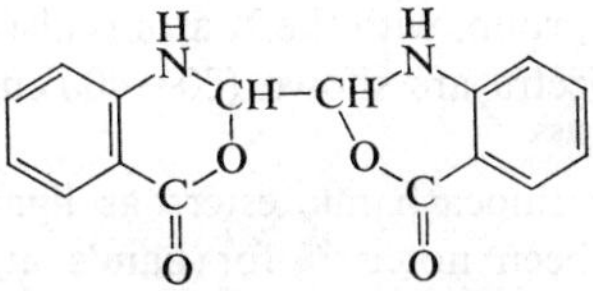

(102)

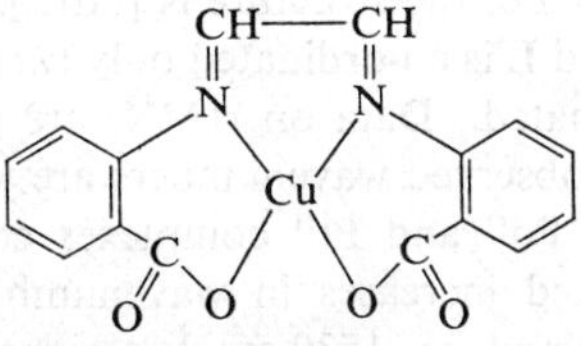

(103)

propane have ν(C=O) at 1670 cm^{-1} and ν(C=N) at 1635—1628 cm^{-1}.[400] Addition of ethylenediamine or trimethylethylenediamine to these complexes afforded the macrocyclic complexes (104; *n*,*m* = 2 or 3; M = Cu or Ni). These have no ν(C=O) band, but show two bands in the region 1645—1625 cm^{-1} attributed to the co-ordinated and unco-ordinated azomethine groups.

Some ligand vibrations in glycine- and leucine-peptide complexes of Cu, Ni, and Zn have been assigned.[401]

The i.r. spectrum of $Et_2Ga(ox)$ (oxH = 8-quinolinol) has been recorded,[402] the absence of any ν(OH) band being used as evidence for the formulation (105).

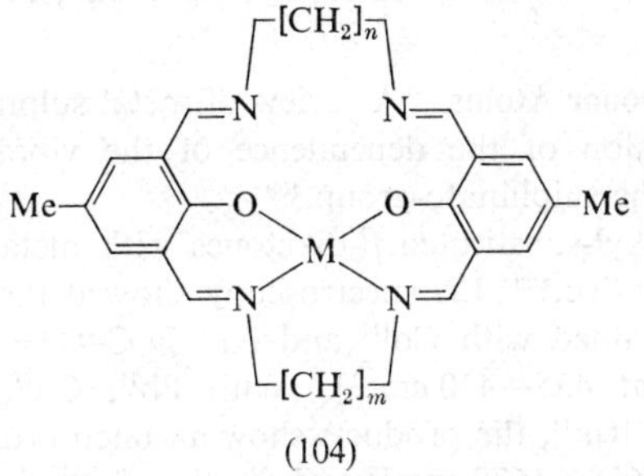

(104)

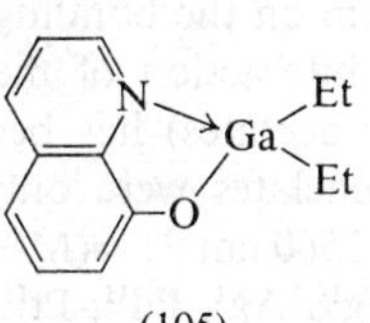

(105)

Ligands containing N and S Donor Atoms.—I.r. frequencies and qualitative assignments have been given for some cobalt complexes of benzothiocarbazide ($PhCONHNHCSNHNH_2$), acetyl thiocarbazide, nicotinoyl thiocarbazide, and 2-chlorobenzoyl thiocarbazide.[403] The ligands are bidentate, co-ordinating *via* S and a hydrazine N.

400 H. Okawa and S. Kida, *Inorg. Nuclear Chem. Letters*, 1971, **7**, 751.
401 M. L. Blair and E. M. Larsen, *J. Amer. Chem. Soc.*, 1971, **93**, 1140.
402 B. Sen and G. White, *Inorg. Nuclear Chem. Letters*, 1971, **7**, 79.
403 N. K. Dutt and N. C. Chakder, *J. Inorg. Nuclear Chem.*, 1971, **33**, 393.

A partial assignment of ligand bands has been attempted [404] for some nitrato-complexes of Ni^{II} with aryl or acyl thiocarbazides RCONHNHC-SNHNH$_2$ (R = PhCO, MeCO, etc.).

Far-i.r. studies on complexes ML$_2$X$_2$ (M = Pd^{II}, Pt^{II}, Ni^{II}, or Cu^{II}; L = 2-methylthioaniline; X = Cl or Br) show that (a) for M = Ni or Cu, M is octahedrally co-ordinated (L bidentate, X co-ordinated); (b) for M = Pd, the structure is [PdL$_2$]X$_2$ (L bidentate), and (c) for M = Pt, the ligand L is co-ordinated only via the NH$_2$ group, with the X atoms also co-ordinated. Data on ν(MX) are given, spectra are shown (80—400 cm^{-1}), and observed wavenumbers are tabulated.[405]

In Pd^{II} and Pt^{II} complexes containing thiocarbamic esters as ligands, marked increases in wavenumber have been noted [406] for bands at ca. 1200 and ca. 1530 cm^{-1}, assigned as ν_{as}(OCS) and ν(CN) [with δ(NH)], respectively.

This has been interpreted in terms of resonance forms such as (106a) and (106b) for the complexed ligands, compared with the free ligand form

$$\begin{array}{cc}
\overset{\nwarrow S^-}{\underset{H}{R-\overset{+}{N}=C-OR}} & \overset{\nwarrow S^-}{\underset{H}{R-N-\overset{+}{C}=OR}} \\
(a) & (b)
\end{array}$$

(106)

R—NH—CS—OR. I.r. spectra (mainly of the solids) have been used [407] to confirm structural predictions for dithizonates of Ag^I, Hg^{II}, Cu^{II}, Pt^{II}, and Pd^{II}.

Ligands containing O and S Donor Atoms.—A review of metal sulphinato-complexes includes a discussion of the dependence of the vibrational spectrum on the bonding of the sulphinato-group.[408]

The interaction of meso-alkyl-substituted β-diketones with metal salts (chiefly acetates) has been studied.[409] I.r. spectroscopy showed that true metal chelates were only obtained with Co^{III} and Ni^{II} [ν(C$\cdots$O—M) at 1530—1560 cm^{-1}; ν(M—O) at 455—470 cm^{-1}]. With Pb^{II}, Cd^{II}, Zn^{II}, Cu^I, Hg^{II}, Ag^I, Pd^{II}, Pt^{II}, and Ru^{III}, the products show an unco-ordinated carbonyl group [ν(C=O) at 1655—1670 cm^{-1}] and are thus formulated as mercaptides.

Assignments have been proposed [410] for the i.r. spectra of nine thiosulphate complexes of known crystal structure: Na$_2$S$_2$O$_3$ (ionic S$_2$O$_3^{2-}$);

[404] N. K. Dutt and N. C. Chakder, *Inorg. Chim. Acta*, 1971, **5**, 188.
[405] M. Ikram and D. B. Powell, *Spectrochim. Acta*, 1971, **27A**, 1845.
[406] T. Tarantelli and C. Furlani, *J. Chem. Soc. (A)*, 1971, 1213.
[407] W. Kemula and T. Ganko, *Chem. Comm.*, 1971, 1063.
[408] G. Vitzthum and E. Lindner, *Angew. Chem. Internat. Edn.*, 1971, **10**, 315.
[409] E. Uhlemann and U. Eckelmann, *Z. anorg. Chem.*, 1971, **383**, 321.
[410] A. N. Freedman and B. P. Straughan, *Spectrochim. Acta*, 1971, **27A**, 1455.

$Na_2S_2O_3,5H_2O$, $MgS_2O_3,6H_2O$ (H-bonded $S_2O_3^{2-}$); $[Zn(tu)_3(S_2O_3)],H_2O$ (tu = thiourea), $BaSe(S_2O_3)_2,2H_2O$, $(NH_4)_2Te(S_2O_3)_2$ (singly S-bonded); $Na_{4n}[Cu(NH_3)_4]_n[Cu(S_2O_3)_2]_n$ (bridging S); $[Ni(tu)_4S_2O_3],H_2O$ (bidentate; bonding *via* S and O); $Ba_2S_2O_3,H_2O$ (both bidentate SO- and OO-bonding). The following general ranges were found for $\nu_{as}(SO_2)$ and $\nu_s(SO_2)$, respectively: S-bridged (type A), > 1175, > 1000 cm^{-1}; M—S co-ordination (type B), 1130—1175, > 1000 cm^{-1}; M—O co-ordination (type C), < 1130, < 1000 cm^{-1}; ionic $S_2O_3^{2-}$, ~ 1130 cm^{-1}. Using these criteria, the structures of the following compounds have been proposed: type A; $Na[Cu(S_2O_3)_2],2H_2O$; $Na_3[Ag(S_2O_3)_2],2H_2O$; $Na[Ag(S_2O_3)],2H_2O$; $Na_3[Au(S_2O_3)_2],2H_2O$: type B; $Na[Cu(S_2O_3)],H_2O$; $(NH_4)_2[Cu_2(S_2O_3)I_2]$; $(NH_4)_2[Ag_2(S_2O_3)I_2]$: type C; PbS_2O_3: type A + B; $Na_4[Cu_2(S_2O_3)_3],3H_2O$, $K_4[Cd(S_2O_3)_2]$. In addition, it was suggested that $UO_2(S_2O_3),H_2O$ and $ZrO(S_2O_3),8H_2O$ contained O-bonded thiosulphate only, being the first examples of such complexes.

Two series of complexes of monothiothenoyltrifluoroacetone (ttas) with Group VI zerovalent metal carbonyls have been obtained.[411] The structures proposed were: (107; M = Cr, Mo, or W); $\nu_{terminal}(CO)$ characteristic of C_{4v} symmetry was found; the β-diketone $\nu(C{=}O)$ mode (1638—1643 cm^{-1}) is at higher wavenumber than in the free ligand; and (108; M = Cr, Mo,

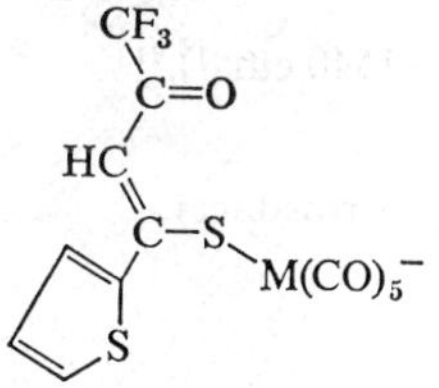

(107)

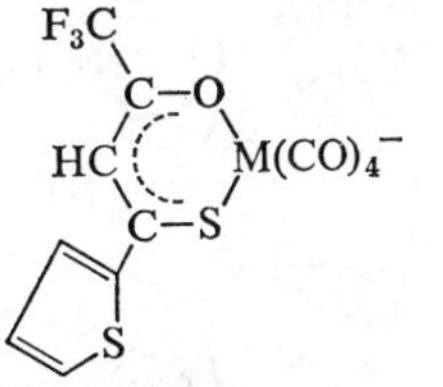

(108)

or W); $\nu_{terminal}(CO)$ characteristic of *cis*-$M(CO)_4L_2$ was found; the β-diketone $\nu(CO)$, at 1612—1600 cm^{-1}, was of lower wavenumber than in the free ligand.

Insertion of SO_3 into the metal–carbon bonds of $RRe(CO)_5$ (R = Me, Ph, or *p*-Me·C_6H_4) has been reported by Lindner and Grimmer.[412] Vibrations arising from the sulphonato-group are assigned to $\nu_{as}(SO_2)$ in the 1290 cm^{-1} region, to $\nu_s(SO_2)$ in the 1160 cm^{-1} region, and to $\nu(SO)$ at *ca.* 1000 cm^{-1}.

I.r. spectroscopy has allowed O- and S-bonding in bipyridyl–iron–sulphinato-complexes to be distinguished, by observation of ν(Fe—O) and $\nu(SO)$ in the former, and ν(Fe—S) and $\nu(SO_2)$ in the latter.[413]

[411] G. H. Barnett and M. K. Cooper, *Chem. Comm.*, 1971, 1082.
[412] E. Lindner and R. Grimmer, *Chem. Ber.*, 1971, **104**, 544.
[413] E. Lindner, I. Lorenz, and G. Vitzthum, *Angew. Chem. Internat. Edn.*, 1971, **10**, 193.

Dorange and Guerchais have studied [414] the i.r. spectra of a number of complexes of β-ketonic dithio-esters $RCOCH_2SSMe$ with Co^{III}, Ni^{II}, Zn^{II}, and Cu^{II}. From the positions of bands assigned to ν(C┄C) (1575—1500 cm^{-1}), ν(C┄O) (1472—1442 cm^{-1}), and ν(C┄S) (1227—1222 cm^{-1}), they conclude that extensive π-electron delocalization takes place in the chelate rings (109).

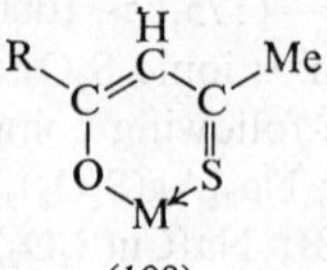

(109)

I.r. data for some Ir sulphinates [415] indicate that $Ir(O_2S$-*p*-tolyl)(CO)-$(PPh_3)_2$ is O-bonded, having ν(SO) at 1085 cm^{-1}, and ν_{as}(SOIr) at 855, 840 cm^{-1}, but that the linkage isomer (S-bonded), $Ir(SO_2$-*p*-tolyl)(CO)-$(PPh_3)_2$, can also be prepared [$\nu_{as}(SO_2)$ at 1155, 1140 cm^{-1}; $\nu_s(SO_2)$ at 1020, 1010 cm^{-1}]. In addition, $Ir(Me)I(O_2S$-*p*-tolyl)(CO)$(PPh_3)_2$ is O-bonded, but $Ir(L)(SO_2$-*p*-tolyl)(CO)$(PPh_3)_2$ (I = O_2 or SO_4) are S-bonded.

Two (monothiocarbamato)nickel(II) complexes (110) and (111) have been prepared. Both have bands characteristic of the 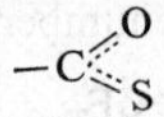 group (*ca.* 1615 and *ca.* 670 cm^{-1}) and of ν(C┄N) (1500—1540 cm^{-1}).[416]

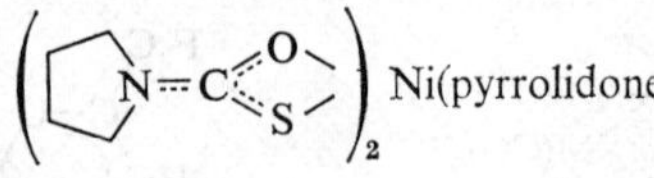

(110)

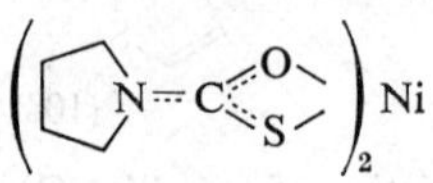

(111)

The i.r. spectra of $M(O_2SPh)_2$ complexes (M = Zn or Cd) and some other phenyl-substituted derivatives have been reported.[417] All are O-bonded, and give $\nu_{as}(SO_2)$ in the range 965—1048 cm^{-1}, with $\nu_s(SO_2)$ between 916 and 962 cm^{-1}. For the Hg analogues $Hg(O_2SR)_2$ (R = Ph, *p*-Br·C_6H_4, 2-naphthyl, *etc.*), however, an S-bonded structure is indicated [418] by the i.r. spectra [$\nu_{as}(SO_2)$ 1180—1246 cm^{-1}; $\nu_s(SO_2)$ 1009—1082 cm^{-1}], *i.e.* (112), while for the arylmercury sulphinates $RHg(O_2SR)$ (R = *p*-

[414] G. Dorange and J. E. Guerchais, *Bull. Soc. chim. France*, 1971, 43.
[415] C. A. Reed and W. R. Roper, *Chem. Comm.*, 1971, 1556.
[416] E. M. Krankovits, R. J. Magee, and M. J. O'Connor, *Inorg. Nuclear Chem. Letters*, 1971, 7, 541.
[417] P. G. Cookson and G. B. Deacon, *Austral. J. Chem.*, 1971, **24**, 935.
[418] P. G. Cookson and G. B. Deacon, *Austral. J. Chem.*, 1971, **24**, 1599.

Br·C_6H_4 or 2,4,6-$Me_3C_6H_2$), O-bonded sulphinate groups are present [$\nu_{as}(SO_2)$ 1020—1090 cm^{-1}; $\nu_s(SO_2)$ 922—907 cm^{-1}], *i.e.* (113).

Stretching and bending frequencies of SO_2 in some Tl^I and Tl^{III} arene sulphinates have been listed.[419]

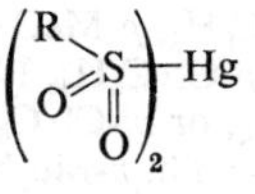

(112)

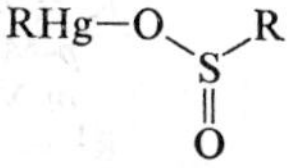

(113)

Appendix: Additional References to Metal Carbonyl Complexes

Titanium Carbonyl Complex

		Ref.
$Ti(CO)_2(C_5Me_5)_2$		420*a*

Vanadium, Niobium, and Tantalum Carbonyl Complexes

$[M(CO)_5(PPh_3)]^-$	M = V, Nb, or Ta	420*b*
$[V(CO)_5(diphos)]^-$	diphos = 1,2-bis(diphenylphosphino)-ethane	420*b*
$[V(CO)_5(EPh_3)]^-$	E = As or Sb	420*b*
$[V(CO)_5\{P(C_4H_9)_3\}]^-$		420*b*
$[M(CO)_4(diphos)]^-$	M = V, Nb, or Ta	420*b*
$[V(CO)_4(fdpa)]^-$	fdpa = ferrocene-1,1′-bisdiphenylarsine	420*b*

Chromium Carbonyl Complexes

$Cr(CO)_5BiPh_3$		420*c*
$Cr(CO)_5L$	L = benzimidazole, benzoxazole, benzothiazole, or benzoselenazole	421
$Cr(CO)_5L$	L = PF_2Bu^t or $PFBu^t_2$	422
$Cr(CO)_5\{Sb(MMe_3)_3\}$	M = Ge or Sn	423
$Cr(CO)_5(C_5H_8N_2)$	$C_5H_8N_2$ = 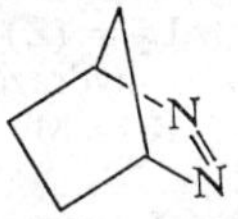	424
$(CO)_5Cr{=}C(X)(Y)$	X = $O^-NMe_4^+$ or OMe; Y = vinyl:	425
	X = $O^-NMe_4^+$, OEt, or NH_2; Y = 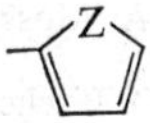 Z = O, S, or NMe	425

[419] A. G. Lee, *Inorg. Chim. Acta*, 1971, **5**, 346.
[420a] J. E. Bercaw and H. H. Brintzinger, *J. Amer. Chem. Soc.*, 1971, **93**, 2045.
[420b] A. Davison and J. S. Ellis, *J. Organometallic Chem.*, 1971, **31**, 239.
[420c] R. A. Brown and G. R. Dobson, *J. Inorg. Nuclear Chem.*, 1971, **33**, 892.
[421] W. Beck, J. C. Weis, and J. Wieczorek, *J. Organometallic Chem.*, 1971, **30**, 89.
[422] O. Stelzer and R. Schmutzler, *J. Chem. Soc.* (*A*), 1971, 2867.
[423] H. Schumann and H. J. Breunig, *J. Organometallic Chem.*, 1971, **27**, C28.
[424] M. Herberhold and W. Golla, *J. Organometallic Chem.*, 1971, **26**, C27.
[425] J. A. Connor and E. M. Jones, *J. Chem. Soc.* (*A*), 1971, 1974.

Chromium Carbonyl Complexes (*cont.*)

		Ref.
$(CO)_5Cr{=}C(N{=}CHMe)(Ph)$		426
$Cr(CO)_5{:}C(NH_2)R$	R = *p*-$Me_2N{\cdot}C_6H_4$, *p*-$MeO{\cdot}C_6H_4$, *p*-$Me{\cdot}C_6H_4$, *p*-$Cl{\cdot}C_6H_4$, Ph, *p*-$Br{\cdot}C_6H_4$, *m*-$MeO{\cdot}C_6H_4$, or *m*-$Cl{\cdot}C_6H_4$	427
$Cr(CO)_5{:}C(OR^1)(R^2)$	R^1 = Me; R^2 = Ph, *p*-$Me_2N{\cdot}C_6H_4$, C_6Cl_5, or *p*-$Cl{\cdot}C_6H_4$: R^1 = Et; R^2 = $C_{10}H_9Fe$ or C_6Cl_5	428
$Cr(CO)_5{:}C(OMe)(R)$	R = *p*-$Me_2N{\cdot}C_6H_4$, *p*-$MeO{\cdot}C_6H_4$, *p*-$Me{\cdot}C_6H_4$, Ph, *p*-$F{\cdot}C_6H_4$, *p*-$Cl{\cdot}C_6H_4$, *p*-$Br{\cdot}C_6H_4$, *p*-$CF_3{\cdot}C_6H_4$, *m*-$Me_2N{\cdot}C_6H_4$, *m*-$MeO{\cdot}C_6H_4$, *m*-$Cl{\cdot}C_6H_4$, *m*-$CF_3{\cdot}C_6H_4$, *o*-$MeO{\cdot}C_6H_4$, *o*-$CF_3{\cdot}C_6H_4$, 2,4,6-$Me_3C_6H_2$, or 2,6-$(MeO)_2C_6H_3$	429
$Cr(CO)_5X$	X = tdp, tris(dimethylamino)phosphine or tdas, tris(dimethylamino)arsine	430
$Cr(CO)_5{:}C(X)Ph$	X = OMe, $NH(CH_2Ph)$, NH(n-C_4H_9), NHPh, or NHC_6H_{11}	431
$Cr(CO)_5(MePX_2)$	X = OMe or NMe_2	432
$Cr(CO)_5C(fu)R$	fu = 2-furyl; R = $O^-NMe_4^+$, OMe, OEt, OPh, SEt, SPh, SCOMe, $C{\equiv}CPh$, NH_2, or $N(CH_2)_4$	433
trans-$Cr(CO)_4L_2$	L = tdp or tdas	430
$[Cr(CO)_4L]_2$, $Cr(CO)_4L_2$	L = Ph_2PCN, $PhP(CN)_2$, Me_2PCN, or $(EtO)_2PCN$	434
cis-$Cr(CO)_4\{MeP(OMe)_2\}_2$ *trans*-$Cr(CO)_4\{MeP(NMe_2)_2\}_2$		432
$Cr(CO)_4(das)$	das = *o*-phenylenebisdimethylarsine	435
$Cr(CO)_4(MeSCH_2CH_2SMe)$ $Cr(CO)_4(Ph_2PCH_2CH_2PPh_2)M(SMe)_2$	M = Pd or Pt	436
$Cr(CO)_4L$	L = ferrocene-1,1′-bisdimethylarsine or ferrocene-1,1′-bisdiphenylarsine	437
$Cr(CO)_4L_2$	L_2 = (*Z*)-pyridine-2-aldehyde 2′-pyridylhydrazone	438
$Cr(CO)_4(SR)_2Ti(\pi\text{-}Cp)_2$	R = Me or Ph	439

[426] L. Knauss and E. O. Fischer, *J. Organometallic Chem.*, 1971, **31**, C68.
[427] E. O. Fischer and H. J. Kollmeier, *Chem. Ber.*, 1971, **104**, 1339.
[428] G. A. Moser, E. O. Fischer, and M. D. Rausch, *J. Organometallic Chem.*, 1971, **27**, 379.
[429] E. O. Fischer, C. G. Kreiter, H. J. Kollmeier, J. Müller, and R. D. Fischer, *J. Organometallic Chem.*, 1971, **28**, 237.
[430] R. B. King and T. F. Korenowski, *Inorg. Chem.*, 1971, **10**, 1188.
[431] E. O. Fischer, B. Heckl, and H. Werner, *J. Organometallic Chem.*, 1971, **28**, 359.
[432] C. E. Jones and K. J. Coskran, *Inorg. Chem.*, 1971, **10**, 55.
[433] J. A. Connor and E. M. Jones, *J. Chem. Soc.* (*A*), 1971, 3368.
[434] C. E. Jones and K. J. Coskran, *Inorg. Chem.*, 1971, **10**, 1664.
[435] H. G. Metzger and R. D. Feltham, *Inorg. Chem.*, 1971, **10**, 951.
[436] P. S. Braterman, V. A. Wilson, and K. K. Joshi, *J. Organometallic Chem.*, 1971, **31**, 123.
[437] J. J. Bishop and A. J. Davison, *Inorg. Chem.*, 1971, **10**, 826.
[438] J. G. Dunn and D. A. Edwards, *J. Chem. Soc.* (*A*), 1971, 988.
[439] P. S. Braterman, V. A. Wilson, and K. K. Joshi, *J. Chem. Soc.* (*A*), 1971, 191.

Chromium Carbonyl Complexes (*cont.*)

		Ref.
$Cr(CO)_3L$ L = N(C$_2$H$_4$X)(C$_2$H$_4$Y)(C$_2$H$_4$Z)	X = Y = Z = PPh_2; X = Y = PPh_2, Z = NEt_2; X = Y = PPh_2, Z = OMe; X = Y = Z = $SCHMe_2$; or X = Y = SMe, Z = NEt_2	440
$Cr(CO)_3(\pi\text{-}Cp)X$	X = Br, I, HgCl, HgBr, or HgI	441
$Cr(CO)_3(\pi\text{-}Cp)SiH_3$		442
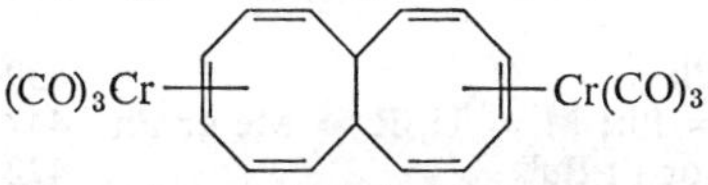	*i.e.* hexacarbonyl-*trans*-6a,12a-dihydro-octalenechromium(0)	443
fac- and *mer*-$Cr(CO)_3\{MeP(OMe)_2\}_3$		432
$Cr_2(CO)_6(C_5H_8N_2)_3$	$C_5H_8N_2$ = 	424
$Cr(CO)_2(C_6H_3Me_3)(C_5H_8N_2)$		424
$Cr(CO)_2(das)_2$	das = *o*-phenylenebisdimethylarsine	435
$\{(\pi\text{-}Cp)Cr(CO)_2L\}_2Hg$	L = $P(OMe)_3$, $P(OPh)_3$, or PPh_3	441
$Cr(CO)_2\{Ph(\pi\text{-}Ph)AsCH_2AsPh_2\}$		444
cis- and *trans*-$Cr(CO)_2(diphos)_2$		445
trans-$[Cr(CO)_2(diphos)_2]^+X^-$	X^- = ClO_4^-, I^-, or BPh_4^-	445
$Cr(CO)_2(SiCl_3)H$		446

Molybdenum Carbonyl Complexes

$Mo(CO)_5BiPh_3$		420*c*
$Mo(CO)_5L$	L = benzimidazole, benzoxazole, benzothiazole, or benzoselenazole	421
$Mo(CO)_5L$	L = PF_2Bu^t or $PFBu^t_2$	422
$Mo(CO)_5Sb[MMe_3]_3$	M = Sn or Ge	423
$Mo(CO)_5L$	L = tdp or tdas	430
$Mo(CO)_5(MePX_2)$	X = OMe or NMe_2	432
$Mo(CO)_4L_2$	L = benzimidazole, benzoxazole, benzothiazole, or benzoselenazole	421
cis-$Mo(CO)_4L_2$	L = PF_2Bu^t or $PFBu^t_2$	422
trans-$Mo(CO)_4(tdp)_2$		430
cis-$Mo\{MeP(OMe)_2\}_2(CO)_4$		432
trans-$Mo\{MeP(NMe_2)_2\}_2(CO)_4$		432
$[Mo(CO)_4L]_2$	L = Ph_2PCN, $PhP(CN)_2$, Me_2PCN, or $(EtO)_2PCN$	434
$Mo(CO)_4L_2$	L = Ph_2PCN, Me_2PCN, $(EtO)_2PCN$, or $(Me_2N)_2PCN$	434

[440] M. Bacci and S. Midollini, *Inorg. Chim. Acta*, 1971, **5**, 220.
[441] A. R. Manning and D. J. Thornhill, *J. Chem. Soc.* (*A*), 1971, 637.
[442] A. P. Hagen, C. R. Higgins, and P. J. Russo, *Inorg. Chem.*, 1971, **10**, 1657.
[443] K. Stöckel, F. Sondheimer, T. A. Clarke, M. Guss, and R. Mason, *J. Amer. Chem. Soc.*, 1971, **93**, 2571.
[444] G. B. Robertson, P. O. Whimp, R. Colton, and C. J. Rix, *Chem. Comm.*, 1971, 573.
[445] P. F. Crossing, and M. R. Snow, *J. Chem. Soc.* (*A*), 1971, 610.
[446] W. Jetz and W. A. G. Graham, *Inorg. Chem.*, 1971, **10**, 4.

Molybdenum Carbonyl Complexes (*cont.*)

		Ref.
$(Ph_2PCH_2CH_2PPh_2)M(SMe)_2Mo(CO)_4$	M = Pd or Pt	436
$(MeSCH_2CH_2SMe)Mo(CO)_4$		436
$Mo(CO)_4L$	L = ferrocene-1,1′-bisdimethylarsine or ferrocene-1,1′-bisdiphenylarsine	437
$Mo(CO)_4L_2$	L_2 = (*Z*)-pyridine-2-aldehyde 2′-pyridylhydrazone	438
$Mo(CO)_4(\pi\text{-}Cp)_2Ti(SR)_2$	R = Me or Ph	439
$(\pi\text{-}Cp)M(SR)_2Mo(CO)_4$	M = W, R = Ph; M = Ti, R = Me or Ph	447
fac-$L_3Mo(CO)_3$	L = PF_2Bu^t or $PFBu^t_2$	422
fac-$Mo(CO)_3[MeP(OMe)_2]_3$		432
μ-L-$[LMo(CO)_3]_2$, $Mo(CO)_3L_2$	L = ferrocene-1,1′-bisdimethylarsine or ferrocene-1,1′-bisdiphenylarsine (uni- and bi-dentate ligands)	437
$Mo(CO)_3L_2Y$	L_2 = (*Z*)-pyridine-2-aldehyde 2′-pyridylhydrazone; Y = PPh_3, $AsPh_3$, $SbPh_3$, Ph_3PSe, or SO_2	438
$Mo(CO)_3L$ L = $N(C_2H_4X)(C_2H_4Y)(C_2H_4Z)$	X = Y = Z = PPh_2; X = Y = PPh_2, Z = NEt_2; X = Y = PPh_2, Z = OMe; X = Y = Z = $SCHMe_2$; or X = Y = SMe, Z = NEt_2	440
$Mo(CO)_3(\pi\text{-}Cp)(SiH_3)$		442
$(\pi\text{-}Cp)_2TiMo(CO)_3(\pi\text{-}Cp)$		448
$Mo(CO)_3ClR$	R = $(\pi\text{-}C_9H_{11})$, $(\pi\text{-}Cp)$, or $(\pi\text{-}C_9H_7)$	449
$[LMo(CO)_3]_2$	L = $\pi\text{-}Me_3SiC_5H_4$	450
$Mo(CO)_3(CH_2)_3Br(h^5\text{-}Cp)$		451
$Mo(CO)_3(Cp)(GeX_3)$	X = F or Cl	452
$Mo(CO)_3(Cp)(GePh_2X)$	X = F or Cl	452
$C_6F_5HgMo(CO)_3(Cp)$		453
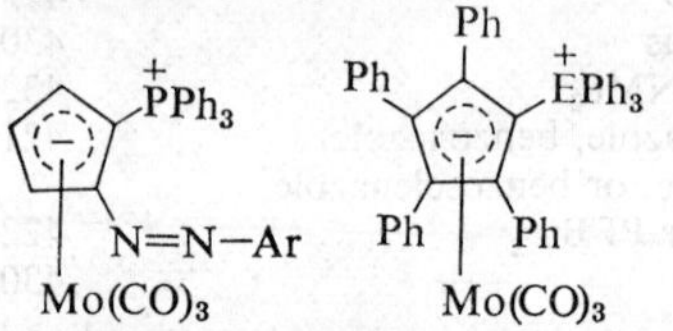	E = P, As, or Sb	454
$(\pi\text{-}C_9H_{11})Mo(CO)_2(PX_3)Cl$	X = Ph, OPh, or OMe	449
$[Mo(CO)_2(PPh_3)(h^5\text{-}Cp)(C_4H_6O)]A$	A = Br or BPh_4	451
$C_6H_5HgMo(CO)_2PX_3(Cp)$	X = Ph or OPh	453

[447] T. S. Cameron, C. K. Prout, G. V. Rees, M. L. H. Green, K. K. Joshi, G. R. Davies, B. T. Kilbourn, P. S. Braterman, and V. A. Wilson, *Chem. Comm.*, 1971, 14.
[448] M. F. Lappert and A. R. Sanger, *J. Chem. Soc.* (*A*), 1971, 1314.
[449] C. White and R. J. Mawby, *J. Chem. Soc.* (*A*), 1971, 940.
[450] E. W. Abel and S. Moorhouse, *J. Organometallic Chem.*, 1971, **28**, 211.
[451] F. A. Cotton and C. M. Lukehart, *J. Amer. Chem. Soc.*, 1971, **93**, 2672.
[452] T. J. Marks and A. M. Seyan, *J. Organometallic Chem.*, 1971, **31**, C62.
[453] T. A. George, *J. Organometallic Chem.*, 1971, **33**, C13.
[454] D. Cashman and F. J. Lalor, *J. Organometallic Chem.*, 1971, **32**, 351.

Molybdenum Carbonyl Complexes (cont.)

		Ref.
(π-Cp)Mo(CO)$_2$(N:CBu^{t}Ph)		455
(π-Cp)Mo(CO)$_2$(HN:CBu^{t}Ph)Cl		
Mo(CO)$_2$(π-Cp)[(p-Me·C$_6$H$_4$)$_2$CNC(p-Me·C$_6$H$_4$)$_2$]		456
R$_2$AlMo(CO)$_2$(Cp)(PX$_3$)	X$_3$ = Ph$_3$ or MePh$_2$	457
(π-Cp)Mo(CO)$_2$LL'	L = H, Cl, Br, I, or Me; L' = PMe$_2$Ph; or L = H, Cl, Br, I, Me, or PhCH$_2$; L' = P(OMe)$_2$Ph	458
Mo(CO)$_2$(π-Cp)(N:CBu$^t{}_2$)		459
[Mo(CO)$_2$(π-Cp){P(OC$_3$H$_5$)$_3$}$_2$]$^+$X$^-$	X$^-$ = BPh$_4{}^-$ or [(π-Cp)Mo(CO)$_3$]$^-$	
[Mo(CO)$_2$(π-Cp){P(OC$_3$H$_5$)$_2$Ph}$_2$]$^+$X$^-$		460
[Mo(CO)$_2$(π-Cp){P(OC$_3$H$_5$)Ph$_2$}$_2$]$^+$X$^-$		
[(π-Cp)Mo(CO)$_2${X'}]$_2$	X' = P(OC$_3$H$_5$)$_3$, P(OC$_3$H$_5$)$_2$Ph, or P(OC$_3$H$_5$)Ph$_2$	
cis- and trans-[Mo(CO)$_2$(Ph$_2$PCH$_2$CH$_2$PPh$_2$)$_2$]$^+$		461
M:N·NH·C(CO$_2$Et)·C·OH		
[M:NH·NH·C(CO$_2$Et)·C·OH]$^+$		
[M:N·N·C(CO$_2$Et)·C·OH]$^-$	M = (π-Cp)(CO)$_2$Mo	462
M:N·NMe·C(CO$_2$Et)·C·OH		
[M:NH·NMe·C(CO$_2$Et)·C·OH]$^+$		
Mo(CO)(π-Cp)I$_2$(N:CBu^{t}Ph)		
Mo(CO)(π-Cp)(PPh$_3$)(N:CBu^{t}Ph)		455
Mo(CO)(π-Cp)I$_2$(N:CBu$^t{}_2$)		459

Tungsten Carbonyl Complexes

W(CO)$_5$L	L = Ph$_3$Bi, Ph$_3$Sb, or Ph$_3$As	420c
W(CO)$_5$L	L = benzimidazole, benzoxazole, benzothiazole, or benzoselenazole	421
W(CO)$_5$L	L = PF$_2$Bu^t or PFBu$^t{}_2$	422
W(CO)$_5$Sb[MMe$_3$]$_3$	M = Ge or Sn	423
W(CO)$_5$X	X = tdp or tdas	430
W(CO)$_5$(MePX$_2$)	X = OMe or NMe$_2$	432
trans-W(CO)$_4$(tdp)$_2$		430
cis-W(CO)$_4$[MeP(OMe)$_2$]$_2$		
trans-W(CO)$_4$[MeP(NMe$_2$)$_2$]$_2$		432
W(CO)$_4$(SMe)$_2$M(Ph$_2$PCH$_2$CH$_2$PPh$_2$)	M = Pd or Pt	436
W(CO)$_4$(MeSCH$_2$CH$_2$SMe)		

[455] M. Kilner and J. N. Pinkney, *J. Chem. Soc.* (*A*), 1971, 2887.
[456] H. R. Keable and M. Kilner, *Chem. Comm.*, 1971, 349.
[457] W. R. Kroll and G. B. McVicker, *Chem. Comm.*, 1971, 591.
[458] J. W. Fuller, A. S. Anderson, and A. Jakubowski, *J. Organometallic Chem.*, 1971, 27, C47.
[459] M. Kilner and C. Midcalf, *J. Chem. Soc.* (*A*), 1971, 292.
[460] R. J. Haines, A. L. Du Preez, and I. L. Marais, *J. Organometallic Chem.*, 1971, 28, 97.
[461] R. H. Reimann and E. Singleton, *J. Organometallic Chem.*, 1971, 32, C44.
[462] M. L. H. Green and J. R. Sanders, *J. Chem. Soc.* (*A*), 1971, 1947.

Tungsten Carbonyl Complexes (cont.)

		Ref.
$W(CO)_4L$	L = ferrocene-1,1'-bisdimethylarsine or ferrocene-1,1-bisdiphenylarsine	437
$W(CO)_4L$	L = (Z)-pyridine-2-aldehyde 2'-pyridylhydrazone	438
$W(CO)_4(SR)_2Ti(\pi\text{-Cp})_2$	R = Me or Ph	439
$W(CO)_3L_3$	L = ethyl cyanoformate	463
fac- and mer-$W(CO)_3[MeP(OMe)_2]_3$		432
$W(CO)_3L$ $L=N{\Large\substack{\diagup C_2H_4X \\ -C_2H_4Y \\ \diagdown C_2H_4Z}}$	X = Y = Z = PPh_2; X = Y = PPh_2, Z = NEt_2; X = Y = PPh_2, Z = OMe; X = Y = Z = $SCHMe_2$; or X = Y = SMe, Z = NEt_2	440
$W(CO)_3(\pi\text{-Cp})(SiH_3)$		442
$[W(CO)_3L]_2$	L = $\pi\text{-Me}_3SiC_5H_4$	450
$W(CO)_2(das)_2$	das = o-phenylenebisdimethylarsine	435
$W(CO)_2(\pi\text{-Cp})(N\!:\!CBu^tPh)$		455
$W(CO)_2(\pi\text{-Cp})\{(p\text{-Me}\cdot C_6H_4)_2CNC(p\text{-Me}\cdot C_6H_4)_2\}$		456
$W(CO)_2(\pi\text{-Cp})(N\!:\!CBu^t{}_2)$		459
$\left.\begin{array}{l} M\!:\!N\cdot NH\cdot C(CO_2Et)\cdot C\cdot OH \\ {[M\!:\!NH\cdot NH\cdot C(CO_2Et)\cdot C\cdot OH]^+} \\ {[M\!:\!N\cdot N\cdot C(CO_2Et)\cdot C\cdot OH]^-} \\ M\!:\!N\cdot NMe\cdot C(CO_2Et)\cdot C\cdot OH \\ {[M\!:\!NH\cdot NMe\cdot C(CO_2Et)\cdot C\cdot OH]^+} \end{array}\right\}$ M = $W(CO)_2(\pi\text{-Cp})$		462
$W(CO)I_2(\pi\text{-Cp})(N\!:\!CBu^t{}_2)$		459
$W(CO)I_2(\pi\text{-Cp})(N\!:\!CBu^tPh)$		455

Manganese Carbonyl Complexes

$Mn(CO)_5GePh_2X$	X = F or Cl	452
$Mn(CO)_5HgC_6F_5$		453
$Mn(CO)_5SiMe_3$		
$Mn(CO)_5Me_2Si(SiMe_3)$		464
$Mn(CO)_5MeSi(SiMe_3)_2$		
$Mn(CO)_5Si(SiMe_3)_3$		
$(CO)_5Mn{\Large\substack{\diagup S \diagdown \\ \diagdown S \diagup}}P{\Large\substack{\diagup R \\ \diagdown R}}$	R = Et or Ph	465
$Mn(CO)_5COCOMe$		466
1-$\{Mn(CO)_5\}$-10-Me-1,10-$(\sigma\text{-}B_8C_2H_8)$		467
1-$\{Mn(CO)_5\}$-2-Me-1,2-$(\sigma\text{-}B_{10}C_2H_{10})$		

[463] J. Y. Chenard, D. Commereuc, and Y. Chauvin, *Compt. rend.*, 1971, **273**, C, 1469.
[464] B. K. Nicholson and J. Simpson, *J. Organometallic Chem.*, 1971, **32**, C29.
[465] E. Lindner and K.-M. Matejcek, *J. Organometallic Chem.*, 1971, **29**, 283.
[466] C. P. Casey and C. A. Bunnell, *J. Amer. Chem. Soc.*, 1971, **93**, 4077.
[467] D. A. Owen, J. C. Smart, P. M. Garrett, and M. F. Hawthorne, *J. Amer. Chem. Soc.*, 1971, **93**, 1362.

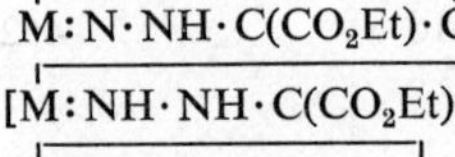

Manganese Carbonyl Complexes (*cont.*)

Ref.

$Mn(CO)_5C_5F_7$ C_5F_7 = 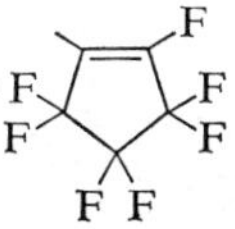 468

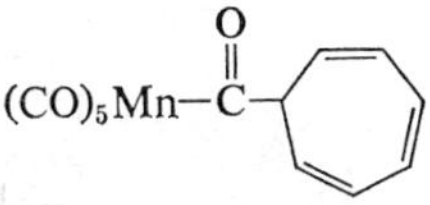 469

trans-$Mn(CO)_4(C_5F_7)(PPh_3)$ C_5F_7 = 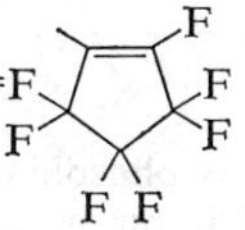 468

cis-$Mn(CO)_4(C_5F_7)(PPh_3)$ 468

trans-$Mn(CO)_4H(PPh_3)$ 470

$Mn(CO)_4(PPh_3)CH_2\cdot C{\equiv}CH$ 471

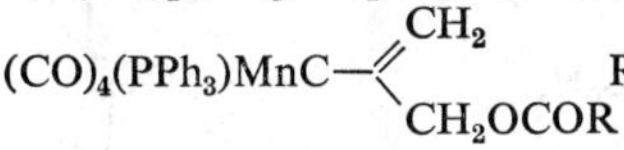 R = Me, Et, or CH_2Cl 471

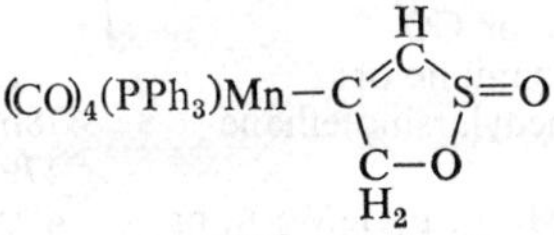 471

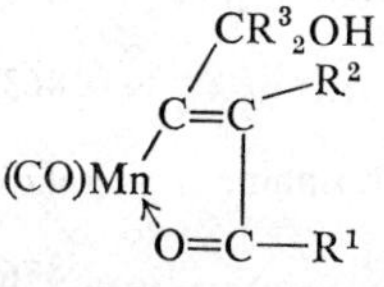 R^1 = Me or Ph, R^2 = H or CH_2OH, R^3 = H or Me 471

$Mn(CO)_4H(H_2C{=}CH\cdot C_6H_4\cdot PPh_2\text{-}o)$ 472

$Mn(CO)_3\pi\text{-}Me_3SiC_5H_4$ 450

fac-$[Mn(CO)_3(XMe_2Ph)_2Br]^+$ X = P or As, 461

fac-$[Mn(CO)_3(dpe)Br]^+$ dpe = $Ph_2PCH_2CH_2PPh_2$, 461

fac-$[Mn(CO)_3(dpm)Br]^+$ dpm = $Ph_2PCH_2PPh_2$ 461

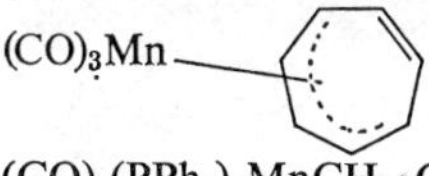 469

$(CO)_3(PPh_3)_2MnCH_2\cdot C{\equiv}CH$ 471

[468] R. E. Banks, R. N. Haszeldine, M. Lappin, and A. B. P. Lever, *J. Organometallic Chem.*, 1971, **29**, 427.
[469] T. H. Whitesides and R. A. Budnik, *Chem. Comm.*, 1971, 1514.
[470] J. R. Miller and D. H. Myers, *Inorg. Chim. Acta*, 1971, **5**, 215.
[471] W. D. Bannister, B. L. Booth, R. N. Haszeldine, and P. L. Loader, *J. Chem. Soc.* (*A*), 1971, 930.
[472] M. A. Bennett and R. Watt, *Chem. Comm.*, 1971, 94.

Manganese Carbonyl Complexes (*cont.*)

		Ref.
$Mn(CO)_3(B_{10}H_{10}PPh)NMe_4$		473
$Mn(CO)_3(pyrazole)\{H_2B(N_2C_3H_3)_2\}$		474
$Mn(CO)_3(\pi\text{-}Me_3SiC_5H_4)$		
$Mn(CO)_3(\pi\text{-}Me_5Si_2C_5H_4)$		475
$\{Mn(CO)_3(\pi\text{-}C_5H_4)\}_2SiMe_2$		
$Mn(CO)_3\{(\pi\text{-}Me_3Si)_2C_5H_3\}$		
$Mn(CO)_2(C_5H_5)(C_5H_8N_2)$	$C_5H_8N_2$ = 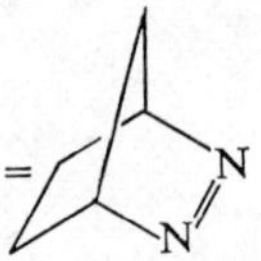	424
$Mn(CO)_2(H)(\pi\text{-}C_5H_5)(Cl_3Si)$		446
$Mn(CO)_2(H)(\pi\text{-}C_5H_4Me)(SiPh_3)$		
$Mn(CO)_2\{H_2B(N_2C_3H_3)_2\}L_2$	L = phosphine or phosphite	474
$Mn(CO)_2(Cp)MCl_3^-$	M = Sn or Ge	476
$Mn(CO)_2(MeC_5H_4)MCl_3^-$	M = Sn or Ge	
$[(Cp)Mn(CO)_2]_2(L')$	L′ = 1,4-butylenebisdiphenylarsine,	
$[(Cp)Mn(CO)_2]_2(L'')$	L″ = 1,2-ethylenebisdiphenylarsine	477
$[(MeCp)Mn(CO)_2]_2(L')$		
$Mn(CO)_2CN(Ar)$	Ar = Ph or $Me_3C_6H_3$	
$[Mn(CO)_2(CNH)(Ar)]^+$		478*a*
$Mn(CO)_2(CNBF_3)(C_6H_3Me_3)$		
$[Mn(CO)_2(C_6H_3Me_3)L]^+$	L = CNEt, $CNCPh_3$, or CO	
$Mn(CO)_2BrL_2$	L = *o*-dimethylarsinoaniline or 1-amino-2-(diphenylarsino)ethane	478*b*
$Mn(CO)_2(Cp)(PhPH_2)$		479
$Mn(CO)_2(\pi\text{-}Cp)L$	L = $P(CMe_3)_3$, $P(SiMe_3)_3$, $P(GeMe_3)_3$, or $P(SnMe_3)_3$	480
$Mn(CO)L_2$	L = ethyl cyanoformate	463
$Mn(CO)L(\pi\text{-}C_5H_5)$		
$Mn(CO)(Cp)L$	L = 1,4-butylenebisdiphenylphosphine	477
$Mn(CO)(MeCp)L$		
$Mn_2(CO)_9MCl_3^-$	M = Sn or Ge	476
$Mn_2(CO)_8(AsMe)_5$		481
$Mn_2(CO)_8L$	L = $AsMe_2I$, $AsMe_2SCF_3$, $AsMe_2SMe$, $AsMe_2H$, $AsMe_2P(CF_3)_2$, or $As(CF_3)_2PMe_2$	482
$Mn_2(CO)_6(AsMe)_8$		481

[473] J. L. Little and A. C. Wong, *J. Amer. Chem. Soc.*, 1971, **93**, 522.
[474] A. Bond and M. Green, *J. Chem. Soc.* (*A*), 1971, 682.
[475] E. W. Abel and S. Moorhouse, *J. Organometallic Chem.*, 1971, **29**, 227.
[476] J. K. Ruff, *Inorg. Chem.*, 1971, **10**, 409.
[477] S. S. Sandhu and A. K. Mehta, *Inorg. Nuclear Chem. Letters*, 1971, **7**, 891.
[478a] P. J. C. Walker and R. J. Mawby, *J. Chem. Soc.* (*A*), 1971, 3006.
[478b] B. Chiswell, R. A. Plowman, and K. Verall, *Inorg. Chim. Acta*, 1971, **5**, 579.
[479] M. Höfler and M. Schnitzler, *Chem. Ber.*, 1971, **104**, 3117.
[480] H. Schumann, O. Stelzer, J. Kuhlmey, and U. Nieder-Reuther, *J. Organometallic Chem.*, 1971, **28**, 105.
[481] P. S. Elmes and B. O. West, *J. Organometallic Chem.*, 1971, **32**, 365.
[482] J. Grobe and F. Kober, *J. Organometallic Chem.*, 1971, **29**, 295.

Rhenium Carbonyl Complexes

Compound		*Ref.*
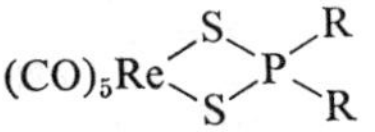	R = Et or Ph	465
$Re(CO)_5(PMe_2Ph)$		483
cis- and *trans*-$Re(CO)_4Cl(PMePh_2)$		483
$Re(CO)_3L$	L = π-$Me_3Si\cdot C_5H_4$	450
$Re(CO)_3$(pyrazole)$\{H_2B(N_2C_3H_3)_2\}$		474
$Re(CO)_3(\pi\text{-}Me_3Si\cdot C_5H_4)$		475
$Re(CO)_3(\pi\text{-}Me_5Si_2\cdot C_5H_4)$		475
$\{Re(CO)_3(\pi\text{-}C_5H_4)\}_2SiMe_2$		475
$Re(CO)_3\{(\pi\text{-}Me_3Si)_2C_5H_3\}$		475
$Re(CO)_3(PMePh_2)_2$		483
$Re(CO)_3Cl(PMePh_2)$		483
$\{Re^I(CO)_3\}_2$-μ-(mesoporphyrin IX dimethylesterato)		484*a*
$ReX(CO)_2(PPh_3)_2$; $ReX(CO)_3(PPh_3)_2$	X = H, Cl, Br, or I	484*b*
$Re(CO)_2L_2\{H_2B(N_2C_3H_3)_2\}$	L = phosphine or phosphite	474
$Re(CO)_2Cl(PMePh_2)_3$		483
$[Re(CO)_2(NO)X_2]_2$	X = Cl, Br, or I	485
$Re(CO)_2(NO)X_2(C_4H_8S)$	X = Cl, Br, or I	485
$Re(CO)_2(NO)X_2(C_5H_5N)$	X = Cl, Br, or I	485
$Re(CO)_2(NO)Cl(acac)$		486
$[Re(CO)(NO)(C_8H_{14})Cl_2]_2$		486
$Re_4(CO)_{10}(PMePh_2)_6$		483
$Re_2(CO)_9(PMePh_2)$		483
$Re_2(CO)_9(PMe_2Ph)$		483
$Re_2(CO)_8(PMePh_2)_2$		483
$Re_2(CO)_7(PMePh_2)_3$		483

Iron Carbonyl Complexes

Compound		Ref.
$Fe(CO)_4X$	X = tris(dimethylamino)phosphine or tris(dimethylamino)arsine	430
$Fe(CO)_4(H)(SiCl_3)$		446
$Fe(CO)_4(H)(SiPh_3)$		446
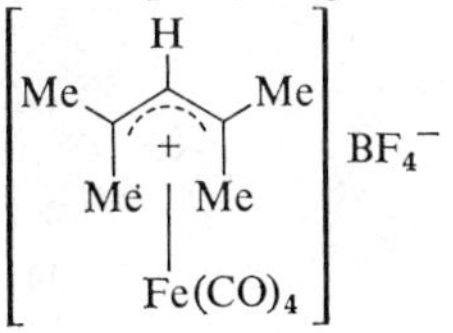		487

[483] J. T. Moelwyn-Hughes, A. W. B. Garner, and N. Gordon, *J. Organometallic Chem.*, 1971, **26**, 373.
[484a] D. Ostfeld, M. Tsutsui, C. P. Hiung, and D. C. Conway, *J. Amer. Chem. Soc.*, 1971, **93**, 2548.
[484b] M. Freni, D. Giusto, and P. Romiti, *J. Inorg. Nuclear Chem.*, 1971, **33**, 4093.
[485] F. Zingales, A. Trovati, F. Cariati, and P. Uguagliati, *Inorg. Chem.*, 1971, **10**, 507.
[486] A. Trovati, P. Uguagliati, and F. Zingales, *Inorg. Chem.*, 1971, **10**, 85.
[487] D. H. Gibson, R. L. Vonnahme, and J. E. McKiernan, *Chem. Comm.*, 1971, 720.

Iron Carbonyl Complexes (*cont.*)

		Ref.
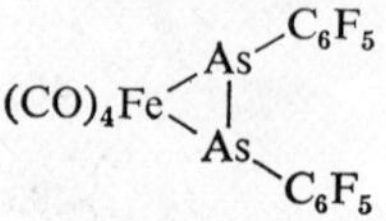		488
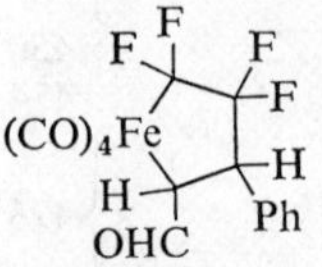		489
$(CO)_4FeCd$		490
$(CO)_4FeZn(NH_3)_3$		490
$(CO)_4FeCd(NH_3)_2$		490
$(CO)_4FeCdL$	L = en, bipy, phen, $(py)_2$, or $(\gamma\text{-pic})_2$	490
$Fe(CO)_4(Pf)$	Pf = $C_5H_{10}NPF_2$, Et_2NPF_2, $(Et_2N)_2PF$, $ClPF_2$, $BrPF_2$, Cl_2PF, $PhPF_2$, $(SCN)PF_2$, $(SCN)_2PF$, or N_3PF_2	491
$Fe(CO)_4HSiPh_3$		492
$[Fe(CO)_4HSiPh_3](NEt_3)$		492
$[Fe(CO)_4SiPh_3]^- Et_4N^+$		492
$Fe(CO)_4HSiCl_3$		492
$Fe(CO)_4(SiCl_3)_2$		492
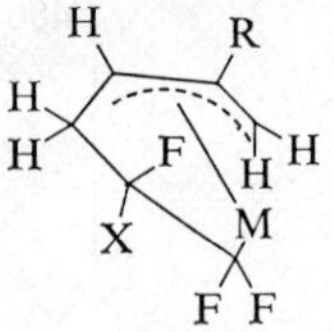	R = H, X = F, M = $Fe(CO)_3$; or R = Me, X = CF_3, M = $Fe(CO)_3$	489
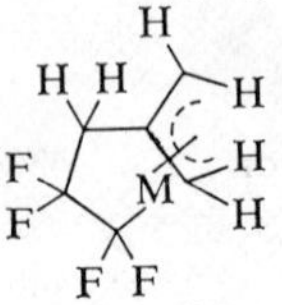	M = $Fe(CO)_3$	489
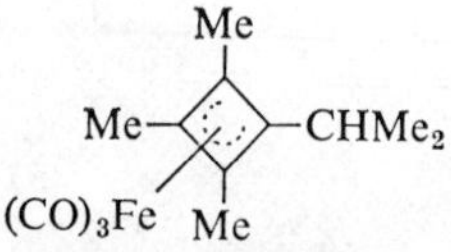		493
$[o\text{-}Me\overline{CHC_6H_4PPh_2Fe}(CO)_3X]$	X = Cl or Br	494

[488] P. S. Elmes, P. Leverett, and B. O. West, *Chem. Comm.*, 1971, 747.
[489] A. Bond, M. Green, B. Lewis, and S. F. W. Lowrie, *Chem. Comm.*, 1971, 1230.
[490] A. T. T. Hsieh, M. J. Mays, and R. H. Platt, *J. Chem. Soc.* (*A*), 1971, 3296.
[491] W. M. Douglas and J. K. Ruff, *J. Chem. Soc.* (*A*), 1971, 3558.
[492] W. Jetz and W. A. G. Graham, *Inorg. Chem.*, 1971, **10**, 1647.
[493] H. A. Brune, H. P. Wolff, and H. Hüther, *Z. Naturforsch.*, 1971, **26b**, 765.
[494] M. A. Bennett, G. B. Robertson, I. B. Tomkins, and P. O. Whimp, *J. Organometallic Chem.*, 1971, **32**, C19.

Iron Carbonyl Complexes (*cont.*)

Ref.

$[Fe(CO)_3X]_2$ X = SePh or TePh 495

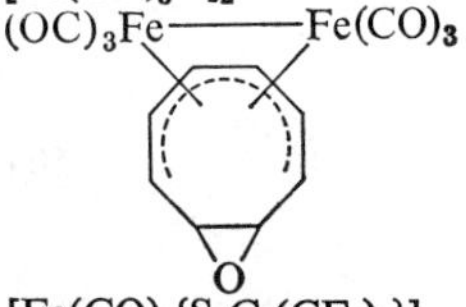

496

$[Fe(CO)_3\{S_2C_2(CF_3)_2\}]_2$ 497

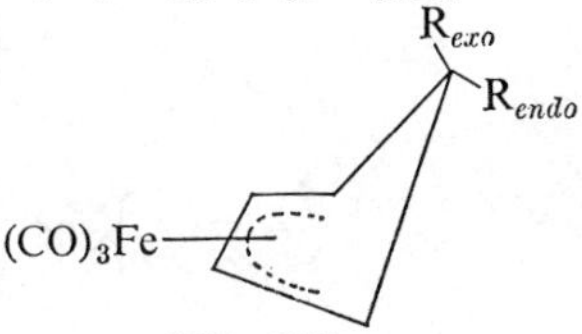

R_{exo}	R_{endo}
Me	CO_2Et
Me	CH_2OH
CH_2OH	Me
Me	CH_2OTs
CH_2OTs	Me

498

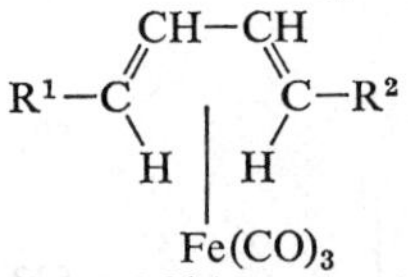

R^1 = H or Me;
R^2 = CHO, CO_2H, or COMe 499

(2,3,4,6-h^4-bullvalene)$Fe(CO)_3$
(7,8,9,10-h^4-bullvalene)$Fe(CO)_3$ 500

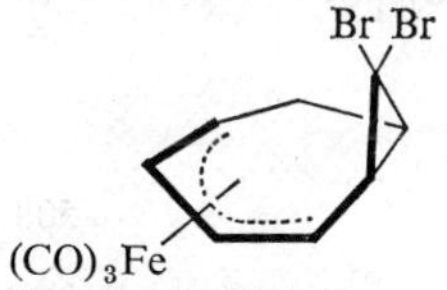

501

$[Fe(CO)_3NO]MPh_3$ M = Ge, Sn, or Pb
$[Fe(CO)_3NO]_3PhSn$
$[Fe(CO)_4NO]_4Sn$ 502

$(RC_8H_7)Fe(CO)_3$ R = CHO, CH_2OH, CH_2OMe, CH_2CN, CH_2N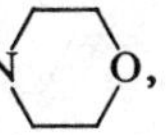O, COMe, CH(Me)COMe, or CH(Me)OH 503

[495] E. D. Schermer and W. H. Baddley, *J. Organometallic Chem.*, 1971, **30**, 67.
[496] H. Maltz and G. Deganello, *J. Organometallic Chem.*, 1971, **27**, 383.
[497] C. J. Jones, J. A. McCleverty, and D. G. Orchard, *J. Organometallic Chem.*, 1971, **26**, C19.
[498] H. Müller and G. E. Herberich, *Chem. Ber.*, 1971, **104**, 2772.
[499] A. Musco, R. Palumbo, and G. Paiaro, *Inorg. Chim. Acta*, 1971, **5**, 157.
[500] R. Aumann, *Angew. Chem. Internat. Edn.*, 1971, **10**, 188.
[501] P. J. van Vuuren, R. J. Fletterick, J. Meinwald, and R. E. Hughes, *J. Amer. Chem. Soc.*, 1971, **93**, 4394.
[502] A. J. Cleland, S. A. Fieldhouse, B. H. Freeland, C. D. M. Mann, and R. J. O'Brien, *J. Chem. Soc.* (*A*), 1971, 736.
[503] B. F. G. Johnson, J. Lewis, and G. L. P. Randall, *J. Chem. Soc.* (*A*), 1971, 422.

Iron Carbonyl Complexes (*cont.*)

Ref.

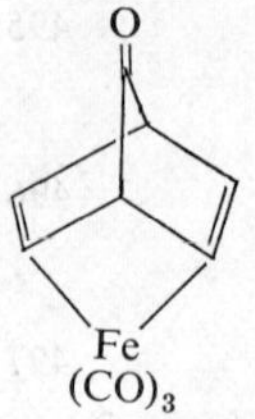

504

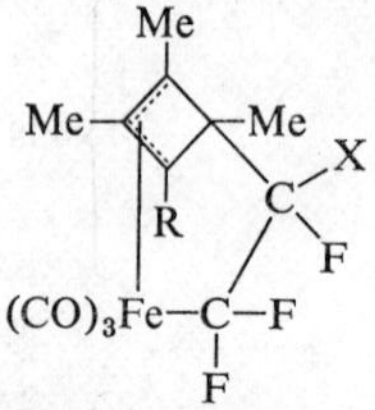

R = Me, X = F; or
R = H, X = CF_3 } 505

$Fe(CO)_3(sp)$ 506

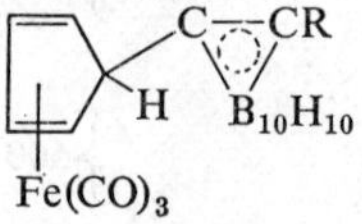

[R = H, Me, Ph, or 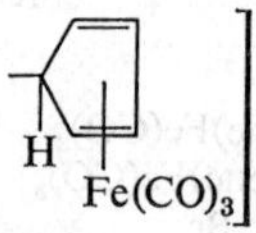] 507

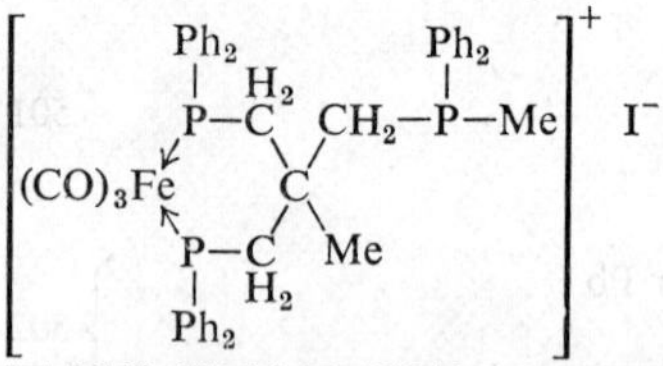

508

$[Fe(CO)_3(SMe)_2Fe(CO)(Ph_2PCH_2PPh_2)]$ 509

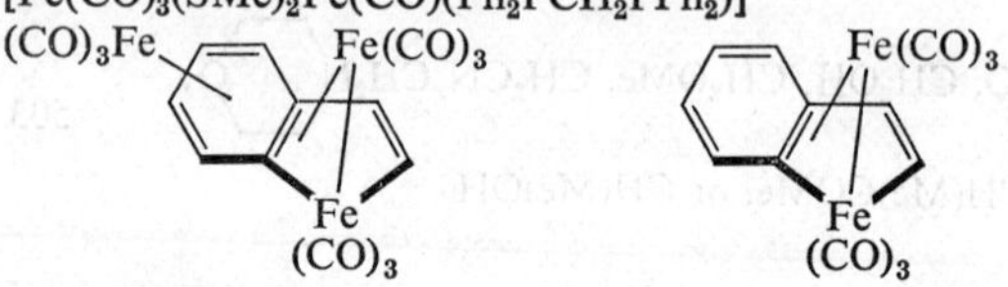

510

$Fe(CO)_2(sp)_2$ 506

$(\pi\text{-Cp})Fe(CO)_2SiCl_2Me$ 511

504 J. M. Landesberg and J. Sieczkowski, *J. Amer. Chem. Soc.*, 1971, **93**, 972.

505 A. Bond and M. Green, *Chem. Comm.*, 1971, 12.

506 M. A. Bennett, G. B. Robertson, I. B. Tomkins, and P. O. Whimp, *Chem. Comm.*, 1971, 341.

507 L. I. Zakharkin, L. V. Orlova, A. I. Kovredov, L. A. Fedorov, and B. V. Lokshin, *J. Organometallic Chem.*, 1971, **27**, 95.

508 H. Behrens, H.-D. Feilner, and E. Lindner, *Z. anorg. Chem.*, 1971, **385**, 321.

509 J. A. De Beer, R. J. Haines, R. Greatrex, and N. N. Greenwood, *J. Organometallic Chem.*, 1971, **27**, C33.

510 R. Victor, R. Ben-Shoshan, and S. Sarel, *Chem. Comm.*, 1971, 1241.

511 J. Dalton, *Inorg. Chem.*, 1971, **10**, 1882.

Iron Carbonyl Complexes (*cont.*)

Compound		*Ref.*
$CpFe(CO)_2(SO_2C_6F_5)$		512
$(\pi\text{-}Cp)Fe(CO)_2(EPh)$	E = S, Se, or Te	513
$(\pi\text{-}Cp)_2Fe_2(CO)_2(EPh)_2$	E = S, Se, or Te	513
$[Fe(CO)_2(NO)(PPh_3)_2]^+$		514
$[(\pi\text{-}Me_3Si\cdot C_5H_4)Fe(CO)_2]_2$		450
$(\pi\text{-}Cp)Fe(CO)_2C_5F_7$	C_5F_7 = 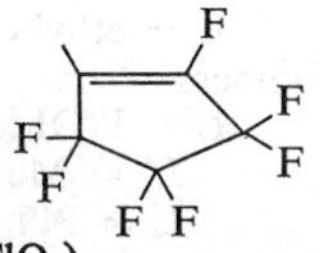	468
$[(\pi\text{-}Cp)Fe(CO)_2(Ph_2PCH_2CH_2PPh_2)](ClO_4)$		515
$[(\pi\text{-}Cp)Fe(CO)_2(Ph_2PCH_2CH_2PPh_2)Fe(CO)_2(\pi\text{-}Cp)](ClO_4)_2$		515
$[(\pi\text{-}Cp)Fe(CO)_2(MeSCH_2CH_2SMe)Fe(CO)_2(\pi\text{-}Cp)](ClO_4)_2$		515
$[(\pi\text{-}Cp)Fe(CO)_2(NC_4H_4N)Fe(CO)_2(\pi\text{-}Cp)](ClO_4)_2$		515
$[Fe(CO)_2(PPh_3)SMe]_2$		509
$[\{Fe(CO)_2SMe\}_2(Ph_2PCH_2PPh_2)]$		509
$[Fe(CO)_2\{P(OMe)_3\}SPh]_2$		509
$(\pi\text{-}Cp)Fe(CO)_2SiX_3$	X = F or Cl	452
$PhHgFe(CO)_2Cp$		453
$(Ph_3M)Fe(CO)_2(NO)\{P(OPh)_3\}$	M = Ge, Sn, or Pb	502
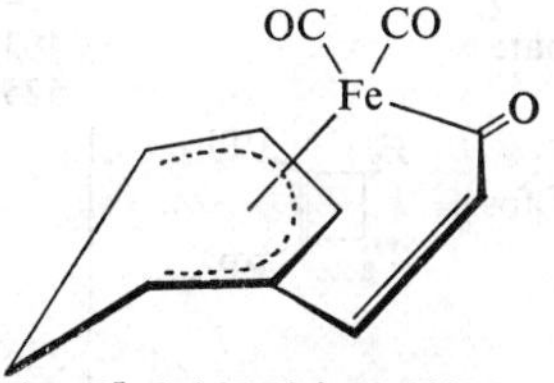		501
$[\{Fe(h^5\text{-}Cp)(CO)_2\}_3SbCl^+]_2[FeCl_4^{2-}],CH_2Cl_2$		516
$(\pi\text{-}Cp)Fe(CO)_2(\sigma\text{-}CH_2CR_2CH{=}CH_2)$	R = H or Me	517
$[(\pi\text{-}Cp)Fe(CO)_2(C_4H_6)]^+PF_6^-$		517
$(\pi\text{-}Cp)Fe(CO)_2(CH_2CMe_2H)$		517
$Fe(CO)_2(2,2',2''\text{-terpyridyl})$		508
$Fe(CO)_2X$	$X = Me{-}C(CH_2PPh_2)_3$	508
$1\text{-}\{(\pi\text{-}Cp)Fe(CO)_2\}\text{-}1,10\text{-}(\sigma\text{-}B_8C_2H_9)$		467
$1\text{-}\{(\pi\text{-}Cp)Fe(CO)_2\}\text{-}10\text{-}Me\text{-}1,10\text{-}(\sigma\text{-}B_8C_2H_8)$		467
$1,10\text{-}\{(\pi\text{-}Cp)Fe(CO)_2\}_2\text{-}1,10\text{-}(\sigma\text{-}B_8C_2H_8)$		467
$1\text{-}\{(\pi\text{-}Cp)Fe(CO)_2\}\text{-}2\text{-}Me\text{-}1,2\text{-}(\sigma\text{-}B_{10}C_2H_{10})$		467
$Fe(CO)(\pi\text{-}Cp)(L)SnR_3$	R = Ph, Me, or Cl; L = PPh_3, $AsPh_3$, or $SbPh_3$	518

[512] M. I. Bruce and A. D. Redhouse, *J. Organometallic Chem.*, 1971, **30**, C78.
[513] E. D. Schermer and W. H. Baddley, *J. Organometallic Chem.*, 1971, **27**, 83.
[514] B. F. G. Johnson and J. A. Segal, *J. Organometallic Chem.*, 1971, **31**, C79.
[515] M. L. Brown, T. J. Meyer, and N. Winterton, *Chem. Comm.*, 1971, 309.
[516] Trinh-Toan and L. F. Dahl, *J. Amer. Chem. Soc.*, 1971, **93**, 2654.
[517] M. L. H. Green and M. J. Smith, *J. Chem. Soc.* (*A*), 1971, 3220.
[518] W. R. Cullen, J. R. Sams, and J. A. J. Thompson, *Inorg. Chem.*, 1971, **10**, 843.

Iron Carbonyl Complexes (*cont.*)

Compound		*Ref.*
$Fe(CO)RCO(L)(\pi\text{-}Cp)$	R = Me, Et, or Pr^i; L = PPh_3, $PPhMe_2$, PPh_2Me, or $P(OPh)_3$	519
$[Fe(CO)(\pi\text{-}Cp)]_4{}^+PF_6{}^-$		520
$[Fe(CO)(\pi\text{-}Cp)]_4{}^-$		520
$Fe(CO)(\pi\text{-}Cp)(PPh_3)NCO$		521
$Fe(CO)L_2$	L = ethyl cyanoformate	463
$[Fe(CO)(Cp)]_4(SO_2)_3$ two isomers		522*a*
$CpFe(CO)LL'$	L = $P(OMe)_3$, $P(OBu^n{}_3)$, or $P(OPh)_3$; L′ = Me or SO_2Me: L = PPh_3; L′ = COEt, Et, or H	523
$Fe(CO)(NO)(PPh_3)_2CO_2Me$		514
$[Fe(CO)(\pi\text{-}Cp)(PPh_2CH_2CH_2PPh_2)](ClO_4)$		515
$[Fe(CO)(SR)\{S_2C_2(CF_3)_2\}]_4$	R = Me, Et, or Ph	497
$[Fe(CO)SePh\{S_2C_2(CF_3)_2\}]_4$		497
$[Fe(CO)(SMe)_2Fe(CO)_3(Ph_2PCH_2PPh_2)]$		509
$(\pi\text{-}Cp)Fe(CO)CH_2\cdot CMe_2CH{=}CH_2$		517
$(\pi\text{-}Cp)Fe(CO)CO\cdot CH_2\cdot CR_2CH{=}CH_2$	R = H or Me	517
$Fe(CO)(\pi\text{-}Cp)(SiCl_3)_2H$		446
$[Fe(CO)(\pi\text{-}Cp)(Me_2PCH_2CH_2PMe_2)]^+BF_4{}^-$		524
$Fe_2(CO)_4L_3$	L = ethyl cyanoformate	463
$Fe_2(CO)_4$(2,2,5,5-tetramethylhex-3-yne)$_2$		525
$Fe_2(CO)_4(SMe)_2\{f_4fos(chelated)\}$		526
$Fe_2(CO)_4(SMe)_2\{f_8fos(chelated)\}$		526
$Fe_2(CO)_4(SMe)_2X$	X = $f_4$fos or $f_4$fars	526
$Fe_2(CO)_4(SMe)_2(Me_2PPh)_2$		526
$Fe_2(CO)_5PPh_3\{(NC_6H_{11})_3C\}$		527
syn- and *anti*-$(SMe)_2Fe_2(CO)_5(PMe_2Ph)$		526
syn- and *anti*-$(SMe)_2Fe(CO)_5(PPh_3)$		526
$(SPh)_2Fe_2(CO)_5(PPh_3)$		526
$Fe_2(CO)_6(AsR)_4$	R = Me or Et	481
$Fe_2(CO)_6(PC_6F_5)_4$		488

$f_4$fos =
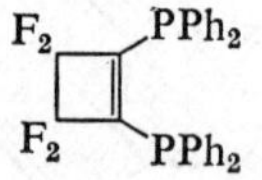

$f_4$fars =
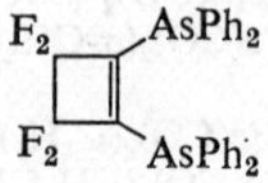

519 M. Green and D. J. Westlake, *J. Chem. Soc.* (*A*), 1971, 367.
520 J. A. Ferguson and T. J. Meyer, *Chem. Comm.*, 1971, 623.
521 M. Graciani, L. Busetto, and A. Palazzi, *J. Organometallic Chem.*, 1971, **26**, 261.
522a D. S. Field and M. J. Newlands, *J. Organometallic Chem.*, 1971, **27**, 221.
522b L. Maresca, F. Greggio, G. Sbrignadello, and G. Bor, *Inorg. Chim. Acta*, 1971, **5**, 667.
523 S. R. Su and A. Wojcicki, *J. Organometallic Chem.*, 1971, **27**, 231.
524 W. E. Silverthorn, *Chem. Comm.*, 1971, 1310.
525 K. Nicholas, L. S. Bray, R. E. Davis, and R. Pettit, *Chem. Comm.*, 1971, 608.
526 J. P. Crow and W. R. Cullen, *Canad. J. Chem.*, 1971, **49**, 2948.
527 N. J. Bremer, A. B. Cutcliffe, M. F. Farona, and W. G. Kofron, *J. Chem. Soc.* (*A*), 1971, 3264.

Iron Carbonyl Complexes (*cont.*)

		Ref.
$Fe_2(CO)_6I(N{:}CR^1R^2)$	$R^1 = R^2 = H$; $R^1 = R^2 = p$-tolyl; $R^1 = Ph$, $R^2 = Bu^t$	528
$Fe_2(CO)_6(C_5H_8N_2)$	$C_5H_8N_2 =$ 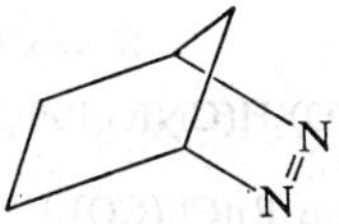	424
$Fe_2(CO)_6\{C(NR)_3\}$	$R = C_6H_{11}$ or C_3H_7	527
syn- and *anti*-$(SMe)_2Fe_2(CO)_6$		526
$(SPh)_2Fe_2(CO)_6$		526
$Fe_2(CO)_8SO_2$		522*a*
$[Fe(CO)_3(\mu\text{-}SR)]_2$		522*b*
$(CO)_3Fe(\mu\text{-}SR^1)_2Fe(CO)_2PR^2_3$		522*b*
$S_2Fe_3(CO)_8(PPh_3)$		526
$S_2Fe_3(CO)_8\{P(n\text{-}C_4H_9)_3\}$		526
$Fe_3(CO)_9(PC_6F_5)_2$		488
$S_2Fe_3(CO)_9$		526
$[Fe_6(CO)_{16}C][Me_4N]_2$		529

Ruthenium Carbonyl Complexes

$Ru(CO)_4(Me_3Ge)_2$		530
$[Ru(CO)_4(Me_3Ge)]_2$		530
$Ru(CO)_4(Me_3Ge)(Me_3Sn)$		530
$Ru(CO)_4(Me_3Ge)(AuPPh_3)$		530
$Hg[Ru(CO)_4(GeMe_3)]_2$		530
$Ru(CO)_4X(MMe_3)$	M = Si, Ge, or Sn; X = Br or I	531
$[Ru(CO)_3X_2]_2$	X = Cl, Br, or I	532
$[Ru(CO)_3X_3]^-$	X = Cl, Br, or I	532
$[o\text{-}MeCHPhPPh_2MRu(CO)_3X]$	X = Cl or Br	494
$Ru(CO)_3(sp)$		506
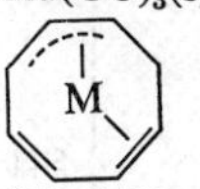	$M = Ru(CO)_3^+\,BF_4^-$	533
$[Me_3MRu(CO)_3X]_2$	M = Si, Ge, or Sn; X = Br or I	531
$[Me_3MRu(CO)_3X(PPh_3)]$	M = Si, Ge, or Sn; X = Br or I	531
$[Ru(CO)_3X]_2$	X = SePh or TePh;	495
$[Ru(CO)_3(EPh)_2]_n$	E = Se or Te	495
$[Ru(CO)_3(Me_3Si)(\mu\text{-}SiMe_3)]_2$		534
$Ru(CO)_2(sp)_2$		506

[528] M. Kilner and C. Midcalf, *Chem. Comm.*, 1971, 944.
[529] M. A. Churchill, J. Wormald, J. Knight, and M. J. Mays, *J. Amer. Chem. Soc.*, 1971, **93**, 3073.
[530] S. A. R. Knox and F. G. A. Stone, *J. Chem. Soc.* (*A*), 1971, 2874.
[531] M. J. Ash, A. Brookes, S. A. R. Knox, and F. G. A. Stone, *J. Chem. Soc.* (*A*), 1971, 458.
[532] R. Colton and R. H. Farthing, *Austral. J. Chem.*, 1971, **24**, 903.
[533] M. Cooke, P. T. Draggett, M. Green, B. F. G. Johnson, J. Lewis, and D. J. Yarrow *Chem. Comm.*, 1971, 621.
[534] A. Brookes, S. A. R. Knox, and F. G. A. Stone, *J. Chem. Soc.* (*A*), 1971, 3469.

Ruthenium Carbonyl Complexes (*cont.*)

		Ref.
$[Ru(CO)_2(NO)(PPh_3)_2]^+$		514
$Ru(CO)_2(PPh_3)_2(Me_3Ge)_2$		530
$[Ru(CO)_2X_2]_n$ } $[Ru(CO)_2X_4]^{2-}$ }	X = Cl, Br, or I	532
cis- and *trans-*$[Ru(CO)_2H(CNR)(PPh_3)_2]^+$ $Ru(CO)_2(CNR)(PPh_3)_2$		} 535
$[Ph_4As]_2$[*cis-* and *trans-*$RuCl_4(CO)_2$]		536
Ru(CO)(mp), Ru(CO)(tpp) Ru(CO)(mp)L Ru(CO)(tpp)L′	mp = mesoporphyrin IX dimethyl ester; tpp = tetraphenylporphin; L′ = imidazole; L = imidazole, 4,5-dimethylimidazole, or 3,5-dimethylpyrazole	} 537
$Ru(CO)Cl_2(Ph_3P)_2$(4-vinylcyclohexene)		538
Ru(CO)(imidazole)(tetraphenylporphin)		539
$Ru(CO)(dtt)Cl(PPh_3)_2$ } $Ru(CO)(dtt)H(PPh_3)_2$ }	dtt = 1,3-di-*p*-tolyltriazenido	540*a*
$Ru(CO)X_3$ $Ru(CO)X_5^{2-}$	X = Cl, Br, or I X = Cl or Br	} 532
$Ru(CO)(Cl)H(CNR)(PPh_3)_2$ $Ru(CO)H(OClO_3)(CNR)(PPh_3)_2$ $[Ru(CO)H(CNR)(PPh_3)_3]^+$ $Ru(CO)(CNR)(PPh_3)_3$ $Ru(CO)O_2(CNR)(PPh_3)_2$ $Ru(CO)(CO_3)(CNR)(PPh_3)_2$		} 535
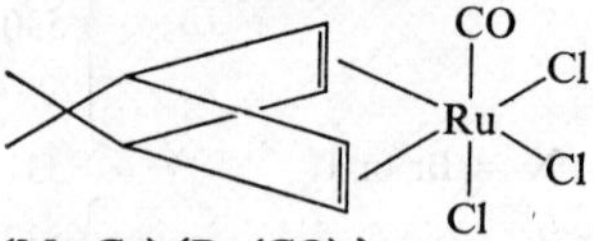		536
$(Me_2Ge)_3\{Ru(CO)_3\}_2$ $[(Me_3Ge)Ru(CO)_3(GeMe_2)]_2$		} 530
$(sem)Ru_2(CO)_6$	sem = 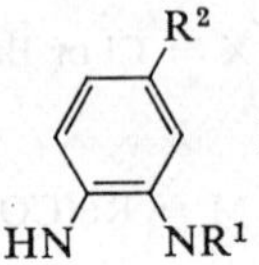 R^1 = Ph, R^2 = H	} 541
$[(azb)Ru(CO)_3]_2$	azb = 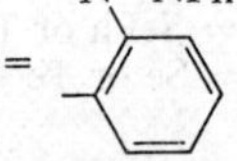	
$[(Me_3M)Ru(CO)_3X]_2$	M = Si, Ge, or Sn; X = Br or I	531

[535] D. F. Christian and W. R. Roper, *Chem. Comm.*, 1971, 1271.
[536] T. A. Stephenson and E. Switkes, *Inorg. Nuclear Chem. Letters*, 1971, **7**, 805.
[537] M. Tsutsui, D. Ostfeld, J. N. Francis, and L. M. Hoffman, *J. Co-ordination Chem.*, 1971, **1**, 115.
[538] J. E. Lyons, *Chem. Comm.*, 1971, 562.
[539] M. Tsutsui, D. Ostfeld, and L. M. Hoffman, *J. Amer. Chem. Soc.*, 1971, **93**, 1821.
[540a] S. D. Robinson and M. F. Uttley, *Chem. Comm.*, 1971, 1315.
[540b] J. V. Kingston and G. R. Scollary, *J. Inorg. Nuclear Chem.*, 1971, **33**, 4373.
[541] M. I. Bruce, M. Z. Iqbac, and F. G. A. Stone, *J. Organometallic Chem.*, 1971, **31**, 275.

Ruthenium Carbonyl Complexes (*cont.*)

		Ref.
$(sem)Ru_3(CO)_9$		
$(H_2sem)Ru_3(CO)_8$	H_2sem = 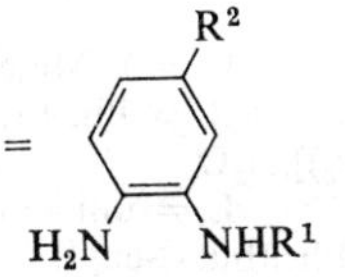 R^1 = Ph, R^2 = H	541
$(H_2sem)_3Ru_2(CO)_4$		
$(H_2sem)_2Ru_2(CO)_4$		
$[(Me_2Ge)Ru(CO)_3]_3$		530
$H_4Ru_4(CO)_{12}$		542
$H_2D_2Ru_4(CO)_{12}$		
$Ru_3(CO)_7(C_{10}H_8)$		543
$H_4Ru_4(CO)_8\{P(OMe)_3\}_4$		
$H_4Ru_4(CO)_9\{P(OMe)_3\}_3$		544
$H_4Ru_4(CO)_{10}\{P(OMe)_3\}_2$		
$H_4Ru_4(CO)_{11}\{P(OMe)_3\}$		

Osmium Carbonyl Complexes

$(Me_3Ge)Os(CO)_4H$		530
$(Me_3Ge)Os(CO)_4$		
$(Me_3M)Os(CO)_4X$	M = Si, Ge, or Sn; X = Br or I	531
$(Me_3M)Os(CO)_3X(PPh_3)$		
$[(Me_3Si)Os(CO)_3(\mu\text{-}SiMe_2)]_2$		534
$[Os(CO)_2(NO)(PPh_3)_2]^+$		514
$Os(CO)_2\{(MeO)_3P\}_2X$	X = C_2F_4, C_2F_3H, or C_2F_3Cl	
$Os(CO)_2(PhPMe_2)_2C_2F_3H$		545
$Os(CO)_2C(CF_3)_2O(PhPMe_2)_2$		
$Os(CO)_2Cl\{C(CF_3){=}C(CF_3)H\}\{(MeO)_3P\}_2$		
$(Me_3Ge)_2Os(CO)_2(PPh_3)_2$		530
$Os(CO)(NO)(PPh_3)_2CO_2Me$		514
$Os(CO)(dtt)H(PPh_3)_2$	dtt = 1,3-di-*p*-tolyltriazenido	540*a*
$[(Me_3M)Os(CO)_3X]_2$	M = Si, Ge, or Sn; X = Br or I	531
$[(Me_2Ge)Os(CO)_3]_3$		
$(Me_2Ge)_3\{Os(CO)_3\}_2$		530
$[(Me_3Ge)Os(CO)_3(GeMe_2)]_2$		
$[(Me_2Ge)Os(CO)_3]_3$		
$H_4Os(CO)_{12}$		542

Cobalt Carbonyl Complexes

$Co(CO)_4(GePh_2X)$		
$Co(CO)_4(GeX_3)$	X = F or Cl	452
$Co(CO)_4(SiX_3)$		
$Co(CO)_4NH_4$		546
$Er[Co(CO)_4]_3,4THF$		547

542 H. D. Kaesz, S. A. R. Knox, J. W. Koepke, and R. B. Saillant, *Chem. Comm.*, 1971, 477.
543 M. R. Churchill, F. R. Scholer, and J. Wormald, *J. Organometallic Chem.*, 1971, **28**, C21.
544 S. A. R. Knox and H. D. Kaesz, *J. Amer. Chem. Soc.*, 1971, **94**, 4594.
545 M. Cooke, M. Green, and T. A. Kuc, *J. Chem. Soc. (A)*, 1971, 1200.
546 I. Rhee, M. Ryang, and S. Tsutsumi, *Bull. Chem. Soc. Japan*, 1971, **44**, 2552.
547 R. S. Marianelli and M. T. Durney, *J. Organometallic Chem.*, 1971, **32**, C41.

Cobalt Carbonyl Complexes (*cont.*)

		Ref.
$Co(CO)_3H(PBu^n{}_3)$		548
$Co(CO)_2L$	L = π-$Me_3SiC_5H_4$	450
$Co(CO)_2(CN)(PR_3)_2$	R = Ph, Et, or Cy; Cy = C_6H_{11}	549
$Et_4N[Co(CO)_2(CN)_2(PCy_3)],H_2O$		549
$Co(CO)_2(OClO_3)L_2$	L = unidentate neutral ligand	550*a*
anti- and *syn*-$Co(CO)_2(PPh_3)$(but-2-enyl)		551
$Co(CO)_2(PPh_3)$(3-methylbut-2-enyl)		551
$[Co(CO)_2(PR_3)]_3$	R = Ph or n-C_4H_9	552
$Co(CO)_2X\{P(n\text{-}C_4H_9)_3\}_2$	X = H, D, or I	553
Co(CO)(NO)(das)	das = *o*-phenylenebisdimethylarsine	435
$Co(CO)(H)(\pi\text{-}C_5H_5)(Cl_3Si)$		446
$Et_4N[Co(CO)(CN)_2(PPh_3)_2],2H_2O$		549
$K[Co(CO)(CN)_2(PMePh_2)_2],3H_2O$		549
$K[Co(CO)(CN)_2(PMe_2Ph)_2],3H_2O$		549
$K[Co(CO)(CN)_2(PEt_3)_2],MeCOMe,H_2O$		549
$K[Co(CO)(CN)_2(Ph_2PCH_2CH_2PPh_2)],H_2O$		549
$Na[Co(CO)(PEt_2Ph)_3]$		554
$Co(CO)(H)(PEt_2Ph)_3$		554
$Zn_2Co_4(CO)_{15}$		555
$R_2SiCo_4(CO)_{14}$	R = Et or Ph	556
$PhCCo_3(CO)_8L$, $MeCCo_3(CO)_8L$	L = Ph_3P, Ph_3As, $Bu^n{}_3P$, $Bu^n{}_3As$, $PhEt_2P$, or Ph_2MeAs	557
$XCCo_3(CO)_8L$	X = F, Cl, or Br; L = Ph_3P, Ph_3As, or $(C_6H_{11})_3P$	557
$RCCo_3(CO)_7L_2$	R = Me or Ph; L = Ph_3P, Ph_3As, or $(C_6H_{11})_3P$	557
$FCCo_3(CO)_7L_2$	L = Et_2PhP or PPh_3	557
$RCCo_3(CO)_6(Et_2PhP)_3$	R = Me or Ph	557
$YCCo_3(CO)_6(\pi\text{-arene})$	Y = Me, Ph, or F; arene = mesitylene, *o*-, *m*-, or *p*-xylene, benzene, or toluene	558

Rhodium Carbonyl Complexes

$[Rh(CO)_3(PPh_3)_2]^+$		559
$Rh(CO)_2(\pi\text{-}Me_3Si\cdot C_5H_4)$		475
$[Rh(CO)_2(PPh_3)_3]^+A^-$	A = ClO_4^-, PF_6^-, or BPh_4^-	559

548 R. Whyman, *J. Organometallic Chem.*, 1971, **29**, C36.
549 J. Bercaw, G. Guastalla, and J. Halpern, *Chem. Comm.*, 1971, 1594.
550*a* J. Peone and L. Vaska, *Angew. Chem. Internat. Edn.*, 1971, **10**, 511.
550*b* D. N. Cash and R. O. Harris, *Canad. J. Chem.*, 1971, **49**, 3821.
551 G. Vitulli, L. Porri, and A. L. Segre, *J. Chem. Soc.* (*A*), 1971, 3246.
552 G. F. Pregaglia, A. Andreetta, G. F. Ferrari, G. Montrasi, and R. Ugo, *J. Organometallic Chem.*, 1971, **33**, 73.
553 G. F. Pregaglia, A. Andreetta, G. F. Ferrari, and R. Ugo, *J. Organometallic Chem.*, 1971, **30**, 387.
554 M. Aresta, C. F. Nobile, M. Rossi, and A. Sacco, *Chem. Comm.*, 1971, 781.
555 J. M. Borlitch and S. E. Hayes, *J. Organometallic Chem.*, 1971, **29**, C1.
556 S. A. Fieldhouse, A. J. Cleland, B. H. Freeland, C. D. M. Mann, and R. J. O'Brien, *J. Chem. Soc.* (*A*), 1971, 2536.
557 T. W. Matheson, B. H. Robinson, and W. S. Tham, *J. Chem. Soc.* (*A*), 1971, 1457.
558 B. H. Robinson and J. L. Spencer, *J. Chem. Soc.* (*A*), 1971, 2045.
559 R. R. Schrock and J. A. Osborn, *J. Amer. Chem. Soc.*, 1971, **93**, 2397.

Rhodium Carbonyl Complexes (*cont.*)

		Ref.
$Rh(CO)_2ClL$	L = PMe_3, PMe_2Ph, PPh_3, or $P(NMe_2)_3$	560
$Rh(CO)_2(PPh_3)_2(C_6F_4R)$	R = F, NC or Et_2OC	
$Rh(CO)_2(PPh_3)_2\{(NC)_2C_6F_3\}$		
$Rh(CO)_2(PPh_3)_2(C_5F_4N)$		561
4-$\{2\text{-}(NC)C_5F_3N\}Rh(CO)_2(PPh_3)_2$		
3-$\{6\text{-}(NC)C_5F_3N\}Rh(CO)_2(PPh_3)_2$		
cis-$[Rh(CO)_2Cl(diaz)]$	diaz = 3,5,7-triphenyl-4*H*-1,2-diazepine, co-ordinated *via* one N atom only	562
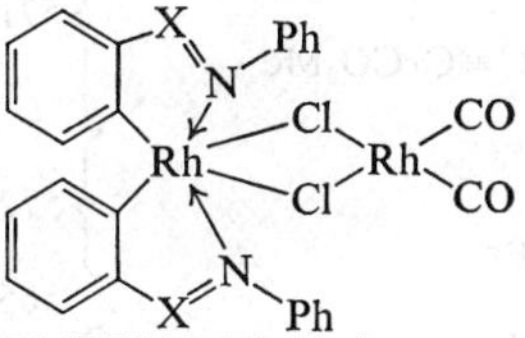	X = N or CH	563
$Rh(CO)(PPh_3)_2(dpt)$	dpt = 1,3-diphenyltriazenido	540*a*
$[Rh(CO)_2(SPh)]_2$		540*b*
$[Rh(CO)(SR)X]$	X = Cl or Br; R = Et or Pr	
$[Rh(CO)(1,3\text{-butadiene})(PPh_3)_2]^+$		559
$Rh(CO)(OClO_3)(L_2)$	L = unidentate neutral ligand	550*a*
$Rh(CO)(\pi\text{-}Cp)(H)SiX_3$	X = Ph or CH_2Ph	
$Rh(CO)(\pi\text{-}Cp)X_3$	X = $SiCl_3$ or $SiCl_2Me$	564
$Rh(CO)(\pi\text{-}Cp)(X)MX_3$	M = Ge or Sn; X = Br, Cl, or I	
$[Rh(CO)(PEt_2Ph)_3]BPh_4$		
$[Rh(CO)(PMePh_2)_3]PF_6$		
$[Rh(CO)(PMe_2Ph)_4]BPh_4$		565
$[Rh(CO)(AsMe_2Ph)_4]PF_6$		
$Rh_2(CO)_3Cl_2L$	L = PMe_3, PMe_2Ph, PPh_3, or $P(NMe_2)_3$	560
$Rh_4(CO)_{12-n}(X)_n$ (n = 2, 3, or 4)	X = PPh_3 or etpo (4-ethyl-2,6,7-trioxa-1-phosphabicyclo[2,2,2]octane)	
$Rh_4(CO)_{10}(PhC_2Ph)$		566
$Rh_4(CO)_{10}(CF_3C_2CF_3)$		
$Rh_6(CO)_{10}R_6$	R = PPh_3, etpo, or $P(OMe)_3$	

Iridium Carbonyl Complexes

C_5Me_2Ir carbonyl complexes		567
$Ir(CO)_3(H)R$	R = PPh_3, PEt_3, $PPr^i{}_3$, $PBu^n{}_3$, or $P(p\text{-}Me\cdot C_6H_4)_3$	548
$Ir(CO)_2Cl_2LR$	R = Et or COEt; L = $AsMe_2Ph$	568
$Ir(CO)_2(PPh_3)_2(O_2S\text{-}p\text{-tolyl})$		569
$Ir(CO)Cl_2RL_2$	R = Et or COEt; L = $AsMe_2Ph$	568

[560] J. Galloy, D. de Montauzon, and R. Poilblanc, *Compt. rend.*, 1971, **273**, *C*, 988.
[561] B. L. Booth, R. N. Haszeldine, and I. Perkins, *J. Chem. Soc.*(*A*), 1971, 927.
[562] R. A. Smith, D. P. Madden, A. J. Carty, and G. J. Palenik, *Chem. Comm.*, 1971, 427.
[563] M. I. Bruce, B. L. Goodall, M. Z. Iqbal, and F. G. A. Stone, *Chem. Comm.*, 1971, 661.
[564] A. J. Oliver and W. A. G. Graham, *Inorg. Chem.*, 1971, **10**, 1.
[565] L. M. Haines and E. Singleton, *J. Organometallic Chem.*, 1971, **30**, C81.
[566] B. J. Booth, M. J. Else, R. Field, and R. N. Haszeldine, *J. Organometallic Chem.*, 1971, **27**, 119.
[567] J. W. Kang and P. M. Maitlis, *J. Organometallic Chem.*, 1971, **26**, 393.
[568] R. W. Glyde and R. J. Mawby, *Inorg. Chim. Acta*, 1971, **5**, 317.
[569] C. A. Reed and W. R. Roper, *Chem. Comm.*, 1971, 1556.

Iridium Carbonyl Complexes (*cont.*)

		Ref.
$[Ir(CO)L_2R]^+A$	L = PMe_2Ph; R = C_2H_4, PhC≡CPh, C_4H_6, C_5H_8, or C_6H_{10}; A = BPh_4^-, ClO_4^-, or BF_4^-	570
$Ir(CO)(OClO_3)L_2$	L = unidentate neutral ligand	550*a*
$Ir(CO)(PPh_3)_2(NO_3)L$	L = $(NO_3)Ag$, O_2, SO_4, SO_2, tcne, Cl_2, Br_2, I_2, or Cl(HgCl)	550*b*
$Ir(CO)(PPh)_2(C{\equiv}CR)$	R = Me or Bu^t	
$Ir(CO)O_2(PPh_3)_2(C{\equiv}CEt)$		
$Ir(CO)(SO_2)(PPh_3)_2(C{\equiv}CMe)$		571
$Ir(CO)(PPh_3)_2(C{\equiv}CMe)X$	X = tcne or $MeO_2C{\cdot}C{\equiv}C{\cdot}CO_2Me$	
$Ir(CO)(PPh_3)_2(MeI)(C{\equiv}CBu^n)$		
$Ir(CO)(Et_3P)_2(MMe_3)_2(HgMMe_3)$	M = Si or Ge	572
$Ir(CO)(Et_3P)_2(MMe_3)_2(HgX)$	M = Ge, X = Cl or Br	
$[Ir(CO)(PR)_3]PF_6$	R = Et_2Ph or $MePh_2$	565
$[Ir(CO)(PPh_3)_2X]^+$	X = Me_2SO or MeCN	
$[Ir(CO)(PPh_2R)_3]^+$	R = Ph, Me, or Et	
$[Ir(CO)(O_2)(PPh_2R)_3]^+$	R = Me or Et	573
$[Ir(CO)(PPh_2Me)_3L]^+$	L = SO_4 or NO_3	
$[Ir(CO)(H_2)(MeCN)(PPh_3)_2]^+$		
$[Ir(CO)(PPh_3)_2X]$	X = OH, F, or CN	
$Ir(CO)(O_2S$-*p*-tolyl$)(PPh_3)_2$		
$Ir(CO)O_2(O_2S$-*p*-tolyl$)(PPh_3)_2$		569
$Ir(CO)(SO_4)(O_2S$-*p*-tolyl$)(PPh_3)_2$		
$Ir(Me)I(CO)(O_2S$-*p*-tolyl$)(PPh_3)_2$		
$Ir_2(CO)_6(PPh_3)_2$		548

Nickel Carbonyl Complexes

$Ni(CO)_3L$	L = PF_2Bu^t or $PFBu^t_2$	422
$Ni(CO)_2(das)$	das = *o*-phenylenebisdimethylarsine	435
$Ni(CO)_2\{P(OPh)_3\}_2$		574
$Ni(CO)\{P(OPh)_3\}_3$		

Platinum Carbonyl Complexes

trans-$[Pt(CO)Me(AsMe_3)_2]PF_6$		575
$[Pt\{(CO)(C_{10}H_{12}{\cdot}OMe)\}(PPh_3)X(CO)]$	X = Cl, Ac, or Ph	576
$\{Pt(CO)X_2\}L$	L = en, $enMe_4$, (4,4′-bipy), or 8-hydroxyquinolinato	577

Copper Carbonyl Complexes

Cu(CO)Cl, Cu(en)(CO)Cl		578
$[(en)Cu(CO)_2Cu(en)]Cl_2$		

570 A. J. Deeming and B. L. Shaw, *J. Chem. Soc.* (*A*), 1971, 376.
571 C. K. Brown and G. Wilkinson, *Chem. Comm.*, 1971, 70.
572 K. A. Hooton, *J. Chem. Soc.* (*A*), 1971, 1251.
573 G. R. Clark, C. A. Reed, W. R. Roper, B. W. Skelton, and T. N. Waters, *Chem. Comm.*, 1971, 758.
574 J. R. Olechowski, *J. Organometallic Chem.*, 1971, **32**, 269.
575 M. H. Chisholm, H. C. Clark, L. E. Manzer, and J. B. Stothers, *Chem. Comm.*, 1971, 1627.
576 G. Carturan, M. Graziani, and U. Belluco, *J. Chem. Soc.* (*A*), 1971, 2509.
577 T. Theophanides and P. C. Kong, *Inorg. Chim. Acta*, 1971, **5**, 485.
578 G. Rucci, C. Zanzottera, M. P. Lachi, and M. Camia, *Chem. Comm.*, 1971, 652.

Mixed Transition-metal Carbonyls

		Ref.
$M_2Ni_3(CO)_{16}\{(PPh_3)_2N\}_2$	M = Cr, Mo, or W	579
$(CO)_5M{-}EMe_2{-}E'Me_2{-}M'(CO)_5$	E,E′ = P or As; M,M′ = Cr or W	580
$(CO)_5M{-}EMe_2{-}M'(CO)_5$	M = Cr, Mo, or W; E = P or As; M′ = Mn or Re	581
$(C_5H_5)(CO)Mo\cdot\mu{-}C_5H_4\cdot Mn(CO)_4$		582
$(CO)_5M{-}M'(CO)_4{-}M''(CO)_5$	M = M″ = Mn, M′ = Fe; M = M″ = Re, M′ = Fe; M = Mn, M′ = Fe, M″ = Re	583
$[(\pi\text{-ring})Fe(CO)_2Co(CO)_4]$	ring = C_5H_5, MeC_5H_4, or indenyl	584
$[(\pi\text{-Cp})Ru(CO)_2Co(CO)_4]$		584
$H_4FeRu_3(CO)_{12}$		542
$(Me_3Si)Ru(CO)_4Re(CO)_5$		531
$[Me_4N][MOs_2(CO)_{12}]$	M = Mn or Re	585
$HMOs_2(CO)_{12}$	M = Mn or Re	585
$HMnOs_3(CO)_{13}$		585
$HReOs_3(CO)_{15}$		585
$H_3ReOs_3(CO)_{13}$		585

Ref. 586:

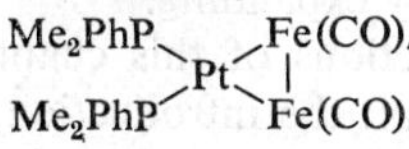

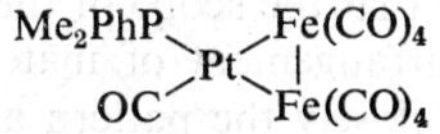

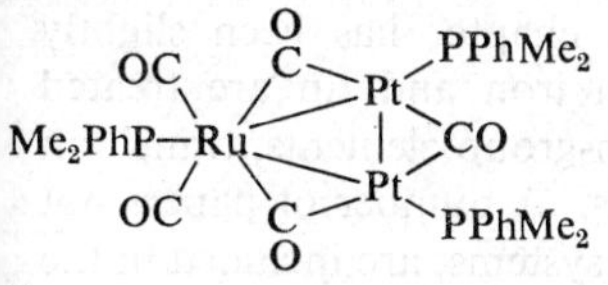

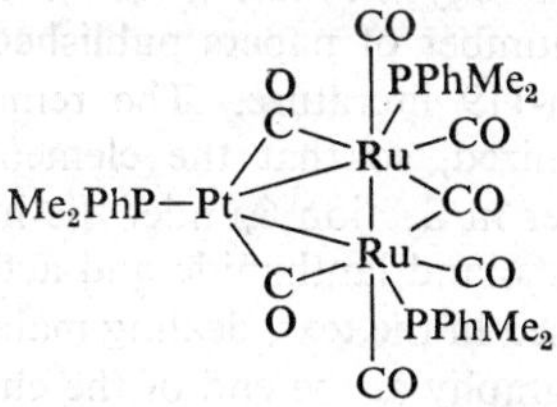

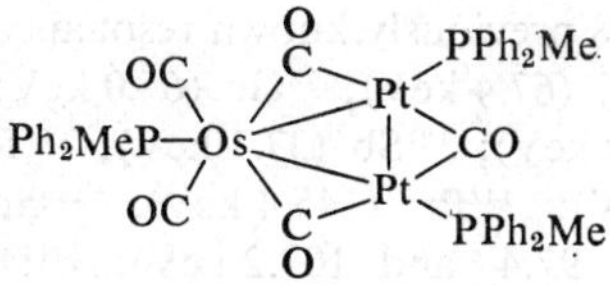

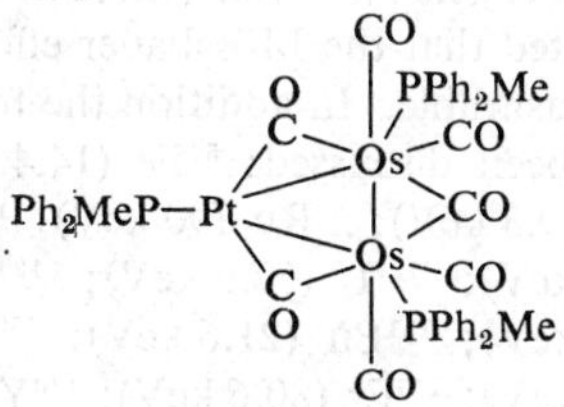

579 J. K. Ruff, R. P. White, and L. F. Dahl, *J. Amer. Chem. Soc.*, 1971, **93**, 2159.
580 H. Vahrenkamp and W. Ehrl, *Angew. Chem. Internat. Edn.*, 1971, **10**, 513.
581 W. Ehrl and H. Vahrenkamp, *Chem. Ber.*, 1971, **104**, 3261.
582 R. Hoxmeier, B. Deubzer, and H. D. Kaesz, *J. Amer. Chem. Soc.*, 1971, **93**, 536.
583 G. O. Evans and R. K. Sheline, *Inorg. Chem.*, 1971, **10**, 1598.
584 A. R. Manning, *J. Chem. Soc.* (*A*), 1971, 2321.
585 J. Knight and M. J. Mays, *Chem. Comm.*, 1971, 62.
586 M. I. Bruce, G. Shaw, and F. G. A. Stone, *Chem. Comm.*, 1971, 1288.

8
Mössbauer Spectroscopy

BY R. GREATREX

1 Introduction

The number of papers on Mössbauer spectroscopy reported here is greater than for any previous year. New resonances and new phenomena associated with the technique (such as the effects of radiofrequency fields on ferromagnetic Mössbauer absorbers described on p. 503) are still being discovered, and reports of applications outside the fields of pure physics and chemistry (*e.g.* the study of ancient Greek pottery discussed on p. 554) indicate that the scope of the technique is continually expanding.

The arrangement of material in the first five sections of this chapter follows closely the pattern adopted in previous years: the introduction is followed by sections on theoretical aspects and on instrumentation and methodology; Section 4 deals with iron-57 and refers to nearly half of the total number of papers published during the year, and Section 5 reviews the tin-119 literature. The remainder of the chapter has been slightly reorganized, so that the elements other than iron and tin are treated together in Section 6 under the headings: main-group elements, transition elements, and lanthanide and actinide elements. A number of papers not discussed in the text, dealing mainly with alloy systems, are included in the Bibliography at the end of the chapter.

Three new resonances have been reported during the year, namely ^{101}Ru (127 keV), ^{147}Sm (121.2 keV), and ^{180}W (104 keV), and it has been predicted that the Mössbauer effect on the 13.34 keV level of ^{73}Ge should be measurable. In addition the following 38 previously known resonances have been discussed: ^{57}Fe (14.4 keV), ^{61}Ni (67.4 keV); ^{73}Ge (67.0 keV); ^{83}Kr (9.3 keV); ^{99}Ru (90 keV); ^{119}Sn (23.9 keV); ^{121}Sb (37.1 keV); ^{125}Te (35.5 keV); ^{127}I (57.6 keV); ^{129}I (27.8 keV); ^{141}Pr (145.4 keV); ^{149}Sm (22.5 keV); ^{151}Eu (21.6 keV); ^{153}Eu (83.4, 97.4, and 103.2 keV); ^{161}Dy (25.7 keV); ^{166}Er (80.6 keV); ^{170}Yb (84.3 keV), ^{171}Yb (66.7 and 75.9 keV); ^{172}Yb (78.7 keV); ^{174}Yb (76.5 keV); ^{177}Hf (113.0 keV); ^{178}Hf (93.2 keV); ^{180}Hf (93.3 keV); ^{181}Ta (6.3 keV); ^{182}W (100.1 keV); ^{183}W (46.5 keV); ^{184}W (111.2 keV); ^{186}W (122.6 keV); ^{193}Ir (73.0 keV); ^{195}Pt (98.8 and 129.4 keV); ^{197}Au (77.3 keV); ^{201}Hg (32.2 keV); ^{232}Th (49.8 keV); and ^{237}Np (59.5 keV).

Two books on Mössbauer spectroscopy have been published during the year. One is a definitive monograph, which records the achievements of the technique at the end of its first decade of applications to chemical problems.[1] The other comprises a series of articles by experts, which introduce and explain the basic concepts of the Mössbauer effect:[2] an introductory chapter [3] is followed by sections on instrumentation,[4] nuclear properties,[5] and the electric field gradient tensor,[6] and on new applications of the Mössbauer effect in such areas as solid-state physics,[7] co-ordination chemistry,[8] organometallic chemistry,[9] physical metallurgy,[10] and biochemical systems.[11]

An important conference proceedings has appeared.[12] The volume contains some 90 papers and covers all aspects of Mössbauer spectroscopy. Much of the material has already been published in primary journals and has been discussed in previous volumes of this series; such papers are therefore merely listed in the Bibliography for completeness. However, the proceedings also contain useful reviews on the application of Mössbauer spectroscopy to the following topics: structural solid-state chemistry,[13] magnetism and crystal chemistry of garnets,[14] extraterrestrial materials,[15] crystal microdynamics,[16] surface phenomena,[17] Goldanskii–Karyagin effect,[18] and biological studies.[19] The past year has also seen the publication of two volumes in the well-established Mössbauer Effect Methodology series. Volume 6 [20] contains chapters on the current status of standards for Mössbauer spectroscopy [21] and on coincidence Mössbauer spectroscopy.[22] There is also a request by S. V. Karyagin that the phenomenon of integral asymmetry of Mössbauer quadrupole doublets, caused by anisotropy of atomic thermal vibrations, be called the Goldanskii–Karyagin effect and not simply the Karyagin effect.[23] A good quarter of

[1] N. N. Greenwood and T. C. Gibb, 'Mössbauer Spectroscopy', Chapman and Hall, London, 1971.
[2] 'An Introduction to Mössbauer Spectroscopy', ed. L. May, Hilger, London, 1971.
[3] P. G. Debrunner and H. Frauenfelder, 'Introduction to the Mössbauer Effect', ref. 2, p. 1.
[4] J. J. Spijkerman, 'Instrumentation', ref. 2, p. 23.
[5] D. W. Hafemeister, 'Nuclear properties determined from Mössbauer measurements', ref. 2, p. 45.
[6] J. C. Travis, 'The electric-field-gradient', ref. 2, p. 75.
[7] R. L. Ingalls, 'Application to solid-state physics', ref. 2, p. 104.
[8] J. Danon, 'Application to co-ordination chemistry', ref. 2, p. 120.
[9] R. H. Herber, 'Application to organometallic compounds', ref. 2, p. 138.
[10] U. Gonser, 'Mössbauer spectroscopy and physical metallurgy', ref. 2, p. 155.
[11] L. May, 'Application to biological systems', ref. 2, p. 180.
[12] 'Proceedings of the Conference on the Application of the Mössbauer Effect—Tihany (Hungary)', 1969, ed. I. Dézsi, Akadémiai Kiado, Budapest, 1971.
[13] C. E. Johnson, ref. 12, p. 663.
[14] I. S. Lyubutin, ref. 12, p. 469.
[15] A. H. Muir and M. Blander, ref. 12, p. 61.
[16] J. J. Bara, ref. 12, p. 93.
[17] V. I. Goldanskii and I. P. Suzdalev, ref. 12, p. 269.
[18] I. P. Suzdalev and E. F. Makarov, ref. 12, p. 201.
[19] G. Lang, ref. 12, p. 637.
[20] 'Mössbauer Effect Methodology', ed. I. J. Gruverman, Plenum Press, 1971, Vol. 6.
[21] R. H. Herber, ref. 20, p. 3.
[22] G. R. Hoy, D. W. Hamill, and P. P. Wintersteiner, ref. 20, p. 109.
[23] S. V. Karyagin, ref. 20, p. 17.

this volume is devoted to studies of Apollo 11 lunar samples;[24–27] however, all of this work was published elsewhere in 1970 and fully reviewed in last year's Specialist Periodical Report; it is therefore not discussed further in this chapter. Volume 7 [28] reviews electric field gradient calculations in ionic crystals,[29] relativistic effects in hyperfine interactions,[30] radiation damage studies,[31] and Mössbauer studies of polynuclear iron(III) compounds.[32] Other papers from refs. 12, 20, and 28 are mentioned at appropriate points later in this chapter.

Applications of Mössbauer spectroscopy to various aspects of solid-state chemistry have been authoratitively reviewed. The areas covered include the electronic structure of point defects,[33] the crystal chemistry and magnetic structures of substituted $Ca_2Fe_2O_5$,[34] and metal fluoride and metal oxide systems.[35] This last review includes many new results on the disordered rutile phases Fe_2MgF_6, $FeMg_2F_6$, and $FeCoNiF_6$, on the trirutile phase $Li^+Fe^{2+}Fe^{3+}F_6$, and on the non-stoicheiometric wüstite phase $Fe_{1-x}O$. The application of large external magnetic fields to diamagnetic compounds of tin and the results of Mössbauer studies of lunar samples returned by the Apollo 11 and 12 missions are also covered.[35] Several other applications to the solid-state have been reviewed.[36, 37]

Rudolf Mössbauer has discussed interpretations of chemical isomer shift and quadrupole splitting for active transition-metal isotopes,[38] and Mössbauer spectroscopy of ^{195}Pt has been surveyed.[39] There has been an extensive review of the applications of ^{119}Sn Mössbauer spectroscopy to the study of organotin compounds.[40] Applications to the study of co-ordination compounds,[40a] mixed-valence compounds,[41] and macro-

[24] A. J. Muir, ref. 20, p. 163.
[25] C. L. Herzenberg and D. L. Riley, ref. 20, p. 177.
[26] S. S. Hafner, B. Janik, and D. Virgo, ref. 20, p. 193.
[27] G. P. Huffman, G. R. Dunmyre, R. M. Fisher, P. J. Wasilewski, and T. Nagata, ref. 20, p. 209.
[28] 'Mössbauer Effect Methodology', ed. I. J. Gruverman, Plenum Press, 1971, Vol. 7.
[29] J. O. Artman, ref. 28, p. 187.
[30] B. D. Dunlap, ref. 28, p. 123.
[31] D. Schroeer, R. L. Lambe, and C. D. Spencer, ref. 28, p. 3.
[32] W. M. Reiff, ref. 28, p. 213.
[33] G. K. Wertheim, A. Hausmann, and W. Sander, 'The electronic structure of point defects as determined by Mössbauer spectroscopy and by spin resonance', in 'Defects in Crystalline Solids', ed. S. Amelinckx, R. Gevers, and J. Nihoul, North-Holland, Amsterdam, 1971, Vol. 4.
[34] R. Geller, R. W. Grant, and V. Gonser, 'Crystal chemistry and magnetic structures of substituted $Ca_2Fe_2O_5$' in *Progr. Solid-state Chem.*, 1971, Vol. 5.
[35] N. N. Greenwood, *Angew. Chem.*, 1971, **10**, 716.
[36] V. G. Bhide, *J. Sci. Ind. Res., India*, 1971, **30**, 160.
[37] D. Feldman, H. R. Kirchmayr, A. Schmolz, and M. Velicescu, *IEEE Trans. Magn.*, 1971, **7**, 61.
[38] R. L. Mössbauer, *Angew. Chem.*, 1971, **10**, 462.
[39] N. Benczer-Koller, *Adv. Chem. Ser.*, 1971, No. 98, pp. 135—149.
[40] J. J. Zuckerman, 'Applications of ^{119}Sn Mössbauer spectroscopy to the study of organotin compounds' in 'Advances in Organometallic Chemistry', ed. F. G. A. Stone and R. West, Academic Press, 1971.
[40a] M. L. Good and C. A. Clausen, 'Application of Mössbauer spectroscopy in the study of co-ordination compounds' in A.C.S. Monograph No. 168, ed. A. E. Martell, 1971.
[41] M. B. Robin, *Kagaku* (*Tokyo*), 1971, **41**, 121.

molecular compounds,[42] and semiconducting compounds of iron[43] have also been reviewed.

Other topics surveyed include chemical after-effects of nuclear reactions,[44, 45] applications to the study of relaxation in magnetic resonance,[45a] applications to the determination of time- and space-modified electron-shell parameters,[46] applications to biophysics,[47, 48] metal physics,[49] high pressure physics,[50] and chemical engineering,[51] and Mössbauer effect studies on magnetic thin films.[52] A number of reviews of a more general nature have also appeared.[53–56] The Mössbauer effect data index covering the 1969 literature has been published.[57]

2 Theoretical

The equivalence of the gravitational red shift and the temperature shift in the Mössbauer effect has been discussed further.[58] The Debye integrals, which enter into the theoretical expressions for the thermal shift and for the Mössbauer fraction, can be evaluated easily by means of infinite series.[59]

A dynamical theory for the propagation of Mössbauer radiation in magnetically ordered crystals has been developed. The results can be used to distinguish between ferromagnetic, weak ferromagnetic, and anti-ferromagnetic interactions, and between canted and helical structures.[60, 61]

General formulae have been given for direct calculations of the transition probabilities of polarized γ-rays in the case of coupled electric and magnetic hyperfine interactions. The expressions are applied to the specific case of the $\frac{3}{2} \rightarrow \frac{1}{2}$ transition.[62]

[42] V. I. Goldanskii, *Vysokomol. Soedineniya, (A)*, 1971, **13**, 311.
[43] J. P. Suchet and J. P. Senateur, 'Semin. Chim. Etat Solide', 1969–70, No. 4, pp. 67—81 (pub. 1971).
[44] S. I. Bondarevskii, A. N. Murin, and P. P. Seregin, *Uspekhi Khim.*, 1971, **40**, 95.
[45] G. K. Wertheim, *Accounts Chem. Res.*, 1971, **4**, 373.
[45a] C. P. Poole and H. A. Farach, 'Relaxation in Magnetic Resonance: Dielectric and Mössbauer Applications', Academic Press, New York and London, 1971.
[46] W. Meisel, *Z. Chem.*, 1971, **11**, 371.
[47] C. E. Johnson, *Phys. Today*, 1971, **24**, 35.
[48] C. E. Johnson, *J. Appl. Phys.*, 1971, **42**, 1325.
[49] W. Keune and A. Trautwein, *Metall.*, 1971, **25**, 138.
[50] A. Sawaoka, *Nippon Kinzoku Gakkai Kaiho*, 1971, **10**, 358.
[51] P. Gütlich and H. Prange, *Chem.-Ing.-Tech.*, 1971, **43**, 1049.
[52] W. Zinn, *Czech. J. Phys.*, 1971, **21**, 391.
[53] R. Wappling, *Kem. Tidskr.*, 1971, **83**, 38.
[54] F. W. Karasek, *Res. Develop.*, 1971, **22**, 18.
[55] R. Wappling, *Kem. Tidskr.*, 1971, **83**, 28.
[56] R. H. Herber, *Sci. Amer.*, 1971, **225**, 86.
[57] 'Mössbauer Effect Data Index covering the 1969 Literature' ed. J. G. Stevens and V. E. Stevens, IFI/Plenum, New York, 1971.
[58] G. P. Gupta and K. C. Lal, *Phys. Letters*, 1971, **36A**, 421.
[59] J. Herberle, ref. 28, p. 299.
[60] Y. M. Aivazyan and V. A. Belyakov, *Fiz. tverd. Tela*, 1971, **13**, 968.
[61] V. A. Belyakov, *Fiz. tverd. Tela*, 1971, **13**, 2170.
[62] D. Barb, S. Constantinescu, and D. Tarina, *Rev. Roumaine Phys.*, 1971, **16**, 263.

The theory for the intensity asymmetry of the ^{57}Fe quadrupole doublet in iron-bearing solutes in particular liquid-crystal systems has been extended. The order parameter of the Mössbauer impurity and the asymmetry parameter of the electric field gradient at the site of the resonant nucleus can be determined.[63] The effect of Brownian motion on the Mössbauer lineshape has also been treated theoretically.[64]

The effect of vacancy diffusion on Mössbauer line-broadening has been considered. The Mössbauer effect can be useful for determining the diffusion mechanism and for providing information about host-vacancy and impurity-vacancy exchange rates.[65] A theoretical expression has been derived for the change in the Debye–Waller factor for a cubic crystal disordered by the introduction of a vacancy-interstitial pair.[66] Other aspects of impurity atoms in crystals have been discussed.[67, 68]

The effect of a magnetic field on the hyperfine structure of a Mössbauer line when there is spin relaxation has been considered generally, and for the specific case of the high-spin iron(II) ion ($S = \frac{5}{2}$).[69]

A general analytical method has been outlined for determining the magnetic and quadrupole hyperfine parameters directly from an observed Mössbauer spectrum for the majority of the common nuclear spins. To illustrate the method, the hyperfine parameters for ^{129}I and ^{237}Np were obtained from spectra of IBr and $NpCl_4$ reported in the literature.[70] The determination of local field parameters from nuclear γ-resonance spectra has been discussed elsewhere.[71]

The concept of additivity of electric field gradients and the correlation of Mössbauer quadrupole splitting with stereochemistry has received theoretical treatment. The correlation is associated with a situation known as intermediate symmetry, in which the electric field gradient has higher symmetry than that required by the point symmetry of the complex. Underlying symmetry features responsible for the intermediate symmetry are elucidated and particular models (*e.g.* the 'point-charge model') for rationalizing the correlations are seen as manifestations of these symmetries.[72]

It has been shown that in general there is more than one value for each of the electric field gradient parameters, which give the same spectrum for a powder sample. The ambiguity disappears in certain situations, for

63 J. M. Wilson and D. L. Uhrich, *Mol. Crystals Liquid Crystals*, 1971, **13**, 85.
64 V. G. Bhide, R. Sundaram, H. C. Bhasin, and T. Bonchev, *Phys. Rev.* (*B*), 1971, **3**, 673.
65 R. C. Knauer, jun., *Phys. Rev.* (*B*), 1971, **3**, 567.
66 B. P. Srivastava, H. N. K. Sharma, and D. L. Bhattacharya, *Phys. Stat. Sol.* (*B*), 1971, **45**, 647.
67 A. Szczepánski, *Phys. Stat. Sol.* (*B*), 1971, **46**, K103.
68 A. Widom, *Solid State Comm.*, 1971, **9**, 1033.
69 V. V. Svetozarov, *Fiz. tverd. Tela*, 1971, **13**, 1263.
70 P. G. L. Williams and G. M. Bancroft, ref. 28, p. 39.
71 G. N. Belozerskii, V. N. Gittsovich, A. N. Murin, and Y. P. Khimich, *Fiz. tverd. Tela*, 1971, **13**, 2687.
72 M. G. Clark, *Mol. Phys.*, 1971, **20**, 257.

example when the magnetic field is either parallel or perpendicular to V_{zz}.[73] Elsewhere it has been pointed out that the commonly held belief that magnetically perturbed ^{57}Fe and ^{119}Sn Mössbauer spectra are far easier to calculate in the parallel arrangement than in the case where the γ-ray is incident perpendicular to the direction of the applied field need not be the case, provided the proper choice of co-ordinates and method of averaging is used.[74]

In normal curve-fitting procedures it is assumed that the velocity is known exactly for a particular channel position. In practice this situation is rarely achieved and each data point corresponds to a range of velocities. In addition there are often external factors (*e.g.* drive circuit drifts and non-linearities) which limit the precision to which the velocity is known. The effect of this 'velocity smearing' on a Lorentzian lineshape has been considered theoretically and it is recommended that the effects be taken into account when fitting Mössbauer spectra.[75] Techniques for correcting for the so-called 'blackness distortion' (*i.e.* thickness broadening) caused by finite absorber thickness have been described.[76, 77] Some statistical considerations relevant to reducing the counting-time necessary to locate Mössbauer lines have also been presented.[78] Two general-purpose computer programs for fitting Mössbauer spectra have been described,[79, 80] and another has been developed specifically for the analysis of spectra of heat-treated stainless steels.[81]

3 Instrumentation and Methodology

Two all-purpose spectrometers,[82, 83] an electronic system for Mössbauer spectrometry,[84] and a highly stabilized time-mode spectrometer employing crystal-controlled digital circuits[85] have been described. Electromechanical feedback systems may become regenerative after a sudden mechanical disturbance. This 'conditionally stable' behaviour is found to be due to a time lag in operational amplifiers when they are saturated. The systems can be made unconditionally stable by 'bounding' the output signal of the amplifier.[86] A Mössbauer drive which has bearings rather than springs

[73] L. Dabrowski, J. Piekoszewski, and J. Suwalski, *Nuclear Instr. and Meth.*, 1971, **91**, 93.
[74] G. Lang, *J. Chem. Soc.* (*A*), 1971, 3245.
[75] S. A. Wender and N. Hershkowitz, *Nuclear Instr. and Meth.*, 1971, **93**, 571.
[76] M. C. D. Ure and P. A. Flinn, ref. 28, p. 245.
[77] B. T. Cleveland and J. Herberle, *Phys. Letters*, 1971, **36A**, 33.
[78] A. J. C. Wilson, *Nuclear Instr. and Meth.*, 1971, **94**, 225.
[79] A. J. Stone, H. J. Aagaard, and J. Fenger, Danish Atomic Energy Commission, Risoe Report, 1971, RISO-M-1348.
[80] B. L. Chrisman and T. A. Tumolillo, *Computer Phys. Comm.*, 1971, **2**, 322.
[81] O. W. Albritton and J. M. Lewis, *Weld. Res.* (*New York*), 1971, 327.
[82] A. V. Dolenko, B. G. Egiazarov, and A. I. Shamov, *Pribory Tekhn. Eksp.*, 1971, 43.
[83] V. S. Indurkar and P. K. Patwardhan, Proceedings of the 15th Nuclear Physics and Solid State Physics Symposium, 1970 (pub. 1971), Vol. 2, p. 551.
[84] J. Thenard and G. Victor, *Nuclear Instr. and Meth.*, 1971, **93**, 311.
[85] H. Miyajima, *Japan J. Appl. Phys.*, 1971, **10**, 1405.
[86] H. de Waard, B. A. Pijpker, and P. F. Greve, *Nuclear Instr. and Meth.*, 1971, **94**, 195.

has been designed. This new drive avoids the use of high-gain feedback electronics required to reduce the inherent non-linearity of systems using springs or diaphragms.[87]

Mössbauer velocity calibration *via* laser interferometry has been discussed; a computer program for handling the data is given.[88] A system has been described in which the width of the black region of an ^{57}Fe black absorber is tripled by moving two absorbers with opposite velocities while keeping a third absorber stationary. Possible applications of the system, such as the precise determination of *f*-measurements, were discussed.[89]

Some general considerations on the effects of detectors, sources, and absorbers on the widths of spectral lines observed in Mössbauer experiments have been reviewed. The derivation of the observed lineshape for both conventional and resonant detectors was outlined, and the underlying reasons for the superior resolution of the resonant detector was explained.[90] The relative merits of transmission and back-scatter geometry have been discussed.[91] Various detection techniques can be employed in back-scattering experiments, namely: detection of re-emitted 14.4 keV radiation, detection of the 6.3 keV conversion *X*-rays, or detection of conversion electrons. The design of special detectors required for these techniques and specific applications of the conversion electron detector have been described.[92] Aspects of this work have been published previously and were detailed in last year's Report.

A method for cooling a moving source [93] and an economical method for varying the temperature of both source and absorber independently inside a 4.2 K helium bath [94] have been reported. An improved indium seal for low-temperature cryostats has been described.[95] A high-pressure cryostat has been constructed for *X*-ray scattering and Mössbauer measurements on solids at temperatures in the range 4.2—350 K and at pressures of 0—180 kbar.[96] An improved technique for measuring phase transitions of solids under such conditions has been developed. Results of Curie point studies for a dilute alloy of iron in palladium at 0 and 60 kbar were given.[97] A detailed description has appeared of the apparatus and operating procedures for studying the Mössbauer effect of nuclei embedded in rare-gas matrices at 4.2 K. The apparatus incorporated a furnace to create the

[87] J. Van Overbeeke, *J. Phys. (E)*, 1971, **4**, 548.
[88] J. G. Cosgrove and R. L. Collins, *Nuclear Instr. and Meth.*, 1971, **95**, 269.
[89] B. Kolk and B. Harwig, *Nuclear Instr. and Meth.*, 1971, **94**, 211.
[90] S. A. Schmidt, *Amer. J. Phys.*, 1971, **39**, 1003.
[91] R. N. Ord and C. L. Christensen, *Nuclear Instr. and Meth.*, 1971, **91**, 293.
[92] J. J. Spijkerman, ref. 28, p. 85.
[93] I. Ortalli, *Rev. Sci. Instr.*, 1971, **42**, 1739.
[94] J. Gal and J. Hess, *Rev. Sci. Instr.*, 1971, **42**, 543.
[95] M. Kuchnir, M. F. Adam, J. B. Ketterson, and P. Roach, *Rev. Sci. Instr.*, 1971, **42**, 536.
[96] K. Syassen, C. W. Christoe, and W. B. Holzapfel, *Z. angew. Phys.*, 1971, **31**, 261.
[97] C. W. Christoe, A. Forster, and W. B. Holzapfel, *Z. angew. Phys.*, 1971, **31**, 263.

necessary flux of atoms or molecules for injection into the rare-gas matrix.[98] A preliminary account of this apparatus and some accompanying data on ^{57}Fe in solid argon were mentioned in last year's Report. These results have also been fully discussed in a separate publication which is covered in a later section (see ref. 102). A miniature vacuum furnace for Mössbauer spectroscopy has been described.[99]

4 Iron-57

The plan of this section is very similar to that adopted in previous years. The first section contains some general topics. This is followed by a section on compounds of iron, which contains sub-sections on high-spin iron(II) compounds, high-spin iron(III) compounds, spin crossover and biological compounds, and low-spin and covalent compounds. Finally there is a section on oxide and sulphide systems containing iron, which has sub-sections on binary oxides, spinel oxides and garnets, other oxide systems, and minerals (including the lunar samples returned by the Apollo 12 mission). Metals and alloys are not discussed, but the relevant papers are listed in the Bibliography at the end of the chapter.

General Topics.—The large quantity of material relevant to this section is discussed in five sub-sections which deal in turn with nuclear parameters, hyperfine interactions, and new effects; pressure-dependence studies; lattice dynamics and relaxation phenomena; alloy-type systems and impurity studies; and finally, new ^{57}Co sources and decay after-effect phenomena.

Nuclear Parameters, Hyperfine Interactions, and New Effects. Previously reported values of the nuclear quadrupole moment, Q, of the first excited state of ^{57}Fe vary from 0.1 to 0.59 b in iron(III) compounds, with a concensus of *ca.* 0.3 b, whereas the Q values obtained from iron(II) data are close to 0.18 b. On the basis of new estimates of Q from the calculated electric field gradients and the quadrupole splitting data in Fe_2O_3 and Al_2O_3:Fe^{3+}, it is now suggested that this well-known iron(II)–iron(III) anomaly can be disposed of. The values of Q obtained in the two cases are 0.180 ± 0.015 and 0.204 ± 0.018 b, which are consistent with the values previously reported in iron(II) compounds.[100] An independent estimate of $Q = 0.16 \pm 0.03$ b has been arrived at by comparing the values of e^2qQ for *trans*-$[Co(NH_3)_4Cl_2]^+$, $[Co(NH_3)_5Cl]^{2+}$, and $[Co(NH_3)_5CN]^{2+}$, measured by nuclear quadrupole resonance, with those calculated from partial quadrupole splitting values for the corresponding hypothetical iron(II) compounds[101] (see also p. 520).

Studies on ^{57}Fe atoms in rare-gas matrices at temperatures 1.45—20.5 K have also led to an estimate of Q. The spectra show two features –

[98] T. K. McNab and P. H. Barrett, ref. 28, p. 59.
[99] V. S. Sundaram, V. P. Gupta, and E. C. Subbarao, *Rev. Sci. Instr.*, 1971, **42**, 1616.
[100] R. R. Sharma, *Phys. Rev. Letters*, 1971, **26**, 563.
[101] G. M. Bancroft, *Chem. Phys. Letters*, 1971, **10**, 449.

a single line ascribed to an isolated ^{57}Fe atom (monomer) with an atomic configuration $3d^64s^2$, and a doublet believed to arise from an iron dimer. From the values of the quadrupole splitting of the latter a value of $Q = 0.213$ b was deduced. In addition, the measured isomer shift for the monomer relative to metallic iron ($\delta = -0.75 \pm 0.03$ mm s^{-1}) gives a new calibration point in the isomer shift *vs.* electron density plot for ^{57}Fe (see also ref. 98) and leads to a value for the fractional change in the nuclear charge radius upon excitation, $\Delta R/R = -(10.2 \pm 0.3) \times 10^{-4}$, in good agreement with previous estimates.[102]

α-Iron is cubic only in the paramagnetic state and undergoes a lattice distortion (magnetostriction) upon ferromagnetic ordering. The quadrupole coupling predicted to be associated with this distortion has now been established as a result of some very precise experimental work.[103, 104] The line positions in the spectrum of N.B.S. standard reference material 1541 α-iron foil determined in this study are given in Table 1, as they may

Table 1

Absorber temperature/K	Line 1	2	3	4	5	6
298	− 5.4823 (8)	− 3.2473 (8)	− 1.0132 (10)	0.6624 (7)	2.8967 (7)	5.1338 (10)
4.3	− 5.5006 (20)	− 3.2099 (9)	− 0.9175 (8)	0.8012 (8)	3.0948 (12)	5.3938 (13)

prove useful for spectrometer calibration. The values are quoted in mm s^{-1} with respect to a ^{57}Co/Pd source at 298 K and may be converted to the sodium nitroprusside scale by adding 0.4361 ± 0.0009 mm s^{-1}. The figures in parentheses are standard deviations and are associated with the final decimal places. The magnetic hyperfine field at the iron nucleus is -330 ± 0.3 kG at 298 K and -339.0 ± 0.3 kG at 4.3 K and the excited-state magnetic moment is -0.1547 ± 0.001 nm. At 298 K the electric quadrupole coupling, e^2qQ, is either greater than $+0.0023 \pm 0.0015$ mm s^{-1} or less than -0.0046 ± 0.0030 mm s^{-1}, and at 4.3 K is either greater than $+0.0088 \pm 0.0025$ mm s^{-1} or less than -0.0176 ± 0.0050 mm s^{-1}.

Hartree–Fock calculations on a number of ionic states of iron form the basis of a new interpretation of the ^{57}Fe isomer shift. Experimental isomer shifts were compared with estimated configurations and these were compared with available MO calculations for a few cases. Specific species considered were K_2FeO_4, $FeCl_4^-$, $FeCl_4^{2-}$, FeF_6^{3-}, $Fe(CO)_5$, $Fe(CO)_4^{2-}$, $Fe(\pi\text{-}C_5H_5)_2$, and haemoglobin and porphin complexes.[105] The ^{57}Fe

[102] T. K. McNab, H. Micklitz, and P. H. Barret, *Phys. Rev.* (*B*), 1971, **4**, 3787.
[103] J. J. Spijkerman, J. C. Travis, D. N. Pipkorn, and C. E. Violet, *Phys. Rev. Letters*, 1971, **26**, 323.
[104] C. E. Violet and D. N. Pipkorn, *J. Appl. Phys.*, 1971, **42**, 4339.
[105] J. Blomquist, B. Roos, and M. Sundbom, *J. Chem. Phys.*, 1971, **55**, 141.

isomer shift has also been discussed elsewhere and a modified Walker–Wertheim–Jaccarino plot given.[106] The effective electronic population of the iron ion in the complexes $K_3[FeF_6]$, $K_3[Fe(CN)_6]$, $K_4[Fe(CN)_6]$, and $Na_2[Fe(CN)_5NO]$ have been estimated by combining *X*-ray emission K_α shifts with Mössbauer isomer shifts. It was suggested that the effect of $4p$ electrons on the isomer shift cannot be safely neglected, although it is probably small for most iron complexes.[107]

Previous correlations between quadrupole splitting and isomer shift have been extended to highly ionic and highly covalent materials. The former (*e.g.* $FeCO_3$) exhibit a small range of isomer shift whereas the latter [*e.g.* $Fe(CN)_6^{3-}$] show little variation in quadrupole splitting. For intermediate compounds (*e.g.* FeS) the quadrupole splitting values increase linearly with increase in isomer shift, forming a continuous progression between the two extremes. These effects are discussed in terms of the expansion and contraction of the $3d(t_{2g})$ and $3d(e_g)$ orbitals of iron.[108–110] In addition, correlations of the Fermi contact interaction (H_f) and the isomer shift in the compounds $K_3Fe(CN)_6$, FeS, FeI_2, $FeBr_2$, $FeCl_2$, $FeTiO_3$, $FeCO_3$, and FeF_2 extrapolate to a value of $H_f = -550$ kG for the free Fe^{2+} ion, in agreement with theory, and also confirm the calculated value $\langle r^{-3} \rangle = 5.1$ a.u. for this species.[108, 110]

It has been pointed out that subtle effects of structure and bonding need not necessarily be reflected by changes in the observed quadrupole splitting. However, V_{zz} and η can vary in a complex fashion and their behaviour can be examined using magnetic perturbation or oriented-single-crystal techniques. The distorted octahedral compound $[Fe(phen)_2(N_3)_2]$, which is known to have a negative V_{zz} and small η, was discussed in relation to this theoretical study.[111]

The relative orientations of the electric field gradient and susceptibility tensors in monoclinic symmetry have been discussed. Specific calculations were carried out for the $d\varepsilon^4 d\gamma^2$ and $d\varepsilon^5$ configurations. $K_3Fe(CN)_6$, $Na_2Fe(CN)_5NH_3$ diluted in DMF, and $FeCl_2,4H_2O$ were presented as examples and Mössbauer spectra were calculated to show the effects of monoclinic symmetry.[112]

Experiments to study the effects of radiofrequency fields on ferromagnetic Mössbauer absorbers have led to the discovery of two new physical phenomena,[113–115] both of which are displayed in Figure 1. When an iron absorber is subjected to an r.f. field of a few gauss, extra lines are

[106] L. Korecz, ref. 12, p. 725.
[107] J. Blomquist, B. Roos, and M. Sundbom, *Chem. Phys. Letters*, 1971, **9**, 160.
[108] Y. Hazony, *Phys. Rev.* (*B*), 1971, **3**, 711.
[109] Y. Hazony and R. C. Axtmann, *Chem. Phys. Letters*, 1971, **8**, 571.
[110] Y. Hazony, ref. 28, p. 147.
[111] J. G. Cosgrove and R. L. Collins, *J. Chem. Phys.*, 1971, **55**, 4238.
[112] W. T. Oosterhuis, *Phys. Rev.* (*B*), 1971, **3**, 546.
[113] N. D. Heiman, J. C. Walker, and L. Pfeiffer, ref. 20, p. 123.
[114] L. Pfeiffer, ref. 28, p. 263.
[115] L. Pfeiffer, *J. Appl. Phys.*, 1971, **42**, 1725.

produced, displaced from the main lines by multiples of the applied frequency. These are thought to be frequency-modulated sidebands caused by magnetostrictively produced vibrations.[113, 114] The second effect involves extending the technique of r.f. hyperfine enhancement to the

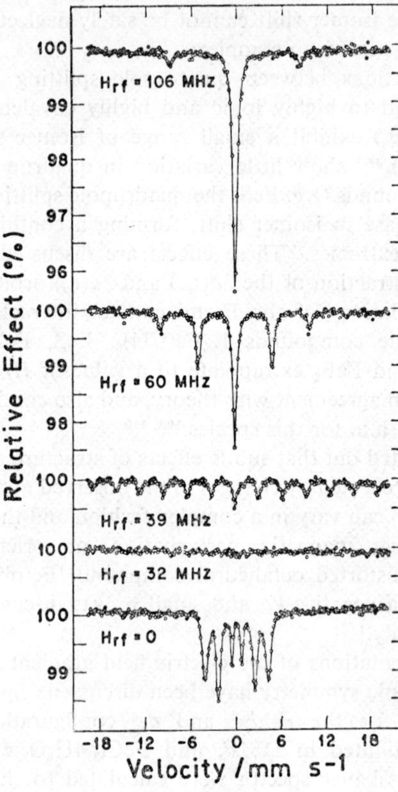

Figure 1 *Mössbauer spectra showing* rf *collapse as a function of the frequency of the applied field. In the lowest spectrum the* rf *applied field is zero; for all others* H_{rf} = 15 G

(Reproduced by permission from *J. Appl. Phys.*, 1971, **42**, 1725)

extreme, so that the entire magnetic hyperfine field at the iron nucleus is forced to oscillate synchronously through large angles in response to the r.f. driving field. The result is that the time-averaged field as seen by the iron nuclei is reduced to zero and the six-line spectrum collapses to a singlet.[114, 115]

The feasibility of using Mössbauer scattering of γ-radiation in order to measure directly the phase of the structure factor which is required in *X*-ray crystal-structure determination has been successfully tested experimentally. The phases of the (020) and the (040) reflections of a $K_3Fe(CN)_6$

single crystal were determined by measuring the interference between nuclear resonance scattering and electronic scattering.[116]

Pressure-dependence Studies. There have been a number of studies in which the pressure dependence of the electronic structure of iron ions has been followed by Mössbauer spectroscopy. For convenience these are now discussed together. The first clear-cut measurement of a high- to low-spin conversion as a function of pressure is shown in Figure 2, which shows spectra of iron as a dilute substitutional impurity in MnS_2. The transition begins near 40 kbar and is complete by *ca.* 120 kbar. (1 atm = 1.013 bar; 1 bar = 10^5 N m^{-2} = 10^5 kg m^{-1} s^{-2}). MnS_2 has the cubic structure of pyrites, FeS_2, in which the iron is low-spin at all pressures.[117]

The mononuclear low-spin iron(III) compound $Fe(phen)_3(ClO_4)_3,2H_2O$ is rapidly and irreversibly reduced to low-spin iron(II) with increasing pressure, the reduction being complete at 15 kbar. By contrast, the single oxo-bridged dimer $Fe_2O(phen)_4Cl_4,4H_2O$, which contains high-spin iron(III) at atmospheric pressure, is only partially reduced to an iron(II) species, even at 167 kbar; the reduction is reversible with some hysteresis. However, this is the first time such a reduction with pressure has been observed for interacting iron(III) ions in a single molecule.[118] The behaviour of bis- and tris-iron(II) phenanthroline complexes depends on the degree of back-bonding to the non-phenanthroline ligands. For example, the complexes $Fe(phen)_2X_2$ (X = Cl or Br) are converted from high- to low-spin with increasing pressure, the rate of conversion decreasing above 80 kbar, but the bis-cyanide undergoes a low- to high-spin conversion. Bis-complexes with ligands of intermediate back-bonding character [*e.g.* $Fe(phen)_2X_2$ (X = NCS, NCSe, NCO, or N_3)] initially undergo a high- to low-spin conversion, but at *ca.* 30—40 kbar there is a net low- to high-spin conversion. All tris-complexes are initially low-spin and exhibit measurable conversion to high-spin with increasing pressure. These results are explained in terms of reduced back-donation owing to thermal occupation of the ligand π^*-orbitals by ligand π-electrons. Bisphenanthrolineiron(II) oxalate, which is an intermediate spin complex, and several bipyridyl complexes were also studied.[119] Similar trends are noted in a subsequent report which discusses the relationship between the amount of spin conversion and the electronic properties of the ligands in some substituted phenanthroline complexes.[120] Other spin crossover studies are discussed on p. 525.

Iron(II) in its phthalocyanine complex is in an intermediate spin state at all pressures up to 175 kbar, but its axially co-ordinated adducts with pyridine, 3-picoline, 4-picoline, and piperidine show interesting pressure

[116] F. Parak, R. L. Mössbauer, U. Biebl, H. Formanek, and W. Hoppe, *Z. Physik.*, 1971, **244**, 456.
[117] C. B. Bargeron, M. Avinor, and H. G. Drickamer, *Inorg. Chem.*, 1971, **10**, 1338.
[118] C. B. Bargeron and H. G. Drickamer, *J. Chem. Phys.*, 1971, **54**, 2288.
[119] D. C. Fisher and H. G. Drickamer, *J. Chem. Phys.*, 1971, **54**, 4825.
[120] C. B. Bargeron and H. G. Drickamer, *J. Chem. Phys.*, 1971, **55**, 3471.

dependencies. These contain low-spin iron(II) at 1 bar, which undergoes partial conversion to an intermediate spin-state as the pressure is raised. The extent of conversion is greatest in the picoline adducts, in which π-back-bonding is most pronounced, and least in the piperidine adduct, which has no π-back-bonding character.[121]

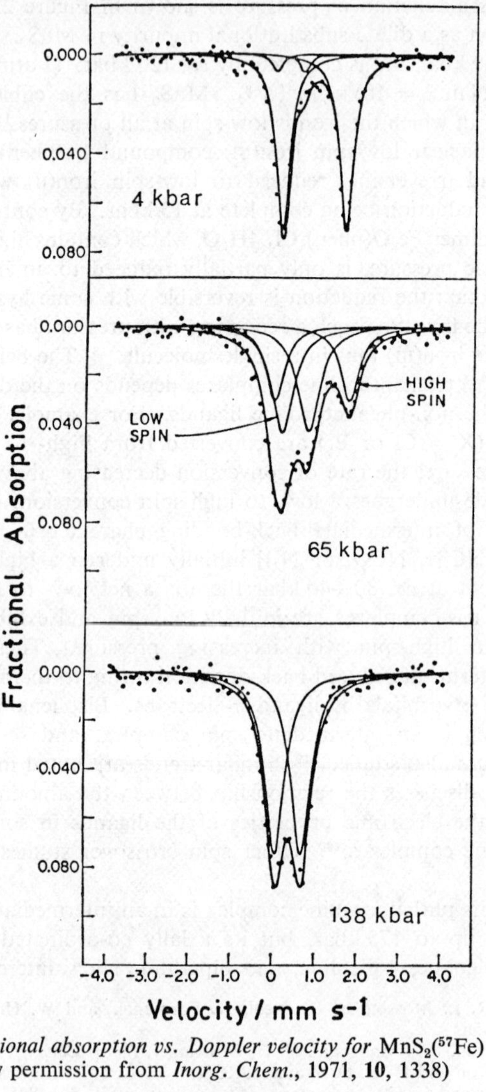

Figure 2 *Fractional absorption vs. Doppler velocity for* $MnS_2(^{57}Fe)$ (Reproduced by permission from *Inorg. Chem.*, 1971, **10**, 1338)

[121] D. C. Grenoble and H. G. Drickamer, *J. Chem. Phys.*, 1971, **55**, 1624.

Some biological compounds have also been studied at high pressures. Imidazole protohemichrome is a low-spin iron(III) compound. At room temperature and pressures above *ca.* 45—50 kbar it is reduced to an intermediate- (or mixed-)spin iron(II) state. At 383 K it is reduced at moderate pressure to a low-spin iron(II) state, but with increasing pressure the iron(II) is transformed to an intermediate spin. Haemin and haematin are high-spin iron(III) compounds which are reduced to intermediate- or mixed-spin iron(II) compounds with increasing pressure. The two iron(III) compounds monochlorobis-(*NN*-diethyldithiocarbamato)iron(III) and monohydroxybis(salicyl)iron(III) – which have square-pyramidal symmetry like haemin and haematin, but differ in having planar neighbours of sulphur or oxygen instead of nitrogen – are both reduced with increasing pressure, in the latter case with a probable change of spin-state from intermediate- to low-spin. All reductions are reversible.[122] Other studies of biological compounds are considered on p. 525.

The pressure dependencies of both the velocity shift and the Mössbauer fraction of a dilute solution of ^{57}Fe in copper have been studied to 130 kbar at 298 and 94 K; the results are consistent with a Debye temperature of *ca.* 317 K for copper.[123] The theoretically calculated pressure dependence of the recoilless fraction of the 14.4 keV γ-ray of ^{57}Fe in natural iron has been compared with that determined experimentally.[124]

Lattice Dynamics and Relaxation Phenomena. In the past, very few Mössbauer fraction measurements have been fitted to lattice-dynamical models. A paper has now appeared which describes measurements of the Mössbauer fraction and thermal shift in the spectra of a metallic iron absorber and three powdered absorbers – sodium nitroprusside, sodium ferrocyanide, and potassium ferrocyanide – over the temperature range 78—293 K. The data are fitted numerically to Einstein and Debye models.[125]

Previous treatments of the Mössbauer lineshapes produced in the spectra of Fe^{2+} compounds in which there is relaxation have yielded poor agreement with experiment and have also yielded large values for the relaxation time of Fe^{2+} in some paramagnetic systems. It has now been shown that the predicted relaxation is considerably reduced if the crystal-field and spin–orbit interactions are considered.[126] Spin–spin relaxation has been studied in the spectra of ^{57}Fe in the organic polymers polyacrylic acid, poly(vinylpyridine), polytetracyanobenzene, and sulphonated polystyrene–divinylbenzene cation-exchange resin, and also in polycrystalline ZnO and single crystals of TiO_2 and Al_2O_3.[127] New evidence has

[122] D. C. Grenoble, C. W. Frank, C. B. Bargeron, and H. G. Drickamer, *J. Chem. Phys.*, 1971, **55**, 1633.
[123] S. S. Nandwani, D. Raj, and S. P. Puri, *Phys. Letters*, 1971, **34A**, 365.
[124] D. L. Williamson and R. Ingalls, *Phys. Letters*, 1971, **34A**, 33.
[125] L. D. Lafleur and C. Goodman, *Phys. Rev.* (*B*), 1971, **4**, 2915.
[126] J. K. Srivastava, *Phys. Stat. Sol.* (*B*), 1971, **46**, K93.
[127] V. P. Korneev, I. P. Suzdalev, V. I. Goldanskii, T. Y. Kureneva, E. F. Makarov, and A. S. Plachinda, *Fiz. tverd. Tela*, 1971, **13**, 354.

been presented which favours the suggestion of Wertheim and Herber that the non-appearance of a magnetic hyperfine pattern in spectra obtained with a source of $Al_2O_3/^{57}Co$ is due to the presence of Fe^{2+} impurities which increase the rate of spin relaxation of Fe^{3+} ions in the source.[128]

Alloy-type Systems and Impurity Studies. Many references to Mössbauer effect studies in alloy-type systems are included in the Bibliography at the end of this chapter.

There has been a continuing interest in the application of Mössbauer spectroscopy to the study of iron carbide systems.[129–134] The nitride γ'-Fe_4N has also been studied and its magnetic easy axis shown to lie along $\langle 100 \rangle$.[135] Previous Mössbauer and *X*-ray diffraction data for β-$FeSi_2$ have been interpreted in terms of a single iron environment and *Cmmm* space group symmetry, whereas a structural study using a vapour-deposited single crystal favoured a *Cmca* space group with two crystallographically different iron atoms. The Mössbauer spectrum has now been re-examined in an attempt to resolve the apparent discrepancy. β-$FeSi_2$ does in fact give only a two-line spectrum, but the lines are broad and asymmetric. This is therefore taken as evidence for space group symmetry with two iron sites.[136] Elsewhere it is reported that the room-temperature spectrum is an almost symmetric doublet, and that this becomes asymmetric with increasing temperature owing to the formation of small polarons.[137] There has been an extensive study of binary and ternary metal phosphides containing iron, *e.g.* FeP, Fe_2P, Fe_3P, FeNiP, FeNbP, FeRuP, FeCoP, and $Fe_3P_{1-x}B_x$: in most cases complex magnetic behaviour is observed.[138] The semi-metallic, ferrimagnetic compound, Fe_3Se_4, has been studied.[139] Spectra have also been obtained for some iron–tin phases and for the products of their reactions with carbon and nitrogen. The chemical isomer shift increases markedly on going from Fe_3Sn to Fe_3SnC, owing to a large increase in the electron population of the 3*d* band; at the same time the hyperfine field induced at the tin nucleus in Fe_3Sn disappears in the carbide.[140]

128 S. K. Date, *J. Phys. Soc. Japan*, 1971, **30**, 1203.
129 Z. Mathalone, M. Ron, J. Pipman, and S. Niedzwiedz, *J. Appl. Phys.*, 1971, **42**, 687.
130 G. Le Caer, A. Simon, A. Lorenzo, and J. M. Genin, *Phys. Stat. Sol.* (*A*), 1971, **6**, K97.
131 E. E. Yurchikov and A. Z. Men'shikov, *Fiz. Metal. Metalloved.*, 1971, **32**, 168.
132 A. Mathalone, M. Ron, and S. Niedzwiedz, *J. Mater, Sci.*, 1971, **6**, 957.
133 J. Polakova, T. Zemcik, and M. Hnlicka, *Cesk. Casopis Fys.*, 1971, **21**, 126.
134 M. Ron and Z. Mathalone, *Phys. Rev.* (*B*), 1971, **4**, 774.
135 J. C. Wood and A. J. Nozik, *Phys. Rev.* (*B*), 1971, **4**, 2224.
136 R. Wandij, C. Le Corre, J. M. Genin, and B. Roques, *Phys. Stat. Sol.* (*B*), 1971, **45**, K123.
137 I. A. Dubovtsev and F. A. Sidorenko, *Pis'ma Zhur. eksp. teor. Fiz.*, 1971, **14**, 205.
138 R. Wappling, L. Hoggstrom, S. Rundqvist, and E. Karlsson, *J. Solid State Chem.*, 1971, **3**, 276.
139 J. R. Regnard and J. C. Hocquenghem, *J. Phys.* (*Paris*) *Colloq.*, 1971, 1 (Pt. 1), C1–268.
140 T. Gibb, N. N. Greenwood, B. Mortimer, and I. J. McColm, *J. Inorg. Nuclear Chem.*, 1971, **33**, 2819.

Mössbauer experiments which involve the doping of non-iron compounds with ^{57}Fe (and in some case ^{57}Co) are gaining in popularity. A variety of hosts including metals, oxides, sulphides, halides, and other materials have been used. These papers are now discussed, roughly in the order of this classification.

The recoil-free fractions in the emission of the 14.4 keV γ-ray of ^{57}Fe in 30 different metallic host lattices have been determined using the 'black' absorber technique, and force constants between the impurity–host and host–host atoms determined.[141] The dynamic properties of an iron impurity in nickel have also been studied.[142] Relaxation effects of ^{57}Fe in thin films, consisting of natural iron + 1% ^{57}Co deposited at 10^{-9} Torr, have been studied at 10^{-11} Torr [1 Torr = 1 mmHg = 133.3 N m^{-2}]. Samples thinner than 7 Å (70 pm) showed complex lineshapes due to superparamagnetic phenomena, whereas samples thicker than 10 Å gave spectra characteristic of bulk iron.[143]

The effect of electron acoustic resonance on the shape of the Mössbauer spectrum has been discussed, Fe^{2+} in MgO being used as an example.[144] The effect of a small external magnetic field on the hyperfine structure of ^{57}Fe in Al_2O_3 has been studied.[145] The spectrum of Ti_3O_5 doped with 1.8% ^{57}Fe has been studied at temperatures above and below the first-order phase transition from the monoclinic to the pseudobrookite structure at 448 K. The spectra are identical and show that the iron is present in two non-equivalent sites, both of which give quadrupole-split resonances. Similar behaviour was observed in 1.8% ^{57}Fe-doped VO_2, although in this case only a single iron(III) doublet is present at temperatures above and below the phase transition at 372 K. The latter result disagrees with previous reports of two different iron sites in ^{57}Fe-doped VO_2.[146] The Fe^{2+} : Fe^{3+} ratio as determined by Mössbauer spectroscopy has been used to give information about the semiconducting properties of pure and lithium-, chromium-, or gallium-doped cobalt(II) and nickel(II) oxides.[147] The magnetic and magneto-optical properties of iron-doped europium oxide films have been studied.[148] An antiferroelectric mode in $PbZrO_3$ has been shown to lead to a drop of 40 ± 8% in the recoilless fraction at the antiferroelectric transition temperature (503 K).[149]

141 S. M. Qaim, *J. Phys.* (*F*), 1971, **1**, 320.
142 C. Janot and H. Scherrer, *J. Phys. and Chem. Solids*, 1971, **32**, 191.
143 M. N. Varma and R. W. Hoffman, *J. Appl. Phys.*, 1971, **42**, 1727.
144 S. S. Bashkirov, N. M. Galeeva, V. I. Goldanskii, and E. K. Sadykov, *Fiz. tverd. Tela*, 1971, **13**, 1775.
145 V. D. Gorobchenko, I. I. Lukhashevich, V. V. Sklyarevskii, K. F. Tzitzkishvili, and N. I. Filippov, ref. 12, p. 143.
146 C. N. R. Rao, S. Ramdas, R. E. Loehman, and J. M. Honig, *J. Solid State Chem.*, 1971, **3**, 83.
147 W. R. Helms and J. G. Mullen, *Phys. Rev.* (*B*), 1971, **4**, 750.
148 T. R. McGuire, G. F. Petrich, B. L. Olson, V. L. Moruzzi, and K. Y. Ahn, *J. Appl. Phys.*, 1971, **42**, 1775.
149 K. K. Mani and S. N. Shringi, *Phys. Stat. Sol.* (*B*), 1971, **44**, K49

Spectra have been obtained for ^{57}Fe as an impurity in CdS [150] and ZnS.[150, 151] There is no evidence for a static Jahn–Teller distortion of the tetrahedral surroundings of the Fe^{2+} ions in the latter. Quadrupole splitting is observed when the impurities fall together in pairs.[55]

Measurements of the hyperfine interaction in Fe^{2+} in MnF_2 in external fields above the spin–flop field at 4.2 K have yielded a value of $g_{\perp} = 2.13$, in good agreement with theory.[152] The room-temperature spectrum of a single crystal doped with *ca.* 0.05 at.% ^{57}Fe indicates the presence of iron(II) ($\delta = 1.40 \pm 0.02$ mm s^{-1}) and iron(III) ($\delta = 0.80 \pm 0.03$ mm s^{-1}) ions. These shifts have been used to re-evaluate the experimental value of α, the constant of proportionality between the isomer shift and the *s*-electron density at the iron nucleus. The new value is significantly smaller than that proposed by Walker *et al.* in 1961 ($-$ 0.28 compared with $-$ $0.5a_0{}^3$ mm s^{-1}).[153] The iron(II) resonance is quadrupole-split below 12 K and from the magnitude of the splitting the first-excited spin-orbit level was estimated to lie at about 120 cm^{-1} above the Γ_{5g} ground-state.[154] The Mössbauer lineshape for iron(II) ions in silver chloride crystals has been discussed.[155]

Spectra have been obtained in the temperature range 73—473 K for the semiconductor compounds $A^{II}B^{IV}C_2{}^{V}$ doped with 1 atom % ^{57}Fe. From a comparison with e.p.r. data it was concluded that in all cases the impurity iron atoms have the electronic configuration $3d^{6-x}4s^{y}$ ($0 < x, y < 1$) and substitute the Group II elements. The isomer shift data incidate that the electron density at the iron nucleus decreases as C is changed from phosphorus to arsenic to antimony, and as B is changed from tin to germanium to silicon in ternary phosphides.[156]

Magnetic hyperfine splitting is observed in the spectra of ^{57}Fe-doped InSb, GaSb, and Ge, and in the emission spectra of ^{57}Co in InSb. The temperature dependence of the spectra and their relaxation character suggest the presence of regions of *ca.* 10 Å (1 nm) in size possessing short-range order. The small value of H_{eff} (100—150 kG) indicates that these are not aggregates of metallic iron. In GaSb the iron is thought to exist as $FeGa_{1.3}$, and in Ge as $FeGe_2$. The nature of the species in InSb is not clear.[157] It was found almost impossible to dissolve an appreciable amount of ^{57}Fe in MnSb, so that previous reports of the absence of a hyperfine field in such a system could not be confirmed. However, a static

150 A. Gerard, P. Imbert, H. Prange, F. Varret, and M. Wintenberger, *J. Phys. and Chem. Solids*, 1971, **32**, 2091.
151 M. Bancie-Grillot and P. Bourtayre, *Compt. rend.*, 1971, **272**, *B*, 804.
152 C. R. Abeledo, R. B. Frankel, A. Misetich, and N. A. Blum, *J. Appl. Phys.*, 1971, **42**, 1723.
153 J. Chappert, J. R. Regnard, and J. Danon, *Compt. rend.*, 1971, **272**, *B*, 1070.
154 J. Chappert, J. R. Regnard, R. B. Frankel, A. Misetich, and N. A. Blum, *J. Phys. (Paris) Colloq.*, 1971, **1** (Pt. 2), C1–941.
155 T. A. Tumolillo, *Phys. Stat. Sol.* (*B*), 1971, **45**, 515.
156 V. K. Yarmarkin, V. D. Prochukhan, and V. S. Grigoreva, *Phys. Stat. Sol.* (*B*), 1971, **48**, 129.
157 G. V. Il'menkov, I. F. Mironov, D. N. Nasledov, Y. S. Smetannikova, B. A. Shustrov, and V. K. Yarmarkin, *Fiz. tverd. Tela*, 1971, **13**, 1407.

internal field has been found at the iron impurity produced by the electron-capture decay of ^{57}Co in MnSb. The hyperfine constants (H_{eff} = 144 ± 8 kG, e^2qQ = + 0.75 ± 0.05 mm s^{-1}) extracted from the spectrum were compared with those of the ^{55}Mn measured previously. It was concluded that the internal fields originate mainly from the contact term and that both the manganese and iron *d*-orbitals are significantly delocalized.[158]

New ^{57}Co Sources and Decay After-effect Phenomena. The performances of sources of ^{57}Co in various metallic matrices have been compared and methods outlined for the production of sources of nearly natural line-width (< 0.21 mm s^{-1}). Chromium is considered to be the best matrix, provided one does not wish to use the source below 70 K. For applications which require an unsplit source at 4.2 K rhodium is the better choice. Palladium is a useful matrix for sources of reasonable strength and has the advantage of being the easiest matrix in which to prepare the source. Electrostatic broadening and the change of isomer shift with age are relatively unimportant for these sources, provided the density of activity is < 150 mCi cm^{-2}. Copper, platinum, and gold are markedly inferior because of the broadening due to electrostatic interactions for all but the weakest sources.[159] A single-line source for use at room-temperature can be made from cobalt(II) oxide doped with lithium. It has a Mössbauer temperature of θ_m = 410 K, and a linewidth of Γ = 0.22 mm s^{-1} (compared with 0.20 for $^{57}Co/Pd$). It gives a larger percentage effect than that of $^{57}Co/Pd$ and can be fabricated with extremely intense activity (> 1 Ci).[160] The metallurgical preparation of non-ferromagnetic sources of CoAl, CoBe, Co + 25 atom % Cr, and Co + 80 atom % Rh, have been discussed,[161] and details given of a simple monochromatic polarized iron-57 source.[162]

There has been continuing discussion on the relative importance of factors which determine the nature of the ionic species formed after ^{57}Co decay in cobalt compounds containing simple ligands. The emission spectrum of $CoBr_2$ consists of a quadrupole-split doublet characteristic of Fe^{2+}, whereas the spectrum of β-$Co(OH)_2$ indicates that a small quantity of Fe^{3+} is present also. Since these two compounds both have the same type of structure, size effects are considered to be unimportant in the stabilization of charge states different from those of the parent cobalt(II) ions (*i.e.* aliovalent charge states). Instead the results suggest that thermodynamic factors are important, and that the stabilization of aliovalent species increases with the lattice energy of the host lattice.[163] Similar arguments have been used to account for the fact that Fe^{3+} ions are produced in the electron-capture decay of $^{57}Co^{2+}$ in MgF_2, CaF_2, and K_2CoF_4, whereas only Fe^{2+} ions are produced in $MgCl_2$.[164] The observation of

158 M. Avinor and M. Pasternak, *J. Phys. and Chem. Solids*, 1971, **32**, 1395.
159 G. Longworth and B. Window, *J. Phys. (D)*, 1971, **4**, 835.
160 W. R. Helms and J. G. Mullen, *Nuclear Instr. and Meth.*, 1971, **91**, 291.
161 S. K. Godovikov and R. N. Kuz'min, *Vestn. Mosk. Univ., Fiz., Astron.*, 1971, **12**, 220.
162 J. P. Stampfel, ref. 20, p. 95.
163 A. Cruset and J. M. Friedt, *Phys. Stat. Sol. (B)*, 1971, **44**, 633.
164 A. Cruset and J. M. Friedt, *Phys. Stat. Sol. (B)*, 1971, **47**, 655.

Fe^{3+} in the emission spectra of $^{57}Co : CoFe_2O_4$ has also been attributed to the thermodynamic stability of this charge state in the host matrix. From the isomer shift it appears that the Fe^{3+} occupies the octahedral B sites. The fact that the magnetic hyperfine field is 3% less than the field acting on B sites in a $CoFe_2O_4$ absorber is thought to reflect the decreased overlap of the electron wavefunctions of Fe^{3+} and O^{2-} because of the smaller ionic radius of Fe^{3+} compared with Co^{2+}.[165] The concentration of Fe^{3+} ions formed in paramagnetic CoO has been found to depend on the stoicheiometry of the samples as well as on the morphology of the crystallites and their defect structures.[166]

It has been suggested in the past (see last year's Report) that the stabilization of anomalous ionic species could be understood on the basis of a process of autoradiolysis by the Auger electrons and *X*-rays. This conclusion was based on the observation that the emission spectra of ^{57}Co-labelled $Co(acac)_3$ are similar to the absorption spectra of electron-irradiated $Fe(acac)_3$. However, the comparison of these spectra was made on the ^{57}Fe species produced in the two different environmental matrices, $Co(acac)_3$ and $Fe(acac)_3$. In order to prove the radiolytic process and to check whether the host material is involved in the process, the effects of γ-irradiation of ^{57}Fe-doped $Mn(acac)_3$, $Fe(acac)_3$, and $Co(acac)_3$, and the effects of ^{57}Co electron-capture decay in these compounds doped with ^{57}Co have been compared. The results are shown in Figure 3. The spectra consist of two doublets ascribed to $^{57}Fe^{2+}$ and $^{57}Fe^{3+}$ species and one broad singlet ascribed to the original $^{57}Fe(acac)_3$. The shapes of the two spectra are quite similar when the host material, $M(acac)_3$, is the same, whereas they are not always similar when the host material is different. The similarities indicate that the radiolytic process initiated by the Auger effect determines the final oxidation state of the ^{57}Fe in a time of 10^{-7} s. The increased peak intensity and quadrupole splitting of the Fe^{2+} species observed in Figure 3(e) and 3(f) suggest that the surrounding neighbours of an electron-capture decayed $^{57}Co(acac)_3$ or a γ-ray absorbed $^{57}Fe(acac)_3$ moiety are involved in the local radiolytic processes and that the $Fe(acac)_3$ matrix is more radiosensitive than are $Mn(acac)_3$ or $Co(acac)_3$.[167, 168] Emission spectra for ^{57}Co-labelled cobaltocene, cobaltocinium, and $CoCl_2$–C_5H_5N complexes have also been accounted for on the basis of the autoradiolysis model. Aliovalent ions are observed only in cobaltocinium.[169] The chemical state of ^{57}Fe in ^{57}Co-labelled $Co(py)_xCl_2$ ($x = \frac{2}{3}$, 1, 2, or 4) is similar to that in $Fe(py)_xCl_2$; there is no evidence for the formation of a tetrahedral monomer.[170]

165 A. Cruset and J. M. Friedt, *Phys. Stat. Sol.* (*B*), 1971, **45**, 189.
166 J. Blomquist, S. Grapengiesser, and R. Söderquist, *Phys. Stat. Sol.* (*A*), 1971, **4**, 435.
167 H. Sano, K. Sato, and H. Iwagami, *Bull. Chem. Soc. Japan*, 1971, **44**, 2570.
168 H. Sano and H. Iwagami, *Chem. Comm.*, 1971, 1637.
169 J. M. Friedt, R. Poinsot, and J. P. Sanchez, *Radiochem. Radioanalyt. Letters*, 1971, **7**, 193.
170 N. Saito, M. Takada, and T. Tominaga, *Radiochem. Radioanalyt. Letters*, 1971, **6**, 169.

The spectrum of ^{57}Co-doped silicon consists of two resonances: a singlet from iron atoms with the electronic configuration $3d^74s^1$, located in cubic surroundings, and a weakly resolved doublet from iron atoms with the electronic configuration $3d^54s^0$ (Fe^{3+}) located in a distorted cubic

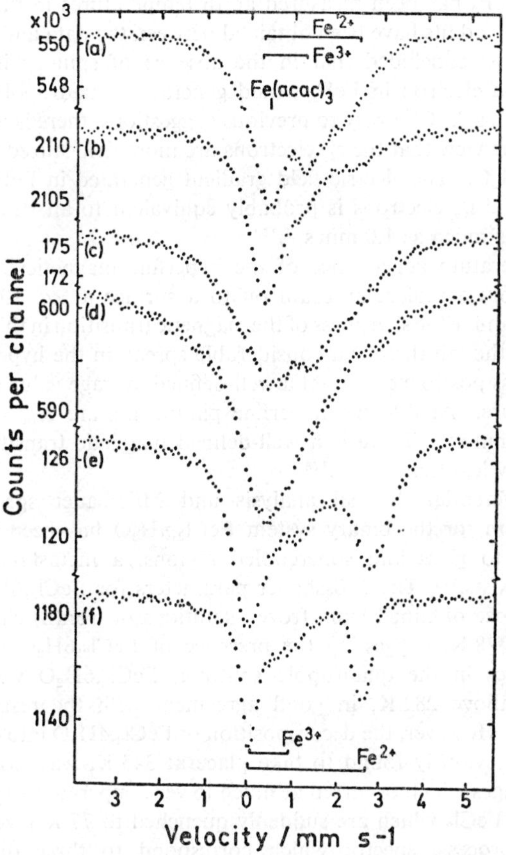

Figure 3 *Mössbauer spectra at* 78 K *of* (a) ^{57}Co-*labelled* Co(acac)$_3$ *source,* (b) *γ-irradiated* (3.0 × 10^9 r) (Co,^{57}Fe)(acac)$_3$ *absorber,* (c) (Mn,^{57}Co)(acac)$_3$ *source,* (d) *γ-irradiated* (3.0 × 10^9 r) (Mn,^{57}Fe)(acac)$_3$ *absorber,* (e) (^{57}Fe,Co)(acac)$_3$ *source, and* (f) *γ-irradiated* (3.0 × 10^9 r) Fe(acac)$_3$ *absorber. All velocity scales normalized with respect to iron*

environment. The form of the spectrum does not depend on the density of dislocations, type of conductivity, or concentration of current carriers.[171]

Compounds of Iron.—The topics covered in this section are: high-spin iron(II) compounds; high-spin iron(III) compounds, spin-crossover systems

[171] B. I. Boltaks, M. K. Bakhadyrkhanov, and P. P. Seregin, *Fiz. tverd. Tela*, 1971, **13**, 2810.

and biological compounds; and low-spin and covalent complexes of iron. Oxide and sulphide systems containing iron are considered in the following section.

High-spin Iron(II) *Compounds.* The quadrupole interaction at the ^{57}Fe nucleus in FeF_2 has been measured at 16 temperatures in the range 78—1023 K and good fits have been obtained between the data and a theoretical model. It was concluded that in the absence of spin–orbit coupling a single iron 3*d* electron in FeF_2 would generate a quadrupole splitting of 3.8 ± 0.3 mm s^{-1}. Contrary to previous suggestions, there is little evidence to support the view that the t_{2g} electrons are more delocalized in FeF_2 than in $FeSiF_6,6H_2O$. The electric field gradient generated in FeF_2 by sources other than the t_{2g} electrons is probably equivalent to an ^{57}Fe quadrupole splitting of as much as 1.0 mm s^{-1}.[172]

The temperature dependence of the hyperfine magnetic field of $FeCl_2$ deposited from a molecular beam on to a surface at *ca.* 10 K has been studied to examine the sharpness of the magnetic transition in this amorphous material. Although there is a considerable spread in the hyperfine field, it is nevertheless possible to extract a well-defined average value as a function of temperature. At 8 K the hyperfine parameters are H_{hf} = 85 kG and e^2qQ = 5.3 mm s^{-1}. There is a well-defined magnetic transition at 22 K, broadened by less than 0.5 K.[173]

Using differential thermal analysis and Mössbauer spectroscopy, a phase diagram for the binary system $FeCl_2,nH_2O$ has been constructed. In addition to glass and supercooled regions, a metastable crystalline phase also exists.[174] The Mössbauer parameters for $FeCl_2,6H_2O$ are very similar to those obtained from frozen solutions of iron(II) chloride in the range 221—228 K, suggesting the presence of $FeCl_2,6H_2O$ in the latter. From changes in the quadrupole splitting, $FeCl_2,6H_2O$ was shown to decompose above 282 K, in good agreement with the results of other experiments. However, the decomposition of $FeCl_2,4H_2O$ into $FeCl_2,2H_2O$, which was previously found to take place at 343 K, was associated with spectral changes which occurred at much lower temperatures (> 293 K).[175] Solutions of $FeCl_2$ which are suddenly quenched to 77 K give, during the rewarming process, spectra which correspond to three distinct states connected by irreversible transitions.[176] The first state, which is reached after quenching and is observed between 93 and about 183 K, has been studied in greater detail and shown to exhibit a glass transition at about 163 K.[177] For slowly cooled solutions only two different phases, connected by an irreversible transition, are observed. The phase which persists until

[172] B. W. Dale, *J. Phys.* (*C*), *Solid State Phys.*, 1971, **4**, 2705.
[173] A. J. F. Boyle, G. M. Kalvius, D. M. Gruen, J. R. Clifton, and R. L. McBeth, *J. Phys.* (*Paris*) *Colloq.*, 1971, **1** (Pt. 1), C1-224.
[174] S. L. Ruby, B. J. Zabransky, and J. G. Stevens, *J. Chem. Phys.*, 1971, **54**, 4559.
[175] I. Dézsi, P. J. Ouseph, and P. M. Thomas, *Chem. Phys. Letters*, 1971, **9**, 390.
[176] B. Brunot, U. Hauser, W. Neuwirth, and J. Bolz, *Z. Physik*, 1971, **249**, 125.
[177] B. Brunot, U. Hauser, and W. Neuwirth, *Z. Physik*, 1971, **249**, 134.

the melting point is the same for both the quenched and the slowly cooled solutions.[176] A number of other studies on frozen solutions of iron(II) compounds have been reported.[178–181]

Hydrated iron(II) formate, $Fe(HCO_2)_2,2H_2O$, has been shown to contain two distinct iron sites, whereas the anhydrous α-$Fe(HCO_2)_2$ contains only a single iron environment. The spectra have been used to show that, even after 30 days' exposure to the atmosphere, only partial hydration of α-$Fe(HCO_2)_2$ occurs.[182] Iron(II) fumarate, $Fe(HCCO_2)_2$, is thought to have a polymeric structure containing iron in a distorted octahedral environment.[183]

Lattice-sum calculations have been carried out for iron(II) sulphate and compared with experimental data collected over a wide temperature range (up to 600 K). The principal electric field gradient axis was calculated to be parallel to the crystallographic *b* axis at low temperatures, and this is consistent with both the Mössbauer measurements and neutron diffraction data. The ground orbital state is a singlet separated by $7|\lambda|$ from the first excited state. The calculated values for the quadrupole splitting and the asymmetry parameter are 3.38 mm s^{-1} and 0.48 respectively, in good agreement with the low-temperature experimental values.[184] A detailed angular orientation study on a single crystal of $FeSO_4,4H_2O$ has yielded the principal-axis values, the asymmetry parameter, and the direction cosines of the electric field gradient axes with respect to the crystal axes. The following t_{2g} orbital populations of the odd electron in the high-spin iron(II) ion were estimated from the quadrupole splitting: $n_{xy} = 0.89$, $(n_{yz} + n_{xz}) = 0.11$. A slight asymmetry in the peak areas of the spectrum for the polycrystalline powder was attributed to the Goldanskii–Karyagin effect.[185] The products of proton irradiation of $FeSO_4,7H_2O$ have been studied by Mössbauer spectroscopy. In addition to the known products of dehydration, a number of new species were also present. Prolonged irradiation leads to complete decomposition of the compound; among the products identified were Fe_2O_3 and FeS_2.[186]

Further Mössbauer studies of vivianite, $Fe_3(PO_4)_2,8H_2O$, have confirmed that the spin directions of the two iron sites are not collinear, the angle between them being 42°. Although the temperature dependences of the two sublattices are greatly different, both internal fields disappear at the same temperature, 90 K. From measurements made at 4.2 K in the

[178] P. M. Thomas, M. Sanders, I. Dezsi, and P. J. Ouseph, *Chem. Phys. Letters*, 1971, **11**, 42.
[179] M. Kaplan, A. J. Nozik, J. V. DiLorenzo, and T. X. Carroll, ref. 12, p. 597.
[180] J. A. Cameron, L. Keszthelyi, H. Nagy, and L. Kacsoh, *Chem. Phys. Letters*, 1971, **8**, 628.
[181] I. Dézsi, L. Keszthelyi, G. Nagy, and D. L. Nagy, ref. 12, p. 607.
[182] J. Pipman and M. Ron, *Solid State Comm.*, 1971, **9**, 241.
[183] A. N. Garg, P. N. Shukla, and P. S. Goel, *Inorg. Chim. Acta*, 1971, **5**, 520.
[184] H. N. Ok, *Phys. Rev. (B)*, 1971, **4**, 3870.
[185] V. K. Garg and S. P. Puri, *J. Chem. Phys.*, 1971, **54**, 209.
[186] M. Kopcewicz and A. Kotlicki, *Phys. Stat. Sol. (B)*, 1971, **44**, K127.

presence of a large external field (up to 50 kG), applied in a direction perpendicular to the crystal *ac* plane, it was shown that ferromagnetic spin flipping occurs only at the sublattice of one iron site, the field dependence at the other site being very small.[187]

The occurrence or absence of cation ordering and the extent of magnetic ordering in complex phases derived from the rutile structure (FeF_2) have been studied over a wide range of temperature. The Fe^{2+} compounds $FeCoNiF_6$, Mg_2FeF_6, and $MgFe_2F_6$ are disordered rutile phases with

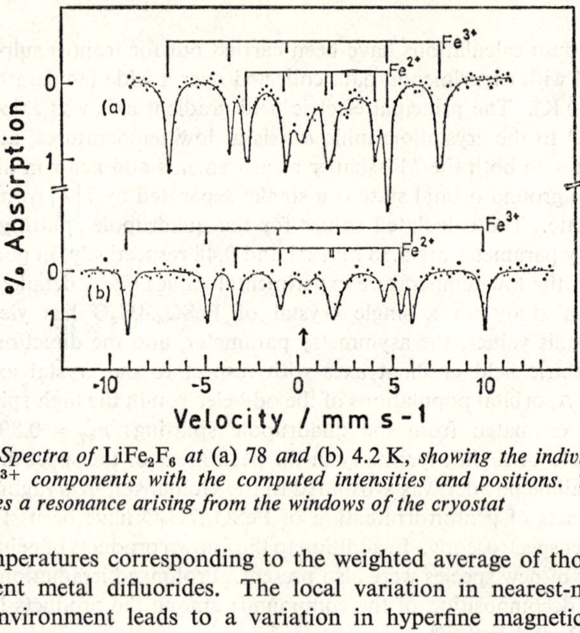

Figure 4 *Spectra of* $LiFe_2F_6$ *at* (a) 78 *and* (b) 4.2 K, *showing the individual* Fe^{2+} *and* Fe^{3+} *components with the computed intensities and positions. The arrow indicates a resonance arising from the windows of the cryostat*

Néel temperatures corresponding to the weighted average of those of the component metal difluorides. The local variation in nearest-neighbour cation environment leads to a variation in hyperfine magnetic field of $\pm$ 30 kG about the mean for each compound. The altervalent fluoride $LiFe_2F_6$ shows separate spectra from Fe^{2+} and Fe^{3+} in both the paramagnetic and the magnetically ordered regions [Figure 4 (a) and (b)], thus demonstrating the integrity of the individual oxidation states and the absence of electron hopping. The narrow linewidths indicate complete geometric ordering of all cations in the trirutile structure and the repeating cation sequence in the *c* direction was shown to be $Li^+(Fe^{2+}\uparrow)(Fe^{3+}\downarrow)$-$Li^+(Fe^{2+}\downarrow)(Fe^{3+}\uparrow)$ leading to overall antiferromagnetic behaviour. By contrast the trirutile compound $LiMgFeF_6$ shows a range of quadrupole splittings consistent with the ordering of Fe^{3+} into every third layer along the *c* axis and a randomization of Li^+ and Mg^{2+} in the intervening two layers. This model is dictated by the absence of magnetic ordering even at 1.4 K, and the spectra at this temperature and at 4.2 K show instead the

[187] T. Shinjo and H. Forstat, *J. Phys. Soc. Japan*, 1971, **31**, 1399.

effects of slow spin–spin relaxation characteristic of magnetically dilute Fe^{3+} ions (Figure 5).[188]

Charge hopping does not occur in the mixed-valence compound $K_{0.5}FeF_3$, which contains equal amounts of Fe^{2+} and Fe^{3+}. The compound orders magnetically at 20 K. At this temperature the Fe^{3+} pattern exhibits narrow lines, suggesting that the ions occupy equivalent sites, and reveals a

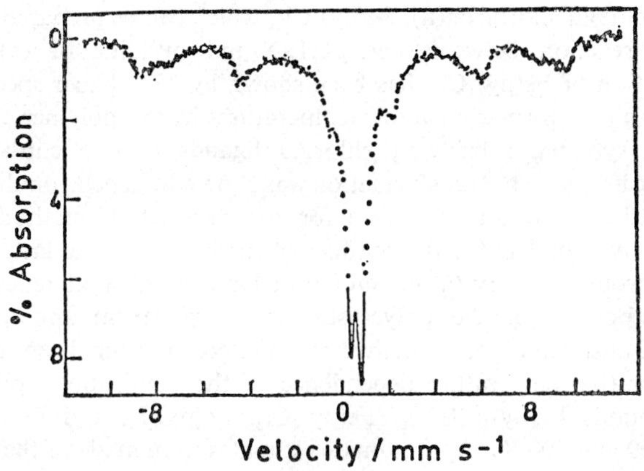

Figure 5 *Spectrum of* $LiMgFeF_6$ *at* 4.2 K

large hyperfine field (590 kG) at the iron nucleus. The Fe^{2+} pattern is broadened, presumably because of the effect of both the random distribution of the potassium neighbours and the distribution of the valence electrons of the iron neighbours on the electric field gradient at the nucleus in the Fe^{2+} ions.[189] Very recently it has been shown that in some cubic compounds, even in the absence of a static Jahn–Teller effect, magnetic interactions can induce an electric field gradient at the ^{57}Fe nucleus. Attempts have now been made to calculate the magnitude of this effect for the case of a high-spin iron(II) ion in an axially distorted cubic crystal field. Two different cases were considered, (*a*) when the magnetic field is internal to the crystal, and (*b*) when the magnetic field is applied externally. Experimental results were discussed for $RbFeF_3$ in its antiferromagnetic phase.[190]

The four compounds $(RNH_3)_2FeCl_4$ (R = Me, Et, Pr, or $C_6H_5CH_2$) are considered to have structures featuring two-dimensional metal–halogen sheets whose spacing depends on the length of the substituted ammonium cation. Single-crystal spectra showed that the principal axis of the electric field gradient is perpendicular to the sheet and that e^2qQ is positive. This is consistent with the $^5B_{2g}$ electronic ground-state predicted

[188] N. N. Greenwood, A. T. Howe, and F. Ménil, *J. Chem. Soc.* (*A*), 1971, 2218.
[189] D. N. E. Buchanan, M. Robbins, H. J. Guggenheim, G. K. Wertheim, and V. G. Lambrecht, jun., *Solid State Comm.*, 1971, **9**, 583.
[190] R. Kamal and R. G. Mendiratta, *Phys. Rev.* (*B*), 1971, **3**, 1649.

by crystal field theory for compressed octahedral geometry. The methyl, ethyl, and propyl compounds are antiferromagnetically ordered at 77 K, with internal fields of 240, 220, and 220 kG respectively, lying in the planes of the sheets.[191] The unsubstituted and phenyl-substituted dithiolium salts of $[FeCl_4]^{2-}$ have been shown to exhibit a strongly temperature-dependent quadrupole splitting, which arises directly from the lifting of the orbital degeneracy of the ^5E ground state owing to distortion of the tetrahedron.[192]

Bis(pyridine)dichloroiron(II), $Fe(py)_2Cl_2$, which can be prepared by either (*a*) direct reaction between $FeCl_2,4H_2O$ and pyridine or (*b*) thermal decomposition of $Fe(py)_4Cl_2$, has been shown by Mössbauer spectroscopy to undergo a transformation at low temperatures from a polymeric structure containing symmetric bridging chloride ligands to one containing an asymmetric bridge. The transformation was shown to depend on the method of preparation, and was complete for the product of method (*a*), but only partially complete for the product of method (*b*). The latter method probably produces a crystal containing a large number of defects which prevent a portion of the polycrystalline sample from undergoing the proposed transformation.[193] A theoretical expression has been derived to account for the temperature dependence of the quadrupole splitting for the compounds $Fe(py)_4(SCN)_2$, $Fe(py)_4Cl_2$, $Fe(py)_4I_2$, and $Fe(py)_4I_2,2py$ between 100 and 300 K. Approximate estimates were made of the displacements of the d_{yz} and d_{xz} orbitals from the ground orbital d_{xy}. $Fe(py)_4Cl_2$ has a *cis* configuration whereas $Fe(py)_4I_2,2py$ has a *trans* configuration.[194] Intermediates formed in the thermal decomposition of the compounds $Fe(py)_4X_2$ (X = Cl, Br, I, or SCN) have been studied.[195] The six-co-ordinate iron(II) complexes $Fe(pic)_2,4H_2O$, $Fe(nic)_2,4H_2O$, and $Fe(isonic)_2$,-$4H_2O$ (picH = picolinic acid, nicH = nicotinic acid, isonicH = isonicotinic acid)[196] and the complex $FeSO_4(CH_2NH_3)_2SO_4,4H_2O$[197] have also been studied. It was suggested that the four water molecules in the latter form an octahedron similar to that in $FeSO_4,4H_2O$, and that the ethylene diammonium ion $(H_3NCH_2CH_2NH_3)^{2+}$ plays no role in the co-ordination around the high-spin iron(II) ion.[197]

The observed temperature dependence of the quadrupole splitting for the distorted tetrahedral iron(II) compound $Fe(quinoline)_2Cl_2$ gives only qualitative agreement with calculations based on Ingall's model. The agreement is poor possibly because of neglect of the lattice contribution to the quadrupole splitting or because of a breakdown in the perturbation

191 M. F. Mostafa and R. D. Willet, *Phys. Rev.* (*B*), 1971, **4**, 2213.
192 R. L. Martin and I. A. G. Roos, *Austral. J. Chem.*, 1971, **24**, 2231.
193 G. J. Long, D. L. Whitney, and J. E. Kennedy, *Inorg. Chem.*, 1971, **10**, 1406.
194 P. B. Merrithew, P. G. Rasmussen, and D. H. Vincent, *Inorg. Chem.*, 1971, **10**, 1401.
195 G. Liptay, L. Korecz, I. Kiraly, and E. Papp-Molnar, *Magyar Kém. Folyóirat*, 1971, **77**, 90.
196 B. W. Fitzsimmons, A. Klenstein, N. J. Seeley, and G. A. Webb, *Rev. Roumaine Chim.*, 1971, **16**, 1197.
197 A. N. Garg, *Phys. Stat. Sol.* (*B*), 1971, **48**, K143.

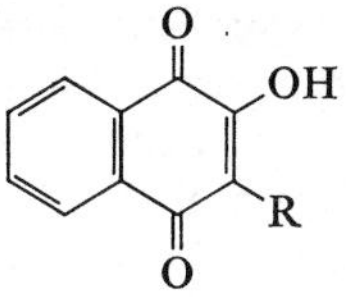

(1)

techniques in the presence of large distortions from tetrahedral symmetry.[198]

Preliminary data for the chelates $Fe^{II}L_2$ and $Fe^{III}L_3$ [where L = (1; R = H, Me, or $CH_2CH{=}CMe_2$)] have confirmed that the iron(II) complexes are high-spin. The mean-square vibrational amplitude of the iron nucleus is less in the iron(II) compounds than in the iron(III) compounds and is slightly dependent on the nature of R. The chemical isomer shift is independent of the nature of R.[199]

The chemical isomer shifts of the complexes $FeX_2(Hthsc)_2,nH_2O$ and $FeX_2(Hsc)_2$ (where Hthsc = thiosemicarbazide, Hsc = semicarbazide, and X = $\frac{1}{2}SO_4^{2-}$, Cl^-, NCS^-, or NO_3^-) confirm that they are octahedral. The hydrated sulphate complex is claimed to have the largest quadrupole splitting (4.33—4.38 mm s^{-1}) so far reported.[200] Spectra have also been obtained for compounds of iron with thio- and seleno-semicarbazones.[201]

The spectra of the complexes $(\pi\text{-}C_5H_5)_2Mo(SR)_2FeCl_2$ (R = Me or Bu^n) are characteristic of tetrahedral high-spin iron(II) and this has been confirmed by *X*-ray analysis for the butyl complex. The low value of the chemical isomer shift suggests a high covalency of the metal–sulphur bond.[202]

The cubic ferrimagnet $FeCr_2S_4$ and the derived system $Cd_{1-x}{}^{57}Fe_xCr_2S_4$ (x = 0.02) have been shown to contain tetrahedrally co-ordinated iron(II).[203]

High-spin Iron(III) *Compounds.* Antiferromagnetism can be induced in FeF_3 above the Néel point (T_N = 363.2 K) by the application of an external magnetic field of $\leqslant$ 20 kG. At $T = T_N$ an external field of 20 kG, applied perpendicular to the direction of the γ-ray beam, induces an effective magnetic field of 100 kG on the ^{57}Fe nuclei.[204] Relaxation effects have been studied in this compound.[205] The Néel temperature of $FeCl_3$ has been shown to be 9 ± 1 K, which is consistent with recent neutron diffraction data. Experiments carried out in applied fields of between 10 and 50 kG

198 G. L. Long and D. L. Whitney, *J. Inorg. Nuclear Chem.*, 1971, **33**, 1196.
199 C. G. deLima, F. F. T. deAraújo, A. DuFresne, and J. M. Knudsen, *Inorg. Nuclear Chem. Letters*, 1971, **7**, 513.
200 A. V. Ablov and N. V. Gerbeleu, *Zhur. neorg. Khim.*, 1971, **16**, 184.
201 K. I. Turta, A. V. Ablov, V. I. Goldanskii, N. V. Gerbeleu, and R. A. Stukan, *Doklady Akad. Nauk S.S.S.R.*, 1971, **196**, 1383.
202 A. R. Dias and M. L. H. Green, *J. Chem. Soc.* (*A*), 1971, 2807.
203 R. P. Van Stapele, J. S. Van Wieringen, and P. F. Bongers, *J. Phys.* (*Paris*), *Colloq.*, 1971, **1** (pt. 1), C1–53.
204 V. Gamlitskii, S. S. Yakimov, V. I. Nikolaev, and N. F. Simonov, *Pis'ma Zhur. eksp. teor. Fiz.*, 1971, **13**, 129.
205 F. van der Woude, C. Blaauw, and A. J. Dekker, ref. 12, p. 133.

at 4.2 K gave no evidence for a spiral structure; instead there was a uniform decrease of the internal hyperfine field with increasing applied field.[206]

From magnetic susceptibility, neutron diffraction, and Mössbauer measurements it has been shown that $KFeF_4$ exhibits two-dimensional antiferromagnetic ordering above $T_N = 137$ K. Below this temperature there is a three-dimensional magnetic structure, with the moments directed along the *c* axis.[207, 208] The magnetic behaviour of the layer-type antiferromagnet $RbFeF_4$ ($T_N = 133.4$ K) has also been studied in detail over the temperature range 4.2—300 K.[209]

Spectra of frozen aqueous solutions of $Fe(NO_3)_3$ have revealed the presence of an oxo-bridged species, a dihydroxo-bridged species, and a magnetic component. The latter increases in importance relative to the other two as the pH is lowered from 2.3 to 0.7.[210]

Several groups have studied iron(III) oxychloride. Experiments with single-crystal and polarized sources have shown that V_{zz} is negative and directed along the crystallographic *a* axis, and that V_{xx} is parallel to *c*; the asymmetry parameter, η, was found to be 0.32 ± 0.03. A paramagnetic to antiferromagnetic transition occurs at 92 ± 3 K and near liquid helium temperature the antiferromagnetic structure is non-collinear; at least two sets of non-equivalent Fe^{3+} sites and a doubling of the unit cell are required. The internal magnetic field is perpendicular to V_{zz} at all of the Fe^{3+} sites. The most probable magnetic space group is *P*2*cm*'2*m*', which has two antiferromagnetic sublattices with spins parallel to *b* and *c* respectively.[211, 212] Similar results have been obtained in an independent study.[213] Attempts to fit a lattice-sum calculation to the experimental data yielded unsatisfactory values for the anion polarizabilities and for the quadrupole moment of the ^{57}Fe nucleus ($Q \approx 0.33$ b).[211] However, it has been shown elsewhere that, by taking into account contributions to the electric field gradient components from overlap of the Fe^{3+} 2*p* and 3*p* orbitals with the oxygen 2*p* and the chlorine 3*p* orbitals, it is possible to match the experimental asymmetry parameter value of $\eta = 0.32$ and to get a Q value of 0.19 b, in good agreement with the iron(II) concensus (see p. 501).[214]

Mössbauer and Raman spectra have shown that the distortion of the octahedral environment in $Fe(O_2PF_2)_3$ is greater than that in $Fe(O_2PCl_2)_3$.[215]

206 J. P. Stampfel, W. T. Oosterhuis, and F. de S. Barros, *J. Appl. Phys.*, 1971, **42**, 1721.
207 G. Heger, R. Geller, and D. Babel, *Solid State Comm.*, 1971, **9**, 335.
208 G. Heger and R. Geller, *Z. Angew. Phys.*, 1971, **32**, 63.
209 M. Eibschütz, H. J. Guggenheim, and L. Holmes, *J. Appl. Phys.*, 1971, **42**, 1485.
210 J. M. Knudsen, F. T. DeAraújo, A. Dufresne, and C. G. de Lima, *Chem. Phys., Letters*, 1971, **11**, 134.
211 R. W. Grant, H. Wiedersich, R. M. Housley, P. G. Espinosa, and J. O. Artman, *Phys. Rev. (B)*, 1971, **3**, 678.
212 R. W. Grant, *J. Appl. Phys.*, 1971, **42**, 1619.
213 E. Kostiner and J. Steger, *J. Solid State Chem.*, 1971, **3**, 273.
214 D. Sengupta, J. O. Artman, and G. A. Sawatzky, *Phys. Rev. (B)*, 1971, **4**, 1484.
215 J. Pebler and K. Dehnicke, *Z. Naturforsch.*, 1971, **26b**, 747.

It has been established that iron plays an important role in the establishment of ferroelectricity in $(NH_4)Fe(SO_4)_2,12H_2O$.[216] Electronic spin–flip processes in this material have also been studied elsewhere.[217]

The Mössbauer spectrum of jarosite, $KFe_3(OH)_6(SO_4)_2$, has been analysed in detail in order to estimate the spin arrangement. This information is of interest because jarosite is antiferromagnetic below 60 K and yet still possesses a 'kagome' lattice of iron atoms. Previous work has shown that usual collinear antiferromagnetic spin arrangements are not stable in a 'kagome' lattice, because it is simply not possible to distribute equal numbers of 'up' and 'down' spins over the lattice points in any ordered manner. The suggested spin arrangement is shown in Figure 6.[218]

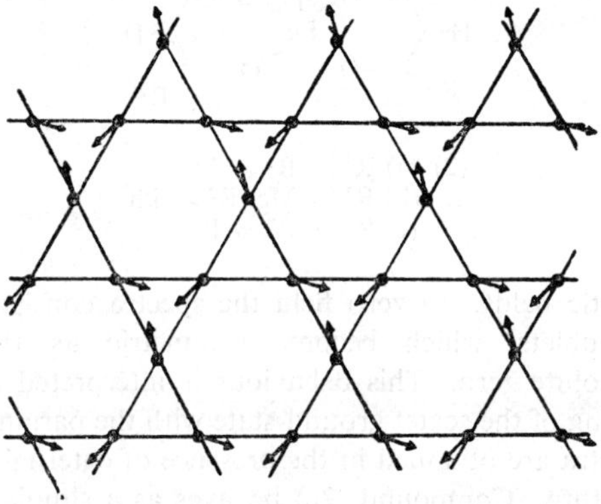

Figure 6 *Spin arrangement in* $KFe_3(OH)_6(SO_4)_2$
(Reproduced by permission from *J. Phys. Soc. Japan*, 1971, **30**, 1049)

Mössbauer spectra at 298, 77, and 4.2 K and magnetic data have been obtained for the extraction products [*e.g.* $(C_9H_{19}NH_3)_2FeOH(SO_4)_2$] from the equilibration of an aqueous iron(II) sulphate solution with a benzene solution of a long-chain primary or secondary amine. In general, the spectra are asymmetric doublets at 298 K, becoming symmetrical at 4.2 K. This behaviour was rationalized in terms of spin–spin relaxation between equivalent iron(III) sites in a trinuclear hydroxy-bridged system.[219]

The compounds $Fe(pic)_2$, $Fe(nic)_2,2H_2O$, and $Fe(isonic)_2,H_2O$, which are products of dehydration of the iron(II) compounds mentioned on p. 518, are probably polymeric iron(III)-containing species.[195] Preliminary data have also been given for the iron(III) analogues of the 2-hydroxy-1,4-napthoquinone iron(II) species mentioned on p. 519.[198]

[216] P. A. Montano, H. Schechter, and A. Biran, *Solid State Comm.*, 1971, **9**, 2029.
[217] S. Mørup and N. Thrane, *Phys. Rev. (B)*, 1971, **4**, 2087.
[218] M. Takano, T. Shinjo, and T. Takada, *J. Phys. Soc. Japan*, 1971, **30**, 1049.
[219] R. Prados and M. L. Good, *J. Inorg. Nuclear Chem.*, 1971, **33**, 3733.

It has been shown that the very large linewidths (*ca.* 2 mm s^{-1}) in the spectrum of $Fe(acac)_3$ are reduced to *ca.* 0.6 mm s^{-1} by γ-irradiation. To explain this it was suggested that a small proportion (< 1%) of the material is converted into $Fe(acac)^-$, and that the subsequent hopping of the odd electrons from reduced to non-reduced species provides an efficient relaxation mechanism.[220] $Fe(acac)_3$ has also been discussed in another paper.[221]

The spectra of the five-co-ordinate trigonal-bipyramidal compounds (2a—c) have been measured at temperatures between 300 and 1.7 K and in

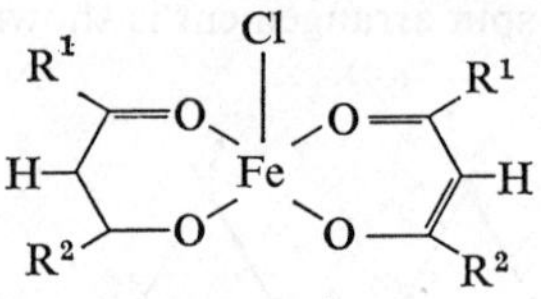

(2). (a) $R^1 = R^2 = Me$
(b) $R^1 = Me, R^2 = Ph$
(c) $R^1 = R^2 = Ph$

external magnetic fields. In zero field the spectra consist of asymmetric quadrupole doublets, which become symmetric as the temperature approaches absolute zero. This behaviour is interpreted as arising from zero-field splitting of the sextet ground-state with the parameter D positive. Interesting spectra are obtained in the presence of external magnetic fields at low temperature. Compound (2c) behaves as a simple paramagnet at 4.2—1.7 K, a six-line spectrum with intensities in the ratio 3 : 4 : 1 : 1 : 4 : 3 and a total spread of *ca.* ± 7 mm s^{-1} being obtained (Figure 7). The other two compounds give only poorly resolved patterns with a very small overall spread (*ca.* ± 3 mm s^{-1}) in a field of 30 kG (Figures 8 and 9). This is taken as evidence of intramolecular antiferromagnetism involving a binary interaction. An unfortunate printing error in the original paper resulted in the figures being presented in an incorrect order and with the wrong captions. All of the figures are reproduced here to remove this possible source of misunderstanding.[222]

Spectra have been reported for the high-spin iron(III) tris-(*OO'*-sulphinate) complexes $Fe(MeSO_2)_3$, $Fe(PhSO_2)_3$, and $Fe(p\text{-}MeC_6H_4SO_2)_3$.[223]

Magnetically perturbed spectra have been obtained for the three strong-exchange-coupled intramolecular antiferromagnets, $[Fe(salen)]_2O$, $[Fe(phen)_2Cl]_2OCl_2,5H_2O$, and $[FeB(H_2O)]_2O(ClO_4)_4$ {B = the macrocyclic

[220] G. M. Bancroft, K. G. Dharmawardena, and A. J. Stone, *Chem. Comm.*, 1971, 6.
[221] D. Hanzel, M. Schara, I. Levstik, and A. Moljk, *Fizika*, 1971, **3**, 117.
[222] M. Cox, B. W. Fitzsimmons, A. W. Smith, L. F. Larkworthy, and K. A. Rogers, *J. Chem. Soc.* (*A*), 1971, 2158.
[223] E. König, E. Lindner, I. P. Lorenz, G. Ritter, and H. Gausmann, *J. Inorg. Nuclear Chem.*, 1971, **33**, 3305.

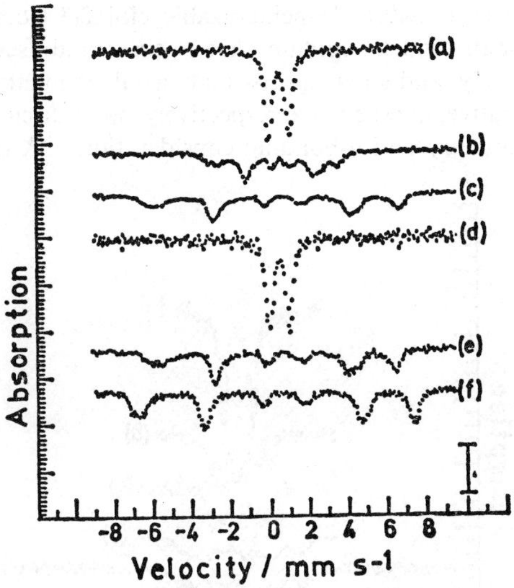

Figure 7 *Mössbauer spectra* (*absorption scale* 10%) *of compound* (2c) *at* (a) 4.2 K; (b) 4.2 K, 15 kG *magnetic field;* (c) 4.2 K, 30 kG; (d) 1.8 K; (e) 1.8 K, 15 kG; *and* (f) 1.8 K, 30 kG

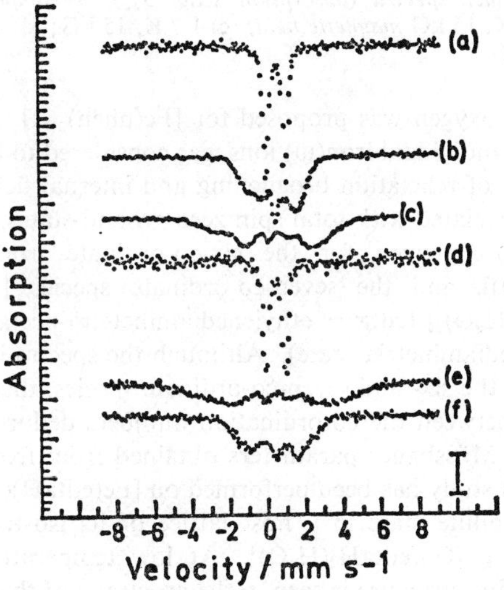

Figure 8 *Mössbauer spectra* (*absorption scale* 5%) *of compound* (2b) *at* (a) 4.2 K; (b) 4.2 K, 15 kG *magnetic field;* (c) 4.2 K, 30 kG; (d) 1.8 K; (e) 1.8 K, 30 kG; *and* (f) 1.8 K, 15 kG

ligand 2,13-dimethyl-3,6,9,12,18-penta-azabicyclo[12,3,1]octadeca-1(18),2,-12,14,16-pentaene}, which contain five-, six-, and seven-co-ordinate iron(III) respectively, and show approximate axial symmetry. The signs of V_{zz} (positive, negative, and negative respectively) were discussed in terms of point-charge geometry and σ-bonding considerations. A *cis* arrangement

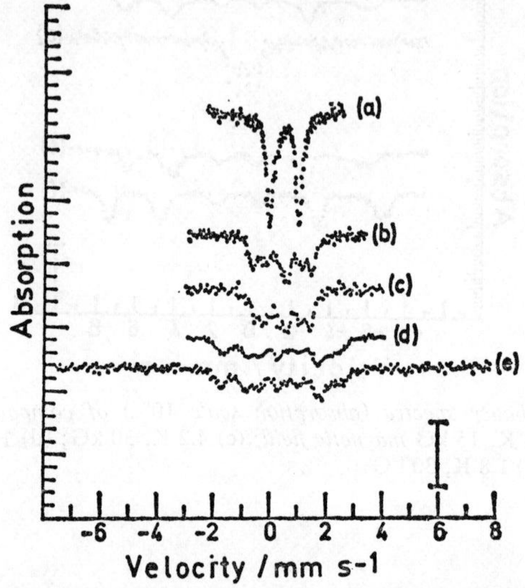

Figure 9 *Mössbauer spectra (absorption scale* 5%) *of compound* (2a) *at* (a) 1.7 K; (b) 4.2 K, 15 kG *magnetic field;* (c) 1.7 K, 15 kG; (d) 4.2 K, 30 kG; *and* (e) 1.7 K, 30 kG

of chlorine and oxygen was proposed for $[Fe(phen)_2Cl]_2OCl_2,5H_2O$. The spin state of the individual iron(III) ions was considered to be $\frac{5}{2}$ in each case and the absence of relaxation broadening and internal field augmentation at 4.2 K was correlated with total spin zero ground-states.[224]

Data have been given for the six-co-ordinate iron(III) compound $[Fe(edtaH)(H_2O)]$, and the seven-co-ordinate species $[Fe(edta)(H_2O)]^-$ and $[Fe(hedta)(H_2O)_2]$ (edta = ethylenediaminetetra-acetate and hedta = hydroxyethylenediaminetriacetate). Although the spectra in the solid state are different for the six- and seven-co-ordinate species, there appears to be no correlation between the co-ordination numbers deduced from the *d–d* spectra and the Mössbauer parameters obtained from frozen solutions.[225] A more detailed study has been performed on $[Fe(edtaH)(H_2O)]$ present in a magnetically dilute state in a host lattice of its isostructural and isomorphous analog $[Ga(edtaH)(H_2O)]$. At low temperature, complicated magnetic hyperfine structure is seen in the spectra, and this is considerably

[224] W. M. Reiff, *J. Chem. Phys.*, 1971, **54**, 4718.
[225] K. Garbett, G. Lang, and R. J. P. Williams, *J. Chem. Soc.* (*A*), 1971, 3433.

sharpened by application of spin-decoupling fields of *ca.* 100 G. The measurements agree well with calculated spectra based on crystal field splittings derived from the e.s.r. spectra.[226]

The Mössbauer effect has been used to study solvation effects of five-co-ordinate iron(III) complexes with sulphur-containing ligands. Bis-(*NN*-diethyldithiocarbamato)iron(III) chloride and thiocyanate, the tetrabutylammonium salts of bis(maleonitriledithiolato)iron(III), and bis(toluene-dithiolato)iron(III) all give large reductions in quadrupole splitting on going from the solid state to the frozen solution. This is thought to be due to solvation changing the co-ordination number from five to six. Some four-line spectra observed for the solid state are explained as superposition spectra of partly five- and partly six-co-ordinated iron. For example, this situation is thought to arise from the dimerization:

$$2Fe(dtc)_2(Cl)(DMF) \rightleftharpoons [Fe(dtc)_2Cl]_2 + 2DMF$$

which is thought to occur on evaporation of the solvent DMF.[227]

The spectrum of the bis(carbollyl)iron(III) complex, $Me_4N[Fe(C_2B_9H_{11})_2]$ has been re-examined and found to be an asymmetric quadrupole-split doublet, rather than a broad singlet as reported earlier. The asymmetry was thought to be due to relaxation effects, although a small asymmetry due to the Goldanskii–Karyagin effect was not completely ruled out. Line-broadening in the single-line absorption spectra of the related ferrocinium salts $[Fe(\pi\text{-}C_5H_5)_2^+][X^-]$ ($X = FeCl_4^-$, picrate, BF_4^-, PF_6^-, Cl^-, or BPh_4^-) was also attributed to spin-relaxation processes, rather than to unresolved quadrupole splitting.[228]

Spin-crossover Systems and Biological Compounds. There have been only a few reports during the year on spin-crossover and biological systems, and no other systems containing iron in unusual oxidation states have been studied.

Magnetic and Mössbauer measurements have shown that the iron(II) complexes $[Fe(ppa)_2]X_2,nH_2O$ [ppa = *N'*-(2-pyridylmethyl)picolinamidine, 3(a), X = univalent anion] exist in a temperature-dependent high-spin ⇌ low-spin equilibrium in accordance with a prediction based on the ligand field spectra of the nickel(II) complexes. However, in the Mössbauer

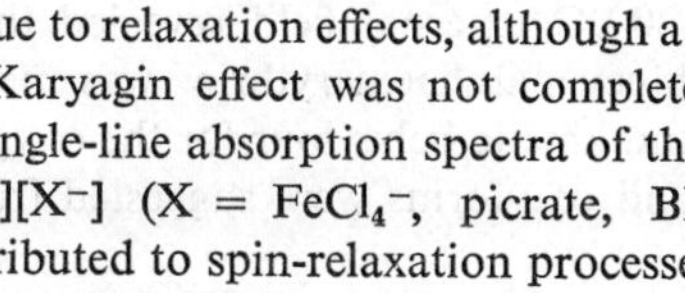

(3) (a) $R^1 = R^2 = H$ (ppa)
(b) $R^1 = Me$, $R^2 = H$ (Meppa)
(c) $R^1 = R^2 = Me$ (Me_2ppa)

[226] G. Lang, R. Aasa, K. Garbett, and R. J. P. Williams, *J. Chem. Phys.*, 1971, **55**, 4539.
[227] J. L. K. F. de Vries, J. M. Trooster, and E. de Boer, *Inorg. Chem.*, 1971, **10**, 81.
[228] T. Birchall and I. Drummond, *Inorg. Chem.*, 1971, **10**, 399.

spectrum at 293 K, no signals attributable to the high-spin ($S = 2$) form could be discerned, because < 6% of the PF_6^- salt and < 3% of the ClO_4^- salt is present in the high-spin form at this temperature. By contrast, the iron complexes of the substituted ligands [3(b) and (c)] are fully high-spin over the temperature range investigated (4.2—300 K).[229] The temperature dependence of the $^5T_2$—$^1A_1$ energy separation at the ground-state crossover in $[Fe^{II}–N_6]$ complexes has also been discussed, but no new data were given.[230]

The haeme-iron in metmyoglobin of horse heart has been shown to be in the high-spin iron(III) oxidation state, whereas the H_2O_2 complex of metmyoglobin is thought to contain quadrivalent iron. Magnetic hyperfine splitting is observed at 4.2 K in the spectrum of the former.[231] Investigation of the mean-square displacement of the iron atoms in crystals of myoglobin, using Mössbauer and *X*-ray methods, has given information about the vibrational portion and the crystal lattice portion of the temperature factors in this material.[232]

Oxyhaemerythrin and various methaemerythrin derivatives have been shown to contain antiferromagnetically coupled iron(III) ions, which are in non-equivalent environments in the former but are equivalent in the latter. Experiments with a 30 kG external field revealed that the electric field gradient in metisothiocyanatohaemerythrin was negative in sign. A possible mechanism and kinetic behaviour for the oxygenation equilibrium between the deoxy- and oxy-forms were suggested from a comparison of their structures.[233]

The protein *Scenedesmus* ferredoxin enriched in ^{57}Fe has been shown to give spectra identical with those of the ferredoxins from spinach and *Euglena*. The oxidized state contains two antiferromagnetically coupled high-spin iron(III) ions in slightly different environments, and the addition of one electron to the system reduces one of these to high-spin iron(II).[234] Similar results have been obtained for the two-iron ferridoxins from spinach, parsley, *Azotobacter vinelandii*, *Clostridium pasteurianum*, and pig adrenal cortex,[235] for putidaredoxin,[236] and for adrenodoxin purified from porcine adrenal glands.[236, 237] A model iron–sulphur protein, formed by reduction of S—S links in albumin by means of $HOCH_2{\cdot}CH_2SH$ in the presence of iron and sulphide ions, has been shown to contain high-spin iron(III). The full hyperfine pattern was not observed at 4.2 K, owing to

229 M. J. Boylan, S. M. Nelson, and F. A. Deeney, *J. Chem. Soc.* (*A*), 1971, 976.
230 E. König and S. Kremer, *Chem. Phys. Letters*, 1971, **8**, 312.
231 Y. Morita, Y. Maeda, and C. Yoshida, *J. Biochem.* (*Tokyo*), 1971, **70**, 509.
232 F. Parak and H. Formanek, *Acta Cryst.*, 1971, **A27**, 573.
233 K. Garbett, C. E. Johnson, I. M. Klotz, M. Y. Okamura, and R. J. P. Williams, *Arch. Biochem. Biophys.*, 1971, **142**, 574.
234 K. K. Rao, R. Cammack, D. O. Hall, and C. E. Johnson, *Biochem. J.*, 1971, **122**, 257.
235 W. R. Dunham, A. J. Bearden, I. T. Salmeen, G. Palmer, R. H. Sands, W. H. Orme-Johnson, and H. Beinert, *Biochim. Biophys. Acta*, 1971, **253**, 134.
236 R. Cammack, C. E. Johnson, D. O. Hall, and K. K. Rao, *Biochem. J.*, 1971, **125**, 18p.
237 R. Cammack, K. K. Rao, D. O. Hall, and C. E. Johnson, *Biochem. J.*, 1971, **125**, 849.

spin relaxation, but the high-velocity component of the quadrupole doublet was broadened. Some nitrosyl derivatives of the protein were also discussed.[238]

The paramagnetic Mössbauer spectra of a number of other biological molecules such as ferritin,[239] mixed-spin haemes, peroxidase complexes, ferridoxins, and nitrogen-fixing bacterial proteins have been interpreted in terms of their electronic behaviour.[240] The nitrogen-fixing protein *Ectothiorhodospira shaposhnikovii* has also been studied.[241]

Low-spin and Covalent Compounds. The vacuum pyrolysis of iron(III) ferrocyanide has been followed by Mössbauer spectroscopy and shown to yield iron(II) ferricyanide, $Fe_3[Fe(CN)_6]_2$.[242] Redox reactions in various mixtures of $FeCl_2$, $FeCl_3$, $K_3[Fe(CN)_6]$, and $K_4[Fe(CN)_6]$ have also been studied.[243] Isotopic labelling with Mössbauer active (^{57}Fe) and inactive (^{56}Fe) iron isotopes has been used to simplify the spectra of Turnbulls' and Prussian blues and confirm their identity as $K^+Fe^{3+}[Fe^{II}(CN)_6{}^{4-}]$.[244] The Lamb–Mössbauer factor in $Na_4[Fe(CN)_6]$ has been estimated to be 0.28 ± 0.03.[245] The ferroelectric phase transition in $K_3Fe(CN)_6,3H_2O$ has been investigated.[246]

Spectra have been reported for several structurally related alkali-metal nitroprussides $M_2[Fe(CN)_5NO],xH_2O$ (M = Li, Na, K, Rb, Cs, NH_4, Ag, or Cu), and bivalent transition-metal nitroprussides $M[Fe(CN)_5NO],xH_2O$ (M = Mn, Fe, Co, Ni, Cu, Zn, or Cd). $Fe[Mn(CN)_5NO],2H_2O$ was also studied. The spectra suggest that the nitroprusside complex ion is present in all of the compounds and that the electronic environment of the iron atom is insensitive to change in the outer cation.[247] The positive sign for the electric field gradient at the iron nucleus in this anion has been confirmed from experiments using a single crystal and a magnetic field of 23.5 kG.[248] From a consideration of the known physical (including Mössbauer data) and chemical properties of the nitroprusside ion, it has been suggested that its formulation as an Fe^{2+}–NO^+ complex is misleading; rather it should be grouped with ions known to contain Fe^{3+}.[249] A reinterpretation has been given of the chemical isomer shifts and of the signs and magnitudes of the electric field gradients at the iron nuclei in

[238] G. V. Novikov, E. N. Frolov, L. Cher, G. I. Likhtenstein, L. A. Syrtsova, V. A. Trukhtanov, and V. I. Goldanskii, *Doklady Akad. Nauk S.S.S.R.*, 1971, **196**, 390.
[239] J. F. Boas and G. J. Troup, *Biochim. Biophys. Acta*, 1971, **229**, 68.
[240] G. Lang, *J. Phys. (Paris), Colloq.*, 1971, **1** (Pt. 2), C1–822.
[241] Y. S. Moshkovskii, N. Y. Uspenskaya, and S. S. Mardanyan, *Biofizika*, 1971, **16**, 933.
[242] D. S. Murty, J. G. Cosgrove, and R. L. Collins, *Nature*, 1971, **231**, 311.
[243] B. V. Bonshagovski, V. I. Goldanskii, G. B. Seifer, and R. A. Stukan, *Izvest Akad. Nauk S.S.S.R., Ser. khim.*, 1971, 1016.
[244] A. K. Bonnette, jun., and J. F. Allen, *Inorg. Chem.*, 1971, **10**, 1613.
[245] D. L. Decker and L. E. Lortz, *J. Appl. Phys.*, 1971, **42**, 830.
[246] P. A. Montano, H. Schechter, and U. Shimony, *Phys. Rev. (B)*, 1971, **3**, 858.
[247] A. N. Garg and P. S. Goel, *Inorg. Chem.*, 1971, **10**, 1344.
[248] M. H. Clear, J. F. Duncan, and D. McConchie, *J. Roy. Soc. New Zealand*, 1971, **1**, 71.
[249] D. Brown, *Inorg. Chim. Acta*, 1971, **5**, 314.

the reduced forms of sodium nitroprusside $[Fe(CN)_5NO]^{3-}$ and $[Fe(CN)_5NO_2]^{n-}$ (n = 3, 4, or 5). $Fe(NO)(S_2CNEt_2)_2$ was also discussed.[250]

The i.r. and Mössbauer spectra of the adducts $[Fe(CNR)_4(CN,BX_3)_2]$ (R = H, Me, or Et; X = F or Cl) have been interpreted in terms of structures in which the Lewis acids are linked to the nitrogen atoms of the cyanide group. The spectra are thought to be consistent with local octahedral symmetry at the iron atom and to indicate that the groups CNR and CN,BX_3 have similar π-acid ligand properties. From the chemical isomer shift it is concluded that $:C{\equiv}N{-}BF_3$ is a better π-acceptor from iron(II) than is $:C{\equiv}N{:}^-$, and that $:C{\equiv}N{-}R$ and $:C{\equiv}N{-}H$ are similar to one another in this respect.[251]

An unusually large asymmetry in the room-temperature spectrum of $[(PhCH_2NC)_4Fe(CN)_2]$ has been shown to disappear at 4.2 K and is therefor attributed to the Goldanskii–Karyagin effect. Other possible factors, such as magnetic relaxation and particle orientation effects, were ruled out on the basis of this material being both diamagnetic and randomly orientated. The compounds $[(PhCH_2NC)_5Fe(CN)]Br,H_2O$ and $[(PhCH_2NC)_5Fe(CN)]SCN$ were also studied, but the results were not discussed in detail.[252]

The signs of V_{zz} (and e^2qQ) have been shown to be positive in *trans*-$[FeCl_2(p\text{-}MeO{\cdot}C_6H_4{\cdot}NC)_4]$, *trans*-$[FeCl_2(depe)_2]$ (depe = 1,2-diethylphosphinoethane), $Na_3[Fe(CN)_5NH_3],H_2O$, and $K_3[Fe(CN)_5H_2O],7H_2O$, and negative in *cis*-$[FeCl_2(p\text{-}MeO{\cdot}C_6H_4{\cdot}NC)_4]$ and *trans*-$[Fe(CN)_2(EtNC)_4]$. The 2 : 1 *trans* : *cis* ratio was confirmed and partial quadrupole splittings (p.q.s.) assigned for sixteen ligands. The p.q.s. values become more positive as the π-acceptor function of the ligand increases (see Table 2)

Table 2

NO^+	+ 0.02	N_3^-	− 0.38	NCO^-	− 0.50	½dmpe	− 0.67
Br^-	− 0.28	$SnCl_3^-$	− 0.43	NH_3	− 0.51	RNC	− 0.69
I^-	− 0.29	H_2O	− 0.44	½depb	− 0.58	CN^-	− 0.84
Cl^-	− 0.30	NCS^-	− 0.49	½depe	− 0.62	H^-	− 1.04

and as the σ-donor ability decreases [*i.e.* p.q.s. $\propto (\pi - \sigma)$].[253] V_{zz} has been shown to have a positive sign in both $Fe(bipy)_3(ClO_4)_3$ and $Fe(en)_3Cl_3$ and the results have been interpreted in terms of negative trigonal fields and orbital doublet ground levels.[254] Spectra have been reported for 24 low-spin iron(II) and iron(III) derivatives with 1,10-phenanthroline and 2,2′-bipyridyl ligands.[255]

[250] J. B. Raynor, *J. Inorg. Nuclear Chem.*, 1971, **33**, 735.
[251] D. Hall, J. H. Slater, B. W. Fitzsimmons, and K. Wade, *J. Chem. Soc.* (*A*), 1971, 800.
[252] N. Malathi and S. P. Puri, *Chem. Phys. Letters*, 1971, **10**, 154.
[253] G. M. Bancroft, R. E. B. Garrod, and A. G. Maddock, *J. Chem. Soc.* (*A*), 1971, 3165.
[254] W. M. Reiff, *Chem. Phys. Letters*, 1971, **8**, 297.
[255] V. K. Garg, N. Malathi, and S. P. Puri, *Chem. Phys. Letters*, 1971, **11**, 393.

The temperature dependence of the Mössbauer spectra of some iron-containing compounds of dithioacetylacetone has been studied between 4 and 293 K. The quadrupole splitting of [Fe(sacsac)$_3$] is larger than that usually found for low-spin six-co-ordinated iron(III) and is ascribed to the lifting of the orbital degeneracy of the $^2T_2$ ground state by small axial and rhombic distortions.[209]

The compounds [Fe(ttd)$_2$(dtt)], [Fe(ttd)(dtt)$_2$], and [Fe(dtt)$_3$,CHCl$_3$] (ttd = trithioperoxy-*p*-toluate, dtt = dithio-*p*-toluate) have been shown to have Mössbauer parameters which are similar to one another and characteristic of low-spin iron(III). There is a slight increase in chemical isomer shift along the series because of the better π-acceptor ability of ttd compared with dtt. The sign of V_{zz} is negative for the first two compounds and positive for the third. Taken in conjunction with e.p.r. data, these results suggest that the ground state of the first two compounds is a d_{xy} hole, whereas for the third compound, for which no e.p.r. spectrum was detectable, the ground state is probably an equal mixture of d_{xz} and d_{yz} holes.[256]

The low-spin iron(III) complex (Ph$_4$P)$_3$[Fe{S$_2$C$_2$(CN)$_2$}$_3$] has been studied over a wide temperature range (1.3—300 K) and in various applied magnetic fields. At 300 K the spectrum consists of a symmetrical doublet with a large quadrupole splitting [Figure 10(a)]. As the temperature is lowered, magnetic hyperfine broadening of the left-hand peak, due to increasing electron spin relaxation times, is observed [Figure 10(b) and (c)]. The presence of a small (5 kG) external magnetic field appears to increase the relaxation rate [Figure 10(d) and (f)], and a mechanism for this effect is suggested. The large value of the quadrupole splitting ($\frac{1}{2}e^2qQ = -1.85$ mm s^{-1} at 77 K) indicates a strong distortion from cubic symmetry. The negative sign was deduced from spectra [Figure 10(g) and (f)] obtained in the presence of a large external magnetic field. The spectra also reveal that V_{zz} is parallel to the major axis of the magnetization tensor. These results indicate that the ground state of the low-spin iron(III) ion is a d_{xy} hole, well separated from higher states; this conclusion is in disagreement with previous speculations of Cotton and Gibson which were based solely on e.p.r. data.[257]

The interpretation of magnetically split Mössbauer spectra of $S = \frac{1}{2}$ iron compounds has been treated in detail. The treatment was limited to those paramagnetic systems in the limiting cases of extremely long ($\tau \gg 10^{-7}$ s) or extremely short ($\tau \ll 10^{-7}$ s) electron spin relaxation times. Calculated spectra were presented for each case and compared with experimental spectra for powders and in some cases single crystals in both zero and small applied magnetic fields.[258]

256 R. Rickards, C. E. Johnson, and H. A. O. Hill, *J. Chem. Soc. (A)*, 1971, 1755.
257 R. Rickards, C. E. Johnson, and H. A. O. Hill, *J. Chem. Soc. (A)*, 1971, 797.
258 W. T. Oosterhuis, ref. 28, p. 97.

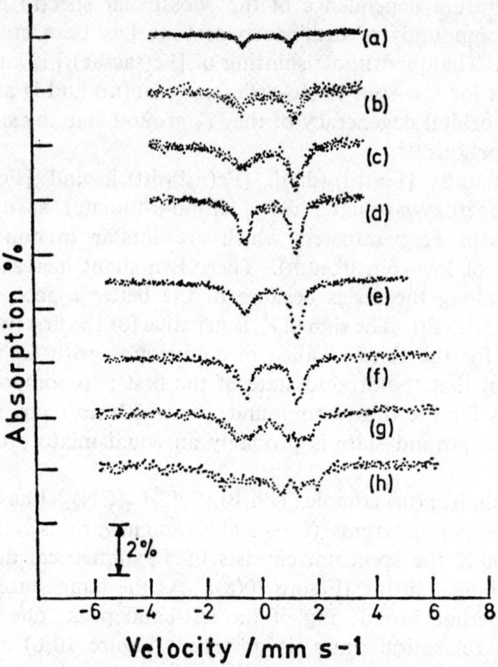

Figure 10 *Mössbauer spectra of* $(Ph_4P)_3[Fe\{S_2C_2(CN)_2\}_3]$: (a) *at* 300 K; (b) *at* 195 K; (c) *at* 77 K; (d) *at* 77 K *in a* 5 kG *external field;* (e) *at* 4.2 K; (f) *at* 4.2 K *in a* 5 kG *external field;* (g) *at* 4.2 K *in a* 30 kG *external field;* (h) *at* 1.6 K *in a* 30 kG *external field*

From i.r. and Mössbauer data it has been suggested that the compounds $MFe(CO)_4$ (M = Cd or Hg) consist of infinite zig-zag chains in which the M atoms link $Fe(CO)_4$ units (4). The co-ordination at the iron is *cis*-octahedral so that only a very small quadrupole splitting results. The Cd—Fe and Hg—Fe bonds are thought to be covalent, rather than ionic as in $Na_2Fe(CO)_4$. The isomorphism of the two compounds was confirmed by *X*-ray diffraction.[259] The i.r. and Mössbauer spectra of eight

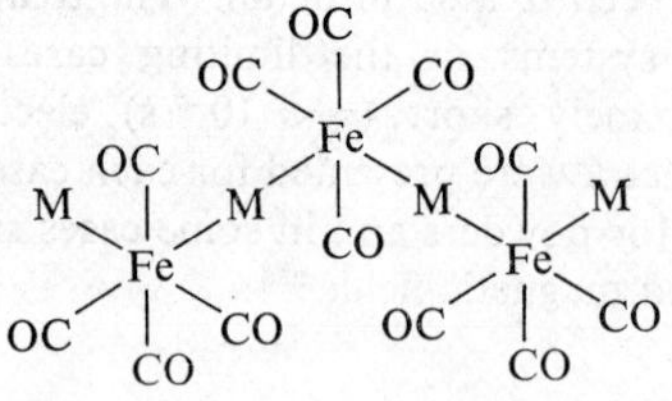

(4)

[259] T. Takano and Y. Sasaki, *Bull. Chem. Soc. Japan*, 1971, **44**, 431.

complexes of the types $L_2MFe(CO)_4$ (M = Zn or Cd; L = nitrogen donor ligand) also favour polymeric structures, containing octahedrally co-ordinated low-spin iron(II), rather than monomeric trigonal-bipyramidal or ionic $[L_2M^{2+}][Fe(CO)_4{}^{2-}]$ arrangements.[260]

Data have been reported for compounds of the types (L–L)$Fe(NO)_2$, $LFe(NO)_2X$, $[Fe(NO)_2X]_2$, and $LFe(CO)(NO)_2$ [where L is a tertiary phosphine, phosphite, or arsine and L–L = $Me_2AsCCF_3{=}CCF_3AsMe_2$ or $Ph_2PC{=}CPPh_2(CF_2)_n$ (*n* = 2, 3, or 4)]. The π-acceptor strengths of L were shown to increase in the order PPh_3 < Ph_2PMe ~ Ph_3As < $(PhO)_3P$ < CO < NO, and the ditertiary phosphines in the compounds (L–L)$Fe(NO)_2$ were thought to be similar to PPh_3. The sign of V_{zz} in the fluorocarbon-bridged derivatives is predicted to be positive.[261]

Data for the π-complexes of 1*H*-1,2-diazepines (5)—(7) have been interpreted in terms of a quasi-octahedral configuration for the iron,

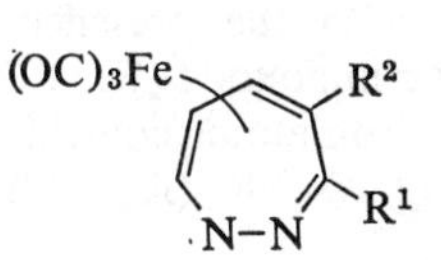

(5) (a) $R^1 = R^2 = H$
(b) R^1 = Me, R^2 = H
(c) R^1 = H, R^2 = Me

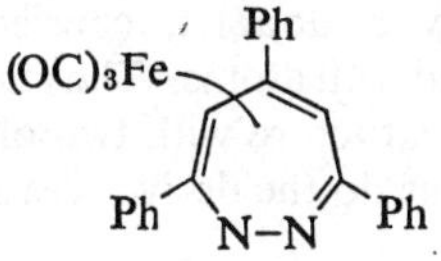

(6)

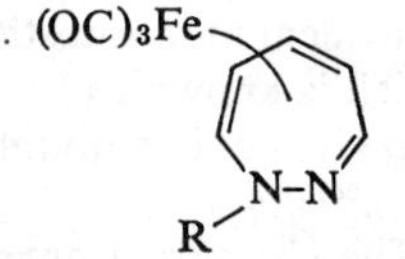

(7) (a) R = $SO_2C_7H_7$
(b) R = COPh

involving two σ bonds and one μ bond from the diene system to the iron atom.[262]

The increase in chemical isomer shift from 0.03 to 0.16 mm s^{-1}, which occurs when a nitrogen donor ligand in $[Fe_2(CO)_6(N{:}CPh_2)_2]$ is replaced by iodine to give $[Fe_2(CO)_6I(N{:}CPh_2)]$, has been discussed in terms of the poorer σ-donor and π-acceptor properties of iodine compared with nitrogen. The observed increase in the quadrupole splitting is thought to reflect a tendency towards five-co-ordination at iron.[263]

The presence of non-equivalent iron environments has been confirmed in the complex $[Fe_2(CO)_5NEt_2(CONEt_2)_2]$, which contains $CONEt_2$ as a novel bridging ligand.[264]

Mössbauer spectroscopy has helped to elucidate the structures of products of the reactions between $[Fe(CO)_3SR]_2$ (R = Me, Et, or Ph) and the ditertiary phosphines L = $Ph_2PCH_2PPh_2$, *cis*-$Ph_2PC_2H_2PPh_2$, and $Ph_2PNEtPPh_2$, and the ditertiary arsine L = $Ph_2AsCH_2AsPh_2$. Five

260 A. T. T. Hsieh, M. J. Mays, and R. H. Platt, *J. Chem. Soc.* (*A*), 1971, 3296.
261 J. P. Crow, W. R. Cullen, F. G. Herring, J. R. Sams, and R. L. Tapping, *Inorg. Chem.*, 1971, **10**, 1616.
262 A. J. Carty, G. Kan, D. P. Madden, V. Snieckus, M. Stanton, and T. Birchall, *J. Organometallic Chem.*, 1971, **32**, 241.
263 M. Kilner and C. Midcalf, *Chem. Comm.*, 1971, 944.
264 E. O. Fischer and V. Kiener, *J. Organometallic Chem.*, 1971, **27**, C56.

different structural types were distinguished. These include $[Fe_2(CO)_5L(SR)_2]$ and $[Fe(CO)_2LSR]_2$ in which the ligands are unidentate, $[\{Fe(CO)_2SR\}_2L]$ in which the ligand bridges the two iron atoms, $[Fe(CO)_3(SR)_2Fe(CO)L]$ in which the ligand chelates a single iron atom, and $[Fe_2(CO)_3L_2(SR)_2]$ in which one ligand is unidentate and the other bidentate. The monosubstituted $[Fe_2(CO)_5L^1(SR)_2]$ (R = Me; L^1 = $AsPh_3$ or $SbPh_3$), bis-substituted $[Fe(CO)_2L^1SR]_2$ [R = Me; L^1 = PPh_3, $AsPh_3$, or $SbPh_3$; R = Ph, L^1 = $P(OMe)_3$] and tris-substituted $[Fe_2(CO_3$-$L_3{}^1(SR)_2]$ [R = Me, L^1 = $P(OPh)_3$] derivatives were also studied, and new data recorded for the parent compounds $[Fe(CO)_3SR]_2$ (R = Me, Et, or Ph). In addition to these structural aspects, systematic variations in the chemical isomer shift and quadrupole splitting were discussed in relation to the bonding in these compounds.[265, 266] In a related study it has been shown that differences noted in the spectra of the *syn*- and *anti*-isomers of $[Fe(CO)_3SMe]_2$ are preserved in its ditertiary phosphine and arsine derivatives. Derivatives of the type $[\{Fe(CO)_2SMe\}_2(L\text{–}L)]$ (L–L = $f_4$fars or $f_4$fos) give only a doublet, consistent with the presence of equivalent monosubstituted iron atoms. The cluster compound $S_2Fe_3(CO)_9$, which is known to have a structure with two of the iron atoms equivalent, also gives only a doublet, as do the derivatives $S_2Fe_3(CO)_8PR_3$ (R = Ph or Bu^n).[267]

The chemical isomer shift has been shown to increase along the series $[Fe(\pi\text{-}C_5H_5)(CO)_2X]$ (X = CS, CO, CN, PPh_3, NCO, NCS, or Br), indicating a decrease in the π-acceptor ability of the ligands in the order given. Some attempt was made to use the data to derive 'partial isomer shifts' for the various ligands.[268] Data have also been given for the complex salts $[Fe(\pi\text{-}C_5H_5)PhX]$ (X = H, OMe, NHAc, F, Me, NH_2, CO_2H, $CONH_2$, CN, NH_2, or Cl), $[Fe(\pi\text{-}C_5H_5)(p\text{-}MeC_6H_4X)]PF_6$ (X = CO_2H, Et, OMe, NHAc, Cl, or NH_2), and $[Fe(\pi\text{-}C_5H_4X)(\pi\text{-}C_6H_6)]PF_6$ (X = NHAc, Et, MeS, Me, CN, Cl, NH_2, F, or MeO). The chemical isomer shift is practically independent of the type of substituent, but the quadrupole splitting shows a more sensitive dependence: the latter decreases as the electronegativity of the substituent in the benzene ring is increased but increases as the electronegativity of the substituent in the cyclopentadienyl ring is increased. From MO calculations and the observed linear correlation between the quadrupole splitting and the Hammett–Taft inductive constant, an electron occupying a d_{xy} or $d_{x^2-y^2}$ orbital was calculated to contribute 5.4 mm s^{-1} to the quadrupole splitting, and an electron occupying a d_{xz} or d_{yz} orbital to contribute $-$ 2.7 mm s^{-1}.[269]

[265] J. A. de Beer, R. J. Haines, R. Greatrex, and N. N. Greenwood, *J. Organometallic Chem.*, 1971, **27**, C33; *ibid.*, **29**, C22.
[266] J. A. de Beer, R. J. Haines, R. Greatrex, and N. N. Greenwood, *J. Chem. Soc.* (*A*), 1971, 3271.
[267] J. P. Crow and W. R. Cullen, *Canad. J. Chem.*, 1971, **49**, 2948.
[268] K. Burger, L. Korecz, P. Mag, U. Belluco, and L. Busetto, *Inorg. Chim. Acta*, 1971, **5**, 362.
[269] K. I. Turta, R. A. Stukan, V. I. Goldanskii, N. A. Vol'kenau, E. I. Sirotkina, I. N. Bolesova, L. S. Isaeva, and A. N. Nesmeyanov, *Teor. i eksp. Khim.*, 1971, **7**, 486.

The nature of the iron–tin bond in the complexes [$Fe(\pi\text{-}C_5H_5)(CO)$-$LSnR_3$] (R = Me, Ph, or Cl; L = phosphine, arsine, or stibine) and [$Fe(\pi\text{-}C_5H_5)L_2SnR_3$] has been studied with the aid of Mössbauer, i.r., and n.m.r. spectra. The fact that the chemical isomer shift for both the ^{57}Fe and ^{119}Sn nuclei increases upon introduction of L is thought to indicate that π-bonding is of little importance. For the methyl compounds the chemical isomer shift increases along the series: L = CO, $(PhO)_3P$, Ph_2PCF_3, $f_6$fos, Ph_3P, Ph_3As, Ph_2AsCF_3, or Ph_3Sb, and it is suggested that this ordering reflects the decreasing ability of the ligands to act as π-acceptors.[270] The sign of the quadrupole coupling constant e^2qQ (and hence the sign of V_{zz}) has been shown to be positive for the ^{57}Fe nucleus in the compounds [$Fe(\pi\text{-}C_5H_5)(CO)_2SnCl_3$], [$Fe(\pi\text{-}C_5H_5)(CO)_2SnBu_3$], [{Fe-($\pi\text{-}C_5H_5)(CO)_2\}_2SnCl_2$], and [$\{Fe(\pi\text{-}C_5H_5)(CO)_2\}_2Sn(NCS)_2$]. The significance of these results was not discussed as the ^{57}Fe resonance was only of secondary interest in this study. The paper was mainly concerned with the electric field gradient at the ^{119}Sn nucleus and this is discussed in a later section.[271] The positive sign of e^2qQ in [$Fe(\pi\text{-}C_5H_5)(CO)_2SnCl_3$] has been confirmed elsewhere.[272]

The spectra of the complex cations [$\{Fe(\pi\text{-}C_5H_5)(CO)_2\}_2X]^+$ (X = H or Cl) consist of sharp doublets, consistent with the presence of equivalent iron atoms and a symmetrically bridging X atom in these species.[273] The spectra of the novel iron carbonyl derivatives (8) and (9) confirm the

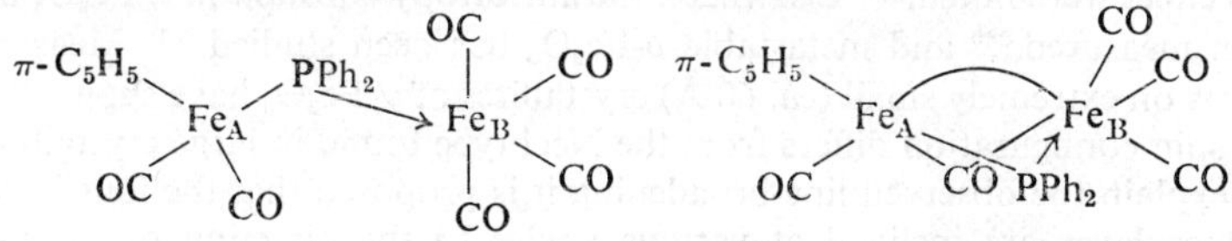

(8) (9)

non-equivalence of the iron atoms and indicate five-co-ordination for Fe_B in (8). The cyclopentadienyl-bearing iron atom gives the larger chemical isomer shift of the two iron atoms.[274] Although more than one isomeric form of the polynuclear iron–nickel carbonyl compound [$FeNi_2(\pi\text{-}C_5H_5)_2(CO)_5$] may be present in solution, the spectrum of the crystalline solid shows a single sharp quadrupole-split doublet, which suggests that perhaps only one form is present in the solid.[275]

270 W. R. Cullen, J. R. Sams, and J. A. J. Thompson, *Inorg. Chem.*, 1971, **10**, 843.
271 B. A. Goodman, R. Greatrex, and N. N. Greenwood, *J. Chem. Soc.* (*A*), 1971, 1868.
272 S. R. A. Bird, J. D. Donaldson, A. F. Le C. Holding, B. J. Senior, and M. J. Tricker, *J. Chem. Soc.* (*A*), 1971, 1616.
273 D. A. Symon and T. C. Waddington, *J. Chem. Soc.* (*A*), 1971, 953.
274 R. J. Haines, C. R. Nolte, R. Greatrex, and N. N. Greenwood, *J. Organometallic Chem.*, 1971, **26**, C45.
275 A. T. T. Hsieh and J. Knight, *J. Organometallic Chem.*, 1971, **26**, 125.

A temperature-dependence study of the compound $[Fe(\pi\text{-}C_5H_5)PhF]PF_6$ in the temperature range 80—300 K has given evidence for a phase transition at 215 K.[276]

X-Ray photoelectron spectra have been measured for the following six ferrocene compounds: ferrocene, biferrocene, ferrocene Fe^{III} picrate, ferrocene Fe^{III} fluoroborate, biferrocene $Fe^{II}Fe^{III}$ picrate, and biferrocene $Fe^{III}Fe^{III}$ fluoroborate, and the data correlated with the Mössbauer quadrupole splitting.[277]

Oxide and Sulphide Systems containing Iron.—This section is subdivided into four parts; the opening section deals with the binary oxides and this is followed by sections on spinel oxides and garnets, other oxide systems, and minerals. Only a few papers have reported work on sulphide materials and these fall conveniently into the section on minerals (see also ref. 203).

Binary Oxides. Spectra of ultrafine particles of α-Fe_2O_3 with differing mean size have been measured at various temperatures and applied external magnetic fields. Surface ions were found to have a faster electron spin relaxation and a larger mean-square displacement than the inner ions.[278] Mössbauer data have been shown to be incompatible with a recent proposal by Searle and Dean, which ascribes the anomalous temperature dependence of the weak ferromagnetic moment of haematite to a large temperature-dependent inclination of the antiferromagnetic axis out of the basal plane above the Morin temperature. Some possible alternative explanations of this effect were noted.[279] Elsewhere, the anisotropy constant in α-Fe_2O_3 has been measured,[280] and metastable δ-Fe_2O_3 has been studied.[281] Measurements on extremely small (*ca.* 60 Å) crystallites of γ-Fe_2O_3 have shown that the spin configuration differs from the Néel type found in large crystallites. To explain the observed line-broadening it is proposed that the ions in the surface layer are inclined at various angles to the direction of the nett moment.[282]

New low-temperature n.m.r. and Mössbauer data have been interpreted as further evidence against the validity of the Verwey model of magnetite, Fe_3O_4. Details are not given here, as the results parallel very closely those of Hargrove and Kündig reported last year – except that spectral resolution is improved.[283] A possible method for the evaluation of the electron-hopping time between the Fe^{2+} and Fe^{3+} sites in Fe_3O_4 has been discussed. The intensity of a forbidden line related to the B sublattice varies with relaxation time; measurement of the intensity of this line therefore

[276] R. A. Stukan, K. I. Turta, V. I. Goldanskii, A. M. Kaplan, N. A. Vol'kenau, and E. I. Sirotkina, *Teor. i eksp. Khim.*, 1971, **7**, 74.
[277] D. O. Cowan, J. Park, M. Barber, and P. Swift, *Chem. Comm.*, 1971, 1444.
[278] A. M. Van de Kraan, *J. Phys.* (*Paris*), *Colloq.*, 1971, **1** (Pt. 2), C1–1034.
[279] L. M. Levinson, *Phys. Rev.* (*B*), 1971, **3**, 3965.
[280] V. Beckmann, G. Ritter, H. Spiering, and H. Wegener, ref. 12, p. 31.
[281] G. V. Loseva and N. V. Murashko, *Izvest. Akad. Nauk S.S.S.R., Neorg. Materialy*, 1971, **7**, 1467.
[282] J. M. D. Coey, *Phys. Rev. Letters*, 1971, **27**, 1140.
[283] M. Rubinstein and D. W. Forester, *Solid State Comm.*, 1971, **9**, 1675.

offers a method for the determination of such relaxation times.[284] The Mössbauer spectrum of magnetite has been discussed in two other reports.[285, 286]

The controlled thermal degradation of ferricinium nitrate produces a material composed of iron oxide (mainly magnetite) particles embedded in an inert carbonaceous matrix. The small particle size causes the material to be superparamagnetic down to and below 77 K; at 4.2 K the material is ferromagnetic and exhibits exchange anistropy in the form of a shifted hysteresis loop. In the ferromagnetic state, local disorder around the magnetic ions manifests itself in a broadening of the spectra.[287] The state of iron resulting from the thermal decomposition of iron salts on carriers such as γ-Al_2O_3, SiO_2, MgO, Cr_2O_3, and ZnO has been studied. Spectra were also obtained for the reduction products of the finely divided iron oxides. The larger particles are reduced to ferromagnetic iron whereas the smaller particles are transformed into Fe^{2+}ions.[288] The properties of $Fe(OH)_2$ and Fe_3O_4 precipitated from alkaline solutions and of their thermal decomposition products have been studied in detail.[289]

The use of Mössbauer spectroscopy for the non-destructive analysis of corrosion products has been further demonstrated by the identification of γ-Fe_2O_3 and γ-FeOOH formed in the corrosion of cubic α-Fe in air at 25 and 100% humidity; Fe_3O_4, α-Fe_2O_3, and α- and β-FeOOH were not found.[290] The nature of the oxide films formed on iron after exposure to dry air and to chromate solutions at pH 4 has been studied. The first film contained > 75% Fe_3O_4–Fe_2O_3, whereas the chromate-formed film had only *ca.* 25% Fe_3O_4–Fe_2O_3. Fe^{3+} was also present in either the paramagnetic or superparamagnetic high-spin state and in sixfold co-ordination.[291] Iron deposited at 4.2 K on to a thin aluminium foil in the presence of residual oxygen and water at 10^{-6} Torr has been shown to give a six-line pattern (H_i = 460 kG, δ = + 0.42 mm s^{-1}), whereas a cooled beryllium foil substrate induces only a quadrupole splitting. When the latter is warmed and then recooled to 4.2 K, an unresolved magnetic hyperfine pattern (H_i = 475 kG) appears, which is due to an amorphous-to-crystalline transition leading to γ-FeOOH.[292]

Spinel Oxides and Garnets. The effect on the Mössbauer spectrum of doping lithium into α-Fe_2O_3 has been studied. At first the magnetic hyperfine field decreases as the lithium spinel $LiFe_5O_8$ is formed. However, above a mole fraction of 0.2 Li, a paramagnetic component appears in the spectrum.

[284] L. Cser and D. L. Nagy, ref. 12, p. 165.
[285] J. M. D. Coey, A. H. Morrish, and G. A. Sawatzky, *J. Phys.* (*Paris*), *Colloq.*, 1971, **1** (Pt. 1), C1–271.
[286] L. Cser, I. A. Gladkih, L. Keszthelyi, D. L. Nagy, and I. Vincze, ref. 12, p. 553.
[287] S. M. Aharoni and M. H. Litt, *J. Appl. Phys.*, 1971, **42**, 352.
[288] H. Hobert and D. Arnold, ref. 12, p. 325.
[289] A. M. Pritchard and B. T. Mould, *Corrosion Sci.*, 1971, **11**, 1.
[290] W. Meisel, *Z. Chem.*, 1971, **11**, 238.
[291] G. M. Bancroft, J. E. O. Mayne, and P. Ridgway, *Brit. Corrosion J.*, 1971, **6**, 119.
[292] W. Keune and U. Gonser, *Thin Solid Films*, 1971, **7**, R7.

This increases in intensity until at a mole fraction of 0.8 the magnetic hyperfine component is almost absent. The magnetic ordering caused by the superexchange interaction between iron atoms in adjacent octahedral sites is reduced as the probability of an iron having a next-near iron neighbour is decreased.[293]

A number of spinel systems derived from $LiFe_5O_8$ have been studied and are now discussed individually; these include $Li_{0.5}Fe_{2.5-x}Cr_xO_4$,[294] $Li^+_{1-x}Fe^{3+}_x[Li^+_{x-0.5}Fe^{3+}_{2.5-x-a}Cr^{3+}_a]O_4$,[295] $Zn_xLi_{0.5-x/2}Fe_{2.5-x/2}O_4$,[296] and $Li_{1.2}Fe_{4.6}Sb_{0.2}O_8$.

The compounds $Li_{0.5}Fe_{2.5-x}Cr_xO_4$ (x = 0, 1.4, or 2.0) undergo order–disorder transitions at temperatures (T_{o-d} = 1008—1028 K) higher than the Curie temperature (T_C = 940 K). It was therefore possible for the influence of the order–disorder transition on the quadrupole splitting to be investigated in the absence of magnetic hyperfine splitting, while retaining the symmetry of the lattice in its original form. It was found that the destruction of the superstructure in lithium ferrite, owing to high temperatures and partial substitution of the iron with chromium (x = 1.4), caused considerable changes in the chemical isomer shift and quadrupole splitting for the tetrahedral and octahedral positions, as a result of changes in the degree of covalence of the Fe^{3+}—O^{2-} bonds.[294]

The ferrites $Li^+_{1-x}Fe^{3+}_x[Li^+_{x-0.5}Fe^{3+}_{2.5-x-a}Cr^{3+}_a]O_4$ (a = 1.15, 1.25, or 1.50) exhibit only a single six-line hyperfine pattern, whereas when a = 1.70 two internal fields are observed. To explain these results it was argued that the strongest influence on H_{eff} in the A-sublattice is from Li^+ ions on the other A-sites. Calculations were then performed, for different values of a, of the probability that a given Fe^{3+} ion on an A-site is adjacent to n = 4, 3, 2, or 1 Fe^{3+} ions of the same sublattice. These calculations showed that for a = 1.15, 1.25, and to a lesser extent 1.50 there is a maximum probability for one particular contribution (*i.e.*, n = 4) and thus accounted for the fact that similar spectra, containing only one six-line pattern, are observed for these three cases. On the other hand, the calculations showed that for a = 1.70 no single contribution is dominant, and that it is therefore possible for more than one hyperfine pattern to appear in the spectrum.[295]

Spectra of the ferrites $Zn_xLi_{0.5-x/2}Fe_{2.5-x/2}O_4$ ($0 < x < 1$) at 22 and 330 K can be curve-fitted by a superposition of components from ions on tetrahedral and octahedral sites with different numbers of magnetic neighbours. Values of the saturation magnetization calculated from the Mössbauer results, assuming no spin-canting, are in good agreement with published experimental data.[296]

293 M. Ramachandran, J. Ghose, and A. B. Biswas, *J. Inorg. Nuclear Chem.*, 1971, **33**, 3175.
294 V. I. Nikolaev, F. I. Popov, V. M. Cherepanov, and S. S. Yakimov, *Fiz. tverd. Tela*, 1971, **13**, 1145.
295 A. Z. Hrynkiewicz, D. S. Kulgawczuk, and K. Tomala, ref. 12, p. 549.
296 J. W. Young and J. Smit, *J. Appl. Phys.*, 1971, **42**, 2344.

From the near-natural linewidths for both the *A*- and *B*-site patterns in $Li_{1.2}Fe_{4.6}Sb_{0.2}O_8$ in which *B*-site disorder is known to predominate over *A*-site disorder, it was concluded that in this material *B*-site cation disorder is not nearly as effective as *A*-site cation disorder in producing inhomogeneities in H_{eff} at either site. The area ratio of the two hyperfine patterns, which are well resolved in an applied field, yields equal recoilless-fractions for the two sites.[297]

The line-broadening in the spectra of $MgFe_2^{3+}O_4$ and $MnFe_2^{3+}O_4$ has been analysed in detail on the basis of a model proposed earlier for $CoFe_2O_4$. The ratio of the Mn(*A*)–Fe(*B*) to the Fe(*A*)–Fe(*B*) superexchange interactions was found to be 0.66 ± 0.04.[298] Scandium-substituted magnesium–manganese ferrites,[299] and magnesium–aluminium ferrites with square hysteresis loops have also been studied.[300]

The magnetic properties and site distributions in the system $FeCr_2O_4$–Fe_3O_4 ($Fe^{2+}Cr_{2-x}Fe_x^{3+}O_4$) have been studied in detail in order to obtain information on the transition, with increasing x, between the normal spinel $FeCr_2O_4$ with conical spiral spin structure and the inverse spinel Fe_3O_4 with a Néel antiferromagnetic structure. It was shown that Fe^{3+}, substituted for Cr^{3+} in $FeCr_2O_4$, goes into the *B*-sites up to $x = 0.68$ (region 1) and has a spin which is canted relative to the *A*-site moment. In the region $0.68 < x < 1.38$ (region 2) the Fe^{3+} goes into the *A*-site, displacing corresponding amounts of Fe^{2+} to the *B*-sites and leading to the beginning of charge hopping. From $x = 1.38$ to magnetite (region 3) the *B*-sites tend to have equal amounts of Fe^{2+} and Fe^{3+}, presumably because of charge hopping between the two ions. These conclusions are summarized in the site distribution diagram in Figure 11. The Verwey transition present in Fe_3O_4 and the attendant cubic to orthorhombic lattice distortion are suppressed by the presence of Cr^{3+} in the *B*-site. This is illustrated for $Fe_{2.75}Cr_{0.25}O_4$ in Figure 12; the increasing linewidth shows that the charge hopping slows down, but even at 1.4 K (not shown) the hopping is still sufficiently fast to prevent the *B*-site Fe^{2+} and Fe^{3+} resonances from being separately resolved.[301]

The spectrum of $MnFe_2O_4$ has been remeasured and explained on the basis of lattice symmetry arguments. The spectra obtained at temperatures above the Curie point (565 K) were interpreted in terms of a singlet from iron atoms located in the *A*-sites and a doublet from iron atoms in *B*-sites; the intensity ratio of the two resonances was 0.15 : 0.85. From spectra taken below the Curie point the domain magnetization was found to lie

[297] B. J. Evans and L. J. Swartzendruber, *J. Appl. Phys.*, 1971, **42**, 1628.
[298] G. A. Sawatzky, F. van der Woude, and A. H. Morrish, ref. 12, p. 573.
[299] V. F. Belov, T. A. Khimich, L. A. Alekseyuk, V. V. Korovushkin, E. V. Korneev, L. N. Sharanevich, M. N. Shipko, and G. S. Podval'nykh, *Porosh. Met.*, 1971, **11**, 46.
[300] V. F. Belov, T. A. Khimich, M. N. Shipko, E. V. Korneev, and L. Letyuk, *Izvest. V.U.Z. Fiz.*, 1971, **14**, 105.
[301] M. Robbins, G. K. Wertheim, R. C. Sherwood, and D. N. E. Buchanan, *J. Phys. and Chem. Solids*, 1971, **32**, 717.

along [111].[302] In the scandium-substituted manganese ferrites, $MnFe_{2-x}Sc_xO_4$ (x = 0—0.7), scandium replaces iron in octahedral *B*-sites only. Whereas the parent manganese ferrite has an inversion of 15%, the cation distribution in $MnFe_{1.3}Sc_{0.7}O_4$ is that of a normal spinel.[303, 304] The use of

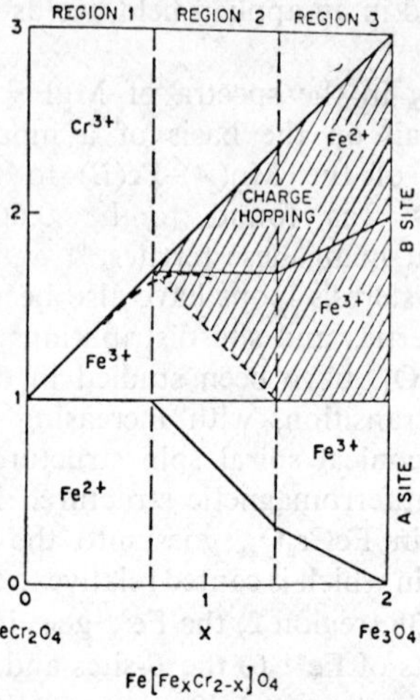

Figure 11 *Site distribution diagram*
(Reproduced by permission from *J. Phys. and Chem. Solids*, 1971, **32**, 717)

local molecular field theory in the interpretation of the Mössbauer spectra of substituted systems has been discussed generally and with particular reference to the compound $(Zn_{0.34}Mn_{0.66})Fe_2O_4$.[305]

The system $Co_{1-x}Zn_xFe_2O_4$ has been studied in detail at 4 K in magnetic fields from 0 to 80 kG. The cation distributions and the hyperfine fields at ^{57}Fe nuclei in *A*- and *B*-sites were determined as a function of the zinc concentration. For samples with x = 0.4 and 0.6 the Néel temperatures were found to be 513 ± 5 and 322 ± 5 K, respectively. The results showed that, except near x = 0 where a small fraction of Co^{2+} ions appear to be in *A*-sites, the structure of $Co_{1-x}Zn_xFe_2O_4$ is inverse. In the presence of a magnetic field, applied parallel to the γ-ray direction, $\Delta m = 0$ lines are present in the spectra of all samples, but their intensity increases with increasing x. These results are explained on the basis of a localized distribution of canting angles among the spins of Fe^{3+} ions at octahedral *B*-sites. For large values of x there is localized spin-canting with a distribution of canting angles determined by the statistical distribution of

302 U. König, *Solid State Comm.*, 1971, **9**, 425.
303 K. V. Pocholok, B. I. Pocrovskii, A. M. Babeshkin, A. K. Gapeev, and L. N. Komissarova, *Solid State Comm.*, 1971, **9**, 1727.
304 A. M. Babeshkin, K. V. Pocholok, Z. P. Yershova, D. Yu. Perfilev, I. B. Pokrovsky, and L. N. Komissarova, ref. 12, p. 541.
305 J. M. D. Coey and G. A. Sawatzky, *Phys. Stat. Sol.* (*B*), 1971, **44**, 673.

non-magnetic Zn^{2+} neighbouring ions. For low concentrations of non-magnetic ions, the distribution of hyperfine fields is, to a good approximation, independent of the distribution of canting angles, and these spectra can be analysed in terms of supertransferred hyperfine fields from neighbouring ions.[306] The dependence of hyperfine fields at ^{57}Fe nuclei

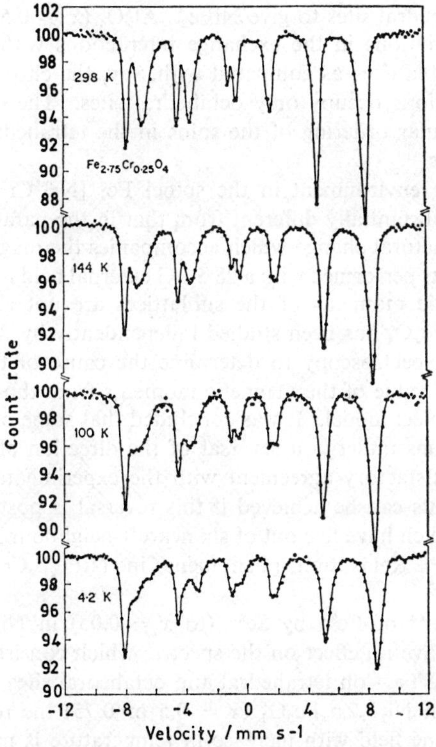

Figure 12 *Mössbauer spectra of* $Fe_{2.75}Cr_{0.25}O_4$ *below room temperature showing a slowing down of the charge hopping but giving no indication of the Verwey transition which occurs at* 119 K *in* Fe_3O_4
(Reproduced by permission from *J. Phys. and Chem. Solids*, 1971, **32**, 71)

in octahedral *B*- and tetrahedral *A*-sites in ferrites on the kind of nearest neighbours has been discussed elsewhere in terms of supertransferred hyperfine fields.[307] $Co_{0.5}Zn_{0.5}Fe_2O_4$, a single member of the cobalt–zinc ferrite series discussed above, has been studied independently over the temperature range 90—455 K, and shown to give spectra characteristic of Fe^{3+} ions undergoing relaxation effects. A stochastic model for spin fluctuations was used to calculate theoretical spectra and these agree satisfactorily with those obtained experimentally. The effects of relaxation

[306] G. A. Pettit and D. W. Forester, *Phys. Rev.* (*B*), 1971, **4**, 3912.
[307] F. Van der Woude and G. A. Sawatzky, *Phys. Rev.* (*B*), 1971, **4**, 3159.

on the shapes of the spectra led to significant differences in the Néel temperature determined by Mössbauer spectroscopy (448 K) and by neutron diffraction (400 ± 5 K).[308]

From spectra for the system $Ni_xFe_{3-x}O_4$ it has been concluded that the iron 3*d* electrons are essentially non-localized for $x < 0.7$ and localized for $x > 0.7$.[309] It has been shown that Al^{3+} replaces iron in both tetrahedral and octahedral sites to give $NiFe_{2-x}Al_xO_4$ ($x \geqslant 0.5$). This leads to considerable variations in the exchange interactions within and between the magnetic sublattices as compared with, say, the chromium ferrites in which the Cr^{3+} ions occupy only octahedral sites. The complex spectra confirm the angular ordering of the spins in the tetrahedral sublattice of these compounds.[310]

The electronic environment in the spinel $Fe^{3+}[Ni^{2+}Cr^{3+}]O_4$ below the Curie point is substantially different from that in the paramagnetic region because of a structural change which accompanies the magnetic transition. From experiments performed with a 28.5 kG external field it has been shown that the magnetic moments of the sublattices are not collinear.[311] The system $NiCr_{2-x}Fe_xO_4$ has been studied independently by *X*-ray diffraction and Mössbauer spectroscopy to determine the cause of the deviation of the experimental value of the magnetic moments from those calculated on the basis of the Néel model. It was concluded that some of the octahedral Fe^{3+} and Cr^{3+} ions undergo a reversal of the direction of their magnetic moments. A satisfactory agreement with the experimental values of the magnetic moments can be achieved if this reversal is postulated to occur for those ions which have five out of six nearest-neighbour *B*-sites occupied by Cr^{3+} ions.[312] Relaxation phenomena in $NiFe_{2-x}Cr_xO_4$ have been studied.[313]

The replacement of Fe^{3+} by Sc^{3+} (to $x = 0.05$) in $NiFe_{2-x}Sc_xO_4$ has been shown to have no effect on the spectra, which consist of two six-line patterns from the Fe^{3+} on tetrahedral and octahedral sites.[299]

For the system $Ni_{1-x}Zn_xFe_2O_4$ ($x = 0.5$ or 0.75) the reduction of the magnetic hyperfine field with increase in temperature is much more than that predicted by the corresponding decrease in sub-lattice magnetization. At temperatures well below the reported Néel points well-resolved nuclear Zeeman splitting is not observed, only relaxation spectra. This behaviour is explained by the assumption of collective spin-flipping due to the

[308] P. K. Iyengar and S. C. Bhargava, *Phys. Stat. Sol.* (*B*), 1971, **46**, 117.
[309] I. Bunget, C. Nistor, and M. Rosenberg, *J. Phys.* (*Paris*) *Colloq.*, 1971, **1** (Pt. 1), C1–274.
[310] V. F. Belov, M. N. Shipko, T. A. Khimich, V. V. Korovushkin, and L. N. Korablin, *Fiz. tverd. Tela*, 1971, **13**, 2018.
[311] V. I. Nikolaev, S. S. Yakimov, F. I. Popov, and V. S. Rusakov, *Fiz. tverd. Tela*, 1971, **13**, 388.
[312] V. F. Belov, T. A. Khimich, M. N. Shipko, P. P. Kirichok, V. V. Korovushkin, N. S. Oranesyan, G. S. Podval'nykh, and V. A. Trukhtanov, *Fiz. tverd. Tela*, 1971, **13**, 900.
[313] V. I. Nikolaev, F. I. Popov, and S. S. Yakimov, *Pis'ma Zhur. eksp. teor. Fiz.*, 1971, **13**, 143.

relatively small spin-flip barrier of these soft ferrites.[314] A well-resolved magnetic hyperfine structure is observed in the spectrum of $Zn_{0.8}Ni_{0.1}$-$Fe_{2.1}O_4$, even though neutron diffraction indicates that no long-range order is present. This phenomenon is explained on the basis of the existence of isolated magnetic clusters in which the electronic spins are ordered.[315] Spectra of $Ni_{0.65}Zn_{0.02}Fe^{2+}_{0.33}Fe^{3+}_{2}O_4$ and $Ni_{0.55}Zn_{0.11}Fe^{2+}_{0.28}Fe^{3+}_{2.07}O_4$ indicate the change in the degree of inversion with changing zinc content of the solid solutions $Ni_{1-x-y}Zn_yFe^{2+}_xFe^{3+}_2O_4$.[316]

The signs of the quadrupole coupling constants of ^{57}Fe in the normal spinels $ZnFe_2O_4$ and $CdFe_2O_4$ have been shown to be negative by the magnetic perturbation technique. The field gradients at the Fe^{3+} ions in octahedral *B*-sites were interpreted in terms of the ionic point-multipole model, modified by some charge transfer between oxygen and the cations. The point-charge contribution to the electric field gradient is positive in the case of $ZnFe_2O_4$ and nearly zero in the case of $CdFe_2O_4$; the predominant contribution is from the electric dipole moments of the oxygen ions and is negative. The dipole polarizability of the oxygen ions which gives the best fit is $\alpha_D = 0.8$ Å^3. The effect of charge transfer on the field gradient is small. The isomer shifts and Fe—O distances indicate that Fe^{3+} in $ZnFe_2O_4$ is somewhat more covalently bonded than in $CdFe_2O_4$.[317] A method has been developed for analysing Mössbauer spectra broadened by hyperfine field distributions and applied to the spectra of $ZnFe_2O_4$. In principle it is possible to follow separately the thermal variation of each internal field present in the sample.[318] Clustering behaviour similar to that described above for $Zn_{0.8}Ni_{0.1}Fe_{2.1}O_4$ has been observed for $Zn_{0.88}Fe_{2.12}O_4$; even though the Mössbauer spectrum shows a six-line pattern, no long-range order is detected by neutron diffraction at 4.2 K.[319]

The generation of a double resonance between Mössbauer resonance and an electron-paramagnetic resonance in paramagnetic materials requires such a high microwave power that it is hard to avoid heating the sample. However, magnetically ordered systems in which a large number of spins are strongly coupled to each other by the exchange interaction should require much less power for pumping. A line-broadening due to the effect of microwave radiation ($\lambda = 3$ cm) on the Mössbauer spectrum of ^{57}Fe in polycrystalline yttrium iron garnet (YIG) has now been observed. The Mössbauer line-broadening is most pronounced under conditions for ferromagnetic resonance in YIG. Experiments with oriented samples show that the microwave radiation has an anisotropic effect. A mechanism for the broadening was discussed in terms of the spin-wave spectrum and the

[314] P. Raj and K. S. Kulshreshtha, *Phys. Stat. Sol.* (*A*), 1971, **4**, 501.
[315] E. Pekoshevski, Y. Sural'ski, and L. Dombrovski, *Fiz. tverd. Tela*, 1971, **13**, 393.
[316] V. A. Potakova, V. P. Romanov, N. D. Zverev, O. I. Fromovenko, and E. A. Rubalskaya, *Phys. Stat. Sol.* (*A*), 1971, **4**, 327.
[317] B. J. Evans, S. S. Hafner, and H. P. Weber, *J. Chem. Phys.*, 1971, **55**, 5282.
[318] F. Varret, A. Gerard, and P. Imbert, *Phys. Stat. Sol.* (*B*), 1971, **43**, 723.
[319] A. G. Kocharov, J. Leciejewicz, M. K. Fayek, and A. Murasik, *Phys. Stat. Sol.* (*A*), 1971, **4**, 53.

related set of effective fields at the ^{57}Fe nuclei.[320] The temperature dependence of the combined electric and magnetic hyperfine interaction of the Fe^{3+} ion in YIG has been studied in the range 85—570 K. Calculations show that the results can be explained in terms of a temperature-dependent change in the direction of spontaneous magnetization.[321] It has been shown that in the high-pressure phase of YIG the Fe^{3+} ions are situated in two different lattice sites (dodecahedral and octahedral) and that the hyperfine fields at 4.2 K are 491 and 542 kG respectively. A small amount (*ca.* 2 wt. %) of Fe^{2+} is also present at the dodecahedral sites.[322] YIG has been studied elsewhere,[323–325] and the substituted garnets $Y_3Fe_{5-x}Al_xO_{12}$,[325, 326] $Y_{3-2x}Ca_{2x}Fe_{5-x}V_xO_{12}$,[327, 328] $Y_{3-x}Ca_xFe_{5-x}Si_xO_{12}$,[327] $Y_3Fe_{5-2x}Cu_xGe_xO_{12}$,[329] and $Y_{2.9}Ca_{0.1}Sn_{0.1}Ga_xFe_{4.9-x}O_{12}$ [330] (x = 0, 0.5, 1.0, or 1.5) have also been studied. In the last-mentioned compounds, for $x < 1$ it was found that less than 3% of the gallium ions occupy the *a*-site, and for x = 1.5 it was found that 8 ± 2% occupy the *a*-site (see p. 559 for a discussion of the ^{119}Sn resonance in these garnets).[330]

The hyperfine fields at the ^{57}Fe nuclei in dysprosium iron garnet have been measured over the temperature range 4.2—750 K. The values extrapolated to 0 K are H_a = 551 kG and H_d = 480 kG for the *a*- and *d*-sites respectively. There is a slight discontinuity in the temperature dependence of the isomer shift at the Néel point (T_N = 547 K).[331]

Other Oxide Systems. Evidence has been obtained for clustering and vacancy formation in MgO–$Fe_{1-x}O$ solid solutions.[332]

Mössbauer spectra of ten members of the solid solution series $(Fe_{1-x}Rh_x)_2O_3$, taken over a wide range of temperature, have been analysed in terms of the different magnetic environments possible for the ^{57}Fe nuclei. The interactions of a central ion with its neighbours were determined on the basis of molecular field theory. There is some evidence that magnetic interactions extend beyond nearest neighbours. The spectra at 20 K can be explained by a supertransferred hyperfine field of 7.8 kG

[320] Y. V. Baldokhin, E. F. Makarov, V. A. Povitskii, and A. I. Pristupa, *Fiz. tverd. Tela*, 1971, **13**, 1349.
[321] S. S. Bashkirov, N. G. Ivoilov, and V. A. Chistyakov, *Fiz. tverd. Tela*, 1971, **13**, 689.
[322] M. Shimada, *Phys. Letters*, 1971, **37A**, 341.
[323] G. N. Belozerskii, V. N. Gittzovich, A. N. Murin, L. A. Marshak, A. I. Shapiro, and Yu. M. Yakovlev, ref. 12, p. 527.
[324] A. M. van der Kraan and J. J. van Leof, ref. 12, p. 519.
[325] L. A. Alekseev, P. L. Gruzin, M. N. Uspenskii, and M. R. Gryaznov, *Pis'ma Zhur. eksp. teor. Fiz.*, 1971, **14**, 292.
[326] J. Piekoszewski and J. Suwalski, ref. 12, p. 499.
[327] V. A. Bokov, S. I. Yushchuk, G. V. Popov, and N. N. Parfenova, ref. 12, p. 531.
[328] L. Dabrowski, J. Piekoszewski, J. Suwalski, and S. Makolagwa, *Inst. Nuclear Res. (Warsaw), Rep.*, 1971, No. 1301/II/PS.
[329] L. Brossard, R. Krishnan, and G. A. Fatseas, ref. 12, p. 491.
[330] I. Nowik, E. R. Bauminger, J. Hess, A. Mustachi, and S. Ofer, *Phys. Letters*, 1971, **34A**, 155.
[331] G. Crecelius, S. Hüfner, and D. Quitmann, ref. 12, p. 507.
[332] V. V. Kurash, V. I. Goldanskii, T. V. Maylsheva, V. S. Urusov, L. M. Kuzynetsov, and L. A. Moskovkina, *Izvest. Akad. Nauk S.S.S.R., Neorg. Materialy*, 1971, **7**, 1574.

for each strongly interacting iron neighbour. A comparison of experiment and theory also yielded information about the Fe—O bond in α-Fe_2O_3. The charges on the Fe^{3+} and O^{2-} ions are approximately + 2.43 and − 1.62 respectively, so that the bond is 80% ionic. The spin of pure 3*d* character is reduced from 2.5 for the free iron ion to 2.32 in α-Fe_2O_3, the corresponding reduction in hyperfine field being *ca.* 100 kG. Variations in chemical isomer shift and quadrupole splitting were also discussed and it was estimated that the electron charge density at the nucleus is 5.6 au^{-3} higher than it is for the free ion.[333]

$Sr(FeTa)_{\frac{1}{2}}O_3$, which is a disordered perovskite tetragonal at room temperature, and $Sr(CrTa)_{\frac{1}{2}}O_3$, which is an ordered perovskite cubic at room-temperature, form solid solutions over the whole range of composition. The chemical isomer shift is constant throughout the series and indicates the presence of Fe^{3+}. The quadrupole splitting shows a complicated behaviour, explained in terms of local distortion arising from the disordered distribution of the Ta, xFe, and $(1-x)$Cr ions. Spectra were only taken at room temperature and no magnetic hyperfine interactions were observed.[334] By careful control of the oxygen partial pressure the ordered perovskite $(SrLa)(FeTa)O_6$ has been obtained free from any major contaminant. The Mössbauer spectrum at room temperature indicates the presence of Fe^{2+} and a small amount of Fe^{3+} as impurity. It is also suggested on the basis of magnetic susceptibility that the larger cations Sr^{2+} and La^{3+} are distributed randomly, whereas the smaller ones, high-spin Fe^{2+} and Ta^{5+}, are ordered.[335]

New data have been reported for the perovskite $BiFeO_3$. The internal field in this compound follows a Brillouin curve of $S = \frac{5}{2}$ with a Néel temperature of 640 ± 5 K and H_{eff} (0 K) of 555 ± 10 kG. There are discontinuities in the chemical isomer shift and quadrupole splitting at 400 and 500 K which correspond to crystallographic phase changes. The series $(BiFeO_3)_x$–$(PbZrO_3)_{1-x}$ (x = 0.2, 0.3, 0.4, or 0.5) was also studied; no magnetic ordering was detected even at 78 K, despite the fact that ordering has been reported for the system $BiFeO_3$–$PbTiO_3$. The difference in behaviour may be due to the fact that the latter system has a tetragonal structure up to 70% $BiFeO_3$, whereas the former is pseudocubic. Changes in the quadrupole splitting were interpreted in terms of the induced dipole moment at the Fe^{3+} site. Both the quadrupole splitting and the cell volume increase with increasing mole percentage of $BiFeO_3$, but there are no significant changes in the chemical isomer shift.[336] By contrast, the chemical isomer shift is particularly sensitive to the unit cell volume for the solid solutions $(BiFeO_3)_x$–$(PbTiO_3)_{1-x}$. The similarity and continuity in internal structure between the ferroelectrics $PbTiO_3$ and $BiFeO_3$ have been

[333] J. M. D. Coey and G. A. Sawatzky, *J. Phys.* (*C*), 1971, **4**, 2386.
[334] S. Nomura and T. Nakagawa, *J. Phys. Soc. Japan*, 1971, **30**, 491.
[335] T. Nakamura and T. Sata, *J. Phys. Soc. Japan*, 1971, **30**, 1501.
[336] A. Biran, P. A. Montano, and U. Shimony, *J. Phys. and Chem. Solids*, 1971, **32**, 327.

demonstrated by studies on this system at temperatures between 295 and 600 K. The quadrupole splitting increases and the chemical isomer shift decreases as x is changed from 0.05 to 0.70, a range in which the solutions are tetragonal and paramagnetic, and in which the unit cell volume increases with increasing x.[337] Phase transitions in antiferroelectric $PbZrO_3$ and ferroelectric $PbTi_{0.2}Zr_{0.8}O_3$ have been studied by means of the Mössbauer effect in $BiFeO_3$ admixed to the extent of 5 mol % into each crystal. The compounds give similar chemical isomer shifts whether they are in the ferroelectric, antiferroelectric, or cubic state. However, there is a pronounced and approximately equivalent dip in the recoil-free fractions at the Curie points of both materials, associated with a low-frequency lattice mode at the Brillouin zone boundary. Discontinuities in the quadrupole splitting also occur.[338]

The reorientation of the magnetic moment in $SmFeO_3$ has been shown to occur over the temperature range 423 ± 5—483 ± 5 K.[339]

It has been shown that iron occupies three different magnetic sites in $Na_3Fe_5O_9$, with hyperfine fields (extrapolated to 0 K) of $H_{Fe(1)}$ 485, $H_{Fe(2)}$ 506, and $H_{Fe(3)}$ 539 kG. Two of the environments [Fe(1) and Fe(2)] are tetrahedral and have the same temperature dependence of the hyperfine field; the third environment is octahedral and has a hyperfine field which decreases more rapidly with increase in temperature. A spin arrangement which agrees with this behaviour was proposed and it was concluded that the magnetic field lies in the (0.03, 0.82, − 0.53) direction.[340]

The calcium-magnetites $Ca_xFe_{3-x}O_4$ ($0 \leqslant x \leqslant 0.55$) have been studied at room temperature and at 77 K. A single calcium magnetite phase, giving spectra with magnetite-like Zeeman splitting, was observed up to $x \simeq 0.28$, but for higher calcium concentrations a second phase, represented by an additional quadrupole doublet and thought to arise from $CaFe_2O_4$, was detected. The increasing substitution of Fe^{2+} ions by Ca^{2+} ions leads to a slight but systematic decrease in the effective magnetic fields at the iron nuclei in both octahedral and tetrahedral sites, but the isomer shifts are invariant and remain the same as in Fe_3O_4. Surprisingly, even a large calcium substitution does not remove the Verwey mechanism of electron exchange between Fe^{2+} and Fe^{3+} ions in octahedral sites, but the Verwey temperature may be affected.[341]

From measurements on a single crystal, the sign of the quadrupole coupling constant, e^2qQ, has been shown to be positive for ilmenite, $FeTiO_3$, corresponding to an $|xy\rangle$ orbital ground state for the Fe^{2+} ion.

[337] C. M. Yagnik, R. Gerson, and W. J. James, *J. Appl. Phys.*, 1971, **42**, 395.
[338] J. P. Canner, C. M. Yagnik, R. Gerson, and W. J. James, *J. Appl. Phys.*, 1971, **42**, 4708.
[339] A. M. Balbashov, V. M. Golubev, E. F. Makarov, V. A. Povitskii, and A. Y. Chervonenkis, *Fiz. tverd. Tela*, 1971, **13**, 685.
[340] P. J. Schurer, *J. Phys. (Paris) Colloq.*, 1971, **1** (Pt. 1), C1–278.
[341] A. Z. Hrynkiewicz, D. S. Kulgawczuk, E. S. Mazanek, A. J. Pustówka, J. A. Sawicki, and M. E. Wyderko, *Phys. Stat. Sol. (B)*, 1971, **43**, 401.

V_{zz} lies along the $\langle 111 \rangle$ direction, which is the spin axis, and the asymmetry parameter, η, is zero. From measurements on the powder it was shown that there is no significant Goldanskii–Karyagin effect in this material.[342] Both Fe_2TiO_5 and $FeTi_2O_5$ have been shown to be antiferromagnetic at 4.2 K. The crystal structure of these compounds is orthorhombic and they are isomorphous with Ti_3O_5. Fe_2TiO_5 gives six broad lines which correspond to an internal magnetic field of 310 kG. The line-broadening is due to the fact that some of the Fe^{3+} ions are located in Ti^{4+} sites. The spectrum of $FeTi_2O_5$ is rather complex and indicates that the internal field of 190 kG is parallel to the major axis of the electric field gradient.[343]

$FeVO_4$ has been shown to undergo a transition to an ordered anti-ferromagnetic state at $T_N = 22 \pm 1$ K. The magnitude of the internal field extrapolated to 0 K is 470 ± 10 kG, which is rather low for Fe^{3+} in a $^6S_{\frac{5}{2}}$ configuration.[344]

$CuFeO_2$ (delafossite) has been studied in detail between 10 and 14 K. It orders antiferromagnetically at *ca.* 14 K and undergoes a crystallographic transition in the range 10—12 K to a probable orthorhombic phase which is also antiferromagnetic. These transitions are very sensitive to sample imperfections, which can cause the two structures to coexist over a temperature range of several degrees.[345]

Mössbauer spectroscopy has been used to study the phase composition of catalysts based on MoO_3–Fe_2O_3, which are used for the oxidation of methanol to formaldehyde.[346]

Modifications in the magnetic ordering in $BaFe_{12}O_9$, in which small amounts of Fe are replaced by Al, Cr, ZnTi, ZnGe, ZnSn, ZnZr, CuTi, or NiTi, have been studied.[347]

The spectra at 295 K of $Bi_2Fe_4O_9$ [348, 349] and of its solid solutions $Bi_2Fe_{3.4}In_{0.6}O_9$ [349] and $Bi_2Fe_3GaO_9$ [349] have been shown to consist of three components, resolvable into two doublets corresponding to Fe^{3+} in octahedral and tetrahedral sites. The cation distributions in the substituted compounds were estimated to be $Bi(Fe_{1.44}Ga_{0.56})[Fe_{1.56}Ga_{0.44}]O_9$ and $Bi_2(Fe_{1.7}In_{0.3})[Fe_{1.7}In_{0.3}]O_9$. The Ga^{3+} ions are therefore shown to prefer tetrahedral to octahedral sites, whereas the In^{3+} ions are evenly distributed between the two.[349] Chromium on the other hand prefers the octahedral sites.[348] At 85 K magnetic hyperfine splitting is observed for $Bi_2Fe_4O_9$ and its indium derivative, but not for its gallium derivative.[349]

[342] V. K. Garg and S. P. Puri, *Phys. Stat. Sol. (B)*, 1971, **44**, K45.
[343] S. Muranaka, T. Shinjo, Y. Bando, and T. Takada, *J. Phys. Soc. Japan*, 1971, **30**, 890.
[344] L. M. Levinson and B. M. Wanklyn, *J. Solid State Chem.*, 1971, **3**, 131.
[345] A. H. Muir, jun., R. W. Grant, and H. Wiedersich, ref. 12, p. 557.
[346] A. Popescu, A. Szabo, and H. Hobert, *Rev. Roumaine Chim.*, 1971, **16**, 869.
[347] J. G. Rensen, J. A. Schulkes, and J. S. Van Wieringen, *J. Phys. (Paris), Colloq.*, 1971, **1** (Pt. 2), C1–924.
[348] E. Kostiner and G. L. Shoemaker, *J. Solid State Chem.*, 1971, **3**, 186.
[349] V. A. Bokov, S. I. Yushchuk, G. V. Popov, N. N. Parfenova, and A. G. Tutov, *Fiz. tverd. Tela*, 1971, **13**, 1590.

Minerals. This section begins with a discussion of the Mössbauer spectra obtained from the lunar samples returned by the Apollo 12 mission from Oceanus Procellarum. These results are contained principally in six papers,[350–355] published simultaneously in the Proceedings of the Second Lunar Science Conference which was held in Houston. The subject matter and conclusions in three of the reports are such that it is convenient to discuss them together, but first of all the scope of each individual investigation and the particular samples studied are outlined. The sample numbers are those assigned by the NASA Lunar Receiving Laboratory in Houston.

Gay *et al.* used the spectra of the three fines samples (soils) 12001.58, 12057.53, and 12070.22 to identify the iron-bearing minerals present and to estimate their relative proportions.[350] Herzenberg's group also studied three fines samples, 12032.63, 12033.93, and 12037.12, to obtain similar information. In addition this group analysed two microbreccias, 12013.7 and 12013.17, and ten rock samples, 12002.168, 12004.52, 12004.48, 12004.35, 12020.17, 12020.56, 12020.44, 12063.100, 12063.114, and 12063.58.[351] Housley *et al.* studied fines samples 12042.38 and 12001.117, core-tube samples 12025.15, 12025.42, 12028.88, and 12028.113, and rocks 12038.47, 12052.16, and 12063.59. They also obtained spectra from heavy and light fractions separated from sample 12042.38 using methylene iodide.[352]

Typical spectra of lunar fines, taken from the work of Gay *et al.*, are shown in Figure 13. The peaks are assigned as follows: 1 and 1′ to ilmenite ($FeTiO_3$); 2 and 2′ to pyroxenes [$Ca(Mg,Fe)Si_2O_6$]; 3 and 3′ to olivine [$(Mg,Fe)_2SiO_4$]; 4 and 4′ to an overlap of the pyroxene and olivine peaks; 5, 5′, and 5″ to metallic iron; and 6 and 6′ to troilite (FeS). The relative percentages of the various iron-bearing minerals in the fines studied by Gay *et al.*, based on a computer analysis of their spectra, are compared in Table 3 with the values for an Apollo 11 soil (10084).[350] The Apollo 12 fines contain more olivine but less ilmenite, metallic iron, and troilite than the Apollo 11 fines.

A comparison of the spectrum in Figure 13 with that of lunar soil 12037[351] in Figure 14 reveals an encouraging agreement between the

[350] P. Gay, M. G. Bown, I. D. Muir, G. M. Bancroft, and P. G. L. Williams, Proc. 2nd Lunar Science Conference, vol. 1, *Geochim. Cosmochim. Acta*, Supplement 2, MIT Press, 1971, p. 377.
[351] C. L. Herzenberg, R. B. Moler, and D. L. Riley, Proc. 2nd Lunar Science Conference, vol. 3, *Geochim. Cosmochim. Acta*, Supplement 2, MIT Press, 1971, p. 2103.
[352] R. M. Housley, R. W. Grant, A. H. Muir, jun., M. Blander, and M. Abdel-Gawad, Proc. 2nd Lunar Science Conference, vol. 3, *Geochim. Cosmochim. Acta*, Supplement 2, MIT Press, 1971, p. 2125.
[353] S. S. Hafner, D. Virgo, and D. Warburton, Proc. 2nd Lunar Science Conference, vol. 1, *Geochim. Cosmochim. Acta*, Supplement 2, MIT Press, 1971, p. 91.
[354] D. E. Appleman, H. U. Nissen, D. B. Stewart, J. R. Clark, E. Dowty, and J. S. Huebner, Proc. 2nd Lunar Science Conference, vol. 1, *Geochim. Cosmochim. Acta*, Supplement 2, MIT Press, 1971, p. 117.
[355] S. Sullivan, A. N. Thorpe, C. C. Alexander, F. E. Senftle, and E. Dwornik, Proc. 2nd Lunar Science Conference, vol. 3, *Geochim. Cosmochim. Acta*, Supplement 2, MIT Press, 1971, 2433.

Table 3

Mineral	10084	12001	12057	12070
	Percentage of total Mössbauer area			
Ilmenite	19.7	4.9	9.1	6.1
Pyroxene + glass	67.6	⩾ 71.8	⩾ 72.7	⩾ 64.9
Olivine	4.4	⩽ 19.2	⩽ 23.0	⩽ 25.0
Iron	5.8	3.4	3.4	3.3
Troilite	⩽ 1.1	0.7	0.9	0.7

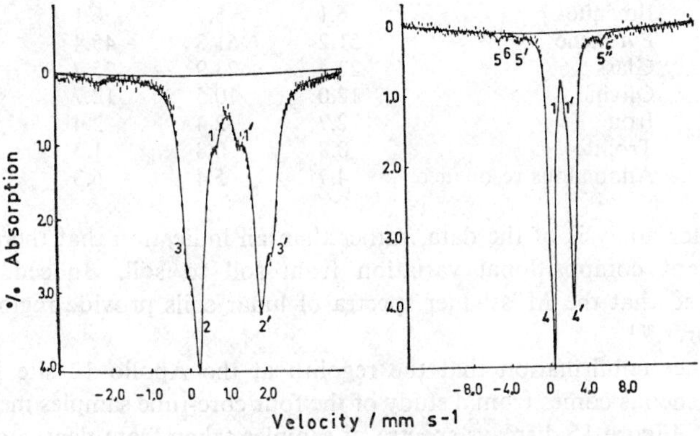

Figure 13 (a) *Low- and* (b) *high-velocity scans of fines sample* 12001 *at room temperature*
(Reproduced by permission from Proc. 2nd Lunar Science Conference, vol. 1, *Geochim. Cosmochim. Acta,* Supplement 2, MIT Press, 1971, p. 377)

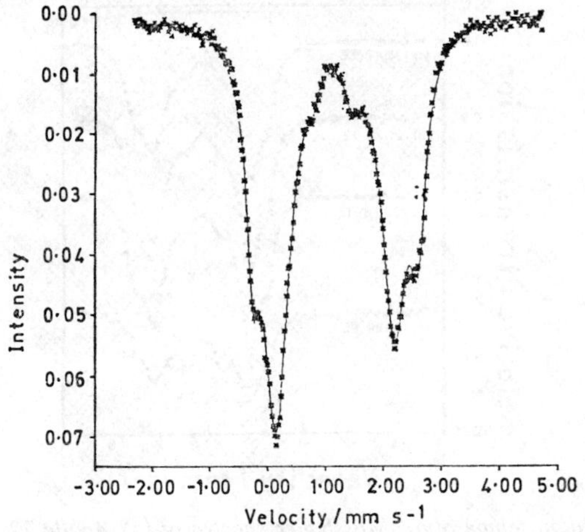

Figure 14 *Mössbauer hyperfine spectrum for lunar soil* 12037 (*measured at room temperature*) *with least-squares computer fit*
(Reproduced by permission from Proc. 2nd Lunar Science Conference, vol. 3, *Geochim. Cosmochim. Acta,* Supplement 2, MIT Press, 1971, p. 2103)

results from different laboratories. The phase distribution of iron in this soil and in other soils studied by Herzenberg *et al.* is shown in Table 4. The only major discrepancy between this set of figures and those given earlier for three different samples is in the olivine content, and this is probably a reflection of the fact that different models were adopted for the

Table 4

Mineral	12032.63	12033.93	12037.12
Ilmenite	6.1	5.7	6.1
Pyroxene	51.2	51.3	45.4
Glass	22.4	23.9	25.7
Olivine	12.0	10.5	12.7
Iron	2.7	2.4	2.4
Troilite	0.8	0.8	1.3
Anomalous resonance	4.7	5.4	6.3

computer analysis of the data, rather than an indication that there is any significant compositional variation from soil to soil. Indeed, it was suggested that the Mössbauer spectra of lunar soils provide regional soil signatures.[351]

Further confirmation that the regolith at the Apollo 12 site is fairly homogeneous comes from a study of the four core-tube samples mentioned earlier. Figure 15 displays spectra of samples taken from depths of 5 and 30 cm respectively, together with the spectrum of the fines sample 12042.38, and it can be seen that they are all very similar to one another. A typical Apollo 11 fines spectrum is also shown for comparison.[352]

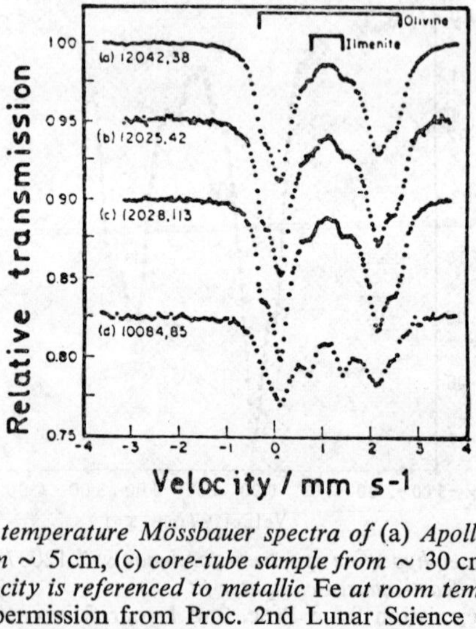

Figure 15 *Room temperature Mössbauer spectra of* (a) *Apollo* 12 *dust,* (b) *core-tube sample from* ~ 5 cm, (c) *core-tube sample from* ~ 30 cm, *and* (d) *Apollo* 11 *dust.* (*Zero velocity is referenced to metallic* Fe *at room temperature*)
(Reproduced by permission from Proc. 2nd Lunar Science Conference, vol. 3, *Geochim. Cosmochim. Acta*, Supplement 2, MIT Press, 1971, p. 2125)

In most of the spectra of the fines the peak at lower velocity is invariably more intense than its partner. This 'anomalous absorption' (see Table 4) has been attributed in two of the reports to the presence of extremely finely divided, superparamagnetic iron.[351, 352] It is further suggested that the absorption is associated with an admixed foreign component having a very different composition from that observed among the larger igneous rock samples, and known as the 'kreep' component.[351]

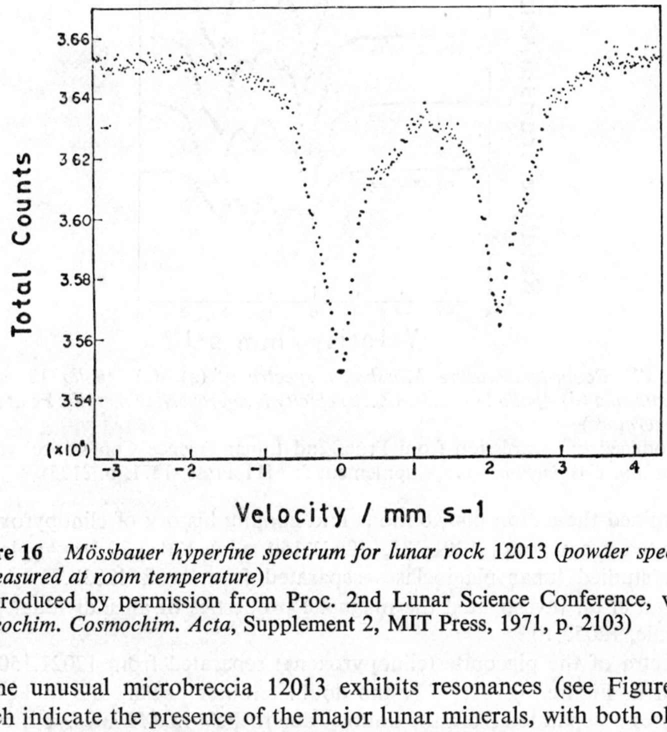

Figure 16 *Mössbauer hyperfine spectrum for lunar rock* 12013 (*powder specimen measured at room temperature*)
(Reproduced by permission from Proc. 2nd Lunar Science Conference, vol. 3 *Geochim. Cosmochim. Acta*, Supplement 2, MIT Press, 1971, p. 2103)

The unusual microbreccia 12013 exhibits resonances (see Figure 16) which indicate the presence of the major lunar minerals, with both olivine and ilmenite less abundant than in any other Apollo 12 sample examined. In addition there is an absorption excess in the low-velocity region thought to be associated with the 'kreep' component, and a diffuse absorption contribution extending between about -1 and $+3$ mm s^{-1} which is thought to be associated with a history of intense shock.[351]

Typical room-temperature spectra of the rocks, taken from the work of Housley *et al.*, are shown in Figure 17, where they are compared with the spectrum of an Apollo 11 rock. There are substantial differences among the spectra of the individual rocks, which indicates that there is considerable diversity in the phase distribution of iron and in the modal mineralogy. Furthermore the rocks are substantially higher in ilmenite and lower in olivine than the mineral grains in the fines. Compared with the Apollo 11 rocks the Apollo 12 samples contain less ilmenite and more olivine.[352] These general trends were also confirmed by the work of Herzenberg *et al.*[351]

The three remaining papers all describe work on materials separated carefully from the bulk sample prior to the Mössbauer study. Hafner *et al.*

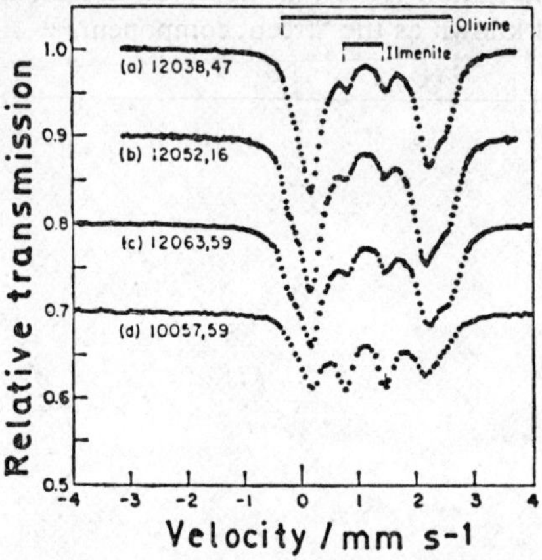

Figure 17 *Room-temperature Mössbauer spectra of* (a)—(c) *Apollo* 12 *olivine basalts, and* (d) *Apollo* 11 *basalt.* (*Zero velocity is referenced to metallic* Fe *at room temperature*)

Reproduced by permission from Proc. 2nd Lunar Science Conference, vol. 3, *Geochim. Cosmochim. Acta*, Supplement 2, MIT Press, 1971, p. 2125)

determined the cation distributions and cooling history of clinopyroxenes separated from rocks 12018.35, 12021.150, and 12053.79;[353] Appleman *et al.* studied lunar plagioclases separated from rock 12038.72;[354] and Sullivan *et al.* investigated the magnetic properties of an individual glass spherule, 12057.10.[355]

Spectra of the pigeonite (clinopyroxene) separated from 12021.150 are shown in Figure 18. At 77 K the doublet (outer peaks) due to Fe^{2+} in $M1$ sites is separated from the doublet due to Fe^{2+} in $M2$ sites. However, the two doublets differ in their temperature dependences with the result that at 295 K they are superimposed on one another. From these spectra the $Fe^{3+} : (Fe^{2+} + Fe^{3+})$ ratio was estimated to be smaller than 0.003.[353]

The equilibrium distribution of Mg^{2+} and Fe^{2+} over the $M1$ and $M2$ sites was approximately determined, for some samples, from trends in the Fe^{2+} distribution number obtained after heating at various temperatures. In this way an equilibrium temperature of *ca.* 843 K was estimated for the pigeonite from 12021, and it was suggested that this sample had been subjected to long-range ordering over an exceedingly long period of time in the temperature range 1083—753 K.[353, 356] The equilibrium temperature

[356] S. S. Hafner, D. Virgo, D. Warburton, H. Fernández-Morán, M. Ohtsuki, and A. Hibino, *Nature, Phys. Sci.*, 1971, **231**, 79.

for the phenocrysts from basalt 12053 is much higher (*ca.* 943 K), and it was concluded that this rock had cooled more rapidly through this range. The pigeonites from basalt 12018 exhibit apparently high degrees of Mg^{2+} Fe^{2+} disorder, whereas the site occupancy in the calcium-rich clinopyroxenes

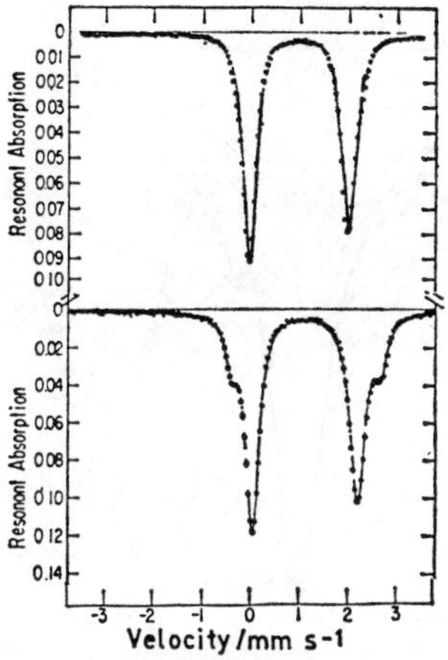

Figure 18 *Mössbauer spectra of* ^{57}Fe *in clinopyroxene* 12021, 150-P1 *measured at room temperature* (*upper spectrum*) *and at* 77 K (*lower*)
(Reproduced by permission from Proc. 2nd Lunar Science Conference, vol. 1, *Geochim. Cosmochim. Acta*, Supplement 2, MIT Press, 1971, p. 91)

of 12021 is indicative of nearly complete ordering, magnesium occurring at the *M*1 sites.

Figure 19 shows the spectrum of 530 mg of crushed plagioclase feldspar separated from a sample of rock 12038.72 weighing 3.90 g. The doublet with the smaller chemical isomer shift and quadrupole splitting is assigned to Fe^{2+} in tetrahedral sites, usually occupied by aluminium or silicon ions, and the other doublet to Fe^{2+} ions in the larger, irregularly co-ordinated sites, normally occupied by calcium or sodium cations. The weak resonance between the major peaks is due to ilmenite impurity. No Fe^{3+} was detected in the spectrum but, because of interference from ilmenite, the limit of detection is as high as *ca.* 5% of the total iron content in this case.[354]

A glass bubble (designated 12057.10) from the rock chip 12057, 6 mm in diameter and weighing 44.4 mg, has been studied. The spectrum contained only two broad lines characteristic of Fe^{2+} but there was no indication of the six-line spectrum of metallic iron or other inclusions such as ilmenite.[355]

Further discussion of the analysis of some Apollo 11 samples has also appeared[357] (see also refs. 24—27).

[357] C. L. Herzenberg and D. L. Riley, *Phys. Earth Planet Interiors*, 1971, **4**, 204.

Papers have appeared on a wide range of terrestrial minerals and these are now discussed in the following order: pyroxenes, olivines, amphiboles, clay minerals, other minerals, and sulphides.

Apparent anomalies in the spectra of *C2/c* clinopyroxenes, observed by a number of investigators in the past, have been shown to be due to the

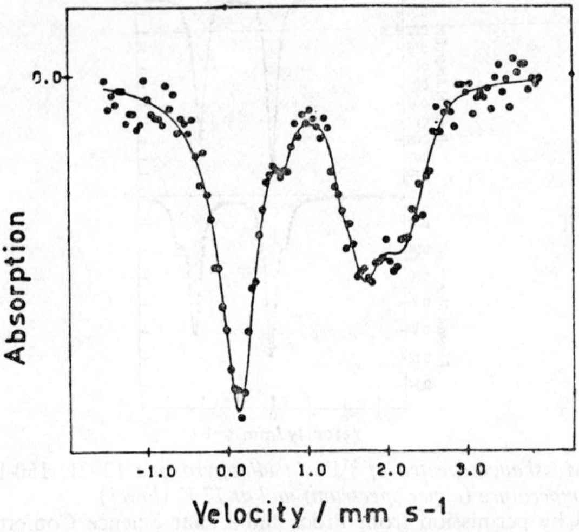

Figure 19 *Mössbauer spectrum of feldspar concentrate from rock* 12038.72. *There are two major doublets present; the left-hand peaks are superposed, but the right-hand peaks are partially resolved. These doublets are due to ferous iron in the* Ca/Na *and tetrahedral sites. A much less intense doublet due to ilmenite impurity can also be seen between the major peaks*

Reproduced by permission from Proc. 2nd Lunar Science Conference. vol. 1, *Geochim. Cosmochim. Acta*, Supplement 2, MIT Press, 1971, p. 117

presence of more than two doublets as a result of the existence of two or more exsolved *C2/c* phases with markedly different *M*1 quadrupole splittings. The usual assignment of the inner doublet to Fe^{2+} in *M*2 sites and the outer doublet to Fe^{2+} in *M*1 sites was confirmed. A three-doublet fit to the spectrum of a Skaergaard intrusion sample [$(Ca_{0.82}Fe_{1.02}Mg_{0.16})$-$Si_2O_6$] suggested that 0.20 Fe^{2+} per formula unit enters the *M*2 sites, in reasonable agreement with the concept of Fe^{2+} entirely filling the Ca^{2+} discrepancy.[358] Spectra have been obtained for five *C2/c* clinopyroxenes lying on, or close to, the diopside–hedenbergite tie line. Particular attention was directed towards the temperature dependences of the quadrupole splitting of Fe^{2+} in *M*1 sites and towards the increase of the *M*1 quadrupole splitting with increase in the ratio Fe : (Fe + Mg).[359]

358 P. G. L. Williams, G. M. Bancroft, M. G. Bown, and A. C. Turncock, *Nature, Phys. Sci.*, 1971, **230**, 149.
359 G. M. Bancroft and P. G. L. Williams, *Amer. Mineral*, 1971, **56**, 1617.

A series of solid solutions of diopside ($CaMgSi_2O_6$) and the ferri-Tschermak 'molecule' ($CaFe^{3+}SiO_6$) has been studied. The spectrum of one of these, $CaMg_{0.74}Fe^{3+}_{0.52}Si_{1.74}O_6$, is shown in Figure 20. It consists of two overlapping doublets which can be assigned to Fe^{3+} at the octahedral ($M1$) and the tetrahedral (Si) sites and a weak resonance from 5 to 10% of

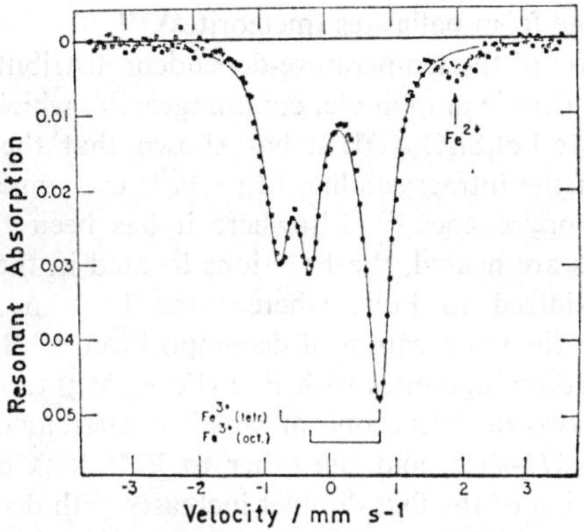

Figure 20 *Mössbauer absorption spectrum of* ^{57}Fe *and assignment of the* Fe^{3+} *doublets in a synthetic ferri-diopside* $CaMg_{0.74}Fe^{3+}_{0.26}(Fe^{3+}_{0.26}Si_{1.74})O_6$. *The solid line is a least-squares fit to the data assuming three Lorentzian lines. The sample shows a small substitution of* Fe^{2+} *for* Mg^{2+}
(Reproduced by permission from *Nature, Phys. Sci.*, 1971, **233**, 9)

iron in the Fe^{2+} state. The quadrupole splitting (1.49 mm s^{-1}) at the tetrahedral position in this ferri-diopside is one of the largest observed for Fe^{3+} in a four-co-ordinated configuration. Throughout the series there is very little dependence of the Mössbauer parameters on the amount of ferri-Tschermak's molecule present.[360]

The spectrum of ilvaite, a natural silicate mineral of composition $Ca^{2+}Fe_2^{2+}Fe^{3+}(Si_2O_8)^{7-}OH^-$, has been studied as a function of temperature and the results interpreted in terms of an electron-hopping process producing a motional narrowing of the lines, similar to the case of magnetite.[361] Isomorphous substitution of manganese for iron in ilvaite has produced the first example of magnetic order in iron silicate minerals at 80 K. At low manganese content ($< 4.5\%$) the spectrum consists of only two doublets corresponding to Fe^{2+} and Fe^{3+} in octahedral oxygen environments. As the degree of substitution increases, the spectra become more complex and, above 12% manganese, hyperfine structure is observed. The hyperfine fields at the Fe^{2+} and Fe^{3+} sites are 210 ± 7 and $- 340 \pm 7$ kG respectively at 80 K. The value of the quadrupole splitting derived from the magnetic spectra is less than half of the value derived from paramagnetic samples.

[360] S. S. Hafner and H. G. Huckenholz, *Nature, Phys. Sci.*, 1971, **233**, 9.
[361] A. Gerard and F. Grandjean, *Solid State Comm.*, 1971, **9**, 1845.

This is attributed to the existence of an angle between the direction of the magnetic field and the principal axis of the electric field gradient tensor of the iron atoms.[362]

The Fe^{2+} distribution for a pyroxene from eulysite of the Mariupol iron ore deposit has been studied.[363]

Mössbauer spectroscopy has been used to determine the Fe^{2+},Mg^{2+} ordering in olivine from pallasites (meteorites).[364]

An investigation of the temperature-dependent distributions of Fe^{2+} and Mg^{2+} in the monoclinic amphibole, cummingtonite, which has a composition close to $(Mg,Fe)_7Si_8O_{22}(OH)_2$, has shown that the thermodynamic characteristics of the intracrystalline Mg^{2+},Fe^{2+} exchange are very similar to those of orthopyroxenes.[365] Elsewhere it has been shown that when these amphiboles are heated, the Fe^{2+} ions located in the *M*1-, *M*2-, and *M*3-sites are oxidized to Fe^{3+}, whereas the Fe^{2+} in *M*4-sites remain unchanged up to the temperature of decomposition.[304] Spectra have been obtained for 26 cummingtonites with Fe : (Fe + Mg) ratios of 0.98—0.18. They consist of two doublets, one of which is attributed to Fe^{2+} ions in *M*1-, *M*2-, and *M*3-sites, and the other to Fe^{2+} ions in *M*4-sites. The quadrupole splitting of the first doublet increases with decreasing temperature between 300 and 77 K, whereas the quadrupole splitting of the second doublet is nearly invariant. Both the quadrupole splitting and the chemical isomer shifts exhibit a correlation with the radial distortion of the oxygen octahedra at the two sets of sites. The chemical isomer shift at *M*4 is indicative of enhanced covalent participation in the Fe^{2+}—O bonding compared with the other sites. The Fe^{2+} distribution over the two sets of sites is generally quite ordered.[366] A volcanic cummingtonite has been studied at 4.2, 55, 78, and 298 K.[367]

Resolution of three quadrupole-split Fe^{2+} doublets in the spectra of minerals of the actinolite series $[Ca_2(Mg,Fe^{2+})_5Si_8O_{22}(OH)_2]$ has been achieved for the first time. Fe^{3+} doublets were also resolved for the first time. The results support earlier deductions of relative enrichments of Fe^{2+} ions in actinolites of $M1 > M3 \geqslant M2$ and suggest that Mn^{2+} ions, present in small amounts, are enriched in *M*2 positions.[368, 369]

A convincing example of the versatility of the Mössbauer technique is provided by a recent analytical study of 16 samples of ancient Greek pottery

[362] B. V. Borshagovskii, A. S. Marfunin, A. R. Mkrtchyan, G. N. Nagyaryan, and R. A. Stukan, *Phys. Stat. Sol. (B)*, 1971, **43**, 479.
[363] N. D. Zverev, A. A. Valter, V. P. Romanov, and L. I. Gorogotskaya, *Lithos*, 1971, **4**, 17.
[364] T. V. Malysheva, V. V. Kurash, A. N. Ermakov, and A. K. Lavrukhima, *Geokhimiya*, 1971, 355.
[365] K. Shürmann and S. S. Hafner, *Nature, Phys. Sci.*, 1971, **231**, 155.
[366] S. S. Hafner and S. Ghose, *Z. Kristallogr., Kristallgeometrie, Kristallphys., Kristallchem.*, 1971, **133**, 301.
[367] A. N. Buckley and R. W. T. Wilkins, *Amer. Mineral.*, 1971, **56**, 90.
[368] C. Greaves, R. G. Burns, and G. M. Bancroft, *Nature, Phys. Sci.*, 1971, **229**, 60.
[369] R. G. Burns and C. Greaves, *Amer. Mineral.*, 1971, **56**, 2010.

from the Vitsa Zgorion (north-western Greece) excavation. Two typical spectra of samples of two different groups are shown in Figure 21. The spectra can be analysed as a superposition of components of paramagnetic iron(II) and iron(III) ions similar to those found in silicate minerals and ferromagnetic iron oxides (Fe_2O_3 or Fe_3O_4). Differences in the spectra,

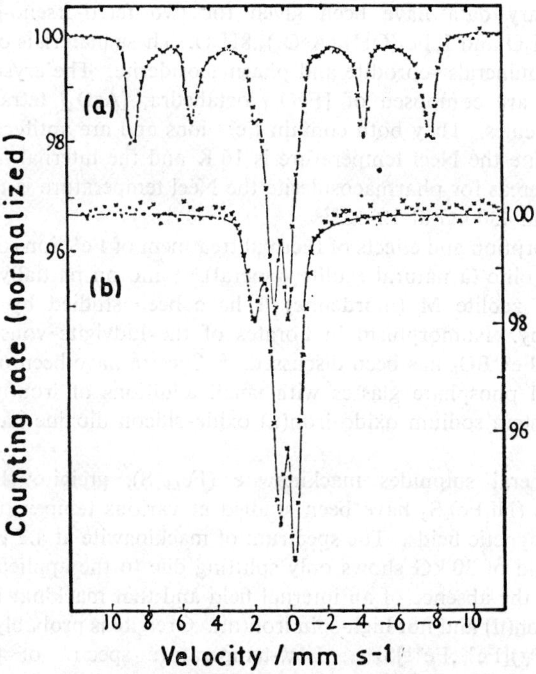

Figure 21 *Representative Mössbauer spectra from two groups of ancient ware from the Vitsa Zgorion (north-western Greece) excavation.* (a) *Local hand-made ware about* 800 B.C.; (b) *Korinthian like ware about* 750 B.C. (*Note that the velocity scale runs from positive to negative*)
(Reproduced by permission from *Nature*, 1971, **229**, 485)

which can arise from such factors as (*i*) the constitution of the original clay, (*ii*) the constitution of tempering additives, and (*iii*) the firing conditions, were found to correlate well with a classification of the samples based on archaeological criteria. The technique can therefore be used to provide at least supporting evidence for the classification of the findings from an archaeological site.[370]

The exact Mössbauer parameters for tetrahedral Fe^{3+} in micas have been evaluated using a synthetic ferrophlogophite, $KMg_3Fe^{3+}Si_3O_{10}(OH)_2$, in

[370] N. H. J. Gangas, A. Kostikas, A. Simopoulos, and J. Vocotopoulou, *Nature*, 1971, **229**, 485.

which the octahedral sites are not populated. As expected, there is a marked decrease in chemical isomer shift for Fe^{3+} in tetrahedral co-ordination compared to Fe^{3+} in octahedral co-ordination, owing to the increased covalent bonding character in the former case.[371] The thermal stability of the three-layer type iron-bearing clay minerals illite and montmorillonite has been studied after various chemical treatments.[372]

Preliminary data have been given for two ferro-arseno-compounds $FeAsO_4,2H_2O$ and $KFe_4(OH)_4(AsO_4)_3,8H_2O$. These materials occur naturally as the minerals scorodite and pharmacosiderite. The crystals of each compound are composed of $[FeO_6]$ octahedra, $[AsO_4]$ tetrahedra, and water molecules. They both contain Fe^{3+} ions and are antiferromagnetic. For scorodite the Néel temperature is 16 K and the internal field 520 kG at 4 K, whereas for pharmacosiderite the Néel temperature is 6 K and the internal field 430 kG at 1.5 K.[373]

The adsorption and effects of thermal treatment of Fe^{3+} ions on silica-gel, Al_2O_3, natrolite (a natural zeolite mineral),[374] and on partially exchanged samples of zeolite M (mordenite)[375] have been studied by Mössbauer spectroscopy. Isomorphism in borates of the ludvigite–vonsenite series $(Fe^{2+},Mg)_2Fe^{3+}BO_5$ has been discussed.[376] Spectra have been obtained for silicate and phosphate glasses with small additions of iron(III) oxide,[377] and the system sodium oxide–iron(II) oxide–silicon dioxide has also been studied.[378]

The mineral sulphides mackinawite ($Fe_{1+x}S$), greigite (Fe_3S_4), and pentlandite $(Ni,Fe)_9S_8$ have been studied at various temperatures and in external magnetic fields. The spectrum of mackinawite at 4.2 K and in an external field of 30 kG shows only splitting due to the applied field, thus confirming the absence of an internal field and that mackinawite contains low-spin iron(II) and not high-spin iron(III). Greigite is probably an inverse spinel $(Fe^{3+})[Fe^{2+},Fe^{3+}]S_4^{2-}$. Low-temperature spectra of this phase, prepared by different methods, suggest the presence of differing amounts of high-spin iron(II) and iron(III) ions. Pentlandite is cubic with a unit cell containing four octahedral sites and 32 tetrahedral sites. The spectrum at room temperature comprises an unsplit peak of intensity 0.134 due to octahedral iron, with a superimposed quadrupole doublet of relative intensity 1 from iron in the distorted tetrahedral sites. The spectrum in an external field of 30 kG at 4.2 K confirms the absence of an internal field and

[371] H. Annersten, S. Devanarayanan, L. Häggström, and R. Wäppling, *Phys. Stat. Sol.* (*B*), 1971, **48**, K137.
[372] N. Malathi, S. P. Puri, and I. P. Saraswat, *J. Phys. Soc. Japan*, 1971, **31**, 117.
[373] M. Takano, T. Takada, T. Wada, and K. Okada, *J. Phys. Soc. Japan*, 1971, **31**, 298.
[374] N. Malathi and S. P. Puri, *J. Phys. Soc. Japan*, 1971, **31**, 1418.
[375] W. Meisel, A. S. Platschinda, and I. P. Susdalev, *Z. anorg. Chem.*, 1971, **382**, 188.
[376] T. V. Malisheva, A. N. Yermakov, S. M. Alexandrov, and V. V. Kurash, ref. 12, p. 745.
[377] G. H. Frischat and G. Tomandl, *Glastech. Ber.*, 1971, **44**, 173.
[378] J. A. Boon, *Chem. Geol.*, 1971, **7**, 153.

suggests the presence of low-spin iron.[379] The vacancy distribution and types of magnetic ordering in natural pyrrhotite, $Fe_{1-x}\square_xS$ ($0.091 \leqslant x \leqslant 0.119$), has been studied. $Fe_{0.881}S$ is ferromagnetic with long-range order in the vacancy distribution.[380] The temperature reorientation of the spins in this material has been discussed elsewhere.[381] $Fe_{0.900}S$ is hexagonal and has ferromagnetic properties which are associated with the co-existence of ordered and disordered states within the limits of a single phase. Below some minimum vacancy concentration ($x \leqslant 0.096$) the ordered state is destroyed completely and the materials become anti-ferromagnetic.[380]

Troilite from the Sardis meteorite gives a six-line pattern with intensities in the ratio 3 : 2 : 1 : 1 : 2 : 3. The linewidths are much narrower than those of synthetic FeS, suggesting that the deviation from a 1 : 1 stoicheiometry is very slight.[382]

Spectra of synthetic iron-bearing sphalerite grown from pyrite and (or) pyrrhotite indicate that at least 95% of the iron is present as Fe^{2+}, and is randomly distributed over the tetrahedral sites. The study showed that the state of iron does not change with the conditions of growth of sphalerite, and therefore provides no useful information as regards geothermometry.[383]

5 Tin-119

The discussion of the tin-119 literature is divided into three parts; the first section deals with some general topics and ^{119}Sn impurity studies, and this is followed by sections on tin(II) and tin(IV) compounds.

General Topics.—The lifetime of the 23.8 keV first excited state of ^{119}Sn has been measured by the delayed coincidence technique and found to be $t_{\frac{1}{2}} = 17.75 \pm 0.12$ ns, which corresponds to a natural linewidth of $\Gamma = 0.325 \pm 0.002$ mm s^{-1}.[384] This value appears to be the best of all the recently reported estimates.[385]

The spectrum of the ferromagnetic compound Fe_3Sn, which is shown in Figure 22, has yielded a value of $\mu_{ex} = (0.685 \pm 0.003)$ n.m. for the magnetic moment of the 23.8 keV excited state of ^{119}Sn. Furthermore the compound is proposed as a velocity calibration standard for ^{119}Sn Mössbauer spectroscopy. It is easily prepared, has a high Curie temperature ($T_C = 743$ K), gives a well-resolved magnetic hyperfine splitting even at room temperature, and gives a high percentage effect. The peak positions in mm s^{-1} relative to a $Ba^{119}SnO_3$ source are reproduced in Table 5.[386]

379 D. J. Vaughan and M. S. Ridout, *J. Inorg. Nuclear Chem.*, 1971, **33**, 741.
380 N. S. Ovanesyan, V. A. Trukhtanov, G. Odinets, and G. V. Novikov, *Zhur. eksp. teor. Fiz.*, 1971, **60**, 2220.
381 N. S. Ovanesyan, V. A. Trukhtanov, G. Y. Odinets, G. V. Novikov, and L. Cser, *Pis'ma Zhur. eksp. teor. Fiz.*, 1971, **13**, 488.
382 A. Okada, H. Yabuki, and M. Shima, *Ganseki Kobutsu Kosho Gakkaishi*, 1971, **66**, 76.
383 S. D. Scott, *Canad. Mineral.*, 1971, **10**, 882.
384 N. Benczer-Koller and T. Fink, *Nuclear Phys. (A)*, 1971, **161**, 123.
385 R. L. Cohen and N. Benczer-Koller, *Phys. Letters*, 1971, **35A**, 50.
386 E. Both, G. Trumpy, and C. Djéga-Mariadassou, *Phys. Letters*, 1971, **35A**, 27.

The concept of internal standardization has been applied to Mössbauer spectroscopy, with SnO_2 and β-Sn being used as analyte and internal standard absorbers respectively. One of the difficulties encountered in using the Mössbauer technique for assay work is that the γ-ray is

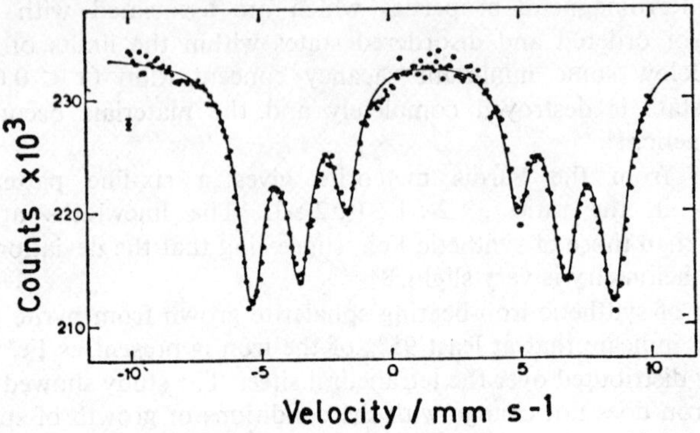

Figure 22 *Mössbauer absorption spectrum of* Fe_3Sn (24.0 at % Sn) *at* 78 K, *fitted with six Lorentzian peaks*
(Reproduced by permission from *Phys. Letters*, 1971, **35A**, 27)

Table 5

Absorber temperature 78 K	300 K
− 5.44 ± 0.02	− 3.65 ± 0.02
− 3.57 ± 0.02	− 2.22 ± 0.02
− 1.70 ± 0.02	− 0.85 ± 0.02
+ 4.95 ± 0.02	+ 4.09 ± 0.02
+ 6.82 ± 0.02	+ 5.47 ± 0.02
+ 8.71 ± 0.02	+ 6.83 ± 0.02

accompanied by background radiation. However, it is claimed that by simultaneous measurement of the analyte and internal standard absorbers, the background contribution can be eliminated.[387]

The alloy $Ni_{21}Sn_2B_6$ has been shown to have properties which make it suitable as a source for both ^{119}Sn and ^{121}Sb (see p. 577) Mössbauer experiments. It is not recommended for use at room temperature, but at 80 K its recoilless fraction is comparable to that of other sources (*e.g.* $BaSnO_3$).

The isomer shift conversions are set out in Table 6.[388]

The after-effects of the ^{119m}Sn isomeric transition have been studied in $SnSO_4$ at 100 K. The spectrum shows the simultaneous presence of both Sn^{2+} and Sn^{4+} ions. The stabilization of the latter is attributed to the consequences of the autoradiolysis process accompanying the Auger

[387] P. A. Pella and J. R. Devoe, *Appl. Spectroscopy*, 1971, **25**, 472.
[388] L. H. Bowen, K. A. Taylor, H. Z. Dokuzoguz, and H. H. Stadelmaier, ref. 28, p. 233.

cascade, rather than to the thermodynamic stability of the aliovalent species.[389]

The hyperfine structure of $^{119}Sn^{4+}$ doped into α-Fe_2O_3 has been studied at temperatures above and below the Morin temperature (286 K) of the latter. The hyperfine fields at the tin nucleus are $H_0 = 123 \pm 4$ kG at 293 K and $H_0 = 132 \pm 4$ kG at 77 K. On the basis of a value of 1.0 mm s^{-1} for the quadrupole coupling e^2qQ (taken from the spectrum of Sn^{4+} in Al_2O_3), the angle θ between the direction of the internal field and the direction of the principal axis of the electric field gradient was calculated to be $31° \pm 8°$ and $55° \pm 8°$ for $T > T_M$ and $T < T_M$ respectively. The difference was thought to be consistent with the reorientation of the Fe^{3+} magnetic moments at the Morin transition.[390]

Table 6

Absorber	*Shift*/mm s^{-1} (at 80 K)
$BaSnO_3$	$-\ 1.500 \pm 0.005$
$Ni_{21}Sn_2B_6$	$-\ 0.003 \pm 0.005$
β-Sn	$+\ 1.050 \pm 0.005$

Impurity nuclei of ^{119}Sn have been studied in magnesium ferrites in which some of the iron ions are replaced by either gallium or indium ions. Only those tin nuclei which are adjacent to iron atoms give a six-line pattern, resulting from a supertransferred magnetic hyperfine interaction; the rest give only single lines.[391]

A very well-defined magnetic hyperfine field exists at the tin nuclei in $\{Ca_xY_{3-x}\}[Sn_xFe_{2-x}](Fe_3)O_{12}$ at 4.2 K for $x = 0.1$. As the extent of substitution increases, the spectra show a distribution of magnetic hyperfine fields acting on the tin nuclei. The spectra can be simulated on the basis of a recent model which takes into account canting and flipping of iron spins due to diamagnetic substitution.[392] The origins of supertransferred hyperfine fields acting on ^{119}Sn nuclei in iron garnets has been studied further in the closely related series of compounds $Y_{2.9}Ca_{0.1}Sn_{0.1}$-$Ga_xFe_{4.9-x}O_{12}$ ($x = 0$, 0.5, 1.0, and 1.5). The spectra were analysed on the assumption that the major contribution to the hyperfine field acting on the ^{119}Sn nuclei comes from the six nearest neighbours. It is clear from Figure 23 that good agreement was obtained between the theoretical and experimental spectra. The supertransferred hyperfine field of 209.5 ± 0.5 kG acting on ^{119}Sn in YIG (*i.e.*, $x = 0$) is short-ranged and originates mainly from the tetrahedral iron first-nearest neighbours (the tin is known from previous measurements to enter the *a*-sites of the garnets exclusively).

[389] Y. Llabador and J. M. Friedt, *Chem. Phys. Letters*, 1971, **8**, 592.
[390] P. B. Fabritchnyi, A. M. Babechkin, and A. N. Nesmeianov, *J. Phys. and Chem. Solids*, 1971, **32**, 1701.
[391] B. I. Pokrovskii, P. B. Fabrichnyi, A. M. Babeshkin, A. K. Gapeev, L. N. Komissarova, and A. N. Nesmeyanov, *Teor. i eksp. Khim.*, 1971, **7**, 419.
[392] E. R. Bauminger, I. Nowik, and S. Ofer, *J. Phys. (Paris), Colloq.*, 1971, **1** (Pt. 2), C1–913.

The respective contributions of farther octahedral and tetrahedral iron shells are − 2 ± 2 kG and 13 ± 2 kG. The ratio of the nuclear *g* factors g_{ex}/g_o obtained from the measurements is − 0.2150 ± 0.0005.[330] Usually the hyperfine fields acting on nuclei of impurities in ferromagnetic crystals do not follow the temperature dependence of the host magnetization, and

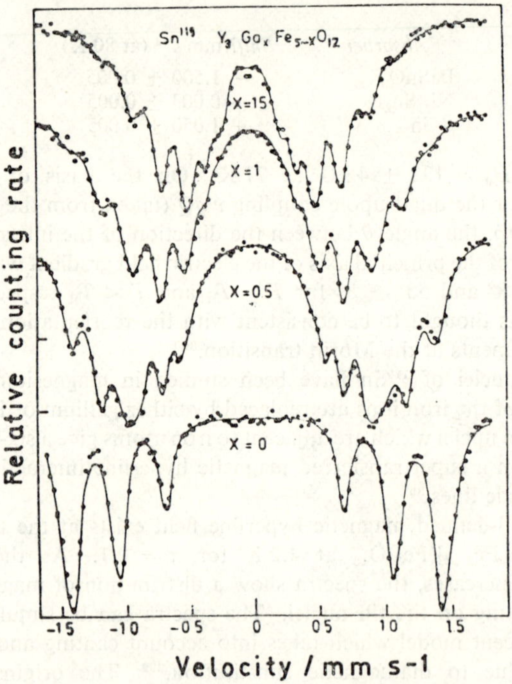

Figure 23 *Recoilless absorption spectrum of the* 23.9 keV *gamma ray of* ^{119}Sn *in* $Y_{2.9}Ca_{0.1}Sn_{0.1}Ga_xFe_{4.9-x}O_{12}$ *compounds at* 4.2 K. *The solid lines are the theoretical curves obtained from least-square computer fits to the experimental spectra*

(Reproduced by permission from *Phys. Letters*, 1971, **34A**, 155)

several mechanisms have been suggested to explain this. However, it has now been shown conclusively that the temperature dependence of the hyperfine field at ^{119}Sn nuclei in $\{Ca_{0.1}Y_{2.9}\}[Sn_{0.1}Fe_{1.9}](Fe_3)O_{12}$ is determined by the magnetic behaviour of the ions in the immediate vicinity of the impurity. A simple parameterless molecular field calculation reproduces the experimental observations perfectly.[393]

Tin(II) Compounds.—The signs of the quadrupole coupling constants e^2qQ, in the compounds $NaSnF_3$, $NaSn_2F_5$, $SnSO_4$, $Sn(HCO_2)_2$, $Sn(AcO)_2$,

[393] D. Lebenbaum, I. Nowik, E. R. Bauminger, and S. Ofer, *Solid State Comm.*, 1971, **9**, 1885.

and $K_2Sn(C_2O_4)_2,H_2O$ have been shown to be positive. The crystal structures of two of the compounds, $NaSn_2F_5$ and $SnSO_4$, are accurately known. Both covalent and point-charge approaches to the interpretation of the origin of the quadrupole splitting in these materials predict the correct sign for e^2qQ, but neither is able to predict the magnitude of the quadrupole splitting.[394]

The Mössbauer and electronic reflectance spectra of the cubic tin(II) trihalogenostannates $CsSnBr_{3-x}Cl_x$ (x = 0—3) and their behaviour on heating suggests that the non-bonding tin(II) electrons are involved in a band system, and not located in non-bonding molecular orbitals. The monoclinic modifications of certain of these species have smaller chemical isomer shifts than the cubic analogues, and this is thought to reflect the shorter Sn—X bond lengths in the former.[395] All ten of the binary and mixed trihalogeno-complexes of tin(II) $[SnXYZ]^-$ (X, Y, or Z = Cl, Br, or I) have been prepared as salts of one or more of the cations $[Et_4N]^+$, $[Bu^n{}_4N]^+$, and $[Ph_4As]^+$ and studied by ^{119}Sn Mössbauer spectroscopy. For salts of a given cation the chemical isomer shifts show a regular variation with the sums of the Mulliken electronegativities of the halide ligands. However, the shifts are influenced more by changing the cation than by changing the ligands. In addition there is an inverse relationship between the quadrupole splitting and the chemical isomer shift within each series.[396] Asymmetric quadrupole-split spectra have been reported for the salts $MSnCl_3$ (M = Na, K, Rb, or Cs); the asymmetry decreases along the series and is attributed to the Goldanskii–Karyagin effect.[397]

On the basis of vibrational spectroscopy and ^{119}Sn Mössbauer spectroscopy the 1 : 1 adducts of SnF_2 with SbF_5, AsF_5, and BF_3 have been formulated as salts of the fluorine-bridged $(Sn\text{–}F)_n{}^{n+}$ cation, with the anions SbF_6^-, AsF_6^-, and BF_4^-. The 1 : 2 adduct $SnF_2,2SbF_5$, which is better regarded as $Sn(SbF_6)_2$, has the highest reported ^{119}Sn chemical isomer shift (4.40 mm s^{-1} rel. $BaSnO_3$) but the value is still below that estimated for a bare Sn^{2+} ion and this, together with the vibrational spectral evidence, is thought to indicate considerable cation–anion interaction. The same applies to $Sn(SO_3F)_2$.[398]

Data have been reported for 12 complexes of the type SnX_2,L and $SnX_2,2L$ (X = Cl, Br, or I). In all cases complex formation decreases the chemical isomer shift and induces a substantial quadrupole interaction, as expected. On the basis of the shifts for these compounds and for others in the literature, the order of ligand strengths for secondary and tertiary amines towards tin(II) follows the sequence: *N*-methylmorpholine >

394 J. D. Donaldson, E. J. Filmore, and M. J. Tricker, *J. Chem. Soc.* (*A*), 1971, 1109.
395 J. Barrett, S. R. A. Bird, J. D. Donaldson, and J. Silver, *J. Chem. Soc.* (*A*), 1971, 3105.
396 M. Goldstein and G. C. Tok, *J. Chem. Soc.* (*A*), 1971, 2303.
397 A. N. Murin, S. I. Bondarevskii, and P. P. Seregin, *Vestn. Leningrad Univ., Fiz. Khim.*, 1971, 138.
398 T. Birchall, P. A. W. Dean, and R. J. Gillespie, *J. Chem. Soc.* (*A*), 1971, 1777.

morpholine ~ piperidine ~ piperazine > pyridine ~ α-picoline > γ-picoline > 1,10-phenanthroline ~ 2,2′-bipyridyl.[399]

The stannous chloride and bromide complexes of thiourea (tu) and tetramethylthiourea (tmtu), of formula $Sn(tu)Cl_2$, $Sn(tu)_2Cl_2$, $Sn(tu)_2Br_2$, $Sn(tmtu)Cl_2$, and $Sn(tmtu)Br_2$ have been studied, both as solids and as frozen solutions. I.r. and n.m.r. data were also given, and the relative merits of the various techniques in determining their structural and bonding properties were compared.[400] Spectra have also been obtained for some chelates of tin(II) with 2-mercapto-, 5-bromo-8-hydroxy-, 8-hydroxy-, and 8-mercaptoquinoline,[401] for a rhenium–thiourea–tin complex,[402] and for tin(II) sensitizer deposits on kapton.[403] Frozen solutions of tin(II) and tin(IV) halides,[404] and the electron-exchange reaction between tin(II) and tin(IV) have also been studied.[405]

Tin(IV) Compounds.—The causes of discrepancies between the published electronegativity *vs.* chemical isomer shift correlations for the ^{119}Sn spectra of SnX_4 and $SnX_4Y_2^{2-}$ (X, Y = F, Cl, Br, or I) have been examined and it has been suggested that certain chemical isomer shift values used in earlier treatments were inaccurate. On the basis of new data for $SnBr_4$ and SnI_4 it is now claimed that the sensitivity of the chemical isomer shift to changes in the electronegativity of the halogen is essentially independent of the co-ordination number of the tin atom. Values of the effective electronegativity for various groups, calculated from previous correlations, have been revised to take account of the new results and are as follows: NCO, 3.8 (Pauling scale—see also p. 570); O, 3.7; N_3, 3.3; NEt_2, 3.1; NMe_2, 3.1; C_6F_5, 2.9; Me, Et, and Ph, 2.8; $Co(CO)_4$, 2.3; $Fe(\pi\text{-}C_5H_5)(CO)_2$, 2.1.[406]

The claim by Clausen and Good that Mulliken ligand-electronegativities correlate significantly better than Pauling electronegativities with the chemical isomer shifts of tris-halogeno- complexes has been contested. Moreover, it is shown that the parameter of most relevance is not the inherent electronegativity of the substituents, but the charge which the substituents can induce on the tin atom. Indeed, for both tetrahedral and octahedral species there is a linear correlation between induced charge and chemical isomer shift (see Figure 24). Even the hexafluorostannate anion, which has proved troublesome in previous correlations, falls on the octahedral line.

399 N. N. Greenwood and A. Timnick, *J. Chem. Soc. (A)*, 1971, 676.
400 J. F. Lefelhocz and C. Curran, ref. 20, p. 55.
401 V. N. Belogurov, J. Bankovskis, E. Luksa, and P. E. Senkov, *Latv. PSR Zinat. Akad. Vestis, Fiz. Teh. Zinat. Ser.*, 1971, 13.
402 V. A. Dolgopolov, T. V. Malysheva, L. V. Borisova, and A. N. Ermakov, *Zhur. strukt. Khim.*, 1971, **12**, 543.
403 R. L. Cohen, J. F. D'Amico, and K. W. West, *J. Electrochem. Soc.*, 1971, **118**, 2042.
404 A. Vertes, E. Szilagyi-Gyori, M. Gal, and L. Suba, *Magyar Kém. Folyóirat*, 1971, **77**, 189.
405 A. Vertes, *Magyar Kém. Folyóirat*, 1971, **77**, 344.
406 R. V. Parish and P. J. Rowbotham, *Chem. Phys. Letters*, 1971, **11**, 137.

The slopes for the octahedral and tetrahedral correlations are clearly different in this treatment.[407]

The chemical isomer shift has been shown to increase with ligand electronegativity for the tin(IV) oxyhalides $SnOX_2$ (X = Cl, Br, or I). This trend is in contrast with that well established for compounds of the types

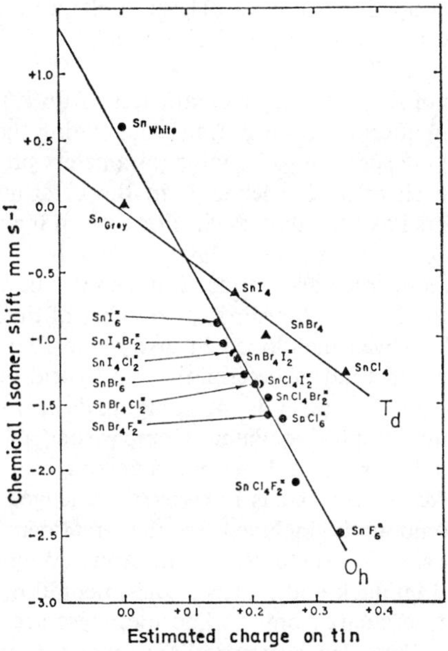

Figure 24 *Chemical isomer shift as a function of estimated charge on the tin atom in tetrahedral and octahedral species*
(Reproduced by permission from *Inorg. Chem.*, 1971, **10**, 1553)

SnX_4, SnX_6^{2-}, and R_3SnX (X = halogen). The results can be explained on the basis of an sp^2 hybridized metal atom with a lone pair occupying the third (non-bonding) p-orbital. As the halogen becomes more electronegative, charge is transferred from the metal to the halogen *via* a bond which increases in p character. To offset this charge transfer, electron density flows from the oxygen atom to the metal *via* an interaction which increases in s character. Both of these processes serve to increase the 5s electron density at the tin nucleus; hence the chemical isomer shift increases as the halogen electronegativity increases. There may also be an additional effect owing to the change in the size of the halogen ligands.[408]

Data have been given for the adducts SnX_4,L (L = NMe_3, X = Cl or Br; L = PPh_3 or $Ph_2PCH_2PPh_2$, X = Cl) and SnX_4,2L (L = NMe_3,

[407] J. E. Huheey and J. C. Watts, *Inorg. Chem.*, 1971, **10**, 1553.
[408] H. S. Cheng and R. H. Herber, *Inorg. Chem.*, 1971, **10**, 1315.

X = Cl, Br, or I; L = NEt_3, X = F; L = py, PBu^n_3, $AsPh_3$, or $AsEt_3$, X = Cl or Br; L = MeCN, PhCN, PPh_3, PPh_2Me, $PPhEt_2$, or PEt_3, X = Cl). The nitriles are the only nitrogen donors to give a resolved quadrupole splitting. Triphenyl-phosphine and -arsine adducts give single-line spectra but quadrupole splitting is induced by replacing phenyl groups by alkyl groups. The five-co-ordinate adducts $SnCl_4,NMe_3$ and $SnBr_4,NMe_3$ give very sharp singlets. These results were discussed in terms of a point-charge model for trigonal-bipyramidal and distorted trigonal-bipyramidal geometry.[409]

Interpretations of the Mössbauer parameters of tin(IV) compounds have been reviewed and alternative suggestions for relating the chemical isomer shift and quadrupole splitting to bonding parameters presented.[410]

The use of the chemical isomer shift in the determination of tin co-ordination numbers has been discussed. For a given ligand, a smaller shift is characteristic of a higher co-ordination number. The temperature dependence of the percentage absorption cannot be used as a direct method for determining the co-ordination number of tin(IV) compounds.[411]

In a paper which was mentioned in last year's Report, Williams and Kocher described a method of determining tin hybrid orbital populations for tetrahedral R_nSnX_{4-n} (n = 1—4) compounds from chemical isomer shift and quadrupole splitting data. Comparisons of Mössbauer- and n.q.r.-derived populations were then used to postulate ($p \rightarrow sp^3$) π-bonding. The validity of this treatment has now been challenged on the basis that many of the compounds included in the treatment may have five-co-ordinate, associated structures.[412] In reply, Williams and Kocher pointed out that Maddock and Platt's criticism could not apply to the tin tetrahalides, because, apart from the fluoride, these are now known to be four-co-ordinate. They also questioned the reliability of the point-charge model, as used by Maddock and Platt, to determine the co-ordination number of tin. They did, however, concede that it is hazardous to use Mössbauer data for orbital population analyses unless the tin atom is known to be four-co-ordinate. Finally, they presented new populations analyses of compounds for which strong evidence of four-co-ordination does exist.[413]

The ratios of the quadrupole splittings to the chemical isomer shifts for the triorganotin fluorides R_3SnF (R = Me, Et, Pr, Bu, or Ph) lie in the range 2.62—2.79, which indicates that the tin has a co-ordination number greater than four in all of these compounds.[414]

The vibrational anisotropy of the tin atom in Me_3SnF has been studied in the temperature range 7—296 K. The intensity ratio of the high-

[409] D. Cunningham, M. J. Frazer, and J. D. Donaldson, *J. Chem. Soc.* (*A*), 1971, 2049.
[410] A. J. Carty and H. D. Sharma, ref. 28, p. 167.
[411] T. V. Malysheva and V. A. Dolgopolov, *Zhur. strukt. Khim.*, 1971, **12**, 52.
[412] A. G. Maddock and R. H. Platt, *J. Chem. Phys.*, 1971, **55**, 1490.
[413] D. E. Williams and C. W. Kocher, *J. Chem. Phys.*, 1971, **55**, 1491.
[414] K. Licht, H. Geissler, P. Köhler, K. Hottman, H. Schnorr, and H. Kriegsmann, *Z. anorg. Chem.*, 1971, **385**, 271.

velocity to the low-velocity peak decreases steadily from *ca.* 0.98 ± 0.02 as the temperature is raised. In contradiction to previous results this ratio does not level off at 170 K. At room temperature the difference in the mean-square amplitudes of vibration parallel and perpendicular to the (nearly) threefold symmetry axis defined by the ···F···Sn···F framework is given by $\langle x_{\parallel}^2 - x_{\perp}^2 \rangle = -3.4 \times 10^{-18}$ cm^2. The Mössbauer temperature for the π and σ transitions were estimated to be 105 ± 2 K and 114 ± 2 K respectively.[415] The temperature dependences in the range 85—160 K of the resonance effect magnitude and of the Goldanskii–Karyagin asymmetry have been used in conjunction with chemical isomer shift and quadrupole splitting data to elucidate the structures of some butyltin compounds. A monomeric dinuclear structure is inferred for $(Bu_3Sn)_2SO_4$, in agreement with i.r., mass spectroscopy, molecular weight, and solubility data. Polymeric chain structures with bridging fluorines are favoured for Bu_3SnF and Bu_2SnF_2, the tin being five-co-ordinate and six-co-ordinate respectively. The data for Bu_3SnOAc were interpreted in terms of a low-molecular weight polymeric form for this compound, similar to that reported for the 'soluble' form of trimethyltin acetate. $Bu_2Sn(OAc)_2$, like the difluorides, is assumed to have a *trans*-octahedral configuration of the two alkyl groups, but its physical properties suggest that, unlike the fluoride, it is monomeric, with the two acetate ligands occupying bidentate positions in the equatorial plane.[416] The origin of the asymmetric quadrupole splitting in Ph_2SnCl_2 (both hydrated and dehydrated),[417] Me_2SnF_2, Et_2SnCl_2, Me_3SnF, and Ph_3SnCl,[418] and in frozen solutions of Me_3SnCl [419] have also been discussed.

Results have been discussed for 35 organotin compounds of the type R_3SnX. The trialkyltin iodides R_3SnI (R = Me, Et, Pr^i, Pr^n, Bu^i, or Bu^n) have chemical isomer shifts in the range 1.48—1.63 mm s^{-1} (except for the isopropyl derivative, which has δ = 1.75 mm s^{-1}) and on the basis of their large quadrupole splittings (Δ = 2.71—3.15 mm s^{-1}) are considered to be five-co-ordinate. Data were also given for 20 compounds containing tin–metal bonds, but these are more conveniently discussed later in the section (see p. 574). Other compounds included in the study were: Me_3SnX (X = C_5H_5, C≡CPh, or NMe_2), Ph_3SnNMe_2, and SnX_4 (X = *o*-MeC_6H_4 or *p*-MeC_6H_4). The tetrahedral trimethyl and triphenyl compounds were found to have chemical isomer shifts within the narrow ranges 1.33—1.45 mm s^{-1} and 1.36—1.46 mm s^{-1} respectively. This insensitivity of δ to the nature of the group X demonstrates the domination of the use of the tin *s*-orbitals by the organic groups. Deviations from shifts in these narrow ranges occur for Me_3SnX (X = C_5H_5, C≡CPh, or

[415] R. H. Herber and S. Chandra, *J. Chem. Phys.*, 1971, **54**, 1847.
[416] R. H. Herber, *J. Chem. Phys.*, 1971, **54**, 3755.
[417] N. Balabanov, B. A. Komissarova, A. A. Sorokin, and V. S. Shpinel, ref. 12, p. 235.
[418] V. Ya. Rochev, V. I. Goldanskii, and R. A. Stukan, ref. 12, p. 227.
[419] H. Schnorr and H. Kriegsmann, ref. 12, p. 629.

NMe_2) and for Ph_3SnNMe_2, all of which have a weak Sn—X bond; significantly these species also show a resolvable quadrupole splitting.[420]

The trigonal-bipyramidal compounds $Me_3SnCl(C_5H_5N)$ and Me_3SnF give quadrupole splittings of 3.35 and 3.47 mm s^{-1} respectively, whereas the tetrahedral compounds $Me_3Sn{\cdot}S{\cdot}CS{\cdot}NMe_2$ and $Ph_3Sn{\cdot}S{\cdot}CS{\cdot}NEt_2$ give quadrupole splittings of 2.25 and 1.85 mm s^{-1} respectively. This drop in quadrupole splitting from *ca.* 3.5 to *ca.* 2.0 mm s^{-1} is therefore taken as an indication of a change-over in geometry from (10) to (11), and on this basis the compounds Ph_3SnX (X = Cl, Br, or NCS); Ph_3SnXL [X = Cl, Br, or NCS; L = (C_5H_5NO), Me_2SO, or Ph_3PO]; $[Ph_3SnX_2]^-[Ph_3PC_{10}H_{21}]^+$

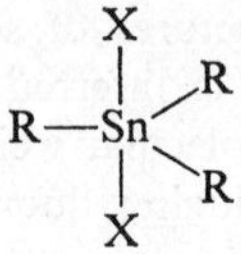

(10)

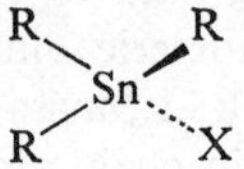

(11)

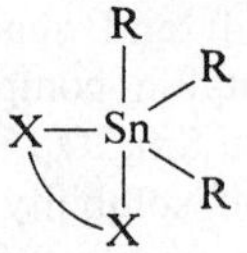

(12)

(X_2 = Cl_2, ClBr, or Br_2), $Me_3SnBr[OP(morph)_3]$, $Me_3SnNCS[OP(pip)_3]$, $Me_3SnCl(C_5H_5N)$, and Me_3SnF were assigned trigonal-bipyramidal structures, and the compounds $Me_3Sn{\cdot}S{\cdot}CS{\cdot}NEt_2$ and $Ph_3Sn{\cdot}S{\cdot}CS{\cdot}NMe_2$ tetrahedral structures. For the compound Ph_3Sn(oxin) it was not possible to distinguish between structure (11) and a third structure (12) in which the organotin unit is constrained by the chelating oxinate ligand. A general conclusion to be drawn from the data is that a range of quadrupole splitting of 1.6—2.5 mm s^{-1} is appropriate for tetrahedral organotin compounds (see, however, refs. 426, 439, and 445). The values predicted on the basis of the point-charge model for these three geometries are (10) 3.12, (11) 1.66, and (12) 1.57 mm s^{-1}.[421]

The determination of the sign of the quadrupole interaction in organotin(IV) compounds by the magnetic perturbation method has been the subject of five papers. The results constitute a significant advance in our understanding of the origins of the electric field gradient in tin compounds, and have added greatly to our knowledge of many structural aspects of tin chemistry.

In one report a study of three compounds of rigorously established stereochemistry was described. The *trans*-octahedral compound Me_2SnF_2 was found to have a positive e^2qQ, whereas the trigonal-bipyramidal compounds $Me_3Sn(NCS)$ and $Me_3Sn(OH)$ were shown to have a negative e^2qQ. Based on this evidence, trigonal-bipyramidal structures were then deduced for Ph_3SnF and Ph_3SnCl (e^2qQ negative), and a distorted tetrahedral structure for Bu_2SnO (e^2qQ positive), with bond angles decreasing in the order C—Sn—C > C—Sn—O > O—Sn—O.[422]

[420] S. R. A. Bird, J. D. Donaldson, S. A. Keppie, and M. F. Lappert, *J. Chem. Soc.* (*A*), 1971, 1311.

[421] J. Ensling, P. Gütlich, K. M. Hasselbach, and B. W. Fitzsimmons, *J. Chem. Soc.* (*A*), 1971, 1940.

[422] B. A. Goodman and N. N. Greenwood, *J. Chem. Soc.* (*A*), 1971, 1862.

In a consecutive paper in the same journal, complexes formed between the compounds Pr^nSnCl_3, $Pr^n_2SnCl_2$, and Ph_xSnCl_{4-x} (x = 1—3) and the Lewis bases piperidine, morpholine, β-picoline, γ-picoline, and isoquinoline were studied. The complexes of Ph_3SnCl have quadrupole splittings of *ca.* 3.0 mm s^{-1} of negative sign, and on this basis were assigned trigonal-bipyramidal structures with C—Sn—C bond angles of *ca.* 120°. The adducts of Ph_2SnCl_2 and $Pr^n_2SnCl_2$ with piperidine, γ-picoline, β-picoline, and isoquinoline give quadrupole splittings of *ca.* 4.0 mm s^{-1} of positive sign, and are therefore thought to be six-co-ordinate with *trans*-organo-ligands. By contrast, the morpholine adducts have a much smaller quadrupole splitting (*ca.* 2.0 mm s^{-1}) suggestive of a *cis* structure, but the sign of e^2qQ is still positive, rather than negative as expected for a *cis* structure. However, the compounds are thought to be *cis*, and the apparent inconsistency can be explained by an increase in the C—Sn—C bond angle to *ca.* 120°, which necessitates redefining the axis system of the electric field gradient tensor. The adducts of Pr^nSnCl_3 and $PhSnCl_3$ have Δ-values of < 2 mm s^{-1} and of positive sign, consistent with a six-co-ordinate polymeric structure containing tin surrounded by five chlorines and one organo-group.[423]

Elsewhere a systematic study of some four-, five-, and six-co-ordinate organotin(IV) compounds, covering seven different structural types, has been reported. The *trans*-octahedral species $(Me_4N)_2[EtSnCl_5]$, $Cs_2[Me_2SnCl_4]$, $K_2[Me_2SnF_4]$, and $Bu_2Sn(O_2CCH{:}CHO_2)_2$ all have positive coupling constants; the trigonal-bipyramidal complexes $Me_4N[Me_3SnCl_2]$, $Ph_3PCH_2Ph[Et_3SnCl_2]$, $Me_4N[Ph_3SnCl_2]$, and $[Et_3SnCN]_n$, in which the electronegative groups lie along the z-axis, have negative coupling constants; the trigonal-bipyramidal $Et_4N[Me_2SnBr_3]$, in which the organo-groups lie along the z-axis, has a positive coupling constant; and the four-co-ordinate complexes $(Bu_3Sn)_2$, $Me_3SnC_6F_5$, $Ph_3SnC_6F_5$, $Ph_3SnSnPh_3$, and Ph_3SnMe, in which the more electronegative groups lie along the z-axis, have negative coupling constants. All of these results agree with the predictions of point-charge theory.[424] However, the four *cis*-R_2SnX_4 complexes $Me_2Sn(oxin)_2$, $Ph_2Sn(oxin)_2$, $Ph_2Sn(S_2CNEt_2)_2$, and $Ph_2Sn(NCS)_2$phen all have positive coupling constants, which are at first sight anomalous, but which can be explained by arguments similar to those used to account for the positive coupling constants of the morpholine adducts discussed above.[424, 425]

A conclusion common to all four studies is that the electric field gradient arises primarily from imbalances in the σ-bonding network and is dominated by the disposition of the tin–carbon bonds. Related studies on compounds containing tin–transition-metal bonds are described on p. 572.

[423] B. A. Goodman, N. N. Greenwood, K. L. Jaura, and K. K. Sharma, *J. Chem. Soc.* (*A*), 1971, 1865.
[424] R. V. Parish and C. E. Johnson, *J. Chem. Soc.* (*A*), 1971, 1906.
[425] R. V. Parish, *Chem. Phys. Letters*, 1971, **10**, 224.

The full paper dealing with the spectra of cyclohexyltin derivatives and of the compounds $Ph_2Sn(X)[CH_2]_4Sn(X)Ph_2$ (X = I, OH, O·COMe, or Ph), $Bu_nSn(O{\cdot}COMe)_{4-n}$ (n = 2 or 3), and Bu_2Sn(maleate) has now appeared. The data were used to define an upper limit of 2.9 mm s^{-1} for the quadrupole splitting associated with tetrahedral geometry (see also refs. 421, 439, and 445) and to discuss intermolecular association in organotin compounds. In addition it was shown that the simple point-charge model does not account satisfactorily for the quadrupole splittings of compounds with differing co-ordination numbers.[426]

^{119}Sn Mössbauer quadrupole splitting and $J(^{119}Sn\text{–}C\text{–}^{19}F)$ data have been used to establish the existence of intramolecular co-ordination in the perfluoroethyltin compounds $SnEt_2(C_2F_5)_2$ and $SnEt_2(C_2F_5)(I)$.[427]

The 1 : 1 adducts $MeSnCl_3$(acen), R_2SnCl_2(acen) [R = Me, Bu, or Ph; acen = bis(acetylacetone)ethylenedi-imine], Me_2SnBr_2(acen), and Me_3SnCl-(acen) have been studied by i.r. and Mössbauer spectroscopy. The diorgano-complexes exhibit large quadrupole splittings (*ca.* 4.0 mm s^{-1}) and are thought to have structures of the type shown, in which the organo-ligands are *trans* and the acen ligands are bonded through nitrogen (13) or oxygen (14). Measurements of the temperature dependence of the Debye–Waller factor for the Me_3SnCl adduct indicate a polymeric structure. An

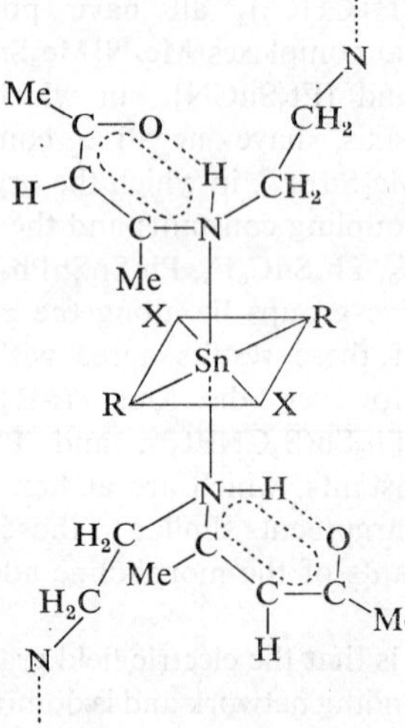

(13)

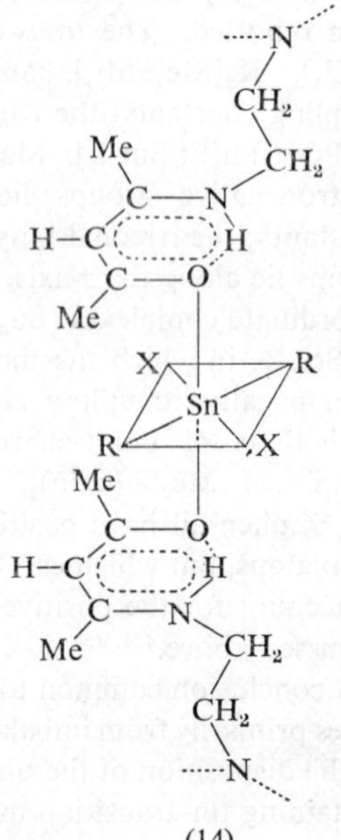

(14)

[426] A. G. Maddock and R. H. Platt, *J. Chem. Soc.* (*A*), 1971, 1191.
[427] P. G. Harrison, S. E. Ulrich, and J. J. Zuckerman, *J. Amer. Chem. Soc.*, 1971, **93**, 5398.

octahedral-type configuration in which the planar organotin(IV) halide moieties are bridged by the ligand, axially co-ordinating tin(IV), is proposed.[428] The large quadrupole splittings for the 1 : 1 adducts R_2SnCl_2(acen) (R = Me, Et, Bu, or Ph) have been confirmed elsewhere and interpreted in terms of the *trans*-R_2SnX_4 octahedral configuration (15). Structure (16)

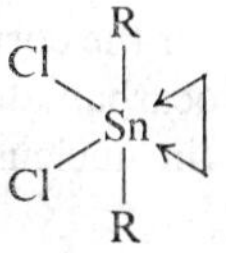

(15)

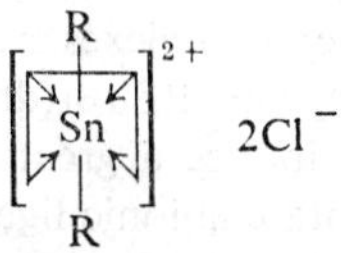

(16)

was eliminated on the basis of conductivity measurements. The possibility of these compounds adopting structures (14) or (15) was not considered.[429]

trans-Arrangements of C—Sn—C bonds have also been proposed for the complexes X_2Sn(picolinato)$_2$ (X = Bu^n or C_2H_3), Bu^n_2Sn(dipicolinato), and Ph_2Sn(dipicolinato),H_2O, the quadrupole splittings being *ca.* 4.0 mm s^{-1} in each case. The quadrupole splitting for Ph_2Sn(picolinato)$_2$ is *ca.* 1.94 mm s^{-1}, so this presumably has a *cis*-C—Sn—C arrangement. Other compounds studied included X_2Sn(picolinato)$_2$ (X = I, Br, or Cl), XSn(picolinato)$_3$ (X = Cl or Br), and Sn(dipicolinato)$_2$,H_2O.[430]

The chelates of tin(IV) with 2-mercapto-, 5-bromo-8-hydroxy-, 8-hydroxy-, and 8-mercapto-quinoline, and the complex $SnCl_2$(4-methyl-8-mercapto-quinoline) have been studied.[401]

Mössbauer spectroscopy has been used to confirm the presence of new adducts in the products of the reaction between Ph_2SnCl_2 and the sulphoxide ligands, L = Me_2SO, $(CH_2)_4SO$, Pr_2SO, or Bu_2SO. The quadrupole splittings (*ca.* 3.0 mm s^{-1}) of the (presumed) trigonal-bipyramidal 1 : 1 adducts Ph_2SnCl_2,L lie between those for Ph_2SnCl_2 (Δ = 2.83 mm s^{-1}) and the 1 : 2 adducts (Δ *ca.* 3.8 mm s^{-1}), which are thought to be octahedral with *trans*-phenyl groups. In all cases the chemical isomer shift decreases on complex formation from 1.37 to values lying between 1.21 and 1.30 mm s^{-1}.[431]

In an attempt to broaden the correlation between the Mössbauer and electric dipole moment studies, previously undertaken on R_2SnX_2 and $RSnX_3$ complexes, data have now been reported for complexes of R_2SnCl_2 and $RSnCl_3$ (R = Bu or Ph) with the ligands L = Ph_3PO, Ph_3AsO, Bu_3PO, or Ph_3PO. The trigonal-bipyramidal complexes, Ph_2SnCl_2L, are considered to have equatorial phenyl groups, whereas the organo-groups in the

428 R. Barbieri, R. Cefalù, S. S. Chandra, and R. H. Herber, *J. Organometallic Chem.*, 1971, **32**, 97.
429 P. J. Smith and D. Dodd, *J. Organometallic Chem.*, 1971, **32**, 195.
430 D. V. Naik and C. Curran, *Inorg. Chem.*, 1971, **10**, 1017.
431 R. S. Randall, R. W. J. Wedd, and J. R. Sams, *J. Organometallic Chem.*, 1971, **30**, C19.

$R_2SnCl_2Y_2$ complexes are probably *trans* to one another. Data for the $RSnCl_3Y_2$ complexes suggest that in $BuSnCl_3(Ph_3AsO)_2$ and $BuSnCl_3(Ph_3P)_2$ two chlorines are *trans* and that in $BuSnCl_3(Ph_3PO)_2$ and $PhSnCl_3(Ph_3PO)_2$ the three chlorines are *cis*.[432]

Data have been presented for 32 tin(IV) compounds containing bidentate *NN′*-dialkyldithiocarbamate (dtc) ligands. The chemical isomer shifts for the $RSn(dtc)_3$ complexes are more positive than those for the corresponding $RSnCl(dtc)_2$ and $RSnBr(dtc)_2$ species, and the dithiocarbamate ligand is claimed to induce a greater electron density at the tin nucleus than any other bidentate anionic ligand studied so far.[433]

The effective Pauling electronegativity of a non-bridging cyanate group has been estimated to be *ca.* 3.4 from the Mössbauer results for $[Sn(NCO)_6]^{2-}$ and related octahedral hexahalogen complexes (see also p. 562). The trialkyltin and dialkyltin cyanates R_3SnNCO (R = Et, Pr, Bu^n, Ph, or $C_6H_5CH_2$) and $R_2Sn(NCO)_2$ (R = Me, Bu^n, Bu^i, or $C_6H_5CH_2$) are all thought to have tetrahedral structures with non-bridging NCO groups, whereas Me_3SnNCO probably has a distorted pyramidal structure.[434]

The compounds Me_3SnMO_4 (M = S, Se, or Cr) have been shown to give similar Mössbauer parameters to Me_3SnCl, and are therefore thought to have similar polymeric, trigonal-bipyramidal structures. Based on a comparison with the data for Me_2SnF_2, the compounds Me_2SnMO_4 (M = S or Se) are also thought to be polymeric, but with six-co-ordination at the tin atom.[435]

The structure reported last year for $(Bu^n_3Sn)_2SO_4$, in which the sulphate is quadridentate and joined through pairs of oxygen atoms to two Bu_3Sn moieties, has been confirmed.[436] On the basis of Mössbauer and vibrational spectra, the new trimethyltin(IV) sulphonates Me_3SnSO_3X (X = F or Me) and the previously reported $Me_2SnSO_3CF_3$ are thought to have polymeric structures with bridging SO_3X groups and five-co-ordinate tin.[437]

Data have been discussed for some trigonal-bipyramidal diorganotin(IV) derivatives containing terdentate planar ligands with ONO and SNO donor groups.[438]

Mössbauer and i.r. spectra indicate that the compounds $Ph_3SnOCOR$ (R = Me, CH_2I, CH_2Br, CH_2Cl, $CHCl_2$, or CF_3) are five-co-ordinate polymers in the solid state with O—C—O bridges between tin atoms. For R = CCl_3 a four-co-ordinate monomeric species is suggested. The magnitude of the quadrupole splitting (Δ = 2.97 mm s^{-1}) for the latter indicates that previous upper limits set on the magnitude of this parameter

432 F. P. Mullins, *Canad. J. Chem.*, 1971, **49**, 2719.
433 J. C. May, D. Petridis, and C. Curran, *Inorg. Chim. Acta*, 1971, **5**, 511.
434 K. L. Leung and R. H. Herber, *Inorg. Chem.*, 1971, **10**, 1020.
435 B. F. E. Ford, J. R. Sams, R. G. Goel, and D. R. Ridley, *J. Inorg. Nuclear Chem.*, 1971, **33**, 23.
436 R. H. Herber, *Inorg. Nuclear Chem. Letters*, 1971, **7**, 617.
437 P. A. Yeats, J. R. Sams, and F. Aubke, *Inorg. Chem.*, 1971, **10**, 1877.
438 R. H. Herber and R. Barbieri, *Gazzetta*, 1971, **101**, 149.

for four-co-ordinate tin (refs. 421 and 426) may be too low (see also ref. 445).[439]

Homologous series of trialkyltin acetates in which the methyl group of the acetate is successively chlorinated and fluorinated have been studied by i.r., ^{1}H and ^{19}F n.m.r., and Mössbauer spectroscopy. Substitution by groups of increasing electronegativity was found to produce a steady increase in the chemical isomer shift, and, for the methyl series, in the quadrupole splitting. In the solid state the tin atom is usually five-co-ordinate with bridging carboxylate groups; however, five-co-ordinate monomeric species may also be present in frozen solutions.[440]

The temperature dependence of the mean-square displacement of the ^{119}Sn nucleus in the polymeric triphenyltin acetate and triphenyltin levulinate [$Ph_3SnOCO(CH_2)_2COMe$], and in the monomer triphenyltin methacrylate ($Ph_3SnOCOCMeCH_2$) has been determined. The results indicate that the helical chain polymeric structure reduces the inter-molecular interaction to the same extent as does the monomer structure. In a single-crystal study the sign of V_{zz} was shown to be positive ($e^2qQ < 0$) for the acetate, suggesting that the σ-polarization in the Sn—C bond predominates over π-bonding effects in determining the sign of the electric field gradient.[441]

The compound $Ph_4Sn_2(OCOPh)_2$, which is known to have a structure (17) with five-co-ordinate tin, has been found to give a quadrupole splitting to chemical isomer shift ratio of $\rho = 2.12$ which is close to that expected (> 2.1) on the basis of Herber's suggestion. $Ph_4Sn_2(OOCR)_2$ (R = Me, CH_2Cl,CCl_3, or CF_3) give similar ρ-values and are therefore thought to have structures analogous to (17).[442]

The 1,3-difunctional tetra-alkyldistannoxanes $XBu_2SnOSnBu_2X$ (X = F, Cl, Br, NCS, OCOMe, OPh, OC_6H_4Me-4, or OC_6H_4Cl-4) give symmetrical doublets with quadrupole splittings in the range 2.74—3.36 mm s^{-1}, thought to be consistent with the ladder structure (18). When X = $OSiMe_3$ the quadrupole splitting is anomalously low (2.46 mm s^{-1}); it is suggested

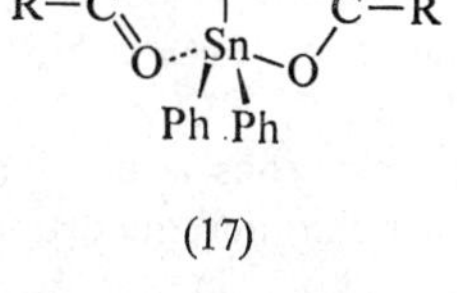

(17)

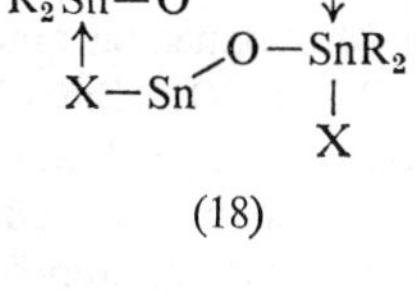

(18)

[439] B. F. E. Ford and J. R. Sams, *J. Organometallic Chem.*, 1971, **31**, 47.
[440] N. W. G. Debye, D. E. Fenton, S. E. Ulrich, and J. J. Zuckerman, *J. Organometallic Chem.*, 1971, **28**, 339.
[441] H. Sano and R. Kuroda, *Chem. Phys. Letters*, 1971, **11**, 512.
[442] M. Delmas, J. C. Maire, Y. Richard, G. Plazzogna, V. Peruzzo, and G. Tagliavini, *J. Organometallic Chem.*, 1971, **30**, C101.

that this is caused by the bulkiness of the trimethylsiloxy-groups, which in turn results in a lengthening of the Sn—O and Sn—X bonds.[443]

Mössbauer, i.r., and ^1H and ^{13}C n.m.r. spectroscopy have been used to search for donor–acceptor interactions between the tin atom and olefinic double bonds in some norbornenyl and norbornadienyl derivatives of tin. The chemical isomer shifts were similar for all of the compounds and in no case was a quadrupole splitting observed. It was concluded that, in the ground state at least, no appreciable bonding occurs between tin and the olefinic moiety in these compounds.[444]

Magnetic perturbation studies have been performed on a number of compounds containing tin–transition-metal bonds, in order to obtain information about the origin of the electric field gradient at the tin nucleus. The quadrupole coupling constant, e^2qQ, was found to have a large positive value (V_{zz} negative) in the compounds [$Mn(CO)_5SnCl_3$], [$Mn(CO)_5$-$SnCl_2Me$], [$Fe(\pi$-$C_5H_5)(CO)_2SnCl_3$], [$\{Fe(\pi$-$C_5H_5)(CO)_2\}_2SnCl_2$], and [$\{Fe(\pi$-$C_5H_5)(CO)_2\}_2Sn(NCS)_2$] and a small negative value (V_{zz} positive) in the new compound [$Fe(\pi$-$C_5H_5)(CO)_2SnBu_3$]. The zero-field spectrum of this butyl compound is reproduced in Figure 25(a) to show the extent of resolution of the quadrupole splitting. The slight asymmetry in the spectrum in a transverse field of 50 kG [Figure 25(b)] is seen to match that present in the lineshape simulated for $\frac{1}{2}e^2qQ = -0.59$ mm s^{-1} [Figure 25(c)]. These results can be adequately explained on the basis of a point-charge approach to the bonding, taking into account the bond angles and the relative contributions to V_{zz} from the Sn—M and Sn—X bonds (M = Mn or Fe, X = Cl or Bu^n). In these compounds the Sn—M bonds make a negative contribution to V_{zz}, whereas the Sn—X bonds make a positive contribution. Two of the compounds studied ([$Fe(\pi$-$C_5H_5)$-$(CO)_2SnCl_3$] and [$\{Fe(\pi$-$C_5H_5)(CO)_2\}_2SnCl_2$]) are of accurately known crystal structure and from the results for these compounds it is apparent that the $\langle r^{-3} \rangle$ contribution to V_{zz} from an Sn—Fe bond is much larger than that from an Sn—Cl bond. The change in sign of V_{zz} from negative to positive in going from [$Fe(\pi$-$C_5H_5)(CO)_2SnCl_3$] to [$Fe(\pi$-$C_5H_5)(CO)_2SnBu_3$] is a reflection of the enhanced σ-donor ability of the butyl group compared with chlorine, which in turn increases the positive contribution to V_{zz} from the Sn—X bonds.[271]

The ^{119}Sn Mössbauer parameters for the series [$Fe(\pi$-$C_5H_5)(CO)_2SnCl_3$] and [$\{Fe(\pi$-$C_5H_5)(CO)_2\}_2SnCl_2$] (X = Cl, Br, I, NCS, HCO_2, or AcO) also show that the use of the tin bonding electrons is dominated by the Sn—Fe bond. The trends in chemical isomer shift and quadrupole splitting were explained with reference to the very short Sn—Fe bond in these materials. The positive sign for e^2qQ for [$Fe(\pi$-$C_5H_5)(CO)_2SnCl_3$] was also

443 A. G. Davies, L. Smith, and P. J. Smith, *J. Organometallic Chem.*, 1971, **29**, 245.

444 C. H. W. Jones, R. G. Jones, P. Partington, and R. M. G. Roberts, *J. Organometallic Chem.*, 1971, **32**, 201.

confirmed and was explained in terms of a high *p*-electron density on the tin atom in the direction of the Sn—Fe bond.[272]

The point-charge model has been applied elsewhere to four-co-ordinate compounds of this type. The partial quadrupole splitting value of + 0.63 for Cl, derived by Parish and Platt in 1970, was used to calculate quadrupole

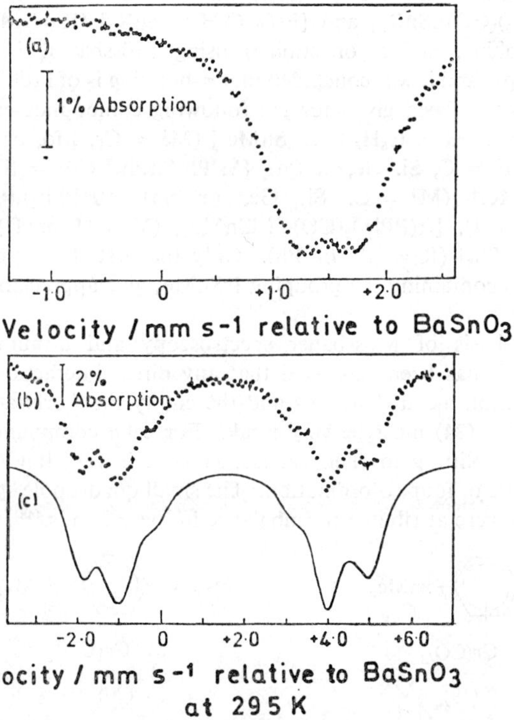

Figure 25 (a) *Zero-field* ^{119}Sn *spectrum for* [Fe(π-C_5H_5)$(CO)_2$SnBu_3] *at* 77 K. (b) *Spectrum in a transverse field of* 50 kG *at* 4.2 K. (c) *Lineshape simulated for* δ 1.47 mm s^{-1}, $\frac{1}{2}e^2qQ$ − 0.59 mm s^{-1}, Γ 0.8 mm s^{-1}, η 0, *H* 50 kG ⊥

splitting values. In general these calculated values were in fair agreement with known observed values, both in magnitude and sign. The splitting for [Mn$(CO)_5$SnPh$(OSOPh)_2$] (Δ = 3.06 mm s^{-1}) agrees with the calculated value and is the largest yet reported for a tetrahedral tin(IV) compound (see also discussion on pp. 566, 568, and 570).[445]

In the series [Mn$(CO)_5$Sn$R_{3-x}X_x$] (R = Me or Ph; X = halogen) the ^{119}Sn chemical isomer shift has been found to correlate with the ^{1}H and ^{55}Mn n.m.r. chemical shifts, and with the coupling constant J[^{119}Sn–^{1}H(Me)]. On the basis of these results it was claimed that the

[445] R. V. Liengme, M. J. Newlands, and J. R. Sams, *Inorg. Nuclear Chem. Letters*, 1971, 7, 1223.

$Mn(CO)_5$ group is a stronger electron donor than the methyl, phenyl, or halogen groups and that the Sn—Mn bond is predominantly $s\sigma$ in character, whereas the Sn—halogen bond is predominantly $p\sigma$ in character.[446, 447] These conclusions regarding the distribution of the tin bonding electrons apparently conflict with those [271, 272] discussed earlier.

The nature of the Sn—Fe bond has also been studied in the series $[Fe(\pi\text{-}C_5H_5)(CO)LSnR_3]$ and $[Fe(\pi\text{-}C_5H_5)L_2SnR_3]$ (R = Me, Ph, or Cl; L = phosphine, arsine, or stibine) using Mössbauer, i.r., and n.m.r. spectroscopy, and it was concluded that π-bonding is of little importance.[270] Data have also been given for the following compounds containing tin–metal bonds: $[M^1(\pi\text{-}C_5H_5)(CO)_3SnMe_3]$ (M^1 = Cr, Mo, or W), $[M^2Me_3SnMe_3]$ (M^2 = C, Si, Ge, or Sn), $[M^3Ph_3SnPh_3]$ (M^3 = Si, Ge, or Sn), $[M^4Ph_3SnMe_3]$ (M^4 = C, Si, Ge, or Sn), $[Ir(PPh_3)_2(CO)HXSnPh_3]$ (X = Cl or I), $[Ir(PPh_3)_2(CO)YClSnMe_3]$ (Y = H or D), $[Ir(PPh_2R)(CO)HClSnPh_3]$ (R = Me or Ph). Only the first three compounds and compounds containing the grouping $IrSnMe_3$ give appreciable quadrupole splitting.[420]

On the basis of Mössbauer spectroscopy and a number of other techniques it has been suggested that any direct interaction between the filled transition-metal *d*-orbitals and the empty tin *d*-orbitals in the compounds (19)—(24) must be very weak. For each compound the ratio of quadrupole splitting to chemical isomer shift is very low (*ca.* 0.50) and characteristic of four-co-ordination. The small quadrupole splittings which are present were attributed to imbalance in the σ-bonds.[448]

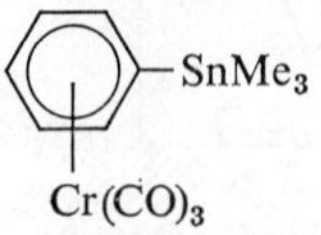

(19)

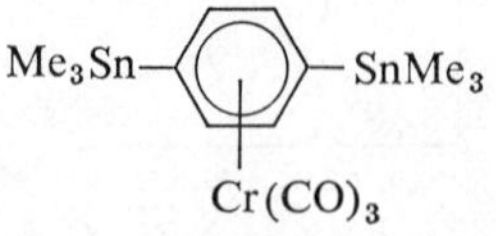

(20)

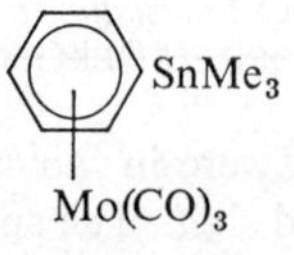

(21)

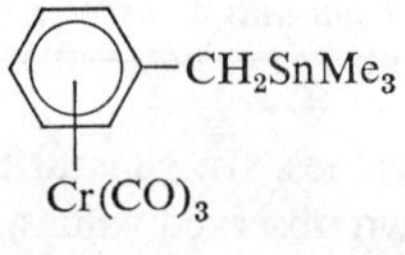

(22)

The reaction of dicyclopentadienyltin(II) with phenylmagnesium bromide has been investigated. The product $Ph(\pi\text{-}C_5H_5)_2SnMgBr$ was found to contain tin(IV), favouring carbenoid insertion into the Mg—C bond with formation of a covalent Sn^{IV}—Mg bond, rather than the formation of a

[446] S. Onaka, Y. Sasaki, and H. Sano, *Bull. Chem. Soc. Japan*, 1971, **44**, 726.
[447] S. Onaka, T. Mijamoto, and Y. Sasaki, *Bull. Chem. Soc. Japan*, 1971, **44**, 1851.
[448] T. P. Poeth, P. G. Harrison, T. V. Long, jun., B. R. Willeford, and J. J. Zuckerman, *Inorg. Chem.*, 1971, **10**, 522.

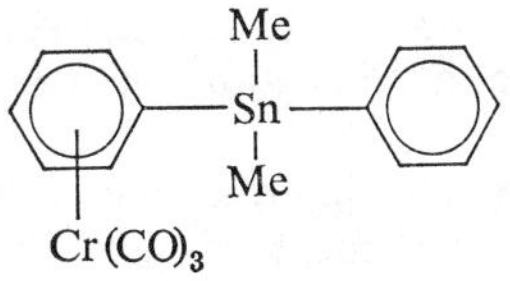

(23)

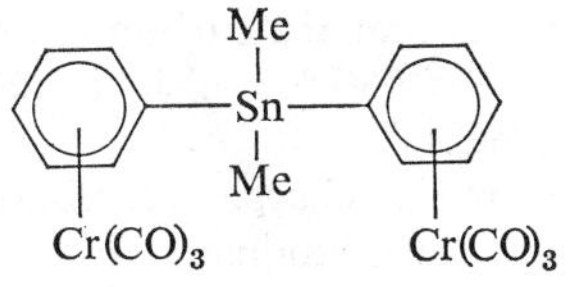

(24)

donor–acceptor adduct containing a co-ordinated $Sn^{II} \rightarrow Mg$ bond. The related compounds $Ph_3SnMgBr$, $Ph_3SnZnCl$, and $Ph_3SnZnCl$,(tmed) (tmed = *NNNN*-tetramethylethylenediamine) also contain tin(IV) and are thought to have the cyclic structure (25).[449]

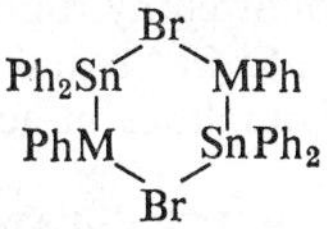

(25)

The tin–sulphur system has been re-investigated and, in contrast to previous reports, materials of composition Sn_2S_3 and Sn_3S_4 apparently contain only SnS and SnS_2. Tin–selenium mixtures with 10—33.3 at. % Sn were shown to contain only $SnSe_2$, and alloys with 35.4 at. % Sn contained SnSe but not Sn_2Se_3.[450] The solid solution tin–sulphur–selenium systems, $SnS_{1-x}Se_x$ and $SnS_{2-x}Se_x$ [451] and the tin–tellurium system [452] have also been studied. Tin(IV) could not be detected in the latter and it was concluded that $SnTe_2$ does not exist in the solid phase.

Data from a number of lanthanide–tin and iron–tin phases have been used in conjunction with other studies to help clarify the phase relationships in these systems. In the binary alloys $CeSn_3$, Ce_2Sn_3, Ce_2Sn, and Ce_3Sn a decrease in the tin content is accompanied by a decrease in the chemical isomer shift, which probably reflects an increasing tendency of the tin to donate electrons to the *f*-band of cerium. The relatively sharp lines in the quadrupole split spectrum of $CeSn_3$ suggest that there is a higher degree of order in this alloy than in the other phases in this system. There is very little difference between the chemical isomer shifts for the alloys and those for the carbides Ce_3SnC_x (x = 0.20, 0.61, 1.05, or 1.97),

449 P. G. Harrison, J. J. Zuckerman, and J. G. Noltes, *J. Organometallic Chem.*, 1971, **31**, C23.
450 G. M. Bartenev, A. D. Tsyganov, S. Dembovskii, and V. I. Mikhailov, *Izvest. Akad. Nauk S.S.S.R., Neorg. Materialy*, 1971, 7, 1442.
451 B. T. Melekh, P. P. Seregin, and V. T. Shipatov, *Izvest. Akad. Nauk S.S.S.R., Neorg. Materialy*, 1971, 7, 502.
452 B. T. Melekh, P. P. Seregin, and V. T. Shipatov, *Izvest. Akad. Nauk S.S.S.R., Neorg. Materialy*, 1971, 7, 693.

which suggests that the carbon is probably bonded to the lanthanide and not to tin.[140, 453] (See also p. 508 for a discussion of the system Fe_3Sn–Fe_3SnC.)

The electronic charge distribution in the intermetallic compound $ZnSnSb_2$ has been studied. This compound is more covalent than both $ZnSnP_2$ and $ZnSnAs_2$.[454]

The polymerization of formaldehyde by $SnCl_4$ in PhMe has been shown to occur *via* the initial formation of a symmetrical six-co-ordinate complex. In the presence of a two-fold or greater excess of $CH_2(OMe)_2$ a product is obtained which gives a quadrupole splitting characteristic of a *trans*-Sn–$CH_2(OMe)_2$ complex. Polyformaldehyde exhibits a new line in the spectrum, which facilitates monitoring of the polymerization.[455] The mechanism of the reaction of a triethylstannyl derivative of polyvinyl chloride with HCl has been shown to proceed by initial substitution of chlorine for an ethyl group, followed by rupture of the tin–polymer bond, as further HCl is added, to give Et_2SnCl_2 and $(-CH_2CH_2CH_2CHCl-)_n$.[456] The role of tin tetrachloride complexes in acetal bond-breaking reactions has been discussed.[457]

Frozen solutions of tin(IV) in HCl and $HClO_4$ have been examined,[404] and phase transitions in frozen HCl solutions doped with tin(IV) ions have been studied.[458]

6 Other Elements

The elements in this section are discussed under three main headings: main-group elements, transition elements, and lanthanide and actinide elements. In each section the elements are treated in order of increasing atomic number.

Main-group Elements.—In addition to tin, papers have appeared during the year on five other main-group elements – germanium, krypton, antimony, tellurium, and iodine.

Germanium (^{73}Ge). A calculation based on the experimentally determined value of the total conversion coefficient of the 13.34 keV level in ^{73}Ge, $\alpha_{tot} = 1310 \pm 60$, has shown that the Mössbauer effect on this level should be measurable.[459] A value of $\delta R/R \simeq +(2 \times 10^{-3})$ has been estimated for the 67 keV level of ^{73}Ge from a comparison of *X*-ray and Mössbauer spectra of some germanium, tin, and antimony oxides.[460]

453 I. J. McColm, N. J. Clark, and B. Mortimer, *J. Inorg. Nuclear Chem.*, 1971, **33**, 49.
454 L. V. Kradinova, V. D. Prochukhan, and V. K. Yarmarkin, *Phys. Stat. Sol. (A)*, 1971, **5**, K137.
455 I. F. Sanaya, N. F. Kedrina, R. A. Stukan, V. I. Goldanskii, and N. S. Enikolopyan, *Doklady Akad. Nauk S.S.S.R.*, 1971, **197**, 869.
456 A. Y. Alexandrov, V. I. Goldanskii, T. B. Zavarova, L. A. Korytko, and N. A. Plate, *Vysokomol. Soedineniya*, 1971, **13B**, 76.
457 I. S. Morozova, G. M. Tarasova, V. V. Ivanov, E. V. Bryuchova, and N. S. Enikolopyan, *Doklady Akad. Nauk S.S.S.R.*, 1971, **199**, 654.
458 S. L. Ruby and I. Pelah, ref. 20, p. 21.
459 D. Riegel and K. P. Vester, *Z. Physik*, 1971, **241**, 188.
460 L. M. Dautov and V. V. Nemoshkalenko, ref. 12, p. 53.

Krypton (^{83}Kr). The Mössbauer effect has been observed in the past in ^{83m}Kr using hydroquinone–krypton clathrates, solid krypton, and KrF_2, the latter being the only *bona fide* chemical compound studied. Work on this isotope has now been extended to study the extent of chemical bonding between the krypton atom and its nearest-neighbour environment in metallic selenium, SeO_2, and $(NH_4)_2SeO_4$ (*N.B.* the 9.3 keV transition in ^{83}Kr is populated from ^{83}Se by a two-β-decay sequence), using a standard hydroquinone–krypton clathrate absorber. The largest resonance effect (*ca.* 8%) was observed for the SeO_2 source, which also gave a broadened spectrum. In contrast, only a small room-temperature resonance was observed for $(NH_4)_2SeO_4$. These results are thought to indicate the existence of a definite Kr—O bond in both the SeO_2 and the Kr–hydroquinone clathrate matrices.[461] Several papers have dealt with the recoilless fraction of solid krypton.[462–465]

Antimony (^{121}Sb). As mentioned on p. 558, the alloy $Ni_{21}Sn_2B_6$ has been shown to have properties which make it suitable as a source for both ^{119}Sn and ^{121}Sb Mössbauer experiments. Its chief advantage with regard to the latter isotope is that its chemical isomer shift is intermediate between those of antimony(III) and antimony(V), thus avoiding the need for wide velocity scans and the attendant difficulties of velocity calibration. The chemical isomer shift conversions are shown in Table 7.[388]

Table 7

Absorber	*Chemical isomer shift*/mm s^{-1} (at 80 K)
$SnO_2(^{121}Sb)$	$+ 6.91 \pm 0.05$
$Ni_{21}Sb_2B_6$	$- 0.03 \pm 0.05$
InSb	$- 1.70 \pm 0.05$
Sb	$- 4.72 \pm 0.05$

^{121}Sb Mössbauer data have been presented for the fluoroantimonate anions $NaSbF_4$, $KSbF_4$, KSb_2F_7, K_2SbF_5, $KSbF_6$, $CsSb_2F_{11}$, and for $(NH_4)_2SbCl_5$ in order to provide a catalogue of data to aid the analysis of more complex antimony–fluorine systems in future. The antimony(III) compounds possess a lone pair of 5*s* electrons and therefore resonate at more negative velocities than the antimony(V) compounds. The known and often contradictory information on the structures of these species was reviewed and attempts were then made to rationalize the Mössbauer data in terms of this structural information. The sign of the quadrupole coupling constant, e^2qQ, is positive for all of the compounds, and this implies a

[461] Y. Hazony and R. H. Herber, *J. Inorg. Nuclear Chem.*, 1971, **33**, 961.
[462] J. S. Brown, *Phys. Rev. (B)*, 1971, **3**, 21.
[463] B. Kolk, *Phys. Letters*, 1971, **35A**, 83.
[464] K. Mahesh and S. V. Kapil, *Phys. Stat. Sol. (B)*, 1971, **47**, 397.
[465] K. Mahesh and N. D. Sharma, *Phys. Stat. Sol. (B)*, 1971, **44**, 167.

concentration of negative charge along the symmetry axis rather than in the equatorial plane.[466]

Spectra have been obtained for the chloroantimony salts $Rb_4[Sb_2Cl_{12}]$, $Rb_{16}[Sb_6Cl_{36}]$, $[Co(NH_3)_6]SbCl_6$, $K_3[SbCl_6]$, $Cs_3[SbCl_6]$, $[NH_4]_3[SbCl_6]$, $[NH_4]_2[SbCl_5]$, $Rb[SbCl_6]$, and for $[NH_4]_4[Sb_2Br_{12}]$. The presence of both antimony(III) and antimony(V) in Rb_2SbCl_6 was confirmed and the salt should be formulated as $Rb_4[Sb^{III}Sb^{V}Cl_{12}]$. The spectrum of $Rb_{16}[Sb_6Cl_{36}]$ has an antimony(III) resonance which is much more intense than that in the antimony(V) region, the ratio being 2 : 1. However, the data are poor and the formulation of this species as $Rb_{16}[Sb_5{}^{III}Sb^{V}Cl_{36}]$ may not be correct. A particularly interesting aspect of this study was the observation that the $[Sb^{III}Cl_6]^{3-}$ anion gives the most negative chemical isomer shift yet reported, in agreement with a recent estimate by Stevens and Bowen of the chemical isomer shift for a bare Sb^{3+} ion ($-$ 10.6 $\pm$ 0.3 mm s^{-1} rel. InSb at 80 K). This result was interpreted in terms of a stereochemically inactive lone pair of electrons.[467] Similar results have been reported independently for Rb_2SbCl_6 and Cs_2SbCl_6. The Sb^{3+} resonance was broadened by a quadrupole interaction, attributed to deformation of the $[SbCl_6]^{3-}$ octahedra by a 'statistical Jahn–Teller pseudoeffect'.[468]

The compounds SbSI, SbSBr, SbSeBr, and SbTeBr have been shown to give single quadrupole-broadened resonances with very similar chemical isomer shifts. The degree of ionicity calculated from the latter for the SbSBr and SbSeBr crystals was estimated to be 0.23 $\pm$ 0.10, indicating that these compounds are characterized by an essentially covalent bond between the antimony atoms and the lattice. As iodine is replaced by bromine the ionic character of the bond increases.[469]

A number of complex sodium, silver, and potassium antimony oxides (*e.g.* $NaSb_{2.48}O_{6.10}$, $AgSb_{1.02}O_{2.92}$, and $KSb_{4.96}O_{11.46}$) have been studied, and the chemical isomer shift for antimony(V) has been correlated with structural changes in the oxides. Most of the oxides with the pyrochlore structure contain some antimony(III), the percentage of which is readily obtained from the Mössbauer spectrum. In no case is this amount greater than 25%.[470]

The Mössbauer effect and lattice parameters have been studied for silicon doped with antimony,[471] and spectra have been obtained for the ternary semiconductor compounds $M^{I}SbX_2{}^{VI}$ (M^{I} = Li, Na, K, Rb, or Cs; X^{VI} = S, Se, or Te), showing them to contain antimony(III).[472] Other

466 T. Birchall and B. Della Valle, *Canad. J. Chem.*, 1971, **49**, 2808.
467 T. Birchall, B. Della Valle, E. Martineau, and J. B. Milne, *J. Chem. Soc.* (*A*), 1971, 1855.
468 A. Y. Alexandrov, S. P. Ionov, A. M. Pritchard, and V. I. Goldanskii, *Pis'ma Zhur. eksp. teor. Fiz.*, 1971, **13**, 13.
469 T. A. Khimich, V. F. Belov, O. K. Zhukov, V. A. Yurin, L. N. Korablin, M. N. Shipko, A. N. Lobachev, and V. I. Popolitov, *Fiz. tverd. Tela*, 1971, **13**, 1507.
470 L. H. Bowen, P. E. Garrou, and G. G. Long, *J. Inorg. Nuclear Chem.*, 1971, **33**, 953.
471 J. R. Teague, C. M. Yagnik, G. J. Long, R. Gerson, and L. D. Lafleur, *Solid State Comm.*, 1971, **9**, 1695.
472 B. N. Veitis, S. Berulis, V. Grigalis, and Y. D. Lisin, *Latv. P.S.R. Zinat Akad. Vestis, Fiz. Teh. Zinat. Ser.*, 1971, 48.

compounds studied include $ZnSnSb_2$,[454] the ternary nickel boride $Ni_{21}Sb_2B_6$,[473] and the Heusler alloy Ni_2CuSb.[473]

Tellurium (^{125}Te). The stereochemistry of the complexes [Te(tu)$_4$Cl$_2$], [Te(tu)$_2$Cl$_2$], and [Te(tu)$_2$Cl$_2$],nH$_2$O (tu = thiourea) has been investigated by ^{125}Te Mössbauer spectroscopy and i.r. spectroscopy; the results favour a tetrahedral geometry rather than octahedral or square-planar configurations.[474]

The Mössbauer effects associated with the 35.6 keV and 27.8 keV transitions in ^{125}Te and ^{129}I respectively have been used to measure the quadrupole splitting in paramagnetic Ni_2Te_3. The quadrupole coupling constant was found to be 0.0 ± 4.0 mm s^{-1} at the telluride sites, and 10.03 ± 0.14 mm s^{-1} at the sites of the iodine impurities. From the chemical isomer shift data it was concluded that the telluride ions and iodine impurities have the same electronic structures, which indicates that almost no change occurs in the electronic structure of tellurium in its β-decay to iodine. The ^{129}I data indicate that 0.27 electrons are transferred from nickel to iodine and that the configuration of the latter is $5s^25p^{5.27}$. This configuration, coupled with the positive e^2qQ for iodine suggests that the bonding in Ni_2Te_3 is mainly through p_x and p_y iodine electrons.[475]

Iodine (^{127}I, ^{129}I). (See also previous section.) The iodine bonding in iodobenzene, iodobenzene dichloride, iodosobenzene, iodoxybenzene, and an iodosodilactone (26) has been investigated using the 57.6 keV

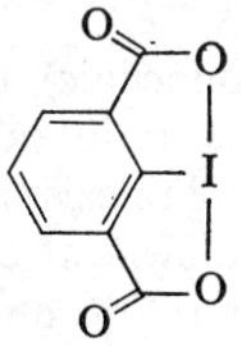

(26)

resonance in ^{127}I. The data are consistent with pure p-bonding in PhI, $PhICl_2$, PhIO, and (26), and this provides an opportunity for recalibrating the correlation between the chemical isomer shift and the number of p electron holes, h_p (see Figure 26). A small amount of s hybridization in the iodine p-orbitals is necessary to explain the results for $PhIO_2$. The bonding in $PhICl_2$, PhIO, and (26) is adequately described on the basis of an MO model, a localized hybridization model being unsatisfactory. The C—I—O angle in PhIO is shown to be near 90°, rather than linear, and (26) has an approximately planar configuration with a linear IO_2 group.

[473] H. Dokuzoguz, H. H. Stadelmaier, and L. H. Bowen, *J. Less-Common metals*, 1971, **23**, 245.
[474] A. Y. Alexandrov, A. M. Babeshkin, V. I. Goldanskii, S. P. Ionov, V. A. Lebedev, and R. A. Lebedev, *Zhur. strukt. Khim.*, 1971, **12**, 328.
[475] J. Granot, *Phys. Letters*, 1971, **36A**, 391.

$PhIO_2$ has iodine orbital populations which are significantly different from those in other compounds and which can be associated with an O—I—O bond angle of about 95° in this molecule.[476]

Investigations by different groups into the consequences of the β-decay of ^{129}Te in some oxo-tellurium compounds have yielded some conflicting results. In one study it was found that the β-decay of the ^{129}Te nucleus in

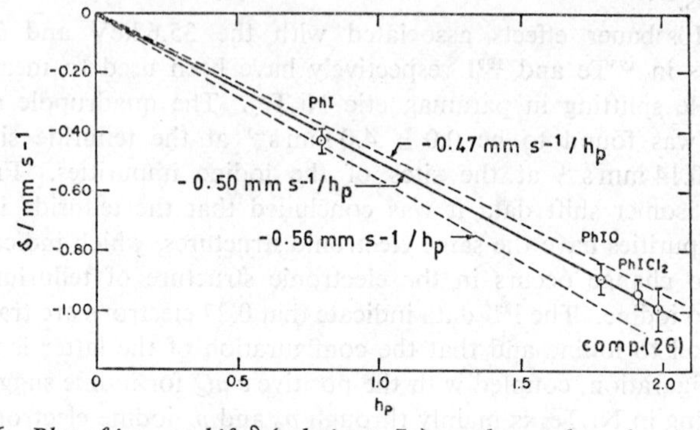

Figure 26 *Plot of isomer shift δ (relative to* I^-*) vs. the number of* p*-electron holes,* h_p*, for molecules with pure* p*-bonding of iodine. The solid line is the best fit to the present data, and yields an isomer shift calibration of* − 0.50 (mm s^{-1})/h_p*. The two dashed lines correspond to previously reported calibrations*
(Reproduced by permission from *J. Chem. Phys.*, 1971, **54**, 612)

TeO_2 (tetragonal), H_6TeO_6 (monoclinic), $(NH_4)_2H_4TeO_6$, and $Na_2H_4TeO_6$ was not accompanied by changes in its valence shell, whereas in tellurium metal the daughter ^{129}I was found to exist in an environment different from that of the ^{129}I parent at the moment of emission.[477] Elsewhere, chemical after-effects of the isomeric transition and of the $^{128}Te(n,\gamma)^{129}Te$ nuclear reaction were clearly observed in H_6TeO_6 and $(H_2TeO_4)_n$, but not in TeO_2, H_2TeO_3, or tellurium metal. Where molecular decomposition was observed, only a single product, $TeO_3{}^{2-}$, was identified.[478] The emission spectrum of β-$^{129m}TeO_3$ at 80 K consists of two resonances, one with an isomer shift of + 2.59 ± 0.03, corresponding to an octahedral tellurium(VI) parent, and another with an isomer shift of − 7.96 ± 0.05, which corresponds to an unusual oxidation state and confirms an earlier result of Lebedev. No ^{129}I emission was detected in the 80 K spectrum of α-TeO_3.[479] The state of ^{129}I atoms formed after β-decay in GeTe, SnTe, and PbTe, and in PbTe–SnTe solid solutions has also been studied.[480]

[476] B. S. Ehrlich and M. Kaplan, *J. Chem. Phys.*, 1971, **54**, 612.
[477] R. A. Lebedev, A. Babeshkin, and A. N. Nesmeyanov, *Vestn. Mosk. Univ. Khim.*, 1971, **12**, 113.
[478] C. H. W. Jones, J. L. Warren, and P. Vasudev, *J. Phys. Chem.*, 1971, **75**, 2867.
[479] J. L. Warren and C. H. W. Jones, *Radiochem. Radioanalyt. Letters*, 1971, **7**, 97.
[480] P. P. Seregin and E. P. Savin, *Fiz. tverd. Tela*, 1971, **13**, 2790.

The ^{129}I resonance has been used to investigate the electronic structure of ICl and IBr molecules, isolated in an argon matrix at 20.4 K. From the quadrupole coupling constants and chemical isomer shifts the iodine configurations in these two molecules were found to be $5s^{1.98}5p^{4.70}$ and $5s^{1.97}5p^{4.81}$ respectively. The ionic character of the bond in these isolated molecules is therefore considerably reduced relative to the solid phase.[481] Several charge-transfer complexes of iodine have been studied and spectra for two of these are reproduced in Figure 27. The complex I_2,phenazine

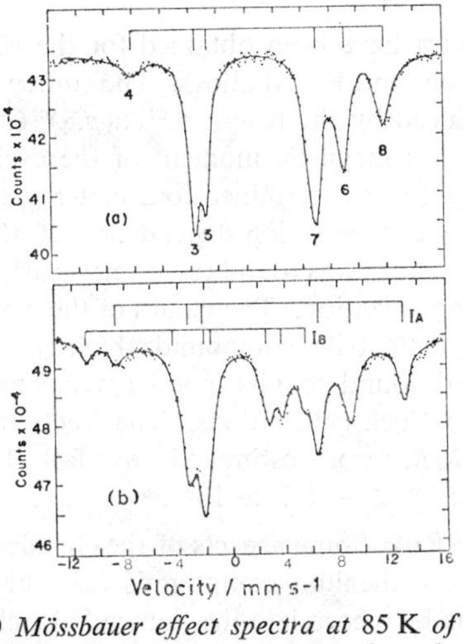

Figure 27 *Iodine*-129 *Mössbauer effect spectra at* 85 K *of* (a) I_2,phenazine, *and* (b) I_2,hexamethylenetetramine. *The curves drawn through the points are computer fits to the experimental data*
(Reproduced by permission from *J. Chem. Phys.*, 1971, **54**, 1627)

contains only a single iodine environment [Figure 27(a)] and is thought to have the structure shown in (27), which agrees with the results of *X*-ray diffraction. By contrast, both I_2,acridine and I_2,hexamethylenetetramine contain two types of iodine, bridging and terminal [see Figure 27(b) and

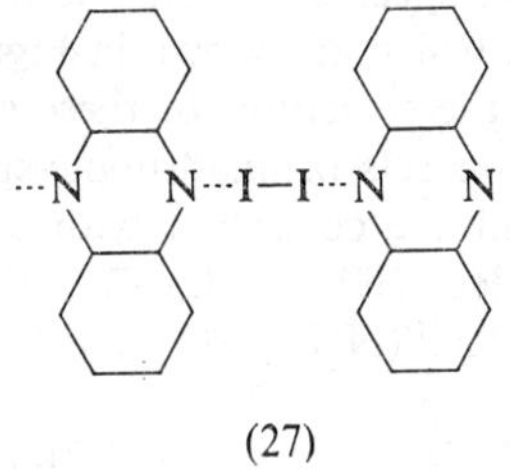

(27)

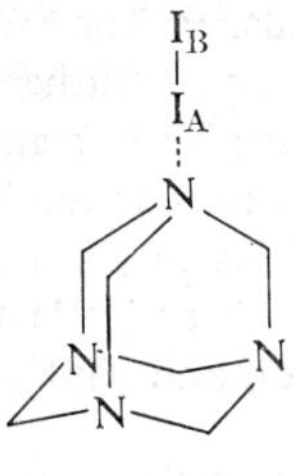

(28)

[481] C. Goldstein and T. Barnoi, *Chem. Phys. Letters*, 1971, **10**, 136.

structure (28)]. In the iodine–phenazine complex 0.09 electron is transferred to the iodine atom from the phenazine molecule. In the acridine and hexamethylenetetramine complexes the negative charges on the bridging iodine atoms are decreased and those on the terminal atoms are increased. A nett charge of 0.08 electron is transferred to the iodine molecule through the nitrogen atom from the hexamethylenetetramine molecule.[482]

Transition Elements.—In addition to iron, papers have appeared on eight other transition metals: nickel, ruthenium, hafnium, tungsten, tantalum, iridium, platinum, and gold.

Nickel (^{61}Ni). Spectra have been obtained for the 67.4 keV resonance of ^{61}Ni in Cu–Ni, Co–Ni, and Fe–Ni alloys. The source contained the parent isotope, ^{61}Co, produced by the reaction ^{64}Ni(p,α)^{61}Co, in a non-magnetic $^{64}Ni_{0.86}V_{0.14}$ alloy. The magnetic moment of the excited state, $\mu(67.4) = +(0.481 \pm 0.006)\mu_N$, was determined to a better precision than in earlier measurements. The concentration-dependence of the average magnetic hyperfine field in the alloys was found to be very nearly proportional to that of the bulk alloy magnetization. The values of the magnetic hyperfine field in the antiferromagnetic Ni^{2+} compounds $KNiF_3$, NiO, and NiF_2 were also determined and found to give good agreement with the results of unrestricted Hartree–Fock calculations. The fractional change in nuclear charge radius, $\delta R/R$, was estimated to fall between the limits, $-1.8 \times 10^{-4} \leqslant \delta R/R \leqslant -0.7 \times 10^{-4}$.[483]

Ruthenium (^{99}Ru, ^{101}Ru). Some aspects of the chemical applications of the Mössbauer effect in ruthenium compounds have been reviewed and an attempt has been made to correlate the known ^{99}Ru chemical isomer shifts with the ruthenium oxidation state and with the *s*-electron density at the ruthenium nucleus, calculated on the basis of restricted Hartree–Fock wave-functions.[484]

The quadrupole splittings and chemical isomer shifts of the anhydrous ruthenium trihalides have been shown to decrease as the Mulliken electronegativity of the halide decreases. This trend was interpreted in terms of an increase in 4*d* electron delocalization which results from increased overlap of populated π-type ligand orbitals with partially occupied t_{2g} metal orbitals.[485] The ^{99}Ru Mössbauer spectra of 'ruthenium red' and its oxidation product 'ruthenium brown', which are shown in Figure 28, have been used to give information about the structures of these species. The spectra were interpreted in terms of two sets of quadrupole-split doublets with intensities in the ratio 2 : 1, and are consistent with a linear configuration such as $[(NH_3)_5Ru^{III}—O—Ru^{IV}(NH_3)_4—O—Ru^{III}(NH_3)_5]^{6+}$ for the red cation and $[(NH_3)_5Ru^{IV}—O—Ru^{III}(NH_3)_4—O—Ru^{IV}(NH_3)_5]^{7+}$ for

482 S. Ichiba, H. Sakai, H. Negita, and Y. Maeda, *J. Chem. Phys.*, 1971, **54**, 1627.
483 J. C. Love, F. E. Obenshain, and G. Czjzek, *Phys. Rev.* (*B*), 1971, **3**, 2827.
484 C. A. Clausen, R. A. Prados, and M. L. Good, ref. 20, p. 31.
485 C. A. Clausen, R. A. Prados, and M. L. Good, *Chem. Phys. Letters*, 1971, **8**, 565.

the brown cation.[486] ^{99}Ru spectra have also been obtained at 4.2 K for a series of octahedral nitrosyl ruthenium(II) complexes $[RuL_5(NO)]^{n\pm}$. The chemical isomer shift decreases steadily as the ligand field strength of L decreases ($CN^- > NH_3 > NCS^- > Cl^- > Br^-$), implying a positive sign

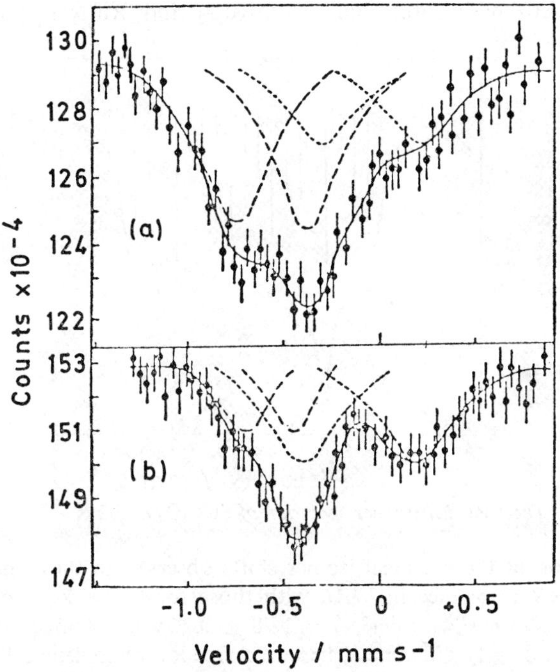

Figure 28 (a) *Mössbauer spectrum of 'ruthenium red' at* 4.2 K. (b) *Mössbauer spectrum of 'ruthenium brown' at* 4.2 K
(Reproduced by permission from *Inorg. Nuclear Chem. Letters*, 1971, **7**, 485)

for $\delta R/R$ for ^{99}Ru. The quadrupole splitting also decreases along the series and this can be rationalized in terms of the relative σ-donor and π-acceptor abilities of the ligands.[487]

The first observation of a hyperfine magnetic field in a ruthenium compound, $SrRuO_3$, has been reported. The spectrum at 4.2 K shown in Figure 29 yields a value for the internal field of 352 kG (315 kG at 77 K), which corresponds to 1.0 unpaired electrons on the ruthenium. This value compares with values of 1.4 ± 0.4 BM mol^{-1} from neutron diffraction data and 0.85 BM mol^{-1} at 77 K from the bulk magnetization, and confirms the collective-electron magnetism model for this compound. It is clear

[486] C. A. Clausen, R. A. Prados, and M. L. Good, *Inorg. Nuclear Chem. Letters*, 1971, **7**, 485.
[487] R. Greatrex, N. N. Greenwood, and P. Kaspi, *J. Chem. Soc.* (*A*), 1971, 1873.

that ^{99}Ru Mössbauer spectroscopy provides the basis for a detailed study of related magnetic ternary and quaternary ruthenium oxides and should play an important role in the understanding of the magnetic properties of the 4*d* transition series.[488]

The Mössbauer effect of the 127 keV γ-rays of ^{101}Ru has been observed for the first time in ruthenium metal, RuO_2, and $[Ru(NH_3)_4(HSO_3)_2]$. By

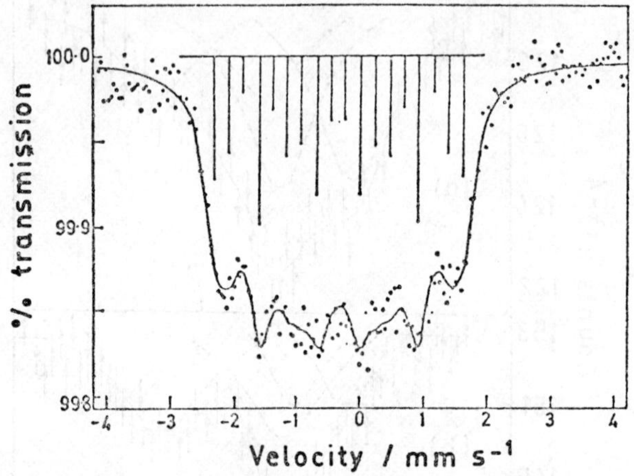

Figure 29 *The* ^{99}Ru *Mössbauer spectrum of* $SrRuO_3$ *at* 4.2 K

comparison of the chemical isomer shifts observed in these materials for the 127 keV resonance in ^{101}Ru with those for the 90 keV resonance in ^{99}Ru, the ratio $\delta\langle r^2\rangle_{101} : \delta\langle r^2\rangle_{99} = 1.78 \pm 0.26$ was obtained. An estimate of electron density differences, based on free-ion relativistic self-consistent field calculations, yielded $\delta\langle r^2\rangle : \langle r^2\rangle_{99} \doteqdot + 1.4 \times 10^{-3}$ and $\delta\langle r^2\rangle : \langle r^2\rangle_{101} \doteqdot + 2.4 \times 10^{-3}$. These results were discussed in terms of the core excitation model.[489]

Hafnium (^{177}Hf, ^{178}Hf, ^{180}Hf). The magnetic hyperfine field in $(Hf_{0.1}Zr_{0.9})Fe_2$ has been measured by means of Mössbauer experiments on the first excited (93.3 keV) 2^+ state of ^{180}Hf and found to be $H_{hf} = -200 \pm 20$ kG. This value was then combined with the known hyperfine splitting of the 1142 keV 8^- isomeric state to yield a value of $\mu(8^-) = +(8.6 \pm 1.0)\mu_N$ for the magnetic moment of this level. The result identifies the 8^- state as a virtually pure two-proton configuration.[490]

The use of the Mössbauer effect in ^{177}Hf, ^{178}Hf, ^{180}Hf, ^{182}W, ^{183}W, ^{184}W, and ^{186}W to study recoil radiation damage following Coulomb excitation in several refractory hafnium compounds and in tungsten metal has been

488 T. C. Gibb, R. Greatrex, N. N. Greenwood, and P. Kaspi, *Chem. Comm.*, 1971, 319.
489 W. Potzel, F. E. Wagner, R. L. Mössbauer, G. Kaindl, and H. E. Seltzer, *Z. Physik*, 1971, **241**, 179.
490 H. J. Körner, F. E. Wagner, and B. D. Dunlap, *Phys. Rev. Letters*, 1971, **27**, 1593.

discussed further. Much of the work was published elsewhere in 1970, and was included in last year's Report.[491]

Tungsten (^{180}W, ^{182}W, ^{183}W, ^{184}W, ^{186}W). (See also previous section.) Experiments following Coulomb excitation with WO_3, WC, and WS_2 as absorbers have yielded consistent results for the relative quadrupole moments for three of the tungsten isotopes. The ratios are $Q(^{182}W) : Q(^{184}W) : Q(^{186}W) = 1 : 0.965 \pm 0.008 : 0.906 \pm 0.018$.[492]

The Mössbauer effect in the 104 keV 2^+-0^+ transition in ^{180}W has been observed for the first time. The source was ^{180}Ta, produced by irradiation of a tantalum foil with fast neutrons, the nuclear reaction being $^{181}Ta(n,2n)^{180}Ta$. The width of the resonance line was 3.1 ± 0.8 mm s^{-1}, which corresponds to the natural linewidth.[493]

The 100 keV transition of ^{182}W, which occurs between a 2^+ excited state and a 0^+ ground state, has been used to study the tungsten compounds WO_3, $FeWO_4$, Na_2WO_4, and $W(CO)_6$. Na_2WO_4 and $W(CO)_6$, which are known from *X*-ray crystallography to contain tungsten in symmetrical environments, accordingly give single-line spectra. By contrast, WO_3 and $FeWO_4$ show quadrupole interactions, which, because tungsten(VI) has a d^0 configuration, must arise from lattice effects. The crystal structures of both WO_3 and $FeWO_4$ do indeed confirm that the tungsten atoms are in distorted octahedral environments. However, the results are unusual in that the electric field gradients are much larger than those normally associated with lattice effects. No chemical isomer shifts were detectable.[494]

Tantalum (^{181}Ta). In the past, great difficulty has been encountered in observing resonance for the 6.25 keV γ-ray of ^{181}Ta, because of the very small natural linewidth, and resonance absorption has only been obtained in metallic tungsten and tantalum lattices. A major improvement towards high-resolution Mössbauer spectroscopy with this isotope has now been reported. Chemical isomer shifts covering a range of 25 mm s^{-1} (8000 times the natural linewidth) have been observed between sources of ^{181}W diffused into single crystals of the 4*d* transition metals molybdenum and niobium, and the 5*d* transition metals iridium, platinum, tungsten, and tantalum, against absorbers of tantalum metal and $KTaO_3$. Single crystals were used to ensure high-purity hosts and to preclude preferential diffusion along grain boundaries. Representative spectra are shown in Figure 30, where it is apparent that the shifts for tantalum in the 4*d* metals are much greater than those for tantalum in the 5*d* metals. In all cases except $^{181}Ta(Pt)$ the product $\delta\langle r^2\rangle\,[|\psi^2(0)|_{host} - |\psi^2(0)|_{Ta}]$ is negative. Interpretations based on electronegativity arguments fail to explain the results

491 N. Hershkowitz and C. G. Jacobs, jun., ref. 20, p. 143.
492 L. W. Oberley, N. Hershkowitz, S. A. Wender, and A. B. Carpenter, *Phys. Rev.* (*C*), 1971, **3**, 1585.
493 K. Zioutas and B. Wolbeck, *Z. Naturforsch.*, 1971, **26a**, 1088.
494 G. M. Bancroft, R. E. B. Garrod, and A. G. Maddock, *Inorg. Nuclear Chem. Letters*, 1971, **7**, 1157.

consistently, instead it is necessary to consider the band structures of the host metals. In this way $\delta\langle r^2\rangle$ for the 6.25 keV γ-transition of ^{181}Ta was deduced to be positive.[495] The observation of the resonance in ^{181}Ta diffused into a host matrix of platinum has been reported independently.[496]

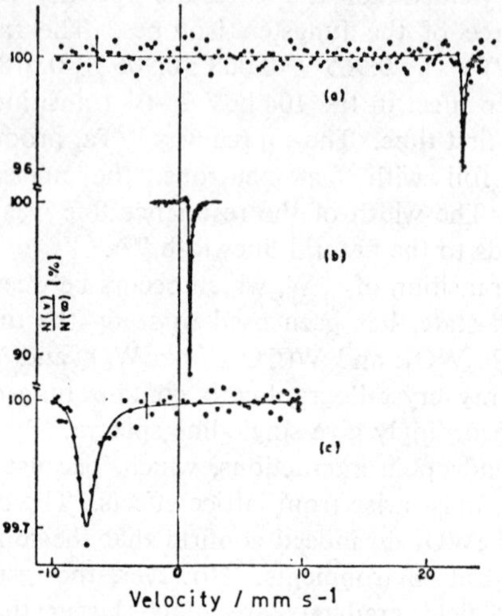

Figure 30 *Absorption spectra of the* 6.25 keV *γ-rays:* (a) *source* $^{181}W(Mo)$, (b) *source* $^{181}W(W)$, *both with* Ta *metal as absorber;* (c) *source* $^{181}W(W)$, *absorber* $KTaO_3$. *Adjacent points were added for plots, and the solid curves represent the results of least-squares fits*
(Reproduced by permission from *Phys. Letters*, 1971, **36A**, 457)

A pulsed *X*-ray beam has been used to obtain resonance excitation on Mössbauer levels and the results on ^{181}Ta have been discussed briefly.[497] The sort of information which can be obtained about dynamical effects on ^{181}Ta nuclei in frozen aqueous solutions from the Mössbauer effect and from the technique of perturbed angular correlation has been compared.[498]

Iridium (^{193}Ir). The 73 keV resonance in ^{193}Ir, which occurs between a spin $\frac{3}{2}$ + ground state and a spin $\frac{1}{2}$ + excited state has been used to study the electronic state of iridium in $IrCl(CO)(PPh_3)_2$ and $IrBr(CO)(PPh_3)_2$ and in the molecular adducts formed between the former compound and O_2, H_2, HCl, MeI, I_2, and Cl_2. Four-peak spectra were observed in each case, because of quadrupole splitting in both the source and absorber. The O_2 adduct was found to adopt a five-co-ordinate structure in which the O—O bond is retained, whereas the other adducts adopt a six-co-ordinate structure in which the X—Y bond of the ligand XY is broken. For the

495 D. Salomon, G. Kaindl, and D. A. Shirley, *Phys. Letters*, 1971, **36A**, 457.
496 G. Wortmann, *Phys. Letters*, 1971, **35A**, 391.
497 P. S. Kamenov, *Doklady Bolg. Akad. Nauk*, 1971, **24**, 713.
498 J. A. Cameron, P. R. Gardner, L. Keszthelyi, and W. V. Prestwich, ref. 12, p. 613.

latter series, the chemical isomer shift and, therefore, the *s*-electron density at the iridium nucleus, decrease and ν(C=O) increases in going from the H_2 to the Cl_2 adduct. These trends were discussed in terms of the usual arguments of the relative σ-donor and π-acceptor abilities of the ligands.[499]

Platinum (^{195}Pt). The nuclear magnetic moment of the 130 keV level of ^{195}Pt has been found to be $\mu_{130} = (0.81^{+0.13}_{-0.25})\mu_N$ from a measurement on

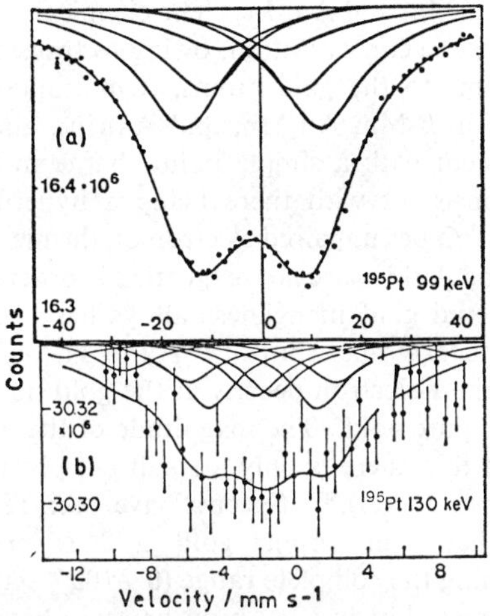

Figure 31 *The Mössbauer spectrum of* (a) *the* 99 keV *and* (b) *the* 130 keV *transition in* ^{195}Pt *obtained with an* Fe–Pt *absorber at* 4.2 K
(Reproduced by permission from *Phys. Letters*, 1971, **37A**, 355)

a ^{130}Pt–Fe alloy. Chemical isomer shift measurements for the 130 keV and 99 keV transitions in platinum alloys yielded $\delta\langle r^2\rangle_{130} : \delta\langle r^2\rangle_{99} = 1.5 \pm 0.2$, and from the chemical isomer shifts of the 99 keV transition in compounds the fractional change in nuclear radius for this level was found to be $\delta\langle r^2\rangle_{99} : \langle r^2\rangle_{99} = -(1.6^{+4.4}_{-0.0}) \times 10^{-4}$.[500]

Subsequently a more accurate value of the magnetic moment of the 130 keV state of ^{195}Pt, $\mu_{130} = (0.90 \pm 0.08)\,\mu_N$, was determined, again from measurements on an Fe–Pt alloy absorber, and a source of ^{195}Au embedded in a platinum matrix. The magnetic hyperfine field at the platinum nucleus was first determined from the spectrum of the 99 keV ^{195}Pt resonance shown in Figure 31(a) to be $H_{int} = 1.34 \pm 0.09$ MG. This value was then used as a fixed parameter in the computer analysis of the 130 keV spectrum shown in Figure 31(b). The ratio $\delta\langle r^2\rangle_{130} : \delta\langle r^2\rangle_{99} = 1.3 \pm 0.2$ derived from this study was in good agreement with the value reported earlier (ref. 500). In addition new and improved chemical isomer

[499] H. H. Wickman and W. E. Silverthorn, *Inorg. Chem.*, 1971, **10**, 2333.
[500] D. Walcher, *Z. Physik*, 1971, **246**, 123.

shift data were obtained for the 99 keV resonance in FePt, PtO_2, $PtCl_2$, $PtCl_4$, and PtI_2.[501]

Gold (^{197}Au). As a basis for future studies on gold alloys, high-precision measurements have been made of the recoilless fraction for pure gold in the temperature range 4.2—100 K.[502] The 77 keV ^{197}Au resonance has been investigated at 4.2 K for all the phases of the gold–manganese system which could be retained at room temperature, namely: Au_4Mn, Au_3Mn, Au_5Mn_2, Au_2Mn, AuMn, and $AuMn_2$. The chemical isomer shift increases linearly with manganese concentration, owing to transfer of electrons from the manganese atoms to the gold atoms. Quadrupole interactions were observed for ^{197}Au in β-Mn, Au_2Mn, and $AuMn_2$, and the values of the splitting are consistent with a simple point-charge model. The magnetic interactions are consistent with there being a hyperfine field at a gold nucleus of about 55 kG per unpaired electron on the manganese neighbours. The phases β-Mn and $AuMn_2$ are not magnetically ordered.[503] The magnetic structure of disordered gold–manganese alloys has been studied independently.[504] The chemical isomer shift of ^{197}Au in $TbAu_2$ has been measured and indicates that the *s*-electron density at the gold nucleus in this alloy is greater than that in pure gold. The magnitude of the magnetic field at the gold nucleus shows that there is only a small net polarization of the conduction electrons (*ca.* 50 kG).[505] Spectra have been recorded for ordered and disordered phases of copper–gold and silver–gold alloys with compositions spanning the complete range (0—100% Au). The dependence of the chemical isomer shift in Cu_3Au on pressure between 0 and 65 kbar was found to be in good agreement with a theoretical model.[506] The change in the chemical isomer shift, produced by the introduction of up to 12% mercury in mercury–gold alloys, is five times smaller than that produced by any solute studied previously.[507]

Mercury (^{201}Hg). The linewidth of the spectrum of the 32.2 keV transition in ^{201}Hg has been measured and gives a lower limit of 0.1 ns for the half-life of the second excited state.[500]

Lanthanide and Actinide Elements.—The eight elements discussed in this section are the six lanthanides praseodymium, samarium, europium, dysprosium, erbium, and ytterbium, and the two actinides thorium and neptunium.

Praseodymium (^{141}Pr). Nuclear hyperfine interactions in the 145.4 keV state of ^{141}Pr have been investigated in the following 14 compounds: PrB_6, $PrAl_2$,

501 W. Ruegg and J. P. Launaz, *Phys. Letters*, 1971, **37A**, 355.
502 D. J. Erickson, L. D. Roberts, J. W. Burton, and J. O. Thomson, *Phys. Rev.* (*B*), 1971, **3**, 2180.
503 G. Longworth and B. Window, *J. Phys.* (*F*), 1971, **1**, 217.
504 R. L. Cohen and K. W. West, *J. Phys.* (*Paris*), *Colloq.*, 1971, **1** (Pt. 2), C1–781.
505 L. R. Sill, A. J. Fedro, and C. W. Kimball, *Internat. J. Magn.*, 1971, **1**, 319.
506 P. G. Huray, L. D. Roberts, and J. O. Thomson, *Phys. Rev.* (*B*), 1971, **4**, 2147.
507 R. L. Cohen, Y. Yafet, and K. W. West, *Phys. Rev.* (*B*), 1971, **3**, 2872.

PrO_2, Pr_6O_{11}, Pr_2O_3, PrF_3, $PrPO_4$, PrC_2, $PrFeO_3$, PrS, Pr_2S_3, $Pr_{0.5}Gd_{0.5}S$, PrN, and $Pr_{0.5}Dy_{0.5}PO_4$. The praseodymium is tervalent ($4f^25s^25p^6$) in all of the compounds, except for PrO_2 and Pr_6O_{11} in which it is tetravalent ($4f^15s^25p^6$). The extra $4f$ electron in Pr^{3+} increases the shielding of the $5s$ and $5p_{\frac{1}{2}}$ electrons, thereby decreasing the total electron density at the nucleus. The fact that the quadrivalent compounds have chemical isomer shifts which are more positive than those of the tervalent compounds indicates that the change in nuclear charge radius accompanying the transition is positive; the numerical value of $\langle r_{ex}^2\rangle - \langle r_{gr}^2\rangle$ was estimated to be $(6.3 \pm 2.5) \times 10^{-3}$ fm^2. It was pointed out that Groves, Debrunner, and De Pasquali, who in 1970 reported a negative value for $\delta R/R$ have now revised their opinion and agree with the present results. Although evidence of a magnetic hyperfine interaction at 4.2 K was observed for a number of compounds, only in the case of PrB_6, PrO_2, and $PrAl_2$ were the spectra reasonably well resolved. The magnetic fields were estimated to be 2.07 ± 0.08, 1.7 ± 0.4, and 0.85 ± 0.04 MG respectively, and the excited-state magnetic moment to be $(2.78^{+0.12}_{-0.06})\ \mu_N$, a value which is considerably larger than predicted by Kieslinger and Sorenson.[508]

In an independent study, Mössbauer scattering spectra were obtained for the following 12 compounds: PrF_3, K_3PrF_6, Cs_2PrF_6, $PrCl_3$, $Pr_2(SO_4)_3$, $Pr(OH)_3$, PrO_3, Pr_6O_{11}, PrO_2, PrC_2, $CsPrF_5$, and $PrFeO_3$ at 1.8—24 K. From the chemical shift difference between the Pr^{3+} and Pr^{4+} fluorine compounds the change of nuclear charge radius was determined to be 12×10^{-3} fm^2, and from the magnetic hyperfine pattern of antiferromagnetic PrO_2 the magnetic moment of the 145.4 keV state was found to be $(2.8 \pm 0.2)\ \mu_N$.[509]

Samarium (^{147}Sm, ^{149}Sm). The Mössbauer effect has been observed for the first time in the decay of the $\frac{5}{2}-$ 121.1 keV state of ^{147}Sm. The sources used were ^{147}Pm in a mixed promethium–samarium oxide matrix and ^{147}Eu in Sm_2O_3. Both sources give single Lorentzian lines when used with an Sm_2O_3 absorber. The ^{147}Sm source was used with a samarium iron garnet absorber to obtain the excited-state nuclear moments. The magnetic dipole moment was found to be $-0.445 \pm 0.025\mu_N$ and, using an atomic-beam measurement of the ground-state quadrupole splitting, an estimate of -0.31 ± 0.12 b was obtained for the excited-state quadrupole moment.[510]

The hyperfine structure of the 22.5 keV level of ^{149}Sm has been studied in the canted-antiferromagnetic, distorted perovskites $SmFeO_3$ and $SmCrO_3$ at 1.3—300 K. At 1.3 K a well-resolved hyperfine pattern is observed, revealing a unique magnetic field and electric field gradient at all samarium

[508] M. F. Bent, D. D. Cook, and B. I. Persson, *Phys. Rev. (C)*, 1971, **3**, 1419.
[509] W. H. Kapfhammer, W. Maurer, F. Wagner, and P. Kienle, *Z. Naturforsch.*, 1971, **26**, 357.
[510] F. T. Parker, K. A. Hardy, and J. C. Walker, *Phys. Rev. (C)*, 1971, **3**, 841.

nuclei, with H_{eff} = 2.09 MG for $SmFeO_3$ and 2.28 MG for $SmCrO_3$. At $T \leqslant 20$ K the temperature dependence of H_{eff} follows the equation $H_{eff} = H_{eff}(0)\tanh(\Delta kT/2)$, with Δ = 4.8 and 8 K form $SmFeO_3$ and $SmCrO_3$ respectively.[511] On the basis of new transport measurements in combination with analyses of the magnetic susceptibility and lattice energetics, a model for the electronic structure of SmB_6 has been proposed, which successfully describes the resistivity, Hall effect, magnetic susceptibility, and Mössbauer chemical isomer shift data.[512]

Europium (^{151}Eu, ^{153}Eu). The interference of the photoeffect and nuclear resonance absorption followed by internal conversion has now been observed for the 21 keV *M*1 transition in ^{151}Eu. The measured interference parameter is $\beta_{M1}^{exp} = (0.92 \pm 0.18) \times 10^{-2}$.[513] The after-effects of the electron-capture decay of $^{151}Gd^{3+}$ in $Eu_2(C_2O_4),10H_2O$ have been studied and $^{151}Eu^{2+}$ has been detected as an anomalous charge state. It is proposed that the Auger cascade produces, by radiolysis, a sufficient number of reducing species in the vicinity of the $^{151}Eu^{3+}$ daughter atom to reduce the Eu^{3+} to Eu^{2+} prior to emission of the 21.7 keV γ-ray. The radiolytic mechanism is complete in $< 10^{-9}$ s.[514]

The static and, especially, the dynamic magnetic properties of the nearly Heisenberg ferromagnet EuO have been studied near its Curie point, T_C = 69.2 K. Just below T_C the spectra exhibit relaxation effects, characteristic of critical superparamagnetism which was shown to have its origins in the non-ideal composition of the crystal.[515] Systematics of the ^{151}Eu and ^{153}Eu Mössbauer data for the compounds EuX (X = O, S, Se, or Te) have been discussed.[516] The direction of easy magnetization in a number of mixed compounds of rare-earth iron garnets $R_xSm_{\frac{1}{2}-x}Eu_{\frac{1}{2}}$ IG ($0 \leqslant x \leqslant \frac{1}{2}$) at 4.2 K has been found to depend on the relative amounts of R and Sm. It was found to coincide with [111] in most of the RIG and [110] in the SmIG. For ErIG the easy axis at 4.2 K is in the [100] direction, a result about which there is some debate in the literature.[517]

^{151}Eu Mössbauer studies of the spontaneous magnetization of the cubic perovskite $EuLiH_3$ have shown that this compound can be regarded as a nearly ideal ferromagnet with only nearest-neighbour exchange interactions. The extrapolated hyperfine field at 0 K is 251 ± 3 kG – considerably smaller than that expected for a free Eu^{2+} ion.[518]

[511] M. Eibschütz, R. L. Cohen, and L. G. Van Uitert, *J. Phys.* (*Paris*), *Colloq.*, 1971, **1** (Pt. 2), C1–922.
[512] J. C. Nickerson, R. M. White, K. N. Lee, R. Bachman, T. H. Geballe, and G. W. Hull, jun., *Phys. Rev.* (*B*), 1971, **3**, 2030.
[513] P. Steiner and G. Weyer, *Phys. Letters*, 1971, **36A**, 201.
[514] P. Glentworth, A. L. Nichols, N. R. Large, and R. J. Bullock, *Chem. Comm.*, 1971, 206.
[515] G. Groll, *Z. Physik*, 1971, **243**, 60.
[516] W. Zinn, *J. Phys.* (*Paris*), *Colloq.*, 1971, **1** (Pt. 2), C1–724.
[517] U. Atzmony, K. Hardy, J. C. Walker, and E. Loh, *J. Phys.* (*Paris*) *Colloq.*, 1971, **1** (Pt. 2), C1–920.
[518] C. L. Chien and J. E. Greedan, *Phys. Letters*, 1971, **36A**, 197.

The nature of the bonding between chlorine and europium in aquochloro-complexes of europium(II) has been investigated using ^{151}Eu Mössbauer spectroscopy on frozen solutions and u.v.–visible spectrophotometry. The chemical isomer shift decreases with increasing chloride ion concentration, suggesting that the chloride resides in the inner co-ordination sphere of the europium and that the interaction is not purely ionic. The complexing of chloride ions with europium(III) ions is dramatically dependent on pH and this is thought to reflect a substantial change in the activity coefficients at high ionic strength.[519] The chemical isomer shifts for europium(III) chloride, acetate, perchlorate, and sulphate and their adducts with nitrogen and oxygen donor ligands all lie between 0 and 1 mm s^{-1} (rel. Eu_2O_3), even though the co-ordination numbers within the series range from 6 to 10.[410]

The chemical isomer shifts of the chelates of europium(III) and dysprosium(III) with the ligands edta, bsedi (bis-salicylaldehyde-ethylenedi-imine) and oxine (8-hydroxyquinoline) are slightly negative with respect to Eu_2O_3 and Dy_2O_3 respectively, corresponding to lower *s*-electron densities at the metal nuclei in the chelate compounds. These small differences are attributed to a 4% transfer of electrons from the ligands to the partly filled 4*f* shell of the metal.[520]

The ^{151}Eu resonance above the superconducting transition (T_C = 3.4 K) in EuLa is a single line, owing to the rapid conduction electron relaxation. Below T_C there is a line-broadening followed by a resolved hyperfine splitting as the relaxation rate decreases.[521] The solid-solution alloys Pd_5Eu and Pd_3Eu have been shown to contain europium(III) whereas Pd_2Eu, PdEu, $PdEu_2$, and $PdEu_3$ contain europium(II). The latter group of compounds are magnetically ordered at 4.2 K, with effective magnetic fields at the europium nuclei of 17 ± 10, 162 ± 5, 148 ± 10, and 244 ± 15 kG respectively.[522]

Dysprosium (^{161}Dy). (See also the discussion on dysprosium chelates in previous section.) The interference of the photoeffect and internal conversion following nuclear resonance absorption of the 26 keV *E*1 gamma quanta in ^{161}Dy has been detected.[523] The *L*-shell conversion coefficient has been found to be $\alpha_L = 2.4^{+0.3}_{-0.2}$.[524] Calculations of the conduction electron density at the nucleus in dysprosium metal have been shown to agree with the value obtained from Mössbauer isomer shift measurements, provided proper account is taken of the relativistic effect.[525]

519 N. N. Greenwood, G. E. Turner, and A. Vertes, *Inorg. Nuclear Chem. Letters*, 1971, 7, 389.
520 A. Z. Hrynkiewicz, D. S. Kulgawczuk, A. M. Pustówka, I. Stronski, and K. Tomala, *J. Inorg. Nuclear Chem.*, 1971, **33**, 3707.
521 S. Hüfner, J. Brinkman, and G. Crecelius, *Phys. Letters*, 1971, **36A**, 367.
522 I. R. Harris and G. Longworth, *J. Less-Common Metals*, 1971, **23**, 281.
523 P. Steiner and G. Weyer, *Z. Physik*, 1971, **248**, 362.
524 P. Steiner and G. Weyer, *Z. Physik*, 1971, **248**, 370.
525 K. C. Das and D. K. Ray, *Solid State Comm.*, 1971, **9**, 1061.

A magnetic moment of 9.7 ± 0.3 BM has been deduced from the spectrum of Dy^{3+} in antiferromagnetic DyOOH.[526] The magnetic properties of Dy_2O_2S have also been studied. The nuclear parameters are $g_1/g_0 = -1.243$ and $Q_1/Q_0 = +1.007$. The disappearance of the $\Delta m = 0$ lines in a single-crystal experiment showed that the antiferromagnetic axis is parallel to the *c*-axis, contrary to the previous interpretation

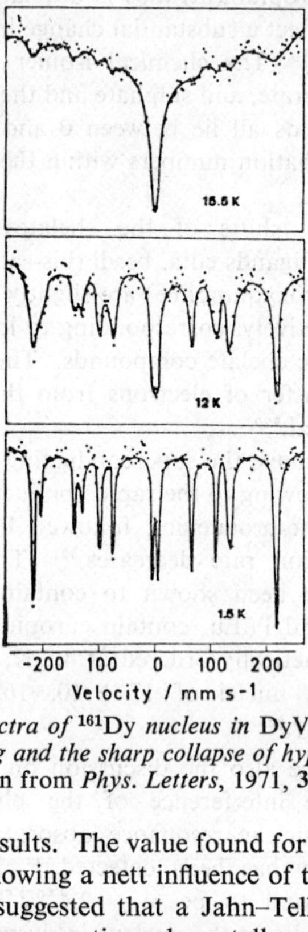

Figure 32 *Mössbauer spectra of* ^{161}Dy *nucleus in* $DyVO_4$ *at* 1.5, 12, *and* 15.5 K, *exhibiting the broadening and the sharp collapse of hyperfine splitting*
(Reproduced by permission from *Phys. Letters*, 1971, **34A**, 361)

of neutron diffraction results. The value found for the saturation magnetic moment was 8.0 BM, showing a nett influence of the crystal field.[527]

It has recently been suggested that a Jahn–Teller distortion is present in $DyVO_4$ at 14 K. The magnetic and crystallographic transitions in this compound have now been studied further using low-temperature *X*-ray, Mössbauer, and neutron diffraction measurements. Direct evidence of the distortion is given by the *X*-ray experiment, but it is hard to say whether the spin–lattice relaxation effects in the Mössbauer spectra around 14 K (see Figure 32) are connected with the distortion. The neutron diffraction measurements indicate a Néel temperature of $T_N = 3.1 \pm 0.1$ K, but there is no evidence for a crystallographic change accompanying the magnetic transition, from the Mössbauer or *X*-ray data. The magnetic moment at

[526] A. N. Christensen, *Solid State Comm.*, 1971, **9**, 925.
[527] M. Belakhovsky, *J. Phys. (Paris), Colloq.*, 1971, **1** (Pt. 2), C1–915.

1.5 K was found to be 8.6 ± 0.5 BM from the neutron measurements and 9.1 ± 0.2 BM from the Mössbauer measurements.[528]

DyH_2, which has a Néel point at 8 K, has been studied at 4.12, 1.1, and 0.23 K, and it has been shown that even at 0.23 K two Dy^{3+} levels are still populated. The magnetic moments of the two levels, the splitting between them, and relaxation times are estimated.[529]

The magnetic moments of Dy^{3+} ions in the compounds DyP, DyAs, and DySb have been determined at temperatures below that of magnetic ordering and are respectively: $\langle\mu\rangle = 9.52 \pm 0.20$, $\langle\mu\rangle = 9.6 \pm 0.20$, and $\langle\mu\rangle \simeq 10$ BM. These values are close to the maximum possible magnetic moment of Dy^{3+} in a magnetically ordered compound.[530] A systematic study of the hyperfine parameters in the intermetallic cubic compounds of dysprosium with copper, silver, zinc, and rhodium has appeared. The ground state was found to be a $|\frac{15}{2}\rangle$ state along a three-fold direction. This was consistent with point-charge calculations and supports the hypothesis of a non-collinear spin structure in DyCu. All hyperfine parameters reveal the influence of conduction electrons, the *s* character of the conduction band decreasing on passing from DyCu to DyZn.[531] Point-charge model calculations have also been tested in the intermetallic compound $DyNi_3$, and the magnetic structure obtained by neutron diffraction was discussed in the light of the Mössbauer data. The magnetic moments derived from the Mössbauer data were identical (10 BM) for the two inequivalent sites in this compound, but the values obtained from neutron diffraction differed: 9.8 ± 0.05 and 8 ± 0.5 BM.[532]

Erbium (^{166}Er). Systematics of the ^{166}Er Mössbauer data have been discussed for the Laves phases $(Er_{1-x}X_y)Al_2$ (X = Y or Gd) and $ErNi_{2-x}Co_x$.[516]

Ytterbium (^{170}Yb, ^{171}Yb, ^{172}Yb, ^{174}Yb). Measurements have been made of the chemical isomer shift of ^{170}Yb in $Yb_{0.218}Ca_{0.782}O$ at 5 K and the value obtained ($\delta = -0.14 \pm 0.05$ mm s^{-1} rel. $YbAl_2$) is thought to be consistent with the presence of Yb^{2+}.[533] Quadrupole and magnetic hyperfine interactions in Yb_2O_3 have been studied by use of the Mössbauer effect in the isotopes ^{170}Yb, ^{171}Yb, ^{172}Yb, and ^{174}Yb. There are two different sites in the sesquioxide and these have temperature-dependent hyperfine fields but temperature-independent electric field gradients. By using the same asborber for all measurements, it was possible to obtain very accurate values for the quadrupole moments of the first excited states of the four

528 F. Sayetat, J. X. Boucherle, M. Belakhovsky, A. Kallel, F. Tcheou, and H. Fuess, *Phys. Letters*, 1971, **34A**, 361.

529 J. Hess, E. R. Bauminger, A. Mustachi, I. Nowik, and S. Ofer, *Phys. Letters*, 1971, **37A**, 185.

530 V. D. Gorobchenko, I. I. Lukashevich, V. G. Stankevich, N. I. Filippov, V. I. Chukalin, and E. Y. Yarembash, *Fiz. tverd. Tela*, 1971, **13**, 1085.

531 M. Belakhovsky and J. Pierre, *Solid State Comm.*, 1971, **9**, 1409.

532 J. Yakinthos, J. Rossat-Mignod, and M. Belakhovsky, *Phys. Stat. Sol.* (*B*), 1971, **47**, 247.

533 J. C. Achard, O. Gorochov, F. Gonzalez, and P. Imbert, *Compt. rend.*, 1971, **272**, *C*, 868.

isotopes:[534]

$$\frac{^{172}Q_0(2\,+)}{^{170}Q_0(2\,+)} = 1.020 \pm 0.012 \quad \frac{^{174}Q_0(2\,+)}{^{170}Q_0(2\,-)} = 1.000 \pm 0.019$$

$$\frac{^{171}Q_0(\frac{3}{2}\,-)}{^{170}Q_0(2\,+)} = 1.074 \pm 0.023 \quad \frac{^{171}Q_0(\frac{5}{2}\,-)}{^{170}Q_0(2\,+)} = 1.020 \pm 0.027$$

Chemical isomer shifts and quadrupole splitting of the 2 + rotational state in ^{174}Yb have been measured. As can be seen in Figure 33, Yb,

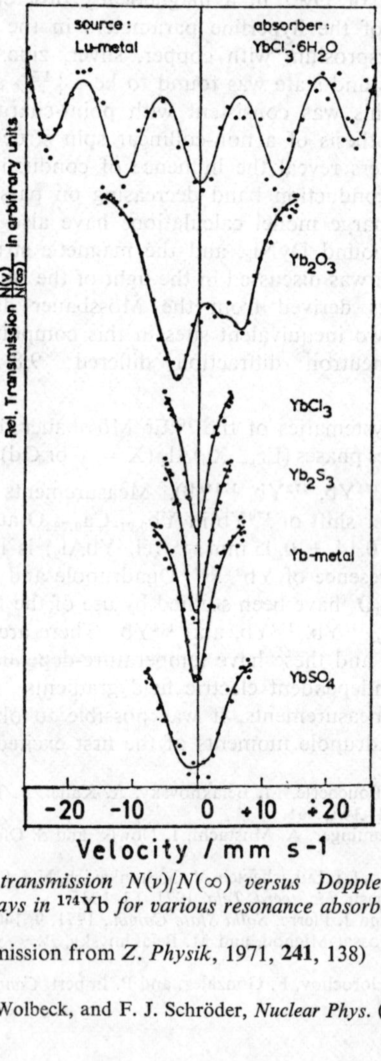

Figure 33 *Relative transmission $N(v)/N(\infty)$ versus Doppler velocity v of the* 76.5 keV *gamma rays in* ^{174}Yb *for various resonance absorbers. The source was lutetium metal*
(Reproduced by permission from *Z. Physik*, 1971, **241**, 138)

[534] K. G. Plingen, B. Wolbeck, and F. J. Schröder, *Nuclear Phys. (A)*, 1971, **165**, 97.

Yb_2S_3, and $YbCl_3$ give single-line spectra, Yb_2O_3 and $YbSO_4$ give quadrupole splitting with a negative V_{zz}, and paramagnetic $YbCl_3,6H_2O$ gives magnetic hyperfine splitting due to long spin–spin and spin–lattice relaxation times. YbOOH and paramagnetic $Yb(SO_4)_3,8H_2O$ were also studied but their spectra are not shown; both give quadrupole splitting, with a positive V_{zz}, and in addition $Yb(SO_4)_3,8H_2O$ shows magnetic hyperfine splitting. There is an approximately linear correlation (see Figure 34) between the shifts of

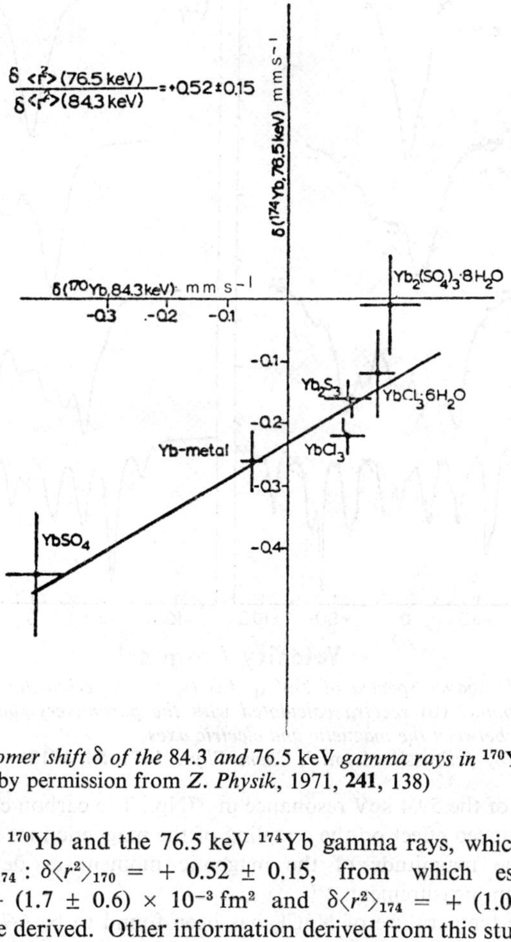

Figure 34 *Isomer shift δ of the* 84.3 *and* 76.5 keV *gamma rays in* ^{170}Yb *and* ^{174}Yb (Reproduced by permission from *Z. Physik*, 1971, **241**, 138)

the 84.3 keV ^{170}Yb and the 76.5 keV ^{174}Yb gamma rays, which yields the ratio $\delta\langle r^2\rangle_{174} : \delta\langle r^2\rangle_{170} = +\ 0.52 \pm 0.15$, from which estimates of $\delta\langle r^2\rangle_{170} = +\ (1.7 \pm 0.6) \times 10^{-3}\ \text{fm}^2$ and $\delta\langle r^2\rangle_{174} = +\ (1.0 \pm 0.45) \times 10^{-3}\ \text{fm}^2$ were derived. Other information derived from this study included an estimate of the quadrupole moment ratio $^{174}Q_0 : {}^{170}Q_0 = +\ 1.001 \pm 0.021$, and a value for the effective magnetic field in $Yb(SO_4)_3,8H_2O$ of $H_{\text{eff}} = (2.60 \pm 0.10)$ MG.[535]

[535] W. Henning, G. Bähre, and P. Kienle, *Z. Physik*, 1971, **241**, 138.

Thorium (^{232}Th). Resonant absorption following Coulomb excitation has been observed for γ-rays from the first excited state of ^{232}Th in thorium metal and ThC targets with ThO_2, thorium metal, ThN, and ThC_2 absorbers.[536]

Neptunium (^{237}Np). The neptunium monocarbide alloys $NpC_{0.96}$, $NpC_{0.89}$, and $NpC_{0.82}$, which have the sodium chloride structure, have been studied

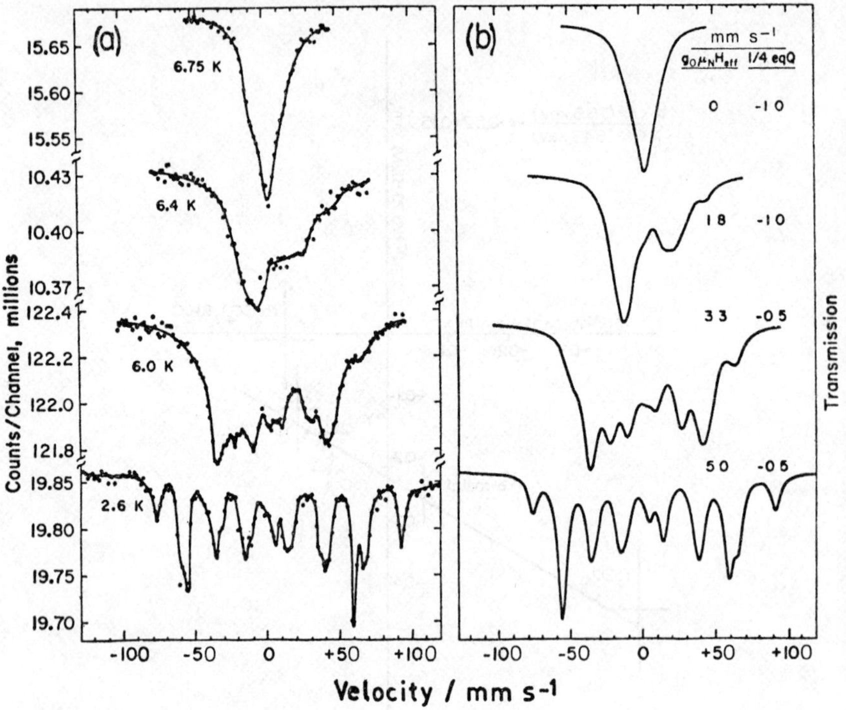

Figure 35 *Mössbauer spectra of* $NpCl_4$: (a) *typical experimental spectra below the Curie point;* (b) *spectra calculated with the parameters shown, assuming collinearity between the magnetic and electric axes*

(Reproduced by permission from *J. Chem. Phys.*, 1971, **54**, 1713)

with the aid of the 59.4 keV resonance in ^{237}Np. The carbon concentration has a pronounced effect on the coupling of the magnetic ions but has little effect on the magnitude of the magnetic moment as determined by magnetization measurements.[537]

The Curie temperature of $NpCl_4$ has been found to be 6.7 ± 0.1 K by magnetic susceptibility measurements, and this is confirmed by the Mössbauer measurements (see Figure 35). The experimental spectra can

[536] P. Durkee and N. Hershkowitz, *Phys. Rev.* (*B*), 1971, **3**, 3607.
[537] D. J. Lam, M. H. Mueller, A. P. Paulikas, and G. H. Lander, *J. Phys.* (*Paris*), *Colloq.*, 1971, **1** (Pt. 2), C1–917.

be satisfactorily simulated on the basis of the parameters shown and on the assumption of collinearity between the magnetic and electronic axes. There is no need to invoke relaxation effects to explain the spectra. The temperature dependence of the hyperfine field visually resembles a Brillouin function, but no Brillouin function from $J = \frac{1}{2}$ to $J = \infty$ will fit the data.[538]

Neptunium has been shown to exist in the + 7 oxidation state in Li_5NpO_6, $[Co(en)_3]NpO_5,xH_2O$, $Ba_3(NpO_5)_2,xH_2O$, and $Ca_3(NpO_5)_2,xH_2O$. Li_5NpO_6 does not have O_h symmetry; the $[NpO_6]^{5-}$ anion is probably compressed in the z-direction and rhombically distorted in the xy-plane. The other anions are presumed to contain a tetragonal-bipyramidal configuration, with the neptunyl group oriented in the z-direction and two oxygen atoms, as well as two OH groups, both in *trans*-positions in the xy-plane.[539]

A number of americium compounds have been used as sources of the 59.6 keV resonance in ^{237}Np. The compounds AmSb, Am_3Se_4, and Am_3Te_4 were shown to be tervalent on the basis of their chemical isomer shifts, but were not magnetically ordered at $T \geqslant 4.2$ K. Similarly, no magnetic hyperfine interaction was observed for Am_2C_3 even though Np_2C_3 has a ferromagnetic transition at 109 K. Conducting americium compounds tend to form a non-magnetic tervalent state. The results were discussed in terms of a crystal-field calculation, which includes J-mixing caused by the large crystal field interaction present in actinide compounds.[540]

7 Bibliography

The following list contains references on Mössbauer spectroscopy which have not been discussed in the text. Most of the papers are concerned with alloy systems containing iron or tin. A few papers, from relatively inaccessible journals, came to the attention of the reviewer at too late a date to be discussed in the main text and are therefore included here for completeness. The references are ordered alphabetically with respect to the first author.

M. A. Abidov, G. S. Zhdanov, and R. N. Kuz'min. Mössbauer spectra in copper–manganese–tin alloys, *Metalloved., Mater. Simp.*, 1968 (pub. 1971), 470.

K. Adachi, K. Sato, M. Matsui, and Y. Fujio. Band antiferromagnetism in fcc $(CoMn)_{1-x}Fe_x$, *J. Phys. Soc. Japan*, 1971, **30**, 1201.

A. Afanas'ev and V. D. Gorobchenko. Theory of degenerate spectra of Mössbauer line hyperfine structure in paramagnetic substances, *Zhur. eksp. teor. Fiz.*, 1971, **60**, 283.

A. F. Afanasiev, I. P. Suzdalev, and E. A. Manikin. Study of the superparamagnetism of ferromagnetic particles by Mössbauer spectroscopy, ref. 12, p. 183.

V. K. Agarval and R. N. Kuz'min. Mössbauer effect in $Fe_{17}Gd_2$ and $Fe_{17}Nd_2$, *Kristallografiya*, 1971, **16**, 774.

G. M. Aivazyan, M. Y. Aivazyan, V. I. Goldanskii, R. N. Kocharyan, E. F. Makarov, and A. R. Mkrtchyan. Effect of ultrasonic excitation in crystals on Mössbauer effect probability. *Pis'ma Zhur. eksp. teor. Fiz.*, 1971, **13**, 543.

538 J. A. Stone and E. R. Jones, jun., *J. Chem. Phys.*, 1971, **54**, 1713.
539 K. Fröhlich, P. Gütlich, and C. Keller, *Angew. Chem.*, 1971, **10**, 829.
540 B. D. Dunlap, D. J. Lam, G. M. Kalvius, and G. K. Shenoy, *J. Appl. Phys.*, 1971, **42**, 1719.

Z. Alexan, M. Bornaz, A. Calusaru, G. Filoti, and A. Gelberg. Structure of ferric compounds formed in electrolytical aluminium oxide layers, ref. 12, p. 349.

M. A. Andreeva and R. N. Kuzmin. Narrowing of the Mössbauer lines, ref. 12, p. 77.

Yu. M. Ayvazan and V. A. Beliakov. On the theory of Mössbauer diffraction in crystals with electric field gradients of complicated structure, ref. 12, p. 127.

A. M. Babeshkin, P. B. Fabrichny, and A. N. Nesmeianov. The study of the transitions on the surface of stannic acid by *X*-ray diffraction, Mössbauer and proton magnetic resonance spectroscopy, ref. 12, p. 319.

A. M. Babeshkin, Yu. D. Perfilev, V. A. Lebedev, E. V. Lomikin, L. A. Kulikov, R. A. Lebedev, and A. N. Nesmeianov. Investigation of local changes in valence and structure caused by nuclear transformation in Co, Fe, I, and Te compounds by gamma resonance spectroscopic method, ref. 12, p. 697.

R. A. Baker. Evaluation of pyrite oxidation by nuclear methods, *Water Pollut. Contr. Res. Ser.*, 1971, No. 14010F1103/71.

E. Banks, O. Berkooz, and J. A. de Luca. Magnetic interactions in the complex fluorides $AM^{II}M^{III}F_6$, *J. Appl. Phys.*, 1971, **42**, 1722.

J. J. Bara, K. Krolas, and T. Matlak, Investigation of organometallic iron compounds by Mössbauer effect, ref. 12, p. 769.

O. A. Bayukov, V. N. Ikonnikov, M. I. Petrov, V. N. Seleznev, R. N. Smolin, and V. V. Uskov. Mössbauer study of iron borate, *Pis'ma Zhur. eksp. teor. Fiz.*, 1971, **14**, 49.

V. A. Beliakov, Yu. M. Ayvazan, and V. P. Orlov. On the theory of the Mössbauer diffraction of magnetically ordered crystals, ref. 12, p. 129.

V. M. Belova, V. I. Nikolaev, and E. P. Stepanov. Resonance scattering of γ quanta by the ^{57}Fe nuclei in a ticonal-type alloy, *Fiz. tverd. Tela*, 1971, **13**, 934.

G. N. Belozerskii, V. N. Gittsovich, V. N. Kramer, O. G. Sokolov, and Y. P. Khimich. Mössbauer effect in some invar alloys, *Fiz. tverd. Tela*, 1971, **13**, 562.

G. N. Belozerskii, Y. N. Grinblat, and A. I. Shapiro. Dynamics of the thermomagnetic treatment of YuNDK35T5 alloy studied by the Mössbauer effect, *Fiz. Metal Metalloved.*, 1971, **32**, 301.

G. N. Belozerskii, O. G. Sokolov, and V. N. Gittzovich. Mössbauer study of the γ-FeMn alloys, ref. 12, p. 441.

L. H. Bennet, L. J. Swartzendruber, and R. E. Watson. Interactions between small magnetic clusters in copper-rich Cu–Ni–Fe alloys, *J. Appl. Phys.*, 1971, **42**, 1547.

B. S. Bokshtein and Y. B. Voitkovskii. Titanium-iron alloys studied by the Mössbauer method, *Metalloved., Mater. Simp.*, 1968 (pub. 1971), 486.

Zw. Bonchev, A. Jordanov, and A. Minkova. Method of analysis of thin surface layers by the Mössbauer effect, ref. 12, p. 333.

E. Both, G. Trumpy, and C. Djéga-Mariadassou. Magnetic structure and electron properties of iron–tin alloys, ref. 12, p. 375.

L. Brossard, G. A. Fatseas, J. L. Dormann, and P. Lecocq. Mössbauer study in α-phase solid solutions of an FeGe system, *J. Appl. Phys.*, 1971, **42**, 1306.

G. Bueche, H. Appel, and W. Renz. Residual nucleus of the reaction $^{56}Fe(d,p)^{57}Fe^*$ (14 keV ^{57}Fe), *Nuclear Phys.* (*A*), 1971, **170**, 55.

K. Burger, L. Korecz, and A. Vértes. The effect of back-co-ordination on the Mössbauer parameters of high-spin iron(II) complexes, ref. 12, p. 721.

D. C. Champeney and D. F. Sedgwick. The molecular dynamics of solid cyclohexane by Mössbauer scattering, *J. Phys.* (*C*), *Solid State Phys.*, 1971, **4**, 2220.

D. Chandra and L. H. Schwartz. Mössbauer effect study of the 475 °C decomposition of iron–chromium, *Met. Trans.*, 1971, **2**, 511.

D. Chandra and L. H. Schwartz. Mössbauer effect study of the 475 °C decomposition of iron–chromium. Reply to comments, *Met. Trans.*, 1971, **2**, 2294.

D. Chandra and L. H. Schwartz. Mössbauer effect studies of spinodal decomposition in Fe–Cr, ref. 20, p. 79.

G. Chandra and T. S. Radhakrishna. Hyperfine fields at ^{57}Fe in nickel–manganase alloys, *Phys. Stat. Sol.* (*B*), 1971, **45**, 639.

C. C. Chao, P. Duwez, and C. C. Tsuei. Metastable fcc Fe–Rh alloys and the Fe–Rh phase diagram, *J. Appl. Phys.*, 1971, **42**, 4282.

D. Christov, Z. Bonchev, B. Manouchev, D. Dimov, and N. Nenov. Application of the Mössbauer effect for the investigation of the resin vulcanization of butylrubber, ref. 12, p. 795.

L. Cser, G. Konczos, D. L. Nagy, Yu. M. Ostanevich, and L. Pál. Investigation of the local behaviour of the DO_3 type ordered Fe–Al alloys, ref. 12, p. 419.

L. Cser and I. Vincze. Mössbauer experiments on dilute iron-based alloys with non-transition elements, *J. Phys.* (*Paris*), *Colloq.*, 1971, **1** (Pt. 2), C1-787.

L. M. Dautov, M. M. Kadykenov, D. K. Kaipov, and S. P. Ionov. Use of the Fermi–Segre–Goudsmit formula for detecting electron density of non-alkaline atoms in the area of the nucleus, *Izvest. Akad. Nauk Kaz. S.S.R., Fiz. Mat.*, 1971, **9**, 6.

I. Y. Dekhtyar, M. M. Nistchenko, and V. P. Pomashko. Positron annihilation and Mössbauer effect in dilute solid solutions of Ni, Pd, and Pt, with iron, *Phys. Stat. Sol.* (*B*), 1971, **48**, K51.

I. Ya. Dehtiar, P. S. Nizin, M. M. Nishchenko, and K. V. Chuistov. Mössbauer effect in the segregating iron–beryllium alloy, ref. 12, p. 435.

I. Ya. Dehtiar, P. S. Nizin, M. M. Nishchenko, and R. G. Fedchenko. Investigation of the magnetic transformations in dilute palladium–iron solid solutions by Mössbauer effect, ref. 12, p. 395.

N. N. Delyagin and E. N. Kornienko. Hyperfine magnetic fields for ^{119}Sn atoms in ferromagnetic CoFe, *Fiz. tverd. Tela*, 1971, **13**, 1497.

I. Dézsi, K. Kulcsár, D. L. Nagy, and L. Pócs. On the Mössbauer spectra of $FeCO_3$ (siderite), ref. 12, p. 247.

Y. Endoh and Y. Ishikawa. Antiferromagnetism of γ-iron–manganese alloys, *J. Phys. Soc. Japan*, 1971, **30**, 1614.

V. Fano and I. Ortalli. A study on some Co–Te compositions by Mössbauer spectroscopy, *J. Phys. and Chem. Solids*, 1971, **32**, 2305.

G. A. Fatseas, J. L. Dormann, and L. Brossard. Analysis and interpretation of the Mössbauer spectrum of $B8_2$ phase iron germanide, *J. Phys.* (*Paris*), *Colloq.*, 1971, **1** (Pt. 2), C1–785.

A. Ferro, G. Griffa, and I. Ortalli. Quench enhanced ordering of iron 50-nickel 50 alloy, *Nuovo Cimento Soc. Ital. Fis. B*, 1971, **3**, 269.

G. Filoti, A. Gelberg, V. Gomolea, and M. Rosenberg. Mössbauer study of the Jahn–Teller effect in spinels, ref. 12, p. 593.

A. A. Firsova, N. N. Khovanskaya, A. D. Tsyganov, I. P. Suzdalev, and L. Y. Margolis. Surface compounds of propylene, acrolein, oxygen, and water on a tin–molybdenum catalyst studied by γ-resonance spectroscopy, *Kinetika i Kataliz*, 1971, **12**, 792.

J. Flechon, C. Varnier, G. Le Caer, and J. M. Genin. Structure of chemical iron–nickel–boron deposits studied by the Mössbauer effect, *Compt. rend.*, 1971, **272**, *B*, 437.

J. M. Friedt. Mössbauer spectra of ^{57}Fe in $CoF_2(^{57}Co)$: application to the re-evaluation of the quadrupole moment of ^{57m}Fe, ref. 12, p. 121.

J. M. Friedt. Mössbauer spectra of ^{57}Fe in K_3CoF_6: A possible explanation for anomalous charge states observed after E.C. in ^{57}Fe compounds, ref. 12, p. 709.

I. I. Georgescu, C. Nistor, M. Rusu, M. Andreescu, and R. Mihu. Mössbauer effect in some iron derivative salts of stilbene acids, ref. 12, p. 781.

V. I. Goldanskii, E. F. Makarov, I. P. Suzdalev, and I. A. Vinogradov. On the polarization phenomena, absolute probabilities, and anisotropy of the Mössbauer effect in siderite, ref. 12, p. 239.

I. E. Grey, H. Hong, and H. Steinfink. The crystal structure of $Ba_7Fe_6S_{14}$, a trinuclear complex, *Inorg. Chem.*, 1971, **10**, 340.

M. L. Grigor'ev, Relative intensities of Mössbauer lines in magnets with spiral structure, *Fiz. Metal. Metalloved.*, 1971, **31**, 685.

G. Gruener and I. Vincze. Hyperfine field distribution and average hyperfine field in dilute iron–cobalt alloys, *Kozp. Fiz. Kut. Intez.*, 1971, 71.

Yu. S. Grushko, B. G. Lurie, and A. N. Murin. Investigation of I_2O_4 by Mössbauer effect, ref. 12, p. 681.

D. Gumprecht, P. Steiner, G. Crecelius, and S. Heufner. Critical phenomena in Fe–Ni, *Phys. Letters*, 1971, **34A**, 79.

M. P. Gupta and H. B. Mathur. Nature of precipitated tin oxide in internally oxidized tin–silver alloy, *Indian J. Chem.*, 1971, **9**, 864.

P. Gütlich. Physical methods in chemistry: Mössbauer spectroscopy II, *Chemie in unserer Zeit*, 1971, **5**, 131.

P. Gütlich and H. Prange. Mössbauer spectroscopy and its applications to chemical problems, *Chem.-Ing.-Tech.*, 1971, **43**, 1049.

D. Hanzel, M. Schara, and N. Trsan. Characterization of high-alloy ferritic steels by Mössbauer spectroscopy, *Zelezarski Zb.*, 1971, **5**, 111.

M. Hayase, M. Shiga, and Y. Nakamura. Thermal expansion of ordered Fe_3Pt alloy, *Phys. Stat. Sol.* (*B*), 1971, **46**, K117.

R. H. Herber and Subhas Chandra. The Goldanskii–Karyagin effect in dimethyltin difluoride, ref. 12, p. 253.

E. Hermon, R. D. Nolan, and S. Shtrikman. Mössbauer effect and resistivity studies in $Fe_{1.11}Te$, *Israel J. Chem.*, 1971, **9**, 1.

C. Hohenemser. Comparison of the Mössbauer effect to time-differential perturbed angular correlations: The case of ^{57}Fe in a nickel host, ref. 20, p. 43.

A. Z. Hrynkiewicz, B. Sawicka, and J. Sawicki. Temperature dependence of the electric field gradient in ferricyanides, ref. 12, p. 731.

A. Z. Hrynkiewicz, B. Sawicka, and J. Sawicki. The effective magnetic fields in cubic ferricyanides, ref. 12, p. 739.

G. P. Huffman. Mössbauer study and molecular field theory of the magnetic properties of iron–aluminium alloys, *J. Appl. Phys.*, 1971, **42**, 1606.

G. P. Huffman and G. R. Dunmyre. Localized polarization on tin atoms in ferromagnetic transition metals and alloys, *J. Appl. Phys.*, 1971, **42**, 1613.

B. C. Huguelet, W. C. Harper, C. L. Hummel, C. W. Kimball, A. T. Aldred, and A. P. Paulikas. Mössbauer investigation of iron-rich iron+arsenic and iron+antimony solid solutions, *J. Appl. Phys.*, 1971, **42**, 1312.

P. S. Huyen. Theory of γ-quanta transition through a resonant medium, *Zhur. eksp. teor. Fiz.*, 1971, **61**, 359.

P. K. Iyengar and S. C. Bhargava. Spin relaxation of Fe^{3+} ion in ferrites using the Mössbauer effect. Proceedings of the 15th Nuclear Physics and Solid State Physics Symposium, 1970 (pub. 1971), **3**, 709.

J. Jach, R. J. Borg, and D. Y. F. Lai. Clustering and giant moments in copper–nickel alloys, *J. Appl. Phys.*, 1971, **42**, 1611.

C. Janot, G. Marchal, M. Piecuch, and H. Scherrer. Study of iron thin film by the Mössbauer effect, *Compt. rend.*, 1971, **273**, *B*, 399.

C. Janot and M. Piecuch. Mössbauer effect studies of beryllium–iron compounds with the same structure as Be_5Fe, *Acta Met.*, 1971, **19**, 713.

K. Kai and T. Nakamichi. Magnetic moment of the iron atom in the Laves phase compounds, *J. Phys. Soc. Japan*, 1971, **30**, 1755.

M. Kaplan and T. X. Carroll. Anomalous Mössbauer spectra of ferric chloride hexahydrate, ref. 12, p. 169.

M. Kaplan and B. S. Ehrlich. Mössbauer effect in CsI_3 and tri-iodide complexes of benzamide and amylose, ref. 12, p. 689.

N. Kazama and H. Watanabe. Magnetic and crystallographic studies of the $Cr_{1-x}Fe_xAs$ system, *J. Phys. Soc. Japan*, 1971, **30**, 578.

A. P. Kirianov and N. E. Alekseevskii. Anistropy of the gamma radiation probability of ^{119}Sn nuclei in single crystals of active tin, ref. 12, p. 261.

A. P. Kirianov, Yu. A. Samarskii, N. E. Alekseevskii, and V. I. Tzebro. Mössbauer effect on ^{119}Sn nuclei in tin layers condensed at 4.2 K, ref. 12, p. 313.

M. Komor, A. Vértes, I. Dézsi, and I. Ruff. Investigation of the electron-exchange reaction between iron(II) and iron(III) in solution by Mössbauer effect, ref. 12, p. 713.

M. Krishnan and N. D. Sharma. Mössbauer effect studies of the 9.3 keV transition in solid krypton, Proceedings of the 15th Nuclear Physics and Solid State Physics Symposium, 1970 (pub. 1971), **3**, 363.

K. Krop and J. M. Williams. Mössbauer studies of the relaxation behaviour of β-Co fine particles, *J. Phys.* (*F*), 1971, **1**, 938.

D. S. Kulgawczuk, E. Nowicka, B. Sawicka, J. Sawicki, I. Stroński, and K. Tomala. Iron chelate complexes investigated by Mössbauer effect, ref. 12, p. 773.

R. N. Kuzmin, A. A. Opalenko, and V. S. Shpinel. Mössbauer effect in Te single crystals, ref. 12, p. 785.

G. Lang, T. Asakura, and T. Yonetani. Mössbauer spectroscopy of mesohaem and protohaem myoglobins and their fluoride complexes, ref. 12, p. 655.

J. K. Lees and P. A. Flinn. Mössbauer effect of ^{119}Sn in alpha-brass and related alloy systems, *Phys. Rev.* (*B*), 1971, **3**, 591.

W. Leiper, D. J. W. Geldart, and P. J. Pothier. Hyperfine field at the tin sites in the Heusler alloy Ni_2MnSn, *Phys. Rev.* (*B*), 1971, **3**, 1637.

S. Ligenza. The Mössbauer study of ^{57}Fe and ^{119}Sn in FeSn at 5 K, *Phys. Stat. Sol.* (*B*), 1971, **44**, 775.

S. Ligenza. A spin-flip effect in FeSn, *Phys. Stat. Sol.* (*B*), 1971, **45**, 721.

S. Ligenza. Investigations of internal magnetic and electric fields on iron and tin nuclei in the antiferromagnetic compound FeSn by the Mössbauer effect and neutron diffraction, *Inst. Nuclear Res.* (*Warsaw*), *Rep.*, 1971, No. 1290/II/PS.

V. I. Lisichenko, S. L. Korduk, O. L. Orlov, and A. N. Smoilovskii. Mössbauer study of dispersed systems, ref. 12, p. 339.

I. I. Lukashevich, V. D. Gorobchenko, V. V. Sklyarevskii, and N. I. Filippov. Direct observation of the interference between internal conversion and photoeffect in ^{161}Dy, ref. 12, p. 27.

V. A. Makarov, E. B. Granovskii, E. F. Makarov, V. A. Povitskii, and A. A. Fridman. Nature of the single line in Mössbauer spectra of highly coercive ticonal alloys, *Fiz. Metal Metalloved.*, 1971, **32**, 438.

T. V. Malisheva and V. A. Dolgopolov. On the determination of the co-ordination numbers in complex compounds of tin by Mössbauer effect, ref. 12, p. 763.

P. R. Marshall and D. Rutherford. Physical investigations on colloidal iron–dextran complexes, *J. Colloid and Interface Sci.*, 1971, **37**, 390.

W. Meisel. Inhomogeneities in solids measured by approximate defolding of Mössbauer spectra, *Phys. Stat. Sol.* (*B*), 1971, **43**, K129.

W. Meisel, K. Hennig, and H. Schnorr. The state of Sn impurities in AgCl crystals, ref. 12, p. 761.

R. Mishima, Y. Ishida, and M. Kato. Mössbauer analysis of the ageing process of splat-cooled aluminium–iron alloys, *Seisan-Kenkyu*, 1971, **23**, 85.

H. S. Moller and H. G. Drickamer. Pressure dependence of the Curie temperature and magnetization in ferromagnetic PdFe alloys, *J. Phys. and Chem. Solids*, 1971, **32**, 745.

S. Mørup. Hyperfine structure of ^{161}Dy in Gd_2O_3 at 300 K, ref. 12, p. 87.

S. S. Nandwani, D. Raj, and S. P. Puri. Anharmonic contribution to the Mössbauer fraction of krypton-83 in solid krypton, Proceedings of the 15th Nuclear Physics and Solid State Physics Symposium, 1970 (Publ. 1971), **3**, 353.

S. Nasu and Y. Murakami. Second-order Doppler shift of ^{57}Fe Mössbauer resonance in Al–0.01% ^{57}Fe alloy solid solution, *Phys. Stat. Sol.* (*B*), 1971, **46**, 711.

S. Nasu, Y. Murakami, and R. Katano. Dilute ^{57}Co Mössbauer study on a ferromagnetic Ni–Cu alloy near the transition temperature, *Phys. Letters*, 1971, **36A**, 117.

L. R. Newkirk and C. C. Tsuei. Mössbauer study of hyperfine magnetic interactions in Fe–Ga solid solutions, *Phys. Rev.* (*B*), 1971, **4**, 4046.

S. V. Nikitina, S. B. Zezin, R. N. Kuz'min, and N. M. Matveeva. Mössbauer spectra and interatomic interactions in $MnSn_2$–$FeSn_2$, $MnSn_2$–$CoSn_2$, and $FeSn_2$–$CoSn_2$, quasibinary systems, *Metalloved., Mater. Simp.*, 1968 (pub. 1971), 394.

I. N. Nikolaiev, V. V. Svetozarov, and V. Ya. Gamlitskii. On the sign of the *s–d* exchange interaction in iron, ref. 12, p. 367.

C. Nistor, I. I. Georgescu, M. Rusu, and L. Macarie. Mössbauer effect in some iron derivative salts of substituted benzoic acids, reg. 12, p. 779.

G. V. Novikov, A. I. Mihailov, V. A. Truchtanov, E. F. Abdrashitov, and V. I. Goldanskii. The effect of external magnetic fields on the hyperfine split Mössbauer spectra of ^{57}Fe in $FeCl_3$ n-butanol solution, ref. 12, p. 621.

H. Ohno and M. Mekata. Antiferromagnetism in hcp iron–manganese alloys, *J. Phys. Soc. Japan*, 1971, **31**, 102.

P. Peretto, H. Rechenberg, L. Billard, and A. Chamberod. Influence of short-range order on non-dilute alloys, *Phys. Stat. Sol.* (*B*), 1971, **44**, K77.

V. A. Povitskii, Yu. V. Baldohin, E. B. Granovsky, V. A. Makarov, and E. F. Makarov. Mössbauer study of Fe–Co–Ni–Al alloys, ref. 12, p. 427.

R. S. Preston and R. Gerlach. Mössbauer effect in dilute alloys of iron in aluminium, *Phys. Rev.* (*B*), 1971, **3**, 1519.

T. M. Quick, R. J. Borg, D. N. Pipkorn, and C. E. Violet. Spin relaxation in magnetically ordered Pd–Au (Fe 2.8 at. %) alloys, *Phys. Letters*, 1971, **36A**, 42.

P. Raj and S. K. Kulshreshtha. The effect of microstructure on Mössbauer line shapes in nickel zinc ferrite, Proceedings of the 15th Nuclear Physics and Solid State Physics Symposium, 1970 (Pub. 1971), **3**, 703.

R. Raudsepp and E. Realo. Mössbauer spectra of the methionine–iron complex, *Eesti NSV Tead. Akad. Toim. Fuus., Mat.*, 1971, **20**, 225.

W. E. Sauer and R. J. Reynik. Electronic and magnetic structure of dilute iron–base alloys. *J. Appl. Phys.*, 1971, **42**, 1604.
D. Schroeer. Rare-earth ions in colour centres, *Govt. Rep. Announce.* (*U.S.A.*), 1971, **71**, 262.
P. J. Schurer, G. A. Sawatzky, and F. van der Woude. Localized *versus* itinerant description of ferromagnetism in iron, *Phys. Rev. Letters*, 1971, **27**, 586.
H. Schuster and J. Bostock. Explanation of the temperature dependence of the Debye–Waller factor in Nb_3Sn, *Phys. Letters*, 1971, **35A**, 31.
L. H. Schwartz and D. Chandra. Hyperfine fields in concentrated Fe–Cr alloys, *Phys. Stat. Sol.* (*B*), 1971, **45**, 201.
R. B. Schwartz and R. B. Frankel. Local magnetic moments and the Mössbauer effect, ref. 12, p. 21.
T. E. Sharon and C. C. Tsuei. Mössbauer effect study of amorphous Fe–Pd–Si alloys, *Solid State Comm.*, 1971, **9**, 1923.
M. Shiga. The Mössbauer evidence of antiferromagnetic spin-ordering in fcc iron–iridium alloys, *Phys. Stat. Sol.* (*B*), 1971, **43**, K37.
T. Shinjo, T. Matsuzawa, T. Takada, S. Nasu, and Y. Murakami. Mössbauer evidence against the existence of magnetically 'dead' layers, *Phys. Letters*, 1971, **36A**, 489.
V. S. Shpinel and S. E. Gukasian. Isomer shifts and quadrupole splittings in the compounds of atoms with 5*s*5*p* valence electrons (^{119}Sn, ^{121}Sb, ^{125}Te and ^{127}I, ^{129}I, ^{129}Xe, ^{131}Xe), ref. 12, p. 41.
G. V. Smirnov, V. V. Sklyarevskii, R. A. Voskanian, and A. N. Artemiev. Diffraction of resonant γ-radiation by α-Fe_2O_3 antiferromagnetic crystals, ref. 12, p. 73.
J. K. Srivastava. Crystal field effects on time-dependent hyperfine interactions, Proceedings of the 15th Nuclear Physics and Solid State Physics Symposium, 1970 (pub. 1971), **3**, 719.
C. F. Steen, D. G. Howard, and R. H. Nussbaum. Evidence for stable interstitial and substitutional sites of cobalt in gold from Mössbauer studies, *Solid State Comm.*, 1971, **9**, 865.
P. Stetsenko, S. Antipov, and V. Satbaev. Hyperfine interactions in the invar alloys on iron–platinum base, *J. Phys.* (*Paris*), *Colloq.*, 1971, **1** (Pt. 2), C1–1117.
C. A. Stickels and R. H. Bush. Precipitation in the system aluminium − 0.05 wt. % iron, *Met. Trans.*, 1971, **2**, 2031.
R. A. Stukan, V. A. Prusakov, Y. N. Novikov, M. E. Vol'pin, and V. I. Goldanskii. γ-Resonance spectroscopic study of products of the reduction of laminated compounds of graphite with iron chlorides, *Zhur. strukt. Khim.*, 1971, **12**, 622.
J. Suwalski, J. Pickoszewski, and J. Leciejewicz. Mössbauer study of $Fe_{1.11}Te$, ref. 12, p. 385.
I. P. Suzdalev. Superparamagnetism of antiferromagnetic particles of ultramicro-size, ref. 12, p. 193.
I. P. Suzdalev, V. P. Korneev, and Yu. F. Krupansky. Investigation of the effect of weak magnetic fields on the hyperfine splitting in Fe^{3+} paramagnets with a crystal field of non-axial symmetry, ref. 12, p. 147.
L. J. Swartzendruber. Localized moments on iron impurities in Nb–Mo alloys: Mössbauer effect absorber study, *J. Appl. Phys.*, 1971, **42**, 1549.
N. Thrane. On the anomalous quadrupole splitting in $FeCl_3,6H_2O$, ref. 12, p. 175.
S. Tomiyoshi, H. Yamamoto, and H. Watanabe. Temperature-dependent distribution of internal magnetic fields at ^{57}Fe nuclei in fcc iron–nickel alloys, *J. Phys. Soc. Japan*, 1971, **30**, 1605.
T. Tomov, T. Ruskov, and S. A. Georgiev. An industrial application of the Mössbauer effect, ref. 12, p. 793.
F. Varret, A. Gerard, F. Hartmann-Boutron, P. Imbert, and R. Kleinberger. Observation of a first-order antiferro-paramagnetic transition in $ZnCr_2O_4$ and $MgCr_2O_4$, ref. 12, p. 581.
A. Vértes and B. Zsoldos. A study of thermal decomposition of iron(II) salt hydrates by Mössbauer spectroscopy, ref. 12, p. 751.
P. J. Viccaro and W. T. Oosterhuis. Back–Goudsmit effect in Mössbauer spectra, *J. Appl. Phys.*, 1971, **42**, 1723.
L. J. Vieland. *X*-Ray study of the Debye–Waller factor in Nb_3Sn, *Phys. Rev.* (*B*), 1971, **3**, 1804.
I. Vincze and L. Cser. Mössbauer spectra of Fe–Ga alloys, ref. 12, p. 371.

I. Vincze, L. Cser, and D. L. Nagy. Conduction electron polarization investigated by the temperature dependence of the hyperfine magnetic field in Fe–Mn alloys, ref. 12, p. 389.
Y. B. Voitkovskii, A. N. Pokhvisnev, Y. S. Yusfin, V. V. Dan'shin, N. F. Pashkov, T. N. Bazilevich, O. N. Generalov, and Y. A. Litvinenko. Use of the Mössbauer effect for controlling the production of metallized pellets, *Izvest. V.U.Z., Chern., Met.*, 1971, **14**, 23.
J. C. Walker and B. Cleveland. Magnetic ordering in very thin foils using the Mössbauer effect, ref. 12, p. 307.
J. C. Walker, F. Munley, and E. Loh. Relaxation effects near the Néel point in $FeCO_3$, ref. 12, p. 153.
G. K. Wertheim, D. N. E. Buchanan, and J. H. Wernick. Iron-57 hyperfine fields in iron–cobalt alloys, *J. Appl. Phys.*, 1971, **42**, 1602.
J. M. Williams. The sign of the internal field at ^{119}Sn nuclei in Co_2MnSn, ref. 12, p. 353.
R. O. Williams. Mössbauer effect study of the 475 °C decomposition of iron–chromium. Comments, *Met. Trans.*, 1971, **2**, 2294.
B. Window. Hyperfine field distributions from Mössbauer spectra, *J. Phys.* (*E*), 1971, **4**, 401.
B. Window. Mössbauer studies of iron in copper alloys, *J. Phys.* (*F*), 1971, **1**, 533.
B. Window. Electric field gradients at iron-57 nuclei in gold alloys, *J. Phys. and Chem. Solids*, 1971, **32**, 1059.
B. Window and G. Longworth. Electrostatic interactions at ^{57}Fe nuclei in nickel, palladium, and platinum alloys, *J. Phys.* (*F*), 1971, **1**, 718.
B. Window, G. Longworth, and C. E. Johnson. Mössbauer effect in rhodium–iron alloys, *J. Phys.* (*Paris*), *Colloq.*, 1971, **1** (Pt. 2), C1–863.
K. Yamaguchi, H. Watanabe, H. Yamamoto, and Y. Yamaguchi. Magnetic properties of the systems (CrFe)Sb and (CrCo)Sb, *J. Phys. Soc. Japan*, 1971, **31**, 1042.
K. Yamanaka, H. Ino, R. Oshima, and F. E. Fujita. Study of ordering of iron–19–25 at. % aluminium alloys by the Mössbauer effect and electron microscopy, *Nippon Kinzoku Gakkaishi*, 1971, **35**, 566.
E. N. Yefremov, A. M. Babeshkin, and A. N. Nesmeianov. The valance states of europium and dysprosium in the Laves phases and the isomer shifts in the Mössbauer spectra, ref. 12, p. 459.
H. Yoshida and C. Yonezawa. Determination of the chemical state of tin in copper–tin alloy by using the Mössbauer effect, *Bunseki Kagaku*, 1971, **20**, 970.
E. E. Yurchikov, A. Z. Menshikov, and V. A. Tzurin. Mössbauer study of magnetic transformation in γ-Fe–Ni alloys, ref. 12, p. 405.
E. E. Yurchikov, A. Z. Menshikov, and V. A. Tzurin. Mössbauer study of the $\alpha \rightleftharpoons \gamma$-transitions in Fe–Ni alloys, ref. 12, p. 413.
T. Zemcik and J. Vrestal. Mössbauer effect study of the Ni_3Co alloy, *Czech. J. Phys.*, 1971, **21**, 104.

Author Index